Life Science

LIFE SCIENCE

SECOND EDITION

BY **Gerard J. Tortora** *Bergen Community College*
Joseph F. Becker *Montclair State College*

IN CONSULTATION WITH

Guy L. Bush *University of Texas*
P. J. Docter *Eastern Illinois University*
John L. Howland *Bowdoin College*
J. R. McClintic *California State College, Fresno*
Ernst S. Reese *University of Hawaii*

MACMILLAN PUBLISHING CO., INC.
New York

Collier Macmillan Publishers
London

PRINTED IN THE UNITED STATES OF AMERICA

Macmillan Publishing Co., Inc.
866 Third Avenue, New York, New York 10022

Collier Macmillan Canada, Ltd.

Library of Congress Cataloging in Publication Data

Tortora, Gerard J.
Life science.
Includes index.

1. Biology. I. Becker, Joseph F., joint author.
II. Title
QH308.2.T67 1978 574 77-1231
ISBN 0-02-420920-1

Printing: 1 2 3 4 5 6 7 8 Year: 8 9 0 1 2 3 4

Preface

THE second edition of *Life Science,* like the first edition, is designed for use in introductory biology courses and presumes no previous study of biology or chemistry. It has been the purpose of the authors to introduce the material at an elementary level and in a simplified manner. From the comments and suggestions of those who have used the first edition the authors have reorganized and expanded certain fundamental concepts. There is a logical development of biological principles as the student continues through the text. Where previously learned concepts are helpful in the development of new material, a brief review or text reference is given.

The authors, aware of the expanding number of students entering the allied health professions, have expanded those areas that will better prepare these students for advanced courses as well as enrich the background for all students of biology. The text is adaptable to the various approaches to college biology and can be used for science as well as nonscience majors. Through planning by the instructor, it is possible to proceed sequentially through the text emphasizing those concepts and areas most applicable to individual needs while giving a broad overview of the science of biology.

Continuing in the second edition with the current approach to the study of biology, the various concepts are developed by placing emphasis on the structure and function of organisms at all levels of organization, interrelationships among organisms, and interrelationships between organisms and their environments. To develop logically these concepts, *Life Science* is organized into six major areas: (1) the chemical organization, origin, and diversity of life; (2) the cellular level of biological organization; (3) biological processes related to maintenance at the organismic level; (4) biological processes related to continuity at the organismic level; (5) the biology of organisms as components of ecosystems; and (6) behavior and its adaptive nature.

In addition to overall updating, accompanied by reorganization and expansion in many areas, we have made certain major changes in this edition. While the text still adheres to the now widely accepted five-kingdom classification of R. H. Whittaker, the detailed classification and specific examples have been placed in the Appendix. Without in any way reflecting on the importance of this material, the authors believe that the new arrangement improves the continuity of the text. In addition, the content of the first edition's Section VII, "The Challenges: Present and Future," has been revised and relocated elsewhere in the text. For example, exobiology has been integrated with the origin of life in Section One; human population and pollution are discussed in Chapter 26, "Human Populations and the Biosphere," in Section Six; and microbes, diseases, and transplantation have been placed in the new Chapter 15 in Section Three. With this reorganization, these important topics are integrated with the discussion of the concepts they represent, thus giving them greater practical application.

Section One, "The Organization, Origin, and Diversity of Life" (Chapters 1–4), is designed to provide the student with an understanding of the relationships of matter and energy to living systems, the basic concepts of chemistry needed to comprehend modern biological principles, the principal molecules of living systems, the hypothetical sequence of events related to the chemical origin and evolution of life, and the phylogenetic relationships and the diversity of organisms.

In Chapter 1, "Matter, Energy, Organization, and Life," the student is introduced to the concepts of matter and energy and how these relate to living systems. The essentials of atomic structure, chemical bonding, and chemical reactions prepare the student for an analysis of molecules of biological importance. In Chapter 2, "The Molecules of Life," a study is made of the chemical constituents of biological systems. At the molecular level, the physical, chemical, and biological properties of the building units of life are examined. The authors feel that such a full, chapter-length treatment not only reinforces previously learned chemical principles but also provides a new learning experience for those who have never taken chemistry. In Chapter 3, "The Origin of Life," the various hypotheses concerning the origin of life and the presumed events associated with chemical origin and evolution are developed. This chapter concludes with a discussion of exobiology and the search for the universality of life processes. Chapter 4, "Diversity of Living Organisms," presents a broad overview of the diversity that is the product of evolutionary development, organized according to the five-kingdom system of classification developed by Dr. R. H. Whittaker of Cornell University. This system, which groups organisms on the basis of levels of structural organization and patterns of nutrition, is in conformity with recent biochemical and evolutionary evidence and has gained wide acceptance. The Appendix presents a detailed analysis of this system, with descriptions and examples.

In Section Two, "The Cellular Level of Biological Organization" (Chapters 5–8), the structural and functional aspects of cells are presented. Initially, a generalized cell is discussed as the basic unit of structure and function. Next, cells are treated as units of energy procurement and transduction. Then the role of cells in energy utilization is presented. Finally, the mechanisms involved in the control of cellular metabolism are examined.

In Chapter 5, "Cells: Basic Units of Structure and Function," a generalized cell is analyzed with emphasis on newly acquired knowledge of cellular ultrastructure as revealed by the electron microscope and recent biochemical data. In this context, illustrations of most cell parts are designed to show the location of a specific part under consideration in a composite cell, detailed ultrastructure of the part by means of an electron micrograph, an interpretation of the micrograph by an adjacent labeled diagram, and the presumed chemical composition of the part. The structure and chemistry of cell parts are related to function. Chapter 6, "Cells and Energy: Procurement and Transduction," examines the flow of energy through cells, that is, the transduction of environmental energy into a biologically usable form. Photosynthesis—the acquisition of energy—and respiration—the transformation of energy—are the two major concerns of this chapter. In Chapter 7, "Cells and Energy: Utilization," an examination is made of some of the ways in which cells expend energy. Among the uses discussed are bioluminescence, bioelectricity, active transport, motion, and biosynthesis. Chapter 8, "The Control of Cellular Metabolism," deals with various mechanisms that regulate chemical activities of cells. After a discussion of DNA and its relationship to enzymes, gene function is studied with regard to the control of enzymatic activity and the control of enzymatic synthesis.

Section Three, "Organismic Biology: Maintenance" (Chapters 9–16), deals with metabolism at the organismic level. Various metabolic activities are considered for representative organisms on a comparative basis and for the human organism. In addition, the student is introduced to some fundamentals of microbiology, disease, and the responses of the human organism to disruption of the body's homeostasis.

In Chapter 9, "Multicellular Patterns of Organization," the transition between cellular biology and organismic biology is facilitated by a presentation of the organization of cells into tissues, organs, and systems in selected organisms. Next are two chapters on the concept of homeostasis; Chapter 10, "Homeostasis: Chemical Control," considers the regulatory mechanisms of organisms that are primarily hormonal in nature, and Chapter 11, "Homeostasis: Nervous Control," examines the nervous control in representative organisms. Further aspects of the biology of organisms are considered in Chapter 12, "Protection, Support, and Locomotion"; Chapter 13, "The Procurement and Digestion of Nutrients"; Chapter 14, "Transportation Mechanisms"; Chapter 15, "Microbes, Disease, and Immunity"; and Chapter 16, "Breathing and Waste Disposal."

Section Four, "Organismic Biology: Continuity" (Chapters 17–21), takes up the propagation of living forms at the organismic level. After a survey of reproductive patterns among selected living forms, developmental processes are considered. This is followed by an analysis of the principles of inheritance from Mendelian to molecular genetics, a study of human heredity, the evolution of life, and finally, population genetics.

Chapter 17, "Patterns of Reproduction," surveys various modes of reproduction among representative groups of organisms. In Chapter 18, "Developmental Processes," embryonic and postembryonic morphogenesis is discussed, with special attention to the factors controlling development. Chapter 19, "Principles of Inheritance," is an analysis of the work of Mendel and its modern interpretation, the chromosome theory of heredity, gene interactions, mutations, and environmental effects on heredity. Chapter 20, "Human Heredity," deals with the application of the principles of heredity to humans. Chapter 21, "The Evolution of Life," presents theories and mechanisms of evolution and discusses population genetics.

Section Five, "Ecosystems" (Chapters 22–26), examines individual organisms at higher levels of organization (populations, communities, and ecosystems). The various forces and processes that determine the existence of living forms within the biosphere are analyzed.

Chapter 22, "The Abiotic Environment," describes the physical world in which organisms live. The physical factors of the lithosphere, hydrosphere, and atmosphere that help to shape the characteristic forms and patterns of living forms are presented. Chapter 23, "Populations Within Ecosystems," discusses the characteristics of populations and their various intraspecific, interspecific, and symbiotic interactions. Chapter 24, "Dynamics of Ecosystems," introduces the various components of ecosystems. There is then a discussion of energy and the biogeochemical cycles. These concepts are brought together in the development of ecological pyramids and factors affecting ecosystems. Chapter 25, "Types of Ecosystems," introduces the student to terrestrial and aquatic ecosystems and their various components. The major biomes and biogeographical regions are then described. Chapter 26, "Human Populations and the Biosphere," brings all the various concepts of ecology together and discusses the effects on human beings through population growth and pollution on the total biosphere.

The study of ecosystems logically leads to Section Six, "Behavior." Behavior is basically adaptive and has evolved in much the same way as other aspects of living organisms. Unlike other areas of biology, however, behavior generally cannot be described by well-defined laws because of the difficulties encountered in its study. In many instances, there are different interpretations of the same set of observations and scientific data.

Chapter 27, "Behavior: Its Nature and Study," discusses three different approaches to the study of behavior—psychological, ethological, and neurophysiological—and various levels of behavior, from the simple taxis and tropism to the highest level of abstract reasoning and thought. Chapter 28, "Behavior as an Individual Adaptation," shows how individual behavior is adapted to specific modes of living in various organisms. In addition, the basic responses of all organisms are studied. Chapter 29, "Social Behavior," deals with behavior within and among populations. These behavior patterns are interpreted from the ethological point of view since this approach emphasizes the behavior of organisms within their natural populations. In Chapter 30, "Human Behavior," human behavior in relation to the total world of living organisms and human social behavior are discussed.

Supplementary learning aids are employed throughout the book. These include a considerable number of electron micrographs with accompanying diagrammatic representation; numerous photographs, line drawings, and tables; a list of suggested readings at the end of each section; and guide questions at the beginning of each chapter instead of at the end. The rationale for these guide questions, instead of review questions at the ends of chapters, is explained in "Note to the Student" on page 2. Finally, an extensive Glossary follows the Appendix.

For this revision, as for the preparation of the first edition of this text, Macmillan Publishing Co., Inc., has made available to us the assistance of five consultants. These consultants, specialists in their respective academic disciplines, have read their individual areas of expertise as well as the complete manuscript and have provided pointed comments, helpful suggestions, and new insights. These five biologists, to whom we express our grateful appreciation, are Guy L. Bush, University of Texas; Patrick J. Docter, Eastern Illinois University; John L. Howland, Bowdoin College; J. Robert McClintic, California State University, Fresno; and Ernst S. Reese, University of Hawaii.

Finally, we wish to express our deepest appreciation to Mr. Woodrow Chapman, Biology Editor, and Mrs. Elisabeth Belfer, Production Supervisor, at Macmillan. Mr. Chapman offered us all the resources of Macmillan to successfully complete the project and personally supervised and became involved in all phases of the project. Mrs. Belfer, who also worked with us on the first edition, brought copyediting consistency and organization to the final manuscript and provided us with many insights.

We would like to invite readers of this textbook to send their reactions and suggestions concerning the book to us at our academic addresses. These responses will be very helpful in formulating plans for subsequent editions.

GERARD J. TORTORA
JOSEPH F. BECKER

Contents

SECTION ONE

The Organization, Origin, and Diversity of Life

1 Matter, Energy, Organization, and Life 3

2 The Molecules of Life 21

3 The Origin of Life 51

4 Diversity of Living Organisms 73

SECTION TWO

The Cellular Level of Biological Organization

5 Cells: Basic Units of Structure and Function 85

6 Cells and Energy: Procurement and Transduction 129

7 Cells and Energy: Utilization 158

8 The Control of Cellular Metabolism 174

SECTION THREE

Organismic Biology: Maintenance

9 Multicellular Patterns of Organization 203

10 Homeostasis: Chemical Control 231

11 Homeostasis: Nervous Control 259

12 Protection, Support, and Locomotion 298

13 The Procurement and Digestion of Nutrients 321

14 Transportation Mechanisms 349

15 Microbes, Disease, and Immunity 373

16 Breathing and Waste Disposal 392

SECTION FOUR

Organismic Biology: Continuity

17 Patterns of Reproduction 419

18 Developmental Processes 454

19 Principles of Inheritance 486

20 Human Heredity 519

21 The Evolution of Life 530

SECTION FIVE

Ecosystems

22 The Abiotic Environment 559

23 Populations Within Ecosystems 582

24 Dynamics of Ecosystems 598

25 Types of Ecosystems 631

26 Human Populations and the Biosphere 650

SECTION SIX

Behavior

27 Behavior: Its Nature and Study 689

28 Behavior as an Individual Adaptation 712

29 Social Behavior 723

30 Human Behavior 737

APPENDIX

The Five-Kingdom System of Classification According to R. H. Whittaker 749

Glossary 773

Index 787

ONE
The Organization, Origin, and Diversity of Life

1 Matter, Energy, Organization, and Life

2 The Molecules of Life

3 The Origin of Life

4 Diversity of Living Organisms

NOTE TO THE STUDENT

The authors have departed from the convention of placing guide questions at the *end* of each chapter; instead, we have decided to place the questions *before* each chapter. The questions are designed to guide your learning efforts by providing you with an indication of the materials that you should understand when you have finished reading each chapter. We hope that you will review the questions before reading each chapter, and after you have completely read the chapter answer the questions to see if you have met your goals.

1 Matter, Energy, Organization, and Life

GUIDE QUESTIONS

1 Differentiate between matter and energy. What are the kinds of energy? How are they related to each other?
2 List and define the forms of energy. Relate each of these to a living and nonliving system. What is an energy transduction? Give several examples in a living system.
3 Relate the first and second laws of thermodynamics to the stability of matter. Describe the structural organization of matter in living systems.
4 List and define the components of an atom. What is an isotope? Explain what is meant by an electronic configuration.
5 Relate valence to chemical bonding. Define and give examples of ionic, covalent, and hydrogen bonding.
6 What is a chemical reaction? Describe five principal types. How does the collision theory explain the mechanism of a chemical reaction? How is activation energy related to chemical reactions? What factors influence the rates of chemical reactions?
7 Discuss the relationships between energy and chemical reactions. Why do some exergonic reactions absorb energy? Why do some endergonic reactions release energy?
8 What is a reversible reaction? Relate this to steady state.
9 Discuss the sequential nature of biochemical reactions. Of what importance is the sequential arrangement of biochemical reactions to a living system?

Matter and Energy Relationships

Matter

ALL biological systems represent an efficiently organized complex of matter. **Matter** may be defined as anything that occupies space and possesses *mass*. Essentially, then, any substance or object within the universe, either living or nonliving, represents one of several forms of matter. Matter may be measured by conventional units of mass, such as milligrams, grams, ounces, pounds, and tons, and by conventional units of volume, such as milliliter, liter, cubic centimeters, fluid ounces, and cubic feet. Atmospheric gases, mineral elements comprising the earth's surface, bodies of water, and all forms of life consist of matter. Matter may exist in the solid, liquid, or gaseous state, and most materials may be converted from one of these states to either of the others by the addition or removal of heat or pressure. Consider the conversion of water from a solid (ice) to a liquid to a gas (steam) under the influence of increasing temperatures. Inasmuch as such a change alters only the physical state of the substance and not the chemical composition, it is termed a *physical change.*

Energy

Energy, by contrast, does not possess mass and therefore does not occupy space. It is not measured in such units as grams, pounds, or quarts but in units such as footpounds, calories, amperes, coulombs, or ergs. **Energy** may be defined as the capacity to do work and by its effects on matter; it is the ability to move a body of matter. The greater the mass of matter that has to be moved, the more energy must be expended.

KINDS OF ENERGY. It is convenient to classify energy into two principal kinds—potential energy and kinetic energy. *Potential energy* is inactive or stored energy. When work is not actually being done, although a system has the capacity to do work, it is said to contain potential energy. A steel ball, for example, placed at the edge of a table contains potential energy (Figure 1-1). If the ball is dropped, the potential energy is capable of causing an effect on matter. Therefore the energy in the system is no longer potential energy; it is now kinetic energy. When work is being done, the energy used is kinetic energy. *Kinetic energy* is possessed by any moving object. As the ball descends to the floor, potential energy is converted to kinetic energy. In order to reestablish the system, the ball must be lifted back to the table top. This requires an expenditure of energy, that is, energy put back into the system equal in amount to the kinetic energy given off when the ball fell to the floor. In the conversion of potential energy to kinetic energy and kinetic energy to potential energy, the sum of the energies remains the same. The kinetic energy, as is true of most forms of energy, is finally degraded to heat energy.

Forms of Energy. In addition to kinds of energy—potential and kinetic—scientists also recognize a number of different forms of energy. The commonest forms of energy that are directly related to biological systems are chemical, mechanical, radiant, and electrical energy. All of these forms may exist as either potential or kinetic energy.

Chemical energy is the energy released or absorbed in the breaking or forming of bonds between atoms. In living systems, this energy can be released for various activities when bonds are broken (as when molecules are broken down), or it can be stored by molecules when bonds are formed (as when molecules are built up). For example, in a process called cellular respiration (Chapter 7), molecules of sugars and fats, which contain large amounts of chemical energy in their atomic configurations, are degraded, and the energy is made more readily available to the cell in the form of energy-rich compounds. In photosynthesis, the process by which chlorophyll-bearing plants synthesize complex organic compounds from simple inorganic compounds, a great deal of chemical energy is built up and stored in the bonds of the sugars. Ultimately, every life process carried on by living organisms is based on the release of chemical energy.

Mechanical energy, as illustrated in Figure 1-1, is energy directly involved in moving matter. Movements

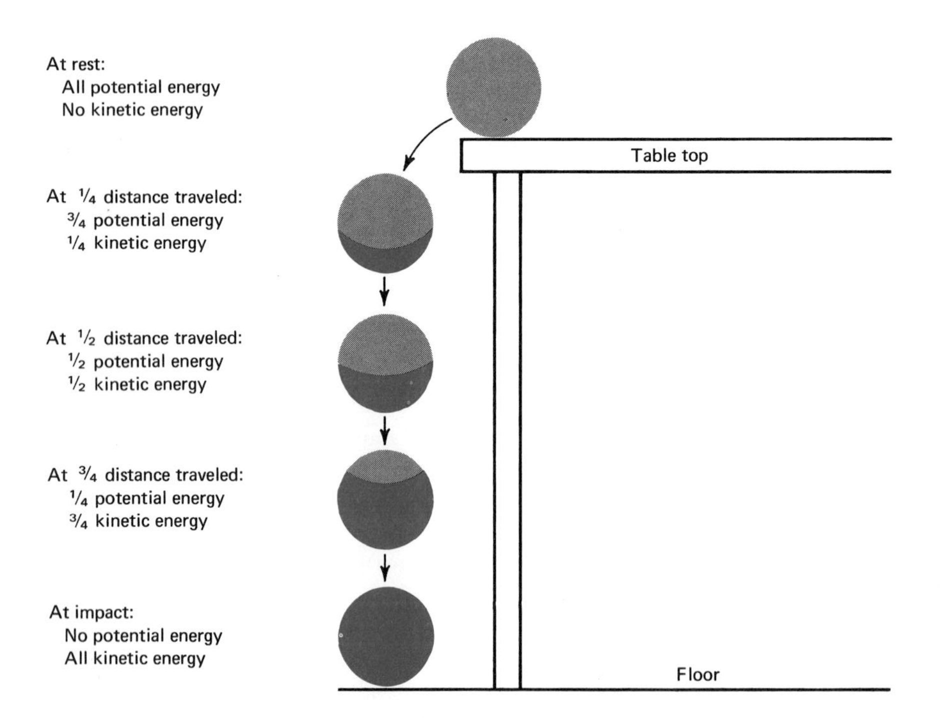

Figure 1-1. Illustration of the transduction of potential energy into kinetic energy. The ball sitting on the edge of the table represents potential energy, or stored energy. If the ball is dropped, however, potential energy becomes kinetic energy, or energy in action. Note that as the ball falls to the floor and potential energy is converted to kinetic energy, the sum of the energies remains the same.

of many organisms, especially those involving muscles and bones, demonstrate the action of mechanical energy. *Radiant energy,* on the other hand, is electromagnetic energy, such as heat and light, which travels in waves. The principal effect of heat energy on living systems is to accelerate the movements of atoms and molecules, an effect that assumes significance in their chemical reactions. Light energy, which will be discussed in Chapter 6, is the primary source of energy for all organisms, since, through the agency of photosynthesis, green (chlorophyllous) plants convert light energy into chemical energy. The final form of energy to be considered, *electrical energy,* is the result of the flow of charges (either electrons or ions) along or through a conductor. Although electrons do not flow through cells as they do in nonliving conductors, they play a prominent role in cellular energy changes. Electrical currents associated with living cells represent ionic flows, not electron flows, except in the electron transport system involved in energy metabolism. The conduction of nerve impulses along a nerve fiber, for example, is the result of ionic flow between the inside and outside of a nerve cell.

Energy Transductions. In the preceding discussion of forms of energy, it was noted that energy **transductions** (conversions of energy from one form to another) took place in living systems (e.g., photosynthesis). Inasmuch as a number of these transductions will be noted in subsequent discussions, only one common example from the nonliving world will be given here. Consider one type of power plant. In such a complex, fuel (chemical energy) is transduced to heat energy by combustion. Some of this energy is further transduced into mechanical energy by steam turbines. This mechanical energy is then transduced into electrical energy by a generator. Finally, by passing an electric current along a metal wire, electrical energy can be transduced to light energy. Thermal energy, formed at all stages, represents inefficiencies in the operating processes since there is a loss of useful energy.

Although energy can be transduced from one form into another, both living and nonliving systems illustrate the *first law of thermodynamics,* which states that energy cannot be created or destroyed but is only changed in form. This law applies to all systems (physical, chemical, and biological) in which energy changes occur. The implication of this law is that the amount of energy present at the beginning of a reaction is also present, perhaps in a different form, at the termination of a reaction. It should be noted that although the energy of a system may change form in the course of a chemical reaction, there are limitations on the type of transduction that different forms of energy may undergo. Some examples of energy transductions that occur in living systems are listed in Table 1-1.

Living matter, like all other matter, is composed of chemical elements. If you were to combine the elements required to make living matter in their exact proportions in a test tube, do you suppose that the mixture would yield a living organism? The answer to this question is obviously no, since the resulting compounds would not be organized into a living complex. Living matter contains certain complex organic compounds that are produced by living things; because organic compounds are generally more complex and highly organized than inorganic ones, they are less stable, that is, organic compounds tend to break down to a more randomly ordered state. This is especially true because there are living systems present which can degrade

Table 1-1. Representative Energy Transductions That Occur in Living Systems

Site of Transduction	Type of Transduction
Nerve cells	Chemical energy ⟶ electrical energy
Inner ear	Sound energy ⟶ electrical energy
Kidney	Chemical energy ⟶ osmotic energy
Retina of eye	Light energy ⟶ chemical energy ⟶ electrical energy
Chloroplast of a photosynthetic cell	Light energy ⟶ chemical energy
Muscle cell	Chemical energy ⟶ mechanical energy
Luminescent organ of a firefly	Chemical energy ⟶ radiant energy
Organs of taste and smell	Mechanical energy ⟶ chemical energy ⟶ electrical energy

them. Logically, then, in order to maintain the organic compounds it produces, a living system must constantly expend energy to rebuild them.

A further analysis of order and energy can be made by considering the *second law of thermodynamics,* which states that the *entropy* (disorder) of the universe is constantly increasing. It says, in effect, that the natural tendency of a system is to break down to the most randomly ordered state—the state that is the least organized. The green plant, in producing complex organic compounds, expends energy. The energy expenditure is necessary in order to create a high degree of organization. Conversely, the breakdown of these compounds by respiration brings about the release of energy. In this regard, the application of the first law of thermodynamics to living systems becomes even more important. To a living system, the first law means that to carry out any action or movement there must be an input of energy. A further interpretation of the second law is also quite basic; even if a living system carries out no observable action or movement, it must have a constant supply of energy just to maintain its high degree of organization. It will be seen in later chapters that living systems derive energy for maintenance from the breakdown of organic compounds and that these substances are formed initially using energy from the sun. One of the principal criteria for a living system, from the standpoint of energy and organization, is that it must perpetuate its organization by chemically processing matter for energy and by manufacturing complex matter from stable raw materials.

The Structural Organization of Matter

The concepts of energy and order can be applied to all levels of organization that exist in the living world. The continuum of life consists of basic levels of biological organization, which range from an interaction of exceedingly small atoms to the entire living world, the biosphere. All these levels of biological organization are dynamic and interacting. Any given level is a part of all higher ones and contains all lower levels as part of itself.

The smallest structural units of matter are **atoms** which are composed of subatomic particles—protons, neutrons, and electrons. Atoms interact in certain combinations to form **molecules,** which may be joined to form more elaborate units called **macromolecules.** Some macromolecules exhibit many properties characteristic of living forms. Aggregations of macromolecules result in structural entities termed **organelles,** specialized components of a cell that perform different life activities. The next higher level of organization, the **cell,** is the smallest structural and functional unit that may exhibit all properties of life.

If one considers the entire living world, other more complex levels of organization become apparent. It is obvious, for example, that not all organisms are single-celled units; in fact, most are not. Cells of multicellular organisms are organized into more complex levels called **tissues.** Tissues are composed of cells, either similar or different types, that usually perform a single function, such as muscle tissue and nervous tissue. In the vast majority of organisms, tissues are intimately associated at a still higher level of organization into structures called **organs.** Organs are composed of structurally and functionally integrated tissues. Common examples of plant organs are roots, seeds, stems, leaves, and flowers; familiar animal organs are the heart, lungs, kidneys, stomach, liver, and brain. The further integration of organs constitutes an even higher level of organization, called **systems.** Systems are groups of functionally correlated organs that typically perform a specialized activity. In most multicellular forms, the various systems are interconnected and constitute the total functioning **organism.**

The organismic level does not represent the highest organizational complex in the world of living things. The next higher grouping is the **population,** a relatively permanent association of organisms of the same kind that occur in a particular habitat. Populations may be local units such as families, tribes, and herds. Familiar examples of local populations are daisies in a field, earthworms in a garden, spruce trees in a forest, frogs in a pond, and people in a city. The sum of all populations of the same kind, and therefore, the sum of all organisms of the same kind, represents a species. Specifically, a species is a group of organisms with certain characteristics in common, which breed freely among themselves but are prevented from breeding indiscriminately with other species by one or several isolating mechanisms.

Populations of several different species associate locally to form a more complex level of organization, the **community.** Communities always contain a variety of organisms. Representative communities would be all forms of life in a pond, in an evergreen forest, and in a city.

Communities possess unique characteristics such as numbers, kinds of species, and patterns of behavior. A community is a dynamic organization and may be replaced by other communities over periods of time.

The **ecosystem** represents the highest level of organization and presents the greatest degree of complexity. It is a broad environmental unit consisting of a community of organisms interacting with each other and with the physical environment through interchanges of chemical nutrients and energy.

It has been shown that changes in the environment can affect more than the level of a single ecosystem or group of ecosystems within a given geographical area. There are certain relationships that more or less affect all forms of life on earth. The term **biosphere** is a collective designation for all forms of life that inhabit the globe.

Although the foregoing discussion has dealt only with levels of biological organization, it should be noted that none of these levels can be separated from the physical environment of the organism.

As one moves up the scale toward higher and higher levels of organization, each level is more complex than the last. In terms of energy and organization, each higher level represents a unit that is more unstable than the preceding level. Each level requires a constant supply of energy to maintain its degree of organization. If a cell, for example, could no longer obtain or utilize energy in order to maintain its organization, it would break down into the molecules of which it is composed and these molecules would eventually break down further to the most randomly ordered state. Without a constant expenditure of energy, each level in the hierarchy would break down into successively less complex units until the least organized state was reached.

Matter and Biological Systems

Elements and Life

All forms of matter that comprise the earth—water, air, soil, rocks, and all living organisms—consist of a limited number of building blocks referred to as **chemical elements.** These fundamental substances of matter are unique in that they cannot be decomposed into simpler substances by ordinary chemical reactions. Presently, there are 106 different chemical elements, of which 92 are naturally occurring and 14 are synthetic, or man-made elements. For consistency and convenience, elements are designated by letter abbreviations called *chemical symbols.* The symbols are usually derived from the first letter or two of the Latin or English name of the element. Table 1-2 contains a listing of some representative chemical elements and their appropriate chemical symbols.

Of the 92 naturally occurring elements, only about one-quarter are found as constituents of living organisms. Various chemical analyses of the living material of the human body indicate that carbon, hydrogen, oxygen, and nitrogen comprise about 96% by weight. These elements, together with calcium and phosphorus, make up approximately 99% of the total body weight. A various assortment of some eighteen other elements

TABLE 1-2. Some Chemical Elements Found in Living Organisms, Their Chemical Symbols, and Respective Atomic Numbers

Element	Chemical Symbol	Atomic Number
Hydrogen	H	1
Boron	B	5
Carbon	C	6
Nitrogen	N	7
Oxygen	O	8
Fluorine	F	9
Sodium	Na	11
Magnesium	Mg	12
Silicon	Si	14
Phosphorus	P	15
Sulfur	S	16
Chlorine	Cl	17
Potassium	K	19
Calcium	Ca	20
Manganese	Mn	25
Iron	Fe	26
Cobalt	Co	27
Copper	Cu	29
Zinc	Zn	30
Molybdenum	Mo	42
Iodine	I	53

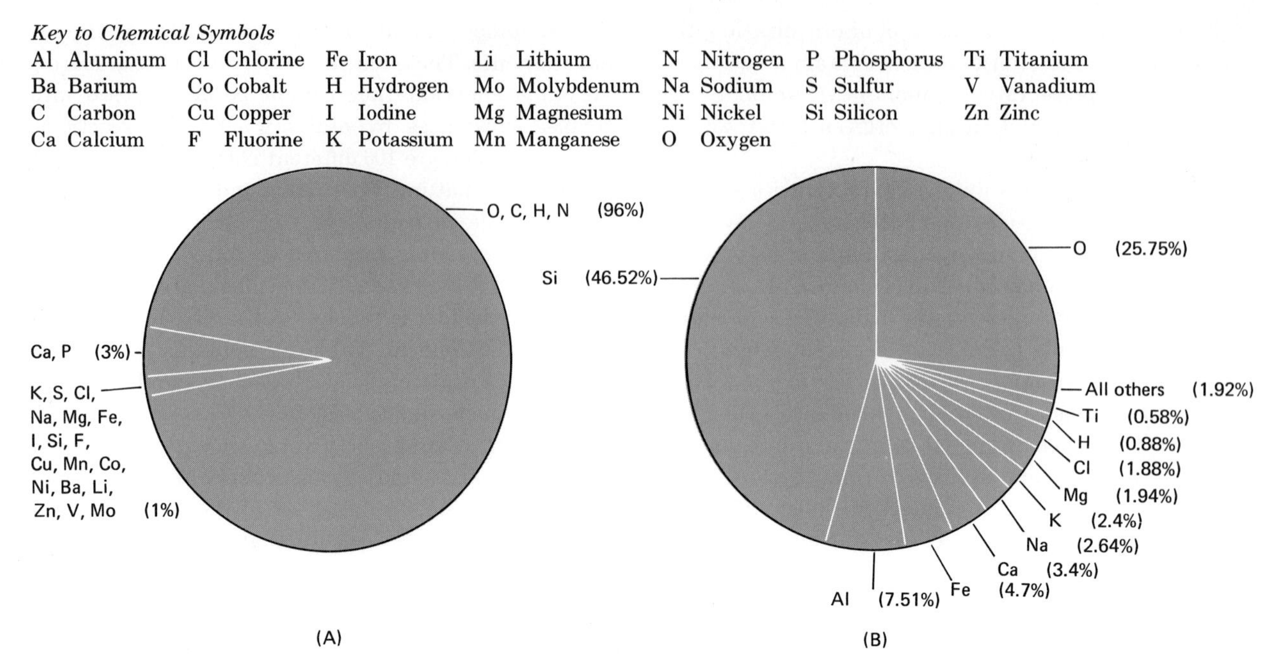

Figure 1-2. Comparison of the elemental composition of humans with that of the earth's crust. (**A**) Percentages of various elements relative to total body weight of the human organism. (**B**) Percentages of chemical elements found in the crust of the earth.

constitutes the remaining 1% of living matter in the human organism (Figure 1-2A). The earth's crust contains elements not found in living systems and, more significantly, has a different proportion of elements than does the human living system (Figure 1-2B). For example, aluminum, found abundantly in the physical environment, assumes no function in living systems; carbon, hydrogen, and nitrogen appear in relatively high proportions in living organisms but are only sparsely represented in the earth's materials. A further consideration of the chemical elements that constitute the hydrosphere (water) and atmosphere (air) of the earth would reveal that all protoplasmic elements can be found in nature. In other words, no single chemical element is characteristic of living things alone.

The Fundamental Particles of Elements: Atoms

Each of the chemical elements found in nature consists of distinct particles called **atoms,** the smallest particles of elements that enter into chemical reactions. In this context, then, a chemical element is a form of matter of which all the atoms contain the same number of protons and electrons. For example, each hydrogen atom contains one proton and one electron, and each oxygen atom contains eight protons and eight electrons. It has been estimated that the largest atoms in nature are less than $\frac{1}{50,000,000}$ of an inch in diameter, while the smallest, those of element hydrogen, are less than $\frac{1}{250,000,000}$ of an inch in diameter. In other words, a line of 50 million atoms placed end to end would be, at most, one inch long.

The Structure of Atoms. It was once believed that atoms of elements were the ultimate, smallest particles of matter. Physicists, however, have demonstrated that atoms can be decomposed into even smaller particles by nuclear reactions and related high-energy processes. It would seem that the concept of the indivisibility of the atom was an incorrect hypothesis. However, at the chemical reaction level, the concept is still applicable.

Studies of atomic structure have revealed that atoms

consist of two basic portions—a centrally located *nucleus* and particles called *electrons* located around the nucleus in possible arrangements often called electronic configurations (Figure 1-3). The nucleus of an atom is fixed, or nonreactive, in that it does not take part in chemical reactions. Within it are contained positively (+) charged particles called *protons* and uncharged particles termed *neutrons.* The nucleus as a unit, therefore, bears a net positive charge. Both neutrons and protons have approximately the same mass (weight), being about 1840 times heavier than electrons. The electrons are in more or less definite regions called *shells* or *energy levels* at varying distances from the nucleus. The charge on electrons is negative (−) and in all atoms the number of protons is equal to the number of electrons. Since the total positive charge on the nucleus equals the total negative charge of the electrons, each atom is an electrically neutral unit. The number of electrons ranges from one, in the case of hydrogen atoms, to more than 100 in some of the largest known atoms such as mendelevium, which has 101 electrons. Atoms of elements are often listed in terms of **atomic number,** that is, the number of positive charges on the nucleus (Table 1-2).

An element may consist of several atomic species differing in mass because of a difference in the number of neutrons in the nucleus although the number of protons is the same. Atomic species that have the same atomic number and different atomic mass are called **isotopes,** and an element is the total of all its isotopes. For example, in a given natural sample of oxygen, all atoms will contain 8 protons. However, 99.759% of the oxygen atoms will have 8 neutrons, 0.037% will contain 9 neutrons, and the remaining 0.204% will contain 10 neutrons. Therefore, the three isotopes comprising a natural sample of oxygen will have atomic masses of 16, 17, and 18, respectively, and all will have the atomic number of 8. These may be designated as $^{16}_{8}O$, $^{17}_{8}O$, and $^{18}_{8}O$.

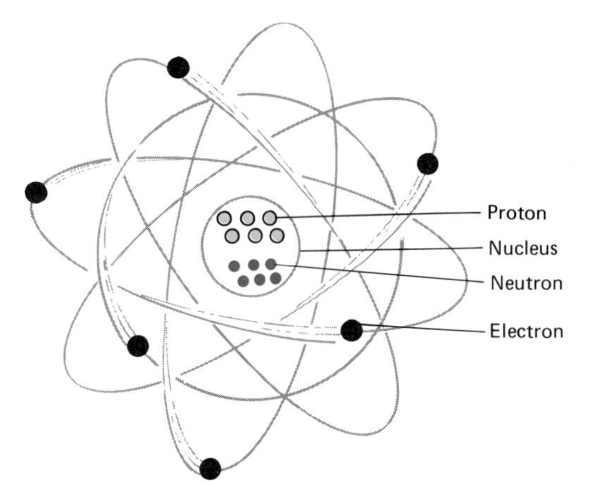

FIGURE 1-3. Diagrammatic representation of atomic structure. An atom of an element consists of a relatively heavy, compact, centrally located nucleus and lighter particles called electrons located in configurations at varying distances from the nucleus.

ELECTRONIC CONFIGURATIONS. For purposes of simplicity, somewhat at the expense of accuracy, it will be assumed that electrons occupy a definite orbital at a fixed distance from the nucleus and these orbitals form shells or energy levels. It will be further assumed, in the same context, that the innermost shell, or *first energy level,* designated as the *K shell,* can accommodate a maximum of two electrons; the next shell, the *L shell,* or *second energy level,* a maximum of eight electrons; and the third shell, the *M shell,* or *third energy level,* a maximum of eight electrons if there are no more energy levels within a given atom. The *N, O,* and *P* shells or fourth, fifth, and sixth energy levels, respectively, can each accommodate eighteen electrons, although some variations exist with respect to this generalization. In terms of the positions of these shells relative to the nucleus, the *K* shell is closest and the *P* shell is farthest away. The distribution, or arrangement of the electrons of atoms in various shells, is referred to as an **electronic configuration.** Table 1-3 shows the electronic configurations for the atoms of a few elements found in living organisms.

The various chemical properties of atoms are largely a function of the number of electrons in the outermost shell. Chemical reactions take place in such a manner as to produce the most stable electronic configurations in the outermost shells of the reacting atoms. When this shell is completely filled in terms of the maximum number of electrons that it can hold, the atom does not generally react with other atoms. Helium (atomic number 2) and neon (atomic number 10) are examples of atoms that have filled outer shells. Helium has two electrons in the *K* shell, the outer shell for helium, and

TABLE 1-3. **Electronic Configurations for the Atoms of Some Elements Found in Living Systems**

Element	Electron Distribution			Diagrammatic Representation of Electron Distribution	Number of Outer Shell Electrons
	K Shell	*L* Shell	*M* Shell		
Hydrogen	1				1
Carbon	2	4		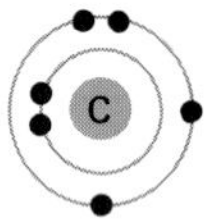	4
Nitrogen	2	5		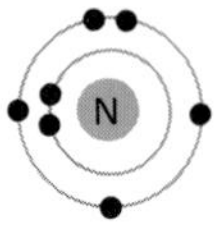	5
Oxygen	2	6		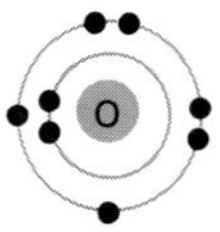	6
Magnesium	2	8	2	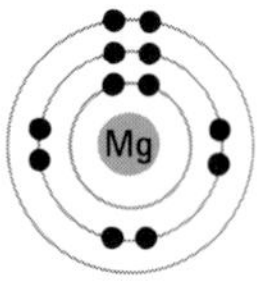	2
Phosphorus	2	8	5	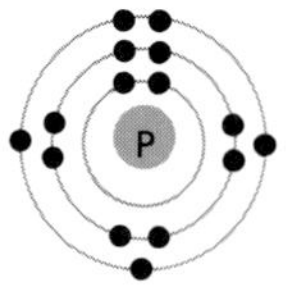	5
Sulfur	2	8	6	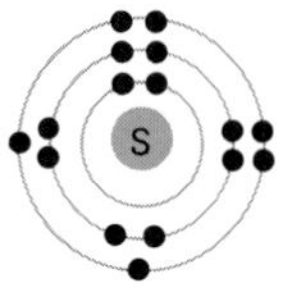	6

neon has two electrons in the *K* shell and eight electrons in the *L* shell, the outer shell for neon.

Most atoms contain outer shells that are only partially filled. As a result, they have unstable electronic configurations and they will tend to react with each other in order to attain stability. The degree of reactivity is dependent, at least in part, on the degree to which the outer shells are filled. Note the number of electrons in the outer shells for the atoms represented in Table 1-3.

Molecule Formation

VALENCE AND CHEMICAL BONDING. When the outermost shell of an atom is not completely filled, it may be visualized as having either unfilled spaces or as having extra electrons in that shell. For example, an atom of oxygen (*K*-2, *L*-6) has two unfilled spaces in the *L* shell, while an atom of potassium (*K*-2, *L*-8, *M*-8, *N*-1) has one extra electron in the *N* shell. For these two atoms to attain the most stable electronic configuration, oxygen has to gain two electrons and potassium has to lose one electron. Atoms of elements combine so that the extra electrons in the outermost shell of some atoms can fill the spaces of the outermost shell of other atoms. The *valence,* or combining capacity, of an element is a numerical expression which represents the number of extra electrons or deficient electron spaces in the outermost shell. For some elements the valence is a fixed value, while for others it may vary.

The atoms of most elements are capable of chemically combining with each other to form aggregates called **molecules.** In the process of combination the valence electrons in the outer shells of the combining atoms interact with each other in such a way that attractive forces are set up between the atomic nuclei involved. These attractive forces, which act to bind the nuclei together, are referred to as **chemical bonds.** Energy is required for bond formation. As a result, each chemical bond possesses a certain amount of potential chemical energy. Atoms of elements form only a specific number of chemical bonds and, in this regard, valence may also be viewed as the bonding capacity of elements. For example, hydrogen has a valence of 1, oxygen has a valence of 2, and carbon has a valence of 4. Essentially, this means that hydrogen can form one chemical bond, oxygen can form two, and carbon can form four chemical bonds with various atoms.

As stated previously, atoms combine chemically to form molecules. Basically, atoms form molecules and gain stability by completing the full complement of electrons in the outermost shells. For reasons far beyond the scope of this textbook, most, but not all, such molecules thus formed are stable. In general, atoms may form molecules in two ways: (1) by gaining or losing outer shell electrons or (2) by sharing outer shell electrons. In both cases attractive forces are set up and chemical bonds are formed. When atoms gain or lose outer shell electrons, the resulting chemical bond is referred to as an ionic or electrovalent bond. By contrast, when outer electrons are shared, the bond formed is called a covalent bond. It should be noted that although ionic and covalent bonds will be discussed separately, the kinds of bonds actually found in molecules do not fall clearly into ionic or covalent categories. Different molecules range from being highly ionic to highly covalent.

IONIC BONDING. In order to understand the concept of ionic bonding, consider first an atom of sodium. It has an atomic number of 11 and a *K*-2, *L*-8, *M*-1 electronic configuration. In order for sodium to attain the stable electronic configuration, *K*-2, *L*-8, the single electron in the *M* shell is passed on to an atom that would serve as an electron acceptor. Once it has lost the electron, the sodium atom is left with one more proton than electrons, and therefore has a net charge of $+1$ (Figure 1-4A). Any such charged atom, or groups of atoms, is called an *ion.*

Now consider an atom of chlorine. It has an atomic number of 17 and a *K*-2, *L*-8, *M*-7 electronic configuration. In order for it to attain the stable electronic distribution *K*-2, *L*-8, *M*-8, the chlorine atom accepts a single electron from an electron donor atom. In the process, as in the case of sodium, an ion is formed (Figure 1-4B).

The nucleus of the positively charged sodium ion and the nucleus of the negatively charged chloride ion tend to attract each other. This phenomenon may be explained on the basis that the unlike charges of the ions attract each other. The resulting association, NaCl, or common table salt, is thus formed by the transfer of an electron from sodium to chlorine and the ions are held together by an electron affinity or attraction called an **ionic bond** (Figure 1-4C). Ionic compounds, such as sodium chloride, play an important role in the living economy and, as will be seen, many of the substances entering living systems are ions.

COVALENT BONDING. Within organisms, most atoms are bonded together to form molecules by sharing valence electrons. Once again, the most stable configuration is one in which the outer shell contains the maximum number of electrons. This type of bonding, in which pairs of outer electrons of atoms are shared, is

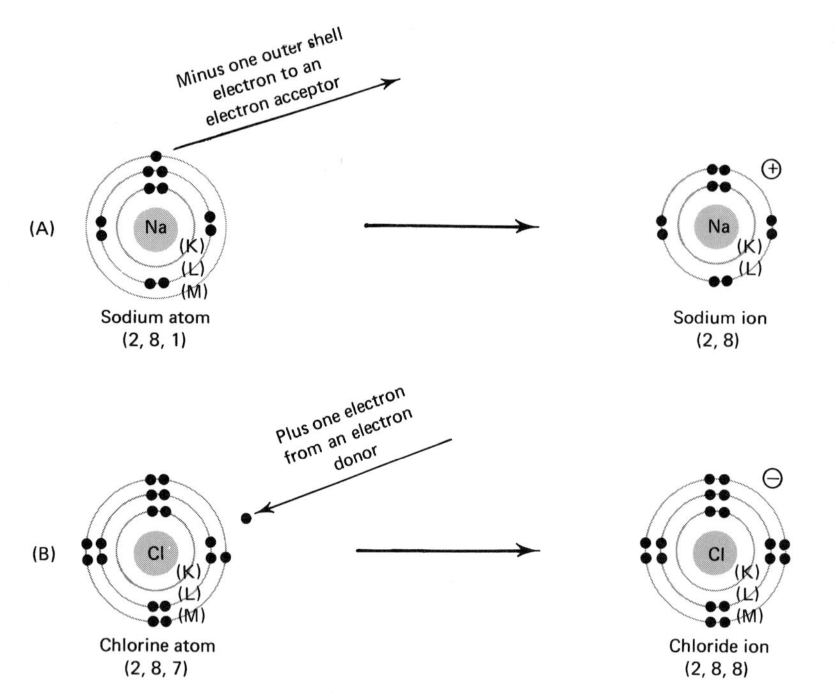

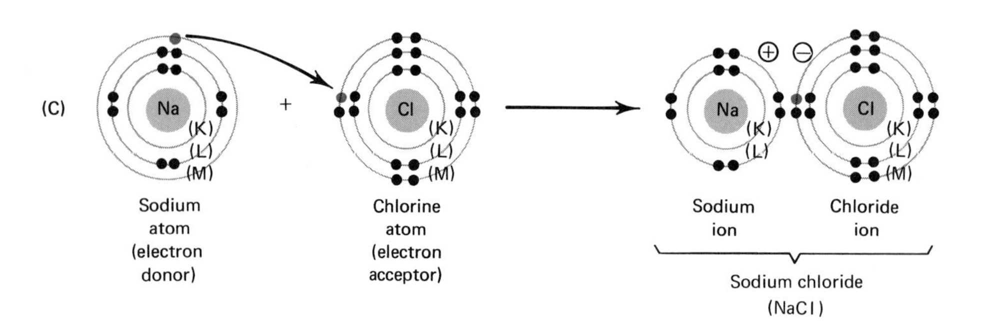

Figure 1-4. Ion formation and ionic bonding. (**A**) Sodium attains a stable configuration by acting as an electron donor, passing a single outer shell electron to an electron acceptor. (**B**) Chlorine becomes stable by acting as an electron acceptor. (**C**) When a sodium ion and a chloride ion combine, they are held together by an electrostatic attraction called an ionic bond. The association of ions formed from this attraction is sodium chloride (NaCl). The transferred electron in (**C**) is shown in color.

termed covalent bonding, and the bond shared between such atoms is called a **covalent bond.** The shared pair of electrons involves one electron from each of the bonded atoms. Covalent bonds may be formed by atoms of the same element or by atoms of different elements. For example, two hydrogen atoms may be covalently bonded to form a molecule of hydrogen (H_2) in which each atom shares its outer electron with the other atom so that both have two electrons in the outer shell (Figure 1-5A).

In addition to the single covalent bond formed between one pair of electrons, atoms may also form double and triple covalent bonds. For example, when two atoms of oxygen bond together to form an oxygen molecule (O_2), they form a double bond, that is, a sharing of two pairs of electrons, since each oxygen atom needs two electrons to complete its outer shell (Figure 1-5B). Similarly, when two nitrogen atoms bond together to form a molecule of nitrogen (N_2), they form a triple covalent bond because each atom needs three electrons in order to fill its outer shell (Figure 1-5C).

In the discussion of covalent bonds so far, single,

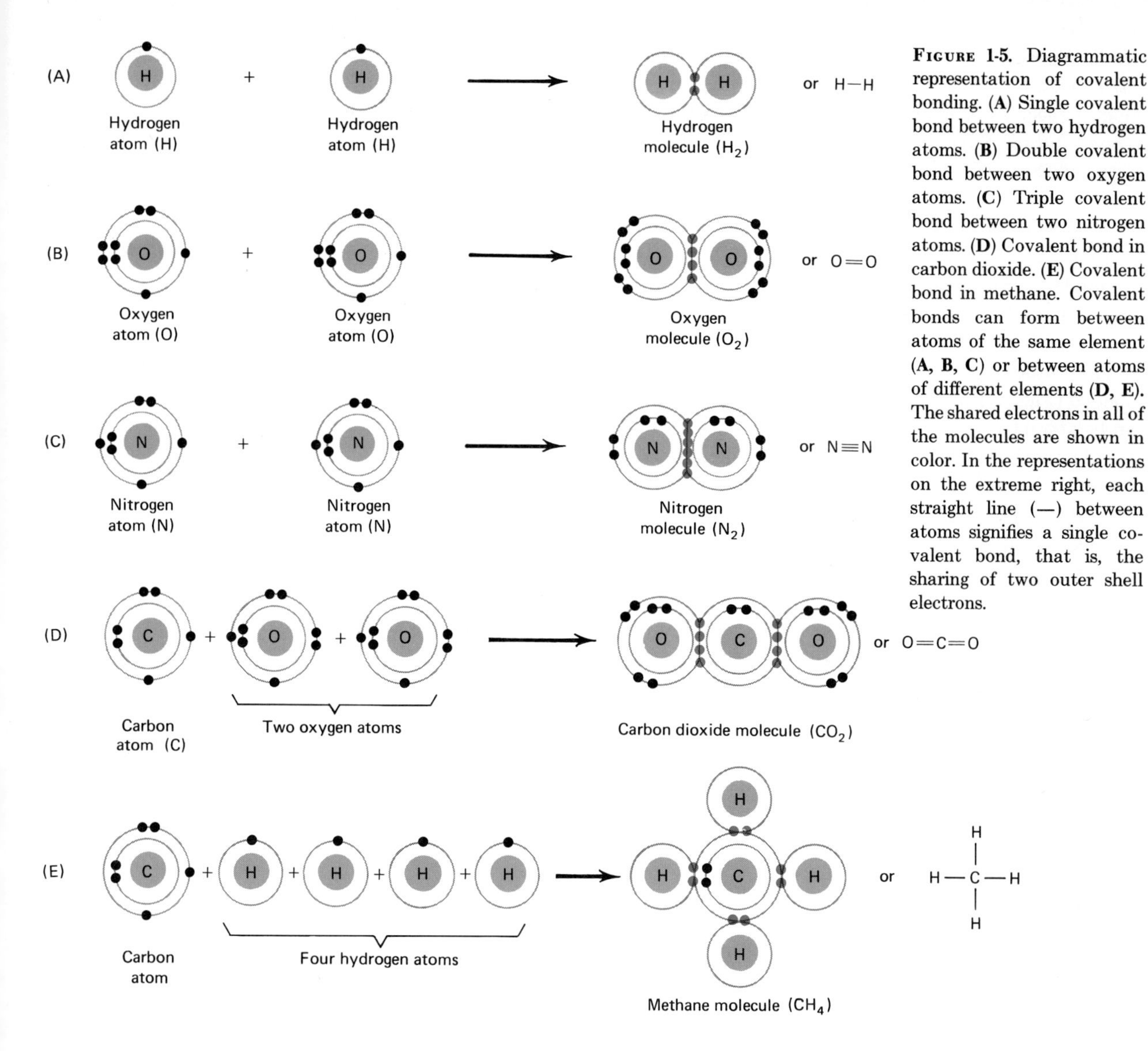

Figure 1-5. Diagrammatic representation of covalent bonding. (**A**) Single covalent bond between two hydrogen atoms. (**B**) Double covalent bond between two oxygen atoms. (**C**) Triple covalent bond between two nitrogen atoms. (**D**) Covalent bond in carbon dioxide. (**E**) Covalent bond in methane. Covalent bonds can form between atoms of the same element (**A, B, C**) or between atoms of different elements (**D, E**). The shared electrons in all of the molecules are shown in color. In the representations on the extreme right, each straight line (—) between atoms signifies a single covalent bond, that is, the sharing of two outer shell electrons.

double, and triple covalent bonds have been shown between atoms of the same element. Atoms of different elements also form covalent bonds. Consider the nature of covalent bonds formed between atoms in molecules of carbon dioxide (CO_2) and methane (CH_4). In the formation of carbon dioxide (Figure 1-5D), note that the carbon atom forms a double covalent bond with each oxygen atom. Methane (Figure 1-5E) represents a molecule in which the carbon atom forms a single covalent bond with each hydrogen atom.

In carbon dioxide and methane, as well as most other covalently bonded substances, the individual atoms forming the molecule have stable electronic configurations. From these examples it becomes apparent that the number of covalent bonds formed by hydrogen is one, by carbon is four, and by oxygen is two. In a similar manner, sulfur may form two covalent bonds, nitrogen three, and phosphorus three. Essentially, it is the numbers and types of atoms involved in molecule formation that determine whether a single, double, or triple covalent bond, or any combination of these, may be formed.

Hydrogen Bonding. A third kind of chemical bond of special importance to all organisms is the **hydrogen bond.** Such bonds do not bind atoms into molecules, but rather serve as bridges between different molecules or between various portions of the same molecule. When hydrogen combines with certain atoms, such as oxygen and nitrogen, the relatively large positive nuclei of these atoms attract the electron of hydrogen. As a result of the single electron located closer to the atom with which hydrogen is bonding, the nucleus (single proton) of the hydrogen atom is relatively unshielded (Figure 1-6A). This positively charged nucleus can be attracted to an unshared pair of electrons of an atom in another molecule. Since the hydrogen remains covalently bonded to the original molecule, this attraction draws the two molecules together, forming the hydrogen bond (Figure 1-6B). If the molecule containing hydrogen is relatively large, the attraction may occur between the hydrogen and other atoms within the same molecule. Inasmuch as nitrogen and oxygen have unshared pairs of electrons, they are most frequently involved in hydrogen bonding.

Hydrogen bonds are considerably weaker than either ionic or covalent bonds; they are only about 5% as strong. As a consequence, they may be formed and broken relatively easily. This property is of special significance as it accounts for the temporary bonding between certain atoms of large complex molecules such as proteins and nucleic acids (Chapter 2). Even though hydrogen bonds are relatively weak bonds, some protein molecules may contain several hundred of them and this results in considerable strength and stability.

Chemical Reactions: Interaction of Matter

The term **chemical reaction** may be viewed as any process in which some chemical bonds are broken and others are formed. For purposes of this textbook, the five principal types of reactions are:

1. Rearrangement reactions.
2. Synthetic reactions.
3. Decomposition reactions.
4. Displacement reactions.
5. Double displacement reactions.

Rearrangement Reactions. A *rearrangement reaction* involves a reorientation of atoms within

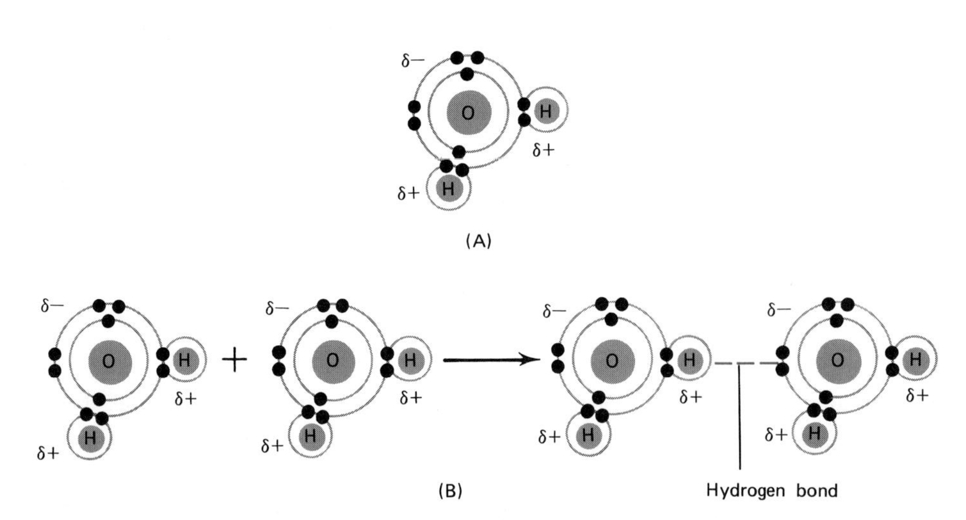

Figure 1-6. Hydrogen bonding. (**A**) In a water molecule, the electrons of the hydrogen atoms are attracted to the oxygen atom. That portion of the water molecule containing the hydrogen atoms is partially positively charged and is symbolized $\delta+$. The portion of the water molecule containing oxygen is partially negatively charged and is symbolized $\delta-$. (**B**) Hydrogen bonding between water molecules. The hydrogen of one water molecule is attracted to a pair of unshared electrons on the oxygen atom of another water molecule by way of a hydrogen bond (- - -). In a similar manner, many water molecules may be bonded to each other by way of hydrogen bonds.

a molecule; the number and kinds of atoms remain the same. Such rearrangements are very common in living systems, especially in chemical reactions that break down food molecules. For example, in the first series of chemical reactions (glycolysis) in which sugars are prepared for later breakdown to produce energy, there is a rearrangement of atoms in a phosphorylated glucose molecule. The product of this reaction is another sugar molecule, phosphorylated fructose. This reaction may be represented as follows.

$$\underset{\text{glucose-6-phosphate}}{\begin{array}{c} \mathrm{H} \\ | \\ \mathrm{C{=}O} \\ | \\ \mathrm{H{-}C{-}OH} \\ | \\ \mathrm{HO{-}C{-}H} \\ | \\ \mathrm{H{-}C{-}OH} \\ | \\ \mathrm{H{-}C{-}OH} \\ | \\ \mathrm{H{-}C{-}O{-}P} \\ | \\ \mathrm{H} \end{array}} \rightleftharpoons \underset{\text{fructose-6-phosphate}}{\begin{array}{c} \mathrm{H} \\ | \\ \mathrm{H{-}C{-}OH} \\ | \\ \mathrm{C{=}O} \\ | \\ \mathrm{HO{-}C{-}H} \\ | \\ \mathrm{H{-}C{-}OH} \\ | \\ \mathrm{H{-}C{-}OH} \\ | \\ \mathrm{H{-}C{-}O{-}P} \\ | \\ \mathrm{H} \end{array}}$$

At this point emphasis should not be placed on how, when, or why such a reaction occurs. The important aspect to note is that atoms within a molecule are rearranged; a molecule of phosphorylated glucose is converted to a molecule of phosphorylated fructose. Count the number of carbon, hydrogen, and oxygen atoms in each molecule. Are they the same? This type of reaction may be represented symbolically as

$$A \rightleftharpoons B$$

(The double arrows mean that the reaction is reversible.)

Synthetic Reactions. A *synthetic reaction* involves the interaction of two substances to form a third. In living systems, many reactions occur in which larger molecules are synthesized from smaller ones. A typical synthetic reaction may be represented as

$$\underset{\text{hydrogen}}{2\,H_2} + \underset{\text{oxygen}}{O_2} \longrightarrow \underset{\text{water}}{2\,H_2O}$$

This reaction type is symbolized

$$A + B \longrightarrow AB$$

Decomposition Reactions. A *decomposition reaction* involves the breakdown of one molecule into two or more smaller ones, the opposite of synthesis. Decomposition reactions are vital to living organisms in such processes as digestion and respiration. Consider the following.

$$\underset{\text{water}}{2\,H_2O} \longrightarrow \underset{\text{hydrogen}}{2\,H_2} + \underset{\text{oxygen}}{O_2}$$

$$AB \longrightarrow A + B$$

Displacement Reactions. A *displacement,* or *exchange, reaction* involves the replacement of an atom or group of atoms in one substance by an atom or group of atoms from another. For example,

$$A + BC \longrightarrow AC + B$$

Displacement reactions are essential for the normal functioning of living systems in which various substances are exchanged between body fluids and tissues. Hemoglobin, an iron-containing molecule, picks up CO_2 from cells and tissues of the body and transports it via the blood to the lungs. In the lungs, inhaled O_2 replaces the CO_2 and the CO_2 is exhaled. This displacement reaction may be represented as

$$\underset{\substack{\text{carbamino-}\\\text{hemoglobin}}}{HbCO_2} + \underset{\text{oxygen}}{O_2} \longrightarrow \underset{\substack{\text{hemoglobin–}\\\text{oxygen complex}}}{HbO_2} + \underset{\substack{\text{carbon}\\\text{dioxide}}}{CO_2}$$

Double Displacement Reactions. Finally, a *double displacement reaction* is a mutual exchange of atoms between molecules. An example is

$$\underset{\substack{\text{silver}\\\text{nitrate}}}{AgNO_3} + \underset{\substack{\text{hydrochloric}\\\text{acid}}}{HCl} \longrightarrow \underset{\substack{\text{silver}\\\text{chloride}}}{AgCl} + \underset{\substack{\text{nitric}\\\text{acid}}}{HNO_3}$$

$$AB + CD \longrightarrow AD + CB$$

Neutralization reactions, one of the ways by which the acid-base environment of an organism is maintained, are biologically important double displacement reactions.

How Chemical Reactions Occur

There is no totally satisfactory explanation of the phenomenon by which atoms and molecules interact to produce chemical reactions. However, the **collision theory** does offer a mechanism for explaining how reactions might occur, and it also helps to explain the involvement of certain factors that affect the rates of chemical reactions. The basic assumption of the collision theory is that all atoms, ions, and molecules are constantly moving. It is further assumed that in order for reaction to take place, the substances involved must make contact (collide) with each other. Among the factors that determine whether collisions will cause a chemical reaction are the velocity of the colliding particles, the energy of the colliding substances, and the specific chemical configurations of the reacting substances. The more rapidly the particles travel, the greater the probability that they will collide. Particles of various elements and compounds also have specific minimum energy requirements that must be met before a reaction will occur. Even if the particles collide and they possess the minimum energy for reaction, no reaction will take place unless the colliding particles are properly oriented (joined at their surfaces). In living systems there are specific substances, the enzymes, that hold colliding substances in definite positions in order to facilitate chemical reactions.

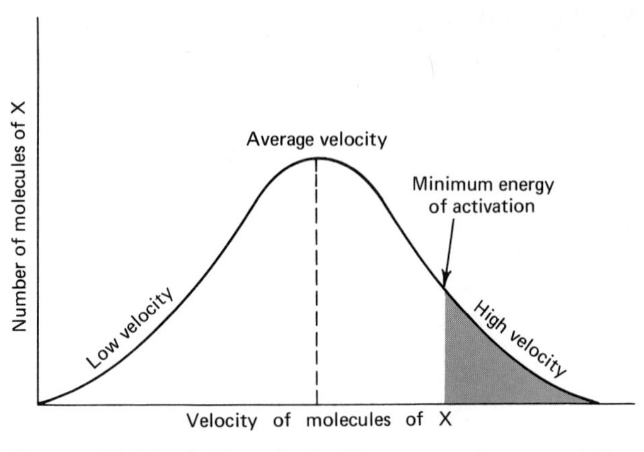

Figure 1-7. Distribution of velocities among a large population of molecules. The bell-shaped curve indicates that some molecules of X have a low velocity, some have a very high velocity, and a major fraction of the population possesses average velocity. In order for a chemical reaction to occur, molecules of X must have sufficient activation energy. That portion with the minimum energy of activation is represented by the colored area. Only this distribution of molecules of X is capable of being converted to molecules of Y.

Assume that molecules of substance X, the reactant, are to be converted to molecules of substance Y, the product. In a given population of molecules of substance X, at a specific temperature, there will be some that possess relatively little energy; a large number of molecules, indeed the major fraction of the population, will possess average energy; and a small fraction of the population will have quite a high energy. Only the energy-rich X molecules are able to react and be converted to Y molecules. Therefore, only a relatively few molecules at any one time possess the reaction level of energy. The minimum kinetic energy of collision required for chemical reaction is termed **activation energy** (Figure 1-7). The reaction rate (the frequency of collisions of sufficient energy) depends upon the relative number of reacting molecules at the activation energy level. One way to increase the reaction rate is to raise the temperature of the system. This addition of heat energy causes the molecules to move faster, thereby increasing the frequency of collisions and also increasing the number of molecules at the necessary activation energy (Figure 1-8A). Heat is a nonenzymatic agent of acceleration. Other nonenzymatic agents of acceleration are high pressure and high concentrations of reactants. Increases in pressure and concentrations of reactants reduce the distance between molecules and thereby increase the number of collisions between reacting molecules. Logically, if temperature, pressure, and reactant concentrations are reduced, the reaction will proceed at a slower rate.

If, on the other hand, the appropriate enzyme is present in the same reaction mixture, the difference is significant (Figure 1-8B). The activation energy of the reaction is lowered because the enzyme joins (forms an enzyme–substrate complex) with molecules that have a broader range of energies, not just those that are energy-rich. The enzyme–substrate complex enables the collisions to be more effective by lowering the activation energy required for the reaction. In this way many more molecules of X can participate in the reaction than if the enzyme were absent. The presence of an enzyme, therefore, speeds up the reaction since more of the X molecules attain sufficient activation energy.

An **enzyme** is capable of accelerating a reaction without an increase in temperature. This is of considerable importance to living systems, since all enzymes are

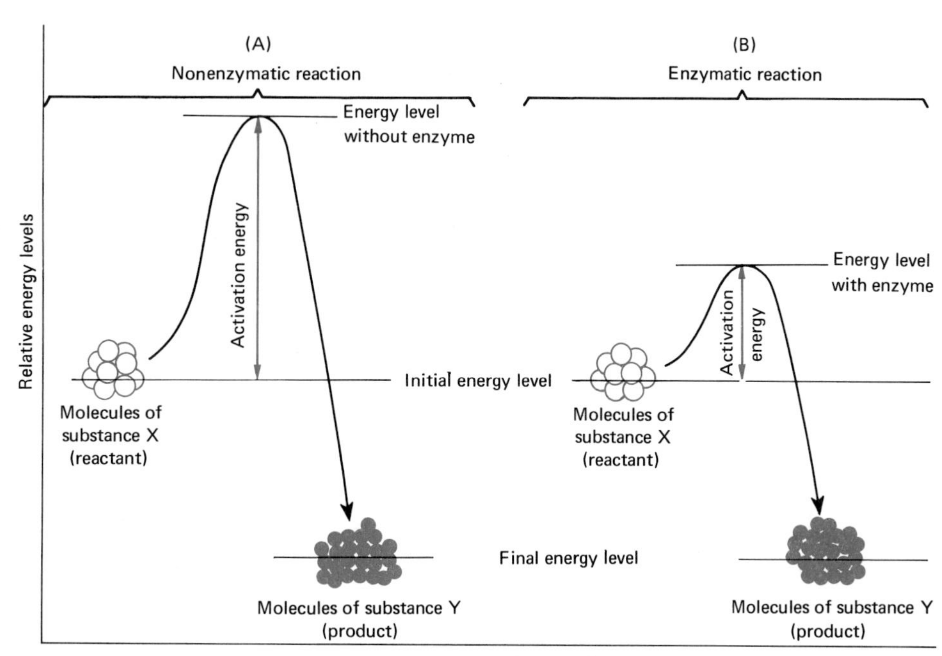

Figure 1-8. Energy requirements of a chemical reaction with and without an enzyme. Before the reaction can occur, molecules of X must collide with one another with sufficient energy (activation energy) to form molecules of Y. In reaction (**A**), the frequency of collisions may be increased by raising the temperature. In reaction (**B**), an enzyme-catalyzed reaction, the presence of the enzyme lowers the required activation energy. In this way more molecules of X are converted to Y (at a lower temperature) since a larger fraction of molecules of substance X possess the required activation energy for reaction. [*Modified from Thomas P. Bennett and Earl Frieden,* Modern Topics in Biochemistry. *New York: Macmillan Publishing Co., Inc. Copyright © 1966.*]

protein in composition and proteins are susceptible to breakdown by heat. Consequently, a significant temperature increase would destroy the enzymes as well as other cellular proteins. The essential function of enzymes, therefore, is to speed up biochemical reactions at a temperature that is compatible with the normal functioning of the cell.

Factors Influencing Chemical Reactions

The rate of a given chemical reaction can be determined by measuring the amount of product that is produced, or the amount of reactant that remains, after a given period of time. For example, consider the reaction

$$\underset{\text{reactants}}{A + B} \longrightarrow \underset{\text{products}}{C + D}$$

If the reactants A and B are mixed at the beginning of the reaction, the rate of reaction can be determined by measuring how quickly A and B disappear and are converted into products. As time passes, the amounts of A and B decrease since they are being converted to products. Simultaneously, the amounts of products C and D increase, and the rate of their increase is directly dependent upon the rate at which the reactants interact and are converted to products.

One of the most important influences on the rate of chemical reactions is that of temperature. Within limits, as there is an increase in temperature, there is an increase in reaction rate. Temperature increases the velocity of interacting particles and thus increases the number of collisions between particles per unit of time and increases the kinetic energy of the particles so that they can more readily approach the activation state. Temperature decreases produce opposite effects. Pressure increases have basically the same influences as temperature increases.

Catalysts are substances that affect the rate of a chemical reaction. Catalytic agents, such as enzymes, have profound influences on the rate of chemical reactions. They increase reaction rates by lowering the activation energy requirements of interacting molecules, thus allowing the reaction to take place with less elevation of temperature or, in some instances, without any change in temperature. Since catalysts are recovered after the reaction, they may be used again to accelerate a similar reaction.

Energy Exchanges and Chemical Reactions

Biochemical reactions involve exchanges of energy that are of great importance in maintaining biological organization and continuity of the life processes. Through various chemical reactions cells obtain energy to carry on their activities.

Three concepts from thermodynamics are useful in discussing energy exchanges and their relations to biological activity. The change in *free energy,* ΔG, is a measure of the energy available for doing useful work. The change in *enthalpy,* ΔH, is a measure of the total energy change in a system during reaction. The change in *entropy,* ΔS, is a measure of changes in the degree of order of a system. An increase in entropy indicates that a system has changed to a more random, more probable, and more stable condition—a situation that means that there is a loss of available useful energy.

At constant temperature and pressure, these three characteristics of a reacting system are related by

$$\Delta G = \Delta H - T\Delta S$$

where T is the absolute temperature ($C^\circ + 273^\circ$) of the system. Thus the equation shows that the free energy available for doing useful work is equal to the total energy change of the system minus any energy lost as reacting molecules achieve a more stable state. It should be noted that most of the energy lost is in the form of heat energy.

When heat is given off during a reaction, the reaction is *exothermic;* when heat is absorbed, the reaction is *endothermic.* Similarly, when a reaction proceeds with the release of free energy, the reaction is **exergonic;** a reaction that absorbs free energy is **endergonic.**

Catabolic reactions, that is, those reactions in which molecules are broken down, generally are exergonic reactions and the energy made available by these reactions is utilized by the cell in performing work. For example, although it is an oversimplification, the process of respiration, the breakdown of larger molecules to smaller ones, is an exergonic reaction from which the cell obtains useful energy.

Anabolic (synthetic) reactions, that is, those reactions in which molecules are synthesized, are generally endergonic and require energy from another source. Again, although an oversimplification, the process of photosynthesis, by which plants synthesize complex molecules from simple precursors, is an endergonic reaction that proceeds only by using the energy of sunlight.

Generally speaking, the buildup of molecules is an energy-absorbing process. According to the first law of thermodynamics, energy cannot be created or destroyed in a given system, and thus such synthetic reactions require an outside source of free energy.

Many chemical reactions within living organisms proceed on the basis of a coupling of endergonic and exergonic reactions. The energy released by the exergonic reaction is used to drive an endergonic reaction. Figure 1-9 shows the interrelationships of photosynthe-

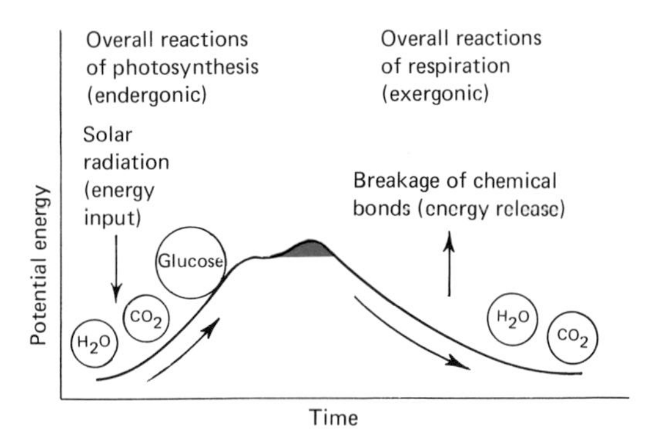

FIGURE 1-9. Comparison of endergonic and exergonic reactions. The initial reactants for photosynthesis are CO_2 and H_2O. Through a series of intermediate reactions, utilizing the energy of sunlight, food molecules (glucose) are produced. Since the overall photosynthetic reactions require energy from sunlight, they are endergonic. Energy is put into the system. Respiration, by contrast, involves a series of reactions in which food molecules (usually glucose) are broken down. In the process, energy is released as chemical bonds are broken. The products of the reaction are CO_2 and H_2O. Inasmuch as energy is released from the overall reactions, respiration is exergonic. The colored area of the curve represents the amount of activation energy necessary to initiate the process of respiration.

sis (endergonic) and respiration (exergonic). Cells trap the free energy resulting from the breakdown of molecules and store it in molecules of ATP which can then pass this energy on to other reaction systems that are endergonic.

Before leaving the topic of energy exchanges and chemical reactions, it is important to emphasize the concept of entropy. A majority of reactions proceed in such a manner that the distribution and arrangement of molecules become more random or less ordered. Molecules tend to reach lower free energy levels and a more stable condition. This is represented by an increase in their entropy content and a loss of available energy for doing work.

A complex macromolecule such as a protein has a lower entropy content than a solution of its constituent amino acids. When a protein molecule breaks down, part of the chemical energy of the protein is converted into vibrational energies of the resulting amino acids and this energy is lost insofar as doing useful work is concerned. The entropy content of the amino acids is higher than the entropy content of the complex highly organized protein molecule.

Cells exist only because they directly or indirectly utilize a continuous inflow of energy from the sun. During the time of their existence, they lower the entropy of the system and maintain a highly ordered state. However, overall, all reactions tend to an increase in entropy—an idea which leads to the concept of the heat death of the universe. Briefly, this states that at some far distant time all energy will have changed into a form of entropy, with a loss of all useful energy and a resulting inability of any system to do work.

Chemical Equilibria and Steady States

A *reversible chemical reaction* is one in which the products can react to form the original reactants; that is, the reaction can go in either direction. All reactions of this type continue as long as reactants and products are present in the system. In a closed system, one in which there is no inflow or outflow of material, a reversible reaction will reach a point at which the rate of the forward reaction is equal to the rate of the reverse reaction. This is the point of *dynamic equilibrium.* The nature of a reversible reaction and its ultimate equilibrium can be illustrated by considering the reaction

$$A + B \rightleftharpoons C + D$$

The arrows pointing in different directions indicate reversibility; the difference in the lengths of the arrows indicates that the conversion

$$A + B \longrightarrow C + D$$

occurs more readily than the reverse reaction

$$C + D \longrightarrow A + B$$

Given equal concentrations initially, the energy of activation is lower for the interaction of A + B than it is for the interaction of C + D.

If initially only the reactants A and B are present, there will be a relatively rapid conversion of these molecules to C + D. When enough molecules of C + D have been produced, they will be reconverted to A + B. Since the rate of a reaction depends partially on the concentrations of reacting molecules, as C + D build up in concentrations, the reverse rate of reaction will increase. Similarly, as A + B are converted to C + D, the concentrations of the former decrease and so does the forward rate of reaction. Dynamic equilibrium is reached when the two rates become equal. At this point, although there will be no net change in the concentrations of either reactants or products, each will be reacting to form the other.

Figure 1-10 shows the relationships between the reactants and products of the above reaction. At the start of the reaction (zero time), only molecules of A + B are present in the system, and molecules of A + B are converted to C + D. The reaction occurs to the right. As C + D accumulate, reaction to the left begins and proceeds at an increasing rate (the possibility for collisions between C + D increases) while the rate of the forward reaction decreases (the rate of collisions between A + B decreases as their concentrations become less). When the rate of the forward reaction equals the rate of the reverse reaction, no further changes in concentration take place. This is the point of dynamic equilibrium. Note that at equilibrium the concentrations of reactants do not have to be equal to the concentrations of products. In the reaction just described,

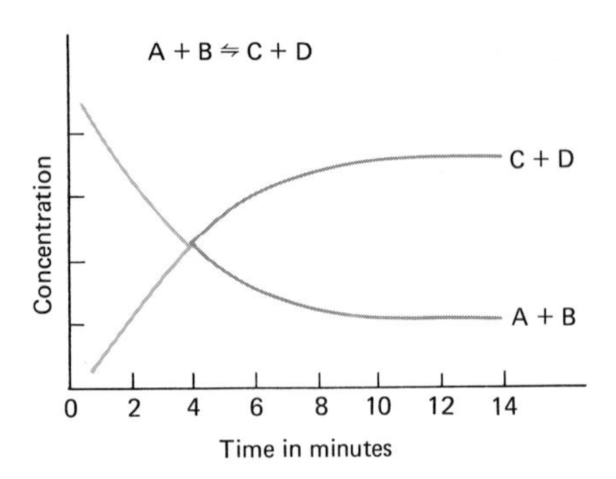

FIGURE 1-10. Reversibility and equilibrium. At the start, only molecules of A and B are present. When the reaction begins, molecules of A and B are converted to molecules of C and D. For a while there is a higher concentration of A and B. At the end of 4 minutes the concentrations of reactants and products are equal. From that point on, however, the concentration of C and D increases until an equilibrium is established. Once equilibrium is established, at about the tenth minute, the rate of forward reaction is equal to the rate of reverse reaction. Note that the concentrations of reactants and products do not have to be equal for an equilibrium to be established.

the concentrations of C + D are substantially higher than those of A + B. While the proportions of A + B and C + D remain constant, there is always more C + D present than A + B.

Such a chemical equilibrium system can be made to proceed further in one direction by any of several mechanisms. If one of the products is a gas that escapes from solution, there will be a continuous conversion of A + B to C + D until all of the reactants are used up. Similarly, if one of the products is insoluble, the reaction will continue to the right. In living systems, although most individual reactions are chemically reversible, such reversibility does not lead to an equilibrium condition because products of one reaction are used as reactants (substrates) in further metabolic reactions.

Living systems are not closed systems. At all levels these systems are open, with a continuous inflow of matter and energy, a continuous transformation of material, and a continuous outflow of material. At the cellular level, there are a continuous inflow of nutrients, salts, vitamins, and oxygen into the cell, a continuous metabolism within the cell, and a continuous outflow of waste products such as CO_2 and other materials from the cell.

Such open systems cannot reach chemical equilibrium but they can arrive at a *steady state* condition in which concentrations and proportions of materials within the system tend to remain at constant levels although the rates of inflow and outflow of materials may change.

Again, although individual biochemical reactions are reversible, they are arranged in sequential fashion in the cell so that the product of a reaction is immediately used as a substrate in the next enzyme reaction.

Consider a sequence of enzyme-catalyzed reactions:

$$A \rightleftharpoons B \rightleftharpoons C \rightleftharpoons D \rightleftharpoons E \longrightarrow F$$

Although each reaction, except the last, is reversible, the system cannot arrive at equilibrium because the product of each reaction is used in the next step of the sequence. Thus, as fast as A is converted to B, the latter is removed by its conversion to C, and so on. The entire reaction system proceeds to the right. Usually the final product (represented by F above) is removed from the system either as a diffusible gas or by some other mechanism. This series of enzyme reactions, which is typical of the arrangement of metabolic reactions, is an example of an open system that can arrive at a steady state so that the concentrations of all materials remain at constant levels. If, for some reason, the concentration of A coming into the system should increase, all reactions increase their rates to the right or forward direction and the final product is formed faster.

Again, cells and their subsystems are closer to steady state than equilibrium conditions. In addition, although certain individual reactions or entire reaction sequences may be chemically reversible, cells generally utilize alternate pathways to reverse reactions. For example, glucose is broken down to pyruvic acid by one set of enzymes in the initial reaction sequence in respiration, while the conversion of pyruvic acid back to glucose involves an entirely different set of enzymes.

The steady state is, as already mentioned, a regulating type of system. Such systems play an important role in the cell in regulating cellular metabolism and preventing the buildup of large concentrations of unnecessary materials. The open systems of the cells are made possible by the presence of enzymes that govern the flow of materials and energy in the cell.

2 The Molecules of Life

GUIDE QUESTIONS

1 Define a compound. Differentiate between an inorganic compound and an organic compound.
2 Describe the structure of the water molecule and account for its polarity. Relate the function of each of the unique properties of water to living organisms.
3 Define an acid and a base according to the Brønsted–Lowry, the Lewis, and the classical theories. What is meant by a strong and weak acid and a strong and weak base?
4 What is pH? How is pH expressed mathematically?
5 Define a buffer. By means of equations show what occurs when alkali (OH^-) is added to the carbonic acid–bicarbonate buffer system; when acid (H^+) is added.
6 State five physiological functions of inorganic compounds.
7 What property of the carbon atom accounts for the large number of organic compounds? Give specific examples.
8 Give the structural formulas for the following functional groups: hydroxy, carbonyl, amine, sulfhydryl, carboxyl, ether, and ester.
9 Define, classify, and state the physiological importance of each of the following: carbohydrates, lipids, and proteins.
10 Name three monosaccharides and discuss the importance of each. How are disaccharides formed? Describe three polysaccharides of biological importance.
11 Describe the hydrolysis of a simple fat. What is an ester linkage?
12 Describe the formation of the peptide link. How do amino acids function as buffers? Describe in detail the primary, secondary, tertiary, and quaternary structures of protein organization.
13 What is the sequence of events believed to occur in an enzymatic reaction? What is an active site? Relate the concept of active site to enzyme specificity. Describe the four exceptional catalytic properties of enzymes.
14 Differentiate among apoenzyme, holoenzyme, and prosthetic group. What is an activator? What is a coenzyme?
15 What is a nucleotide? Compare DNA and RNA with regard to structure, function, and location in the cell.
16 What is ATP? How is it related to ADP? What is the essential role of ATP in a living system?

IT HAS been shown that atoms may combine chemically to form molecules. Any substance formed by the chemical combination of the atoms of two or more elements is called a **compound,** and it will have characteristics different from those of the individual atoms forming it. Whereas the fundamental unit of an element is an atom, the fundamental unit of a compound is a **molecule.** The compounds found in organisms may be subdivided into two general types: (1) inorganic compounds and (2) organic compounds. **Inorganic compounds** typically lack carbon, dissolve readily in water, generally resist decomposition (breakdown), undergo rapid chemical reactions, and are usually more ionic in character. **Organic compounds** always contain the element carbon and typically contain hydrogen as their essential constituents.

Chemical Constituents of Biological Systems

The principal inorganic substances that comprise living systems are water, gases, and certain acids, bases, and salts. Some important organic compounds are carbohydrates, lipids, proteins, and nucleic acids.

Inorganic Constituents

WATER. Any living organism requires a wide variety of inorganic compounds for growth, repair, maintenance, and reproduction. Water is one of the most important, as well as one of the most abundant, of these. In fact, with few exceptions, such as the enamel of teeth and bone tissue, water is by far the most abundant cellular component. It is present in amounts ranging from 5 to 95% or more. In humans, about 60% of the red blood cells, 75% of muscle tissue, and 92% of blood plasma consists of water. Spores and seeds of certain plants may contain as little as 5–10%. Although there is no definite amount of water that must be present in living matter, the general average is about 65–75%. All organisms die if their water supply is either inadequate or discontinued.

The importance of water to all life resides in unique properties directly related to the structure of the water molecule. It will be recalled from Chapter 1 that in the water molecule two hydrogen atoms each form a covalent bond with oxygen. In the formation of these two bonds, four electrons are involved. The electrons involved in the covalent bonds between hydrogen and oxygen are attracted more strongly toward the oxygen nucleus, with the result that all the electrons are closer to the oxygen nucleus than to the hydrogen proton. While the total charge on the water molecule is neutral, the positive and negative charges are not equally distributed within the molecule. The oxygen region of the molecule is slightly electronegative, and the hydrogen region of the molecule is slightly electropositive (see Figure 1-6A).

Any molecule in which there is unequal distribution of charges is known as a **polar molecule.** While the water molecule is one of the outstanding examples of a polar molecule, there are many other compounds made up of polar molecules. The importance of polar molecules to living systems is two-fold. First, because of the attraction of unlike charges, polar molecules tend to be arranged in specific configurations with respect to other molecules. Second, the polarity accounts for some of the physical and chemical characteristics of molecules.

For example, the polarity of the water molecule accounts for the fact that it is an excellent suspending medium and solvent. Many substances will dissolve or dissociate in water because the negative part of the water molecule is attracted to the positive part of the *solute* (that which is dissolved) molecule while the positive part of the water molecule is attracted to the negative part of the solute molecule (Figure 2-1). The attractive forces of the water molecules for the solute molecules reduce the attractive forces between the solute molecules, allowing them to separate and become surrounded by water molecules. The polarity of the water molecule also accounts for its chemical properties. Water is often a reactant or product of chemical reactions. For example, water is a reactant in the digestive process of organisms—a process in which larger molecules are broken down into smaller ones for utilization by cells. Conversely, water molecules are involved in synthetic reactions. In a chemical sense, water is a principal source of the hydrogen and one of the sources of the oxygen that are incorporated into numerous organic compounds in living systems as a component of protoplasm, organelles, and membranes.

In the liquid state, molecules of water are randomly arranged. As water freezes, the partially negatively and partially positively charged parts of the molecules become oriented in such a manner that the resulting ice crystal has an ordered geometrical structure. The arrangement of the molecules is such that they are less closely spaced than in the liquid state. This accounts for the fact that the density of ice is less than that of water. As a result, ice floats on top of water. This is a most important property from an ecological point of view and will be discussed further in Chapter 22.

The hydrogen bonding between water molecules accounts for some of the unique physical properties of water. A given quantity of water, when compared to the same quantity of many other substances, requires a much greater amount of heat to increase its temperature by a given number of degrees and a greater loss of heat to decrease its temperature. Heat absorption by molecules increases their kinetic energy and thus their rate of motion (this is measured as an increase in temperature). However, heat absorption by water is used first to break hydrogen bonds rather than increasing molecular motion, and thus much more heat must be applied to raise its temperature than to a nonhydrogen-bonded liquid. The reverse is true on cooling. In this

Figure 2-1. Solvating property of water. The $\delta-$ portion of the water molecule is attracted to the positive sodium ion, while the $\delta+$ portion of the water molecule is attracted to the negative chloride ion. The water molecules reduce the attractive forces between the sodium and chloride ions in the crystal and aid in dissolving the sodium chloride molecule.

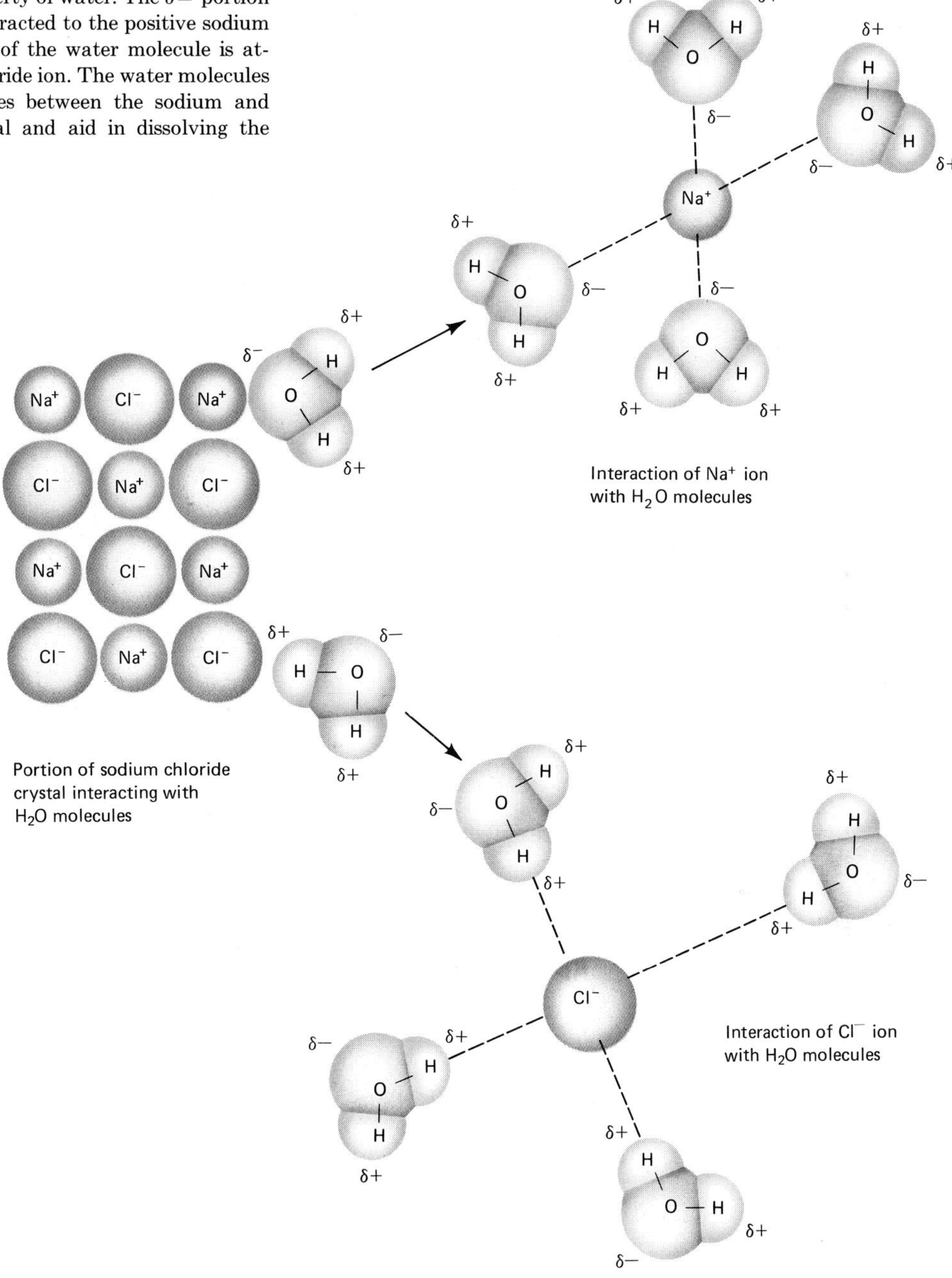

regard, water acts as a temperature buffer by preventing rapid heating and rapid cooling. By doing so, water maintains a more constant temperature than other solvents, even though there may be fluctuations of environmental temperatures, thereby helping to maintain a constant body temperature. These and other properties will be discussed further in Chapter 22.

Gases. Three of the gases of the atmosphere are of prime importance to living organisms: nitrogen, carbon dioxide, and oxygen. Atmospheric nitrogen (N_2) comprises about 78% of the gaseous envelope around the earth. Gaseous carbon dioxide (CO_2), which constitutes about 0.03% of the earth's atmosphere, is the source of carbon, a major component of living matter. Molecular oxygen (O_2), which comprises about 20% of the atmospheric gases, is essential for the metabolism of most cells. The cyclic patterns followed by these gases in the biological and physical worlds are treated in more detail in Chapter 24.

Acids, Bases, and Salts. Within living organisms are certain chemical substances (acids, bases, and salts) that are considered as inorganic compounds because they lack the typical carbon-to-hydrogen bonds characteristic of organic compounds. It might be noted at this point that there are also organic acids, bases, and salts, which, of course, contain carbon-to-hydrogen bonds. The inorganic substances are important to living organisms because they provide the ions for many essential biochemical reactions.

There are several different points of view with regard to classifying a substance as an acid, base, or salt. For example, an **acid** may be defined as a proton donor (Brønsted–Lowry), an electron pair acceptor (Lewis), or a hydrogen ion (H^+) donor (Arrhenius, or classical). Conversely, a **base** may be defined as a proton acceptor (Brønsted–Lowry), and electron pair donor (Lewis), or a hydroxyl ion (OH^-) donor (Arrhenius, or classical). **Salts** may be regarded, from a classical point of view, as compounds made up of positive ions (except hydrogen ions) and negative ions (except hydroxyl ions). While all three of the above definitions are important for a complete understanding of acid-base reactions, the definitions proposed by Arrhenius will be applied in the following discussion. These definitions will be used because the inorganic substances of living organisms occur in an aqueous system (dissolved in water) for which these definitions provide a simple explanation.

When certain molecules dissolve in water, regardless of whether they are more or less covalent or ionic in character, they are further acted upon by water molecules. The water molecules reduce the attraction of the atoms within an ionic molecule and the ionic substance breaks apart or *dissociates.* If the molecule is covalently bonded, then the term *ionization* rather than dissociation is used. While there is a slight difference in the meaning of these terms, they are frequently used interchangeably. When the molecule breaks apart in solution, the electrons forming the original bond will remain with one of the parts (called *species*), and each species will be free in solution as an **ion** (charged atom or group of atoms). If an ion has more protons than electrons, it is positively charged; if it has more electrons than protons, it is negatively charged. Solutions that contain ions are capable of conducting an electric current; the substances forming the ions are called **electrolytes.** In any given solution of an electrolyte, the total number of plus charges will be equal to the total number of negative charges.

When an acid is dissolved in water, it dissociates into one or more hydrogen ions and one or more negative ions (Figure 2-2A). When a base is dissolved in water, it forms one or more positive ions and one or more hydroxyl ions (Figure 2-2B). A salt, when dissolved in water, dissociates into one or more positive ions and one or more negative ions, neither of which is an H^+ or an OH^- (Figure 2-2C). Most salts are made up of a metallic ion and a nonmetallic ion. Salts are the source of the essential metallic ions as well as certain nonmetallic ions, such as phosphate (PO_4^{3-}), sulfate (SO_4^{2-}), and nitrate (NO_3^-), needed for proper metabolism. The inorganic acids and bases are important in maintaining the proper pH (which will be discussed subsequently), the electrolytic environment in which biochemical reactions occur. For example, reactions of proteins, one of the major groups of biochemicals, are greatly affected by the number of hydrogen ions present. Relatively small changes in the pH (the amount or concentration of hydrogen ions present) can alter the structure of protein molecules as well as change the direction of a biochemical reaction.

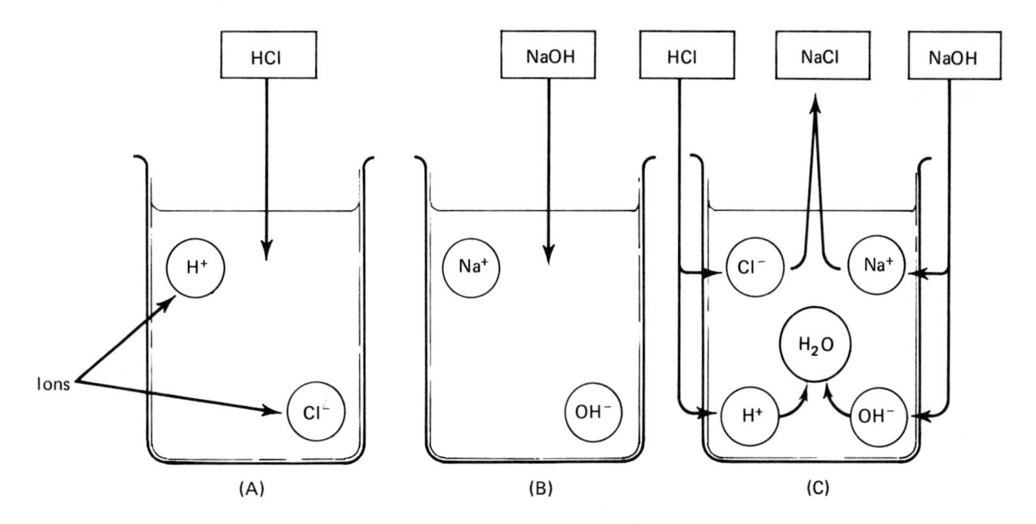

Figure 2-2. Acids, bases, and salts. (**A**) An acid, upon ionization, yields H^+ ions. (**B**) A base, by contrast, yields OH^- ions upon ionization. (**C**) A salt is a compound resulting from the chemical combination of an acid and a base and, upon ionization, yields ions that are neither H^+ nor OH^-. In this case the salt is NaCl. Water is also produced in this reaction of an acid and base.

Not all electrolytes ionize or dissociate to the same degree in aqueous (water) solution. For example, a one-tenth molar solution of hydrochloric acid (HCl) will dissociate almost 100%; that is, when pure HCl is placed in water, few if any molecules of HCl are present. On the other hand, a one-tenth molar solution of acetic acid will only ionize about 4.26%; that is, most of the acetic acid will remain as molecules with only a small portion dissociating into hydrogen ions and acetate ions. The degree or percentage of ionization or dissociation varies from electrolyte to electrolyte. In addition, the percentage of dissociation of an electrolyte will vary with its concentration.

It is to be noted that concentration of solutions is most frequently expressed as molarity. The standard is a one molar solution where the molecular weight of the solute (that which dissolves) expressed in grams is added to enough solvent to make one liter of solution. For example, the molecular weight of hydrochloric acid is 36.5 atomic mass units. Therefore 36.5 grams of hydrochloric acid dissolved in enough water to make one liter makes a one molar (1.0 *M*) solution. If a liter of solution contained 18.25 grams of HCl, its concentration would be 0.5 *M*; if it contained 3.65 grams, its concentration would be 0.1 *M*; and so on.

In the following examples showing degree of dissociation, specific small numbers of species are used for simplification. These numbers would not hold true in nature because even the smallest visible amount of a chemical substance contains millions of molecules and also all electrolytes will dissociate completely in extremely small concentrations.

Suppose 100 molecules of hydrochloric acid (HCl) are placed in water. The hydrochloric acid molecules acted upon by the water molecules will produce 100 chloride (Cl^-) and 100 hydronium (H_3O^+) ions.

$$\underset{\text{100 molecules of hydrochloric acid}}{HCl} \xrightarrow{+H_2O} \underset{\text{100 hydronium ions}}{H_3O^+} + \underset{\text{100 chloride ions}}{Cl^-}$$

It has been found that the H^+ (a single proton) from an acid is not free when in aqueous (water) solution. It attaches to a water molecule, forming the hydronium ion.

$$\underset{\text{hydrogen ion}}{H^+} + \underset{\text{water}}{H_2O} \longrightarrow \underset{\text{hydronium ion}}{H_3O^+}$$

For simplicity, the ionization equation of an acid may be represented as

$$\underset{\text{hydrochloric acid}}{HCl} \longrightarrow \underset{\text{hydrogen ion}}{H^+} + \underset{\text{chloride ion}}{Cl^-}$$

which is in keeping with the classical definition of an acid (H^+ donor). It must be remembered, however, that the water molecule is intimately involved in the reaction and that the hydrogen ion is not present as such but is in the form of the hydronium ion.

As the equation shows, hydrochloric acid molecules completely ionize. An acid that completely ionizes in solution is said to be a *strong acid.* Sulfuric acid is a strong acid; therefore, in aqueous solution sulfuric acid reacts as follows.

$$\underset{\text{sulfuric acid}}{H_2SO_4} \longrightarrow \underset{\text{hydrogen ions}}{2\,H^+} + \underset{\text{sulfate ion}}{SO_4^{2-}}$$

Now suppose that 100 molecules of carbonic acid (H_2CO_3) are added to water. The result might be written

$$\underset{\substack{\text{98 molecules}\\ \text{of carbonic acid}}}{H_2CO_3} \longrightarrow \underset{\text{2 hydrogen ions}}{H^+} + \underset{\text{2 bicarbonate ions}}{HCO_3^-}$$

Most of the carbonic acid molecules do not ionize, but remain in the molecular form. While there are potentially 200 hydrogen ions possible (the bicarbonate that contains a hydrogen atom may form a hydrogen ion under certain circumstances), only two are formed. Acids such as carbonic acid, in which ionization is slight, are known as *weak acids.* In comparing hydrochloric acid and sulfuric acid with carbonic acid, it can be seen that many more hydrogen ions are produced by strong acids than weak acids. The stronger acids are therefore considered to be more acidic than the weaker acids. The degree of ionization of weak acids is dependent upon the specific acid and its concentration.

The ionization of weak acids or bases more often is represented as

$$H_2CO_3 \rightleftharpoons H^+ + HCO_3^-$$

The arrows going in both directions mean that carbonic acid molecules in solution dissociate into hydrogen (H^+) and bicarbonate (HCO_3^-) ions and that hydrogen and bicarbonate ions react to form H_2CO_3. The shorter upper arrow indicates that H_2CO_3 ionizes to a relatively small extent. At equilibrium, the concentrations of the ions and molecules are constant although not necessarily equal.

Strong and weak bases are comparable to strong and weak acids. The same principles of ionization and ionization equilibrium apply. Generally, the inorganic bases do not have great application within living organisms. In addition, strong acids and bases are not usually found throughout a living structure. If they do occur, they are limited to specific regions of the body such as hydrochloric acid in the stomach. The strength or weakness of an acid or base can be determined experimentally by testing the electrical conductivity of two solutions of the specific compound at different concentrations. The greater the degree of ionization, the greater the degree of conductivity.

pH. Biochemical reactions are extremely sensitive to small changes in the electrolytic environment in which they take place. Concentrations of ions, especially the hydrogen ion, when expressed in molar concentration or absolute amounts, are unwieldly for the biologist. The concept of pH provides a convenient way to express the hydrogen ion concentration. By definition, the **pH** of a given solution is equal to the negative logarithm of the hydrogen ion concentration. Expressed mathematically,

$$pH = -\log\,[H^+]$$

In the ionic equilibrium of normal blood plasma, the concentration of hydrogen ions is 0.0000000395 M or $3.95 \times 10^{-8}\,M$. It is converted to pH as follows.

$$\begin{aligned} pH &= -\log\,[H^+] \\ &= -\log\,(3.95 \times 10^{-8}) \\ &= -(\log 3.95 + \log 10^{-8}) \\ &= -(0.5966 - 8) \\ &= -(-7.4034) \\ &= 7.4034 \text{ or } 7.4 \end{aligned}$$

and the pH of blood plasma is about 7.4.

Water ionizes to an extremely small extent.

$$H_2O \rightleftharpoons H^+ + OH^-$$

Since it produces equal numbers of hydrogen and hydroxyl ions, it can be considered as much acidic as basic (alkaline) according to the classical concept and, therefore, is considered neutral. It has been found that the concentration of hydrogen ions of water is $1 \times 10^{-7}\,M$.

If we calculate the pH of water,

$$\begin{aligned} pH &= -\log[H^+] \\ &= -\log(1 \times 10^{-7}) \\ &= -\log(\log 1 + \log 10^{-7}) \\ &= -(0 - 7) \\ &= -(-7) \\ &= 7 \end{aligned}$$

we find it to be 7. A solution of pH 7 is considered neutral, with equal amounts of hydrogen and hydroxyl ions. Through the use of pH, a scale can be set up whereby a pH of less than 7 is acidic because there are more hydrogen ions and fewer hydroxyl ions than are found in water (a neutral solution). Conversely, a pH of greater than 7 is basic or alkaline because it has fewer hydrogen ions and more hydroxyl ions than water. From the pH of 7.4, it can be quickly determined that normal blood plasma is slightly basic, or alkaline. Within living organisms slight variations in pH will greatly affect life activities. It is important to realize that a change in one whole number of the pH scale, while numerically small, represents a ten-fold change from the previous concentration (Figure 2-3).

Buffers. It was previously stated that inorganic substances play an important part in maintaining an electrolytic balance within an organism. Although there is variation in pH in different parts of an organism, the limits are generally quite specific. For example, within a range of 0.3 above or below the 7.4 pH of normal blood plasma, severe acidosis or alkalosis results, with disastrous effects. In order to maintain a relatively constant pH, certain of the inorganic electrolytes function as **buffer systems.** A buffer resists a change of pH on the addition of alkali or acid. The most notable inorganic buffer system of the blood plasma is the carbonic acid–bicarbonate buffer system. It will be used to demonstrate the way in which buffers function.

Carbonic acid is a weak acid and ionizes according to the following equation:

$$H_2CO_3 \rightleftharpoons H^+ + HCO_3^-$$

In solution, and at a given concentration, an equilibrium is established between the molecular form of the acid and its ions. Now, if the salt sodium bicarbonate ($NaHCO_3$) is added to a solution of carbonic acid, it ionizes completely (all salts, if soluble, are dissociated completely).

$$NaHCO_3 \longrightarrow Na^+ + HCO_3^-$$

The effect of this salt on the carbonic acid equilibrium is that it greatly increases the concentration of the bicarbonate ion in solution. According to the ionic equilibrium equation for carbonic acid, this will force the reaction in the direction to form more carbonic acid molecules. New equilibrium concentrations will be established according to the combined equations:

$$\begin{aligned} NaHCO_3 &\longrightarrow Na^+ + HCO_3^- \\ HCO_3^- + H^+ &\rightleftharpoons H_2CO_3 \end{aligned}$$

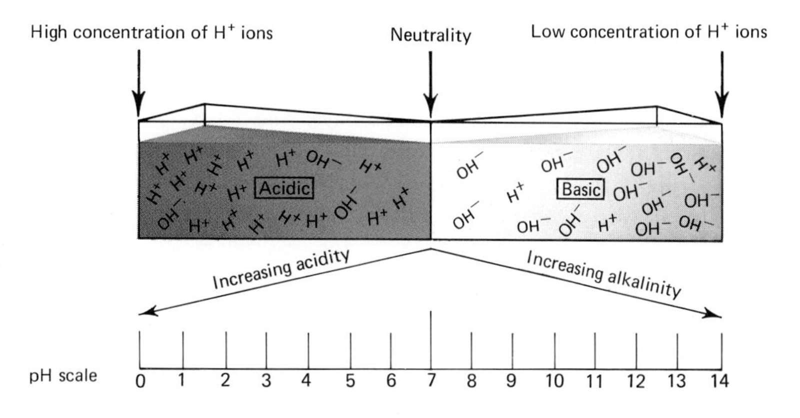

Figure 2-3. The pH scale. A pH below 7 indicates an acid solution, that is, a solution in which there are more H^+ ions than OH^- ions. The lower the numerical value, the higher the H^+ ion concentration. The opposite holds true for the concentration of OH^- ions. As the numerical expression increases, the OH^- ion concentration increases and the solution becomes increasingly basic. At pH 7, the concentrations of H^+ ions and OH^- ions are equal and the solution is neutral. A change in one whole number of the pH scale represents a ten-fold change from the previous concentration. See text for explanation.

It might be noted here that the sodium ion plays no part in re-establishing the ionic equilibrium but serves to balance the electrical charges. At equilibrium, the following situation exists: there are many more carbonic acid molecules and bicarbonate ions than hydrogen ions.

If hydrogen ions are added to this buffer system, they combine with bicarbonate ions, forming molecules of carbonic acid. The net result is that although hydrogen ions are added to the system, they do not increase the total hydrogen ion concentration greatly. On the other hand, if hydroxyl ions are added to the buffer system, they combine with hydrogen ions to form water. The removal of hydrogen ions causes more carbonic acid molecules to ionize, forming more hydrogen ions, replacing those removed in order to re-establish the equilibrium. The formation of these hydrogen ions tends to restore the original concentration of hydrogen ions. It can be seen from this that the carbonic acid acts as a reservoir for additional hydrogen ions to replace those removed by alkali or base and the bicarbonate ions act as a reservoir to combine with hydrogen ions resulting from the addition of an acid. Thus, within limits, a buffer system tends to maintain a relatively constant pH. If large quantities of acid or base are added, there may not be sufficient acid or negative ions to react, and there will be a sharp change in pH.

The carbonic acid–bicarbonate buffer system just discussed was used as an example. There are many other inorganic as well as organic buffer systems. Generally, inorganic buffer systems may be of two types:

1. A weak acid and a salt of the weak acid, which contains the same negative ion as is found in the acid.
2. A weak base and a salt of the weak base, which contains the same positive ion as in the base.

There are other inorganic buffer systems based upon compounds that are classically defined as salts, such as the phosphate buffer system of the kidneys. The mechanisms of the various types of inorganic buffer systems are essentially the same; that is, one component ionizes slightly and the other component ionizes almost completely.

Physiological Functions of the Inorganic Components. Only a few of the important reactions of the inorganic constituents of living forms have been mentioned. A number of other inorganic substances have important roles in life processes. Table 2-1 gives a few of the elements found in living organisms. In the functions listed, most do not react in their elementary forms but as ions or constituents of molecules. In addition, it is important to realize that while these substances may be considered as inorganic, within the living organism they are closely interrelated with the organic components. In general, inorganic constituents perform such functions as helping to maintain electrolytic balance (buffer system), play an

Table 2-1. Relative Abundance and Functions of Some Elements Found in Living Matter

Element	Weight (%)	Functions
Oxygen	62	Constituent of water and many organic molecules; used in cellular respiration
Carbon	20	Constituent of all organic molecules
Hydrogen	10	Constituent of water and organic molecules
Nitrogen	3	Constituent of many organic molecules (e.g., proteins and nucleic acids)
Calcium	2.50	Structural component of bone and teeth; required for blood clotting and hormone synthesis, membrane stability, muscle contraction
Phosphorus	1.14	Constituent of some proteins, ATP, and nucleic acids; component of nerve tissue and bone; an important component of certain lipids
Chlorine	0.16	Component of salt (NaCl)
Sulfur	0.14	Constituent of many proteins
Potassium	0.11	Needed for growth and the conduction of nerve impulses
Sodium	0.10	Structural component of bone and other tissues; component of blood that maintains osmotic balance; needed for nerve conduction
Magnesium	0.07	Component of chlorophyll and many enzymes
Iodine	0.014	Vital to functioning of thyroid gland
Iron	0.101	Essential part of hemoglobin and cytochromes
Others	0.756	

important part in the formation of teeth and bones, provide the ions for nerve and other cells to respond to stimuli, and furnish specific activators for enzymatic reactions.

Organic Constituents

Inorganic compounds, excluding water, constitute about 1–1.5% of living material; the remainder consists of organic compounds. Inorganic molecules serve as the raw materials for the synthesis of organic compounds, and both are interrelated through various chemical processes. Organic compounds are considerably larger and structurally more complex than inorganic molecules. Many organic molecules contain hundreds and even thousands of component atoms.

The element carbon, perhaps more than any other, is characteristic of organic matter. It will be recalled from Chapter 1 that a covalent bond is formed by the sharing of a pair of electrons between two atoms. Carbon is unusual in that its four outer orbital electrons can form up to four covalent bonds, and carbon atoms can bond to each other, forming chain, branched, or ring structures.

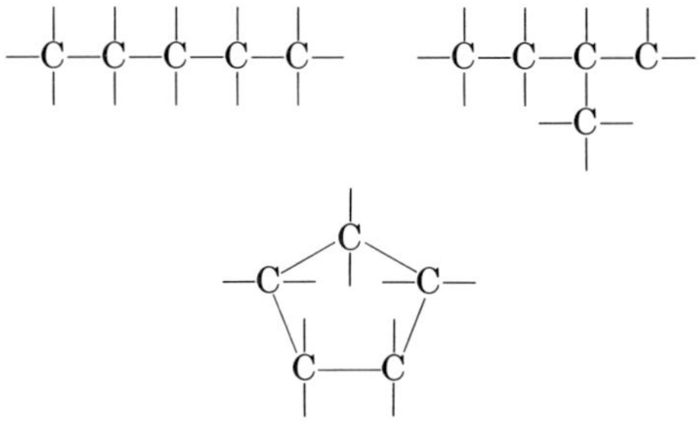

When a carbon is bonded to four other atoms, the four bonds form a tetrahedron shape, which gives a characteristic three-dimensional structure to the molecules that are formed (Figure 2-4). Carbon atoms can also bond with double or triple bonds:

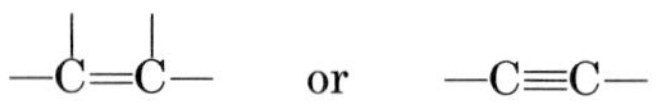

In these cases, the molecular shape is not tetrahedral. Carbon can also bond to other elements.

In addition to carbon, the most frequently found elements in organic compounds are hydrogen (one bond), oxygen (two bonds), and nitrogen (three bonds).

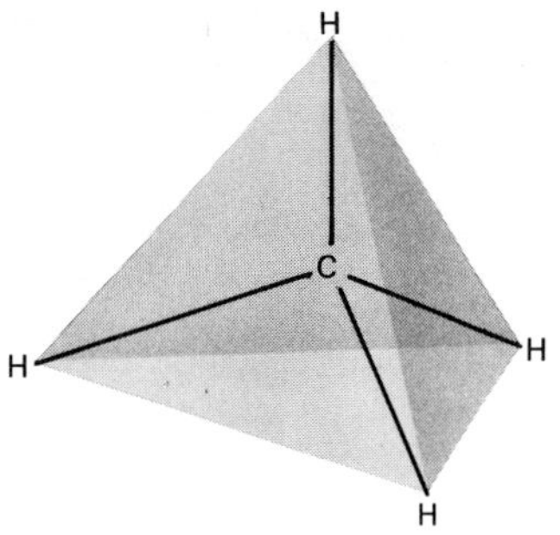

FIGURE 2-4. The tetrahedral carbon atom. The four electrons forming single covalent bonds of the carbon atom result in a tetrahedron shape. The directions of these bonds give a characteristic structure to organic molecules containing carbon atoms with four single bonds.

To a lesser extent are found sulfur (two bonds) and phosphorus (five bonds). Other elements are found, but only in a relatively few organic compounds.

The basic straight or branched chain, or the ring structure, of carbon atoms in the molecule is called the *carbon skeleton*. Most of the free bonds of the carbon skeleton are satisfied by hydrogen atoms. The additional bonding of other elements with carbon and hydrogen form characteristic groupings called *functional groups* (Table 2-2). These functional groups contribute to most of the characteristic chemical properties and many of the physical properties of the given organic compound. Since there is an almost infinite number of combinations possible in carbon skeletons, a common characteristic grouping or bonding helps to classify organic compounds. One method of classification is based upon the type of functional group. For example, in the following structures

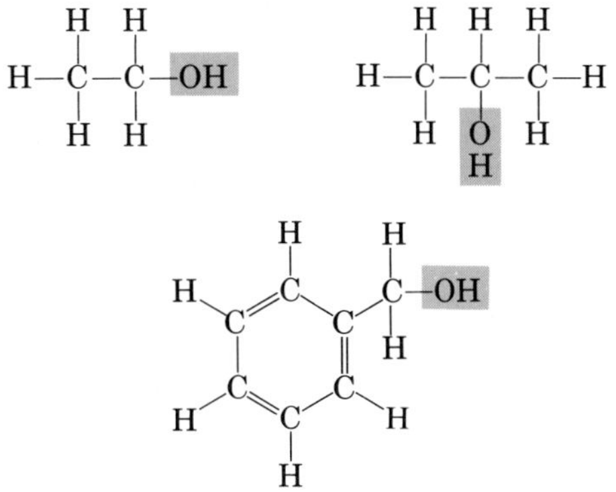

TABLE 2-2. Representative Functional Groups and Compounds in Which They are Found

Functional Group	Name	Class of Compounds
$R{-}O{-}H$	Hydroxy	Alcohol
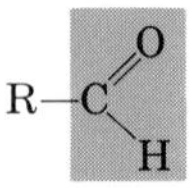 $R{-}C({=}O){-}H$	Carbonyl (terminal)	Aldehyde
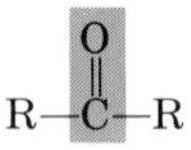 $R{-}\overset{O}{\overset{\Vert}{C}}{-}R$	Carbonyl (internal)	Ketone
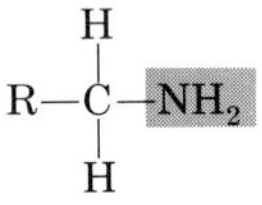 $R{-}\overset{H}{\underset{H}{C}}{-}NH_2$	Amine	Amine
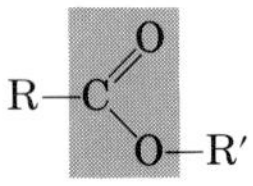 $R{-}C({=}O){-}O{-}R'$	Ester	Ester
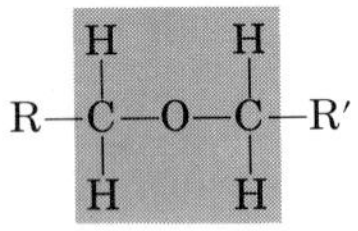 $R{-}\overset{H}{\underset{H}{C}}{-}O{-}\overset{H}{\underset{H}{C}}{-}R'$	Ether	Ether
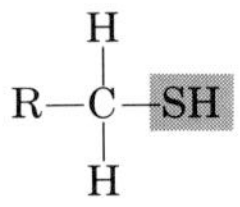 $R{-}\overset{H}{\underset{H}{C}}{-}SH$	Sulfhydryl	Sulfhydryl
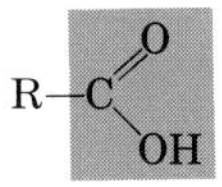 $R{-}C({=}O){-}OH$	Carboxyl	Organic acid

the —OH group is present in each of the molecules. Since the characteristic reactivity of the molecules is based upon the —OH group, they are all classified as alcohols. The —OH group is called the *hydroxy group* and is not to be confused with the hydroxyl ion (OH^-) of bases. The hydroxy group of alcohols does not ionize.

Since the functional group is the classifying characteristic, frequently the letter R is substituted for a carbon skeleton. For example, if one refers to alcohols in general, instead of listing a specific alcohol, such as:

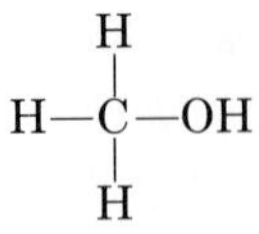

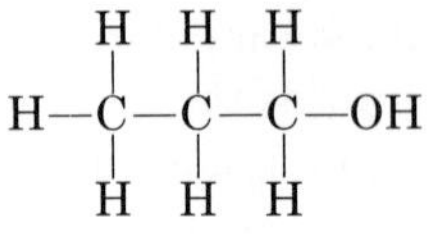

$$\underset{\text{methanol}}{H{-}\overset{H}{\underset{H}{C}}{-}OH} \qquad \underset{\text{ethanol}}{H{-}\overset{H}{\underset{H}{C}}{-}\overset{H}{\underset{H}{C}}{-}OH} \qquad \underset{n\text{-propanol}}{H{-}\overset{H}{\underset{H}{C}}{-}\overset{H}{\underset{H}{C}}{-}\overset{H}{\underset{H}{C}}{-}OH}$$

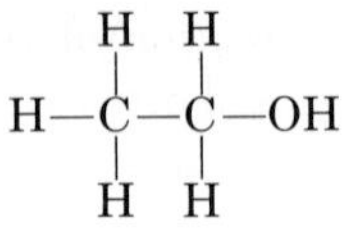

the R may be substituted for a carbon chain as follows: R—OH. Frequently, more than one functional group will be found in one molecule. For example,

$$\text{(amine)}\quad H_2N{-}\overset{H}{\underset{H}{C}}{-}C({=}O){-}OH \quad\text{(carboxyl)}$$

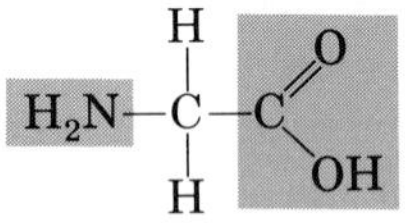

If a molecule contains amine and carboxyl groups, it is called an amino acid. Most of the organic compounds found in living organisms are quite complex, with a large number of carbon atoms forming the skeleton and many functional groups attached. In considering representations of organic and biochemical molecules it is important to note that each of the four bonds of carbon is satisfied (attached to another atom) and each of the attaching atoms has its characteristic number of bonds satisfied. The principal organic compounds found in living matter are carbohydrates, lipids, proteins, and nucleic acids.

CARBOHYDRATES. **Carbohydrates** represent a large group of organic compounds that assume important roles in all living systems. First, they play a pivotal role in the energy relations of all organisms. Through the agency of photosynthesis, green plants synthesize carbohydrates which contain huge quantities of stored chemical energy. The breakdown of these carbohydrates releases energy required by both plants and animals for their everyday activities. Second, carbohydrates enter into the formation of a number of structural units of organisms. For example, chitin forms the bulk of the exoskeleton of insects and crustaceans (crayfish, crabs, etc.) and cellulose is organized into the rigid structural organization of plant cell walls. Third, some carbohydrates serve as storage or reserve energy materials in organisms. Starch, formed principally in the organs of higher plants, may be broken down into smaller units that can be used by the plant during periods when the photosynthetic process is operating inefficiently. In animals, the corresponding carbohydrate storage product is glycogen, which is stored in the liver and muscles of higher animals. Finally, carbohydrates, through a series of intermediate reactions, may be used for the synthesis of other organic substances such as fats and proteins.

Carbohydrates all contain carbon, hydrogen, and oxygen as constituent chemical elements. Typically, the hydrogen and oxygen atoms are present in a 2:1 ratio, the same as in water (H_2O). The general molecular formula for carbohydrates, although there may be exceptions, is $(CH_2O)_n$, where n may represent three or more such units. A *molecular formula* is a symbolic expression of the specific number and types of atoms or groups of atoms present in a given molecule.

Monosaccharides. The basic building blocks of carbohydrates are simple sugars, or **monosaccharides.** This group of carbohydrates consists of compounds containing three, four, five, six, or seven carbon atoms, the most important being the monosaccharides containing either five or six carbon atoms. Collectively considered, the names of monosaccharides end in *-ose* (meaning a sugar), and the prefix designates the number of carbon atoms present (see Table 2-3).

It is often convenient to designate chemical compounds by a *structural formula,* that is, the arrangement of atoms within the molecule. In addition, for simplicity and from a reference standpoint, the carbon atoms of organic compounds are numbered. The structural formulas for the monosaccharides ribose (a pentose), glucose (a hexose), and fructose (a hexose) are shown in Figure 2-5. By comparing the molecular formulas for glucose and fructose, it should be noted that both monosaccharides contain six carbon atoms, six oxygen atoms, and twelve hydrogen atoms. The chemical properties of these compounds, however, are quite

TABLE 2-3. Classification of Monosaccharides by Number of Carbon Atoms and Molecular Formulas

Sugar	Number of Carbon Atoms	Molecular Formula $(CH_2O)_n$	Value for n
Triose	3	$C_3H_6O_3$	3
Tetrose	4	$C_4H_8O_4$	4
Pentose	5	$C_5H_{10}O_5$	5
Hexose	6	$C_6H_{12}O_6$	6
Heptose	7	$C_7H_{14}O_7$	7

```
        H                    H                    H
        |                    |                    |
       1C=O                 1C=O             H — 1C — OH
        |                    |                    |
  H — 2C — OH          H — 2C — OH               2C=O
        |                    |                    |
  H — 3C — OH         HO — 3C — H          HO — 3C — H
        |                    |                    |
  H — 4C — OH          H — 4C — OH          H — 4C — OH
        |                    |                    |
  H — 5C — OH          H — 5C — OH          H — 5C — OH
        |                    |                    |
        H              H — 6C — OH          H — 6C — OH
                             |                    |
                             H                    H

       (A)                  (B)                  (C)
```

FIGURE 2-5. Structural formulas for three monosaccharides. (**A**) Ribose. (**B**) Glucose. (**C**) Fructose.

different. This is in part because the carbonyl is an aldehyde in glucose and a ketone in fructose. Any such molecules with similar chemical compositions but with different chemical properties due to different atomic arrangements are referred to as *isomers,* and the phenomenon is called *isomerism.*

Some monosaccharides, such as glucose and fructose, assume a molecular configuration forming a ring-shaped arrangement of atoms. This form, called a *ring form,* is predominant when these compounds are in solution in cells. By comparison, the *straight chain,* or linear arrangement of atoms, for molecules of glucose and fructose along with ribose are shown in Figure 2-5. When atoms of carbohydrate molecules assume a ring form, specific carbon atoms are joined to each other via an "oxygen bridge." In the case of glucose, the first and fifth carbon atoms are bridged together, while in fructose, the bridge is between the second and sixth carbon atoms (Figure 2-6).

Among the biologically important monosaccharides are the triose *glyceraldehyde,* the pentose *ribose* and *deoxyribose,* and the hexoses *glucose, fructose,* and *galactose.* Glyceraldehyde, as will be noted later, is fundamental in the process of glycolysis and lipid formation. Ribose is a constituent of RNA (a nucleic acid) and of several important coenzymes (substances that assist enzymes in their activities). Deoxyribose is a constituent of the nucleic acid DNA and is distinguished from

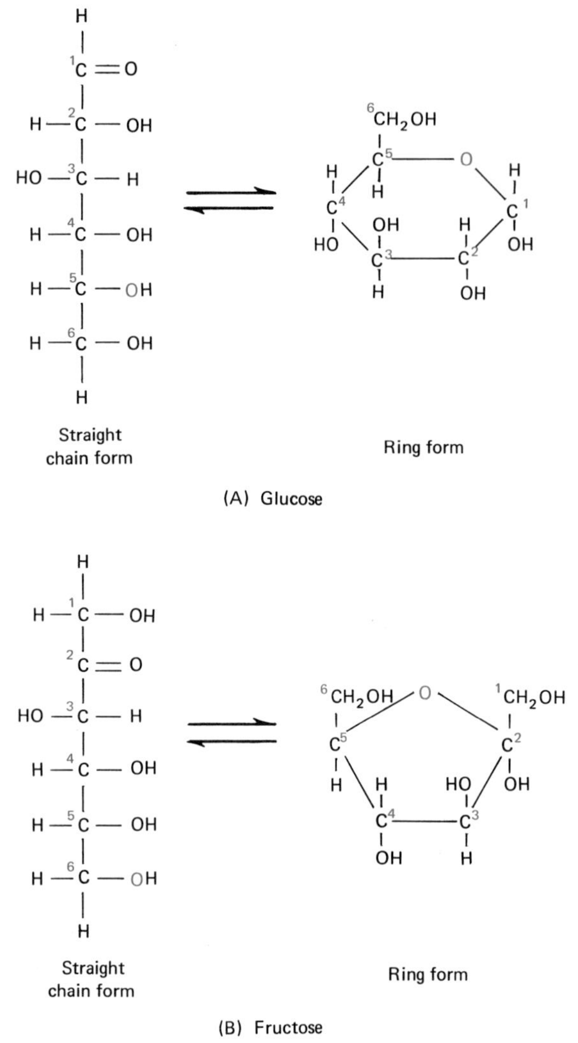

Straight chain form

Ring form

(A) Glucose

Straight chain form

Ring form

(B) Fructose

Figure 2-6. Ring forms for glucose (**A**) and fructose (**B**). Note the relative positions of the numbered carbon atoms as each molecule assumes a ring configuration as well as the oxygen bridge between carbon atoms.

ribose by the fact that its number two carbon atom lacks an atom of oxygen. The most important monosaccharide is glucose. In addition to being one of the major products of photosynthesis, it is the principal energy-supplying molecule of all living systems. Glucose is also a constituent of an important structural carbohydrate, cellulose.

Disaccharides. Within living cells two monosaccharides may combine with each other; the resulting molecule is called a **disaccharide.** In the process of disaccharide formation, a molecule of water is lost and the reaction is termed **dehydration synthesis.** An example of this type of reaction in which the disaccharide maltose is formed from two glucose molecules may be represented as follows:

$$\underset{\text{glucose}}{C_6H_{12}O_6} + \underset{\text{glucose}}{C_6H_{12}O_6} \longrightarrow \underset{\text{maltose}}{C_{12}H_{22}O_{11}} + \underset{\text{water}}{H_2O}$$

Note that the molecular formula for disaccharides, as represented by maltose, is $C_{12}H_{22}O_{11}$. In a similar manner, the chemical combination (dehydration synthesis) of molecules of glucose and fructose will produce the disaccharide, *sucrose* (common table sugar). The disaccharides may also be placed in a category called *oligosaccharides,* carbohydrates that contain more than two linked monosaccharide units.

The basic mechanism of disaccharide formation involves the union of smaller molecules, with the loss of water, to form a larger, more complex molecule. Many other organic compounds in cells, besides carbohydrates, are synthesized in a similar manner. It should be noted that dehydration synthesis is only one mechanism by which large molecules are made from smaller ones.

It is also possible for large molecules to be broken down into smaller, simpler ones with the addition of water. This reverse chemical process is called **hydrolysis,** or **digestion.** A molecule of sucrose, for example, may be hydrolyzed into glucose and fructose while a molecule of maltose may undergo hydrolysis into two glucose molecules. Figure 2-7 represents the dehydration synthesis and hydrolysis of the disaccharides maltose and sucrose.

Polysaccharides. The dehydration synthesis of eight or more monosaccharides results in the formation of a third group of carbohydrates called **polysaccharides.** Polysaccharides have the molecular formula $(C_6H_{10}O_5)_n$. Like disaccharides, they can be broken down into their constituent sugars by hydrolysis. Unlike either monosaccharides or disaccharides, they are usually not soluble in water and lack the characteristic sweetness of sugars. Chief among the important polysaccharides are starches, cellulose, chitin, and glycogen.

Starches are found in most green plants, particu-

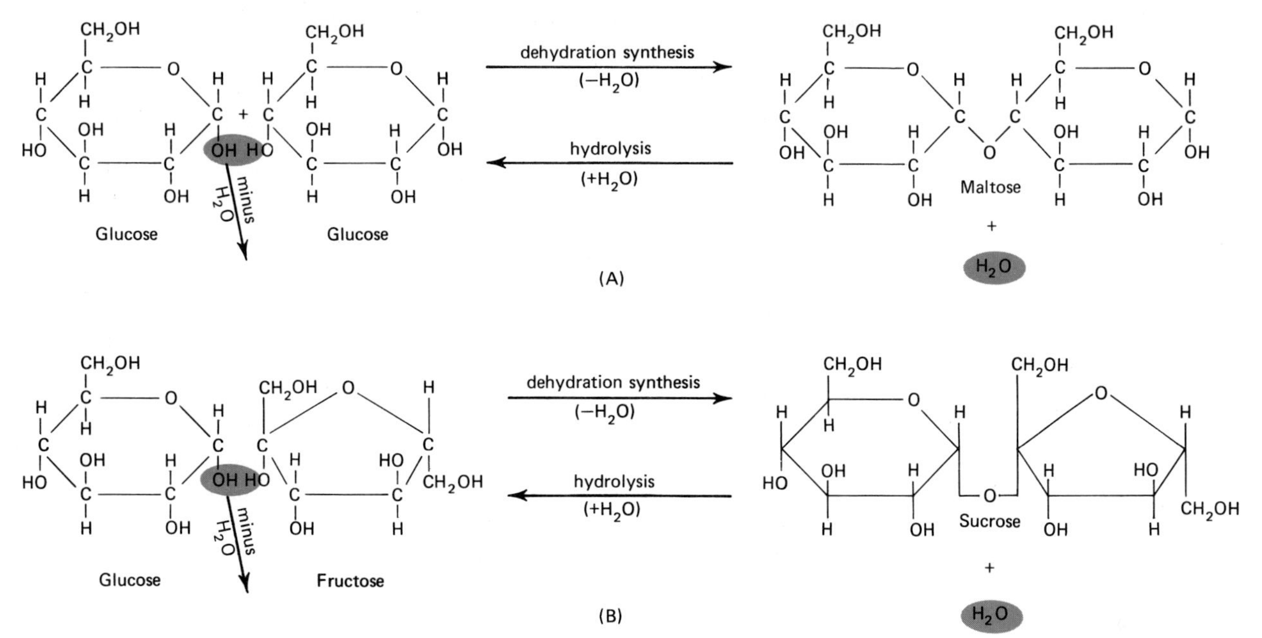

FIGURE 2-7. Dehydration synthesis and hydrolysis of maltose (**A**) and sucrose (**B**). In dehydration synthesis smaller molecules are built up into larger ones and there is a loss of water. In the reverse reaction, hydrolysis, there is an addition of water to a larger molecule and smaller molecules are formed. Although not indicated here, these reactions, like most other biochemical reactions, are enzymatically controlled.

larly more complex land plants, as the principal food reserve or storage product. Typically, starches are stored in seeds, stems, leaves, and particularly the roots of plants in the form of *starch grains.* Starch grains consist of concentrically oriented layers of starch molecules. The concentric ring arrangement of starch molecules is owed to the fact that during the day when the plant is most active, the starch is produced and deposited more densely than at night when the plant is less active. Starch grains vary greatly in size and number in different plants, but the consistency of form and shape is adequately diagnostic for identifying certain plants (see Figure 5-4).

Chemically, starches consist of long series of glucose molecules bonded to each other. As a result of the manner in which this bonding occurs, starch molecules reveal two distinguishable components: amylose and amylopectin. The *amylose* portion consists of unbranched coiled chains of glucose molecules and may contain from 300 to 1000 such glucose units in a single, long molecule. *Amylopectin,* on the other hand, exhibits a fairly high degree of branching. Although the percentages of amylose and amylopectin differ widely in starches from different kinds of plants, most plants contain from 20 to 30% amylose. The repetition of many simple, smaller molecules such as glucose, into larger, chain-like molecules such as starch, is called *polymerization* and the resulting molecule thus formed is termed a *polymer.* In all cases, polymerized molecules are fairly large.

Cellulose is an important component of the cell walls of higher plants. Generally, it is found in combination with other compounds in plants, such as lignin; however, one source of pure cellulose is cotton fibers, a product of cotton plants. Other varieties of pure cellulose include flax and hemp, the raw materials of textiles. Chemically, cellulose is a polymer of 1000 or more glucose molecules. Like amylose, it is unbranched but, unlike amylose, it is not coiled. Thousands of cellulose molecules are chemically bonded to each other in an

TABLE 2-4. Representative Lipid Derivatives of Biological Importance

Derivative	Remarks and Importance
I. Phospholipids	Glycerol joined to two fatty acid molecules, the third place is occupied by a phosphate group; major lipid constituent of cell membranes; found in human plasma; abundant in seeds and egg yolk; found in high concentrations in brain, nerves, and neural tissues generally.
II. Waxes	Lipids composed of fatty acids with long-chain alcohols, not glycerol.
A. Suberin	Waterproofing material found in the cell walls of cork.
B. Cutin	Waterproofing substance covering the epidermis of leaves, stems, and fruits.
III. Carotenoids	Plant and animal pigments.
A. Carotenes	Occur in all leaves; responsible for the red, orange, yellow, and cream-white colors of tomatoes, carrots, egg yolk, milk, and other plant and animal products; *vitamin A* is a derivative of carotenes; a further derivative of vitamin A, *retinene,* acts as a photoreceptor in the retina of the eye.
B. Xanthophylls	Examples include *lutein,* a yellow substance in autumn leaves, and *fucoxanthin,* a pigment in brown and other algae.
IV. Steroids	All are related in that they possess a structure similar to that of cholesterol; regulate sexual development, and influence numerous aspects of metabolism.
A. Cholesterol	Present in all animal cells, blood, and particularly nervous tissues.
B. Vitamin D	Synthesized in human skin upon exposure to ultraviolet light.
C. Cortisone	Hormone produced by the vertebrate adrenal glands.
D. Androgens	Male sex hormones; the most influential is *testosterone.*
E. Estrogens	Female sex hormones, of which *progesterone* is the most powerful.
F. Bile salts	Substances produced by vertebrates which emulsify fats prior to digestion.
V. Porphyrins	Pigment compounds distantly related to lipids which are combined with protein.
A. Chlorophyll	Green pigment of photosynthetic organisms.
B. Hemoglobin	Oxygen-transporting substance in the blood or tissues of many animals.
C. Cytochromes	Electron-carrying proteins in the respiration of all cells.
D. Bile pigments	*Bilirubin* (reddish) and *biliverdin* (green) have their origin in hematin, a derivative of hemoglobin; responsible for brown color of feces.
VI. Other lipoid substances	Among other lipid-like compounds in living matter are *xanthocyanins,* plant pigments; *vitamin E,* the "antisterility" vitamin; *vitamin K,* the antihemorrhagic vitamin; and *coenzyme Q,* a substance that functions in respiration.

intricate, organized network to form plant cell walls. The basic physical properties of cell walls—tensile strength, elasticity, and plasticity—are directly dependent upon the amount and arrangement of the cellulose molecules. Figure 5-14 shows the organization of cellulose into plant cell walls.

Chitin, like cellulose, is also a structural polysaccharide. It is prevalent among certain invertebrates such as crustaceans and insects, and constitutes part of the exoskeletal structure of these organisms. *Glycogen,* in animals, is the counterpart of starches; it serves as a storage form of glucose. It is considerably more branched than starches and is more soluble in water. Glycogen, like starches, is capable of being hydrolyzed into glucose molecules as the need arises.

Lipids. **Lipids** constitute a second major group of organic compounds found in living matter. Like carbohydrates, they are composed of atoms of carbon, hydrogen, and oxygen, without the 2:1 ratio between H and O atoms. Most are insoluble in water, but dissolve readily in solvents such as ether, chloroform, or alcohol. Lipids represent a very diverse group of organic compounds that function in living organisms as structural components of cells, fuels for energy, and storage materials. Inasmuch as the diversity of lipids is so great, only one kind of lipid, the fats, will be discussed in detail. Many lipid derivatives, however, are shown in Table 2-4 in order to emphasize their importance in living matter.

Chemically, a molecule of fat is composed of two structural components: an alcohol called *glycerol* and a group of compounds known as *fatty acids.* A single molecule of glycerol consists of three carbon atoms to which there are attached three hydroxy (—OH) groups. The structural formula for glycerol is shown in Figure 2-8A. Fatty acids consist of a hydrocarbon chain (composed only of carbon and hydrogen atoms) plus a carboxyl (—COOH) or acid group. Fatty acids differ in length of the hydrocarbon chain they possess, that is, in the number of carbon atoms making up the length of the chain. The most common fatty acids contain an even number of carbon atoms (Figure 2-8B). A single molecule of fat is formed when a molecule of glycerol combines with one to three fatty acid molecules. In the reaction, one to three molecules of water are formed (dehydration synthesis), depending upon the number of fatty acid molecules reacting. The linkage formed where the water molecule is removed is the *ester linkage.* In the reverse reaction, hydrolysis, a fat molecule may be broken down into its component fatty acid and glycerol molecules (Figure 2-8C).

The three fatty acids involved in the formation of fats may be similar or different. For example, three molecules of fatty acid A may combine with a glycerol molecule, or one molecule each of fatty acids A, B, and C may unite with a molecule of glycerol. As a result, there is variation in fat molecules.

Fats are quite common in the bodies of animals but are found to a lesser degree in plants. They represent an organism's most highly concentrated source of biologically usable energy. Inasmuch as fats contain more hydrogen than carbohydrates and since they are capable of a greater degree of oxidation, they provide twice as many calories per gram as carbohydrates.

The oxidation of a fat yields water. This factor, coupled with a high energy yield, assumes importance in considering the storage of fats prior to hibernation of certain animals. It provides them with large quantities of usable energy and allows them to survive for long periods of time without a supply of drinking water.

Plants synthesize some fats that, because of their relative insolubility, are stored in the watery contents of cells in the form of clear droplets or globules. Animals store fat in large clear globules within fatty tissue called *adipose* tissue. Adipose serves not only as storage for fats but also as a source of heat insulation for the body.

Proteins. The **proteins** are far more complex in structure than either carbohydrates or lipids. As a group, they play a unique role in living matter since the complexity and diversity of life itself are directly related to the complexity and diversity of proteins. In addition to comprising a major portion of the structure of cells, they assume a function in its vital activities. As enzymes, or functional proteins, they catalyze most biochemical reactions that constitute cell activities. Other proteins, such as myosin, function in muscular contraction; several act as hormones which regulate metabolic processes; and still others function as antibodies, one of the organism's defenses against invasion by bacteria and viruses. Important structural proteins include: keratin in the skin, fingernails, and hair, and collagen, a protein of bone and connective tissue. Moreover, they are found in every essential component of the cell.

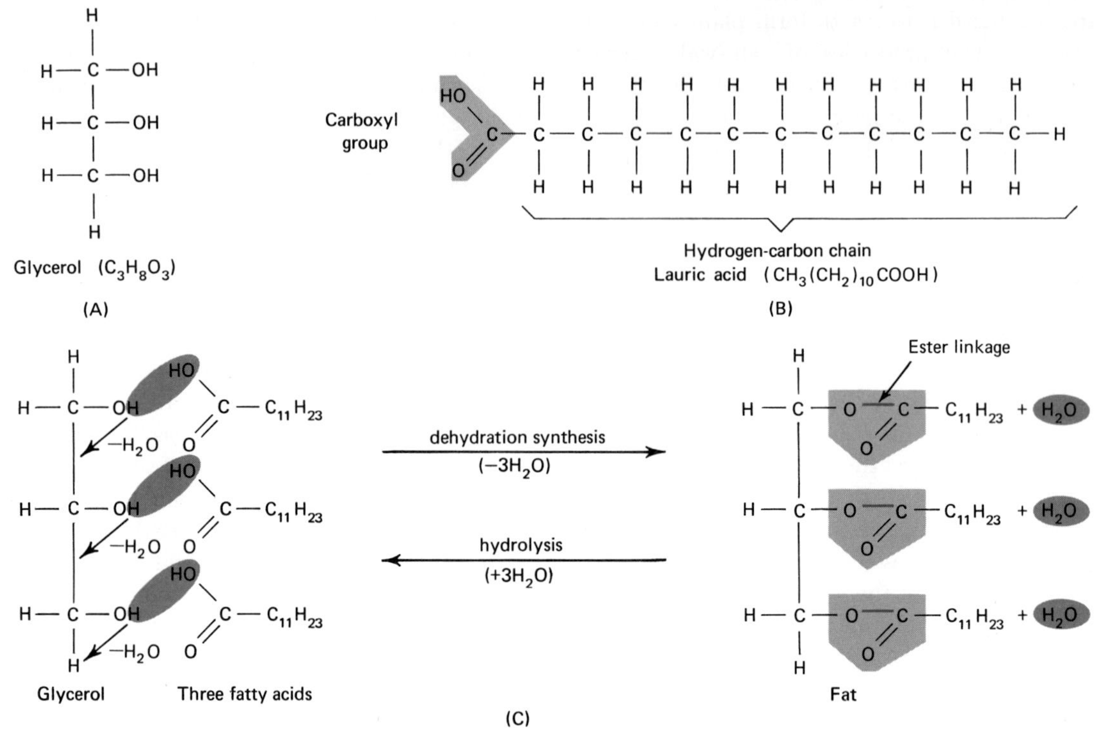

Figure 2-8. Structural formulas for glycerol (**A**) and lauric acid, a fatty acid (**B**). (**C**) The chemical combination of a molecule of glycerol with three fatty acid (lauric) molecules forms a molecule of fat and three molecules of water (dehydration synthesis). The addition of three water molecules to a fat forms glycerol and three fatty acid molecules (hydrolysis). Note the position of the ester linkage.

Proteins, like carbohydrates and fats, may also be broken down to provide energy for cellular use.

All proteins contain carbon, hydrogen, oxygen, and nitrogen. In addition, some contain sulfur and/or phosphorus. Just as the monosaccharides are the building blocks of larger carbohydrate molecules, and fatty acids and glycerol are the building blocks of fats, **amino acids** are the fundamental units of proteins. Amino acids that are found in living organisms contain at least one carboxyl (—COOH) group and one amine (—NH_2) group attached to the same carbon (alpha carbon) atom (Figure 2-9A). Such amino acids are called *alpha-amino acids.* Also attached to the alpha carbon is a side group (R group), which distinguishes the various amino acids. The side group may be an unbranched or branched chain or a ring structure that may be cyclic (all carbons in the ring) or heterocyclic (an atom other than carbon is included in the ring). The side group may contain functional groups such as the sulfhydryl group (—SH), the hydroxy group (—OH), or additional carboxyl or amine groups. These side groups and the carboxyl and alpha-amine groups of the basic structure of amino acids play an important part in the total structure of a protein. One of the innumerable possibilities in the structures of amino acids, only about twenty different amino acids occur in proteins (Table 2-5).

Proteins are important buffers. The Brønsted–Lowry definitions of an acid and a base whereby an acid is a proton donor and a base is a proton acceptor were briefly stated previously and can now be applied to amino acids. The hydrogen of the carboxyl group is slightly ionized in aqueous solution. Since it contributes a hydrogen ion or single proton (when the hydrogen atom loses its electron, a single proton remains), it fits

Figure 2-9. Building units of proteins. (**A**) Structural formula for glycine, a simple amino acid. (**B**) Chemical reaction of glycine and alanine to form a dipeptide, glycylalanine. In this reaction, the carboxyl group of one amino acid (glycine) attaches to the amine group of another (alanine) and water is lost. The carboxyl group yields the OH while the amine group supplies the H for the formation of water. The newly formed bond between the carbon atom of glycine and the nitrogen atom of alanine is called a peptide link.

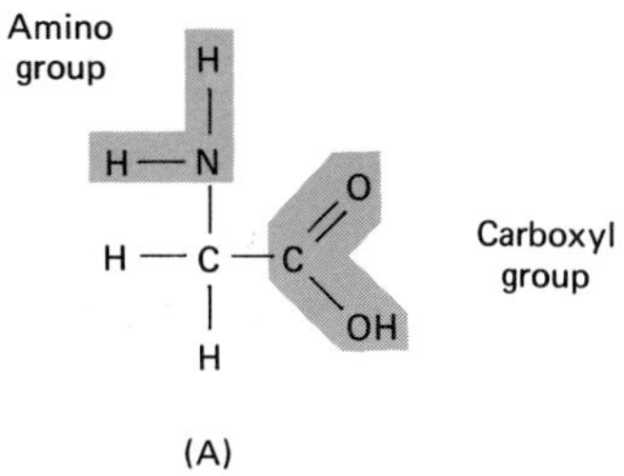

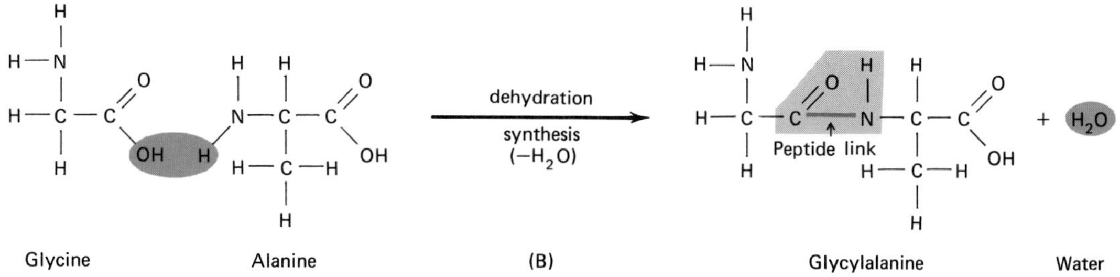

both the classical and Brønsted–Lowry definitions of an acid. On the other hand, there are no hydroxyl ions in the amino acids and therefore, classically they cannot be considered a base. However, according to the Brønsted–Lowry definition of a proton acceptor, it can be shown that the amine group is basic. The electronic configuration for the amine group is as follows:

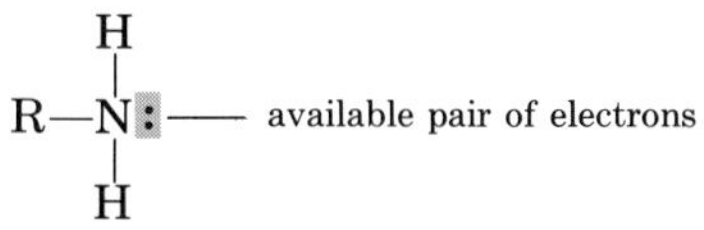

$$\mathrm{R{-}\overset{\displaystyle H}{\underset{\displaystyle H}{N}}{:}} \text{ — available pair of electrons}$$

The free pair of electrons on the nitrogen is capable of accepting a proton.

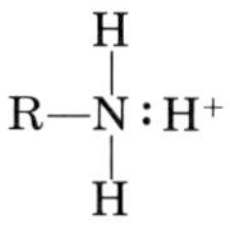

$$\mathrm{R{-}\overset{\displaystyle H}{\underset{\displaystyle H}{N}}{:}H^+}$$

Therefore, if there is an excess of acid (H^+ ions) present, the free amine groups of the protein will remove them from solution; and if there is an excess of alkali (OH^- ions) added, they will combine (be removed from solution) with available hydrogen ions from the carboxyl group to form water. Thus, the protein molecule is able to assist in maintaining a stable pH. A substance which is able to react with both acids and alkalies is called *amphoteric.* This ionization quality of amino acids accounts for many of the unique and important chemical and physical properties of proteins.

In the formation of proteins, the carboxyl group of one amino acid combines with the alpha amine group of another amino acid with the removal of a water molecule. The resulting bonding between amino acids is known as a **peptide linkage** and provides the basic protein structure (Figure 2-9B).

In the example shown in Figure 2-9B, two amino acids undergo dehydration synthesis. The carboxyl group of one amino acid yields the OH^-, while the amine group of the other supplies the H^+ for the formation of water. The peptide link is formed between the carbon atom of the carboxyl group of one amino acid and the nitrogen atom of the amine group of another amino acid. The resulting compound, because it consists of two amino acids joined together via a peptide link, is referred to as a *dipeptide.* The addition of another amino acid to a dipeptide would form a *tripeptide.* Further addition of amino acids would result in the formation of a long chain-like molecule referred to as a *polypeptide,* or a large protein molecule.

It should be obvious that an extremely large number of proteins is possible, owing to the number and sequences of amino acid linkages that can occur. The twenty different amino acids found in proteins may be arranged

TABLE 2-5. Naturally Occurring Amino Acids[a]

Amino Acid	Structural Formula	Standard Abbreviation	Characteristic of R Group
Glycine	$H-\overset{H}{\underset{NH_2}{C}}-COOH$	Gly	Unbranched chain
Alanine	$CH_3-\overset{H}{\underset{NH_2}{C}}-COOH$	Ala	Unbranched chain
Valine	$H_3C-\underset{CH_3}{CH}-\overset{H}{\underset{NH_2}{C}}-COOH$	Val	Branched chain
Leucine	$H_3C-\underset{CH_3}{CH}-CH_2-\overset{H}{\underset{NH_2}{C}}-COOH$	Leu	Branched chain
Isoleucine	$H_3C-CH_2-\underset{CH_3}{CH}-\overset{H}{\underset{NH_2}{C}}-COOH$	Ileu	Branched chain
Serine	$HO-CH_2-\overset{H}{\underset{NH_2}{C}}-COOH$	Ser	Hydroxy (—OH) containing
Threonine	$H_3C-\underset{OH}{CH_2}-\overset{H}{\underset{NH_2}{C}}-COOH$	Thr	Hydroxy (—OH) containing
Cysteine	$HS-CH_2-\overset{H}{\underset{NH_2}{C}}-COOH$	Cys	Sulfur containing

[a] Note that all the amino acids are similar in that each contains at least one carboxyl group and one alpha amine group (except proline). These similarities are indicated by the colored area. They differ in the chemical composition of R group. The R group is the unique number and arrangement of atoms that give each amino acid its specific structure and characteristic chemical properties.

TABLE 2-5. Naturally Occurring Amino Acids[a] [*Continued*]

Amino Acid	Structural Formula	Standard Abbreviation	Characteristic of R group
Cystine	$S{-}CH(NH_2){-}COOH$ / $S{-}CH(NH_2){-}COOH$ (the two S atoms joined)	Cy	Sulfur containing
Methionine	$H_3C{-}S{-}CH_2{-}CH_2{-}CH(NH_2){-}COOH$	Met	Sulfur containing
Glutamic acid	$HOOC{-}CH_2{-}CH_2{-}CH(NH_2){-}COOH$	Glu	Additional carboxyl (—COOH) group, acidic
Aspartic acid	$HOOC{-}CH_2{-}CH(NH_2){-}COOH$	Asp	Additional carboxyl (—COOH) group, acidic
Lysine	$H_2N{-}CH_2{-}CH_2{-}CH_2{-}CH_2{-}CH(NH_2){-}COOH$	Lys	Additional amine (—NH_2) group, basic
Arginine	$H_2N{-}C({=}NH){-}NH{-}CH_2{-}CH_2{-}CH_2{-}CH(NH_2){-}COOH$	Arg	Additional amine (—NH_2) group, basic
Asparagine	$H_2N{-}C({=}O){-}CH_2{-}CH(NH_2){-}COOH$	Asn	Additional amine (—NH_2) group, basic
Glutamine	$H_2N{-}C({=}O){-}CH_2{-}CH_2{-}CH(NH_2){-}COOH$	Gln	Additional amine (—NH_2) group, basic

(continued)

TABLE 2-5. Naturally Occurring Amino Acids[a] [*Concluded*]

Amino Acid	Structural Formula	Standard Abbreviation	Characteristic of R Group
Phenylalanine	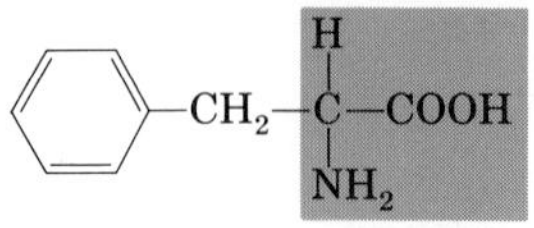 (benzene ring)$-CH_2-C(H)(NH_2)-COOH$	Phe	Cyclic
Tyrosine	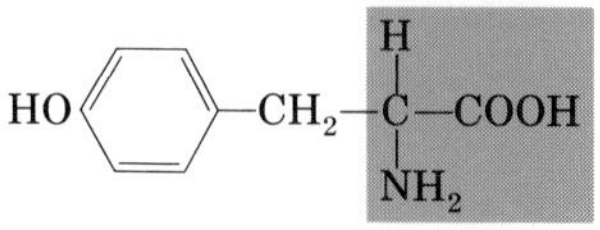 HO(benzene ring)$-CH_2-C(H)(NH_2)-COOH$	Tyr	Cyclic
Histidine	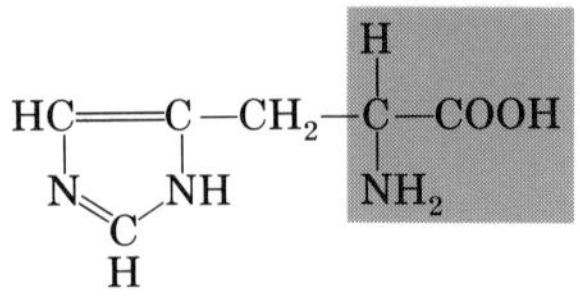 $HC=C-CH_2-C(H)(NH_2)-COOH$; ring: $HC=C$, N, NH, C–H	His	Heterocyclic
Tryptophan	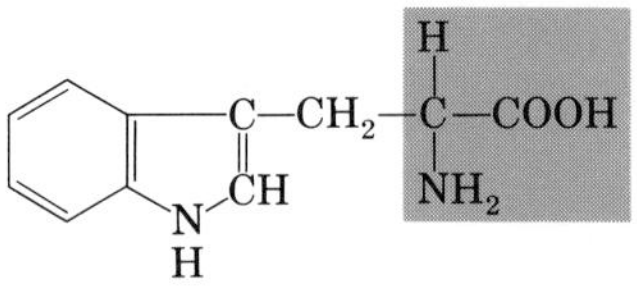 (indole ring: benzene, C, CH, $N-H$)$C-CH_2-C(H)(NH_2)-COOH$	Trp	Heterocyclic
Proline	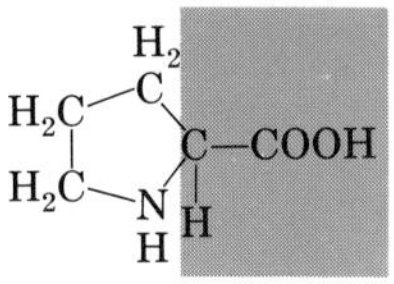 ring: H_2C, H_2C, $C(H_2)$, $N(H)$; $C(H)-COOH$	Pro	Heterocyclic (—NH—) imine group

in any order to form a sequence of hundreds of units in the formation of a single protein molecule. This situation is quite similar to having an alphabet of twenty letters with which to form words. Assuming that no prescribed sequence of letters exists to form these words, the number of possible words is exceedingly large.

Most protein molecules are so large and complex that, aside from their size, and in some cases their molecular formulas, it has not yet been possible to learn everything about their structure. One outstanding recent development, however, has been the discovery of methods for determining the amino acid sequence of a complete protein molecule. Sanger established for the first time the exact amino acid sequence of a protein, insulin (Figure 2-10A). This hormone, formed by the pancreas, is required for the proper utilization of sugar by the body. If it is secreted in insufficient quantities, the disease diabetes mellitus results. Relative to other proteins, insulin is fairly small, having a molecular weight of about 6000. In general, proteins range in molecular weight from 6000 to 40 million.

Levels of Structural Organization. It is generally recognized that proteins assume at least four levels of structural organization. These are referred to as primary, secondary, tertiary, and quaternary. The *primary structure* of a protein is determined by the type and sequence of amino acids comprising the polypeptide chain or chains (Figure 2-10A). The definite sequence of the amino acids results in a specific protein. It has been found that alterations in the sequence of amino acids can have profound metabolic effects. For example, the

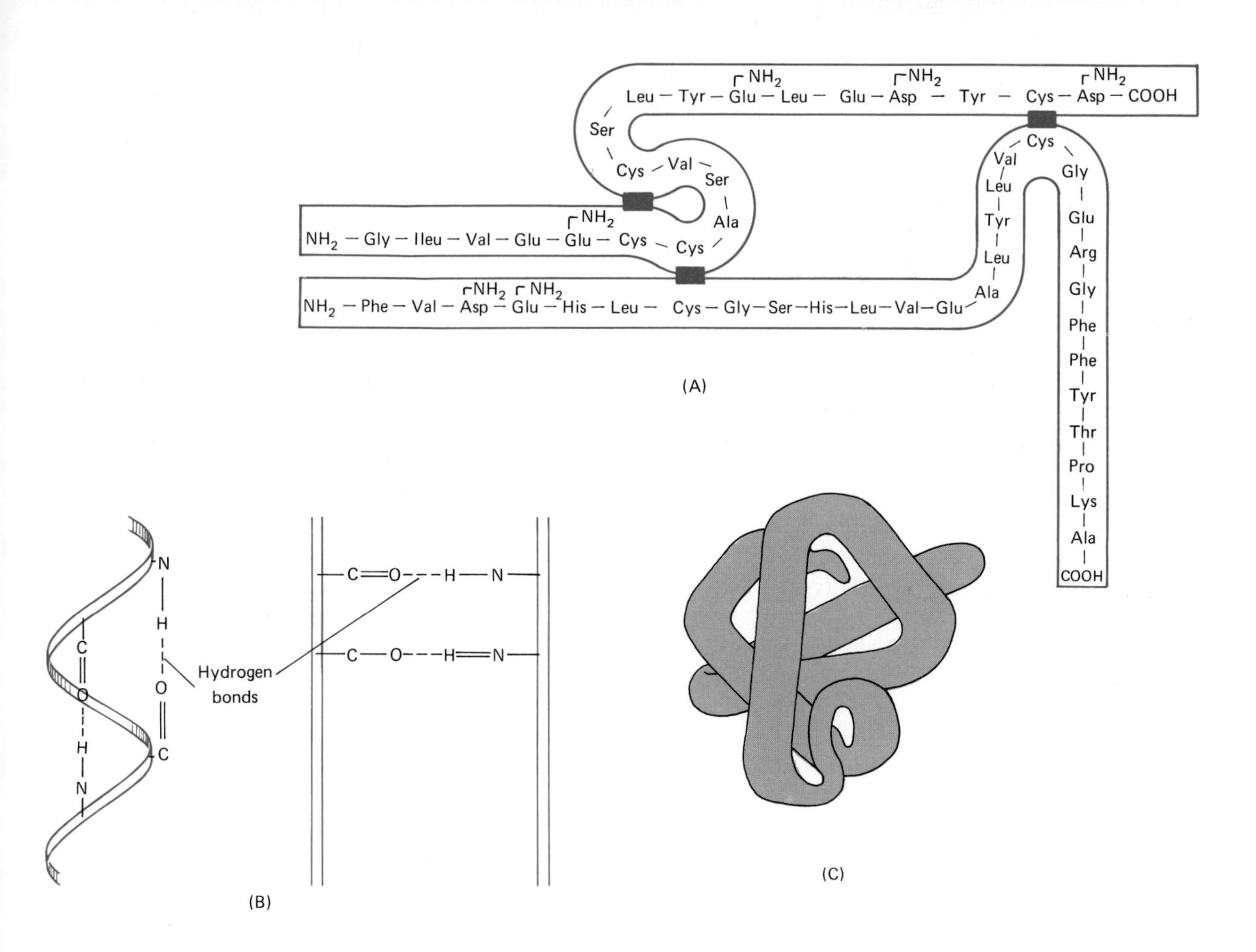

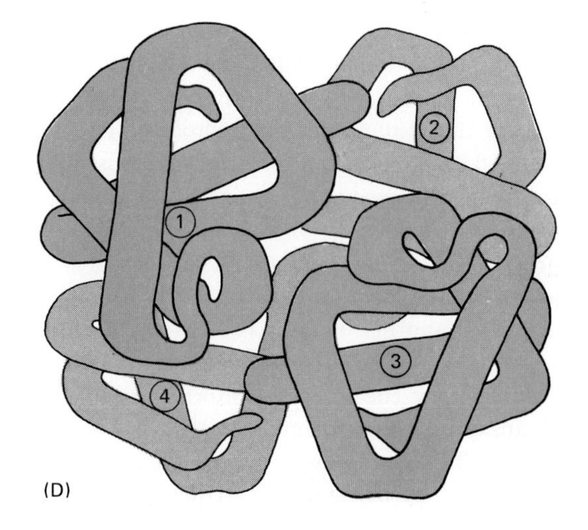

FIGURE 2-10. Structural levels of protein organization. (**A**) Amino acid composition and sequence of beef insulin illustrating the primary level of organization. Secondary structure: The helix of a polypeptide chain. (**B**) (*left*) the pairing of chains side by side via hydrogen bonding (*right*). (**C**) Tertiary structure—the bending or folding of the polypeptide chain over itself. (**D**) Quaternary structure—the relationship of four polypeptide units (numbered) of a complex protein to one another. The increasing levels of complexity are due to hydrogen bondings, as well as other chemical attractions and reactions between the side groups of certain amino acids within the polypeptide chains.

replacement of glutamic acid by valine in a specific position in the hemoglobin molecule accounts for the metabolic hereditary disease sickle cell anemia.

Analysis through physical and chemical investigations indicates that few proteins exist in living matter as straight-chain sequences of amino acids at the primary level of organization. In fact, the available data suggest that most proteins are coiled or twisted in a number of ways at the *secondary level* of organization. The most common form is a spiral-shaped structure called a *helix* (Figure 2-10B, left). These clockwise spirals or helixes of varying degrees are maintained by weak hydrogen bonds, generally between the hydrogen atom attached to the nitrogen atom of the peptide link at one location on the molecule and the oxygen atom of the peptide link at another location. These hydrogen bonds are formed at given distances, giving a regular repeating spiral structure. Another type of secondary structure occurs in certain proteins when series of chains line up side by side via hydrogen bonding, forming what is called a *pleated sheet structure.* This partially accounts for the structure of such protein fibers as silk, wool, and hair.

In addition to assuming straight-chain and helical configurations, globular proteins are also arranged into a third structural level of organization called the *tertiary level* (Figure 2-10C). The tertiary structure is the nonregular coiling, looping, and twisting of a polypeptide chain as a result of interactions among the various functional groups and branches on the specific amino acids. For example, an ester linkage can be formed by the action of a carboxyl group and a hydroxy group on two different amino acids. The sulfhydryl groups on different amino acids can form a disulfide (—S—S—) link by the oxidation of the hydrogens. The negatively charged carboxyl group formed by the loss of a hydrogen ion is attracted to the amine group when it becomes positively charged by picking up a hydrogen ion. Hydrogen bonds may be formed between the hydrogen atoms at one location on a molecule with nitrogen or oxygen at another. Finally, there is a type of attraction which is caused by a *hydrophobic interaction* (mutual repulsion) by the solvent (water). Certain of the side chains or branches are made up of carbon and hydrogen atoms. These chains are relatively nonpolar, that is, the charges are rather equally distributed throughout the group. The polarity of the water molecule repulses these groups, causing them to come together. These various types of interactions within the protein molecule resulting in tertiary structure are shown in Figure 2-11.

The fourth level of organization is the *quaternary structure* resulting from an aggregation or polymerization of individual polypeptide units (Figure 2-10D). For example, the enzyme phosphorylase *a* is made up of four identical protein subunits. Each of these units alone does not exhibit enzymatic activity but becomes active when associated in groups of four. If the subunits making up the protein are identical, it is described as a homogeneous quaternary structure, and if the units are not alike, it is described as a heterogeneous quaternary structure. For example, tobacco mosaic virus is an association of ribonucleic acid with protein units that together form the virus particle (see Figure 8-3). Not all proteins exhibit quaternary structure. It may also be noted that the amount of secondary structuring varies among different proteins or even may be lacking. For example, the globular proteins such as hemoglobin have primary, tertiary, and quaternary structure, but little in the way of secondary structure.

Classification of Proteins. The large size and complexity of proteins precludes any simply unambiguous classification scheme. Not enough is known chemically about most proteins to base classification on their amino acid composition or on their secondary or tertiary structure. Proteins are usually classified on the basis of their functions, shape, or physicochemical properties such as solubility and coagulability. Such a classification scheme is shown in Table 2-6.

In recent years the structure of various proteins has been elucidated by various methods such as sedimentation velocity in centrifuging, osmotic pressure, radioactive tracers, and X-ray diffraction. X-ray diffraction has been extremely useful in determining the shape of the protein molecule. When an X-ray beam is directed toward a substance, generally a crystal, the planes formed by the atoms or molecules composing the structure of the matter diffract the beam to form a pattern on photographic paper. This diffraction is caused by a modification of the X-ray beams as they appear to be deflected by the arrangement of the atoms or groups of atoms in the molecule. By carefully studying series of these X-ray diffraction patterns, it has been possible to

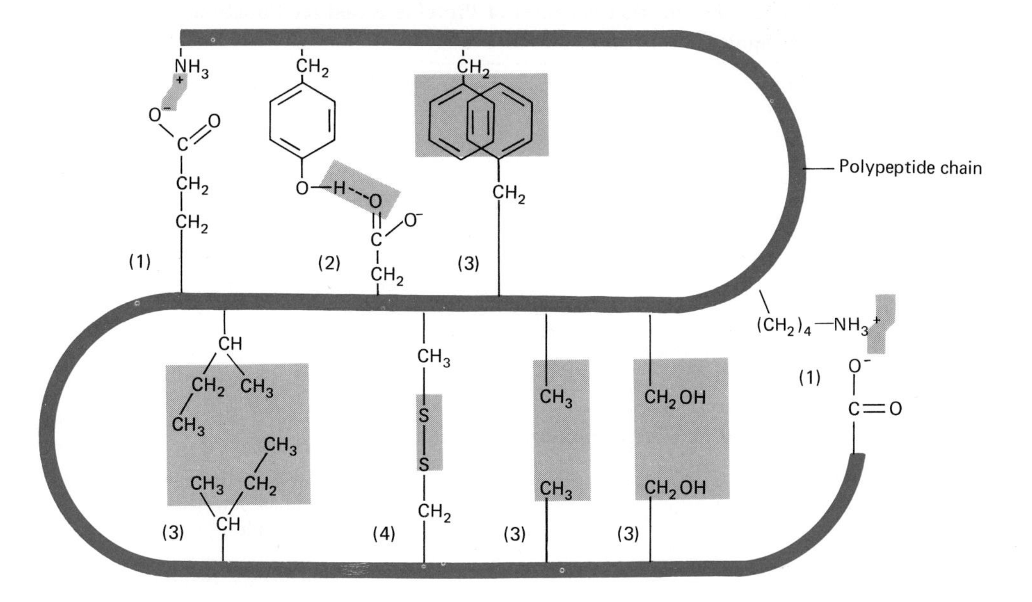

FIGURE 2-11. Stabilization of the tertiary structure of a polypeptide chain. The reactions of various side groups on the amino acids result in a folding or looping of the chain. (1) Attraction of unlike charged groups. (2) Hydrogen bonding. (3) Hydrophobic symmetrical groups. (4) Disulfide linkage. [*Modified from C. B. Anfinsen,* The Molecular Basis of Evolution. *New York: John Wiley and Sons, Inc. Copyright © 1959.*]

establish the tertiary structure of some proteins. In general, fibrous proteins produce fairly regular diffraction patterns showing the peptide chain to be orderly and rigidly oriented. Globular proteins exhibit a more random arrangement in X-ray diffraction patterns, showing a nonlinear, unordered arrangement of peptide chains.

Some Chemical Reactions of Proteins. The buffering action of amino acids has already been explained. Since proteins are composed of amino acids, they react to some extent as buffers. However, it is to be noted that only free acidic and basic groups found in the R group (see Table 2-5) can buffer. The alpha-amine group and carboxyl group form the peptide link and, therefore, cannot buffer. The various functional groups attached to the amino acids, in addition to reacting with each other as previously mentioned in the formation of tertiary structure, also react with other substances. These groups will react with acids, alkalies, and metallic ions. Physical forces such as temperature, agitation, and electrical current may also alter the chemical bonding within protein molecules. The results of an alteration in any of the bonds of the protein molecule may result in denaturation. Generally, **denaturation** is an alteration of the tertiary structure of the molecule which may greatly change the properties of the specific protein. Once a protein is denatured, the reaction sometimes cannot be reversed. Denaturation can be easily demonstrated with the white of an egg which contains large amounts of protein. If the white of an egg is rapidly agitated, it forms meringue; if it is heated it solidifies; and, if acid, alkali, or salt of a heavy metal is added, a precipitate forms. Since protein molecules are formed by the removal of water molecules, they can be broken down to amino acids by the addition of water.

Functional Proteins—Enzymes. **Enzymes** are generally globular proteins ranging in molecular weight from about 10,000 up into the millions. Of the 1000 or more known enzymes, each has a three-dimensional characteristic shape with a specific surface configuration due to its primary, secondary, and tertiary structures.

The manner in which an enzyme actually lowers activation energy is not completely understood. However, from what has been learned from the study of enzyme behavior, the following sequence of events is believed to occur.

1. The surface of the **substrate** molecule(s)—that is, the molecule or molecules with which the enzyme reacts—comes into contact with the surface of the enzyme molecule.
2. A temporary intermediate compound called an *enzyme-substrate complex* forms.
3. The substrate molecule is transformed (by rear-

TABLE 2-6. Different Classification Schemes of Proteins Based on Function, Shape, and Physicochemical Properties

Kinds of Proteins	Description and/or *Examples*
I. CLASSIFICATION BASED UPON FUNCTION	
A. Enzymes (catalytic proteins)	*Lactase, ribonuclease, pyruvic dehydrogenase, fumarase.*
B. Structural proteins	*Collagen, elastin, keratin.*
C. Regulatory or hormonal proteins	*Insulin, adrenaline.*
D. Transport proteins	*Hemoglobin, myoglobin.*
E. Genetic proteins	*Nucleoproteins, histones.*
F. Immune proteins	*Gamma globulin.*
G. Contractile proteins	*Actin, myosin.*
H. Storage proteins	*Zein, ovalbumin, casein.*
II. CLASSIFICATION BASED UPON SHAPE	
A. Fibrous proteins	Long thread-like molecules whose helical strands often form fibers or sheets; often insoluble in water. Examples: *collagen, elastin, keratin.*
B. Globular proteins	Generally soluble in aqueous media; spheroid or ovoid shape; further classified on the basis of solubility—water soluble, heat coagulable include proteins in blood serum, egg white, milk, and in certain plants (pea, wheat); soluble in dilute salt solutions include proteins in blood serum, egg white, and in plants such as peanuts.
III. CLASSIFICATION BASED UPON PHYSICOCHEMICAL PROPERTIES	
A. Simple proteins	Yield only amino acids on hydrolysis.
1. Albumins, globulins, glutelins	Soluble in dilute acids and alkalies.
2. Scleroproteins	Nonsoluble proteins such as *collagen* or *keratin.*
3. Prolamines	Soluble in alcohol but insoluble in water; found in the seeds of plants.
4. Histones	Soluble in water, dilute acids, and alkalies; contain a large portion of basic amino acids such as the *globin* of hemoglobin.
5. Protamines	Basic proteins which are essentially large polypeptides.
B. Conjugated proteins	Yield amino acids and nonprotein products on hydrolysis.
1. Glyco- or mucoproteins	Proteins plus carbohydrates. Example: *mucin* of saliva.
2. Lipoproteins	Proteins plus a lipid. Example: *liporutellin* of egg yolk.
3. Chromoproteins	Proteins plus a pigmented prosthetic group. Examples: *hemoglobin, myoglobin.*
4. Metalloproteins	Protein plus a metal element such as iron, magnesium, copper, or zinc. Examples: *ferritin* (Fe), *tyrosine oxidase* (Cu), *alcohol dehydrogenase* (Zn).
5. Nucleoproteins	Protein plus nucleic acid. Example: *nucleohistone.*

rangement of existing atoms, a breakdown of the substrate molecule, or a combination of several substrate molecules).

4. The transformed substrate molecules (products) move away from the surface of the enzyme molecule.

5. The recovered enzyme, now freed, reacts with other substrate molecules (Figure 2-12).

An outstanding characteristic feature of an enzyme reaction is its specificity with regard to the substrate with which the enzyme will react and the type of reaction it will catalyze. The fate of a given reactant (substrate) depends upon the specific enzyme that reacts upon it. For example, glucose-6-phosphate may be acted upon by at least four different enzymes, each one of which will give a different product. Similarly, a specific enzyme may be capable of hydrolyzing a peptide link only between two specific amino acids, and only those amino acids. Moreover, there are other enzymes that are capable of hydrolyzing starch and are not capable of hydrolyzing glycogen, even though the two substrates are quite similar.

The specificity of enzymes is probably related to what are called **active sites.** It is at these active sites that the transformation of the substrate is believed to occur. It has been found that the enzyme is able to distinguish between fairly similar compounds. In most instances the substrate is much smaller than the enzyme. Therefore, in consideration of size, it would appear that only a small fraction of the enzyme surface is in contact with the substrate surface. From this it can be assumed that some of the amino acids of the protein are in direct contact with the substrate (forming an active site), while others play an indirect role and still others play no role at all. For example, it has been found that removal of one portion of the enzyme molecule may not affect its ability to catalyze a reaction, while removal of another portion will.

Previously, the enzyme was thought to be somewhat like a template (pattern) that interacted with its substrate because the two molecules fitted together much like pieces of a jigsaw puzzle. According to modifications of this concept, the active site of the enzyme consists of contact amino acids whose fit with the substrate determines specificity. As a result of the typical protein three-dimensional folding, these specific contact amino acids, which are relatively far apart in primary structure (sequence), are fairly close together. This arrangement of the amino acids forming a specific active site, rather than rigidly fitting the enzyme–substrate together in jigsaw puzzle fashion, allows for flexibility, resulting in what is called an *induced fit.* The induced fit of the active site provides a flexibility that has been found to be needed for the proper alignment of the catalytic groups. This explanation helps to account for the possible function of large segments of the enzyme molecule that do not take place in the active site. The reactions between various functional groups on the amino acids in these segments maintain the specific three-dimensional structure of the molecule and probably properly orient and hold the active site together.

The major difficulty in hypothesizing the mechanisms of enzymatic action is accounting for the exceptional catalytic properties of enzymes. While it is above the level of this text to discuss quantitatively and qualitatively the problems in determining these mechanisms, several properties of enzymes will be mentioned.

1. Enzymes are exceedingly efficient. Under optimum conditions they catalyze reactions at rates that are 10^8 to 10^{10} times more rapid than those of comparable reactions without enzymes. The *turnover number* (number of substrate molecules metabolized per enzyme molecule per minute) is generally 1000 and in some instances may be as high as one million.

2. As previously mentioned, enzymes are specific in the reactions they catalyze as well as in the substrates utilized.

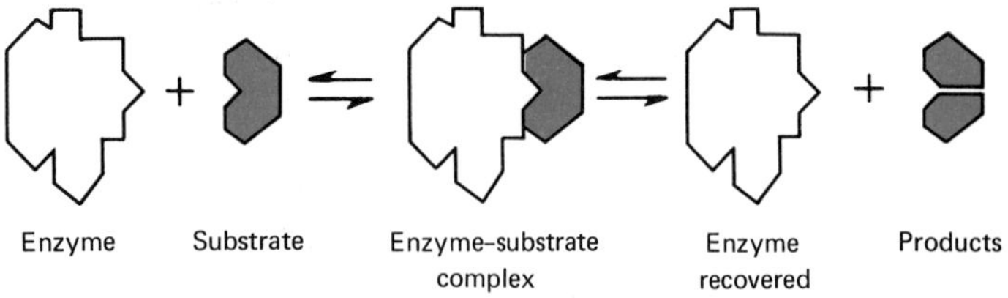

FIGURE 2-12. Enzyme action. During a biochemical reaction, an enzyme and a substrate molecule combine to form an unstable enzyme–substrate complex. In the process of combination, the substrate is activated and transformed into products. Following transformation, the enzyme is recovered and may be used again to catalyze a similar biochemical reaction.

3. The reactions that enzymes catalyze take place in aqueous solution and at relatively low temperatures.

4. Enzymes are subject to various cellular controls. Their rate of synthesis and their concentration are under genetic control and are influenced by various other molecules present in the cell (Chapter 8). Enzymes are present in the cell in both active and inactive forms. The rate at which the inactive form becomes active or the active form becomes inactive is determined by the cellular environment.

Enzymes share with other proteins the ability to become denatured. In the denatured state, enzymes lose their native configuration (normal biological form) at the tertiary level and sometimes at the secondary level. As might be expected, this alteration in structure changes the arrangement of the amino acids in the active site with a loss in catalytic ability and of vital biological activity. In addition to heat, other agents that may denature enzymes include high concentrations of heavy metal ions (e.g., copper, zinc, silver, arsenic, and mercury), alcohol, ultraviolet radiation, and concentrated acidic and basic solutions.

Some enzymes appear to consist solely of proteins. Many other enzymes, however, consist of two distinct portions. The first part, called an **apoenzyme,** is the protein portion composed of only amino acids. The other part is a nonamino acid constituent, the **prosthetic group,** which assists the enzyme in catalysis. The apoenzyme and its related nonprotein portion are collectively designated the **holoenzyme.** The relationship between the apoenzyme and prosthetic group is such that, if they are dissociated, the holoenzyme is rendered ineffective. If, however, they are recombined, the catalytic activity of the holoenzyme is re-established (Figure 2-13).

Prosthetic groups associated with enzymes are generally of two types: activators and cofactors (coenzymes). **Activators** are metals, usually ions of salts such as iron, copper, magnesium, manganese, zinc, calcium, and cobalt. Although the exact function of these activators is not completely understood, it is believed that they form a bridge between the enzyme and the substrate, thus binding them together, in order to facilitate substrate transformation. For example, Mg^{2+} is required by many phosphorylating enzymes that act together with ATP. The Mg^{2+} may act to form a bond between the

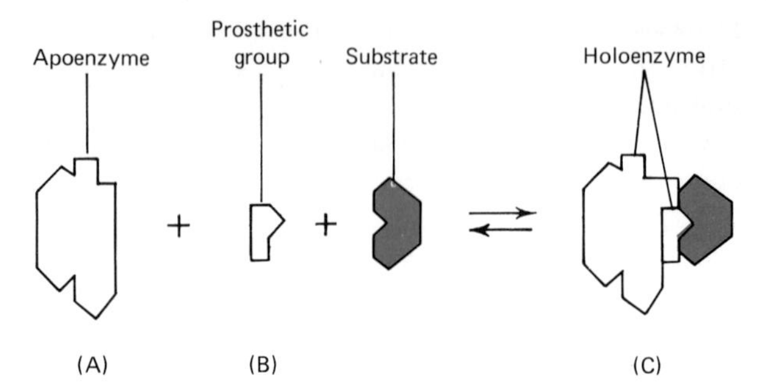

FIGURE 2-13. Components of an enzyme. Most enzymes are composed of an apoenzyme (protein portion) (**A**) and a prosthetic group (nonprotein portion) (**B**). Both portions together constitute the holoenzyme, or complete enzyme (**C**). If the apoenzyme and prosthetic group are separated, the holoenzyme becomes inactive and will not catalyze a chemical reaction.

enzyme and ATP. In general, it may be stated that most trace elements are probably required as activators for cellular enzymes.

In contrast to enzymes requiring metals, some enzymes require organic prosthetic groups called **coenzymes.** Generally a coenzyme assists the enzyme in transforming the substrate by acting as an acceptor of atoms being removed from the substrate or as a donor of atoms by contributing atoms in a synthetic reaction.

Many coenzymes are vitamins. Two of the most important coenzymes in cellular metabolism are *NAD*$^+$ (*nicotinamide adenine dinucleotide*) and *NADP*$^+$ (*nicotinamide adenine dinucleotide phosphate*) (see Figure 6-6). Both compounds contain derivatives of the B vitamin *nicotinic acid* (*niacin*) and function with their respective enzymes in the removal and transfer of hydrogen ions and electrons from substrate molecules. Enzymes that participate in reactions in which hydrogen atoms are removed (dehydrogenation) are collectively called **dehydrogenases.**

The flavin coenzymes, such as *FMN* (*flavin mononucleotide*) and *FAD* (*flavin adenine dinucleotide*), contain derivatives of the B vitamin riboflavin and are important in hydrogen transfer reactions as well as photosynthetic reactions (see Figure 6-8). Like NAD^+ and $NADP^+$, these coenzymes function in hydrogen transfer reactions with dehydrogenase enzymes.

Cytochromes are a group of proteins that function as electron carriers in photosynthesis and respiration (see Figure 6-8). These compounds are metalloprotein pigments which contain iron and are structurally related to hemoglobin and chlorophyll. Cytochromes function in serial succession in reactions in which electrons are systematically passed from one cytochrome to another. This methodical, synchronized passage of electrons from one cytochrome to another forms the *cytochrome system.* This system plays an important role in both respiration and photosynthesis.

Another important coenzyme, called *coenzyme A* (*CoA*) contains a derivative of *pantothenic acid,* another B-vitamin complex. This coenzyme assumes important roles in the synthesis and breakdown of fats and in a series of oxidizing reactions of respiration called the *Krebs cycle.* Coenzyme A is used in decarboxylation (removal of CO_2) reactions and is associated with a useful fragment of cellular catabolism, the acetyl $\left(CH_3{-}\overset{O}{\overset{\|}{C}}{-}\right)$ group.

Nucleic Acids. Either directly or indirectly, **nucleic acids** direct the synthesis of proteins. While some nucleic acids have extremely high molecular weights, most, like the proteins, are of medium molecular weight—about 50,000—and they contain carbon, hydrogen, oxygen, nitrogen, and phosphorus. Nucleic acids are found in all living forms. On the basis of chemical composition, there are two classes of nucleic acids: **DNA** or **deoxyribonucleic acid** and **RNA** or **ribonucleic acid.** DNA serves as the primary carrier of genetic information; it is the hereditary material. The function of RNA is to build proteins as specified by DNA.

Nucleic acids are long chain-like molecules composed of simpler units joined together by the removal of water at points of linkage between the units. These basic structural units of nucleic acids are called **nucleotides.** Since nucleic acids are composed of repeating nucleotide units, they may also be categorized as polymers.

Each nucleotide consists of three distinguishable components: one nitrogenous base, one pentose sugar, and one phosphoric acid or phosphate group. *Nitrogenous bases* are ring-shaped structures consisting of carbon, hydrogen, oxygen, and nitrogen atoms. They are further classified into two general types—*purines* and *pyrimidines.* The specific purines found in DNA are *adenine* and *guanine;* the pyrimidines are *thymine* and *cytosine.* The structures for these nitrogenous bases are shown in Figure 2-14. The pentose (five-carbon sugar) found in DNA is called *deoxyribose.* This sugar is similar to ribose (Figure 2-5A) except that it has no oxygen attached to the second carbon atom. The structural formula of deoxyribose is represented in Figure 2-15 along with the structural formula of *phosphoric acid,* the third constituent of nucleotides.

If one of the purines or one of the pyrimidines is chemically bonded to a sugar such as deoxyribose, and the sugar is then attached to a phosphate group, a nucleotide is formed. Each nucleotide thus formed is named for the constituent nitrogenous base. For example, a nucleotide containing adenine would be *adenine nucleotide;* thymine, a *thymine nucleotide;* cytosine, a *cytosine nucleotide;* and guanine, a *guanine nucleotide.* These four nucleotides are the ones commonly found in DNA.

Although the chemical components of DNA were known before 1900, it was not until 1953 that J. D. Watson and F. H. C. Crick presented a model for the arrangement of these components. Using data from various investigations, the following information concerning the structure of DNA was recorded:

1. Most DNA molecules consist of two strands, with cross members, twisted about each other in the form of a double helix (similar to a twisted ladder).
2. The uprights of the ladder are made of alternating phosphate and deoxyribose portions of a nucleotide.
3. The cross members, or rungs of the ladder, consist of paired purines and pyrimidines linked together by way of weak hydrogen bonds.
4. In the pairing of purines and pyrimidines, adenine always pairs off with thymine, and cytosine always pairs off with guanine.

From these data and those of other investigators, Watson and Crick constructed a model of the probable structure for DNA (Figure 2-16). The significance of this structure will be treated in Chapter 8.

The exact structure of RNA is much less clear than that of DNA. In some cases RNA consists of only a single strand of nucleotides, in others it is doubly stranded like DNA, and in still others it appears as a combination of both. Some data concerning RNA, however, have been ascertained. For example, it has been determined that RNA contains the pentose sugar ribose, not deoxyribose as found in DNA. Second, RNA

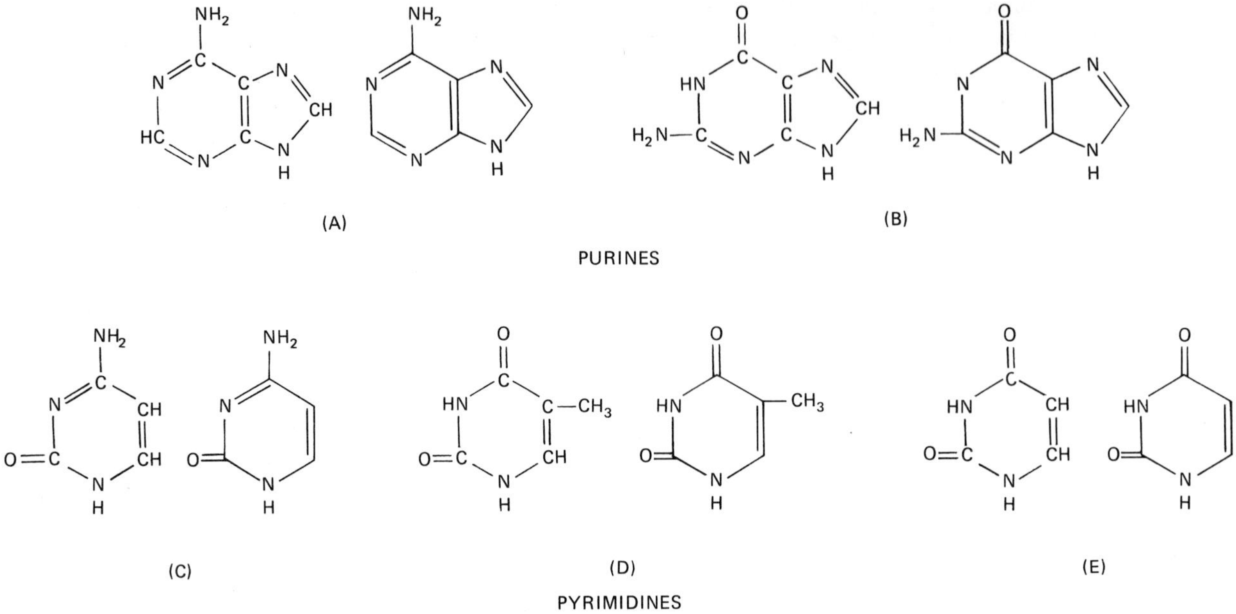

(A) (B)

PURINES

(C) (D) (E)

PYRIMIDINES

Figure 2-14. Structural formulas for the nitrogenous bases found in DNA and RNA. *Above:* The purines adenine (**A**) and guanine (**B**). *Below:* The pyrimidines cytosine (**C**), thymine (**D**), and uracil (**E**). To the right of each of the complete structural formulas for each of the nitrogenous bases is a simplified version of the structure from which some of the carbon and hydrogen symbols have been omitted. The simpler versions of these structures are used in later chapters.

does not contain the nitrogenous base thymine. Instead, the thymine is replaced by a pyrimidine called *uracil,* (Figure 2-14E). A final comparative aspect between DNA and RNA is that DNA is generally found in the nuclei of cells while RNA may be found in both the nuclei and the cytoplasm, being more abundant in the cytoplasm. Moreover, cytological evidence indicates that both DNA and RNA are also found in cellular organelles. The DNA and RNA unique to these organelles are discussed in Chapter 8.

(A) (B)

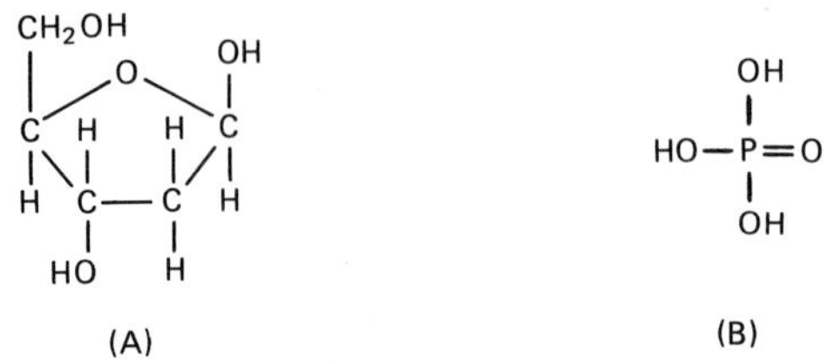

Figure 2-15. Structural formulas for deoxyribose (**A**) and phosphoric acid (phosphate) (**B**). The second carbon atom of deoxyribose lacks one oxygen atom as compared with ribose; thus, it is designated as deoxyribose.

RNA is divided into three subgroups based upon certain structural and functional aspects: *ribosomal RNA, messenger RNA,* and *transfer RNA.* The importance of these RNA types in the process of protein synthesis will also be noted later (Chapter 8).

Adenosine Triphosphate (ATP). In addition to the nucleotides that are found as components of DNA and RNA, one other compound should be mentioned at this point. This substance is **adenosine triphosphate (ATP).** ATP is a key organic compound that appears universally in plant and animal cells. The essential function of ATP is to store and carry energy for various cellular activities. Figure 2-17 shows the structure of this basically simple but extremely important molecule.

The major significance of ATP lies in the nature of the phosphate groups. The two terminal (end) phosphate groups are called *high-energy phosphate groups.*

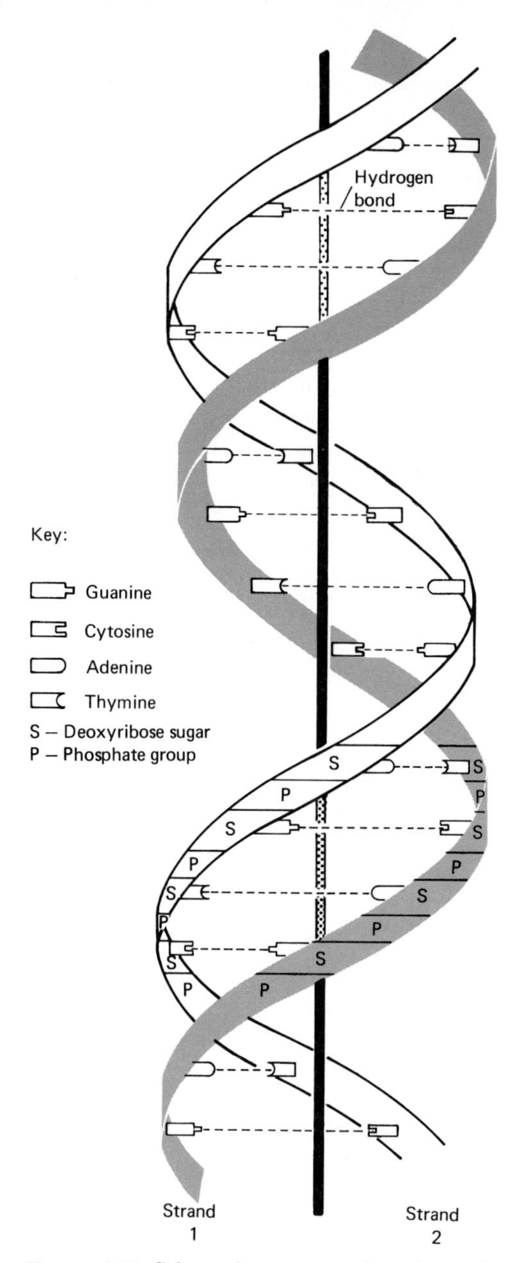

Figure 2-16. Schematic representation of a portion of a DNA molecule according to Watson and Crick. The entire molecule assumes a double-stranded helical configuration. The cross members represent paired purine and pyrimidine nitrogenous bases, while the spiral backbone consists of alternating phosphate and sugar units. Note that adenine always pairs off with thymine, cytosine always pairs off with guanine, and the nitrogenous bases are attached to the sugar molecules. In each case a purine nucleotide and its complementary pyrimidine nucleotide are held together by hydrogen bonds.

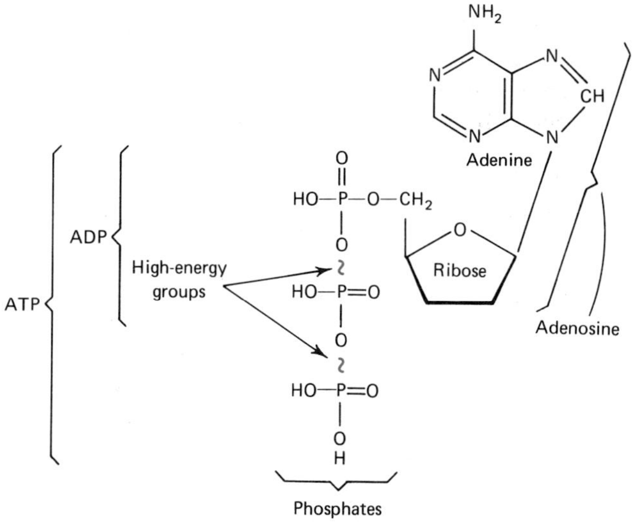

ATP
ADP
High-energy groups
NH_2
Adenine
Ribose
Adenosine
Phosphates

Figure 2-17. The structure of ATP. Adenosine triphosphate is composed of an adenine unit (adenine + ribose) combined with three phosphate units. The first phosphate attached to the ribose sugar is a low-energy group. The second and third phosphates are high-energy groups. These high-energy groups provide most of the chemical energy utilized by living cells.

When these groups are broken off they liberate four to six times as much energy as other chemical groups. The attachment of such high-energy phosphates is indicated by the wavy bond lines (~). This larger than normal amount of energy liberated by splitting off a high-energy phosphate is used to carry out basic cell activities. If one phosphate group is split from ATP, the compound *adenosine diphosphate* (*ADP*) results and energy is released. This reaction is represented as follows.

ATP	$\longrightarrow$	ADP	+	P	$\Delta G = -$
adenosine triphosphate		adenosine diphosphate		phosphate	(exergonic)

(*Note:* A negative ΔG signifies that the reaction is exergonic, while a positive ΔG signifies that the reaction is endergonic.) Since cells are constantly using energy supplied by the conversion of ATP into ADP, and since the supply of ATP is small and limited at any given time, provision is made for its constant replenishment or the constant rephosphorylation of any ADP produced. This is accomplished when a phosphate group is added to ADP and a molecule of ATP is formed. Remember that when ATP is broken down, the reaction liberates energy. Conversely, when ATP is built up, the reaction requires the same amount of energy.

ADP	+	P	$\longrightarrow$	ATP	$\Delta G = +$
adenosine diphosphate		phosphate		adenosine triphosphate	(endergonic)

The energy needed to convert ADP and phosphate into ATP is derived from a long series of chemical reactions that will be considered in later chapters.

Cyclic Adenosine Monophosphate. **Cyclic AMP (cAMP)** is a compound of great biological importance. This compound was first isolated in 1959, and it has been determined that cyclic AMP is involved in many different biochemical processes as a major regulator of metabolic and physiological activities in all organisms. In mammals, many hormones activate an enzyme, adenylate cyclase, which stimulates the production of cyclic AMP. The cyclic AMP then acts to control the activities of other enzymes, most frequently by an allosteric action (see page 195); the cyclic AMP acts as a secondary regulator, carrying the message of a primary regulator, which is typically a hormone.

In this chapter, an attempt has been made to provide an understanding of the organization of matter in living systems and to introduce some of the biologically important molecules. This background will be assumed in subsequent discussions dealing with the origin of life, cellular structure and physiology, energy relationships, genetics, development, and evolution. In this regard, reference should be made to this chapter frequently, as needed.

3 The Origin of Life

GUIDE QUESTIONS

1 What relationship exists between chemicals and living forms? How are organismic diversity and unity related to the concept of evolution?
2 Define abiogenesis. List and discuss the contributions of those who supported abiogenesis. What is biogenesis? How was this concept eventually substantiated? Discuss some of the major steps in the development of the cell theory. How did this theory lend support to biogenesis?
3 What is Oparin's hypothesis on the origin of life? How was the earth probably formed? What were the chemical constituents of the ancient atmosphere? How may simple substances of the ancient atmosphere have been converted into organic compounds?
4 What was S. Miller's contribution to the heterotroph hypothesis? What is a macromolecule? How may it be formed?
5 What is the importance of HCN in the development of the heterotroph hypothesis? Outline the principal tenets of the coacervate theory. How has this theory been substantiated? What are microspheres?
6 Discuss some of the events which may have occurred in the organization of procells into the first living forms.
7 Outline the major advances that may have taken place as part of early biological evolution.
8 Do you believe that the data are sufficient to substantiate the heterotroph hypothesis? Explain and defend your response.
9 What is the autotroph hypothesis? What are some of the objections raised against this hypothesis?
10 What are the two lines of evolutionary development that may have been followed by the first cells? Discuss in detail.
11 What are the principal concerns of the biologist regarding space exploration? Define exobiology.
12 What are the alternative possibilities with regard to the existence of extraterrestrial life?
13 Briefly describe the characteristics of the major and minor planets that may be related to the existence or nonexistence of life.
14 What data substantiate the probability that life is not limited to the solar system of the sun?

THE advent of better instruments and techniques of investigation has enabled scientists to study living systems at smaller and smaller levels of organization. Perhaps the greatest influence on the biological sciences has been the integration and implementation of the principles of physics and chemistry with those of the life sciences. As a result of these advances, the biologist now interprets the structures and activities of organisms in terms of their constituent chemicals. Biologists clearly recognize that all living forms consist solely of chemicals in various organizational and functional patterns.

An Introduction to Chemical Evolution

A cursory examination of a large number of organisms would reveal a considerable degree of structural diversity. The diversity of living forms becomes readily apparent when one considers the numerous variations exhibited by the millions of species of organisms that have once lived and become extinct and by those that are a conspicuous part of today's world. Among some of the outstanding examples of diversity are the sizes of organisms, ranging from the exceedingly small bacteria to the giant redwood trees; the varied colors of organisms as exemplified by brilliant hues of fruits, flowers, and leaves, skin pigmentation in man and other animals, multicolored feathers of birds, and color patterns of fish; and the forms and shapes of life as represented by the amorphous amoeba, the radial starfish, the highly modified insectivorous plants, and the very specialized insects.

Although structurally diverse, organisms are functionally similar. Unity among organisms is also apparent, and even limited contacts with a variety of organisms would suggest certain commonalities. For example, growth is an integral process of all life; even though it is difficult to define, it is an obvious phase in the life cycle of all organisms. Various types of energy transformation and utilization in living systems represent another example of unity. The vast majority of life processes in all organisms are dependent upon the continuous procurement, transformation, and utilization of energy. This energy is expended for numerous activities, including locomotion, circulation of materials, movements of substances into and out of cells, synthesis of food, and the further transformation and utilization of energy. A final example of unity among organisms, and perhaps the most outstanding, is the ability of living systems to reproduce. The continuity of all life is directly related to this basic and essential activity; without it, life would cease.

The concepts of diversity and unity are integral components of the theory of evolution, which, more than any other, is the unifying theme of our study of life. Evolution provides many answers concerning the directions that living forms have taken in the course of their histories, the forces that control extinction and the origin of new forms, and the influences that act upon organisms to establish the myriad patterns of diversity and unity. Logically, then, if evolution is considered as a plausible explanation for the commonalities and variations that exist among organisms, the basic questions that must be asked are: How did the process of evolution begin? Where did it originate? When did evolution become operative? In other words: How did life originate? The basic enigma of life resides in this one simple question. To be sure, the answers to this question are exceedingly complex and at best only tentative. In this chapter an analysis will be made of the major hypotheses that have been proposed to account for the origin of life.

Hypotheses on the Origin of Life

Early Ideas: Abiogenesis (Spontaneous Generation)

Until the second half of the nineteenth century, it was generally believed that life could arise spontaneously from nonliving substances. According to this belief called **abiogenesis,** or **spontaneous generation,** nonliving matter was assumed to have a capacity within itself to turn into certain types of organisms. For example, toads, snakes, and mice could be formed from moist soil; flies could arise spontaneously from decaying matter, manure, and dirt; molds could originate from fruits; and lice could suddenly appear on decaying human bodies.

Greek Ideas—Aristotle's Hypothesis. Belief in spontaneous generation can be traced as far back as 600 B.C., to the early Ionian philosophers. They maintained that living organisms originated in sea slime under the influence of factors in the environment such as heat, air, and the sun. Once this rationale was established, it received considerable support from Aristotle (384–322 B.C.), whose teachings did much to reinforce the belief. The essence of his philosophy concerning the origin of life is summed up in the following passage quoted from *Historia Animalium:*

> Such are the facts, everything comes into being, not only from the mating of animals but from the decay of earth. . . . And among plants the matter proceeds in the same way; some develop from seed, others, as it were, by spontaneous generation by natural forces; they arise from decaying earth and from certain parts of the plants.

Aristotle believed that there was some vital force in the elements (air, water, fire, and earth) that could transform lifeless matter into a living form. The observations of Aristotle, added to the earlier Ionian speculations, provided strong support for the doctrine of spontaneous generation which persisted for nearly 2000 years.

Helmont's Support for Abiogenesis. In

the middle of the seventeenth century, Jean Baptiste van Helmont, a Belgian physician, published an account of a 21-day "experiment" that gave added impetus to the concept of abiogenesis. He stated that if wheat grains are placed in a dirty shirt for 21 days, mice would originate. He believed that the active principle for such spontaneous generation was human sweat, which acted to convert the wheat grains into mice.

Analysis of his experiment would show that his experimental design was not scientific. If, for example, he had set up a controlled experiment in which the dirty shirt and wheat grains were placed in a closed container, his results would have been quite different. Such a controlled experiment would have shown that the mice came from the outside and not from the kernels of wheat. A few years later, in 1668, the idea of spontaneous generation was subjected to a carefully controlled experiment by Francesco Redi, an Italian physician and biologist.

Spontaneous Generation Is Questioned: The Rise of Biogenesis

Redi's Experiments. The experimental work of Redi offered the first strong scientific opposition to the belief in abiogenesis. Observations made by the abiogenesists led to the conclusion that "worms" arose spontaneously from decaying meat. (These worms are now known to be maggots—the larvae of developing insects.) It was this conclusion that Redi sought to disprove. He too observed that worms appeared in decaying meat. But he also noted that flies hovered over the meat that became wormy and so he decided to ascertain what relationship, if any, existed between the flies and the worms in the meat. In his experimental setup he filled three jars with the decaying flesh of various animals (Figure 3-1A). This set of jars was sealed tightly. He then filled three similar jars with the same types and amounts of decaying flesh. These, how-

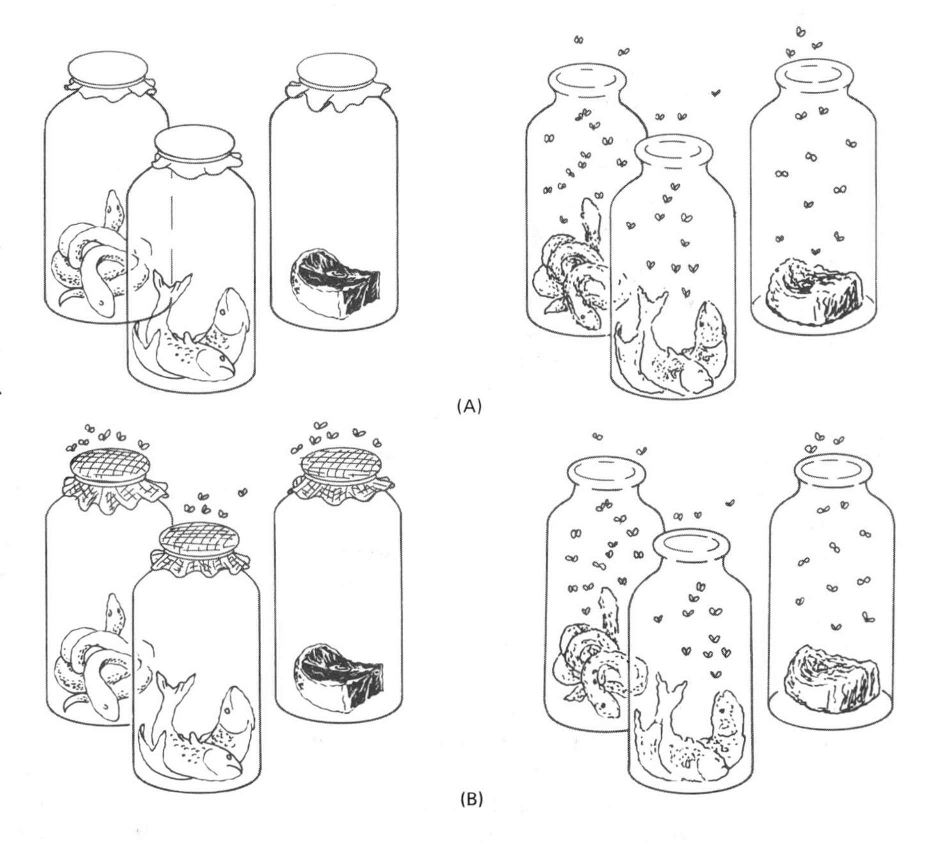

Figure 3-1. Redi's challenge to spontaneous generation. In the experiment shown in (**A**) Redi filled two sets of jars with decaying flesh. One set (left) was made airtight; the other set (right) was left open. In a few days, the uncovered set became contaminated with larvae, and flies entered and exited each of the contaminated jars. The covered jars showed no evidence of flies. Those who believed in abiogenesis claimed that the sealed jars were deprived of the vital life force (air) required for spontaneous generation. Redi answered this charge by setting up another experiment (**B**) in which half the jars were left uncovered (right) and half were covered (left) with gauze *to permit the entrance and free flow of air.* The result was the same: the uncovered jars became contaminated, while the covered jars did not.

ever, were left open. It was not long before the decaying meat in the open vessels became wormy and flies were continuously entering and leaving each of the contaminated jars. The sealed containers showed no macroscopic forms of life in the decaying meat, and there were no flies associated with these vessels. These observations suggested a relationship between the flies and the worms and thus the study of **biogenesis,** the origin of life from pre-existing life, had its experimental beginnings.

It should be noted that although Redi's experiment was more scientific than that of van Helmont, there is still an additional factor that must be considered. Inasmuch as the flasks in which no worms appeared had been closed, the entrance and flow of fresh air were prevented. The abiogenesists claimed that fresh air and not the absence of flies was the factor (vital force) required for spontaneous generation.

In order to overcome this difficulty, Redi set up a similar experiment in which three of the jars were covered with a fine net (instead of being sealed) and the others were left open (see Figure 3-1B). His results were identical to those observed in the first experiment. No worms appeared in the jars covered with gauze, even though air was permitted to enter and flow freely inside the vessels. Worms appeared only when the eggs were deposited by the flies on the decaying meat.

Redi's experiments, convincing as they were, did not completely disprove the idea of spontaneous generation since the abiogenesis of worms might have been possible under a different set of circumstances. Furthermore, Redi did not experimentally determine that other forms of life could not be generated spontaneously. His work, however, as well as contributions of others, seriously questioned the long-standing concept which held that life could arise from lifeless material.

Leeuwenhoek's Entry into the Controversy. At approximately the same time that Redi was conducting his experiments, Anton van Leeuwenhoek introduced the microscope. This instrument enabled scientists to make observations of exceedingly small organisms that no one had ever seen before. Leeuwenhoek believed that the microorganisms he observed arose from pre-existing organisms. He did not support abiogenesis. Other scientists, however, were convinced that although larger animals could not arise from spontaneous generation, microorganisms could. One observation in support of this conclusion was that when a small amount of organic matter (e.g., hay) was combined with pure rain water, a tremendous number of microorganisms appeared. It is known today that such a hay infusion contains numerous spores (inactive forms) of microscopic animals and bacteria that, when given suitable growth conditions, will become active forms capable of reproduction.

The battle between experimentalists over the question of spontaneous generation continued for two hundred years. Conflicting results were obtained in experiments using hay infusions and other nutrient solutions. Prominent among those who disputed the issue were John T. Needham, an abiogenesist, and Lazzarro Spallanzani, a biogenesist.

Needham's Support for Abiogenesis. In 1745, Needham lent considerable support to the concept of abiogenesis with a series of experiments in which he subjected various nutrient fluids to elevated temperatures. He heated the fluids (chicken broth, mutton gravy, corn infusions, and wheat infusions) in corked containers to boiling and noted that after the solutions were allowed to cool they were teeming with microorganisms. Needham claimed that he had shown experimentally that microorganisms developed spontaneously out of nonliving organic matter.

Spallanzani's Support for Biogenesis. Some 25 years after Needham's results and conclusions were reported, Spallanzani offered another serious challenge to abiogenesis. He performed experiments similar to those conducted by Needham but was unable to substantiate the spontaneous generation concept. Essentially, he boiled a series of test tubes containing nutrient fluids for long periods of time. In addition to prolonging the heating process, he used airtight seals instead of corks. Half of the test tubes were sealed with corks; the other half were made airtight with nonporous coverings. After boiling both sets of test tubes, it was found that the corked vessels contained microorganisms; the airtight ones were free of microorganisms. Although it might appear that this experiment settled the argument, in fact, it did not. Supporters of spontaneous generation replied that prolonged heating had destroyed not only the microbes in the broth but also the vital life force present in the broth and in the air.

The dispute remained unsettled for another hundred years, until Louis Pasteur designed a series of ingenious and convincing experiments that finally resolved the debate.

Pasteur Resolves the Issue. Pasteur approached the problem in three basic steps. First, he demonstrated that microorganisms exist in the air. Second, he showed that when solutions of broth are boiled, they do not lose their capacity to support life. Finally, using these data, he showed that life could not arise from nonliving materials. He began his experiments with the working hypothesis that life comes from preexisting life.

The initial phase of his experimental work was concerned with demonstrating that microorganisms are present in the air and that these microorganisms caused the appearance of life in broth cultures. He filled several flasks with beef broth and subjected them to boiling. Some were left open and allowed to cool, and in a few days they were contaminated with microbes. Other flasks were sealed after sterilization; upon examination, they were found to be free of microorganisms. Based upon these results, Pasteur hypothesized that microbes in the air were the agents responsible for contaminating nonliving matter. Despite the apparent conclusiveness of these experiments, the argument that the heat of sterilization destroyed the active principle in the flask had to be answered.

In order to answer this objection, Pasteur placed broth as before in a specially designed flask, the neck of which was drawn out to give it an S-shaped curvature (Figure 3-2). The contents of the flask were boiled and then allowed to cool. Examination revealed that the broth did not decay; there was no sign of microbes after days, weeks, and months. (A number of these original flasks are still in existence and are on display at the Pasteur Institute in Paris. None of them shows any signs of contamination even after over one hundred years.) The significant aspect of the design of these flasks is that air is permitted to pass down the neck of the flask to make contact with the broth while the curved portion traps microorganisms and dust and prevents their passage into the broth.

Pasteur provided further experimental proof by tipping one of the S-shaped flasks so that the broth made contact with the trapped material in the curved portion

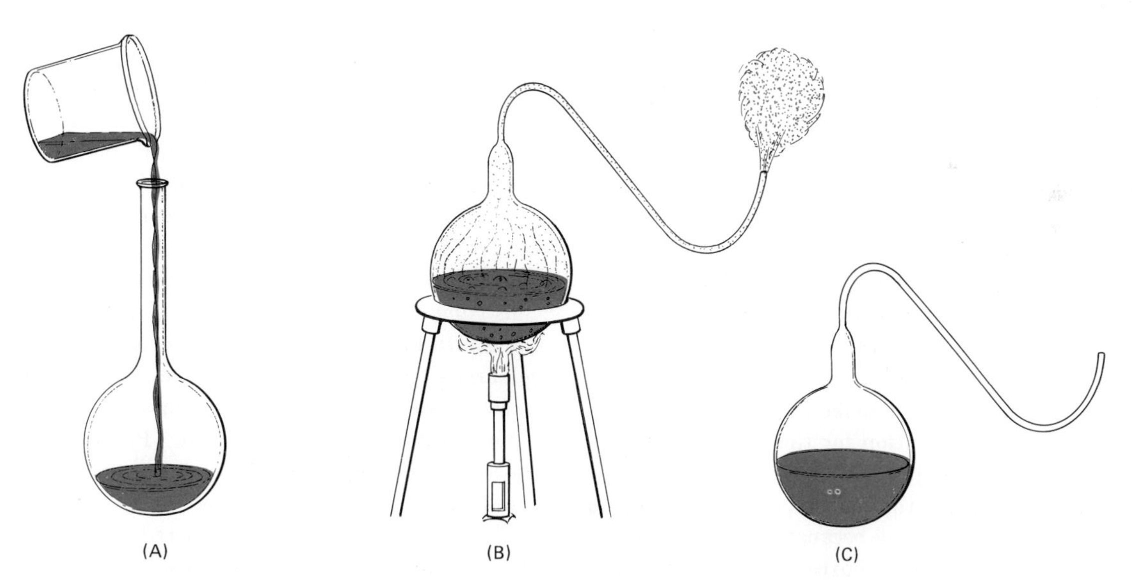

Figure 3-2. Pasteur's classic experiment that disproved spontaneous generation. In this simple but ingenious experiment, Pasteur first added a nutrient solution to an ordinary flask (**A**). He then contoured the neck of the flask by heating to give it an S-shaped curvature. (**B**) The solution was boiled for several minutes and cooled. (**C**) Even after the passage of long periods of time, there were no signs of decay; no microorganisms appeared.

of the flask. After the flask was placed upright again so that the mixture of broth and contents of the curved neck flowed back into the flask, it was found that the originally clear broth became contaminated in a few days.

Pasteur's experiments provided some evidence that life did not arise from mystical forces present in nonliving materials. Any appearance of life was due to the presence in the air of microorganisms or their spores, which became active when placed in a suitable environment. Nevertheless, his work did not completely rule out the possibility of spontaneous generation; although it did not occur in his experiments, there was still the possibility that it could occur in other systems under a different set of circumstances. In other words, Pasteur did not prove that abiogenesis never occurs; he showed only that it did not take place within the limits of his experimental design. Since the time of Pasteur, however, biologists generally have agreed that spontaneous generation cannot occur under the environmental conditions that now exist on earth.

Further Support for Biogenesis

Although Pasteur disproved the doctrine of spontaneous generation, he did not explain how the first living form originated. His experiments showed that living organisms came only from living organisms. Other scientists, pursuing different lines of investigation, were also seeking to provide an answer to the question of life's origin.

The Discovery of Cells. The beginning of this chain of discoveries was quite accidental. Robert Hooke, an English physician, published a book entitled *Micrographia* in which he presented his microscopic analyses of the structure of cork. His research was directed primarily toward an explanation of the buoyancy of cork. At the time, Hooke was not interested in searching for an explanation for the origin of life. He found the cork cells empty and bounded by strong cell walls and that gave him the solution to his problem. He concluded that cork floated because it was held up by tiny air-filled cells or "little boxes." Although he went on to discover the cells of other plants, he never fully realized the true significance of his discovery. As microscopes were improved, however, other biologists began to suspect that living forms consisted entirely of cells.

Cells Come from Cells. In 1809 Jean Lamarck, a French zoologist, called attention to the jelly-like substance inside living cells. He believed that this internal fluid was more important than the outside appearance of a cell; Hooke, by contrast, observed the cell wall. In this regard, Lamarck wrote:

> Every living thing is essentially a mass of cellular tissue in which more or less complex fluids move more or less rapidly.

Conclusions along similar lines were also reached by René H. Dutrochet about fifteen years later. His findings, based upon careful research of the constituents of plant and animal tissues were more precise than those of Lamarck. He wrote:

> The globular corpuscles which by their assemblage make up all the organic tissue of animals are actually cells of exceedingly smallness. . . . Thus, all tissue, all animal organs, are actually only cellular tissue variously modified. . . .

Two German scientists, Matthias Schleiden, a botanist, and Theodor Schwann, a zoologist, consolidated the discoveries of many other investigators and independently posited a coherent theory on cells. This enunciation, one of the most important generalizations of the biological sciences, called the cell theory, was published in 1838–39.

According to the present concept of the **cell theory,** all plants and animals are composed of cells and cell products. After this theory was accepted, a very important extension of it was offered in 1858 by Rudolf Virchow, a German physician. He stated that all living cells come from pre-existing cells; that is, they do not come into being by spontaneous generation. "When a cell arises," Virchow wrote, "there must be a cell before it; in the same way, an animal can arise only from an animal, a plant only from a plant." Proof of Virchow's theory of biogenesis came only a few years later as a result of the experiments of Pasteur.

The chain of discoveries started accidentally by Hooke ended in a startling revelation—all living things are composed of cells and all cells arise from other cells before them. Ultimately, the theory of biogenesis is based upon the assumption that it takes a living organism to produce another living organism. A logical question, then, that arises from the concept of biogenesis is: If life comes only from life, how did the first living form arise?

Abandonment of the concept of abiogenesis left scientists with no plausible hypothesis on the origin of life. The dilemma with which they were faced was that they had rejected abiogenesis as a result of Pasteur's work and they were unwilling to accept the origin of life by supernatural forces. A small group of naturalists, however, maintained that the origin of living things involved not a sudden and spontaneous genesis from organic matter, but rather a gradual evolution of chemicals that led to the emergence of the first primitive life form.

Chemosynthesis: The Heterotroph Hypothesis

This evolutionary concept of the origin of life extended into the twentieth century and received considerable support from an increasing number of experimentally derived facts. This new view, called *chemosynthesis,* or the *heterotroph hypothesis,* held that conditions on earth billions of years ago were markedly different from those of the present and that the first form of life arose spontaneously from chemicals under these conditions over a long period of time. The basic tenets of this hypothesis were introduced by a Russian biochemist, A. I. Oparin, in 1936, in a publication titled *The Origin of Life.*

In this book he detailed an account of the origin of life. What Oparin suggested was a return to spontaneous generation, but his conception of the process was quite different from the original ancient doctrine. Oparin based his hypothesis on the most recent data gathered from all fields of science, not on a vague "life force" that could not be defined. Geological studies contributed to Oparin's hypothesis by providing data about conditions on the earth billions of years ago; physics offered information about the intensity and effect of solar radiation; chemistry provided facts concerning the properties and behavior of atoms and molecules; and the life sciences afforded important information related to the processes involved in evolutionary change. Synthesis of all of these data enabled Oparin to formulate a unified concept of the origin of life. Before examining the details of the heterotroph hypothesis as proposed by Oparin (as well as supportive data provided by others), the nature of the prelife conditions of the primitive earth will be examined.

Prelife Conditions. Since it is reasonably certain that before life originated there were only chemicals and that living things originated from chemicals, the starting point to determine the origin of life should be an analysis of the prebiotic chemicals. Such an analysis must first consider the origin of the earth.

The Origin of the Earth. There is no totally satisfactory explanation to account for the origin of the earth, despite the numerous hypotheses and theories that have been advanced. The circumstances of its origin present one of the most fascinating mysteries still before science. Nevertheless, the age of the earth has been estimated by various methods to be about $4\frac{1}{2}$ billion years. The hypothesis most widely accepted today is that the sun and its planets were formed within a massive cloud of cosmic dust and gas in an empty region of our galaxy. It has been approximated that initially the cosmic cloud was one light year in diameter and had a temperature far below 0°C. Most of this material began to condense rapidly into a more compact mass as a result of gravitational forces. The great heat and enormous pressure resulting from condensation initiated thermonuclear reactions that converted the main condensed mass into a luminous body, the sun. Within the remainder of the dust and gas cloud, lesser centers of condensation were produced, and these became the planets.

Once formed, the earth also began to condense. As a result of condensation, the temperature of the solid mass increased to 1000–3000°C and a selective melting and stratification of its chemicals took place. The heavier molten metals, such as iron and nickel, sank to the core while the lighter substances migrated nearer the surface. Included in these lighter materials were basalt, granite, and various dehydrated minerals. As the earth cooled, its surface eventually solidified.

The Ancient Atmosphere. It has been assumed, and widely accepted, that when the earth was first formed, there was no atmosphere. This is because the gases produced as part of the earth's formation probably escaped into outer space immediately. However, as the earth cooled, an ancient atmosphere evolved as a result of volcanic eruptions. The newly formed atmosphere probably consisted of an envelope of methane (CH_4), water vapor (H_2O), ammonia (NH_3), hydrogen (H_2), and hydrogen cyanide (HCN) (Figure 3-3A) and was highly reducing. The ancient atmosphere also probably contained carbon dioxide (CO_2), carbon monoxide (CO), hydrochloric acid (HCl), and hydrogen sulfide (H_2S). In contrast, the present atmosphere, which is relatively oxidizing, contains about 78% molecular nitrogen (N_2), 20% molecular oxygen (O_2), and less than 1% (0.03%) carbon dioxide (CO_2), with the remainder distributed as rarer gases such as neon and helium. It is believed that free nitrogen and oxygen were not present in the primitive atmosphere because of the large amounts of hydrogen and the exceedingly high temperatures. Under these conditions, free nitrogen would have combined with hydrogen to form ammonia; carbon would have united with hydrogen to produce methane and with oxygen to produce carbon monoxide and carbon dioxide; and oxygen would also have combined with iron, silicon, and aluminum to form assorted minerals of the earth's crust and with hydrogen to form water vapor. The principal essential chemical elements found in all living forms today are carbon, hydrogen, oxygen, and nitrogen, the same substances that are believed to have been constituents of the ancient atmosphere.

Energy and the Chemical Compounds of the Primitive Earth. Most certainly, the simple inorganic gases comprising the ancient atmosphere were not the chemicals from which living forms were directly produced. Any mixture of these compounds is fairly stable, and such gases do not react with each other to form other compounds. It is reasonable to assume that the primary molecules from which life arose would be organic substances. Therefore, certain conditions must have existed in order to transform the simple compounds of the primitive atmosphere into more complex organic substances, the precursors of the components of modern life forms.

As the earth cooled, temperatures became sufficiently low so that the water vapor of the atmosphere liquefied and began to collect in the basins on the crust. Basins and shallows filled as torrential rains fell, the low places on the crust slowly filled with water, and the first oceans formed. Dissolved in the seas were atmospheric ammonia and methane deposited by the rains and salts and minerals carried by rivers flowing down mountain slopes. In addition, volcanic eruptions supplemented the mineral contents of the seas.

If the primitive seas and the ancient atmosphere contained the basic materials (simple inorganic compounds) that eventually gave rise to life, some external source of energy must have acted upon the mixtures to rearrange the atoms into organic molecules. One possible source may have been heat energy from the interior of the earth, which could have increased the motion of the molecules of the gases, making them collide with sufficient energy to form new combinations of molecules. A second possibility may have been the energy of solar radiation in the form of visible light, ultraviolet light, and X-rays bombarding the existing molecules to produce new combinations. A third form of energy that may have rearranged the molecules was electrical energy from lightning. Finally, the radioactive materials in the earth itself, more abundant billions of years ago than now, were another source of energy that may have acted on the mixtures of inorganic molecules (see Figure 3-3B).

The Formation of Simple Organic Compounds. Given the earth conditions just described, the atoms of the inorganic gases were continuously breaking apart and recombining to form new substances. Regarding the importance of the reactions, Oparin wrote

> It is absolutely unthinkable that such complex structures like organisms could have been generated spontaneously, directly from carbon dioxide, water, oxygen, nitrogen, and mineral salts. The generation of living things must have been inevitably preceded by a primary development on the Earth's surface of those organic substances of which the organisms are constructed.

As part of his general working hypothesis, then, Oparin suggested that one of the first steps necessary for the origin of life was the transformation of the atmospheric gases into simple organic compounds. What types of evidence exist that such an event could have taken place? And what organic compounds could have been formed under such conditions?

The first significant experimental data to demonstrate that organic molecules could be produced from a mixture of ammonia, methane, water vapor, and hydro-

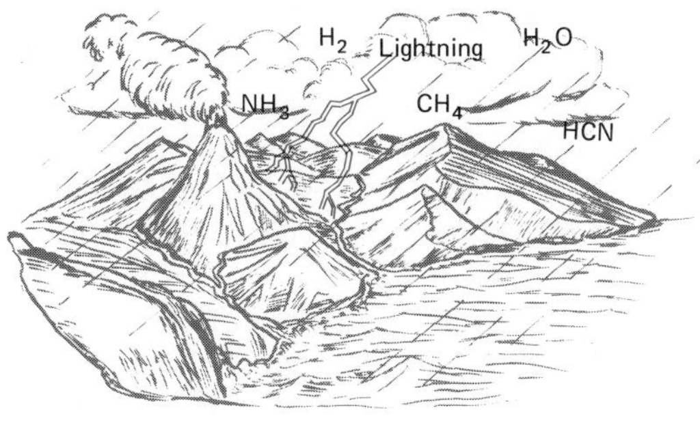

Formation of the primitive atmosphere consisting of H_2O, NH_3, CH_4, H_2, and HCN
(A)

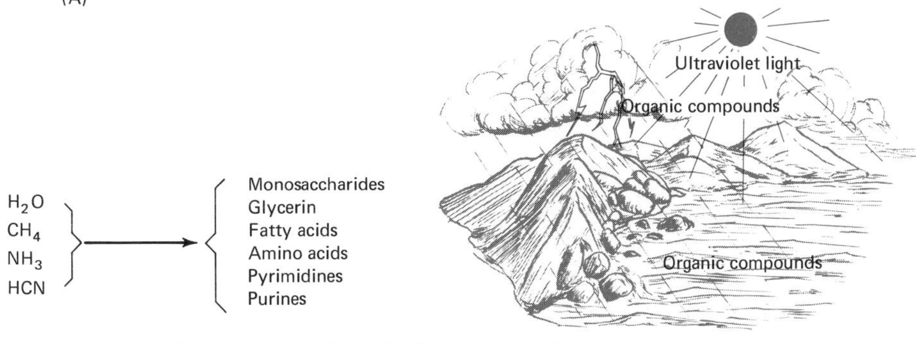

Organic compound formation from the atmospheric gases in the ancient seas
(B)

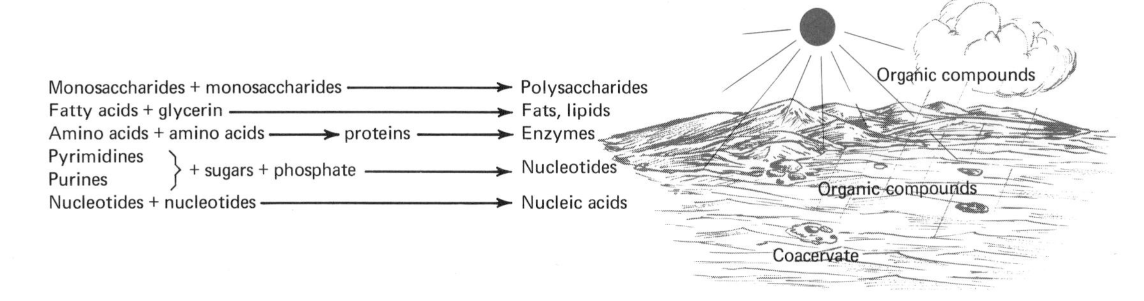

Formation of complex organic compounds and coacervates.
(C)

Figure 3-3. Diagrammatic representation of some of the principal assumptions of the heterotroph hypothesis. (**A**) Presumed conditions on the newly formed earth and the hypothesized constituents of the primitive atmosphere. (**B**) Various types of energy may have converted the inorganic atmospheric gases into simple organic compounds. (**C**) The aggregation of complex organic compounds into procells (coacervates), the structures from which life may have originated over long periods of time.

gen were published in 1953 by Stanley Miller, a graduate student of Dr. Harold Urey at The University of Chicago. In these experiments, the conditions assumed to have existed on the primitive earth two or three billion years ago were simulated. In the airtight apparatus shown in Figure 3-4, inert methane, ammonia, hydrogen, and water vapor were circulated past electrical discharges from tungsten electrodes. The boiling and condensation of the water kept the substances circulating through the system. After one week, the contents of the system, which appeared as a residue collected from a trap, were analyzed. It was found that a variety of organic compounds had been synthesized. Among the compounds identified by Miller were a number of amino acids, including some present in proteins, and a variety of simple organic acids known to occur in living systems. The amino acids included glycine, aspartic acid, and glutamic acid; among the organic acids were acetic, formic, propionic, lactic, and succinic acids.

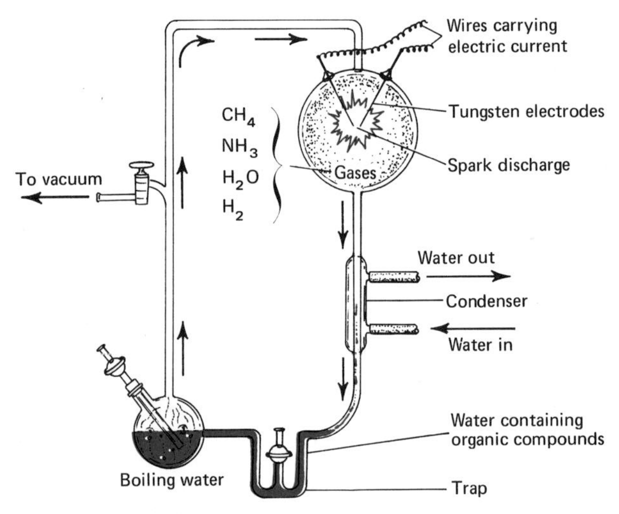

Figure 3-4. Miller's apparatus for simulating primitive earth conditions. The gases in the large vessel are methane (CH_4), ammonia (NH_3), hydrogen (H_2), and water vapor (H_2O). The water vapor was continuously supplied by boiling water in the smaller vessel on the left. Energy was supplied by the artificial lightning of electrical discharges across the two tungsten electrodes. As the inorganic gases were broken down into atoms, new combinations of molecules were formed. These heavier newly formed substances drifted down into the trap. There appeared in this solution several types of organic compounds, including amino acids, which have the same atomic components as the gases, but which possess a more complex atomic structure. [*From S. L. Miller,* Journal of the American Chemical Society, *vol. 77, p. 2351, 1955. Copyright © 1955 by the American Chemical Society.*]

Since this classic experiment, other investigators, using a variety of energy sources and regulating the types and amounts of other suspected constituents of the primitive atmosphere, have shown that still other organic compounds could be produced. In more recent experiments, mixtures consisting of N_2, H_2, CO, and CO_2 (no preformed ammonia or methane), have yielded amino acids and other organic compounds. In this regard, a compound central to most of the reaction pathways leading to the formation of simple organic compounds is hydrogen cyanide (HCN). It has been suggested that HCN may have been formed from primitive atmospheric constituents as follows.

$$\underset{\text{methane}}{CH_4} + \underset{\text{ammonia}}{NH_3} \longrightarrow \underset{\text{hydrogen cyanide}}{HCN} + \underset{\text{hydrogen}}{3\,H_2}$$

$$\underset{\text{carbon monoxide}}{CO} + \underset{\text{ammonia}}{NH_3} \longrightarrow \underset{\text{hydrogen cyanide}}{HCN} + \underset{\text{water}}{H_2O}$$

In analyzing the contents of his spark discharge apparatus, Miller postulated that the various organic products arose by various sequences of reactions. For example, HCN reacts with substances such as ethylene (C_2H_4) to produce substances called nitriles (precursors of amino acids). Other investigations have shown that HCN is an important precursor of purines and pyrimidines, the building blocks of nucleic acids. The central role of HCN as a starting substance leading to the production of organic compounds is substantiated by the fact that over fifty different derivatives of HCN may be produced under simulated primitive earth conditions. Some of these derivatives are shown in Figure 3-5.

Macromolecule Formation. Once chemical evolution progressed to the point where amino acids, sugars, fatty acids, and other simple organic compounds were present, it is probable that they began to join together to produce larger molecules. Recall from Chapter 2 that the covalent bonds between building blocks of proteins, polysaccharides, and lipids are the result of the removal of water (dehydration synthesis) from successive building block units. Under what con-

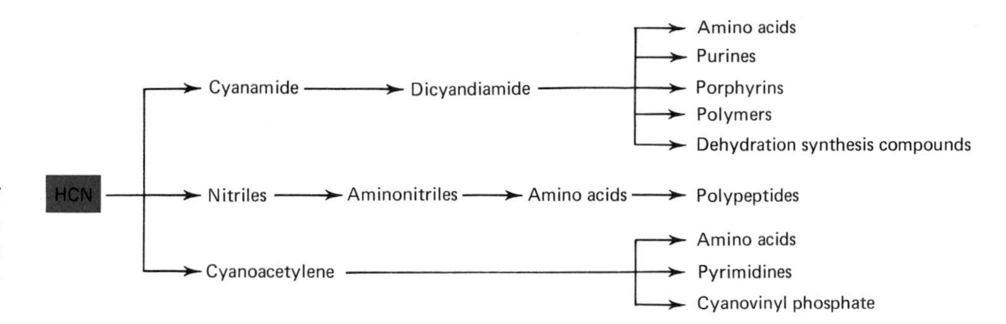

FIGURE 3-5. Central role of hydrogen cyanide as a starting substance in the formation of organic compounds.

ditions, then, could larger molecules such as proteins, lipids, polysaccharides, and nucleic acids be formed?

In general, dehydration synthesis reactions occur in one of two ways. The first of these is when the temperature is above the boiling point of water. Under such thermal conditions, over a period of time, water is removed from the reactants. For example, thermal formation of polymers of amino acids has been demonstrated by Dr. Sidney W. Fox and his associates at the University of Miami. The conditions under which amino acids might be bonded together to form proteins were simulated in a number of Fox's experiments by heating mixtures of different amino acids at temperatures over 100°C for varying periods of time. It was found that, in addition to low molecular weight proteins, a number of large and complex proteins were produced. Some of the proteins assumed highly complex structures, thus indicating a certain level of order and organization. Elaboration of these experiments by other investigators has shown that other macromolecules may also be formed under the influence of thermal energy. Keep in mind that even though the earth was cooling slowly, its surface would still have retained heat. Moreover, the torrential rains must have become less frequent. Water could have evaporated between periods of rain, thus leaving dry amino acids on the hot surface. Under these conditions, amino acids may have bonded together, and later rains could have washed the resulting proteins into the oceans. It seems reasonable to assume that thermal energy may have been responsible for forming the first macromolecules in the primitive seas.

A second general way in which dehydration synthesis occurs is by the action of chemical compounds that excise molecules of water. Living cells of today utilize the pyrophosphate group of ATP to effect the removal of water during certain dehydration synthesis reactions. Melvin Calvin of the University of California, Berkeley, and his associates have shown that HCN is the precursor of chemical substances capable of removing water during such reactions. These substances, derivatives of carbodiimides, are capable of removing water from amino acids to form simple peptides and also promote the formation of compounds that have been the forerunners of ATP.

PROCELLS: COACERVATE AND MICROSPHERE DEVELOPMENT. Assuming that the waters of the primitive earth now contained all the compounds necessary for life, there still remains the problem of organizing the various components into a structural pattern characteristic of even the simplest organism. As far back as 1938, Oparin suggested what was called the *coacervate theory*. According to this theory, protein-like substances in the prehistoric oceans aggregated and developed enveloping membranes. Since the original proposal by Oparin, it has been shown experimentally that protein molecules tend to self-assemble to form more complex units, **coacervates** or **proteinoid microspheres.** These units are clusters of proteins or protein-like substances held together in small droplets within a surrounding liquid. Each coacervate is surrounded by a shell of water so that there is a definite separation between the coacervate droplet and the medium in which it floats. In essence, the water molecules form a membrane around the droplet. Experimental data have also shown that coacervates are able selectively to absorb substances from the surrounding medium. Although coacervates are not living, they do exhibit certain structural and functional properties ordinarily associated with living organisms.

In 1959, Dr. Sidney W. Fox elaborated on Oparin's coacervate theory and demonstrated that thermally produced proteins will orient into spherical aggregates called microspheres after the solution has cooled (Figure 3-6). These microspheres not only resemble spherical bacteria but they also show certain structural properties of cells. For example, in addition to having diameters about the same size as bacterial cells, they have surfaces that function in certain aspects like those of living cells and they react to chemicals much as living cells do. Some microspheres even contain internal components similar in appearance to structures found in living cells. A few have also been observed to split apart by a process (fission) similar to that which occurs in bacteria. However, the similarities between microspheres and living forms are superficial and highly speculative.

Despite structural and functional similarities to cells, it must be stressed that neither coacervates nor microspheres are living. They both lack well-developed structural organization and the capacity to reproduce. They do, however, reinforce the plausibility of a nonbiological origin for the first living forms (see Figure 3-3C).

The Organization of Procells into Living Forms. The time at which procells evolved into the first living forms and the mechanisms by which the process occurred are purely matters of speculation. Certain assumptions, however, can be made. For example, it is reasonable to assume that an evolving procell would have required a controllable and constant source of energy to make large complex molecules, to organize such molecules into definite structural patterns, and to maintain their structural organization. It can also be assumed that the organic molecules in the primitive seas contained energy. This energy, mainly from solar radiation, was converted into chemical energy and stored in the chemical bonds that held the atoms in the organic molecules together. Since carbohydrates are the compounds that supply most of the energy to living cells today, it can be further assumed that similar compounds in the ancient oceans also supplied procells with energy.

Large quantities of ATP are produced by present-day organisms that use oxygen as part of the reaction system that breaks down organic molecules. Such a process is called *aerobic respiration*. Since it has been assumed

(A)

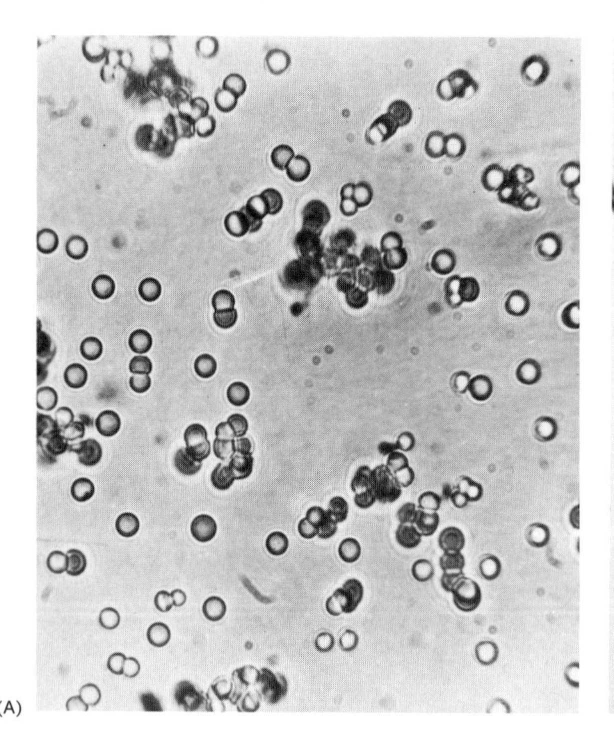

(B)

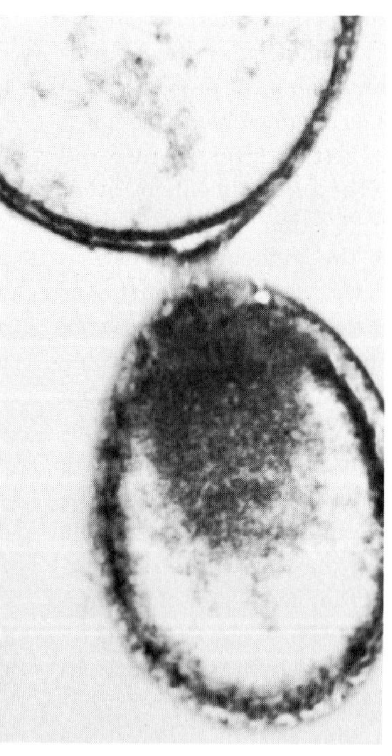

Figure 3-6. Proteinoid microspheres. (**A**) In this photomicrograph note the regular appearance of the spheres and their similarity to spherical bacteria. (**B**) This electron micrograph shows a microsphere containing a double-walled membrane in the process of "fission." [*Courtesy of S. W. Fox, Institute of Molecular Evolution, University of Miami.*]

that the primitive atmosphere did not contain free molecular oxygen, the evolving procells must have utilized another type of process that forms ATP without using oxygen gas. An example of such a process, which will be described in Chapter 6, is *fermentation.* Basically, fermentation is a series of enzyme-catalyzed reactions that release energy from organic molecules and store this energy in the chemical bonds of ATP.

Recall that the release of energy from compounds requires chemical bonds to be broken. In living systems the energy must be released in small amounts in a controlled manner, trapped, and converted into a form that is usable by the cell. If such a mechanism did not exist, much of the energy released by organic compounds would escape as heat and would be of little value to the cell. In this regard, it is believed that as procells continued their development, they eventually acquired the means to harness the chemical energy of the compounds in the primitive seas. More specifically, enzymes and ATP became involved in these energy-harnessing reactions. Enzymes split the bonds of the organic molecules and released the energy in a stepwise series of reactions. The chemical energy stored in ATP molecules was released when needed for various activities.

No mention has yet been made of the replication of life, that is, a continuity from one living individual to another (heredity). While some of the procells that have been discussed are capable of splitting by fission (simple division), there is no guarantee that the resulting divisions will each contain all the components necessary for life. Without some system of transmitting genetic information, life would not have evolved beyond the level of the proteinoid microspheres. Therefore, a most significant evolutionary step in the origin of life was the origin and transmission of the genetic code. However, it is most difficult to prove definitely or explain this particular step. Perhaps, during the billion or so years of chemical evolution preceding the evolution of life, different systems for the transmission of information arose and were selectively eliminated.

The buildup of a genetic code requires energy. Inasmuch as the present genetic code has ATP (high-energy compound) as a building block, it is reasonable to speculate that this and similar high-energy molecules originally polymerized to form the nucleic acids (genetic code). Selective pressure would favor the polymerization of a reproducible system rather than a random one because a reproducible system, once established, could perpetuate itself. While the evolution from a nonliving ancestral procell to the first living ancestral bacterium may seem difficult to envision, this aspect of evolution is no more difficult to conceive than the evolution of the simple bacterium to the complex multicellular organizations of the higher forms of life.

It is impossible to say at what stage nucleic acid control became operative, but it was probably quite early. Once this control was established, procells became primitive living organisms. These first living forms are believed to have been heterotrophs. A **heterotroph** is an organism that depends upon the environment for its nutrient requirements; inasmuch as primitive life forms probably derived organic molecules from the ancient oceans, they are assumed to have been heterotrophs. With the appearance of the first living organisms, chemical evolution gave rise to biological evolution. The major events involved in the chemosynthesis of life are summarized in Figure 3-3.

The Beginnings of Biological Evolution. Once continuity was established, the primitive heterotrophs began to compete for available food supplies in the ancient oceans and as the first cells multiplied and flourished, they caused significant changes in the environment. For example, competition was necessary because the surrounding organic compounds were used not only as foods, but also as building materials for the nucleic acids. But, such competition caused a depletion of the existing nutrients; they were probably being utilized faster than they could be produced. Moreover, as the primitive heterotrophs carried on fermentation-like activities, great quantities of carbon dioxide were added to the water and atmosphere. The production of more nutrients would have been more efficient without the increasing quantities of carbon dioxide since such concentrations of carbon dioxide would have absorbed some of the available solar radiation.

In the evolutionary history of the first heterotrophs, then, competition assumed a prominent role. Probable adaptive changes in the original life forms enhanced the chances for survival because they did not become extinct as the supply of organic compounds decreased. Any changes that would enable an organism to manufacture its own food, and not depend on the dwindling supply in the oceans, would be considered highly favorable. Among favorable changes that did occur were those that made possible the synthesis of light-absorbing pigments and the essential enzymes for a new process called photosynthesis. Those organisms that could carry on photosynthesis are called autotrophs. **Photosynthesis** is a process by which chlorophyll-bearing organisms utilize the raw materials of water and carbon dioxide in the presence of light to manufacture their own food. Such photosynthetic organisms are not dependent on foods available in the environment and for this reason they are referred to as **autotrophs.** The appearance of this new process not only provided a rapid and efficient mechanism for producing food molecules, it also brought about significant alterations of the physical environment. One of the by-products of photosynthetic reactions is free molecular oxygen, a highly reactive gas that combines readily with other substances. In the presence of large quantities of oxygen in the primitive atmosphere, most of the hydrogen, ammonia, and methane were converted to other substances. Oxygen probably reacted with methane to produce carbon dioxide and water, and with ammonia, to produce nitrogen and water. Thus, the new atmosphere consisted of water vapor, carbon dioxide, molecular nitrogen, and large quantities of free molecular oxygen itself. In addition, at higher altitudes, oxygen molecules combined with each other under the influence of high-energy radiation to form an ozone (O_3) layer. Such a layer absorbs solar radiation and protects present-day organisms from ultraviolet radiation, which, even in small doses, is lethal. Once biological evolution had progressed to this point, the physical environment was altered to such an extent that the conditions which had made possible the origin of life were destroyed.

The free molecular oxygen supplied through photosynthesis, along with the appearance of certain mutations, made possible a new form of respiration more efficient than fermentation. These mutations provided the organisms with the ability to carry on **aerobic respiration,** a process by which far more energy could be obtained from organic molecules.

CHEMICAL EVOLUTION: EXPLANATION OR SPECULATION? The account of the origin of life just presented is, of course, only hypothetical. In this regard, however, scientists often are required to hypothesize in order to offer solutions to problems. They must offer the best possible explanation based upon the available data. Any account of the origin of life is highly speculative and subject to continual and often drastic revision. At present, there is no crucial test to ascertain the accuracy of the heterotroph hypothesis, or for that matter, any other proposed account of the origin of life.

Alternate Hypotheses

A large portion of this chapter has dealt with the origin of life as posited by the basic assumptions of the heterotroph hypothesis. Essentially, it holds that life has arisen slowly from chemicals on a world-wide basis over a vast span of time. The plausibility of chemosynthesis is substantiated because it is a concept in keeping not only with present ideas regarding the evolution of life but also with probabilities concerning the origin of the earth and the solar system. The heterotroph hypothesis is by no means the sole account of the origin of life. There have been many others, and, it is quite likely that other concepts will emerge as more data become available. Among the many other hypotheses that have been proposed, two more will be examined here.

THE AUTOTROPH HYPOTHESIS. Some biologists believe that the first living form may have been an autotroph. Proponents of this hypothesis, the *autotroph hypothesis,* maintain that the first living organisms possessed all the necessary reactions for making food from simple inorganic materials using either chemicals or solar radiation as a source of energy. Those who wish to refute this hypothesis reason that since the chemical reactions involved in food making are very complex and require a structurally and functionally complex organism, such a possibility is not likely. In other words, the first autotrophs would have been complex organisms from the very beginning. Another major objection to the autotroph hypothesis is that complex organisms are typically the result of an accumulation of gradual changes over long periods of time. The changes leading to complexity are evolved; they are not acquired all at once. This reasoning is derived principally from fossil evidence. Moreover, no one has succeeded in constructing a model or providing sufficient data to show how autotrophs may have originated from molecular precursors.

The Cosmozoic Hypothesis. Unlike either the heterotroph hypothesis or the autotroph hypothesis, the *cosmozoic hypothesis* holds that organisms from various parts of the universe were transported to earth, perhaps on dust or meteorites, and were the source of life on this planet. Among the objections to this hypothesis are the following.

1. The ultimate source of these organisms is not explained—the problem is merely removed from earth and placed at some other portion of the universe.
2. It is improbable that living forms, as they are known, could have survived the extremes of heat and cold and the intense cosmic radiation of interplanetary travel.

Despite these objections, the cosmozoic hypothesis is being subjected to closer scrutiny by astronomers, geologists, and biologists. With the advent of artificial satellites, man is able to collect materials from outer space and analyze them for signs of life. In addition, the present commitment to space exploration may uncover some significant clues concerning the origin of life. There is hope of finding traces of chemicals on the moon where evolutionary processes halted as atmospheric water leaked away. Rocks four to five billion years old are not found on the earth, but are present on the moon in accessible condition.

Evolution and Organismic Diversity

It would appear that the first cells may have followed two lines of evolutionary development. In one direction, the nucleic acids were freely suspended within the cytoplasm of the cell. These nucleic acids found their way out of the cells and into the environment, most probably at the death of the cell, accidental or otherwise. Within the environment they remained in a relatively inactive state. When these nucleic acids came into contact with other cells, they entered and became active again. These ancient nucleic acids were possibly the evolutionary ancestors of present-day viruses and functioned in much the same way. Viruses consist mainly of a core of nucleic acid surrounded by an external protein coat (see Figure 8-2). When a virus comes into contact with a cell, it attaches itself to the cell membrane and the nucleic acid enters the cell, leaving the protein coat behind. Within the cell, the nucleic acid reorganizes and directs (parasitizes) the cell's resources, producing more nucleic acids with corresponding protein coats. It has been found that even fragments of viruses (nucleic acids) are capable of performing the same function.

In evolutionary development, it is probable that these virus-like nucleic acids came into existence at about the same time as cells. Since nucleic acids are the biochemicals of the hereditary substance (genes), their transfer from cell to cell could have had an important effect upon the evolutionary development of cells. As evolutionary development proceeded, the nucleic acids may have come together into filaments and clumps within the cells. These loose aggregates of nucleic materials were passed on as the cells divided, altered during cell division, and possibly aggregated with additional nucleic acids entering from the environment. All of these factors probably played an important part in the cellular evolution of a primitive cell. The present-day bacteria and blue-green algae are possibly the evolutionary descendents of these primitive cells.

In a second possible direction, nucleic acids were associated in cellular evolution in much the same way as that just mentioned. However, within the cell, instead of being freely suspended aggregates, they became associated with proteins, forming nucleoproteins and thus were precursors of chromosomes. As this type of cell evolved, the nucleoproteins came together and a membrane formed around them, resulting in a clearly defined nucleus (eucaryotic).

There is a further hypothesis supported by some modern evolutionary biologists that the development of the more complex eucaryotic cell began with a symbiotic (two species living together) relationship with procaryotic cell ancestors. A procaryotic cell is one in which a membrane-bound nucleus and membranous organelles such as mitochondria are lacking. During the evolution of the eucaryotic cell, a bacteria-like or blue-green algal ancestor entered the eucaryotic cell and evolved into the chloroplast. Aerobic bacteria living within the eucaryotic cell evolved into mitochondria, and the advanced flagellum, centriole, and spindle were the evolutionary result of a symbiotic relationship of a spirochete-like organism. While this proposed evolutionary development of the eucaryotic cell is only a hypothesis, there are phylogenetic and biochemical data to support it. Some of the organelles of the cell are highly developed intracellular symbionts with their own DNA, reproduction, and mode of existence. It also helps to account for the possible development of the autotrophic cell from the heterotrophic cell.

A few of the most widely accepted explanations of the origin of life based upon evidence that can be physically studied have been presented. We expand now to encompass theories of the origin of life which include the whole universe.

Exobiology

For many centuries man has been intrigued with the possibility of the existence of life outside of the earth. One of the most far-reaching and admirable achievements of modern man has been his exploration of outer space. While the technological accomplishments tend to get the publicity, another aspect of great importance, and of particular interest to biologists, is the possible origin and existence of life outside the earth, or **exobiology.**

In the search for extraterrestrial life man hopes to find a greater understanding of life on earth and its origins as well as a possible proof of its universality. Unlike chemistry and physics, biology has no universal laws or principles since the only life that is known is that which exists on earth. Chemists and physicists can observe and make measurements of far distant planets and stars and can substantiate the universality of their laws. While it is possible to plan and perform experiments concerning the origin of life on earth and the possibility of extraterrestrial life, only through space exploration will the existence of extraterrestrial life be established or refuted.

While there are economic, political, intellectual, and technological reasons for exploring space and placing man on other planets, the most important reason for the biologist is that it provides the only adequate means of determining the existence of extraterrestrial life. Although unmanned spacecraft are being developed to bring back samples from distant planets, these probes will provide only partial answers at best (Figure 3-7). The most satisfactory means of study is to place man in space. Therefore, sustaining man and other forms of terrestrial life in space becomes a major goal for the biologist. This involves a consideration of the degree to which terrestrial life is adapted to the physical environment here on earth and the degree to which these adaptations can be modified to enable terrestrial life to exist in another environment. In certain instances experimentation has provided actual and theoretical answers. For example, in determining the temperature limits of terrestrial life, it is also possible to theorize about the ability of life as we know it to exist on a distant planet.

Today, the interrelationships among the various sciences are well known. However, *space biology,* as a new science, is unique in its inherent connection with the other natural sciences. It makes extensive use of the achievement of chemistry, astronomy, physics, geophysics, geology, radioengineering, psychology, and many other disciplines. The unique quality of this new science has resulted in bold ideas in research, inventiveness, and methodology. The most important aspect of space biology is that it poses far more questions than will be readily answered.

The Nature of Exobiology

Establishment of the existence of extraterrestrial life depends to a degree upon the understanding of the dynamic state of terrestrial life. A most important aspect of extraterrestrial life would be its chemical composition, since this would indicate the nature and mechanisms of life functions. Therefore, in considering the chemical composition of extraterrestrial life, there are alternative possibilities. First, there may be some form of life with a chemical basis totally different from that found on earth. For example, the structure and metabolism of a living creature could be based upon silicon rather than carbon, or ammonia rather than water. Even with these examples, speculations are based upon similarities in the properties of the chemical substances substituted. With present knowledge, it is difficult to conceive of other possibilities, yet they may exist. Second, the forms of extraterrestrial life may be almost identical with those found on earth. If an origin independent of earth could be established, it would have a significant effect on our present theory of evolution. A

FIGURE 3-7. Unmanned spacecraft. A view of the unmanned Surveyor III spacecraft taken during the Apollo XII second extravehicular activity on the surface of the moon. The Apollo XII lunar module, with astronauts Charles Conrad, Jr., and Alan L. Bean aboard, landed within 600 feet of Surveyor III in the Ocean of Storms. The television camera and several other pieces were taken from Surveyor III and brought back to earth for scientific examination. Surveyor III landed on the side of this small crater in the Ocean of Storms on April 19, 1967. [*National Aeronautics and Space Administration photo.*]

third possibility is that extraterrestrial life is generally similar to terrestrial life but differs in certain important aspects. Of the various possibilities, this one is considered the most probable. Extraterrestrial life of this kind would have carbon and water as the dominant life chemicals, but there would be differences such as not having cellulose or proteins as structural materials. The genetic material might be some substance other than nucleic acids. Enzymes would not be protein in structure. There are other chemical possibilities than these just cited; those mentioned simply serve as examples. Fourth, evidence of extinct life may be found. While this would not provide the biologist with as important information as living forms, fossil remains might provide a means of reconstructing previous life forms. Fifth, there is the possibility that life never existed elsewhere. The difficulty with this alternative is that it would apply only to the specific extraterrestrial bodies explored. Although man's exploration of the moon has shown no evidence of life, this does not mean that life does not or did not exist elsewhere. There is a statistical probability that there are in other solar systems undetected planets possessing a moon and having an environment similar to that of the earth.

Extraterrestrial Life

In considering the existence of extraterrestrial life, two different possibilities have been entertained. First, there is the possibility of the existence of life within our solar system besides that on earth, and/or second, its existence on a distant planet in a different solar system within the universe. An additional consideration concerns the possible types of living forms, and whether or not these possess intelligence—an area in which speculation reaches the level of science fiction. Yet, interestingly enough, the science fiction of yesterday has in many instances become the reality of today.

PLANETS. The *planets* of our solar system lie within the realm of vehicular space exploration on the basis of present technology. For this reason, the greatest attention has been devoted to investigating the possibility of life on these planets. Of course, as has been previously stated, the search for life is not the only purpose for space exploration, but it is the most intriguing, and for the biologists, the most significant. In considering the possibility of life on the planets, a knowledge of their physical characteristics is important.

Our solar system has nine planets, thirty-one *natural satellites* (*moons*), and innumerable *asteroids* and *comets.* The planets all revolve around the sun in the same direction in elliptical paths, with their orbital planes slightly inclined to one another. The physical conditions that exist on a planet are dependent upon its distance from the sun and its size. Mercury, Venus, Earth, and Mars, which are closest to the sun, and Pluto, which is farthest from the sun, all have average densities four or five times that of water. The four other planets (Jupiter, Saturn, Uranus, and Neptune), called the *major planets,* have average densities from less than one to almost three times that of water. This would indicate that the actual structure of the four major planets is much different than that of the earth and the other *minor planets.*

In order for a planet to have an atmosphere, it must be large enough to exert a gravitational force sufficient to hold gaseous molecules. For example, the moon, which is much smaller than the earth and has about one-sixth the gravitational field, has no atmosphere. A knowledge of the size and mass of a planet makes it possible to determine if the planet has an atmosphere and to deduce something about the components of that atmosphere. Since hydrogen is the lightest element, it occurs only in the atmosphere of the larger planets. Oxygen, which is sixteen times heavier than hydrogen, has sufficient mass to be held in the earth's atmosphere but not in that of Mars, which is a still smaller planet. Carbon dioxide, which is forty-four times heavier than hydrogen, is probably present in the atmospheres of all planets except Mercury. Some of the components of the planets and their atmospheres have been determined through spectrographic analysis. Space vehicles (Figure 3-8) to the moon and those passing within a relatively short distance of Venus and Mars have sent back information about the composition of these extraterrestrial bodies. Of the minor planets, Venus, although slightly smaller than the earth, has an atmosphere. In fact, Venus is perpetually enveloped by a thick cloud cover. The Mariner II space probe indicated that the surface of Venus was about 600°F. Radar roughness measurements show enough relief to indicate that its surface is not liquid. However, there may be lakes of molten metals or other fluids. Mercury is so close to the sun that it is difficult to observe. Since Mercury does not rotate on its axis as does the earth, one side always faces the sun and probably has a surface temperature greater than 1000°F, while the "dark side" is probably close to absolute zero (−459.4°F).

Mars (Figure 3-9) has been probably the most intriguing of all planets. It has an atmosphere sufficiently thick to give it a climate. The changes in its white polar caps suggest an apparent difference in seasons. These earth-like properties have often suggested the possibility of some type of life. However, the likelihood of life on Mars is not based on these apparent properties, for in reality Mars differs greatly from the earth. The diameter of Mars (4220 miles) is slightly more than half that of the earth. Its atmosphere consists largely of nitrogen, carbon dioxide, water vapor, and oxygen, but in much different proportions than the atmosphere of the earth. Since Mars is tilted to the sun at about the same angle as the earth, it has a summer and winter. As the seasons change, different regions of the planet undergo color change. The temperatures of various areas of the planet have been measured by infrared radiation. Near the equator the temperature may rise to 80°F during the day, but it drops to −120°F at night. Other areas are colder. It was originally believed that there was only a very small amount of water present. Despite the cold temperature and small amount of water, recent evidence shows considerable water erosion, indicating that more water is present than was originally suspected. In addition, present water levels and temperatures are within a range to support certain types of bacteria.

The major planets all have similar characteristics. Gases such as hydrogen, helium, methane, and ammonia have been detected above the surface of these planets. The chemical and physical nature of the planets below their atmospheres is most intriguing. The "greenhouse effect" (see Chapter 22) of an atmosphere suggests that these planets may not be as cold as was

Figure 3-8. First space bounce. This is an artist's conception of the first United States multiplanet flight—the 1973 Mariner Venus-Mercury project. The 1973 Mariner trajectory employed the interplanetary billiard technique, using Venus' gravity to change speed and direction to shorten the flight time to Mercury. Both planets were photographed by TV cameras and probed by various scientific instruments—Venus in February 1974 and Mercury a month later. Earth is in the foreground of the drawing. Mercury is the planet closest to the sun. [*National Aeronautics and Space Administration photo.*]

previously believed. There has been recent speculation that conditions may not be so alien to life as was formerly thought.

Although similar, each of the major planets has unique characteristics. Jupiter possesses a giant red spot that varies in size. Recent evidence shows this to be a huge storm of gases. Saturn has spectacular rings that astronomers believe to be fine particles of ice. Uranus is tilted so far on its axis that its rotation is actually retrograde; that is, it has a direction contrary to that of the other planets. Neptune has a very dense moon much larger than the earth's moon.

While the physical aspects of space explorations are most important, the search for extraterrestrial life has been a driving force in the space effort. Any clues or hints as to the possibility of life on another planet

(A)

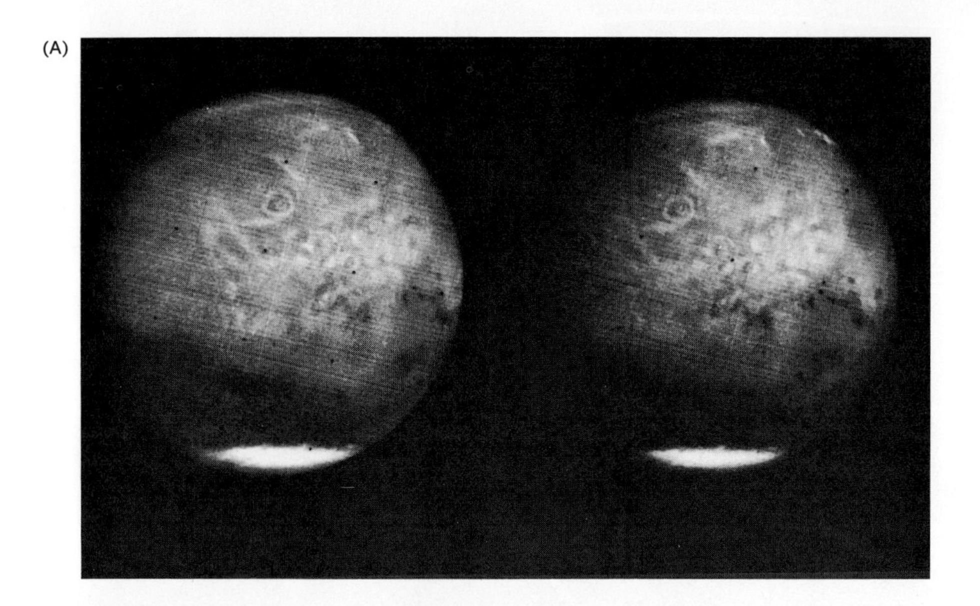

(B)

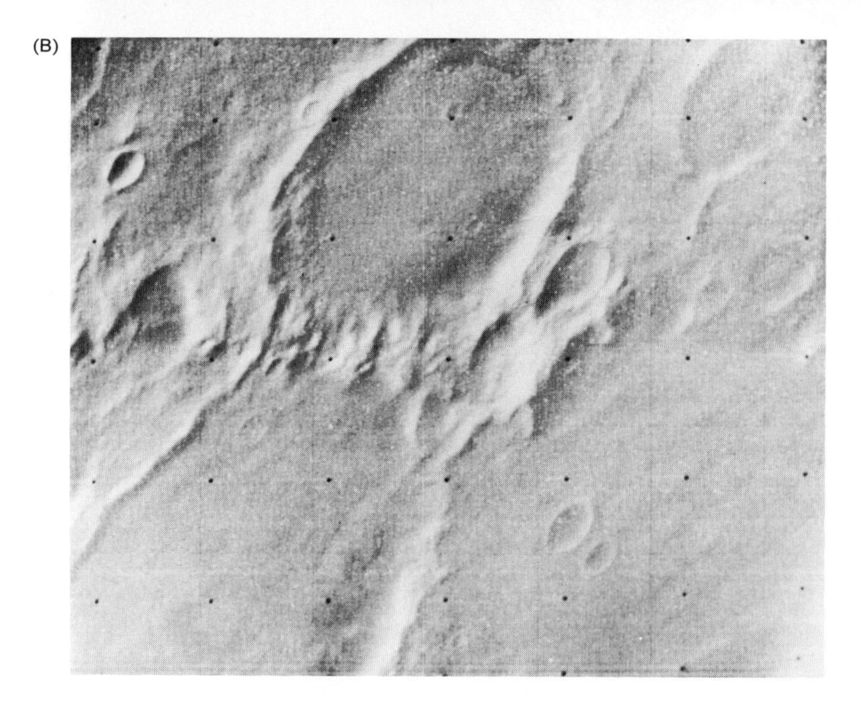

Figure 3-9. Mars. (**A**) These photographs are Nos. 73 and 74 of the Mariner VII far-encounter sequence. They were taken at an interval of 47 minutes, during which time Mars rotated through an angle of 12 degrees. Prominent in the picture are the bright ring-shaped Nix Olympica and the complex bright streaks of the Tharsis-Candor region. At the right edge the "canal" Agathadaemon disappears beyond the horizon. The dark, diffuse area at the lower left is Mare Sirenum. (**B**) The "Giant Footprint" consists of two adjacent craters, which are foreshortened by oblique viewing of the south polar cap of Mars. This high-resolution photograph was taken by Mariner VII and shows an area approximately 85 × 200 miles located about 75 degrees south latitude. [*National Aeronautics and Space Administration photos.*]

create excitement and interest. At the present time, no evidence from astronomy, biology, or space exploration conclusively indicates the existence or nonexistence of life outside the earth. However, speculation about the existence of such life continues. An important and encouraging hypothesis is that, given the right conditions, life may arise spontaneously and evolve. These "right conditions" for life as we know it include the presence of water, oxygen, and carbon, a source of radiant energy, and proper temperatures. Considerable experimentation

Figure 3-10. The 85-foot Howard E. Tatel radiotelescope. Observation instruments such as this are being used to search for signals from an intelligent source in the solar system. [*National Radio Astronomy Observatory photo.*]

has been done on the creation of organic molecules. Evidence also indicates that simple terrestrial life forms are able to survive the rigors of outer space. In the early life of the earth, conditions were far different than they are today. In addition, the presence of life on the earth is known to have altered considerably the physical conditions of the earth's surface and atmosphere.

Even if life is not found on the other planets, various stages of chemical evolution may be found. These stages may provide substantial information about the origins of life on earth. While the conditions on the other planets are different from those on earth, scientists do not dismiss the possibility of the existence of some form of life. For example, at the dawn of terrestrial life the atmosphere of the earth was probably similar to that found on Jupiter. Conditions may be such on any of the planets that the evolution of life is in its earliest stages. Over millions of years alteration in the environment of the planets by the presence of life could lead to the evolution of higher living forms. One of the interesting aspects of space biology is that it allows great freedom of imagination and considerable speculation.

Stars. In discussing the possibility of extraterrestrial life, one cannot be limited to the solar system of the sun. In fact, there is a substantial statistical probability that within the universe there are stars similar to our sun with an associated planet possessing an environment similar to that of the earth. Harlow Shapley of Harvard University suggests that one star in a million has a planet that meets the necessary conditions for the existence of life. In our galaxy there are 10^{11} stars; therefore, there may be 100,000 planets similar to the earth. In the entire universe there are more than 10^{23} stars. Shapley's estimate would therefore be increased to 10^{16} planets similar in size to the earth and its distance from their respective suns. The probability exists that these planets support life. Su Shu Huange of the Princeton Institute for Advanced Study concluded that 1 or 2% of these planets may have at one time supported intelligent life. Of course, the preceding estimates disregard the fact that the earth-moon system is often considered a binary planet. This consideration substantially reduces the probability of the existence of life as we know it because astronomers have yet to discover another planet-moon system similar to ours.

Astronomical distances are so great that the chances of two life forms meeting are exceedingly small. Interstellar trips with present space vehicles are impossible. Most of the nearest stars are more than 20 light years away. The distance of a light year is approximately 6×10^{12} miles. Even traveling close to the speed of light (186,000 miles per second), it would take almost an earth lifetime for man to make such a trip.

One of the most practical means of possibly determining the existence of life in the universe would be by means of *electromagnetic radiation* (radio waves). Electromagnetic radiation travels at the velocity of light to distances 1000 light years or more. The National Astronomical Observatory has already made an initial search for signals from an intelligent source (Figure 3-10). For a period of 2 months, an antenna was directed toward two stars. No signals were heard. However, it will take many more years with larger radiotelescopes and more sensitive receivers to detect such signals if they do exist.

To date, major considerations of exobiology have been the development of life detection instruments to be sent to the planets and instruments for possible radio contact. However, there is another approach to the detection of extraterrestrial life. Through laboratory experimentation it may be possible to determine which materials and conditions in the universe might give rise to life. Tracing the possible patterns by which life appeared and evolved on earth would give strong evidence of the existence of life elsewhere. This experimental approach has a rational basis in the Darwinian theory of evolution and recent advances in biochemistry. If the evolution of higher forms from lower forms is projected back to its logical beginnings, it is essential to postulate chemical evolution.

By the twenty-first century manned and unmanned space exploration of our solar system, radiotelescopes scanning the universe, and laboratory experimentation tracing the steps in chemical evolution may establish the existence of extraterrestrial life.

4 Diversity of Living Organisms

GUIDE QUESTIONS

1 What is taxonomy? Why is classification necessary? How does a natural classification scheme differ from an artificial one?
2 Define binomial nomenclature. Why are these designations useful? How are scientific names derived? Give several examples.
3 In what respects do homology and analogy serve as taxonomic criteria? How are vestigial structures and the appearance of the nucleus related to taxonomy?
4 What is a species? Define each of the principal taxons and give an example of each.
5 Discuss the roles of cellularity, symmetry, body cavities, and embryonic layers in taxonomic schemes.
6 Of what importance are nutritional patterns in classifying organisms?
7 What is the basis for Whittaker's five-kingdom classification scheme used in this text?
8 What are the outstanding characteristics of the Kingdom Monera? Protista? Fungi? Animalia? Plantae? Give specific examples to illustrate your response (see the Appendix).

According to the heterotroph hypothesis outlined in Chapter 3, it is believed that all life evolved from individual cells and from simple aggregates of cells. As new forms evolved from these beginnings, they diverged in several distinct directions. Cumulatively considered, there are more than 1½ million classified species of organisms. Moreover, several thousand species are identified and added to the list each year, and it is reasonable to assume that a large number of species are yet to be recognized. It becomes apparent, then, that some type of logical and meaningful system of cataloguing such vast numbers of organisms is essential.

The Classification of Organisms

The science of classification, called **taxonomy,** was one of the first specialized disciplines to evolve from the study of organisms. Historically, the earliest recorded systems of classification date back to the time of Aristotle. Since then, numerous systems have been devised, abandoned, and modified as newly acquired data have become available. In general, two principal types of classification schemes may be distinguished—artificial and natural.

The first satisfactory system of classification was proposed in 1735 by Carolus Linnaeus. Early systems of classification, including the one proposed by Linnaeus, were artificial. An *artificial system of classification* is one that is based upon structural characteristics without regard for the ancestry or evolutionary development of organisms. Such artificial systems are no more than a catalogue of terms. For example, the classification of plants devised by Linnaeus was based essentially on the number of stamens and styles (reproductive structures) in each flower. The difficulty with his system was that plants having the same number of stamens are not necessarily related to each other. Conversely, plants related to each other may differ in the number of stamens present. For the most part, artificial classification systems have been greatly revised or replaced by natural ones.

A *natural classification system* attempts to classify organisms on the basis of their genetic relationships to each other. Such a system is based upon the similarities and differences of a great many external and internal characteristics and the relationship of these structures as organisms have evolved from relatively simple to more complex forms. It is a system of descent, in which organisms are grouped according to the presence of certain primitive or advanced characteristics.

Although preferable to artificial systems because they demonstrate evolutionary relationships, natural systems also contain inherent weaknesses.

1. The number and diversity of organisms in terms of form and function are very great.
2. Numerous organisms have appeared at various times during evolutionary history, while others have become extinct.
3. Inasmuch as the fossil record is discontinuous, many data concerning the evolutionary origin and development of various groups of organisms are fragmentary, vague, or not available.

In spite of these difficulties, as well as others, taxonomists have generally attempted to devise classification systems that are more natural than artificial since, even though subject to change, natural schemes do provide fairly accurate systems based upon genetic relationships and evolutionary patterns.

Classification Units (Levels)

In order to be useful, a natural classification scheme must place a particular species in a given position among all other organisms. An individual organism is given two names, usually derived from either Latin or Greek, a convention designated as *bionomial nomenclature.* The obvious advantage of employing scientific names for organisms, and not common names, is that these names are universally accepted and convey specific meanings that show natural relationships. A scientific name consists of the *genus* and the *species* designation; both terms are underlined or italicized, and the genus name is capitalized. The scientific name for the black locust, for example, is *Robinia pseudoacacia.* The generic, or group, name is always a noun, whereas the specific, or individual, name is usually an adjective.

The species designation for organisms may be derived according to a number of criteria. Some are chosen on the basis of habitat. *Ranunculus aquaticus,* the water buttercup, is so designated because it inhabits a water environment. *Pyrus americana,* American mountain ash, is named on the basis of location. Still other plants, such as *Rosa alba,* the white rose, are designated on the basis of a principal characteristic of the plant. Finally, some scientific names are chosen from the name of the discoverer of the plant; *Aster drummondii,* Drummond's aster, is an example.

The species name of an organism represents the lowest category of taxonomic grouping. Occasionally, lower levels of classification such as the variety, subspecies, and others are used; but, for purposes of this textbook, the species will be considered the lowest taxonomic category. Despite the significance of the species as a taxonomic grouping, disagreement still exists about its definition. Recall from Chapter 1 that a species is defined as a group of organisms, with certain characteristics in common, that breed freely among themselves but are prevented from breeding indiscriminately with other species by one or several isolating mechanisms. Logically, some species show more variation than others. When these variations are quite conspicuous, some species are subdivided into *subspecies* and/or *varieties.* Taxonomists generally agree, however, that the species is the fundamental natural unit of classification.

A species, then, is a group of closely related individuals. A *genus* consists of a number of species different from each in certain respects but related to each other by descent. For example, *Quercus,* the oak genus, consists of all types of oak trees (white, red, bur, velvet, etc.); while each species of oak differs from all others, they are all related genetically and therefore constitute the genus *Quercus.*

Just as a number of species are combined into a larger unit, the genus, related genera constitute a *family.* A group of similar families constitutes an *order,* and a group of similar orders makes up a *class.* Related classes in turn comprise a *phylum.* In general, the higher the unit, the more individuals it contains, and the fewer the number of units at the same level. By contrast, the lower the unit, the fewer the number of individuals in it, and the larger the number of units at that level. Consider the classification of the organisms shown in Figure 4-1.

Taxonomic Criteria

Quite obviously, the multitude of features that characterize organisms is almost limitless. In attempting to classify organisms, therefore, some decisions must be

Taxonomic Category	Common Name				
	Pneumonia bacterium	Amoeba	Bread mold	White oak	Man
Kingdom	Monera	Protista	Fungi	Plantae	Animalia
Phylum	Eubacteriae	Sarcodina	Zygomycota	Tracheophyta	Chordata
Class	Eubacteria	Lobosa	Zygomycetes	Angiospermae	Mammalia
Order	Eubacteriales	Amoebina	Muscorales	Fagales	Primates
Family	Enterobacteriaceae	Amoebidae	Mucoraceae	Fagaceae	Hominidae
Genus	*Klebsiella*	*Amoeba*	*Rhizopus*	*Quercus*	*Homo*
Species	*K. pneumoniae*	*A. proteus*	*R. nigricans*	*Q. alba*	*H. sapiens*

FIGURE 4-1. The classification of representative organisms from the five separate kingdoms of living forms. The basis for the separation of organisms into these five kingdoms is discussed later in this chapter.

made regarding which characteristics should be employed as taxonomic criteria. Most of the criteria used in classifying organisms are based on structural similarities, since organisms that share similarities in body structure also share a common ancestry or are closely related in their evolution. Taxonomists are often faced with the problem of deciding which of these structural characteristics are most significant in ascertaining natural relationships.

HOMOLOGY. Homologous structures represent one of the principal criteria for classifying organisms. **Homologous structures** are structures in different organisms that have a common origin or that correspond to one another in terms of inheritance from a common ancestor. Such body parts do not necessarily perform similar functions. Comparison of the vertebrate forelimbs (see Figure 20-1) shows that even though they are built on the same pattern, each has been adapted to serve a particular function. The differences have arisen only since the separation of the lines of descent leading from a common ancestor to each respective organism. In general, the closer the structural similarities, the more closely related the organisms are to a common ancestor.

ANALOGY. In contrast to homologous structures, **analogous structures** are body parts of different organisms that have similar functions but different origins. The wing of a bird and the wing of an insect illustrate the concept of analogy (Figure 4-2). Both are functionally similar, yet their structure and embryological origins are quite distinct. The wing of a bird has a bony framework; an insect wing has a membranous framework.

The evolutionary route of homologous and analogous structures is quite different. Homologous structures have an evolutionary origin in the same structure (a common ancestor), and in time the homologous structures of different organisms evolve in different ways; one structure is modified into several. The formation of analogous structures is just the opposite; two different structures, as they evolve, come to resemble each other.

APPEARANCE OF THE NUCLEUS. The cells that comprise the vast majority of organisms are referred to as **eucaryotic cells.** This means that the DNA of such cells is organized into discrete bodies, the chromosomes, which are contained in a membrane-bounded nucleus. Eucaryotic cells also contain specialized organelles such as the Golgi apparatus and mitochondria. These structures will be discussed in Chapter 5. Other cells, called **procaryotic cells,** possess dispersed clumps of genetic

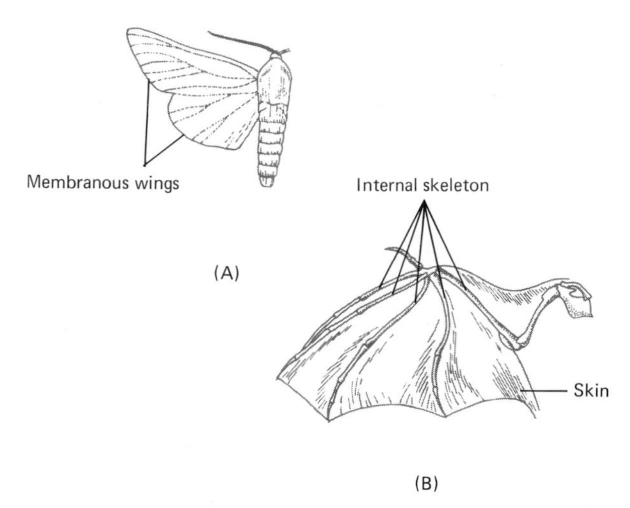

FIGURE 4-2. Analogy between the wing of an insect (**A**) and the wing of a bat (**B**). Although these structures have similar functions, they differ in embryological origin.

material and lack the membrane-enveloped nucleus. These cells also lack the aforementioned organelles.

As Figure 4-1 indicates, biologists do not classify all organisms into either plants or animals. In recent years there has been an increasing dissatisfaction with this simple division into two main kingdoms. Some taxonomists recognize three, four, and even five separate kingdoms. In all of these recently devised schemes the appearance of the nucleus of various cells has been employed as an important criterion in classification. In an evolutionary context, it is believed that certain major groups (kingdoms) of organisms evolved from procaryotic ancestors and others from eucaryotic ancestors.

Cellularity. Large numbers of organisms exist that consist of only a single cell. In such *unicellular* organisms there are no specialized tissues or organs. Each individual cell carries on all activities—digestion, conduction of impulses, movement, excretion, reproduction, etc.—essential to the survival of the species. In *multicellular* organisms, which are composed of many cells that are differentiated into tissues and organs, there is a clear division of labor among the component cells. In humans, for example, particular cells are specialized for activities such as reproduction, digestion, movement, conduction of impulses, and excretion.

Intermediate between unicellular and multicellular forms there are some groups of organisms that exhibit a third type of cellularity. They consist of individual cells that are loosely, though interdependently, associated in filaments or colonies. Although many such intermediate forms do contain cells that are specialized for various functions, the cells are not differentiated into tissues and organs, so these forms are not considered to be multicellular. At the same time, since individual cells of a colony or a filament cannot exist independently, such forms are not truly unicellular. It is reasonable to assume that these intermediate forms represent the transition from unicellularity to multicellularity that occurred during evolutionary history. In the taxonomic schemes that have been devised in recent years, cellularity is a principal criterion for classification of organisms.

Symmetry (Body Form). The component parts of an organism have a definite relationship to one another that results in a characteristic body form or shape. Such an orientation is conveniently described in terms of *symmetry*. Some organisms, such as amoebae and many sponges, have no symmetry and are referred to as *asymmetric* (Figure 4-3A). Others, indeed the vast majority, have a definite form and are symmetrical.

Organisms with *spherical symmetry* may be divided into two equal parts by any plane passing through the diameter of the body (Figure 4-3B). Spherically symmetrical organisms are not differentiated into anterior (front end) or posterior (back end) areas, dorsal (back or upper surface) or ventral (front or lower surface) regions, or sides; they have no head, no tail, and no appendages.

Radial symmetry is best demonstrated by coelenterates (hydra, jellyfish, anemones, etc.) and echinoderms (starfish, sand dollar). In these organisms, a number of similar parts radiate outward from a central axis like spokes of a wheel (Figure 4-3C). This type of symmetry allows a stationary animal to encounter food and enemies from all sides.

The final type of symmetry is *bilateral symmetry* (Figure 4-3D). In this particular arrangement, only one plane of division separates the organisms into two similar parts. Such a plane must pass through the longitudinal axis, through the center of the back, and through the center of the front; thus the organism is divided into mirror images. Organisms with bilateral symmetry have a definite right and left side, conspicuous dorsal and ventral surfaces, and prominent anterior and posterior ends. The anterior end of such organisms typically contains a concentration of nervous tissue and sense organs. All vertebrates and most invertebrates have this type of symmetry.

Body Cavity (Coelom). Another characteristic considered in grouping animals is the type of body cavity, or coelom, present in the adult organism. A **coelom** may be defined as a cavity within the mesoderm, between the alimentary canal and the body wall, that is lined by a peritoneum. (The mesoderm is one of the three embryonic layers from which tissues and organs develop in most animals; the peritoneum is a derivative or modification of the mesoderm.) Some animals, such as the flatworms (Figure 4-4A), coelenterates, and sponges, do not possess a body cavity between the internal organs and the body wall. These are designated **acoelomates.** As the body of an animal

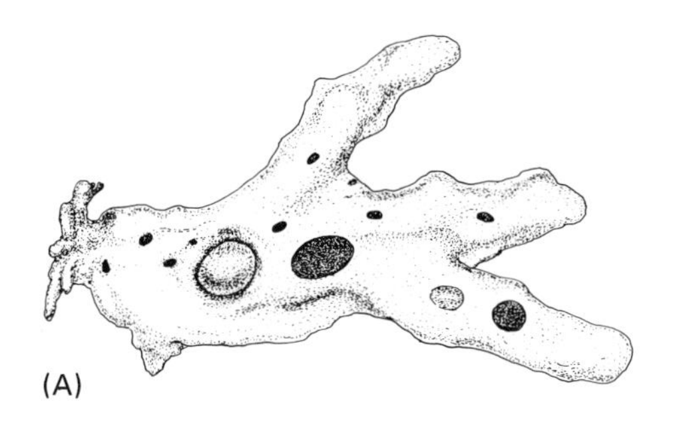

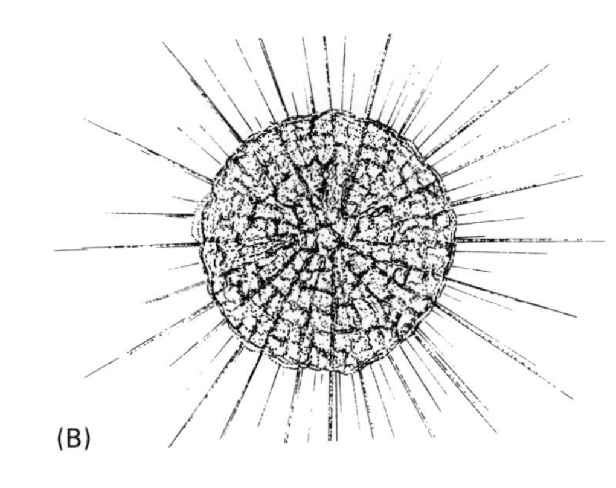

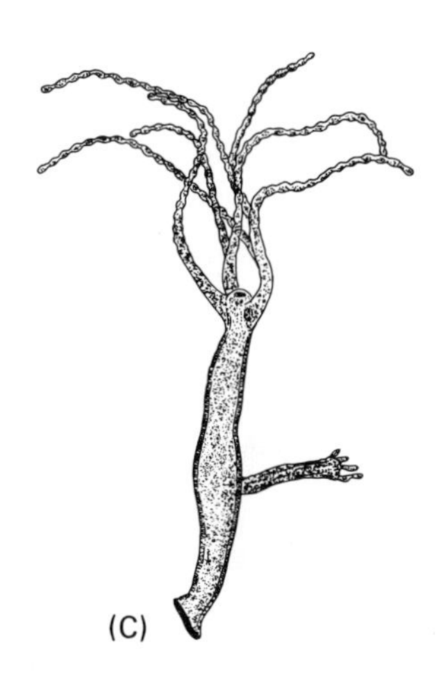

FIGURE 4-3. Kinds of symmetry. (**A**) *Amoeba* (amoeba)—asymmetry. (**B**) *Actinosphaerium* (radiolarian)—spherical symmetry. (**C**) *Hydra* (hydra)—radial symmetry. (**D**) *Perca* (perch)—bilateral symmetry.

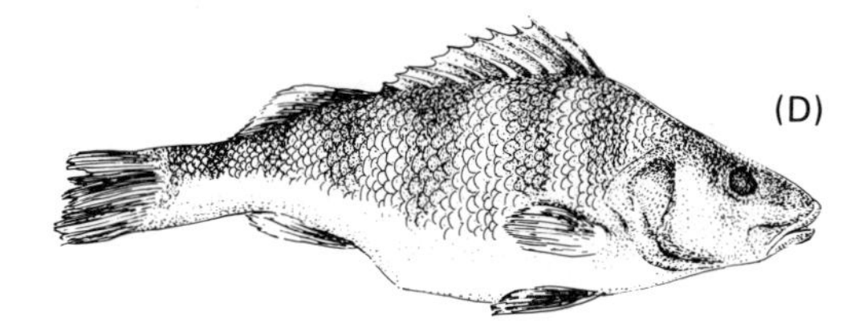

increases in size, the acoelomate condition would not be physiologically efficient. A fluid-lubricated cavity allows for looped intestines and aids in circulation of food and oxygen, as well as the removal of wastes. It is for this reason that most animals have a coelom.

Those organisms possessing a true coelom are referred to as **coelomates** (Figure 4-4B). The coelom develops within the mesoderm of the embryo and has a lining of specialized covering cells, the peritoneum. This lining surrounds the digestive tract and covers the inner surface of the body wall. In humans the coelom is subdivided into two principal cavities, the abdominal cavity and the thoracic cavity.

Between the animals with a true coelom and those that are acoelomate there are a number of organisms possessing a *pseudocoel,* or false coelom (Figure 4-4C). These organisms (roundworms), referred to as **pseudocoelomates,** have a body cavity, but it is not lined with peritoneum. Thus, the internal organs are not suspended in the coelom, but rather lie freely within it.

EMBRYONIC LAYERS. Following fertilization of the eggs of multicellular animals, there is a differentiation of cells into **germ layers** (see Figure 18-15). These specialized layers of cells have the potential to develop into all tissues and organs that form the adult animal. In the vast majority of organisms, three germ layers develop. These are called the ectoderm, mesoderm, and endoderm, and such animals are designated as **triploblastic.** Other organisms, such as sponges and coelenterates, do not develop mesoderm and are therefore referred to as **diploblastic.**

NUTRITIONAL PATTERNS. In recently devised classification schemes, the mode of nutrition utilized by an organism has assumed a prominent role. In general, two major nutritional types are recognized, autotrophic and heterotrophic. **Autotrophs,** represented by green (photosynthetic) plants and certain bacteria (chemosynthetic), are capable of synthesizing their own food molecules from energy and raw materials obtained from the physical environment. **Heterotrophs,** by contrast, cannot manufacture nutrients and must feed on complex materials that are produced by other organisms. Some heterotrophs (the animals) take solid food into the body by a process called **ingestion.** Organisms that eat in this manner are termed **holotrophs.** Other heterotrophs, referred to as **saprophytes,** are dependent

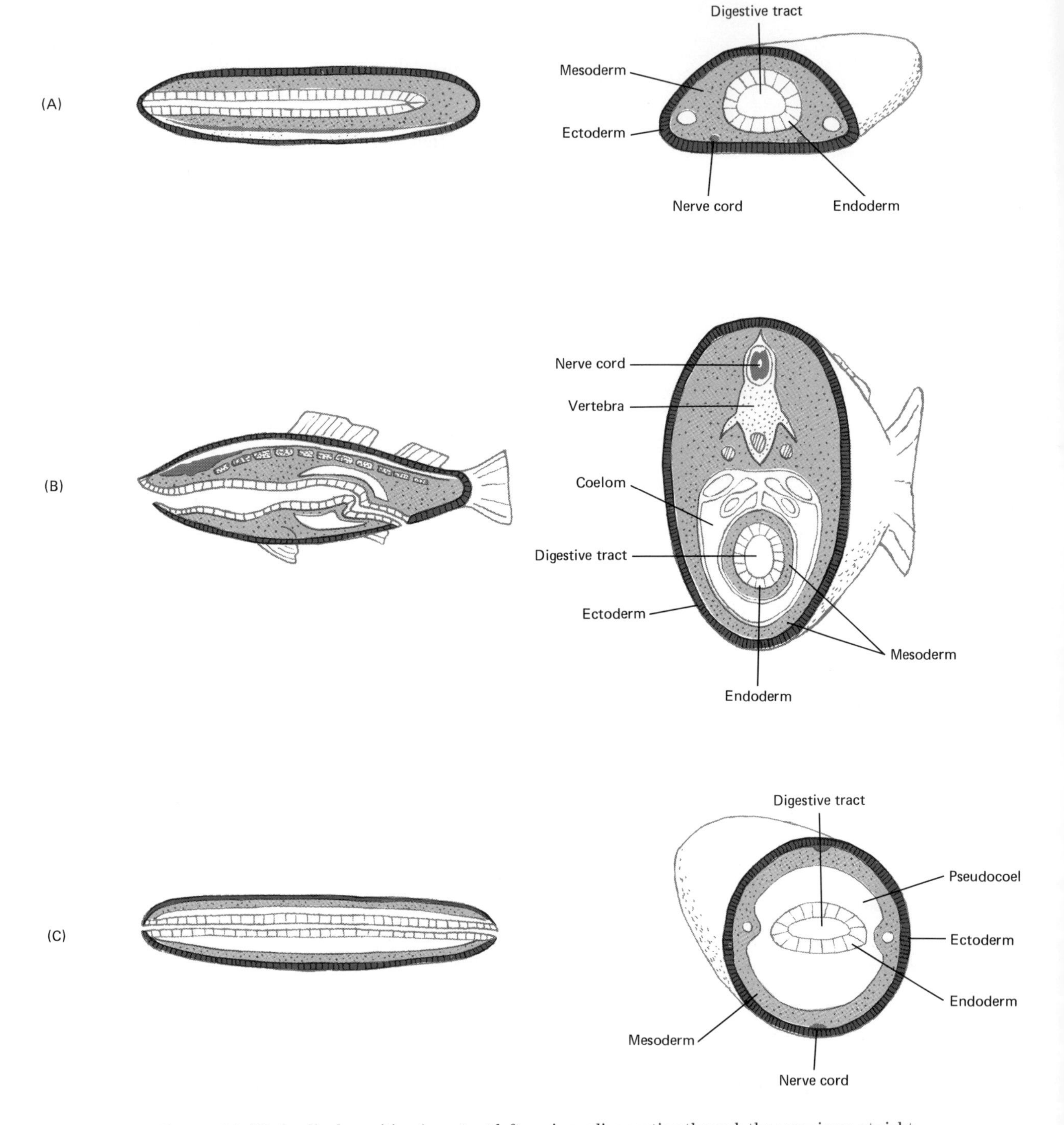

Figure 4-4. Kinds of body cavities. Aspects at left are in median section through the organisms; at right are cross sections showing the embryonic germ layers and body cavities, where present. (**A**) Since the flatworm is an acoelomate, it has no body cavity. (**B**) True coelom of a vertebrate. (**C**) Pseudocoel of a roundworm.

upon the absorption of organic food molecules from nonliving matter. In the process of absorption the molecules of the nutrients are of a relatively small size and are generally soluble. They pass through a membrane and are then utilized by the living form. Some heterotrophs (holotrophs) ingest food and have special systems (digestive) for preparing the nutrients for absorption. Nutritionally, all animals are holotrophs. Other heterotrophs (saprophytes) generally absorb their nutrients directly from the external environment. Most fungus organisms are of this type. A third form of heterotroph is the **parasite,** an organism that lives on or in an organism of a different species and derives nutrients from it. Many parasites are quite specific with regard to the hosts they will attack and also to the part of the host's body on or in which they will live.

It will be seen later in this chapter that three modes of nutrition—photosynthesis, absorption, and ingestion—lead to the very different types of organization that characterize the three higher kingdoms—the plants, fungi, and animals, respectively.

OTHER TAXONOMIC CRITERIA. In addition to the criteria discussed above, biologists rely on a number of others to place organisms in a natural taxonomic framework. Among these are

1. The type of skeletal system used for support, protection, and locomotion. Some animals have an internal skeleton (endoskeleton), others have only shells or protective tubes (exoskeletons).
2. Variations in organ systems, specifically those relating to locomotion, circulation, excretion, digestion, and reproduction.
3. The appendages of animals that are adapted for functions such as protection, obtaining food, locomotion, and respiration.
4. The organs and methods of reproduction.
5. Behavioral patterns.
6. Habitats occupied by different organisms.
7. Degree of body segmentation.
8. Physiological and biochemical similarities.
9. Genetic makeup.

The last is the most conclusive evidence of relationships among organisms, since all other taxonomic criteria are determined by genes.

Organismic Diversity

Since the time of Linnaeus, the formal classification of organisms into two kingdoms—plants and animals—has been used extensively. Early taxonomists differentiated between these two major groups of organisms on the basis of gross, macroscopic examinations. They had little difficulty in grouping higher, photosynthetic, rooted organisms as plants and higher, motile, food-ingesting forms as animals. To the early naturalists, such a dichotomous subdivision was obvious. Biologists still tend to categorize organisms as animals or plants. Throughout this text, reference is occasionally made to plants and/or animals; yet, to be absolutely correct, in some instances other major classifications of living forms might be included.

With the development of biological instrumentation, especially the microscope, the problem of classification became more difficult. For the first time, biologists could examine unicellular organisms that were motile and ingested food; naturally these were classified as animals (protozoans). Other organisms were nonmotile and photosynthetic and were designated as one-celled plants. A wide variety of unicellular organisms, however, exhibited other combinations of nonmotility or motility and nutritional pattern. They were neither solely plant-like nor solely animal-like. Thus, the designation of all organisms as either plants or animals was inappropriate.

In view of this problem, as well as many others, biologists began to devise new taxonomic schemes that attempted to incorporate the myriad differences among organisms. While there are relative strengths to a number of these classification systems currently in use, this text employs one that groups organisms into five principal kingdoms (Figure 4-5) on the basis of levels of structural organization and patterns of nutrition that conform to current biochemical and evolutionary evidence.

1. Monera
2. Protista
3. Fungi
4. Plantae
5. Animalia

This classification has been set forth by R. H. Whittaker of Cornell University. Details of the Whittaker

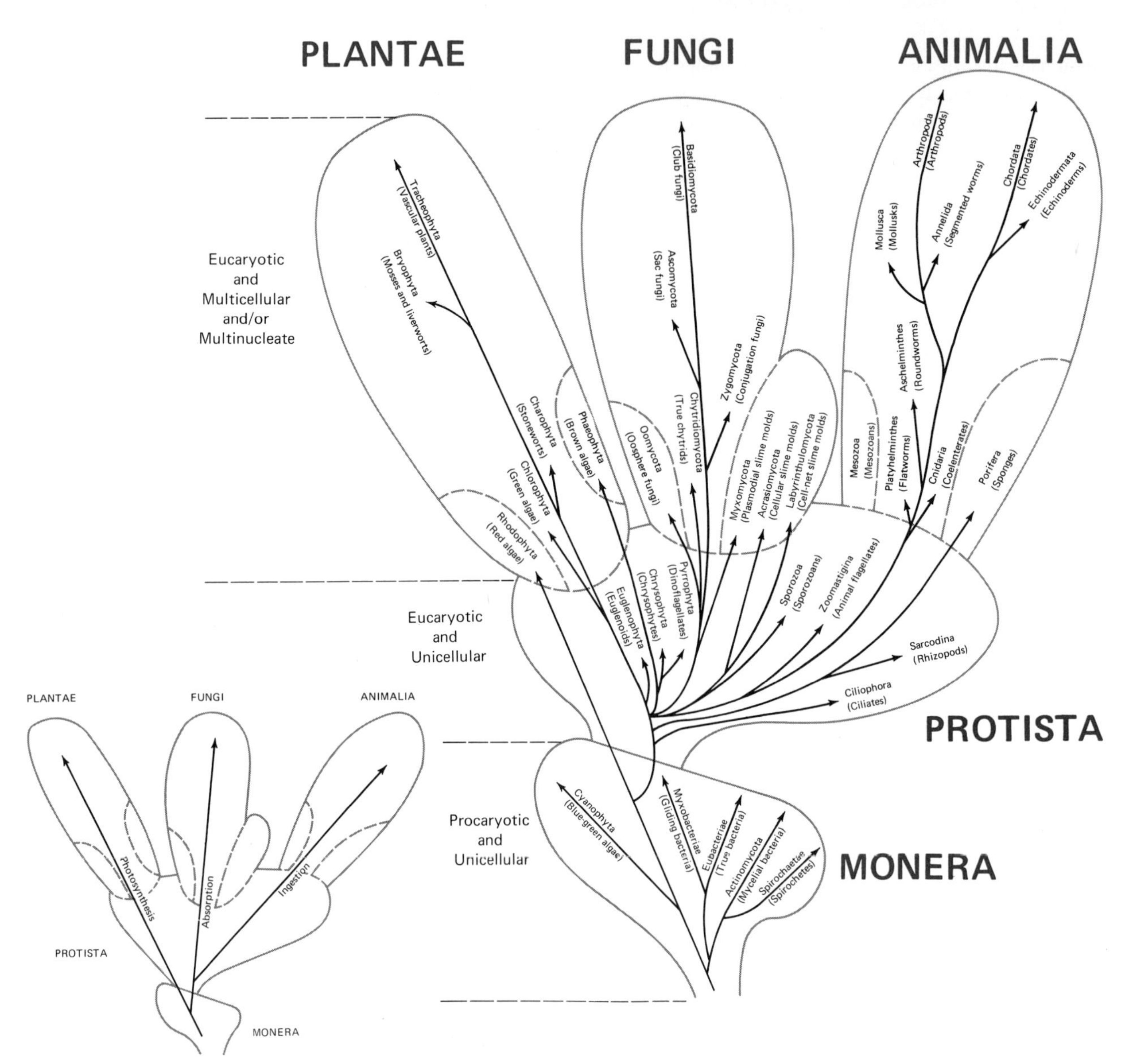

FIGURE 4-5. The five-kingdom system of classification as proposed by R. H. Whittaker. In this taxonomic scheme, levels of organization and nutritional patterns are the principal criteria employed in grouping organisms into natural categories. With regard to levels of organization, note that members of the Kingdom Monera are procaryotic and unicellular; organisms of the Kingdom Protista are eucaryotic and unicellular; and representatives of the Kingdoms Plantae, Fungi, and Animalia are eucaryotic, multicellular, and/or multinucleate. On each level there is a divergence with respect to the three principal nutritional patterns—photosynthesis, absorption, and ingestion. Ingestion is lacking in the Monera; all three patterns are present in the Protista; and on the multicellular and/or multinucleate level, the three patterns are used to classify the widely divergent groups of organisms that make up the Plantae, Fungi, and Animalia. In other words, members of the Plantae are primarily photosynthetic, Fungi are typically absorptive, and most Animalia are ingestive. The larger diagram shows the relationships of selected phyla to each other (these phyla are discussed in the Appendix), and the smaller diagram to the left illustrates the principal modes of nutrition that characterize the five kingdoms. [*Modified from H. R. Whittaker,* Science, *vol. 163, Jan. 10, 1969, pp. 150–60, Fig. 3. Copyright © 1969 by the American Association for the Advancement of Science. By permission of Dr. Whittaker and the AAAS.*]

classification scheme are presented in the Appendix.

The distinction between the procaryotic cells represented by the *Kingdom Monera* (monerans) and the eucaryotic cells of other organisms provides the clearest separation of organismic levels in the world of living things.

The *Kingdom Protista* (protistans) represents a wide diversity of organisms with a highly organized cellular structure. However, the basis for their classification into a separate kingdom is their lack of tissue formation and an absence of integration between associating cells. The Protista represent the evolutionary ancestors of the three kingdoms—Plantae, Fungi, and Animalia—in which there is tissue formation and integration between the cells making up the organisms. It must be noted, however, that certain unicellular organisms are integrated into these kingdoms.

The placement of *Kingdom Fungi* (fungi) in a separate kingdom represents one of the most recent proposals of this broad classification scheme. It is probable that the lower fungi (Chytridiomycota, Oomycota, and Zygomycota) were evolved from nonphotosynthetic flagellate ancestors and the higher fungi (Basidiomycota and Ascomycota) were evolved from the lower fungi. Their mode of nutrition is absorptive, with the absorbing tissue typically embedded in the food source. While the Fungi are frequently classified with the plants, they have a different origin, a different direction of evolution, and a different nutritional basis.

The slime molds possess the organization and nutritional distinctions of both the Kingdoms Animalia and Fungi. Their placement in the Kingdom Fungi is a matter of choice. However, the fruiting body in the reproduction of the slime molds is similar in structure to certain stages in the life cycle of the Fungi.

The *Kingdom Plantae* (plants) probably evolved from several different protistan organisms; however, plants represent an autotrophic nutritional organization. The phylum Chlorophyta (green algae) has certain unicellular forms and thus does not conform to the multicellular characteristic of the Kingdom Plantae. These unicellular forms, however, show a close relationship to certain of the multicellular Chlorophyta. The diversity and number of living forms are so extensive within known means of present-day classification that certain inconsistencies are inevitable. Within the Plantae, distinguishing differences between the Rhodophyta (red algae) and Phaeophyta (brown algae) appear to indicate that neither phylum had followed the line of development of the higher green plants. It appears that they have evolved from a different unicellular Protista and their resemblance to other groups of the Plantae is probably that of evolutionary convergence.

The *Kingdom Animalia* (animals) is believed to have descended from several protistan ancestors. Within this kingdom, the Porifera (sponges) and Mesozoa (mesozoans) probably evolved as separate groups from different ancestral protistans. The main line of evolutionary development follows from the Cnidaria (coelenterates). The Animalia are multicellular with varying degrees of integration among the cells. The nutritional pattern is predominantly ingestion with some forms obtaining nutrition by absorption from the external environment.

The system of classification used in this text attempts to establish a broad classification expressing the major structural and nutritional relationships. In the discussion of these kingdoms in the Appendix, only the principal characteristics of the major groups and certain representative species are included.

SUGGESTED SUPPLEMENTARY READINGS FOR SECTION ONE: THE ORGANIZATION, ORIGIN, AND DIVERSITY OF LIFE

ALEXOPOULOS, C. T., and H. C. BOLD, *Algae and Fungi.* New York: Macmillan, 1967.

BAKER, J. J. W., and G. E. ALLEN, *Matter, Energy and Life,* 2nd ed. Reading, Mass: Addison-Wesley, 1970.

BALDWIN, ERNEST, *The Nature of Biochemistry.* New York: Cambridge University Press, 1967.

BENNETT, T. P., and EARL FREIDEN, *Modern Topics in Biochemistry,* 2nd ed. New York: Macmillan, 1969.

CALVIN, MELVIN, *Chemical Evolution.* New York: Oxford University Press, 1969.

CAMPBELL, J. A., *Why Do Chemical Reactions Occur?* Englewood Cliffs, N.J.: Prentice-Hall, 1965.

CARSON, H. S., "Chromosome Tracer of the Origin of Species," *Science,* vol. 168, June 19, 1970, pp. 1414–18.

EISENBERG, DAVID, and W. KOUZMANN, *The Structure and Properties of Water.* New York: Oxford University Press, 1969.

FRIEDEN, EARL, "The Chemical Elements of Life," *Scientific American,* July 1972.

FOX, SIDNEY W., *Molecular Evolution and the Origin of Life.* San Francisco: Freeman, 1975.

FOX, SIDNEY W. (ed.), *The Origin of Prebiological Systems and Their Molecular Matrices.* New York: Academic Press, 1965.

GAMOW, GEORGE, *The Creation of the Universe.* New York: New American Library, 1965.

GOLDSBY, R. A., *Cells and Energy,* 2nd ed. New York: Macmillan, 1977.

LEHNINGER, A. L., *Biochemistry,* 2nd ed. New York: Worth Publishers, 1975.

MILLER, S. L., "The Origin of Life," in W. H. Johnson and W. C. Steere (eds.), *This Is Life.* New York: Holt, Rinehart and Winston, 1962.

OPARIN, A. I., *Life: Its Nature, Origin, and Development,* 2nd ed. New York Academic Press, 1962.

ORR, R. T., *Vertebrate Biology.* Philadelphia: Saunders, 1976.

PHILLIPS, DAVID C., "The Three-Dimensional Structure of an Enzyme Molecule," *Scientific American,* November, 1966.

SCAGEL, ROBERT, et al., *An Evolutionary Survey of the Plant Kingdom.* Belmont, Cal.: Wadsworth, 1966.

WATSON, JAMES D., *The Double Helix.* New York: Atheneum, 1968.

WHITE, E. H., *Chemical Background for the Biological Sciences.* Englewood Cliffs, N.J.: Prentice-Hall, 1970.

WHITTAKER, R. H., "New Concepts in Kingdoms of Organisms," *Science,* vol. 163, January 10, 1969, pp. 150–60.

TWO
The Cellular Level of Biological Organization

5 Cells: Basic Units of Structure and Function

6 Cells and Energy: Procurement and Transduction

7 Cells and Energy: Utilization

8 The Control of Cellular Metabolism

5 Cells: Basic Units of Structure and Function

GUIDE QUESTIONS

1 Discuss some of the major discoveries that have contributed to the development of the cell theory.
2 Compare the magnification possibilities of a light microscope and an electron microscope. What principles are involved in each type of microscope?
3 Define a cell. List and briefly discuss the principal subdivisions of a cell. In what ways are all cells unified? In what ways are cells diversified?
4 How has electron microscopy advanced knowledge of the structure of the plasma membrane? Discuss the following characteristics of the plasma membrane: molecular composition, permeability, and structural modifications.
5 Of what importance to the cell are diffusion and osmosis? What is meant by diffusion pressure and osmotic pressure? What factors may affect these processes?
6 Define passive absorption and active absorption and discuss several examples of each. What is meant by a concentration gradient? How are phagocytosis and pinocytosis similar processes? How do they differ?
7 Discuss the functions of the following cell wall components: middle lamella, primary cell wall, secondary cell wall, pits, and plasmodesmata. How do the primary cell walls and the secondary cell walls differ in structure?
8 What functions do intercellular substances serve? Discuss two such materials in detail.
9 Describe the general chemical composition and physical nature of the cytoplasm. Explain sol–gel transformations with respect to cytoplasmic functions. Discuss cyclosis.
10 Describe the structure and composition of a typical cell nucleus. What are the chief functions of the nucleus? the nucleolus? the nuclear envelope? the chromatin? Discuss several experiments that suggest the importance of the nucleus.
11 Outline in detail the structure and function of the following cell organelles: centrioles, cilia and flagella, endoplasmic reticulum, ribosomes, Golgi apparatus, mitochondria, plastids, lysosomes, and microtubules.
12 What constitutes cell inclusions? Of what importance are cell inclusions to metabolic activities of the cell?
13 Define mitosis and cytokinesis. Name and describe in detail the various stages of cell division. Are all the stages of mitosis of equal duration? Explain.
14 What is the origin of the spindle? What are the similarities that may indicate that various types of biological motility have a common biochemical basis? Describe one of the most widely held theories as to how chromosomes move.
15 List nine modifications of the typical mitotic pattern and state where they occur.
16 What are three significant outcomes of cell division?

PRIOR to the seventeenth century, little was known about the structure of cells. Today, it is universally accepted that cells are the units of life and that the forms and functions of these basic units are as diverse and as complex as life itself. This concept did not develop spontaneously, however; it evolved from the results of a long series of investigations that used improved tools and techniques of observation and analysis.

Development of the Cell Concept

Probably the most significant discovery leading to the development of the cell concept was the invention of

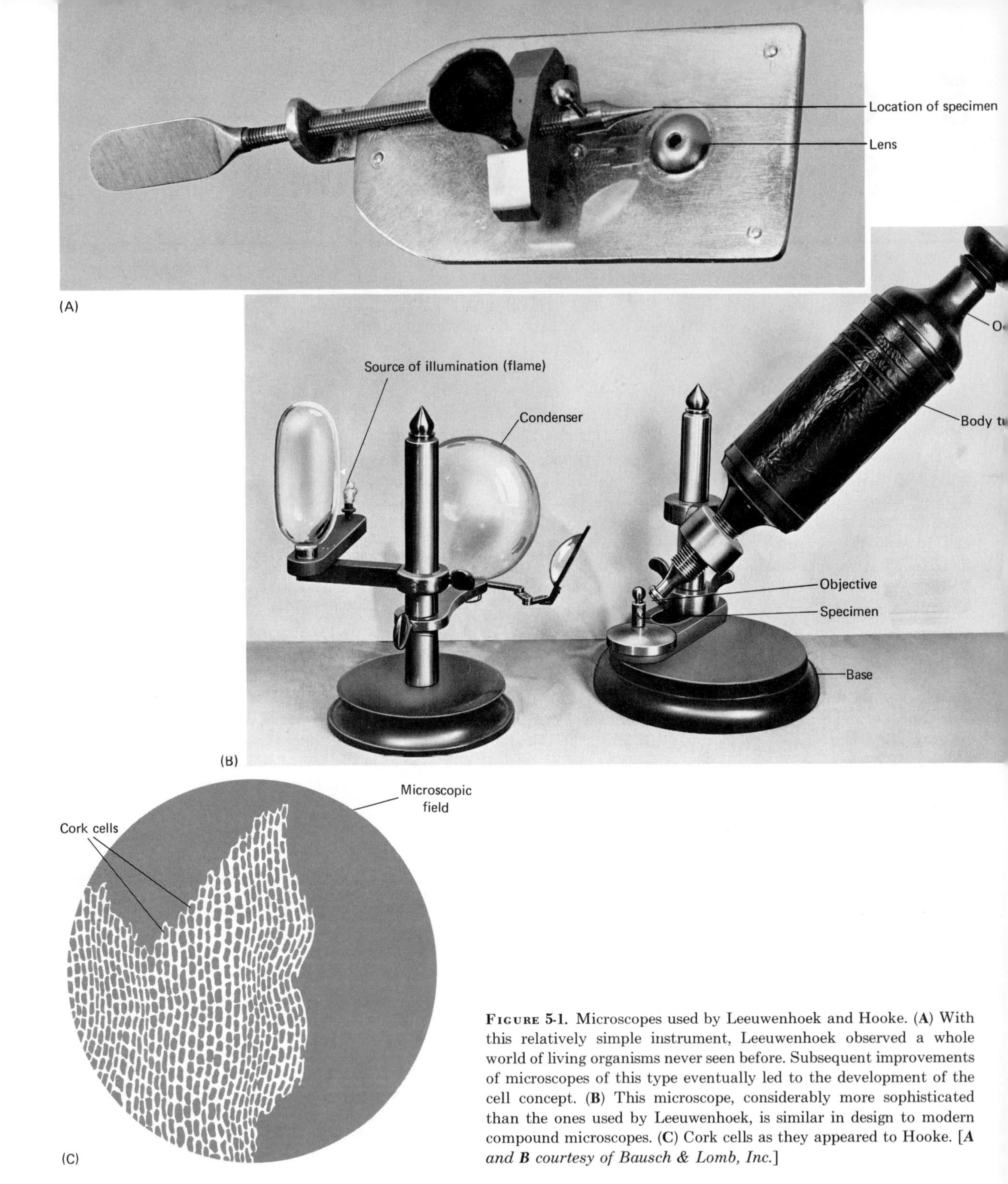

FIGURE 5-1. Microscopes used by Leeuwenhoek and Hooke. (**A**) With this relatively simple instrument, Leeuwenhoek observed a whole world of living organisms never seen before. Subsequent improvements of microscopes of this type eventually led to the development of the cell concept. (**B**) This microscope, considerably more sophisticated than the ones used by Leeuwenhoek, is similar in design to modern compound microscopes. (**C**) Cork cells as they appeared to Hooke. [**A** *and* **B** *courtesy of Bausch & Lomb, Inc.*]

the compound microscope by the Dutch lens maker Zacharias Janssen at the close of the sixteenth century. Through the efforts of Anton van Leeuwenhoek (1632–1723) the science of microscopy advanced further. His instrument was a very simple microscope consisting of a single lens of minute size but high magnification (Figure 5-1A). In the years following Leeuwenhoek, better microscopes were developed and more scientists began to use them. Robert Hooke (1636–1700) developed an instrument that was the true forerunner of the compound microscope used today (Figure 5-1B). In one of his publications there appeared a sketch of a thin slice of cork that he had examined under the microscope (Figure 5-1C). He described the cork as being composed of "little boxes" or "cells." The term *cell,* originally applied to dead plant material, is still in use.

In 1675 Marcello Malpighi, an Italian physician, published an account of his microscopic investigations of plants. This account consisted of a detailed description of the internal structure of plants and, despite minor errors in observation, he provided subsequent investigators with many clear and surprisingly accurate illustrations. The English microscopist Nehemiah Grew also made detailed studies of sections through roots and stems, the results of which he published in 1682.

In 1823 Robert Brown, a Scottish botanist, recognized for the first time the presence of a small spherical transparent object within each cell. He called this body the nucleus but did not concern himself with its exact nature or function. He also described the haphazard movement of pollen grains suspended in water. This agitation of the pollen grains, called *Brownian movement,* is caused by the constant motion of water molecules too small to be seen, even by microscopic observation. Such molecular motion, which is common to all matter, is extremely significant in bringing about chemical reactions and in activities such as diffusion.

As the magnifying power of the microscope was increased, investigators penetrated deeper into the structure of cells. A liquid observed by Hooke was subjected to closer scrutiny by Dujardin in 1835, at which time he described it as a jelly-like mass. In 1839 Purkinje called this viscous, jelly-like substance protoplasm, the material now regarded as the physical and chemical basis of all life.

All of these observations were consolidated by the German biologists Matthias Schleiden and Theodor Schwann in the early part of the nineteenth century. In 1838, Schleiden, a botanist, announced that the cell was the basic structural unit of all plants. One year later, Schwann, a zoologist, announced that all plants and animals consist of cells and that each cell consists of a fluid substance (protoplasm) and a solid, spherical inclusion, the nucleus. Just as an atom is the smallest particle of an element capable of retaining its identity, the cell is the basic unit of life. All subsequent investigations into the structure and functions of cells were based upon the cell theory as experimentally established by Schleiden and Schwann.

Until a relatively few years ago, the concept of the cell was limited to what could be seen through a *compound microscope* employing visible light (Figure 5-2A). Any optical instrument is limited in the structural details it can make visible by its *resolution.* The limit of resolution, defined as the smallest distance between two points on an object which can still be seen as two distinct points, is determined, among other factors, by the wavelength of radiation used to illuminate the object. The smaller the wavelength, the smaller the limit of resolution and the more structural detail is visible. With wavelengths in the visible range, the smallest object that can be seen in detail is about 0.3 μ and parts of a cell smaller than this cannot be resolved in the light microscope. A **micron,** symbolized μ, is a unit of measurement equal to 0.001 millimeter or $^1/_{25,000}$ of an inch and is more convenient when discussing dimensions of cells and cellular constituents. Please refer to Table 5-1 for an explanation of commonly used metric units of length and their English equivalents.

Many cell components are much smaller than 0.3 μ, and another instrument was needed before their structures could be seen in any detail. Even with special staining techniques, smaller structures of the cell, such as mitochondria, appear only as dots or rods in the light microscope; many structures remain invisible. It may be noted that resolution, not magnification, is the important criterion in determining whether a structure can be seen in detail in a microscope. Normal maximum magnifications with the light microscope are about 1000–2000 times and the use of greater magnifications would reveal no further structural details, but only a blurrier picture.

TABLE 5-1. Metric Units of Length and Some English Equivalents

Metric Unit (abbreviation, symbol)	Meaning of Prefix	Metric Equivalent	English Equivalent
1 kilometer (km)	kilo = 1000	1000 m	3280.84 ft or 0.62 mi; 1 mi = 1.61 km
1 hectometer (hm)	hecto = 100	100 m	328 ft
1 dekameter (dkm)	deka = 10	10 m	32.8 ft
1 meter (m)	standard unit of length		39.37 in. or 3.28 ft or 1.09 yd
1 decimeter (dm)	deci = 1/10	0.1 m	3.94 in.
1 centimeter (cm)	centi = 1/100	0.01 m	0.394 in; 1 in. = 2.54 cm
1 millimeter (mm)	milli = 1/1,000 = 10^{-3}	0.001 m	0.0394 in.
1 micrometer (μm) or micron (μ)	micro = 1/1,000,000 = 10^{-6}	1×10^{-6} m	3.94×10^{-5} in.
1 nanometer (nm) [formerly millimicron (mμ)]	nano = 10^{-9}	1×10^{-9} m	3.94×10^{-8} in.
1 Ångström (Å)		1×10^{-10} m	3.94×10^{-9} in.
1 picometer (pm)	pico = 10^{-12}	1×10^{-12} m	3.94×10^{-11} in.

An important breakthrough in the study of submicroscopic (below the resolution of the light microscope) structures came in the 1940's with the advent of the first practical *electron microscope* (Figure 5-2B). This instrument employs high electrical voltages to drive a beam of electrons. Such a beam can be focused by magnetic lenses and can produce magnified images of objects through which it passes. The electron microscope increases resolution because the wavelength of electrons is much smaller than that of visible light (about $\frac{1}{100,000}$ of visible light). The present limit of resolution of the electron microscope is about 3 Å. One **Ångström unit,** symbolized Å, is equal to $\frac{1}{10,000}$ of a micron, or $\frac{1}{254,000,000}$ of an inch. Magnifications of 200,000 times (200,000 ×) are therefore possible.

The observer looks directly at the object when using a light microscope. In the electron microscope, however, either the observer views a fluorescent screen which lights up as the beam of electrons hits it, or the image is directed onto a photographic emulsion (a film or a glass plate) which is then developed, enlarged, and examined. The resulting photograph is called an *electron micrograph* (Figure 5-3).

A beam of electrons can only be stabilized and focused in a vacuum where the electrons will not collide with air particles. Therefore, specimens to be examined in the electron microscope are placed in a vacuum. This means that they must be completely dehydrated; otherwise they would boil as the electron beam struck them. The electron microscope has elucidated many of the mysteries associated with the cell. It has revealed cellular structures not even known to exist previously and added a new dimension to the knowledge of the structural organization of the cell. Undoubtedly, the electron microscope, coupled with other tools and techniques of investigation, will reveal smaller and smaller components of cellular organization.

Cellular Structure

Despite the myriad differences that exist among cells, there is an underlying unity. First, with relatively few exceptions, every cell possesses a nucleus embedded in a cytoplasmic matrix. This may be viewed as unity of form. Second, there is a unity of function among cells since the metabolism of all cells is basically similar. Finally, cells exhibit a unity of composition, for the main macromolecules of living things are built up from the same small molecules; in other words, cells are of similar chemical composition.

In a very broad sense, a **cell** may be defined as the structural and functional unit of living organisms. Each is a highly complex unit composed of molecules organized to form a series of component cell structures.

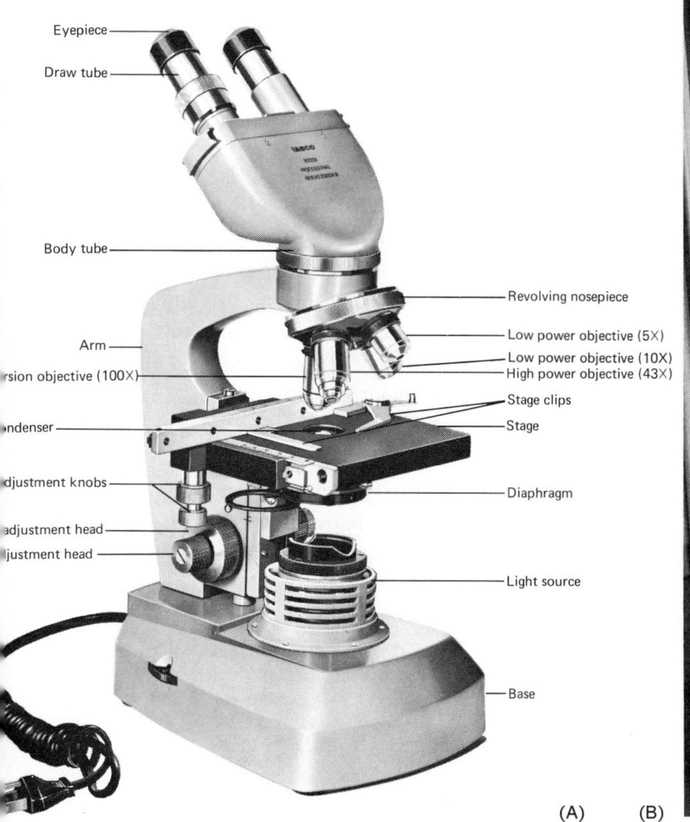

Figure 5-2. (**A**) Compound light microscope. Inasmuch as microscopes of this type use light, they cannot resolve cellular structures smaller than 0.3 μ. This limitation is the result of the relatively long length of visible light waves employed in the instrument. (**B**) Electron microscope. Instead of light, this instrument uses a beam of electrons. Since the wavelengths of electrons are about $\frac{1}{100,000}$ that of the normal wavelength of white light, an electron microscope can magnify objects up to 200,000 times. [**A** *courtesy of TASCO Sales, Inc.* **B** *courtesy of RCA Laboratories, Princeton, N.J.*]

Although cellular components constantly change shape, sometimes assume different functions, and may even change in chemical composition, for purposes of our discussion they will be treated as they appear in electron micrographs. That is, they will be viewed as having a given structure, chemical composition, and function.

In the discussion of most cell parts that follows, a series of illustrations has been designed to show (1) the location of the specific part under consideration and its anatomic relationship to other structures in a composite cell, (2) detailed ultrastructure of the part by means of an accompanying electron micrograph, (3) an interpretation of the electron micrograph by means of an adjacent labeled diagram, and (4) the presumed chemical composition and arrangement of molecules for the structure based upon recent research. In this way, cell components are seen as highly organized molecular units at various levels of structural organization. The functions of many of these structures are discussed only briefly since there will be more detailed coverage in other chapters.

For convenience, a generalized cell (composite of

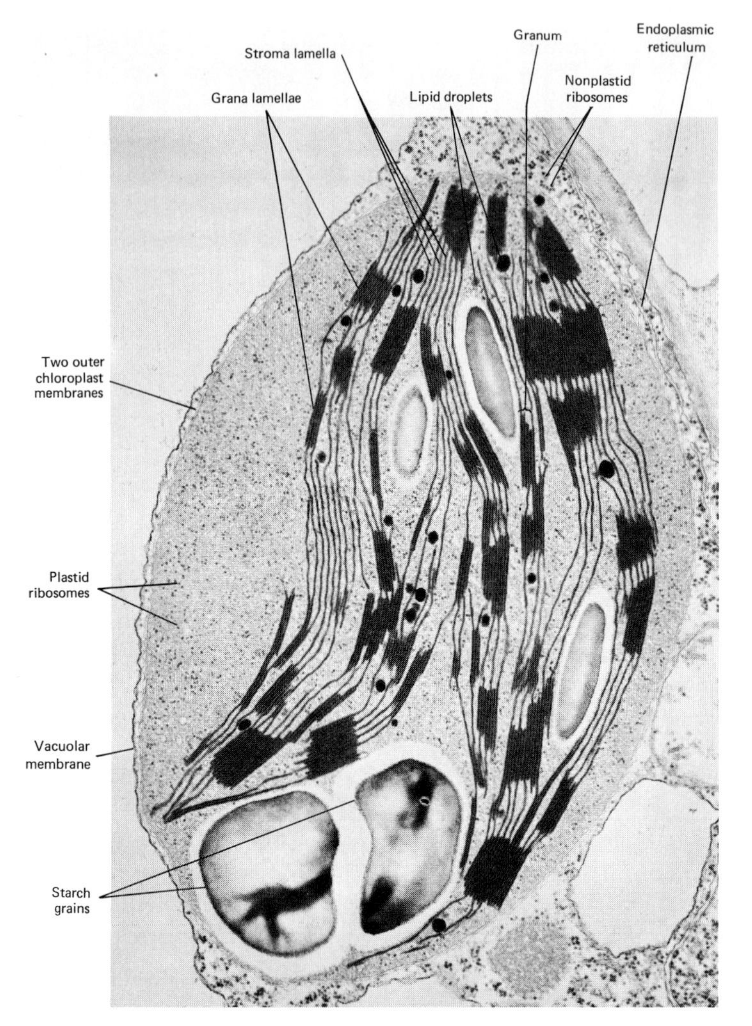

Figure 5-3. Electron micrograph of a single chloroplast from the leaf of *Phleum pratense* (timothy grass). 20,600×. In micrographs such as this, the resolution of the electron microscope becomes readily apparent. Note the numerous detailed structural features. [*Courtesy of Dr. Myron C. Ledbetter, Brookhaven National Laboratory.*]

many cells—Figure 5-4) may be considered to consist of five basic portions.

1. The *cell surface,* an outer limiting membrane that separates one cell from another and from the environment.

2. *Extracellular materials,* substances external to the membrane.

3. *Cytoplasm,* the ground substance of the cell.

4. *Organelles,* highly organized and specialized portions of the cytoplasm that assume various roles in cellular maintenance, growth, repair, and continuity.

5. *Cell inclusions,* the secretory products and storage areas of cells.

The Cell Surface

Structure. Prior to the advance of electron microscopes, biologists assumed that the protoplasm of cells was bounded by a **plasma (cell) membrane.** Recent electron microscopic studies have confirmed the universal existence of this membrane and have shown that it is invisible in the light microscope since it ranges from 65 to 100 Å in thickness, a dimension below the limits of resolution of the light microscope. Prior to this discovery with the electron microscope, the general chemical nature of the membrane was determined primarily on the basis of permeability studies. First, the low permeability of the cell to ions and high permeability to lipids and many substances soluble in lipids suggested that the membrane must contain a lipid layer. Further measurements led to the conclusion that the lipid must be in the form of a bimolecular (double) layer. Second, since it was observed that a number of water-soluble substances also move freely through the membrane, it was theorized that the membrane contained pores (openings). From later studies of the membrane it was noted that ions moved across the membrane at varying rates. This observation led to the conclusion that the membrane is ionic and is capable of attracting some ions while repelling others. Finally, based upon investigations of the surface tension and elasticity of the membrane, it was concluded that a layer of protein was attached to the outer surfaces of the bimolecular lipid layer. These various conclusions were finally synthesized by H. Davson and J. F. Danielli in 1940. According to their hypothetical model, the cell membrane consists of a double layer of lipid molecules covered by two protein layers (Figure 5-5C).

J. D. Robertson at Harvard, among others, was able to demonstrate the structure of the membrane in the electron microscope and to assign it a thickness of about 75 Å. He called it the **unit membrane** to emphasize its cellular universality and gave support to the earlier model proposed by Davson and Danielli. In electron micrographs (Figure 5-5A), the membrane appears

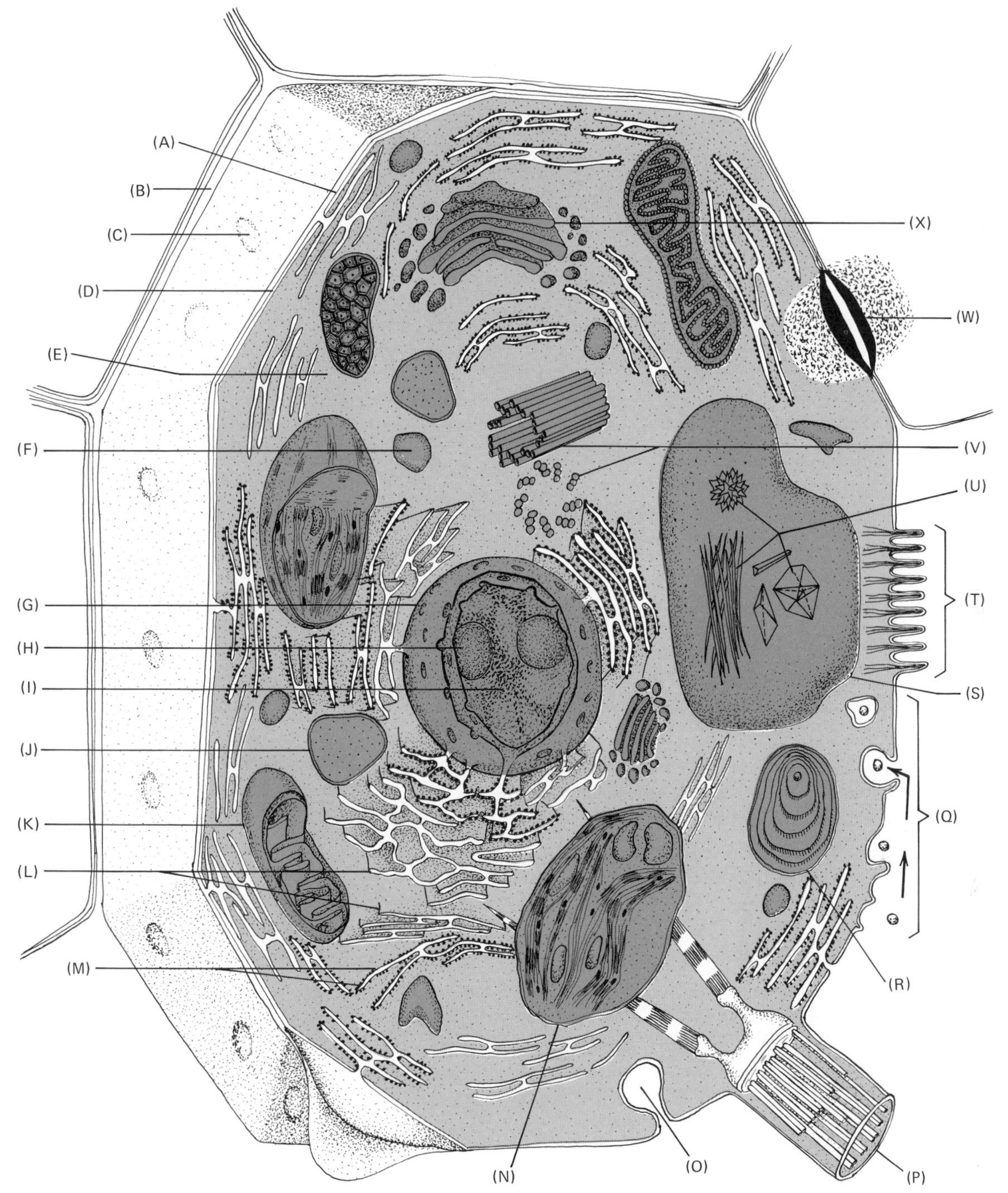

Figure 5-4. Highly diagrammatic representation of a generalized cell based upon electron microscope studies. This composite cell, prepared to illustrate the principal features of cells, combines components of a wide variety of cells. It should not be interpreted as a "typical" cell of any particular organism or group of organisms. For purposes of viewing structural relationships, a reduced copy of this generalized cell is included in the figure that accompanies discussion of each cell structure later in this chapter.

A Primary cell wall
B Middle lamella
C Primary pit-field
D Trilamellar plasma membrane
E Cytoplasm
F Chromoplast
G Nucleus
H Nucleolus
I Chromatin
J Lysosome
K Mitochondrion
L Endoplasmic reticulum
M Ribosomes
N Chloroplast
O Pinocytotic vesicle
P Cilium or flagellum
Q Pinocytosis
R Leucoplast (starch grain)
S Vacuole
T Microvilli
U Crystals
V Centrioles
W Desmosome
X Golgi apparatus

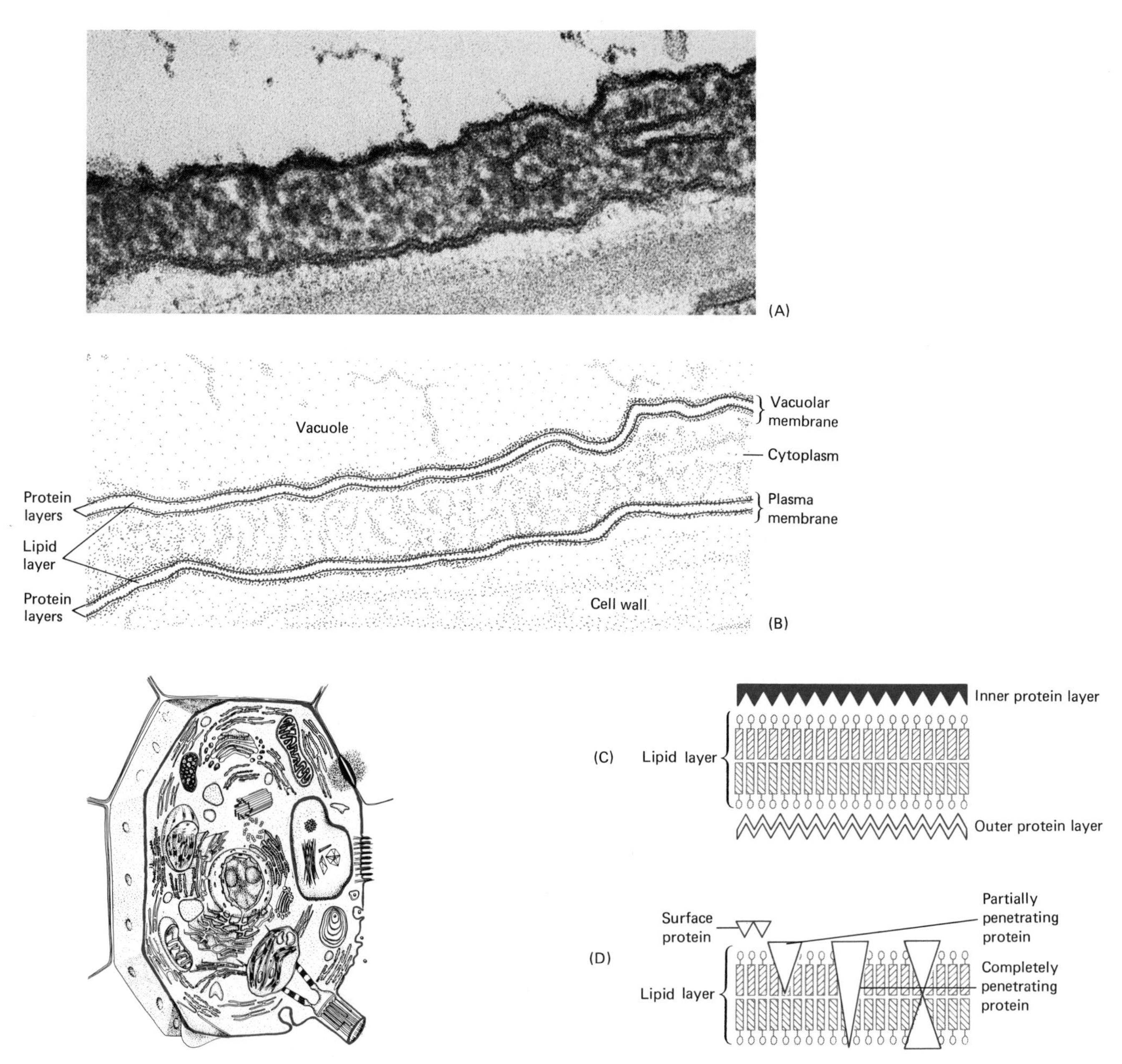

FIGURE 5-5. The plasma membrane. (**A**) Electron micrograph of the plasma membrane and vaciolar membrane of *Statice sinuata* (statice) showing trilamellar structure. 105,800×. (**B**) Diagrammatic interpretation of the electron micrograph. (**C**) A proposed model of the orientation of molecules in the plasma membrane showing the unit membrane as a bimolecular lipid layer covered on each side by a protein layer. The outer and inner protein layers are probably not identical. (**D**) A newer proposed model of the structure of the plasma membrane. [**A** *courtesy of Dr. Myron C. Ledbetter, Brookhaven National Laboratory.* **C** *modified from J. D. Robertson,* Biochemical Society Symposia, *no. 16, 1959, pp. 3–43. Copyright 1959 by The American Chemical Society.*]

as a thin linear profile, 80–100 Å thick, consisting of two dark lines each approximately 30 Å thick and separated by an intermediate light line of about the same thickness. These three layers have usually been interpreted as corresponding in a general way to the one lipid and two protein layers postulated by Davson and Danielli (Figure 5-5B). It should be noted that it is the universality of a triple-layered structure of membranes seen in the electron microscope that is the unit membrane concept. The unit membrane hypothesis specifically states that the plasma (or other) membrane is a triple-layered structure as seen in the electron microscope under special staining conditions.

It should be recognized that alternative interpretations of the organization of cell membranes have also been proposed. Moreover, even though the widespread occurrence of the trilaminar appearance in membranes is an indication that all membranes have the same basic pattern of molecular organization, it would be fallacious to imply that all cell membranes have a similar origin and similar physiological characteristics. E. A. Korn points out that the unit membrane concept provides reasonable but not conclusive support for extending the concept to biological membranes in general. Furthermore, he states that membranes differ widely in chemical composition, metabolism, function, and enzymatic composition and suggests that until more is known about the chemistry of electron microscopy, evidence obtained from electron micrographs alone cannot be interpreted with confidence.

Other studies of membrane structure seem to indicate that the lipids, which account for about one-half of the mass of the membrane, are organized into a thin bilayer. Moreover, it seems that the proteins of the membrane are of two kinds. Some lie at or near the surface of the membrane. Others penetrate the membrane part way, completely, singly, or in pairs (Figure 5-5D). Recent studies suggest that membranes are not static. In other words, both the lipids and proteins exhibit a considerable degree of movement. It is quite possible that many key functions of membranes may be explained on the basis of such lipid and protein movements.

Functions of the Membrane. The cell membrane serves as a boundary between the external and the internal environment of the cell. As such, it maintains the integrity of the cell in relation to the surrounding environment. All substances entering or leaving the cell must pass through this barrier. Cell membranes are described as **differentially permeable;** they permit the passage of certain ions and molecules while excluding others, and they allow substances to pass through at different rates. Before analyzing the role of the plasma membrane in more detail, we must first examine some of the factors involved in the movement of materials from one area to another.

The movement of many dissolved inorganic and organic materials into and out of cells (or through a solution) occurs by a process called diffusion. **Diffusion** may be defined as the net movement of molecules, or ions, from a region of higher concentration to a region of lower concentration. Consider the following example. If a crystal of copper sulfate (blue) is placed in the bottom of a beaker filled with water, it will be noted that a dark blue color quickly appears around the crys-

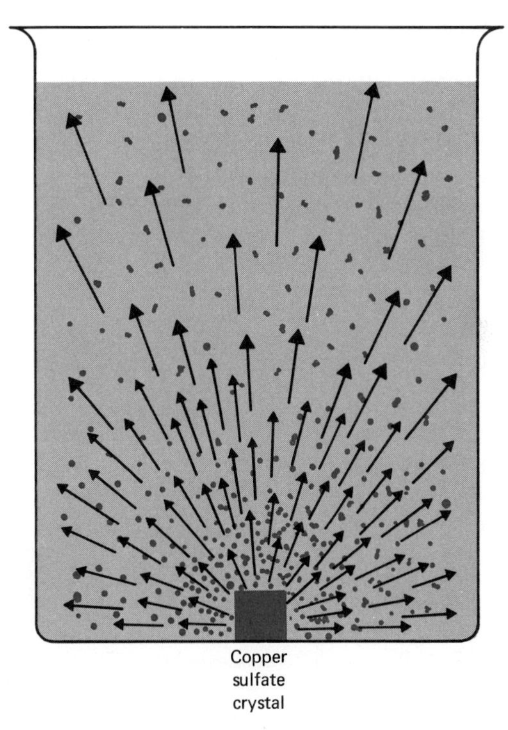

Figure 5-6. Diffusion. Copper sulfate ions move from a more concentrated area to a less concentrated region. Similarly, water molecules move from more concentrated to less concentrated regions. Diffusion ceases at equilibrium when the crystal and water particles become evenly dispersed.

tal. As the distance from the crystal increases, the color becomes less and less intense (Figure 5-6). However, as diffusion of molecules and ions of the crystal progresses, it will be seen, in a few days, that the entire water solution becomes uniform in color. This is because the dissolved crystal particles have been diffused by their kinetic energy and have become evenly distributed among the water molecules. Similarly, water molecules, because of their kinetic energy, have migrated from more concentrated to less concentrated regions. When water and crystal particles have become evenly distributed, an *equilibrium* is attained and net diffusion ceases, although random molecular movements still continue. In this regard, diffusion may be viewed as a tendency toward reaching an equilibrium, because ultimately it results in the equal distribution of molecules or ions within a given system.

As particles of a substance diffuse, they exert a pressure called **diffusion pressure.** This pressure is caused by the high molecular activity of the diffusing molecules or ions and is proportional to the concentration, or number of diffusing particles. In the copper sulfate system, the diffusion pressure is greater near the dissolving crystal since the concentration of copper sulfate particles is greater at that point. Similarly, the diffusion pressure is least at a point furthest from the dissolving crystal since there the concentration of particles is lowest. The behavior of diffusing molecules may be described in terms of diffusion pressures by stating that they diffuse from areas of higher diffusion pressure to areas of lower diffusion pressure.

A process of special importance to living cells is the migration of water and dissolved substances into and between cells. The term **osmosis** refers to the migration of water through a differentially permeable membrane from a region containing a higher concentration of water to one containing a lower concentration of water.

Osmosis can be demonstrated quite readily by reference to an apparatus called an *osmometer* (Figure 5-7). The apparatus may consist of a sac made of parchment, cellophane, dialysis tubing, or other differentially permeable membrane that is filled with a concentrated sugar solution. For ease of observation, the sugar solution may be colored with a dye such as Congo red. The top portion of the sac is stoppered with a rubber stopper through which a glass tube is fitted and the entire apparatus is placed in a distilled water solution.

The relative concentrations of water on the two sides of the membrane are different. There is a lower concentration of water inside the sac than in the beaker. As a result of these unequal concentrations, there is a tendency toward equalization of the concentrations of water on both sides of the membrane. Since the sugar molecules cannot diffuse to reduce the concentration differ-

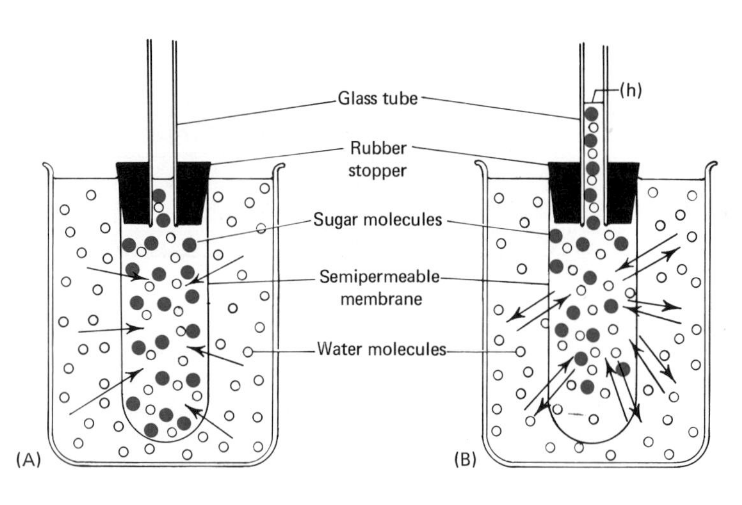

FIGURE 5-7. Demonstration of osmosis in an osmometer. (**A**) Osmometer at start of experiment. (**B**) Osmometer at equilibrium. The setup consists of a sugar solution enclosed within a semipermeable membrane and immersed in a water solution. In **A**, arrows indicate the direction of movement of water molecules, which pass freely across the membrane, whereas the sugar molecules cannot. As more water molecules move into the sac than leave it, the sugar solution is diluted and the volume of liquid in the sac increases, as illustrated in **B**. The increasing volume forces the sugar solution into the glass tube. The final height (h) attained by the column of solution is directly related to the concentration of the sugar solution at the start of the experiment. When the final height is reached, an osmotic equilibrium is established and the same number of water molecules pass across the membrane in both directions. The head of hydrostatic pressure, h, exerts a force that at equilibrium is exactly equal but opposite to the osmotic pressure.

ence owing to the presence of a differentially permeable membrane, water must move from the beaker into the sac. Actually water moves into and out of the sac, but the net movement inward is much greater. The greatest difference in water concentration exists at the moment the sac is submerged in the beaker containing pure water. As a result, it is at this time that the rate of water movement into the sac is highest and the rate of water movement out of the sac is lowest. As the volume of water in the sac increases, the sugar solution becomes more and more diluted and moves up the glass tube. After a period of time the rate of water moving into the sac slows down. As the water builds up in the sac and the tube, the weight of the column of solution begins to exert a pressure. This pressure continues to increase until an equilibrium is reached in which the pressure exerted by this water forces water out of the sac at the same rate as water molecules enter from the beaker. The maximum amount of pressure which can be developed in a solution separated from pure water by a differentially permeable membrane is called the *osmotic pressure* of the solution. Inasmuch as living cells in their environment are not immersed in pure water, the theoretical maximum pressure never develops. In this regard, the term osmotic pressure simply refers to a potential pressure, if a cell were placed under prescribed conditions.

In order to demonstrate that living cells are subject to pressure changes associated with osmosis, let us consider two examples. The first of these might be any of a number of filamentous green algal forms that inhabit freshwater ponds. The surrounding medium in which these organisms live is said to be a **hypotonic solution,** that is, a solution in which there is a lesser concentration of solutes and therefore a higher concentration of water. According to the principles of osmosis previously discussed, water should enter the cells, cause them to swell, and if the medium concentration is high enough, the cells should burst. The cells, however, have rigid walls of cellulose that prevent drastic swelling. In hypotonic solutions, water enters the cells and builds up an internal pressure known as **turgor pressure.** The cellulose wall, being rigid, exerts an equal and opposite pressure called *wall pressure.* As a result of the equal and opposite forces of turgor pressure and wall pressure, the cell is said to be *turgid* (firm or stiff). These relationships are shown in Figure 5-8A. Organs of higher plants whose cells are void of turgor undergo a partial collapse, as externally exhibited by wilting.

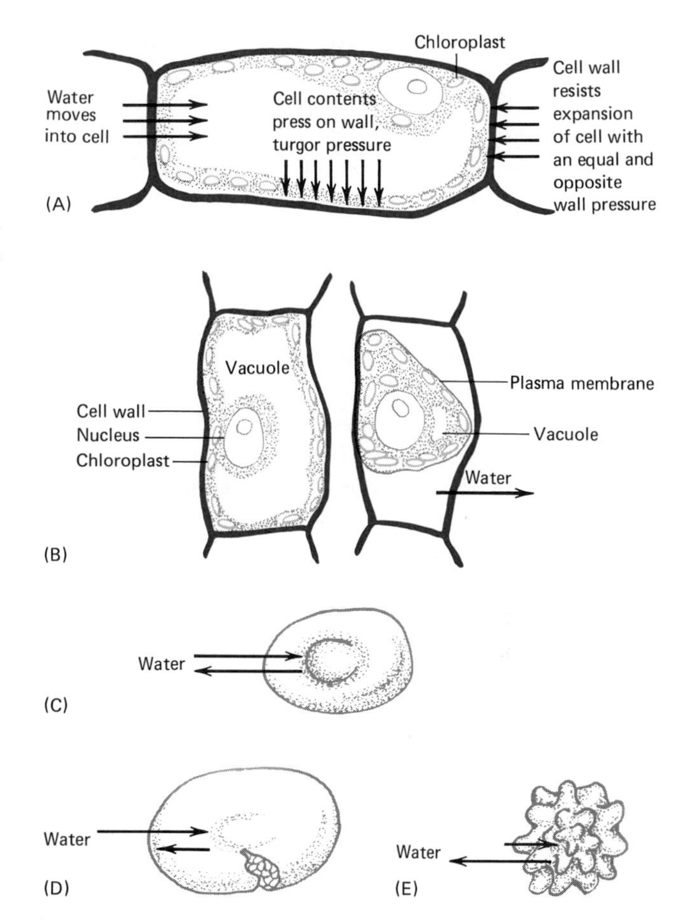

Figure 5-8. Osmotic relationships of cells. (**A**) As water moves into a plant cell, the cell contents are pushed against the wall by pressure called turgor pressure. The wall resists this pressure with an equal and opposite force called wall pressure. As long as these forces are equal and opposite, the turgidity (firmness) of the cell is maintained. (**B**) If the cell is placed in a hypertonic solution, water will diffuse out of the cell and the cytoplasm shrinks away from the cell wall. *Left:* Before plasmolysis. *Right:* After plasmolysis. (**C–E**) Effects of pressure changes associated with osmosis in red blood cells: (**C**) erythrocyte in an isotonic solution (normal), (**D**) erythrocyte in a hypotonic solution (cell swells and may burst), (**E**) erythrocyte in a hypertonic solution (cell shrinks).

When the cells are placed in **hypertonic solutions,** the situation is quite the reverse. In these solutions, there is a higher concentration of solutes in the medium and a lesser concentration of water. The effect of these conditions is that water will move out of the cells, and as such movement proceeds, the cytoplasm contracts and shrinks away from the cell wall. Such shrinkage of plant cells through loss of water is termed **plasmolysis** (Figure 5-8B). Continued plasmolysis results in the death of a cell; if however, plasmolysis has been only slight, the cell will normally regain turgidity when placed in a hypotonic solution.

A second example of the effects of pressure changes associated with osmosis may be considered by noting the effects of these changes on red blood cells. In order to maintain the shape of human or other mammalian erythrocytes, they must be placed in an **isotonic solution.** This is a solution in which the total water and solute concentration is the same as in the red blood cells and which does not cause changes in cell volume. For ordinary purposes, a 0.85% NaCl solution is isotonic for these cells (Figure 5-8C). If red blood cells are placed in distilled water, a hypotonic solution, water enters the cells, causing the membrane to swell and eventually burst (Figure 5-8D). Conversely, if the erythrocytes are placed in a 10% NaCl solution, a hypertonic solution, water leaves the cells and they shrink (Figure 5-8E). It should be apparent from the above statements that cells, such as red blood cells, may be destroyed or killed by solutions that deviate significantly from the isotonic state.

The osmotic relationship between a cell and its surrounding medium assumes critical importance in the life of a biological system. In higher animals, cells are surrounded by fluids that are isotonic to the cytoplasm. Among freshwater plants, the presence of rigid cell walls enables these organisms to withstand osmotic pressure by developing wall pressure. Freshwater animals, such as certain protozoans (*Amoeba, Paramecium*), which do not have cell walls, are protected against excessive pressures by employing energy and some contractile structure to discharge the excess water back into the environment. Marine organisms are also adapted to their environment in that the cytoplasm of their cells is isotonic to the sea water.

The processes of diffusion and osmosis and the semipermeable nature of membranes are critical components in the life of a cell. Although membrane permeabilities vary among different types of cells, a number of broad generalizations can be made. Membranes are relatively permeable to water, some simple sugars, amino acids, and lipid-soluble materials, but are relatively impermeable to very large molecules such as proteins and polysaccharides. The lipid permeability of the membrane is probably due to the presence of lipid layers in the membrane itself; lipid-soluble substances probably move across the membrane by first dissolving in it. With regard to the passage of smaller molecules and the exclusion of larger ones, it is postulated that there are small pores in the membrane through which smaller molecules can move. Another generalization that can be made is that negatively charged ions traverse membranes more readily than positively charged ones, though neither passes as readily as nonelectrolytes.

The movement of many substances into and out of cells in most cases is due simply to differences in diffusion pressure. Such movement of materials into or out of a cell as a result of diffusion only is called **passive transport.** In passive transport, the molecules or ions of a dissolved substance diffuse from a region of greater concentration to a region of lesser concentration. The continuation of this type of movement depends upon the maintenance of a concentration difference (*gradient*) of the particles of the diffusing substance. In other words, passive transport is operational only when the diffusing material is present in a greater concentration on one side of the membrane than on the other. As substances enter cells, they may become incorporated into various chemical compounds, transformed into other molecules, or they may be passed on to other cells. As a cell uses a certain substance, the cellular concentration of that material decreases and more of it diffuses inward as long as the external concentration is greater than the internal concentration.

Passive transport, a result of simple diffusion, is not the only method by which materials enter or leave cells. There is evidence that dissolved substances, especially mineral ions, continue to move into cells even though there is a greater concentration of them within the cells than outside. Such a movement of materials is directly opposite to what would occur by diffusion alone since the movement is from a region of lesser concentration to a region of greater concentration. The ability of cells to accumulate ions in concentrations higher than those of the surrounding medium was first recognized in certain types of aquatic unicells. In the freshwater protist *Nitella,* potassium ions were found to be 1000 times more concentrated inside the cells than in the surrounding medium. Similarly, iodine may accumulate in kelp, at least 100 times more concentrated than in the sea. In many cell types, the sodium ion concentration is far less than in the surrounding medium. Under these conditions sodium should move from the medium into the cell until an equilibrium is reached. This does not occur; rather, as quickly as sodium diffuses into the cell, it is pumped out against the concentration gradient. In a similar, though reverse situation, many of these same cells maintain a high potassium ion concentration. In other words, potassium ions, even though more concentrated inside the cell than out, continue to move into the cell against a concentration gradient.

The movement of materials against a concentration gradient is called **active transport.** Such a movement of materials requires an expenditure of energy by the cells. Although the exact mechanism of active transport is not known, many data suggest that this energy comes from respiration, specifically energy supplied by ATP. Studies of ion intake and accumulation have been carried out with erythrocytes, liver mitochondria, squid axons and other nerve cells, intestinal cells, and gall bladder cells. In these studies, it has been demonstrated that an increase in the rate of respiration is accompanied by an increase in ion accumulation, and that if certain chemicals are used to block respiration, the ion intake by active transport is stopped. The way in which energy from ATP is utilized is not completely understood, but various cellular membranes and enzymes probably assume an active role. It is believed that carriers, which are enzyme-like molecules located in or on cell membranes, combine with an ion and form a compound that is soluble in the lipid portion of the membrane. This complex then moves across the membrane where it is split by enzymes. Next, the transported ion is released and the carrier diffuses back through the membrane to pick up more of the transported substance (Figure 5-9).

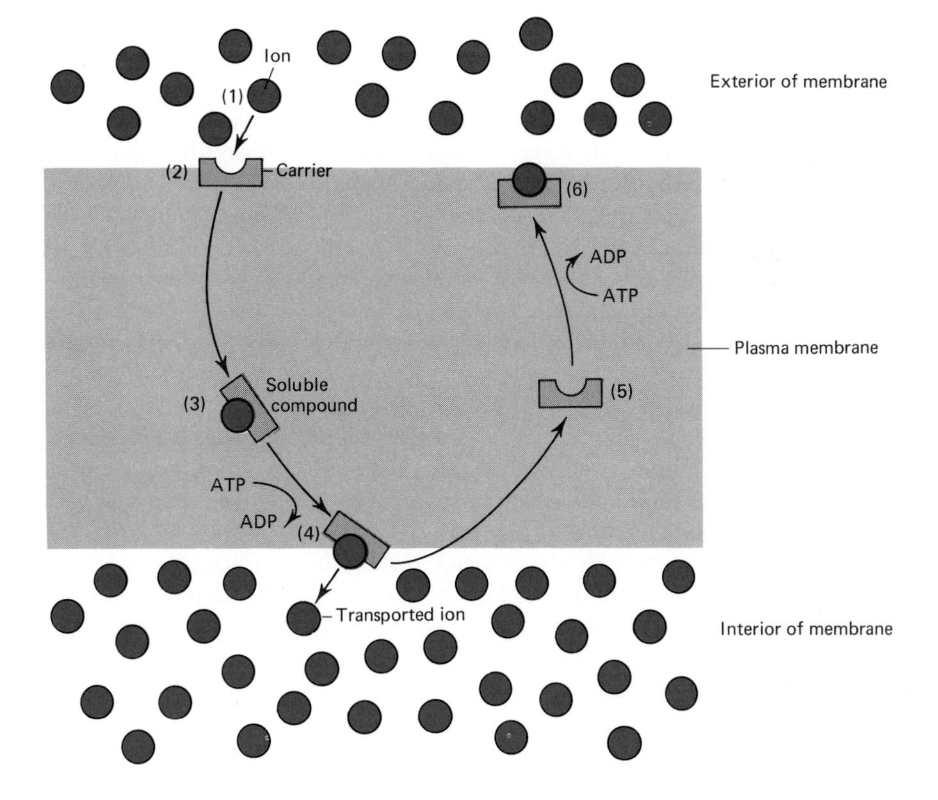

FIGURE 5-9. Proposed model of the mechanism of active transport. The suggested sequence of events is as follows. An ion (1) external to the membrane is picked up by a carrier molecule (2). A compound soluble in the lipid portion of the membrane is formed (3). The compound migrates toward the interior portion of the membrane, moves across the membrane where it is split by enzymes, and the transported molecule is released (4). The carrier diffuses back through the membrane (5) and picks up another ion (6). Note that the energy for the process is supplied by ATP.

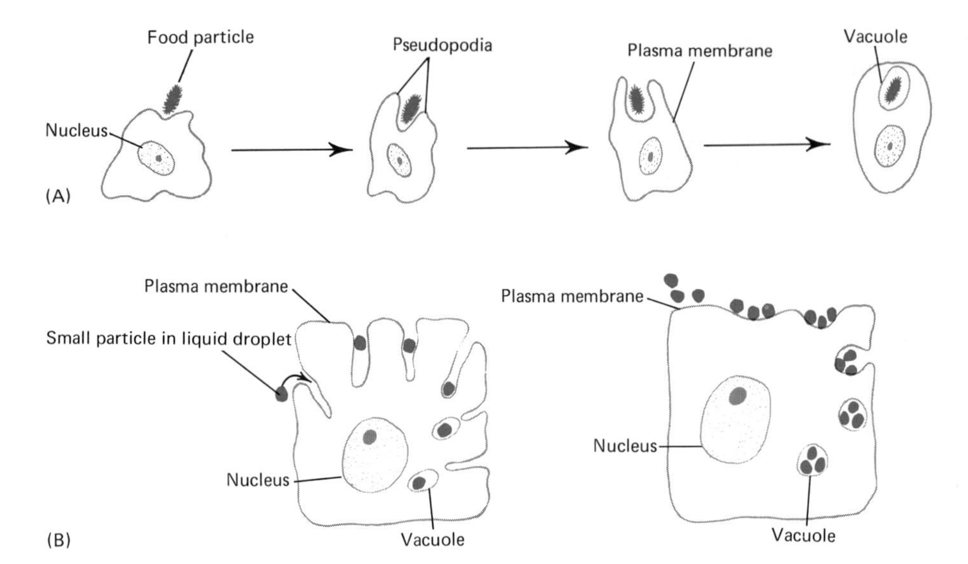

Figure 5-10. Phagocytosis and pinocytosis. (**A**) In phagocytosis, cytoplasmic projections (pseudopodia) flow around a food particle, forming a vacuole that becomes incorporated into the cytoplasm. (**B**) In pinocytosis, a liquid droplet containing one or more very small particles is adsorbed on the cell surface; no pseudopodia are formed. Two alternative forms of pinocytosis are shown. *Left:* The particle moves into a deep channel at the end of which it becomes enclosed in a vacuole. *Right:* After adsorption, particles may be enclosed in vacuoles that form and detach directly from the cell surface.

Some cells utilize still other active methods of absorbing materials by an inward folding of a portion of the plasma membrane. In one of these processes, called **phagocytosis** ("cell eating"), projections of cytoplasm engulf solid particles and invaginations of the membrane pinch off and form a vacuole that floats freely in the cytoplasm (Figure 5-10A). Once inside the vacuole, the solid material is digested. Ingestion of food in the *Amoeba* is facilitated through phagocytosis. Phagocytic cells (white blood cells) in the blood of many animals represent an important defensive mechanism in protecting these organisms against invasion by bacteria and other foreign substances. Phagocytosis also permits molecules to which the membrane is normally impermeable to enter cells along with ingested particles.

The ability to engulf solid materials has been noted in only a few kinds of cells. Many cells, however, are capable of a related process referred to as **pinocytosis** ("cell drinking"). In pinocytosis, the engulfed material consists of a liquid containing small particles. It also differs from phagocytosis in that the material is not surrounded by projections of the cell but instead is adsorbed on the cell surface (Figure 5-10B). Moreover, once adsorption occurs, either the membrane flows inward and vacuoles are formed at the end, or the vacuoles are detached directly from the membrane surface.

Modifications of Plasma Membrane. Electron microscopy has made possible a quite detailed analysis of the fine structure of the plasma membrane and has revealed a number of structural modifications in different cell types. These studies indicate that the membrane is not simply an envelope enclosing the cell contents. For example, the cells lining the intestines (epithelium) of higher animals contain membrane specializations at their surfaces. These modifications consist of minute, cylindrical, surface processes called **microvilli** (Figure 5-11A). Microvilli enormously increase the absorbing area of the cell surface. A single cell may have as many as 3000 microvilli, and in a square millimeter of intestine, there may be as many as 200 million.

Cells in contact with each other also give rise to several different types of specialization. One such specialization is represented by local thickenings of the opposing surfaces from which fine filaments radiate into the cytoplasm. Such structures, found also in epithelial tissue, are referred to as **desmosomes** (Figure 5-11B.)

Another modification of the membrane may be seen in rods and cones, cells in the eye that serve as photoreceptors. Figure 5-12 shows an electron micrograph and a diagrammatic representation of rod cells. The upper portion of each rod cell consists of disc-shaped, double-layered membranes (sacs), which contain the pigments involved in vision.

A final type of specialized membrane is illustrated by a nerve fiber (axon) and its surrounding cell, the Schwann cell. As the Schwann cell grows, its cytoplasm

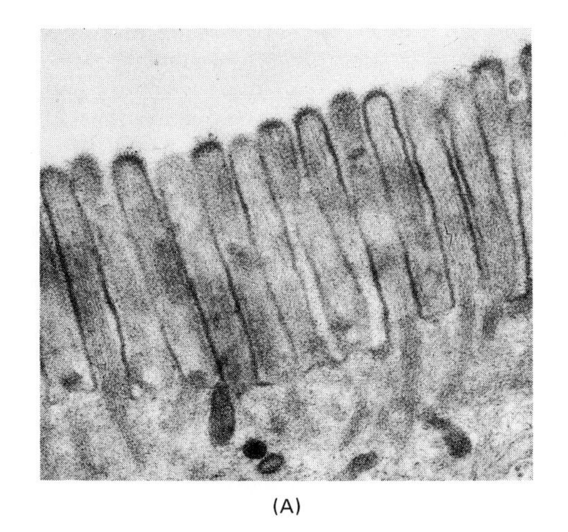

(A)

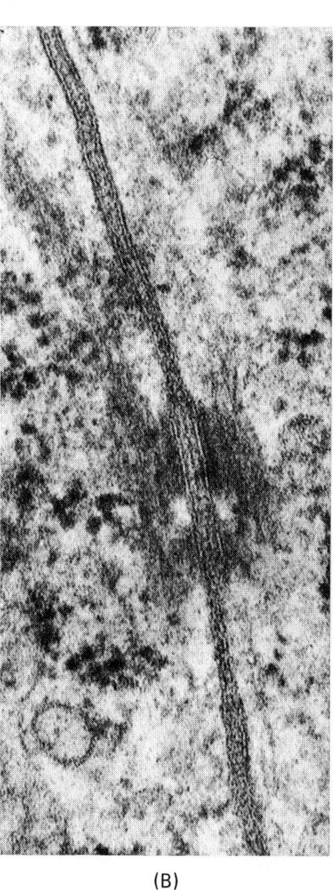

(B)

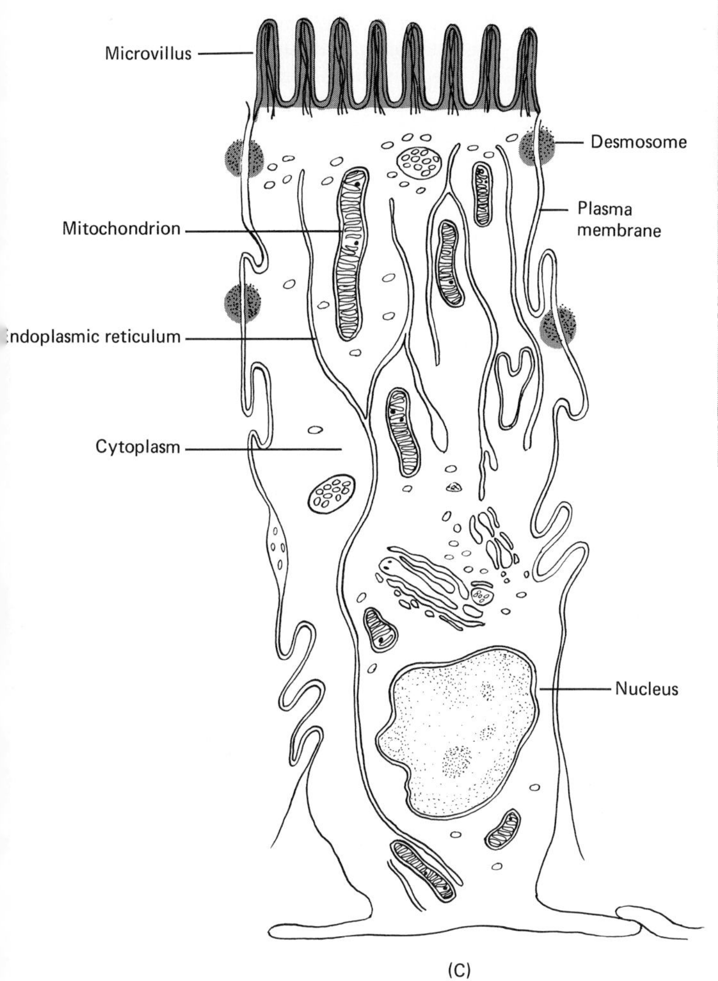

(C)

Figure 5-11. Structural modifications of the plasma membrane: microvilli and desmosomes. (**A**) Electron micrograph of microvilli from the duodenum. 26,700×. (**B**) Electron micrograph of a desmosome from the duodenum. 22,500×. (**C**) Diagrammatic representation of an epithelial cell from the small intestine showing microvilli and desmosomes. [***A*** *and* ***B*** *courtesy of E. B. Sandborn, M.D., Université de Montréal.* **C** *after Zetterqvist.*]

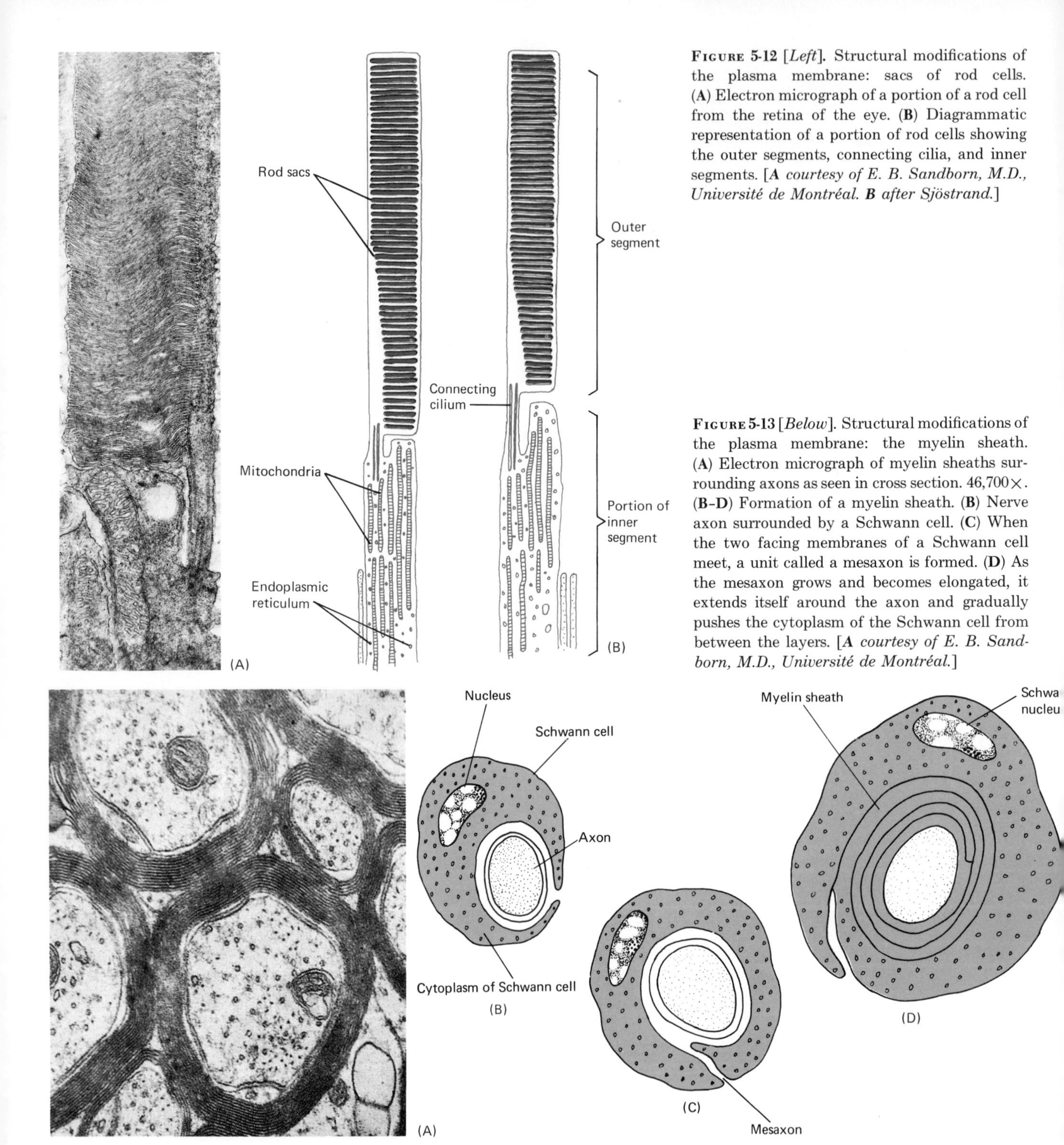

FIGURE 5-12 [*Left*]. Structural modifications of the plasma membrane: sacs of rod cells. (**A**) Electron micrograph of a portion of a rod cell from the retina of the eye. (**B**) Diagrammatic representation of a portion of rod cells showing the outer segments, connecting cilia, and inner segments. [**A** *courtesy of E. B. Sandborn, M.D., Université de Montréal.* **B** *after Sjöstrand.*]

FIGURE 5-13 [*Below*]. Structural modifications of the plasma membrane: the myelin sheath. (**A**) Electron micrograph of myelin sheaths surrounding axons as seen in cross section. 46,700×. (**B–D**) Formation of a myelin sheath. (**B**) Nerve axon surrounded by a Schwann cell. (**C**) When the two facing membranes of a Schwann cell meet, a unit called a mesaxon is formed. (**D**) As the mesaxon grows and becomes elongated, it extends itself around the axon and gradually pushes the cytoplasm of the Schwann cell from between the layers. [**A** *courtesy of E. B. Sandborn, M.D., Université de Montréal.*]

is squeezed to the outside and moves around the axon, depositing a membrane with each revolution (Figure 5-13C). Collectively, these layers are referred to as a **myelin sheath.** An electron micrograph of myelin sheaths is also shown in Figure 5-13A. The multilayered sheath probably protects the nerve, assists in the transmission of impulses, and assumes a role in nutrition.

Extracellular Materials

In general, the plasma membrane is regarded as the outer boundary of the cell. It is not, however, the outermost structure of all cells. In plant cells, for example, the outer boundary is typically the cell wall and among animal cells and unicellular organisms there is a variety of comparable external substances. These include the chitinous covering of insects for protection, collagen which provides support in bone and cartilage, gelatinous layers which prevent desiccation in many algae, siliceous shells of diatoms, and elastic fibers of skin or arterial walls. The discussion of extracellular materials that follows will be limited to two principal categories: (1) plant cell walls and (2) the intercellular substances of cells.

THE CELL WALL. The outer boundary of the plant cell, produced by the living protoplast, consists of a nonliving structure called the **cell wall.** With relatively few exceptions, all plant cells possess walls. These walls have protective and supportive functions, and they also determine the shape and the texture of the cells. On the basis of its development and structure, the cell wall consists of three distinguishable parts: an intercellular substance called the middle lamella, a primary cell wall, and a secondary cell wall.

The **middle lamella** is a relatively thin layer of intercellular material formed between adjacent plant cells during division and which persists after division (Figure 5-14C).The lamella is composed of viscous pectin, calcium, cellulose, and other polymers. In terms of consistency, the middle lamella, owing to the presence of pectin, is a viscous, jelly-like substance. Commercial preparations of pectin are obtained from the cell walls of various fruits and are added to jellies to ensure the "jelling" of these materials. In plant cells, the middle lamella serves as an intercellular cementing material.

The **primary cell wall,** found in all plant cells, is formed during the early stages of growth and development (Figure 5-14C). It is relatively thin, being about 1–3 μm thick. It is also quite plastic and capable of considerable extension as the cell increases in size. The cells of many fruits, roots, fleshy stems, and leaves possess only a primary cell wall and middle lamella.

The **secondary cell wall,** as the name implies, follows the primary wall in order of appearance and is found only in certain mature cells (Figure 5-14B, C). It forms after the primary cell wall ceases to grow and, in many cell types after secondary wall formation is completed, the living components of the cell die and disappear. The secondary wall may be about 5–10 μm thick, and lends considerable strength and mechanical support to the cell. Plant products or derivatives containing these thickened cell walls, such as lumber and textiles, are much stronger and more durable than those containing only primary walls. An electron micrograph showing the primary and secondary walls from the cells of the black locust is shown in Figure 5-14A.

For the most part, the cell wall is a continuous covering. In some plants, however, especially multicellular ones, secondary walls are characterized by the presence of cavities or depressions called *pits* (Figure 5-15C). Cells with only primary walls contain somewhat similar depressions called *primary pit fields* (Figure 5-15B, D). Pits and primary pit fields originate when the wall is formed unevenly, leaving depressions or extremely thin portions of wall material. These pitted areas greatly facilitate the movement of materials from one cell to another.

Pits are usually found in nonliving cells that are concerned with conduction and support such as fibers and tracheids (Chapter 9). In some cases, thickened overhangs form around the pits. These pits are referred to as *bordered pits* (Figure 5-15E). In other cases, the overhanging borders are absent and the pits are known as *simple pits* (Figure 5-15F). With special staining techniques and high magnification many primary pit fields reveal exceedingly small openings (pores) in the membrane of living cells through which fine, delicate strands of cytoplasm connect adjacent cells. These cytoplasmic projections are termed **plasmodesmata** (Figure 5-15B, D). Figure 5-15A shows an electron micrograph of plasmodesmata formed between two adjacent cells of a corn plant.

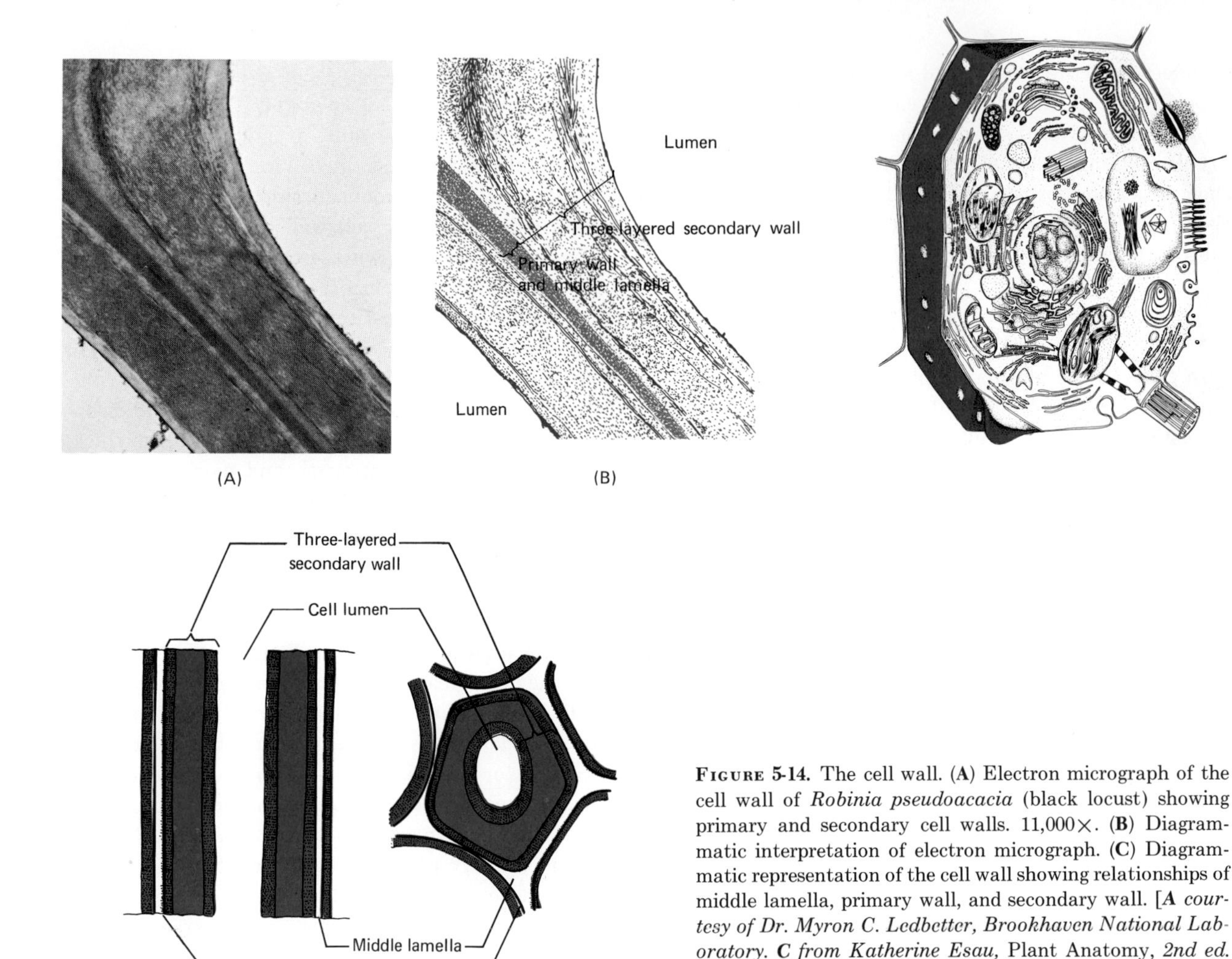

Figure 5-14. The cell wall. (**A**) Electron micrograph of the cell wall of *Robinia pseudoacacia* (black locust) showing primary and secondary cell walls. 11,000×. (**B**) Diagrammatic interpretation of electron micrograph. (**C**) Diagrammatic representation of the cell wall showing relationships of middle lamella, primary wall, and secondary wall. [*A courtesy of Dr. Myron C. Ledbetter, Brookhaven National Laboratory. C from Katherine Esau,* Plant Anatomy, *2nd ed. New York: John Wiley & Sons, Inc. Copyright © 1965.*]

The most abundant organic compound in most plant cell walls is the polysaccharide *cellulose.* Other compounds include *hemicelluloses* and *pectic compounds,* substances structurally designated as polysaccharides also. At maturity, if a secondary wall develops, another wall component called lignin appears. *Lignin,* a constituent of the cell walls of materials such as wood, affords a high degree of rigidity and strength. In addition, lipids are found as structural components of cell walls. Representative lipids include suberin, cutin, and waxes.

Figure 5-15 [opposite]. Pits, primary pit-fields, and plasmodesmata. (**A**) Electron micrograph of primary pit-fields and plasmodesmata of *Zea mays* (corn). 12,100×. (**B**) Diagrammatic interpretation of electron micrograph. (**C**) Pits in the secondary wall of *Pyrus* (apple) wood. (**D**) Primary pit-fields in a parenchyma cell from the stem of *Nicotiana* (tobacco). (**E**) Diagrammatic representation of a bordered pit. (**F**) Diagrammatic representation of a simple pit. [*A courtesy of Dr. Myron C. Ledbetter, Brookhaven National Laboratory.* **C** *and* **D** *adapted from L. G. Livingston, American Journal of Botany, vol. 22, 1935, pp. 75–87.*]

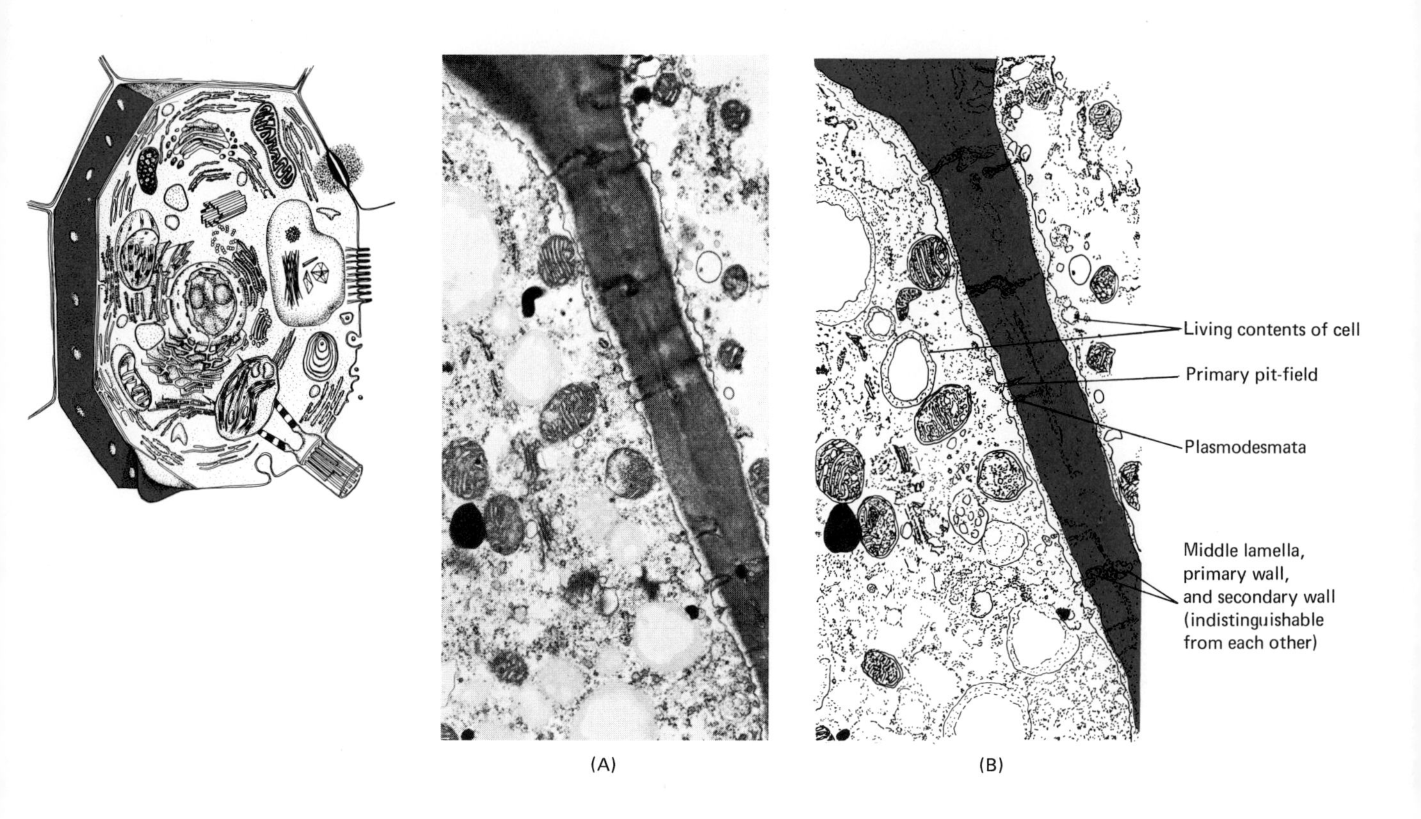
Living contents of cell
Primary pit-field
Plasmodesmata
Middle lamella,
primary wall,
and secondary wall
(indistinguishable
from each other)
(A)
(B)

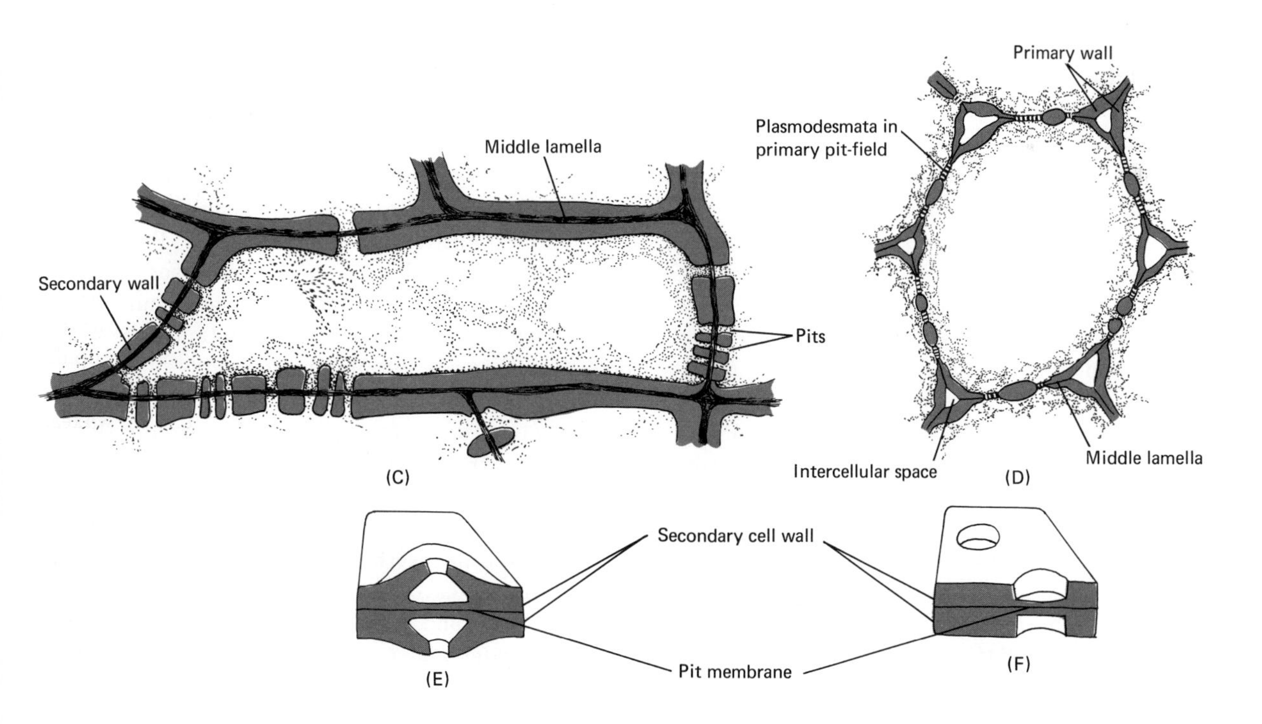
Middle lamella
Secondary wall
Pits
(C)
Primary wall
Plasmodesmata in
primary pit-field
Intercellular space
Middle lamella
(D)
Secondary cell wall
Pit membrane
(E)
(F)

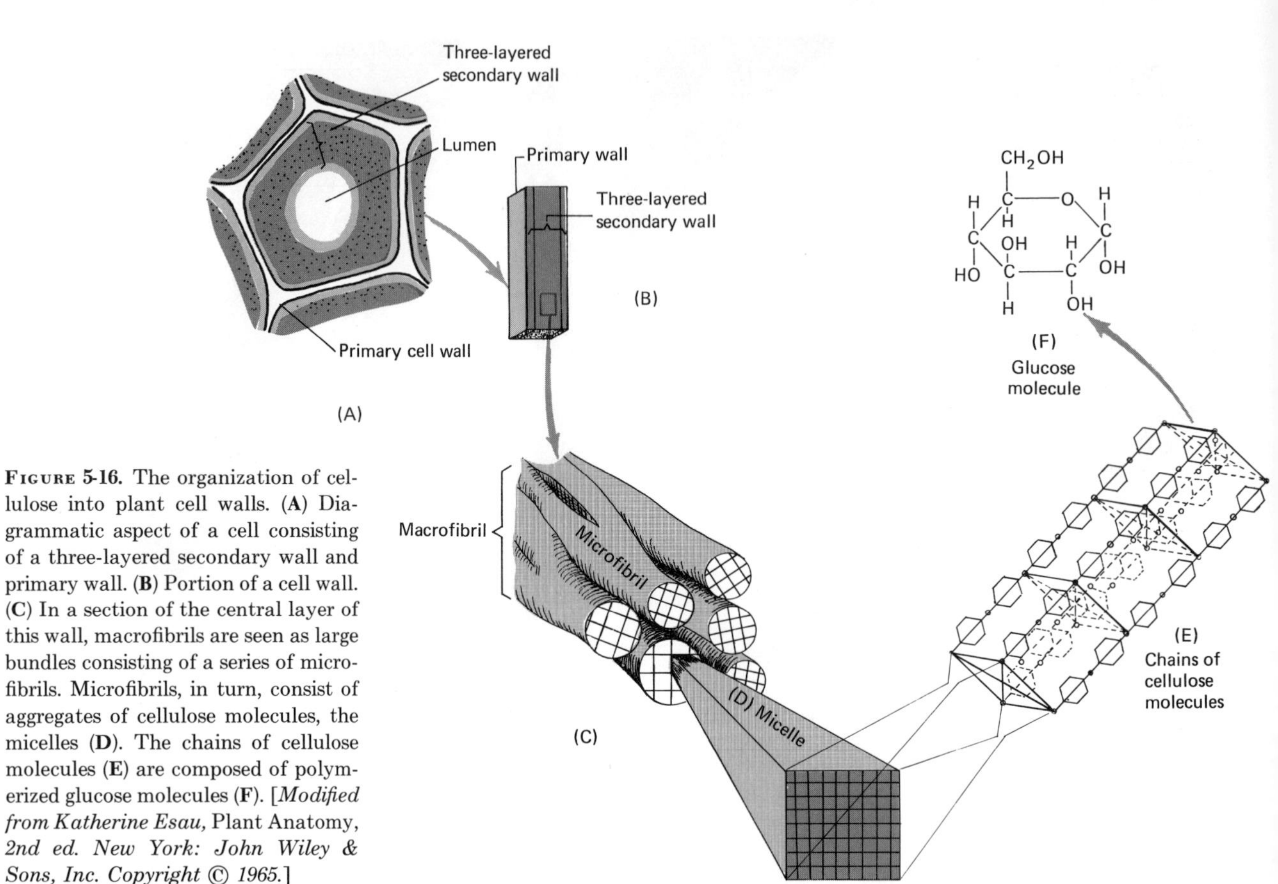

FIGURE 5-16. The organization of cellulose into plant cell walls. (**A**) Diagrammatic aspect of a cell consisting of a three-layered secondary wall and primary wall. (**B**) Portion of a cell wall. (**C**) In a section of the central layer of this wall, macrofibrils are seen as large bundles consisting of a series of microfibrils. Microfibrils, in turn, consist of aggregates of cellulose molecules, the micelles (**D**). The chains of cellulose molecules (**E**) are composed of polymerized glucose molecules (**F**). [*Modified from Katherine Esau,* Plant Anatomy, *2nd ed. New York: John Wiley & Sons, Inc. Copyright © 1965.*]

These substances are usually abundant in cell walls found at the periphery of various plant organs.

The primary wall is composed chiefly of cellulose, hemicellulose, and pectic substances, whereas the secondary cell wall contains cellulose, hemicellulose, and lignin as the chief constituents. The middle lamella, by comparison, contains pectic compounds as its main component.

The organizational pattern of cellulose in the formation of cell walls is shown in Figure 5-16. As previously stated, cellulose is a polymer of glucose molecules and together with other cellulose molecules, forms a structure called a *micelle.* Micelles are in turn organized into larger bundles called *microfibrils.* The microfibrils are then arranged into layers forming the cell wall.

INTERCELLULAR SUBSTANCES. *Intercellular substances* comprise a rather large and varied group of materials that form the matrix or mold in which cells live. They are nonliving, cellular-produced materials and occupy the spaces between cells. Essentially, they provide strength and support for tissues and act as a medium for the diffusion of nutrients and wastes. The two main types of intercellular substances are amorphous (nonformed) and fibrous (formed).

Among the principal amorphous intercellular substances are *hyaluronic acid* and *chondroitin sulfate* (collectively, the *ground substances*). The former is a viscous, fluid-like mucopolysaccharide of high molecular weight that is capable of binding readily. Functionally, it binds cells together (as well as affording flexibility), in the fluids of joints it serves as a lubricant (synovial fluid), and in the humors of the eye it may act to retain water and maintain the shape of the eye. Chondroitin sulfate is more jelly-like than hyaluronic acid and is also a mucopolysaccharide. It is found primarily in cartilage and bone, in the aorta and heart valves, in the cornea of the eye, and in the umbilical cord. In conjunction with other substances, chondroitin sulfate provides good support and adhesiveness while affording a degree of flexibility.

The function of providing strength and support for tissues is performed mainly by the fibrous intercellular substances. The fibrous elements embedded in the ground substance are *collagen* (collagenous fibers), *elastin* (elastic fibers), and *reticulin* (reticular fibers), all of which are complex proteins. Collagen is found in all types of connective tissue, principally in tendons, muscles, and bones. Elastin is a long thin fiber that has the ability to be stretched and relaxed. It is prevalent in areas that require elasticity (skin and tissues around blood vessels). Reticulin consists of fibers of small diameter and occurs as a network around blood vessels, muscle fibers, nerve fibers, and fat cells.

Cytoplasm

Cytoplasm is a rather general term for the living matter or physical substance of the cell inside the plasma membrane and external to the nucleus (Figure 5-17A, B). It is the ground substance, contains organelles, and usually comprises the bulk of cell contents. It is best described as a semifluid, semitransparent, viscous, elastic substance.

Chemically, cytoplasm consists of 75–90% water. The remainder is solid material in the form of assorted chemical compounds. Of the solid materials, more than one-half is protein with lipids, carbohydrates, acids, and other inorganic and organic compounds comprising the remaining portions. It should not be thought that cytoplasm is a random mixture of water and assorted chemical substances. On the contrary, it represents a complex and highly organized chemical and physical system of detailed ultrastructure capable of myriad energy transfer reactions.

In a physical context, cytoplasm is neither a solid nor a liquid, but a system composed of both solid and liquid components. The inorganic constituents of cytoplasm, as well as most carbohydrates, are soluble in water and are present as a true solution. The vast majority of organic compounds, specifically proteins and lipids, are present as colloids. True solutions and colloids may be differentiated on the basis of particle size.

If all the particles are small, such as those of inorganic salts or small organic molecules, the system is termed a *true solution.* A true solution is a homogeneous mixture of two or more components, that is, the particle sizes of the components are very small and cannot be distinguished from each other. In fact, the molecules are so small that they cannot be detected even with the most powerful microscope. No matter how long a true solution stands, the dispersed particles will not settle. Two examples of true solutions are salt (NaCl) dissolved in water and glucose ($C_6H_{12}O_6$) dissolved in water and in the cytoplasm of cells.

If the particles are somewhat larger than those in a true solution, ranging from 0.001 to 0.1 μm in diameter, the system is termed a *colloid.* As in a true solution, the particles of a colloid remain suspended. In the case of colloids, this is because colloidal particles bear positive and/or negative electric charges; like charges repel each other, keeping the particles suspended as well as apart from each other.

Cytoplasmic colloids undergo reversible *sol–gel* transformations. In the *sol state,* the colloidal particles are dispersed in a random fashion. In the *gel state,* the particles interact and form a network (Figure 5-17C). Interconversions between sol and gel may be brought about by changes in temperature, concentration of various cytoplasmic components, and salt concentration, among others. Consider gelatin as an example. At high temperatures it is a sol (fluid); when the temperature is lowered, it is transformed into a gel (semisolid).

Cytoplasm is, then, both a colloid and a true solution. It differs from ordinary colloids, however, in that the colloidal components are not haphazardly arranged but rather systematically organized. It also differs from an ordinary colloid in that it contains specialized bodies called organelles.

One distinctive cytoplasmic characteristic is **cytoplasmic streaming (cyclosis),** the flowing or streaming of cytoplasm within a cell in a continuous movement. Though there is no movement from one cell to another, many of the distinct bodies found in the cytoplasm, such as chloroplasts, are carried about much like logs in a stream. As shown in Figure 5-17D, the movement of one portion of cytoplasm may be opposite to

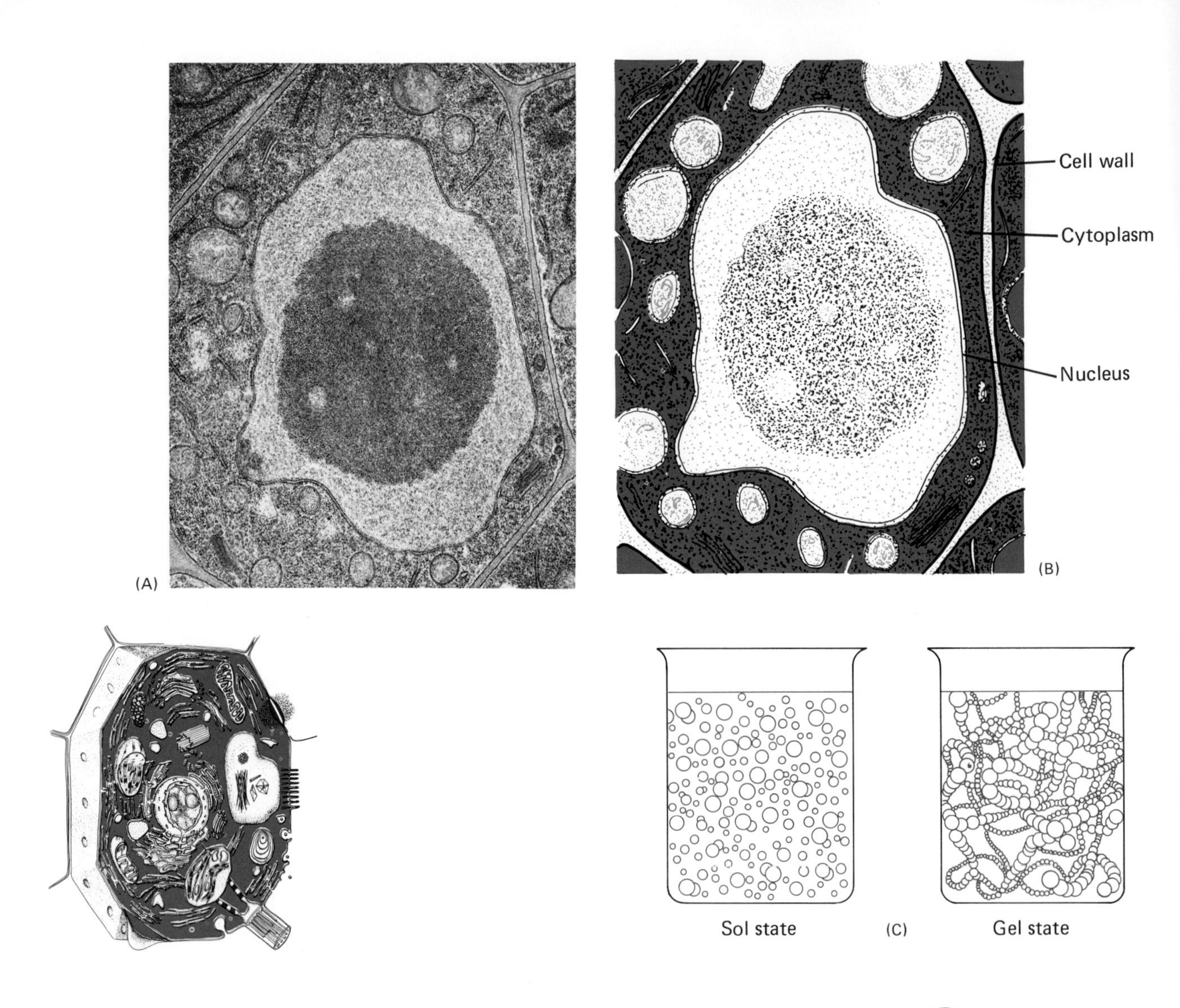

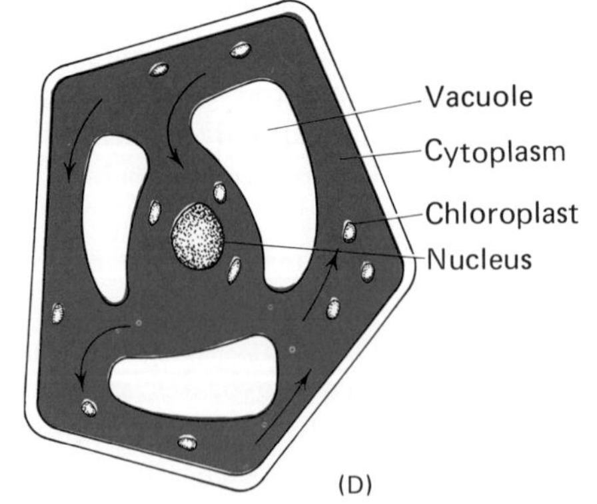

Figure 5-17. Cytoplasm. (**A**) Electron micrograph of cytoplasm as seen in a cell of *Arabidopsis thaliana* (mouse ear cress). 7800×. (**B**) Diagrammatic interpretation of electron micrograph. (**C**) Diagrammatic representation of the sol and gel states of a colloid. In the sol state (*left*) the colloidal particles are randomly dispersed. The particles in a gel (*right*) interact and form a more orderly network. (**D**) Diagrammatic representation of cyclosis in a plant cell. [**A** *courtesy of Dr. Myron C. Ledbetter, Brookhaven National Laboratory.* **C** *modified from M. S. Gardiner and S. C. Flemister,* The Principles of General Biology. *New York: Macmillan Publishing Co., Inc. Copyright © 1967.*]

that of another; lower layers of the cytoplasm may flow in one direction, while the upper layers flow in the opposite direction. Such movements indicate great chemical activity and may be important in the transport of food and other materials.

Organelles

Organelles, or "little organs," constitute a fourth principal subdivision of the parts of a cell. These are highly organized and specialized portions of the cytoplasm that assume various roles in cellular maintenance, growth, repair, and continuity. The chemical machinery by which cells perform various activities is localized within these organelles.

NUCLEUS. The **nucleus** is a constant and essential component of nearly all cells. Its importance in the life of the cell is readily apparent if one considers a few exceptions to this generalization. For example, enucleate mammalian red blood cells and the cells comprising the center of the lens of the eye are somewhat restricted in their metabolic activities and are not capable of growth or division.

Experimental enucleation by microsurgery also provides proof of the prominence of the nucleus in the cell's activities. In one of these experiments, performed by Joachim Hämmerling of the Max Planck Institute for Marine Biology in Berlin, it was shown that the nucleus is the source of information that controls cell morphology. His experimental organisms were two species of *Acetabularia,* photosynthetic protists of warm marine waters (Figure 5-18). These organisms consist of a cap, a chlorophyllous stalk, and a root-like structure that contains the single nucleus. The two species differ primarily in the morphology of their caps; *A. mediterranea* has a compact umbrella-shaped cap, whereas *A. crenulata* possesses a cap composed of loose petal-like structures. If the cap is removed from either species, a new one resembling the decapitated one is produced (Figure 5-18A). However, if a piece of the stalk of one species is grafted onto the nucleus-containing root-like structure of the other species, the cap that forms is characteristic of the species that contributed the nucleus, not of the species that contributed the stalk (Figure 5-18B). It was further shown that if two root-like structures (one from each species), both containing nuclei, are grafted together, the cap produced is intermediate in form, a product of the influence of both nuclei (Figure 5-18C).

The nucleus not only exercises control over structure, it also influences cell function. If a unicellular form, such as *Amoeba,* is cut in two, the nucleated half functions. The enucleate portion, however, does not form pseudopodia and therefore cannot ingest food or undergo locomotion. It cannot metabolize, grow, or reproduce. Transplantation of the nucleus from the formerly nucleated half into the formerly enucleate half shows the opposite effects. The half containing the nucleus is now capable of motility, phagocytosis, growth, and division. The nucleus is an important center of control; it directs the many aspects of cellular activity and contains the hereditary factors (genes) responsible for the traits of the organism.

Microscopically, the nucleus of eucaryotic cells is the most prominent organelle. It assumes a spherical or oval shape and its contents are more viscous than those of the cytoplasm. A nucleus of characteristic appearance is lacking in procaryotic cells of the Kingdom Monera and a few nonmoneran types, such as erythrocytes and mature sieve cells in phloem tissue. Although most cells contain only a single nucleus, some, such as those of certain species of algae and fungi, contain two or more nuclei. This condition is referred to as *multinucleate* or *coenocytic.*

The nucleus is delimited from the cytoplasm in which it is embedded by a double unit membrane, the **nuclear membrane (nuclear envelope),** with a cavity, the *perinuclear cisterna,* between them (Figure 5-19). This structure resembles other cellular membranes both in appearance and chemistry. Moreover, electron micrographs reveal that the nuclear membrane communicates with an internal membranous network, the endoplasmic reticulum, by minute pores (see Figure 5-4). These pores may facilitate an exchange of materials between the nucleus and the cytoplasm.

The body of the nucleus within the nuclear membrane is composed of a gel-like nuclear cytoplasm called the *nuclear matrix,* one or more spherical bodies called nucleoli (nucleolus, singular), and a thread-like network referred to as chromatin.

Nucleoli are composed chiefly of protein and RNA and are believed to assume a role in protein synthesis. **Chromatin,** which is the genetic material, consists of

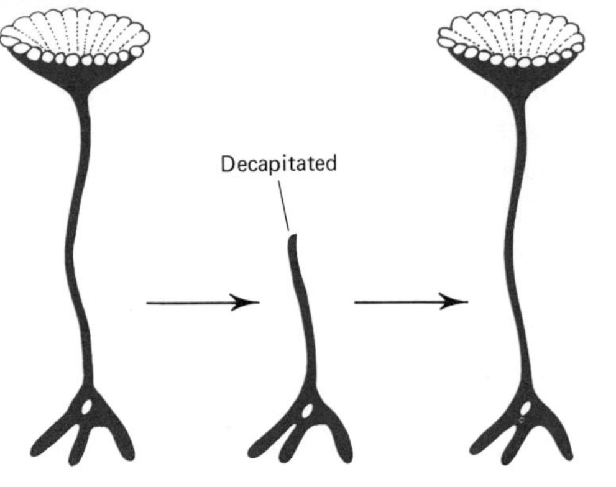

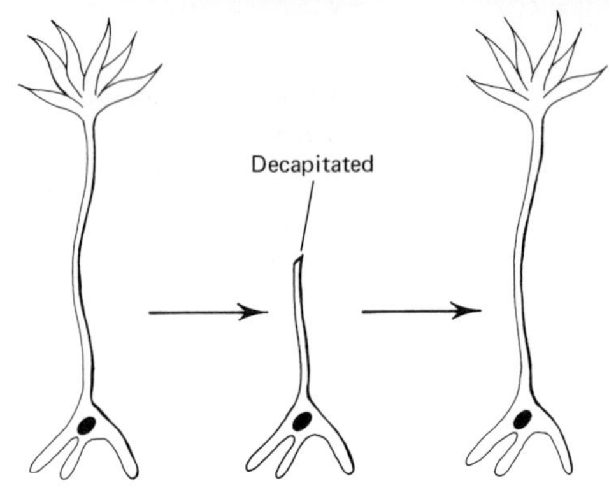

(A)

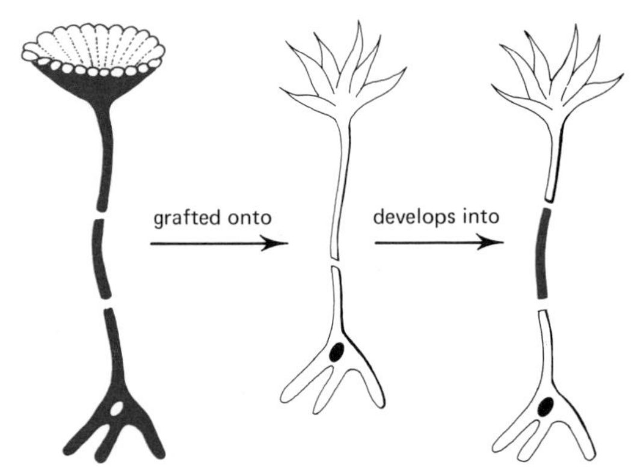

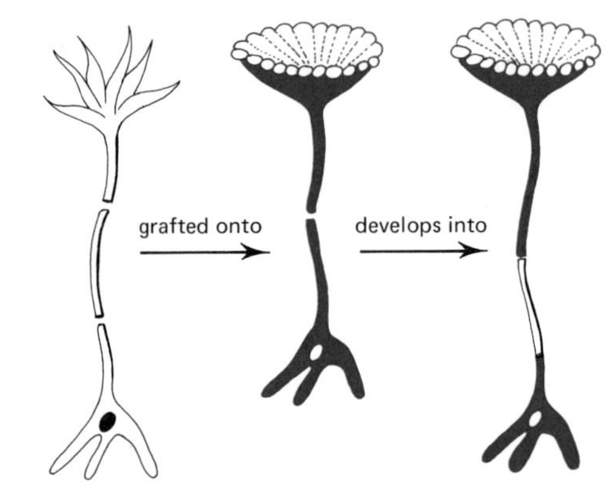

(B)

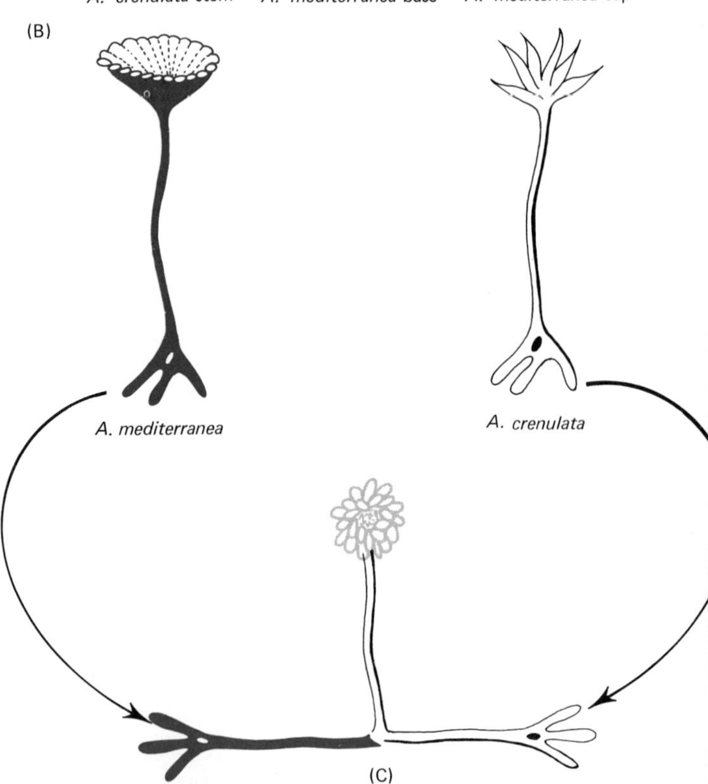

(C)

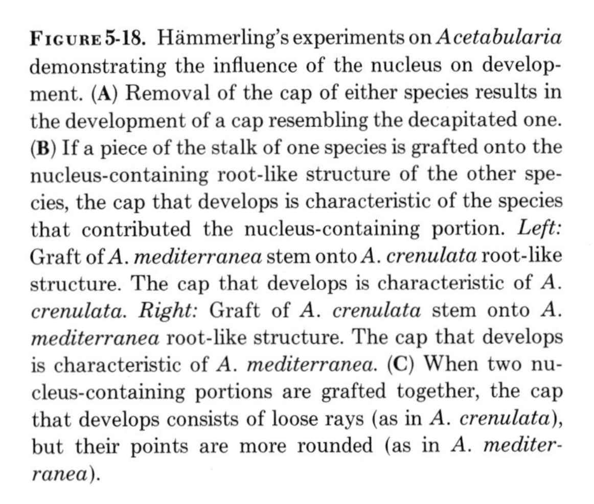

Figure 5-18. Hämmerling's experiments on *Acetabularia* demonstrating the influence of the nucleus on development. (**A**) Removal of the cap of either species results in the development of a cap resembling the decapitated one. (**B**) If a piece of the stalk of one species is grafted onto the nucleus-containing root-like structure of the other species, the cap that develops is characteristic of the species that contributed the nucleus-containing portion. *Left:* Graft of *A. mediterranea* stem onto *A. crenulata* root-like structure. The cap that develops is characteristic of *A. crenulata*. *Right:* Graft of *A. crenulata* stem onto *A. mediterranea* root-like structure. The cap that develops is characteristic of *A. mediterranea*. (**C**) When two nucleus-containing portions are grafted together, the cap that develops consists of loose rays (as in *A. crenulata*), but their points are more rounded (as in *A. mediterranea*).

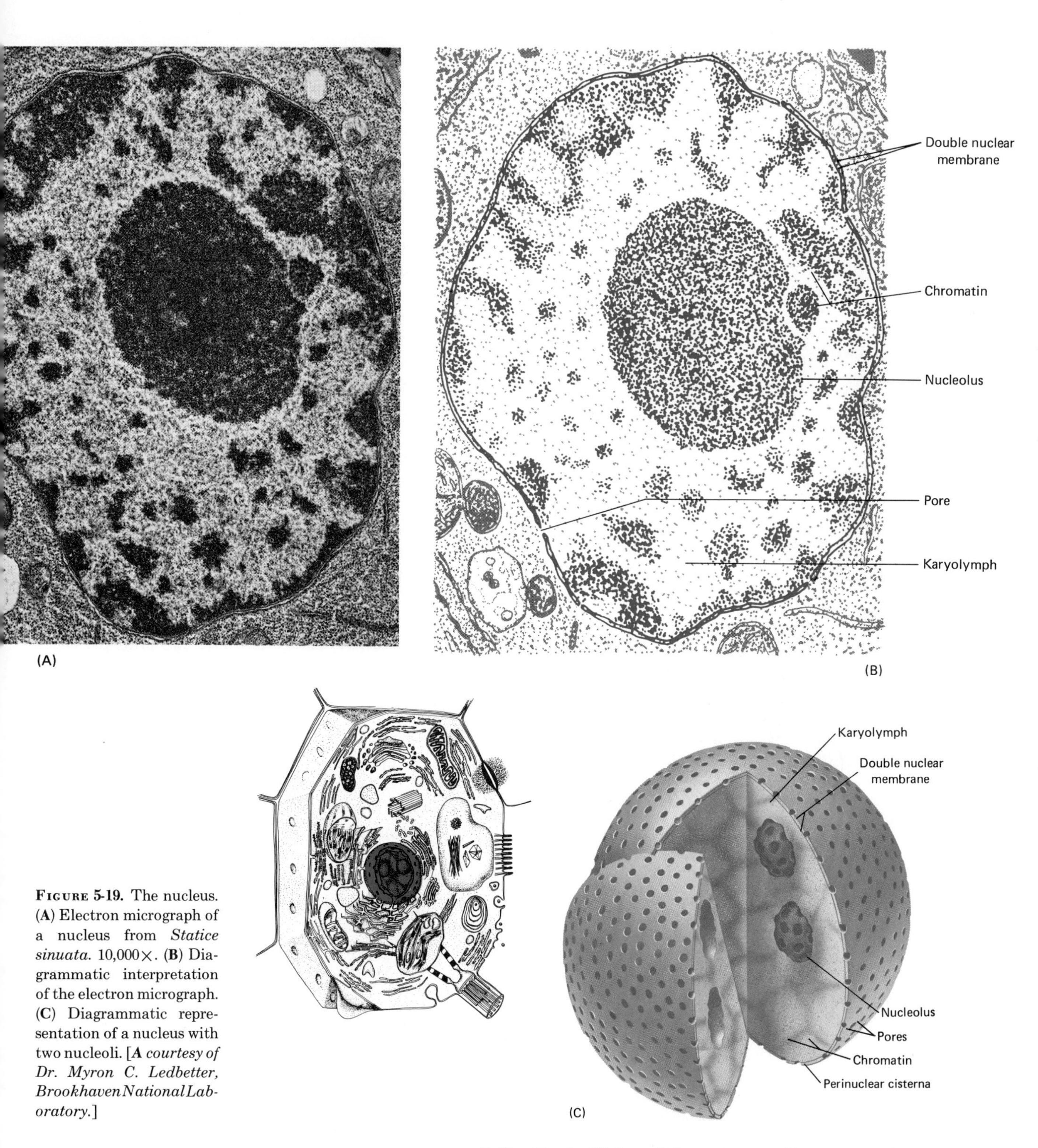

Figure 5-19. The nucleus. (**A**) Electron micrograph of a nucleus from *Statice sinuata*. 10,000×. (**B**) Diagrammatic interpretation of the electron micrograph. (**C**) Diagrammatic representation of a nucleus with two nucleoli. [*A courtesy of Dr. Myron C. Ledbetter, Brookhaven National Laboratory.*]

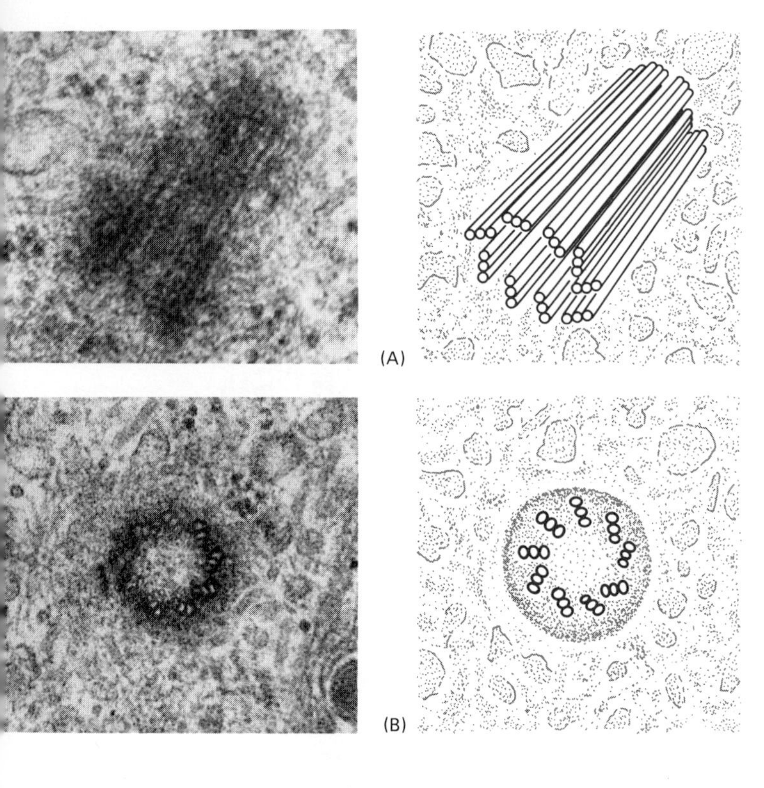

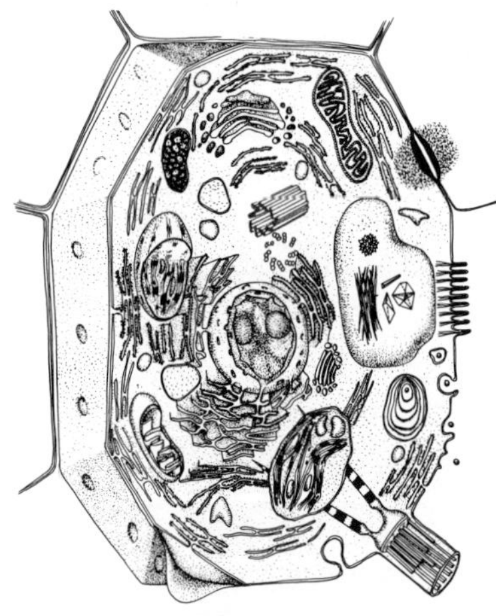

Figure 5-20. Centrioles. Usually there is only one pair of these self-replicating organelles in each cell. They duplicate in the early stages of mitosis so that two are transmitted to each daughter cell. (**A**) Electron micrograph (*left*) and diagrammatic representation (*right*) of a centriole in semilongitudinal section. (**B**) Electron micrograph (*left*) and diagrammatic representation (*right*) of a centriole in cross section. [*Electron micrographs courtesy of E. B. Sandborn, M.D., Université de Montréal.*]

DNA and protein. In a nondividing cell, chromatin appears as a distended thread-like mass. When the cell is about to divide, the chromatin mass shortens and thickens to form rod-shaped bodies called **chromosomes.**

Centrioles. The cells of many animals, as well as those of many protists, typically contain one pair of small cylinder-shaped organelles in the nuclear region called **centrioles.** Each centriole is found to be a hollow cylinder 300–500 nm long and 150 nm in diameter (Figure 5-20). Its wall is composed of nine evenly spaced, triplet, hollow tubules embedded in a rather dense, amorphous matrix; one end of the centriole appears open while the other is closed. Centrioles are often arranged so that the long axis of one is perpendicular to the long axis of the other. Just before a cell divides, its centrioles duplicate so that two are transmitted to each daughter cell. The centrioles appear to have two basic functions: (1) they assume a role in cell division and (2) they may also serve as the basal body from which cilia and flagella originate. Their role in cell division will be described later.

Cilia and Flagella. A large variety of cells, especially animal cells, possess projections that are utilized for various activities. In general, these motile cell processes are categorized into two principal types.

1. If they are few in number and long in proportion to the size of the cell, they are termed **flagella.**
2. If they are numerous and relatively short, they are referred to as **cilia.**

There appears to be no fundamental structural difference between cilia and flagella (Figure 5-21). Each consists of an extension of the typical unit cell membrane containing a cytoplasmic matrix, with eleven groups of fibrils embedded in the matrix. Typically, nine of these groups of fibrils are oriented around the periphery of the cylinder and the other two are centrally located. This arrangement is commonly referred to as the "9 + 2" pattern. At the base of the stalk is a structure, the *basal body,* which is found within the cellular cytoplasm. The peripheral fibrils grow out of this body, a hollow cylinder with the same structure as a centriole. Between the stalk and the basal body is a structure, the *basal plate,* which gives rise to the two central fibrils. Some cilia and flagella also contain rootlet fibers, which project from the basal body into the cellular cytoplasm.

Cilia and flagella normally function by a beating action that results in movement of an entire cell or organism (*Paramecium, Chlamydomonas, Euglena*), of the male reproductive cells of many plants and animals, or of liquids or particles across the cell surface. For example, in clams, the motion of cilia on gills creates a steady current of water containing food particles. In humans, the ciliated cells of the respiratory tract move lubricating fluids over the surface and trap foreign particles.

Studies with isolated cilia and flagella have demonstrated that the beat originates within the structure itself. Current hypotheses on the mechanism of ciliary motion are based on knowledge of the structure as elucidated by electron microscopy. According to one hypothesis, the matrix of the cilium is stiff and the nine peripheral fibrils can contract and propogate waves of contraction from the base to the tip. This hypothesis further postulates that the central pair of fibrils is not contractile, but specialized for rapid conduction, and that the impulse that initiates the beat arises rhythmically from the basal body and spreads sequentially to the different fibrils.

ENDOPLASMIC RETICULUM (ER) AND RIBOSOMES. The **endoplasmic reticulum** is an organelle that was discovered with the electron microscope and since has been found to be present in all nucleated cells. The presence of this structure has discredited the idea that the cytoplasm is merely a viscous homogeneous substance. Although highly variable in extent and configuration, the endoplasmic reticulum in its typical form is a system of pairs of parallel membranes (similar to other cellular membranes) enclosing narrow cavities of varying shapes (Figure 5-22A–C). This appearance is the result of sectioning the cylindrical membranous structure. In some cells, the network appears as fine tubules, 50–100 nm in diameter; in others, the membrane-bound cavities form flattened sac-like structures called *cisternae.* On the surfaces of the tubules and cisternae are areas where biochemical reactions occur. The ER system may be more or less continuous between the plasma membrane and nuclear envelope and is assumed to be part of the cell's complex membrane system (see Figure 5-4). The ER has been described as a type of cytoskeleton that provides surfaces for chemical reactions, pathways for the transportation of cellular molecules, and storage for synthesized molecules.

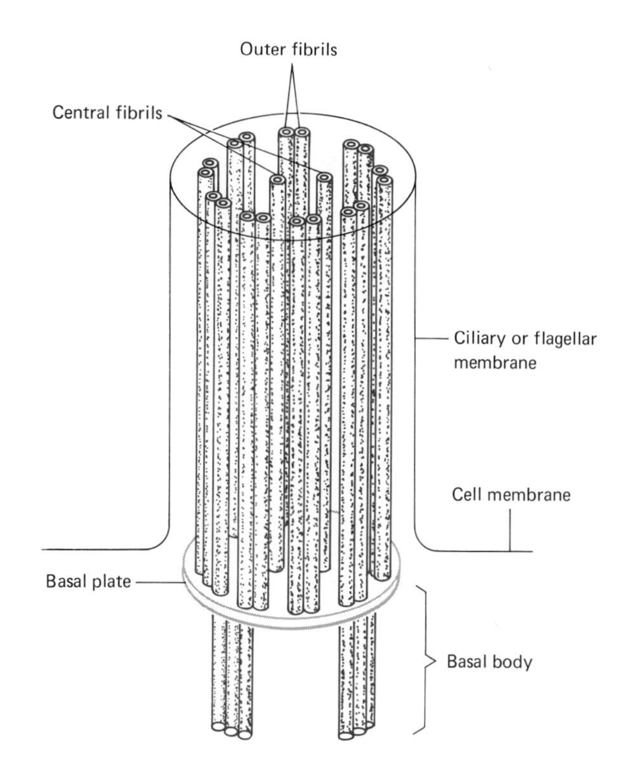

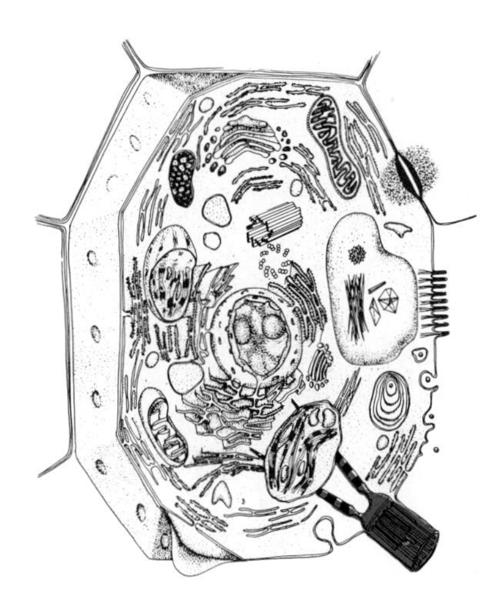

FIGURE 5-21. Flagellar and ciliary structure. Since there appears to be no fundamental difference between flagellar and ciliary structure, a single ultrastructure (common to both) is shown. See text for amplification.

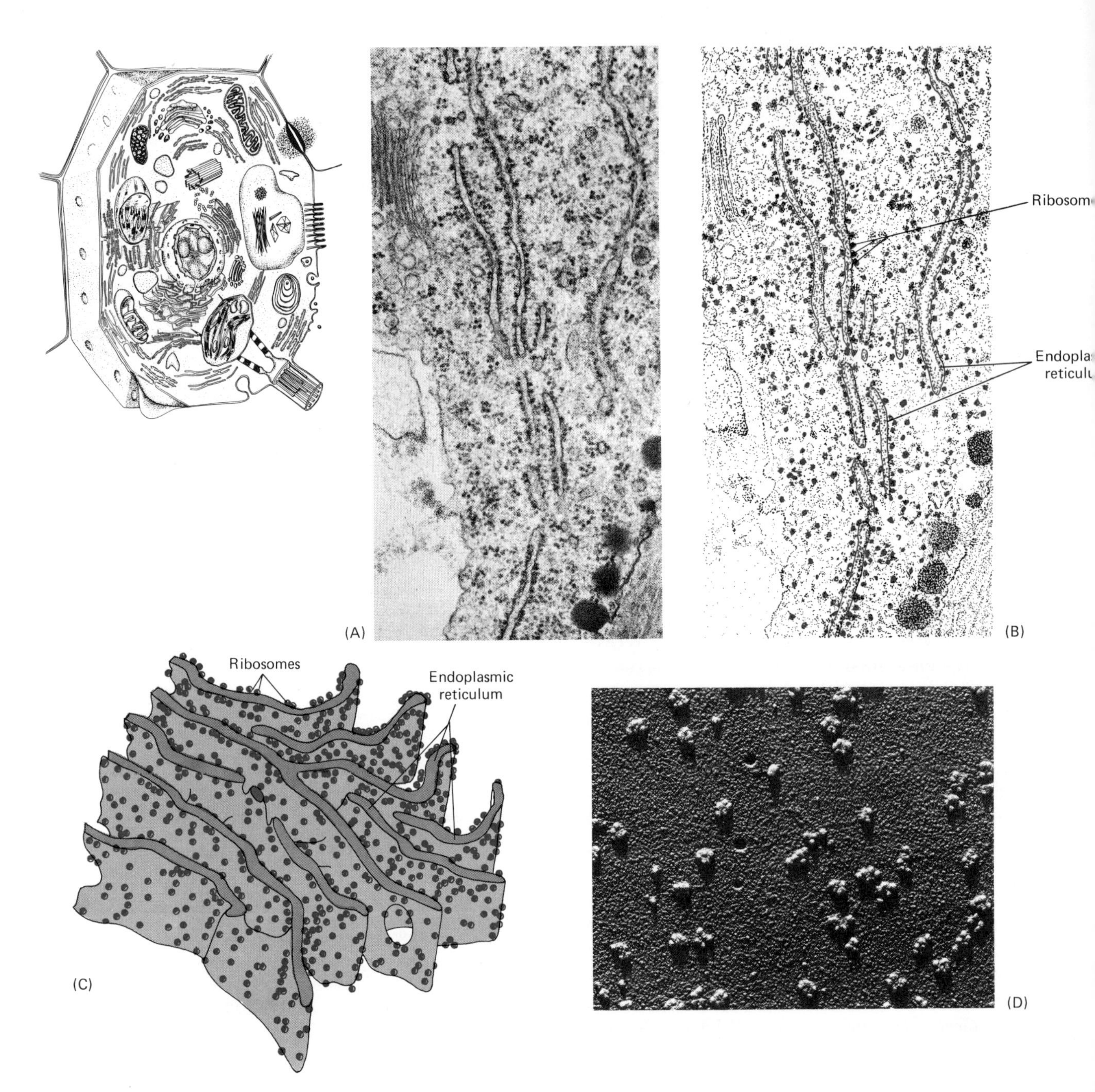

Figure 5-22. The endoplasmic reticulum. (**A**) Electron micrograph of the endoplasmic reticulum with ribosomes from *Phleum pratense* (timothy grass). 120,000×. (**B**) Diagrammatic interpretation of the electron micrograph. (**C**) Diagrammatic representation of the endoplasmic reticulum with ribosomes. (**D**) Electron micrograph of polyribosomes from rabbit reticulocyte cells. [**A** *courtesy of Dr. Myron C. Ledbetter, Brookhaven National Laboratory.* **C** *modified from E. D. P. DeRobertis, W. W. Nowinski, and F. A. Saez,* Cell Biology, *4th ed. Philadelphia: W. B. Saunders Co. Copyright © 1965.* **D** *courtesy of Dr. Alexander Rich, Massachusetts Institute of Technology.*]

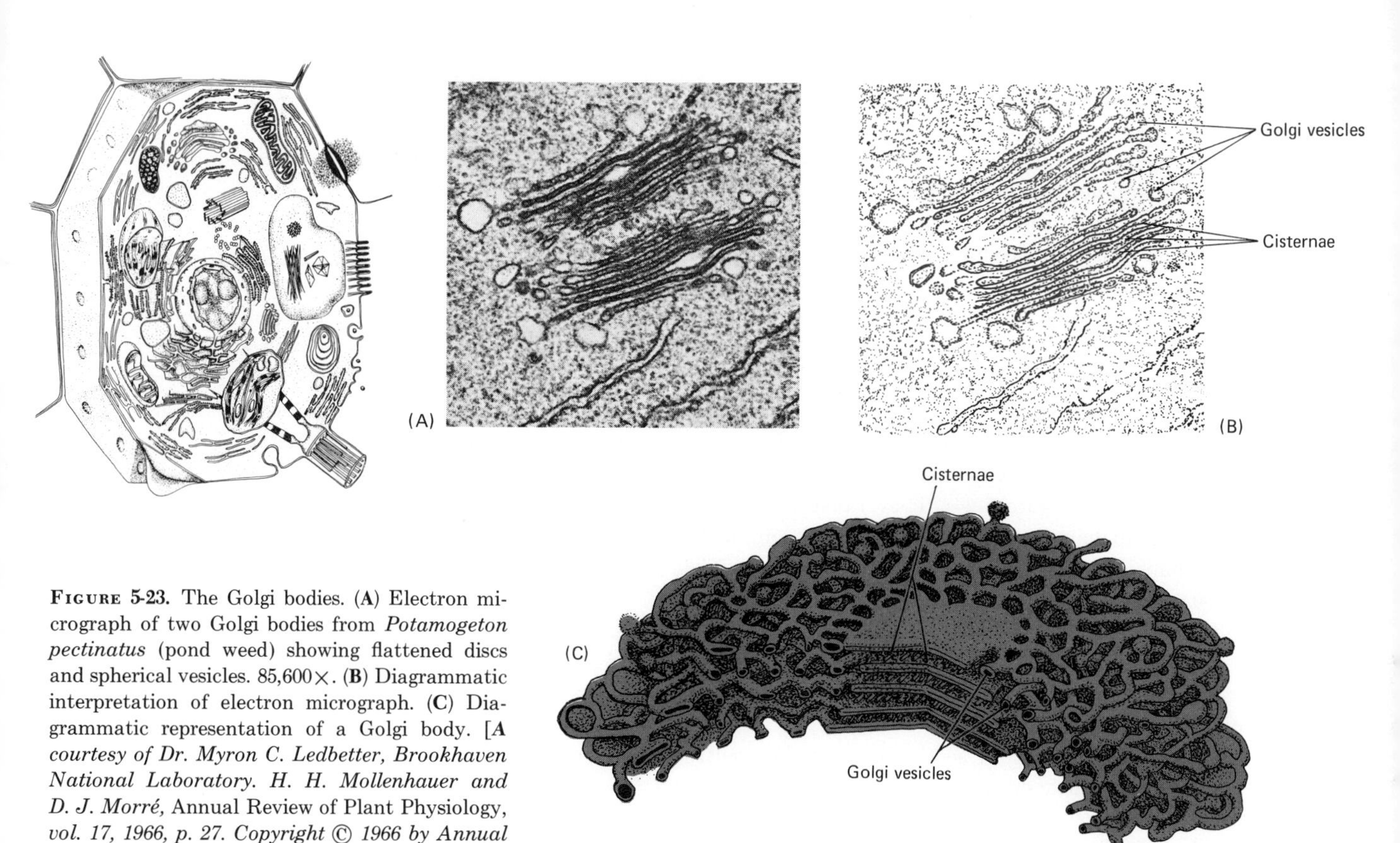

FIGURE 5-23. The Golgi bodies. (**A**) Electron micrograph of two Golgi bodies from *Potamogeton pectinatus* (pond weed) showing flattened discs and spherical vesicles. 85,600×. (**B**) Diagrammatic interpretation of electron micrograph. (**C**) Diagrammatic representation of a Golgi body. [**A** *courtesy of Dr. Myron C. Ledbetter, Brookhaven National Laboratory. H. H. Mollenhauer and D. J. Morré,* Annual Review of Plant Physiology, *vol. 17, 1966, p. 27. Copyright © 1966 by Annual Reviews, Inc.*]

In some instances, the membranes of the ER are lined on their outer surfaces by exceedingly small dense bodies called **ribosomes;** in this case the ER is referred to as a *granular* ("rough") *reticulum* (Figure 5-22A–C). In the absence of ribosomes, the ER is designated as an *agranular* ("smooth") *reticulum.* When ribosomes are organized in groups or clusters, they are referred to as *polyribosomes* (Figure 5-22D). Functionally, ribosomes are the sites of protein synthesis (Chapter 8). Ribosomes are also found in mitochondria and plastids; these ribosomes are smaller than the usual ER ribosomes.

GOLGI APPARATUS. **Golgi bodies,** or **dictyosomes,** were first described by Camillo Golgi, an Italian physician, in 1898. Although once considered to be organelles of animal cells only, electron microscopy has demonstrated their presence in plant cells as well. Structurally, dictyosomes appear as aggregations of membrane-bounded (smooth) elements (Figure 5-23). The most conspicuous and characteristic feature is a system of several parallel stacked saucer-shaped elements called *cisternae.* Four to eight such layers, either flat or curved, are typical. Clustered at their ends are numerous *Golgi vesicles,* which are derived from the stacked elements. Some evidence indicates that the smooth unit membranes of the Golgi complex are continuous with those of the ER.

The functions of this organelle have not yet been completely defined. Most investigators believe that lysosomes (discussed later) may be derived from the Golgi apparatus. The lysosomes store intracellular digestive (hydrolytic) enzymes, and because of this relationship between the Golgi apparatus and lysosomes, the Golgi apparatus has long been associated with secretory activities of cells. In pancreatic cells, the enzyme *zymogen,* presumably synthesized on the ribosomes, appears to move into the channels of the ER and from there into the Golgi apparatus. Here, the enzyme appears to be stored and eventually released by the cell through the ER. These data would seem to indicate that the Golgi apparatus is involved in the storage of secretory products and not their actual synthesis. Some evidence also indicates that the role of the Golgi apparatus is not limited to secretory products; it may also involve the storage and distribution of materials within the cell. Still other functions essential to the economy of the cell, such as packaging synthesized materials, have been proposed.

Mitochondria. Another type of organelle found universally in the cytoplasm of plant and animal cells is the small sphere-, rod-, or filament-shaped bodies called **mitochondria** (Figure 5-24). They range in size from 0.5 to 1 μm in diameter and from 1 to 2 μm in length. The size, shape, and distribution of mitochondria are fairly constant in cells of the same type. However, cells of different types exhibit considerable variations.

Although mitochondria are rendered visible under the ordinary light microscope by special staining techniques, very little detailed structure is revealed. Through the use of thin sectioning and the electron microscope, mitochondria exhibit a fairly elaborate internal organization (Figure 5-24A–C). Electron micrographs have shown that each mitochondrion consists of a double unit membrane similar in structure to the plasma membrane. Chemically, the mitochondrial membrane, like the plasma membrane, is lipid and protein in nature (Figure 5-24C). The outer membrane forms a smooth surface separating the internal portions of the mitochondrion from the cytoplasm. The inner membrane is composed of a series of folds that extend into the *matrix,* or center of the mitochondrion. These folds, or *cristae,* assume various forms in the mitochondria of different cells (Figure 5-24A–C). The cristae contain subunits about 100 Å in diameter (Figure 5-24C,D). These exceedingly small stalked particles contain ATP-synthesizing enzymes.

Like other cellular organelles, mitochondria illustrate a high degree of correlation between structure and function. The cristae are arranged in a specific pattern in the form of uneven and incomplete folds. This pattern affords an enormous surface area for chemical reactions, a great deal of reaction space, and the possibility of molecules orienting themselves in a linear arrangement. Experimental evidence indicates that enzyme molecules concerned with energy-releasing reactions are arranged on the cristae. Organic acids, derived from the breakdown of more complex molecules, are further broken down into carbon dioxide and water as part of a stepwise series of reactions termed *respiration.* Each reaction is catalyzed by a specific enzyme, and many steps in the sequence release energy. Some of this energy is then stored as ATP and passed on to other parts of the cell where it is utilized. Because of the central role mitochondria play in cellular respiration, they are referred to as "powerhouses of the cell" (Chapter 6). Mitochondria contain RNA, ribosomes, and DNA. Mitochondria are self-duplicating organelles, and it is thought that the DNA contains at least some of the hereditary information required for mitochondrial growth and replication.

Plastids. Plastids represent a variety of differentiated organelles embedded in the cytoplasm. These, along with cell walls, are truly distinctive plant structures. Microscopically, plastids reveal a number of sizes and shapes. Various forms exhibited by plastids are plates, discs, and spirals, among others. Plastids are believed to originate from minute, defined precursor structures called *proplastids.* The proplastids, typically found in young cells, are transmitted from one generation to another during cell division and are also capable of duplicating themselves in the mature plant cell. As noted earlier, plastids also contain RNA. Plastids are commonly categorized into three distinct types on the basis of color and function. These include chloroplasts, chromoplasts, and leucoplasts.

Chloroplasts. **Chloroplasts** are green plastids. They owe their color to the presence of the pigment chlorophyll. In higher plants, they are usually disc-shaped and range in size from 2 to 4 μm in diameter and from 0.5 to 1 μm in thickness. About 20% of the volume of leaf cells is occupied by chloroplasts; this is not surprising in view of the fact that leaves are the primary organs of photosynthesis. At high magnification, chloroplasts reveal an elaborate, highly organized internal structure (Figure 5-25A, B). Each consists of a double-layered bounding membrane that separates it from the cytoplasm. The outer membrane frequently appears similar to the plasma membrane of the cell. The inner membrane is similar to the membrane system of the chloroplast itself. One interpretation of this variation is

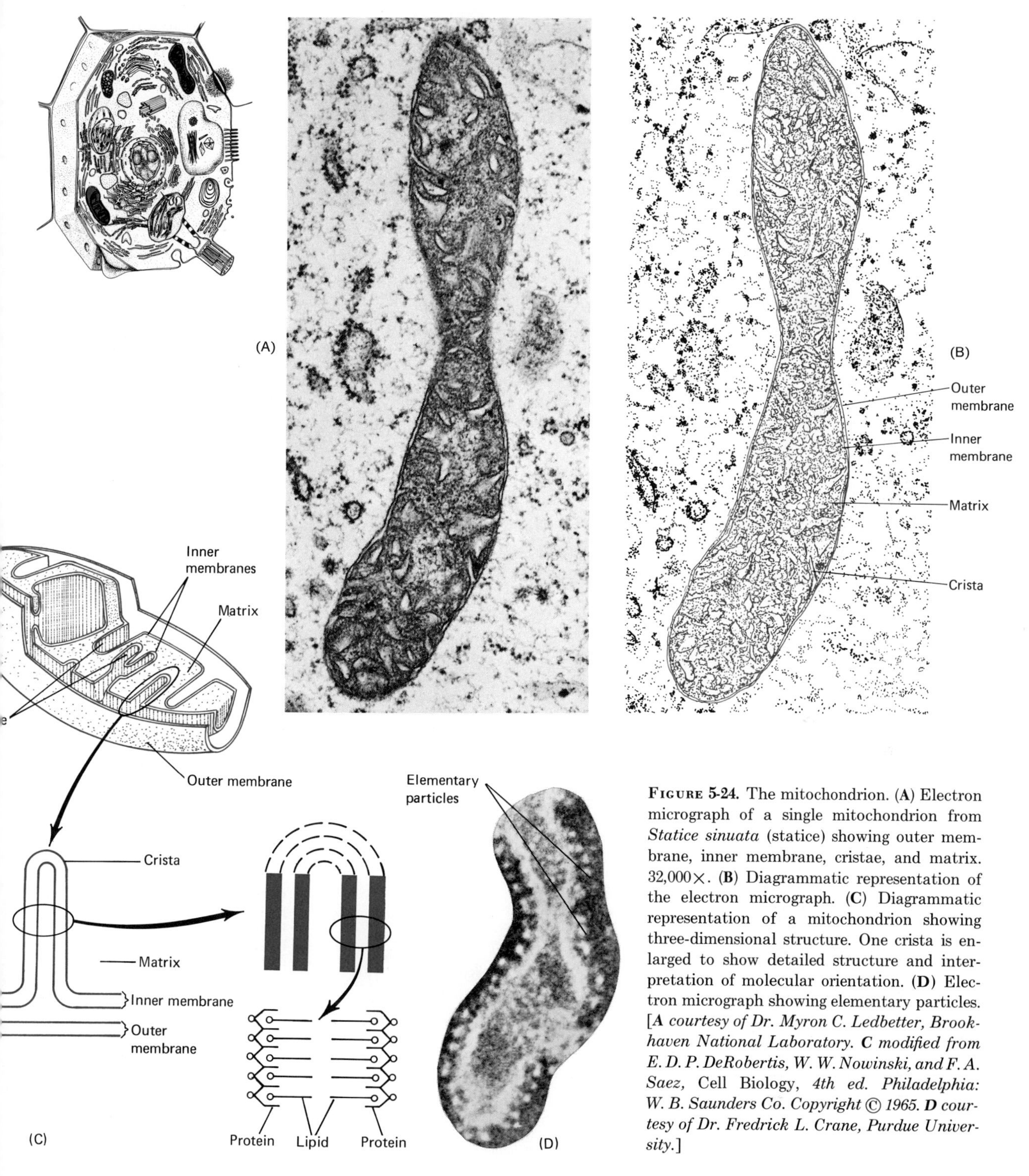

FIGURE 5-24. The mitochondrion. (**A**) Electron micrograph of a single mitochondrion from *Statice sinuata* (statice) showing outer membrane, inner membrane, cristae, and matrix. 32,000×. (**B**) Diagrammatic representation of the electron micrograph. (**C**) Diagrammatic representation of a mitochondrion showing three-dimensional structure. One crista is enlarged to show detailed structure and interpretation of molecular orientation. (**D**) Electron micrograph showing elementary particles. [**A** *courtesy of Dr. Myron C. Ledbetter, Brookhaven National Laboratory.* **C** *modified from E. D. P. DeRobertis, W. W. Nowinski, and F. A. Saez,* Cell Biology, *4th ed. Philadelphia: W. B. Saunders Co. Copyright © 1965.* **D** *courtesy of Dr. Fredrick L. Crane, Purdue University.*]

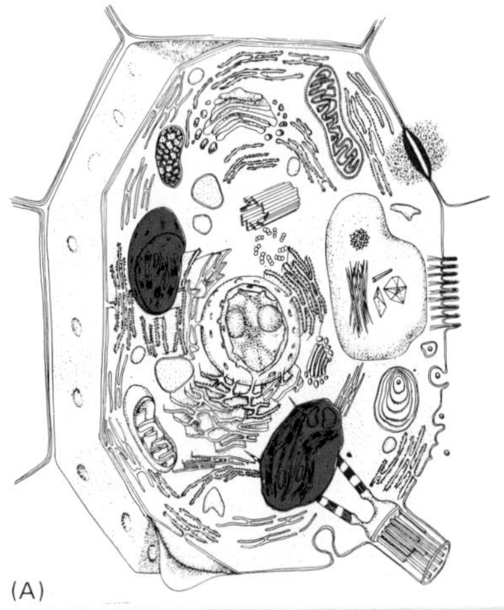

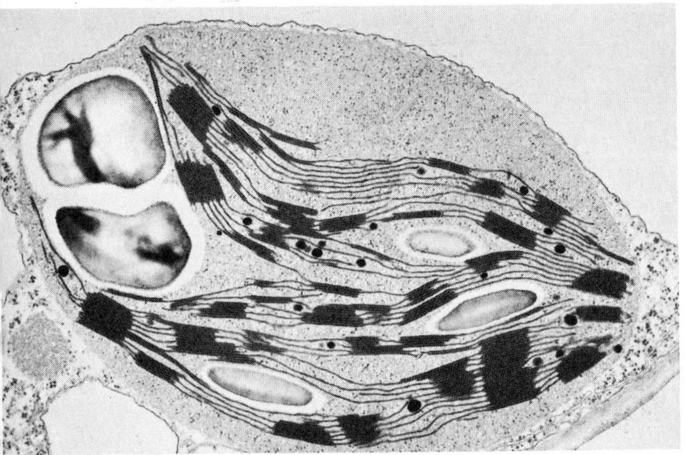

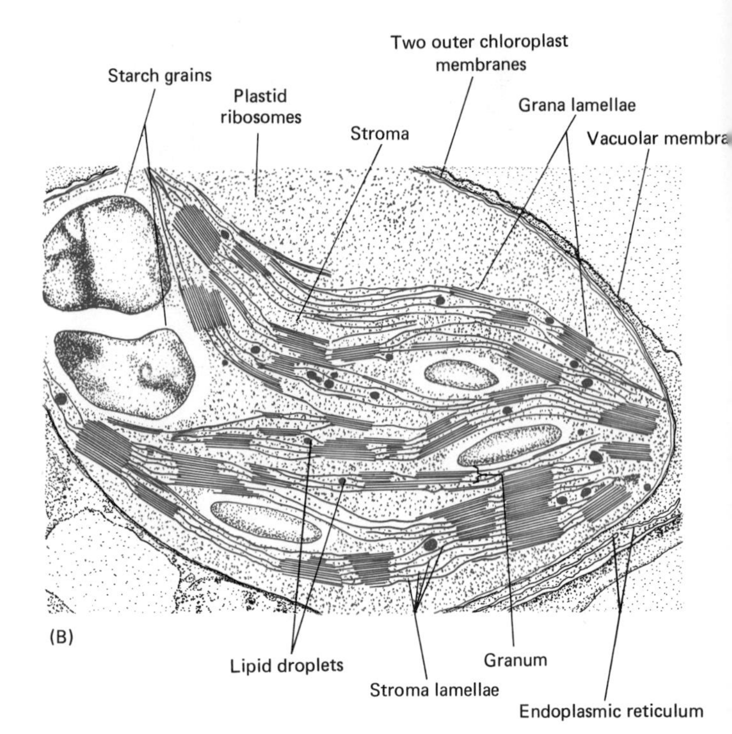

Figure 5-25. The chloroplast. (**A**) Electron micrograph of a chloroplast from the leaf of *Phleum pratense* (timothy grass). 17,200×. (**B**) Diagrammatic interpretation of the electron micrograph. (**C**) Diagrammatic representation of a granum in successive magnifications and in various views. [**A** *courtesy of Dr. Myron C. Ledbetter, Brookhaven National Laboratory.* **C** *modified in part from A. J. Hodge, in J. L. Onceley et al. (eds.),* Biophysical Science—A Study Program. *New York: John Wiley & Sons, Inc. Copyright © 1959.*]

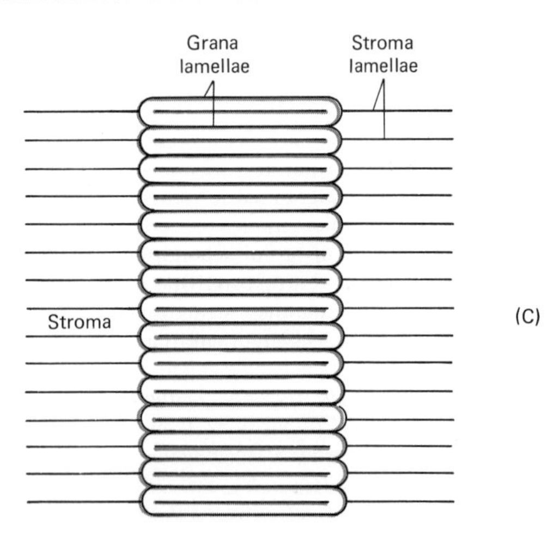

(C)

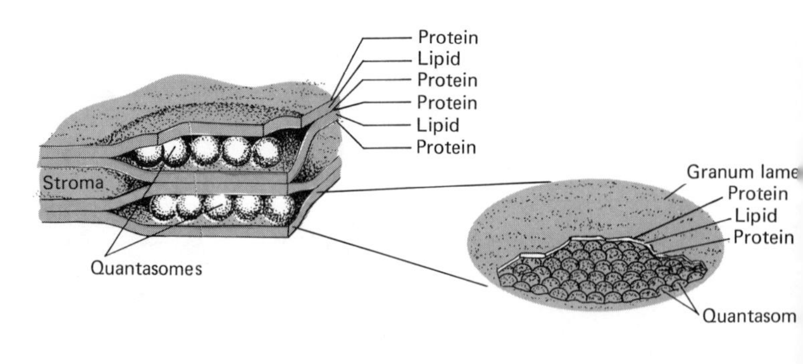

that the outer membrane is derived from the parent cell as a means of isolating itself from the chloroplast; the inner membrane, therefore, serves to envelop the contents of the chloroplast.

In addition to the external double membrane, a number of internal structures are distinguishable. The membrane system consists of closed flattened sacs called *thylakoids.* These membranes contain the chlorophyll and the light conversion apparatus in eucaryotic cells. Thylakoids also represent the sites of oxygen production and photophosphorylation, processes that will be discussed in conjunction with photosynthesis in Chapter 6.

Small thylakoids, or *grana lamellae,* when stacked one on top of the other, produce a structure called a *granum* (Figure 5-25A–C). The membrane structures extending between the stacks of grana are referred to as large thylakoids, or *stroma lamellae* (Figure 5-25A–C). Thylakoids may assume a number of configurations in different types of chloroplasts. The significance of these diverse configurations has yet to be elucidated. Recently, substructures of the grana lamellae, the *quantasomes,* have been described (Figure 5-25C). Chlorophyll and other pigments appear to be localized in a single layer between the lamellae of a granum. Researchers have calculated that each quantasome could contain about 230 chlorophyll molecules.

Figure 5-25 also reveals that internally a chloroplast contains a clear region called the *stroma.* The stroma is a membrane embedded in the matrix which contains the enzymes necessary for the fixation of carbon dioxide into sugar (Chapter 6) and ribosomes involved in protein synthesis (Chapter 8). While chlorophyll and carotenoids are the principal chemical constituents of chloroplasts, the stroma has also been found to contain globules, which may be lipid droplets and starch granules. RNA and DNA of a sort unique to chloroplasts and distinct from nuclear DNA and RNA have also been revealed by chemical analyses. Chloroplasts, like mitochondria, are self-replicating organelles.

Chromoplasts. **Chromoplasts** owe their color to the presence of various carotenoids; these plastids are typically yellow, orange, or red. Although they are generally disc-shaped, some appear as spindle-, angular-, spherical-, or rod-shaped bodies (see Figure 5-4).

Chromoplasts may be plastids in which chlorophyll disappears and becomes less dominant, or they may originate directly from proplastids, never having contained chlorophyll. For example, when bananas, sweet peppers, or tomatoes ripen, chloroplasts become chromoplasts as the chlorophyll breaks down and carotenoids become dominant. The yellow coloration in autumn leaves also results from the takeover by carotenoids, originally present in leaves, as chlorophyll is broken down.

Leucoplasts. **Leucoplasts,** for the most part, are colorless plastids and are difficult to locate and identify unless cells containing them are specially stained (see Figure 5-4). One common stain, iodine, has an affinity for the starch contained in the leucoplasts. These plastids are generally found in colorless leaf cells (variegated leaves), stems, roots, and other storage organs. Usually, tissues not exposed to sunlight and underground portions of plants contain numerous leucoplasts. Functionally, leucoplasts serve as starch-forming and -storing centers in plants.

LYSOSOMES. Lysosomes are fairly recently discovered organelles (1952) that have been conclusively demonstrated in animal cells. When viewed under the electron microscope, a lysosome appears as a spherical membrane-enclosed structure somewhat smaller than a mitochondrion (Figure 5-26). Unlike mitochondria, a lysosome has a single external membrane and lacks internal structure. It apparently functions as a storage vesicle for digestive enzymes. In this regard, the membrane of the lysosome is probably impermeable to the outward movement of these enzymes as well as resistant to their action. It is generally believed that lysosomes store enzymes formed on ribosomes in the ER and then fuse with food vacuoles where digestion takes place. Lysosomes are particularly prominent in phagocytic protozoans.

Presumably lysosomes also function in the digestion of damaged cell parts with the products of digestion being utilized in the mitochondria for energy release or in other organelles as building materials. It is interesting to note that in some cells, such as those of the tail of a tadpole undergoing metamorphosis to a frog, numerous lysosomes are present. They apparently facilitate the orderly destruction (autolysis) of these cells during metamorphosis.

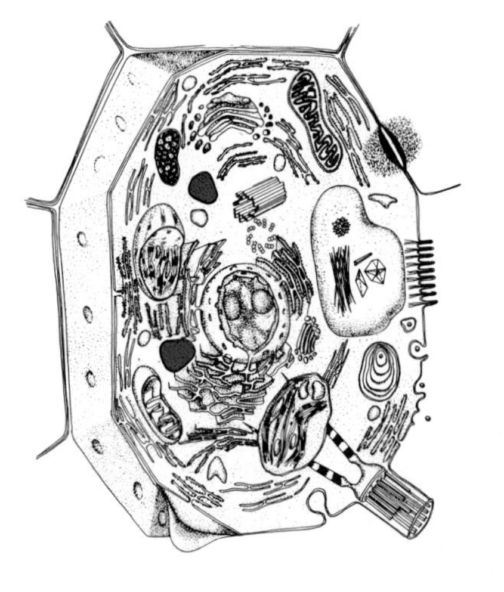

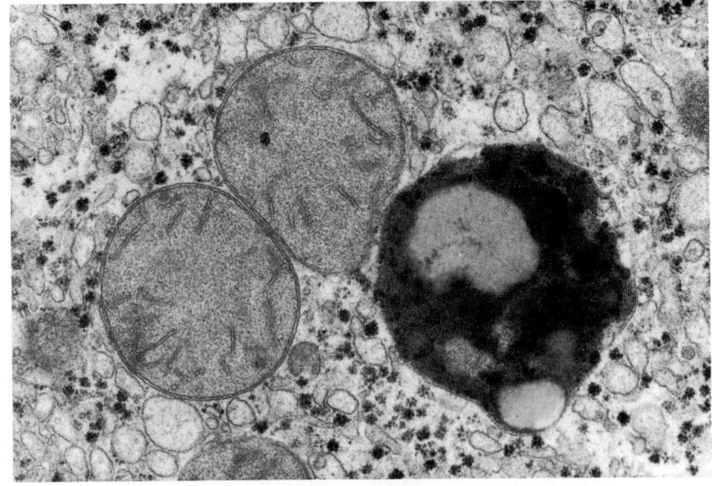

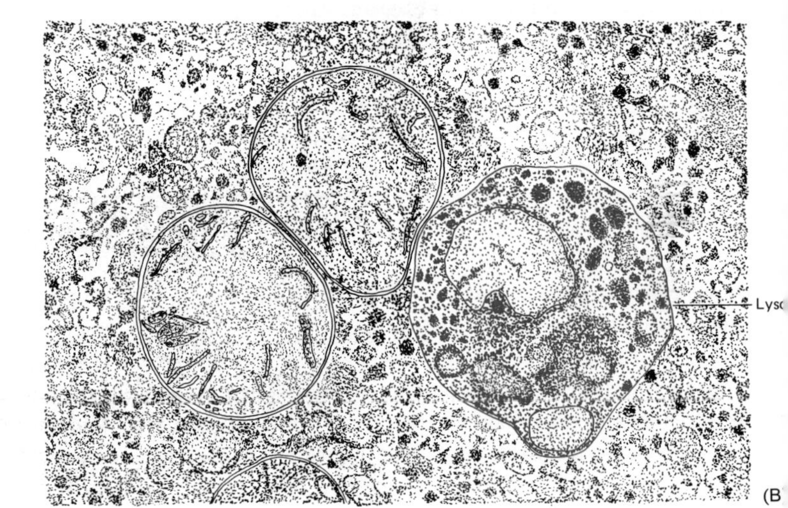

Figure 5-26. The lysosome. (**A**) Electron micrograph of a lysosome in a liver cell of *Homo sapiens* (man). 26,700×. The grey inclusions are probably neutral lipids and the darker ones correspond to lipofuscin. (**B**) Diagrammatic interpretation of a lysosome. [**A** *courtesy of Dr. F. Van Hoof, Université Catholique de Louvain.*]

Microtubules. **Microtubules,** the last organelles to be considered, have recently been recognized as a cytoplasmic constituent of widespread occurrence in both plant and animal cells. In electron micrographs, they appear as thin tube-like structures. They are especially common in the center of the cell where they are often closely related to the centrioles. These structures form the spindle apparatus of dividing cells (Figures 5-27 and 5-28) and have been demonstrated in the tail of many sperm cells, in axons, and in the cytoplasm of many other cell types. They are usually interpreted as cytoskeletal elements that assume a role in maintaining cell shape by imparting stiffness to certain areas of the cell. They may also be involved in the internal movements of the cytoplasm and alterations in a cell shape.

Cell Inclusions

Cell inclusions constitute a rather diverse group of materials that are produced as a result of cellular activities. These substances, much simpler in structure than organelles, are mostly organic in nature and may appear and disappear at different times in the life of a cell. Some are stored as reserve materials, others are products of biochemical processes, and many assume characteristic forms in the cell that are easily distinguished. Among animal cells, inclusions in the form of stored secretory granules, droplets, or crystals appear as zymogen granules (pancreas), mucus (intestinal epithelium), melanin (a pigment of the skin, hair, and eyes of mammals), glycogen (stored polysaccharide in the liver), lipids (fat cells), and hemoglobin crystals (red blood cells). In plant cells, inclusions are quite conspicous. A few of them are described below.

VACUOLES. **Vacuoles** are the most outstanding inclusions of plant cells as well as a distinctive plant cell feature. They are found in the cytoplasm of mature plant cells and to a lesser extent in unicellular organisms. Structurally, vacuoles are membrane-bound (the *tonoplast*), fluid-filled spaces (see Figure 5-4). In protozoans, they are specialized as contractile vacuoles that facilitate the expulsion of excess water and some waste products. Many unicells also possess food vacuoles. Similar vacuoles containing food also formed by a number of cells that take in materials by pinocytosis or phagocytosis.

In many mature plant cells, the vacuole is a region containing a highly dilute solution called *cell sap*. The major component of vacuoles is water. Among the dissolved substances are atmospheric gases, salts, sugars, organic acids, soluble proteins, and certain pigments. The most prevalent pigments are *anthocyanins*. The red color of many flowers, such as the rose, is a result of pigments being concentrated in the vacuoles of the cells in the flower petals. Anthocyanin pigments dissolved in the sap are also responsible for the red, blue, and purple colors of autumn leaves, flowers, fruits, and stems.

STARCH GRAINS AND CRYSTALS. Another important inclusion of plant cells is represented by *starch grains*. Glucose, produced during photosynthesis, is converted into starch and stored in plastids (leucoplasts and chloroplasts) as starch grains (see Figure 5-4).

Crystals are inorganic deposits in plant cells consisting chiefly of calcium salts, the most common crystals being those of *calcium oxalate*. The chemical union of oxalic acid (an organic acid) with calcium produces the calcium oxalate crystals. These crystals assume a number of forms in plants. Some are needle-like and are termed *raphides;* some form clusters and are designated *druses;* others appear as single crystals (see Figure 5-4). In addition to calcium oxalate crystals, crystals of calcium carbonate, calcium sulfate, silica, and protein also occur.

OTHER INCLUSIONS. Proteins, the major constituents of cytoplasm, may also occur as temporary inclusions in plant cells. Two examples are *gluten* and *aleurone grains,* which are stored proteins found in the cells of seeds such as wheat. Lipids also appear as inclusions, being found as *oil globules* in the cytoplasm. Protein, lipid, and starch inclusions are formed when excess foods are produced by the plant cell, and they may be digested and utilized during periods of minimum food synthesis.

Less frequently found inclusions are tannins, resins, gums, and mucilages. It is believed that all are waste products of the physiological activities of plant cells and probably protect the protoplast against desiccation, decay, and injury by animals.

Cellular Reproduction

When a cell divides, the hereditary instructions in the configuration of DNA must be replicated so that a duplicate copy is passed on to cells of succeeding generations. In this way, newly formed cells will inherit the same DNA of the parent cells. This is essential for the continuity of life. In this context, we will examine one of the basic processes of living cells called cell division. Cell division gives rise to cells that are identical to those from which they arose. A special nuclear division, called meiosis, is a process in which egg and sperm cells are formed. Inasmuch as this process is concerned with organismic continuity, it will be discussed along with other reproductive patterns as an organismic process (Chapter 17) and not as a cellular phenomenon.

The vast majority of eucaryotic cells are capable of precisely dividing and distributing their genetic material and dividing their cytoplasm by a process called **cell division.** Although the essentials of cell division are similar in all eucaryotic cells, certain details are conspicuously different in cells of animals and plants. We will therefore discuss cell division as it occurs in animals (Figure 5-27) and note the dissimilar details of plant cell division (Figure 5-28).

Cell division consists of two more or less simultaneous processes; the first of these, **mitosis** or **karyokinesis,** involves the replication and distribution of chro-

Figure 5-27. Photomicrographs and diagrammatic representations of whitefish eggs in the process of cell division. (**A**) Parent cell in interphase. (**B**) Early, middle, and late prophase. (**C**) Metaphase. (**D**) Early and late anaphase. (**E**) Telophase. (**F**) Daughter cells in interphase. [*Photomicrographs courtesy of Carolina Biological Supply Co.*]

(A)
INTERPHASE

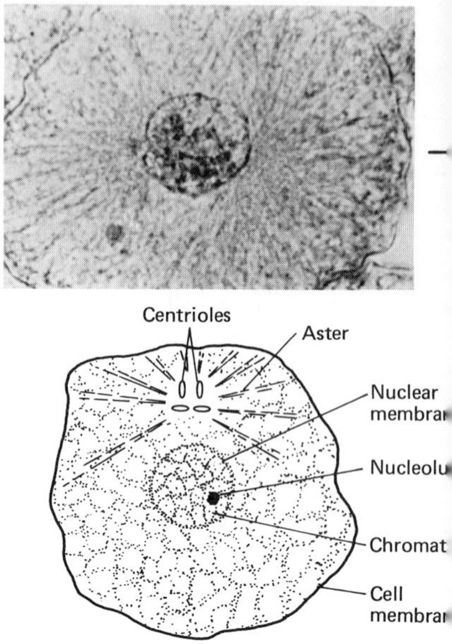

Centrioles
Aster
Nuclear membra
Nucleolu
Chromat
Cell membra

Nuclear membrane and nucleolus visible
Genetic material (chromatin) appears as irregularly shaped granular mass
Centrioles self-replicate

(F)
DAUGHTER CELLS

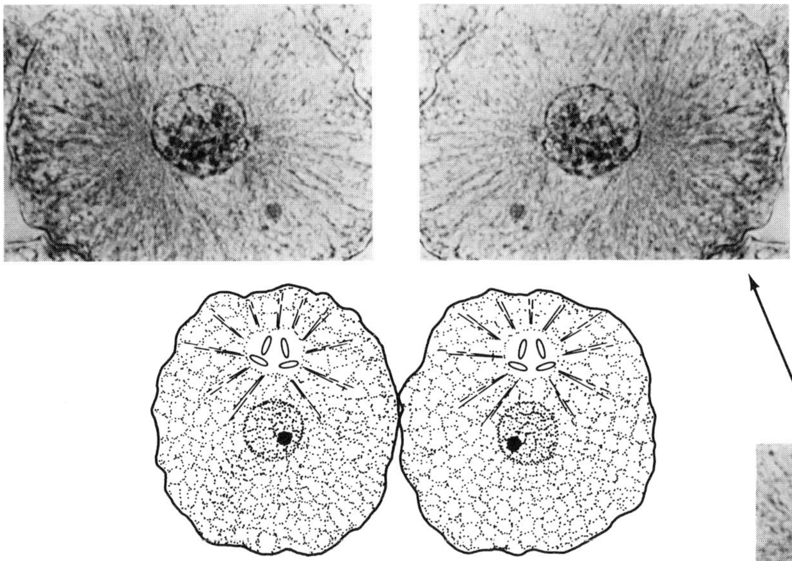

Nuclear membrane and nucleolus clearly visible
Genetic material appears as chromatin
Cytokinesis complete

(E)
TELOPHASE

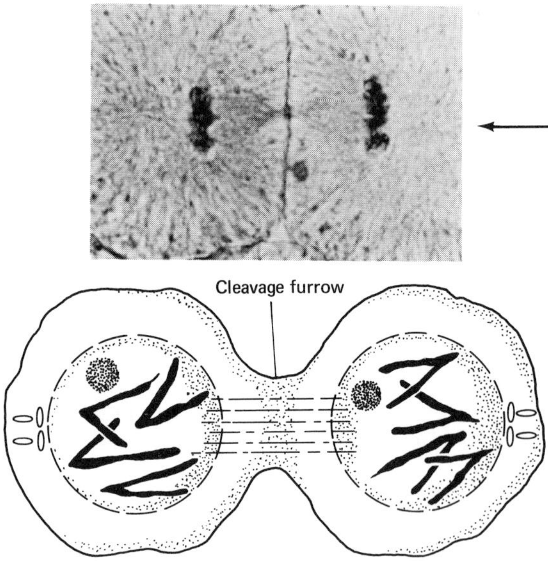

Cleavage furrow

Nuclear membranes reappear and enclose chromosomes
Spindle fibers disappear
Nucleoli reappear
Chromosomes become thread-like and less distinct
Centrioles replicated
Cytokinesis almost complete

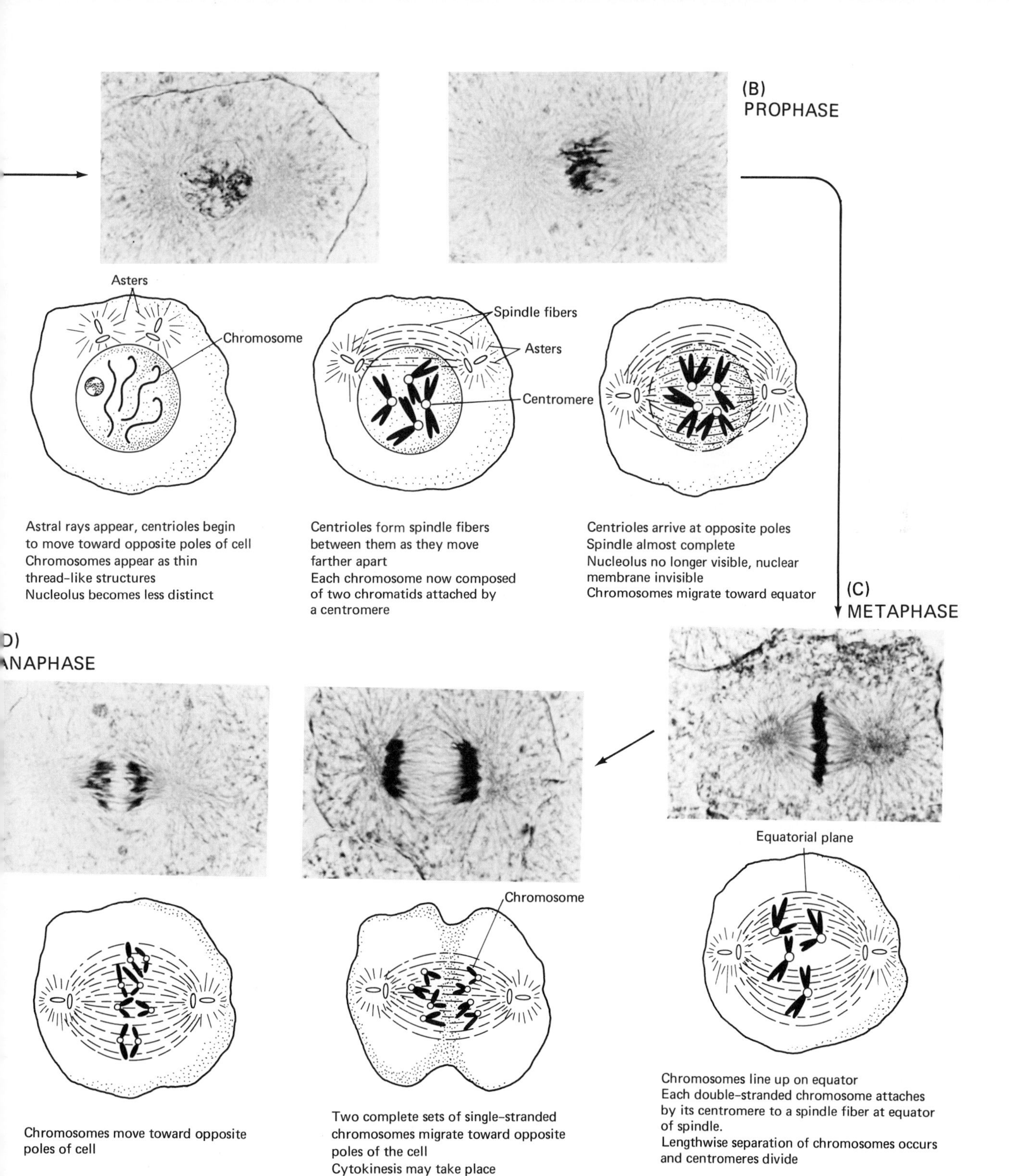
(B)
PROPHASE
Asters
Chromosome
Spindle fibers
Asters
Centromere
Astral rays appear, centrioles begin to move toward opposite poles of cell
Chromosomes appear as thin thread-like structures
Nucleolus becomes less distinct
Centrioles form spindle fibers between them as they move farther apart
Each chromosome now composed of two chromatids attached by a centromere
Centrioles arrive at opposite poles
Spindle almost complete
Nucleolus no longer visible, nuclear membrane invisible
Chromosomes migrate toward equator
(C)
METAPHASE
D)
ANAPHASE
Equatorial plane
Chromosome
Chromosomes line up on equator
Each double-stranded chromosome attaches by its centromere to a spindle fiber at equator of spindle.
Lengthwise separation of chromosomes occurs and centromeres divide
Chromosomes move toward opposite poles of cell
Two complete sets of single-stranded chromosomes migrate toward opposite poles of the cell
Cytokinesis may take place

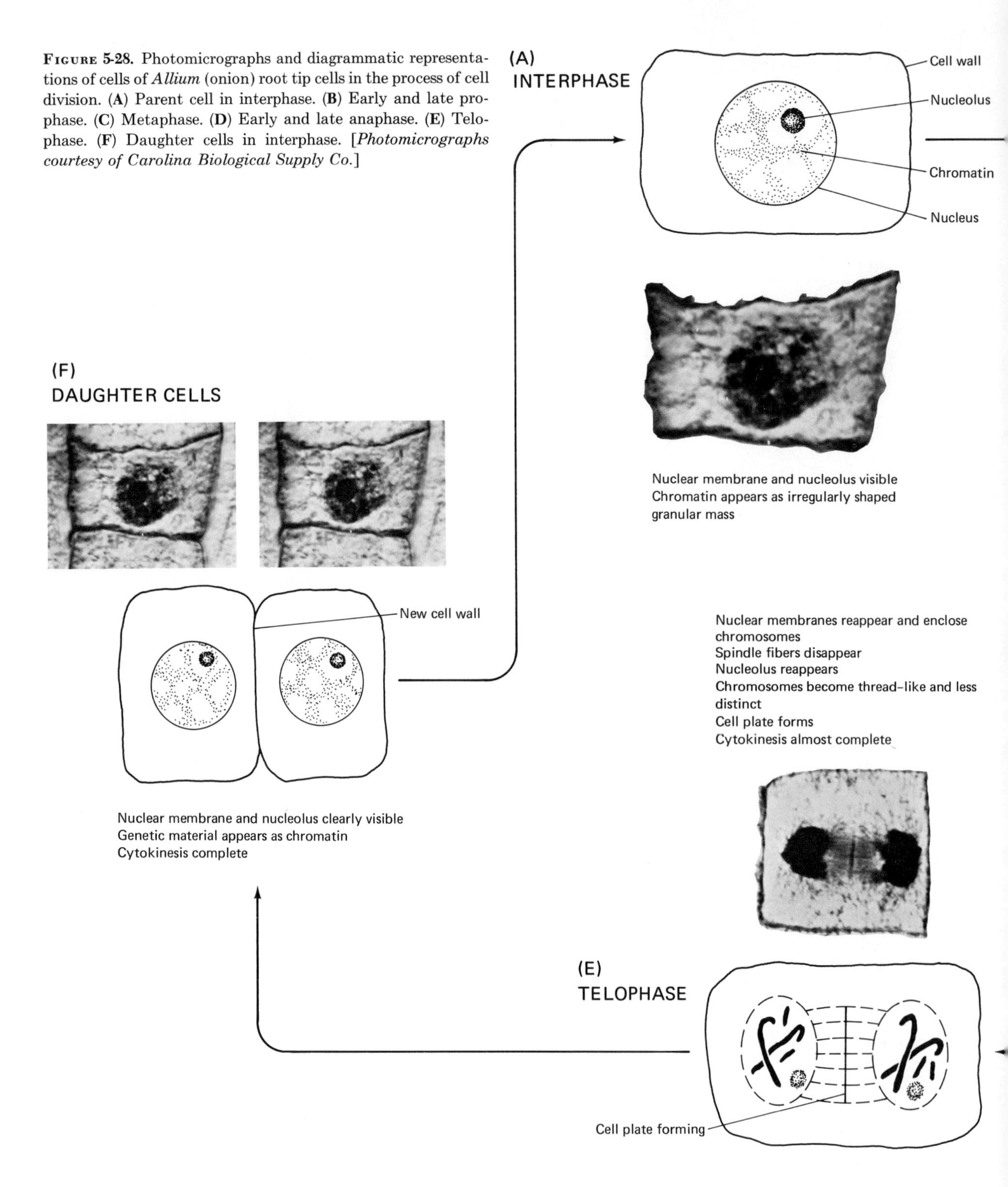

FIGURE 5-28. Photomicrographs and diagrammatic representations of cells of *Allium* (onion) root tip cells in the process of cell division. (**A**) Parent cell in interphase. (**B**) Early and late prophase. (**C**) Metaphase. (**D**) Early and late anaphase. (**E**) Telophase. (**F**) Daughter cells in interphase. [*Photomicrographs courtesy of Carolina Biological Supply Co.*]

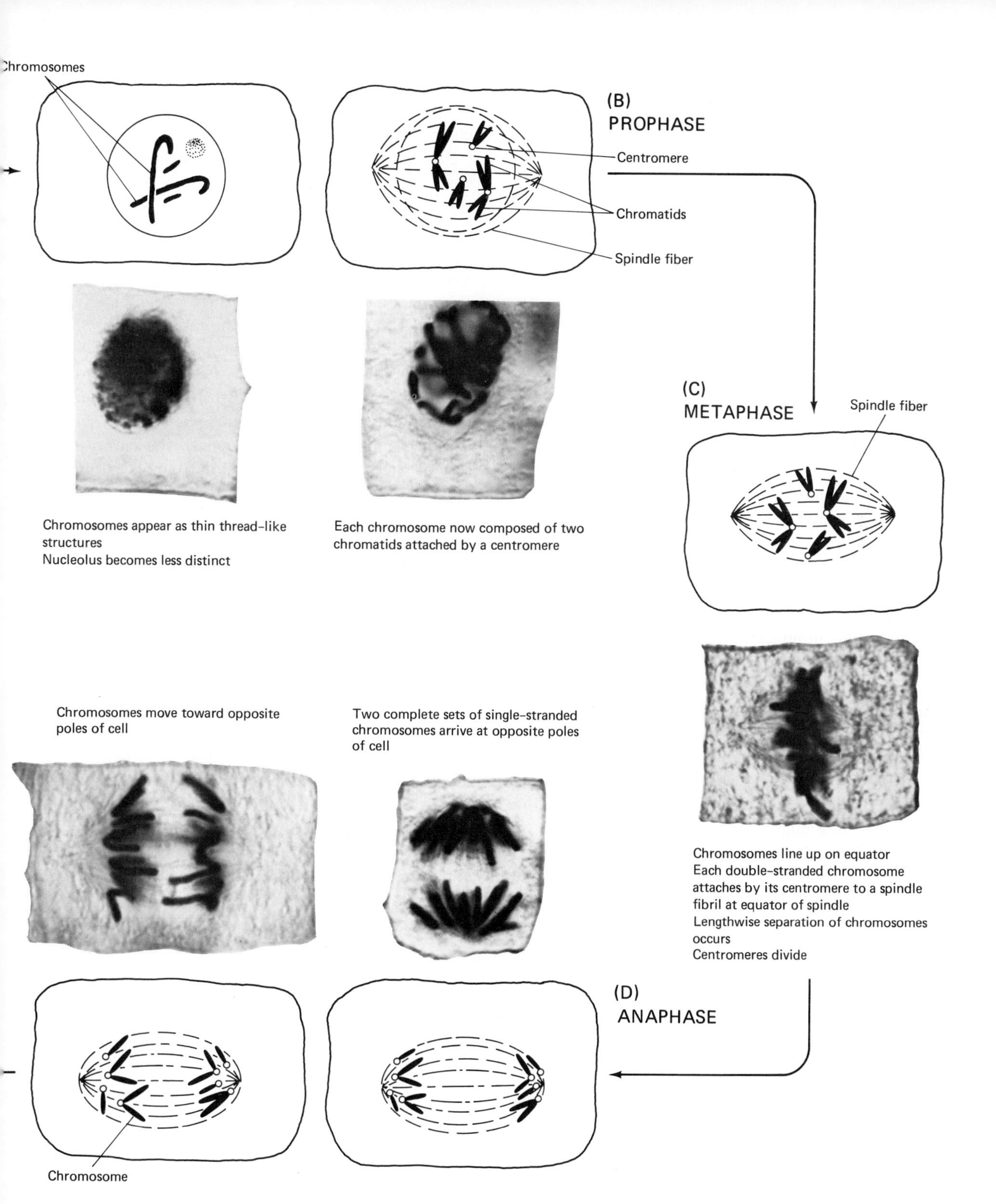

Chromosomes
(B)
PROPHASE
Centromere
Chromatids
Spindle fiber
Chromosomes appear as thin thread-like structures
Nucleolus becomes less distinct
Each chromosome now composed of two chromatids attached by a centromere
(C)
METAPHASE
Spindle fiber
Chromosomes line up on equator
Each double-stranded chromosome attaches by its centromere to a spindle fibril at equator of spindle
Lengthwise separation of chromosomes occurs
Centromeres divide
Chromosomes move toward opposite poles of cell
Two complete sets of single-stranded chromosomes arrive at opposite poles of cell
(D)
ANAPHASE
Chromosome

mosomes into two separate and equal nuclei. The second process, **cytokinesis,** is concerned with the division of the cytoplasm into separate components.

Interphase

When a cell is about ready to divide, it is said to be in the interphase stage (Figures 5-27A and 5-28A). In this condition, the longest phase in the life of a cell, there is a great deal of metabolic activity. Biochemically, the cell is synthesizing most of its RNA and proteins, replicating its DNA, and producing sufficient chemical compounds so that all cellular components can be doubled at a later stage of division. Morphologically, the interphase cell consists of a definitive membrane-bound nucleus containing nucleoli and karyolymph. Distinctive chromosomal bodies are not evident. Rather, the DNA of the nucleus appears as an irregularly shaped granular mass, the chromatin. Chemical and genetic data indicate the presence of chromosomes during interphase, but because of limitations of present optical techniques they can be discerned only as a long, thin, intertwined network.

Animal cells in interphase, but only a relatively few plant cells, contain centrioles, which self-reproduce, move apart, and ultimately organize the mitotic apparatus of a potentially dividing cell. The movement of chromosomes depends on the activities of centrioles, which serve as mitotic centers, or potential poles. In the majority of animal cells, the centrioles separate just prior to the initiation of mitosis; however, in certain other cells their migration takes place well before the onset of mitosis. Both plant and animal cells have mitotic centers, but structures are seen only in animal cells; their presence in cells of higher plants has not been universally demonstrated. However, centrioles are associated with sperm production in many lower plant groups. Once a cell completes its activities associated with interphase, mitosis commences.

Stages of Mitosis

It is customary to subdivide the process of mitosis into four stages or phases. In nature, mitosis is a continuous process so that the arbitrarily designated phases should not be interpreted as a sequence of discontinuous occurrences. These stages, in order of appearance, are prophase, metaphase, anaphase, and telophase.

Prophase. **Prophase** is the earliest recognizable stage of mitosis. During this stage in the animal cell (Figure 5-27B), the mitotic apparatus is formed. The centrioles duplicate, separate, and as they move apart, project a series of radiating fibers referred to as *asters*. The centrioles and associated asters migrate to opposite poles of the cell and the centrioles connect by another system of fibers, forming what is designated the *spindle apparatus*. The cells of most higher plants develop spindles even though they lack centrioles; no asters are formed, however (Figure 5-28B). In the meantime, the formerly indistinct chromosomes shorten, thicken, and become compact. By late prophase the chromosomes assume their characteristic rod-like configurations, the nucleoli become less and less distinct, finally disappearing, and the nuclear membrane disappears.

Close examination of a chromosome in late prophase indicates that it is not a single rod, but rather two separate strands called *chromatids*. The chromatids are attached to each other by a small spherical body, the *centromere*. Chromatids are formed during interphase in which one chromatid serves as a template in producing the other. Essentially, the paired chromatids of a phrophase chromosome are identical.

The last major event of prophase and beginning of metaphase is characterized by the movement of randomly distributed chromosomes within the nucelus to the equator of the spindle. Once this movement has occurred, the chromosomes become oriented midway between the two centrioles and at approximately right angles to the spindle fibrils. This alignment of chromosomes signals the end of prophase.

Metaphase. During **metaphase** (Figures 5-27C and 5-28C), the chromosome pairs are arranged on the equatorial plane of the spindle. Specifically, the centromeres line up along the equatorial plane and each double-stranded chromosome becomes attached by its centromere to a spindle fibril. The lengthwise separation of the chromatid pairs takes place, each centromere divides, and entirely independent chromosomes are produced. Thus, the total number of independent chromosomes in the nucleus is doubled.

Anaphase. **Anaphase** (Figures 5-27D and 5-28D)

is characterized by the movement of one complete set of chromosomes to each pole of the cell. One set of chromosomes migrates away from the metaphase plane toward one spindle pole (centriole), and an identical set migrates in the opposite direction toward the other pole. During the migration, the centromeres attached to spindle fibrils lead, while the arms of the chromosomes trail like streamers. Experimental evidence indicates that a chromosome cannot move during anaphase unless its centromere is connected to a spindle fibril. It is not uncommon for cytokinesis to occur during a late anaphase stage.

TELOPHASE. Telophase (Figures 5-27E and 5-28E) is the last recognizable stage of mitosis. Basically, the events of telophase are described as the opposite of prophase. Once the identical chromosome sets reach their respective poles, new nuclear membranes enclose them and the spindle fibrils disappear. In addition, the chromosomes begin to assume their interphase chromatin form, nucleoli reappear, and the centrioles of each nucleus usually replicate themselves. Cytokinesis, which will be discussed shortly, is also frequently completed during late telophase. The appearance of two nuclei similar to those of cells in interphase terminates telophase and thus completes a mitotic cycle (Figures 5-27F and 5-28F).

The salient features of mitosis may be summarized as follows.

1. It brings about an increase in the number of cells.
2. Each new (daughter) cell has the same number and kind of chromosomes as the original (parent) cell.

As a result, the daughter cells have the same hereditary material and genetic potential as the parent cell.

Cytokinesis

The term cytokinesis refers to the division of the cytoplasm, a process that often begins in late anaphase and terminates during telophase. In the vast majority of cells, mitosis and cytokinesis are more or less simultaneous processes that result in the production of two nucleated cells with approximately equal portions of cytoplasm. In some cells, however, such as those of fungi, slime molds, and vertebrate skeletal and cardiac muscle, mitosis is not accompanied by cytokinesis. As a result, *coenocytic cells* (many separate nuclei without intervening membranes) are formed.

Cytokinesis in animal cells (Figure 5-27E) typically begins with the formation of a cleavage furrow which runs around the cell and appears at the equator, perpendicular to the spindle. As the furrow spreads inward as a constricting ring, it cuts completely through the cell and the spindle apparatus, forming two new cells. In plant cells, the details of cytokinesis are somewhat different (Figure 5-28E). One probable reason for this is that the cell wall of plants, unlike the membrane of animal cells, is a fairly rigid structure that resists constriction. As the chromosomes approach the poles, the continuous spindle fibrils change; on the region of the equatorial plane, they become dense and fibrillar in appearance. This region is termed the *phragmoplast.* Small vesicles, which may be products of the Golgi apparatus, appear to migrate to the phragmoplast where they fuse to form a *cell plate.* As additional materials (cell wall constituents contained in the vesicles) are deposited on the cell plate, it slowly becomes larger until its edges reach the outer surface of the cell and the cell contents are divided in two. Thus, whereas the cleavage furrow of animal cells develops from the periphery to the middle of the cells, the cell plate of plant cells progresses from the middle to the periphery of the cells.

Some Details of Cell Division

In our discussion of cell division thus far, the general features of the process have been described. It should be made clear that many details cannot yet be explained or understood at the molecular level. However, a brief discussion of some of the problems may illustrate the nature of these unsolved problems and some of the techniques that are employed to elucidate them. We will consider (1) the replication of DNA, (2) the origin and nature of the spindle, (3) the movement of chromosomes, and (4) cytokinesis.

THE REPLICATION OF DNA. A cell passes through four phases as part of its complete growth and divisional cycle. Three of these are concerned with activities prerequisite to mitosis, the fourth involves the actual mitotic process. Many cells of both plants and animals studied in tissue culture show the life cycle durations represented in Table 5-2.

The G_1 phase, which lasts from 10 to 20 hours, is a period of growth during which most RNA and protein

TABLE 5-2. Typical Durations of the Phases of Growth and Division

Phase	Duration (hours)
G_1	10–20 (usually less than 50% of the total)
S	6–8
G_2	1–4 or more (usually less than 20% of the total)
M	1
Total	18–33 hours

synthesis takes place. The S period (6–8 hours) is a phase of DNA synthesis; the G_2 period (1–4 hours), like the G_1 period, involves the synthesis of cellular molecules which will be used to double the amount of cellular constituents during mitosis; and the M period, which lasts about 1 hour, is the period of mitosis. Throughout the G_1, S, and G_2 periods, the cell is in interphase. The details of DNA replication, a principal activity of the S phase, are discussed in detail in Chapter 8.

THE SPINDLE: ORIGIN AND NATURE. At or near the beginning of prophase, as the centrioles move apart, a group of fibers radiate from each centriole like the spokes from the hub of a wheel. Some of these fibrils (astral fibrils) radiate away from the center of the cell and end blindly; other fibrils (spindle fibrils) link up with fibrils from the other centrioles to form continuous links between the centrioles. The characteristic spindle, which is best seen in metaphase, seems to provide the machinery for the movement of chromosomes. If the spindle is destroyed, the movement of chromosomes to the poles does not take place.

Various techniques employed in isolating the spindle from metaphase cells of sea urchin eggs have clearly established its physical existence. In addition, recent electron micrographs clearly show the presence of individual spindle fibrils which appear to be microtubules. Chemically, the spindle material appears to consist mostly of protein which can be separated into two components. The major one, about 70%, has a molecular weight of approximately 50,000 and is believed to be the chief structural protein. The other component, a minor fraction, is of large molecular size and is as yet of unknown significance.

The amino acid composition of isolated spindle proteins has also been ascertained. The relative concentrations of the various amino acids rather closely approximate those of the muscle protein actin and the proteins of certain flagella. Moreover, spindle fibers shorten, as do actomyosin and other contractile proteins. These and other similarities may indicate that various types of biological motility have a common biochemical basis at the cellular level.

Spindle protein represents a large amount of the total protein found in the cell (more than 11% in some cells). Studies of radioactive sea urchin eggs suggest that proteins of the spindle are among the first proteins to be synthesized before prophase, and they apparently aggregate to form the spindle which becomes visible later in prophase. The exact mechanism by which spindle fibrils organize and break up is not known.

THE MOVEMENT OF CHROMOSOMES. Despite some of the elaborate physical descriptions that have been written about chromosome movements, the nature of the process is still obscure. Whatever the mechanism, it must account not only for the migration of chromosomes at metaphase, but also for their movement to each pole of the cell at anaphase. Some of the early theories of chromosome movement purported such explanations as magnetic fields, cyclosis, mechanical pushing or pulling forces, and attraction or repulsion by oscillation.

One of the most widely held theories is that chromosomes are pulled toward the poles by the fibrils to which they are attached. During chromosomal migrations, portions of chromosomes may break off and it can be shown that only those segments of chromosomes that contain centromeres and remain attached to fibrils move toward a pole. Fragments without centromeres or fibrils do not take part in mitosis. Such an observation implicates not only fibrils but also centromeres. A further statement of this theory holds that centromeres, the sites at which fibrils are attached to chromosomes, usually move steadily, with the arms of the chromosomes frequently dragging along behind. This situation is analogous to a limp thread being pulled through a liquid medium. The moving chromosomes thus appear as J- or V-shaped structures, depending upon the location of the centromeres. It was hypothesized that if the pulling theory of chromosome movement, or an alternative variation of it was correct, portions of the fibrils between the chromosomes and the poles would contract in a fashion similar to muscle contraction.

Daniel Mazia and co-workers at the University of California have done a great deal of experimentation in attempting to discover the factors involved in chromosome movement. In this connection, he has found an enzyme in the mitotic apparatus that is capable of splitting ATP, just as the myosin in muscle does. Moreover, it has been found that the chromosome-to-pole fibrils do shorten to a fraction of their original length. As the fibrils become shorter or longer, however, they do not become thicker or thinner. Electron micrographs reveal that the fibrils retain a uniform diameter even though they grow shorter or longer. Such an observation is apparently inconsistent with the theory of muscle contraction. As a result, an alternative theory has been proposed by Shinya Inoué of the Dartmouth Medical School. According to his model shortening is due to the actual removal of molecules from the chain of which the fibril is composed and elongation is due to the addition of molecules in the fibril.

Cytokinesis. As with mitosis, the mechanism of cytokinesis is uncertain and there are several current theories to account for the phenomenon. Most are associated with the cleavage furrow of the animal cell. For example, one hypothesis suggests that the cell surface forms a contractile belt around the equator that pinches the cell in two. A second hypothesis holds that the membrane expands at the poles, but not at the equator, thus pushing the equator inward to form a furrow. The presence of asters in animal cell leads to the theory that the astral rays attach to the cell surface and pull the surface inward to form the cleavage furrow. Observed cyclosis has led to the theory that the furrow forms as inner cell contents stream away from the equator.

Any attempt to explain cytokinesis must account for the manner in which the poles of the mitotic apparatus control the formation of a partition between cells, whether it be a cleavage furrow or a cell plate. It has been shown that if the entire mitotic apparatus is removed from sea urchin eggs before the cells begin to divide, no division takes place. If, however, the apparatus is removed at anaphase, then division does occur. The removal or alteration of position of the apparatus before anaphase, thus blocking or altering cytokinesis, suggests its role in the control of cytokinesis. Cytokinesis apparently does not depend on the presence of chromosomes. Experiments have shown that even after chromosomes have been removed, the cell division proceeds normally and cytokinesis occurs. In short, no adequate description of cytokinesis is possible at this time. If, however, cytokinesis is fundamentally the same in plants and animals despite their superficial differences, any unified theory must avoid furrows and asters and rely on less conspicuous internal events.

Duration of the Stages of Mitosis

Some of the technical aspects of measuring the amount of time required for the various stages of mitosis have been alleviated by the use of time-lapse photography. Such observations have demonstrated that the duration of mitosis varies not only among cells of different organisms but among the cells of different tissues of the same organism. The time required for a complete divisional cycle varies with the type of cell, its location, and the effects of certain factors such as temperature. In general, an increase in temperature increases the rate of mitosis. For example, at 59°F the root cells of pea plants undergo a complete divisional cycle in about 25 hours. If the temperature is increased to 77°F, the cycle is completed in about 16 hours.

The various stages of mitosis are not all equal in duration and show considerable variations. Prophase is usually the longest stage, lasting from one to a few hours. Metaphase is much shorter and requires about 5–15 minutes. Another short stage is anaphase, lasting about 2–15 minutes. Finally, the telophase may vary from 10 to 30 minutes. These figures indicate only the relative lengths of time for the various stages of mitosis and should not be interpreted as exact limits.

Whereas some cells require only several minutes for a mitotic cycle, others require several hours. For example, fruit fly egg cells require only about 7 minutes for the entire divisional cycle (prophase, 4 minutes; metaphase, 0.5 minute; anaphase, 1.2 minutes; and telophase, 0.9 minute). The cells in spiderwort staminal hairs, on the other hand, require 340 minutes (prophase, 181 minutes; metaphase, 14 minutes; anaphase, 15 minutes; and telophase, 130 minutes).

Exceptions to the Typical Divisional Pattern

The pattern of cell division described here is characteristic of most eucaryotic cells, both plant and animal. However, in addition to the differences already noted (presence or absence of asters, centrioles, and cleavage furrow and cell plate formation), there are numerous important exceptions to this typical pattern. Indeed, almost every conceivable modification can and does occur in some organism. For example, in many protozoans, the nuclear membrane does not disappear during prophase and the spindle apparatus forms inside the nucleus. The vast majority of chromosomes have a single distinct centromere, but a few plants and animals have chromosomes that lack centromeres; these chromosomes have numerous points of spindle attachment instead.

Significance of Cell Division

The primary significance of cell division is to maintain the continuity of metabolism by transmitting identical copies of the coded instructions in DNA to daughter cells. In a large number of unicellular organisms, cell division is equivalent to reproduction of the entire organism. Newly produced daughter cells generally separate, but in some forms, such as *Volvox* (see Figure A-9, in the Appendix) and *Nostoc* (see Figure A-2), they remain together after division and give rise to colonies.

In multicellular organisms, cell division either contributes to cell replacement or adds to cell number. For example, during each 24-hour period of adult life, the average human loses approximately 500 billion cells from various parts of the body. Cells that have a short life span such as those of the outer layer of skin, the digestive tract, and the cornea of the eye are continually replaced by cell division. Unless lost cells are replaced, a significant decrease in size would become apparent after a period of time.

In those multicellular forms that are produced from the fertilization of a single cell, subsequent divisions represent the means of growth and development for the maturing organism. Essentially, the approximately 100 trillion cells that comprise the body of an adult human are the products of repeated cell divisions of a single egg.

6 Cells and Energy: Procurement and Transduction

GUIDE QUESTIONS

1 What are the major purposes of the procurement and transformation of the chemical constituents of living organisms? What is metabolism? Distinguish between anabolism and catabolism.
2 Define an enzyme. List the six principal classes of enzymes, give an example of each, and tell what the enzyme accomplishes.
3 What is the purpose of photosynthesis? What conditions are necessary for photosynthesis to occur? Discuss the fate of the raw materials in photosynthesis. What is the wave theory of light?
4 What is meant by the visible spectrum? Relate it to photosynthesis. What happens to light striking the surface of a leaf? Define an absorption spectrum.
5 Distinguish between the dark reactions and the light reactions of photosynthesis. What is the role of chlorophyll in photosynthesis? Explain in detail. Define photophosphorylation.
6 Outline the steps of cyclic photophosphorylation and noncyclic photophosphorylation. How are they alike? How do they differ? What is the outcome of each?
7 What is a coenzyme? How are coenzymes related to oxidation-reduction reactions? Give specific examples and describe the function of each in energy metabolism.
8 Discuss the research that demonstrated that PGAL is one of the first stable products of the dark reactions of photosynthesis. What is carbon dioxide fixation? Outline the details of the dark reactions of photosynthesis.
9 What is the role of PGAL in the dark reactions? How are the dark reactions related to the light reactions of photosynthesis? Compare photosynthesis and respiration.
10 How is the ATP cycle related to photosynthesis and respiration? Distinguish between anaerobic and aerobic respiration and discuss how they are related to carbohydrate metabolism.
11 Outline the major steps of glycolysis, alcoholic fermentation, lactic acid formation, and the Krebs cycle.
12 What are the two major functions of the Krebs cycle? What is the cytochrome system? Compare fermentation and aerobic respiration with regard to ATP yield.
13 What is the significance of the hexose monophosphate shunt? How are fats broken down? Relate the lipid metabolic pathways to the Krebs cycle.
14 What are the alternative pathways of pyruvic acid in glycolysis? Define deamination, transamination, intermediary metabolism, and beta oxidation.
15 Summarize the relationship of the Krebs cycle to other phases of energy metabolism.

WE NOW turn to physiological aspects of cells. The primary concern of this chapter is to examine the flow of energy through cells, that is, the transduction of environmental energy into a biologically useful form. Energy consumption is a characteristic of life. Living organisms are highly ordered systems, and they must expend energy to maintain their orderliness. All biological energy, regardless of its form, comes initially from the sun. The chloroplasts in autotrophic cells capture the energy of solar radiation and use it to synthesize sugar molecules from CO_2 (carbon dioxide) and H_2O (water). The mitochondria of autotrophic and heterotrophic cells break down the sugar molecules to form ATP (adenosine triphosphate). The energy of ATP is then utilized by cells for various activities.

The sum total of matter and energy transformations is called cellular **metabolism.** For convenience of study, the many and varied aspects of metabolism may be divided into two categories: **anabolism** and **catabolism.** Anabolic reactions are concerned with the synthesis of a variety of molecules and organelles from simpler well-defined starting materials. Typically, all such reactions absorb energy; they are *endergonic.* Catabolic reactions, by contrast, are involved in the degradation of chemical substances and organelles. The starting materials vary greatly, but ultimately yield simple well-defined products such as carbon dioxide, water, and urea. These reactions are *exergonic;* that is, they release energy. The energy and matter relationships of cellular metabolism are illustrated in Figure 6-1. The process of photosynthesis, a unique anabolic reaction, and respiration, a basic catabolic reaction, will be discussed shortly as they occur at the cellular level.

It was noted in Chapter 2 that enzymes are a special group of proteins that catalyze the biochemical reactions occurring in living matter. In that discussion the proposed mechanism of enzyme action, the distinctive characteristic of enzymes, and the composition of enzymes were emphasized. At this point the classes of enzymes and their actions will be outlined.

Enzymes, whose names typically end in *-ase,* are usually named according to three criteria: (1) the specific substrate with which they react, (2) the type of substrate transformation, and (3) the group of related substrates with which they react. In most cases the root represents the criterion according to which the enzyme is named, and the suffix (-ase) designates the compound as an enzyme. For example, the enzyme *pyruvate decarboxylase* removes the carboxyl group (type of transformation) from pyruvic acid (substrate) by removing the carbon dioxide. Enzymes are classified in various ways. One scheme, which classifies enzymes into six groups, is summarized in Table 6-1.

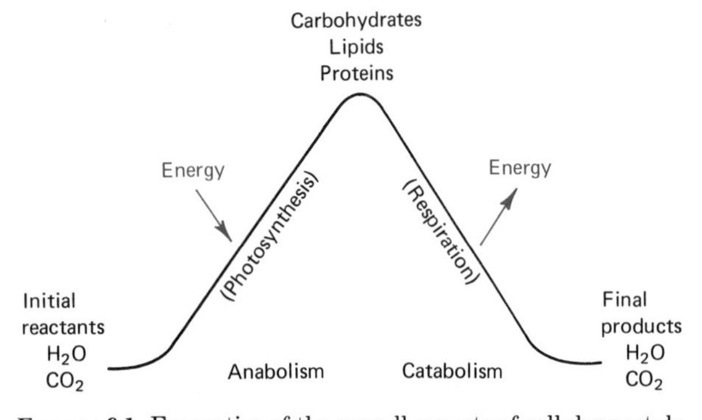

FIGURE 6-1. Energetics of the overall aspects of cellular metabolism. Anabolic reactions, such as photosynthesis, require energy. This energy may become incorporated into the chemical bonds of food molecules (carbohydrates, lipids, and proteins). Catabolic reactions, such as respiration, release the energy of food molecules and convert it into ATP, a form that can be utilized by the cell for all of its metabolic reactions.

Photosynthesis—The Acquisition of Energy

Significance of Photosynthesis

Chlorophyll-containing cells, in the presence of light, are capable of transforming CO_2 and H_2O into carbohydrates and O_2 by a fundamental anabolic process called **photosynthesis.** This process, by which solar energy is transduced into chemical energy, is the single most important chemical process in the world. The carbohydrates produced are not only utilized by plants for various activities but are also used by animals, including humans, for their metabolic processes. In other words, all heterotrophs depend, either directly or indirectly, upon autotrophs for their food. It is this almost unlimited capacity for making their own foods that gives chlorophyllous cells so prominent a place in the physical and biological environments and makes them the connecting link between the inorganic (nonliving) and organic (living) worlds.

Historical Development and General Nature

The discovery of photosynthesis goes back some 200 years, and since then, all data and knowledge of the process have been derived from a long series of tedious, piecemeal researches. The first significant clue regarding photosynthesis came in 1772 when Joseph Priestley observed that oxygen was given off by green plants. Seven years later, Jan Ingen-Housz, a Dutch physician and chemist, re-examined the work of Priestley and further observed that oxygen is evolved by green plants only if they are placed in light. He also noted that plants used CO_2 as a raw material in the process. In 1804 Nicholas Theodore de Saussure, a Swiss botanist,

TABLE 6-1. International Classification of Enzymes Showing Classes, Description of Reactions, and Examples

Class	Reactions Catalyzed	Example
1. Oxidoreductase	Biological oxidations and reductions	Lactate dehydrogenase—converts lactic acid to pyruvic acid
2. Transferase	Transfer of functional groups from one molecule to another	Glucokinase—transfers a phosphate group from ATP to glucose
3. Hydrolase	Hydrolysis (addition of water)	Maltase—adds water to maltose to yield two glucose molecules
4. Lyase	Removal of groups of atoms from a molecule nonhydrolytically	Isocitrate lyase—converts isocitric acid to succinic acid and glyoxylic acid
5. Isomerase	Rearrangement of atoms within a molecule	Glucose phosphate isomerase—converts glucose-6-phosphate to fructose-6-phosphate
6. Ligase	Joining of two molecules in the presence of ATP	Acetyl CoA synthetase—links acetic acid to CoA with a cleavage of ATP

correlated the uptake of CO_2 and the evolution of O_2 to the light requirement and also suggested the role of water as another raw material. Thirty-three years later, René H. Dutrochet, a French scientist, demonstrated the necessity for chlorophyll in the process, and in 1845 the German biochemist J. von Liebig indicated that all organic compounds synthesized by green plants were derived from CO_2. The extensive researches of these scientists, as well as the investigators from many other scientific disciplines, finally reduced the entire photosynthetic process to a generalized chemical formula.

$$\underbrace{\underset{\text{carbon dioxide}}{6\,CO_2} + \underset{\text{water}}{6\,H_2O}}_{\text{Raw materials}} \; \underbrace{\xrightarrow[\text{chlorophyll}]{\text{light}}}_{\text{Necessary conditions}} \; \underbrace{\underset{\text{sugar}}{C_6H_{12}O_6} + \underset{\text{oxygen}}{6\,O_2}}_{\text{Products}}$$

This equation is a mere overall representation of the raw materials, necessary conditions, and final products of photosynthesis in green plants. It does not indicate either the actual amounts of reactants and products or the variety of other molecules that are also synthesized. Nor does it indicate the many intermediate reactions that occur in the formation of sugars. It is known, for example, that carbohydrates are not formed by mixing CO_2 and H_2O; the product of such a reaction would be carbonic acid, H_2CO_3. Obviously, then, the equation does not indicate how sugars are actually synthesized. In order to understand the mechanisms that produce sugars photosynthetically, each aspect of the generalized equation will be considered.

The Fate of the Raw Materials

The chemical composition of carbohydrates (carbon, hydrogen, oxygen) indicates that there must be a supply of compounds containing these elements for synthesis. Water and carbon dioxide are the simple inorganic forms in which these elements are found. Generally, land plants absorb water through the roots; the water is conducted through the stems to the leaves, and photosynthesis typically occurs in the chloroplasts of the leaf cells. Carbon dioxide enters in gaseous form from the atmosphere through openings (stomata) of the leaves by simple diffusion. Because both H_2O and CO_2 are constantly being consumed during photosynthesis, their concentrations are less inside the plant than outside, a condition conducive to continuous diffusion into the photosynthetic sites.

The generalized equation for photosynthesis indicates that the carbon in synthesized sugar comes from CO_2

and that the source of hydrogen is H_2O. The source of oxygen, however, is not self-evident. Experiments to ascertain the source of O_2 released in photosynthesis were conducted by Martin Kamen using an isotope of oxygen (^{18}O). Since isotopic oxygen is heavier than normal oxygen (^{16}O), it can be separated and identified by means of an instrument called a mass spectrometer. In one of his experiments, Kamen supplied a photosynthesizing plant with CO_2 in which all of the oxygen atoms were ^{18}O and found that none of the oxygen released by the plant contained ^{18}O. He then supplied the plant with water molecules in which the oxygen was ^{18}O. Under these conditions, it was found that the ^{18}O was released to the atmosphere.

$$\underset{\text{carbon dioxide}}{6\,CO_2} + \underset{\text{water containing isotopic oxygen}}{6\,H_2{}^{18}O} \xrightarrow[\text{chlorophyll}]{\text{light}} \underset{\text{sugar}}{C_6H_{12}O_6} + \underset{\text{isotopic oxygen}}{6\,{}^{18}O_2}$$

If the oxygen in the water is evolved as gaseous oxygen, the oxygen source for sugar must be the carbon dioxide. But the above equation shows that there are twice as many oxygen atoms (12) in the CO_2 as in the sugar molecule (6). It also shows that twelve oxygen-18 atoms are produced, yet the water supplied only six. To correct this unbalanced situation we write

$$\underset{\text{carbon dioxide}}{6\,CO_2} + \underset{\text{water containing isotopic oxygen}}{12\,H_2{}^{18}O} \xrightarrow[\text{chlorophyll}]{\text{light}} \underset{\text{sugar}}{C_6H_{12}O_6} + \underset{\text{water}}{6\,H_2O} + \underset{\text{isotopic oxygen}}{6\,{}^{18}O_2}$$

Thus the atoms in synthesized sugar ($C_6H_{12}O_6$) are derived as follows: the six carbon atoms from carbon dioxide (6 CO_2), the twelve hydrogen atoms from water (12 $H_2{}^{18}O$), and the six oxygen atoms from carbon dioxide (6 CO_2). The atoms in the six molecules of water as a product (6 H_2O) are derived from the remaining twelve hydrogen atoms in 12 $H_2{}^{18}O$ and six of the oxygen atoms in 6 CO_2. The six molecules of oxygen (6 $^{18}O_2$), another product of the reaction, are derived entirely from water (12 $H_2{}^{18}O$). These relationships are summarized in Figure 6-2.

The Necessary Conditions: Chlorophyll and Light

It has been known for some time that in higher plants photosynthesis occurs in the chloroplasts of leaf cells (see Figure 5-23). Seven different types of chlorophyll pigments are known to exist in plant cells. These are chlorophylls *a, b, c, d,* and *e, bacteriochlorophyll,* and *bacteriovirdin.* Chlorophylls *a* and *b* are the most abundant and are found in all autotrophs except pigment-containing bacteria, which contain bacteriochlorophyll and bacteriovirdin. Chlorophylls *c, d,* and *e* are found only in algae along with chlorophyll *a.*

The two principal chlorophylls found in chloroplasts of higher plants are the blue-green chlorophyll *a* ($C_{55}H_{72}O_5N_4Mg$) and the yellow-green chlorophyll *b* ($C_{55}H_{70}O_6N_4Mg$). It is believed that the lipid tail ($H_{39}C_{20}$) of these chlorophylls, built up of —CH_2— units, is anchored to the lipid and protein layers of the grana of the chloroplast. Although the relative proportions of chlorophyll *a* and chlorophyll *b* vary somewhat in higher plants, on the average there are three molecules of chlorophyll *a* to every molecule of chlorophyll *b.*

Chloroplasts also contain *accessory pigments,* so called because they assist chlorophyll molecules during the initial phase of photosynthesis. The general types of accessory pigments are the orange *carotenes* and yellow *xanthophylls* of higher plants, and the *phycobilins,* red or blue pigments that are found in certain groups of algae.

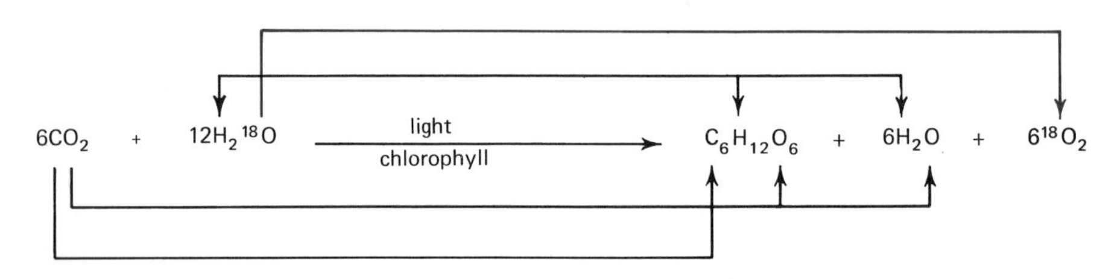

FIGURE 6-2. The fate of carbon dioxide and water during the process of photosynthesis.

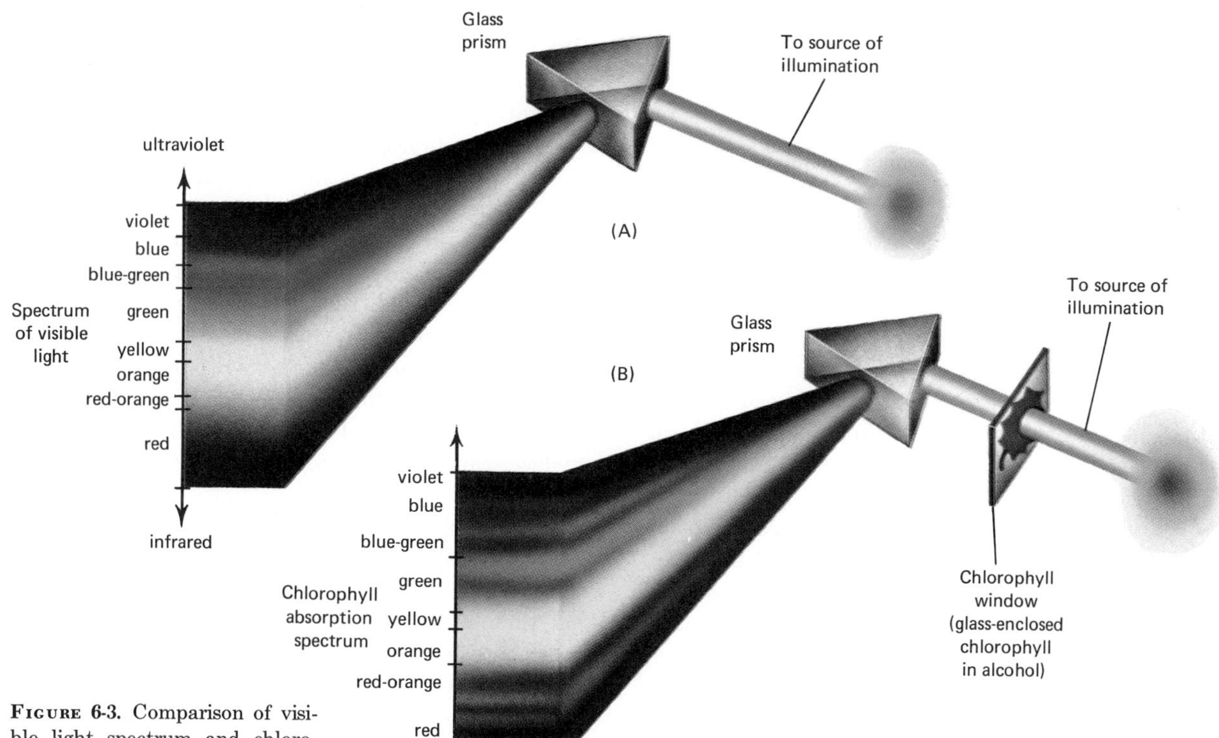

Figure 6-3. Comparison of visible light spectrum and chlorophyll absorption spectrum. (**A**) Visible white light passing through a prism separates into a range of wavelengths from red (longest) to violet shortest. (**B**) The spectrum produced by sunlight passing through a chlorophyll extract in front of the prism has dark bands in certain regions. These bands indicate that chlorophyll has absorbed light of these wavelengths. [*Modified from Biological Science: An Inquiry Into Life. Copyright © 1968 by the Regents of Colorado. By permission of the Biological Sciences Curriculum Study.*]

Light, the second prerequisite photosynthetic condition, provides the energy to drive the photosynthetic process. Two models are in use to describe the behavior of light because it appears to have a dual nature. The *wave theory* states that light is the result of minute, rapid vibrations or waves with characteristic dimensions measured in *wavelengths.* These wavelengths vary from one type of light to another; the longer the wavelength, the smaller is the energy conveyed. Conversely, the shorter the wavelength, the greater is the energy conveyed. Sunlight, although appearing to be white, is a mixture of lights of all colors, the colors being determined by different wavelengths. When white light is passed through a glass prism, the various wavelengths can be seen as different colors. This band of colors, the *visible spectrum,* includes those wavelengths perceived by the eye (Figure 6-3A). The wavelengths of the visible spectrum range from the relatively long waves of red light to the short waves of violet light. Violet light conveys more energy than red light.

The second model is the *photon theory,* which holds that light is composed of a stream of tiny particles called *quanta* or *photons.* Photons are discrete packets of energy given off by any light-emitting object. This conception of the nature of light is particularly helpful since, although light seems to travel as a wave, it apparently interacts with matter, such as chlorophyll, as a particle.

Some solar radiation is absorbed by various gases and dust particles in the atmosphere before striking the surfaces of leaves. Of the light that actually reaches leaves, a portion is reflected, another portion is transmitted, and a larger portion absorbed. Of course, the various portions are dependent upon the physical characteristics of the particular leaf studied. In general, more than two-thirds of the light striking the leaf is absorbed. Most of this is transduced to heat energy, which is dissipated to the environment. Only a small portion, from 1 to 4%, is used in the actual photosynthetic process.

In order to determine which portion of the absorbed visible light makes up the 1–4% utilized in the photosynthetic process, chlorophyll is extracted from a leaf with organic solvents such as carbon tetrachloride or alcohol. The extract is then spread on a clean glass microscope slide and placed between a source of white light and a prism. Examination of the spectrum now shows that certain wavelengths are partly or completely absent, indicating that they have been more or less completely utilized by the chlorophyll molecules on the slide (Figure 6-3B). Compare the two spectra in Figure 6-3 and note the positions of the dark bands in the chlorophyll spectrum. In regions where little absorption occurs, changes in the spectrum will be slight.

As implied previously, only those portions of the spectrum that are absorbed by chlorophyll can be used in photosynthesis. The chlorophyll spectrum indicates that much of the red, orange, blue, indigo, and violet regions is absorbed. Chlorophyll can absorb all wavelengths of the visible spectrum, but the maximum absorption is in the red and blue regions, and the minimum absorption (major reflection) is in the green region (which, incidentally, accounts for the green color of chlorophyll).

The Reactions of Photosynthesis

Many data indicate that photosynthesis is not a single process but consists of at least two main steps. For example, if light is supplied to a green plant in flashes separated by dark periods, a higher photosynthetic yield for a given quantity of light is possible than if the light is given in one continuous period of illumination. Accordingly, the entire process of photosynthesis may be subdivided into two interrelated phases: light reactions and dark reactions. The *light reactions* refer to a group of reactions that depend upon light in order to take place. The *dark reactions* refer to reactions that do not require light energy in order to occur. They will proceed in the light or in thedark.

LIGHT REACTIONS. Although many reactions of photosynthesis are incompletely understood, it is agreed that the initial reaction is the absorption of light by pigments within the chloroplasts. Chlorophyll *a* appears to be the principal light-absorbing pigment, although other chlorophylls, as well as other pigments (carotenes and xanthophylls), may also absorb light energy and transfer it to chlorophyll *a*.

All atoms normally possess a given number of electrons arranged at certain distances from the nucleus called energy levels. Electrons at the most distant energy level are the most weakly bound to the nucleus and are the most chemically reactive. Atoms can frequently absorb energy from an outside source and the added energy raises the outermost electrons into energy levels more distant from the nucleus. Such "activated" or "excited" electrons tend to return to their normal energy levels, and, in doing so, release the energy they had absorbed.

When a photon of light strikes a molecule of chlorophyll, and is absorbed, its energy is transferred to an outer electron of the chlorophyll molecule (Figure 6-4). This "excited" electron is then raised to a higher energy level (moves farther from the nucleus), but because the molecule is now unstable, the electron tends to fall back to its normal energy level, giving up the absorbed energy. It has been estimated that the shift of an "ex-

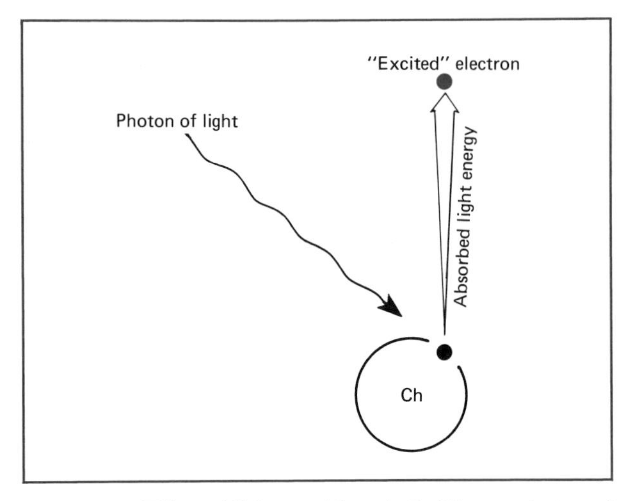

FIGURE 6-4. Effect of light on chlorophyll. When a photon of light strikes a molecule of chlorophyll, energy is transmitted from the light to an outer electron of a chlorophyll molecule. As a result, the chlorophyll electron is "excited" to a higher energy level.

cited" electron to a higher energy level and back again occurs in a millionth of a second. The energy-rich "excited" electrons do not fly off freely into space before falling back to their normal energy levels; they are trapped by various electron acceptors. These acceptors function in serial succession; that is, the electrons are passed from one acceptor to another in a series of chain reactions. The essential role of chlorophyll, therefore, is to absorb photons of light energy and transfer this energy to electron acceptors.

Through various pathways of electron transfer, chloroplasts are able to convert light energy into chemical energy. In other words, light energy (photons) absorbed by chlorophyll electrons is converted into chemical energy (ATP) by the passage of excited electrons from chlorophyll to specific electron acceptors. This conversion of light energy to chemical energy is called **photophosphorylation.** There are two rather different pathways available to the electrons given up by chlorophyll. These are cyclic photophosphorylation and noncyclic photophosphorylation.

In *cyclic photophosphorylation* (Figure 6-5), a chlorophyll molecule temporarily loses an excited electron to some other substance and then regains it, thus setting up a cycle. When chlorophyll donates an electron to an appropriate acceptor molecule, it becomes a powerful electron acceptor. Inasmuch as the chlorophyll molecule has lost an electron, it is left as a positive ion (+1). Thus, chlorophyll serves as both an electron donor and an electron acceptor. The excited electron from chlorophyll is immediately passed to a water-soluble iron–protein cofactor (coenzyme) called **ferredoxin.** Once ferredoxin has gained an electron, it transfers the electron to other cofactors or acceptors, one of which is the flavoprotein *flavin mononucleotide* (*FMN*) (Figure 6-6). Next the electron is passed to a group of acceptors collectively called **cytochromes,** iron-containing pigments structurally related to hemoglobin. Each cytochrome passes the electron to another cytochrome acceptor molecule, and so on. As the electron is cycled in this manner, it progressively loses energy. When the electron left the chlorophyll molecule, it was energy rich, but upon its return it is energy poor. Thus, the electron is eased down an energy gradient from the excited state to the normal state. At several points along the carrier chain of electron acceptors, the released electron energy is converted to chemical energy by being incorporated into molecules of ATP.

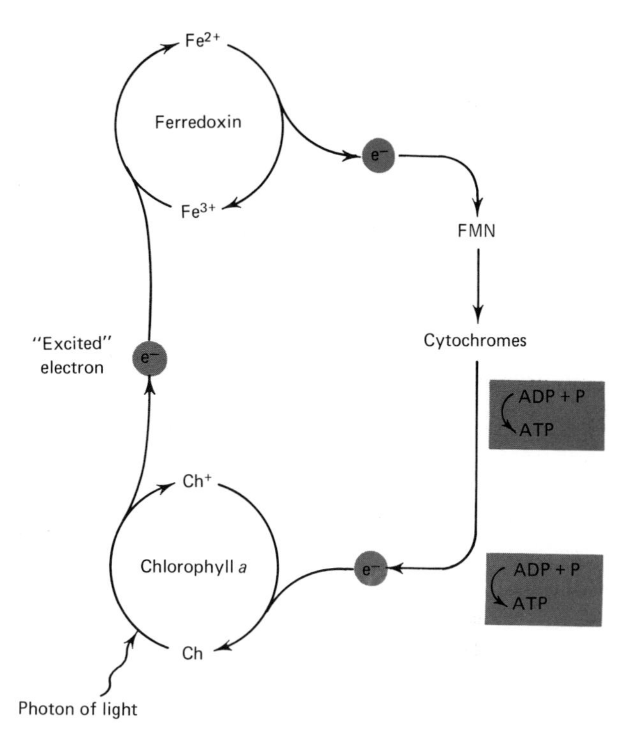

Figure 6-5. Cyclic photophosphorylation. Light energy is absorbed by an outer electron of chlorophyll. The electron becomes energized and is raised to a higher energy level. The energized electron is picked up by cofactors, such as ferredoxin and FMN, and is passed in serial succession through the cytochrome system of electron carriers. With each electron transfer, the electron is systematically de-energized as light energy is converted into chemical energy in the form of ATP. Ultimately, the de-energized electron returns to the chlorophyll molecule.

It will be recalled that when an inorganic (low energy) phosphate group is added to ADP in the presence of energy, a molecule of ATP is formed.

$$\underset{\text{adenosine diphosphate}}{\text{ADP}} + \underset{\text{inorganic phosphate}}{\text{P}} + \underset{\text{energy}}{E} \longrightarrow \underset{\text{adenosine triphosphate}}{\text{ATP}}$$

The ATP produced during cyclic photophosphorylation is now available to the cell as a source of chemical energy. Some of this energy, as will be discussed shortly, is used during the dark reactions for the synthesis of sugar.

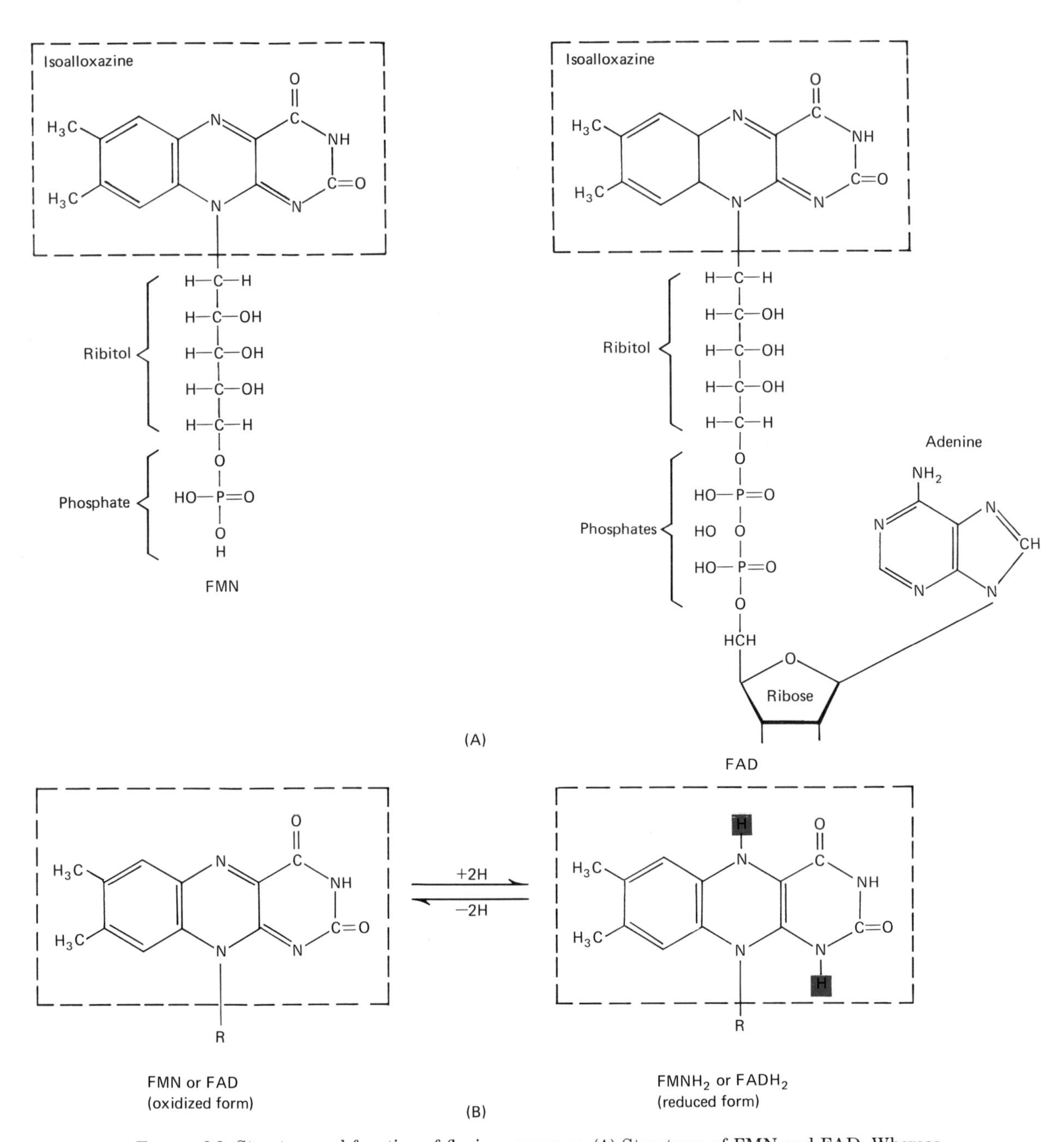

Figure 6-6. Structure and function of flavin coenzymes. (**A**) Structures of FMN and FAD. Whereas FMN is a flavin mononucleotide—isoalloxazine + ribitol (a sugar alcohol) + a phosphate group—FAD is a flavin dinucleotide—isoalloxazine + ribitol + two phosphate groups + ribose + adenine. (**B**) Oxidized and reduced forms of FMN and FAD. As shown, these coenzymes are capable of picking up two hydrogens, usually from NADH + H^+ or NADPH + H^+.

If cyclic photophosphorylation (Figure 6-5) is considered in these terms, it can be seen that the stored energy in a chlorophyll electron, which is light energy, is passed from one compound to another. Since each compound successively accepts and donates an electron, each is successively reduced and oxidized as the energy is transferred. Ultimately, the energy is stored in ATP. Chlorophyll, by losing the electron temporarily, is oxidized, and therefore releases energy. Because of the electron loss, it has acquired a positive charge. Next, the electron is passed to cofactors, which in turn are reduced and therefore store energy. The cofactors then transfer the electron to cytochromes, thus being oxidized. Each cytochrome passes the electron to another cytochrome, being first reduced and then oxidized in the process. The final cytochrome molecule passes the de-energized electron to chlorophyll, and the chlorophyll molecule is reduced. As a result of regaining an electron, the chlorophyll loses its positive charge. Thus cyclic photophosphorylation is completed.

The dark reactions of photosynthesis are concerned with combining the carbon and oxygen of CO_2 with hydrogen in order to synthesize sugars. Intimately related with the dark reactions of photosynthesis is a process called **noncyclic photophosphorylation** (Figure 6-7). It is so named because the excited electron lost by chlorophyll is passed on to a succession of acceptors but not returned to chlorophyll.

In noncyclic photophosphorylation, the electrons lost by chlorophyll are picked up by an acceptor and eventually passed to $NADP^+$. Each $NADP^+$ picks up two electrons. Unlike the electron acceptors in cyclic photophosphorylation, $NADP^+$ does not immediately pass the electrons to another acceptor, but instead retains them. The added electrons make $NADP^+$ unstable, and it is stabilized by the addition of two protons (H^+) from water. The resulting compound is the reduced form, $NADPH + H^+$ (Figure 6-8). It is this compound that furnishes hydrogen atoms for combination with carbon and oxygen during the dark reactions.

Two different aggregations of chlorophyll *a* and accessory pigments have been located in the chloroplast. These aggregations selectively absorb light of slightly different wavelengths and are called Pigment System I

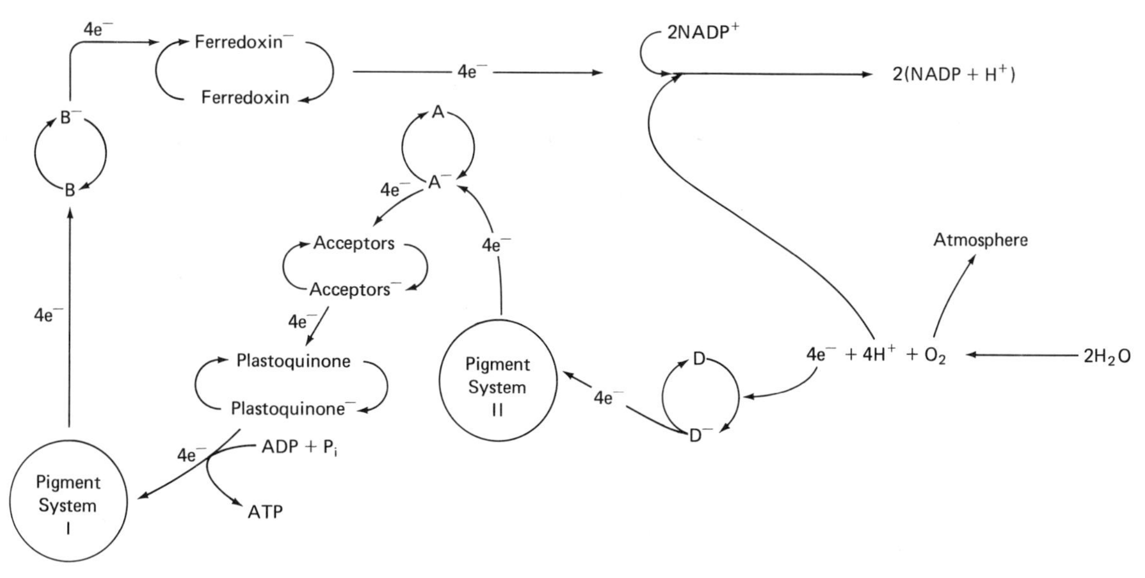

FIGURE 6-7. Noncyclic photophosphorylation. The diagram represents an overall reaction of the transfer of four electrons ($4e^-$). Electrons are actually passed through the scheme one at a time. The structures and mechanisms of electron acceptors, A and B, and electron donor, D, have yet to be determined. The net overall products are 2 ($NADPH + H^+$), ATP's, and oxygen. Consult the text for full description of the process.

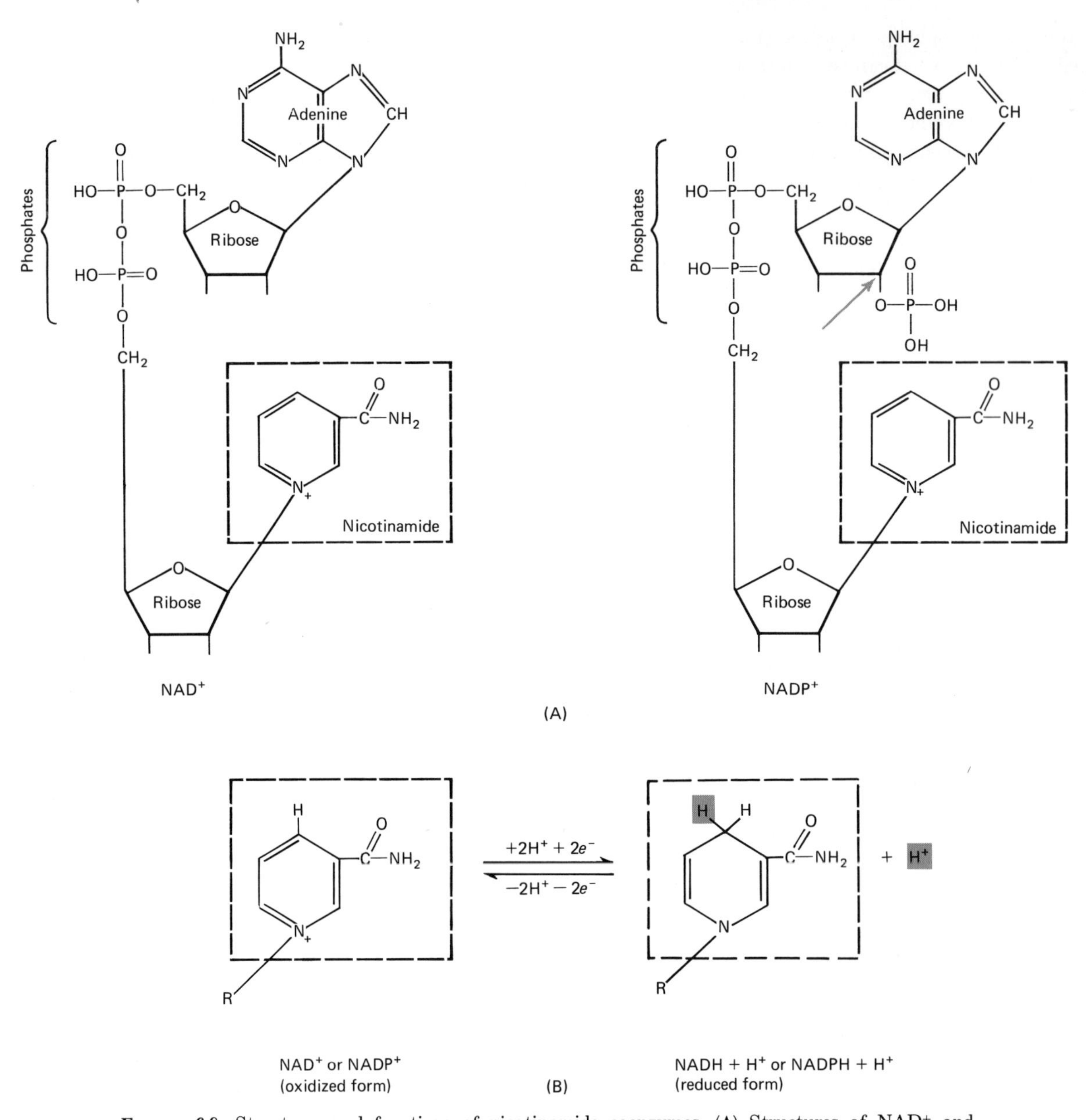

Figure 6-8. Structures and functions of nicotinamide coenzymes. (**A**) Structures of NAD^+ and $NADP^+$. As shown, these coenzymes are dinucleotides (two combined nucleotides, each consisting of an organic ring structure + ribose + phosphate). $NADP^+$ differs from NAD^+ in having a phosphate group attached to carbon-2 of the ribose that is attached to adenine (arrow). (**B**) Oxidized and reduced forms of the coenzymes. In this illustration, only the nicotinamide portion of the two coenzymes is shown (boxed area). Nicotinamide is a derivative of the B-vitamin niacin, and is an example of the usual role played by vitamins in forming important coenzymes in the cell. Only the nicotinamide portion plays a role in oxidation-reduction reactions; the remainder of the molecule is probably necessary for attachment to the enzyme and formation of an enzyme–coenzyme complex that can specifically combine with substrate molecules.

and Pigment System II. It has been implied and somewhat substantiated that, because of the two systems and their close interrelationship, noncyclic photophosphorylation occurs in two stages.

In the first stage a photon of light strikes Pigment System II and results in an electron being transferred from donor (D) to an acceptor (A) (Figure 6-7). The donor is now electron deficient and must receive an electron to re-establish its neutrality or original state. The donor receives an electron from water that has been broken up into electrons (e^-), protons (H^+), and oxygen (O_2). While the process takes place by transferring one electron at a time, overall the first stage reaction involves the transfer of four electrons ($4\,e^-$) from two water molecules and results in the production of one molecule of oxygen (O_2) and four protons ($4\,H^+$). The acceptor (A) of the electrons passes the electrons on to other acceptors, one of which is plastoquinone. From plastoquinone they are passed on through a cytochrome system that ultimately yields ATP. The number of ATP molecules produced has not been ascertained at the present time. The electrons are then passed to Pigment System I, but not before photons of light have promoted electrons from this system to another acceptor (B). The removal of electrons from Pigment System I begins the second stage.

Since electrons have been removed from Pigment System I, the effect of passing electrons from Pigment System II is to restore both systems to their original state. The electrons from Pigment System I, which have been passed to acceptor B, are then passed on to ferredoxin. Two $NADP^+$ molecules are then reduced by accepting electrons enzymatically from ferredoxin. In order to regain stability after acceptance of the electrons, the two $NADP^+$ molecules also pick up four protons ($4\,H^+$) and become reduced

$$\text{NADP} \longrightarrow 2\,(\text{NADPH} + \text{H}^+)$$

Since the reaction takes place in an aqueous environment, the four protons ($4\,H^+$) are picked up from the water. This does not destroy the neutrality of the water environment since, in an earlier step, two water molecules were oxidized to produce four protons ($4\,H^+$).

While the overall scheme has been worked out, there is still much to be studied and determined. The main obstacles to the solution include the rapidity of the reactions and the fact that the various steps can take place only within the living chloroplast.

How are electrons released from the pigment systems upon the absorption of photons? How does the oxidation of water really occur? What is the exact mechanism for the formation of ATP? What are the structures and mechanisms of the various acceptors and donors? These are some of the questions yet to be answered.

In higher plants, the oxygen diffuses through the chlorophyll-containing cells of the leaf into air spaces and through the stomata into the atmosphere. The important outcomes of noncyclic photophosphorylation, then, are the production of NADPH + H^+, the formation of ATP, and the release of oxygen as a waste product.

In all probability both cyclic and noncyclic photophosphorylations occur in living green cells in the light. Thus, in a chloroplast as a whole, some chlorophyll molecules are involved in cyclic photophosphorylation and the production of ATP only, while others are concerned with noncyclic photophosphorylation and the production of both ATP and NADPH + H^+. The formation of NADPH + H^+ signals the end of the light phase of photosynthesis. This compound, together with ATP, now enters the dark phase of photosynthesis.

Dark Reactions. For many years it was believed that a hexose sugar (glucose) was the first stable product of photosynthesis. Extensive research, however, carried on by Melvin Calvin and his associates at the University of California demonstrated that a three-carbon compound, *phosphoglyceraldehyde* (*PGAL*), is one of the first stable substances produced during the dark reactions. For these studies, Calvin received the Nobel Prize in 1961.

Essentially, Calvin sought to trace the path of carbon through the dark reactions by labeling carbon dioxide with ^{14}C. His experimental organism, *Chlorella,* was placed in a culture medium containing labeled carbon dioxide and permitted to carry on photosynthesis. In order to determine which compounds had incorporated the labeled carbon, the reaction was stopped at various times and the products analyzed. In this manner, a number of the intermediate products were identified and a total scheme for the dark reactions determined.

The dark reactions are concerned with combining the

hydrogen atoms from NADPH + H^+ with CO_2 to form sugars. In this series of synthetic reactions, energy is needed to reduce CO_2. ATP and NADPH + H^+ function as energy donor and reducing agent, respectively.

Sugars contain a great deal of stored chemical energy, but CO_2 possesses relatively little. Just as energy is slowly released in a stepwise series of oxidation-reduction reactions during photophosphorylation, so also is energy slowly built up during *CO_2 fixation,* the reduction of CO_2 by NADPH + H^+ to form sugar. In effect, CO_2 is slowly pushed up an energy gradient through a series of intermediate compounds, some of them unstable, until a stable end product is formed. However, since free molecules of carbon dioxide cannot be directly reduced, CO_2 must first be combined with a larger molecule (Figure 6-9).

The first step of the dark reactions, preparatory to the fixation of CO_2, consists of the conversion of a ribulose phosphate (RuP) into a high-energy compound, ribulose diphosphate (RuDP), by the addition of a phosphate, which is obtained from ATP. This conversion is necessary because CO_2 is not able to combine with ribulose phosphate unless energy is added, in this instance in the form of a high-energy phosphate. Ribulose diphosphate is continually being produced in photosynthetic cells and is always present in chloroplasts.

Within a fraction of a second after CO_2 reaches a chloroplast, it combines with the ribulose diphosphate and forms an unstable six-carbon intermediate compound that rapidly breaks down into two three-carbon compounds called phosphoglyceric acid (PGA). Each molecule of PGA is then phosphorylated to a triose sugar by ATP. It is at this juncture that energy from the light reactions, which was incorporated in the formation of ATP and NADPH + H^+, participates in the dark reactions. In the process of phosphorylation, molecules of ATP are broken down into ADP + P. In addition, hydrogen is transferred to the PGA molecules from NADPH +H^+ and NADPH + H^+ is oxidized back to $NADP^+$.

The product of the phosphorylation and the reduction of PGA is an energy-rich triose sugar called phosphoglyceraldehyde (PGAL). The ADP + P and the $NADP^+$ return to the light reactions to perform the functions cited earlier.

For every three molecules of CO_2 that are used, six molecules of PGAL subsequently follow one of two paths. One molecule of PGAL represents the stable end product of photosynthesis and, in chemical combination with single PGAL molecules from similar photosynthetic reactions, is converted into hexoses such as fructose and glucose. For this reason, it is customary to consider a hexose ($C_6H_{12}O^6$) as the end product of photosynthesis, and not PGAL. The remaining five molecules of PGAL are fed back into a self-sustaining cycle to form three molecules of ribulose diphosphate for further action as a CO_2 acceptor. Ribulose diphosphate regeneration consists of a rather complex series of enzymatically controlled reactions involving extensive reshuffling and phosphorylation.

Respiration—The Transformation of Energy

The essential feature of photosynthesis is the transduction of solar energy into potential chemical energy. As a synthetic phase of metabolism, it is an energy-requiring process, the energy being supplied by solar radiation. PGAL, the photosynthetic product, and the sugars and other compounds synthesized from it are the recipients of stored chemical energy. Before any of these compounds can release stored energy, they must be broken down chemically through respiration.

Photosynthesis and respiration are reverse metabolic processes, one storing energy and the other releasing it, and the pattern of energy-storing reactions differs somewhat from the energy-releasing reactions. The processes are alike, however, in consisting of a series of step-by-step reactions, each enzymatically catalyzed. In addition, the energy released in respiration does not

FIGURE 6-9 [OPPOSITE]. Dark reactions of photosynthesis. These reactions, which proceed independently of light, are concerned with combining the hydrogen atoms of NADPH + H^+ with carbon dioxide to form PGAL. ATP and NADPH + H^+ function as energy donor and reducing agent, respectively. [*Modified from J. A. Bassham and M. Calvin,* The Path of Carbon in Photosynthesis. *Copyright © 1957, Prentice-Hall, Inc., Englewood Cliffs, N.J. By permission.*]

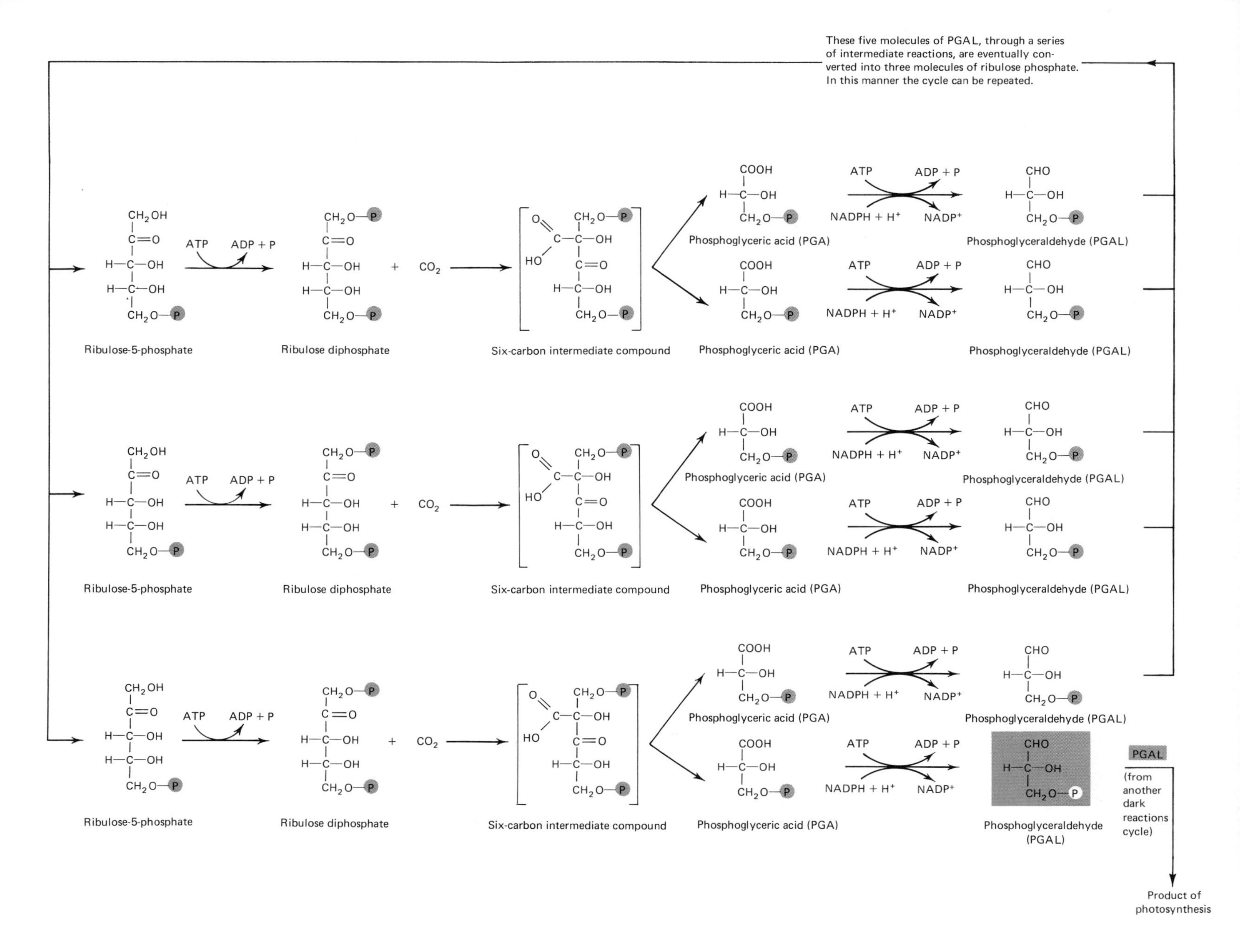
These five molecules of PGAL, through a series of intermediate reactions, are eventually converted into three molecules of ribulose phosphate. In this manner the cycle can be repeated.
Ribulose-5-phosphate
ATP
ADP + P
Ribulose diphosphate
CO_2
Six-carbon intermediate compound
Phosphoglyceric acid (PGA)
NADPH + H+
NADP+
Phosphoglyceraldehyde (PGAL)
PGAL
(from another dark reactions cycle)
Product of photosynthesis

dissipate as heat or light but is conserved and stored in ATP for cellular utilization. The interconnection between photosynthesis and respiration is one example of the interrelationships that exist between anabolic and catabolic processes in cells, all contributing to the total functioning organism. Some aspects of photosynthesis and respiration are compared in Table 6-2.

Respiration, or **biological oxidation,** may be defined as the oxidation of organic compounds in cells, with a release of chemical energy. The overall process in the respiration of glucose can be represented by the following general equation.

$$\underset{}{6\,O_2} + \underset{\text{glucose}}{C_6H_{12}O_6} \xrightarrow{\text{enzymes}} \underset{\substack{\text{carbon}\\\text{dioxide}}}{6\,CO_2} + \underset{\text{water}}{6\,H_2O} + \underset{\text{(ATP)}}{\text{energy}}$$

As in the case of photosynthesis, this equation is a mere representation of the raw materials, necessary conditions, and products of the reaction. It in no way accounts for the myriad intermediate chemical reactions that actually occur.

The energy released by the oxidation of organic molecules is incorporated into the high-energy groups of ATP. In other words, the chemical energy in the bonds of food molecules is transformed into the energy in ATP, a form that can be readily utilized by the cell to do work. If such a transfer and storage of energy from food molecules to ATP did not occur, the energy contained in foods would dissipate as heat in the cell and be unavailable as cellular energy. The energy stored in ATP is used for reactions that require energy; thus, energy from one reaction (respiration) is transferred to drive another reaction (synthesis). In addition to providing energy for synthetic reactions, such as those involved in the production of amino acids, proteins, and lipids, the energy of ATP may also be used in active transport, cyclosis, contraction, and bioluminescence, to mention only a few. These will be described in Chapter 7.

Once ATP is formed, its energy may be utilized at various places in the cell to drive energy-requiring reactions. In the process, one of the three phosphate groups is removed from the ATP molecule, leaving a compound, adenosine diphosphate (ADP), which is less rich in energy than ATP, and an inorganic phosphate (P) group. Molecules of ADP and P may be recombined into ATP by trapping energy released from the oxidation of glucose or other organic molecules. The role of ATP as an intermediate energy-transferring compound between energy-releasing and energy-consuming reactions is shown in Figure 6-10.

It is to be noted that in certain reactions two high-energy phosphate groups may be removed from ATP, yielding adenosine monophosphate (AMP) and inorganic pyrophosphate (PP) with release of energy.

TABLE 6-2. Comparison of Some Aspects of Photosynthesis of Green Plants, and Respiration of Glucose

Photosynthesis	Respiration
1. Occurs only in chlorophyll-containing cells	1. Occurs in all cells
2. Takes place only in the presence of light	2. Takes place continually both in the light and in the dark
3. Sugars, water, and oxygen are products (green plants)	3. CO_2 and H_2O are products
4. CO_2 and H_2O are raw materials	4. O_2 and food molecules are raw materials
5. Synthesizes foods	5. Oxidizes foods
6. Stores energy	6. Releases energy
7. Results in an increase in weight (gain of CO_2 and H_2)	7. Results in a decrease in weight (loss of CO_2)
8. $CO_2 + H_2O \longrightarrow$ glucose	8. Glucose $\longrightarrow CO_2 + H_2O$

The Reactions of Energy Metabolism

The first stage of energy metabolism, called **glycolysis** ("sugar splitting"), consists of a sequence of reactions occurring in the cytoplasm in which one molecule of glucose, the starting substrate, is converted into two molecules of a three-carbon organic acid called *pyruvic acid* (Figure 6-11). The major events of glycolysis are the preparation of glucose for reaction, splitting it into two three-carbon compounds, partial oxidation of the three-carbon compounds, and the formation of ATP. Let us now consider these events in more detail.

Glucose, a product of photosynthesis, is a fairly stable molecule; it resists breakdown. If the energy contained in its molecular configuration is to be released, it must first be made more reactive. "Activation" of glucose is

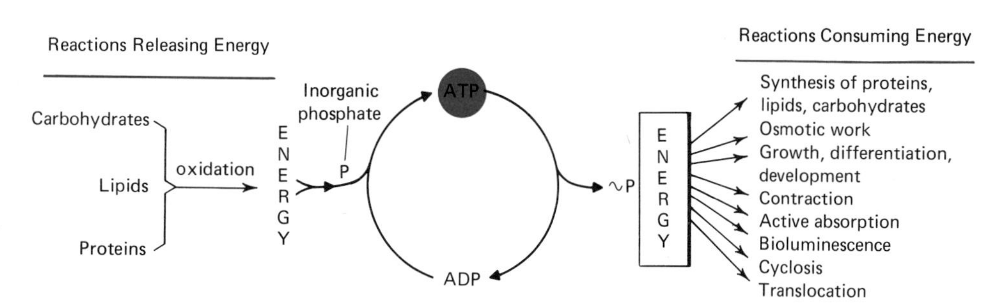

Figure 6-10. The ATP cycle. ATP is an intermediate energy-transferring compound between energy-releasing and energy-consuming reactions. After ATP is broken down to release energy for cellular activities, it is rebuilt by the addition of a phosphate group. The energy for this rebuilding is supplied by the breakdown of food molecules during respiration.

accomplished by the transfer of some of the energy of ATP, derived from previous respiration, to the glucose molecule. This reaction is called **phosphorylation** and involves the conversion of ATP to ADP and the transference of some of the stored energy to glucose. The product of this reaction is *glucose-6-phosphate* (activated glucose), so named because a phosphate group of ATP is attached to the sixth carbon atom of the glucose molecule. Catalysis is facilitated by the enzyme *hexokinase* with Mg^{2+} as an activator (Figure 6-11A).

The glucose-6-phosphate now undergoes a reaction in which atoms are rearranged. In the presence of an *isomerase* enzyme (*phosphoglucoisomerase*), the atoms of glucose-6-phosphate are rearranged to produce a molecule of *fructose-6-phosphate* (Figure 6-11B). Following this, another oxidative phosphorylation reaction, catalyzed by a *kinase* (*phosphofructokinase*), converts fructose-6-phosphate into *fructose-1,6-diphosphate*. In this reaction, a phosphate group is added to the first carbon atom of the fructose molecule (Figure 6-11C).

The fructose-1,6-diphosphate molecule is then split into two three-carbon sugar phosphate molecules in the presence of an *aldolase* enzyme. One of these sugars is *phosphoglyceraldehyde* (*PGAL*); the other is a compound called *dihydroxyacetone phosphate*. In the presence of an appropriate *isomerase* enzyme (*phosphotriose isomerase*), dihydroxyacetone is usually converted into PGAL (Figure 6-11D). In effect, at this juncture, the pathway is an energy-consuming process since two molecules of ATP have been used in order to phosphorylate the starting glucose molecule.

The next step in glycolysis involves the oxidation of PGAL (and dihydroxyacetone phosphate since it is converted to PGAL) to a compound called *1,3-diphosphoglyceric acid* (Figure 6-11E). In this oxidation reaction, a pair of electrons is removed from PGAL by a *dehydrogenase* enzyme (*phosphoglyceraldehyde dehydrogenase*). It should be noted that the removal of electrons presents somewhat of a problem to the cellular machinery since electrons (−) are attracted to the protons (+) of the molecule to which they are attached. Inasmuch as a great deal of energy would have to be expended to remove only the electrons, a proton is removed along with an electron, a process requiring less energy. In effect, then, oxidation of the PGAL molecule involves the removal of hydrogen atoms (an electron and a proton together), a process called *dehydrogenation*. It should be kept in mind, however, that the energy released during oxidation reactions resides in the electrons and not the protons. Two hydrogen atoms are picked up by the coenzyme NAD^+, thus reducing it to $NADH + H^+$. In addition to the oxidation of PGAL, PGAL also picks up an inorganic (low-energy) phosphate, raising the energy level of the inorganic phosphate to a high-energy level equivalent to that found in ATP.

In the presence of ADP and a *kinase* enzyme (*phosphoglyceric kinase* activated by Mg^{2+}), 1,3-diphosphoglyceric acid is converted to 3-phosphoglyceric acid and ATP is formed (Figure 6-11F). This acid, in turn, is transformed to *2-phosphoglyceric acid* by the activity of the *mutase* enzyme *phosphoglyceromutase* (Figure 6-11G). This enzyme transfers the phosphate from the

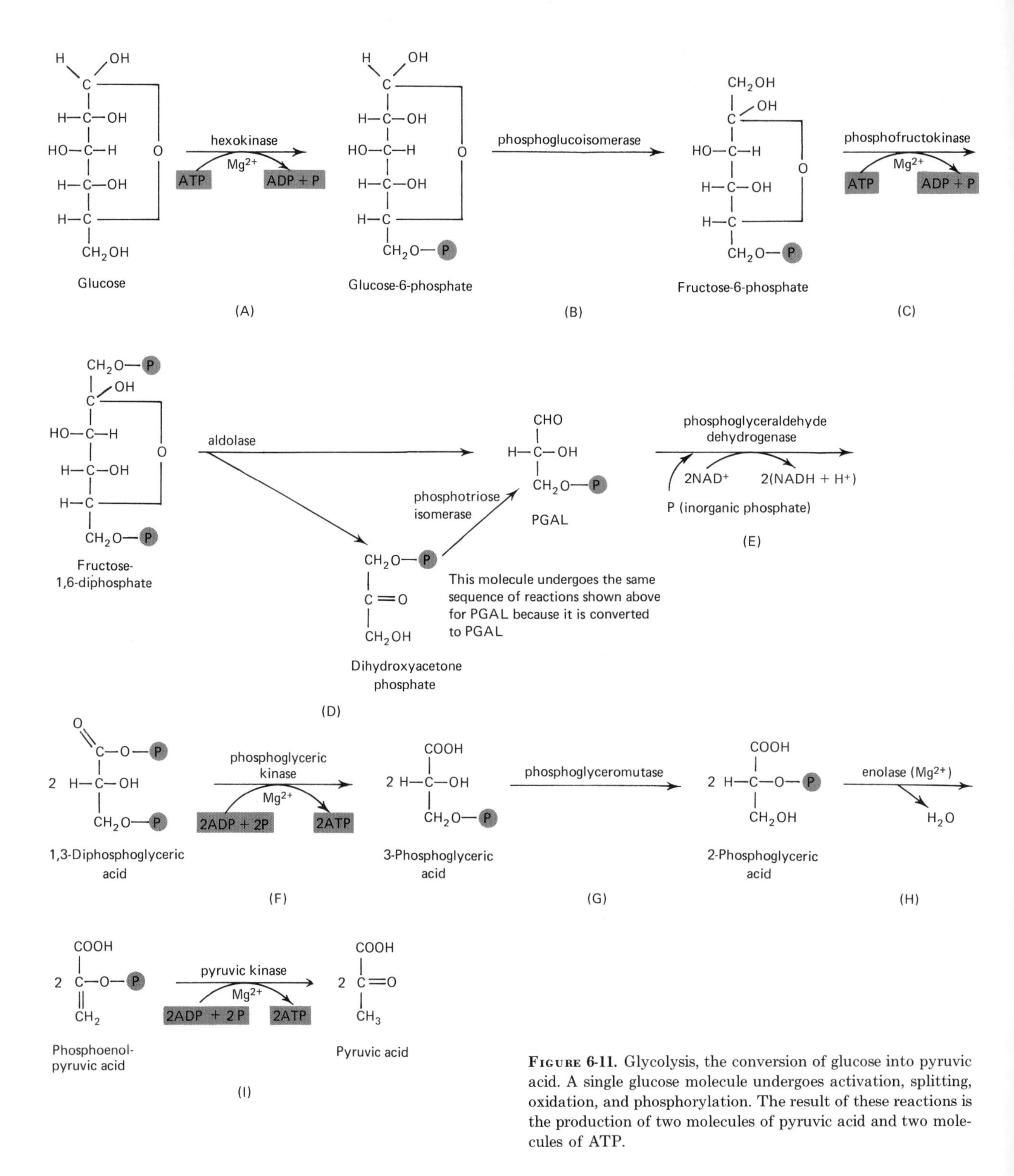

Figure 6-11. Glycolysis, the conversion of glucose into pyruvic acid. A single glucose molecule undergoes activation, splitting, oxidation, and phosphorylation. The result of these reactions is the production of two molecules of pyruvic acid and two molecules of ATP.

third carbon atom of glyceric acid to the second carbon atom. As glycolysis nears completion, a molecule of water is removed from 2-phosphoglyceric acid by the enzyme *enolase* which is activated by Mg^{2+}. The elimination of water from 2-phosphoglyceric acid results in the formation of *phosphoenolpyruvic acid* (Figure 6-11H). Finally, in the presence of ADP and the enzyme *pyruvic kinase,* phosphoenolpyruvic acid is converted to *pyruvic acid.* In this reaction, the phosphate group of phosphoenolpyruvic acid is transferred to ADP to form ATP (Figure 6-11I).

The formation of pyruvic acid signals the end of the glycolysis phase of respiration. It should be noted that whereas two molecules of ATP are converted to ADP in order to activate glucose, for each molecule of glucose that undergoes glycolysis, four are produced, for a net gain of two. Remember that since dihydroxyacetone phosphate is converted to PGAL, it also undergoes the same sequences of reactions as PGAL. In addition, note that glycolysis requires no molecular oxygen for its operation. Of the total energy built into PGAL, about 5% is transferred and stored in the ATP, 15–20% is absorbed by NADH + H^+, and the remainder is left in pyruvic acid. In view of this unequal energy distribution, it would appear that the fate of pyruvic acid, especially, is of great importance to the cell.

At this point it might be well to examine, in somewhat more detail, the mechanism of electron transport since energy transductions in cells result from electron transfers.

The bulk of cellular energy in the form of ATP is finally obtained from oxidation-reduction reactions. In the cell, *oxidation-reduction* may involve the transfer of electrons (e^-), the transfer of hydrogen atoms (H), or the transfer of hydrogen ions (H^+) and electrons. Only in the final step of energy metabolism is oxygen used in an oxidation-reduction reaction. The various oxidation-reduction reactions occurring in the cell are catalyzed by enzymes which require a particular coenzyme.

An important generalization concerning cellular oxidation-reductions is that in the initial step a pair of hydrogen atoms is removed from a substrate molecule (i.e., a food molecule, acting as a substrate in an enzyme reaction, is oxidized by the removal of two hydrogen atoms). This dehydrogenation reaction always requires either the coenzyme *NAD*$^+$ (*nicotinamide adenine dinucleotide*) or the coenzyme *NADP*$^+$ (*nicotinamide adenine dinucleotide phosphate*). The structures of these coenzymes are shown in Figure 6-6A. These hydrogen atoms and their potential energy, especially that of their associated electrons, are used to provide energy for phosphorylating ADP and forming ATP. As with most cellular metabolic schemes, there is a series of reaction steps leading to the procurement of this energy and there is a variety of coenzymes involved in the overall oxidation-reduction scheme. The particular fate of the hydrogens and their associated electrons depends upon which coenzyme in the series is involved.

The coenzymes NAD^+ and $NADP^+$ function in dehydrogenation by accepting one hydrogen atom plus one electron from a second hydrogen atom from a substrate. In the course of the reaction, a hydrogen ion (H^+) is formed and goes into solution. Only the nicotinamide portion of the coenzyme is actually involved in this reaction (Figure 6-6B). Nicotinamide picks up one hydrogen atom and also one electron, leaving an H^+ in solution. In biological oxidation-reduction reactions for energy procurement, the next substrate in the series is usually a *flavin coenzyme,* a derivative of the vitamin riboflavin. This type of coenzyme picks up two hydrogen atoms during its reduction. Examples of such coenzymes are *FMN* (*flavin mononucleotide*) and *FAD* (*flavin adenine dinucleotide*). The general structure of oxidized and reduced forms of flavin coenzymes is shown in Figure 6-7.

A third important type of oxidation-reduction coenzyme includes the cytochromes. These proteins contain heme groups whose iron atom acts in oxidation-reduction reactions by picking up or releasing electrons from $FMNH_2$ or other reduced coenzymes. The cytochromes are reduced only by the electrons, and the resulting hydrogen ions go into solution.

In biological oxidation, the final step is the combining of electrons from cytochromes and hydrogen ions from solution with atomic oxygen to form water (Figure 6-12). Whenever one substance is reduced, another is oxidized. Substances that are oxidized lose energy, while substances that are reduced gain energy.

Types of Respiration

During glycolysis, hydrogen atoms are removed from glucose and, in the process, NAD^+ is reduced to

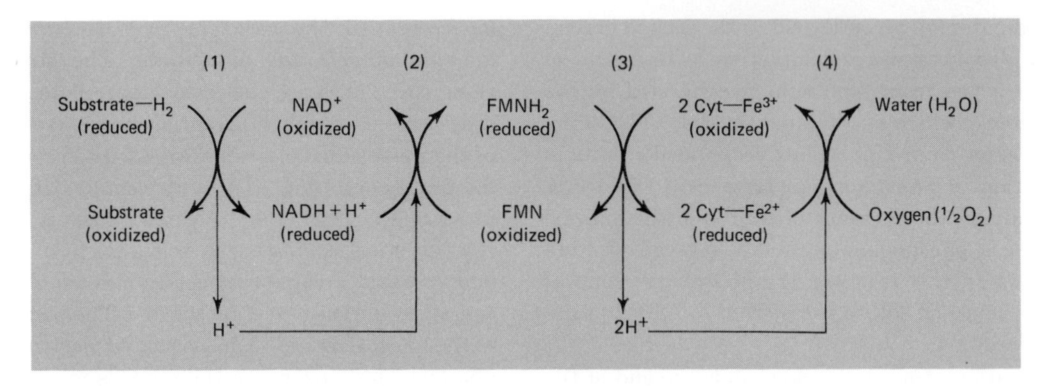

FIGURE 6-12. Generalized stages of biological oxidation-reduction for energy utilization. A substrate molecule is first dehydrogenated (step 1), usually by either NAD^+ or $NADP^+$. During this reaction, the substrate is oxidized and the coenzyme is reduced. Because of the nature of NAD^+, a hydrogen ion goes into solution. The next step in the system is the reduction of the coenzyme FMN or FAD. This molecule picks up two hydrogens—the hydrogen ion from solution and a hydrogen atom (with one electron) from NADH or NADPH. At step 2, the pyridine coenzyme is reoxidized (and made ready to pick up more hydrogens from substrates), while the flavin coenzyme is reduced. At step 3, the flavin coenzyme transfers two electrons to two molecules of cytochrome and releases two hydrogen ions into solution. Each cytochrome contains an atom of iron that can pick up and release electrons. Finally (step 4), the reduced cytochrome transfers an electron to oxygen, which combines with hydrogen ions taken from solution to form water.

NADH + H^+. Organisms possess various mechanisms for disposing of the hydrogen held as NADH + H^+ so that NAD^+ can be freed to pick up more hydrogen. The removal of hydrogen from NADH + H^+ is extremely important to organisms. If NADH + H^+ could not be freed of hydrogen and reused as NAD^+, glycolysis would stop and the organism would die. NAD^+, like other coenzymes, is present in only small quantities in the cell. This is the important reason why any NADH + H^+ formed must be immediately converted back to the oxidized form if glycolysis or other metabolic systems are to function. Based upon the manner in which organisms oxidize NADH + H^+ to NAD^+, two principal types of respiration exist, anaerobic and aerobic respiration. In **anaerobic respiration** the electrons in NADPH + H^+ are not passed through biological oxidation reactions with atmospheric oxygen as the ultimate acceptor but are transferred to an organic compound or inorganic radical. Among the acceptors are acetaldehyde, pyruvic acid, SO_4^{2-}, and NO_3^-. The last two acceptors are employed by certain types of bacteria. If alcohol is the final product of the transfer of hydrogen, the process is **alcoholic fermentation.** If lactic acid is the final product, it is lactic acid fermentation. If the hydrogen from NADPH + H^+ is released, ultimately to be combined with molecular oxygen (O_2) from the air, the process is known as **aerobic respiration.**

FERMENTATION. *Alcoholic fermentation* is a process carried on by yeasts, some bacteria, and even green plants when deprived of oxygen. The first sequence of reactions in alcoholic fermentation is the same as that occurring during glycolysis in which a glucose molecule is split and oxidized into two pyruvic acid molecules. In this part of the process, there is a net gain of two ATP molecules. In alcoholic fermentation, the pyruvic acid resulting from glycolysis is then converted in two steps into alcohol and CO_2. First, CO_2 is removed from pyruvic acid, leaving *acetaldehyde* (CH_3CHO). Second, the acetaldehyde is converted into *ethyl alcohol* (C_2H_5OH). These reactions are shown in Figure 6-13A.

During alcoholic fermentation, there is a net production of only two ATP molecules for each molecule of glucose oxidized. The bulk of the energy originally present in the sugar is now in the alcohol. In view of the fact that the alcohol is an end product of the reactions and undergoes no further oxidation, alcoholic fermentation

FIGURE 6-13. Alcoholic (A) and lactic acid (B) fermentation. These fermentations are only partial oxidations of glucose molecules. Only two ATP molecules (net) are produced by glycolysis. The remaining energy is still stored in the chemical bonds of ethyl alcohol and lactic acid.

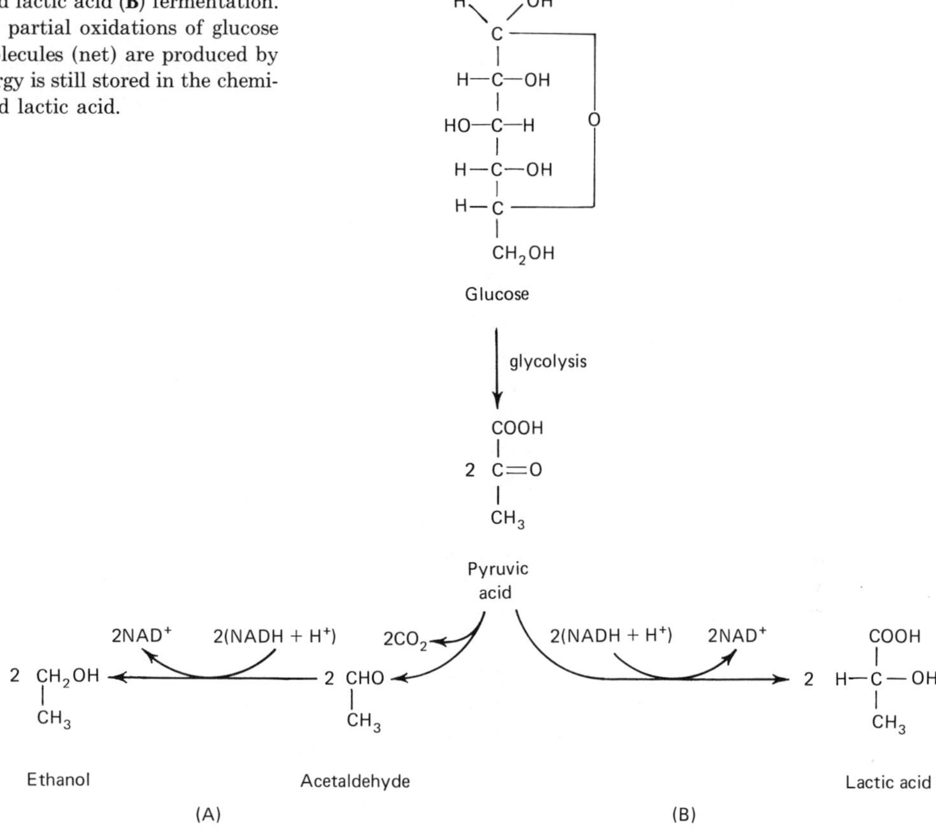

is an inefficient energy-releasing process. Also note that the hydrogen acceptor in alcoholic fermentation is acetaldehyde forming ethyl alcohol.

Lactic acid fermentation is characteristic of animals and some forms of bacteria and is very similar to alcoholic fermentation. The main difference is that pyruvic acid, formed by glycolysis, acts as the hydrogen acceptor and *lactic acid* is the end product (Figure 6-13B). As in alcoholic fermentation, there is a net production of two ATP molecules during lactic acid fermentation.

Different fermentations produce different byproducts, many of which are of economic importance. The two principal commercial uses of yeasts, baking and brewing, are dependent on the two products of alcoholic fermentation, CO_2 and alcohol. The fermentation of fruit juices, malted grains, and molasses by yeasts produces wine, beer, and rum, respectively. Different types of yeast are also used to leaven bread. The CO_2 produced during fermentation makes bubbles in the dough and causes the bread to rise, giving it lightness and a desirable texture when baked. Other bacterial fermentations produce a wide variety of metabolic byproducts. Lactic acid, for example, is used commercially in the manufacture of many dairy products, including cheeses, buttermilks, and yogurts.

AEROBIC RESPIRATION. Glycolysis and fermentation are anaerobic types of respiration; that is, they occur in the absence of free molecular oxygen. In the presence of free oxygen the pathway called *aerobic respiration* is followed. During aerobic respiration, the complete oxidation of glucose into CO_2 and H_2O is brought about by a complex series of reactions in which oxygen serves as the final hydrogen acceptor. Essentially, the two additional mechanisms involved in these complete oxidation reactions are the Krebs cycle and the cytochrome system.

Krebs Cycle. The first phase of the complete breakdown or oxidation of pyruvic acid, an energy-rich compound, into carbon dioxide and water is achieved in the inner mitochondrial membrane and the intracristal space by a complex series of reactions collectively called the **Krebs cycle.** Pyruvic acid, like glucose during glycolysis, must undergo preparation for aerobic respiration. Before entering the Krebs cycle, pyruvic acid is converted to a two-carbon compound by the loss of CO_2 (*decarboxylation*) and two hydrogen atoms (*dehydrogenation*). This compound, bonded to a coenzyme called *coenzyme A* (CoA), is an activated substance known as *acetyl coenzyme A* (Figure 6-14). When a molecule of pyruvic acid is oxidized to acetyl coenzyme A and carbon dioxide, one hydrogen atom is removed from the pyruvic acid and one is removed from coenzyme A, and these are picked up by NAD^+, thus forming $NADH + H^+$ (Figure 6-15, step 1). Since two pyruvic acid molecules are formed from each glucose during glycolysis, two molecules of $NADH + H^+$ are formed, as shown in Figure 6-14.

Acetyl CoA, which contains only two carbons of the original pyruvic acid, is activated and able to enter the Krebs cycle (Figure 6-15). The cycle starts in the mitochondrion when a molecule of acetyl CoA combines with a four-carbon organic acid, *oxaloacetic acid,* already present. The result of this reaction (step 2) is the formation of a six-carbon organic acid called *citric acid* (A) and CoA—SH. Step 3 involves the removal of H_2O from citric acid, which results in the formation of *cis-aconitic acid,* also a six-carbon acid (B). In step 4 H_2O is added to *cis*-aconitic acid to form a different six-carbon acid, *isocitric acid* (C).

At step 5 the first oxidation-reduction reaction takes place. In the presence of $NADP^+$ (or NAD^+) two hydrogen atoms are removed (dehydrogenation) from isocitric acid. The products of this reaction are $NADPH + H^+$ (or $NADH + H^+$) and the six-carbon acid *oxalosuccinic acid* (D). Next (step 6), oxalosuccinic acid is decarboxylated (CO_2 is removed) and a five-carbon acid, *α-ketoglutaric acid,* is formed (E).

The oxidation of α-ketoglutaric acid (step 7), the second oxidation-reduction reaction of the cycle, is very similar to the formation of acetyl CoA from pyruvic acid. The α-ketoglutaric acid is decarboxylated and dehydrogenated, and NAD^+ is reduced to $NADH + H^+$. In addition, CoA enters the reaction and the product is a four-carbon compound called *succinyl coenzyme A* (F).

Step 9, the conversion of succinyl CoA to *succinic acid* (G), is a unique step. The energy released in the formation of succinic acid and coenzyme A (CoA—SH) is used to add an inorganic phosphate group to *guanosine diphosphate* (GDP) to form *guanosine triphosphate* (GTP). GTP is a high-energy compound similar to ATP and can be used as a source of energy for synthesis reactions or to convert ADP to ATP, which in turn can be used as a source of energy. This is the only step of the Krebs cycle in which a high-energy phosphate compound is produced.

A third oxidation occurs at step 9 when succinic acid is dehydrogenated to the four-carbon acid *fumaric*

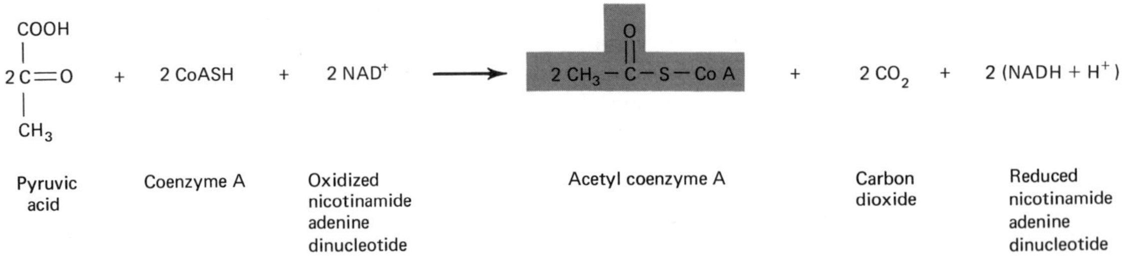

$$2\,\underset{\substack{|\\ CH_3}}{\overset{\substack{COOH\\ |}}{C}}{=}O \;+\; 2\,CoASH \;+\; 2\,NAD^+ \longrightarrow 2\,CH_3-\overset{\substack{O\\ \|}}{C}-S-CoA \;+\; 2\,CO_2 \;+\; 2\,(NADH + H^+)$$

Pyruvic acid | Coenzyme A | Oxidized nicotinamide adenine dinucleotide | Acetyl coenzyme A | Carbon dioxide | Reduced nicotinamide adenine dinucleotide

FIGURE 6-14. Formation of acetyl coenzyme A. Pyruvic acid, which represents the last product formed during glycolysis, is converted to acetyl CoA if sufficient oxygen is present. The principal mechanisms involved in this conversion are decarboxylation (removal of CO_2) and dehydrogenation (removal of hydrogen atoms). Although acetyl CoA formation is shown here as a single reaction, it should be noted that four intermediate reactions are involved.

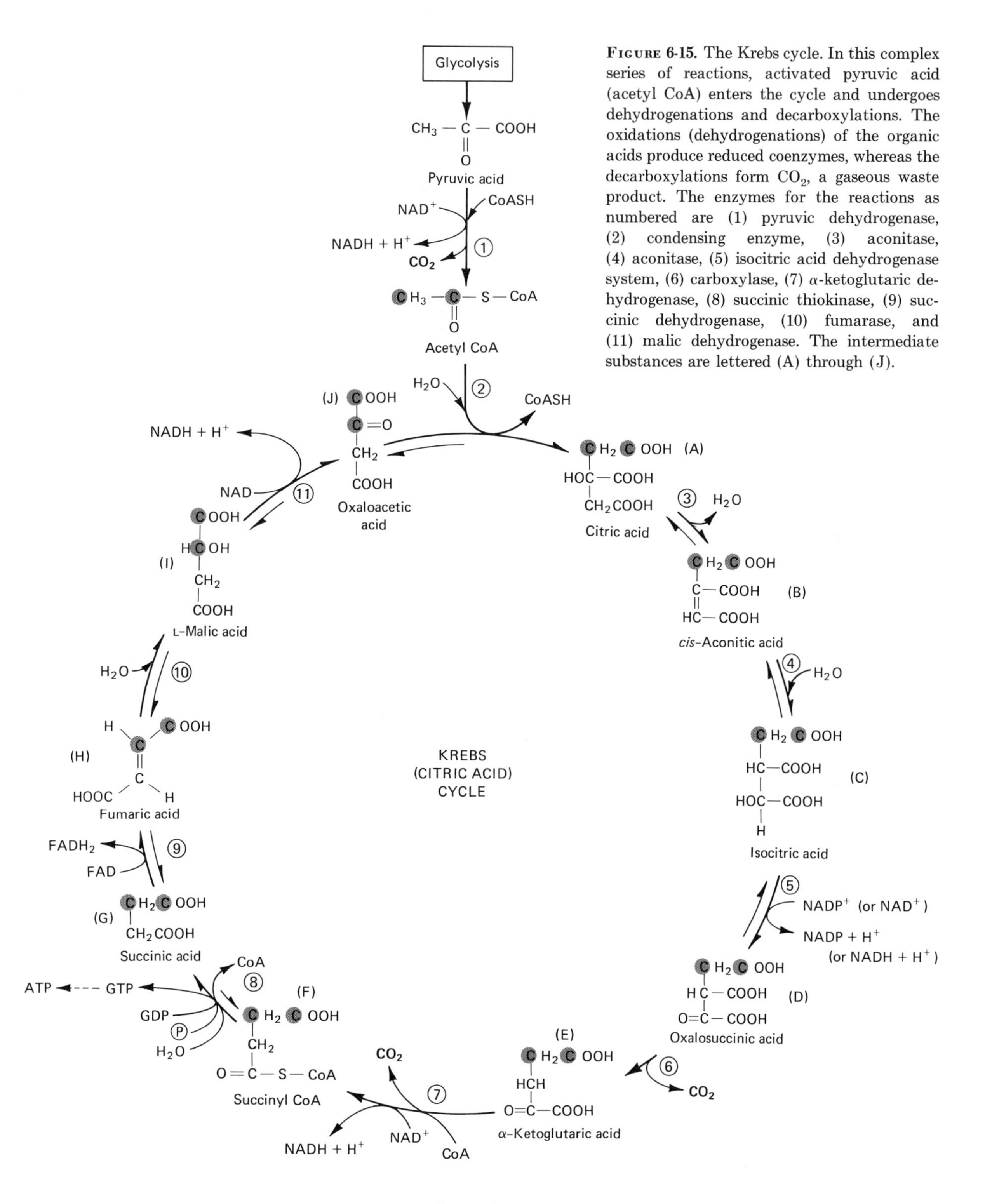

FIGURE 6-15. The Krebs cycle. In this complex series of reactions, activated pyruvic acid (acetyl CoA) enters the cycle and undergoes dehydrogenations and decarboxylations. The oxidations (dehydrogenations) of the organic acids produce reduced coenzymes, whereas the decarboxylations form CO_2, a gaseous waste product. The enzymes for the reactions as numbered are (1) pyruvic dehydrogenase, (2) condensing enzyme, (3) aconitase, (4) aconitase, (5) isocitric acid dehydrogenase system, (6) carboxylase, (7) α-ketoglutaric dehydrogenase, (8) succinic thiokinase, (9) succinic dehydrogenase, (10) fumarase, and (11) malic dehydrogenase. The intermediate substances are lettered (A) through (J).

acid (H). It is interesting to note that this reaction is the only oxidation in the cycle in which a pyridine nucleotide (NAD^+ or $NADP^+$) does not serve as the hydrogen acceptor coenzyme. Instead, *FAD* is reduced to *$FADH_2$*. Fumaric acid, by the addition of water, is then converted to the four-carbon acid *malic acid* (I).

In the fourth oxidation step (11) of the Krebs cycle, malic acid is converted to oxaloacetic acid and, in the process, NAD^+ is reduced to NADH + H^+. This reaction, which results in the regeneration of oxaloacetic acid (J), completes the cycle. The oxaloacetic acid can now combine with more acetyl CoA—SH to form citric acid and another cycle is initiated. Remember that since one molecule of glucose produces two pyruvic acid molecules during glycolysis, two turns of the cycle occur for each molecule of glucose oxidized. The carbon dioxide produced as a result of decarboxylation of the organic acids in the Krebs cycle is the source of the CO_2 given off as a product of respiration. Furthermore, examination of Figure 6-15 shows that the carbon dioxide molecules that are removed from the Krebs cycle come from the oxaloacetic acid and not from the acetyl coenzyme A that entered. Thus, it is only on the third time through the cycle that the carbons from the entering acetyl coenzyme A are removed.

Cytochrome System. Approximately 95% of the ATP produced during aerobic respiration is formed by the transfer of electrons from hydrogen acceptor molecules (NADH + H^+, NADPH + H^+, and $FADH_2$) formed during glycolysis and the Krebs cycle. These electrons are passed along, in serial succession, to electron acceptor molecules. The energy released in the reduction and oxidation (accepting and donating electrons) of the various acceptors at certain points in their pathway is used to convert ADP + P to ATP. The source of electrons in the scheme is hydrogen atoms. The electron is separated from the hydrogen atom, and the resulting hydrogen ion (H^+) is released into the cytoplasm. The entire sequence of electron transfers resulting in the production of ATP involves an organized group of electron acceptor molecules on the inner mitochondrial membrane referred to as the **cytochrome system** (Figure 6-16).

The hydrogen atoms removed from the pyruvic acid in glycolysis and from the various compounds of the Krebs cycle are picked up by NAD^+, which is reduced to

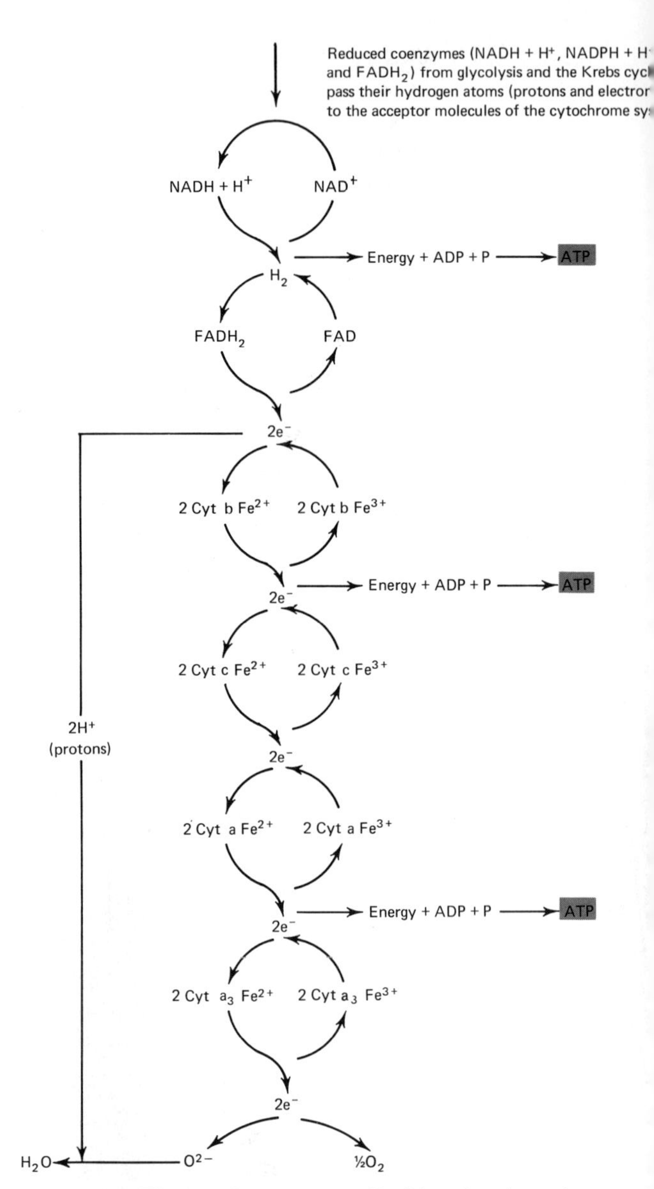

Figure 6-16. The cytochrome system. In this series of reactions, the main line of oxidation-reduction in cells that use oxygen, electrons are passed to various electron acceptors and, in the process, energy is released and stored in ATP. Dehydrogenase enzymes remove hydrogen atoms (protons and electrons) from reduced coenzymes formed during glycolysis and in the Krebs cycle. These hydrogen atoms are passed to NAD^+ and then to FAD. FAD liberates the protons into the mitochondrial matrix and passes the electrons to the cytochromes. The last cytochrome in the series, cytochrome oxidase (cytochrome a_3), passes the electrons to oxygen, and protons are picked up to form water. Note the three junctures at which ATP is formed.

NADH + H^+ ($NADP^+$ is used instead in some species). This compound is the initial hydrogen acceptor in the cytochrome system. The other acceptor molecules are *FAD, cytochrome b, cytochrome c, cytochrome a,* and *cytochrome* a_3 (*cytochrome oxidase*), the terminal acceptor molecule in the chain. Cytochromes are proteins that have a reversible valency charge on their iron-containing group. As components of the cytochrome system, they are alternately oxidized and reduced as they transfer electrons. The iron portions of the molecules can change their valency by going from Fe^{2+} (reduced form) to Fe^{3+} (oxidized form). Entire hydrogen atoms are passed from NADH + H^+ to FAD. The ionization of hydrogen occurs with the oxidation of $FADH_2$ so that from this point on in the system only electrons are transferred through the various cytochromes. When these electrons reach the end of the system, they are passed by cytochrome oxidase to molecular oxygen. The oxygen, now negatively charged, picks up the H^+ ions from the ionization of hydrogen to form water, a product of respiration. In this regard, oxygen serves as the final hydrogen acceptor during aerobic respiration. As electrons are passed along the system, they pass down an energy gradient, and at specific points their energy is transferred into ATP. Three ATP molecules are believed to be formed from three ADP and three inorganic phosphate molecules for each pair of electrons that passes through the complete system. If two hydrogens enter the system at the point of NAD^+ and are passed through the system, three ATP's are formed. However, if two hydrogens enter at the point of FAD, two ATP's are formed.

Fermentation and Aerobic Respiration Compared. Fermentation produces a total of four ATP molecules for each molecule of glucose that is oxidized. Inasmuch as two ATP molecules are utilized to activate glucose for reaction during glycolysis, only two of these four molecules of ATP are made available as stored chemical energy. Much of the energy in the glucose molecule has not been released in its breakdown to two pyruvic acid molecules, and when the pyruvic acid is converted to lactic acid or ethanol and CO_2 in the fermentation process, no energy is released. Thus, much of the original energy of the glucose molecule is not utilized by anaerobic organisms.

In contrast, aerobic respiration, through the agency of the Krebs cycle and the cytochrome system, completely oxidizes glucose into CO_2 and H_2O with the net production of 38 or 39 ATP molecules if GTP is converted to ATP. If glycogen is the starting substrate, 39 molecules of ATP are produced because the glucose unit, when separated from glycogen, is already phosphorylated when it enters the glycolysis pathway. If, on the other hand, glucose is the initial substrate, it must first be phosphorylated by the conversion of ATP to ADP. Thus, if glycogen is utilized, one less ATP is consumed. These 38 (or 39) ATP molecules represent 40–60% of the total energy present in a glucose molecule. The remaining energy not recovered by ATP is lost in the cell as heat.

Both fermentation and aerobic respiration involve a series of oxidation reactions in which the potential energy of glucose is systematically passed to ATP. Two molecules of ATP are initially expended in order to activate glucose for reaction. Once activated, glucose may be visualized as passing down an energy hill, releasing energy at various steps in the formation of ATP. Since fermentation is a partial oxidation of glucose, only two ATP molecules are formed. Aerobic respiration releases an additional 36 ATP molecules from glucose for a net production of 38 ATP's. The end products of fermentation (alcoholic) are CO_2 and ethyl alcohol; CO_2 and H_2O are the end products of aerobic respiration.

Further Details of Carbohydrate Metabolism. For most aerobic cells, the oxidation of glucose occurs according to the series of reactions of glycolysis and the Krebs cycle. However, alternate pathways exist, one of the most important being a sequence of reactions collectively termed the *hexose monophosphate shunt* (Figure 6-17). Although we are not concerned with the details of this mechanism, the entire cycle is shown in Figure 6-17 to illustrate its complexity. The major point to be made regarding this cycle is that one of the intermediates is a five-carbon sugar, *ribulose-5-phosphate,* which is of major importance in the dark reactions of photosynthesis. It is also the source of a five-carbon sugar, *ribose-5-phosphate,* which is a constituent of nucleotides and nucleic acids. In further reactions, rearrangements occur in which the upper two or three carbons of the molecule are transferred as a unit to another sugar. By such reactions, sugars containing three, four, five, six, or seven carbon atoms are metabolized. Many of these sugars also participate in the CO_2 fixation phase of photosynthesis. The hexose monophosphate shunt could ultimately lead to the synthesis of 36 molecules of ATP. Thus, the energy released in the oxidation of glucose via this pathway is almost as efficient as that of the glycolysis–Krebs cycle pathway.

Fatty Acid and Amino Acid Metabolism

The discussion of catabolism thus far has been concerned largely with the oxidation of glucose, the principal energy-supplying carbohydrate of the cell. It should be noted, however, that the Krebs cycle (and to a lesser extent, glycolysis), in addition to representing the major pathways for carbohydrate metabolism, is also interrelated with the metabolism of the other chemical components of living forms. While there are certain essen-

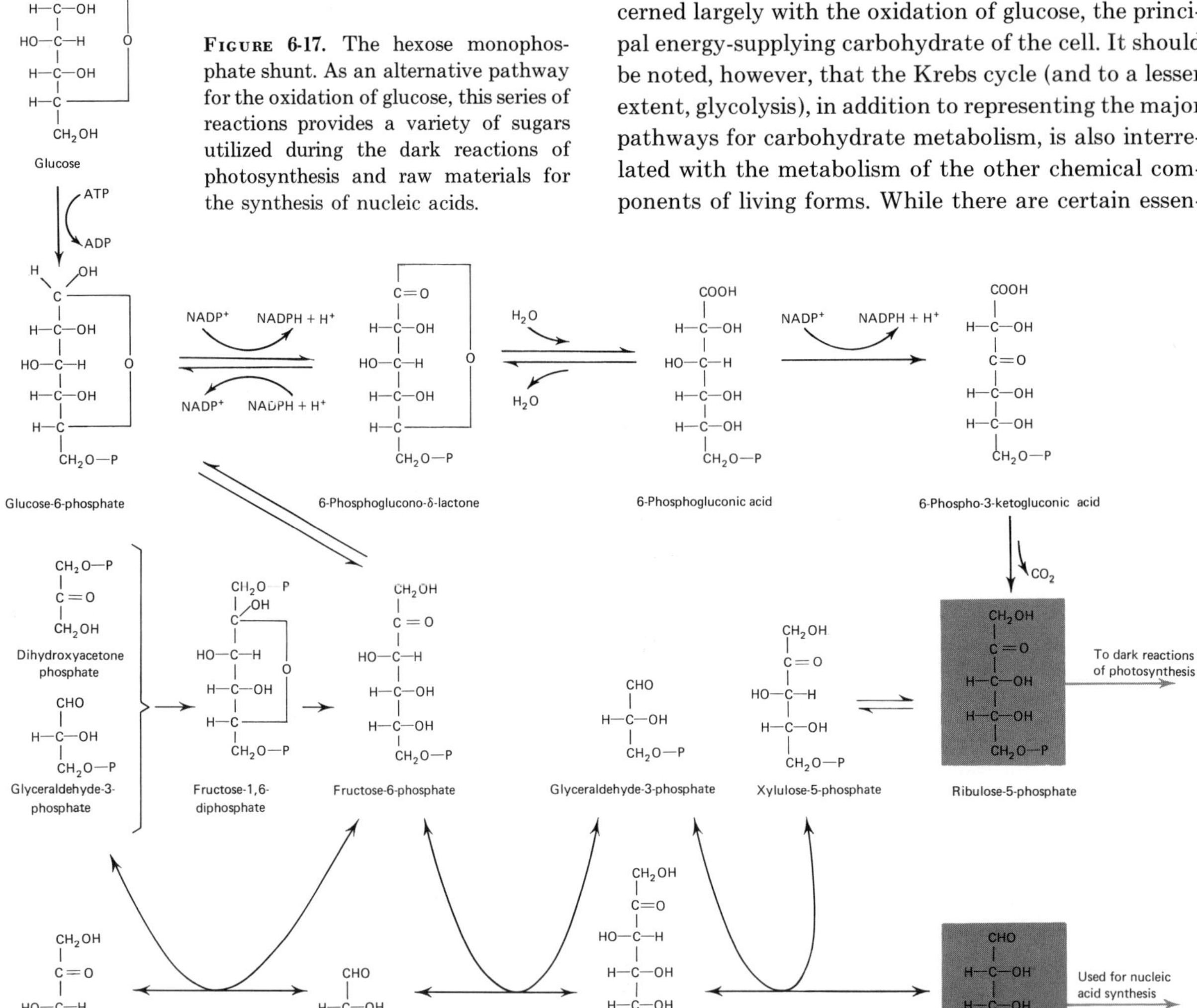

Figure 6-17. The hexose monophosphate shunt. As an alternative pathway for the oxidation of glucose, this series of reactions provides a variety of sugars utilized during the dark reactions of photosynthesis and raw materials for the synthesis of nucleic acids.

tial compounds (i.e., vitamins, minerals, and essential amino acids) that some living forms must obtain from their environment, others can be synthesized within the cell. Most nutrients are reduced to a relatively small molecular size during digestion. These smaller molecules are elaborated in the cell to form the specific chemical requirements of the organism, through various metabolic pathways. In addition, larger molecules produced in a cell for use in another cell are reduced in size before they can pass out of the cell.

Most of the chemical components making up living forms are constantly being renewed through breakdown and buildup processes. Since this involves the utilization of energy, the Krebs cycle becomes intimately involved. The various chemical substances produced within the Krebs cycle and glycolysis also provide some of the precursors for more complex molecules required by the cell. Other substances, such as fats and proteins, can be utilized as a source of energy. Complete breakdown of these substances involves the entrance of some of the intermediate products during breakdown into some step in the carbohydrate metabolic pathway. Several of these interrelationships will now be discussed.

Fatty Acid Catabolism (Beta Oxidation)

The first major attempt to study fat metabolism was made by F. Knoop, a German biochemist, in 1905. In an ingenius experiment he was able to attach a phenyl group to the hydrocarbon end of fatty acids containing odd and even numbers of carbon atoms. He fed these acids to dogs and then analyzed their urine for the end products of oxidation. He found that the end products of the odd-numbered carbon fatty acids yielded derivatives of benzoic acid and that the end products of the even-numbered fatty acids yielded derivatives of phenylacetic acids (Figure 6-18). From this analysis he deduced that the breakdown of fatty acids occurs at the carboxyl or acid end. Subsequent extension of Knoop's findings indicated that there was a sequential removal of two carbon fragments by oxidation at the beta carbon. After many years of investigation the details of the mechanism of beta oxidation of fatty acids were worked out.

In the breakdown of a fatty acid one ATP molecule is introduced to activate the reaction by the formation of a coenzyme A derivative. Through a series of steps

Figure 6-18. Beta oxidation of fatty acids. Consult text for discussion.

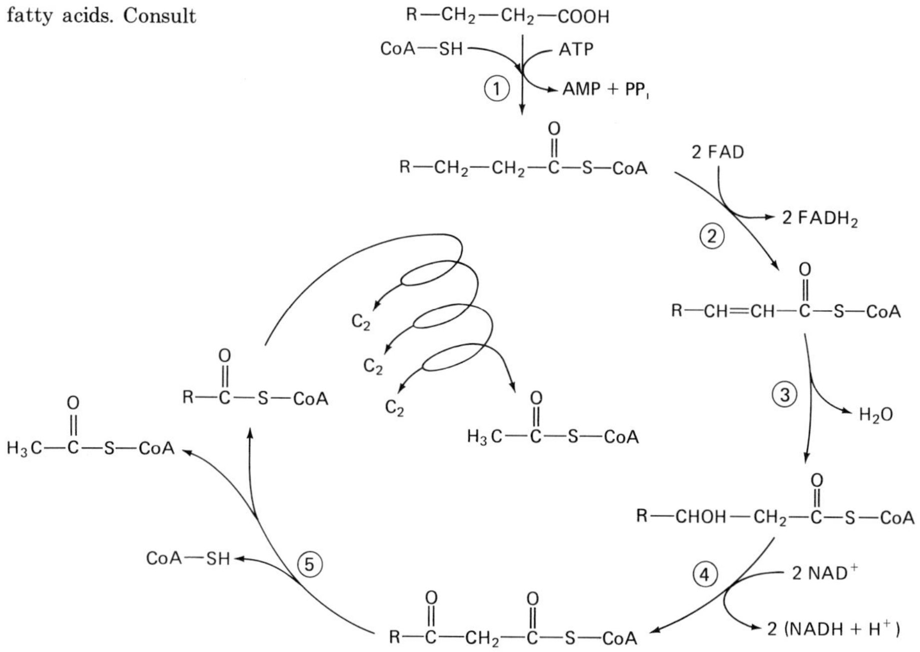

involving removal of four hydrogen atoms and the addition of water and another coenzyme A molecule, this molecule is cleaved into one two-carbon fragment, acetyl coenzyme A, and another coenzyme A derivative of the acid, which is two carbons shorter than the original acid. Since this derivative is already in the coenzyme A form, it can be passed through the pathway again without being activated; the products again are a two-carbon fragment and a shorter coenzyme A derivative. The sequence is repeated until the coenzyme A derivative is completely broken up into two-carbon acetyl coenzyme A fragments. Since this is a catabolic reaction, it is energy yielding. In two steps of the pathway hydrogens are removed. In one step FAD accepts two hydrogens and two ATP molecules are formed. In the other step NAD^+ accepts two hydrogens and three molecules of ATP are formed. Therefore, each time an acetyl coenzyme A is removed from the fatty acid five ATPs are produced. It is possible to calculate the net gain of ATPs from beta oxidation of a fatty acid.

1. Count the carbons in the acid and divide by 2 to find the number of acetyl coenzyme A molecules produced.
2. Subtract 1 to get the number of coenzyme A molecules *removed* (at the final degradation removal of *one* produces *two* coenzyme A molecules).
3. Multiply by 5.
4. Subtract 1 for the ATP used to activate the original acid.

For example, lauric acid (see Figure 2-8) contains twelve carbon atoms. This will produce six acetyl coenzyme A molecules. For each acetyl coenzyme A formed, except for the last one (the sixth is produced when the fifth one is removed), five ATPs are produced. Therefore 25 ATP molecules will be formed. However, since one ATP is needed to start the reaction, there will be a net gain of 24 ATPs in the degradation of lauric acid to six acetyl coenzyme A molecules. The hydrogens accepted by FAD and NAD^+ are passed through the cytochrome system and finally combine with atmospheric oxygen. The reaction, therefore, is aerobic.

Beta oxidation is only part of the complete oxidation of fatty acids. The resulting acetyl coenzyme A molecules may be further degraded in the high-energy-yielding Krebs cycle. Since each fatty acid generally produces a relatively high number of acetyl coenzyme A molecules, it is apparent that fats, consisting of from one to three fatty acids, are exceedingly high energy compounds. Fats are the principal source of fatty acids for organisms; there are relatively few free fatty acids within the cell. Hydrolysis of fats yields fatty acids and glycerol. The glycerol may also be metabolized through the glycolysis pathway, beginning at the point of the three-carbon compounds.

Fatty acids are also synthesized within the cell. However, the buildup is by a different pathway. The process of fatty acid synthesis will be discussed in the next chapter.

Amino Acid Catabolism

Proteins, unlike carbohydrates and simple lipids, are not primarily energy-producing compounds. However, within the living organism all molecules are constantly being metabolized. Amino acids, the building blocks of proteins, have been synthesized from simpler molecules and as such contain energy. When these are catabolized, they release energy that can be utilized in much the same manner as the breakdown products of carbohydrates and fats.

Much is known about the individual metabolic pathways of the twenty amino acids that make up proteins; however, this discussion will be limited to generalized reactions that apply to all amino acids. These reactions are classified as transamination, deamination, and decarboxylation.

TRANSAMINATION. **Transamination,** as the term implies, involves the enzymatic transfer of an amine group from one amino acid to another acid of a keto type (one containing a double-bonded oxygen on the alpha carbon). The result of this transfer is a new amino acid and a keto acid of the original amino acid donor.

$$R_1{-}\underset{NH_2}{\overset{H}{C}}{-}C\!\!\begin{smallmatrix}{}^{\nearrow O}\\{}_{\searrow OH}\end{smallmatrix} + R_2{-}\underset{O}{\overset{\|}{C}}{-}COOH \xrightleftharpoons{\text{transaminase}} R_1{-}\underset{O}{\overset{\|}{C}}{-}COOH + R_2{-}\underset{NH_2}{\overset{H}{C}}{-}COOH$$

Given the proper keto carbon skeleton the transaminases are capable of forming nearly all of the amino acids. These reactions are readily reversible and occur in both the cytoplasm and mitochondria. In many instances the resulting keto acid fits into the carbohydrate catabolism pathway.

Deamination. Deamination, the removal of the amino group, may be oxidative or nonoxidative. Oxidative deamination is further divided into one type of reaction involving NAD^+ or $NADP^+$ and a second type involving FAD. NAD^+ is involved in the deamination of glutamic acid. The glutamic acid is converted to α-ketoglutaric acid, which is an important compound in the Krebs cycle; therefore, the reaction is closely integrated with this metabolic pathway.

$$HOOC{-}CH(NH_2){-}CH_2{-}CH_2{-}COOH + NAD^+ + H_2O \rightleftharpoons$$

$$HOOC{-}C({=}O){-}CH_2{-}CH_2{-}COOH + NADH + H^+ + NH_3$$

The ammonia (NH_3) resulting from this reaction may then be picked up in the urea cycle and excreted. The glutaric acid may be metabolized in the Krebs cycle to yield energy. The deamination of glutamic acid may also be coupled with a transamination reaction and used to synthesize other amino acids.

The deamination reaction involving FAD is somewhat more complicated. Through a series of steps the amine group is removed and the resulting carbon skeleton is further catabolized. This reaction is not reversible and cannot be used to synthesize other amino acids.

Nonoxidative deamination involves specific enzymes for the various amino acids. Many of the compounds produced by this removal of the amine group participate in the Krebs cycle.

$$HOOC{-}CH(NH_2){-}CH_2{-}COOH \xrightleftharpoons{\text{deaminase}} HOOC{-}CH{=}CH{-}COOH + NH_3$$

Decarboxylation. Decarboxylation reactions involve the enzymatic removal of carbon dioxide. Unlike transamination, which is both anabolic and catabolic, decarboxylation is important only in anabolic reactions. The reaction involves specific enzymes in which carbon dioxide is removed from the amino acid leaving an amine.

$$R{-}CH(NH_2){-}COOH \xrightleftharpoons{\text{decarboxylase}} R{-}CH_2{-}NH_2 + CO_2$$

Urea Cycle. Closely associated with amino acid metabolism and the Krebs cycle is the urea cycle of mammals. Although nitrogen is retained within the organism through transamination sufficient quantity is lost through catabolism of amino acid as to require constant replacement. In fact, sufficient replacement through protein consumption is the major food problem of the world. In the breakdown of amino acids and other compounds containing nitrogen, the nitrogen is usually removed as ammonia (NH_3). This substance is toxic and must be removed. Through a series of reactions, the ammonia combines with carbon dioxide and is converted to urea, the nitrogenous waste product of mammals.

Intermediate Metabolism

The photosynthetic process provides the major source of energy either directly or indirectly for most living organisms. The various biochemical pathways discussed in this chapter have been greatly oversimplified. It is beyond the scope of this text to present the complex step-by-step metabolism of the biochemical pathways. However, it should be somewhat apparent that there is considerable interrelationship among all the various schemes (Figure 6-19). Central to metabolism in almost all cells is the Krebs cycle. Not only does the Krebs cycle perform the catabolic function of yielding available energy, but it performs a crucial role in the biosynthesis of important metabolites ranging from the amino acids, purines, and pyrimidines to long-chain fatty acids and steroids.

The various interconversions of metabolites, some of which are shown in Figure 6-19, provide a great deal of flexibility in the cell's economy. While the precursors of

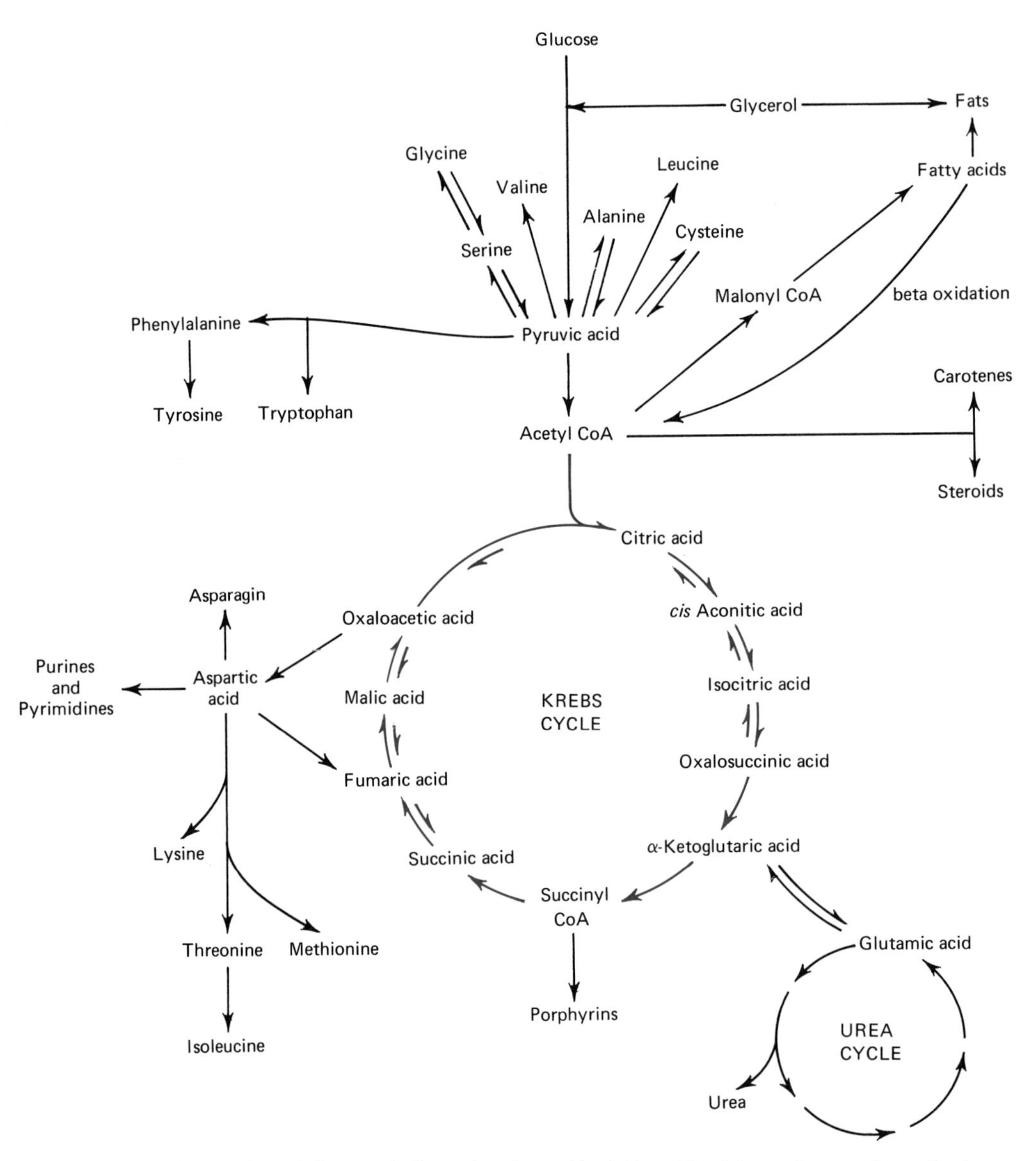

FIGURE 6-19. Integration of the metabolism of amino acids, fatty acids, fats, purines, and porphyrins with the Krebs cycle.

large molecules, as well as energy-rich molecules, are obtained in the consumption of nutrients, the cell need not depend entirely upon obtaining each specific type of molecule from the environment. The combination of glycolysis and the Krebs cycle with other metabolic pathways is capable of manufacturing many of the macromolecules, which then can be altered to meet particular cell requirements. Many of the reactions of this complex scheme, often referred to as *intermediary metabolism,* are reversible. In effect, this means that whereas one set of reactions (catabolic) may produce energy-rich compounds for immediate utilization of energy, the same reactions may, in reverse, lead to synthesis.

7 Cells and Energy: Utilization

GUIDE QUESTIONS

1 List and briefly discuss the forms of energy that perform biological work.
2 What is bioluminescence? How widespread is it among living forms? Describe the bioluminescent reaction in the firefly.
3 How is bioelectricity related to neuronal function? Enumerate the principal kinds of neurons with regard to function. What is the function of the following: dendritic zone, perikaryon, axon, axon telodendria?
4 Outline and diagram the initiation and transmission of a nerve impulse along the surface of a neuronal membrane.
5 List and discuss the major steps involved in impulse transmission from one neuron to another.
6 Describe the structure of a skeletal muscle at various levels of structural organization. How is a skeletal muscle fiber adapted to impulse conduction?
7 By means of sequential illustrations, describe the principal features of the sliding filament hypothesis.
8 Discuss the biochemistry of muscular contraction. What role does creatine phosphate play in the process? What is an oxygen debt?
9 How are flagellary, ciliary, and amoeboid movements accomplished?
10 Define biosynthesis. Relate photosynthesis to polysaccharide synthesis. How are glycolysis and lipid biosynthesis interrelated? How are amino acids synthesized?

In Chapter 6 it was shown that adenosine triphosphate is the immediate source and usable form of energy for cellular work. Through the photosynthetic activities of autotrophs, solar radiation is transduced into chemical energy in the bonds of food molecules. Both autotrophs and heterotrophs oxidize food molecules and transfer the energy within the bonds of these molecules to ATP. The chemical energy of ATP, in turn, is transduced into various forms of energy that are capable of performing useful biological work. Among these forms of energy are bioluminescence (light energy), bioelectricity (electrical energy), active transport (chemical energy), motion (mechanical energy), and biosynthesis (chemical energy). The general scheme for the acquisition, transduction, and utilization of energy is summarized in Figure 7-1.

Bioluminescence (Light Energy)

Bioluminescence is the ability of an organism to emit light. Perhaps the firefly is the most familiar light-emitting organism. Bioluminescent organisms are also found among bacteria, fungi, radiolarians, sponges, flagellates, corals, nemerteans, ctenophores, echinoderms, clams, snails, crustaceans, squids, centipedes, millipedes, insects, and—many fishes are luminous. Bioluminescence has not been observed in amphibians, reptiles, birds, or mammals. No plants are bioluminescent.

It appears that there is no single evolutionary line of luminescent forms; the capacity to produce light has developed independently several times during evolution. The adaptive nature of bioluminescence among advanced multicellular organisms may be illustrated by considering a few examples. In fireworms and fireflies, it serves as a mating signal. The female of a species of marine worm comes to the surface and secretes a glowing circle of green luminous material along with her eggs. The males, which normally stay well below the surface, swim toward the small circle of glowing material, emitting short flashes of light as they travel. Frequently, several males converge on a single female,

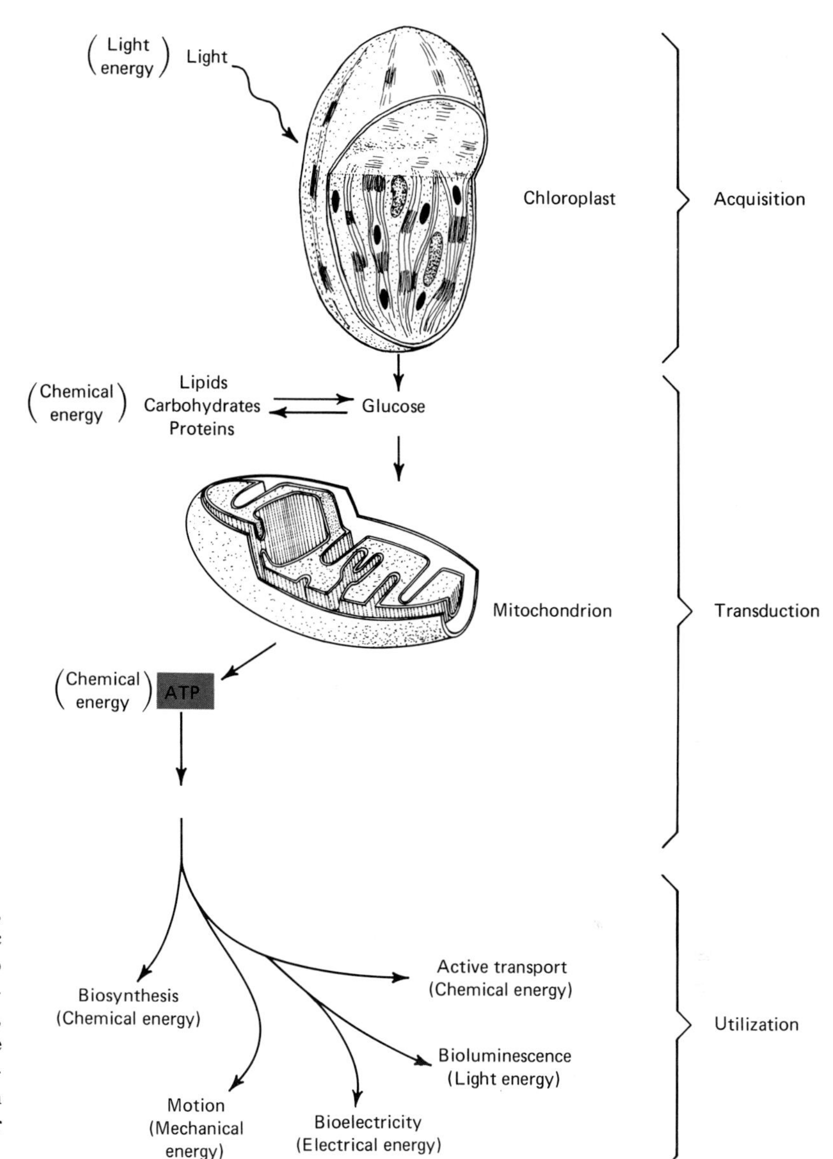

FIGURE 7-1. The acquisition, transduction, and utilization of energy. Photosynthetic autotrophs transduce solar radiation into chemical energy in the bonds of food molecules. Both autotrophs and heterotrophs, through the agency of various oxidative processes, convert the energy of food molecules into the chemical energy of ATP, in which form it can be utilized by the cell for biological work.

rotate in the glowing circle, and discharge sperm into the water. Bioluminescence can also be used to obtain food. A deep-sea angler fish carries a light-emitting organ at the top of a rectractable appendage with which it lures prey into its jaws. A deep-sea squid discharges a protective bioluminescent screen for camouflage. The function of bioluminescence among lower forms such as bacteria, dinoflagellates, and fungi is less obvious; it may only represent a by-product of metabolism.

At the cellular level, bioluminescence is the transduction of chemical energy into light energy in a series of enzyme-controlled reactions. The details of the mechanism differ for each species, but there are similarities among the cases that have been studied. The general terms *luciferin* for the substrate and *luciferase* for the enzyme are used, although these components of bioluminescence and the reactions in which they occur do differ considerably among various bioluminescent organisms. Since the bioluminescence of the firefly has been intensively studied, it will be used here as an example. The luciferin–luciferase reaction of the firefly is a form of oxidation and can occur only in the presence of oxygen. If luciferin and luciferase are extracted from a firefly and mixed with ATP in a test tube, luminescence occurs. The ATP supplies the energy for the reaction as its chemical energy is converted to light energy. The mechanism of light production in the firefly is as follows:

$$\underset{\text{(reduced)}}{\text{Luciferin}} + O_2 \xrightarrow[\text{ATP} \curvearrowright \text{AMP}]{\text{luciferase}} \underset{\text{(oxidized)}}{\text{luciferin}} + CO_2 + \text{light}$$

The bioluminescence of organisms may last for extended periods or may occur in brief flashes. The light emitted by different organisms is entirely in the visible spectrum and is perceived by the human eye as red, green, yellow, or blue. Some organisms are able to produce two different colors. The actual colors that are perceived are probably due to variations in enzyme structure or conformation. Bioluminescence is sometimes referred to as "cold light" since most of the energy is expended in light production and very little is lost as heat.

Bioelectricity (Electrical Energy of Cells)

All cells examined have been found to possess a difference of *electrical potential* across their plasma membrane. That is, the inside of the membrane is at a different electrical potential than the outside. This potential difference is based primarily on active transport mechanisms that establish a difference in ion concentrations between the inside and outside of the cell. The chemical energy of ATP is used to establish ion concentration differences across the membrane and these, in turn, lead to the development of differences in electrical potential. All membrane potentials arise from unequal charge movement.

In nearly all cells, active transport mechanisms pump sodium ions out of the cell as soon as they enter by diffusion down their concentration gradient. Potassium is more diffusible through membranes, but its diffusion outward is retarded because of the presence of other substances (ATP, proteins, etc.) in all cells. Potassium ions are associated with these negatively charged impermeable materials. Although potassium ions tend to move out of the cell in response to the concentration gradient between the inside and outside of the cell, they cannot do so because they must remain associated with the negatively charged impermeable materials in the cell. Such a separation of charges requires work and the cell has no mechanism for accomplishing such work. However, the tendency for potassium ions to move through the membrane in response to the concentration gradient means that a few potassium ions enter the membrane and tend to move outside. This tendency for the outward movement of positive potassium ions establishes a positive charge on the outside of the membrane, while the negative impermeable materials inside tend to make the inside of the membrane negative with respect to the outside. A potential difference is created across the membrane that just balances the force of the potassium ion concentration gradient. The electrical sign is such as to attract potassium ions on the inside and repel them on the outside of the membrane. The magnitude of this transmembrane potential is determined by the actual concentrations of diffusible ions inside and outside the cell and varies from about 1 mV (millivolt) in human erythrocytes to 50–75 mV in larger nerve and muscle cells. The inside of the membrane is always negative with respect to the outside. Any potassium ions that actually escape to the outside are usually returned by active transport mechanisms, which return the potassium ions against their concentration gradient to the inside of the cell.

Excitable cells are those that develop electrical charges as a result of stimulation of their membranes, such as nerve and muscle cells. The transmembrane potential present in the unstimulated excitable cell is called the *resting potential* to distinguish it from other active potentials that arise following stimulation. The basis of the resting potential is that described above for the transmembrane potential of any cell.

Nerve cells, especially, are specialized for the conduction and transmission of electrical signals, and cells of this type possess a membrane whose permeability to sodium and potassium ions can change drastically upon stimulation. Changes in membrane permeability to specific ions at specific times can produce changes in the membrane potential; such changes lie at the basis of the nerve impulse (action potential) or the action potential of other excitable cells such as muscle. Before considering these potential changes, it is necessary to discuss briefly the morphology of the nerve cell.

The unit of nervous tissue is the *neuron,* or nerve cell. Depending upon their functions, neurons may be classified into three types.

1. *Sensory neurons,* located in receptor structures, pick up an impulse and transmit it to the brain or spinal cord.

2. *Association neurons,* found in the brain and spinal cord, form the intermediate link in the pathway of almost all nervous coordination.

3. *Motor neurons* transmit impulses from association (or sensory) neurons to muscles and glands to elicit coordinated responses.

All neurons are specialized to conduct electric impulses from one part of the organism to another—sometimes over long distances. Although at the functional level, neurons operate on similar mechanisms, the gross morphological appearance of neurons is varied. But in general terms (to which there are many exceptions), neurons may be considered to consist of a nucleated *cell body* (which functions in normal cellular metabolism) and two or more cytoplasmic extensions from the cell body, or perikaryon. Two types of morphological processes are recognized: **axons**—thicker, often longer processes, which often have branches (collaterals) that are responsible for conduction of impulses; and **dendrites**—thinner, usually profusely branched processes that are responsible either for receiving sensory information from the environment or for transmitting impulses, i.e., receiving impulses from other neurons.

Figure 7-2 represents a few receptor and effector neu-

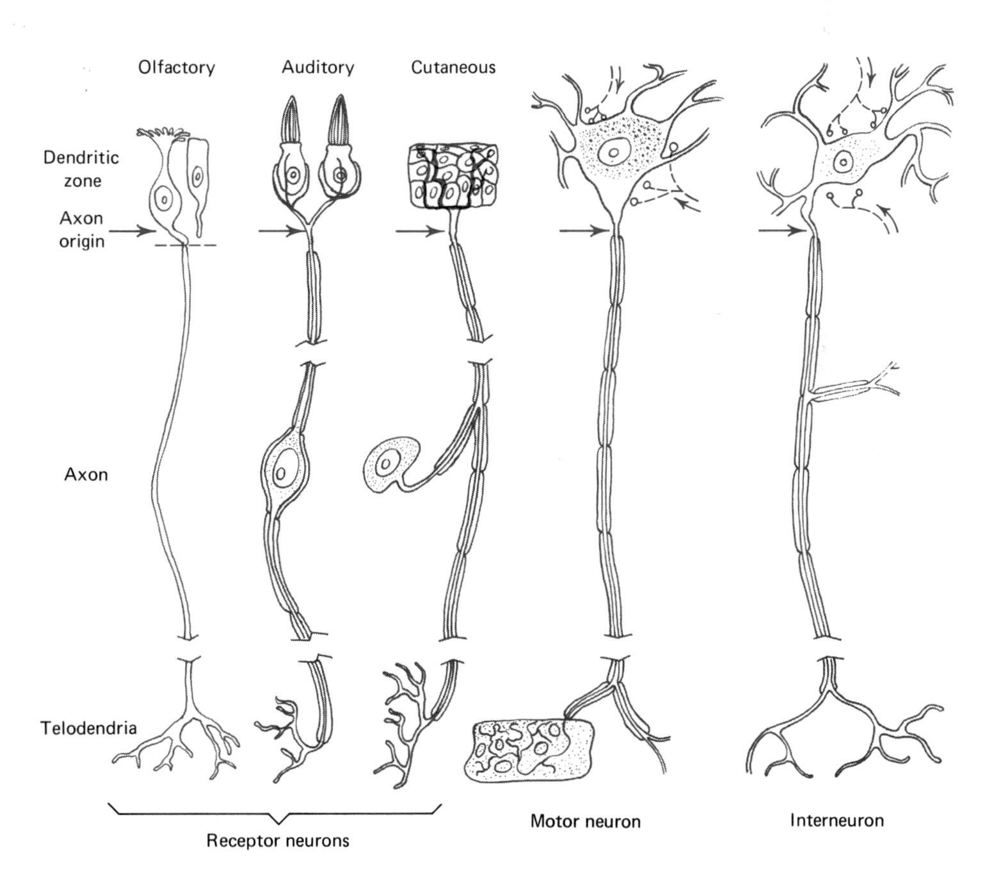

FIGURE 7-2. Receptor and effector neurons of vertebrates arranged to illustrate the idea that impulse origin, rather than cell body position, is the most reasonable focal point for the analysis of neuron structure in functional terms in neurons with an axon process. In all axon-bearing neurons, the four major points of interest are the dendritic (receptor) zone, the area of axon origin, the axon (conducting) zone, and telodendria (transmitting or synaptic) zone. Although the location of the perikaryon does not have a constant relationship to the geometry of neurons in general, its location is fixed in any specified type of neuron. [*From David Bodian,* Science, *vol. 137, pp. 323–26, 1962, Fig. 1. Copyright © 1962 by the American Association for the Advancement of Science. By permission of the publisher and author.*]

rons of vertebrates. Of the nearly 100 different neuronal types, only these have been selected to illustrate neuronal structure in functional terms. The *dendritic zone,* the receptor membrane of a neuron, consists of cytoplasmic extensions (dendrites) that receive synaptic endings of other neurons and are differentiated to transduce environmental stimuli into local response-generating activity. The axon is a single, often branched, and typically elongated cytoplasmic extension that is morphologically and uniquely differentiated to conduct impulses away from the dendritic zone. The cell body, or *perikaryon,* is a nucleated cytoplasmic mass characterized by chromidial substance. It is the center of embryonic growth of dendrites and axons, axon regeneration, and possibly enzymatic synthesis in the differentiated neuron. Morphologically, it may appear in the dendritic zone, within the axon, or attached to the axon. The terminal portions of axons, the *axon telodendria,* are diversely differentiated and exhibit membranic and cytoplasmic differentiations adapted to synaptic transmission or neurosecretory activity. The bulb-like terminals of the telodendria commonly contain concentrations of mitochondria, secretory granules, or synaptic vesicles. Telodendria transmit electrical or chemical signals capable of giving rise to generator potentials in the dendritic zone of other neurons and tissue, stimulatory effects in innervated glandular tissue, or in distant cells via neurohormones.

It is generally believed by most neurophysiologists that a nerve impulse is the result of a series of electrochemical changes over the membrane of a neuron. Although impulses are a function of many other kinds of cells besides neurons, the impulse phenomenon as it relates to neurons only will be discussed here. As noted earlier, the neuron is surrounded by a semipermeable membrane, the inside of which is always negative with respect to the outside. This transmembrane potential (resting potential) exists because of the selective permeability of cell membranes to sodium and potassium ions. Measurements indicate that there are about twenty times as many potassium ions (K^+) inside the cell as outside, and approximately ten times as many sodium ions (Na^+) outside the cell as inside. The energy stored in ATP is used to actively transport Na^+ ions outside and K^+ ions inside. These differences in concentration create a strong tendency for the K^+ ions to diffuse out of and the Na^+ ions to diffuse into the cell. The membrane of the resting (unstimulated) neuron is virtually impermeable to the diffusion of Na^+ ions; it is also very impermeable to K^+ ions. Negatively charged ions of both inorganic chloride and organic forms distribute themselves in such a way as to achieve an electrically neutral state on both sides of the neuronal membrane. However, the complex system of ions involved cannot overcome the selective property of the membrane in concentrating K^+ ions inside the cell and Na^+ ions outside. As a result, the ions cannot distribute themselves so that the same number of both ions and electrical charges appear inside and outside the cell. Thus, in a resting cell, the inside of the cell is electrically negative with regard to the outside; that is, there is a positive charge on the outside of the membrane. Recall that such an electrical potential existing across the membrane of the cell is referred to as the *resting potential* (Figure 7-3A).

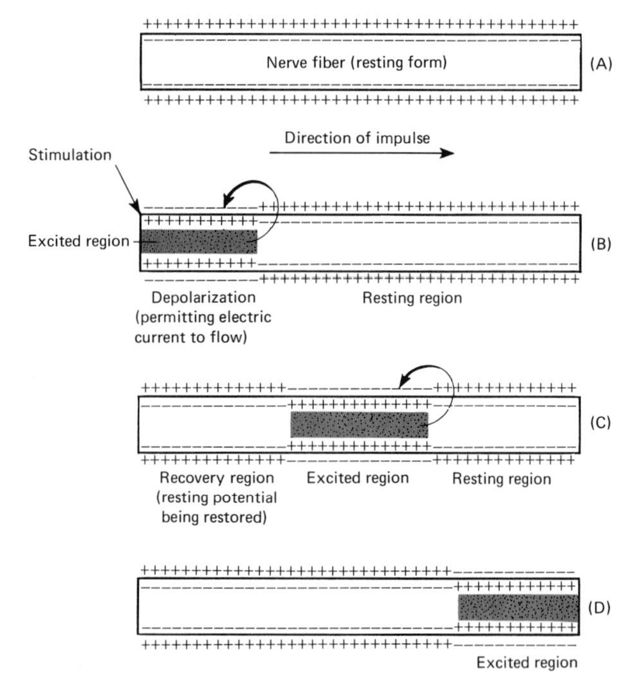

FIGURE 7-3. Diagrammatic representation of impulse conduction along the surface of a neuronal membrane. The colored area represents the excited region of the neuronal membrane; the direction of movement of the impulse is from left to right. See text for discussion of events involved.

If any factor should alter the normal permeability of the membrane, the distribution of ions is affected and the potential across the membrane is also affected. In stimulation of a nerve, the resting potential is increased (and reversed in sign) by changing the permeability of the membrane until it has reached a high enough level to become propagated along the neuronal membrane. This propagated impulse is called the *action potential.*

Stimulation of a nerve cell causes the membrane permeability to ions to change. There is a localized increased permeability to Na^+, while K^+ permeability remains the same. This causes Na^+ to move into the cell, down its concentration gradient. The inward movement of the positively charged Na^+ ions causes *depolarization* of the membrane, i.e., the potential difference across the membrane starts to decrease from the resting potential level. If the stimulus is great enough, the permeability of the membrane is changed enough so that sodium ions rush into the nerve cell. Since the stimulus acts by depolarization, it is complete; the membrane potential falls to zero. In fact, the membrane potential actually reverses sign, and the membrane becomes polarized with the outside negative (in the resting condition, the membrane potential is polarized with the outside positive). This *reversal of membrane potential* is caused by the higher level of Na^+ inside the cell (Figure 7-3B). The reversal of the membrane potential is part of the action potential (nerve impulse).

The action potential is conducted along the length of the axonal membrane because an impulse at one point sends out electrical currents to other regions of the membrane. These local currents act as stimulating agents, causing a slight depolarization of the membrane that again acts as an agent in increasing the membrane permeability to sodium ions in this new region (Figure 7-3C).

When the peak of the action potential is reached, a new series of changes in membrane permeability is initiated. The membrane now becomes permeable to K^+ and impermeable to Na^+. Potassium ions leave the cell down their concentration gradient, and this outward movement of K^+ causes the membrane potential to fall back to its normal resting level of polarization. When the membrane has reached its resting level of potential, a recovery period follows in which the resting state of membrane permeability is restored and any ions that have moved into or out of the nerve cell are restored to their original site (Figure 7-3D). While the ionic movements responsible for producing the changes in potential of the nerve membranes are well known, the mechanisms by which the membrane specifically changes its permeability to given ions are not known.

The conduction of an impulse through a neuron ultimately terminates at the end of a fiber (axon). At this point the impulse must travel across a gap between nerve cells called a **synapse** (Figure 7-4). The mechanism of transmitting a nerve impulse from one neuron to another is different from those involved in conducting an impulse along an axon. Such transmission may be achieved by either chemical or electrical means. At most nerve–nerve functional junctions (synapses), the impulse arriving at an axon terminal causes the release of a chemical transmitter. This chemical diffuses across the small gap between the axon ending and the dendritic portion of the next nerve cell and attaches to the dendritic membrane, causing a change in permeability to ions. In the majority of cases, the permeability change is such that all ions become free to move across the membrane: Na^+ into the cell, K^+ and Cl^- out of the cell. This movement of ions depolarizes the membrane, the potential change being a *synaptic potential.* This potential change sets up an ionic current which moves through adjacent axon membranes and initiates an action potential in them (Figure 7-4).

The best-known chemical transmitter is *acetylcholine,* which functions at nerve–nerve and nerve–muscle junctions in most peripheral vertebrate synapses. After acetylcholine has caused a depolarization of the postsynaptic membrane, the enzyme *acetylcholinesterase* destroys the acetylcholine and thus prevents continuous depolarization of the membrane. Since axons alone produce acetylcholine, impulses can proceed only from the axon of one nerve to the dendritic zone of another nerve cell.

In an increasing number of cases, it has been found that transmission at synapses occurs by direct electrical transmission. A nerve impulse at an axon ending sets up flows of currents that go through the postsynaptic membrane and cause depolarization and the generation of a postsynaptic potential. The postsynaptic potential, in turn, can cause the production of an action potential

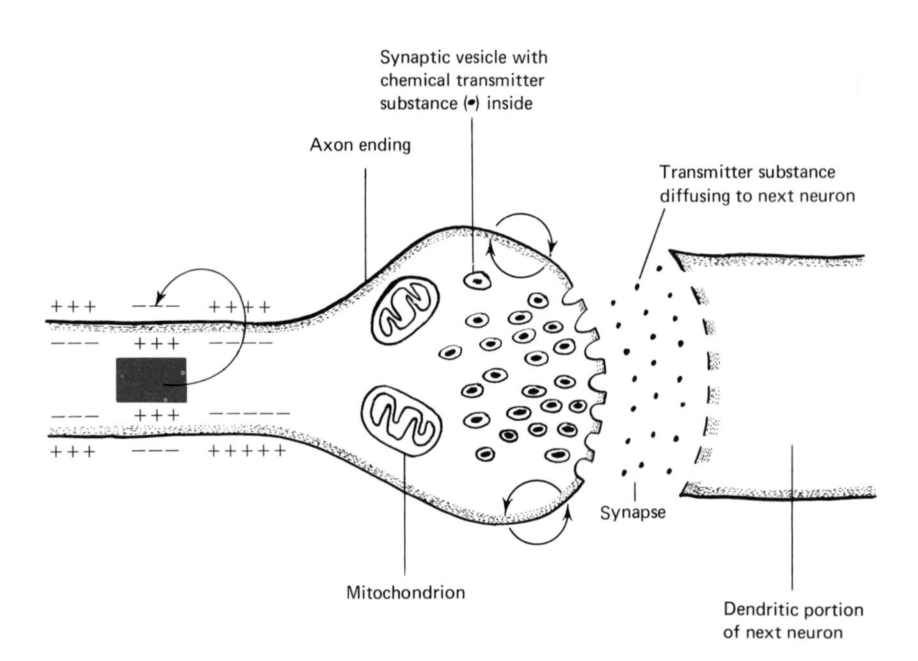

FIGURE 7-4. Diagrammatic representation of impulse conduction from the axon of one neuron to the dendrite of another neuron. Refer to text for details.

in the axon membrane of that nerve cell. Note that dendritic regions (which include the membrane of dendrites and the membrane of the nerve cell body in some cases) can generate only synaptic (receptor) potentials if the dendritic zone is part of a receptor. These potentials establish ionic current flows that, by slight depolarization, can lead to a change in Na^+ permeability of axon membranes. Only axon membranes exhibit action potentials that are conducted. Some electrically transmitting synapses are rectified and impulses can pass in only one direction; others are not polarized, and impulses can pass in either direction.

Although bioelectricity is a by-product of all cellular processes in which ions play a part, certain organisms are highly specialized in their ability to produce electricity. Marine vertebrates such as the ray and the electric eel contain electric organs capable of converting the energy of ATP into an electric current. These electric organs develop from muscle-like cells that are disc-shaped, noncontractile, and arranged into stacks somewhat like storage batteries connected in a series. The generation of electricity in these cells depends on ATP and acetylcholine.

Active Transport (Chemical Energy)

Inasmuch as the proposed mechanism of active transport has already been discussed in Chapter 5, its details will not be examined further. The importance of the process in the transmission of an impulse should now be obvious.

Motion (Mechanical Energy)

One of the most important functions of energy in living organisms is to carry on mechanical work (motion). This movement may be the external movement of the entire organism or the internal movement of parts of the organism. A characteristic of higher animals is muscular movement. The muscular system of these organisms, in addition to facilitating running, swimming, flying, and climbing, also maintains breathing, heart beat, blood pressure, posture, and shape.

Essentially the same mechanism involved in motion is also involved in maintaining tension. For example, if a man lifts a weight and then holds it steady, the initial mechanical movement is then changed to an input of energy in maintaining the weight by tension of his muscles. It is important not to confuse the concepts of work and energy at this point. Inasmuch as work is defined as a force acting through a distance, the man holding a weight is not doing work. However, energy is used for muscle tension in maintaining the weight in position. Similarly, tension is necessary to maintain posture and/or shape.

The movement of most unicells is the result of either the activity of cilia and flagella or amoeboid movement. Many living cells also exhibit varying degrees of cytoplasmic movement.

Muscular Contraction

In the following discussion of motion, only the skeletal muscles of man will be considered. These muscles, also called *striated muscles,* are attached to bones and when stimulated produce most of the general motion of the body. A single skeletal muscle such as the biceps (Figure 7-5A), when viewed microscopically, is seen to consist of thousands of muscle *fibers,* or cells (Figure 7-5B, C). The individual fibers, averaging between 10 and 100 μ, are arranged parallel to one another and represent the structural units of a muscle. Each fiber is surrounded by a thin electrically charged membrane, the *sarcolemma,* and contains *sarcoplasm,* several nuclei, and mitochondria. Seen from the top or bottom, muscle fibers exhibit a pattern of cross-bandings, or striations; this appearance is responsible for designating skeletal muscle as striated muscle (Figure 7-5C).

A highly magnified section of a single muscle fiber reveals that it consists of *myofibrils* (Figure 7-5C, D). These are thread-like structures, each about a micron in diameter, and cross-striated like the fiber in which they lie. Electron micrographs show that the myofibrils are composed of still smaller filaments. These filaments are of two kinds: thin (0.005 μ in diameter and approximately 2 μ long) and thick (0.01 μ in diameter and about 1.5 μ long). Chemically, the *thin filaments* consist of a protein, *actin,* while the *thick filaments* are composed of a second protein, *myosin* (Figure 7-5D, E).

The filaments of a myofibril do not extend the entire length of the myofibril. Rather, they are stacked in definite compartments, the *sarcomeres,* which are partitioned by separations called *Z lines.* Z lines are regions where the ends of thin filaments or other protein fibrils are concentrated and often interwoven.

The overlap between the numerous thick and thin filaments and the Z lines produces the pattern of striations visible through the microscope. These broad crosswise bands (striations) are alternately dark and light (Figure 7-5E). The dark band is referred to as the *A band* (designating its anisotropic or dense condition). The midportion of the A band is interrupted by a less dense stripe, the *H zone,* a region consisting solely of thick filaments. Within the H zone is the *M line,* a dark stripe caused by a bulge in the center of each thick filament and the *pseudo H zone,* a bare region immediately surrounding the bulge. The light band, called the *I band* (designating its isotropic or less dense condition), is composed of thin filaments only.

The exterior of the skeletal muscle fiber is positively charged relative to the interior; as in a neuron, polarization is maintained by the active transport of Na^+ and K^+ ions. Electrical shocks and certain chemicals are capable of depolarizing the membrane so that a wave of depolarization, the **action current,** passes along the length of the fiber. The period of time during which the action current is swept along the fiber is referred to as the *latent period* and lasts for about 0.01 second. No visible change occurs in the muscle fiber during this period. The action current also penetrates the fiber during the latent period. Higher magnifications obtained by electron microscopy reveal a network of channels and vesicles within each fiber in many ways similar to the endoplasmic reticulum of other cells. This network, referred to as the *sarcoplasmic reticulum,* consists of a series of tubules that run parallel to the myofibrils. Running transversely through the fiber and perpendicular to the sarcoplasmic reticulum are *T tubules.* The tubules connect with the sarcoplasmic reticulum and open to the outside of the fiber (Figure 7-5D). This network affords an explanation of how the simultaneous contractions of all myofibrils within the fiber could be accomplished.

Contraction of muscle fibers requires the proteins actin and myosin, ATP as an energy source, and Ca^{2+} ions. Even in test tubes, actin and myosin, which can be extracted from muscle and purified, will combine to form a more complex protein, *actomyosin.* If fibers of actomyosin are supplied with ATP and the necessary ions, they will contract.

In living muscle tissue, the actual contraction of a myofibril is believed to involve the inward movement of thin filaments drawing the Z lines inward toward the A band of each sarcomere. This hypothesis, termed the *sliding-filament hypothesis,* was originally proposed by H. E. Huxley and J. Hanson. As the thin filaments slide together, the H zone disappears. These filaments continue to slide until their ends actually overlap to produce a dense zone in the center of the A band. As a direct result of this sliding, the distance between Z lines is greatly reduced. The mechanism by which the thin filaments slide along the thick filaments is still highly theoretical. One theory holds that sliding is related to projections (cross-bridges) of the thick filaments that extend sideways and touch the thin filaments. These bridges may function in a ratchet-like manner to effect the movement of the thin filaments along the thick filaments.

When a nerve impulse reaches a muscle fiber, the neuron releases acetylcholine, which causes an electrical charge in the sarcolemma. The electrical charge travels over the surface of the sarcolemma and into the T tubules. When the impulse is conveyed from the T tubules to the sarcoplasmic reticulum, the reticulum releases calcium ions into the sarcoplasm surrounding the myofibrils. These calcium ions trigger the contractile process. Muscle contraction lasts only as long as calcium ions are present in the sarcoplasm. When the nerve impulse is over, the calcium ions recombine with the reticulum, and muscle contraction ceases.

When calcium is released by the sarcoplasmic reticulum, the cross-bridges connect with thin filaments (Figure 7-6). The bridges pull the thin filaments of each sarcomere inward toward each other until their approaching ends overlap. As the thin filaments slide inward, the Z lines are drawn toward the A band, and the sarcomeres are shortened. The shortening of the

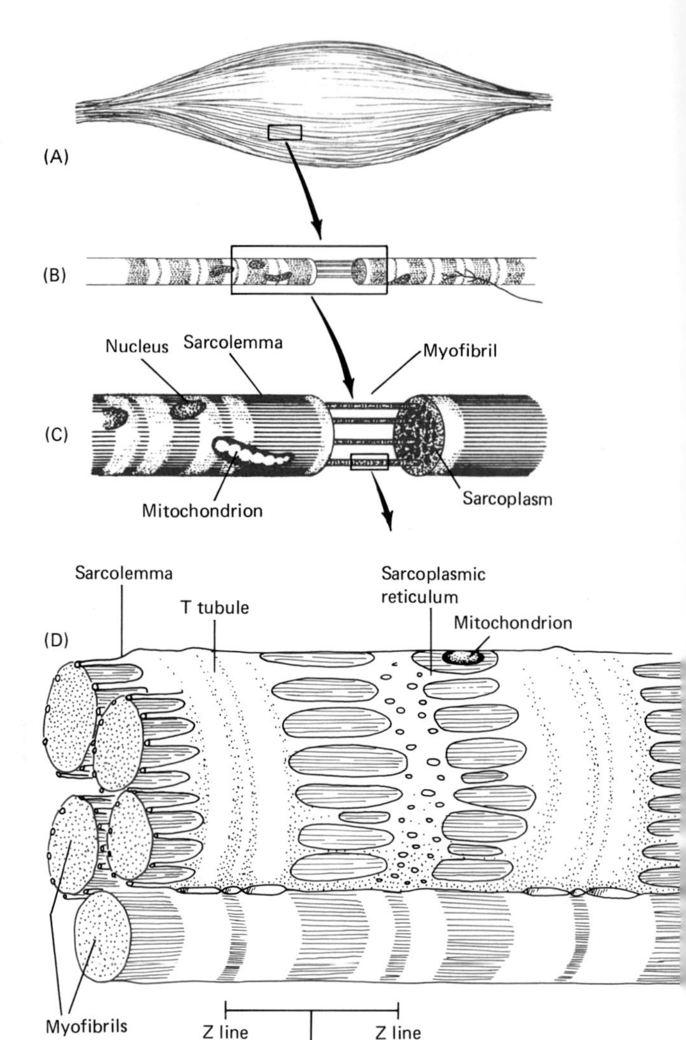

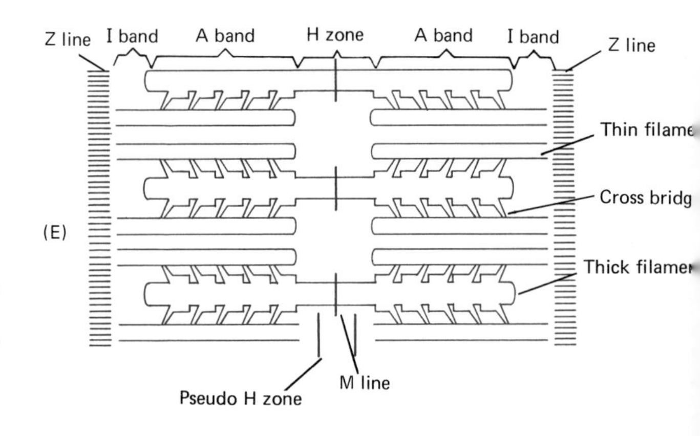

Figure 7-5. Structural organization of striated muscle. (**A**) Entire muscle, such as biceps. (**B**) Enlarged aspect of a single muscle fiber (cell) with central portion removed to show internal structure. (**C**) Further enlargement of a single muscle fiber showing more detail. (**D**) Enlarged aspect of several myofibrils. (**E**) Details of a sarcomere.

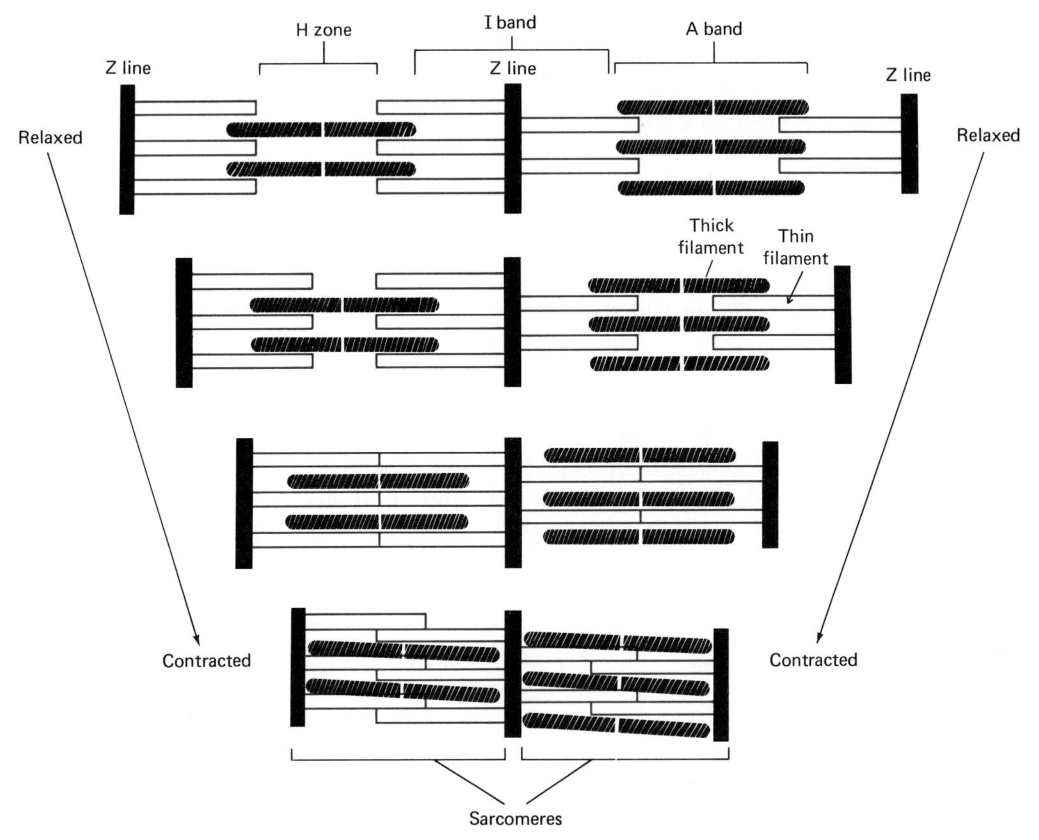

Figure 7-6. Sliding-filament hypothesis of muscular contraction. Positions of various zones in relaxed and contracted sarcomeres and illustration of sliding actin filaments.

sarcomeres and myofibrils causes the shortening of the muscle fibers.

It is believed that the protein, troponin, which is found in the filaments, prevents any interaction between actin and myosin in a noncontracting muscle. When a nerve impulse reaches the sarcoplasmic reticulum and triggers the release of calcium ions, the ions combine with the troponin and deactivate it. The myosin and actin are then free to interact. At the same time, the nerve impulse stimulates the breakdown of ATP. The energy that is released by ATP breakdown is used for the attachment of the cross-bridges and the sliding of the thin filaments. After the impulse ends, the calcium ions return to the sarcoplasmic reticulum. The troponin breaks the cross-bridge attachments, and the thin filaments slip back outward to the original place. The sarcomeres are thereby returned to their resting lengths, and the muscle resumes its resting shape.

Unlike most other types of cells, muscle cells alternate between a high degree of activity and virtual inactivity. Inasmuch as the hypothesized immediate energy source for contraction is ATP, energy utilization in muscle cells occurs only as long as ATP is available. Thus, energy utilization would depend upon the rate of ATP synthesis by oxidative phosphorylation or glycolysis were it not for the presence of other compounds to store high-energy phosphate groups until needed by ATP. Considering that a burst of muscular activity depletes the pool of ATP at a faster rate than it can be replenished, the quantity of ATP available in muscle cells is limited. Each muscle fiber, however, contains a supply of a nitrogen-containing substance called *crea-*

tine. This compound can accept the high-energy phosphate group from ATP to become the high-energy substance *creatine phosphate.* Creatine phosphate is synthesized in the muscle cells when they are not contracting. At such times, muscle cells are synthesizing more ATP than they are breaking down. Thus, during these periods, they convert some of the excess ATP to creatine phosphate, which is stored. During periods of strenuous contraction, when the cell needs more ATP quickly, ADP is recharged to ATP through the creatine phosphate (Figure 7-7).

Other chemical changes also occur in muscle during very strenuous exercise. Soon after vigorous activity begins, muscle fibers manufacture ATP by fermentation. Glycogen (stored glucose) in the muscle is broken down into lactic acid (see Chapter 6). Since most of the potential energy of the glycogen remains trapped in the lactic acid, it yields only two ATP's for each molecule of glucose that undergoes glycolysis. Furthermore, lactic acid starts to accumulate in muscle cells because there is insufficient oxygen to oxidize the pyruvic acid via the Krebs cycle as it forms by glycolysis. In man and a majority of vertebrates, most of the lactic acid is transported to the liver where it is eventually resynthesized to glucose or glycogen. A smaller amount remains behind in the muscle cells. Here, after strenuous exercise has ceased and oxygen is again available, the lactic acid is oxidized back to pyruvic acid, which is then passed on through the Krebs cycle.

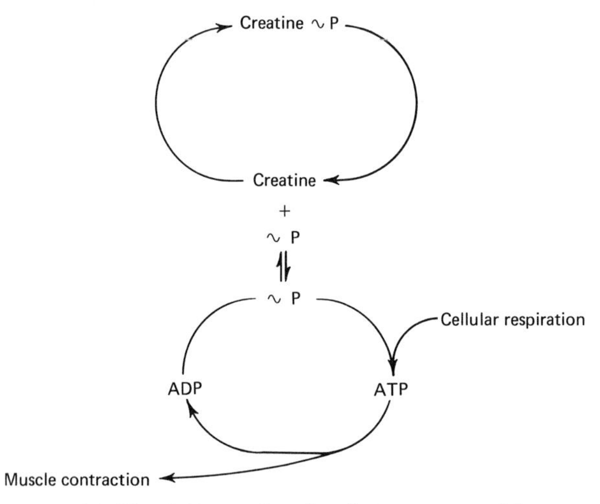

FIGURE 7-7. The ATP–creatine phosphate energy pool for muscular contraction.

The amount of oxygen required for oxidation of the lactic acid accumulated during strenuous exercise is called the *oxygen debt.* This debt is paid by breathing rapidly and deeply for some time after strenuous exercise has stopped. About 20% of the accumulated lactic acid is oxidized to CO_2, H_2O, and ATP via aerobic respiration. The ATP is then used to resynthesize glycogen from the remaining 80% of lactic acid. Muscle glycogen is restored by transport of glucose from the liver to the muscles.

Flagellary, Ciliary, and Amoeboid Movements

The structure of cilia and flagella was presented in Chapter 5. It was shown that cilia and flagella have a common internal ultrastructure and a common ability to effect movement through contraction.

The movement of a cilium involves two separate strokes, a power stroke, and a recovery stroke; the result is a rowing motion (Figure 7-8A). The cilium is fairly stiff during the power stroke, but is relaxed during the recovery stroke. Flagellary movement is somewhat more complicated, involving either rippling or lashing movements. Nevertheless, both types of movement probably result from uneven contraction of the fibril elements. Four or five of the peripheral nine fibril pairs appear to contract, shortening one side of the cilium or flagellum, causing it to bend; the contraction of the remaining peripheral fibrils follows rapidly to restraighten the structure. Chemical investigations reveal the presence of ATP and a protein-like substance in flagella. Experiments have shown that the addition of ATP to a flagellar system initiates beating, or, if they are already beating, appreciably accelerates movement.

In *amoeboid movement,* the characteristic form of locomotion in the *Amoeba,* there is a flow of cytoplasm into one or more projections, the *pseudopodia.* As cytoplasm accumulates in the region of the cell where the movement occurs, it decreases in the opposite end. This net flow results in amoeboid movement (Figure 7-8B). The cytoplasm of the cell is able to assume two consistencies—the sol and the gel. As the sol reaches the top of a pseudopodium, it is pushed to the sides by the cytoplasm behind and is converted to a gel. Simultaneously,

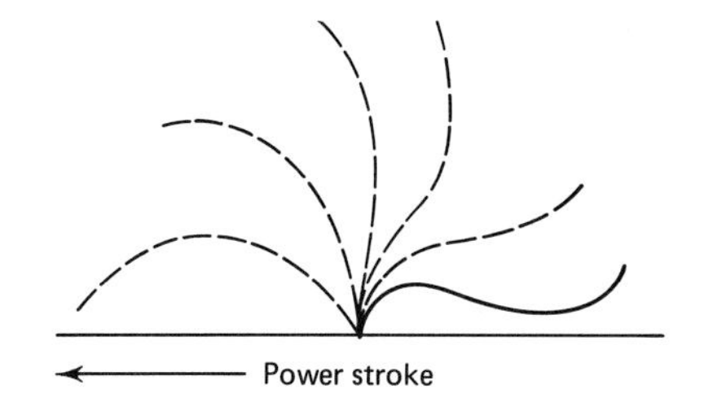

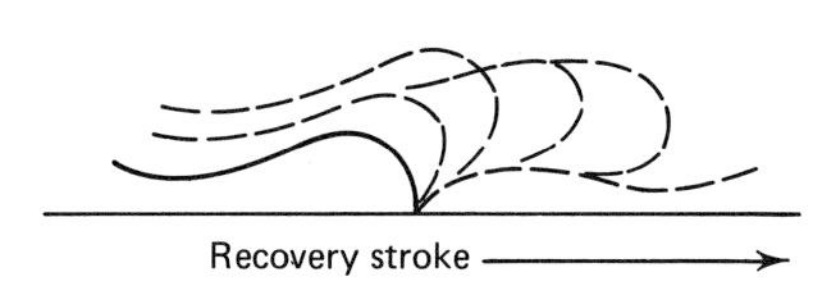

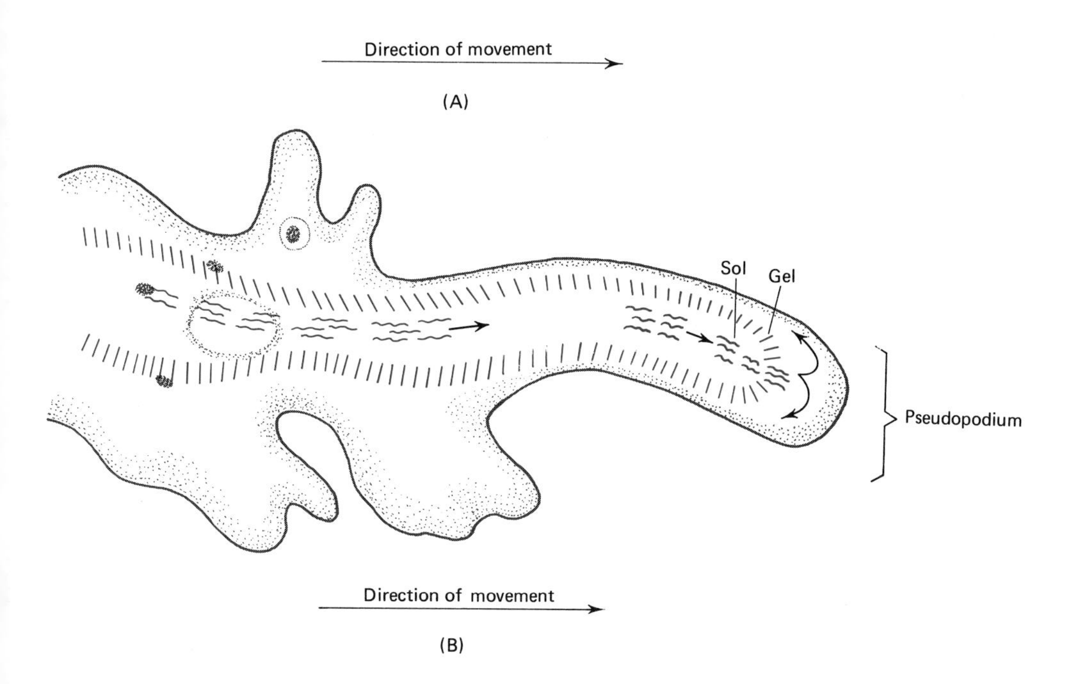

Figure 7-8. Uses of ATP in mechanical energy (motion). **(A)** Ciliary movement. **(B)** Amoeboid movement. See text for discussion.

the gel in the region opposite the pseudopodium becomes a sol and flows forward. In addition to sol–gel transformations, contractile proteins also appear to be involved in amoeboid movement. In slime molds, some of which exhibit amoeboid movement during one phase of their life cycle, a contractile protein, *myxomyosin,* has been isolated. It is chemically similar to the actomyosin involved in muscle contraction. As with other contractile proteins, the energy source for the particular kind of biological work is typically ATP. Although the relationship between contractile proteins and ATP has been established, the mechanism by which they operate in bringing about cytoplasmic movement is still obscure.

Biosynthesis (Chemical Energy)

The formation of chemical substances in living cells is referred to as **biosynthesis.** This process includes the assembly of smaller, simpler molecules into larger, more complex ones. The raw materials from which cells produce larger molecules are smaller molecules composed chiefly of the elements carbon, hydrogen, oxygen, and nitrogen. The ultimate source of carbon is atmospheric CO_2, which is incorporated into molecules through photosynthesis. Hydrogen enters cells chemically combined in the water molecule. Oxygen is supplied either in water or in gaseous form. Nitrogen is taken into plants in the form of ions, and animals must consume plants for their nitrogen. It should be noted that carbon, hydrogen, and oxygen are constituents of sugars, the principal products of photosynthesis.

Once the cell obtains the necessary raw materials, it can synthesize them into the building block molecules. These include amino acids, mononucleotides, monosaccharides, fatty acids, and glycerol. Synthesis of these building block molecules results in the production of the major cell molecules, the macromolecules (DNA, RNA, proteins, polysaccharides, and lipids). Eventually, these macromolecules are incorporated into the various structural elements of the cell (membranes, ribosomes, nuclei, mitochondria, endoplasmic reticulum, chloroplasts, and so on).

Biosynthesis from raw materials to structural elements of the cell is a complex and essential activity of living systems. Living systems are built of, and their functions depend upon, macromolecules. Macromolecules store information for the cell. Such information includes the ability to direct enzyme reactions in specific pathways, to synthesize specific molecules to allow for hereditary variation and species continuity, and to allow for adaptation of organisms. In addition, large molecules are able to store greater quantities of energy than smaller molecules.

Polysaccharide Biosynthesis

The most basic process of carbohydrate synthesis is photosynthesis (Chapter 6). As described earlier, only autotrophic cells can combine CO_2 and H_2O in the presence of chlorophyll and sunlight into carbohydrates. The first stable carbohydrate produced in photosynthesis is PGAL. Once this substance is available, all cells, including nongreen plant cells and animal cells, may synthesize other carbohydrates, as well as other organic compounds from it. The principal monosaccharide and one of the most fundamental raw materials for synthesis is glucose. This compound can be formed within cells from the synthesis of PGAL units. Within any cell, glucose may enter oxidation reactions such as fermentation and aerobic respiration or synthesis reactions to yield more complex carbohydrates.

Our discussion of the biosynthesis will consider carbohydrates first, since the larger ones (polysaccharides), although complex molecules, consist of long chains of identical molecules all joined together by the same kind of linkage. In Chapter 2 the carbohydrates of importance to living systems were considered. At this point, only the biosynthesis of glycogen will be discussed since this mechanism exemplifies the general lines along which most polysaccharides are made in biological systems.

Glycogen, the storage carbohydrate of animal cells, consists of long chains of glucose molecules built up by the condensation of single glucose units. These linkages require the activation of glucose molecules and the entire series of reactions is catalyzed by specific enzymes (Figure 7-9). The initial step in glycogen synthesis consists of the activation of glucose by phosphorylation; that is, a phosphate group from ATP is added to the sixth carbon atom of glucose. This results in the production of glucose-6-phosphate, a compound that raises the energy content of glucose. Next, the glucose-6-phosphate is enzymatically converted to glucose-1-phosphate. In this reaction, the phosphate group has been transferred from the sixth to the first carbon atom of glucose. Glucose-1-phosphate reacts with a high-energy compound, *uridine triphosphate* (*UTP*), a substance similar to ATP that is capable of supplying energy for cellular work. This reaction produces *uridine diphosphate glucose* (*UDPG*), the basic glucose carrier. Once this compound has been formed, an enzyme transfers the glucose component to an existing glycogen chain, thus lengthening it by one unit. The other fraction of UDPG, the *uridine diphosphate* (*UDP*), is regenerated to UTP by the transfer of a phosphate group from ATP. Thus, ATP plays a role in the formation of glucose-1-phosphate and the regeneration of UTP. Inasmuch as glycogen is a branched polymer of glucose units, a special enzyme called a "branching" enzyme transfers the segments of glucose units to the existing glycogen to form a branching pattern every ten or so glucose units.

Fat Biosynthesis

Lipids are much smaller molecules than polysaccharides, but their synthesis is more complex because they contain several different kinds of subunits. In the discussion of the biosynthesis of lipids that follows, only one type will be considered. These are the simple lipids composed of one molecule of glycerol chemically bonded to three long-chain fatty acids.

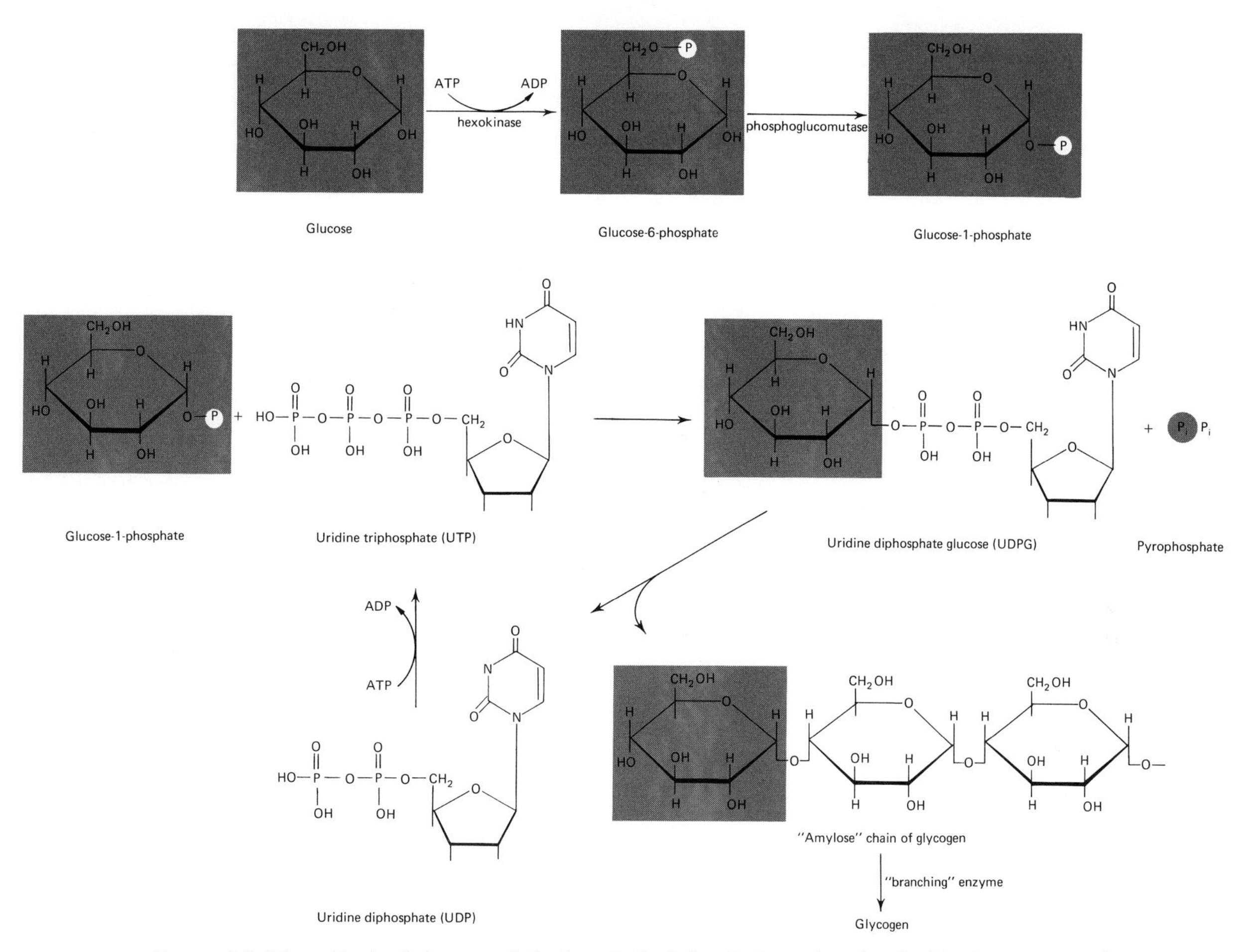

Figure 7-9. Biosynthesis of glycogen. Only the principal chemical reactions involved in the activation of glucose and its synthesis into glycogen are shown.

The energy required for fat biosynthesis is supplied by ATP. Carbohydrate metabolism is the main source of acetyl CoA for fat biosynthesis. During glycolysis, pyruvic acid is formed. Further enzymatic action removes one of the three carbon atoms of pyruvic acid (decarboxylation). The remaining two-carbon fragment eventually reacts with CoA—SH to form acetyl CoA, which can enter either the Krebs cycle or fat biosynthesis.

The use of acetyl CoA in fat synthesis involves a complicated series of reactions requiring ATP and certain cofactors. In these reactions, acetyl CoA reacts with CO_2, biotin, and an enzyme (carboxylase) to synthesize malonyl CoA. The malonyl CoA then acts as the condensing unit to which additional acetyl CoA units are added one at a time to form the long-chain fatty acids (Figure 7-10). The energy required for the formation of these large molecules is supplied by ATP at the first step.

Glycerol, the other component of fats, is obtained from glycolysis. To complete the biosynthesis of fat, fatty acids, still in association with CoA—SH, react with a phosphorylated form of glycerol. The result is the synthesis of fat (Figure 7-11).

Amino Acid Biosynthesis

Plants and microorganisms use ammonia (NH_3) and other inorganic nitrogen-containing compounds to synthesize amino acids, proteins, and nucleic acids. Higher

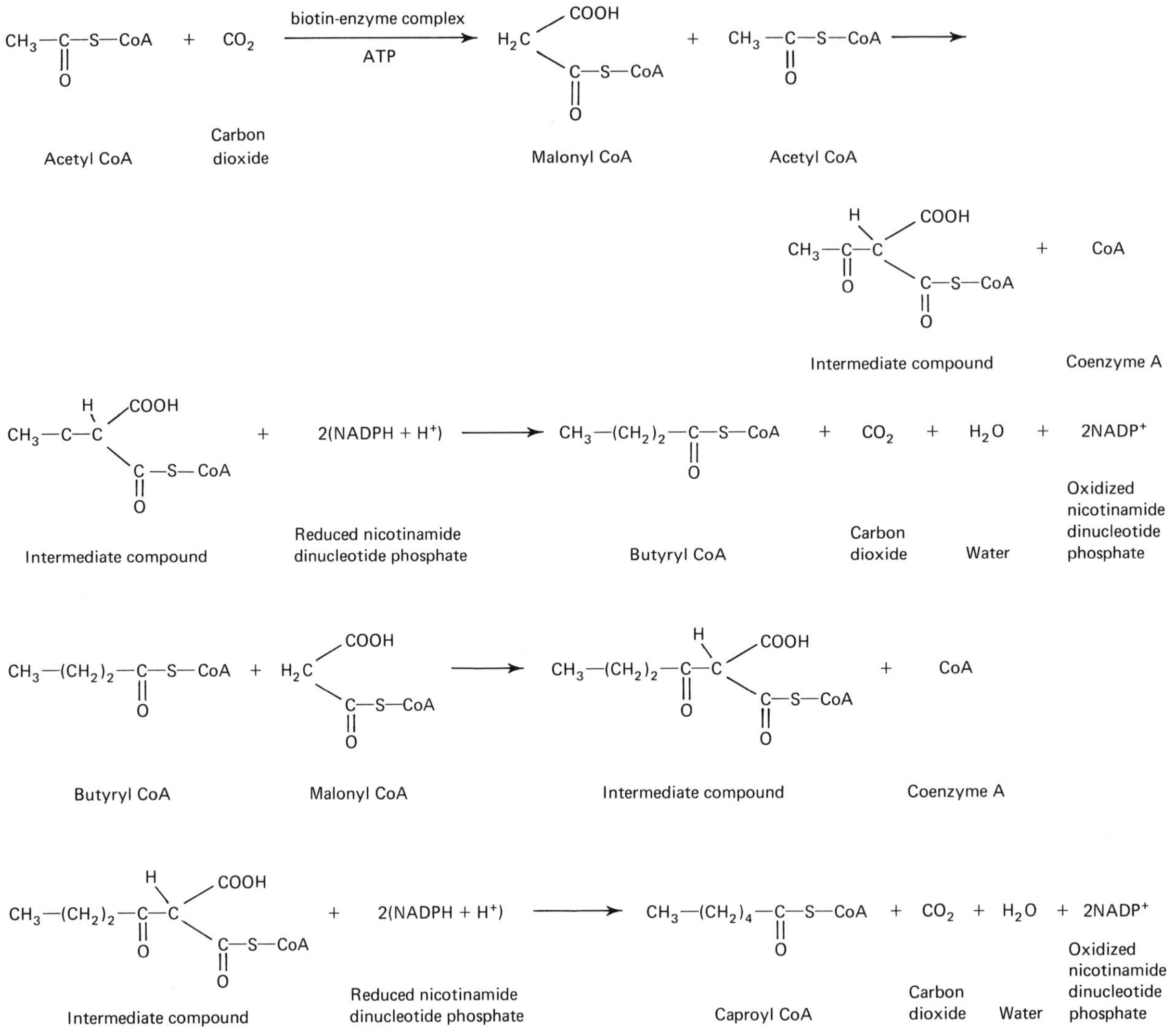

$CH_3-\underset{\underset{O}{\|}}{C}-S-CoA + CO_2 \xrightarrow[\text{ATP}]{\text{biotin-enzyme complex}} H_2C\begin{smallmatrix} COOH \\ \underset{\underset{O}{\|}}{C}-S-CoA \end{smallmatrix} + CH_3-\underset{\underset{O}{\|}}{C}-S-CoA \longrightarrow$

Acetyl CoA | Carbon dioxide | Malonyl CoA | Acetyl CoA

$CH_3-\underset{\underset{O}{\|}}{C}-\overset{H}{C}\begin{smallmatrix} COOH \\ \underset{\underset{O}{\|}}{C}-S-CoA \end{smallmatrix} + CoA$

Intermediate compound | Coenzyme A

$CH_3-C-\overset{H}{C}\begin{smallmatrix} COOH \\ \underset{\underset{O}{\|}}{C}-S-CoA \end{smallmatrix} + 2(NADPH + H^+) \longrightarrow CH_3-(CH_2)_2-\underset{\underset{O}{\|}}{C}-S-CoA + CO_2 + H_2O + 2NADP^+$

Intermediate compound | Reduced nicotinamide dinucleotide phosphate | Butyryl CoA | Carbon dioxide | Water | Oxidized nicotinamide dinucleotide phosphate

$CH_3-(CH_2)_2-\underset{\underset{O}{\|}}{C}-S-CoA + H_2C\begin{smallmatrix} COOH \\ \underset{\underset{O}{\|}}{C}-S-CoA \end{smallmatrix} \longrightarrow CH_3-(CH_2)_2-\underset{\underset{O}{\|}}{C}-\overset{H}{C}\begin{smallmatrix} COOH \\ \underset{\underset{O}{\|}}{C}-S-CoA \end{smallmatrix} + CoA$

Butyryl CoA | Malonyl CoA | Intermediate compound | Coenzyme A

$CH_3-(CH_2)_2-\underset{\underset{O}{\|}}{C}-\overset{H}{C}\begin{smallmatrix} COOH \\ \underset{\underset{O}{\|}}{C}-S-CoA \end{smallmatrix} + 2(NADPH + H^+) \longrightarrow CH_3-(CH_2)_4-\underset{\underset{O}{\|}}{C}-S-CoA + CO_2 + H_2O + 2NADP^+$

Intermediate compound | Reduced nicotinamide dinucleotide phosphate | Caproyl CoA | Carbon dioxide | Water | Oxidized nicotinamide dinucleotide phosphate

Figure 7-10. Simplified scheme of fatty acid synthesis. The formation of malonyl CoA from acetyl CoA and the fixation of carbon dioxide involve a series of complex steps requiring a biotin–enzyme complex and ATP. In the synthesis of the carbon chain, $NADPH^+ + H^+$ and many enzymes participate. Note that each time CoA is removed, it bonds with a hydrogen atom. The diagram shows the addition of two "two-carbon" fragments. Addition of "two-carbon" fragments continues in a like manner until a fatty acid of a given size is synthesized.

animals can also use ammonia for the synthesis of nitrogen-containing compounds, but they obtain most of their nitrogen by consuming protein. The protein consumed in the diet is hydrolyzed to amino acids in the gastrointestinal tract, and the acids are absorbed into the blood and transported to the liver. Within the liver some of the amino acids are reassembled into several blood proteins such as plasma albumin, globulins, fibrinogen, and prothrombin. Other amino acids are passed on to other tissues of the body and are resynthe-

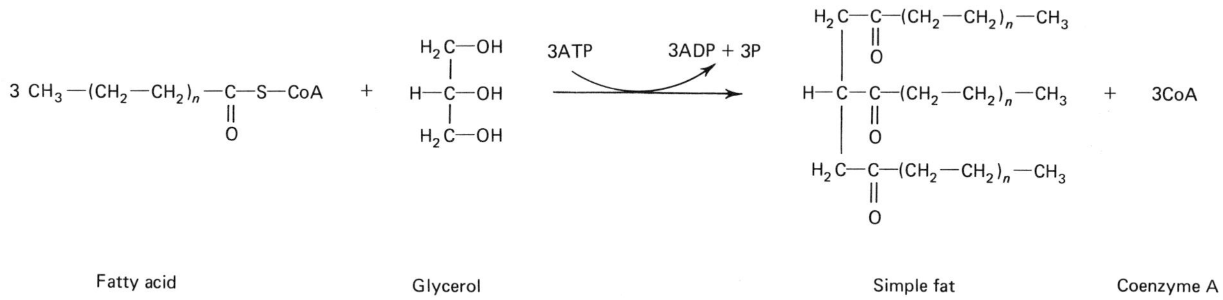

$$3\ CH_3-(CH_2-CH_2)_n-\overset{O}{\overset{\|}{C}}-S-CoA\ +\ \begin{matrix} H_2C-OH \\ | \\ H-C-OH \\ | \\ H_2C-OH \end{matrix}\ \xrightarrow{3ATP\ \rightarrow\ 3ADP + 3P}\ \begin{matrix} H_2C-\overset{O}{\overset{\|}{C}}-(CH_2-CH_2)_n-CH_3 \\ | \\ H-C-\overset{O}{\overset{\|}{C}}-(CH_2-CH_2)_n-CH_3 \\ | \\ H_2C-\overset{O}{\overset{\|}{C}}-(CH_2-CH_2)_n-CH_3 \end{matrix}\ +\ 3CoA$$

Fatty acid — Glycerol — Simple fat — Coenzyme A

Figure 7-11. Simplified scheme of the condensation of fatty acids and glycerol. Note that the glycerol is phosphorylated via ATP and the products of the reaction are a simple fat, CoA, and phosphoric acid.

sized into protein. Amino acids in excess of bodily needs are broken down or converted to other amino acids.

The interconversion of amino acids involves the transamination and deamination reactions described in Chapter 6. Of the two processes transamination is the more important in the synthesis of amino acids.

About twenty α-amino acids occur in nature. Most animals, including man, are able to synthesize only about half of these in amounts sufficient for their needs. There are ten *essential amino acids*—arginine, histidine, isoleucine, leucine, lysine, methionine, tryptophan, phenylalanine, threonine, and valine—that animals must obtain from outside sources. Plants and most microorganisms, on the other hand, are capable of synthesizing most, if not all, of their amino acids.

The synthesis and the breakdown of each of the amino acids involve a complex series of steps. Certain of the transformations are common to all amino acids, yet each acid has its own unique reaction pathways. The ability of an organism to synthesize a given amino acid depends upon the presence of specific enzymes. This ability varies among different organisms.

In the Krebs cycle, there are specific points of entry or exit for the amino acids. However, it must be remembered that not all organisms possess all these pathways. Pyruvic acid is directly or indirectly involved with the metabolism of glycine, serine, valine, leucine, alanine, cysteine, phenylalanine, tryptophan, and tyrosine (see Figure 6-19). α-Ketoglutaric acid is involved with the metabolism of glutamic acid, while fumaric acid and oxaloacetic acid are directly or indirectly involved in the metabolic pathways of aspartic acid, threonine, methionine, isoleucine, and lysine (see Figure 6-19).

In addition to providing the means for synthesizing other amino acids, certain amino acids play important roles as precursors of other nitrogen-containing cellular compounds. Decarboxylation of amino acids is a means of producing important amines (see Chapter 6). Molecules of chlorophyll and hemoglobin contain a complex nitrogen-containing structure called a porphyrin ring. This structure is synthesized via a pathway of reactions originating at succinyl CoA of the Krebs cycle. The purines and pyrimidines of the nucleic acids are involved in a metabolic pathway leading from aspartic acid, which, in turn, leads from oxaloacetic acid and fumaric acid of the Krebs cycle (see Figure 6-19).

Intermediary Metabolism and Cellular Bioenergetics

The metabolic link between photosynthesis and the synthesis of all molecules is represented by myriad reactions collectively termed *intermediary metabolism* (see Figure 6-19). The three major groups of molecules (carbohydrates, lipids, and proteins) function interchangeably as the source of both energy (ATP) and the raw materials (carbon fragments) for biosynthesis. The overall role of intermediary metabolism is the synthesis of all those constituents that a cell does not obtain as prefabricated molecules from the environment or from other cells and the production of useful energy. The breakdown of organic molecules leads to a net buildup of ATP; conversely, the breakdown of ATP to ADP leads to a buildup of organic molecules through biosynthesis.

8 The Control of Cellular Metabolism

GUIDE QUESTIONS

1 Why is it necessary for cells to have some mechanism to control their metabolic activities? How does cellular control differ from organismic control and intraspecific and interspecific control?
2 Outline and discuss the investigations which ultimately led to the elucidation of DNA as the hereditary material. What indirect evidence suggests the role of DNA as the genetic substance?
3 Discuss the events associated with viral replication.
4 What researches provided data for the semiconservative replication of DNA? How do these researches support the Watson-Crick model?
5 Discuss the contributions of Garrod and Beadle and Tatum to ascertaining the relationship between genes and enzymes.
6 What is meant by the genetic code? How is it related to the number of amino acids? Define a codon.
7 Outline the principal stages of protein synthesis. What are the roles of RNA polymerase, mRNA, tRNA, and ribosomes in the process? Define an anticodon. What are polyribosomes?
8 How was the genetic code deciphered? Why is the code said to be degenerate?
9 What are the two principal ways in which gene function is controlled? Define feedback inhibition and allosteric interaction. Cite examples of positive and negative control of enzymatic activity. How may negative and positive control be operative simultaneously?
10 Discuss the components and functions of the Monod–Wyman–Changeux model of enzyme activity.
11 Define the following: inducible enzyme, inducer, enzyme induction, enzyme repression, repressible enzyme, and corepressor. What is the operon model of enzyme synthesis? Be sure to include all components of the model in your discussion.

IN THIS chapter, we will examine the biosynthesis of informational molecules. These are the nucleic acids, which are adapted for the storage and transcription of biological information, and the proteins, which are responsible for the expression of this information. Together, they assume a major role in the control of cellular metabolism.

Control in Living Systems

The two previous chapters dealing with metabolic activities of cells clearly establish the fact that life is characterized by complexity. At any given time in the life of any one organism, hundreds or even thousands of different biochemical reactions may be occurring. Even more amazing is the fact that these reactions do not take place randomly and independently of each other; there is considerable cooperation and interdependence among various metabolic pathways. One group of reactions, such as biological oxidation, produces ATP for cellular utilization, while another set of reactions, such as polysaccharide biosynthesis, will break down the ATP. All processes characteristic of life require that a multiplicity of chemical reactions take place at a rate and to an extent appropriate to the maintenance of the total metabolism of the organism. This implies that the assorted reactions occurring in cells must be controlled.

For many years biologists have been concerned with the problems of control, and a great deal of progress has now been made. It has been determined that the periodic and necessary adjustments of the many individual reactions are not the result of chance. Within any organism, definite control mechanisms direct metabolic adjustments appropriate to the ever-changing demands of the environment. The cell is capable of maintaining a balance of materials within its boundaries and is able to accelerate or decelerate the production of certain key substances.

The principal concern of this chapter is the nature of control at the cellular level of organization, and it will be seen that ultimately cellular control is genetic in origin and operation. At the organismic level of organization, growth regulators, hormones, and nervous coordination comprise the major components of control (see Chapters 10 and 11). Many examples of intraspecific and interspecific control are discussed in Chapter 27.

Essentially, genes have two principal functions. First, they maintain and control cellular metabolism; second, they facilitate self-perpetuation of living systems through successive generations. An understanding of both of these functions requires a basic analysis of the structure of genes and the mechanism by which the genetic material is replicated.

The Elucidation of the Chemistry of the Genetic Material

The work of Gregor Mendel (Chapter 18) clearly demonstrated the presence of a structural entity (gene) related to the reappearance of certain characteristics in successive generations. Much of the first half of the present century was concerned with genetic research involving such phenomena as mutations, gene recombinations, and the construction of chromosome maps. Other researches dealt with chemical analyses of chromosomes. All of these investigations, as well as others, supplied conclusive proof that genes are physical units located on chromosomes and that the chromosomes are composed principally of DNA and protein.

Investigators postulated that whatever substance was the genetic material must satisfy at least two basic requirements. First, it has to have some mechanism for duplicating itself exactly, and second, it must exert control over cellular activities. In the past two decades, intensive research into the chemical nature of the genetic material has demonstrated that DNA is the primary genetic material.

DNA as the Genetic Material

The experiments which ultimately led to the conclusion that DNA is probably the hereditary material were performed by F. Griffith in England in 1928. He was working with two strains of bacteria: one of these, *Diplococcus pneumoniae,* causes pneumonia; the other, an avirulent strain, does not cause disease. The virulent strains are characterized by an enveloping polysaccharide capsule which is missing in the avirulent strains.

Griffith was interested in determining whether injections of heat-killed (60°C) virulent pneumococci could be used to vaccinate mice against pneumonia. He found that the injected mice were not infected by the heat-killed virulent strain. However, when these dead cells were mixed with live cells of the avirulent strain and the mixture injected into the mice, they frequently died. When the blood of the dead mice was analyzed, it was found that it contained living virulent bacteria. It was concluded that something passed from the dead bacteria (virulent) to the live ones (avirulent) which caused the offspring of the harmless bacteria to become transformed into capsulated forms capable of causing pneumonia. Presumably, hereditary material from the dead bacteria had entered the live cells and changed them into capsulated, pathogenic forms (Figure 8-1).

Subsequent investigations based upon Griffith's research revealed that bacterial transformation could also be duplicated using standard microbiological culturing techniques in broth and on agar petri dishes. Dead bacteria with capsules were added to a broth (liquid culture medium) which was inoculated with live bacteria without capsules, and then incubated for some hours. A drop of the culture was smeared over agar (solid culture medium) in petri dishes. After a period of incubation it was found that some of the bacteria had capsules and therefore had been transformed. The bacteria had acquired a new hereditary trait.

The next logical step was to extract various chemical components from the dead cells and to determine whether any of these chemicals would cause transformation. These crucial experiments were performed by O. T. Avery and his associates at the Rockefeller Institute in 1944. After some years of work, they announced that the chemical substance responsible for transforming harmless pneumococci into virulent strains was

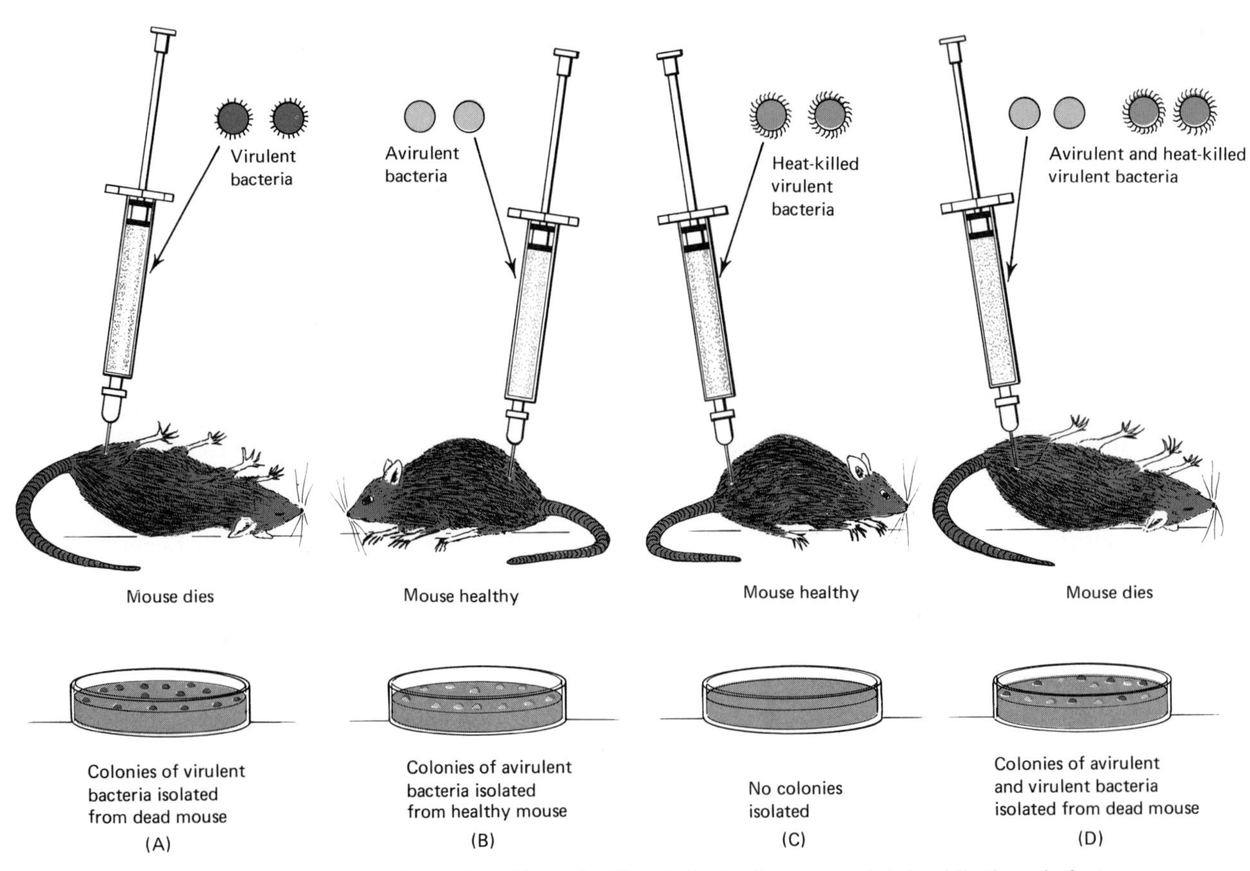

FIGURE 8-1. Genetic transformation of bacteria. The virulent cells are capsulated, while the avirulent ones are not. The virulent principle is transferred from dead cells into live nonvirulent ones. The transformed bacteria become virulent and kill the mouse.

FIGURE 8-2. Structure and life cycle of the T2 bacteriophage. (**A–B**) Diagrammatic representation of the virus in (**A**) external and (**B**) sectional views. The virus consists of a six-sided head with a tail and, like all other viruses, is considered to be a single macromolecule of nucleoprotein. The nucleic acid core of genetic material is enclosed in and protected by a protein coat. The nucleic acid is contained in the head and is prevented from leaking out of the hollow, tubular tail by a tail plug. The tail is the point of attachment of the virus to the bacterial cell wall during the initial stage of infection. The contraction of the tail contributes to the process whereby the phage nucleic acid is injected into the bacterial cell. The baseplate appears to contain some lysozyme, an enzyme that dissolves a hole in the bacterial cell wall. (**C–G**) Viral replication in *E. coli.* (**C**) The tail portion of the phage particle becomes attached to the cell wall of the bacterium. (**D**) The cell wall is penetrated, the nucleic acid is injected, and the protein coat is left behind. (**E**) The viral nucleic acid takes over bacterial cell activities and synthesizes new protein coats. (**F**) Viral nucleic acid is synthesized and becomes incorporated into the protein coats. (**G**) When all phage particles are complete, the bacterial cell bursts (a phenomenon called lysis), and the viral particles are liberated. The entire process of infection, as outlined here, takes about 30 minutes; new viruses appear about 12–15 minutes after infection. In some virus–bacterium systems, as many as 300 viruses are reproduced from a single cell infection. Inasmuch as the infective process occurs at the expense of the host cell (bacterium), it dies.

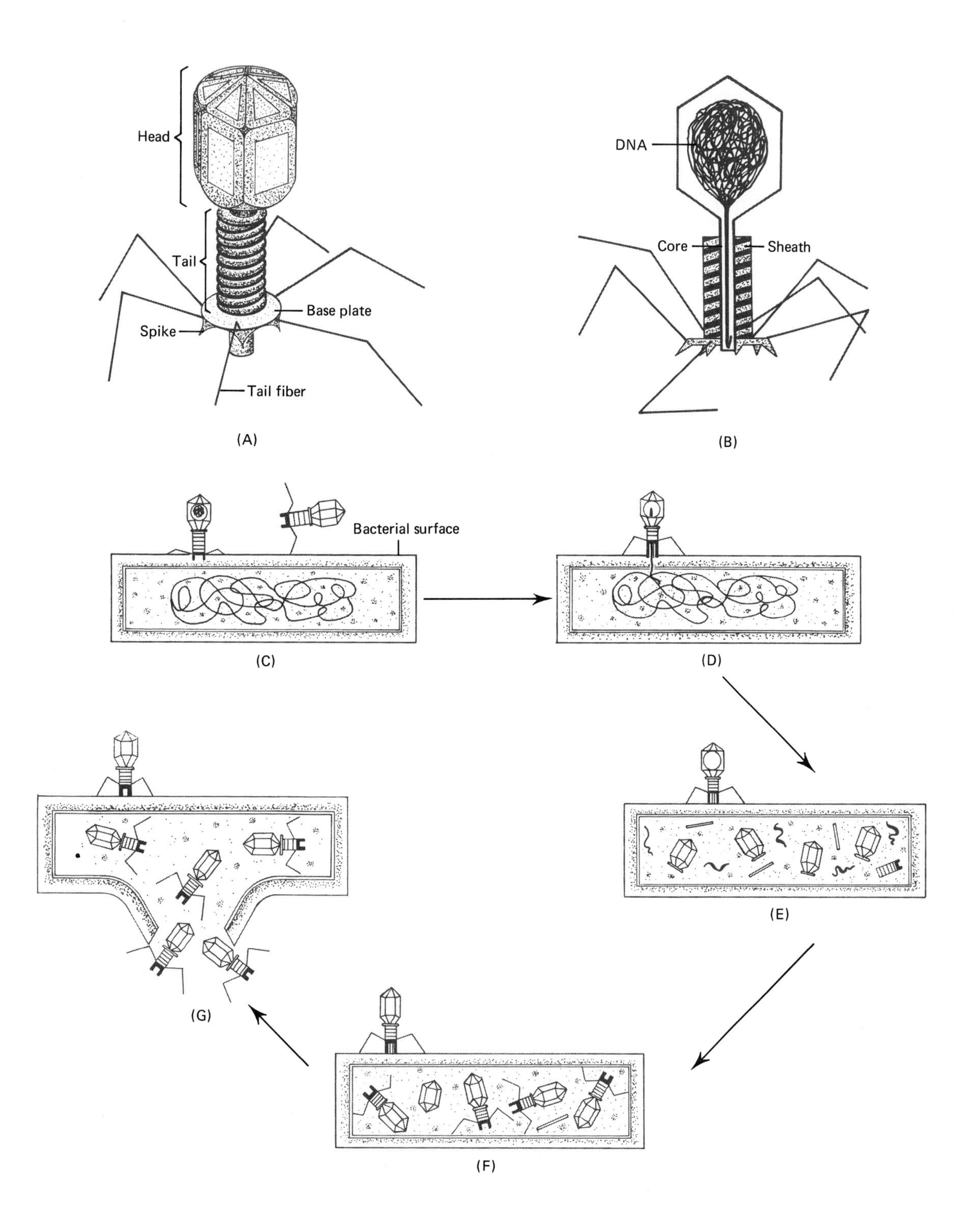
Head
Tail
Base plate
Spike
Tail fiber
(A)
DNA
Core
Sheath
(B)
Bacterial surface
(C)
(D)
(E)
(F)
(G)

DNA. Subsequent experiments showed that other genetic traits could also be passed from one bacterial cell to another and that the substance involved in the transformations was DNA. It was concluded that genes are composed of DNA.

There is also considerable evidence that DNA is the primary genetic material of **bacteriophages** (viruses that attack bacterial cells). The bacterial viruses that have been studied extensively are those that infect the bacterium *Escherichia coli* and are referred to as *T2 bacteriophages.* Electron microscope studies reveal that T2 viruses consist of hexagonal-shaped bodies and a protruding tail (Figure 8-2A, B). When suspensions of *E. coli* and T2 viruses are mixed, the viruses attach themselves by their tails to the surfaces of the bacteria where they remain (Figure 8-2C). After about 20 minutes the bacteria burst and each releases many complete new viruses, while the original infecting viruses still appear in outline attached to the surface of the bacterium (Figure 8-2D–G). The newly produced viruses are identical with those that initiated the infection. Based upon these observations, it is clear that some substance passed from the infecting viruses into the bacteria where it took over the chemical machinery of the bacteria and put it to work manufacturing new viruses. This substance must have contained the genes of the virus.

Chemical analysis of bacteriophages has shown that their structure consists of DNA within a protein coat. The DNA of the viruses is released into the bacteria, while the protein coat remains on the surface. Although these data suggest that the viral genes consist of DNA, the actual proof that the genes of T2 viruses are made of DNA came in 1952.

In that year, A. D. Hershey and M. C. Chase at the Carnegie Institution in Cold Spring Harbor, New York, prepared two kinds of bacterial viruses. One was grown in bacteria containing radioactive sulfur (^{35}S); the other was grown in bacteria containing radioactive phosphorus (^{32}P). After a cycle of multiplication, the newly synthesized viruses contained radioactive sulfur or radioactive phosphorus. Inasmuch as proteins contain sulfur but no phosphorus, all the radioactive sulfur was confined to the protein coats of the viruses. Similarly, since the DNA contains phosphorus while the protein does not, only the DNA contained the radioactive phosphorus. A suspension of the radioactive viruses was then mixed with a suspension of ordinary *E. coli* cells. The viruses attached themselves to the bacteria, but after a few minutes, a brief treatment of the suspension with an electric blender broke the loose attachment between the viruses and the bacterial cells. This treatment had little effect on the number of infected bacteria; when a portion of the suspension was incubated, the bacteria burst, liberating the usual number of bacterial viruses. Thus, during the few minutes of attachment, some substance must have passed from the virus particles into the bacteria and this substance must have contained the genetic material of the viruses.

Hershey and Chase then measured the quantities of radioactive sulfur and radioactive phosphorus to determine whether DNA or protein (or both) was injected into the bacteria. They found that 85% of the radioactive phosphorus of the viruses had passed into the bacteria and that 80% of the radioactive sulfur was still with the viruses. The bulk of the DNA had entered the bacteria while most of the protein had not. It was clear that DNA entered the host cell almost exclusively and that it was sufficient to code for the synthesis of complete virus particles. It was reasoned that nucleic acids, and probably not proteins, constitute the primary genetic material. It should be noted that although protein and DNA always occur together in cells, the reason for this association is not clear.

The role of DNA in bacterial transformation and viral infection, as well as other investigations and lines of research, has demonstrated that DNA is the primary genetic material of most organisms. Exceptions have been found among viruses. All viruses are composed of a nucleic acid core and a protein coat and many of those that infect higher plants contain a core of RNA and not DNA. *Tobacco mosaic virus* (*TMV*) is perhaps the best-studied example of this type (Figure 8-3). Experiments conducted with TMV have shown that the RNA portion of the virus contains all hereditary information required for the construction of new viruses.

A considerable amount of indirect evidence also suggests that DNA is the primary genetic material of higher organisms, including humans. It has been shown by Alfred Mirsky, for example, that all the cells of a given organism contain an equal amount of DNA; the only exceptions are gametes, which typically contain

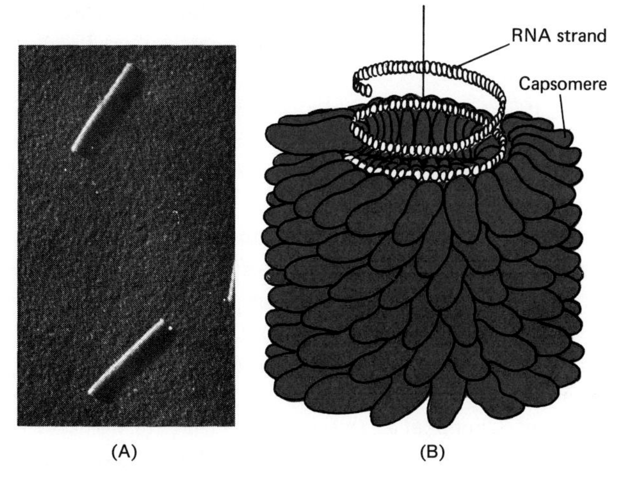

Figure 8-3. The tobacco mosaic virus (TMV). (**A**) Electron micrograph of the needle-shaped tobacco mosaic virus. 70,000×. [*Courtesy of Dr. Robley C. Williams, Virus Laboratory, University of California, Berkeley.*] (**B**) Diagrammatic representation of the arrangement of protein subunits and the RNA of the tobacco mosaic virus. The RNA assumes a helical configuration surrounded by the protein capsomeres. The center of the virus is hollow. [*After Klug in Philip L. Carpenter,* Microbiology, *2nd ed. Philadelphia: W. B. Saunders Company. Copyright © 1967. By permission of the publisher and the author.*]

only half as much DNA. Also, DNA from any cell from any given member of a given species contains the same relative amounts of the four nucleotides. Thus, all cells of similar hereditary origin have DNA of similar composition. Furthermore, there are differences in the composition of DNA and the proportion of nucleotides among different species, and the greater the differences, the greater the hereditary variations between the species.

Additional supportive data come from studies of mutations, which appear in all organisms from viruses to humans. Once a mutant form appears (if it is not lethal), it is continuously inherited according to Mendel's principles. The frequency with which mutants appear can be increased by certain treatments that alter the chemical structure of DNA.

DNA Structure and Replication

The next step in the elucidation of the genetic material was to demonstrate the phenomenon of duplication. Such an explanation is possible based upon the Watson–Crick model of DNA, which was constructed in 1953. At the time Watson and Crick started their investigations, there were a number of clues as to the nature of the DNA molecule. First, many experiments had suggested that DNA was not a highly coiled molecule (e.g., globular proteins), but rather a long, thin, fibrous molecule. Second, some research had suggested that DNA was helical in configuration and that this structure was maintained by hydrogen bonds between turns in the helix. Third, the ratio of nucleotides containing adenine to those containing thymine is 1:1; the ratio of nucleotides containing guanine to those containing cytosine is also 1:1. Finally, DNA molecules in solution have the phosphate groups of each nucleotide exposed to the exterior, while the nitrogenous bases are interiorly oriented.

Based upon these data, Watson and Crick began to construct a model of DNA. After some months of hypothesizing and model building, they postulated the double-stranded helix structure of DNA (see Figure 2-16). In their model, they proposed a mechanism by which replication of DNA might occur based upon the complementary nature of the two chains and base-pairing specificity. They suggested that the two strands separate, thus making available hydrogen bonds for pairing with complementary bases. As the hydrogen bonds are progressively broken, free nucleotides become associated with unpaired complementary ones and a new strand is formed. Each of the old strands serves as a pattern by which a new complementary strand is produced. The net effect of this process is that each new double strand contains the same sequence of nucleotides, and thus the same genetic information, as the old one (Figure 8-4). A current hypothesis is that DNA strands cannot "unzip"; rather, there are enzymes that open up base linkages one at a time, while new complementary bases are added.

If such a model is correct, it would require that each daughter molecule be half old and half new. This type of replication is referred to as **semiconservative;** that is, half of each parent molecule (one strand) is conserved in each daughter molecule, so that the daughter molecule consists of one old strand from the parent and one new strand formed from free nucleotides in the cell.

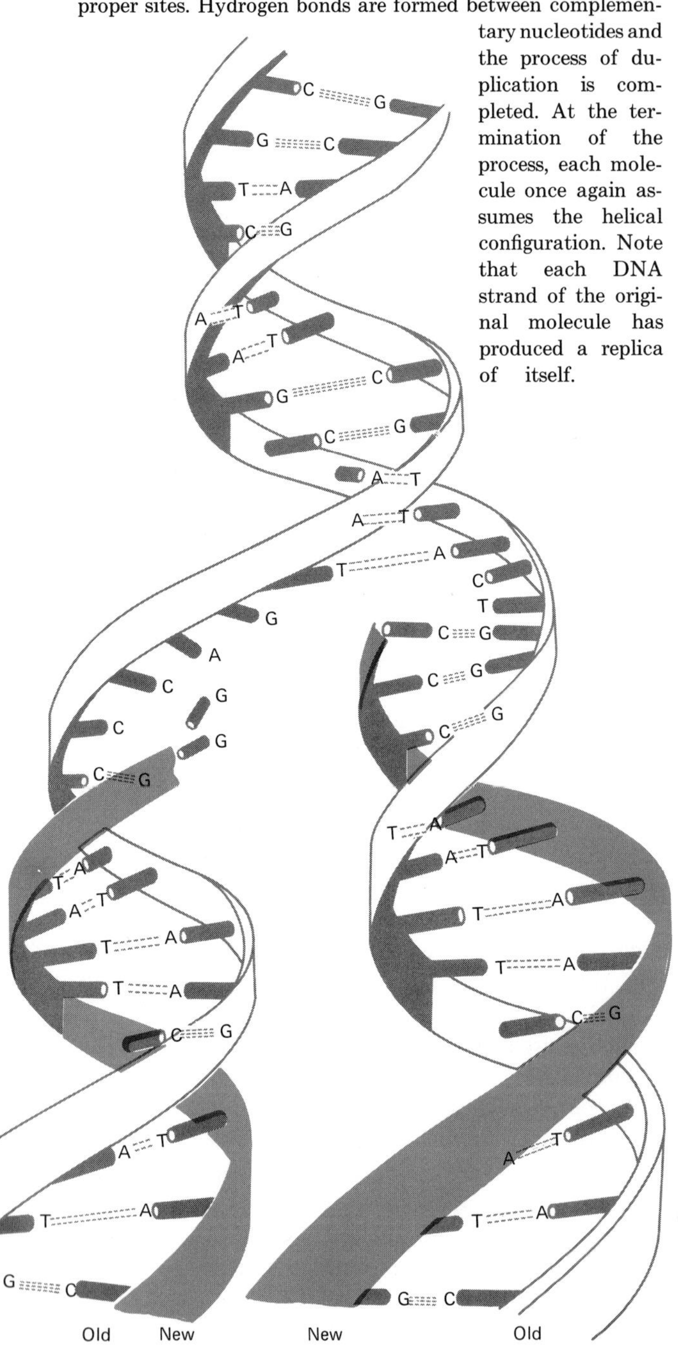

Figure 8-4. Diagrammatic representation of the replication of DNA. The double helical configuration separates, and weak hydrogen bonds between nucleotides of the unspiraled strands break. New complementary nucleotides are attached at the proper sites. Hydrogen bonds are formed between complementary nucleotides and the process of duplication is completed. At the termination of the process, each molecule once again assumes the helical configuration. Note that each DNA strand of the original molecule has produced a replica of itself.

Support for the semiconservative model of replication was immediate and was derived from several lines of investigation.

In 1957, A. Kornberg and his associates isolated an enzyme from living cells that catalyzed the replication of DNA in a test tube. The enzyme, *DNA polymerase,* when placed in a test tube with a sample of DNA, a sufficient supply of nucleotides, and the required energy sources, catalyzed the synthesis of a DNA-like molecule having physical and chemical properties almost identical to those of the DNA sample. Approximately twenty times the amount of the DNA sample originally present was produced until one of the nucleotides was exhausted. Chemical analyses showed that the structures of the sample DNA and the product DNA were almost always identical.

An experiment performed by M. S. Meselson and P. W. Stahl, at the California Institute of Technology, also supported the belief that DNA replicates according to the Watson–Crick model. Using *E. coli* as the experimental organism, they employed a technique called equilibrium density gradient centrifugation, a procedure that makes it possible to distinguish between molecules of the same compound that contain different isotopes of a given element. An isotope of a given element has the same chemical properties but differs in its atomic mass because of a different number of neutrons in its nucleus. By equilibrium density gradient centrifugation Meselson and Stahl were able to distinguish between DNA molecules in which all the nitrogen atoms were normal (^{14}N) and DNA molecules in which all the nitrogen atoms were of the isotope, heavy nitrogen (^{15}N).

In the first part of the experiment they grew *E. coli* in a medium in which the sole source of nitrogen was heavy nitrogen until they were certain that all the nitrogen-containing components of the cells contained the isotope (Figure 8-5A). They then extracted the DNA from the cells containing heavy nitrogen and also from cells grown in a medium containing normal nitrogen. They then put both samples in solution and placed both solutions in a tube and centrifuged for a long period of time. The two different DNA's separated into different levels, which could be distinguished by photographing the tube by ultraviolet light. It was found that the DNA containing heavy nitrogen, ^{15}N, formed a

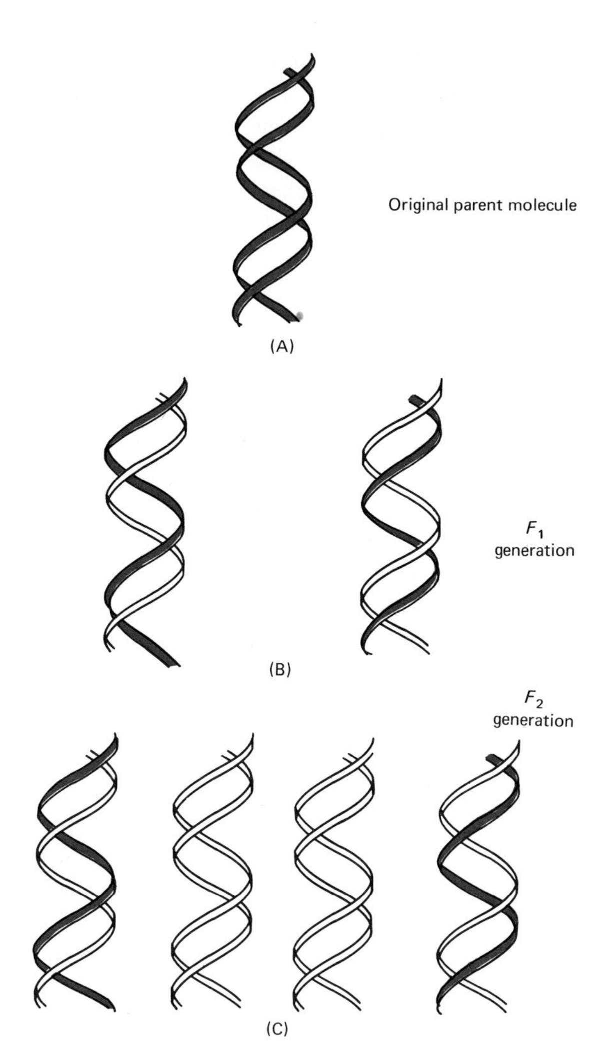

Figure 8-5. The Meselson–Stahl experiment: a demonstration of semiconservative replication of DNA in *E. coli.* **(A)** In the parent DNA molecule, all the nitrogen is ^{15}N or "heavy" nitrogen. **(B)** The F_1 generation molecules contained half ^{14}N, or "light" nitrogen, and half ^{15}N. **(C)** Half of the F_2 generation DNA molecules contained ^{14}N and ^{15}N; the other half contained only ^{14}N. The ^{15}N strands are shown in color. [*From M. S. Meselson and F. W. Stahl,* Proceedings of the National Academy of Sciences, *vol. 44, p. 675, 1958. Copyright © 1958 by the National Academy of Sciences. By permission of the publisher and authors.*]

band lower down the tube than DNA containing normal nitrogen, ^{14}N.

Meselson and Stahl then took a sample of the *E. coli* that had grown in the heavy nitrogen medium and placed it in a medium containing only normal nitrogen. As the *E. coli* colonies grew, normal nitrogen was incorporated in the daughter cells. At successive time intervals they removed some of the *E. coli,* extracted the DNA, and centrifuged it. One of these samples was taken after the bacteria had doubled their numbers (first generation), and therefore had doubled their quantity of DNA. This generation of DNA molecules formed a band in the centrifuge tube that represented a density intermediate between those of ^{14}N and ^{15}N. The DNA consisted of bonded pairs of one heavy (^{15}N) and one light (^{14}N) strand (Figure 8-5B). A sample isolated after the bacteria had quadrupled in number (second generation) showed that their DNA contained particles of two densities; one half had the intermediate density, while the remainder had the density of DNA containing only ^{14}N (Figure 8-5C).

This experiment clearly supported the theory that the two strands of double-helical DNA are complementary. During replication the strands separate and replicate, and the resulting daughter molecules contain one strand from the parental DNA. Later research has shown that this same type of replication of chromosomal DNA also occurs in eucaryotic cells.

The Mechanism of Gene Function

So far, our discussion of DNA has been concerned with the chemistry of the molecule and the mechanisms providing for the replication and conservation of genetic information. The next step is to ascertain how the genetic information contained in DNA finds expression. In other words, how do genes function in the cellular environment? How do they influence cellular metabolism?

Genes and Enzymes. The first clue to the action of genes came around 1909 from investigations by Sir A. E. Garrod in England. In that year he wrote a book titled *Inborn Errors of Metabolism,* in which he discussed human genetic abnormalities that were the result of certain chemical reactions of the body. One of the diseases he discussed was alkaptonuria, in which the urine turns black in air because it contains an abnormal compound called homogentisic acid. *Homogentisic acid,* a normal product of metabolism, is ultimately converted to CO_2 and H_2O in normal individuals by an enzyme that is lacking in individuals with alkaptonuria.

The importance of Garrod's work was that he was the first to postulate that certain diseases were caused by lack of a specific enzyme and were hereditary in nature. In essence, he suggested over sixty years ago that a change in the genetic material could cause a change in an enzyme and that there is a direct relationship between genes and enzymes. It was not until the 1930's, however, that intensive research began to reveal the role of genes in controlling enzyme production.

In 1935, George Beadle began a series of experiments that were designed to discover the chemical reactions by which genes control inherited characteristics. One of his initial experiments was concerned with the inheritance of eye color in *Drosophila* (fruit flies), and he determined that each gene involved controls one chemical reaction by which eye colors are formed. Tremendous progress was made a few years later (1941) when he collaborated with Edward L. Tatum and they reversed their experimental procedure. Instead of selecting a genetic trait and then determining the sequence of chemical reactions that cause it, they worked with known sequences of reactions and determined how genetic traits affected these reactions.

For their experiments they selected the pink bread mold, *Neurospora crassa.* Among the advantages of using this organism are the following.

1. It has a brief life cycle.
2. It exists in the haploid stage through most of its life cycle so that when a mutation occurs, the effects are immediately apparent.
3. It has a simple nutritional requirement consisting only of sugar, a few inorganic salts, and the vitamin biotin.

All other vitamins, amino acids, polysaccharides, and other essential substances are synthesized within the mold from the simple components of the growth medium, and the mold inherits the genes that enable it to do this.

Beadle and Tatum induced mutations in some of these genes with X-rays and ultraviolet light and then studied the effect of these mutations on the reactions by which the chemical compounds were synthesized. As a result of the mutations, they discovered that some of the *Neurospora* strains were unable to grow on a normal minimal medium unless an extra vitamin or a certain amino acid was added to the growth medium. These molds had lost the ability to perform one or more of the synthetic reactions by which these compounds are formed, owing to the mutation of one or more genes. Careful analyses revealed that a number of distinct strains could be isolated that required the same vitamin or amino acid, but had mutations on different genes. This strongly suggested that more than one gene was involved in the synthesis of any one compound, and it was concluded that each gene controlled the formation of an enzyme or enzymes that catalyzed the sequences of reactions.

The evidence for the conclusion was derived in part from the following observations. Three strains of *Neurospora* were isolated which required the amino acid arginine and therefore could not synthesize it from the components of the minimal medium. Each strain had a mutation on a different gene. One of the strains would grow if arginine, or the amino acids citrulline or ornithine, were added to the minimal medium. Another mutant form would grow if either arginine or citrulline was added, and the third mutant type would grow only if arginine was added to the minimal medium. Beadle and Tatum hypothesized that each of these mutant forms involved a different series of enzymes for the synthesis of arginine. From these data it was reasoned that a given gene (gene 1) produces an enzyme (enzyme 1) that converts some prior substance into ornithine; another gene (gene 2) produces an enzyme (enzyme 2) that converts ornithine into citrulline; and still another gene (gene 3) produces an enzyme (enzyme 3) that converts citrulline into arginine (Figure 8-6).

If gene 1 was mutated, the mold could no longer grow and would die unless ornithine was supplied to the minimal medium. If gene 2 was mutated, the mold would die unless citrulline was supplied. Finally, if all three genes had mutated, the mold would die unless arginine was also added. The point here is that by altering the growth medium in various ways, the sequence of biochemical reactions for the synthesis of a specific compound necessary for growth can be arranged in the proper order.

The Beadle–Tatum hypothesis, called the *one-gene, one-polypeptide chain hypothesis,* provided evidence that genes probably exert their influence through the specific enzymes that they produce. Genes exert their control over cellular activities by producing the en-

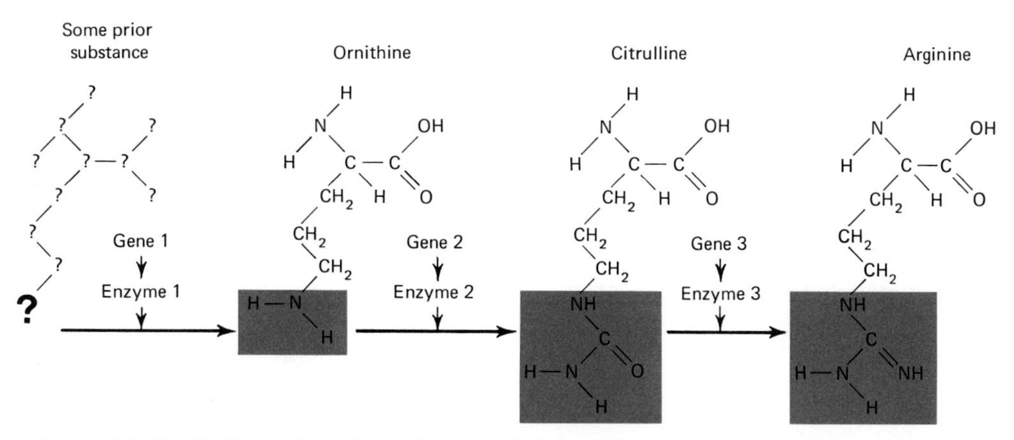

FIGURE 8-6. Beadle–Tatum hypothesis. Based upon their studies of *Neurospora,* Beadle and Tatum reasoned that genes synthesize enzymes that catalyze sequences of biochemical reactions. The action of enzyme 1 is to convert some prior substance into ornithine. Enzyme 2 adds four atoms (carbon, hydrogen, oxygen, and nitrogen) to ornithine, thereby converting it to citrulline. The catalytic activity of enzyme 3 is to replace the oxygen atom of citrulline with a nitrogen and a hydrogen atom. This conversion transforms citrulline into arginine. The colored boxes enclose the differing portions of the amino acids.

zymes that control all the many biochemical reactions of the cell.

THE GENETIC CODE. Once the relationship between genes (DNA) and enzymes (polypeptide chains or proteins) had been established, researchers began to investigate the genetic code, that is, how the information carried in genes determines the structure of proteins. In a later section, we will consider how a group of nucleotides is able to direct the incorporation of a certain amino acid into a protein. At this point, we will consider how many nucleotides of a DNA molecule specify a particular amino acid and the relationship of these nucleotides to each other.

The DNA code is based upon information stored in a nucleotide sequence composed of varying arrangements of the four nitrogenous bases—adenine (A), thymine (T), cytosine (C), and guanine (G). Within the chemical storehouse of the cell there are twenty different amino acids that may be synthesized into proteins of various sizes and levels of complexity. How, then, can a four-letter alphhabet (the nucleotides) specify the twenty different acids? The simplest version would be that each ucleotide in DNA directs one specific amino acid into a otein. According to this reasoning, only four different amino acids of the twenty present could be assembled into proteins. It would likewise be impossible for pairs of nucleotides to direct each kind of amino acid because the corresponding protein could contain a maximum of only sixteen (4^2) amino acids. According to mathematical reasoning, then, at least three nucleotides must specify each amino acid. In this manner, the twenty different amino acids could be arranged in sixty-four (4^3) combinations (Figure 8-7). This concept by which at least three nucleotides specify an amino acid is called the *triplet code* and each group of triplets of nucleotides is referred to as a **codon.**

The simplest possible code, therefore, would be if the sequence of nucleotide triplets from one end of a DNA molecule to the other end specified the sequence of amino acids in a polypeptide. For example, if the gene began with the nucleotides ATCGTACAG, the codon ATC would code for the first amino acid, the triplet GTA would code for the second, and CAG for the third. In this type of code, the triplets are counted off in three's from the first nucleotide in DNA. The members of each triplet are determined solely by their position relative to the first nucleotide. Evidence provided by Crick and others has generally established the concept that three nonoverlapping nucleotides code each amino acid.

	Possible Combinations				Possible Messages (Number of Amino Acids)
Singlet code:	A	G	C	U	4
Doublet code:	AA AG AC AU	GA GG GC GU	CA CG CC CU	UA UG UC UU	16
Triplet code:	AAA AGA ACA AUA GAA GGA GCA GUA CAA CGA CCA CUA UAA UGA UCA UUA	AAG AGG ACG AUG GAG GGG GCG GUG CAG CGG CCG CUG UAG UGG UCG UUG	AAC AGC ACC AUC GAC GGC GCC GUC CAC CGC CCC CUC UAC UGC UCC UUC	AAU AGU ACU AUU GAU GGU GCU GUU CAU CGU CCU CUU UAU UGU UCU UUU	64

FIGURE 8-7. Number of amino acids that can be coded by singlet, doublet, and triplet codes. If each nucleotide coded for a specific amino acid, only four amino acids could be assembled into proteins. If paired nucleotides coded for a specific amino acid, only sixteen amino acids could be assembled into proteins. Since at least twenty codes are needed, one for each of the twenty amino acids, the minimum code length is a sequence of three nucleotides. [*Modified from Thomas D. Brock,* Biology of Microorganisms. *Englewood Cliffs, N.J.: Prentice-Hall, Inc. Copyright © 1970.*]

RIBONUCLEIC ACID AND RIBOSOMES. From our discussion of the generalized cell in Chapter 5, it becomes apparent that genes do not form proteins directly. Recall that proteins are formed on ribosomes of the cytoplasm. In the years following the announcement of the Watson-Crick model of DNA, many researchers sought to ascertain the sequence of reactions leading to protein synthesis. Some of these investigations implicated ribonucleic acid (RNA) as an integral component of polypeptide synthesis. Moreover, it was soon discovered that the genetic information of DNA is passed on to RNA molecules which then serve as the direct templates for the formation of proteins. In other words, DNA → RNA → proteins. Among the evidences for this proposed scheme are

1. Protein synthesis occurs in the cytoplasm (ribosomes) while virtually all the DNA of a cell is confined to the nucleus (probably less than 1% of the cell's DNA is outside the nucleus in mitochondria, and, in cells that contain them, in chloroplasts).
2. Many experiments indicate that proteins can be synthesized in systems lacking DNA.
3. The rate of protein synthesis in a cell is directly correlated with the amount of RNA present.
4. Over half the weight of a ribosome is RNA.

The chemical structures of DNA and RNA differ in a number of ways. First, the sugar found in the nucleotides of RNA is ribose as compared to the deoxyribose of DNA. Second, the four nitrogenous bases of DNA are adenine, thymine, cytosine, and guanine. Those found in RNA are adenine, uracil, cytosine, and guanine; uracil replaces thymine. Third, most RNA exists as a single strand and not as a typical double strand as seen in DNA.

The synthesis of a particular protein requires that the given sequence of amino acids must be assembled in the proper line or order. Such a procedure implies certain problems. First, the linking together of amino acids requires energy. Second, there must be a mechanism for selecting the proper proportions of amino acids from all those available in the cell. Finally, the amino acids must be incorporated into a sequence characteristic of the specific protein being synthesized. RNA plays a critical role in the proper selection and placement of amino acids to form polypeptides.

THE FUNCTION OF A GENE: PROTEIN SYNTHESIS. Ultimately, the genetic information contained in the DNA of a cell is expressed in terms of protein synthesis. The nature of a cell is determined by the quantity and character of its proteins since the enzymes that determine a cell's synthetic and degradative metabolic pathways are proteins. The presence or absence of a relatively few enzymes can completely alter the behavior and function of any cell. Clearly, an understanding of how a gene dictates the hereditary characteristics of cells is linked to an understanding of how they direct the synthesis of proteins. Essentially, it is assumed that genes control the cell's activities by controlling the synthesis of enzymes. Having provided some of the essential background material, we will now examine the details of gene-controlled protein synthesis.

Transcription. In previous sections of this chapter it was shown that DNA is the primary genetic material and that, owing to its molecular configuration, it is capable of storing information. It was also stated that although genes (DNA) actually specify amino acid sequences for proteins, most protein synthesis occurs in the cytoplasm of the cell. Thus, the genetic information coded in the nucleotide sequences of DNA must be transferred to the sites of protein synthesis in the cytoplasm (ribosomes). This transference of information from DNA is referred to as **transcription.**

On the basis of many observations, it appears that the substance responsible for carrying genetic information from DNA to the surface of the ribosomes is a kind of RNA called **messenger RNA (mRNA).** Within the cell nucleus, a portion of the DNA double helix separates, exposing a sequence of nitrogenous bases. In this condition, one strand of DNA serves as a template for the formation of a single strand of mRNA. This transcription of DNA is catalyzed by an enzyme, *RNA polymerase,* apparently found in all cells, in the presence of Mg^{2+}. After a portion of the two DNA's separate, *ribonucleotides* (called nucleotides in DNA) catalytically line up along one of the DNA strands. The adenine of DNA pairs off with uracil of RNA, the thymine of DNA with adenine of RNA, and cytosine of either nucleic acid with guanine of the other. In this way, DNA acts as a template for the transcription of RNA, that is, the RNA molecule is copied on the DNA code.

Figure 8-8 illustrates the transcription of genetic in-

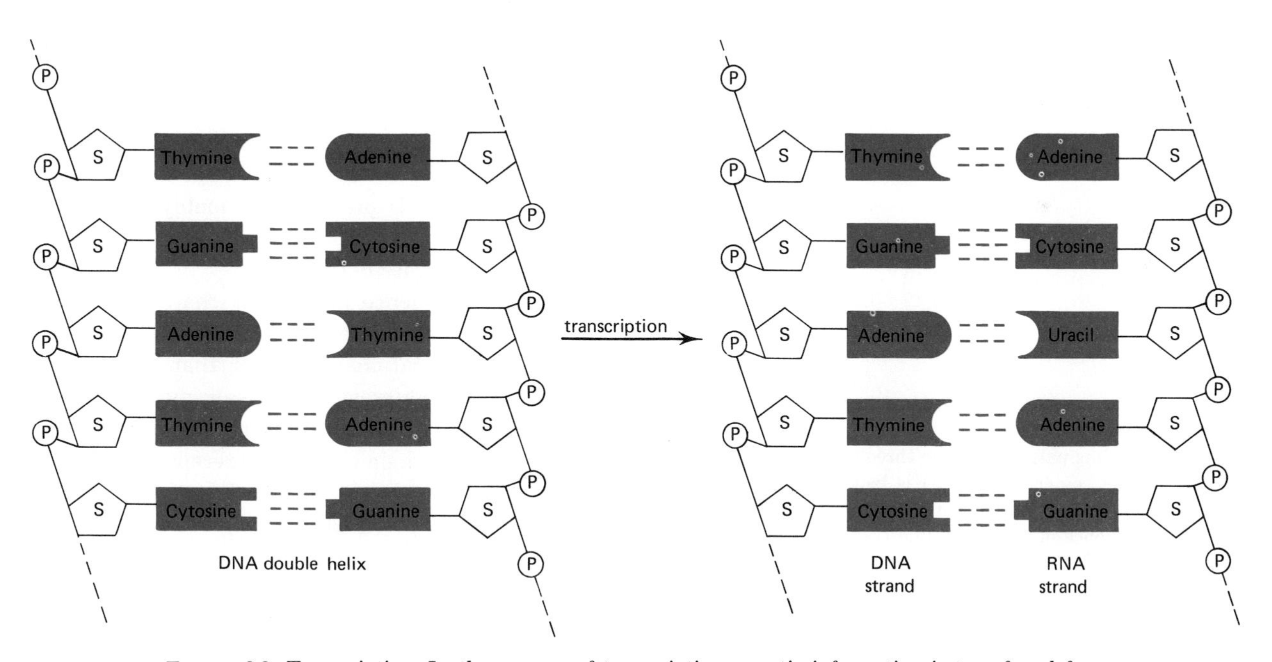

FIGURE 8-8. Transcription. In the process of transcription, genetic information is transferred from DNA to mRNA. *Left:* The DNA double helix consists of the bases T–A, G–C, A–T, T–A, and C–G. *Right:* A portion of one DNA strand (T, G, A, T, C) is transcribed onto a single mRNA strand (A, C, U, A, G).

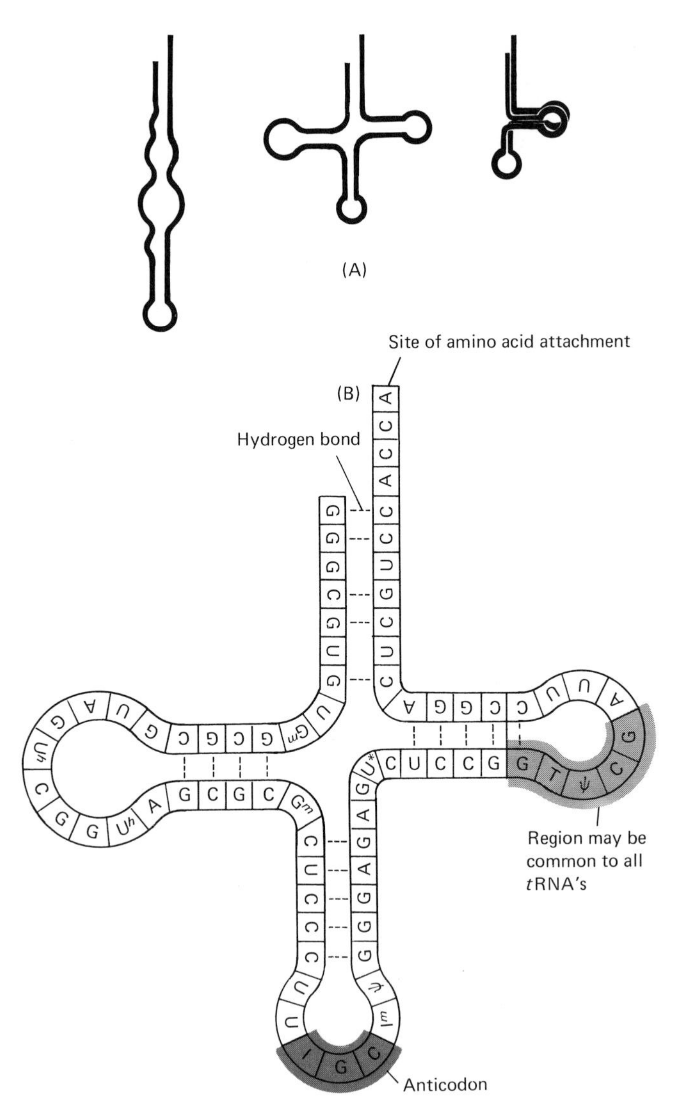

FIGURE 8-9. Structure of tRNA. (**A**) According to Holley, the alanine tRNA isolated for yeast cells can assume any one of a number of folding patterns. Of the three configurations shown, the one in the center (cloverleaf) is believed to be the most likely one. (**B**) In the enlargement of alanine tRNA shown here, the exact sequence of ribonucleotides (letters), of which there are 77, is shown. It is assumed that certain ribonucleotides (A–U and C–G) pair off via hydrogen bonds to form short double-stranded regions. The complementarity, however, is not exact, so that there are some ribonucleotides left over. These unpaired ribonucleotides probably unite with a codon of an mRNA molecule. Note that one end of the tRNA chain terminates with the bases, CCA (the A is the site of amino acid attachment), while the other end of the chain terminates in a guanine ribonucleotide. These terminations are characteristic features of all tRNA's. The triplet at the bottom, IGC, is the presumed anticodon that combines with mRNA. [*From R. W. Holley et al.,* Science, *vol. 147, pp. 1462–64, March 9, 1965. Copyright © 1965 by the American Association for the Advancement of Science.*]

formation from double-stranded DNA into single-stranded mRNA. Once transcription is completed, mRNA has the DNA information to synthesize a protein according to a specific amino acid sequence. In protein synthesis this process is known as **translation.**

Translation. A small group of small RNA molecules exists that serves as acceptors of amino acids, thus playing a key role in protein synthesis. These RNA molecules are called **transfer RNA (tRNA)** or **soluble RNA (sRNA)** molecules. Recently determined evidence indicates that tRNA, like mRNA, is also made on DNA templates. The essential function of tRNA in protein synthesis is to combine with amino acids and transport them to mRNA. It will be seen that although the type and sequence of amino acids in a polypeptide chain are determined by the ribonucleotide sequences of mRNA (through DNA transcription), amino acids do not react directly with the mRNA strand.

Research conducted by Robert W. Holley at Cornell University in 1964 determined the complete ribonucleotide sequence of the tRNA molecule that transfers the amino acid alanine in yeast cells (Figure 8-9B). At least one tRNA molecule occurs for each of the twenty amino acids and each tRNA has a slightly different structure. However, the tRNA's studied have a great similarity in size, each containing seventy-five to eighty ribonucleotides and a total molecular weight of about 25,000. They occur as single strands, but a great deal of hydrogen bonding is present, presumably because the chain folds back on itself and yields the cloverleaf pattern shown in Figure 8-9A.

From base sequence data, some characteristic features of tRNA have been identified. All tRNA molecules have identical chain endings, that is, one chain end always terminates with the base sequence CCA and the A is the site of amino acid attachment (see Figure 8-9B). The other chain end always terminates with the base guanine (G). Moreover, it has been found that one base triplet in the polynucleotide chain is different in all tRNA's studied. This triplet is thought to represent the

anticodon, the base sequence complementary to the triplets (codons) in mRNA (Figure 8-9B). It is assumed that the anticodon triplet has the proper base sequence to form specific hydrogen bonds with the corresponding codon of the mRNA so that an amino acid is properly positioned for transfer to a growing polypeptide chain. For example, assume that one of the codons on mRNA specifying the amino acid phenylalanine is UUU. The anticodon on one type of tRNA that accepts phenylalanine would thus be the complementary triplet AAA.

Amino Acid Activation. The function of tRNA is to pick up amino acids and take them to mRNA for assembly into a protein. Because the linkage of amino acids to tRNA requires energy, the amino acids must first be *activated* by a reaction involving ATP. Amino acids are attached to their corresponding tRNA's by a class of enzymes called *aminoacyl tRNA synthetases.* In these reactions (Figure 8-10A), the carboxyl (—COOH) group of the amino acid reacts with ATP; the products are *aminoacyl adenylate* (activated amino acid), AMP (adenine monophosphate), and PP_i (pyrophosphate). All such reactions have an absolute requirement for Mg^{2+}. The activation for each of the twenty amino acids requires a specific kind of aminoacyl tRNA synthetase enzyme.

Enzyme-bound aminoacyl adenylate reacts immediately with tRNA and produces aminoacyl tRNA (Figure 8-10B). The same enzyme that activates a specific amino acid is also involved in the transfer of that amino acid to a specific tRNA. These enzymes are quite unique since they not only activate the amino acid but are also capable of recognizing a particular tRNA molecule. The anticodons of tRNA are adapted for fitting on to a specific nucleotide sequence (codon) on the mRNA molecule. It is for this reason that tRNA molecules are commonly called *adapter molecules.*

Ribosomes and Protein Synthesis. **Ribosomal RNA (rRNA),** formed by DNA transcription, constitutes the bulk of cellular RNA and occurs in the ribosomes. Inasmuch as ribosomes are the exclusive site of protein synthesis, their conformation or surface arrangement is highly ordered. Chemically, ribosomes are nucleoproteins (RNA + protein) and, in bacteria, ribosomes contain about 60% protein and 40% RNA. Each ribosome, about 180 Å in diameter, has a molecular weight near 3 million. Ribosomes are constructed of two subunits, one having a molecular weight of about 1.8 million and the smaller a molecular weight of approximately 1.0 million. Ribosomes, as well as their subunits, are often described in terms of Svedberg (S) units, that

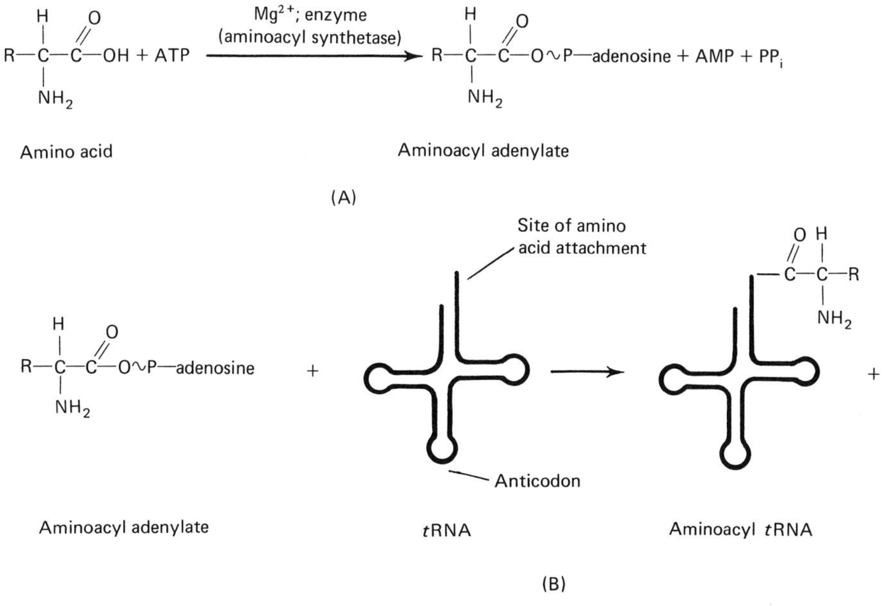

FIGURE 8-10. Amino acid activation and formation of aminoacyl tRNA. (**A**) An amino acid reacts with ATP in the presence of an enzyme and Mg^{2+} to form aminoacyl adenylate (activated amino acid). (**B**) Aminoacyl adenylate, still enzyme-bound, is transferred to a particular amino acid and the product is aminoacyl tRNA.

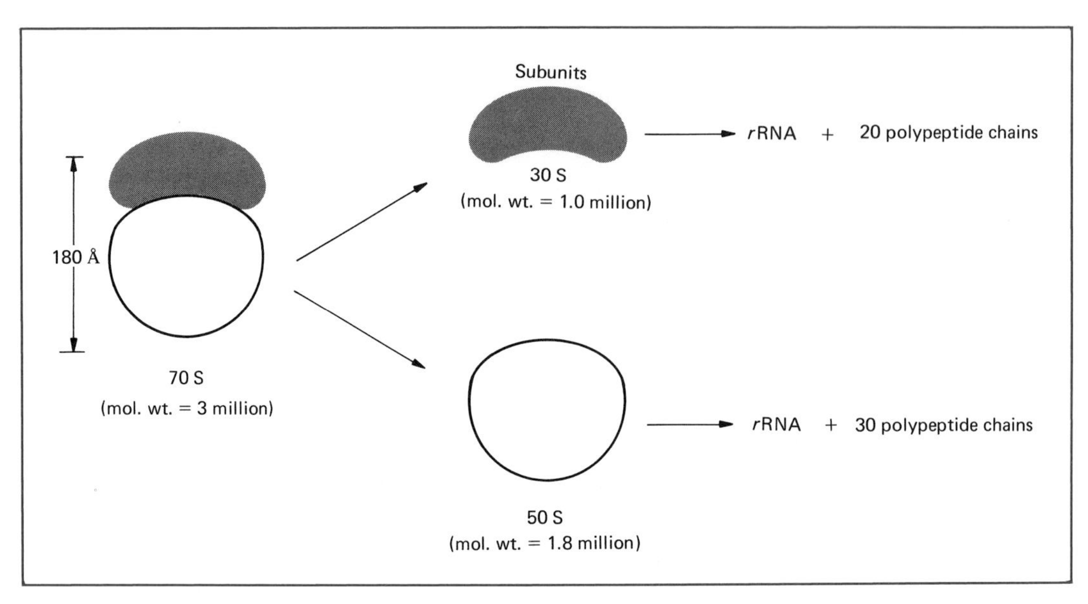

FIGURE 8-11. Components of a ribosome. Shown here is a diagrammatic representation of a ribosome from *E. coli.* Note that each of the subunits consists of both rRNA and polypeptide chains.

is, a measure of the speed at which a particle sediments in the ultracentrifuge (Figure 8-11). The larger subunit has a sedimentation coefficient of 50 S, whereas the smaller one is 30 S; in combination they have a sedimentation constant of 70 S. Both subunits contain rRNA and a large number of separate polypeptide chains. Although some of the proteins of ribosomes probably have a catalytic and structural role, the function of the rRNA is still not clear.

Synthesis of a Polypeptide Chain. In recent years, it has been suspected that the codon for the first amino acid of a polypeptide chain must have some distinctive characteristic enabling the ribosome to recognize it as the starting point for the growth of a polypeptide chain. It has been known that the amine (—NH_2) group of an amino acid forms the beginning of a polypeptide chain. In 1964, F. Sanger and A. Marcker, working with the bacterium *E. coli,* discovered that the synthesis of most, if not all, of *E. coli* proteins begins with the amino acid methionine. They found that the initiating methionine is not simply a methionine–tRNA complex; rather, the methionine of the complex was formylated, that is, a formyl group replaced a hydrogen atom in the amine group of the molecule (Figure 8-12). Normally, when an amino acid enters a polypeptide chain, one of the hydrogen atoms of the amine end of one molecule combines with an —OH from the carboxyl group of another amino acid and a peptide linkage is formed. Formylated methionine (F-Met) prevents this reaction and since an amino group is found at the front end of the protein molecule, F-Met must constitute the initial unit of the molecule. The evidence suggests that methionine is formylated only after the amino acid has become attached to tRNA (F-Met–tRNA).

Since the initial amino acid of some polypeptide chains in *E. coli* is often alanine or serine, a search was made for other formylated amino acids. None were found. Attempts to determine how amino acids other than methionine might become the initial members of a protein chain suggest that the formation of the protein starts with F-Met and that the bacterial cell supplies an enzyme that removes the F-Met later, leaving the second amino acid in the first position.

Although it has been known for some time that ribosomes must bind the mRNA template, the polypeptide chain being formed, and the entering aminoacyl tRNA molecule, only recently have the mechanisms of these reactions been understood. It has been discovered in the

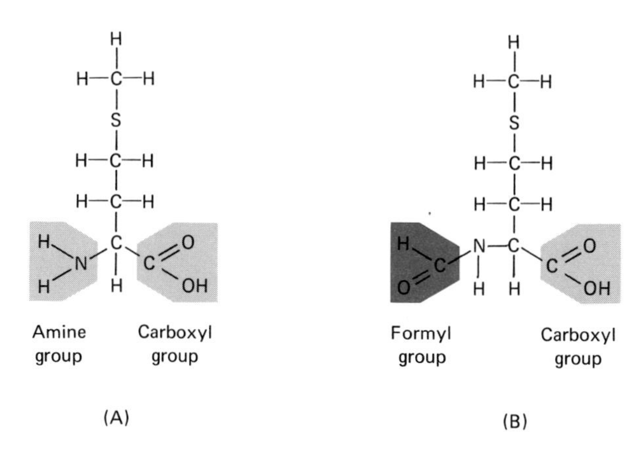

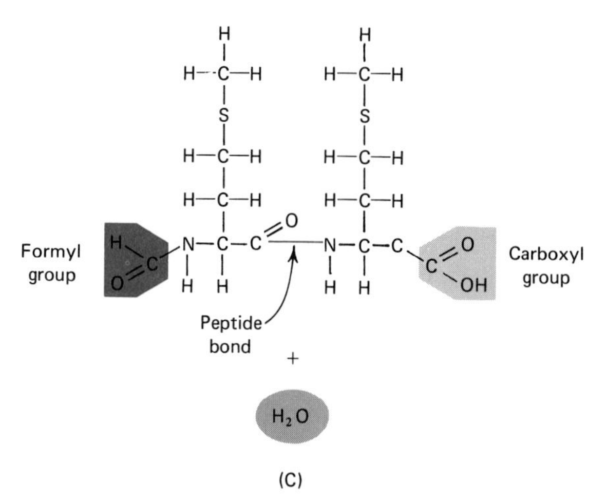

Figure 8-12. The initiating amino acid of a polypeptide chain. (**A**) Methionine has a hydrogen atom as part of its terminal amine group. (**B**) Formylated methionine (F-Met) has a formyl (CHO) group (dark shading) as part of its terminal amine group. Normally, when an amino acid enters a polypeptide chain, one of the H atoms from the amine end of one molecule combines with an —OH group from the carboxyl group of another molecule and a peptide bond is formed between the adjacent amino acids. (**C**) The formyl group of F-Met, however, prevents this reaction; therefore this acid must be initial amino acid in a polypeptide chain.

E. coli system that during protein synthesis, there is a continuous dissociation of 70 S ribosomes into their component subunits and a continuous reassociation of the subunits to form 70 S ribosomes. From this observation it has been postulated that the 70 S ribosomes normally dissociate after completion of synthesis of a protein molecule, and that the subunits recombine to form 70 S ribosomes when they initiate a chain. M. Nomura and his associates have shown that the 70 S ribosomes must first dissociate before either mRNA or aminoacyl tRNA can be bound. They found that the 30 S subunit binds with mRNA and F-Met–tRNA and the resulting complex associates with the 50 S subunit to form the functional 70 S ribosome (Figure 8-13).

Although the nature of the codon–anticodon attraction is not known, experimental evidence indicates that the interaction takes place only in the ribosomes. Alexander Rich at the Massachusetts Institute of Technology has hypothesized that the protein "factories" of the cell are not isolated ribosomes working independently, but are goups of ribosomes called *polyribosomes* (*polysomes*) working together in an orderly fashion. According to this hypothesis, a strand of mRNA attaches to a collection of ribosomes, often five. It is presumed that the ribosomes move along by a ratchet-like mechanism so that they may not reverse direction. Thus, in the illustration they would move from left to right. At each stop along the mRNA route, the properly coded amino acid, attached to a specific tRNA, is added to the growing polypeptide chain. This attachment occurs at a point where the appropriate base sequence (codon–anticodon) occurs. During the sequential condon–anticodon encounter, peptide links form between adjacent amino acids of the growing polypeptide chain (Figure 8-14). Following the complete synthesis of the protein, the ribosome releases the polypeptide chain and itself is liberated from the mRNA strand. Apparently, as each ribosome moves along the mRNA strand, it "reads" the information coded in RNA and builds a polypeptide chain according to that information. The tRNA's uncouple from the mRNA and move away to pick up more amino acids for further protein synthesis. Ribosomes appear to be nonspecific "workbenches" that can function in the production of any one of a number of proteins. Although much work remains to be done before the mechanisms of ribosomal function are fully understood, their central role in protein synthesis is beyond debate.

In the entire sequence (Figure 8-15) of events in protein synthesis, the genes (DNA) act by directing the synthesis of enzymes (proteins) through the intermedi-

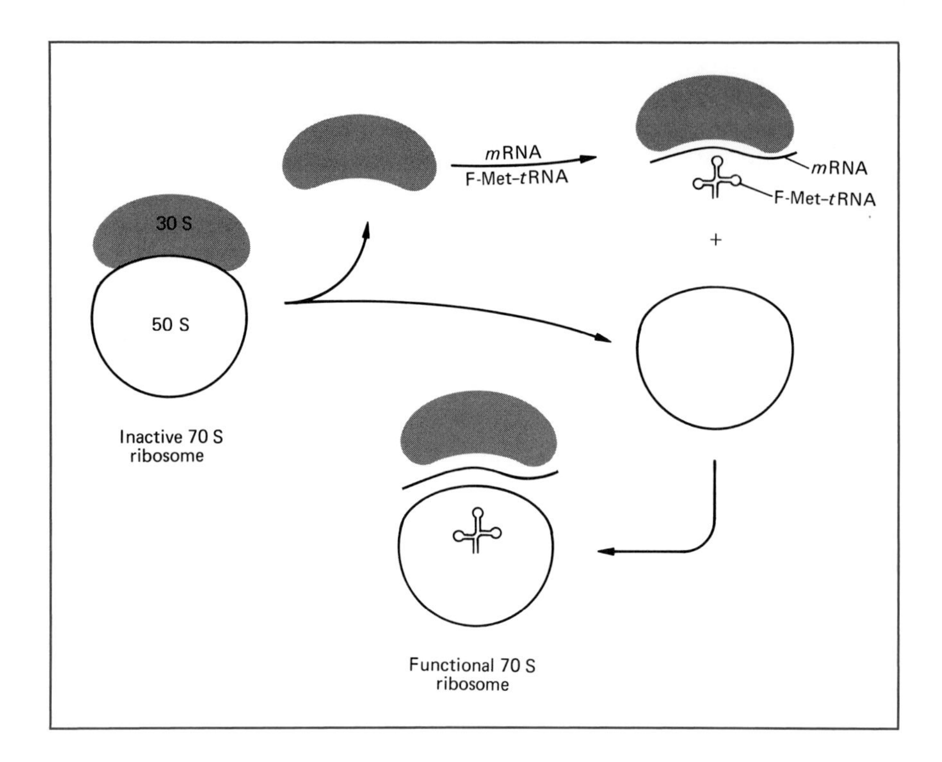

Figure 8-13. Formation of a 70 S functional ribosome. See text for amplification.

ate mRNA. The actual synthesis is facilitated by tRNA and ribosomes in the cytoplasm. The synthesized enzymes then control the various aspects of cellular metabolism. The general outline of protein synthesis just presented is reasonably well established. However, many details are still partly or entirely unknown.

Deciphering the Genetic Code. From our discussion of protein synthesis, it seems reasonable to assume that a given sequence of nucleotides in DNA constitutes a code that determines the sequence of amino acids in a synthesized polypeptide chain. This, once again, is referred to as a genetic code. Earlier in this chapter it was noted that the code consists of nonoverlapping triplets (codons) and is read in an orderly manner from end to end. Research has shown that three codons, UAG, UAA, and UGA, do not code for any of the amino acids. These are called *terminal* (*End*) *codons*. It has been found that one of these codons is located at the end of a nucleotide sequence and signals the termination of the polypeptide chain and its release from the ribosome.

Deciphering the genetic code (assigning triplets to amino acids) was started by Marshall W. Nirenberg, J. Heinrich Matthaei, and their co-workers at the National Institutes of Health, and Severo Ochoa and associates of New York University. In 1964, Nirenberg showed that it was possible for protein synthesis to occur in cell-free extracts of the bacterium *E. coli*. He combined extracts of the bacterial cells (DNA, mRNA, ribosomes, enzymes, and other components) with a source of energy (ATP) and found that the extract would incorporate amino acids into proteins. However, after about 20 minutes, the extracts lost their ability to form proteins. It was proved that this was because DNA had been destroyed during the preparation.

Nirenberg and Matthaei reasoned that protein synthesis stopped because in the absence of DNA, the production of mRNA had also ceased. They then added crude extracts of mRNA from a variety of cell sources to *E. coli* extracts which had lost the ability to form proteins and found that the mRNA extracts stimulated protein synthesis. It occurred to them that if a synthetic mRNA of known nucleotide sequence was added to mixtures of the twenty amino acids, they could de-

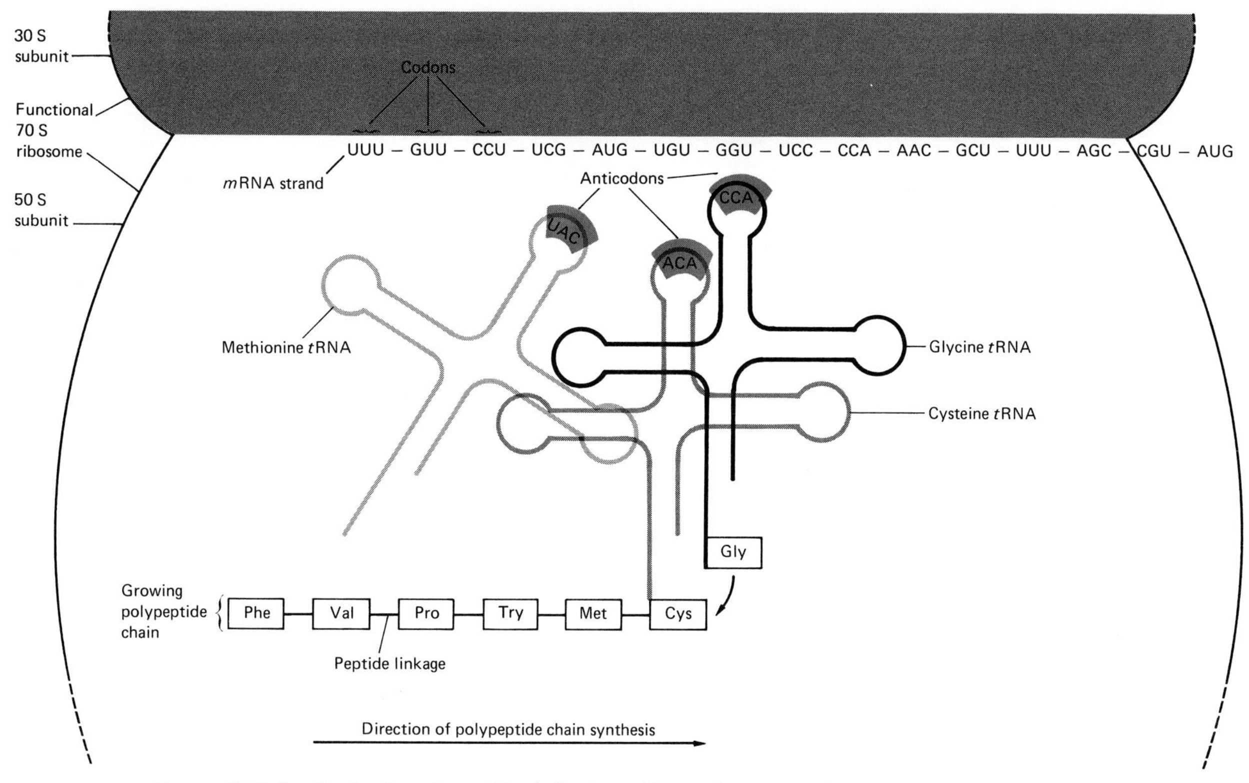

FIGURE 8-14. Synthesis of a polypeptide chain. According to the current view, activated amino acids are transported to the ribosome by tRNA molecules which carry anticodons that can form temporary bonds with the codons of mRNA. In this sequence, read from left to right, the ribosome is shown moving along the chain of mRNA, "reading off" the codons in sequence. The methionine tRNA has added its amino acid to the growing polypeptide chain and is released to pick up more of the same amino acid; the cysteine tRNA holds the growing polypeptide chain; and the arriving molecule of glycine tRNA is delivering its amino acid to be added next to the chain. The completed chain, either alone or in conjunction with other synthesized chains, represents the protein whose specification was originally coded in DNA and transcribed to mRNA. Apparently the ribosome has two bonding sites for tRNA. One site is for positioning a newly arrived tRNA molecule (serine tRNA), and the other is for holding the growing polypeptide chain (leucine tRNA). Note that at any given time three tRNA molecules may be present on the ribosome: one that has just been released, one to which the peptide chain is attached, and one that has just been bound to the new amino acid.

termine which specific sequence of nucleotides coded for specific amino acids.

In 1955, Ochoa discovered an enzyme, *polynucleotide phosphorylase,* that could catalyze the synthesis of RNA ribonucleotides into RNA strands. The enzyme, however, simply linked ribonucleotides together in random order. For Nirenberg's purposes, knowing the order of ribonucleotides was essential. Ochoa also synthesized an mRNA molecule which consisted of only one ribonucleotide (uracil) repeated over and over. Nirenberg and Matthaei prepared the synthetic mRNA polymer containing only uracil (*synthetic poly-U*) and added it to the cell-free extracts together with mixtures of the twenty amino acids. They prepared twenty different test tubes each containing the *E. coli* extracts, ATP, a mixture of amino acids, and poly-U. In addition, in each of the test tubes, only one of the amino acids contained radioactive carbon; the remaining nineteen were not

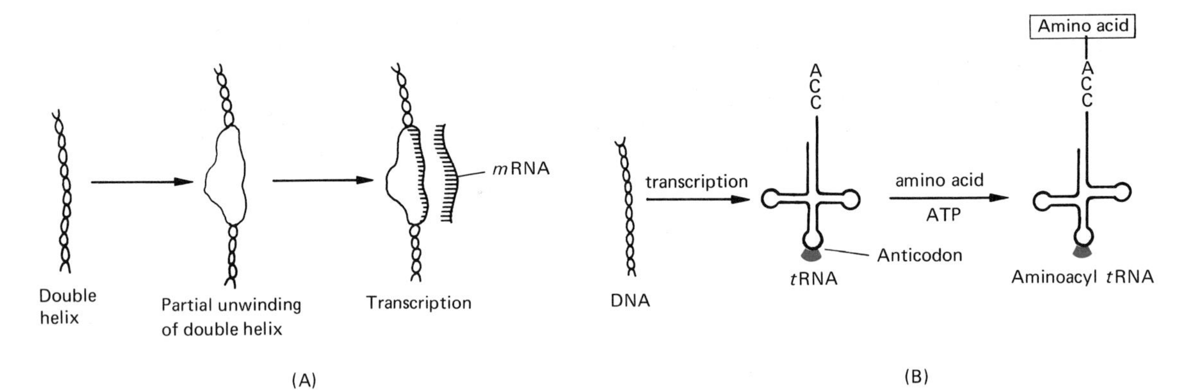

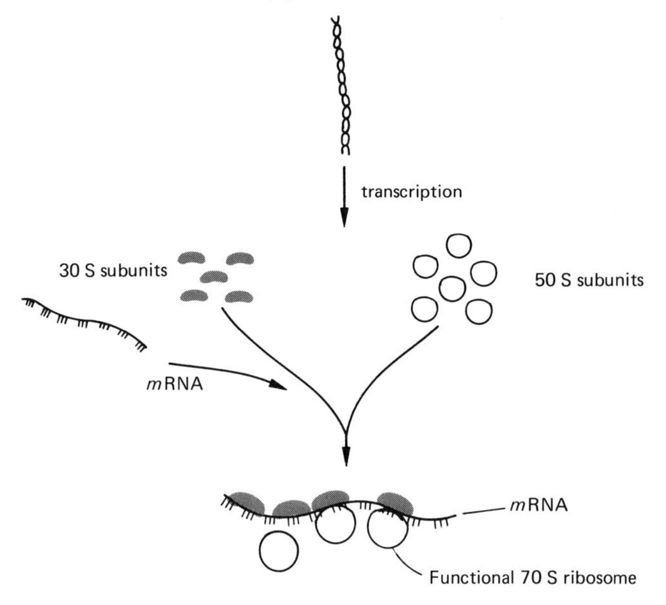

Figure 8-15. Highly schematic representation of protein synthesis. For purposes of simplicity the overall reactions of protein synthesis have been separated into various stages. (**A**) Transcription of mRNA (see Figure 8-8). (**B**) Transcription of tRNA, amino acid activation, and formation of aminoacyl–tRNA (see Figures 8-9 and 8-10). (**C**) Transcription of rRNA and association of 30 S subunits and 50 S subunits to form functional 70 S ribosomes (see Figures 8-11 and 8-13). (**D**) Synthesis of a polypeptide chain (see Figures 8-12 and 8-14).

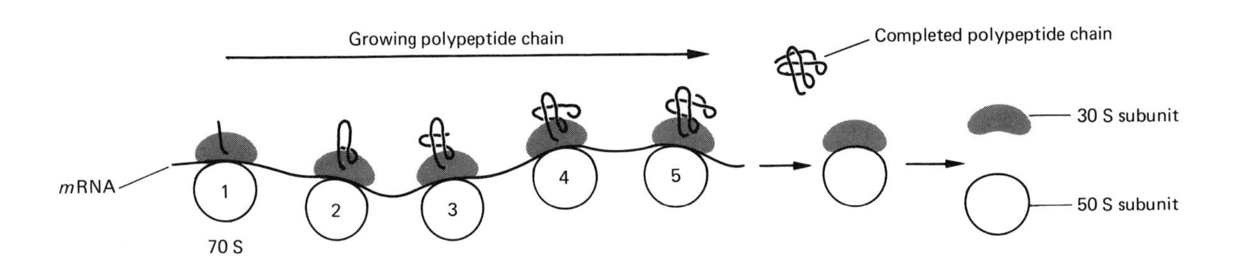

labeled. In this manner, the specific amino acid synthesized into protein by poly-U could be ascertained.

In nineteen of the twenty test tubes, little or no radioactive protein was formed. In the twentieth, however, the one to which radioactive phenylalanine had been added, the protein precipitate was highly radioactive. This protein contained only the amino acid phenylalanine. Thus, the message "UUU" fed into the system in the form of poly-U, came back "phenylalanine" (Figure 8-16). They concluded that the insertion of phenylalanine is determined by three ribonucleotides, each containing uracil. In considering what has been said about protein synthesis, it is possible to trace the sequence of events according to which a gene directs the formation of a protein containing phenylalanine. Thus, one sequence of nucleotides along a DNA molecule (AAA), in the presence of the RNA polymerase, forms mRNA containing the triplet UUU. This triplet on a ribosome facilitates the incorporation of phenylalanine into a protein. The code for phenylalanine is AAA on the gene and UUU on the mRNA.

Since 1961, when Nirenberg and Matthaei worked out the code for phenylalanine, many other experiments by them, Ochoa, and others finally resulted in the deciphering of the basic features of the genetic code. Figure 8-17 reveals that the genetic code is degenerate, that is, more than one triplet codes for most of the amino acids; of the 64 codons, 61 have been shown to specify one or another of the twenty amino acids. The triplets referred to as terminal (End) codons act as starting and stopping signals (punctuations) when polypeptide chains are being synthesized.

The Control of Gene Function

Although the one-gene, one-polypeptide chain hypothesis is compatible with most known data, at least

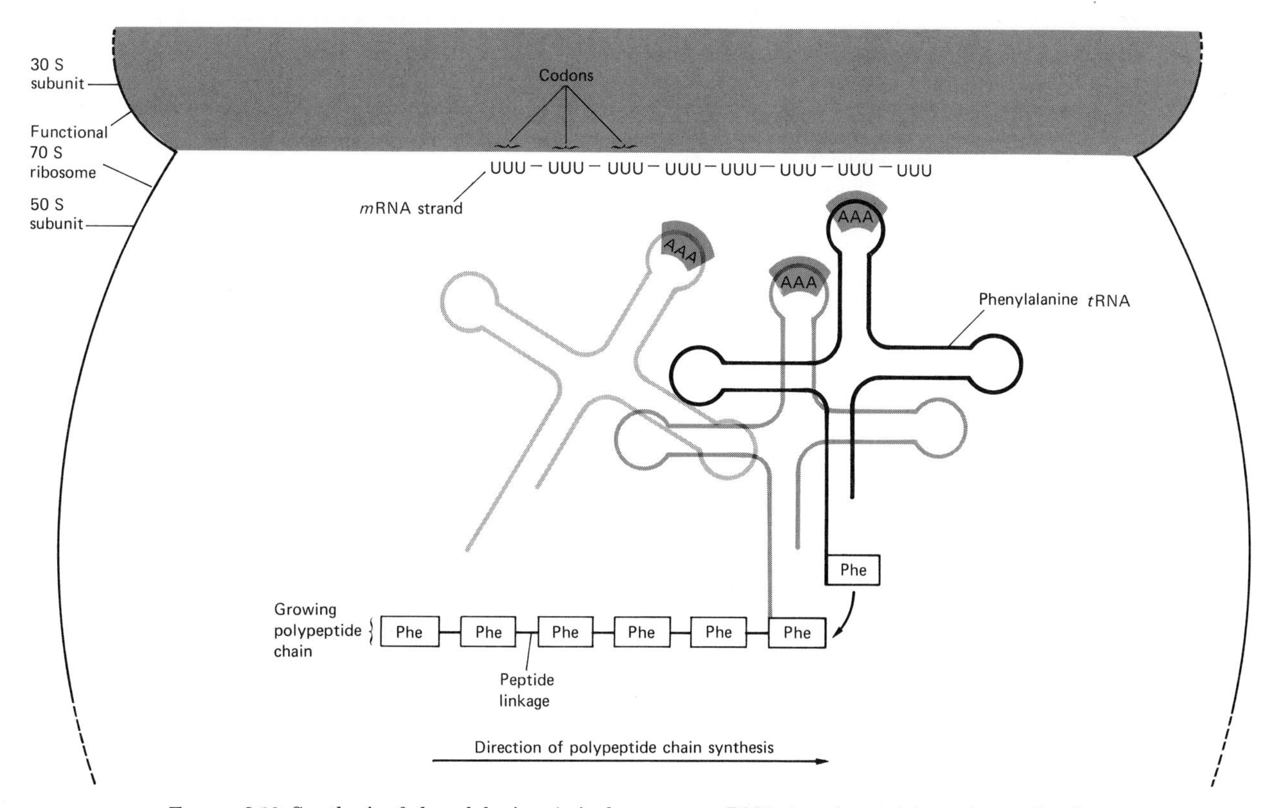

FIGURE 8-16. Synthesis of phenylalanine. A single messenger RNA strand containing only uracil codons will synthesize a polypeptide consisting of only the amino acid phenylalanine repeated over and over.

First Base		Second Base: U	C	A	G	Third Base
	U	Phe	Ser	Tyr	Cys	U
		Phe	Ser	Tyr	Cys	C
		Leu	Ser	Terminal	Terminal	A
		Leu	Ser	Terminal	Trp	G
	C	Leu	Pro	His	Arg	U
		Leu	Pro	His	Arg	C
		Leu	Pro	Gln	Arg	A
		Leu	Pro	Gln	Arg	G
	A	Ileu	Thr	Asn	Ser	U
		Ileu	Thr	Asn	Ser	C
		Ileu	Thr	Lys	Arg	A
		Met	Thr	Lys	Arg	G
	G	Val	Ala	Asp	Gly	U
		Val	Ala	Asp	Gly	C
		Val	Ala	Glu	Gly	A
		Val	Ala	Glu	Gly	G

Figure 8-17. Codon assignments. Represented here are the 64 triplet combinations and their corresponding amino acids. To read the figure, select one first base from the left, a second base from the top, and a third base from the right. Thus, UUU and UUC are codons for the amino acid phenylalanine, UUA for leucine, and so on. Only three combinations (UAA, UAG, and UGA) are not assigned to an amino acid. These codons are called terminal (End) codons and signal the end of a polypeptide chain. Logically, since 61 triplets code for 20 amino acids, a number of triplets code for the same amino acid; for example, GUU, GUC, GUA, and GUG all code for valine). In most of these "synonyms" the first two letters in each triplet are the same. [*Modified from George W. Burns,* The Science of Genetics, *3rd ed. New York: The Macmillan Company. Copyright © 1976, George W. Burns.*]

two important exceptions should be noted. First, some genes do not appear to be primarily concerned with the formation of specific proteins; instead, they regulate other genes. Whether they do so by forming a specific protein is not yet clear. Second, a few genes in each cell appear to direct the formation of the RNA of the ribosomes and of the cell solution, but not the formation of proteins.

In recent years, a great deal of research has demonstrated that genes do not function by forming their products in a continuous and unchanging manner, but that the genes themselves, as well as their products, are subject to some type of regulation. In addition, many other lines of investigation support the view that the environment can directly affect gene activity. Gene regulation appears to involve two principal mechanisms: (1) the control of enzymatic activity and (2) the control of enzyme synthesis.

Control of Enzymatic Activity. A number of investigations performed in the late 1950's were concerned with the control of enzymatic activity. In these investigations, it was noted that when the end product of an enzymatic reaction is added to a culture of bacteria, the enzymatic activity of some earlier step is frequently inhibited. The effect of this addition is immediate and there is no further synthesis of the end product as long as the product exists in high concentrations in the cell. This phenomenon by which the product of an enzymatic reaction may inhibit the activity of one or more enzymes of a metabolic pathway is called *feedback inhibition* (Figure 8-18).

This phenomenon was demonstrated by H. E. Umbarger in strains of *E. coli.* He studied the biosynthetic pathway in which the bacterial cells produced the amino acid isoleucine. In this metabolic pathway, an early precursor, threonine, is enzymatically converted to isoleucine in five separate steps (Figure 8-19). It was found that if isoleucine was added to the *E. coli* culture medium, the bacteria no longer synthesized isoleucine. This condition is maintained until the supply of isoleucine is depleted. The isoleucine control system of *E. coli* is only one kind of control system in the cell. The cell's production of other amino acids, vitamins, and purines and pyrimidines has also been identified.

Feedback inhibition is characterized by its action on enzymes that have already been synthesized; it does not

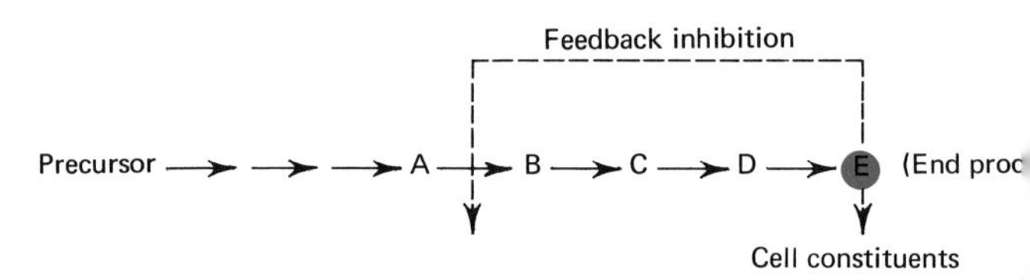

Figure 8-18. The mechanism of feedback inhibition. Accumulation of an end product in the cell results in the combination of the end product with an enzyme catalyzing an earlier step in the metabolic sequence (A → B). This feedback prevents the further synthesis of E from precursor A. Removal of the end product enables the metabolic sequence to proceed normally. [*Reprinted from A. Gib DeBusk,* Molecular Genetics. *New York: The Macmillan Company, Copyright © A. Gib DeBusk 1968.*]

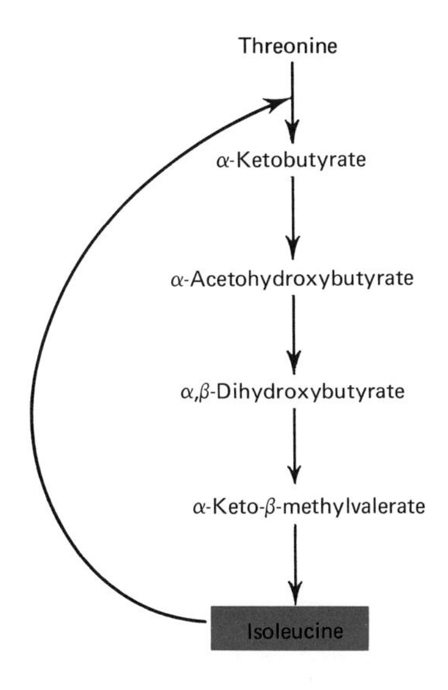

Figure 8-19. Feedback inhibition in *E. coli.* In the sequence of reactions that converts threonine to isoleucine, Umbarger observed that if isoleucine is added to the culture medium, its synthesis is blocked. The isoleucine combines with the enzyme that catalyzes the conversion of threonine into α-ketobutyrate. Thus, further synthesis is blocked.

affect the synthesis of these enzymes. The data suggest that a single enzyme, usually the first one in a metabolic pathway, is affected by feedback inhibition. The chemical interaction between the enzyme and inhibitor (end product) appears to be unique since the inhibitor is structurally different from the normal substrate. The normal substrate and the inhibitor do not compete for the same binding site on the enzyme; it seems probable that the enzyme has two sites—one specific for the substrate and one specific for the inhibitor—and that the binding of the inhibitor results in a change at the catalytic site. This change in the activity of an enzyme that is brought about by the selective binding at a second site on the enzyme (not overlapping the substrate binding site) is called **allosteric interaction** (Figure 8-20). In this regard, the enzyme permits the interaction between inhibitor and substrate.

Allosteric interactions, then, represent another type of system in which regulation may be facilitated by controlling the enzyme activity.

In the examples just cited, the control is *negative* (the inhibition of enzymes). There are other instances in which the control is *positive,* that is, the enzymes are activated. Recall from Chapter 7 (see Figure 7-9) the biosynthesis of glycogen. When the cell has an adequate supply of energy, considerable amounts of glucose-6-phosphate are produced. This serves as a stimulus for the production of glycogen. The presence of excessive amounts of glucose-6-phosphate activates the enzyme which catalyzes the conversion of uridine diphosphate glucose into glycogen. If, by contrast, the energy of the cell falls below a certain critical level, then it is necessary to convert glycogen back into glucose. This requires the activation of the enzyme (glycogen phosphorylase) that splits the glycogen. One stimulus capable of activating this enzyme is adenosine monophosphate (AMP). The AMP activates glycogen phosphorylase, the enzyme splits the glycogen molecule, the splitting releases energy, and the energy is utilized to regenerate more ATP.

Thus, within the cell, two kinds of control of enzyme activity exist. One of these, a negative control, is associated with inhibiting enzyme activity, while the other, positive control, is related to the activation of enzymes. In some cases, both mechanisms operate simultaneously. In the production of nucleic acids in *E. coli,* for example, the synthesis of purine and pyrimidine bases occurs at a rate appropriate to the proportion in which

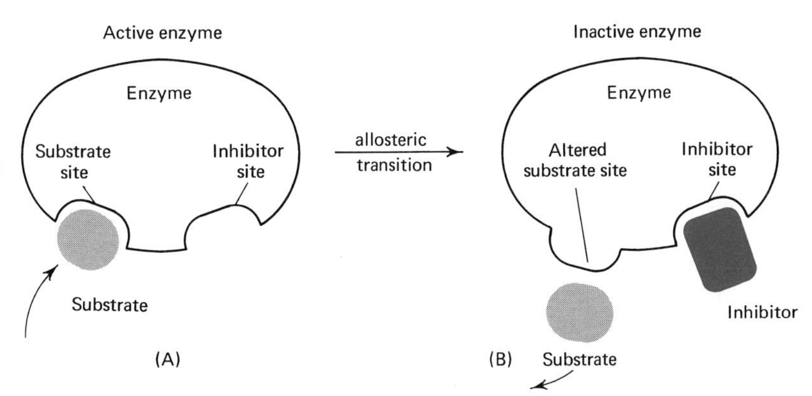

Figure 8-20. Allosteric transition. (**A**) In a normal catalytic reaction, the substrate combines with the enzyme in such a way that the interaction facilitates transformation of the substrate. (**B**) However, when an inhibitor (end product) is attached to its specific site on the enzyme, a structural transition occurs so that the normal substrate can no longer interact with the enzyme. Such a transition inhibits the catalytic activity of an enzyme for a particular biochemical pathway.

they are needed for incorporation into nucleic acids. In this regard, the rate of production of pyrimidines is controlled not only by its own end product, but also by the end product of the purine production line which counteracts the inhibition by the pyrimidine end product. Moreover, the purine end product can activate the pyrimidine production directly when no pyrimidine product is present. Essentially, then, a given enzyme may be inhibited by one signal and activated by another.

There is some experimental evidence favoring the view that specific inhibitors, activators, and substrates fit into separate binding sites on the enzyme and that inhibition or activation takes effect by an interaction between these sites. According to a model proposed by J. Monod, J. Wyman, and J. P. Changeux, an enzyme consists of a set of identical subunits, each containing a specific site for substrate, activator, or inhibitor molecules and the enzyme has an axis of symmetry. In addition, it is assumed that an enzyme can switch back and forth between two states: in the more relaxed state, the enzyme will bind activator and substrate, whereas in the more constrained state, the inhibitor will be bound. Any change in the relative concentrations of substrate, activator, or inhibitor may cause the enzyme to preferentially act in either the relaxed or constrained state. Thus, an enzyme is assumed to have the capacity to recognize and integrate various signals.

In recent years it has become increasingly clear that enzymes once thought to be single, soluble entities may interact with other cellular proteins. The pyruvic acid dehydrogenase complex, for example, which catalyzes the oxidative decarboxylation of pyruvic acid, has been shown to consist of three separate enzymes which perform their activity (transacetylation and electron transfer) better when tightly bound to one another. Individual enzymes of such complexes have also been found to consist of even smaller subunits. Many enzymes are not single-chain protein molecules; rather, they are specific aggregates of polypeptide chains that separately have no activity. If, however, the polypeptide chains are brought together, an active enzyme is formed. Among the bonds that have been implicated in bringing these polypeptide chains together are hydrophobic, disulfide, and hydrogen bonds. An example of such an enzyme, termed an isoenzyme, is lactate dehydrogenase of cardiac and skeletal muscle tissue. This enzyme consists of four polypeptide chain subunits and can exist in five multiple forms. Each subunit has no catalytic activity by itself, but the four chains, in any combination, exhibit lactic dehydrogenase activity. Thus, it appears that enzyme activity may be regulated by an inherent property of many polypeptide chains to form complex proteins, either with similar chains to form a functioning enzyme (e.g., lactic dehydrogenase) or with dissimilar chains to form a multienzyme complex (e.g., pyruvic acid dehydrogenase complex).

In addition to the mechanisms that are involved with the regulation of enzymatic activity, there are other modes of coordination of cellular metabolism related to the regulation of enzyme synthesis.

Control of Enzyme Synthesis. It has been shown from studies of microbial systems, especially those of the bacterium *Escherichia coli,* that certain organisms normally synthesize enzymes only in the presence of specific substrates. For example, *E. coli* cells growing on a glucose medium contain very little of the enzyme β-galactosidase, an enzyme that catalyzes the hydrolysis of lactose into glucose and galactose. If, however, lactose is added to the culture medium, β-galactosidase molecules appear very shortly thereafter and can be easily detected. It has been estimated that the introduction of lactose to the medium causes a thousand-fold increase of the enzyme over the amount present in bacteria growing on a glucose medium. A substrate such as lactose which brings about an increased amount of an enzyme is called an **inducer** and enzymes that are synthesized in the presence of inducers are referred to as **inducible enzymes.** The phenomenon, which is under genetic control, is termed *enzyme induction.*

In addition to enzyme induction, there is a related regulatory phenomenon that results in the alteration of the synthesis of one or often several related enzymes. According to this mechanism, if cells are exposed to a particular end product, there is a relative decrease in the rate of synthesis of enzymes involved in the formation of that end product. This is called *enzyme repression.* For example, cells of *E. coli* grown in a medium lacking any amino acids contain the enzymes necessary for the synthesis of all the amino acids contained in the protein molecules of the bacteria. The introduction of a particular amino acid to the culture medium greatly decreases the synthesis of the biosynthetic enzymes necessary for the production of that amino acid. Enzymes that are reduced in amount by the presence of the end product of a metabolic pathway are called **repressible enzymes** and the molecule (end product) that brings about repression is termed the **corepressor.**

Both induction and repression of enzymes represent types of differential gene activity in which the synthesis of gene products is sensitive to a given set of environmental conditions. In this regard, they are adaptive mechanisms that appear to be of survival value to the organism. When these mechanisms are in operation, cells do not expend large amounts of energy in synthesizing enzymes not immediately required; rather, such mechanisms allow for the synthesis of a limited number of enzymes in response to conditions in the environment. It should be noted that although enzyme repression is similar to feedback inhibition in that the end product acts to prevent its own synthesis, repression differs from feedback inhibition in that the end product in enzyme repression regulates enzyme synthesis, whereas in feedback inhibition the end product regulates enzyme activity.

In 1961, F. Jacob and J. Monod, as a result of their studies on induction and repression, formulated a general model to account for the regulation of enzyme synthesis. This model, called the *operon model,* was based upon studies of enzyme systems in *E. coli.* These studies, as well as other investigations of biochemical and genetic systems of gene regulation, have implicated a number of key components in the control of gene function:

1. Structural genes.
2. Repressor substances.
3. Regulator genes.
4. Operator genes.
5. Cellular metabolites (inducers and corepressors).

Figure 8-21 summarizes these components and their postulated relationships. Their proposed modes of action will now be discussed.

The cells of *E. coli* contain all the genetic information required for metabolic reactions, growth, and reproduction. They can synthesize from glucose and a number of inorganic ions all of the complex molecules that they need. Considering this, it can be assumed that a large number of enzymes is needed to accomplish so many syntheses. Some of these enzymes are present at all times. Among these are the enzymes involved in cellular respiration. Others, however, are produced only when they are needed by the cell. Among these are β-galactosidase, β-galactoside permease (which is involved in the transport of lactose into the cell), and galactoside transacetylase (which is related to lactose utilization). The formation of enzymes only when required by the cell is controlled by a particular **structural gene.** These genes are so designated because they specify the amino acid sequence responsible for the structure of a particular enzyme. These genes, which are closely linked on the bacterial chromosome, produce their respective enzymes rapidly and simultaneously when lactose is introduced into the culture medium.

Data suggesting that all genes are not structural genes have been obtained from studies of mutant strains of *E. coli* which continue to synthesize high levels of an enzyme, regardless of whether its substrate is present. In other words, such mutants are not subject to the usual inductive responses.

Variations in the amount of a specific enzyme in bacteria are generally due to variations in the rate of synthesis of that enzyme. Similarly, the rate of synthesis of an enzyme is governed by the quantity of mRNA present in the cell. According to the operon model, the amount of mRNA template for a given enzyme is controlled by special kinds of molecules, now known to be protein in nature, called **repressors.** Repressors, which block the synthesis of mRNA, are coded for by special genes termed **regulator genes.** Repressors control the rate of mRNA synthesis by combining with specific sites (called operator genes, as discussed subsequently) on DNA and blocking DNA transcription to form mRNA. If regulator genes synthesized repressors all the time, the production of mRNA would always be inhibited. It is therefore necessary to postulate further that repressors exist in two alternative and specific conformations—one active and the other inactive—depending upon whether the repressors are combined with specific small molecules (inducers or corepressors).

The attachment of a specific inducer inactivates a specific repressor. For example, the combination of lactose with a repressor inactivates the repressor and permits the synthesis of β-galactosidase and related en-

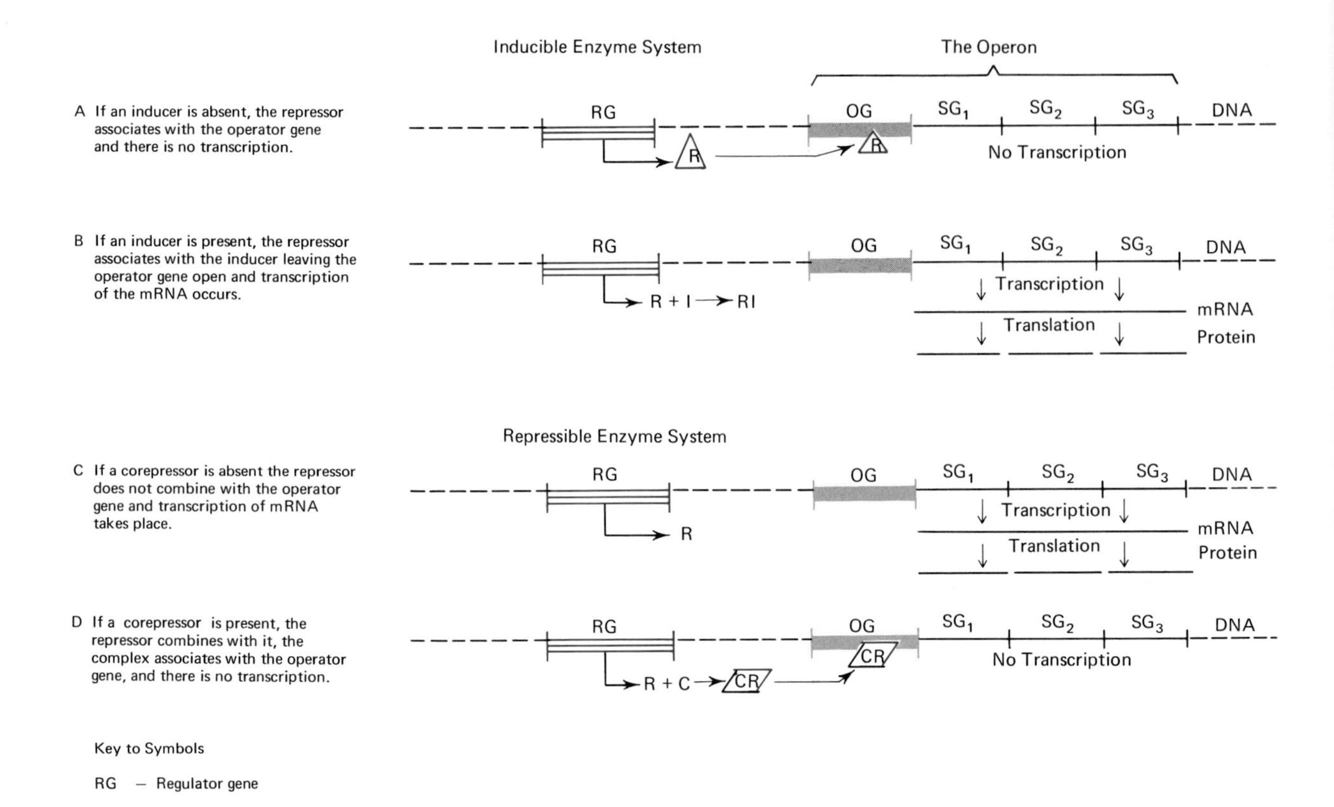

Key to Symbols

RG – Regulator gene

OG – Operator gene

SG_1, SG_2, SG_3 – Structural genes

I – Inducer

C – Corepressor

R – Repressor

FIGURE 8-21. Operon model for gene induction and gene repression according to Jacob and Monod. See text for amplification. [*From F. Jacob and J. Monod,* Journal of Molecular Biology, *vol. 3, p. 318, 1961. Copyright © 1961 by Academic Press Ltd.*]

zymes. The attachment of a corepressor (an amino acid, for example) to a repressor activates the repressor, permitting active repressor molecules to attach to a portion of the DNA molecule and inhibit the transcription of DNA, and therefore the number of RNA molecules that code for specific enzymes. The genes whose codes are transcribed on a single mRNA molecule and that are under the control of a single repressor are collectively termed an *operon.*

One further entity has been postulated in order to account for the control of operons. This structure, called the **operator gene,** is believed to be adjacent to the structural genes in an operon. Operator genes are described as sites in the DNA molecule to which active repressor substances are bound. This association inhibits the synthesis of mRNA by the adjacent structural genes. In the absence of active repressors, the structural genes in the operon are free to be transcribed, mRNA is formed, and enzymes are produced. When repressors are combined with the operator site, the operon is "switched off" and no mRNA is transcribed. This can be the case when either the operon site of the regulator gene is defective through mutation, or when sufficient end product (corepressor) is present. Presumably, combination of a corepressor with a repressor facilitates attachment of the repressor to the operator gene, turning off the operon "switch." In the absence of the corepressor, the free repressor fails to combine with the operator site and the "switch" is open. Inducers also combine with repressor molecules, preventing their attachment to operator sites.

Inducible and repressible systems differ in the form of the repressor substance that will associate with the operator gene. In an inducible system, the repressor substance normally associates with the operator gene and there is no transcription; the presence of a cellular metabolite, which serves as an inducer, inhibits the association and permits transcription to take place. In a repressible system, the presence of a metabolite, which acts as a corepressor, permits the association and prevents transcription.

The overall control mechanisms for the regulation of gene activity may be summarized as follows.

1. Structural genes function by specifying the structure (amino acid sequence) of an enzyme.
2. In addition to structural genes, there are regulatory genes that, under certain unique conditions, code for the synthesis of repressors.
3. Repressors may be active or inactive depending on whether they are combined with inducers or corepressors.
4. Active repressors bind to operator sites and prevent transcription of adjacent structural genes.
5. Either repression or induction often occurs as a coordinate event involving a multigene region, the operon.

It should be noted that the evidence for the existence of control mechanisms involving operons has come almost exclusively from studies of bacterial systems. It is not only possible but quite probable that other types of genetic regulatory systems exist in more complex organisms. Research on regulation in these more complex organisms is currently being conducted in many laboratories and constitutes an important field of study in genetics.

SUGGESTED SUPPLEMENTARY READINGS FOR SECTION TWO: THE CELLULAR LEVEL OF BIOLOGICAL ORGANIZATION

BERNHARD, S., *The Structure and Function of Enzymes.* New York: W. A. Benjamin, 1968.

BOURNE, G. H., *Division of Labor in Cells,* 2nd ed. New York: Academic Press, 1970.

BRACHET, JEAN, "The Living Cell," *Scientific American,* September 1961.

BRETSCHER, M. S., "How Repressor Molecules Function," *Nature,* vol. 217, pp. 509–11, 1968.

BRITTEN, R. J., and E. H. DAVIDSON, "Gene Regulation for Higher Cells: A Theory," *Science,* vol. 165, pp. 349–57, 1969.

CALVIN, MELVIN, "The Path of Carbon in Photosynthesis," *Science,* vol. 136, pp. 879–89, 1962.

CRICK, F. H. C., "The Origin of the Genetic Code," *Journal of Molecular Biology,* vol. 38, pp. 367–79, 1968.

DUPRAW, ERNEST, *Cell and Molecular Biology.* New York: Academic Press, 1968.

EDELMAN, GERALD, "The Structure and Function of Antibodies," *Scientific American,* August 1970.

GIBOR, AHARON, "Acetabularia: A Useful Giant Cell," *Scientific American,* November 1966.

GOLDSBY, RICHARD A., *Cells and Energy,* 2nd ed. New York: Macmillan, 1977.

GREEN, DAVID, and HAROLD BROWN, *Energy and the Mitochondrion.* New York: Academic Press, 1970.

GROS, FRANCOIS, *Messenger RNA.* New York: W. A. Benjamin, 1970.

HERZENBERG, L., G. SWEET, and L. HERZENBERG, "Fluoresence-Activated Cell Sorting," *Scientific American,* March 1976.

HOLLEY, R. W., "The Nucleotide Sequence of a Nucleic Acid," *Scientific American,* February 1966.

LEHNINGER, A. L., *Bioenergetics: The Molecular Basis of Biological Energy Transformations.* New York: W. A. Benjamin, 1965.

LEHNINGER, A. L., *The Mitochondrion.* New York: W. A. Benjamin, 1965.

LERNER, R., and F. DIXON, "The Human Lymphocyte as an Experimental Animal," *Scientific American,* June 1973.

LEVINE, R. P., "The Mechanism of Photosynthesis," *Scientific American,* December 1969.

LIPMAN, F., "Polypeptide Chain Elongation in Protein Biosynthesis," *Science,* vol. 164, pp. 1024–31, 1969.

MANIATIS, T., and M. PTASHNE, "A DNA Operator-Repressor System," *Scientific American,* January 1976.

MAZIA, D., "How Cells Divide," *Scientific American,* September 1961.

MAZIA, D., "The Cell Cycle," *Scientific American,* January 1974.

MCCOSHER, J., "Flashlight Fishes," *Scientific American,* March 1977.

MCELROY, W. D., and H. H. SELIGER, "Biological Luminescence," *Scientific American,* December 1962.

MILLER, O. L., "The Visualization of Genes in Action," *Scientific American,* March 1973.

NEUTRA, MARIAN, and C. P. LEBLOND, "The Golgi Apparatus," *Scientific American,* February 1969.

NIRENBERG, M. W., and P. LEDER, "RNA Codewords and Protein Synthesis," *Science,* vol. 145, pp. 1399–1407, 1964.

NOMURA, MASAYAU, "Ribosomes," *Scientific American,* October 1969.

NOVIKOFF, ALEX B., *Cells and Organelles.* New York: Holt, Rinehart and Winston, 1970.

RABINOWITCH, E. I., and GOVINDJEE, "The Role of Chlorophyll in Photosynthesis," *Scientific American,* July 1965.

ROFF, M., "Cell Surface Immunology," *Scientific American,* May 1976.

SIEKEVITZ, P., "Powerhouse of the Cell," *Scientific American,* July 1967.

STEIN, G., J. S. STEIN, and L. KLEINSMITH, "Chromosomal Protein and Gene Regulation," *Scientific American,* February 1975.

YANOFSKY, C., "Gene Structure and Protein Structure," *Scientific American,* May 1967.

THREE
Organismic Biology: Maintenance

9 Multicellular Patterns of Organization

10 Homeostasis: Chemical Control

11 Homeostasis: Nervous Control

12 Protection, Support, and Locomotion

13 The Procurement and Digestion of Nutrients

14 Transportation Mechanisms

15 Microbes, Disease, and Immunity

16 Breathing and Waste Disposal

9 Multicellular Patterns of Organization

GUIDE QUESTIONS

1 Define division of labor as it is applied to multicellular organisms. How are unicells adapted for maintenance and continuity without such a division of labor?
2 Define the tissue, organ, and organ-system levels of organization. How do these levels illustrate the concept of division of labor?
3 What are the major differences between meristematic and permanent tissues? What are the various types of meristems found in plants? Where are they found and what are their specific functions?
4 List the various simple permanent tissues of plants. Give an example of each type and state its function(s).
5 What different types of cells make up xylem tissue? What are their functions? How do phloem cells differ from xylem with regard to structure and function?
6 List the regions of a typical root tip and discuss the external and internal characteristics of each region. What are the primary tissues of a root? How are they related to the secondary tissues? What is the function of the periderm?
7 What are the major functions of the stem? List the external features of a typical woody stem and explain how each is formed. How do the bud scales and lenticels function in the maintenance of the stem?
8 How do dicot and monocot stems differ? What function does the vascular cambium serve?
9 Describe the internal cell structure of a leaf. How do dicot and monocot leaves differ?
10 On what bases are animal tissues classified? List the major categories and their general functions.
11 Describe the various types of epithelium, where they are located, and their function. What is meant by pseudostratified? What is the function of the basement membrane?
12 In what ways does connective tissue differ from epithelial tissue? Describe the various types of connective tissue, their location, and their function(s).
13 Describe the structure of a typical long bone, including both morphological and microscopic appearances. How do spongy bone and compact bone differ?
14 What are the chief components of vascular tissue? List the various formed elements of blood, their characteristics, and functions.
15 What criteria are employed in classifying muscle tissue? How do cardiac, smooth, and skeletal muscles differ?
16 Describe the structure of a neuron. How are neurons classified? What is the function of the myelin sheath? How is it formed? How do myelin sheath and neurilemma differ?
17 Discuss the skin as an organ. What are the various organ systems of the human body? Give the components and functions of each.

FROM the discussion of cells thus far, it becomes apparent that whereas some organisms are unicellular, others have a multicellular organization. In unicells, the entire spectrum of living activities must be carried out within the limits of the external surface (cell wall or cell membrane). The unicellular organism is adapted in numerous ways to ensure its maintenance and continuity. It must provide for its own nutrition, make most of its own chemical compounds, protect itself, often reproduce itself, and regulate its own activities. Multicellular organisms, by contrast, adapt in a different manner.

Multicellular Levels of Organization

Specialization and Division of Labor

In higher plants and animals, many kinds of cells assume a special function in addition to their basic activities. As part of the maturation process of a multicellular organism, many of its cells *specialize;* that is, they become more or less diversified in form and develop the capacity to perform some function(s) quite well. Normally, once a cell specializes, it cannot revert and respecialize in another way. Multicellular organisms are thus characterized by a *division of labor* in which groups of specialized cells perform a variety of specialized activities. Certain groups of cells are specialized to receive and transmit impulses, contract, transport materials, secrete enzymes, protect, provide support, eliminate wastes, and reproduce.

Tissues, Organs, and Organ Systems

The specialized cells of a multicellular organism are arranged into various organizational patterns. The first of these, above the cellular level of organization, is the tissue level. A **tissue** consists of a mass of cells that are similar in structure and function, together with all associated intercellular material. In some tissues (simple), one type of cell predominates, whereas in others (complex), two or more cell types may be present in significant numbers. The cells of most tissues are usually bound together with varying amounts of intercellular substances to form an organized, compact mass.

In more advanced organisms, even though tissues represent a specialized level of organization, they do not function as isolated entities; they are integrated within a larger framework. This framework is the organ level of organization. An **organ** may be viewed as an integration of two or more tissues that cooperate in the performance of a definite function.

Organs of structurally advanced multicellular organisms are organized into systems. A **system** consists of a collection of interacting organs that specializes as a functional unit to carry out a major process in the body, such as digestion, circulation, and respiration. The various systems, which are structurally and functionally coordinated, collectively constitute the whole **organism.**

A Survey of Plant Tissues

In the following discussion of the tissues and organs of a plant, attention will be directed to those plants with which you are most familiar. These are the *angiosperms,* or flowering plants (Kingdom Plantae). Angiosperms are subdivided into two groups called *Dicotyledonae* (dicots) and *Monocotyledonae* (monocots), which are distinguished on the basis of certain structural characteristics (Figure 9-1).

Plant tissues may be classified in a variety of ways. For example, they may be grouped on the basis of structure and function, stage of development, site and method of origin, and location. Inasmuch as cells of one plant tissue may change from one type to another and may thus share structural and functional characteristics with another tissue, a uniform classification scheme based upon a single criterion has not yet been devised. Moreover, a plant tissue may contain more than one type of cell. For purposes of this textbook, plant tissues will be categorized into two principal types, meristematic and permanent. Meristematic tissue is growth tissue from which all other plant tissues are derived. Permanent tissue, for the most part, consists of specialized cells that do not usually contribute to growth, although under certain conditions, permanent tissue may actively grow.

Meristematic Tissue: Growth Tissue

The continued growth and development of plants depend upon the mitotic activity of **meristematic tissues.** In such tissues the cells are actively dividing and new cells are constantly being produced. Meristematic cells tend to be small, thin-walled, and rich in cytoplasm and to have small, numerous vacuoles. New cells produced by a meristem are initially alike, but as they grow and mature, their characteristics slowly change as they become differentiated into cells of other tissues.

Meristematic tissues located near the tips of roots and in the buds of stems are called **apical meristems** (see Figure 9-5A, B). These meristems bring about an increase in length of roots and stems after they are formed. **Vascular cambium,** or simply **cambium,** is a term applied to meristematic tissue found within the stems and roots of certain plants. This tissue is considered to be a lateral meristem because it causes lateral

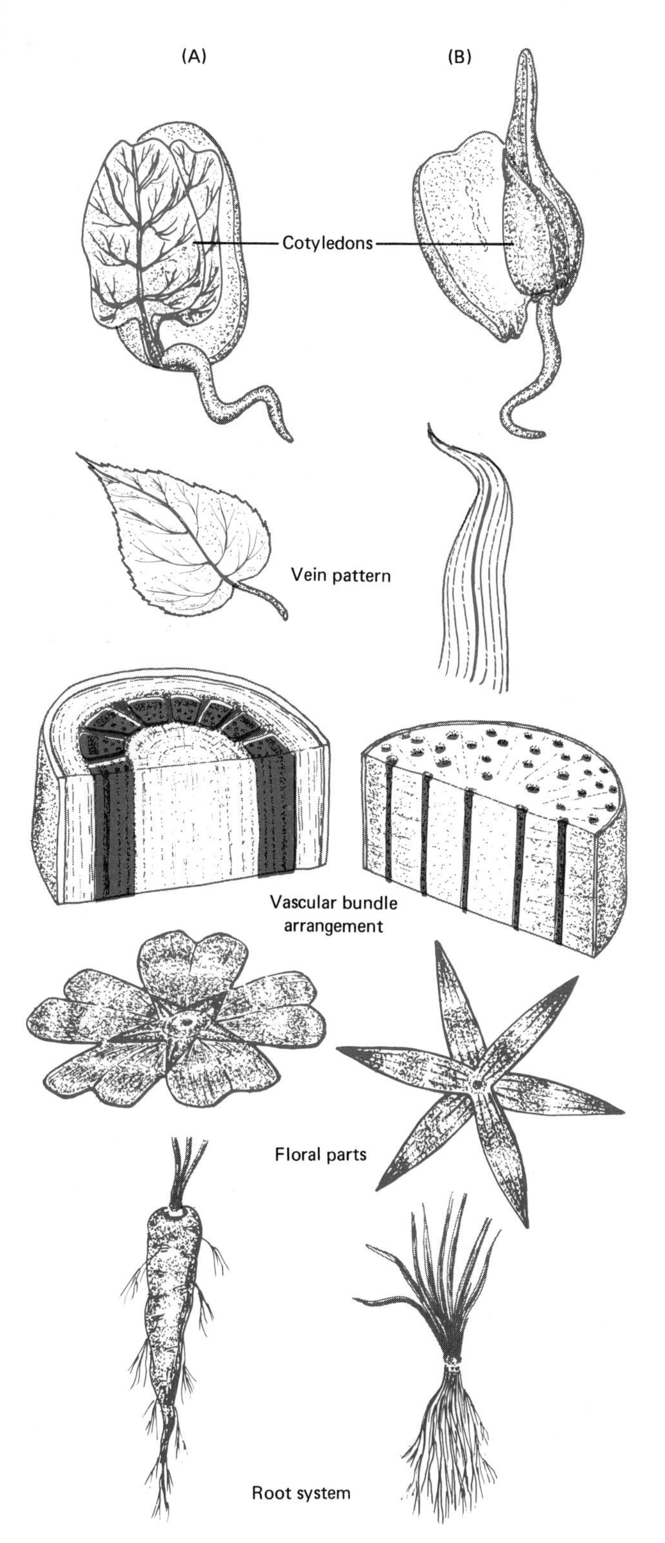

growth, or an increase in diameter, of a plant part. Another type of lateral meristem, called the cork cambium, may develop if a plant is damaged in any way, but it is also quite common in uninjured woody angiosperms as well. It is located in the outer bark of trees and shrubs and produces cork.

Permanent Tissues: Specialized Tissues

Permanent tissues, as contrasted with meristems, are stable in that they no longer actively divide. Although derived from meristematic regions, these tissues develop considerable structural and functional modifications. Each type of permanent tissue is composed of differentiated cells that contribute to an efficient division of labor in the total functioning plant. For example, cells of conducting tissue are usually quite elongated and thick-walled, whereas storage cells are generally isodiametric (length and width are essentially equal) and thin-walled.

Permanent tissues are divided into simple and complex types. Simple permanent tissues are composed of cells that are structurally and functionally similar, while complex permanent tissues are composed of several types of tissues that differ in structure and function. In this latter type of tissue the cells are involved in a group of interrelated activities in which one function usually dominates.

SIMPLE PERMANENT TISSUE. The *simple permanent tissues* of the kingdom Plantae are epidermis, parenchyma, collenchyma, sclerenchyma, and cork (Figure 9-2).

Epidermis: Protective Tissue. The **epidermis** is usually one cell layer thick and forms the surface layers of leaves, flowers, and young stems and roots. In older roots and stems of certain plants, the epidermis is often

FIGURE 9-1. Some structural features compared between dicots and monocots. (**A**) In dicots there are two cotyledons (seed leaves) in the embryo; the veins of the leaves are pinnately or palmately arranged; the vascular bundles in the stem are cylindrically arranged; floral parts are in fours or fives or multiples of these; the root system is typically a taproot; and there is an active cambium. (**B**) In monocots there is only one cotyledon in the embryo; the veins of the leaves are parallel; the vascular bundles are scattered; floral parts are in threes or multiples of three; the root system is typically fibrous; and there is no vascular cambium.

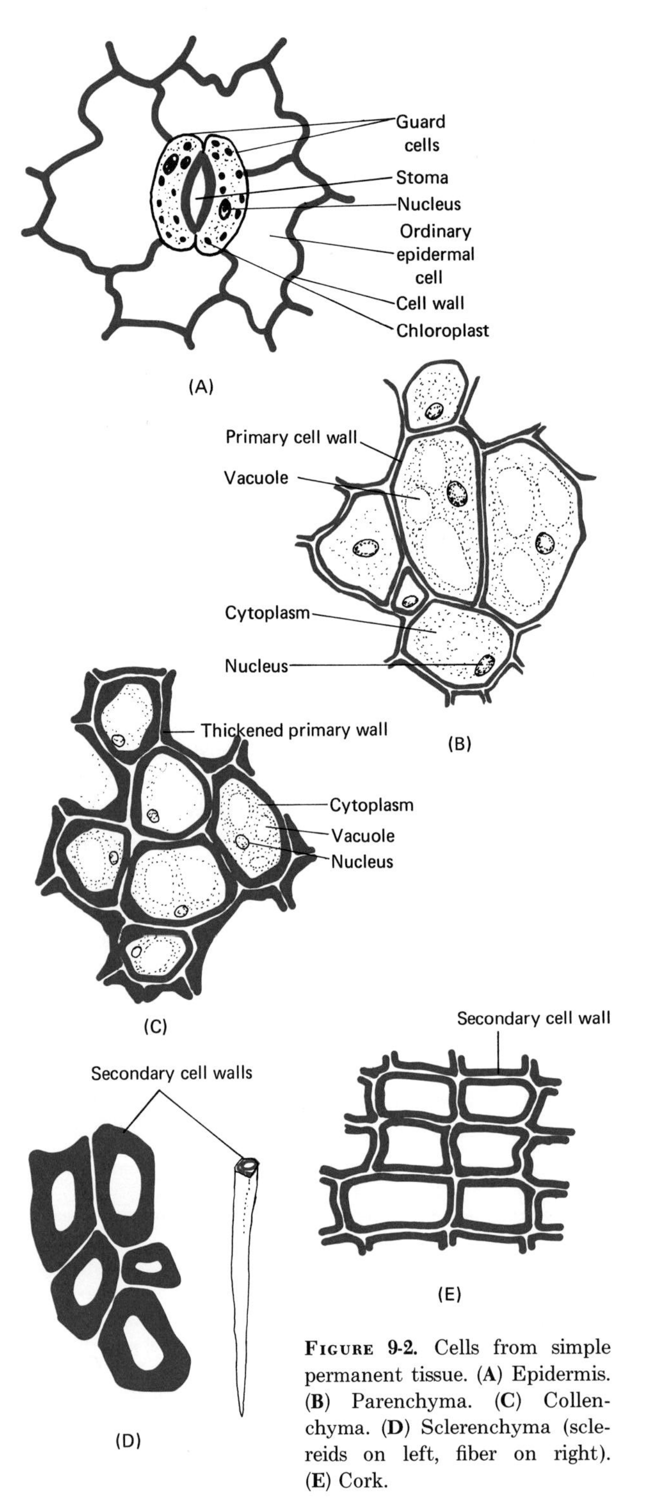

FIGURE 9-2. Cells from simple permanent tissue. (**A**) Epidermis. (**B**) Parenchyma. (**C**) Collenchyma. (**D**) Sclerenchyma (sclereids on left, fiber on right). (**E**) Cork.

replaced by other tissues, such as cork. Epidermal cells are generally relatively flat, with large vacuoles and only a small amount of cytoplasm. Their outer walls, at least on the aerial portions of the plant, are covered by **cutin,** a waxy, waterproofing material. The continuous layer of cutin, termed the *cuticle,* aids in protection against water loss, mechanical injury, and invasion by disease organisms.

Some epidermal cells, especially those covering leaves and green stems, are specialized as guard cells. **Guard cells** are modified epidermal cells, often kidney-shaped, that regulate the size of tiny openings called **stomata.** Each stoma is actually a pore between two adjacent guard cells (Figure 9-2A). Stomata permit the exchange of gases between the atmosphere and internal tissues and also facilitate the evaporation of water.

Parenchyma: Photosynthetic and Storage Tissue. **Parenchyma** tissue occurs in roots, stems, and leaves. The cells of this tissue are characterized by thin primary cell walls, lack of a secondary wall, and well-developed vacuoles surrounded by a peripheral distribution of cytoplasm. In addition, the cells are more or less spherical, although the shape is distorted by the pressure of surrounding cells so that they appear angular or box-shaped (Figure 9-2B). Parenchyma tissue carries on a variety of functions. Most of the chloroplasts of leaves are in the cells of parenchyma tissue (e.g., palisade and spongy mesophyll), and it is in these cells that photosynthesis occurs. Other parenchyma cells, found in roots and to a lesser extent in stems, are colorless and function chiefly in food and water storage.

Collenchyma: Supporting Tissue. **Collenchyma** is composed of cells with unevenly thickened primary walls. The thickened areas are usually most prominent at the angles (corners) of the cells when viewed in cross section (Figure 9-2C). The principal function of collenchyma is mechanical support and strengthening. It is usually the first supportive tissue produced in young stems and leaves and is a permanent tissue in mature plant organs. The combination of tensile strength and flexibility of these cells is ideally adapted to the support of growing organs.

Sclerenchyma: Strengthening Tissue. **Sclerenchyma** is a tissue characterized by thick cell walls. The cells have a definite secondary wall that is often impregnated with lignin. With the appearance of lignin, the protoplast dies; at maturity, therefore, there are no living contents. Sclerenchyma tissue reveals two types of cells that vary greatly in size and shape. These are fibers and sclereids (Figure 9-2D). *Fibers* are elongated sclerenchyma cells with tapered ends. They are tough and strong, but flexible. Because of these properties and because they are aggregated into strands, fibers are important in the manufacture of rope, twine, and textiles from such plants as flax, jute, and hemp. *Sclereids,* or *stone cells,* are similar in strength to fibers but are characteristically different in shape. Typically, they are not elongated, but rather are irregular in form. Sclereids are common in the shells of nuts, the hard parts of seeds, and are scattered in the fleshy portions of certain fruits. The gritty texture of pears, for example, is due to small clusters of stone cells.

Cork: Waterproofing Tissue. **Cork** is composed of cells that have thick walls impregnated with **suberin,** a waxy material (Figure 9-2E). The cells are dead at maturity and form a waterproofing layer of tissue. Cork is found as the outer layer of stems and roots of older woody plants as an epidermal replacement. The essential functions of cork are protection and water conservation. In the case of the cork tree, *Quercus suber,* a profuse amount of cork is produced; this is removed and used commercially in the preparation of cork stoppers, life preservers, and insulating materials. Since the cork cambium produces new cork, it may be again removed from the tree in three to four years.

Complex Permanent Tissue. *Complex permanent tissue* (*vascular tissue*) consists of more than one type of tissue and is a distinctive feature of higher plants. These tissues function as tubes or ducts within which water and dissolved substances (minerals and foods) are conducted from one part of the plant to another. The two principal types of vascular tissue are xylem and phloem.

Xylem: Conducting Tissue. **Xylem,** or wood, is a vascular tissue that functions in the transport of water and dissolved substances, usually minerals, upward in the plant body. In addition, it serves as a supporting tissue. As a conducting tissue, it forms a continuous pathway through the roots, stems, leaves, flowers, and fruits. As in the case of sclerenchyma, a number of types of cells are unique to xylem. These are the tracheids, xylem fibers, and vessel elements.

Tracheids are long, slender cells, tapered at the ends, with well-developed, lignified secondary walls (Figure 9-3A). At maturity, the protoplasm within tracheids dies, leaving a space or cavity called a *lumen.* Xylem fibers and vessel elements are cell types that are dead at maturity and represent two evolutionary modifications of tracheid cells.

In the course of evolution, *xylem fibers* have assumed the function of support at the expense of conduction. Xylem fibers are longer than tracheids, having more extensive and heavier wall thickenings and lignification. In addition, they are more slender and tapering with greatly reduced pits (Figure 9-3B). A wood fiber is a

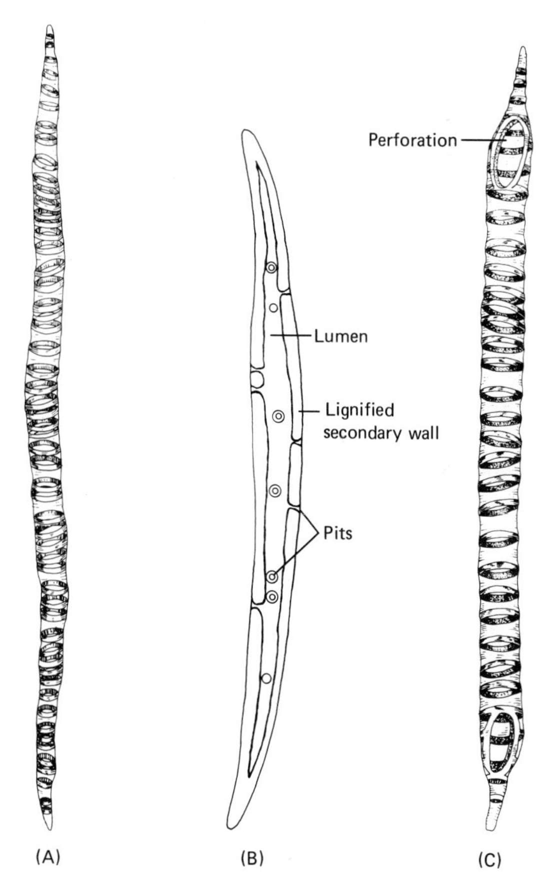

Figure 9-3. Kinds of cells in xylem tissue. (**A**) Annular tracheid. (**B**) Xylem fiber. (**C**) Annular vessel segment. [*A and C from V. A. Greulach and J. E. Adams,* Plants: An Introduction to Modern Botany. *New York: John Wiley & Sons, Inc. Copyright © 1967.*]

good example of an evolutionary derivative of a tracheid in which the supportive function has become predominant.

Vessel elements, by comparison, are derived through evolutionary modifications involving slight shortening and widening of a tracheid with the loss or perforation of the end walls in which the lumina of individual cells become continuous. This latter modification produces vessels that are open and tubular in form (Figure 9-3C). Vessel elements have wall patterns similar to those of tracheids. Vessel elements, in terms of their modifications, have assumed the function of conduction at the expense of support. In the plant, vessel elements are arranged end to end with little or no overlapping. This structural orientation produces a continuous, open, pipe-like structure called a **vessel.**

Phloem: Conducting Tissue. Like xylem, **phloem** is a vascular tissue and is usually associated with xylem in various plant organs. Together they constitute the vascular system. Phloem functions primarily in the transport of dissolved organic substances in a process called **translocation.** Being a complex permanent tissue, phloem contains a variety of cell types. Always present in phloem are sieve cells. In addition, the phloem of higher plants also contains parenchyma cells and fibers.

The fundamental cells in phloem are **sieve cells.** These are long, slender, thin-walled cells, which, at maturity, retain their cytoplasm but lose their nucleus. Just as vessel segments in xylem form a continuous tubular network, the sieve cells in phloem produce a similar structure called a **sieve tube.** The cytoplasm is continuous from one sieve cell to another within the sieve tube through the *sieve plates* on the end walls (Figure 9-4A). These sieve plates are perforated end walls through which cytoplasm passes from one cell to another.

Ordinarily associated with sieve cells are modified parenchyma cells that retain cytoplasm and a functional nucleus. These cells are called *companion cells* (Figure 9-4A). Since there are cytoplasmic connections between sieve cells and adjacent companion cells, it is thought by some that the latter may aid the sieve cells in regulating conduction.

In addition to sieve cells and companion cells, the phloem of higher plants also contains parenchyma cells and fibers. The parenchyma cells function chiefly in

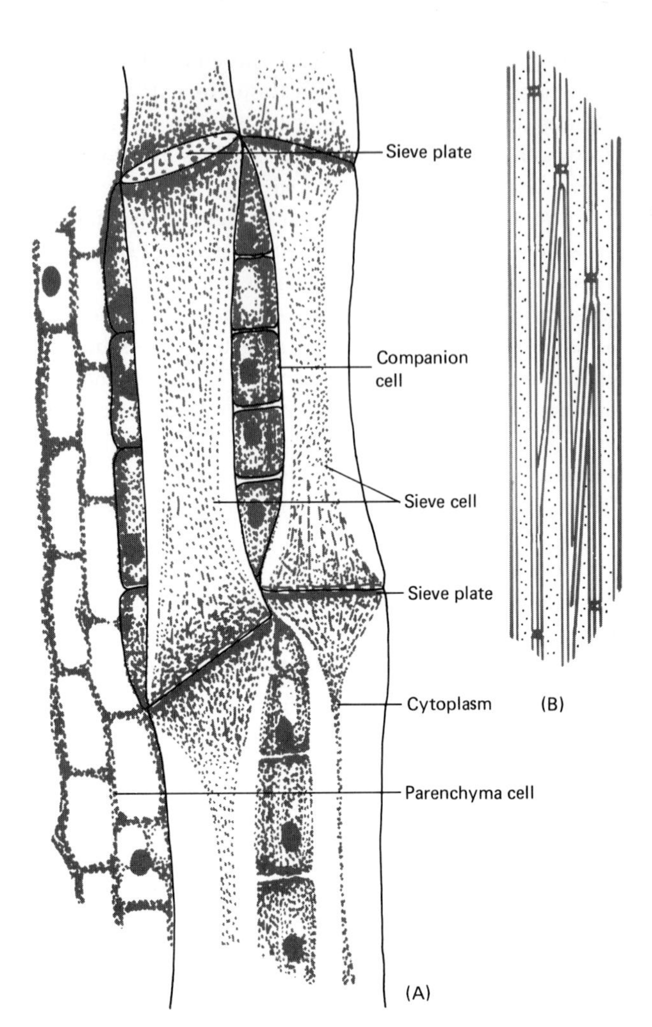

FIGURE 9-4. Kinds of cells in phloem tissue. (**A**) Sieve cells containing cytoplasm. Note the adjacent companion cells with nuclei and the accompanying parenchyma cells. (**B**) Phloem fibers.

food storage while the *phloem fibers* (Figure 9-4B), which are modified sclerenchyma cells, afford strength.

The Organization of Plant Tissues into Organs

Having considered the principal tissues of an angiosperm, we will now discuss their organization into vegetative organs. These organs, which contribute to the maintenance of plant activities, are roots, stems, and leaves.

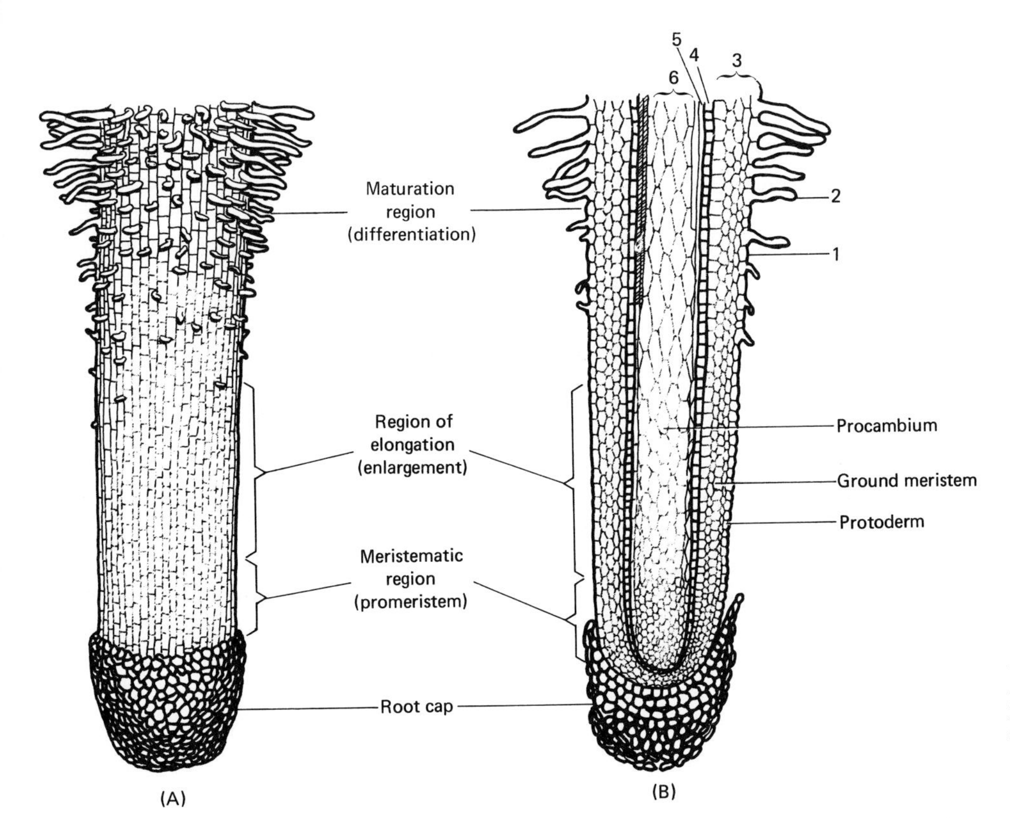

FIGURE 9-5. Various aspects of a young developing root tip. (**A**) External aspect showing root cap and growth regions. (**B**) Internal aspect showing root cap, meristem, region of enlargement with embryonic tissues, and maturation region with primary tissues: (1) epidermis, (2) root hairs, (3) cortex, (4) endodermis, (5) pericycle, and (6) vascular tissue (primary phloem, vascular cambium, and primary xylem). (**C**) Cross-sectional aspect through the region of maturation showing (1) epidermis, (2) cortex, (3) endodermis, (4) pericycle, (5) secondary root formation from pericycle, (6) primary xylem, (7) primary phloem, (8) vascular cambium, and (9) root hair.

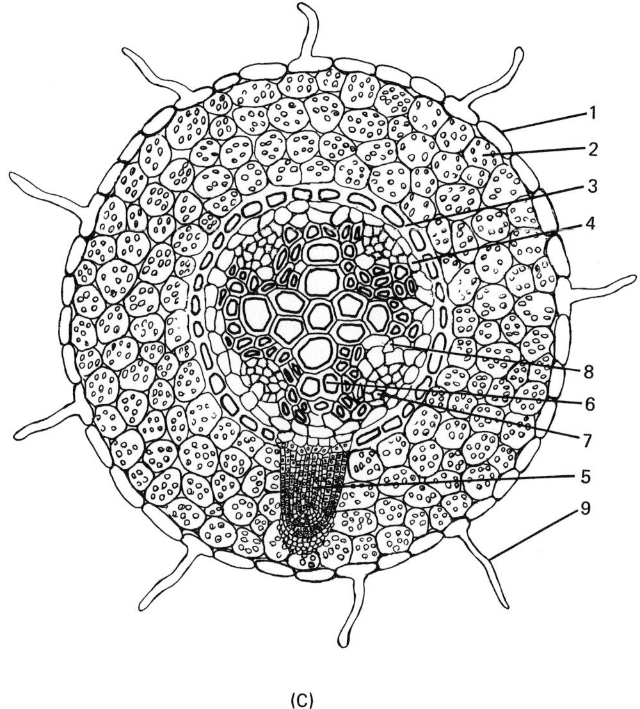

The Root

The **root** of an angiosperm is typically a subterranean organ that functions chiefly to anchor the plant to the substrate and to absorb and conduct water and nutrients from the soil. These absorbed materials are either utilized in the roots themselves or transported to other plant organs where they are used. In addition, some roots such as carrots also serve as food storage organs.

ROOT STRUCTURE. In our discussion of the structure of a root, only the dicot root will be considered. The root system of such plants terminates in delicate white tips, which are usually less than one-sixteenth of an inch in diameter. Close examination of one of these tips reveals four more or less distinct regions: the root cap, meristematic region, enlargement region, and differentiation region.

Root Cap. The *root cap* is found on many roots. Microscopically, it appears as a thimble-shaped mass of parenchyma cells that protects the growing portion of the root tip (Figure 9-5A, B). As the root grows and the tip advances through the soil, the outer cells of the root cap are sloughed or rubbed off and are continually being replaced by divisions of the cells in the region directly behind it. Cells sloughed off the root cap lubricate the passage of the young root through the soil.

Meristematic Region. This region, also called the *promeristem* (Figure 9-5A, B), is an apical meristem and is characterized by active cell division. The cells are small, cube-shaped, parenchyma-type with thin walls, dense cytoplasm, and large nuclei. The division of these cells, and consequently the addition of new ones, brings about lengthening of the root tip. Cells produced by the meristem are destined for one of three possible fates. Those toward the tip of the root become part of the root cap. The cells in the middle remain as cells of the meristem. Those cells at the upper portion enlarge and differentiate into permanent tissues. Such a distribution of cells allows the meristematic zone to remain fairly constant in size even though there is a continual production of new cells.

Enlargement Region. Those cells produced by and located directly above the meristem constitute the *enlargement region,* or region of cell elongation (Figure 9-5A, B). Although the regions are not sharply distinguishable from each other, close microscopic examination reveals two bases for delineation.

1. The cells of the enlargement region tend to be elongated longitudinally as opposed to the cubical form of the meristematic cells.
2. Three distinct immature permanent tissues can be differentiated. These tissues represent the first definitive changes in form from the cells of the meristem and are designated as the *protoderm, ground meristem,* and *procambium* (Figure 9-5B). All these immature meristematic tissues will give rise to *primary plant tissues* (tissues that develop from the apical meristem and bring about an increase in length of a plant part).

Differentiation Region. In the *region of differentiation* (Figure 9-5A, B), cells of the protoderm, ground meristem, and procambium divide, enlarge, and differentiate sufficiently to become recognizable as primary tissues. These primary tissues include the epidermis, which is produced by the protoderm; the cortex and endodermis, which are derived from the ground meristem; and the vascular cylinder, consisting of pericycle, primary phloem, vascular cambium, and primary xylem, all derivatives of the procambium. A cross section of a root tip through the region of differentiation reveals these differentiated primary tissues (Figure 9-5C).

Primary Tissues. The epidermis consists of a single layer of parenchyma cells and forms the outermost tissue of the root. In addition to producing root hairs, it affords protection for the underlying cells. *Root hairs* are the principal structures performing absorption of H_2O and dissolved minerals. The **cortex,** inside the epidermis, is composed of large, thin-walled, angular parenchyma cells. This tissue helps to transfer water and minerals from the root hairs to the xylem and also serves as a food storage tissue. The *endodermis* is a rather conspicuous tissue in roots, consisting of a single layer of thick-walled cells, and forms the inner boundary of the cortex.

Within the endodermis is the **vascular cylinder (stele),** the principal strengthening and conducting portion of the root. The outermost tissue of the vascular cylinder is the *pericycle,* which consists of one or more layers of parenchyma cells. This tissue is extremely important because it produces the secondary roots and the cork cambium. The xylem occupies the center of the stele and is referred to as the *primary xylem.* As viewed in cross section, it is usually arranged in the form of a star, although many alternative arrangements also exist. In each of the angles formed by the star-shaped xylem, there are groups of smaller, thinner-walled cells comprising the *primary phloem.* In dicots, the phloem is separated from the xylem by layers of thin-walled undifferentiated parenchyma cells. One layer of these cells remains meristematic and is known as the vascular cambium. The cambium cells continue to divide and produce the secondary tissues. Most monocots lack a cambium.

Secondary Tissues. The roots of most trees, shrubs, perennial plants, and other dicots have secondary tissues as well as primary tissues. *Secondary tissues* are those that develop from a vascular cambium and cause an increase in diameter, or lateral growth, of a plant part. The two principal types of secondary tissues in roots are those that arise from the vascular cambium and those originating from the pericycle.

Of all the primary tissues, the vascular cambium and pericycle alone remain meristematic and continue to divide. The others are more or less specialized and no longer divide to produce new cells. When the cambium divides, new cells are added to both the primary xylem and the primary phloem. These new secondary tissues, *secondary xylem* and *secondary phloem,* grow at uneven rates, the former growing much more rapidly. As a result, the secondary xylem occupies the space between the radiating arms of the primary xylem, originally occupied by the primary phloem (Figure 9-6). This pushes the phloem cells outward and the xylem region becomes circular. In a similar manner, the phloem also becomes a continuous cylinder of tissue external to the xylem and vascular cambium.

When the cells of the pericycle begin to divide, they produce a second lateral meristem referred to as the *cork cambium,* or *phellogen.* The cork cambium usually remains a single layer of cells, and new layers of cells are added both outwardly toward the epidermis and inwardly toward the phloem as it divides. Those produced outwardly are called cork, or *phellem.* Those developed inwardly are referred to as the *phelloderm.* Collectively considered, this entire new secondary tissue consisting of cork, cork cambium, and phelloderm is designated as the *periderm.*

The cells of the phelloderm are parenchyma-like and remain alive and active. Those of the cork layer become suberized, thus waterproofed, and die at maturity. When the periderm is fully developed, the epidermis (with its root hairs) and the cortex die, so that in older roots the periderm eventually becomes the outermost layer. At maturity, the periderm breaks, cracks, and furrows. Further aspects of growth and development of roots are discussed in Chapter 18.

Stems

The **stem** is that part of the plant that typically rises above the ground and together with the leaves comprises the *shoot.* While most stems are erect, aerial organs, some remain underground (rhizomes) and others creep along the surface of the ground (runners). The stem serves as a mechanical support for leaves, flowers,

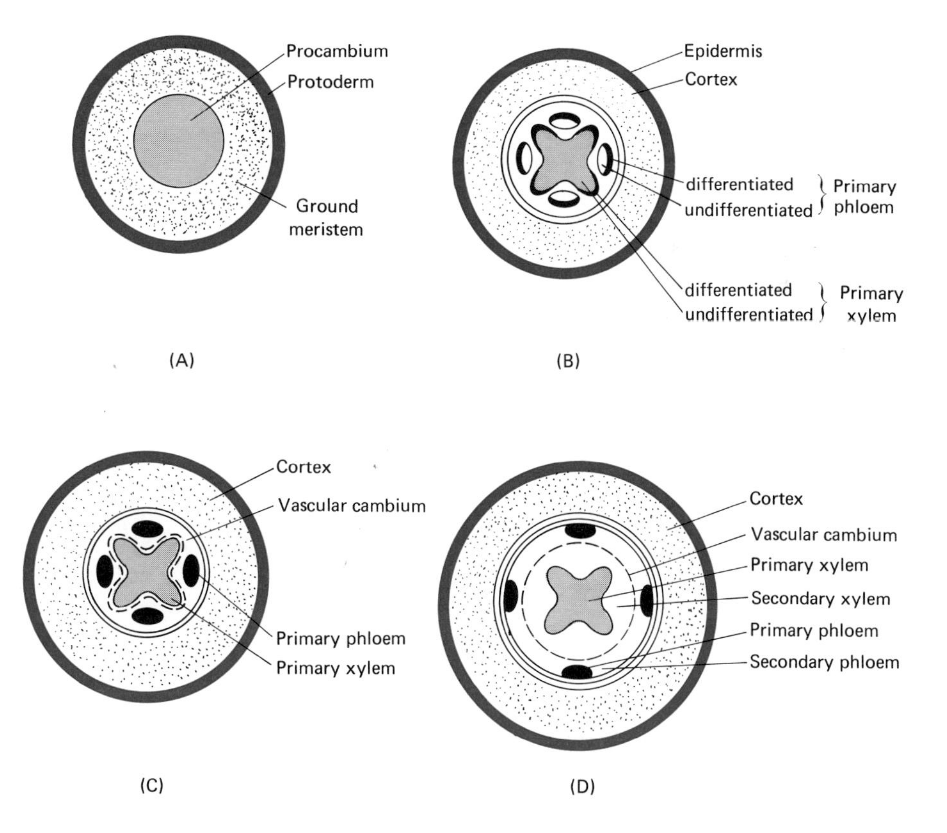

FIGURE 9-6. Secondary growth in a young root as seen in various cross-sectional aspects. (**A**) In a very early stage of development the three embryonic tissues are visible. (**B**) At a later stage of development the embryonic tissues differentiate into primary tissues. (**C**) With the formation of primary tissues the vascular cambium becomes mitotic. (**D**) At an even later stage of growth, the cambium produces secondary xylem and secondary phloem between the primary xylem and primary phloem. Note the relative distribution and spatial arrangement of the primary and secondary tissues in early and later stages of growth.

and fruits; a pathway for the conduction of materials; a site for food manufacture (in green stems); and as a potential reproductive structure.

Stem Structure. As with roots, only the structure of a woody dicot stem will be discussed as a representative type. The twigs of woody plants in the winter condition afford excellent subjects for a study of the external characteristics of stems (Figure 9-7A). New stems and their leaves develop from **buds.** The top of a twig generally bears a large *terminal bud* and occurring at regular intervals along the sides of the stem are *axillary,* or *lateral, buds.* At the base of each axillary bud is a scar that was made when a leaf fell from the twig. This is called the *leaf scar* and is usually circular, oval, or shield-shaped. Since the conducting tissue of the stem is continuous with that of the leaf, this tissue also ruptures when the leaf falls, producing *vascular bundle scars* inside each leaf scar.

Leaves and buds are present at specific locations along the stem, each point of attachment being called a *node;* the distances between nodes are called *internodes.* The immature cells within buds are protected by a series of overlapping structures, the *bud scales,* which drop off as the new stem develops; in doing so they leave scars called *bud-scale scars.* The portion of the stem between sets of terminal bud-scale scars is formed during one growing season. After a portion of a stem is one year old, there is no increase or decrease in the length of that portion. Tiny openings or pores called *lenticels* also appear on woody stems. They facilitate gaseous exchange between the interior of the stem and the atmosphere.

Primary Tissues. The young woody stem consists of primary tissues alone, and secondary tissues are not produced until after the primary tissues are established. All the primary tissues are produced by the apical meristem, after which cell enlargement and differentiation result in these tissues becoming part of the permanent

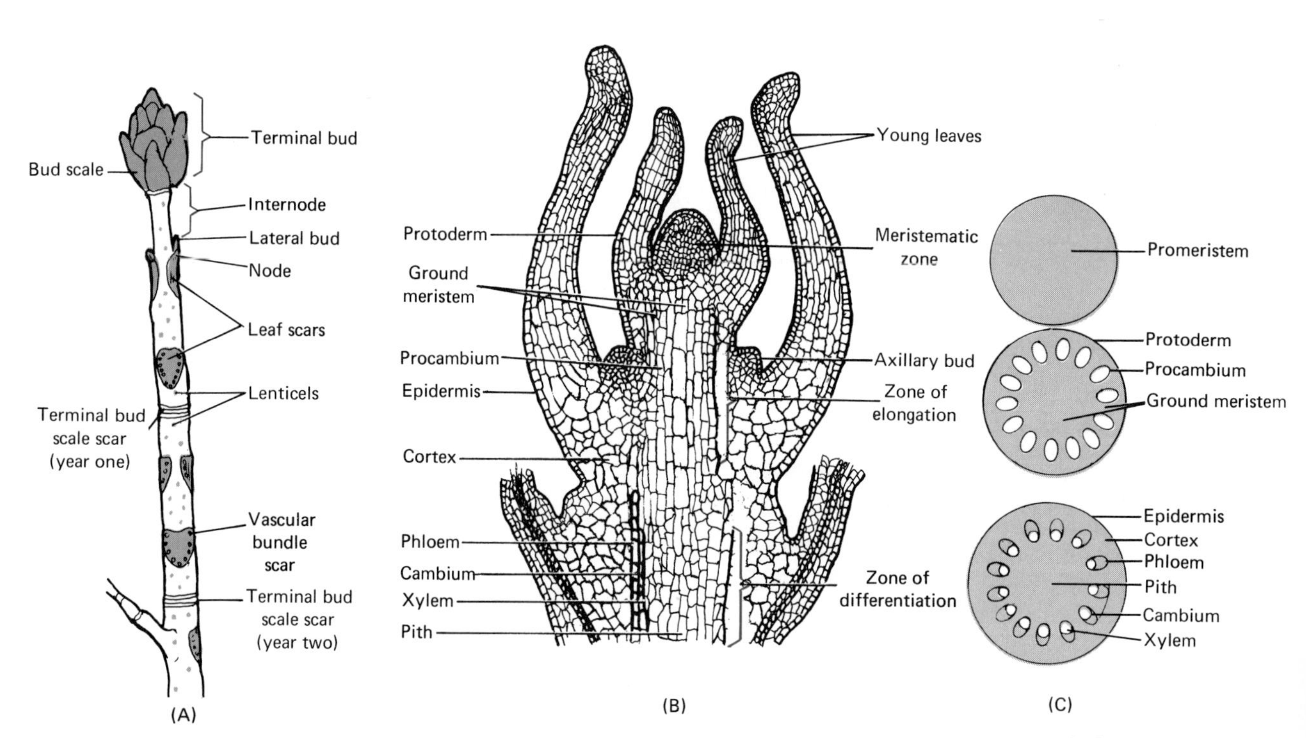

Figure 9-7. External and internal aspects of a stem. (**A**) External characteristics of the stem of a horse chesnut tree in the winter condition. (**B**) Internal aspect of the growth regions and tissues of a stem tip as seen in longitudinal section. (**C**) Transverse view of the same regions and tissues shown in **B.**

tissues. The primary tissues are shown in Figure 9-7B and C in longitudinal and cross-sectional views through various growth regions.

1. Epidermis. This tissue forms the outer surface protective layer and is one cell layer thick. The outer cell walls are frequently thick and covered with cutin. In older stems, the epidermis is replaced by cork.

2. Cortex. This region, just under the epidermis, is composed chiefly of parenchyma cells with some collenchyma and sclerenchyma. The cortex, usually many cells deep, is a food storage tissue which degenerates in older stems.

3. Primary phloem. This complex permanent tissue affords efficient conduction of materials and also assumes the functions of storage and support. The principal conducting cells are sieve cells; the storage cells are companion cells and phloem parenchyma cells; and the strengthening and supporting cells are phloem fibers. These fibers are long, narrow, vertically elongated, thick and dead at maturity. Phloem fibers of flax and hemp plants are made into linen and rope, respectively.

4. Vascular cambium. This tissue is immediately internal to the phloem and consists of a region of meristematic cells.

5. Primary xylem. Like the primary phloem, the primary xylem is a complex permanent tissue and assumes a number of functions. Conduction is facilitated by tracheids and vessel elements. Xylem parenchyma cells and xylem fibers, similar to those found in primary phloem, assume the functions of storage and support.

6. Pith. This region is the central portion, or core, of the stem and contains parenchyma cells that function in storage.

Secondary Tissues. After the primary tissues are formed in a young dicot stem, little or no further growth of the differentiated portion takes place. The terminal bud continues its growth through the growing season, each year forming a new stem segment with its primary tissues. In general, the primary tissues are established in the first year during the first few weeks, and bring about a lengthening of the stem. Also, in the first year, the stem begins to grow in diameter as a result of the continued division of the vascular cambium cells, which produce the secondary tissues.

As the cambium divides, cells produced toward the outer part of the stem develop into the secondary phloem, while those that are produced toward the inner portion of the stem form the secondary xylem. Usually more xylem is produced, and as the stem grows in diameter, more and more xylem is formed. The secondary tissues that reinforce or replace the primary tissues are the secondary phloem, secondary xylem, cork cambium, and cork.

1. Secondary phloem. The cells of this tissue are similar to those of the primary phloem except for two differences: the first is one of origin; the second is the presence of added cells that serve a specific function. As the stem increases in diameter, horizontal as well as vertical conduction is necessary. This specific function is accomplished by *phloem ray cells,* which are not found in primary phloem. These parenchyma-type cells are elongated horizontally (radially in the stem) and function in transporting materials across the stem (Figure 9-8A). The large, simple pits in the walls of phloem ray cells greatly facilitate the horizontal conduction of materials. In the stem, these cells become oriented end

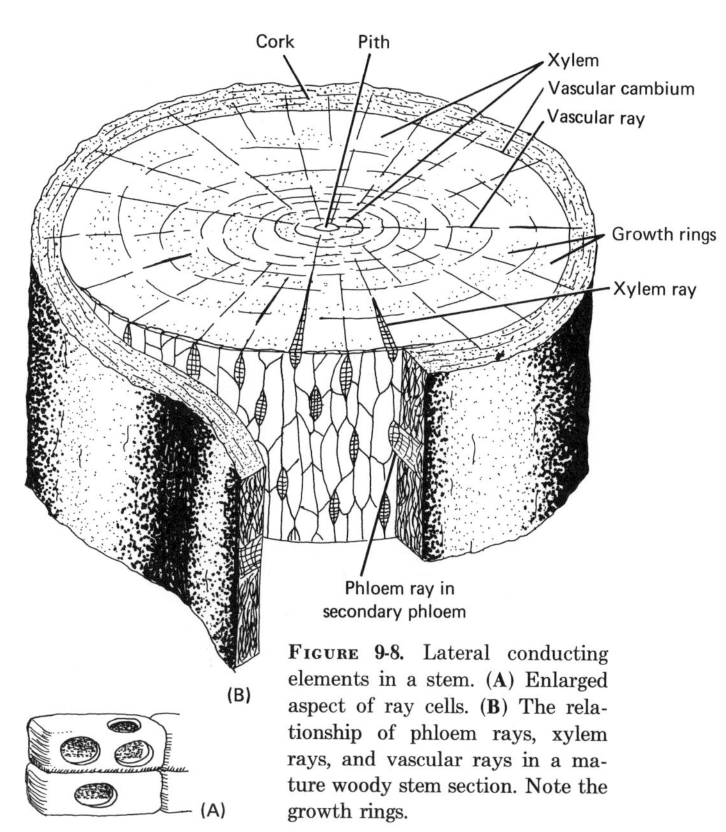

Figure 9-8. Lateral conducting elements in a stem. (**A**) Enlarged aspect of ray cells. (**B**) The relationship of phloem rays, xylem rays, and vascular rays in a mature woody stem section. Note the growth rings.

to end, forming a continuous horizontal conducting structure called a *phloem ray* (Figure 9-8B).

2. Secondary xylem. Here again, the secondary xylem differs from the primary xylem with respect to origin and the presence of new cells. *Xylem ray cells,* similar to those found in the phloem, are attached end to end to the phloem rays and form *vascular rays,* which provide an excellent horizontal conducting system (Figure 9-8B).

The cambium continues to produce secondary xylem and secondary phloem year after year so that the stem progressively increases in diameter. Xylem cells produced in the spring are relatively large but as the season progresses, the cells are considerably smaller. As a result, there is usually an abrupt contrast between the xylem formed during different growth seasons. These annular growth layers that appear as a series of concentric rings when the stem is viewed in cross section are called **annual (growth) rings** (Figure 9-8B).

3. Cork cambium and cork. In addition to secondary tissues produced by the vascular cambium, another group of secondary tissues is produced by the cork cambium (Figure 9-9). The cork cambium is a group of meristematic cells that originates from parenchyma cells of the outer cortex. As the cells of the cork cambium divide, the outer ones develop into cork cells and the inner ones give rise to phelloderm. The cork serves as a protective, waterproofing tissue.

Although cork forms in response to tissue damage, this is not an unusual occurrence. In fact, all woody dicots have external layers of cork by the second year of growth. The reason for cork production is that secondary tissue development brings about an increase in diameter while the epidermis and cortex no longer increase in size. This continued stress causes the epidermis and cortex to tear and rip, and as a result of this damage, cork forms before the damage is apparent. Since cork is impervious to most materials, all cells external to it die because of a lack of nutrition. The **bark** of a tree is a collective designation for the secondary phloem, cork cambium, and cork. Additional aspects of stem development are considered in Chapter 18.

Leaves

Leaves are the characteristic photosynthetic organs of most higher plants. Another important leaf activity is **transpiration,** the loss of water vapor. Leaves, like roots and stems, vary in many respects. For example, leaves differ according to arrangement on the stem, vein patterns, form, structure, and size. Despite these variations, certain basic features are distinguishable and they will be described using a dicot leaf as an example.

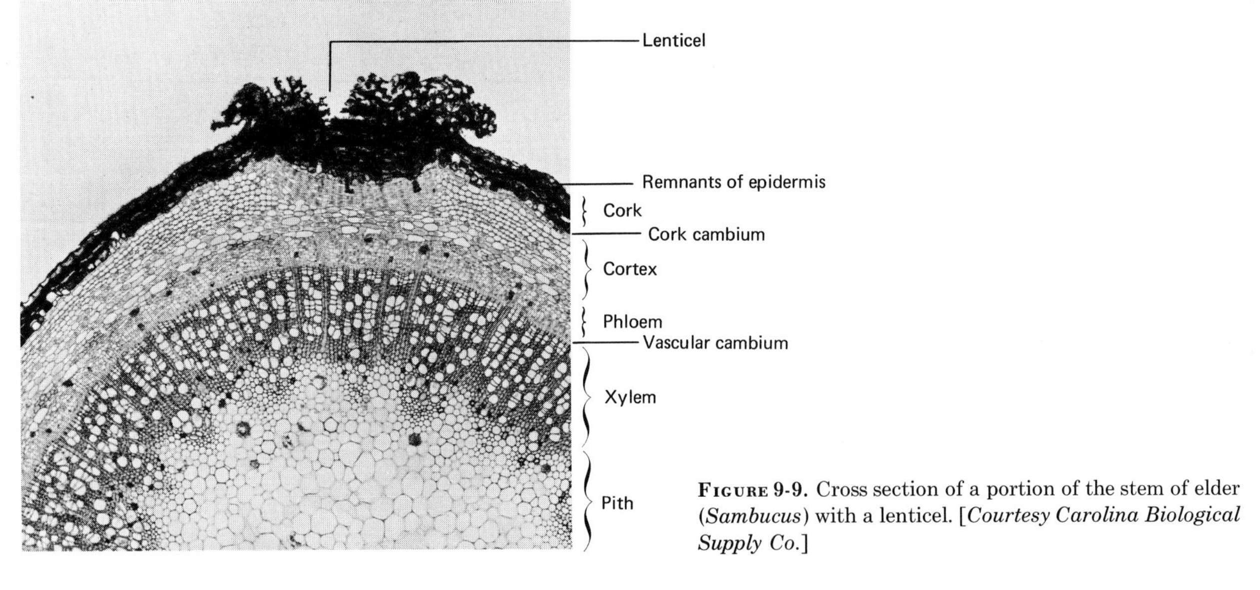

FIGURE 9-9. Cross section of a portion of the stem of elder (*Sambucus*) with a lenticel. [*Courtesy Carolina Biological Supply Co.*]

Leaf Structure. The leaf stalk, or **petiole,** is a continuation of the stem and distributes vascular tissue from the stem to the leaf (Figure 9-10A). The flattened, expanded portion of the leaf is the *lamina* or *blade,* which usually is green because of the presence of chlorophyll-containing cells. At the base of the petiole, small leaf-like structures called *stipules* may be found, along with axillary buds. These axillary buds may produce secondary branches, flowers, or both. The shape and form of the petiole may vary in different species of plants. In some, leaf petioles are absent, and the leaves are said to be *sessile.*

The internal arrangement of the tissues of a leaf is commonly studied by microscopic examination of a cross section cut at right angles to the broad surface of the blade. Such a section reveals three distinguishable areas: (1) epidermis, (2) mesophyll, and (3) vascular tissue (Figure 9-10B).

This epidermis tissue consists of a single layer of more or less rectangular-shaped cells lacking chloroplasts. In surface view, the epidermal cells are fairly irregular in form. The epidermis covers both surfaces of the leaf and is therefore subdivided into the upper and lower epidermis. The *upper epidermis* is covered by a cuticle which prevents drying out of the underlying tissues and protects them against injuries. The *lower epidermis,* and sometimes the upper, is perforated by stomata bordered by guard cells. In contrast to the cells of the epidermis proper, the guard cells contain chloroplasts.

Between the upper and lower epidermal layers, the **mesophyll** is found. On the basis of location and structure, the mesophyll is divided into two portions. The upper, rectangular-shaped, compact rows of cells are called the *palisade mesophyll.* These cells are oriented at right angles to the surface of the leaf. Below these, and extending to the lower epidermis, are irregularly shaped cells with abundant intercellular spaces. These layers of cells are designated the *spongy mesophyll.* Both layers of mesophyll cells are composed of parenchyma with numerous chloroplasts. The term *chlorenchyma* is applied to chlorophyll-bearing cells, whether they are found in the leaf or in other plant organs.

Standing out rather conspicuously in the spongy mesophyll are the veins of the leaf. Veins, or vascular bundles, are specialized structures that function in conduction and support, being similar to the vascular tissue

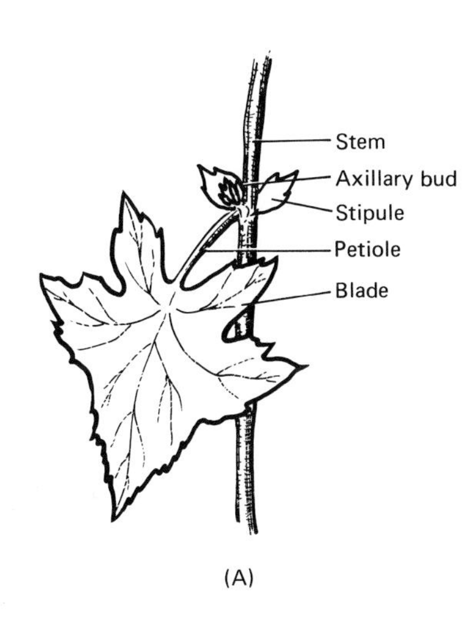

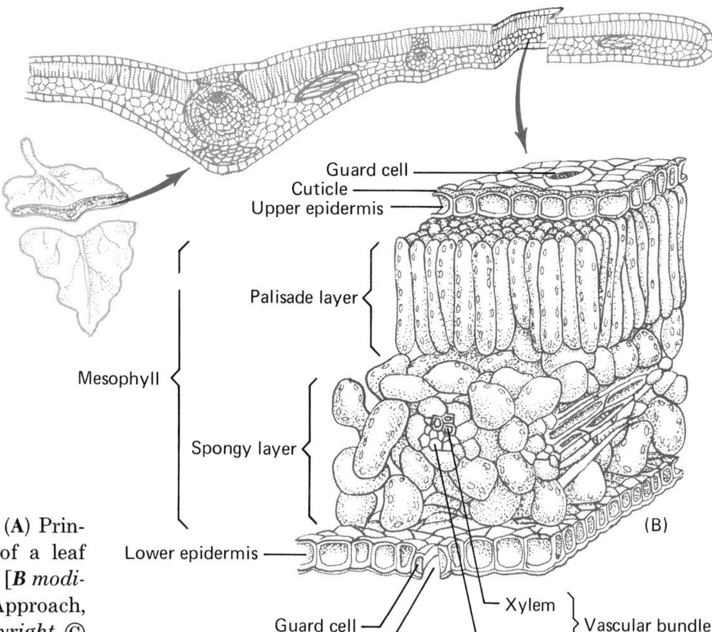

Figure 9-10. External and internal aspects of a leaf. (**A**) Principal external portions of a leaf. (**B**) Section view of a leaf showing internal composition. Consult text for details. [*B modified from Walter H. Muller,* Botany: A Functional Approach, *2nd ed. New York: The Macmillan Company. Copyright © 1969.*]

found in roots and stems. Each vein consists of an outer bundle sheath (parenchyma cells), xylem, and phloem.

The Organ-System Concept

The various organs of an angiosperm are structurally and functionally integrated into the total plant body (Figure 9-11). Vegetative organs are those structures of the plant which are concerned with growth, maintenance, and development, such as roots, stems, and leaves. Those parts of the plant which are concerned with reproduction and the production of seeds are classified as reproductive organs. These would include flowers, which give rise to fruits, which in turn contain seeds.

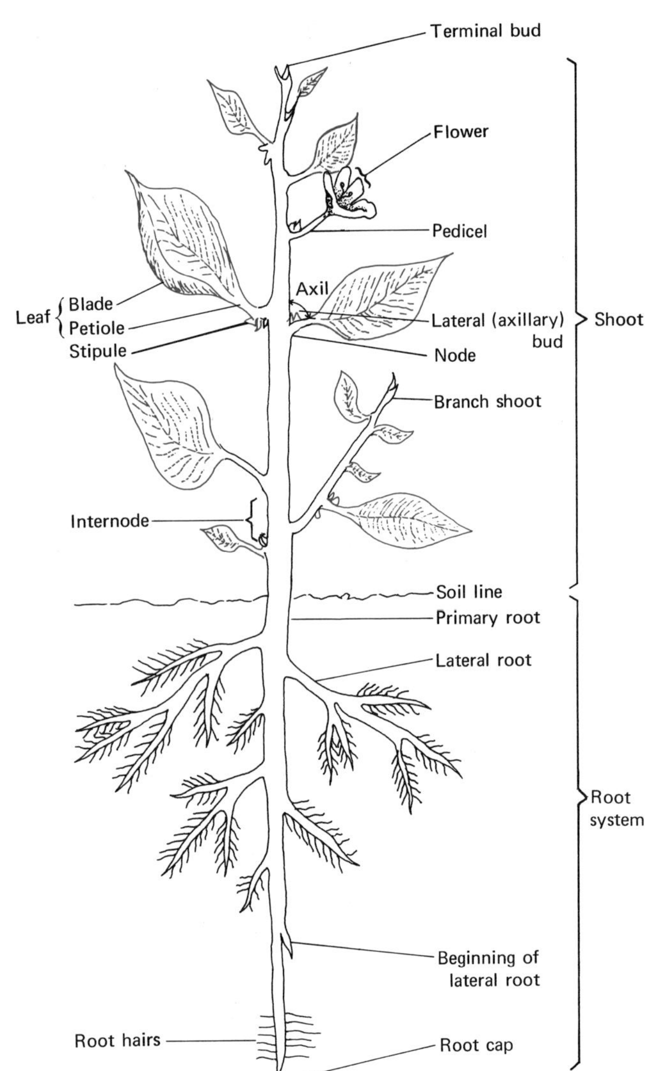

FIGURE 9-11. Diagrammatic representation of the principal external features of a flowering plant.

A Survey of Animal Tissues

As with our discussion of plant tissues, the following sections dealing with animal tissues will be restricted to a particular group of organisms. In this case, we will survey only vertebrate tissues, especially those of the human.

Vertebrate tissues, too, may be classified in a number of ways. For purposes of discussion, they will be subdivided into five principal types: (1) epithelial, (2) connective, (3) vascular (blood), (4) muscular, and (5) nervous. The principal differences among the various types within these major categories are based upon the kinds of cells they contain, the nature and amount of intercellular material present, and the functions they perform. Epithelial tissue covers surfaces, lines cavities, and is formed into tubes; connective tissue forms supporting and binding structures; blood transports materials, provides one form of defensive mechanism against disease, and provides the basis for clotting; muscular tissue contains the contractile elements for movement; and nervous tissue constitutes irritable (excitable) and conducting structures.

Epithelial Tissue: Tissue of Many Activities

Epithelial tissue (epithelium) lines and covers all free surfaces of the body, including cavities of hollow organs and the cavities containing them; makes up the secreting portions of glands; and comprises the sensory areas of sense organs (neuropithelium).

All types of epithelium consist largely or entirely of cells; they contain little intercellular substances. Epithelial cells are also arranged in continuous sheets which may consist of one or many layers. In addition, epithelial cells do not contain blood vessels but may contain nerves; blood vessels are found in the connective tissue which epithelium overlies and adheres to firmly. Since cells arranged in sheets do not provide a great deal of strength, most epithelial tissues are attached to a stronger supporting layer called the *basement membrane.* This structure connects the epithelial membrane to the underlying connective tissue and consists of a thin layer of modified intercellular substance composed of connective tissue permeated with reticular fibers (fine connective tissue fibers). Another common feature of epithelium is that its cells are subjected to at least some wear and tear and injury. Thus, epithelium remains regenerative, that is, its cells retain the capacity to divide and produce new ones to replace old or destroyed ones.

The cells of epithelium are arranged in various ways in different parts of the body (Table 9-1). In those areas of the body in which epithelial tissue is adapted for absorption or filtration, and in which there is little wear and tear, the tissue consists of a single layer of cells. This arrangement is termed *simple epithelium.* If there is considerable wear and tear and no active absorptive filtration or sensory function is performed, the epithelium will then consist of several layers of cells, an arrangement referred to as *stratified epithelium.* A third arrangement, less commonly found, is called pseudostratified epithelium. Although cells of this type of epithelium appear to be stratified, they are not. In *pseudostratified epithelium,* some of the cells in contact with the basement membrane do not reach the surface and because cells cut at right angles to the surface show nuclei at several levels, it is designated as pseudostratified.

In addition to differentiating epithelial tissues on the basis of cellular arrangement, they may also be categorized on the basis of cell shape. Accordingly, there are *squamous cells,* which are flat, and irregular in outline; *cuboidal cells,* which are cube-shaped or in the form of truncated pyramids; *columnar cells,* which are long and cylindrical, appearing as rectangles set on ends; and the cells of *transitional epithelium,* which are large and rounded.

Epithelial tissues exhibit a few modifications of the free surfaces (those opposite the basement membrane). Most epithelial cells are said to be polarized, that is, their proximal (basal) end nearest to the basement membrane is different in structure from the distal (free) end. The distal surface is frequently specialized, and may contain modifications such as deep depressions, cilia, striations, finger-like cytoplasmic projections (microvilli), or coverings of waxy or mucus secretions. In addition, it is not uncommon to find the cellular organelles more concentrated in one end than the other.

The cells of nearly all simple columnar epithelium are adapted to perform some specialized function in addition to providing protection. One of these functions is secretion. In some areas (epithelial lining of stomach), all the cells are modified for secretion. In other areas (epithelium of small intestines), some cells are specialized for secretion while others are modified for absorption. The epithelium of the upper and lower respiratory tract is modified for moving mucus along the free surface by ciliary action, in addition to secretion.

Single mucus-secreting cells of epithelium are referred to as *goblet cells.* These cells are so designated because the portion above the nucleus assumes a goblet (chalice) form, owing to an accumulation of mucus. After discharge of the secretion, the cells repeat the secretory activity. *Absorptive cells* of the intestines contain fine surface striations that approximately parallel their long axes. Collectively, the striations are referred to as a *striated border.* Electron micrographs reveal that the striations consist of minute finger-like projections, the microvilli (see Figure 5-11A). These modifications increase the surface area through which absorption occurs. A modification of pseudostratified epithelium found in the respiratory tract consists of cells that are adapted for the movement of mucus. These cells contain hair-like processes (cilia) that project from their free surfaces. In this type of epithelium, mucus secreted by goblet cells forms a film over the respiratory surface that traps foreign particles and moistens inspired air. The cilia, which beat in unison, move the mucus containing foreign materials to the throat where it can be swallowed or otherwise eliminated.

Other prominent epithelial specializations include *transitional epithelium* (lining of urinary tracts), which is capable of being expanded from within, and *stratified squamous keratinized epithelium* (skin), which is relatively impermeable to bacteria and waterproofs the body.

Most of the epithelial types discussed either cover or line body surfaces. A second general kind of epithelium is that which forms glandular tissue. In parts of the body where a vast number of secretory cells are required, epithelial cells grow into the underlying connective tissue to form highly specialized epithelial structures called *glands*.

Connective Tissue: Binding and Supporting Tissue

Connective tissue performs a variety of functions, all of which are generally related to support, protection,

TABLE 9-1. The Classification of Epithelial Tissue

Name of Tissue	Description	Microscopic Appearance*	Location	Function
I. Simple epithelium	Consists of only a single layer of cells.	See diagrams for simple squamous, simple cuboidal, and simple columnar.	In parts where there is little wear or tear or in areas where absorption or filtration occurs.	See functions for simple squamous; simple cuboidal, and simple columnar.
Squamous	A single layer of thin, flat cells, scale-like in appearance and fitted together like a mosaic; nuclei are large and centrally located.	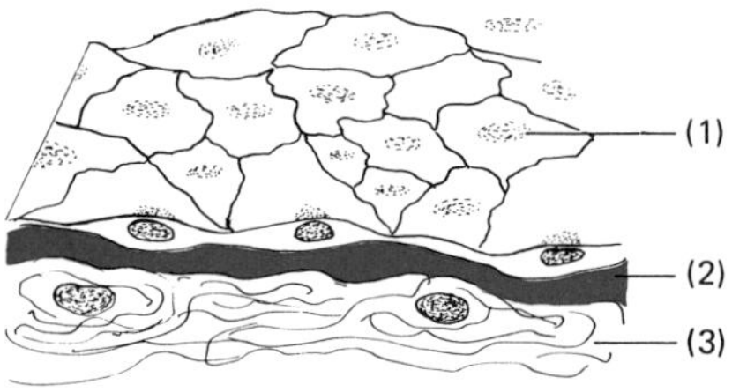 (1) (2) (3)	Lines the blood-vascular and lymphatic systems, where it is called endothelium; lines body cavities and covers visceral organs where it is designated as mesothelium; lines air spaces (alveoli) in lungs, crystalline lens of eye, and membranous labyrinth of inner ear.	Absorption and filtration.
Cuboidal	A single layer of cube-shaped cells; nuclei centrally located.	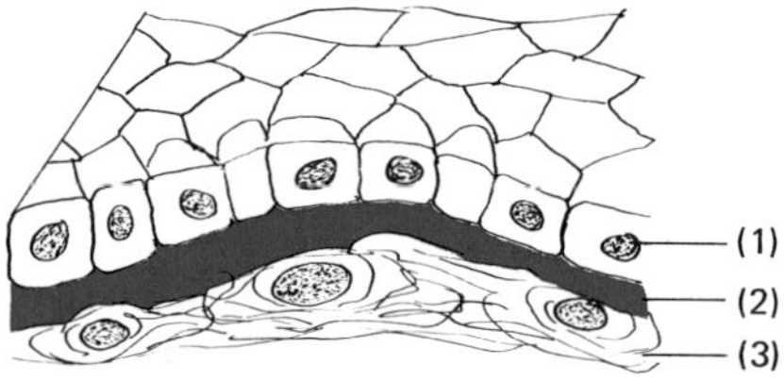 (1) (2) (3)	Thyroid gland, kidney tubules, ducts of glands, smaller bronchi of lungs, ovary.	Secretion and absorption.
Columnar	A single layer of cells which are long and cylindrical, sometimes forming tall, irregular prisms. Among some of the modifications or specializations of columnar epithelium are	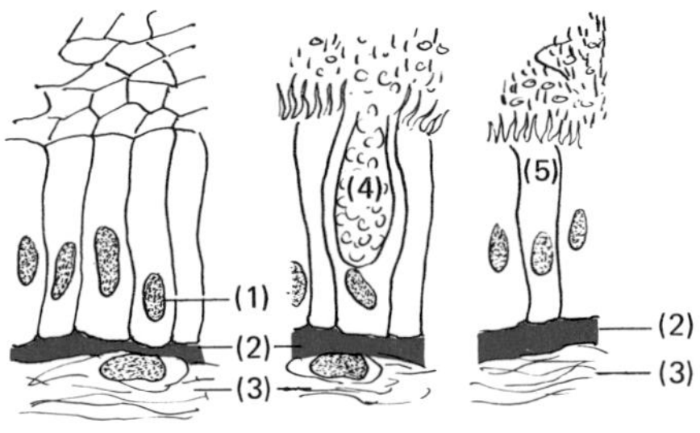 (1) (2) (3) (4) (5) (2) (3)		
	(a) Goblet cells		Surface epithelium of stomach.	Secretion.
	(b) Microvilli		Lining of intestines and kidney tubules; placenta.	Absorption of H_2O and products of digestion.
	(c) Ciliated cells		Upper respiratory tract.	Move mucus by ciliary action.
	(d) Neuroepithelial		Taste buds, rods and cones of retina, lining of nose.	Perception of taste, vision, and smell.

Table 9-1. The Classification of Epithelial Tissue [*Concluded*]

Name of Tissue	Description	Microscopic Appearance*	Location	Function
II. Stratified epithelium	Consists of cells arranged in several layers.	See diagrams for stratified squamous, stratified columnar, and transitional.	In areas where there is considerable wear and tear and little absorptive or filtering functions.	See functions for stratified squamous, stratified columnar, and transitional.
Squamous	Several layers of cells, the deepest tending to be cylindrical and the surface ones flattened and scale-like; as surface cells slough off, they are replaced by cell divisions.	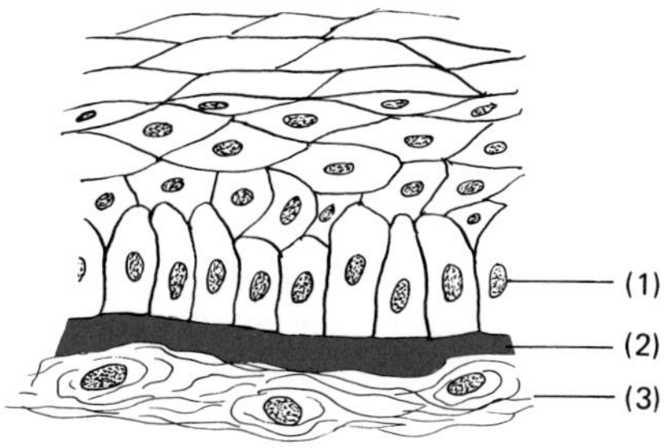 (1) (2) (3)	One type, stratified squamous nonkeratinized epithelium, is found lining mouth, esophagus, and vagina. A second type, stratified squamous keratinized epithelium, makes up the outer waterproofing layer of the skin. Keratin is a fibrous protein found within the surface cells.	Protection and waterproofing.
Columnar	Several layers of elongated cells.	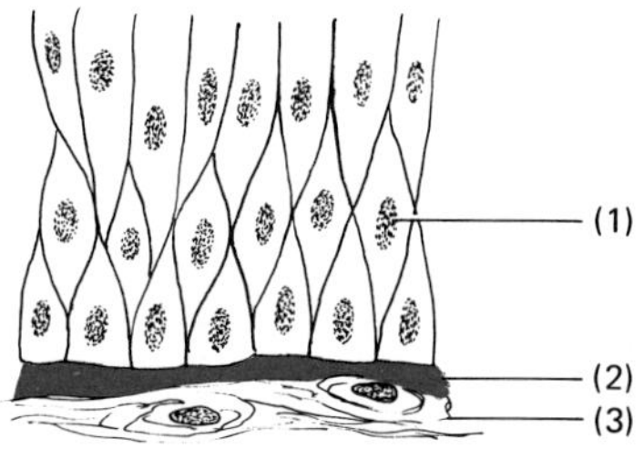 (1) (2) (3)	In pharynx and on epiglottis, larger ducts of glands.	Protection, secretion (in glands).
Transitional	Like stratified squamous nonkeratinized epithelium except that the superficial cells are large and rounded rather than scale-like; generally found in structures that are expanded from within by periodic distention.	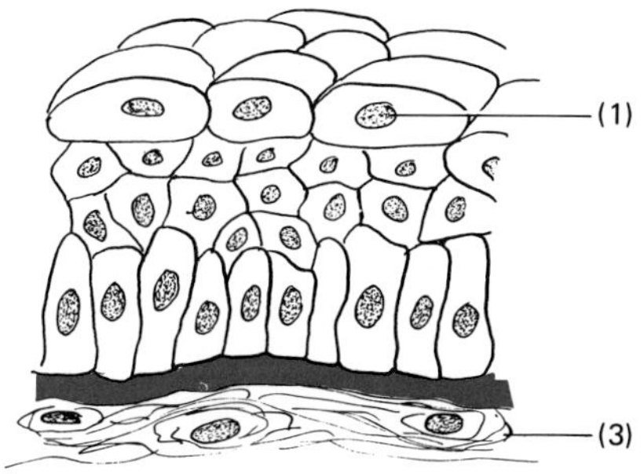 (1) (3)	Pelvis of kidney, ureters, bladder, part of urethra.	Allows distention.
III. Pseudostratified epithelium	Superficially similar to stratified epithelium but not a true stratified tissue. Some of the cells in contact with the basement membrane do not reach the surface.	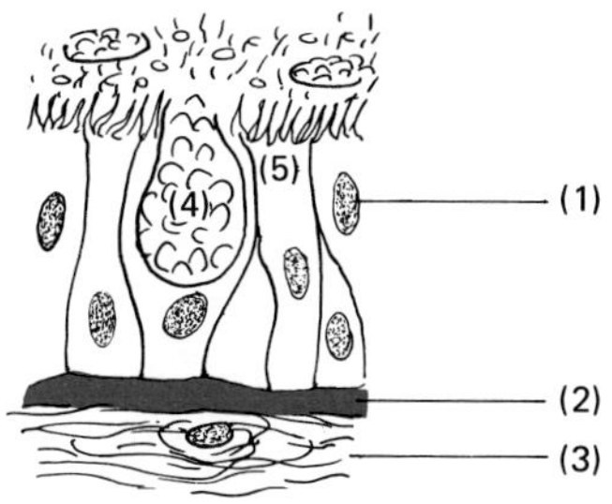 (4) (5) (1) (2) (3)	Lines most of upper respiratory tract.	Secretion and movement of mucus by ciliary action.

* Key to labels: (1) nucleus; (2) basement membrane; (3) connective tissue layer; (4) goblet cell; (5) ciliated cell.

and binding various organs together. It is the most abundant and widely distributed of all animal tissues. In direct contrast to epithelium, connective tissues usually have a rich blood (vascular) supply, typically do not occur on free surfaces, and possess a large amount of intercellular substance with widely scattered cells.

The intercellular substance is a material found between the cells of connective tissue. It varies from a fluid, or semifluid, mucoid (mucus-like) substance to a firm ground substance as in cartilage or the rigid intercellular substance (matrix) of bone. The cells of connective tissue produce the matrix, store fat, produce

TABLE 9-2. Classification of Adult Connective Tissues

Name of Tissue	Microscopic Appearance	Description	Location	Function
I. Connective tissue proper A. Loose connective	 Elastic fiber Fibroblast Plasma cell White fiber Mast cell Histiocyte	Intercellular substance of a fibrous nature. Also called areolar; contains white and yellow fibers, fibroblasts, histiocytes, plasma cells, and mast cells.	Most abundant connective tissue; found under skin, along blood vessels and nerves; forms framework of all organs and is a packing between organs.	Medium through which exchanges between blood and tissue cells occur; water balance; defense against infection; support; attachment.
B. Adipose	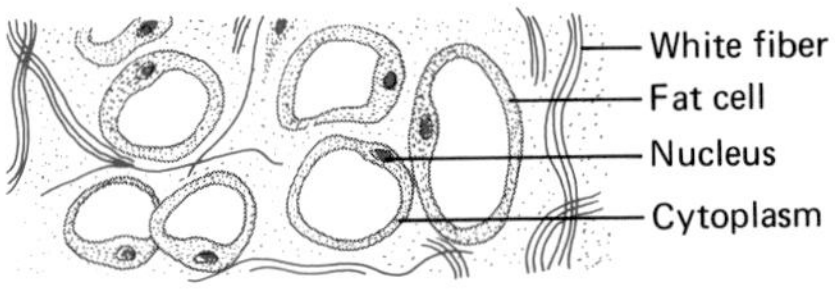 White fiber Fat cell Nucleus Cytoplasm	Closely packed fat-storing cells with elastic and white fibers running in all directions.	Under skin; padding between organs and around joints.	Reserve food supply; insulates against heat loss; support; protection.
C. Reticular	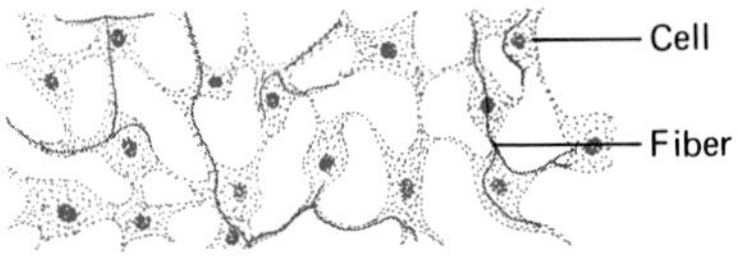 Cell Fiber	Network of cells embedded in a network of intercellular fibers.	Lymph glands; certain endocrine glands; walls of blood vessels; membranes underlying digestive and respiratory passageways.	Supporting framework.
D. Elastic connective in a blood vessel	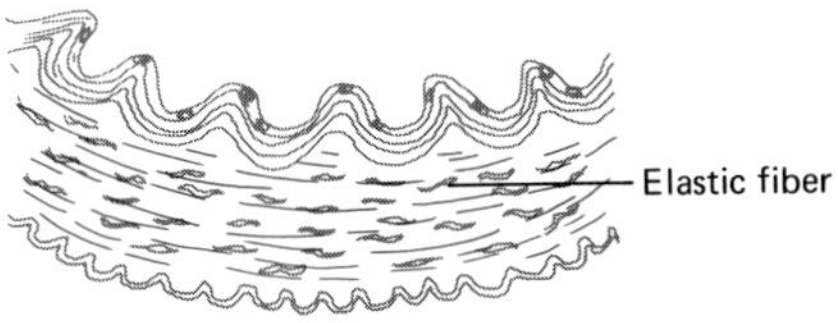 Elastic fiber	Consists mostly of elastic fibers with few of the collagenous type.	Some ligaments; bronchial tubes; walls of arteries.	Support; attachment; elasticity.
E. Fibrous connective	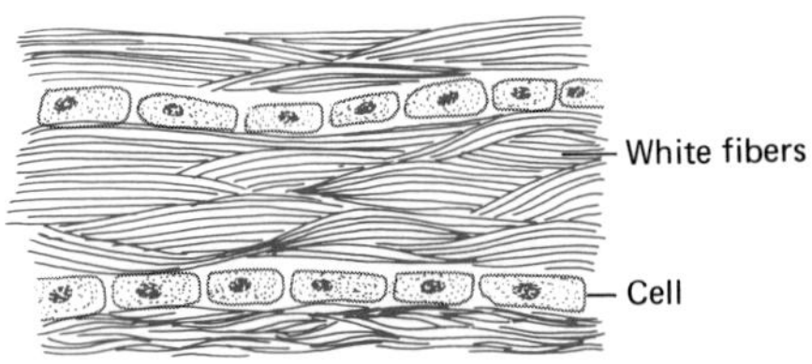 White fibers Cell	Consists mainly of white fibers; few elastic fibers present; relatively few cells; silvery white; tough, strong, yet flexible.	Tendons; ligaments; fasciae; aponeuroses; capsules of organs.	Holds muscle to bone; attaches bone to bone; surrounds muscles; lines various organs.

new blood cells, ingest bacteria and cell debris, form an anticoagulant, and give rise to antibodies, which may afford immunity to various diseases. Of the many kinds of cells found in connective tissue, only a few will be described here.

Loose or areolar connective tissue, the most widely distributed type of connective tissue, is described as an example. *Fibroblasts* are the most numerous cells of areolar tissue (Table 9-2). These are stellate in appearance and give rise to the distinct types of fibers that are found in this connective tissue. *Collagenous* (*white*) *fibers* are very tough and strong, yet flexible and resistant to a

TABLE 9-2. Classification of Adult Connective Tissues [*Concluded*]

Name of Tissue	Microscopic Appearance	Description	Location	Function
II. Cartilage		Also called "gristle"; shapeless intercellular matrix in the form of a gel; fibers form a dense network; cells (chondrocytes) lie in cavities (lacunae); no blood vessels; covered by a fibrous connective tissue membrane (perichondrium).		
A. Hyaline	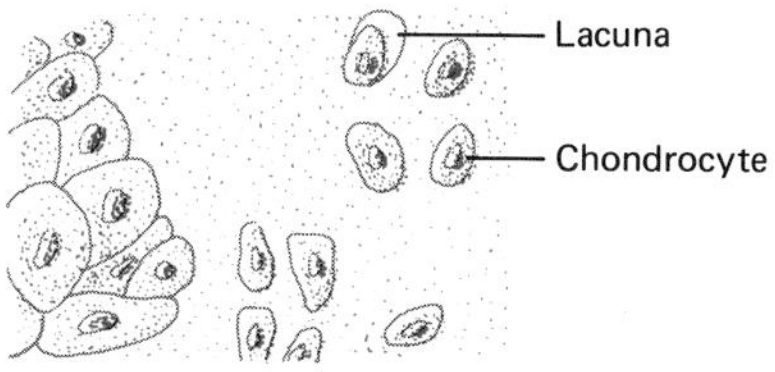	Most common type; pearly white and glassy or translucent.	Embryonic skeleton; articular surfaces of bones; between ribs and sternum; nose; larynx; trachea.	Firm but flexible support.
B. Fibrous	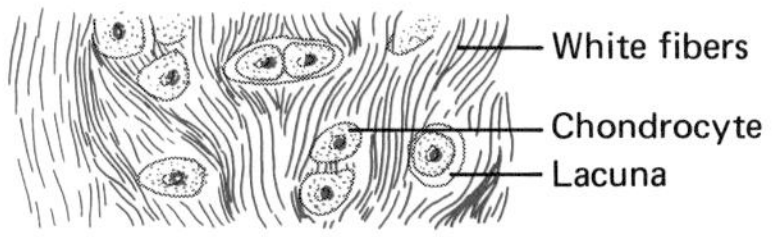	Less firm than hyaline but great strength; many collagenous fibers; few cells.	Discs between vertebrae; pubic symphysis.	Firm but flexible support.
C. Elastic	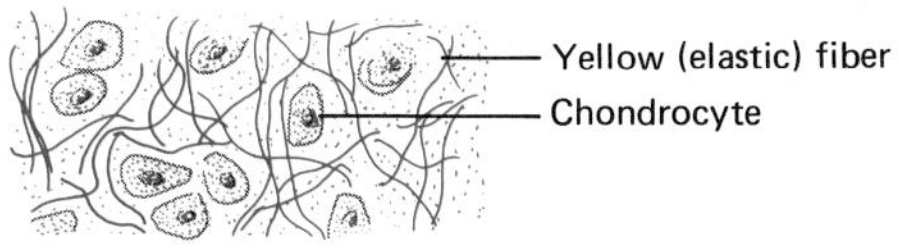	Most flexible type; many elastic fibers.	External ear; eustachian tube; epiglottis.	Firm but flexible support.
III. Bone	See Figure 9-12	Also called osseous tissue; hardest connective tissue; bone cells (osteocytes) are embedded in a matrix impregnated with calcium and phosphorus salts.	Skeleton.	Protection; support; movement; leverage; blood cell formation; calcium storage.

pulling force. These fibers are composed of many minute wavy fibrils lying parallel to each other. They are typically found in relatively large bundles, an arrangement that provides great tensile strength. *Elastic (yellow) fibers* occur singly, are not composed of fibrils, and branch freely; they are highly elastic. *Reticular fibers* are delicate and are arranged in the form of a network that serves to support the cells. They are thought to be the type of fiber that is formed first in a connective tissue.

Macrophages (histiocytes), a second principal kind of cell in loose connective tissue, are irregularly shaped cells with short branching processes. They are derived from monocytes of the blood. Under abnormal conditions, such as inflammation, they exhibit great phagocytic capacity. *Plasma cells,* derived from blood lymphocytes, are small round or irregularly shaped cells that give rise to antibodies. *Mast cells,* found most abundantly near blood vessels, form a substance called heparin that may help to prevent blood from clotting within vessels. Fat cells are specialized fibroblasts that are adapted for fat storage. They resemble signet rings because the fat droplet within each flattens the nucleus and pushes the cytoplasm to the side.

Based upon the consistency of the intercellular substance, adult connective tissues are classified into three principal types (see Table 9-2): (1) connective tissue proper (semisolid intercellular), (2) cartilage, and (3) bone (hard intercellular).

Cartilage, also called gristle, is a tough semisolid tissue containing *chondrocytes,* the cells that produce the intercellular material. The intercellular material, called the *matrix,* contains collagenous and elastic fibers. Cartilage is surrounded by a membrane called the *perichondrium* that contains cells that can develop into chondrocytes and form new cartilage if the tissue is damaged or diseased.

The types of cartilage that are found in the body include

1. *Hyaline cartilage,* which is translucent and forms most of the embryonic skeleton, the cartilage connecting the ribs to the sternum, and the cartilage covering the ends of bones of freely movable joints.
2. *Fibrous cartilage,* a very tough tissue with many collagenous fibers found in the intervertebral discs between the backbones.
3. *Elastic cartilage,* a flexible tissue containing many elastic fibers and found in the external ear and the epiglottis of the larynx.

Before leaving our discussion of connective tissue, the structure of bone (osseous) tissue will be examined in somewhat more detail. Bone, like other connective tissues, contains a predominance of intercellular substance rather than cells, but in bone the intercellular material is calcified. The rigidity of bone is largely a consequence of calcium and phosphorus salts which have been deposited in the cement substance of the intercellular material. Embedded in the calcified matrix are collagenous fibers; these reinforce osseous tissue.

The details of bone structure may be analyzed by considering the morphology and microscopic appearance of a long bone such as the femur, or thigh bone (Figure 9-12A). A typical long bone consists of the *diaphysis,* the shaft or main central portion; the *epiphyses,* the ends or extremities; the *medullary (marrow) cavity* within the diaphysis that contains bone marrow; the *endosteum,* a fibrous membrane lining the medullary cavity; and the *periosteum,* the membrane covering the bone.

Bone is not a solid homogeneous substance; all bone is porous. Based upon the relative degree of porosity, bone tissue is regarded as being one of two types: (1) cancellous (spongy) and (2) dense (compact). Compact bone contains fewer spaces and is always found on the outside of bone and in the diaphysis. Cancellous bone has larger spaces filled with marrow and is found in the interior of a bone and in the epiphyses.

A very thin transverse section of compact osseous tissue reveals its microscopic appearance (Figure 9-12B, C). Here it can be seen that concentric cylindrical layers of calcified matrix (*lamellae*) enclose a central longitudinal canal (*Haversian canal*) which contains blood vessels. The entire unit of canal and enclosing lamellae is termed a *Haversian system,* or *osteon.* The bone cells themselves, called *osteocytes,* occupy small cavities between the lamellae called *lacunae.* Radiating out in all directions from the lacunae are minute canals (*canaliculi*) which branch, and connect with the Haversian canal, forming a network throughout the bone. Minute protoplasmic filaments grow outward from the osteocytes into the canaliculi.

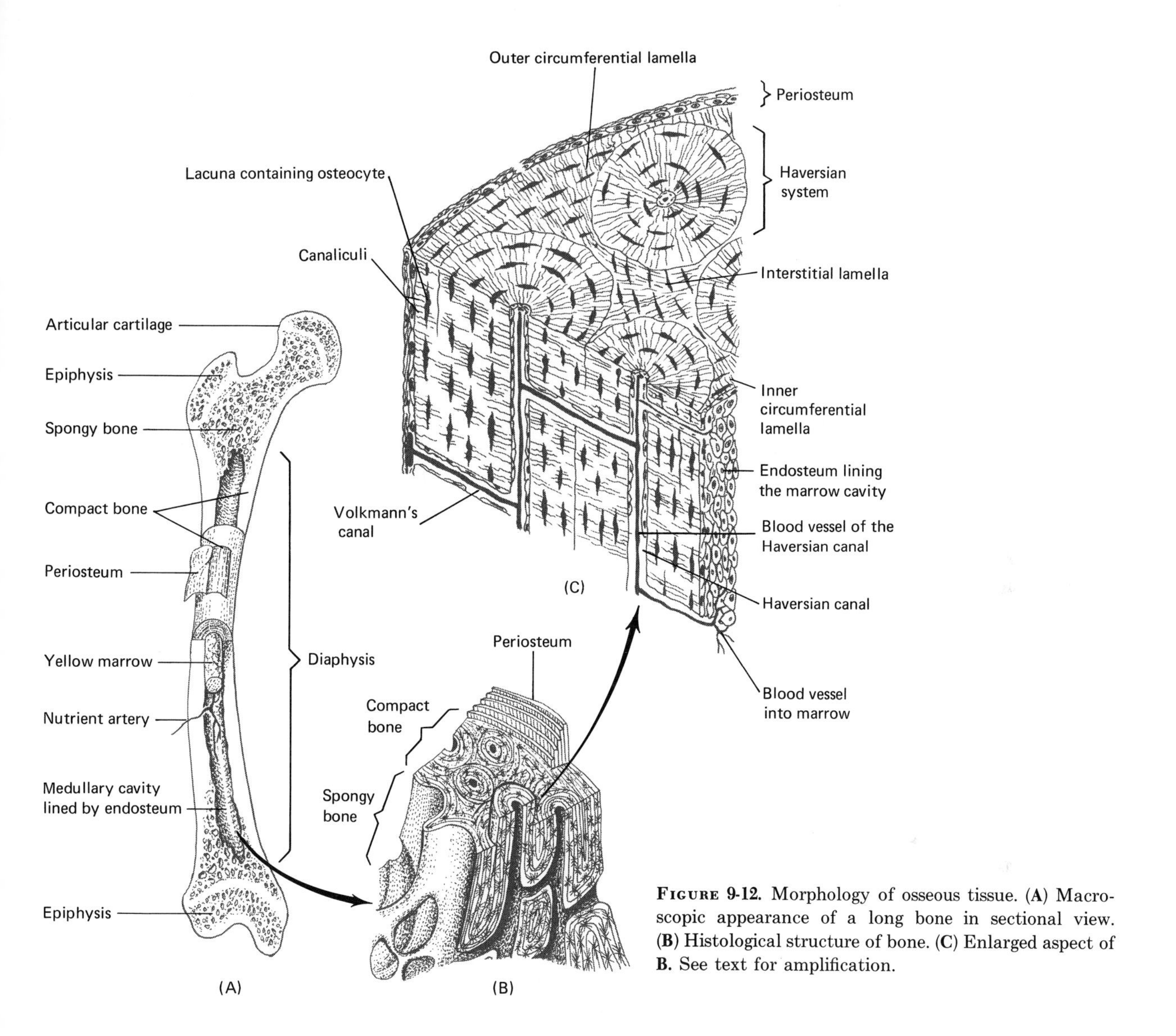

Figure 9-12. Morphology of osseous tissue. (**A**) Macroscopic appearance of a long bone in sectional view. (**B**) Histological structure of bone. (**C**) Enlarged aspect of **B.** See text for amplification.

Thus, the branching network of canaliculi provides numerous routes for nutrients to reach individual osteocytes. The areas between Haversian systems contain lamellae that do not have a concentric arrangement; these are called *interstitial lamellae,* and, like concentric lamellae, they also possess lacunae, osteocytes, and canaliculi, but usually no canals. They are remnants of osteons that have been remodeled as the bone has developed.

The *periosteum* is a tough fibrous membrane that covers bone except at articulating surfaces. It consists of two ill-defined layers: (1) an outer layer (fibrous) composed of connective tissue filled with blood and lymphatic vessels and nerves, and (2) an inner (osteogenic), less vascular layer containing elastic fibers, blood vessels, and cells. The cells of the osteogenic layer, called *osteoblasts,* form new bone during growth and repair. Blood vessels and nerves from the periosteum penetrate into the compact bone through canals (*Volkmann's canals*). Their various passageways traverse the lamellae and some penetrate into Haversian canals.

Vascular Tissue: Transportation Tissue

Although some texts regard **blood** or **vascular tissue** as a type of connective tissue, we will consider it as a separate tissue since the intercellular material is liquid rather than composed of fibers of a more or less dense substance. The chief functions of blood are to keep the internal cellular environment of body tissues in a state of dynamic equilibrium (homeostasis) and to transport materials. Within this context, blood transports oxygen from the lungs to tissues, and carbon dioxide from the tissues to the lungs for elimination; carries foods from the intestines to all parts of the body and returns other wastes to the kidneys for excretion; distributes the heat produced by muscular activity and therefore regulates body temperature; circulates the hormones produced by endocrine glands, assumes a role in maintaining the acid-base balance of tissues; functions in clot formation; and plays an important part in protecting the body against invading bacteria.

Structurally, blood is composed of a liquid, the *plasma,* in which the formed elements are suspended. Plasma is a straw-colored fluid, very complex chemically, containing a wide variety of substances. It is approximately 92% water and 8% solid materials. Among the solids are inorganic ions (calcium, sodium, potassium, magnesium, chloride, bicarbonate, phosphate, and sulfate), and organic compounds (serum albumin, serum globulin, fibrinogen, amino acids, urea, creatine, cholesterol, fats, glucose, enzymes, and hormones). The gases oxygen, carbon dioxide, and nitrogen are also found in plasma. Collectively considered, plasma constitutes 53–58% of whole blood. The formed elements, which comprise 42–47% of the blood, are **erythrocytes** (red blood cells), **leukocytes** (white blood cells), and **thrombocytes** (blood platelets). A few of the important characteristics and functions of these cells are shown in Table 9-3.

Muscle Tissue: Tissue for Movement

Muscle tissue in various locations of the body is specifically adapted for movement by contraction and relaxation of its component cells. These movements, which may be internal or external, are essential for the maintenance of the organism. The movements of muscles may be considered as voluntary (under conscious control) and involuntary (not under conscious control). Among the principal voluntary movements are actions such as the movement of limbs (locomotion), fingers (manipulation), toes (balance), pharynx (swallowing), tongue (vocalization and manipulation of food), and face (expressions). The principal involuntary movements consist of the propulsion of substances through body passages (food through the digestive tract and blood through the vessels), the expulsion of stored materials (bile from the gall bladder), the regulation of the size of openings and tubes (pupil of eye, pyloric valve of stomach, sphincters of anus, blood vessels, and bronchioles), and the beating of the heart.

There are a number of physiological characteristics that adapt muscle tissue for body movements. One of these is *excitability,* the capacity to react to stimuli. Another characteristic is that of *contractility.* A muscle, when stimulated, has the ability to become shorter and thicker, and, in doing so, accomplishes movement. Muscle tissue also exhibits *extensibility,* that is, the ability to be stretched by the application of force. Finally, muscle tissue possesses *elasticity,* the property of returning to its normal length after the force applied to it has been relieved.

Inasmuch as muscle tissues differ from each other on the basis of location, microscopic structure, and nervous control, these criteria are employed to classify them into various types in vertebrates. For example, using the criterion of location, muscle tissue may be classified as: (1) *skeletal* (attached to bones), (2) *visceral* or *smooth* (in the walls of hollow internal structures such as blood vessels, stomach, and intestines), and (3) *cardiac* (wall of the heart). If the criterion of microscopic structures is employed, muscle tissue may be considered either as *striated* or *smooth* (nonstriated). Finally, on the basis of nervous control, there are also two kinds of muscular tissue, *voluntary* and *involuntary.* By combining the three criteria just mentioned, muscle tissue may be

TABLE 9-3. Characteristics and Functions of the Formed Elements in Blood

Name of Cell and Size	Range (no./mm^3)	Description and Microscopic Appearance	Origin, Life Span, and Destruction	Function
I. Erythrocytes (red blood cells) Average 7.7 μ in diameter	5.5–7 million in men, 4.5–6 million in women	Homogeneous circular discs, no nuclei, biconcave in profile. 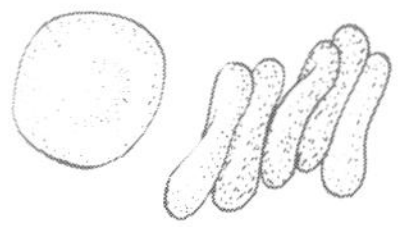	Cells originate in the red bone marrow; average life span about 120 days; destruction by phagocytosis in spleen, liver, and red bone marrow.	The hemoglobin in red blood cells enables them to carry oxygen from the lungs to the tissues, and carbon dioxide from the tissues to the lungs; buffering.
II. Leukocytes* Variable with type.	5000–9000	Larger than red blood cells and nucleated.	Originate in red marrow and lymphatic tissue; life span not definitely known; some probably destroyed by phagocytosis, others by microorganisms.	Defense against microorganisms by phagocytosis; production of antibodies.
A. Granular leukocytes				
(1) Neutrophils 9–12 μ	55–65% of the total number of leukocytes	Nuclei have from 3 to 5 lobes; cytoplasmic granules stained pink by neutral dyes. 	In bone marrow; life span 3–12 days.	High capacity for phagocytosis.
(2) Eosinophils 10–14 μ	1–3% of total	Nuclei usually consist of two oval lobes; cytoplasmic granules stain bright red with acid dye (eosin). 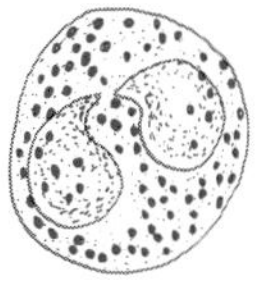	In bone marrow; life span 3–12 days.	No phagocytic action.
(3) Basophils 8–10 μ	0.25–0.7% of total	S-shaped nuclei; cytoplasmic granules stain deep purple with basic dyes. 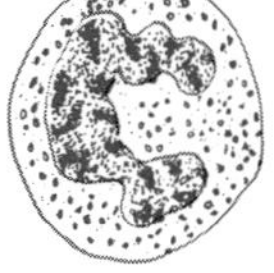	In bone marrow; life span 3–12 days.	Function unknown.

(*continued*)

TABLE 9-3. Characteristics and Functions of the Formed Elements in Blood [*Concluded*]

Name of Cell and Size	Range (no./mm^3)	Description and Microscopic Appearance	Origin, Life Span, and Destruction	Function
B. Agranular leukocytes				
(1) Lymphocytes Small: 10 μ; Large: to 20 μ	Small: 20–25% of total; large: 3% of total	Nucleus is single, large, usually spherical; cytoplasm stains pale blue.	In lymphatic tissue; possibly less than 24 hours—in bloodstream, 200 days.	Forms serum globulin; produces antibodies; good phagocytes.
(2) Monocytes 16–20 μ	6–8% of total	Large oval or kidney-shaped nucleus; cytoplasm stains gray-blue.	Lymph glands and spleen; 3 days in bloodstream, then migrate to tissues and become macrophages.	Phagocytic for microorganisms and debris.
III. Thrombocytes 2–5 μ	250,000–500,000	Oval, granular bodies lacking nuclei; fragments of large cells (megakaryocytes) found in red bone marrow.	In red bone marrow; probably destroyed in spleen; 12 days in blood stream.	Initiate blood clotting.

* There are two major groups of white blood cells: (1) granular (polymorphonuclear) leukocytes, which have granules in the cytoplasm and possess a nucleus with two or more lobes, and (2) agranular leukocytes, which do not have granules in the cytoplasm and contain a nonlobed nucleus that may be spherical or indented.

classified as: (1) skeletal or striated voluntary muscle, (2) cardiac or striated involuntary muscle, and (3) visceral or nonstriated (smooth) involuntary muscle.

Since the structure, physiology, and biochemistry of skeletal muscle have already been considered in Chapter 7, we will discuss some additional aspects of cardiac and visceral muscle tissue only.

Cardiac muscle is a highly specialized tissue found only in the walls of the heart. Individual cells of this tissue are roughly quadrangular in shape and the tissue as a whole is intermediate in structure between skeletal and smooth muscle in that its fibers are striated and contain a single centrally located nucleus (Figure 9-13A). The individual fibers are covered with a thin, ill-defined sarcolemma and the contractile myofibrils within the fibers have alternating light and dark bands, the striations. These striations are less pronounced than those in skeletal muscle. Whereas the fibers of

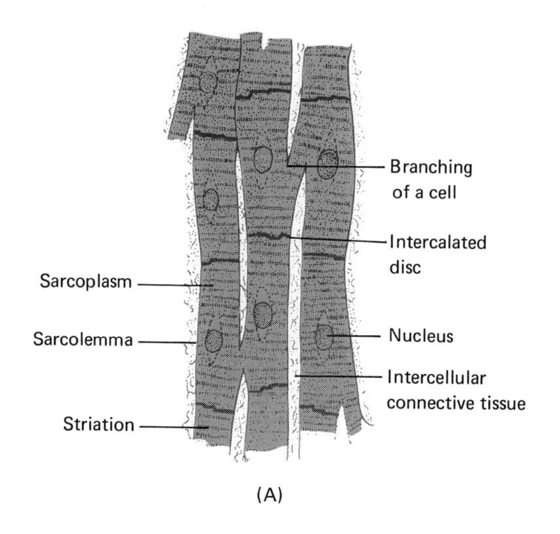

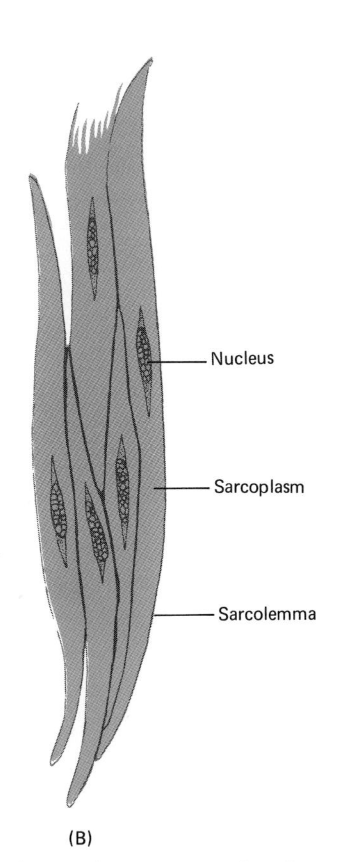

FIGURE 9-13. Microscopic appearance of cardiac (**A**) and smooth (**B**) muscle tissues.

skeletal muscle are arranged in a parallel fashion, those of cardiac muscle branch freely with other fibers, forming a continuous functional network of cells or *syncytium.* Characteristic of cardiac tissue are dense bands that occur at variable intervals across the cardiac fiber. These transverse lines are called *intercalated discs.* Electron micrographs have revealed that these discs are actually places where two cell membranes join at the ends of adjacent cardiac fibers.

Smooth muscle is so designated because it does not exhibit cross striations when viewed microscopically (Figure 9-13B). Isolated smooth muscle fibers are elongate, spindle-shaped, and contain an elongated oval nucleus occupying a central position. The cytoplasm of the fibers (sarcoplasm) contains the contractile elements, or myofibrils. The cells of this tissue are connected by means of reticular connective tissue fibers which act as an intercellular cementing material. In contrast to the quick action of striated muscle, smooth muscle fibers are characterized by slow contraction and relaxation periods.

Nervous Tissue: Communication and Coordination Tissue

The cells and fibers of **nervous tissue** form an interconnecting network that links every part of the organism. The basic function of nervous tissue is to integrate the myriad activities of the organism and to coordinate them with the external (physical) and internal environments. Nervous tissue is composed of two kinds of cells: (1) neuroglia, or supporting cells, and (2) neurons, or nerve cells.

Neuroglia is a term applied to cells comprising the interstitial tissue of the nervous system. They support nerve cells and probably assume an important role in their normal metabolism. In addition, they connect neurons to blood vessels.

Neurons, in direct contrast to neuroglia, are the structural and functional units of nervous tissue (see Chapter 6). Based on function, neurons may be classified as sensory (afferent) neurons, motor (efferent) neurons, and interneurons (internuncial or intercalated). Structurally, neurons are classified as multipolar, bipolar, and unipolar, depending upon the number of processes projecting from the cell body. These relationships are summarized in Table 9-4.

TABLE 9-4. The Classification of Neurons on the Basis of Function and Structure

Criterion	Description
I. Function	
A. Sensory (afferent)	Transmit nerve impulses to brain or spinal cord.
B. Motor (efferent)	Transmit nerve impulses away from brain or spinal cord to muscle or glandular tissue.
C. Interneuron (internuncial or intercalated)	Lie within brain and spinal cord and conduct impulses from sensory to motor neurons.
II. Structure	
A. Multipolar	Have several dendrites but only one axon; most neurons in brain and spinal cord are of this type.
B. Bipolar	Have only one dendrite and one axon; found in retina, inner ear, taste buds, and olfactory area.
C. Unipolar	Have only one process extending from the cell body, but this process divides into a central and a peripheral branch; the central branch is the axon, and the peripheral branch functions as a dendrite; most sensory neurons are unipolar.

The general parts of a neuron are illustrated in Figure 9-14. The *cell body* (*perikaryon*) of a neuron contains a well-defined nucleus and nucleolus, surrounded by granular cytoplasm. Within the cytoplasm are found *neurofibrils,* long, thin fibrils which form a network and are distributed to the processes of the neuron. Scattered throughout the cytoplasm are *Nissl granules,* which consist of a group of flat, membranous sacs and numerous RNA granules scattered between them. These granules may assume a role in the nutrition of the neuron. Cell bodies are essential to the life of a neuron. If they die, the whole neuron dies; they are not capable of mitotic division. After about 16 years of age, neurons lose the ability to divide.

The *cell processes* (nerve fibers) of neurons are elongated extensions of the cell body and are of two types. The first of these are the *dendrites.* They branch extensively and vary greatly in different types of neurons. *Axons,* the second type of projection, are single processes. Frequently one or more side branches (collaterals) emerge from axons at right angles. Each neuron has only one axon, which may be short (fraction of an inch) or long (3 feet or more), depending on the location of its cell body.

Both dendrites and axons may develop sheaths that completely surround them, or they may remain naked.

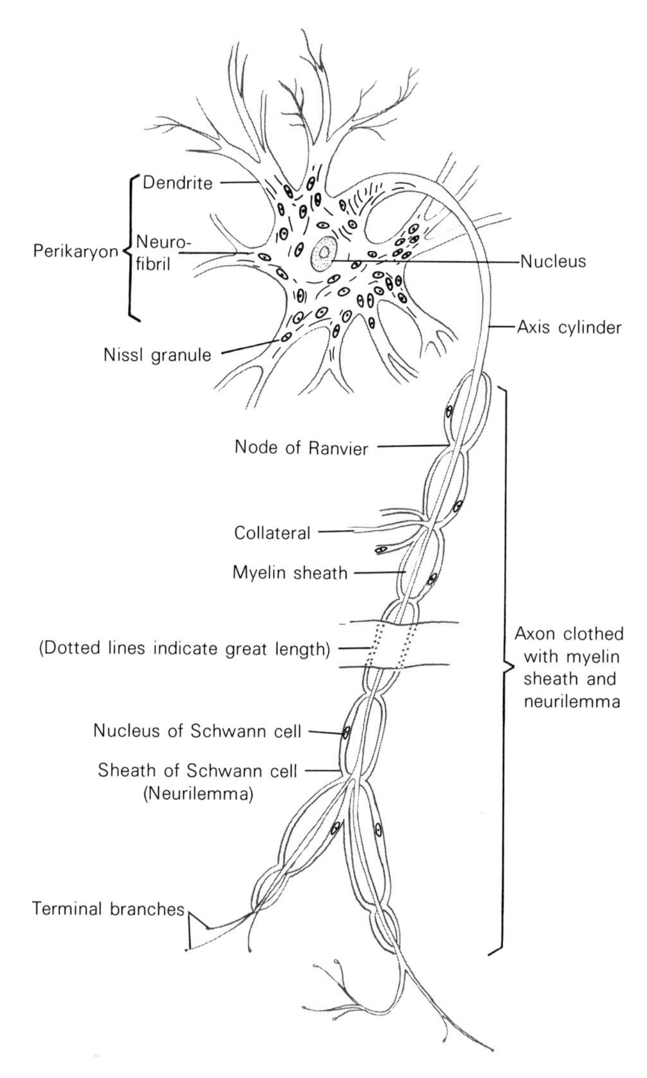

FIGURE 9-14. Diagrammatic representation of a motor neuron from the ventral gray area of the spinal cord.

The sheaths that may develop are of two types, the myelin sheath and the neurilemma. The *myelin sheath* is a segmented wrapping surrounding a nerve fiber. The small gaps between the segments are referred to as *nodes of Ranvier*. The sheath itself is a white, lipid substance that serves both as an insulator and as a source of energy, assisting in the rapid conduction of impulses. The myelin sheath or peripheral nerve fibers are formed by the plasma membrane of *Schwann cells,* located along peripheral nerve fibers. This developmental process is shown in Figure 5-13. Nerve cell processes that possess myelin sheaths are called *myelinated fibers,* while those that lack the sheaths are termed *unmyelinated fibers*. Unmyelinated fibers are also embedded in Schwann cells, and there is no coiling of the membranes around them. The so-called gray matter of the nervous system consists of unmyelinated fibers and cell bodies of neurons; the white matter consists of bundles of myelinated fibers, the myelin produced by a specific glial cell, of the central nervous system.

The *neurilemma* is a delicate, continuous sheath that contains many nuclei. It envelops the myelin sheath of myelinated fibers and the cell processes of unmyelinated fibers. Like the myelin sheath, it is also thought to be derived from Schwann cells. It plays an essential role in nerve fiber regeneration. Although peripheral nerve fibers contain neurilemma sheaths, fibers in the brain and spinal cord do not. Thus, if disease or injury causes degeneration of the fibers in the brain or spinal cord, the destruction is usually permanent, regeneration being very limited.

The Organization of Animal Tissue into Organs

Just as the tissues of a plant are organized into organs (roots, stems, and leaves), animal tissues are similarly organized into higher, more complex levels of organization. Inasmuch as organs of higher organisms (lungs, stomach, heart, kidney, liver, spleen, brain, intestines, etc.) are exceedingly diverse and complex with respect to tissue arrangements, only one example will be cited here to illustrate the organ level of development. A typical example is the human skin, an organ that represents a highly complex integration of various types of cells and tissues. A description of this organ is given in the caption of Figure 9-15.

The Organ-System Concept

No organ of the body operates in isolation. Each is structurally and functionally integrated with a number

TABLE 9-5. Organ Systems of the Human Body in Terms of Structure and Function

Name of Organ System	Representative Organs	Function
Skeletal	All the bones of the body.	Support, protection, movement, leverage, making blood cells, calcium storage.
Muscular	All the muscles of the body attached to the skeleton.	Movement, posture, heat production.
Digestive	Mouth, teeth, esophagus, stomach, intestines, liver, gall bladder, pancreas, salivary glands.	Physical and chemical processing of food for use by the body; elimination; absorption.
Respiratory	Nose, trachea, bronchi, lungs.	Procurement of oxygen and the elimination of carbon dioxide.
Circulatory	Heart, blood vessels, blood.	Internal transport system.
Lymphatic	Lymph nodes and vessels, lymphatic organs, lymph.	Filters lymph and protects body against disease.
Excretory	Kidneys, ureters, urinary bladder, urethra.	Elimination of metabolic wastes; regulation of balance of body fluids.
Endocrine	Pituitary, adrenals, pancreas, ovaries, testes, parathyroids, thyroid.	Internal control through chemical (hormonal) regulation.
Nervous	Brain and spinal cord.	Internal control through nerve impulses.
Reproductive	Ovaries, testes, uterus, penis, vagina.	Propagation of the species.

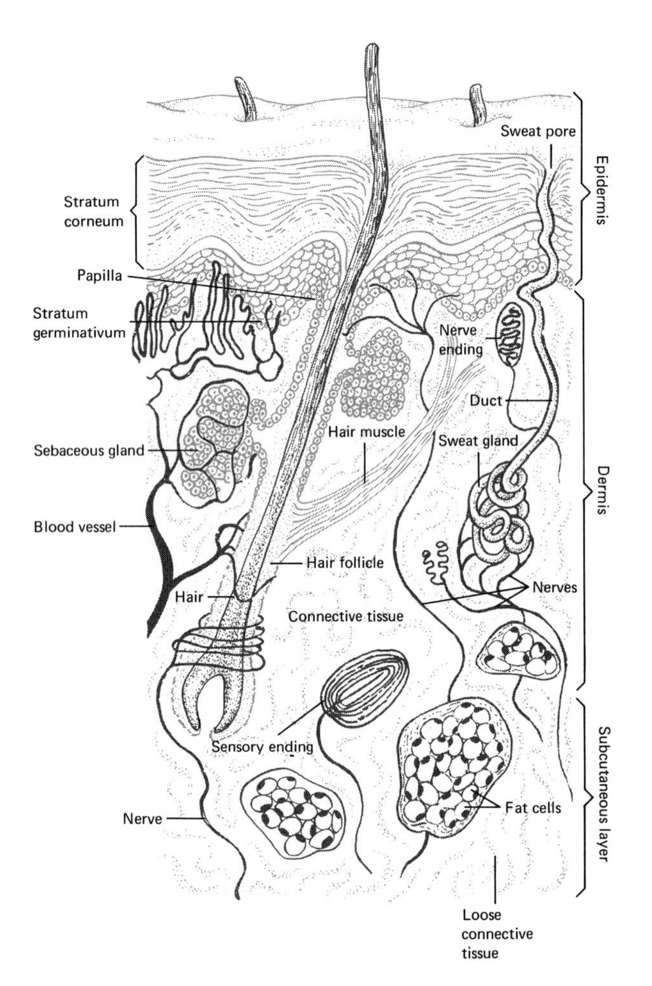

Figure 9-15. The organization of animal tissues into an organ. In this composite diagram of a section of human skin, it can be seen that the organ is composed of a variety of tissues. The skin (integument) consists of two layers: a surface layer, the epidermis, and an underlying thicker layer, the dermis. Beneath the dermis is a subcutaneous layer of loose connective tissue and adipose cells that binds the skin to the body. The epidermis is composed of stratified squamous epithelium, with an outer layer—the stratum corneum—of a tough, keratinized mass of dead cells that are constantly worn away and replaced by cells of the deeper layers. The deepest layer of the epidermis, the stratum germinativum, is actively mitotic and replaces those cells shed at the surface. There are no blood vessels in the epidermis. The dermis constitutes the greater portion of the skin and is subdivided into a superficial papillary layer, which has many blood vessels and brings blood to the stratum germinativum, and a deeper reticular layer, which is a dense mass of white and yellow elastic connective tissue fibers. The uneven surface of the dermis is due to the papillae, finger-like projections that connect the dermis to the epidermis. Some papillae have at their tip a nerve ending that makes the skin sensitive to touch or other stimuli. Among some of the structures that are derived from the epidermis and located in the dermis are sweat glands, whose ducts pass through the epidermis and terminate in pores; sebaceous (oil) glands, whose excretory ducts usually discharge into hair follicles; ceruminous (wax) glands, found only in the skin of the passages leading into the ear; hair follicles; and hair muscles (bundles of smooth muscle).

of other organs. This integration constitutes the *organ-system* level of organization. It is so designated because it is a specialized system, consisting of a series of organs arranged in a particular way, and cooperating to perform one of the general functions of the body. The various systems, all of which contribute to the functioning of the entire organism, will be analyzed in later chapters. Table 9-5 lists the systems of the human body, a few of the organs that comprise them, and some of the functions of each.

10 Homeostasis: Chemical Control

GUIDE QUESTIONS

1 Define homeostasis. What is meant by negative feedback? Give an example of homeostasis and explain how it operates.
2 Describe the three types of growth substances in plants with regard to their effect on plant growth and development.
3 Outline in detail the Boysen-Jensen experiment on coleoptiles. How did Went elaborate on experimentation with coleoptiles?
4 Describe the various influences of auxins, gibberellins, and cytokinins on plant activities. What use does the produce industry make of some of these influences?
5 How may flowering be controlled? What is meant by photoperiodism? Describe Chailakhian's experiment concerning floral hormones.
6 What is the relative importance of the length of the light period as compared with the length of the dark period in short-day plants with respect to flowering? in long-day plants? What are the effects of red and far-red light on both of these plants?
7 Describe the various types of plant movements and reactions. State an example of each.
8 How does nervous control of an animal compare with chemical coordination? What processes are controlled by chemical coordination that are not directly affected by nervous coordination?
9 Explain the chemical nature of hormones. Describe their general roles and interrelationships. How may hormones perform at the molecular level?
10 List some of the hormonal effects on invertebrates. Outline the hormonal control of the metamorphosis of the Cecropia moth.
11 Briefly indicate the processes involved in the transportation of the various hormones to the affected body areas.
12 Outline in detail the endocrine glands, their location, secretion(s), activity(ies), and effects of hypo- and hypersecretion. (This may be conveniently done in table form.)

ONE of the most fascinating problems in modern biology is that of regulation in living systems. When one considers the enormous structural and functional complexities of organisms, it becomes readily apparent that some form of control mechanism must be operative. In Chapter 8 it was noted that the control of cellular metabolism is coded in the base sequences of nucleic acids and operates through the regulation of enzyme production and the regulation of enzyme activity. The coordination of all regulatory functions, as well as the coordination of all other functions of organisms, depends upon special control mechanisms that operate at the organismic level. These are chemical control mechanisms, common to all organisms, and nervous control mechanisms which are typical of multicellular animals.

The Principles of Homeostasis

The term **homeostasis,** in its simplest context, refers to the maintenance of stability (within certain limits) in an organism. It describes a stable internal environment compatible with the continuance of life and implies that dynamic, self-regulating processes are operative. Important to maintenance of homeostasis is the operation of *negative feedback;* that is, if there is a deviation in one direction, there is a reaction in the opposite direction. Thus, homeostasis is the result of self-regulating negative feedback systems that serve to maintain the composition of the internal environment of a living organism within narrow limits.

One of the best examples of homeostasis in animals is the regulation of body temperature. Mammals and birds are referred to as *homeotherms,* or warm-blooded animals. As such, they are capable of maintaining a remarkably constant internal body temperature even though the external temperature varies over a broad range. In man, this constant temperature averages 98.6°F. Assume that a person moves into an external environment with a temperature of 100°F. In response to this temperature increase, a sequence of events opposing the rise in temperature is set into operation. Sensing devices in the skin, called **receptors,** pick up the stimulus (heat), activate nerves, and send the message, in the form of electrical impulses, to the brain. This route is termed a *sensory pathway.* A temperature-regulating center in the brain then acts as an interpreting and response-selecting device, or *modulator.* This structure sends out appropriate impulses over a selected *motor pathway.* The signal leads to an **effector** (sweat glands of the skin). Stimulation of sweat glands causes them to secrete more moisture through the pores of the skin. Evaporation of the moisture lowers the body temperature, thus preventing a drastic temperature increase.

In considering this sequence of events, then, it can be seen that such a pattern of control in living systems consists of (1) a stimulus, (2) a receptor that picks up the stimulus, (3) a sensory pathway that conducts the stimulus from the area of reception to (4) an interpreting and response-selecting device (modulator), (5) a motor pathway that conducts an impulse to an (6) effector, and a (7) response which counteracts the original stimulus (Figure 10-1). It should be made clear that the illustration just presented has been greatly simplified; many other factors are related to temperature homeostasis of humans. Moreover, this is only one kind of control mechanism.

Chemical Control: An Overview

The maintenance of homeostasis by chemical substances is a common feature of all organisms. In unicellular organisms, these chemicals diffuse through the cytoplasm. In multicellular plants and animals lacking elaborate transport systems, chemicals diffuse from one cell to another and may be carried by water of the physical environment in cavities or in the immediate vicinity of the organism. Those organisms with circulatory systems transport chemicals by diffusion through internal fluids.

Within the multicellular pattern of organization, there are certain groups of cells that are specialized for the production of controlling chemicals. Typically, the chemicals produced in one part of the organism elicit a response in an area far removed from the site of synthesis. In vascular plants, these chemicals are usually removed from the site of production to the site of action through the phloem; in higher animals such movement occurs through the circulatory system.

The chemical control substances to which reference has been made in the preceding paragraphs are called **hormones** in animals and **growth regulators** in plants. They are produced on a regular basis, in minute concentrations, by specialized cells in one part of the organism and exert their influences on another part of the organism. In animals, most hormones are synthesized by highly specialized structures (endocrine glands) and transported by the circulating blood, and they regulate a wide variety of activities. In plants, by contrast, the growth regulators are manufactured primarily in meristems and transported in phloem tissue, and they are concerned largely with the regulation of growth patterns.

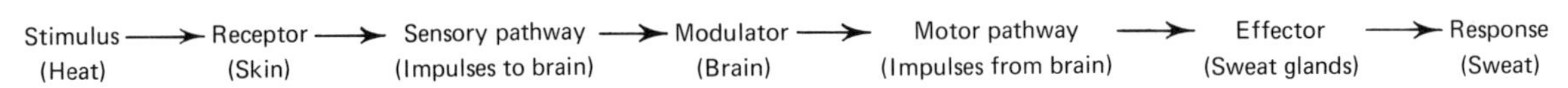

FIGURE 10-1. Homeostasis. In this simplified pattern, it can be seen that body temperature may be regulated by a series of reactions beginning with a stimulus and terminating with a response.

Chemical Control in Plants

Growth Regulators and Plant Growth

Most plant physiologists recognize three principal groups of growth regulators: auxins, gibberellins, and cytokinins. Prominent among these plant growth substances are the auxins, the first group of plant regulators to be discovered (1928) and the most intensively investigated group. The term **auxin,** meaning "to increase," is applied to a group of compounds characterized by their capacity to induce elongation in shoot cells. Although the major physiological activity of auxins is to promote growth, if present in excessive quantities they tend to inhibit growth. Various experimental data have demonstrated that the required auxin concentration for growth promotion varies with different plant organs, being much lower for roots than for stems.

Most data concerning the effects of auxins on plant growth have been obtained from various studies with seedlings of grasses, particularly oats (*Avena sativa*). In very young oat seedlings, the growing shoot is surrounded by a cylindrical sheath called a **coleoptile,** the first structure to rise above the ground after germination. As the growth of the seedling progresses, the coleoptile ceases to elongate and the first foliage leaf of the enclosed shoot breaks through it.

It has long been known that the extreme tip of the coleoptile is necessary for elongation of the coleoptile itself and the enclosed stem below it. As early as 1880, Charles Darwin noted that grass coleoptiles exhibit a characteristic curvature toward light (Figure 10-2A). When he shielded the tips of the coleoptiles, however, no bending occurred (Figure 10-2B) even though the lower portion of the coleoptile was still illuminated from one side. When the upper portion of the coleoptile was illuminated and the lower portion shielded, normal curvature took place (Figure 10-2C). Carrying the experimental procedure one step further, Darwin removed the tip of the coleoptile and found that there was no curvature or elongation (Figure 10-2D). The conclusion drawn from this series of experiments was that some substance was translocated from the tip of the coleoptile downward and caused curvature and elongation.

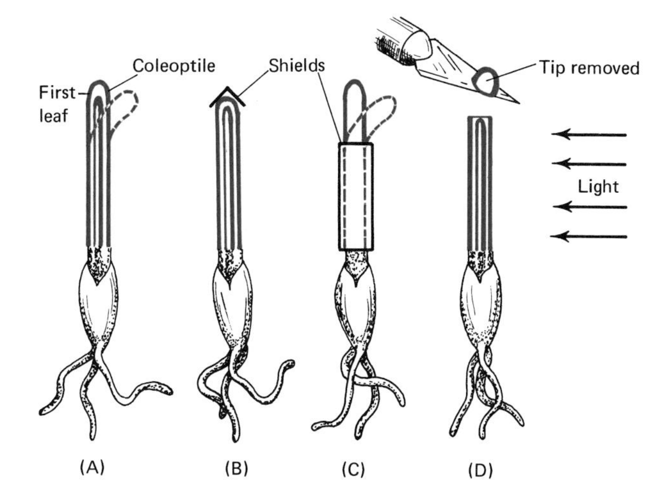

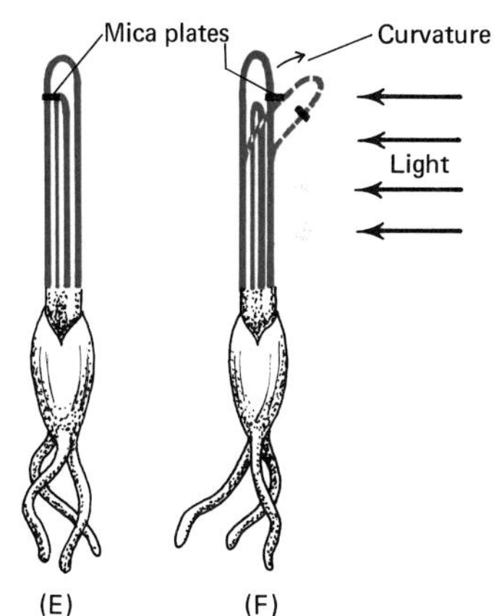

Figure 10-2. Experiments with *Avena* coleoptiles. (**A–D**) In the series of experiments conducted by Darwin it was demonstrated that the stimulus for growth is transmitted from the tip of the coleoptile downward. (**E, F**) Boysen-Jensen demonstrated that the growth stimulus is transmitted down the shaded side of the coleoptile.

Further experimentation on coleoptiles was carried out during 1910–13 by Peter Boysen-Jensen. He demonstrated that if a transverse incision was made on the nonilluminated side of the coleoptile and a plate of mica placed in the incision, no curvature took place (Figure 10-2E). When a similar incision and placement of the mica plate was made on the illuminated side, however, curvature did occur (Figure 10-2F). These responses indicated that the influence translocated from the coleoptile tip downward moved down the shaded side of the coleoptile.

Boysen-Jensen also demonstrated the site of production and diffusible nature of the growth substance in the coleoptile tip. To do this, he removed the tip, placed

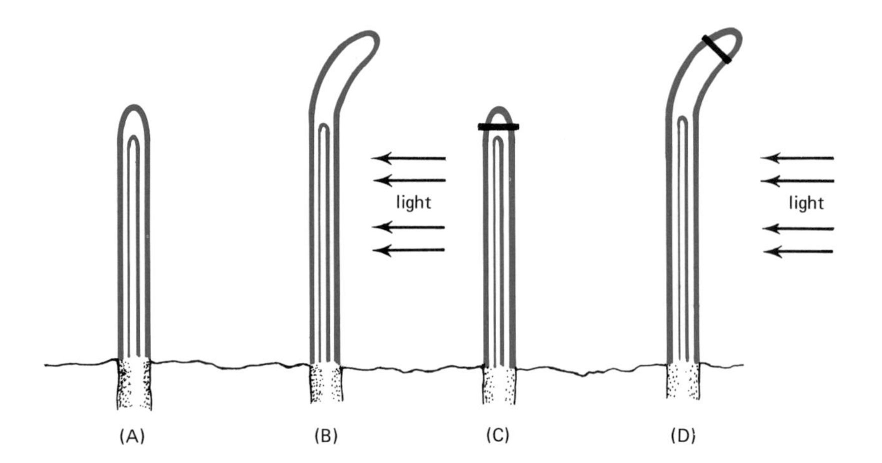

Figure 10-3. Boysen-Jensen's experiment. (**A, B**) The control (untreated) coleoptiles exhibit bending in the presence of light. (**C, D**) The coleoptiles are decapitated, gelatin is placed on the stump, and the tips are replaced on the gelatin. Even after treatment, the coleoptiles continue to bend toward the light. This experiment illustrates the diffusible nature of the growth substance.

gelatin on the stump, replaced the tip on the gelatin, and found that curvature and elongation took place (Figure 10-3). The growth substance diffused from the tip through the gelatin to the stump, thus maintaining the normal growth pattern. If the tip was not replaced, no curvature or elongation took place.

These early experiments, in conjunction with others, eventually led Fritz Went in 1928 to the discovery of the hormonal nature of this substance in coleoptiles. Went showed that the growth substance found in coleoptile tips could diffuse into agar, a gelatin-like polysaccharide, if excised tips are placed on the agar for about 2 hours (Figure 10-4A). If the excised tips are then discarded and small blocks of agar are placed on the decapitated coleoptile stumps, growth and curvature take place (Figure 10-4B). Controls in the form of pure agar show no effect (Figure 10-4C). Went noted that if a treated agar block was placed on one side of a coleoptile stump a curvature resulted, indicating that growth was more rapid along the side containing the agar block. This test, called the *Avena test,* was the basis for determining the amount of growth substance in the block of agar because the degree of curvature was found to be proportional, within limits, to the concentration of auxin in the agar block.

It was not until the mid-1930s that the growth sub-

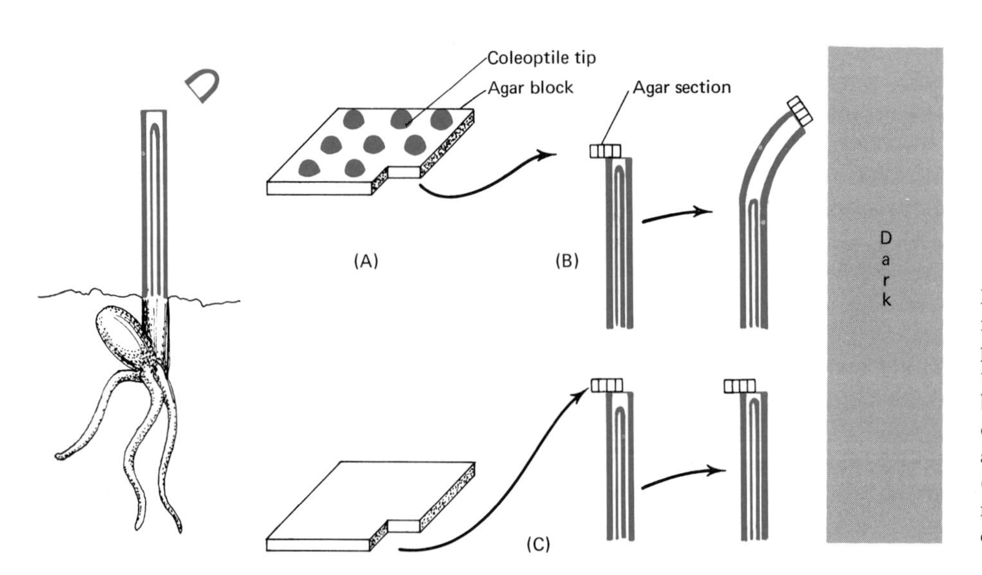

Figure 10-4. Went's experiment. (**A**) Coleoptile tips are placed on agar. (**B**) Excised tips are discarded and agar blocks are placed on the decapitated coleoptiles. Growth and curvature occur. (**C**) Plain (untreated) agar blocks show no effect on either the growth or bending of the coleoptiles.

stance in coleoptile tips was finally characterized as an auxin. The most commonly found naturally occurring auxin is *indole-3-acetic acid* (*IAA*). This compound, as well as other auxins, has been found to be present in growing organs such as foliage leaves, stems, flower stalks, germinating seeds, roots, fruits, and pollen grains.

Influences of Auxins

Among some of the plant activities that are influenced by the action of auxins are

1. Leaf abscission.
2. Fruit drop.
3. Fruit development.
4. Apical bud dominance.
5. Root development.

LEAF ABSCISSION. One of the important functions of auxins is believed to be the control of leaf fall. Leaves separate from most plants as a result of the formation of a weakened region of cells at the base of the petiole called the **abscission layer.** The relationship between abscission and auxins is shown by certain experimental data. If a leaf blade is excised, the petiole stump soon falls off because of the formation of an abscission layer. If, however, an auxin is applied to the cut end of the stump, abscission is delayed or prevented. Inasmuch as the leaf blade is one of the sites of auxin production, it can be concluded that auxin is one controlling factor in abscission.

It has been found that auxin content is high in young leaf blades as compared to the petiole, a condition conducive to leaf retention. As the leaf ages, however, the auxin content of the blade decreases to a point comparable to that found in the petiole, and the leaves abscise. There is evidence to suggest that the most important factor controlling abscission is the nature of the *auxin gradient* across the abscission layer, that is, the relative concentrations of auxin on the stem side and leaf side of the layer. If the gradient is steep, as when the auxin concentration is high on the leaf side of the layer and low on the stem side, no abscission occurs. When the gradient disappears, there is leaf abscission. When the gradient is reversed—that is, when the auxin concentration is high on the stem side of the abscission layer and low on the leaf side—abscission is accelerated.

FRUIT DROP. Fruits also fall from plants at an abscission layer as a result of an auxin gradient on either side of the layer. The fruit-growing industry makes practical application of these relationships. For example, in the spring, at the beginning of the growing season, the stems of fruit-bearing plants are sprayed with auxins, increasing the auxin concentration in the stems. This reversed gradient then causes premature fruit abscission, resulting in fruit "thinning." The fruits that do not drop are usually larger and better because of the availability of more metabolites. At some point later in the season, auxin may again be applied to the remaining fruits, setting up a steep gradient and delaying abscission and permitting longer tree ripening of the fruits (Figure 10-5).

FRUIT DEVELOPMENT. The development, as well as the drop, of fruits is related to auxin control. The growth and development of the ovary of a flower into a fruit are controlled by an adequate supply of auxins produced after a flower has been pollinated. The initial enlargement of the ovary is brought about by auxins supplied by the pollen grains and pollen tubes. Other auxins are produced by the embryo within the seed after fertilization has occurred. Auxins supplied through pollination and fertilization are necessary for fruits to develop fully.

On most plants fruits fail to develop unless flowers are pollinated. In some plants, however, this is not the case. It appears that the ovaries and other floral structures of such plants contain adequate supplies of auxins to promote fruit development. Inasmuch as pollination and fertilization do not occur in such plants, seedless fruits are developed. This phenomenon, called *natural parthenocarpy,* is characteristic of certain varieties of grapes, bananas, pineapples, figs, cucumbers, and oranges. In addition to natural parthenocarpy, man has also developed a commercially important process of *induced* or *artificial parthenocarpy.* In this process, flowers are prevented from pollinating by being sprayed with auxin solutions or treated with auxin pastes. As a result, seedless fruits develop. Common products of artificial parthenocarpy are certain varieties of watermelons, peppers, strawberries, and tomatoes.

APICAL BUD DOMINANCE. In most stems, the terminal bud is the dominant bud; that is, it grows more rapidly and is physiologically more active than lateral buds. The inhibition of lateral bud growth by terminal

(A) (B)

Figure 10-5. The effects of auxins on fruit drop. (**A**) Untreated Jonathan apple tree showing extensive fruit drop. (**B**) A Jonathan apple tree sprayed with an auxin (α-naphthaleneacetic acid, NAA) showing almost no fruit drop. [*USDA photos.*]

buds is called **apical dominance,** and the phenomenon is controlled by auxins. Such a correlation phenomenon may be noted by removing the terminal bud of a sunflower plant (Figure 10-6A). After removal, the plant loses its apical dominance and lateral buds undergo extensive growth into branches (Figure 10-6B). If, however, an auxin preparation is applied to the excised stem tip, lateral bud growth is inhibited and apical dominance is restored.

Root Development. The control of the development of adventitious roots is another important function of auxins. If a stem is removed from a plant and placed into moist soil or water, development of roots is caused by the translocation of IAA from the tip of the stem to the base (Figure 10-7). Such cuttings are commonly used in horticulture as a means of plant propagation. Cuttings of many species of hard-to-root plants form few or no adventitious roots unless supplied with adequate concentrations of auxins. In general, cuttings with leaves react more favorably to rooting than do cuttings devoid of leaves.

Influences of Gibberellins

Gibberellins, a second group of plant growth substances, were first discovered in 1941. These hormones are produced by a fungus, *Gibberella fujikuroi,* that

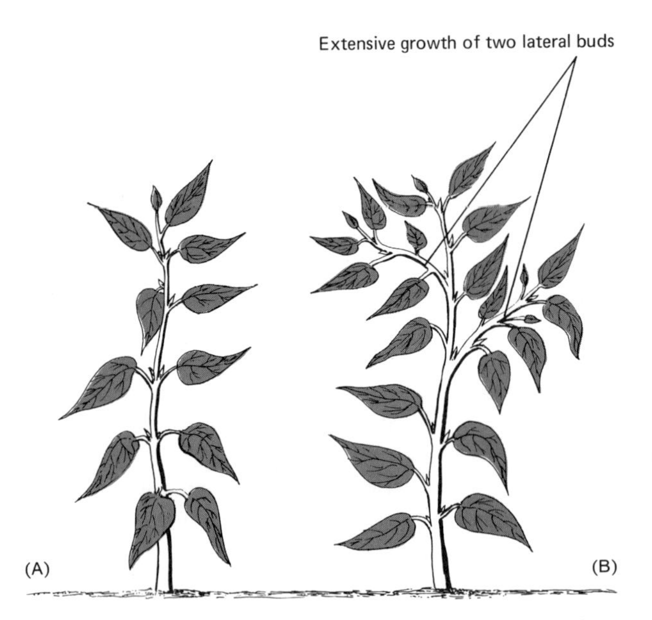

Figure 10-6. Apical dominance. (**A**) Normal sunflower plant exhibiting apical dominance. (**B**) Removal of terminal bud results in loss of apical dominance and extensive growth of two lateral buds into branches.

Not treated

(A)

Treated

(B)

FIGURE 10-7. The effects of auxins on root development of English holly (*Ilex aquifolium*). (**A**) Untreated plants fail to root. (**B**) Plants treated with auxin produce numerous adventitious roots. [*Courtesy of Dr. A. E. Hitchcock, Boyce Thompson Institute for Plant Research, Inc., Yonkers, N.Y.*]

causes a disease in rice grown in the Orient known as "foolish seedling disease." When infected with this fungus, rice seedlings grow rapidly in abnormal, erratic patterns. It was determined that the substance responsible for this abnormal growth was produced by the fungus and was named *gibberellic acid.* Gibberellins, compounds chemically related to gibberellic acid, appear to promote cellular elongation, particularly in stems. In some shoots, auxins and gibberellins interact in their effects on cell enlargement, but details of such interactions are still vague. Both hormones affect aspects of cell elongation, perhaps at different points in the overall process.

GENETIC DWARFISM. *Genetic dwarfism* is usually caused by the mutation of a single gene and is characterized by a shortening of the internodes of a plant. Among the plants that exhibit this type of dwarfism are pea plants (*Pisum sativum*), horse bean plants (*Vicia faba*), and scarlet runner bean plants (*Phaseolus multiflorus*). When gibberellin is applied to such mutant plants, genetic dwarfism is overcome. Two hypotheses have been offered to explain the influence of gibberellin on genetic dwarfism. One is that the application of gibberellin alleviates the shortage of *endogenous* (produced by the organism itself) gibberellin. Such a shortage could be caused by the lack of an enzyme (as a result of the mutation) involved in the synthesis of the hormone. The second hypothesis holds that an excess of a natural inhibitor is present in dwarf plants and that the effects of this inhibitor, which retards growth, are counteracted by the activity of gibberellin. The mechanism still is not fully understood.

OTHER INFLUENCES OF GIBBERELLINS. Aside from their influence on internode elongation, gibberellins are related to other physiological activities such as plant movements, flowering, and increased cambial activity. Many of the responses elicited by gibberellins are probably due to an interaction with auxin as well as other plant hormones. Others appear to be the result of gibberellins acting independently. Whether gibberellins and auxins act independently or together is a function of the species of plant under investigation, the conditions under which the plant is grown, and the particular response being studied. Much work remains to be done on this aspect of plant growth regulation before any conclusions can be reached.

Influences of Cytokinins

Cytokinins, a third group of plant growth substances, appear not only to influence cell enlargement but also to induce cell division. Shortly after the discovery of kinetin, its effects on cell division and enlargement were studied in a number of plant growth systems.

CELL DIVISION. It is believed that cytokinins have a role in the initiation of cell division. It has been demonstrated that isolated plant cells grown in the laboratory and supplied with the necessary minerals, foods, and auxins enlarge enormously but do not divide. If a cytokinin is applied to the cells, not only do the cells continue to enlarge, but they also divide. The mechanism by which cytokinins induce cell division is still unknown. Their interaction with auxin, however, is fairly well established.

Apical Dominance. In an earlier discussion, the influence of auxin on apical dominance was discussed. Removal of the apical bud stimulates the growth of lateral buds because the apical bud inhibits lateral bud growth. In a series of investigations it has been shown that if an intact shoot is soaked in a cytokinin solution, inhibition of lateral buds by the apical bud is to a large extent overcome. Other studies have shown that cytokinins have a stimulatory effect on lateral bud growth. It appears that the phenomenon of apical dominance may be under the influence of a balance between endogenous cytokinins and IAA.

Control of Flowering

The duration of daily exposure to light has significant effects on various developmental processes in plants, especially flowering. The response of plants to length of the daily period of illumination is referred to as **photoperiodism.** Normally, an angiosperm will flower only when the length of day falls within certain limits, a function of the season of the year. Based upon a 24-hour cycle of light and darkness, angiosperms are classified into three main groups.

1. Plants that flower when the day length is less than a certain critical length are called *short-day plants* (Figure 10-8A). Day lengths in excess of this critical value will keep a short-day plant vegetative. Common examples of short-day plants are ragweed, chrysanthemum, violet, aster, cocklebur, and strawberry. The critical day lengths for these, as well as other short-day plants, varies with the species.

2. *Long-day plants* flower after a critical day length is exceeded (Figure 10-8B). This critical day length differs from species to species. Some representative long-day plants are radish, spinach, beet, iris, lettuce, wheat, and clover.

3. *Day-neutral plants* (Figure 10-8C) flower after a period of vegetative growth, regardless of the photoperiod. Some examples of day-neutral plants are to-

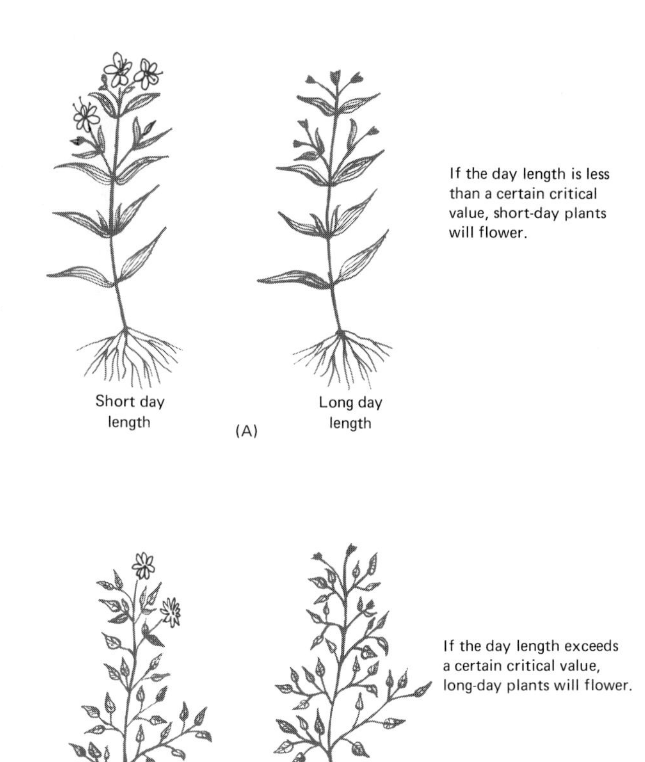

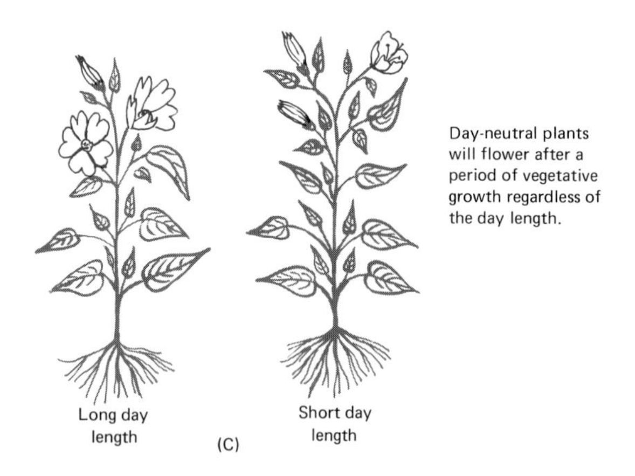

Figure 10-8. Photoperiodism. Based upon a 24-hour cycle of light and darkness, angiosperms may be classified as short-day (**A**), long-day (**B**), and day-neutral (**C**) plants.

mato, four-o'clock, cotton, dandelion, sunflower, and certain varieties of pea.

Attempts to elucidate the mechanism of photoperiodism have been only partially successful. However, several contributory factors are known. First, it has been determined that the leaves of flowering plants are the receptor organs for the transmitted light stimulus (Figure 10-9A). Plants in which leaves have been removed fail to flower, even though the proper photoperiod has been provided (Figure 10-9B).

Second, it has been suggested that hormones might be involved in the flowering process. One line of evidence to suggest the presence of a floral hormone was presented by Chailakhian in 1936. Using chrysanthemums (short-day plants) as his experimental organisms, he excised leaves from the upper part of the plants, leaving those on the lower half intact (Figure 10-9C). He then subjected the lower half to short-day photoperiods while simultaneously exposing the upper half to long-day photoperiods. Under these conditions, the plants

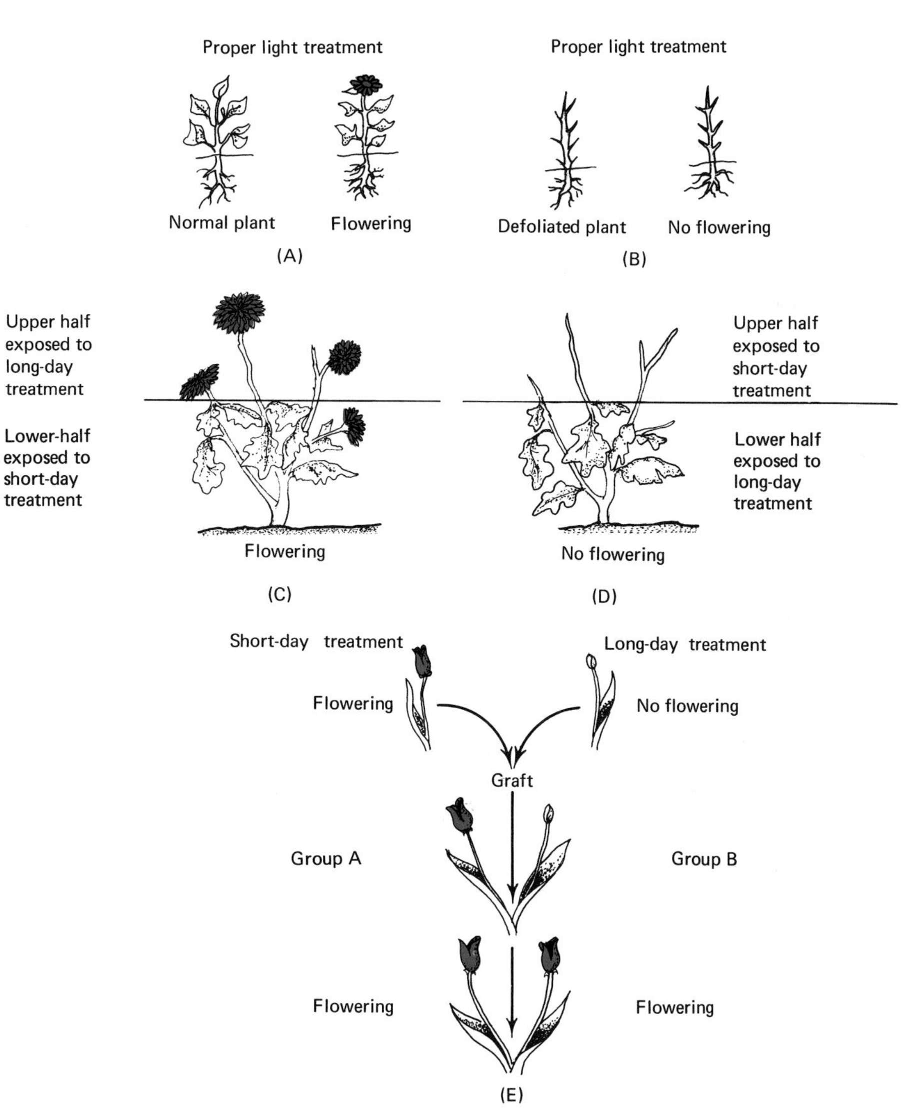

Figure 10-9. Various experiments concerning floral initiation. (**A, B**) Effects of defoliation on flowering: (**A**) plants with normal foliage and proper light treatment produce flowers. (**B**) Plant with excised leaves and proper light treatment fails to flower. (**C, D**) Chailakhian's experiment. (**C**) Chrysanthemums (short-day plants) from which the upper leaves were removed were exposed to long days while the bottom half was exposed to short days. This treatment induced flowering. (**D**) When the experimental procedure was reversed, no flowering took place. (**E**) Experiment showing that florigen may diffuse from one plant to a grafted partner. Both plants are short-day plants. The plant on the left has received a short-day photoperiod, while the one on the right has received a long-day photoperiod. When grafted together, both plants flower.

flowered. Next, he reversed the experimental conditions (Figure 10-9D). This time the lower half was exposed to long-day treatments and the upper portion received short-day treatments. The plants remained vegetative and did not flower. From these data he concluded that day length causes leaves to produce a hormone that moves from the leaves to the buds to induce flowering.

Evidence favoring the concept of a diffusible hormone has also been presented by various grafting techniques. One group of plants (short-day), group A, was grown under short-day photoperiods and produced flowers. A second group of plants (also short-day), group B, was subjected to long-day photoperiods and flowering was inhibited. If plants from both groups are grafted together, and if a short-day photoperiod is maintained for group A and a long-day photoperiod is maintained for group B, the plants exposed to short days (group A) will flower, and soon thereafter the plants exposed to long days (group B) will also flower (Figure 10-9E). The nonflowering (with regard to short- or long-day light requirements) group will produce flowers. These data indicate that a diffusible substance has moved down from its point of origin, the leaf, to another part of the plant, the meristems, where the flowering response is initiated. This flowering hormone, tentatively named *florigen,* has not yet been isolated. Ringing experiments (removal of the phloem) indicate that the flowering hormone is transported in the phloem, as are most organic substances.

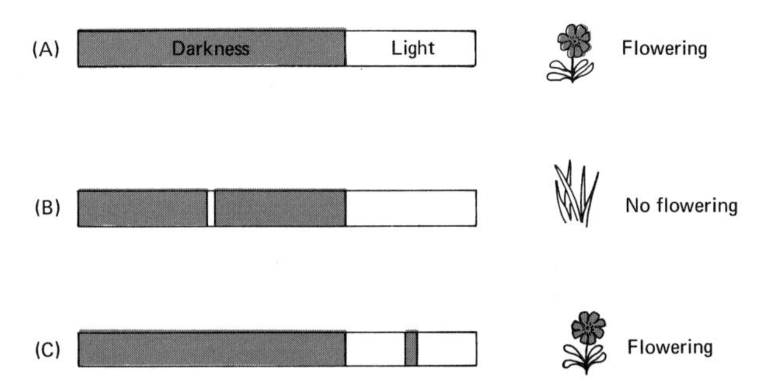

Figure 10-10. The importance of the dark period in short-day plants. (**A**) If a short-day plant is given a long dark period and a short light period, it flowers. (**B**) If the long dark period is interrupted in the middle by a brief light period, the short-day plant does not flower. (**C**) If the plant is given a long night period and a short light period and if the light period is interrupted in the middle by a brief period of darkness, it flowers. By contrast, long-day plants show no requirement for a dark period; they will flower in continuous light.

Intensive investigation of photoperiodism has revealed that the photoperiodic response of short-day plants is more of a response to the dark period than to the light regime. For example, if a long-day plant such as henbane (*Hyoscyamus*) is deprived of light for about 1 hour during the day, it still flowers, whereas a short-day plant such as cockelbur (*Xanthium*), if illuminated for a short period of time during the night, will not flower. It seems then, that a short-day plant is more properly termed a *long-night plant;* what it requires is an uninterrupted dark period of certain minimal length to flower (Figure 10-10). Long-day plants, or *short-night plants,* show no requirement for a dark period and will flower in continuous light. In fact, darkness tends to inhibit flowering in short-night plants.

Data concerning the importance of the dark period to floral initiation and the fact that flowering can be suppressed by a light break suggest that certain chemical reactions, sensitive to minute quantities of light, are in operation within the leaf. One of the initial problems was to determine what wavelengths of light are absorbed. This type of investigation was undertaken by a group of scientists at the United States Department of Agriculture, Beltsville, Maryland, in 1944. They were interested in determining the action spectrum for the inhibitory action of light breaks during the dark period. After exposing Biloxi soybeans (short-day plants) to light of different wavelengths for about 30 seconds in the middle of a long night, it was determined that red light (600 μ) is the most effective inhibitor of flowering in these short-day plants. Apparently, whatever happened during the long dark period was reversed by the flash of red light (Figure 10-11A).

In long-day plants, a brief treatment of red light acts in just the opposite way (Figure 10-11A). Under these conditions, interruption of the dark period stimulated flowering. Thus, both short- and long-day plants are sensitive to red light, but their responses are opposite to each other.

Further experimentation revealed that short-day plants responded to far-red light (730 μ) essentially as they did to darkness. When short-day plants were exposed to brief periods of far-red irradiation during the

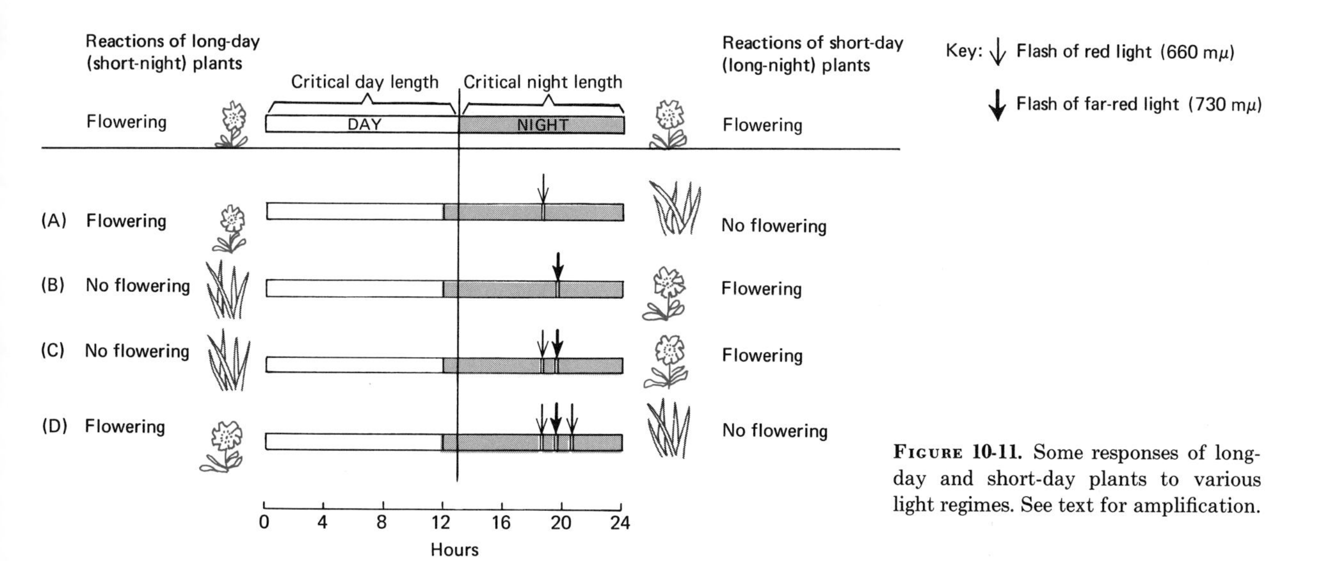

FIGURE 10-11. Some responses of long-day and short-day plants to various light regimes. See text for amplification.

long night, flowering was not affected (Figure 10-11B). Further, if brief far-red light followed red light, the effect of red light was negated and the short-day plants flowered (Figure 10-11C). Almost any number of successive flashes were used (red, far-red, red, far-red, and so on), and the final effect depended solely on whether the last flash given was red or far-red (Figure 10-11D). Not only did red and far-red light have opposite effects, but each reversed the effect of prior exposure to the other. Thus, the short-day plants did not flower if the last flash given was red light, but they did flower if the last flash given was far-red light.

The reactions of long-day plants to red and far-red irradiation were determined in the same manner. In this case, red light promoted flowering, and far-red light alone acted essentially as darkness (Figure 10-11A–D). Here again, red and far-red reversibility was found.

These experiments and many others on light effects have led investigators to conclude that plants contain a receptor pigment capable of existing in two forms. One form of the pigment absorbs red light, the other absorbs far-red light. This photoreceptive protein-like pigment is called **phytochrome.** When the red-absorbing form of phytochrome (P_R) absorbs red light, it is converted into the far-red-absorbing form (P_{FR}). Absorption of far-red light transforms P_{FR} into P_R. These relationships may be symbolized as follows:

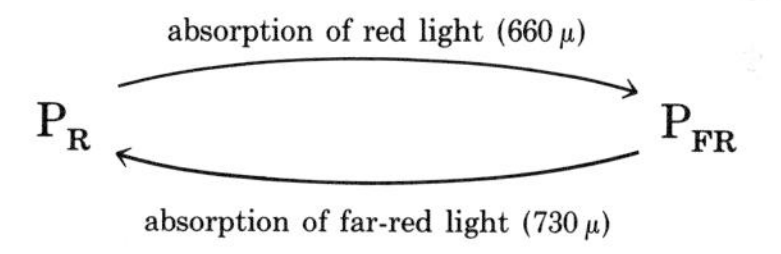

Further investigation showed that if the far-red did not immediately follow the red—that is, if the far-red flash is delayed for about ½ hour or more—the far-red treatment will not reverse the effect of the red light. This observation suggests that the pigment in the P_{FR} form is coupled to biochemical reactions that proceed immediately. P_{FR}, therefore, appears to be the active form of the pigment in the photoperiodic responses. The mechanism of these coupled biochemical reactions is currently under investigation.

It has been shown that if phytochrome is exposed to both red and far-red irradiation simultaneously, the red light dominates and the pigment is converted from the P_R form into the P_{FR} form. Sunlight contains both red and far-red wavelengths, so that during the day phytochrome exists in the plant principally in the P_{FR} form. There is also some evidence to suggest that there is a spontaneous dark conversion of P_{FR} to the P_R form. Based upon half-life reconversion studies, it is postulated that after a period of light, dark conversion of P_{FR} will proceed during the darkness at a rate such that approximately 3% of its initial activity as P_{FR} still remains after 10 hours of darkness. This time period closely approximates the range of critical night lengths for initiation of flowering in many plants.

Hormones and Plant Movements

Reactions and movements of plants usually go unnoticed because the rate of movement is too slow for observation. For this reason it is often thought that plants lack movement. With time-lapse photography, however, many plant reactions and movements may be noted. This special photographic technique makes it possible to observe flowers opening, buds unfolding, organs growing, seeds germinating, stomata opening and closing, stems curving and turning, and organs reacting to various stimuli.

Plant movements may be categorized into two principal types, autonomic and paratonic. *Autonomic movements* are spontaneous movements of a plant brought about by internal activities of the plant that are relatively independent of the environment. *Paratonic movements* are induced movements brought about as a result of stimuli in the environment. Examples of environmental stimuli are gravity, contact, light, heat, shock, and chemicals.

Both autonomic and paratonic movements are the result of two mechanisms, growth movements and turgor movements. *Growth movements* are irreversible movements in which the movement or reaction results from differences in growth rates of cells in different parts of an organ. These growth-rate differences, in turn, arise from unequal distribution of growth substances. *Turgor movements* are temporary and reversible movements that result from changes in the turgor pressure of certain cells. Such movements are usually more rapid than growth movements and may occur over and over again in the same organ.

Autonomic Movements. Aside from growth, which is the principal type of autonomic movement, several other types of autonomic movement may be distinguished. The most common is *nutation.* This is a back-and-forth motion of the tip of a growing stem or other organ caused by alternately changing growth rates on opposite sides of the organ (Figure 10-12A). Although nutation is most obvious in the stems and tendrils of climbing plants, it has also been observed in leaves, roots, flower stalks, and runners. Other autonomic growth movements include *circumnutation,* a

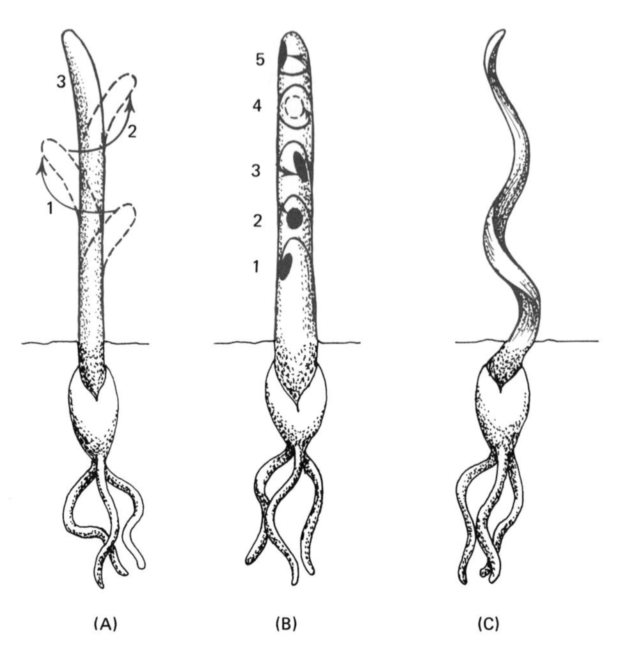

Figure 10-12. Autonomic plant movements: nutation (**A**), circumnutation (**B**), and twining (**C**). These spontaneous movements are brought about by internal activities of the plants, and they occur so slowly that they are not directly observable over a short period of time.

rotational growth of an elongating shoot around its long axis (Figure 10-12B), and *twining,* the spiral orientation of a stem (Figure 10-12C).

Paratonic Movements. Paratonic movements are brought about as a result of external stimuli. Some paratonic movements are such that the direction of movement is independent of the direction from which the stimuli are received. These movements are called **nasties.** Other paratonic movements are such that the direction of movement is determined by the direction from which the stimulus is received. Such paratonic movements are called **tropisms.**

Nasties. Nasties are responses of organs such as leaves, flower petals, bud scales, and other flattened structures of plants. The principal nastic movements are those caused by changes in light intensity (*photonasties*), temperature changes (*thermonasties*), touch (*thigmonasties*), and alteration of day and night (*nyctinasties*).

The fluctuations of light and temperature as well as the alternation of day and night produce various nastic movements, especially in leaves and flowers. Many flowers, such as morning glory, tulip, and dandelion, are either completely or partially opened during the day but close at night. By contrast, flowers of other plants, such as four-o'clock and tobacco, close during the day under the influence of intense illumination and open in light of low intensity. The rapid opening of certain flowers when brought into a warm room from a cold place is an example of a thermonastic response.

One of the most common examples of nyctinastic movements are the "sleep" movements of leaves in which the leaves of certain plants are horizontally oriented during the day and vertically oriented at night (Figure 10-13A). Another prominent and fairly well-

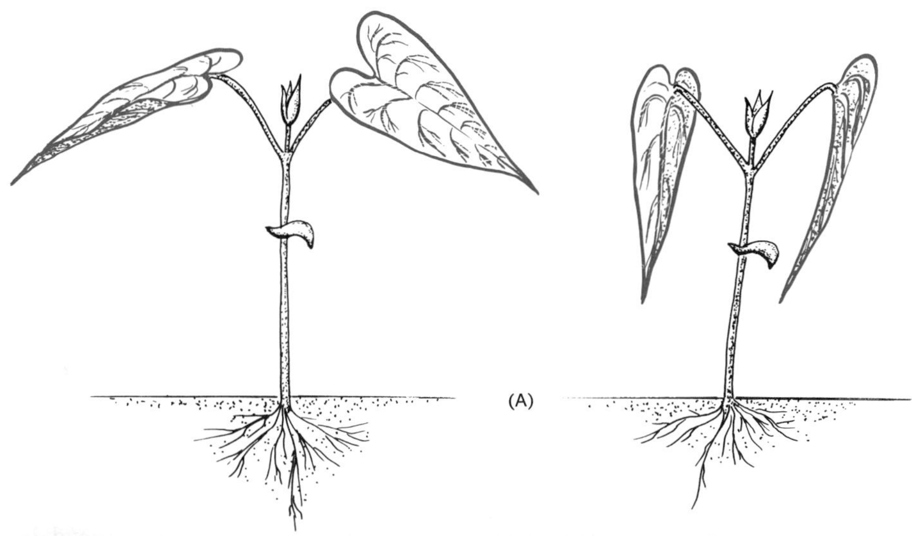

FIGURE 10-13. Paratonic movements. (A) Nyctinastic "sleep" movements of bean (*Phaseolus vulgaris*) leaves. *Left:* Horizontally oriented position of the leaves during the day. *Right:* Vertically oriented leaves at night. (B) Thigmonastic response of the sensitive plant, *Mimosa pudica. Left:* Normal position of the leaflets and the petioles. *Right:* Response of the leaflets and petioles to touch. [*USDA photos.*]

known nastic movement is the thigmonastic response furnished by the sensitive plant (*Mimosa pudica*). This plant, indigenous to tropical and subtropical regions, is extremely responsive to touch and shock. When stimulated, the leaflets fold upward in pairs and the main petiole drops (Figure 10-13B). If a very strong stimulus is applied to one leaflet, it may be transmitted throughout the plant, the leaflets folding and dropping one after another.

Tropisms. Whereas nastic movements may be due to either growth or changes in turgor, tropisms are typically growth movements. Tropisms are plant responses in which the direction of movement is determined by the direction from which the stimulus originates. Depending upon the nature of the stimulus, a tropism may be called phototropism (light), geotropism (gravity), or thigmotropism (contact), and each tropism may be either positive or negative.

The response of a plant organ to light is called *phototropism.* Leaves and stems exhibit positive tropism—that is, they bend toward light—whereas roots show no response to light or exhibit a negative response by bending away. A stem bends because light reduces the auxin concentration on the lighted side, and the shaded side, having a higher auxin concentration, grows faster; this unequal distribution of growth is responsible for the bending toward light (Figure 10-14A). Owing to phototropic movements, the petioles of leaves bend in such a way that leaf blades orient themselves at right angles to the source of illumination. In such an arrangement, called a *leaf mosaic* (Figure 10-14B), few leaves shade others and few spaces between leaf blades are left unfilled.

Geotropism is the response of a plant organ to gravity. Roots exhibit a positive geotropism, whereas stems react negatively to gravity. This can be readily seen if germinating seeds are placed in vertical, horizontal, and inverted positions (Figure 10-14C). In each case, regardless of the orientation of the seeds, stems grow upward and roots grow downward. This growth movement also involves an unequal auxin distribution. Because of gravity auxin accumulates in greater concentration on the lower sides of both stems and roots grown horizontally. The increased auxin concentration on the lower part of the stem causes a greater rate of growth on that side and the stem bends upward. According to this reasoning, roots in the horizontal position should also grow upward. However, this is obviously not the case. It has been found that a high auxin concentration promotes growth on the lower side of stems but inhibits growth on the lower sides of roots. The lower concentration on the upper side of roots is more favorable for

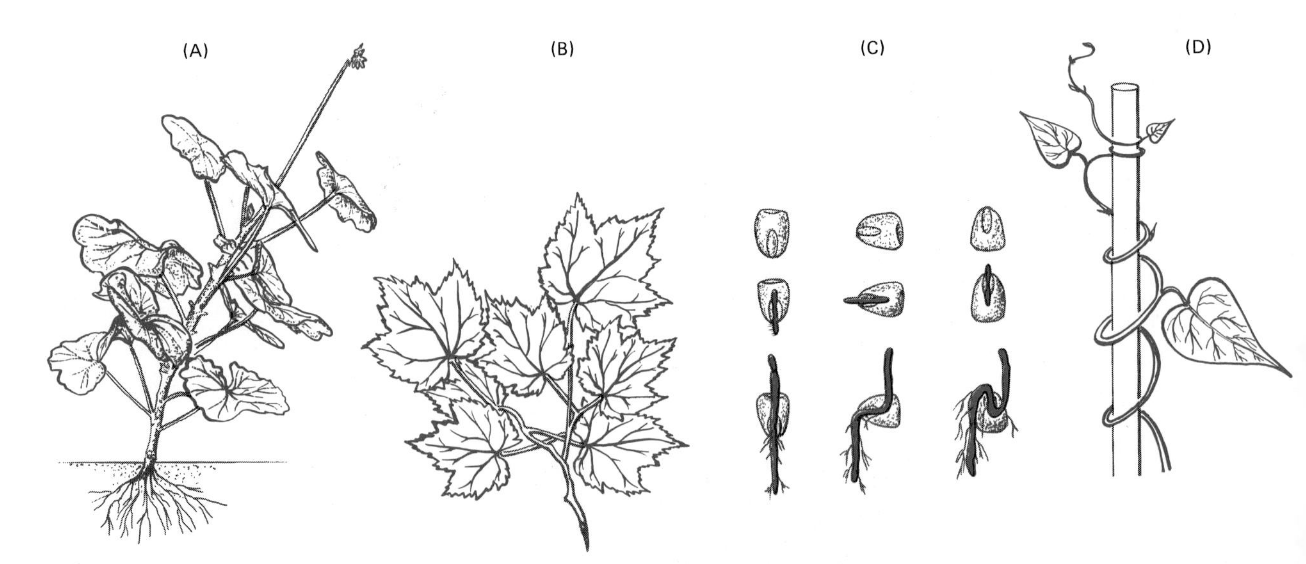

Figure 10-14. Plant tropisms. (**A**) Phototropism in a geranium plant. (**B**) Leaf mosaic formed in response to light. (**C**) Geotropism in germinating corn seeds. (**D**) Tendrils exhibiting thigmotropism.

root growth and the root bends downward. Such a reaction of plant tissues to growth substances is not atypical, for in many instances, high hormone concentrations may inhibit growth.

Thigmotropism is a growth movement in response to contact with a solid object. It is most readily apparent in tendrils of climbing plants such as members of the pea, grape, and passion-flower families. Once a tendril has made contact with a solid object, it bends or spirals because of increased growth on the side opposite to the point of contact (see Figure 10-14D).

Chemical Controls in Animals

Studies on invertebrates and lower vertebrates suggest that chemical integration by hormones and related substances is an overall phenomenon in the entire animal kingdom and that hormonal influences may be exerted during developmental stages as well as in adult organisms. **Hormones** are usually defined as chemical regulators that are synthesized by certain parts of the organism (generally specialized ductless glands) and carried by the blood or circulatory fluids from their site of production to other parts of the body (target tissues or organs) where, in extremely minute quantities, they elicit systemic adjustments. These adjustments typically consist of the stimulation or inhibition of a particular function; in general, hormones do not initiate a specific activity by producing any new cellular reaction or metabolic transformation.

The activities of the organism that require rapid coordination are usually controlled by the nervous system and such coordinations can take place in fractions of a second or minutes. Since hormones are generally conveyed from production sites to target areas via the circulating blood, and since they must also diffuse from the blood through intercellular fluids to reach their targets, they usually regulate processes that require duration rather than speed. Thus, processes such as growth, reproduction, regeneration, blood chemistry, molting, pigmentation, and metabolic rate may require several minutes to several months in order to be completed.

Chemically, hormones are usually steroid, peptide, protein, or amine in composition and based upon their general roles may be viewed as metabolic, trophic (regulatory), and morphogenic. *Metabolic hormones* are those that stimulate or retard certain metabolic activities of the organism. *Trophic hormones* affect the rate and secretion of other glands. *Morphogenic hormones* affect the rate and development of various parts of the organism. The production of these hormones seems to be regulated primarily by certain substances carried in the blood. In some instances, the regulatory substances are other hormones; in others, they are simple compounds such as sugars; while in still others, the stimulatory substance is an acid. It must be stressed that regardless of the generalized roles of hormones or the substances that stimulate their production, hormones within an organism form functionally interlocking systems; one hormone never acts in isolation. It appears that almost every chemical homeostatic adjustment is effected by a balance between hormones acting together or in a sequence.

The mechanism by which a hormone actually performs at the molecular level represents one of the most active areas of current biological investigation. Although the nature of the mechanism has not yet been elucidated, there is considerable evidence that hormones, in one way or another, modify enzymatic reactions within target cells. Since hormone molecules exist in many sizes and shapes, ranging in chemistry from modified amino acids to steroids to complex proteins, it has not been clearly established that all employ similar mechanisms in eliciting their actions, although it is quite possible that they act through allosteric actions. At present, there are three hypotheses to explain the mechanism of hormone action.

1. Hormones exert a direct effect upon intracellular enzymes by interacting with them as coenzymes.
2. Hormones act to control permeability relationships at the cell surface or elsewhere and thus indirectly affect enzymatic reactions.
3. Hormones may produce their effects directly by activating or suppressing the action of particular genes.

A cell responds to a given hormone only after the hormone is bound to a specific receptor site on the cell membrane. If the hormone and the receptor form a new entity that is capable of generating some type of signal, the action of a hormone may be viewed as a specific chemical signal producing a specific adaptive response by the cell. However, more recently, a nonspecific mechanism has been implicated in a large number of metabolic responses to hormones. This mechanism involves the production of one of the adenine nucleotides *called cyclic AMP* (3′,5′-adenosine monophosphate). Cyclic AMP is formed from ATP by the catalytic action of adenyl cyclase.

$$\text{ATP} \xrightarrow{\text{adenyl cyclase}} \text{cyclic AMP} + \text{PP}_i$$

Many hormones appear to act on their target cells by activating the enzyme adenyl cyclase, which is associated with binding sites on the cell membrane that can accept only the specific hormone. Cells respond to increased levels of cyclic AMP with reactions that are characteristic of the particular tissue involved. Thus, specificity of the cellular response to hormones is ensured mainly by the particular binding sites that initially accept the hormone. Several hormones may elicit a similar response in a tissue. For example, lipolysis in fat cells may be induced by ACTH, glucagon, and epinephrine. But the selective sites for binding of different hormones are different in each case.

Invertebrate Hormones

Among invertebrates, evidence of hormones has been found in flatworms, annelids, coelenerates, mollusks, echinoderms, and arthropods. In flatworms, hormones have been isolated and have been shown to affect color change, locomotor activity, water content, and the discharge of gametes. Extracts from starfishes contain a substance that induces the release of gametes from gonads and another compound that inhibits the shedding of eggs. Among annelids, compounds have been found that control the maturation of the gonads, body changes related to reproduction, and regeneration of posterior segments. In arthropods, the largest animal phylum, a large number of hormones involved in myriad activities have been found. For example, in crustaceans, chemical substances have been implicated in the regulation of retinal pigments, coloration, molting (shedding of the exoskeleton), reproduction, heart acceleration, and metabolism. Among insects, hormones control activities such as metamorphosis, molting, heart rate, reproduction, water balance, and color changes.

As an example of hormonal control in invertebrates, we will consider the metamorphosis of the Cecropia (silkworm) moth. Most insects undergo several distinct changes in form during their development from egg to adult; such a series is called **metamorphosis.** Insects such as butterflies, moths, beetles, and flies undergo a *complete metamorphosis;* that is, they pass through four stages in their development.

1. Egg.
2. Larva.
3. Pupa.
4. Adult.

The **larvae** that emerge from eggs are segmented and worm-like and are designated by different names depending upon the kind of insect. For example, the larva of a fly is called a maggot, that of a beetle is termed a grub, that of a moth is designated as a caterpillar, and so on. After a period of feeding and rapid growth, the larva enters the **pupa** stage. In this stage, all the tissues

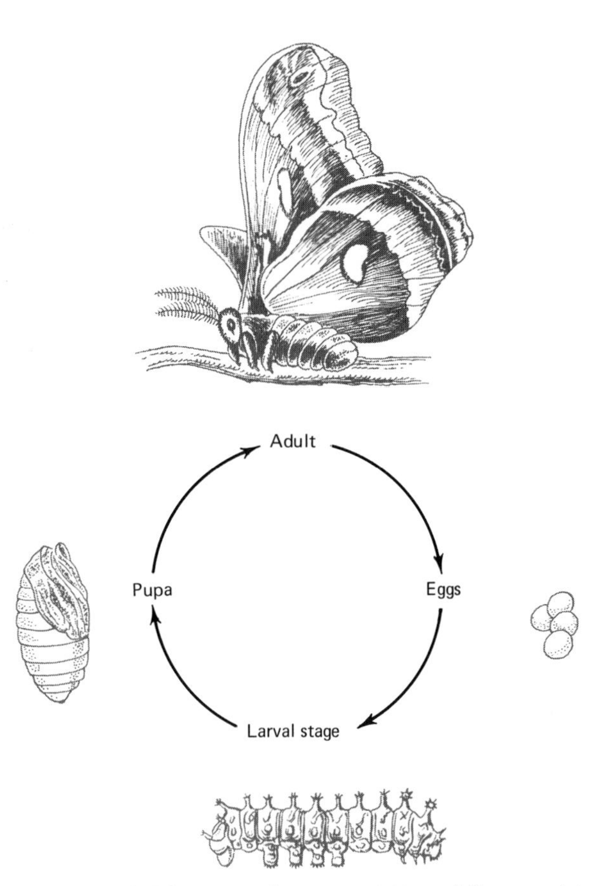

Figure 10-15. Diagrammatic representation of the complete metamorphosis of the Cecropia moth. See text for discussion.

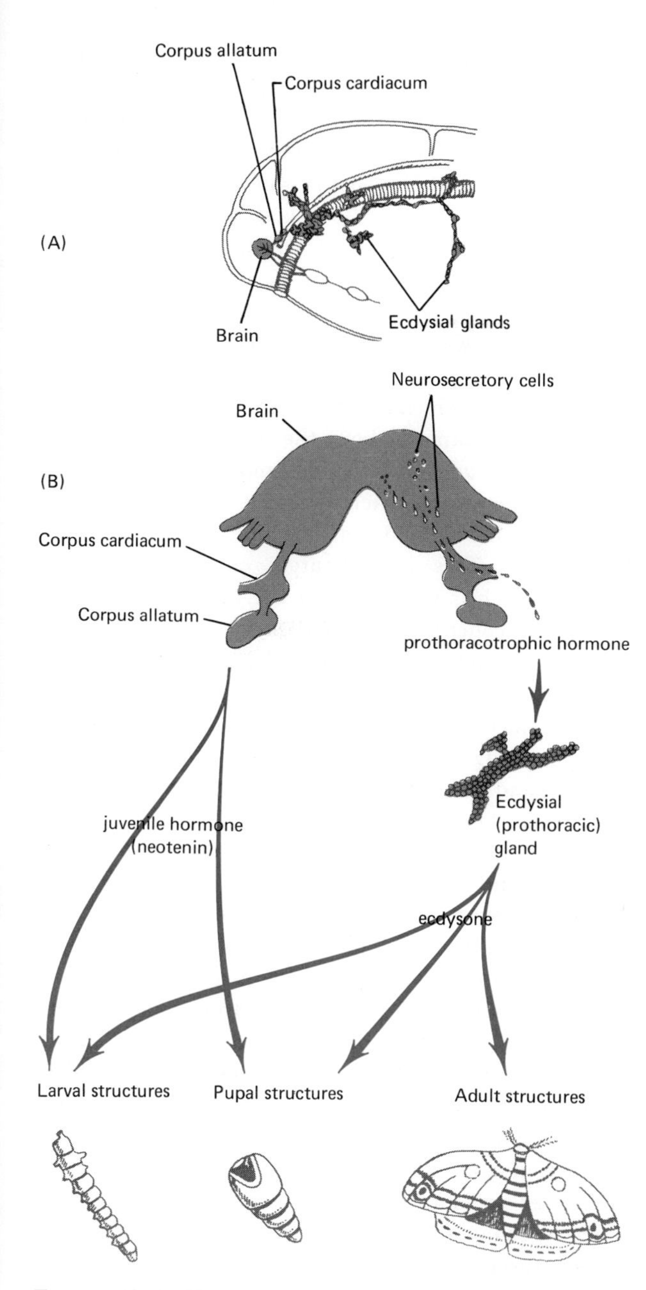

Figure 10-16. Hormonal control of metamorphosis in the Cecropia silkworm. (**A**) Anatomical relationships of the brain, corpus cardiacum, corpus allatum, and ecdysial glands in the head of a larva. (**B**) Interactions of hormones involved in metamorphosis. See text for details. [**A** *modified from C. M. Williams,* Biological Bulletin, *vol. 90, p. 234, 1946; copyright © 1946.* **B** *modified from H. A. Schneiderman and L. I. Gilbert,* Science, *vol. 143, pp. 325–33, Jan. 24, 1964, Fig. 1; copyright © 1964 by the American Association for the Advancement of Science.*]

of the larva are transformed into those of the adult. The complete metamorphosis of the Cecropia moth is shown in Figure 10-15.

Metamorphosis of the pupa into an adult in the silkworm moth results from interaction of two hormones (Figure 10-16). In nature, the low temperatures of the winter are required to terminate the rest period that precedes metamorphosis. After a period of chilling, a growth and differentiating (metamorphosis) hormone, *ecdysone,* is released by the *ecdysial* (*prothoracic*) *glands.* Its production is triggered by another hormone, *prothoracotropic hormone,* produced in the brain. This hormone is carried in nerve axons to the *corpora cardiaced* behind the brain, is released into the blood, and then travels to the ecdysial glands. In early larval life, the *corpora allata* secretes a hormone, *neotenin* (*juvenile hormone*), which inhibits the action of ecdysone. Neotenin favors larval characteristics and tends to delay molting. Removal of the corpora allata from an insect larva can induce molting (or shorten the intermolt period) and metamorphosis (premature), leading to a small adult. Implantation of additional corpora allata delays molting, allows for a longer growth period, and when the next molt takes place, metamorphosis is suppressed and the insect emerges as a giant adult. Apparently neotenin suppresses the action of ecdysone in early larval life; prior to pupation, however, secretion of neotenin decreases so that ecdysone can act.

Vertebrate Control

Among vertebrate organisms especially, the endocrine system is an important adjunct to the nervous system in bringing about coordination of the most complex of physiological activities. The *endocrine system* consists of a series of interrelated glands or body areas that secrete hormones. These hormones are important in growth, development, general metabolic activities, and emotional attitudes. Despite these varied functions, the glands that produce the hormones have several commonalities. For example, all consist of epithelial tissue; most are relatively independent of the nervous system; they possess a rich blood supply; they are provided with nutrients directly from the blood stream; and they pour their secretions directly into the blood or lymph vessels. Inasmuch as they have no specific ducts for the reception of nutrients or for the distribution of their hormones, they are also referred to as *ductless* (*endocrine*) *glands,* or glands of internal secretion.

Having considered some of the principal aspects of hormones and hormone-producing structures, we will now examine the endocrine system of vertebrates, especially that of the human. In this context, attention will be directed to the location and structure of the endocrine glands, their relationships to other glands, and some of the abnormalities that result if the glands are not secreting properly.

Vertebrates have seven clearly recognizable hormone (or hormone-like) producing regions (Figure 10-17). These are the secretory cells of the gonads (ovaries and testes), pancreas, upper gastrointestinal tract (stomach and intestines), thyroid, parathyroids, adrenals, and pituitary. In addition, the placenta secretes hormones during pregnancy, and the pineal body and thymus gland also produce hormone-like substances. The thymus has been shown to produce a chemical (or chemicals) that develops immunological capacity (see Chapter 15).

Hormones of the Gastrointestinal Region: Control of Digestion. Although the stomach and intestines are not ordinarily classified as endocrine glands, they do elaborate a number of chemicals, usually called hormones, that regulate the motor and secretory activities of the digestive organs. These hormones are released into the blood and their actions supplement those of the autonomic nervous system (Chapter 11). Although the activities of other endocrine glands of the body are interrelated to form a system of checks and balances, those of the gastrointestinal tract are not. They are not known to be influenced by other glands or to be integrated among themselves. The secretion of these hormones is conditioned largely by the presence or absence of substances in the gastrointestinal tract.

Among the hormones produced by the gastrointestinal area are *secretin, cholecystokinin–pancreozymin, enterogastrone* (mucosa of the duodenum), and *gastrin* (pyloric mucosa of the stomach). The relationship of these hormones to the digestive system is shown in Figure 10-18 and their functions are outlined in Table 10-1. Further discussion of these hormones is found in Chapter 13 in conjunction with the digestive system.

The Pancreas: Control of Sugar Metabolism. The mammalian **pancreas** is a compound gland containing both exocrine and endocrine tissues. The exocrine portion secretes digestive enzymes, which are poured into the duodenum via the pancreatic duct. The endocrine portion consists of clusters of cells, the *islets of Langerhans,* scattered among the enzyme-secreting cells. Microscopically, the islets reveal at least two types of cells: alpha and beta cells (Figure 10-17). The alpha cells, which are less abundant and tend to be peripherally arranged about the islet, produce a hormone called *glucagon.* This hormone stimulates cyclic AMP production that activates the enzyme *phosphorylase,* which breaks down liver glycogen to glucose-1-phosphate. This is ultimately converted to free glucose, which is secreted into the bloodstream, thus raising the concentration of sugar in the blood. The beta cells, which are more abundant and granular in appearance, secrete the hormone *insulin.* This hormone acts to reduce the concentration of glucose in the blood by stimulating reactions that store it in the liver as glycogen and by increasing cellular uptake of glucose by many cells. Thus, both hormones play a major role in maintaining blood glucose within narrow limits (about 100 mg/100 ml of blood in mammals).

The Gonads: Control of Reproduction. The primary function of the **gonads** (ovaries and testes) is the production of eggs and sperm (see Figure 10-17). In addition, they have important endocrine activities. The testes produce several hormones called **androgens,** the most active being *testosterone,* which stimulates the development and maintenance of secondary sex characteristics, the accessory organs of reproduction, and behavior in all male vertebrates. Since androgens promote anabolic reactions, they are also taken by many athletes to increase muscular strength. The ovaries produce **estrogens,** one of the most important being a compound called *estradiol.* Like testosterone, this hormone stimulates the development and maintenance of secondary sex characteristics and or-

Figure 10-17 [opposite]. Endocrine glands of the human body in terms of anatomical relationships, macroscopic appearance, and histological structure. Consult text for discussion.

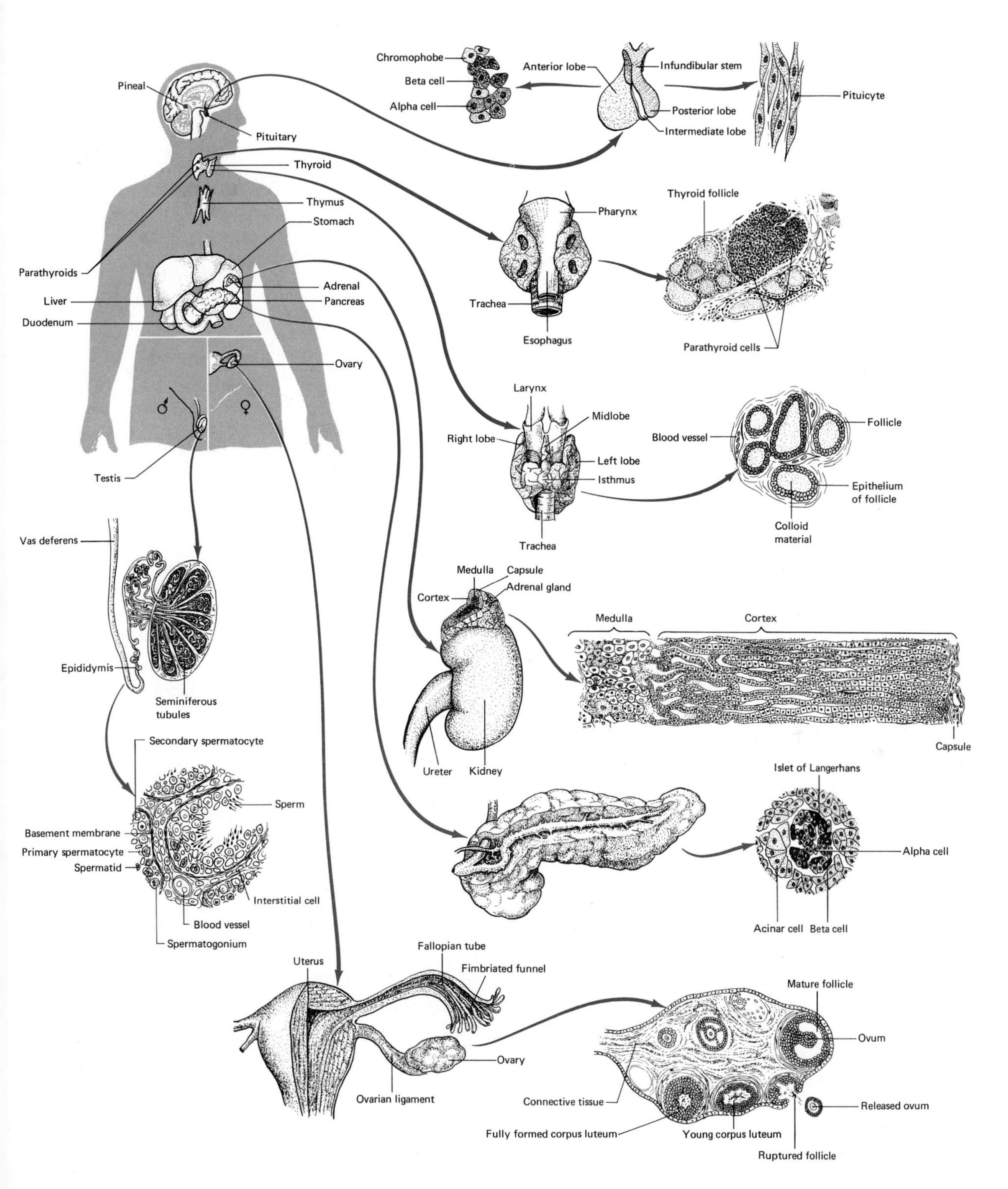

Pineal
Pituitary
Chromophobe
Beta cell
Alpha cell
Anterior lobe
Infundibular stem
Posterior lobe
Intermediate lobe
Pituicyte
Thyroid
Thymus
Stomach
Parathyroids
Liver
Adrenal
Pancreas
Duodenum
Ovary
♂
♀
Testis
Pharynx
Trachea
Esophagus
Thyroid follicle
Parathyroid cells
Larynx
Midlobe
Right lobe
Left lobe
Isthmus
Trachea
Blood vessel
Follicle
Epithelium of follicle
Colloid material
Vas deferens
Epididymis
Seminiferous tubules
Medulla
Capsule
Adrenal gland
Cortex
Medulla
Cortex
Capsule
Ureter
Kidney
Secondary spermatocyte
Sperm
Basement membrane
Primary spermatocyte
Spermatid
Interstitial cell
Blood vessel
Spermatogonium
Islet of Langerhans
Alpha cell
Acinar cell
Beta cell
Uterus
Fallopian tube
Fimbriated funnel
Ovary
Ovarian ligament
Mature follicle
Ovum
Connective tissue
Released ovum
Fully formed corpus luteum
Young corpus luteum
Ruptured follicle

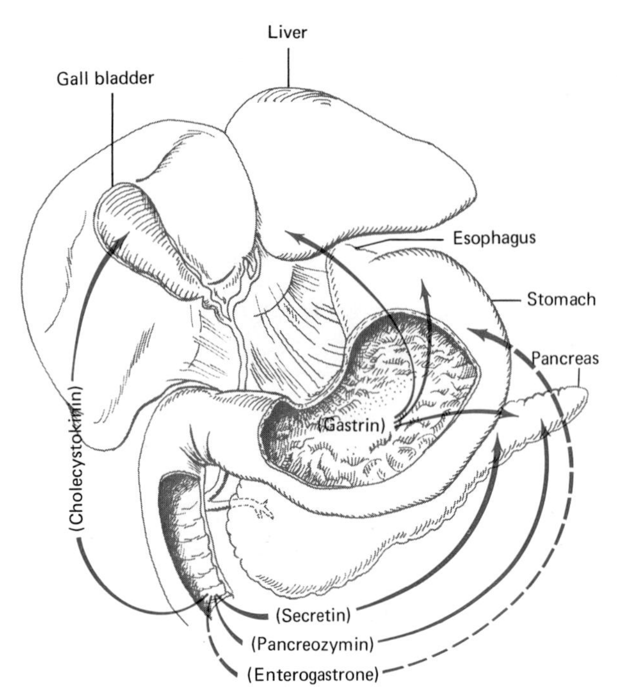

FIGURE 10-18. Source and action of gastrointestinal hormones. The arrows extend from the sources of the hormones to the target organs affected. Solid-line arrows indicate stimulation of the target organ, while the broken-line arrow indicates inhibition. (Pancreozymin and cholecystokinin are apparently the same chemical substance.)

gans in females. A second female hormone, *progesterone,* is necessary for ovulation, the completion of each menstrual cycle, implantation of the fertilized egg in the uterine wall, and development of the breasts during the latter months of pregnancy.

THE THYROID GLAND: CONTROL OF METABOLISM. All vertebrates possess a pair of glands, the **thyroids,** located in the neck. The human thyroid consists of two lobes that lie on either side of the trachea and are usually connected by a narrow isthmus of tissue that passes in front of the trachea (see Figure 10-17). Variations in size of the gland exist with differences in sexual development, diet, and age.

One hormone secreted by the thyroid is *thyroxin.* The synthesis of this hormone involves a series of reactions in which iodine is trapped and removed from the circulatory system and incorporated into the protein *thyroglobulin* which is then hydrolyzed into the active hormone thyroxin. The hormone diffuses into the bloodstream and then to cells where it functions in controlling the oxidative processes of all body tissues. The secretion and production of thyroxin is controlled by a hormone produced by the pituitary gland called *thyroid-stimulating hormone* (*TSH*). The secretion of TSH is in turn regulated by the amount of thyroxin circulating in the blood; it is decreased by a high concentration of thyroxin. If there is a decreased production of thyroxin, this stimulates the pituitary to release TSH which passes to the thyroid and stimulates the secretion of thyroxin. When the concentration of thyroxin reaches its normal level, there is a cutback in the release of TSH. Such a negative feedback system between the thyroid and pituitary controls the output of thyroxin so that it is maintained at a fairly constant level. Through its effect on metabolism, thyroxin has a pronounced effect on growth and development.

Hyposecretion of thyroxin has varied influences in young and adult organisms. In developing frogs (tadpoles), for example, if the thyroid is surgically removed, they will not metamorphose into frogs even though they continue to grow. In humans, if a child has a hypoactive thyroid, a condition called **hypothyroidism (cretinism)** results. Growth and development are seriously retarded, resulting in dwarfism, a protruding abdomen, underdeveloped sex organs, mental deficiency, puffy skin, swollen tongue, occasional deafmutism, and a low metabolic rate. Cretinism is readily treated by the administration of thyroxin if it is diagnosed in its early stages. If the thyroid becomes hypoactive in older children or adults, **myxedema** results. The conspicuous symptoms of this abnormality are a low metabolic rate, reduction in mental activity and physical vigor, increase in weight due to the accumulation of body fat, waxy and puffy skin due to the deposition of mucous fluid in the subcutaneous tissues, lowered heartbeat and body temperature, increased susceptibility to cold, and thinning of hair. Patients with myxedema also respond well to treatment with thyroxin.

In instances where hypothyroidism is caused by a lack of iodine for thyroxin synthesis, the gland tends to compensate for this insufficiency by enlarging. The resulting enlargement, which may be a barely detectable swelling or a conspicuous, disproportionate mass weighing several pounds, is termed a **simple goiter.** This

TABLE 10-1. Vertebrate Hormones and Their Functions

Site of Production	Hormone Produced	Physiological Effect
Gastrointestinal region		
Pyloric mucosa of stomach	Gastrin	Stimulates production of gastric juice.
Duodenal mucosa	Secretin	Stimulates flow of pancreatic juice but not enzymes.
	Cholecystokinin-pancreozymin	Evacuation of bile by gall bladder; stimulates pancreas to produce enzymes.
	Enterogastrone	Inhibits secretion of gastric juice.
Pancreas		
Beta cells	Insulin	Increases metabolism of glucose to CO_2 and H_2O; decreases blood sugar concentration; stimulates formation and storage of glycogen.
Alpha cells	Glucagon	Stimulates conversion of glycogen into glucose.
Testes		
Interstitial cells	Testosterone	Stimulates development and maintenance of secondary sex characteristics in male and regulates sexual behavior.
Ovaries		
Ovarian follicle	Estradiol	Stimulates development and maintenance of female secondary sex characteristics and behavior.
Corpus luteum	Progesterone	Stimulates changes associated with pregnancy and childbirth.
Thyroid	Thyroxine	Stimulates oxidative metabolism in all body tissues.
Parathyroids	Parathyroid hormone (PTH)	Regulates calcium and phosphate metabolism.
Adrenals		
Cortex	Glucocorticoids (cortisone, corticosterone, hydrocortisone)	Stimulate conversion of amino acids and fats into glucose; stimulate formation and storage of glycogen; help regulate normal blood sugar level.
	Mineralocorticoids (aldosterone, deoxycorticosterone)	Regulate metabolism of sodium and potassium.
	Androgenic hormones (adrenosterone, dehydroepiandiosterone)	Stimulate development of secondary sex characteristics, especially in males.
Medulla	Epinephrine	Stimulates a sequence of reactions associated with emergency situations.
	Norepinephrine	Stimulates reactions similar to adrenaline but is more effective in constriction of blood vessels and less effective in carbohydrate metabolism.
Pituitary		
Anterior lobe	Prolactin	Maintains secretion of estrogens and progesterone by ovaries; stimulates milk secretion by mammary glands.
	Thyrotropin (TSH)	Stimulates development and functioning of thyroid.
	Adrenocorticotropin (ACTH)	Stimulates development of adrenals and production of hormones by adrenal cortex.
	Growth hormone	Stimulates growth of tissues, especially bone.
	Follicle-stimulating hormone (FSH)	Stimulates the growth of ovarian follicles and seminiferous tubules of testes.
	Luteinizing hormone (LH)	Maturation of ovarian follicles, release of mature eggs, formation of corpus luteum, stimulates secretion of sex hormones by ovaries and testes.
Intermediate lobe	Melanophore-stimulating hormone (MSH)	Controls skin pigmentation of many vertebrates.
Posterior lobe	Oxytocin	Stimulates contraction of uterine muscles and regulates milk production.
	Vasopressin	Stimulates constriction of arterioles and increased water reabsorption by the kidneys.
		Other Regulatory Substances
Thymus gland	Thymic hormone	Stimulates antibody formation.
Pineal gland	Melatonin	May regulate photoperiodic influences on mammalian gonads.
Placenta	Chorionic gonadotropin	Similar to LH.
	Placental lactogen	Similar to prolactin.
Parasympathetic and skeletal nerve endings	Acetylcholine	Transmission of nerve impulses; may be necessary for rhythmical activities of heart.
Postganglionic sympathetic nerve endings	Norepinephrine	See above.
Damaged tissues	Histamine	Increases the permeability of capillaries.
Mucosa of alimentary canal	Serotonin	Stimulates constriction of blood vessels.

condition is easily remedied by the addition of foods with high iodine content (iodized salt, seafood, etc.) to the diet, and by surgery to remove the excess tissue.

Hypersecretion of thyroxin results either from overactivity of a normal-sized gland or from an increase in the size of the gland itself. In both instances, the metabolic rate may increase as much as 40%. As a consequence, symptoms such as excessive heat production, profuse sweating, increased food intake but loss of weight since little fat is stored, high blood pressure, nervous tension and irritability, and muscular weakness soon follow. Some patients with hyperthyroidism also have protruding eyeballs, a condition called **exophthalmos** (Figure 10-19). The swelling of the thyroid as a result of hyperactivity produces *exophthalmic goiter*. This is distinguished from a simple goiter, which is caused by inadequate iodine intake.

Another hormone produced by the thyroid gland is *thyrocalcitonin* (*calcitonin*). It is involved in the homeostasis of blood calcium level. Part of the remodeling of bone involves the breakdown of bone tissue and the release of calcium into the blood. The other part of the process is the deposition of calcium in the bones and the subsequent laying down of new osseus tissue. Thyrocalcitonin lowers the amount of calcium in the blood by inhibiting bone breakdown and by accelerating the absorption of calcium by the bones.

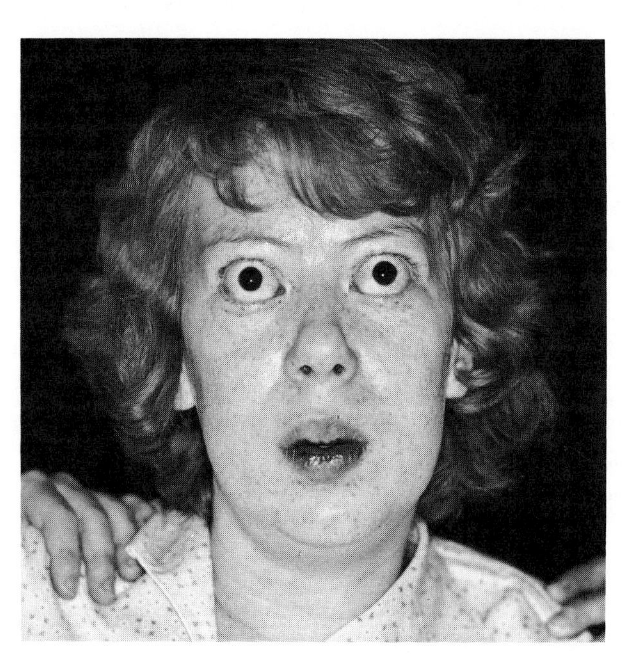

FIGURE 10-19. Results of toxic goiter associated with hyperthyroidism and exophthalmos. [*Armed Forces Institute of Pathology photo.*]

THE PARATHYROIDS: CONTROL OF CALCIUM AND PHOSPHATE METABOLISM. The **parathyroid glands,** usually four in number, are pea-sized masses of tissue that, in the human, are adjacent to or embedded in the posterior thyroid tissue (see Figure 10-17). The active hormone secreted by these glands is *parathyroid hormone* (*PTH*). Its essential function is to control the concentrations of calcium and phosphate in the blood and body tissues, and the calcium content of the blood appears to be the principal factor in regulating the rate of its production. The parathyroid hormone acts by increasing the concentration of calcium in the blood and by decreasing the level of phosphate in the blood. It does so by increasing the absorption of phosphate by the intestines and kidney and by stimulating the bones to release calcium into the blood.

If there is a hyposecretion of parathormone there is a marked decrease in blood calcium concentration and a rise in phosphate concentration, accompanied in a few days by disturbances of nervous and muscular tissue particularly. These tissues become very irritable and respond to the slightest stimuli with severe muscular tremors, cramps and finally convulsions. These symptoms, collectively called *tetany,* may result in death if not treated. Administration of calcium salts by mouth or by injection, or injections of parathyroid gland extracts, relieves the symptoms. Excessive bone tissue loss causes the bones to become weakened so that they are easily bent or fractured.

THE ADRENAL GLANDS: A MULTICONTROL COMPLEX. The **adrenal glands** in humans are paired organs located one above each kidney. They do not, however, have any structural or functional relationship to the kidneys. Morphologically and microscopically (Figure 10-17) the adrenals are composed of two portions: a pale, yellowish pink outer section, the *cortex,* and a dark, reddish brown central core, the *medulla*. In effect, each adrenal gland is a double gland since the medulla and cortex differ in structure, function, and embryological origin. In fish and amphibians the cortex and medulla are two separate glands; in reptiles and birds they are intermingled; and in mammals, although the cortex and medulla are distinct, both lie together.

The adrenal medulla contains cells that produce *epinephrine* (*adrenalin*) by a series of chemical reactions, all of which involve the production of another hormone, *norepinephrine* (*noradrenalin*). This latter hormone is an intermediate of epinephrine synthesis. In humans, 80% of the medullary secretion is epinephrine, 20% is norepinephrine. Both hormones exert similar though not identical effects. They help the individual to meet emergency situations. Of the two, epinephrine functions more closely as the "emergency" hormone of the body. Secretion of epinephrine into the blood is stimulated by the nervous system and brings about the following physiological responses.

1. A more rapid and forceful heartbeat.
2. A rise in blood pressure.
3. Conversion of glycogen into glucose and release of glucose into the blood by the liver.
4. Discharge of red blood cells from the spleen.
5. A greater flow of blood to the muscles, central nervous system, and heart.
6. Relaxation of smooth muscles of the digestive tract and bronchial tubes of the lungs.
7. Erection of hairs.
8. Dilation of the pupils.
9. An increased rate of respiration.

This chain of apparently unrelated events prepares the body to resist stress by mobilizing its resources. The effects of norepinephrine are less extreme than those of epinephrine and, in addition, whereas epinephrine has a more pronounced effect on carbohydrate metabolism, norepinephrine is more effective in constriction of blood vessels.

It should be noted that even though the secretions of the adrenal medulla assume a role in emergency situations, the sympathetic nervous system is believed to be more important in stress situations. In fact, surgical removal of the adrenal medulla causes little noticeable change in the functioning of an organism. Clinically, epinephrine is used to relax the bronchioles in asthmatics; to constrict arterioles of the skin, thus reducing the loss of blood in minor operations; and to revive a heartbeat that has suddenly stopped.

Unlike the adrenal medulla, the adrenal cortex is essential for life. It is far more complex than the medulla and produces myriad hormones, all of which are collectively called *corticoids,* or *cortical hormones.* On the basis of general physiological activities, cortical hormones are of three types.

1. *Mineralocorticoids,* such as aldosterone and deoxycorticosterone, which have a regulatory effect on the relative concentrations of mineral ions (especially sodium and potassium) in the body fluids, and thus on the water content of the tissues.
2. *Glucocorticoids,* such as cortisone, corticosterone, and hydrocortisone, which influence carbohydrate metabolism.
3. Cortical sex hormones, mostly *androgenic hormones,* such as adrenosterone and dehydroepiandrosterone, which regulate the development of sexual characteristics.

The mineralocorticoids stimulate the cells of the kidneys to regulate the concentrations of Na^+, K^+, Cl^-, bicarbonate, and phosphate ions and the water concentrations of the blood and tissue fluids. The loss of water precipitates a decreased blood volume and a lowered blood pressure. It can therefore be seen that the balance of body fluids is regulated by the interactions of parathyroid hormone, insulin, glucagon, epinephrine, and mineralocorticoids.

The glucocorticoids have a number of metabolic functions dealing primarily with carbohydrate metabolism. As a group, they stimulate the conversion of amino acids, and to a more limited extent the fats, into glucose glycogenesis. In addition, they facilitate the formation and storage of glycogen in the liver and other body tissues. Thus, they help to maintain the storage of glycogen and regulate the blood sugar level. In this regard, their action is antagonistic to that of insulin.

The third group of cortical hormones is concerned largely with activities usually associated with sex hormones produced by the gonads. Although both male and female hormones are synthesized by the cortex, the male hormones predominate. Accordingly, they may aid development of secondary sex characteristics in the male, such as deepening of the voice, the growth of a beard, the distribution of body hair, and the development of the genitals.

Hyposecretion of the adrenal cortex is responsible for **Addison's disease.** The principal symptoms of this abnormality are muscular weakness and weight loss, decreased blood pressure, loss of appetite and gastrointestinal distrubances, and an atypical bronzing of the skin. If these symptoms are detected and treated early enough, they can be alleviated by glucocorticoid injections.

Hypersecretion of the adrenal cortex in children can lead to precocious sexual development. A six-year-old male, for example, may exhibit the sexual development of an adult man. In this state, there is a growth of hair on the face and chest, the testes and penis enlarge, and sexual desires may intensify. In young females, if estrogen secretion predominates, menstruation may begin early, the breasts may enlarge, and sexual interest may increase.

Hyperactivity of the cortex in adult females, where androgenic secretion is excessive, causes the development of male characteristics such as the growth of a beard, hair on the chest, a deepened voice, a masculine musculature with a corresponding suppression of menstruation and breast development. Hyperactivity of the cortex in the adult male, with increased androgen secretion, is less common. In some cases, it causes an increase in masculinity; in others, if an excess of female hormone is secreted, the male develops certain female characteristics, such as enlargement of the breasts. In postmenopausal females, hair growth on the face may be increased as the adrenal androgens are "released" to exert their effects by loss of gonadal estrogen.

The development and function of the adrenal cortex are regulated by adrenocorticotropic hormone (ACTH), produced by the anterior lobe of the pituitary gland. ACTH appears to stimulate production of the glucocorticoids of the cortex. Another negative feedback relationship is exhibited by ACTH and glucocorticoids. The ACTH of the pituitary is apparently produced in response to a decreased level of glucocorticoids in the blood. The ACTH, in turn, stimulates the cortex to produce more glucocorticoids until its concentration returns to normal. Conversely, an excess of glucocorticoids in the blood inhibits the pituitary secretion of ACTH. Mineralocorticoid production is controlled mainly by angiotensin and blood sodium levels. Angiotensin is converted from a plasma substance by renin, a chemical produced by the kidneys. Angiotensin raises blood pressure by causing vasoconstriction.

The Pituitary Gland: Control of Other Endocrines. The **pituitary gland (hypophysis),** like the adrenal glands, is structurally, functionally, and embryologically differentiated into two major portions: the anterior lobe, termed the *adenohypophysis,* and the posterior lobe, or *neurohypophysis.* This gland, about the size of a large pea, is attached by a stalk to the hypothalamus portion of the brain on the undersurface of the cerebrum (Figure 10-17).

The anterior pituitary in humans secretes at least six hormones, all of which have profound influences in integrating activities such as growth, metabolism, and the development and functioning of the reproductive systems (Figure 10-20). Because of these varied regulatory roles of the anterior pituitary in controlling the secretions of other glands, it is sometimes referred to as the "master gland" of the body. Among the six hormones secreted by this gland are prolactin (lactogenic hormone), thyrotropin (thyroid-stimulating hormone, TSH), adrenocorticotropic hormone (ACTH), growth hormone (somatotropin), follicle-stimulating hormone (FSH), and luteinizing hormone (LH, also known in the male as interstitial-cell-stimulating hormone, ICSH).

Prolactin maintains the secretion of estrogens and progesterone by the ovary and initiates the secretion of milk by the mammary glands after the birth of a baby. It is effective only after the breasts have been stimulated by the proper amounts of estrogen and progesterone. In the absence of prolactin the secretion of milk by the mammary glands soon stops.

Thyrotropin (TSH) is essential for the normal development and functioning of the thyroid gland. It affects the activity of the thyroid in the production of its hormone, thyroxin. An increase in the concentration of thyroxin in the blood has an inhibitory effect on the rate of secretion of TSH and this negative feedback mechanism prevents hypo- and hyperactivity of the thyroid. A decrease in thyroxin stimulates the secretion of TSH.

The *adrenocorticotropic hormone (ACTH)*, as previously noted, stimulates the adrenal cortex to grow and to synthesize its complex of hormones. The secretions of ACTH and glucocorticoids are mutually affected by a negative feedback mechanism. In the absence of ACTH, the adrenal cortex regresses.

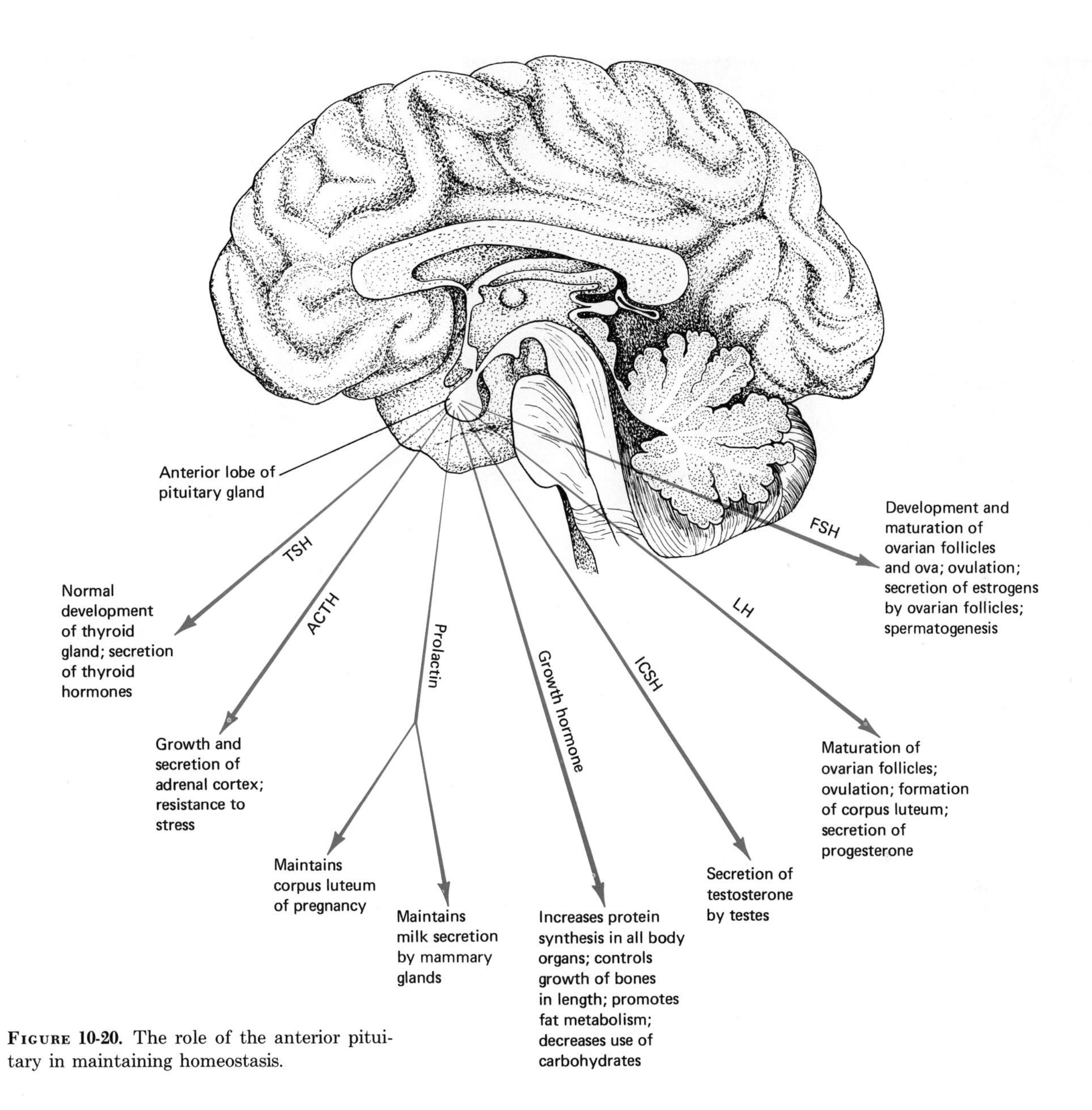

Figure 10-20. The role of the anterior pituitary in maintaining homeostasis.

Somatotropin, or the growth hormone, has a specific influence on the growth of tissues, particularly bone, muscle, and internal organs. As a result, hypersecretion during early life gives rise to an acceleration of growth processes, a condition called **giantism,** in which the individual is very tall and fairly well proportioned. Hypersecretion after the growth period, that is, in adult life, results in a condition termed **acromegaly** (Figure 10-21A). Here, further increase in size is restricted to the regions of the joints and the face so that the individual is characterized by a greatly enlarged and protruding jaw, enlarged cheekbones and eyebrow ridges, a thickened nose, and disproportionately large hands and feet. Hypoproduction of growth hormone during the

(A)

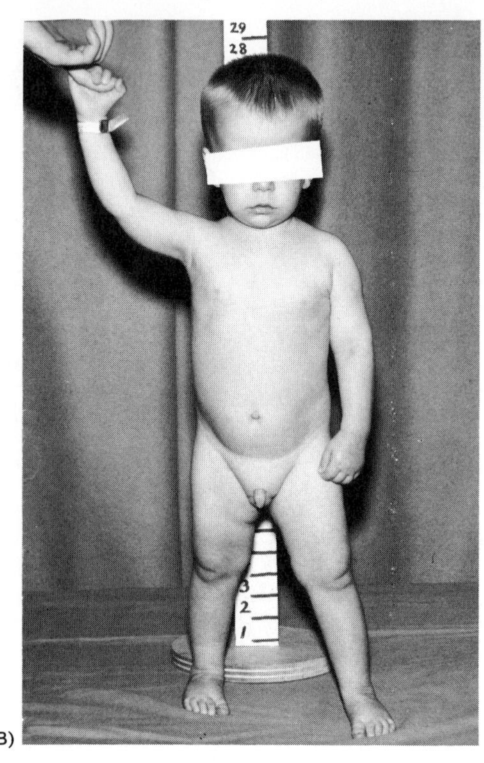

(B)

Figure 10-21. Hormonal abnormalities in man. (**A**) Acromegaly. Note the thick tongue, enlarged facial features, and large hands associated with the hypersecretion of growth hormone during adult life. (**B**) Dwarfism. Note the height of the dwarf in inches and immature sexual development due to hyposecretion of growth hormone during childhood. The dwarf shown is 14 years of age. [*Armed Forces Institute of Pathology photos.*]

growth years results in **pituitary infantilism,** a kind of **dwarfism** (Figure 10-21B), which, in contrast to cretinism, is not accompanied by physical deformities or mental retardation. The adult dwarf, however, usually no more than 3 or 4 feet tall, is usually sexually immature.

In addition to its effect on growth, the growth hormone also influences metabolic processes. For example, it accelerates the rate of protein synthesis, decreases the rate of carbohydrate utilization in muscle and adipose tissue, mobilizes stored fat, and increases the use of fats for energy.

Follicle-stimulating hormone (*FSH*) stimulates the primary sex organs (ovaries and testes) of both sexes. In the female, starting at puberty and continuing throughout the years of sexual maturity, FSH stimulates the development and maturation of ovarian follicles and ova. FSH is also necessary for ovulation, the release of the egg from the follicle. In addition, it has a stimulating effect on the follicles to produce estrogens. In males, starting at puberty, FSH influences spermatogenesis, the production of sperm by the seminiferous tubules of the testes.

The final hormone secreted by the anterior pituitary is the *luteinizing hormone* (*LH*). In the female, this hormone, along with FSH, is responsible for the maturation of ovarian follicles, the release of mature eggs, the formation of the corpus luteum (Chapter 17), and the production of progesterone. In the male, ICSH influences the testes to secrete testosterone. Both LH and FSH regulate the menstrual cycle in the female.

The posterior lobe of the pituitary releases two hormones: the antidiuretic hormone (ADH), or vasopressin, and oxytocin (pitocin). This gland illustrates a very interesting relationship between the endocrine and nervous systems and gives strong support to the concept that nervous and chemical control are part of a unified coordinating system. During embryological development, the posterior pituitary originates as an outgrowth of the hypothalamus of the brain, and once formed, it retains a nervous connection with the hypothalamus by a stalk. There is a considerable amount of evidence to indicate that vasopressin and oxytocin are synthesized in the neurons of the hypothalamus and then migrate via axons to the posterior pituitary where they are stored and subsequently released.

Oxytocin has a stimulatory effect upon smooth muscle, particularly the pregnant uterus. It is sometimes administered during labor, thereby increasing the contraction of the uterus. Whether the posterior lobe normally secretes oxytocin during childbirth is not known, but animals can give birth to their young quite readily after surgical removal of the gland. Oxytocin also functions in regulating lactation (the release of milk).

Vasopressin causes constriction of arterioles, resulting in a significant increase of blood pressure. It is also referred to as the antidiuretic hormone (ADH) because it stimulates the kidneys to reabsorb more water, thus preventing excessive water loss by urination. A deficiency of vasopressin results in a condition called *polyuria insipida* (*diabetes insipidus*) in which the individual is extremely thirsty and excretes large amounts of dilute urine. Injections of vasopressin relieve these symptoms.

The release of pituitary hormones is controlled in part by the concentration of target hormones in the blood. For example, the release of thyrotropin is inhibited by thyroxin, the release of ACTH is inhibited by glucocorticoids, and so on. This regulatory mechanism establishes a feedback system so that the secretions of the pituitary and its target glands are mutually kept in balance. The anterior pituitary is also controlled by the hypothalamus of the brain by way of neurohumors, while the posterior lobe is controlled by nerves. *Neurohumors* are chemical substances produced by nerve cells that cause an endocrine gland to secrete its hormones. One example of a neurohumor is the *somatotropin-releasing factor* (*SRF*) produced by the nerve cells of the hypothalamus. When SRF is released into the blood vessels that connect the hypothalamus with the anterior pituitary, it reaches the anterior lobe of the pituitary and stimulates the lobe to secrete the growth hormone.

Other Regulatory Substances. In addition to endocrine glands that are known to secrete hormones and elicit physiological responses, there are other structures in the body that behave as endocrine-like structures. Chief among these are the thymus gland, the pineal gland, the placenta, and certain tissues of the body.

The **thymus gland** is a fairly large gland that lies in the upper part of the chest, covering the lower end of the trachea (Figure 10-17). It is large in early childhood, but regresses after puberty. Attempts to demonstrate that it secretes a hormone affecting sexual maturity have been unsuccessful. Recently, however, evidence has been provided to show that the thymus of young mice produces a hormone that causes lymphocytes to become plasma cells, the latter producing antibodies in response to antigenic stimulation.

The **pineal gland,** or pineal body, is a small, round structure lying on the dorsal aspect of the brain stem (Figure 10-17). In humans its function as an endocrine gland, like that of the thymus, is obscure. The hormone *melatonin* is produced by the pineal and in lower animals is related to pigment cell change and therefore to skin coloration. In boys tumors of the gland are frequently associated with precocious puberty. Studies in rats indicate that removal of the gland before sexual development causes gonadal retardation. It has been suggested that melatonin may be involved in the timing of human adolescence.

The **placenta,** although an organ primarily for support and nourishment of the developing embryo, also functions as an endocrine gland. It is the source of many hormones, including estrogen, progesterone, chorionic gonadotropin, and placental lactogen. *Chorionic gonadotropin* is similar in action to the luteinizing hormone of the anterior pituitary; *placental lactogen,* to prolactin. The placenta is also a source of a rich supply of immune bodies and of a blood coagulant.

There are a number of parts of the body that secrete chemical substances locally and elicit their responses locally. *Acetylcholine,* for example, secreted at parasympathetic and skeletal nerve endings, not only is a chemical transmitter but may also be necessary for rhythmical activities of the heart. *Norepinephrine* is secreted at most endings of postganglionic sympathetic nerve fibers. Both *histamine* and *heparin* are present in mast cells of connective tissues. Histamine, which is believed to play a role in allergic conditions, causes arteriolar and capillary dilatation, a rise in the temperature of the skin, a fall in diastolic blood pressure, and an increase in heart rate. Heparin is important in the manufacture of certain elements in connective tissue. *Serotonin,* which is found in the mucosa of the alimentary tract, in blood platelets, and in the brain, stimulates the constriction of blood vessels.

11 Homeostasis: Nervous Control

GUIDE QUESTIONS

1 Define a stimulus. List some examples of external and internal stimuli. How do most protistans function without any specialized structures for nervous coordination? Discuss some specialized structures in ciliates; in euglenoids.

2 Discuss nervous coordination in coelenterates. What purpose do the statocysts and ocelli serve in jellyfish?

3 How is cephalization related to bilateral symmetry? Of what advantage is cephalization and centralization of the nervous system to an organism? How may unidirectional impulse conduction benefit an organism's coordination?

4 Discuss the advances of the nervous system in annelids as compared with that of *Hydra*. What activities may be carried on by arthropods that are not possible in lower invertebrates? Explain in detail.

5 Discuss the evolution of the vertebrate brain from fish through mammals. Which activities and/or functions have become more apparent in mammals? Less apparent?

6 Explain in detail the meaning of a reflex and of a reflex arc. What are simple reflexes? conditioned reflexes? State examples of each.

7 Describe the various protective coverings of the brain and spinal cord and explain how each functions. How does gray matter differ from white matter? What are the types of fibers of the cerebral tracts? What is their function(s)?

8 Outline in detail the parts of the brain and their functions. How are the parts coordinated?

9 How is the spinal cord related to the brain? Describe the structure of the spinal cord. Explain the differences between the dorsal and ventral roots.

10 How do sensory, motor, and mixed nerves differ in function, origin, and termination? How are they alike?

11 Outline in detail the distribution of nerves from the sympathetic division of the autonomic nervous system.

12 Outline in detail the twelve cranial nerves, their distribution, and function(s).

13 Summarize the ways in which homeostasis is disrupted by the following conditions: poliomyelitis, syphilis, cerebral palsy, Parkinsonism, epilepsy, and multiple sclerosis. Knowledge of the clinical symptoms will help you to formulate your response.

14 What is the diagnostic usefulness of the EEG for epilepsy?

15 Contrast the various kinds of epilepsy with regard to clinical symptoms.

16 Why are pain receptors important?

17 Distinguish between somatic, visceral, referred, and phantom pain.

18 Define acupuncture. How is the procedure believed to be used to relieve pain?

19 Define a sense organ. How are the various sense organs coordinated with the brain? State specific examples.

20 How do chemical receptors differ from mechanical receptors? Explain how each operates.

21 Describe the structure of the ear and outline the sequence of events that occur after a sound wave strikes the tympanum. How is equilibrium maintained?

22 Explain how a visual image is received by the brain. How do the rods and cones function? What abnormalities in vision may occur because of irregularities in eyeball structures? How may they be corrected?

23 Define a drug. Distinguish a drug from a medicine.

24 Define a hallucinogen. Describe some of the physiological effects of LSD and marijuana.

25 Define a depressant. Distinguish between the types and effects of barbiturates, tranquilizers, and opiates.

26 Define an amphetamine. Describe its action in the body.

As NOTED in the previous chapter, it is becoming increasingly clear that distinctions between the endocrine and nervous systems are artificial. Both probably operate together as a coordinating mechanism. Therefore this discussion of nervous control of homeostasis may be viewed as a continuation of the preceding chapter, except for the shift in emphasis from chemical to nervous control.

The most fundamental activities of a nervous system, regardless of structural complexity or functional diversity, relate to reception of a stimulus, transmission of a stimulus, interpretation and analysis, and response by an effector. Any physical or chemical change in the environment that is capable of eliciting a reaction from a living system may be viewed as a **stimulus.** Some stimuli, such as temperature, moisture, light, pressure, gravity, contact, and chemical substances, arise from the external environment of an organism. Other stimuli related to food, water, oxygen, waste products, pain, fatigue, and disease may originate from the internal environment of an organism. Most organisms possess specialized structures or organs called **receptors** whose primary role is to react to various stimuli. Once the stimulus is detected by a receptor, it is transmitted through the nervous system (in more advanced multicellular organisms) where a particular response to the stimulus is selected. Certain portions of the nervous system then transmit the stimulus to **effectors,** which bring about responses. In the evolution of nervous systems from the simplest to the most complex organisms, four components of nervous coordination are universally operative: (1) stimulus reception, (2) impulse conduction, (3) interpretation and analysis, and (4) response.

Protistan Coordination

Most protistans show no evidence of any structures involved in nervous coordination so that receipt, transmission, and reaction to stimuli are apparently performed entirely within the protoplasm of the single cell. Excitability, the capacity to respond to stimuli, is a universal property of cells. Among ciliates (phylum Ciliophora), however, there is evidence of some specialization in nervous control. In *Paramecia,* for example, there is an interconnected series of fibrils (*neuromotor apparatus*) at the base of the cilia that may function as a simple coordinating mechanism (Figure 11-1A). Many ciliates, such as *Euplotes* (Figure 11-1B), have motionless *bristles* that appear to serve as receptors; euglenoids such as *Euglena* (Figure 11-1C) have a special light receptor organelle called the **stigma.** Thus, among protistans, certain organisms possess receptors (stigma, bristles) and specialized effector organelles (cilia, flagella, contractile fibrils).

Invertebrate Coordination

In sponges (phylum Porifera), as with protistans, there are no definite nerve cells or structures. Sponges are very sluggish in their reactions and it appears that responses to environmental stimuli are limited to and center about the direct response of individual cells. A needle prick near the osculum, for example, provides sufficient stimulus that slowly spreads and elicits responses in nearby cells, bringing about closure of the opening. A similar prick some distance away from the osculum has no effect on its closure (Figure 11-2A).

In the animal kingdom, all groups of organisms above sponges have some kind of nervous system in varying degrees of complexity. *Hydra,* as a representative coelenterate, possesses the simplest form of nervous system (Figure 11-2B). It consists of a network **(nerve net)** of similar nerve-like cells around the body in or under the epithelium. Nerve impulse conduction is slow and can occur in any direction depending on where the stimulus is received. Despite the similar structure of the nerve cells, some (receptors) are more sensitive to certain environmental stimuli, whereas others transmit the stimuli to other parts of the organism. Among the coelenterates, a higher degree of development of the nervous system is illustrated by the jellyfish (Figure 11-2C). Here the nerve network is largely grouped into two parallel bundles of nerve cells or *nerve rings,* and since other nerve cells funnel into these rings, there is a certain degree of centralization for coordinating impulses. The jellyfish also contains the first true sense organs to appear in the animal kingdom. The first of these, *ocelli,* are light-sensitive pigmented areas which serve as lenses. The second sensory structures are the *statocysts* which are special balancing organs that enable the organism to right itself. In the course of evolutionary history, it appears that radial symmetry, a

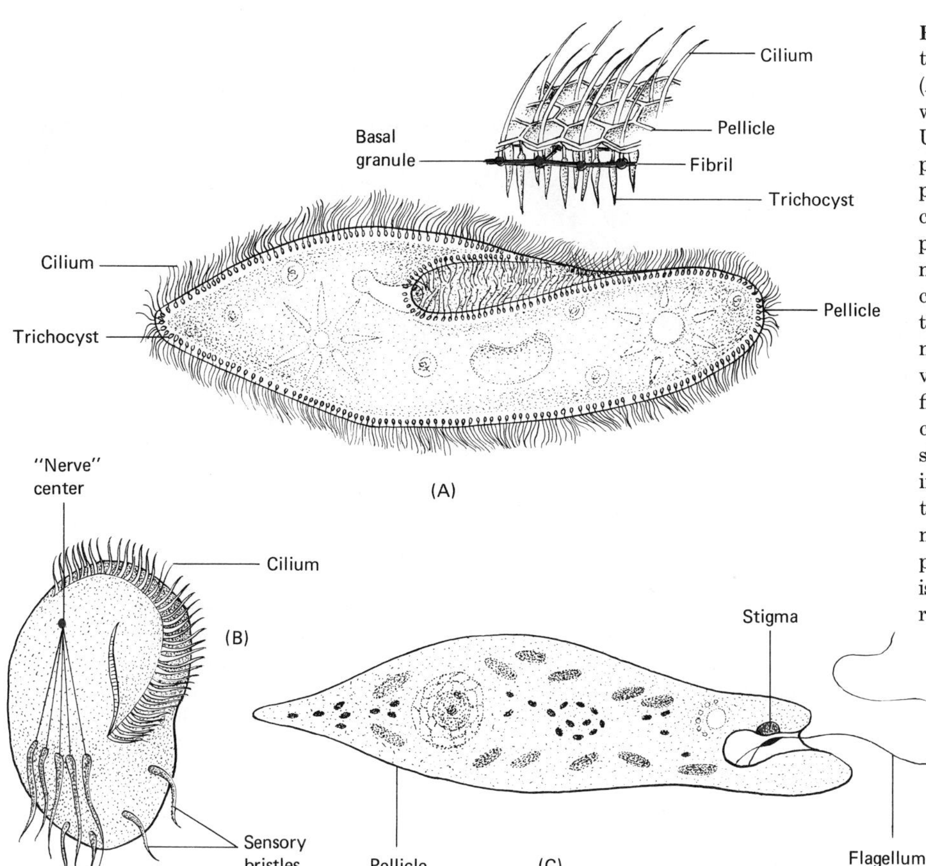

Figure 11-1. Representative protistan coordinating mechanisms. (**A**) Structure of *Paramecium* with enlargement of pellicle. Under high magnification, the pellicle appears as a hexagonal pattern of ridges surrounding a cup-like depression with a cilium projecting from its center. Beneath the pellicle, each cilium connects to a basal granule and the basal granules are interconnected by longitudinal and transverse fibrils. The granules and fibrils are thought to coordinate ciliary action. (**B**) In many ciliates such as *Euplotes,* sensory bristles in conjunction with a very primitive nerve center control the movement of cilia. (**C**) *Euglena* possesses a stigma (eyespot) that is sensitive to light and assumes a role in photosynthesis.

characteristic of coelenterates, has inhibited any high degree of centralization of the nervous system.

In bilaterally symmetrical animals, however, several general patterns of change in the nervous system accompany increasing complexity of structure.

1. Increasing centralization of the nervous system by the evolution of nerve cords that function primarily in the conduction of nerve impulses.
2. Development of many association neurons.
3. Concentration of association neurons in the head region of the animal (*cephalization*), which eventually leads to the evolution of the brain.
4. Imbedding of the centralized nervous system deeper within the protective tissues of the organism.
5. Limitation of impulse conduction to one direction only, so that definite sensory, association, and motor pathways are established.
6. Specialization of cells into receptors for particular stimuli.

These evolutionary trends, which begin in the invertebrates, are progressively more developed in vertebrates. We will now examine the trends as they appear in invertebrates.

The nervous system of most flatworms consists fundamentally of a nerve net. Slightly more advanced flatworms, however, also possess the beginnings of a centralized nervous system in the form of two ventrally located *longitudinal nerve cords* which represent aggregations of nerve cells (Figure 11-3A). The nerve cells in these cords serve as links between stimulus and response. The two nerve cords contain transverse connections as well as lateral projections and connect to an anterior mass of nerve cell bodies called a **ganglion.** The ganglion represents a simple "brain." This "brain,"

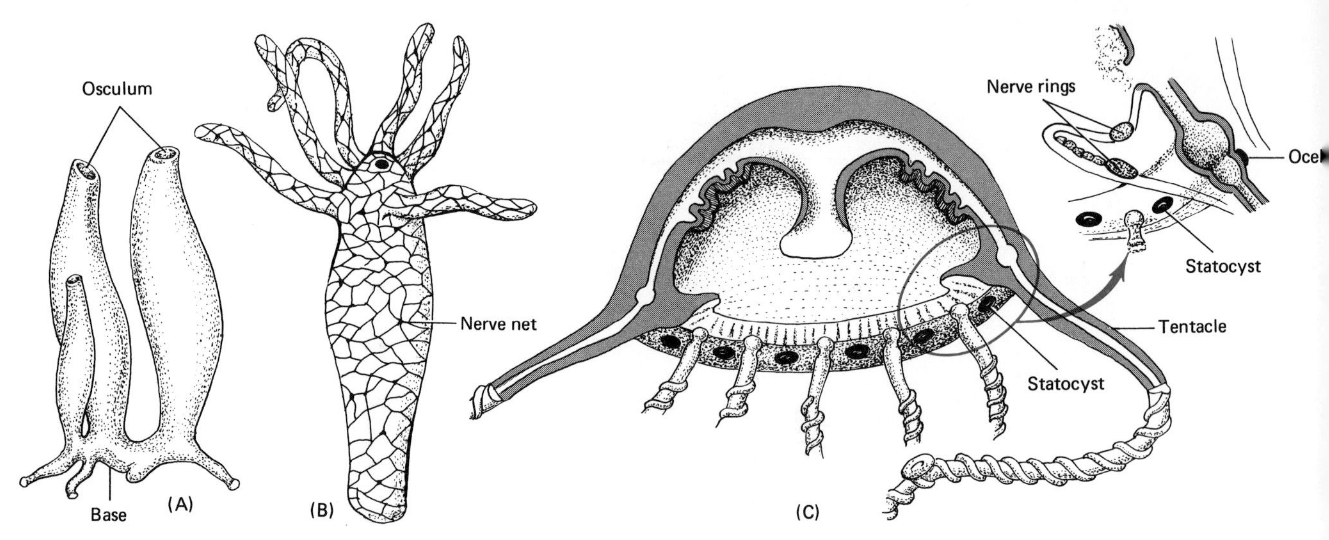

FIGURE 11-2. Nervous coordination in the sponge, hydra, and jellyfish. (**A**) Each cell of the body of a sponge is irritable and reacts to stimuli, but there are no sense cells or nerve cells to enable the organism to react as a unified whole. (**B**) Hydra, a representative coelenterate, possesses the first true nerve cells in the animal kingdom. These nerve cells form an irregular network called a nerve net and connect the sensory cells in the body wall with muscle and gland cells. Since there is no aggregation of nerve cells to form a brain or spinal cord, an impulse received in one part of the body is transmitted more or less equally in all directions. (**C**) In addition to the organization of nerve cells into nerve rings, a condition that establishes a certain degree of coordination, the jellyfish also contains the first true sense organs in the form of statocysts and ocelli.

which exerts only limited control over the rest of the nervous system, probably serves as a relay station between the sense organs and the cords. The *lateral* nerves connecting the two longitudinal cords contain sensory neurons for conducting impulses from receptors to the central system and motor neurons for conducting impulses from the central system to effectors. Among the numerous sensory receptor cells and sense organs present in planarians are *auricles* on the head which are generally sensitive to light, touch, and chemical stimuli, and *eyespots,* concentrations of nerve cells partially covered by pigmented cups, which are especially sensitive to light (Figure 11-3A). These features—the development of nerve cords, a "brain," afferent and efferent neurons, and sense organs—have endowed many of the flatworms with more varied behavior and quicker responses than coelenterates.

The nervous system of oligochaete annelids, such as the earthworm, consists of a large ganglion, or brain, in the head dorsal to the pharynx and a paired *ventral nerve cord* that extends the length of the body (Figure 11-3B). Since the two components of the cord fuse, it appears as a single ventral cord. The cord contains ganglionic masses in each segment with the ganglia connected by fibers running between the segments. Essentially, the brain is an anterior ganglion in the head region and although its dominance over the other ganglia is noticeable, it is not as dominant as the brain of vertebrates. Earthworms have poorly developed sense organs on the head; most receptors are located in the skin around the body wall. In certain polychaete annelids, although there are fewer receptors in the skin, effective chemoreceptors are present on the tentacles and palps and frequently several pairs of eyes are present (Figure 11-3C). Removal of a polychaete brain results in loss of activities such as feeding ability, burrowing, and light sensitivity. These observations suggest a fairly high degree of centralization and integration of the annelid nervous system.

The nervous system of arthropods is one of the most

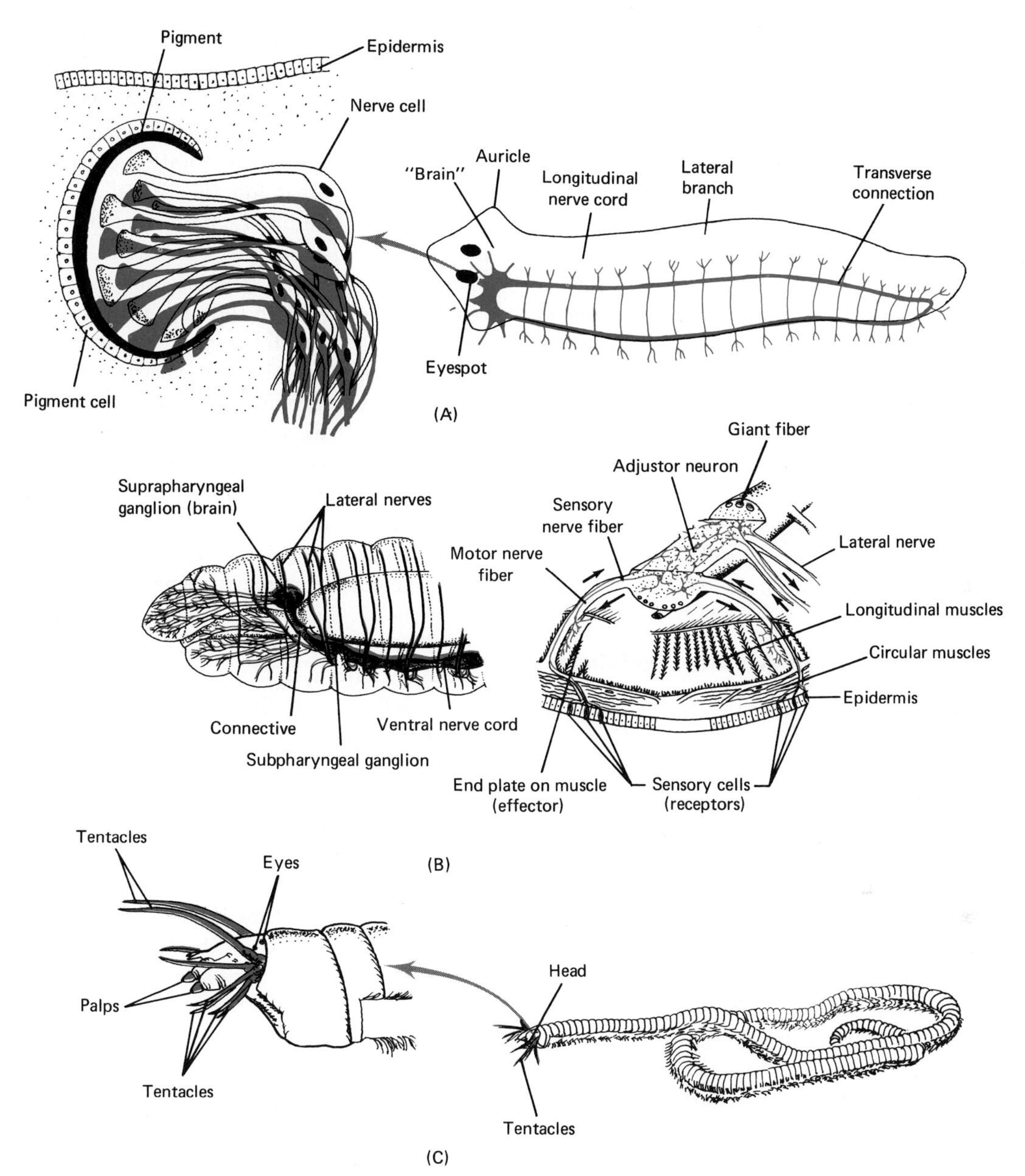

FIGURE 11-3. Nervous systems of representative worms. (**A**) In planaria, two anterior ganglia are joined as a "brain" that distributes nerve fibers to the head and eyespots (enlargement). From the "brain" two longitudinal nerve cords extend posteriorly, one on either side of the body; each has many transverse connections and lateral branches. (**B**) *Left:* The earthworm brain is a ganglion, dorsal to the pharynx, that is connected to a paired ventral nerve cord running the length of the body. There is a ganglion in each segment containing three pairs of lateral nerves. *Right:* The relationship of the ventral nerve cord to muscles and receptors in the skin. (**C**) In *Nereis* (clamworm), the nervous system is basically similar to that of the earthworm. In addition, eyes, palps, and tentacles serve as sensory structures (enlargement).

centralized and complex of all invertebrates. The central nervous system, as represented by that of the grasshopper, consists of two ventrally located solid nerve cords and a brain, which is a bilobed ganglion surrounding the esophagus (Figure 11-4A). In arthropods, activities such as movement, certain behavior, reproduction, and other physiological functions are integrated and controlled by the brain. Sense organs are also highly developed in arthropods. Crayfish, for example, possess organs of equilibrium, the *statocysts,* at the bases of the antennae; *tactile hairs* (bristles) over most of the body that are sensitive to touch; antennules, antennae, and mouth parts that receive chemical stimuli (taste and smell); and *compound eyes* that receive images and transmit them to the brain by the optic nerves (Figure 11-4B). Other arthropods have *simple eyes* (*ocelli*) that consist of a single lens and direct light toward light-sensitive cells, which in turn transmit sensations to the brain.

Nervous Coordination Among Vertebrates

It is among vertebrates that the highest level of structural and functional complexity of the nervous system is reached. In all vertebrates, including humans, the nervous system consists of (1) a **central nervous system** (brain and spinal cord), and (2) a **peripheral nervous system** (network of nerve cells and fibers). Among mammals especially, the brain, and to an extent the spinal cord, is considered to be the most advanced component of the nervous system. Anatomically, the vertebrate brain is a large ganglionic mass at the anterior portion of the spinal cord. The spinal cord in all vertebrates is a dorsally located structure.

The most primitive brain of vertebrates, as represented by that of primitive fishes, and the incompletely developed brains of embryos, consists of three enlargements. These three divisions, called the *forebrain, midbrain,* and *hindbrain* (Figure 11-5A), have undergone considerable modifications in the course of evolution (Figure 11-5B). The main part of the hindbrain, the medulla, has become specialized as a control center for visceral activities and as a connecting tract between the spinal cord and other parts of the brain. Much of the control of equilibrium and muscular movement takes place in the cerebellum, an anterior, dorsal outgrowth of the medulla. Unlike the medulla, the cerebellum, a second portion of the hindbrain, increases immensely in size and importance in the course of evolution of fast-moving birds and mammals. The midbrain has changed somewhat in size with vertebrate evolution, but it has changed even more in function. In fish and amphibians, it mediates the animal's most complex behavior. In more advanced vertebrates, the dorsal portion of the midbrain has become specialized as optic lobes, visual centers associated with the optic nerves, and as a coordinating area between the eyes, ears, and motor system.

The vertebrate forebrain has become divided into two portions, the diencephalon and the telencephalon. The *diencephalon* (thalamus, hypothalamus, and posterior pituitary) is extremely important as a regulatory center, even in primitive vertebrates. This part of the brain has changed little throughout the vertebrates, and is one of the most important structures controlling many forms of complex instinctive behavior. The remainder of the forebrain, the *telencephalon* (cerebrum), has undergone the most significant changes in the course of vertebrate evolution. In fishes, amphibians, and many reptiles, it is little more than an olfactory center, that is, a region concerned with the sense of smell. Modern birds have relatively large cerebrums that have little to do with the sense of smell but are rather significant coordination centers. Beginning with the mammals, the cerebrum progressively increases in size and importance. In general, the principal evolutionary change in the vertebrate brain has been the progressive decrease in size and importance of the midbrain coupled with a corresponding increase in size and importance of the cerebrum.

The Human Nervous System

The human nervous system is subdivided into two principal, though interrelated, components:

1. The *central nervous system* (brain and spinal cord).

2. The *peripheral nervous system* (nerves and ganglia).

The peripheral nervous system consists of twelve pairs of cranial nerves that originate from various areas of the brain and thirty-one pairs of spinal nerves that originate at various points along the length of the spinal

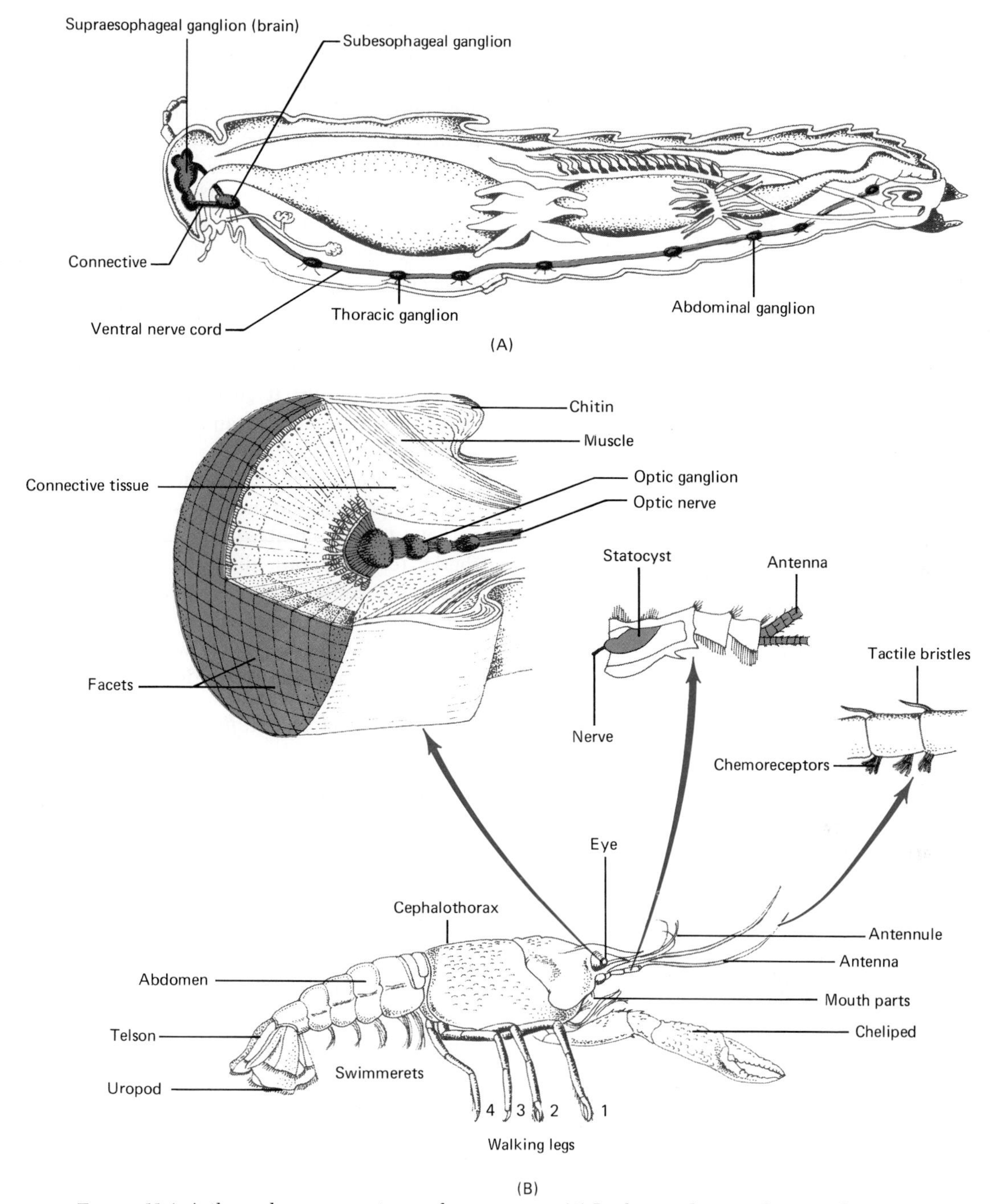

Figure 11-4. Arthropod nervous system and sense organs. (**A**) In the grasshopper, the central nervous system consists of two ventral nerve cords and a brain. Note the presence of ganglia in various body segments. (**B**) Sense organs of the crayfish. *Below:* External structure of the crayfish. *Above* (*left to right*): Compound eye sectioned to show details of internal structure, statocyst at base of antennae, and tactile hairs and chemoreceptors on an antenna.

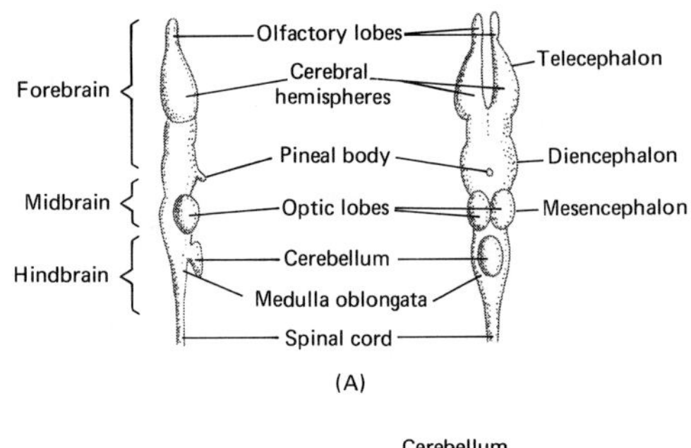

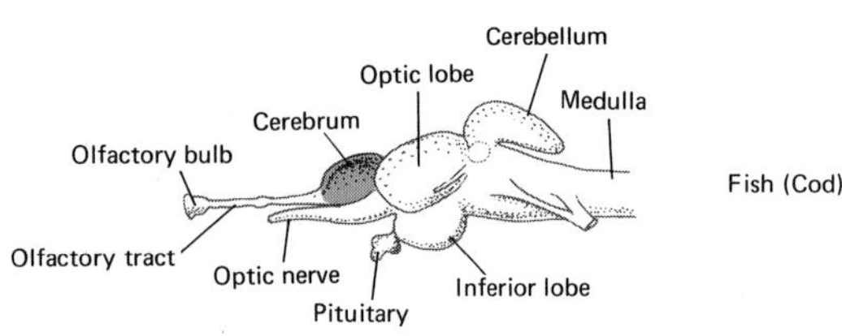

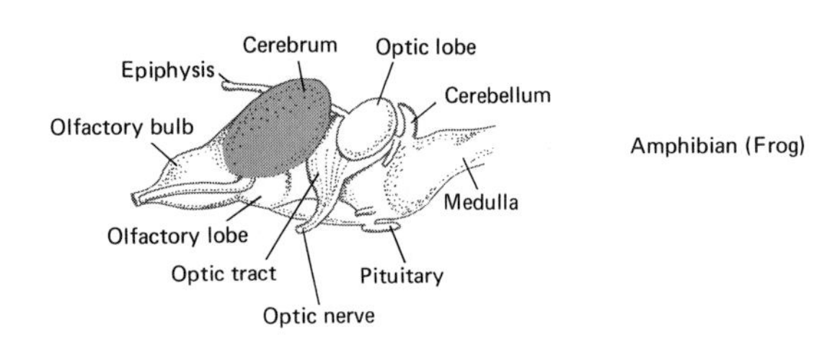

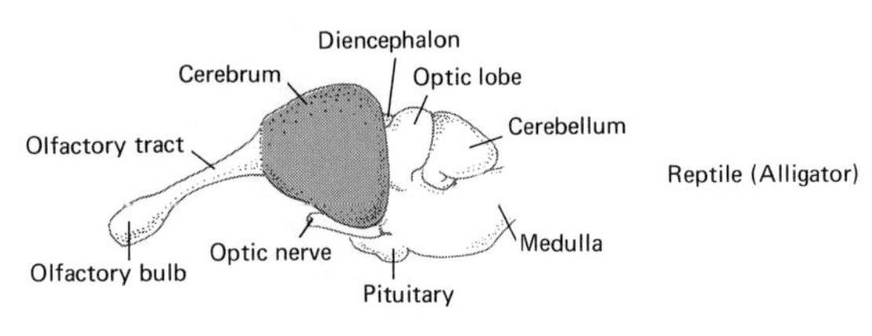

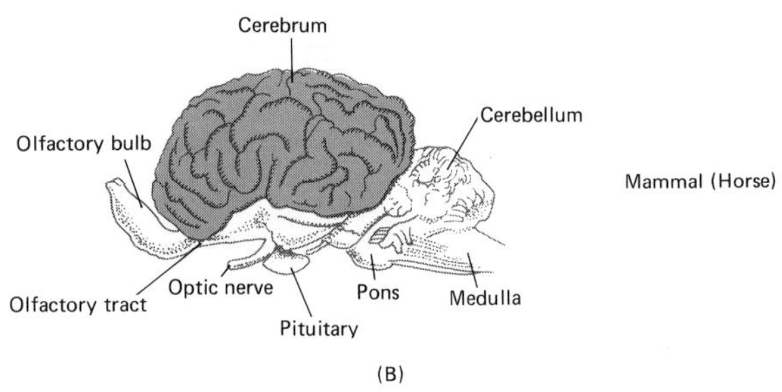

Figure 11-5. Vertebrate brains. (**A**) Generalized diagrams of the vertebrate brain as seen from the left side (*left*) and from the top (*right*). Note the structures that constitute the forebrain, midbrain, and hindbrain. (**B**) Brains of representative vertebrates showing progressive increase in size, especially of the cerebrum and cerebellum.

cord. The cranial nerves are distributed to portions of the head, neck, and internal organs (viscera) of the chest and abdomen; spinal nerves and their branches are distributed to the arms, legs, and trunk of the body. The peripheral nervous system is in turn subdivided into two portions called the somatic nervous system and the autonomic nervous system. The somatic portion usually innervates skeletal muscles and skin and is responsible for movements of various body parts in which there is some conscious control and for transmission of sensory information. The autonomic portion innervates cardiac muscle, some glands, the smooth muscle of digestive, respiratory, excretory, reproductive organs, and blood vessels. Essentially, it governs the functions of these structures automatically and reflexly without any conscious control.

Two anatomically and physiologically separate divisions compose the autonomic nervous system: the sympathetic (thoracolumbar) division and the parasympathetic (craniosacral) division. The structural and functional differences between the two divisions will be considered later in this chapter. A generalized illustration of the overall organization of the human nervous system is shown in Figure 11-6.

Reflexes and the Reflex Arc

Before discussing the structure and functions of the organs comprising the nervous system, we will first examine the meaning of reflexes and the reflex arc. The varied and constant changes in both the external and the internal environments of an organism make necessary some kind of mechanism for automatically adjusting body processes to maintain homeostasis. One of the most important activities of the nervous system is to respond automatically to alterations in the environment that tend to disrupt homeostasis. Such responses, usually occurring below the level of consciousness, are collectively designated as **reflexes** and may be categorized as simple reflexes and conditioned reflexes.

A *simple reflex* is an inborn, unlearned, stereotyped, and unconscious response to a stimulus or change in the environment. Common examples of simple human reflexes are the knee jerk (patellar reflex), in which the leg is involuntarily and momentarily extended in response to a sharp tap below the knee cap (patella); the ankle jerk (Achilles reflex), which is an extension of the

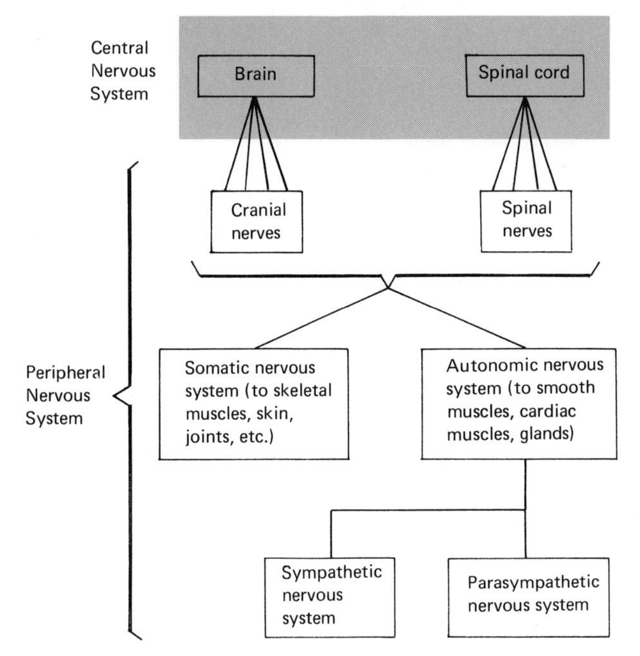

Figure 11-6. Schematic illustration of the overall organization of the human nervous system. See text for amplification.

foot in response to tapping of the Achilles tendon; the quick closing of the eyelid when an object approaches the eye; the rapid withdrawal of the hand in response to being burned or pricked; and the abdominal reflex, which is the drawing in of the abdominal wall in response to stroking the side of the abdomen. In addition to simple reflexes that elicit external responses, innumerable simple reflexes occur inside the body, such as the control of secretions by certain glands, heart rate, movements of the digestive tract, and breathing.

Conditioned reflexes, on the other hand, are responses brought about through learning or experience to a stimulus that originally failed to elicit a reaction; they are not inborn but are acquired and depend upon past training or conditioning. These reactions were first clearly demonstrated by Ivan Pavlov, a Russian physiologist. In the early part of this century, he showed, through experiments with dogs, that conditioned reflexes can be developed to a high degree. Normally, the secretion of saliva when food is ingested is initiated by stimulation of taste buds on the tongue. Working with the dogs, Pavlov supplied an additional stimulus (the ringing of a bell) whenever food was provided. In time, the ringing of the bell in the absence of the sight or smell of food brought about the secretion of saliva and established a conditioned reflex. Note that a previously ineffective stimulus, the ringing of the bell, by association with a stimulus (food) that elicited a reaction, had become effective in evoking a response (the secretion of saliva). The conditioned reflex is considerably more complex than the simple reflex and involves many integrative functions and parts of the brain for complex behavioral activities such as memory, reasoning, and interpretation of sensory impulses.

The structural and functional unit in a simple reflex is termed a **reflex arc.** In its basic form, a reflex arc may be viewed as a simple nervous pathway connecting a receptor and an effector. The simplest reflex arcs in vertebrates involve only two neurons (one sensory and one motor) and since only synapse is involved, they are referred to as *monosynaptic* (Figure 11-7A). As an example, consider the patellar reflex. In this particular reflex, the receptors consist of an association between the dendrites of a sensory neuron in a specialized receptor and a muscle or tendon. Once the stimulus is received, it travels along the sensory neuron to the spinal cord. It then passes to the cell body of the sensory neuron which is located in a structure called the *dorsal root ganglion.* From here, the impulse travels along the axon of the sensory neuron, enters the spinal cord, and traverses the synapse between the axon of the sensory neuron and the dendrites or cell body of a motor neuron in the gray matter of the cord. The axon of the motor neuron, which leaves the cord ventrally through the *ventral root* and passes all the way to the effector cells (muscle fibers in the leg), carries the impulse to the leg where a response is effected. Thus, when the patellar tendon is tapped, the receptor is stimulated, impulses travel through a sensory neuron, the spinal cord, down a motor neuron, and to the muscles of the leg to bring about contraction and a jerking response.

It should be noted that sensory (afferent) neurons transmit nerve impulses toward the central nervous system and their cell bodies are always outside the spinal cord (dorsal root ganglion); the axons of sensory neurons enter the cord dorsally; the axons of motor neurons always leave the cord ventrally; and motor (efferent) neurons transmit impulses away from the central nervous system. These observations apply not

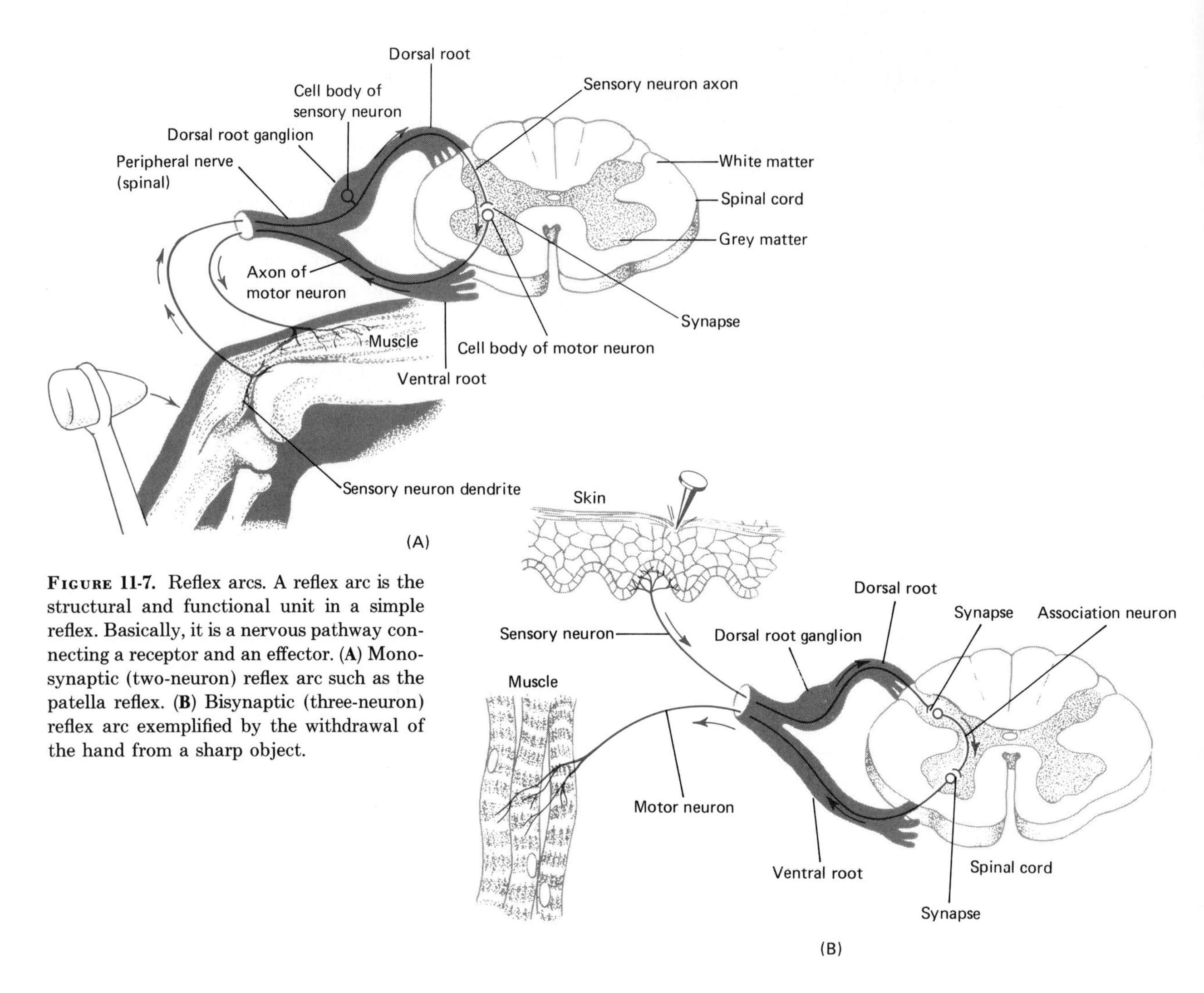

Figure 11-7. Reflex arcs. A reflex arc is the structural and functional unit in a simple reflex. Basically, it is a nervous pathway connecting a receptor and an effector. (**A**) Monosynaptic (two-neuron) reflex arc such as the patella reflex. (**B**) Bisynaptic (three-neuron) reflex arc exemplified by the withdrawal of the hand from a sharp object.

only to the example cited, but also to simple reflexes in general.

It should not be thought that the simple reflex arc just described is the only, or even typical, kind of neural pathway that operates in vertebrates. In fact, very few reflexes involve only two neurons; most also involve one or more association neurons connecting sensory and motor neurons. A common example of a three-neuron reflex arc consisting of a sensory neuron, an association neuron, and a motor neuron is the withdrawal of the hand from a hot object (Figure 11-7B). In this case, the impulse must traverse two synapses. In addition, such a reflex involves a number of reflex arcs, not just one. Consider that when the hand is stimulated by heat, other reflexes also become operative. The person might look in the direction of the stimulus, exhibit a facial and vocal expression of pain, and probably experience a series of emotions such as fear, apprehension, anger, and so on. All of these responses indicate that what started out as a single stimulus has elicited a rather complex reflex behavior involving the integration of a number of components of the nervous system.

This series of responses is explained by the fact that axons of sensory neurons in many cases have a number of branches, each of which may synapse with association neurons at different levels in the spinal cord and brain. Thus, a given sensory neuron has potential attachments with many effectors so that a number of responses can be evoked simultaneously in different parts of the body. Any such linkage of sensory neurons that carries an impulse from lower to higher levels of the cord and to the brain is called an *ascending tract.* By contrast, interconnected motor neurons that carry impulses from the brain to successively lower levels of the spinal cord are termed *descending tracts.*

The Central Nervous System

Protection. Inasmuch as both the brain and spinal cord are extremely delicate and vital organs of the nervous system, they are provided with two kinds of coverings. The first of these, the outer covering, consists of bone. The brain is enclosed by the cranial (skull) bones, while the spinal cord is protected by the vertebrae of the spinal column. The inner covering is formed by three membranes called *meninges,* which are continuous on the brain and cord. The outer membrane, the *dura mater,* is a tough, protective covering comprised of white fibrous tissue. The middle membrane, or *arachnoid,* is a delicate fibrous tissue between the dura mater and the innermost *pia mater,* a transparent layer that adheres to the brain and cord and contains blood vessels (Figure 11-8). Inflammation of the meninges, called *meningitis,* frequently involves the middle and inner layers.

In addition to bony and membranous coverings, the brain and cord are further protected against injury by a lymph-like fluid called *cerebrospinal fluid.* This fluid acts as an internal cushion for the central nervous system. It is found between the arachnoid and pia mater of both the brain and cord, within cavities inside the brain called *ventricles,* and within the central, or spinal, canal inside the cord.

The Brain: Structure and Function. Of all the organs comprising the nervous system, the brain is by far the largest and most complex. In adults, it weighs approximately 3 pounds. The brain grows rap-

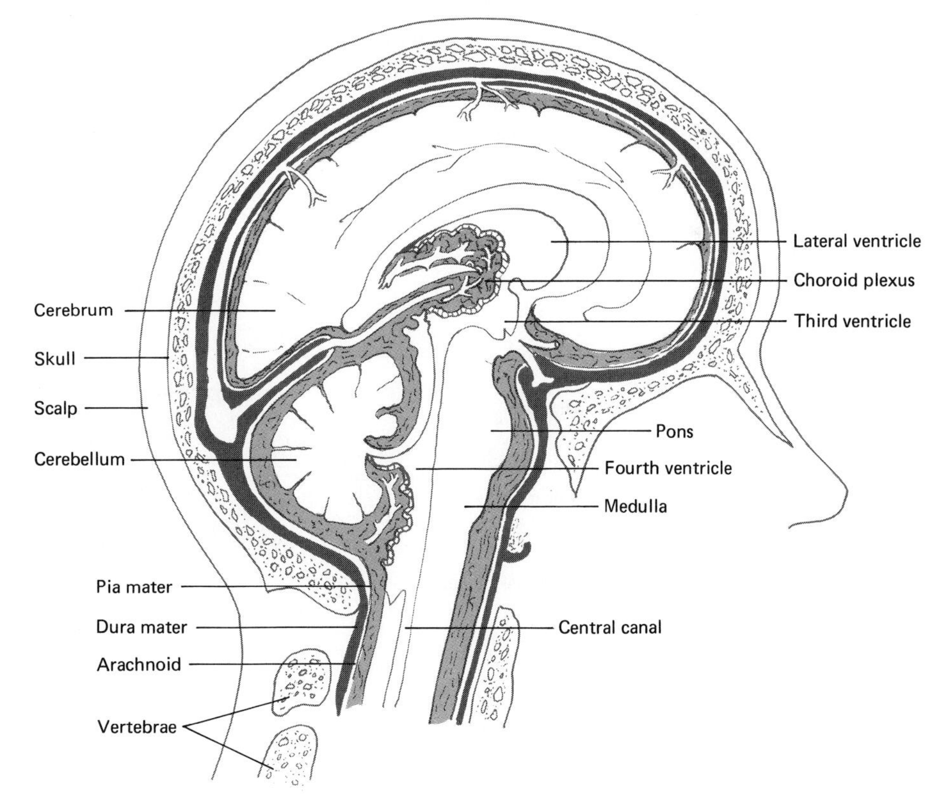

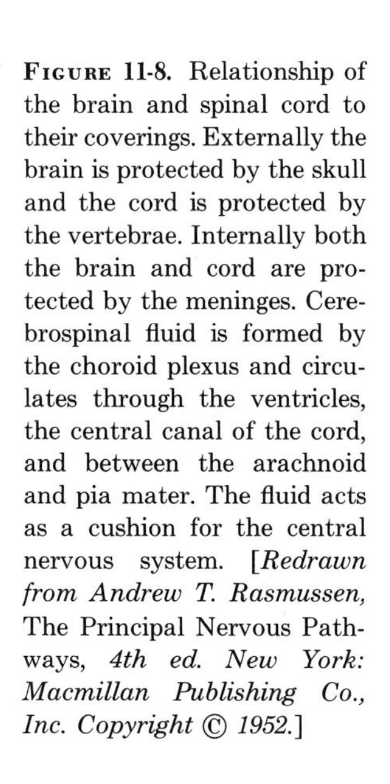
Figure 11-8. Relationship of the brain and spinal cord to their coverings. Externally the brain is protected by the skull and the cord is protected by the vertebrae. Internally both the brain and cord are protected by the meninges. Cerebrospinal fluid is formed by the choroid plexus and circulates through the ventricles, the central canal of the cord, and between the arachnoid and pia mater. The fluid acts as a cushion for the central nervous system. [*Redrawn from Andrew T. Rasmussen,* The Principal Nervous Pathways, *4th ed. New York: Macmillan Publishing Co., Inc. Copyright © 1952.*]

idly only during the first nine years or so, and growth ceases at about the twentieth year. The principal subdivisions of the brain are shown in Figure 11-9.

The Forebrain. The *forebrain* consists of three major areas: cerebrum, thalamus, and hypothalamus. The **cerebrum** is the largest area of the brain and is divided by a *longitudinal fissure* into halves called *hemispheres.* Each hemisphere, in turn, is partitioned into lobes separated by other prominent fissures (Figure 11-10A). Structurally, the hemispheres are made up of three portions: cerebral cortex, tracts of fibers, and basal ganglia. The *cerebral cortex* is a thin surface layer of the cerebrum. *Gray matter* (nerve cell bodies and dendrites) constitutes the cortex, whereas *white matter* (myelinated axons) compose the portion of the cerebrum under the cortex. In early life, the cortex is relatively smooth, but with growth and development, the surface becomes covered with furrows of varying depth. The deeper furrows are called *fissures,* the shallow ones are termed *sulci,* and the ridges between sulci are referred to as *gyri* or *convolutions.*

Cerebral tracts are bundles of axons located inside the brain and spinal cord. The first part of the name of a tract indicates the location of the dendrites and cell bodies of the neurons whose axons compose the tract; the last part of the name indicates the structure in which the axons terminate. For example, axons of the corticospinal tract originate in cell bodies of the cerebral cortex and end in the spinal cord. Thus, the name of a tract indicates in which direction its fibers (axons) conduct impulses. In general, the fibers of the cerebral tracts extend in three directions.

1. *Commissural fibers* transmit impulses from one hemisphere to another.
2. *Projection fibers* connect the cerebrum with other parts of the brain and spinal cord (ascending and descending tracts).
3. *Association fibers* transmit impulses from one part of the cerebral cortex to another within the same hemisphere.

Basal ganglia or *basal nuclei* are masses of gray matter located deep within the cerebral hemispheres (Figure 11-10B). The basal nuclei are all interconnected with many fibers; some are connected as well to the thalamus and hypothalamus. Although the functions of the basal nuclei are not clear, there is some clinical evidence to suggest that they play a role in producing movement.

The cerebrum performs all mental functions (reason, intelligence, will, and memory) and many essential motor, sensory, and visceral activities. As a result of numerous experiments on animals, observations of the

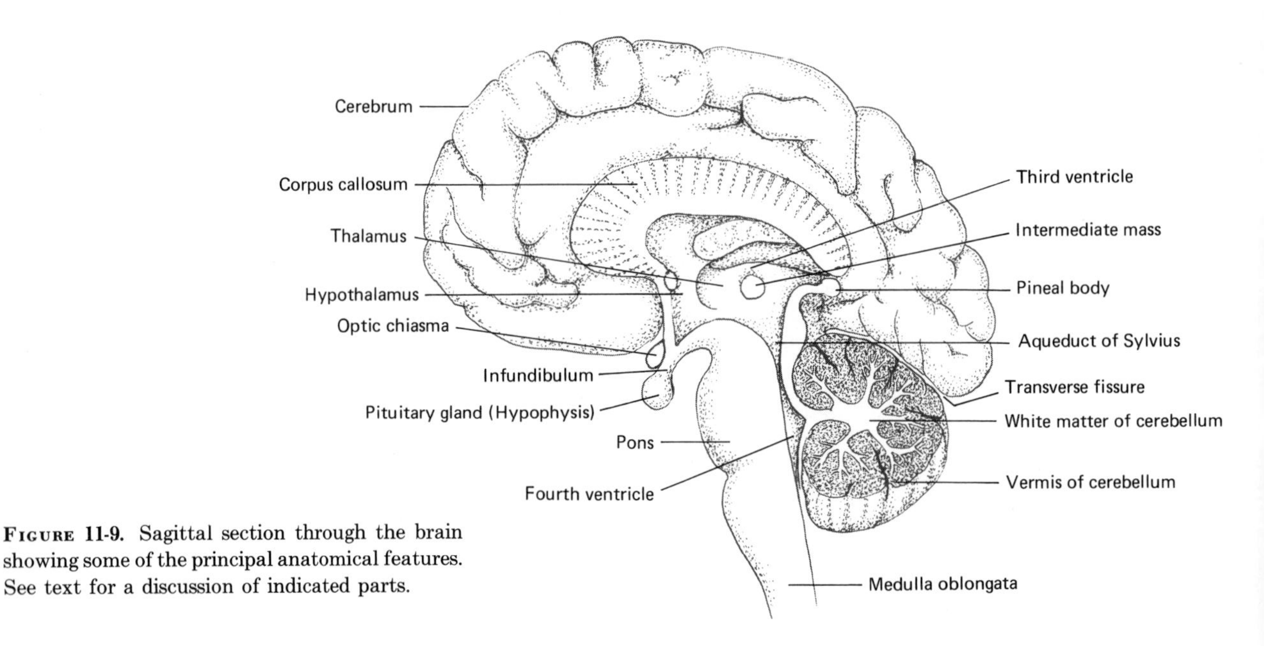

Figure 11-9. Sagittal section through the brain showing some of the principal anatomical features. See text for a discussion of indicated parts.

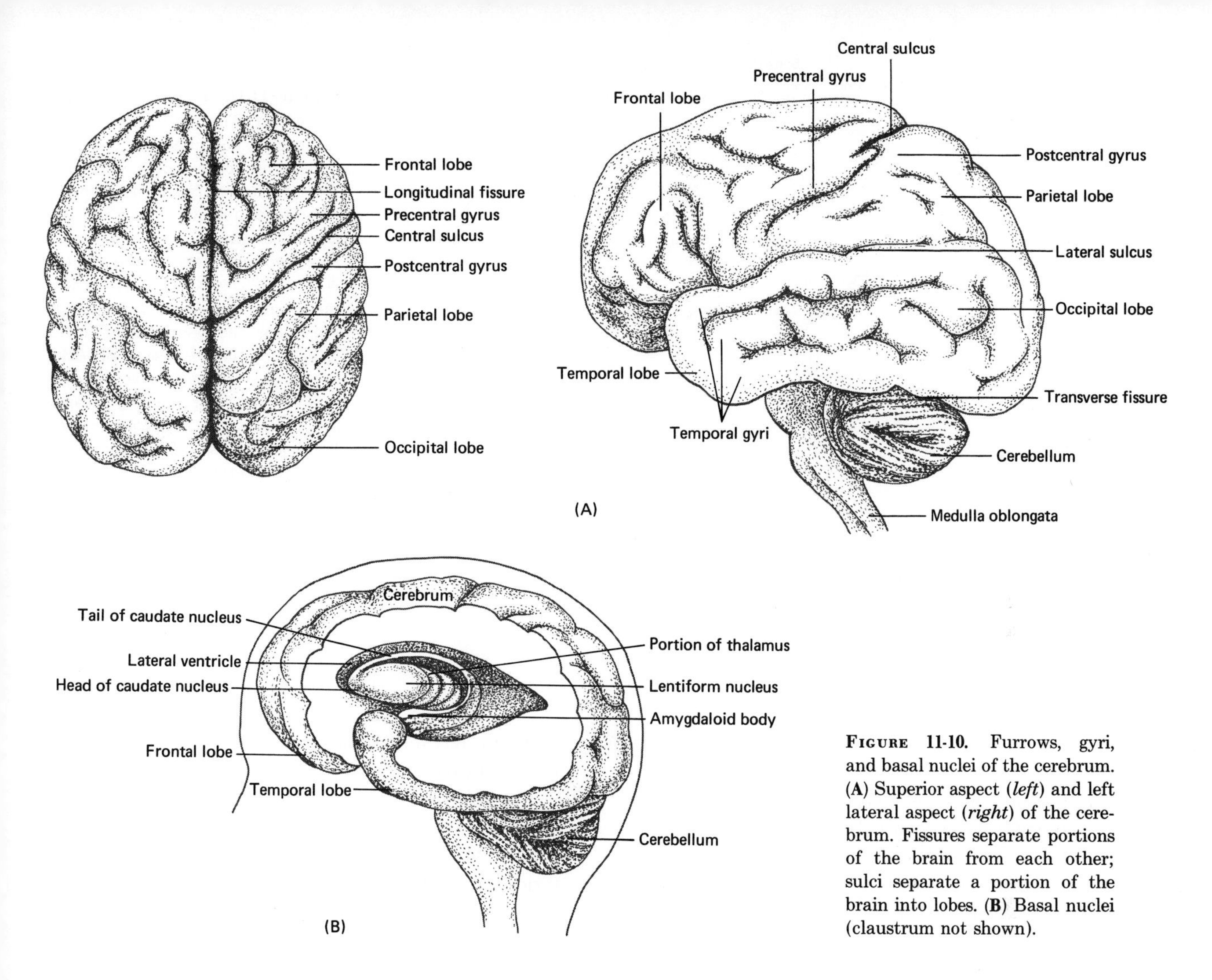

Figure 11-10. Furrows, gyri, and basal nuclei of the cerebrum. (**A**) Superior aspect (*left*) and left lateral aspect (*right*) of the cerebrum. Fissures separate portions of the brain from each other; sulci separate a portion of the brain into lobes. (**B**) Basal nuclei (claustrum not shown).

effects of electrical stimulation of the cerebral cortex, and analyzing symptoms of patients with brain lesions, physiologists have been able to locate certain areas in the brain which control motor, sensory, and other activities. Some knowledge has been gained concerning the areas in the cerebrum which function in higher mental activities.

The portions of the cerebrum that govern muscular movement are known as *motor areas;* those involved in the analysis of sensations are called *sensory areas;* and those concerned with the higher faculties, such as reason and will, are designated as *association areas.* Figure 11-11 illustrates the arrangement of both motor and sensory areas of the cortex.

The second portion of the forebrain, the **thalamus,** is a large, oval, bilateral structure located above the midbrain (Figure 11-9). Each half contains a series of nuclei that are responsible for receiving all sensory impulses either directly or indirectly from all parts of the body with the exception of olfactory sensations. In addition, it plays a part in the mechanism responsible for emotions by associating sensory impulses with feelings of pleasantness and unpleasantness; it functions in the arousal or alert mechanism of the body; and it assumes a role in producing complex reflex movements.

The final portion of the forebrain to be considered is the **hypothalamus** (Figure 11-9). This structure forms the floor and the lower part of the side wall of the third

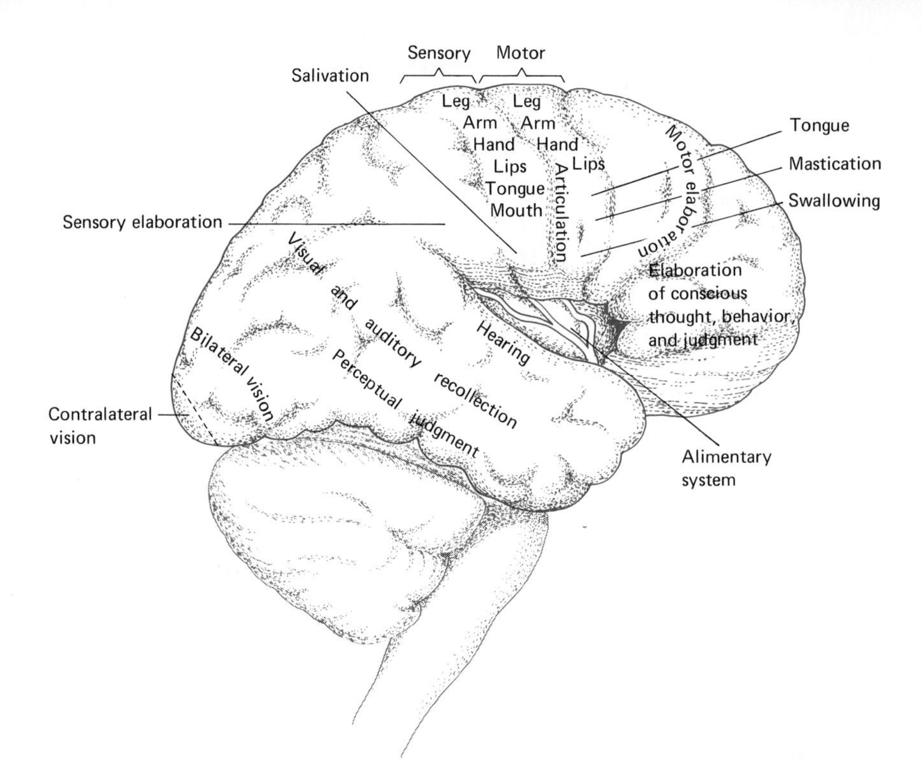

Figure 11-11. Functional areas of the cortex. The motor area is concerned with voluntary movements of skeletal muscles and is located in front of the central sulcus. The sensory structures are behind the central sulcus and serve both cutaneous and deep sensations. Approximately 30% of the cortex is concerned with motor and sensory activities. The remainder, the associational cortex, integrates the activities of other areas and is believed to be related to functions such as memory, reason, and judgment. [*Redrawn from W. Penfield and T. Rasmussen,* The Cerebral Cortex of Man. *New York: Macmillan Publishing Co., Inc. Copyright © 1950.*]

ventricle. Among the functions of the hypothalamus are

1. Partial control of sleep and wakefulness.
2. Regulation of food intake and satiation.
3. Regulation of body temperature.
4. Control of metabolism and water balance through the synthesis of ADH, which is secreted by the posterior pituitary.
5. Regulation of autonomic activities.
6. Control of various reproductive functions.
7. Serving as a major relay station between the cerebral cortex and lower autonomic centers.
8. Producing neurohumors.

The Midbrain. The *midbrain* is a short, constricted mass of projection tracts consisting mostly of white matter surrounding a central cavity. It connects the pons and cerebellum with the cerebral hemispheres. The dorsal part of the midbrain also contains a prominent mass of gray matter, which, in conjunction with the cerebellum, controls muscular coordination. The ventral portion, consisting of two rope-like masses of white matter, serves as the major connection between the hindbrain and the forebrain. The dorsal part of the midbrain has four rounded eminences in which lie certain auditory and visual reflex centers.

The Hindbrain. The three portions that comprise the *hindbrain* are the cerebellum, medulla, and pons (Figure 11-9). The **cerebellum** occupies the lower posterior part of the skull cavity. It is the second largest part of the brain and, like the cerebrum, it consists of an outer layer of gray matter (cortex) and an inner core of white matter (medullary body). Its surface is also grooved, with numerous sulci, but its convolutions are more slender and less prominent than the cerebral furrows. The cerebellum has two large lateral masses, the *cerebellar hemispheres,* and a central portion, the *vermis.* The internal white matter, consisting of both short and long tracts, connects various portions of the cerebellum with each other as well as with other parts of the brain and spinal cord.

Impulses from cellebellar gray matter coordinate the activities of several other brain centers. In this context, the cerebellum performs three general functions, all of which deal with the control of skeletal muscles. First, in conjunction with the cerebral cortex, the cerebellum helps to produce skilled movements by coordinating the activities of groups of muscles. Second, it controls skeletal muscles related to the maintenance of equilibrium. Finally, the cerebellum functions below the level of consciousness to make body movements smooth, timed, precise, and steady. Some recent evidence also suggests that the cerebellum may assume a role in the integration of touch, hearing, and sight sensations.

The **medulla (medulla oblongata),** the second portion of the hindbrain, is an enlargement of the anterior portion of the spinal cord that connects the brain to the cord. It is about 1 inch long and consists principally of ascending and descending tracts of white matter. Nuclei in the medulla contain reflex centers, some of which are so necessary for survival that they are termed *vital centers.* The nuclei serve as centers for various reflexes controlling heart action, blood vessel diameter, and respiration. Other nonvital nuclei contain reflex centers for activities such as vomiting, coughing, sneezing, hiccuping, and swallowing.

The **pons,** or **pons varolii,** lies just above the medulla and is also composed of white matter and a few nuclei. It consists largely of transverse tracts that conduct impulses between the cerebellum, medulla, and cerebrum. Functionally, the pons serves as a reflex center for reflexes mediated by the fifth, sixth, seventh, and eighth cranial nerves (see Table 11-1). In addition, the pons helps to regulate respiration. Finally, it serves as a conduction pathway (projection tracts) between the cord and other parts of the brain.

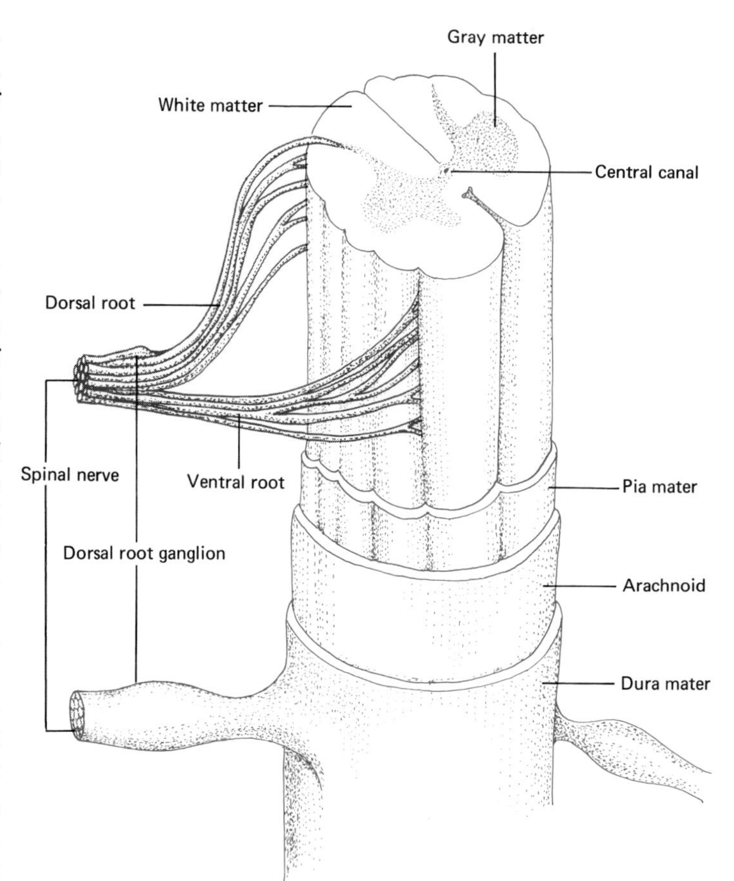

FIGURE 11-12. Section of spinal cord illustrating internal structure, nerves, and meninges.

THE SPINAL CORD. The **spinal cord** in the average adult is an oval-shaped cylinder measuring about 18 inches in length. It extends downward from the medulla through the bony neural arches of the vertebrae of the spinal column, tapering slightly as it descends. In cross section (Figure 11-12), it can be seen that the cord possesses an H-shaped core of grey matter that contains a minute central canal; the portion of the cord external to the grey matter consists of white matter. The transverse bar on the H, the *grey commissure,* connects the two lateral masses of grey matter. The grey matter contains nerve cell bodies and nonmyelinated fibers, whereas numerous myelinated fibers constitute the white matter.

The spinal nerves leave the cord between each pair of vertebrae. Each spinal nerve has two separate connections with the cord, a *dorsal root* and a *ventral root.* The sensory fibers of the nerve enter the cord by way of the dorsal root, while the motor fibers of the nerve emerge from the cord via the ventral root. The swelling on each dorsal root, the *dorsal root ganglion,* is the location of the cell bodies of sensory neurons.

The white matter of the cord is arranged into tracts. Some of these tracts (descending) transfer impulses from the brain to the motor neurons of the spinal nerves; others (ascending) serve as pathways to the brain for impulses entering the cord over afferent fibers of spinal nerves. Both kinds of tracts usually cross over from one part of the spinal cord to the other along their pathways to and from the brain. Thus, the right side of the brain receives impulses that originate on the left side of the body and vice versa.

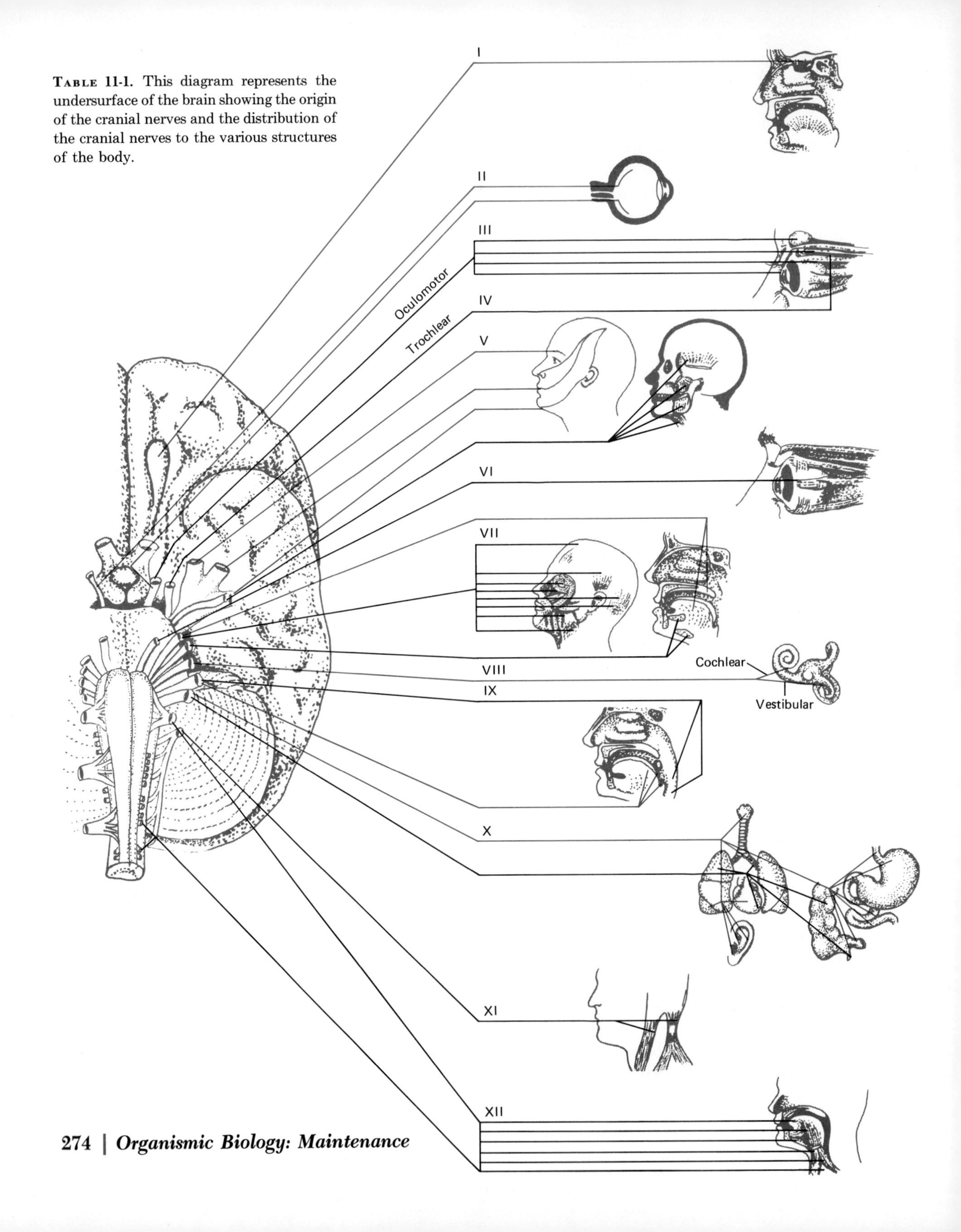

Table 11-1. This diagram represents the undersurface of the brain showing the origin of the cranial nerves and the distribution of the cranial nerves to the various structures of the body.

TABLE 11-1. Cranial Nerves, Distribution, and Functions*

Name	Distribution	Functions
I. Olfactory (sensory)	Nasal mucosa	Smell
II. Optic (sensory)	Retina	Sight
III. Oculomotor (motor)	External eye muscles, except superior oblique and lateral rectus	Accommodation, eye movements, regulation of pupil
IV. Trochlear (motor)	Superior oblique	Eye movements
V. Trigeminal (mixed)	Skin and mucosa of head, teeth, muscles of mastication	Chewing movements, sensations of head and face
VI. Abducens (motor)	Lateral rectus	Movement of eye
VII. Facial (mixed)	Taste buds of anterior two-thirds of tongue, superficial muscles of face and scalp.	Facial expressions, taste, secretion of saliva
VIII. Acoustic (sensory) Vestibular branch	Semicircular canals and vestibule (utricle and saccule)	Equilibrium
Cochlear	Organ of Corti in cochlear duct	Hearing
IX. Glossopharyngeal (mixed)	Pharynx, taste buds and other receptors of posterior one-third of tongue, carotid sinus and carotid body	Taste, secretion of saliva, swallowing, reflex control of blood pressure and respiration
X. Vagus (mixed)	Pharynx, taste buds carotid body, thoracic and abdominal viscera	Movements and sensations of organs supplied (viscera of thorax and abdomen); among functions are decreasing heart rate, increase in peristalsis, and contracting muscles for voice production
XI. Spinal accessory (motor)	Sternocleidomastoid, trapezius	Shoulder and head movements, movements of viscera, voice production
XII. Hypoglossal (motor)	Tongue muscles	Tongue movements

* The diagrams on page 274 represent the undersurface of the brain showing the origin of the cranial nerves and the distribution of the cranial nerves to the various structures of the body.

The principal functions of the spinal cord are to serve as a pathway for the conduction of impulses between the peripheral nervous system and the brain, and to act as a reflex center for numerous local reflexes, particularly simple ones.

The Peripheral Nervous System

As stated earlier, the peripheral nervous system consists of twelve pairs of cranial nerves and thirty-one pairs of symmetrical spinal nerves and their branches. A **nerve** is simply a cord-like structure composed of bundles of nerve fibers enclosed in a connective tissue sheath.

The *cranial nerves* arise from the undersurface of the brain and pass through small openings (foramina) in the skull to their respective designations. They are numbered in the order in which they emerge from front to back, and they are named on the basis of their distribution or function. Some are sensory and consist of afferent fibers only, others consist of efferent fibers and are motor, and still others contain both afferent and efferent fibers and are classified as mixed. Information concerning the twelve pairs of cranial nerves is summarized in Table 11-1.

The thirty-one pairs of *spinal nerves* that originate on the spinal cord have no special names but are merely numbered according to the level of the spinal column at which they emerge. Thus, there are eight cervical, twelve thoracic, five lumbar, five sacral, and one coccygeal pair of spinal nerves. Outside the cord, the dorsal root unites with the ventral root to form a mixed nerve. Unlike cranial nerves, all spinal nerves branch extensively to various portions of the body.

After many of the spinal nerves emerge from the cord, they branch extensively to form networks of nerves called *plexuses.* The first plexus, or network of nerves, is the *cervical plexus* and is formed by the union of the ventral branch of the first four cervical nerves. The second plexus, the *brachial plexus,* is formed by the union of the ventral branches of the last four cervical and the first thoracic nerves. The *lumbar plexus* is formed by a few fibers from the twelfth thoracic and the first four lumbar nerves. The fourth plexus, the *sacral plexus,* is formed by a few fibers from the fourth lumbar nerve, all of the fifth, and the first, second, and third sacral nerves. Refer to Table 11-2 for the nerves of which these plexuses are formed and the body areas that they serve.

TABLE 11-2. Spinal Nerves, Plexuses Formed, and Innervation

Spinal Nerve	Plexus Formed	Innervation (Parts Supplied)
C-1 C-2 C-3 C-4	Cervical	Sensory to back of head, front of neck, and upper part of shoulder; motor to many neck muscles.
C-5 C-6 C-7 C-8 T-1	Brachial	Superficial muscles of scapula, pectoralis major and minor, deltoid, and skin over it, teres minor, muscles of front of arm and skin on outer side of forearm, upper extremities.
T-12 L-1 L-2 L-3 L-4	Lumbar	Skin on lateral half of thigh, flexor muscles of thigh, skin of anterior thigh, hip region, and lower leg.
L-4 L-5 S-1 S-2 S-3	Sacral	Skin on anterior surface of leg and the dorsum of the foot, posterior muscles of the external genitalia, skin of perineum, anal sphincters.

The portion of the peripheral nervous system that innervates skeletal muscles and skin and is responsible for movements of various parts of the body, either reflex or voluntary, is the *somatic nervous system.* The other subdivision of the peripheral nervous system is the autonomic nervous system.

THE AUTONOMIC NERVOUS SYSTEM. The **autonomic nervous system** innervates cardiac muscle, smooth muscle, and glands and controls the functions of the internal organs automatically and unconsciously. The autonomic nervous system is organized into two distinct regions along the central nervous system. These are the sympathetic division and the parasympathetic division. The *sympathetic division* is composed of autonomic fibers from the thoracic and lumbar regions of the spinal cord (Figure 11-13A); *parasympathetic fibers,* by contrast, arise from the brain and sacral portions of the cord (Figure 11-13B). Thus, most organs of the body

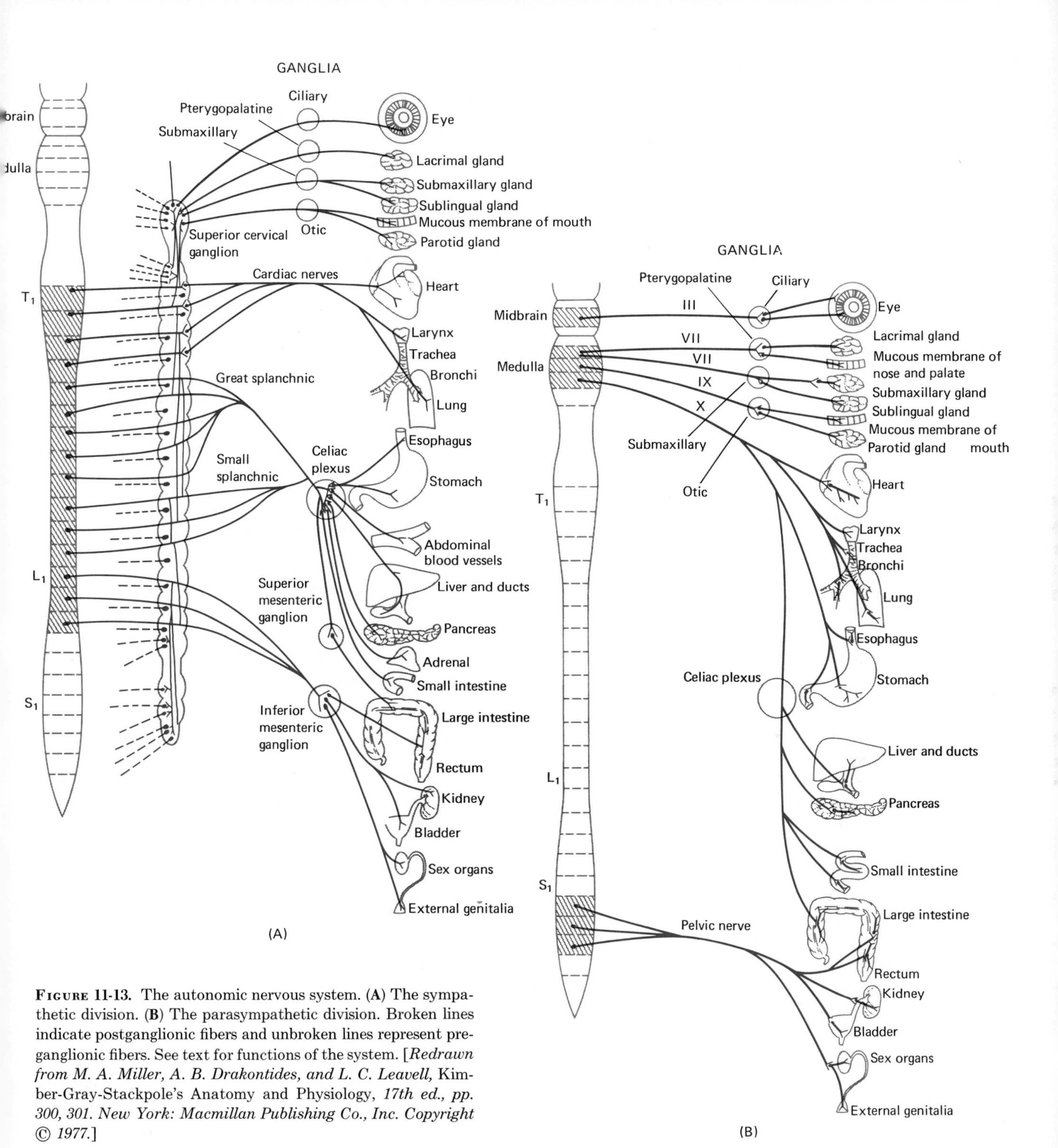

FIGURE 11-13. The autonomic nervous system. **(A)** The sympathetic division. **(B)** The parasympathetic division. Broken lines indicate postganglionic fibers and unbroken lines represent preganglionic fibers. See text for functions of the system. [*Redrawn from M. A. Miller, A. B. Drakontides, and L. C. Leavell,* Kimber-Gray-Stackpole's Anatomy and Physiology, *17th ed., pp. 300, 301. New York: Macmillan Publishing Co., Inc. Copyright © 1977.*]

under control of the autonomic nervous system receive motor stimuli through fibers from some region of the sympathetic division as well as from either the cranial or sacral portion of the parasympathetic division, depending upon the location of the organ concerned. The effects of impulses of the respective sets of nerves are usually antagonistic; that is, if impulses from the sympathetic division increase the heart rate, impulses from the parasympathetic division would decrease heart rate. In this way, a given organ of the body is either inhibited or stimulated in order to maintain homeostasis of the internal environment.

Each neural pathway of motor impulses to organs under autonomic control typically involves two neurons. One of these comes from the spinal cord and one or more passes to the effector concerned. Synapses between the two neurons are generally made in a ganglion external to the cord. The neuron from the cord to the ganglion is termed a *preganglionic fiber,* and the neuron from the ganglion to the organ under control is referred to as a *postganglionic fiber.* The secondary ganglia of the parasympathetic division lie relatively close to the spinal cord, forming two vertical rows or chains (sympathetic chains) with connecting fibers between them, one row on each side of the cord.

By controlling cardiac and smooth muscle and many glands, the autonomic nervous system assumes a critical role in regulating the activity of internal organs of the body. In this regard, it enables the body to adapt to immediate needs by either accelerating or inhibiting the action of a particular organ. For example, nerve impulses of the sympathetic division stimulate heart rate, blood pressure, and the activity of sweat and adrenal glands and in this way enable the body to respond in emergency situations by promoting activities concerned with maximal energy production. Nerve impulses of the parasympathetic division depress these same activities so that the constancy of the internal environment is maintained.

Innervation of visceral organs by autonomic fibers provides rapid and efficient coordination of a given organ under their control. The antagonistic action of the sympathetic and parasympathetic divisions results from responses of the various internal organs to chemical substances produced by the terminal axon branches of postganglionic neurons. These chemicals are nerve-impulse induced. Except in a very few instances, the terminals of postganglionic parasympathetic neurons release acetylcholine. Similarly, sympathetic postganglionic nerve endings secrete another chemical substance, epinephrine. The accumulated evidence indicates that acetylcholine and epinephrine account for the antagonistic effects of the autonomic nervous system.

Before closing our discussion of the autonomic nervous system, it should be made quite clear that the nervous system as a whole (central and peripheral) is, in fact, a unified system. Although each has distinct centers of control and neuron reflex pathways, the two are interrelated structurally and functionally and, in conjunction with the endocrine system, contribute to the overall system of coordination and control. The distinctions we have made in this chapter were made arbitrarily, only for the purposes of discussion.

Selected Disorders of the Nervous System

Many disorders can affect the various parts of the nervous system. Some of the diseases are known to be caused by viruses or bacteria. Other conditions are caused by damage to the nervous system during birth. The origins of many conditions, however, are unknown. We shall now discuss the origins and describe the symptoms of some common nervous system disorders.

Poliomyelitis

Poliomyelitis, also known as **infantile paralysis,** is a viral infection that is most common during childhood. In recent years an immunization against the disease has been used. However, nearly 100% immunization is achieved only if the child is immunized shortly after birth. The onset of the disease is marked by fever, severe headache, a stiff neck and back, deep muscle pain and weakness, and loss of some somatic reflexes. The virus may spread by means of the blood and respiratory passages to the central nervous system. In the central nervous system, it destroys the motor nerve cells, specifically those in the anterior horns of the spinal cord and in the nuclei of some of the cranial nerves. The injury to the spinal gray matter gives the disease its name, *polio,* meaning gray matter, and *myel,* meaning spinal cord. Destruction of the anterior horns produces paralysis of one or more limbs. Poliomyelitis can cause death from respiratory or heart failure if the virus invades the brain cells of the vital medullary centers.

Syphilis

Syphilis is a venereal disease caused by the *Treponema pallidum* bacterium. Venereal diseases are typically infectious disorders that are spread through sexual contact of any sort. The disease process of syphilis goes through several stages—primary, secondary, latent, and sometimes tertiary. During the primary stage, the chief symptom is an open sore, called a chancre, at the point of contact. The chancre eventually heals. About six weeks later, a range of generalized symptoms, such as a skin rash, fever, and aches in the joints and muscles, ushers in the secondary stage. At this stage, syphilis can usually be treated with antibiotics. However, even if the person does not undergo treatment, the symptoms will eventually disappear. Within a few years, he/she will usually cease to be infectious. The signs and symptoms of the disease disappear, but a blood test would show positive results. During this later "symptomless" period, called the latent stage, the bacteria may invade and slowly destroy any of the body organs. This is why untreated syphilis is so dangerous. When symptoms of organ degeneration appear, the disease is said to be in its tertiary stage. If the syphilis bacteria attack the organs of the nervous system, the tertiary stage is called **neurosyphilis.** Neurosyphilis may take any one of several forms, depending on the tissue involved. For instance, about two years after the onset of the disease, the bacteria may attack the meninges, producing meningitis. Or, the blood vessels that supply the brain may become infected. In this case, symptoms vary depending on the parts of the brain that are destroyed by oxygen and glucose starvation. Over the years, the bacteria may spread through the nerve cells of the brain. As one nerve cell after another is destroyed, the patient experiences corresponding symptoms. Cerebellar damage is manifested by uncoordinated movements. For instance, the person may have trouble with skilled activities, such as writing. As the motor areas become extensively damaged, the victim may be unable to control urine and bowel movements, become bedridden, and completely lose control of the limbs. Damage to the cerebral cortex produces memory loss and personality changes that range from irritability to confusion to hallucinations.

A common form of neurosyphilis is **tabes dorsalis,** a progressive degeneration of the posterior columns of the spinal cord. This form affects sensory ganglia and the sensory nerve roots. Tabes dorsalis forms an interesting contrast with polio. The polio virus attacks the anterior columns and destroys motor neurons. The polio victim is unable to move the affected muscles voluntarily, but retains all sensory functions. The person with tabes dorsalis suffers from just the opposite problem. Such an individual retains motor control, but loses many sensory functions. The person often experiences tingling or numbness in the limbs and trunk. Normally, receptors in the joints are able to tell the central nervous system how much a joint is flexed and where one part of the body is in relation to another part. This information is necessary for coordinating movement and maintaining posture and balance. When the sensory nerve roots are destroyed, however, this information cannot pass from the receptors to the brain. Consequently, the victim of tabes dorsalis must rely on sight to carry out motor activities successfully. Walking in the dark is difficult because there is no sense of whether the legs are flexed or where the feet are. In lighted areas, the person may walk with a characteristic shuffle, which consists of jerking the knee up and then letting the leg extend abruptly to the ground.

Syphilis can be treated with antibiotics during the primary, secondary, and latent periods. Some forms of neurosyphilis can also be successfully treated; however, the prognosis for tabes dorsalis is very poor. Unfortunately, not everyone with syphilis shows noticeable symptoms during the first two stages of the disease. But syphilis can usually be diagnosed through a blood test. The importance of these blood tests and follow-up treatment cannot be overemphasized.

Cerebral Palsy

The term **cerebral palsy** refers to a group of motor disorders caused by damage to the motor areas of the brain during fetal life, birth, or infancy. One of the causes is infection of the mother with German measles during the first three months of pregnancy. During early pregnancy, certain cells in the fetus are dividing and changing in order to lay down the basic structures of the brain. These cells can be abnormally changed by toxin from the measles virus. Radiation during fetal life, temporary oxygen starvation during birth, and hydrocephalus during infancy can also damage brain cells.

Cases of cerebral palsy are categorized into three groups depending on whether the cortex, the basal ganglia of the cerebrum, or the cerebellum is affected most severely. Most cerebral palsy victims have at least some damage in all three areas. The location and extent of motor damage determine the symptoms. For instance, the cerebral palsy victim may have a contorted face caused by partial facial paralysis. If the tongue becomes paralyzed, speech is lost and only guttural sounds can be made. Extensive damage to the cerebellum causes very uncoordinated movements. Although cerebral palsy refers only to motor damage, sensory and association areas of the brain may be affected as well. The person may be deaf or partially blind. About 70% of cerebral palsy victims appear to be mentally retarded. However, the apparent mental slowness is often due to the person's inability to speak or hear well. These people are often much brighter than they seem.

Cerebral palsy is not a progressive disease. In other words, it does not get worse as time elapses. But, once the damage is done, it is irreversible.

Parkinsonism

This disorder, also called **Parkinson's disease,** is a progressive malfunction of the basal ganglia of the cerebrum. Recall that the basal ganglia regulate unconscious contractions of skeletal muscles that aid activities desired by the motor areas of the cerebral cortex. Examples of movement produced by the cerebral nuclei are swinging the arms when walking and making facial expressions when talking. In Parkinsonism, the cerebral nuclei produce useless skeletal movements that often interfere with voluntary activity. For instance, the muscles of the arms and hands may alternately contract and relax so that the patient's hands shake. This type of shaking is called a *tremor.* Other muscles may contract continuously and make the involved part of the body rigid. *Rigidity* of the facial muscles gives the face a masklike appearance. The expression is characterized by a wide-eyed, unblinking stare and a slightly open mouth with saliva drooling from the corners. Vision, hearing, and intelligence are unaffected by the disorder, indicating that Parkinsonism does not attack the cerebral cortex.

Parkinsonism may be caused by a malfunction at the neuron synapses. The motor neurons of the basal ganglia release the chemical transmitter acetylcholine. In normal people, the basal ganglia also produce a synaptic transmitter called dopamine. People with Parkinsonism do not manufacture enough dopamine in their basal ganglia. Unfortunately, injections of dopamine are useless because the drug is not able to diffuse from the blood into the brain. However, a few years ago researchers developed a drug called levodopa, which is very similar to dopamine. Levodopa is able to diffuse into the brain and elevate the content of material in the ganglia, thus reducing the severity of the muscular affliction.

Epilepsy

Epilepsy is a brain disorder characterized by short, recurrent, periodic attacks of motor, sensory, and/or psychological malfunction. The attacks, called *epileptic seizures,* are brought on by abnormal and irregular discharges of electricity by millions of neurons in the brain. The discharges stimulate many of the neurons to send impulses over their conduction pathways. As a result, a person undergoing an attack may contract skeletal muscles involuntarily. The victim may sense lights, noise, or smells when the eyes, ears, and nose have not actually been stimulated. The electrical discharges may also inhibit certain centers of the brain. For instance, the waking center may be depressed so that the person loses consciousness.

Many different types of epileptic seizures exist. The particular type of seizure depends on the area of the brain that is electrically stimulated and whether the stimulation is restricted to a small area or spreads throughout the brain. *Grand mal* seizures are brought on by a burst of electrical discharges that travel throughout the motor areas and spread to the areas of consciousness in the brain. The person loses consciousness, experiences spasms of his voluntary muscles, and may also lose urinary and bowel control. Sensory and intellectual areas may also be involved. For instance, just as the attack begins, the person may sense a peculiar taste in his mouth, see flashes of light, or have olfactory hallucinations. The unconsciousness and motor activity last a few minutes. Then the muscles relax and the person awakens. He may be mentally confused for a short period of time afterward. Brain cells have the capacity to generate electrical potentials called *brain waves,* which indicate activity of the cerebral cortex. Brain waves pass easily through the skull, and they can be detected by sensors called electrodes. A record of such waves is called an **electroencephalogram,** or **EEG.** An EEG is obtained by placing electrodes on an individual's head and amplifying the waves by using an instrument called an electroencephalograph. As indicated in Figure 11-14A, four kinds of waves are produced by normal individuals.

1. *Alpha waves.* These rhythmic waves occur at a frequency of about 10–12 cycles per second. They are found in the EEGs of almost all normal individuals when they are awake and in the resting state. These waves disappear entirely during sleep.

2. *Beta waves.* The frequency of these waves is more than 15 cycles per second and may be as high as 60 cycles. Beta waves generally appear when the nervous system is active.

3. *Theta waves.* These waves have frequencies of 5–8 cycles per second. They are normal in children, but also occur in EEGs of adults who are undergoing a great deal of emotional stress.

4. *Delta waves.* The frequency of these waves is usually between 1 and 5 cycles per second and may be as low as 1 cycle every 2–3 seconds. Delta waves appear during deep sleep. They are normal in an awake infant. But when they are produced by an awake adult, they indicate certain types of brain damage.

Studies with EEGs show that grand mal attacks are characterized by rapid brain waves occurring at the rate of 25–30 per second (Figure 11-14B). Many epileptics suffer from electrical discharges that are restricted to one or several relatively small areas of the brain. An example is the *petit mal* form, which apparently involves the thalamus and hypothalamus. Petit mal seizures are characterized by an abnormally slow brain wave pattern occurring at the rate of three waves per second (Figure 11-14B). The person may lose contact with the environment for about 5–30 seconds, but does not undergo the embarrassing loss of motor control that is typical of a grand mal seizure. The victim merely seems to be daydreaming. A few people experience several hundred petit mal seizures each day. For them, the chief problems are a loss of productivity in school or work and periodic inattentiveness while driving a car.

Some epileptics experience motor seizures that are restricted to the precentral motor area of one cerebral hemisphere. These attacks consist of spasms that pass up or down one side of the body. People who suffer from sensory seizures may see lights or distorted objects if the discharge occurs in the occipital lobe. They may hear voices or a roaring in their ears if the discharge is located in the temporal lobe. Or, they may taste or smell unpleasant substances if the discharge is in the parietal lobe. People undergoing attacks of localized motor or sensory disturbances may or may not lose consciousness. A form of epilepsy that is sometimes confused with mental illness is *psychomotor epilepsy.* The electrical outburst occurs in the temporal lobe, where it causes the person to lose contact with reality. It may spread to some of the motor areas and produce mild spasms in some of the voluntary muscles. The person may stare into space and involuntarily smack his lips or clap his hands during an attack. If the motor areas are not involved, he may simply walk aimlessly. When he returns to reality, he is surprised to find himself in a strange or different place.

The causes of epilepsy vary. Many conditions can cause nerve cells to produce periodic bursts of impulses. Head injuries, tumors and abscesses of the brain, and childhood infections, such as mumps, whooping cough, and measles, are some of the causes. Epilepsy may also be *idiopathic,* that is, have no demonstrable cause. It should be noted, however, that epilepsy almost never affects intelligence. If frequent severe seizures are allowed to occur over a long period of time, some cerebral damage may occasionally result. However, damage can be prevented by controlling the seizures with drug therapy.

Epileptic seizures can be eliminated or alleviated by drugs that make neurons more difficult to stimulate. Many of these drugs change the permeability of the nerve cell membrane so that it does not depolarize as easily.

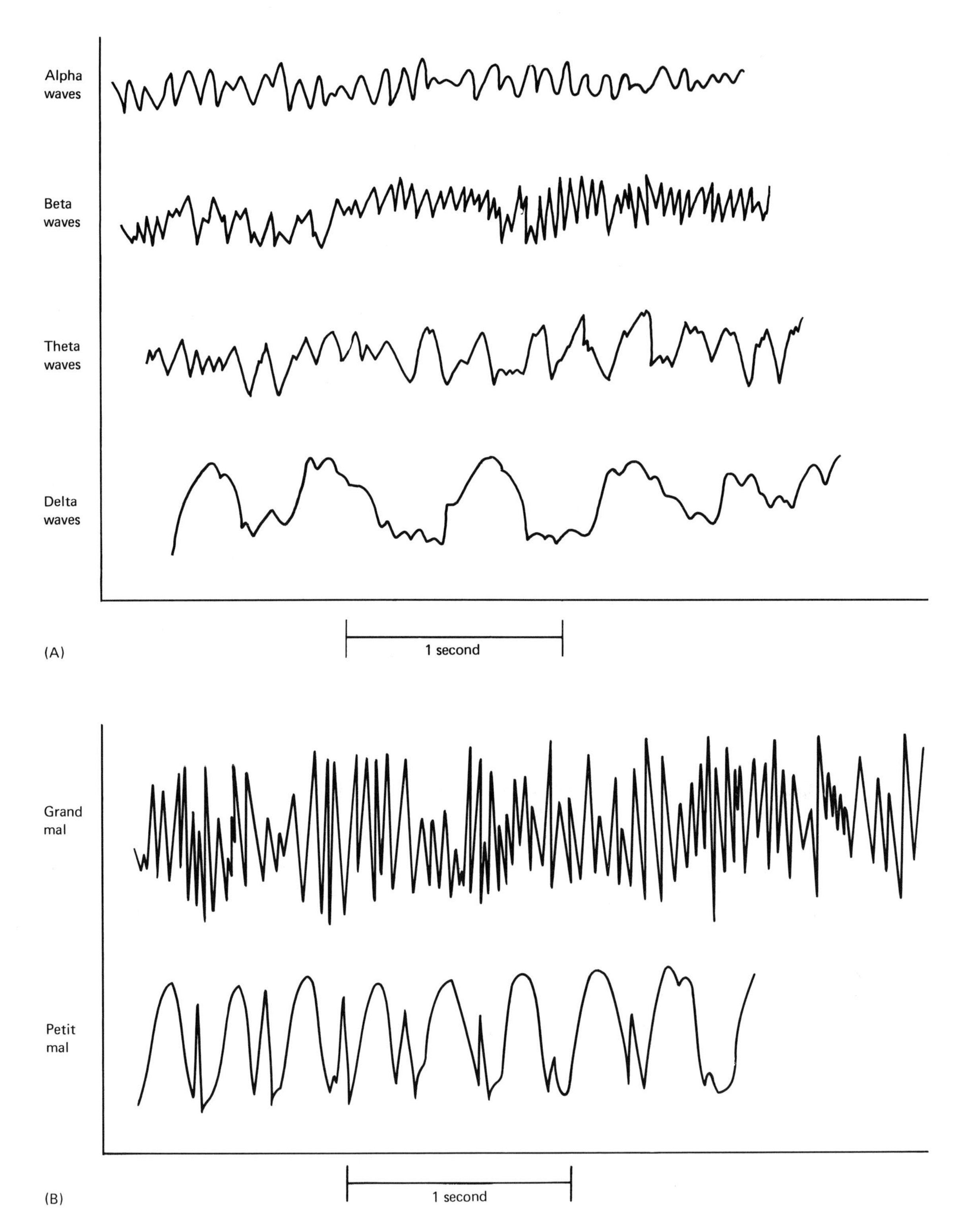
Alpha waves
Beta waves
Theta waves
Delta waves
(A)
1 second
Grand mal
Petit mal
(B)
1 second

Multiple Sclerosis

Multiple sclerosis causes a progressive destruction of the myelin sheaths of neurons in the central nervous system. The sheaths deteriorate to *scleroses,* which are hardened scars or plaques, in multiple regions. Hence, the disorder is given the name **multiple sclerosis.** The destruction of the myelin sheaths interferes with the transmission of impulses from one neuron to another, literally "short-circuiting" conduction pathways. Usually the first symptoms occur between the ages of 20 and 40. The early symptoms are generally produced by the formation of just a few plaques and are, consequently, mild. For instance, plaque formation in the cerebellum may produce some incoordination in one hand. The patient complains that his handwriting has become sloppy. A short-circuiting of some pathways in the corticospinal tract may partially paralyze the leg muscles so that the patient drags his foot when walking. The diseased sheaths often heal, and the symptoms disappear for a while. Later, a new series of plaques develop, and the victim suffers a second attack. One attack follows another over the years. Each time the plaques form, some of the neurons are damaged by the hardening of their sheaths. Other neurons are injured by their plaques. The result is periodic losses of function interspersed with healing periods during which the undamaged neurons regain their ability to transmit impulses.

The etiology of multiple sclerosis is unknown. Many researchers suspect that the disease is caused by a virus that does not affect the myelin sheaths of most people. Other possible causes are metal poisons, accidental injury to the central nervous system, and diseases of the blood vessels that supply the central nervous system.

Sense Organs and the Nervous System

A *sense organ* or sense receptor may be viewed as specialized nervous tissue or other tissue associated with a nerve cell that exhibits greatest sensitivity to a specific stimulus or change in the environment. Regardless of the simplicity or complexity of a given receptor, all are or contain dendrites of sensory neurons or highly specialized sensitive cells in close association with them.

Basically, receptors function, both at conscious and unconscious levels, by informing the body of changes in the external and internal environments so that the body can adjust to these changes in order to maintain homeostasis. As more data have become available, the popular notion of the five senses—touch, taste, smell, sight, and hearing—represents an incomplete list. For example, at least eleven different sensations are recognized among higher animals, including humans. These are touch, taste, smell, vision, hearing, cold, warmth, equilibrium, pain, proprioception (awareness of movement and position), visceral sensations (hunger, nausea, thirst, sexual sensations), and pressure. The sensory receptors of the body that receive these various sensations may be categorized into three principal types:

1. *Chemical receptors,* which pick up chemical stimuli.
2. *Mechanical receptors* for touch and pressure, pain, temperature, sound, and motion, which are sensitive to mechanical stimuli.
3. *Photoreceptors,* such as the eyes, which are sensitive to certain wavelengths of electromagnetic radiation.

Chemical Receptors

Aquatic animals utilize chemoreception in order to locate food, to seek others of their own kind, and to avoid enemies and unfavorable environmental conditions. Among terrestrial organisms, such as arthropods, chemoreceptors for smell are present on the antennae and taste receptors occur at the mouth region. The common housefly has taste receptors on its hindlegs.

The chemical senses of taste and smell in humans are frequently confused since the mouth and nasal cavities are anatomically connected and because a given substance may be smelled and tasted simultaneously. For these reasons the "taste" of many substances is actually a combination of smell and taste. This explains the fact that many foods are tasteless when the nasal passages are congested during a cold or allergic reaction. Taste and smell are associated with two different types of

FIGURE 11-14 [OPPOSITE]. Electroencephalogram. (**A**) Kinds of waves recorded in an electroencephalogram. (**B**) Electroencephalogram of grand mal and petit mal seizures.

receptors that transmit nervous impulses along different pathways to different regions of the brain.

Taste. The receptors for taste, called the *taste buds,* are located principally in the surface layer of the tongue (Figure 11-15A), but they are also distributed in the roof of the mouth, the throat, and the epiglottis. In a sectional view it can be seen that taste buds of the tongue are embedded in mucous membrane elevations called *papillae* (Figure 11-15B). Each receptor cell of a taste bud contains a fine hair-like projection extending toward the minute opening of its taste bud on the surface of the tongue (Figure 11-15C). The basal end of each sensory cell is connected to an axon. In order for the sensory cells to be stimulated, the chemicals must be in solution in saliva so that they can enter the microscopic pores of the taste buds.

Despite the multiplicity of chemicals that are tasted, there are basically only four taste sensations in humans—sour, salt, bitter, and sweet—and each is due to a different response to different chemicals. The different buds are distributed unevenly and in a characteristic pattern over the surface of the tongue. For example, taste buds for sweet and salty substances are concentrated at the tip of the tongue; taste buds for bitter substances are located at the back of the tongue; and receptors for sour substances are found along the edges of the tongue (Figure 11-15A). It appears that the actual taste of a substance is a complex sensation resulting from a blending of all the stimuli.

Smell. The receptors for smell are specialized cells that function as both receptor and conductor. These receptors are located in the mucous membrane that

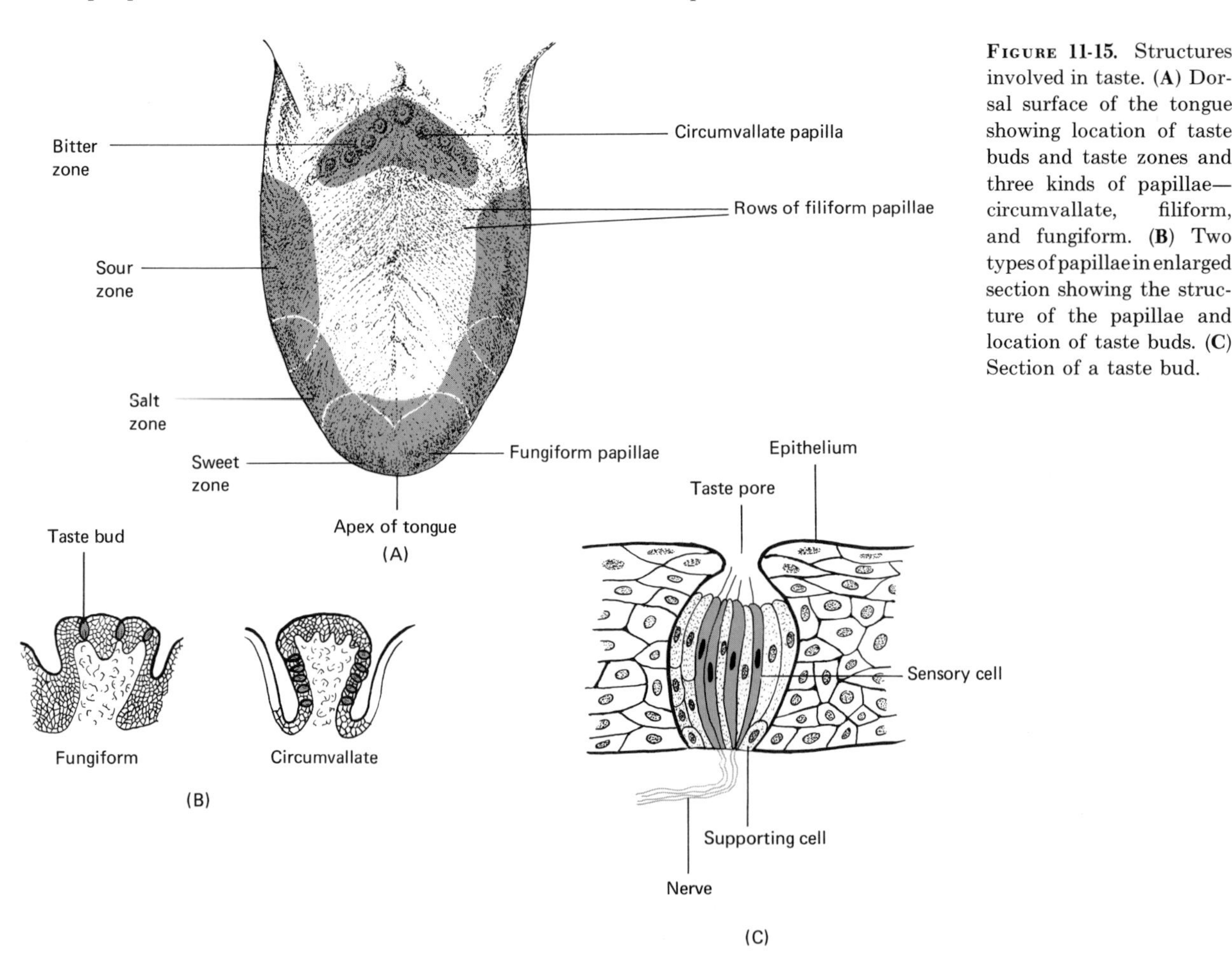

Figure 11-15. Structures involved in taste. (**A**) Dorsal surface of the tongue showing location of taste buds and taste zones and three kinds of papillae—circumvallate, filiform, and fungiform. (**B**) Two types of papillae in enlarged section showing the structure of the papillae and location of taste buds. (**C**) Section of a taste bud.

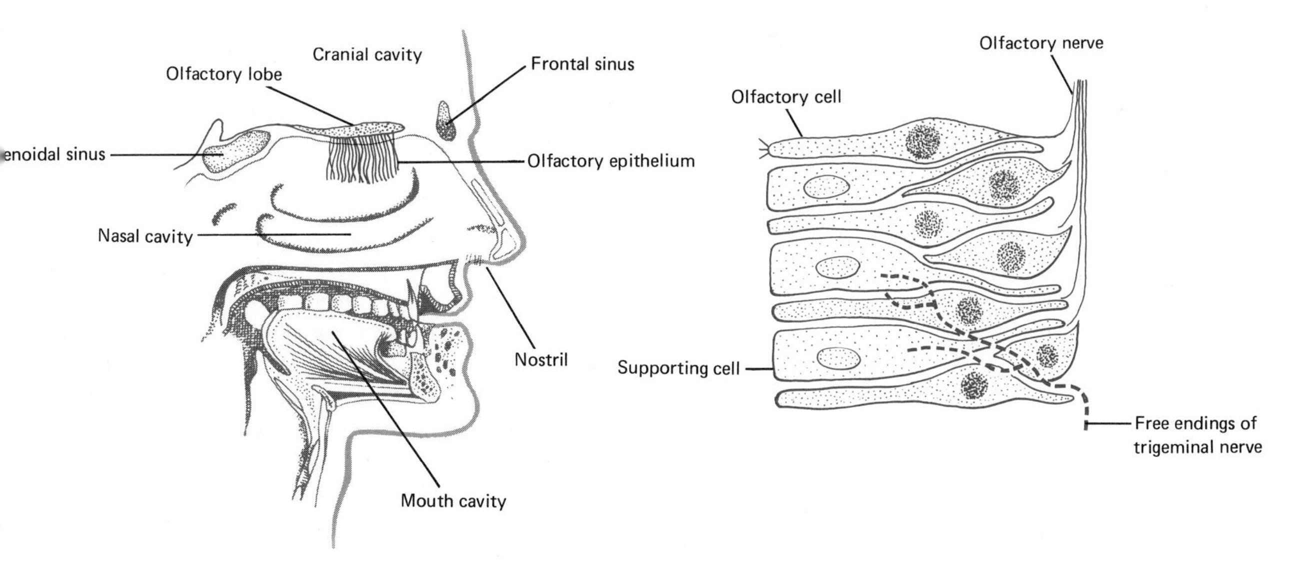

FIGURE 11-16. Structures associated with olfaction (smell). The section through the head shows the relationship of the olfactory lobe and olfactory epithelium to the nasal, mouth, and cranial cavities. The enlargement (*right*) illustrates the details of the olfactory epithelium.

lines the upper portion of the nasal cavities (Figure 11-16). In order for a material to elicit the sensation of smell, it must reach the nasal passages in gaseous form and then become dissolved in the secretions of the mucous membrane. Thus far, it has not been possible to classify odors into categories comparable to those of taste. Compared to taste receptors, the receptors for smell are much more sensitive. The smell receptors are also easily fatigued; odors that are at first quite strong are not sensed at all after a short period of time.

Mechanical Receptors

The types of receptors that are sensitive to mechanical stimulation are extremely diverse in form. They range from simple, free nerve endings in the skin to the highly specialized ear.

CUTANEOUS SENSES. The sensations of heat, cold, touch, and pressure—the *cutaneous senses*—are each dependent upon a special kind of receptor that has specialized endings in the skin (Figure 11-17A–D). These receptors are characterized by a connective tissue capsule surrounding a nerve ending and the capsule varies considerably in shape and thickness with each type of end organ. While the tactile (touch) sense is commonly stimulated by simple nerve endings associated with hair follicles, encapsulated structures called *Meissner's corpuscles* also react to tactile stimuli (Figure 11-17A). These corpuscles are distributed in connective tissue elevations of the skin, the papillae, being most numerous in the fingertips, the palms of the hand, and the soles of the feet. They are also abundant in the tip of the tongue, lips, nipples, clitoris, and glans penis. Located deeply in subcutaneous tissues throughout the body are oval structures called *Pacinian corpuscles* that react to pressure stimuli. Their capsules are composed of layers of connective tissue resembling an onion and each contains a nerve fiber in its central cavity (Figure 11-17B). The outer portion of the dermis and the tip of the tongue contain receptors sensitive to cold, which are referred to as *end bulbs of Krause.* The commonest form of these receptors is an oval capsule with small branching nerve fibers in the cavity (Figure 11-17C). They ordinarily react to temperatures below the normal temperature of the skin or to temperatures lower than those needed to stimulate receptors for warmth. The receptors for warmth are believed to be elongated, encapsulated structures embedded deeply in the dermis that are referred to as *end organs of Ruffini*

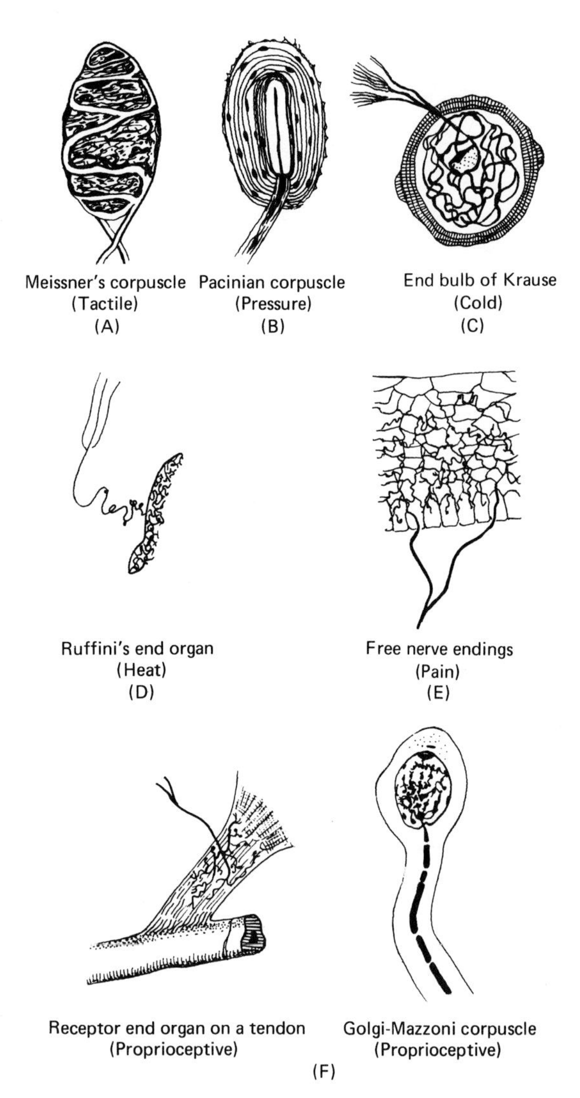

FIGURE 11-17. Cutaneous receptors, pain receptors, and proprioceptors. (**A–D**) Cutaneous receptors. (**E**) Pain receptor. (**F**) Two kinds of proprioceptors. See text for discussion.

(Figure 11-17D). They are not as abundant as the receptors for cold and they ordinarily react to temperatures above normal skin temperature.

RECEPTORS FOR PAIN. The receptors for pain consist of free, naked nerve endings (dendrites) and are found in practically all the tissues of the body (Figure 11-17E). They respond not only to mechanical stimuli but also to any other type of stimulation—chemical, thermal, electrical—that is of sufficient strength. Relative to the receptors discussed thus far, those for pain are far more numerous, occurring in all tissues except the brain and heart, and are located within the deeper organs and tissues of the body.

Types of Pain. Pain receptors may be stimulated by any type of stimulus. Excessive stimulation of any sense organ causes pain. For example, when stimuli for other sensations such as touch, pressure, heat, and cold reach a certain threshold, they stimulate pain receptors as well. Additional stimuli for pain receptors include excessive distention or dilation of an organ, prolonged muscular contractions, muscle spasms, inadequate blood flow to an organ (*ischemia*), or the presence of certain chemical substances. Pain receptors, because of their sensitivity to all stimuli, have a general protective function in that they inform us of changes that could be potentially dangerous to health or life. Adaptation to pain does not readily occur. This is rather important since pain indicates disorder or disease; if we became used to it and ignored it, irreparable damage could result.

Sensory impulses for pain are conducted through spinal and cranial nerves. The lateral spinothalamic tracts of the cord relay impulses to the thalamus. From here, the impulses are relayed to the postcentral convolution of the parietal lobe. Recognition of the kind and intensity of most pain is ultimately localized in the cerebral cortex. Some pain discrimination occurs at subcortical levels.

In general, pain may be divided into two types: somatic and visceral. **Somatic pain** may arise from stimulation of the skin receptors, in which case it is called superficial somatic pain. It may also arise from stimulation of receptors in skeletal muscles, joints, tendons, and fascia, in which case it is called deep somatic pain. **Visceral pain** results from stimulation of receptors in the viscera.

In most cases of somatic pain and in some instances of visceral pain, the cortex accurately projects the pain back to the stimulated area. For example, if you burn your finger, you feel the pain in your finger. Also, an individual with inflammation of the lining of the pleural cavity experiences pain in the affected area. In most instances of visceral pain, however, the sensation of pain is not projected back to the point of stimulation.

The pain may be felt in or just under the skin that overlies the stimulated organ in a surface area of the body that is quite far removed from the stimulated organ. This phenomenon is called **referred pain.** In general, the area to which the pain is referred and the visceral organ involved receive their innervation from the same segment of the spinal cord. For example, afferent fibers from the heart enter segments T1 to T4 of the spinal cord, as do afferent fibers from the skin over the heart and the skin over the medial surface of the left arm. Thus, the pain of a heart attack is typically felt in the skin over the heart and down the left arm.

A kind of pain frequently experienced by amputees is called **phantom pain.** In this instance, the person experiences pain in a limb or part of a limb after it has been amputated. Phantom pain occurs as follows: let us say that a foot has been amputated. A sensory nerve that originally terminated in the foot is severed during the operation but repairs itself and returns to function within the remaining leg. From past experience the brain has always projected stimulation of the neuron back to the foot. So when the distal end of this neuron is now stimulated, the brain continues to project the sensation back to the missing part. Thus, even though the foot has been amputated, the patient still "feels" pain in his toes.

Acupuncture and Pain. **Acupuncture** is a method of inhibiting pain impulses. The word comes from the terms *acus,* meaning needle, and *pungere,* meaning to sting. Needles are inserted through selected areas of the skin and twirled by the acupuncturist or by a battery-operated device. About 20–30 minutes after the twirling starts, pain is deadened, and this condition lasts 6–8 hours. The location of needle insertion varies depending on the part of the body the acupuncturist desires to anesthetize. To pull a tooth, one needle is inserted in the web between the thumb and the index finger. For a tonsillectomy, one needle is inserted about 2 inches above the wrist. For removal of the lung, one needle is placed in the forearm, midway between the wrist and the elbow.

There is no totally satisfactory explanation of how acupuncture works. According to the "gate control" theory, the twirling of the acupuncture needle stimulates two sets of nerves that eventually enter the spinal cord and synapse with the same association neurons. One very fine nerve is the nerve for pain and the other, a much thicker nerve, is the nerve for touch. The speed of the impulse passing along the touch nerve is faster than that passing along the pain nerve. Fibers with larger diameters conduct impulses faster than those with smaller diameters. Because the touch impulse reaches the dorsal horn of the cord first, it has right of way over the pain impulse. It thus "closes the gate" to the brain before the pain impulse reaches the cord. Since the pain impulse does not pass to the brain, no pain is felt. We should mention, however, that additional research and understanding of the process are necessary before acupuncture can be used as a routine procedure by American physicians.

Proprioceptors. Located within skeletal muscles, tendons, connective tissue, and joints are at least three different kinds of receptors that are sensitive to tension or stress and are collectively called **proprioceptors** (Figure 11-17F). Although these receptors do produce well-defined sensations, they help the body to coordinate the position of the limbs and are generally concerned with the kinesthetic sense. Proprioceptive impulses travel to the cerebellum and assist in unifying the contractions of separate muscles into a coordinated motion. Common examples of the kinesthetic sense, mediated in part by proprioceptors, are closing the eyes and touching a particular region of the body, playing a musical instrument, typewriting, and skating. Each of these activities, once learned, is carried on habitually.

Hearing and Equilibrium. The human ear, in addition to containing a receptor for receiving sound waves, also possesses two receptors for equilibrium. Before discussing the physiology of these sensations, however, we will first examine the structure of the ear, a complex mechanoreceptor.

The Human Ear. Anatomically, the ear is subdivided into three principal areas: (1) the outer ear, (2) the middle ear, and (3) the inner ear.

The *outer ear* is structurally designed to collect sound waves and then direct them inward (Figure 11-18A). It consists of a trumpet-shaped flap of elastic cartilage covered by skin called the *pinna* (auricle); a short tube about $1\frac{1}{4}$ inches long leading from the pinna into one of the bones of the head (temporal) that is referred to as the *external auditory canal;* and a very thin, semitransparent membrane, the *tympanic membrane* (ear-

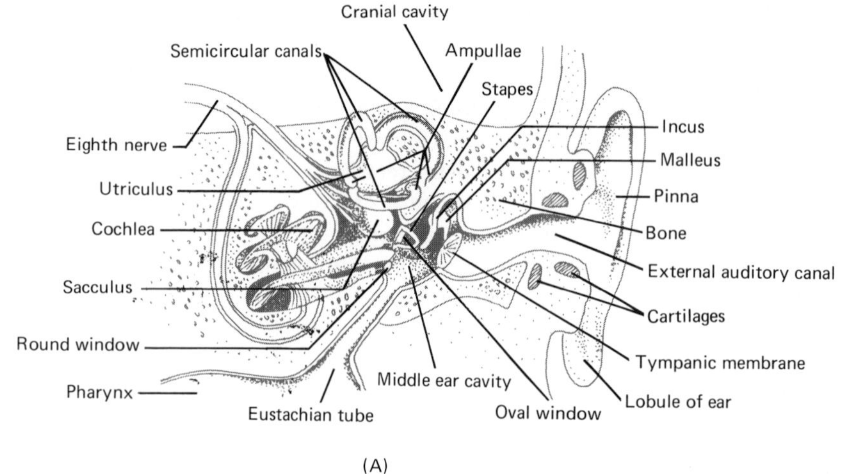

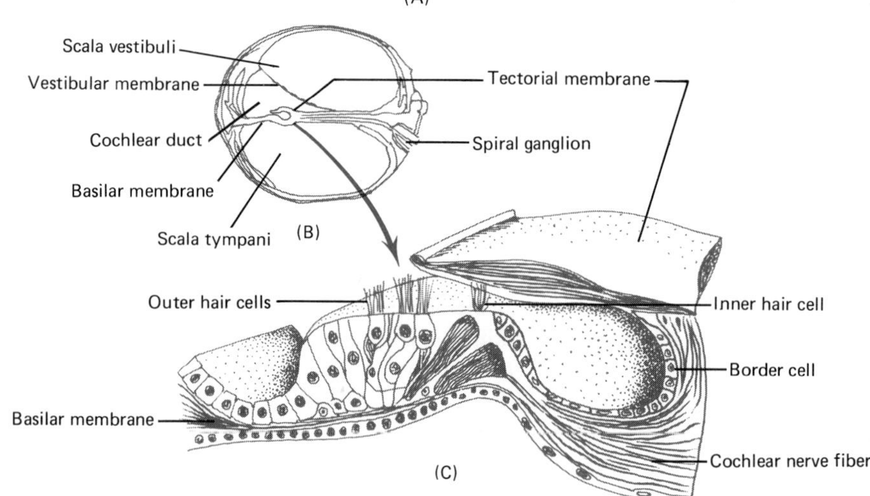

FIGURE 11-18. Auditory apparatus. (**A**) Section of outer, middle, and inner ears and their component parts. (**B**) Cross section of cochlea showing details. (**C**) Enlarged section through the organ of Corti.

drum), that covers the inner extremity of the canal and separates the outer ear from the middle ear.

The *middle ear,* or tympanic cavity, is a small epithelial-lined air-filled cavity hollowed out of the temporal bone (Figure 11-18A). Within it is a lever system of three exceedingly small bones called the *malleus, incus,* and *stapes,* which are named according to their shapes—hammer, anvil, and stirrup—respectively. The "handle" of the malleus is attached to the inner portion of the tympanic membrane, and its rounded end fits into a depression in the incus and is held tightly in place so that these two bones move as one. The incus is the intermediate bone in the series. The stapes fits into the *oval window,* a membrane-covered opening leading into the inner ear. The joint between the incus and stapes is freely movable, the stapes performing a rocking movement at the oval window. A second opening, which is also covered by a thin membrane, connects the middle and inner ears and is referred to as the *round window.* There is also a tube, the *Eustachian tube,* leading from the middle ear to the nasopharynx, the portion of the throat behind the nose. Pathologically, the Eustachian tube is rather important since it offers a passageway by which infections may travel from the throat and nose to the middle ear. It does, however, assume a role in equalizing pressure on both sides of the eardrum. Abrupt changes in external pressure might otherwise cause the eardrum to rupture. Since the tube normally remains closed except during swallowing or yawning, these activities permit a pressure change in the middle ear, thus balancing the pressure on the external surface of the tympanic membrane. Any sudden pressure changes on the eardrum can be offset by deliberately swallowing.

The *inner ear* is also called the labyrinth because of its complicated shape (Figure 11-18A). Structurally, it consists of two main divisions: a bony labyrinth and a membranous labyrinth within the bony structure. The bony labyrinth is composed of the vestibule, cochlea, and semicircular canals. The membranous labyrinth consists of the tissues lining the bony channels—the utricle and saccule inside the vestibule, the cochlear duct inside the cochlea, and the membranous semicircular canals inside the bony ones.

The *vestibule* constitutes the central portion of the bony labyrinth and into it open the oval and round windows from the middle ear, as well as the three semicircular canals of the inner ear. The membranous *utricle* and *saccule* are suspended within the vestibule and are separated from the bony walls of the vestibule by a fluid called *perilymph.* In addition, both structures also contain a fluid termed *endolymph.* Located within the utricle and saccule is a minute structure, the *macula,* that consists principally of hair cells and a gelatinous membrane containing *otoliths* (small particles of calcium carbonate). Changing the position of the head causes a change in the amount of pressure on the gelatinous membrane, which in turn causes the otoliths to pull on the hair cells. This stimulates adjacent receptors of the vestibular nerve, the impulse is carried to the brain, and it produces a sense of the position of the head and also a sensation of a change in the pull of gravity. Stimulation of the macula also gives rise to righting reflexes, muscular movements to restore the body and its parts to their normal position once they have been displaced.

The name *cochlea,* meaning snail, describes the external appearance of this part of the bony labyrinth. In sectional view, the cochlea resembles a tube wound spirally around a cone-shaped core of bone, the *modiolus.* Within the cochlea are three similarly shaped, spiraling membranous channels running its full length (Figure 11-18B). The uppermost channel, the *scala vestibuli,* is attached at its base to the oval window into whose membrane the stapes is inserted. At its other end, located at the apex of the cochlea, the scala vestibuli has a small opening that communicates with the lowermost channel, the *scala tympani.* The base of this channel terminates in the membrane-covered round window. Both the scala vestibuli and the scala tympani are filled with perilymph. The third channel, the *cochlear duct,* is filled with endolymph. It is separated from the upper channel by a partition called *Reissner's* (*vestibular*) *membrane* and from the lower channel by the *basilar membrane.*

The hearing sense organ, the *organ of Corti,* rests on the basilar membrane (Figure 11-18C). Its structure resembles that of the equilibrium sense organ (macula) in that it contains hair cells that project into the endolymph and a gelatinous membrane, the *tectorial membrane.* The cell bodies from which the dendrites originate constitute a ganglion within one of the numerous passageways of the inner ear; the axons constitute the auditory nerve.

Three *semicircular canals,* each at approximately right angles to the others, are found in each temporal bone. Within the bony semicircular canals and separated from them by perilymph are the membranous semicircular canals. Each contains endolymph and attaches to the utricle. One end of each canal, near its junction with the utricle, enlarges into a swelling called the *ampulla* (Figure 11-18A). Each ampulla contains a small elevation of receptor hair cells and nerve endings, collectively designated *cristae,* which project into an overlying gelatinous mass. Inasmuch as the semicircular canals are oriented at right angles to one another, movement of the head in any direction will bring about the movement of endolymph in one or more of the canals. Then, the hair cells of the cristae are stimulated and nervous impulses are transmitted to the brain. These impulses give rise to a sensation of movement and imbalance and to reflexes that tend to right the body. Thus, three structures in the inner ear, the saccule, the utricle, and the semicircular canals, play an important role in maintaining a sense of equilibrium and awareness of position.

The Physiology of Hearing. Sound waves result from the alternate compression and decompression of air. The waves originate from a vibrating object and travel through the air much the same way that waves travel over the surface of water. Sound waves reach the ear and are directed by the pinna into the external auditory canal. When they strike the tympanic membrane, the alternate compression and decompression of the air cause the membrane to vibrate. The central area of the tympanic membrane is connected to the malleus, which also starts to vibrate. The vibration is then picked up by the incus and transmitted to the stapes. As the stapes moves back and forth, it pushes the oval window in and out. The movement of the oval window sets up waves in the perilymph. As the window bulges inward, it pushes the perilymph of the scala vestibuli up into the cochlea. This pressure pushes the vestibular membrane inward and increases the pressure of the endolymph inside the cochlear duct. The basilar membrane gives under the pressure and bulges out into the scala tympani. The sudden pressure in the scala tympani pushes the perilymph toward the round window, causing it to bulge back into the middle ear. Conversely, as the sound wave subsides, the stapes moves backward and the procedure is reversed. That is, the fluid moves in the opposite direction along the same pathway, and the basilar membrane bulges into the cochlear duct. When the basilar membrane vibrates, the hair cells of the organ of Corti are moved against the tectorial membrane. The movement of hairs stimulates the dendrites of neurons at their base, and sound waves are converted into nerve impulses. The impulses are transmitted by way of the auditory nerve to the auditory area of the temporal lobe of the cerebral cortex where the sensation of hearing is perceived.

Photoreception

Sensitivity to light is a common phenomenon of living organisms. Even organisms lacking discrete structures, such as eyes, respond to the changes in light intensity. Recall the eyespot of *Euglena* as an example. Image-forming eyes are found in relatively few animal phyla. These include some mollusks (squid, octopus), a few worms (polychaetes), most arthropods, and vertebrates. Regardless of the nature of the photoreceptive structure, there is apparently a common basic pattern for visual function. As an example, we will consider the human eye.

The adult eye, which measures about 1 inch in diameter, consists of the eyeball and several structures around it. Of the total surface area of the eyeball, only the anterior one-sixth is exposed; the remainder is recessed and protected by the bony socket into which it fits. The accessory structures of the eyeball include the *eyelids* and *eyelashes,* which protect the anterior surface of the eyeball; the *eyebrows,* which protect the eye from falling objects, prevent perspiration from getting into the eye, and tend to shade the eyes from the direct rays of the sun; the *conjunctiva,* a thin, transparent, protective, mucous membrane covering the anterior surface of the eyeball; the *lacrimal apparatus,* which produces tears to flush and cleanse the eye; and *extrinsic eye muscles,* which rotate the eyeball in various directions.

The eyeball itself consists of three layers or coats. The outer coat, called the *fibrous tunic,* consists of the sclera and cornea; the middle coat, the *vascular tunic,* is highly vascularized and contains the choroid, ciliary body, and iris; the inner coat, referred to as the *retina,* is the photosensitive structure of the eye (Figure 11-19A).

The fibrous tunic is a tough white fibrous tissue that is divided into two regions. The first of these, the *sclera,* is the white coat of the eye that covers the entire eyeball except for the cornea. This structure gives shape to the eyeball and affords protection for its inner parts. Extrinsic eye muscles extend from the sclera to the eye socket. The *cornea* is the transparent anterior portion of the fibrous tunic that lies over the colored part of the eye (iris). Over its outer surface there is an epithelial layer that is continuous with the epithelial layer of the conjunctiva.

The middle layer of the eyeball, the *vascular tunic,* or *uvea,* is composed of the choroid, ciliary body, and iris. The *choroid,* or choroid coat, is a thin dark brown membrane containing numerous blood vessels and a large amount of pigment. Since it closely adheres to the undersurface of the sclera, it also lines about five-sixths of the eyeball. The choroid absorbs light rays to prevent their reflection and maintains the nutrition of the retina. The optic nerve passes through the choroid at the back of the eyeball. At its anterior portion, the choroid is modified into the *ciliary body,* which is responsible for changing the shape of the lens and the iris, or colored portion of the eye. The *iris* consists of circular and

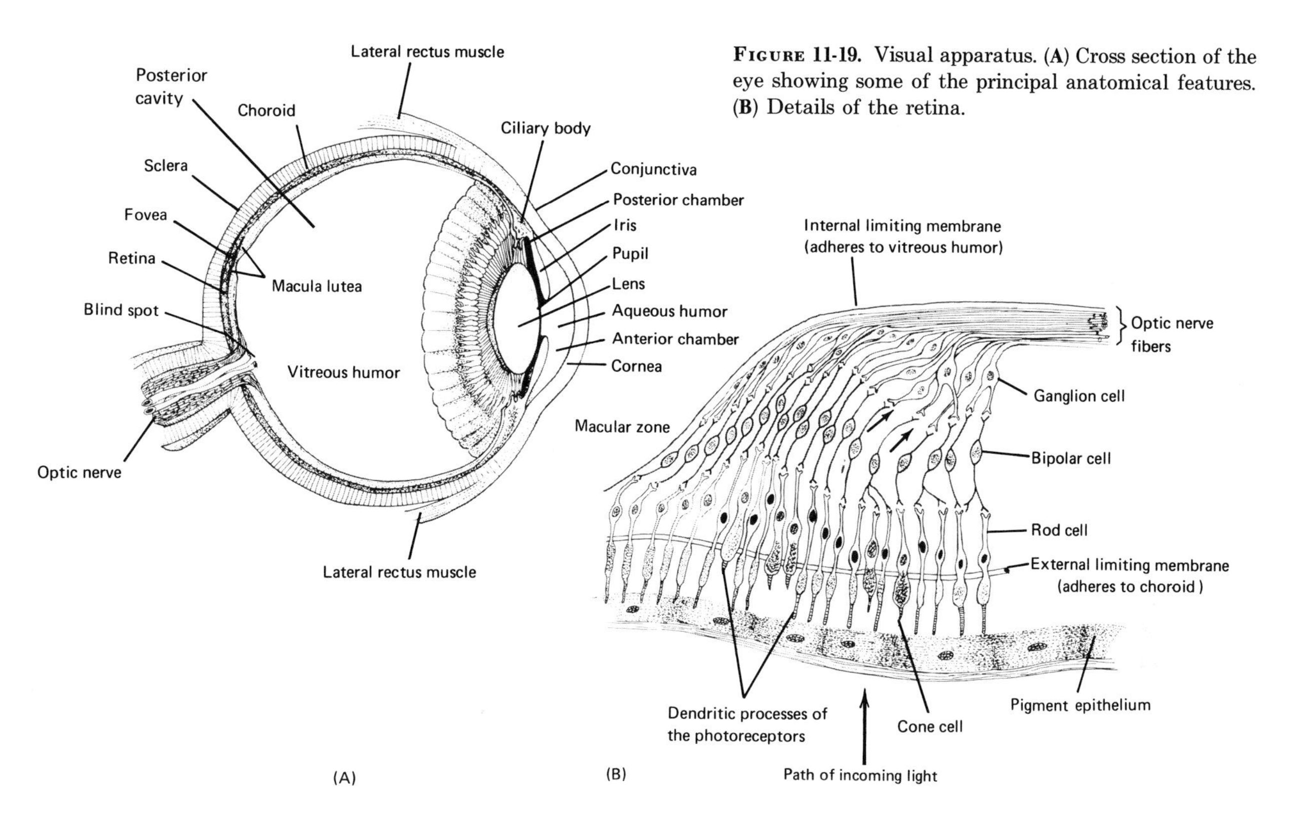

FIGURE 11-19. Visual apparatus. (**A**) Cross section of the eye showing some of the principal anatomical features. (**B**) Details of the retina.

radial smooth muscle fibers arranged so as to form a doughnut-shaped structure, the central opening being called the *pupil.* Anatomically, the iris is suspended between the cornea and the lens and is attached at its outer margin to the ciliary body. One of the chief functions of the iris is to determine the amount of light entering the eye by regulating the size of the pupil. The muscles of the iris also aid in the formation of clear images on the retina.

The third and inner coat of the eye, the *retina,* is an incomplete coat in that it does not extend anteriorly. It consists principally of a nervous tissue layer and a pigmented layer (Figure 11-19B). The outer pigmented layer consists of epithelial cells lying in contact with the choroid coat. The inner sensory layer contains three layers of neurons. Named in the order in which these sublayers conduct impulses, they are photoreceptor neurons, bipolar neurons, and ganglion neurons. The beginning of the dendrites of the photoreceptor cells, the *rods* and *cones,* are so called because of their shapes. The long and narrow rods and the short and thick cones are visual receptors, structures highly specialized for stimulation by light rays. The other two principal sublayers consist of bipolar neurons and ganglion neurons whose axons converge just before leaving the eye to constitute the optic nerve which transmits impulses from the eye to the cerebral cortex.

The estimated number of cones is 7 million and that of rods somewhere between ten and twenty times as many. Cones are densely concentrated in the *fovea centralis,* a small depression in the center of a yellowish area, the *macula lutea,* at about the exact geometric center of the retina (Figure 11-19A). The fovea is the area of sharpest vision and cones become less and less dense from the fovea outward. Rods, by contrast, are entirely absent from the fovea and macula and increase in density toward the periphery of the retina.

All the axons of ganglion neurons extend back to a smaller area in the posterior part of the eyeball known as the *optic disc,* or *blind spot.* This part of the retina contains perforations through which the fibers emerge from the eyeball as the *optic nerve* (Figure 11-19A). The blind spot is so designated because light rays striking this area cannot be seen since it contains no rods or cones, only nerve fibers.

The eyeball is not a solid sphere but contains a large interior cavity that is divided into two cavities, anterior and posterior, separated by the lens (Figure 11-19A). The *anterior cavity* has two subdivisions known as *anterior* and *posterior chambers.* The entire anterior cavity is in front of the lens and is filled with a clear, watery fluid called the *aqueous humor.* This substance often leaks out of the eye when it is injured. The second, larger cavity, the *posterior cavity,* is between the lens and the retina and contains a soft, gelatinous substance called the *vitreous humor.* This semisolid material helps to prevent the eyeball from collapsing.

Also within the eyeball itself, just behind the pupil, is the *lens,* a transparent, biconvex structure. The lens is enclosed by a clear, elastic capsule and just underneath the capsule on the anterior portion of the lens there is a single layer of cells constituting the lens epithelium. The lens itself is constructed of numerous layers of protein fibers that are arranged in concentric lamellae-like layers of onion tissue. The condition called *cataract* manifests itself as an opaque or milk-white appearance of the lens or its capsule. Cataracts commonly occur with aging or after injury to the eye. The lens is connected to the ciliary body, which changes the shape of the lens, permitting the lens to alter its light-focusing properties. It becomes flattened for focusing on distant objects and more convex focusing on closer objects. Changes in the shape of the lens are determined by the degrees of contraction of the ciliary body. It becomes apparent, then, that before light can reach the receptors of the retina, it must first pass through the cornea, aqueous humor, lens, and vitreous humor.

In order for vision to occur, an image must be formed on the retina to stimulate its receptors (rods and cones), and the resulting nerve impulses must be conducted to the visual areas of the cerebral cortex. The formation of an image on the retina requires four processes, all of which are concerned with focusing light rays.

1. Refraction of light rays.
2. Accommodation of the lens.
3. Constriction of the pupil.
4. Convergence of the eyes.

Refraction refers to the bending of light rays and is produced by light rays passing obliquely from one transparent medium into another of different optical density; the more convex the surface of the medium, the greater its refractive capacity. The media of refraction that operate in the eye are the cornea, aqueous humor, lens, and vitreous humor. Light rays are bent at the anterior surface of the cornea as they pass from the air into the denser cornea, and at the anterior surface of the lens as they pass from the lens into the vitreous humor. In a normal eye, the four refracting media together bend light rays sufficiently to focus an object (inverted) on the retina (Figure 11-20A). Many eyes, however, show errors of refraction in that they are not able to focus light rays on the retina properly. Some of these errors of refraction include *myopia* (nearsightedness), *hypermetropia* (farsightedness), and *astigmatism.* These conditions are illustrated and explained in Figure 11-20B–D.

The lens functions chiefly by changing the focusing power of the eye through changes in its curvature. Because of the changes in the lens, a person is able to focus on a distant object at one moment and then a nearby object the next. This change is called *accommodation.* As an example of accommodation, consider the events involved in focusing on a nearby object. Light rays from objects 20 or more feet away are almost parallel, and the normal eye refracts such rays sufficiently to focus them clearly on the retina. Light rays from nearer objects, however, are divergent rather than parallel so that they must be bent to a greater extent in order to focus them on the retina. Accommodation of the lens, that is, an increase in its curvature, takes place to achieve this greater refraction. According to one widely accepted theory, the ciliary muscle contracts, pulling the ciliary body and choroid forward toward the lens. This releases the tension on the suspensory ligament (and therefore on the lens) and, the lens, being elastic, immediately bulges. In near vision, then, the ciliary muscle is contracted and the lens is bulging; for far vision, the ciliary muscle is relaxed and the lens is relatively flat.

The muscles of the iris also play an important func-

Figure 11-20. Normal and abnormal refraction in the eye. (**A**) In the normal eye, light rays from an object are bent sufficiently by the four refracting media, converged on the retinal fovea, and a clear image formed. (**B**) In the myopic eye, the image is focused in front of the retina. This nearsighted condition may be the result of an elongated eyeball or a thickened lens. Correction is by use of a concave lens. (**C**) In the hyperopic eye, the image is focused behind the retina. This farsighted condition may be the result of the eyeball being too short or the lens too thin. Correction is by a convex lens. (**D**) In astigmatism, the curvature of the cornea or lens is uneven, and horizontal and vertical rays are focused at two different points on the retina. Suitable glasses correct the refraction of an astigmatic eye. In the case of an irregular cornea (*left*) the compensating shape of the lens may largely nullify the irregularities of the cornea. With an irregular lens (*right*), the image is not focused on the area of sharpest vision of the retina. The result is blurred or distorted vision.

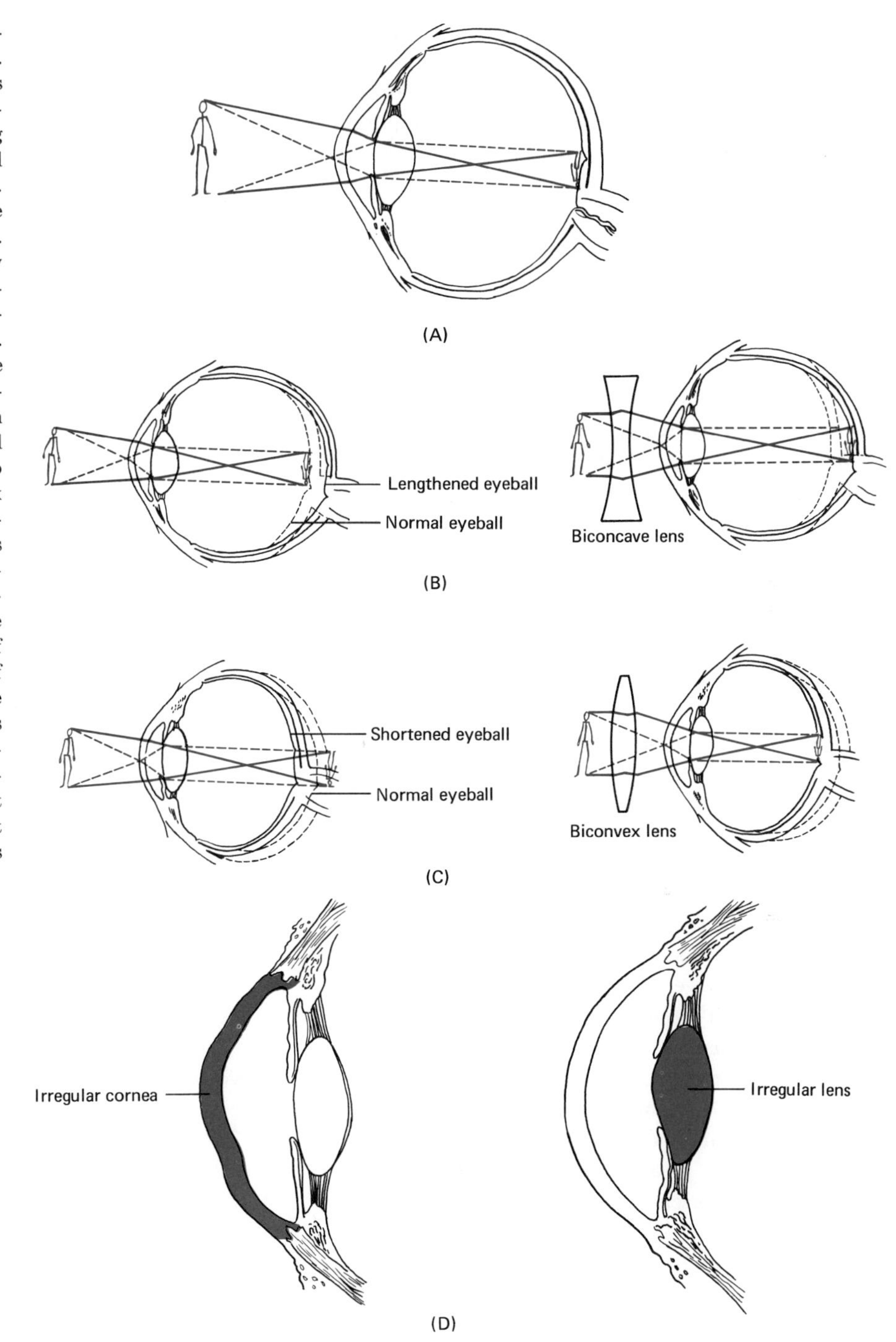

tion in the formation of clear retinal images since part of the accommodation mechanism consists of contraction of the circular fibers of the iris, which constricts the pupil. This constriction of the pupil occurs simultaneously with accommodation of the lens and prevents divergent rays from an object from entering the eye through the periphery of the cornea and lens. The entrance of such peripheral rays would not be brought to focus on the retina and would therefore result in a blurred image. The pupil also constricts in bright light to protect the retina from sudden and intense stimulation.

Single binocular vision—that is, seeing only one object instead of two when both eyes are used—occurs when light rays from an object fall on corresponding points of the two retinas. Whenever the eyeballs move together, either with the visual axes parallel (far objects) or converging on a common point (near objects), light rays strike corresponding points on the two retinas. The movement of the two eyeballs in unison so that they are directed inward toward the viewed object is often referred to as *convergence.* The nearer the object, the greater the degree of convergence necessary to maintain single vision. Convergence is brought about by the balanced action of the external eye muscles.

Once an image is formed on the retina by refraction, accommodation, constriction of the pupil, and convergence, the retina must be stimulated. The exact mechanism by which light falling on the rods and cones of the retina acts as a stimulus to initiate impulses that result in the sensation of sight is not entirely clear. The following data, however, are known. Rods are known to contain a reddish pigmented compound called **rhodopsin (visual purple),** which is a conjugated protein consisting of a protein, *opsin,* and *retinene,* a derivative of vitamin A (retinol). Rhodopsin is highly light-sensitive, so that when the light rays strike a rod, rhodopsin rapidly breaks down (Figure 11-21), and in some manner this chemical change initiates impulse conduction by the rod. If, by contrast, the rod is exposed to darkness for a short time, rhodopsin re-forms from the opsin and retinene and is ready to function again. Although cones also contain photosensitive chemicals, they have yet to be elucidated. It is assumed that the cone compounds require brighter light for their breakdown and are considered to be the receptors for daylight and color vision. Rods, on the other hand, are believed to be the receptors for night vision because their rhodopsin becomes depleted very quickly in bright light owing to its rapid breakdown but slow regeneration. This explains why an individual cannot see for a short period of time after going from brightness to darkness.

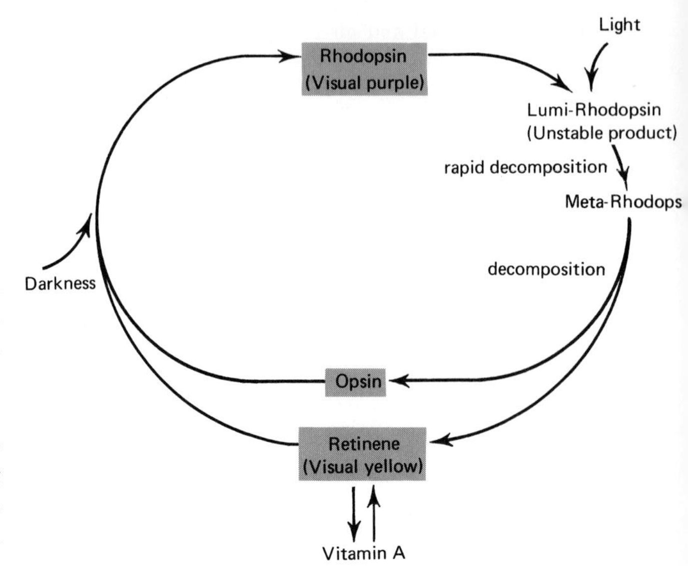

Figure 11-21. The rhodopsin cycle. Consult text for details.

Drugs and the Nervous System

Before leaving our discussion of the nervous system, we will examine some of the relationships between drugs and the functioning of the nervous system. In particular, we will analyze how the homeostasis of the nervous system is altered by certain drugs.

Definition of Drug

According to the Federal Food, Drug, and Cosmetic Act, a **drug** is an "article intended for use in the diagnosis, cure, mitigation, treatment, or prevention of disease in man or other animals," and as an "article intended to affect the structure or any function of the body of man or other animals." In a strict sense, all medicines contain drugs. However, not all drugs are medicines. Examples include marijuana, heroin, and mescaline. There are also substances that are neither medicines nor drugs that are used as if they were drugs such as paint, glue, thinner, and aerosol propellents.

For purposes of our discussion, therefore, we will use the term *drug* to mean any chemical substance capable of altering physiological processes or behavior but without primary medical value. Such compounds may also be viewed as drugs of abuse. The term *medicine* will be defined as a chemical substance administered to a human or animal with the intention of preventing or curing a disease or otherwise enhancing physical and/or mental welfare.

Drugs and Their Effects

HALLUCINOGENS. The *hallucinogens,* also called psychedelic or psychomimetic drugs, comprise a variety of drugs that alter mood, perception, thinking, and ego. In other words, they alter the mind and its functioning. Among the hallucinogens are LSD, mescaline ("mesc"), psilocybin, DMT ("businessman's special"), STP ("serenity, tranquility, and peace"), and marijuana. We will consider here only LSD and marijuana.

LSD (lysergic acid diethylamide) is perhaps the best known of the hallucinogens. It is variously referred to as "acid," "bug D," "sugar," "trips," and "cubes." An experience with LSD or other hallucinogens is called a "trip" or "tripping out." LSD is one of the most powerful drugs known. It is generally used by an individual for insight and exhiliration. The effects of LSD depend on several factors including the user's state of mind and personality, the environment, and the dosage. A dose of 100–250 micrograms can produce the "trip."

LSD is absorbed rapidly into the blood and becomes highly concentrated in the liver, kidneys, and adrenal glands. Among the physiological effects of using LSD are a tingling in the hands and feet, numbness, nausea, lack of appetite, flushed skin, chills, and dilatation of the pupils (mydriasis). There are also increases in heart rate, body temperature, blood pressure, and blood sugar level. It should also be noted that some research indicates that LSD may cause chromosome damage.

Certain psychological dangers are also associated with LSD. One of the first visual effects is ever-changing colors and shapes of objects in a room and the appearance of rainbow-like halos around lights. There may also be a crossing of sensory responses. In other words, the person may "hear" colors and "see" sounds. Time and space perception are lost and a pleasant "trip" may turn into an unpleasant and confused one. In a bad "trip" there is a loss of self-control, judgment, and logic; ego boundaries disappear; and there may be flashbacks characterized by feelings of paranoia, unreality, and estrangement.

Marijuana, also known as "pot," "grass," and "tea," is derived from the flowers and top leaves of the female hemp plant, *Cannabis sativa.* It is generally used to induce euphoria, relaxation, and increased perception. Experimental studies on humans indicate that marijuana can cause nausea, diarrhea, vomiting, and failure of muscular coordination, especially after oral administration. Some data indicate that marijuana also affects visual and time perception, reaction time to complex stimuli, and steadiness of the hands and body. It is generally agreed that marijuana is not physically addicting nor does tolerance build up in the user. To date, the purely physical effects of marijuana have not been found to be permanently damaging, except for a claim by the American Medical Association that inhaled marijuana smoke is irritating and long, continued exposure to it can cause chronic respiratory disorders. Statistics to date seem to indicate that marijuana psychosis is rare. Much research has yet to be done to determine all the short- and long-term effects of marijuana.

DEPRESSANTS. *Depressants* represent a class of drugs that can produce effects ranging from the relief of anxiety and tension to sleep, anesthesia, coma, and death. Common depressants are barbiturates, tranquilizers, and opiates.

Barbiturates ("barbs," "goofballs," "downers") include phenobarbital ("phenos"), secobarbital ("reds," "red devils," "pinks"), amobarbital ("blue," "blue devils," "blue birds"), and nembutal ("nembies"). The depressant action of a barbiturate is directly related to the particular drug used, how the drug is administered (injection produces a reaction faster than oral administration), the health of the central nervous system, the extent to which tolerance has been developed, and most importantly, dosage. Small amounts produce relief of anxiety and tension, and progressively increasing amounts can result in the following sequential effects: drowsiness, sleep, coma, and death. Most users of barbiturates seek euphoria. Physiologically, barbiturates are

TABLE 11-3. Summary of Drugs

Drug	Source	Administration	Possible Long-Term Effects	Dependence Potential	
				Psychological	Physical
I. Hallucinogens					
LSD	Semisynthetic from ergot alkaloids	Swallowed	May intensify existing psychosis; panic reactions	Possible	None
Marijuana	Natural from *Cannabis sativa*	Smoked or swallowed	None determined; possible respiratory disorders and psychosis	Probable	None
Hashish	Natural from *Cannabis sativa*	Smoked or swallowed	Usually none; possible conjunctivitis and psychosis	Probable	None
STP	Synthetic	Swallowed	Not determined	Possible	None
DMT	Synthetic	Injected	Not determined	Possible	None
Mescaline	Natural from Peyote cactus (*Lophophora williamsii*)	Swallowed	Not determined	Possible	None
Psilocybin	Natural from the mushroom, *Psilocybe mexicana*	Swallowed	Not determined	Possible	None
II. Depressants					
Barbiturates	Synthetic from barbituric acid	Swallowed or injected	Severe withdrawal symptoms; possible convulsions; toxic psychosis	Yes	Yes
Tranquilizers	Synthetic	Swallowed or injected	Severe withdrawal symptoms; possible convulsions; toxic psychosis	Yes	Yes
Opiates	Natural from opium; semi-synthetic from morphine; synthetic	Swallowed or injected	Addiction; constipation; loss of appetite	Yes	Yes
III. Amphetamines	Synthetic	Swallowed or injected	Loss of appetite; delusions; hallucinations; toxic psychosis	Yes	Possible

general depressants in that they are capable of inhibiting the normal activity of nerves, skeletal muscle, smooth muscle, and heart muscle. This is accomplished by depression of the activities of certain areas of the brain.

One of the most dangerous aspects of barbiturate abuse is that of physical and psychological dependence. Upon withdrawal, there is extreme hyperexcitability due to the restoration of depressed neural pathways. Among the withdrawal symptoms are apprehension, weakness, tremors, hyperactive reflexes, insomnia, abdominal cramps, nausea, vomiting, dehydration, weight loss, accelerated heart rate and breathing disorientation, and even death.

Tranquilizers, also called "downers," include phenothiazines (Thorazine), meprobamate (Miltown), chlordiazepoxide (Librium), and diazepam (Valium). Although sedative-hypnotics and tranquilizers can produce the same effects, they are classified differently because the difference between doses that produce relief of anxiety and tensions and doses that produce sleep is greater for tranquilizers than for sedative-hypnotics. There is also a greater difference between the dose required to produce sleep and the dose to produce death in tranquilizers.

Opiates include opium and its derivatives such as morphine ("M"), heroin ("H," "junk," "scag," "smack"), and codeine. Although morphine is about ten times more potent than opium, it is one of the least used opiates. Heroin, about two to three times as potent as morphine, is the preferred opiate among those who desire the feeling of euphoria. However, physical dependence results from using as little as one grain every two weeks.

Heroin and other opiates can cause nausea, vomiting, and other untoward effects. Essentially, they are narcotic sedatives that depress the central nervous system, especially the sensory areas of the thalamus and cerebral cortex. A narcotic is an agent that produces insensibility or stupor. Small doses have an analgesic (pain-relieving) effect, large doses cause sleep, and overdoses cause death through depression of the respiratory center in the medulla. Large doses apparently affect the pleasure center of the hypothalamus and other mood control centers, causing euphoria, elevation of mood, and a feeling of peace and contentment. Other effects include constriction of the pupils, chills, and constipation. Withdrawal symptoms include dilatation of the pupils, stomach cramps, and a runny nose.

Amphetamines and Related Stimulants. *Amphetamines* ("drivers") and related drugs are stimulants ("uppers") that are widely used in medicine to treat fatigue, depression, behavioral problems in children, to counteract or prevent sleep, and to suppress appetite. Examples of amphetamines are benzedrine ("benny"), dexedrine ("dex," "hearts"), dexamyl ("Christmas trees"), methamphetamine ("crystal," "speed"), and preludin. Related stimulants include cocaine ("coke," "flake," "snow"), nicotine, and caffeine.

Generally, amphetamines and related stimulants mimic the effects of an activated sympathetic nervous system. Among the responses are constriction of blood vessels, increased heart rate and blood pressure, bronchial dilatation, mydriasis, increased muscle tension, increased blood sugar level, and stimulation of the adrenal glands. Cumulatively, these reactions produce alertness and other characteristics associated with the fight or flight syndrome. Amphetamines are believed to work by releasing norepinephrine stored in axons.

Amphetamine users experience euphoria, a sense of well-being, hyperactivity, and a feeling of increased mental and physical power. Among the reported dangers of amphetamines are

1. Benzedrine—psychological dependence.
2. Dexedrine—nervousness, dizziness, sweating, chills, and headache.
3. Dexamyl—psychological dependence and shock.
4. Methamphetamine—dryness of the mouth, dizziness, and nervousness.
5. Preludin—psychological dependence.

A summary of drugs is presented in Table 11-3.

12 Protection, Support, and Locomotion

GUIDE QUESTIONS

1 Discuss the various types of body coverings of invertebrates. What is meant by ecdysis, and what purpose does it serve?
2 Explain the structure of the skin of a vertebrate. How does its structure compare with the invertebrate coverings? In what ways does skin differ from invertebrate coverings with regard to function? Explain.
3 What are some modifications of skin? Cite examples of each. How do antlers and horns differ from each other? Account for the color variations in human skin.
4 Discuss some skeletal adaptations of protists and invertebrates, and give specific examples of each. Compare the endoskeleton and the exoskeleton with regard to function.
5 State the five basic functions of the vertebrate skeleton. How does the vertebrate skeleton compare with those of invertebrates? Give specific examples.
6 On what basis are bones categorized? What are the four types of bones in the human? Describe the external and internal features of a typical long bone.
7 What are the two main divisions of the human skeleton and which bones comprise each division? Define a fontanel and explain its function. What purpose(s) do foramina serve?
8 Discuss the structure of the vertebral column. What purposes do the intervertebral discs and the various processes of each vertebra serve?
9 Explain the composition and function(s) of the pectoral and pelvic girdles. How are the appendages articulated with each girdle?
10 Describe the various types of articulations between bones and give specific examples of each type.
11 How is bone formed? What is meant by ossification? How do bones increase in length? in diameter?
12 What is a fracture? Describe several kinds.
13 Explain the events involved in the repair of a fracture.
14 Why is locomotion important? How does locomotion occur in protists? in invertebrates? Give examples.
15 How do muscles act? What is meant by origin? insertion? What relationships exist between bones and muscles?
16 Define agonist, antagonist, and synergist as applied to muscles. What criteria are used in naming muscles? State examples.
17 Relate muscle activity to the skeletal and nervous systems. How are the three systems coordinated? In what ways are they dependent? independent?

IN ALL but the lowest animal groups, the external body covering, the supporting framework (skeleton), and the skeletal muscles that facilitate movements are interrelated. In a very broad context, these three components of certain organisms are correlated to the extent that they share several common roles. For example, collectively they serve to protect and support internal structures of the body, they help to give the body shape and form, and they provide the means for locomotion. In this chapter, we will consider the body covering, supporting framework, and locomotive apparatus of representative organisms, with the primary emphasis being on man.

External Coverings

The outer coverings of organisms range in simplicity from a cell wall or membrane to the complexity of skin. In the course of evolution, the external confining layer has become more elaborate in form and function.

Plant Coverings

It should be pointed out that organisms classified as Plantae have protective tissues, although these coverings are unlike the body coverings of animals. The epidermis forms the surface layer of leaves, flowers, and young roots and stems. The outer walls of epidermal cells are covered by cutin so that the epidermis aids in protection against water loss, mechanical injury, and invasion by disease organisms. Cork is found as the outer layer of stems and roots of older woody plants as an epidermal replacement. Inasmuch as the cell walls of cork are impregnated with suberin, cork tissue affords both protection and water conservation.

Moneran and Protistan Coverings

Among monerans, the external covering consists simply of the cell wall. This structure separates the organism from its external environment, affords physical protection, and provides shape. Some protistans also have cell walls. In other protistans (e.g., *Amoeba*), however, the outermost boundary is the plasma membrane which is continuous with a definite nongranular, jelly-like region, the ectoplasm. Still other protistans, such as foraminiferans and radiolarians, have shells consisting of various inorganic and organic compounds secreted to provide protection and to support the internal contents. Ciliates typically possess a dense but flexible external covering, the *pellicle.*

Invertebrate Coverings

All members of the animal kingdom have an outer covering of tissue, the **epidermis.** Among many soft-bodied aquatic animals, such as hydra, jellyfishes, and free-living flatworms, the epidermis is a single layer of cells in direct contact with the environment (Figure 12-1A). In roundworms and annelids the epidermis is covered by a *cuticle,* a layer of noncellular organic material usually secreted by the epidermal cells (Figure 12-1B). The cuticle of roundworms and parasitic flatworms is relatively thick and provides a defense against unfavorable environmental conditions while in annelids, the protective cuticle is relatively thin.

Among most mollusks and other heavily shelled animals, the body covering is made stronger and thicker by the addition of calcium deposits. In clams, for example, the internal body organs are enveloped by the mantle, a tissue formed by a fold of the body wall. The mantle contains the epithelium that secretes the *shell,* a hard protective covering consisting principally of crystals of calcium carbonate, which are laid down in thin sheets (Figure 12-2A). The concentric lines on the surface of the shell represent successive periods of growth.

The body covering of arthropods is a modified cuticular layer composed of chitin secreted by the epithelium of the body surface (Figure 12-2B, left). The cuticle of arthropods serves for the attachment of muscles and as a protective armor plate around the animal, and is, therefore, an *exoskeleton* (external skeleton). The major constituent of the arthropod cuticular layer, *chitin,* is a chemical compound that makes the cuticle very tough and resistant to all but a few very strong chemical substances. The cuticle of certain arthropods (crustaceans) is further hardened by the addition of calcium salts. Unlike mollusk shells, which increase in size by the addition of materials from the epidermis as the animal grows, arthropod exoskeletons are rigid and do not grow as the organism increases in size. The arthropod must periodically shed its exoskeleton (molt) and grow a new and larger exoskeleton.

Close examination of the cuticle reveals that it is composed of three layers (Figure 12-2B, right). The outermost of these is a thin waxy covering, a common feature in terrestrial arthropods, that serves to prevent the loss of body fluids. Immediately below the waxy covering is a hardened layer in which the salts are deposited. A third, flexible layer is adjacent to the epidermis. Inasmuch as arthropods possess jointed appendages, the hardened layer is lacking at the joints, and the flexible layer is relatively thin.

Vertebrate Coverings

The body covering of vertebrates is a **skin,** or **integument,** consisting of an outer epidermis and an underlying dermis that contains blood vessels, nerves, and pigment (see Figure 9-15). Structurally and physiologically it assumes a number of important functions. First, it is concerned with protection. In this regard, it protects all underlying tissues against mechanical injury, bacterial invasion, desiccation, drastic temperature changes in the environment, and aids in preventing

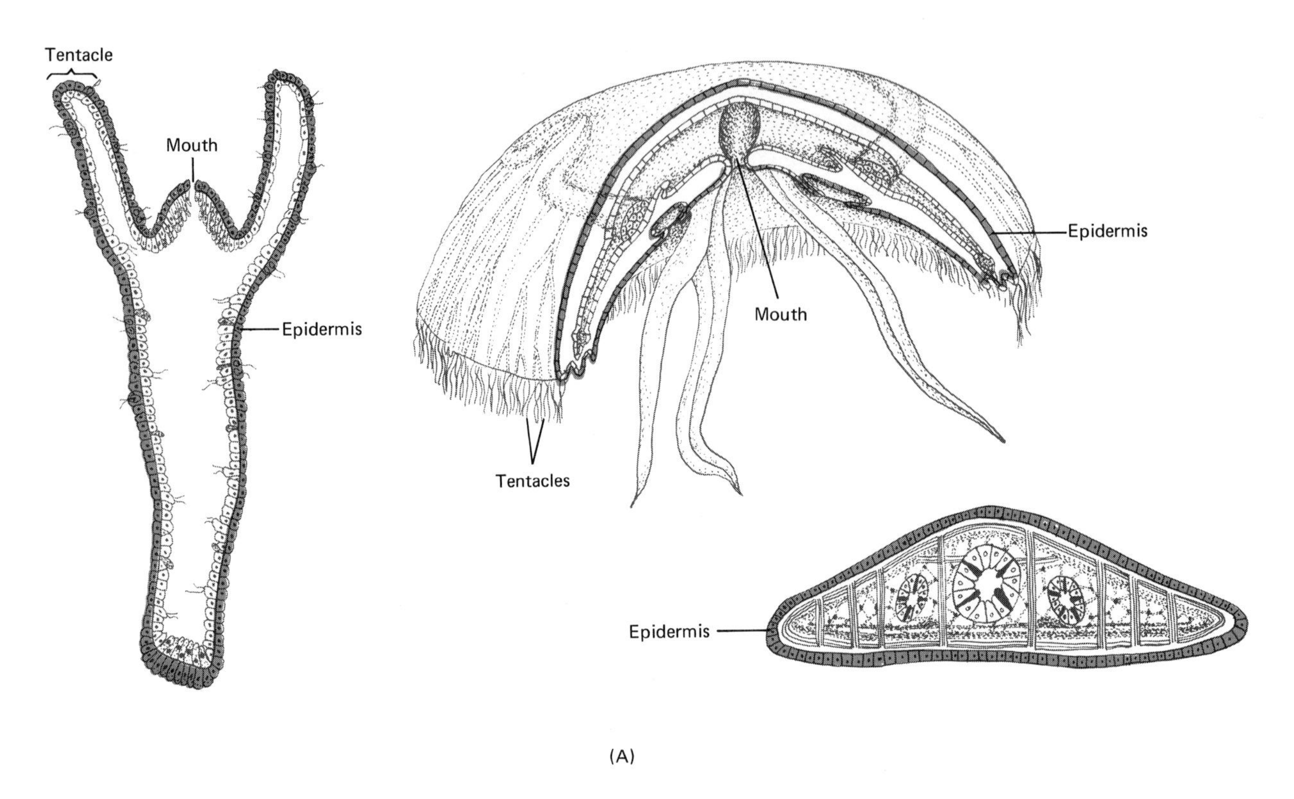

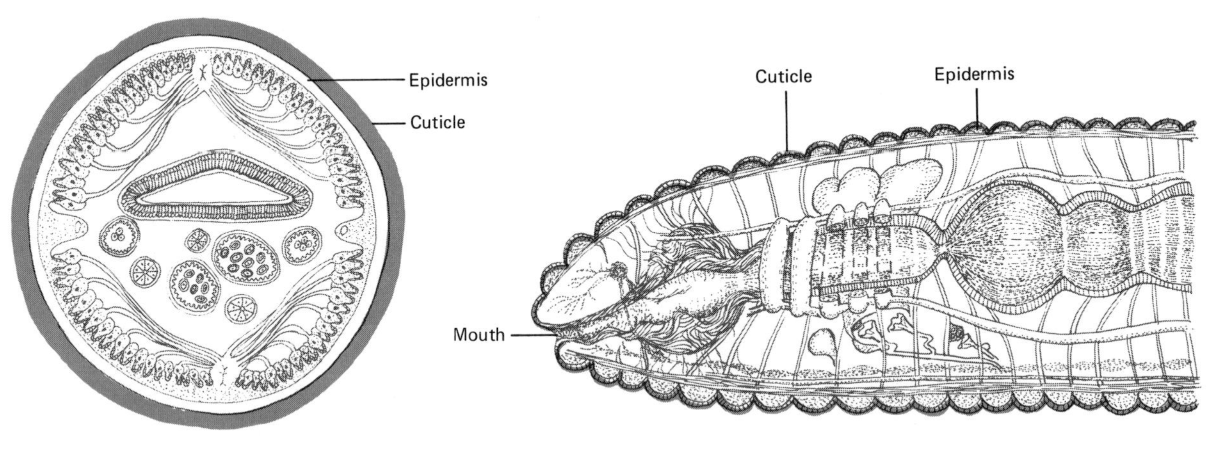

FIGURE 12-1. External coverings of invertebrates. (**A**) Epidermis of soft-bodied and aquatic animals. *Left:* Longitudinal section through *Hydra*. *Center:* Portion of body cut away in *Aurelia* (jellyfish). *Right:* Cross section of a planarian. (**B**) Cuticles of roundworms and annelids. *Left:* Cross section of *Ascaris,* a roundworm. *Right:* Median section of an earthworm, an annelid.

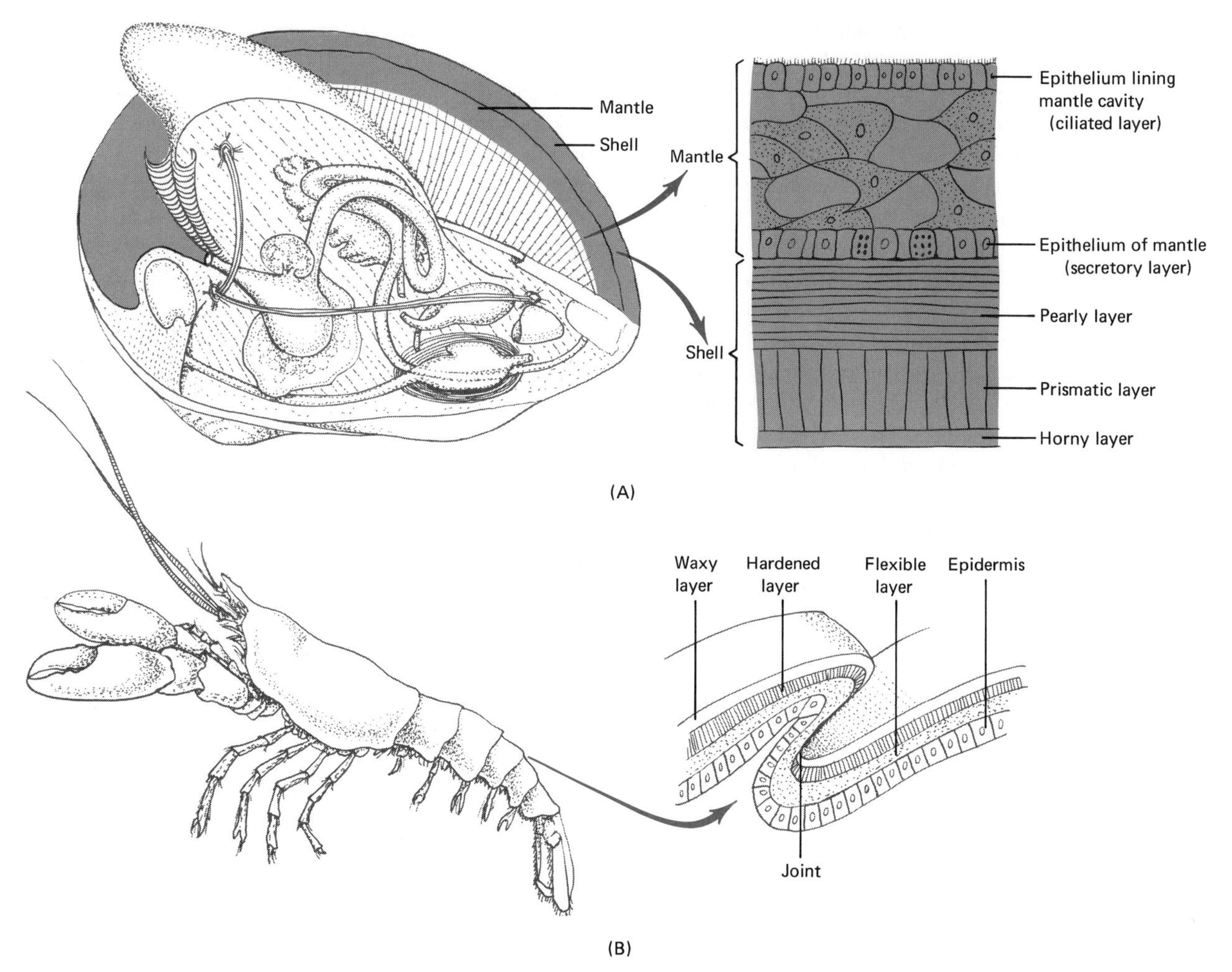

Figure 12-2. Body coverings in mollusks and arthropods. (**A**) Clam (mollusk) shell and mantle in relation to other internal structures (*left*) and enlargement of the shell-mantle relationship (*right*). (**B**) Exoskeleton of a crayfish (arthropod) covering entire body (*left*) and in detail (*right*).

overexposure to ultraviolet radiation from the sun. Second, the skin helps to maintain a constant body temperature in warm-blooded animals (birds and mammals) by serving as a center for the release and evaporation of sweat. Third, the skin has a sensory function since it contains nerve endings or receptors for touch, pressure, pain, and temperature change. The integument also has an excretory function in that sweat, in addition to its large content of water, contains inorganic salts, urea, uric acid, ammonia, and creatinine, substances also excreted by the kidney. Portions of the skin have a secretory role and in mammals, including humans, oil is one of these secretory products. Oil produced by the skin glands helps to keep the skin moist and to protect it against excessive ultraviolet radiation. The mammary glands of mammals, which are highly specialized derivatives of the skin, assume an important secretory function in the nourishment of offspring. Finally, in many vertebrates, especially fishes and amphibians, the skin acts as a route of exchange for a large percentage of the oxygen taken in and the carbon dioxide released during respiration.

The structure of the human skin has already been discussed, and this may be considered as an example of the vertebrate integument. We will now discuss some modifications of the skin in certain other animals. In fishes, the epidermis is quite thin and is composed entirely of living cells. It is conspicuously glandular and is usually underlain by scales derived from the dermis. The glandular secretions, in conjunction with the scales, protect against bacterial invasion and aid in water regulation.

Most land vertebrates possess an epidermis of several layers (stratified epithelium). The surface layer consists of dead cells which are sloughed off and replaced by cells of the basal epithelial layer through mitotic division. Amphibian skin is glandular and moist, as in the frog. In reptiles, birds, and mammals, the outermost epidermal layer is dry and hardened, that is cornified, as a result of the accumulation of a protein-like substance called *keratin.* This anatomical feature is especially prominent in the entire skin of reptiles and the feet of birds in which epidermal scales are also present. Whereas snakes and lizards shed the cornified layer periodically, mammals shed their cornified layer continuously, dandruff being a common result of the accumulation of sloughed cells.

Although the integument itself is a relatively simple structure, its derivatives are numerous and complex. These may be grouped into bony structures, horny structures, glands, and pigments. The bony structures develop within the dermis, though parts of them may be exposed if the overlying epidermis wears off. Thick *bony scales* and plates were prominent in ancestral vertebrates and have been retained and reduced in most groups of living fishes. The shell of a turtle is composed of dermal plates covered by a horny layer and a comparable condition is found in the skin of certain lizards and crocodiles and in the shell of the armadillo. The antlers of deer are also composed of dermal bone which, when the antlers are developing, is covered by skin, the *velvet.* When fully formed, the velvet is shed, exposing the hard bone which affords an excellent defensive (as well as offensive) weapon. Antlers branch, are shed annually and, with the exception of reindeer and caribou, are found only on males. The horns of sheep and cattle, in contrast, do not branch, are not shed, and occur in both sexes. These horns have a core of bone covered by a highly cornified skin.

Horny skin derivatives develop by the accumulation of keratin in the epidermal cells. Reptilian scales and bird feathers are examples of horny structures. Similar horny scales are found on the legs of birds and the tails of certain mammals. Essentially, horny scales reduce water loss through the skin and are periodically shed; bony scales are not shed but increase in size by the addition of new bone. Feathers resemble horny scales in their arrangement but are otherwise quite different in structure. Other horny derivatives of the integument include claws, hoofs, and nails, all of which are homologous structures.

Hair is a different kind of horny skin derivative (see Figure 9-15). A hair lies within a *hair follicle* that is composed of a tubular invagination of the epidermis supported by surrounding fibers of the dermis. A *hair papilla,* containing blood vessels and nerves, protrudes into the base of the follicle and nourishes the adjacent epithelial cells. These cells divide rapidly and add to the base of the hair, which extends up through the follicle as a column of keratinized cells. Associated with each follicle is a small mass of smooth muscle, the *arrector pili.* These muscles are under the control of the autonomic nervous system and when they contract, as a result of cold temperatures or fear, the hairs stand on end, producing "goose flesh." Goose flesh is seen only in humans; in furred animals, erection of the hairs increases the insulative capacity of the fur, thereby warming the animal.

Glands represent another category of skin derivatives and their secretions serve many purposes in vertebrates. Individual mucus-secreting cells are common in the epidermis of fishes, and multicellular mucous glands are abundant in amphibian skin. In addition, many fishes and amphibians also possess poison glands. The substance produced by the epidermal glands of most amphibians is not toxic to humans. Reptiles have lost the epidermal mucous and poison glands, and only a few glands, chiefly scent glands, are found in their dry, horny skin. A similar condition exists among birds, but among mammals glands are quite abundant. For example, oil-secreting sebaceous glands are located near the base of each hair follicle; coiled, tubular sweat glands are abundant in parts of the skin of many mammals; and mammary glands, which are regarded as modified sweat glands, are present and fully developed in most female mammals.

Skin color in vertebrates is determined largely by the kind and amount of pigments present near the base of the epidermis and in the pigment cells of the dermis. In fishes, amphibians, and reptiles, the pigments are concentrated within special cells called **chromatophores.** Chromatophores are rare in mammals, but the brownish pigment *melanin* is present within and between the cells of the epidermis. All humans, in addition to possessing melanin, also have a yellow or yellowish red pigment called **carotene,** and the blood pigment **hemoglobin.** Blacks, Asians, and Caucasoids all have these three pigments, but in different proportions. Blacks have a great deal of melanin; Asians have less melanin and more carotene; Caucasoids have the pink or flesh-colored complexion because the red of the hemoglobin shows through the skin. Evidence of the presence of melanin in white-skinned people can be drawn from the fact that these individuals may tan quite readily. Sunlight has a decided effect, not only on increasing the amount of pigment (tanning), but also on increasing the thickness of the epidermis itself. The purpose of such a response is protection because ultraviolet light is injurious to the tissues of the body beneath the epidermis. It should be noted that a person of any race who lacks the ability to produce pigment is referred to as an *albino*. Albinism is also found in almost every animal species and some plants as well. Human albinos are characterized by very white translucent skin, straw-colored hair, and pink eyes. Obviously, albinos have a low tolerance for light.

Support in Plants

Whereas epidermal and cork tissues of higher plants provide protection, other tissues are concerned largely with support. The principal function of collenchyma tissue (see Figure 9-2C) is for mechanical support and strength. This is usually the first supportive tissue to appear in young stems and leaves and is also a permanent tissue in mature plant organs. Sclerenchyma (see Figure 9-2D) is another plant tissue that affords a considerable degree of support, especially sclerenchyma fibers. This tissue, in addition to having thick secondary walls, is also impregnated with lignin. The cells of xylem tissue (see Figure 9-3) also assume an important role in support.

The Supporting Framework of Organisms

Representative forms within most animal phyla have a firm framework, or skeleton, that gives physical support and protection for the body and often provides surfaces for the attachment of muscles. A skeleton is not absolutely necessary, however, since a number of aquatic invertebrates and even a few land animals have none.

Types of Skeletons

In general, animal skeletons are usually classified into two principal types: (1) **exoskeletons,** or external skeletons, and (2) **endoskeletons,** or internal skeletons. Exoskeletons serving as defensive armor were present on fossil animals such as trilobites, early amphibians, and some ancient reptiles (dinosaurs). Among extant forms they are found on brachiopods, most mollusks, barnacles, some fishes, turtles, and the armadillo. Although the exoskeleton affords a high degree of protection, its limitations are weight and the fact that it must be shed to provide for growth. The internal skeleton of a vertebrate affords far less protection, but provides for attachment of muscles on the outside of the frame so that there is no necessity for the periodic shedding of parts. As a result, the largest animals of land and water are vertebrates.

The Framework of Protista and Invertebrata

Some protozoans (Figure 12-3A) form skeletons of calcium (foraminiferans) or silicon (radiolarians). Some sponges secrete microscopic internal crystalline rods called *spicules* (Figure 12-3B). These structures, which typically consist of lime or a silica-like material, comprise an internal framework. Other sponges, such as the bath sponges, contain fine irregular and interconnected fibers of a sulfur-containing compound called *spongin*. The exoskeletons of corals, brachiopods, mollusks, and echinoderms consist principally of calcium carbonate ($CaCO_3$) and are retained throughout life. All arthro-

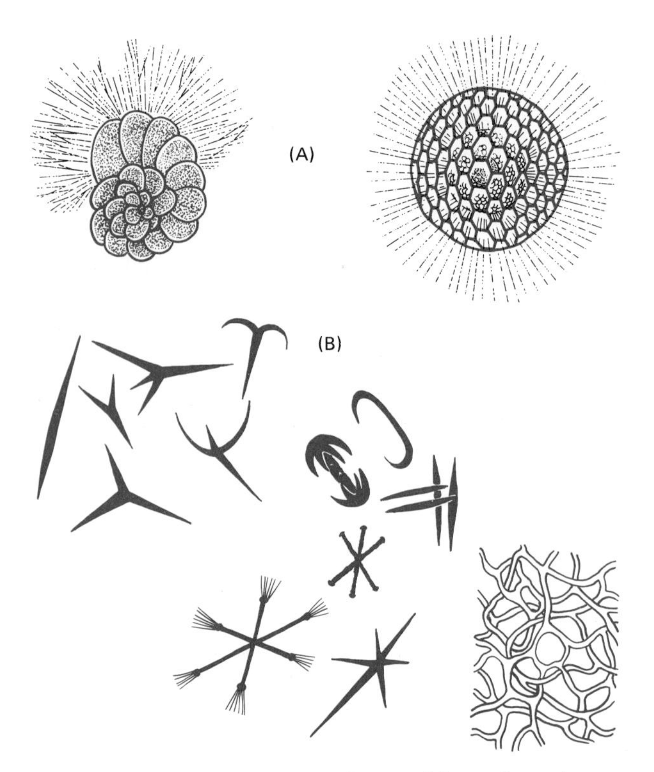

FIGURE 12-3. Representative protistan and invertebrate frameworks. (A) Calcium skeleton of *Rotalia,* a foraminiferan (*left*) and silicon skeleton of *Heliosphaera,* a radiolarian (*right*). (B) Spicules of sponges. Calcium (*left*), siliceous (*center*), and spongin (*right*).

pods, as noted earlier, are completely covered by a jointed exoskeleton consisting mainly of chitin.

Vertebrate Skeletons

Structurally, the vertebrate skeleton consists of bones and/or cartilage. The joints are formed between bones by binding connective tissues (ligaments). The basic substance of the endoskeleton of advanced adult vertebrates is bone, whereas in primitive vertebrates and all embryos of higher vertebrates it is cartilage. The skeletal system performs five basic functions:

1. Support.
2. Protection.
3. Movement and leverage in conjunction with skeletal muscles.
4. Hemopoiesis (blood cell formation) by the bone marrow.
5. Calcium and phosphorus storage in bones.

The human skeleton will now be analyzed as a representative vertebrate skeleton.

The Human Skeletal System

Before proceeding with this section, the student is advised to re-read the description of bone and cartilage tissues in Chapter 9. On the basis of shape, the bones of the body are categorized as *long bones* (femur, tibia, fibula, humerus), *short bones* (wrist and ankle bones), *flat bones* (several cranial bones, ribs, and shoulder blades), and *irregular bones* (vertebrae, lower jaw bone). Long bones (see Figure 9-12) consist of a diaphysis, epiphyses, articular cartilage, periosteum, medullary (marrow) cavity, and endosteum. Short bones consist of a core of cancellous (spongy) bone encased in a thin layer of dense bone. A layer of cancellous bone lies between two plates of compact bone in the flat bones of the body. The cancellous bone of the skull, ribs, and sternum also contains red bone marrow. Irregular bones are similar in structure to short bones.

The human skeleton consists of two main parts: (1) the *axial skeleton,* composed of the bones that form the upright portion or axis of the body (skull, ear bones, hyoid bone, vertebral column, sternum, and ribs), and (2) the *appendicular skeleton,* made up of the bones attached to the axial skeleton as appendages (shoulder, hips, arms, and legs). These relationships are shown in Figure 12-4.

Axial Skeleton

The *skull,* which is formed from twenty-eight irregularly shaped bones, including those of the inner ear, is composed of the bones of the *cranium* (brain case) and the *face* (Figure 12-5). Of the eleven paired and six single bones of the skull, only one, the mandible (lower jaw bone) is movable. The bones of the skull, excepting the articulation of the mandible, are joined together by jagged-edged immovable articulations, the *sutures.*

At birth, several cranial bones are not completely sutured (fused) so that six spaces are left without any bony covering. These spaces, called *fontanels,* allow alteration of the shape of the child's head in passing through the birth canal (vagina) during birth and allow for brain growth. The fontanels are almost completely fused by the second year of life.

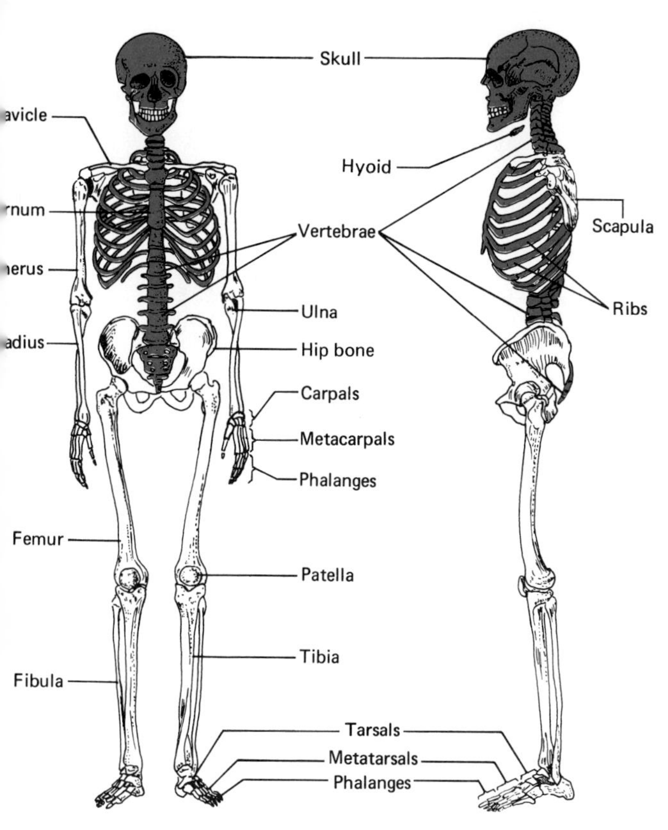

Figure 12-4. Principal subdivisions of the skeletal system. *Left:* Anterior view. *Right:* Lateral view. The axial skeleton is shown in color.

The *hyoid bone* is a single bone in the neck—a part of the axial skeleton. Its U shape may be felt just above the larynx and below the mandible where it is suspended. The hyoid is the only bone in the body which does not articulate (join) directly with any other bone (see Figure 12-4).

The *vertebral column* constitutes the longitudinal axis of the skeleton on which the head is balanced (Figure 12-6A). It consists of twenty-four separate bones called *vertebrae,* so joined to each other to permit forward, backward, and sideways movement of the column. The seven *cervical vertebrae* comprise the skeletal framework of the neck; the next twelve, the *thoracic vertebrae,* lie behind the thoracic (chest) cavity; the next five spinal bones, the *lumbar vertebrae,* form the small of the back; and below the lumbar vertebrae lie the *sacrum* and the *coccyx.* In the adult, the sacrum is a single bone which has resulted from the fusion of five separate vertebrae, whereas the coccyx is a single (or double) bone formed by the fusion of four or five separate vertebrae.

Although the vertebrae exhibit characteristic specializations in different regions of the column, all are constructed on the same basic plan (Figure 12-6B). The *body,* or *centrum,* of a vertebra is the thick, disc-shaped anterior portion. The upper and lower surfaces of the body are roughened for attachment of intervertebral discs of fibrocartilage, whereas the anterior surface is perforated with numerous openings for blood vessels. The *neural arch* forms the dorsal portion of the vertebra. The arch bears three processes for attachment to muscles: two *transverse processes,* one on either side, which serve as points of attachment for the ribs, and a *spinous process* which projects dorsally and serves as a point of attachment for muscles and ligaments. The arch also bears four *articular processes* (two superior and two inferior) for articulation with the vertebrae above and below. The opening formed by the arch, called the *vertebral foramen,* serves as a passageway for the spinal cord. Pairs of small openings, the *intervertebral foramina,* occur between the vertebrae and permit the passage of nerves to and from the spinal cord. The *laminae* form the remainder of the arch and constitute the posterior wall of the vertebral column. Individual vertebrae are held together by *ligaments,* connective tissue masses that attach bone to bone. Between each of the vertebrae are cartilaginous pads, the *intervertebral discs.*

The bony chest cage, or *thorax* (see Figure 12-4), is formed by the twelve thoracic vertebrae, twelve pairs of ribs, and the sternum (breast bone). Each rib articulates with one or more adjacent thoracic vertebrae and curves outward, forward, and downward. Anteriorly, each of the first seven ribs is directly attached to the sternum by the costal cartilages. Each of the cartilages of the next three ribs, however, joins the cartilage of the rib above to be thus indirectly attached to the sternum. Because the cartilage of the eleventh and twelfth ribs does not attach them even indirectly to the sternum, they are referred to as floating ribs.

Appendicular Skeleton. The human appendicular skeleton consists of 126 bones and includes the bones of the arms and legs and the bones comprising the

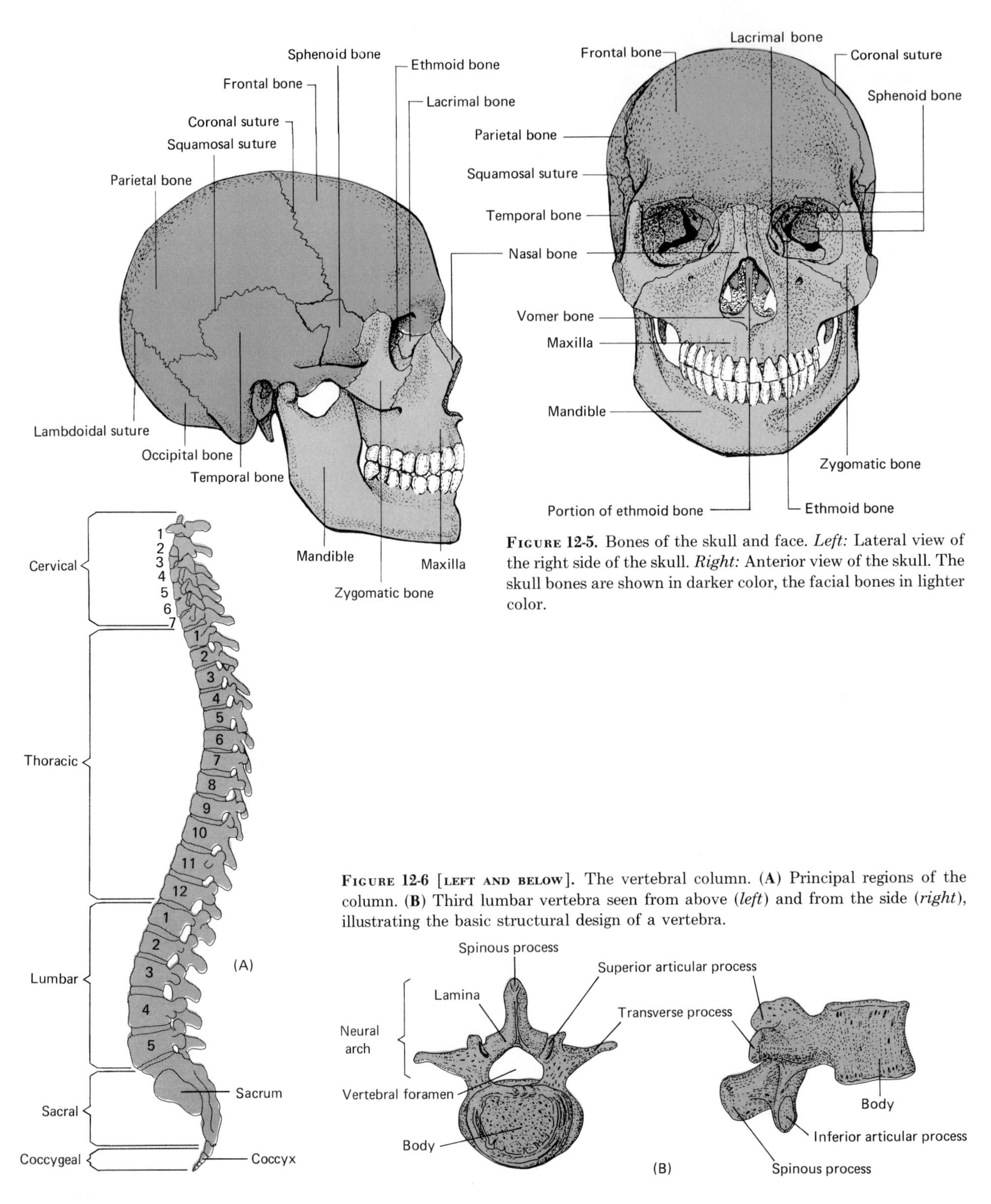

Figure 12-5. Bones of the skull and face. *Left:* Lateral view of the right side of the skull. *Right:* Anterior view of the skull. The skull bones are shown in darker color, the facial bones in lighter color.

Figure 12-6 [left and below]. The vertebral column. (**A**) Principal regions of the column. (**B**) Third lumbar vertebra seen from above (*left*) and from the side (*right*), illustrating the basic structural design of a vertebra.

girdles which attach the appendages to the axial skeleton (see Figure 12-4). The *pectoral girdle* (clavicle and scapula) attaches the arm to the axial skeleton. Each clavicle (collarbone) forms a joint with the breastbone at the inner end and a joint with the scapula (shoulder blade) at the outer end. The scapula is attached to the axial skeleton by muscles and tendons, an anatomical feature which accounts for the flexibility and freedom of movement of the shoulders and arms. A socket on the scapula serves as the point of attachment for the arm. Each arm consists of three long bones: (1) the *humerus,* or upper arm bone which articulates with the shoulder socket, (2) the *ulna,* or lower arm bone on the little finger side, and (3) the *radius,* or lower arm bone on the thumb side. The eight short bones of the wrist, collectively called the *carpals,* are held together in two rows of four by a series of ligaments. The hand consists of five bones of the palm called the *metacarpals* and fourteen finger bones, or *phalanges.* Each finger has three phalanges, the thumb has only two. A thumb is a digit, but not a finger.

The *pelvic girdle* is composed of the two osso coxae or hip bones and with the sacrum constitutes a stable, circular base that supports the trunk and to which the legs attach. Each hip bone is composed of three fused bones. The fused vertebrae of the sacrum complete the circle at the back, leaving a large opening through which offspring pass in the birth process in the female. The dimensions of this opening are one of the clues used in determining the sex of a skeleton since the pelvic girdle of the female is broader than that of the male, a feature that facilitates the child-bearing function.

The upper leg bone, or *femur,* is the longest and heaviest bone of the body. Its *proximal end* (end nearer the trunk) fits into a deep socket in one of the hip bones and, although this joint is much more secure than the arrangement in the shoulder region, it does not have equivalent freedom of movement. This movement is limited by muscles. At the *distal end* (end farther from the trunk) the femur forms the knee joint with one of the two lower leg bones, the *tibia* (shin bone). The other lower leg bone, the *fibula,* is smaller and lies on the outside of the leg. The knee joint is protected anteriorly by the knee cap, or *patella.* The structure of the foot is similar to that of the hand, with certain differences which adapt it for supporting weight. Each foot has twenty-six bones, seven ankle bones, or *tarsals;* five *metatarsals,* which correspond to the metacarpals of the palm; and fourteen *phalanges,* three in each toe except the large one, which has only two.

Articulations (Joints) Between Bones. Bones are joined to one another in several ingenious ways to permit varying degrees of movement. The point at which two or more bones meet is called an **articulation,** or **joint,** and, in general, joints may be classified into three main types on the basis of structure. Functional types will also be included in the discussion of each type.

The first general type of joint is the *fibrous joint.* This class includes all joints in which the adjoining margins of the bones are united by a thin layer of fibrous tissue or by dense fibrous tissue. The two kinds of fibrous joints are the *suture* and the *syndesmosis.* Sutures are found between bones of the skull; they do not permit any appreciable movement. Some sutures consist of jagged, interlocking indentations and projections of the bone margins; others involve an overlapping of the margins. In either case, the bones are slightly separated by a thin layer of fibrous tissue. In the syndesmosis joint, the bone margins are united by dense fibrous tissue membranes. The syndesmosis, as represented by the lower ends of the tibia and the fibula, is a slightly movable joint.

The second principal kind of joint is the *cartilaginous joint* or *amphiarthrosis.* In this joint, the adjoining bony surfaces are connected by cartilage. A *synchondrosis* is a cartilaginous joint in which the connecting cartilage is eventually converted into bone. Thus, it is a temporary joint. An example is the joint between the epiphysis and diaphysis of a long bone. The resulting bony joint is referred to as a *synostosis.* The synchondrosis, like the suture, is an immovable joint. Permanent cartilaginous joints are formed by the rib–sternum junctions. A *symphysis* joint, another type of cartilaginous joint, consists of adjoining bony surfaces connected by broad, flattened discs of fibrocartilage. This joint, like the syndesmosis, is also slightly movable. An example of a symphysis joint is the symphysis pubis of the pelvis.

The third general type of joint is the *diarthrosis,* or *synovial,* joint. This is a joint in which a small space, the joint cavity, exists between the articulating surfaces of the bones. Because of this cavity and because there is no tissue growing between the articulating surfaces, the surfaces are free to move against each other. Functionally, therefore, diarthroid joints are freely movable. Diarthroses are further characterized by a thin layer of hyaline cartilage (articular cartilage) covering the joint surfaces of the articulating bones; the joint is encased within a fibrous capsule lined with a lubricating membrane; and additional ligaments connect the articulating bones. While diarthroses are basically similar in structure, there are distinct variations in the shape of the articulating surfaces, and on this basis, they are subdivided into six types.

1. Gliding.
2. Hinge.
3. Pivot.
4. Condyloid.
5. Saddle.
6. Ball and socket.

Table 12-1 contains a listing of the joints described, a description of each, movements possible, examples, and diagrammatic representation.

Bone Formation and Growth. The embryonic skeleton, when first formed, consists of hyaline cartilage or fibrous membrane structures shaped like bones. The process by which the membranous and cartilaginous precursors of bone are replaced by bone is termed **ossification.** Specifically, *intramembranous ossification* is the process by which bone replaces the membranous predecessors, while *endochondral ossification* replaces cartilaginous structures with bone. The flat skull bones, some bones of the face, and a portion of the clavicle are formed by intramembranous ossification; the remaining bones are formed endochondrally.

The exact mechanism of ossification is not known. However, the data indicate that groups of **osteoblasts** (bone-forming cells) present in the membranous or cartilaginous structures undergoing ossification synthesize organic material of the new bone matrix. Simultaneously with this synthesis, it is believed that the deposition of calcium salts occurs. In long bones, endochondral ossification starts in the diaphysis and develops later in both epiphyses. Increase in bone length occurs by a continual formation of new cartilage at the epiphyseal ends, followed by ossification. Bones grow in diameter by the combined activity of two kinds of cells. One of these, osteoclasts, enlarges the diameter of the medullary cavity by dissolving the bone of its walls. The other cells, the osteoblasts of the periosteum, build new bone around the outside of the bone. Thus, a bone with a large diameter and larger medullary cavity is produced from a smaller bone with a smaller medullary cavity.

The requirements for normal bone growth in the young and bone replacement in the adult are

1. Sufficient quantities of calcium and phosphorus, components of the chief salt that makes bone hard, in the diet.
2. Sufficient amounts of vitamins A, C, and D, which are particularly responsible for the proper utilization of calcium and phosphorus by the body, in the diet.
3. Manufacture by the body of the right amounts of the hormones responsible for bone tissue activity.

Fractures and Bone Repair. In simplest terms, a **fracture** may be defined as any break in a bone. Although fractures of the bones of the extremities may be classified in several different ways, we will use the following scheme for purposes of our discussion.

1. With a *partial* fracture, the break across the bone is incomplete (Figure 12-7A).
2. In a *complete* fracture, the break occurs across the entire bone. The bone is completely broken in two (Figure 12-7B).
3. A *simple* fracture is also called a *closed* fracture. The fractured bone does not break through the skin (Figure 12-7C).
4. In a *compound,* or *open,* fracture, the broken ends of the fractured bone protrude through the skin (Figure 12-7D).
5. With a *comminuted* fracture, the bone is splintered at the site of impact, and smaller fragments of bone are found between the two main fragments (Figure 12-7E).
6. In a *greenstick* fracture, the bone bends, causing the fracture (Figure 12-7F).
7. A *displaced* fracture is a fracture in which the anatomical alignment of the bone fragments is not preserved.
8. A *nondisplaced* fracture is a fracture in which the anatomical alignment of the bone fragments has not been disrupted.

Unlike the skin, which may repair itself in days, or muscle, which may mend in weeks, a bone sometimes

TABLE 12-1. Articulations of the Human Skeleton on the Basis of Structure

I. Fibrous. Adjoining margins of bones are united by a thin layer of fibrous tissue or dense fibrous tissue.

A. Suture. Bones are slightly separated by a thin layer of fibrous tissue.
Movement: immovable
Example: sutures of skull

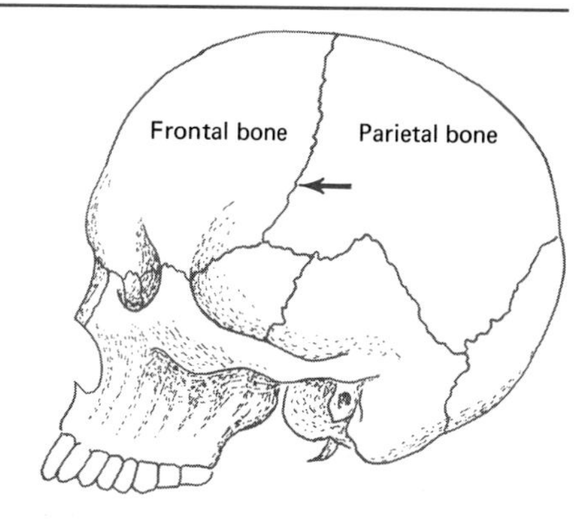

B. Syndesmosis. Bone margins are united by dense fibrous tissue.
Movement: slightly movable
Example: lower ends of tibia and fibula

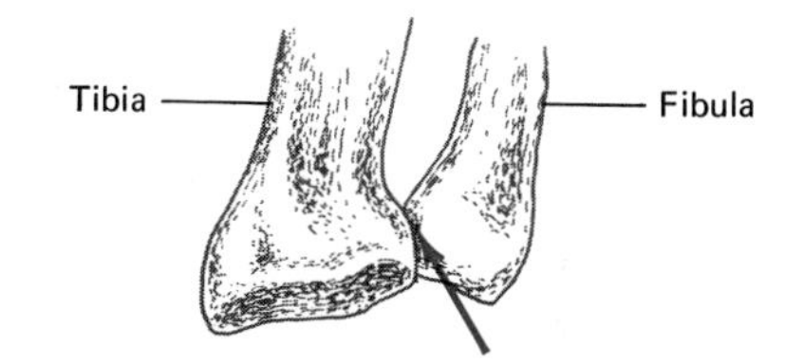

II. Cartilaginous. Adjoining bony surfaces are connected by cartilage.

A. Synchondrosis. Connecting cartilage is eventually converted to bone.
Movement: immovable
Example: between epiphysis and diaphysis of a long bone

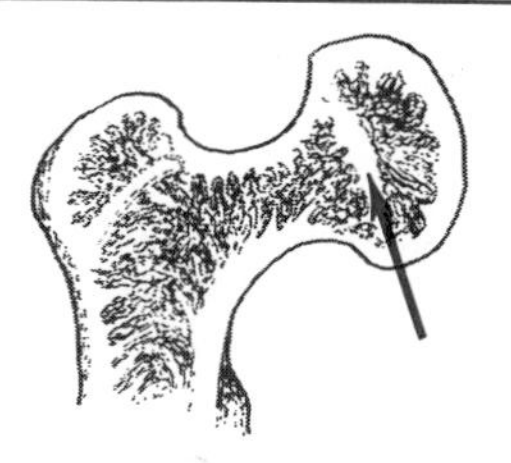

B. Symphysis. Adjoining bony surfaces connected by broad, flattened discs of fibrocartilage.
Movement: slightly movable
Example: symphysis pubis

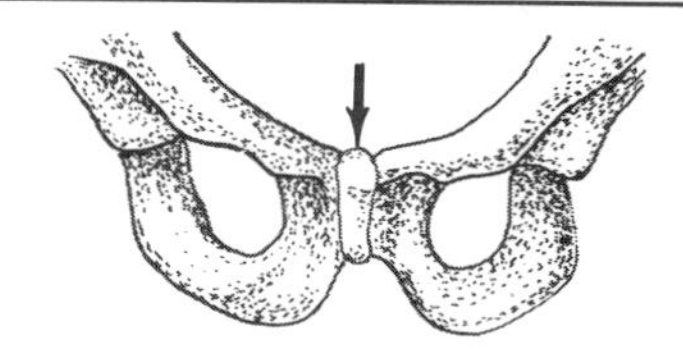

III. Synovial (diarthrosis). A joint cavity exists between the articulating surfaces of bones, articular cartilage covers bones, joint encased in a fibrous capsule lined by a lubricating membrane, additional ligaments connect bones.

A. Gliding (arthrodia). Articulating surfaces usually flat.
Movement: gliding, a nonaxial movement
Example: between carpal bones

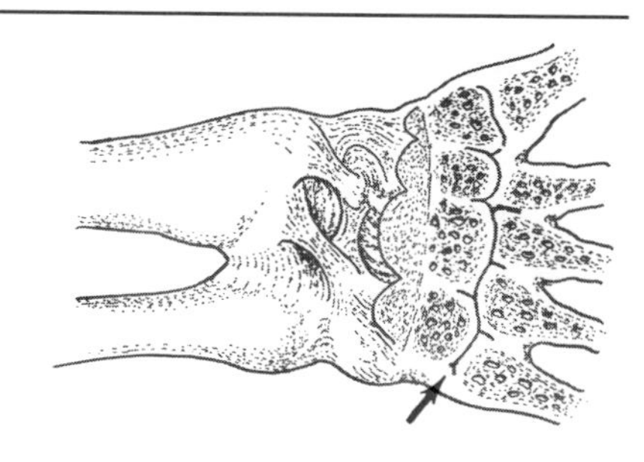

(*continued*)

TABLE 12-1. Articulations of the Human Skeleton on the Basis of Structure [*Continued*]

B. Hinge (ginglymus). Spool-shaped surface fits into a concave surface.
Movement: in one plane about a single axis (uniaxial); like a hinged-door movement
Example: elbow, knee

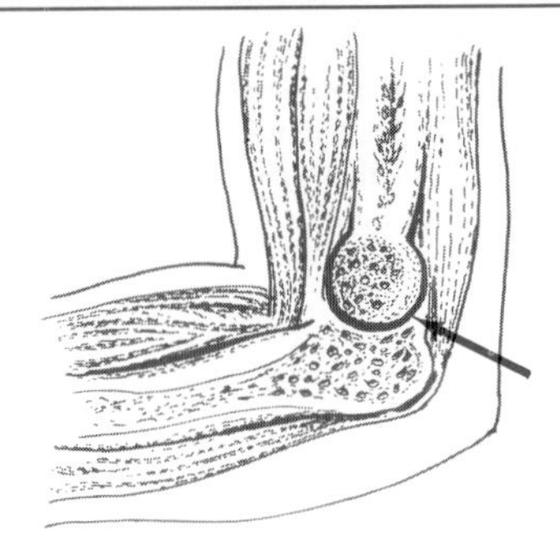

C. Pivot (trochoid). Arch-shaped surface rotates about a rounded or peg-like pivot.
Movement: rotation, uniaxial
Example: between atlas and axis, and between radius and ulna

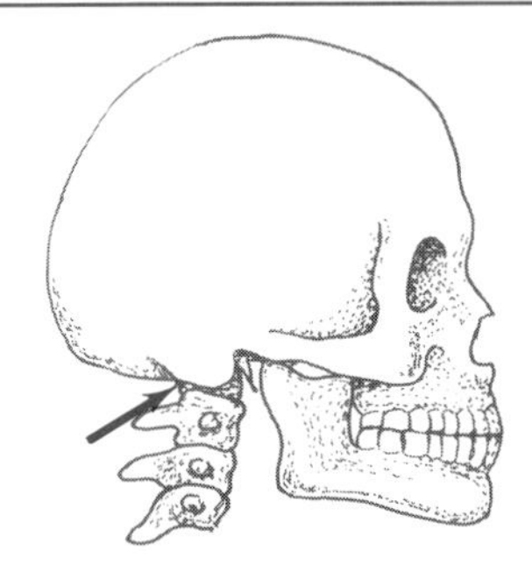

D. Condyloid (ellipsoidal). Oval-shaped condyle fits into an elliptical cavity.
Movement: in two planes at right angles to each other; back and forth, and side to side; biaxial
Example: wrist joint between radius and carpals

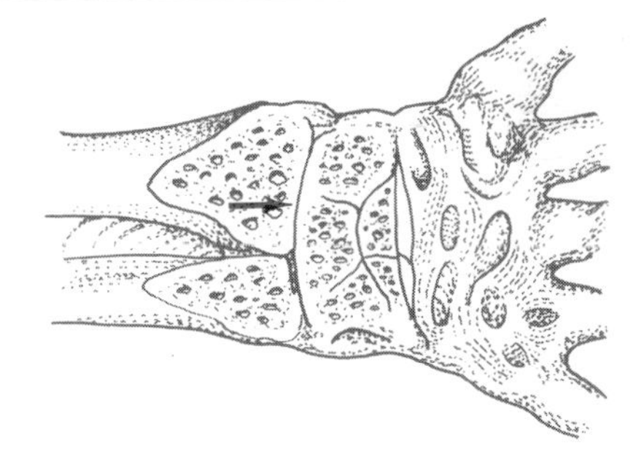

E. Saddle. Saddle-shaped bone fits into socket that is concave–convex in opposite direction; a modified condyloid joint.
Movement: same as condyloid joint but freer
Example: thumb, between metacarpal and carpal bones

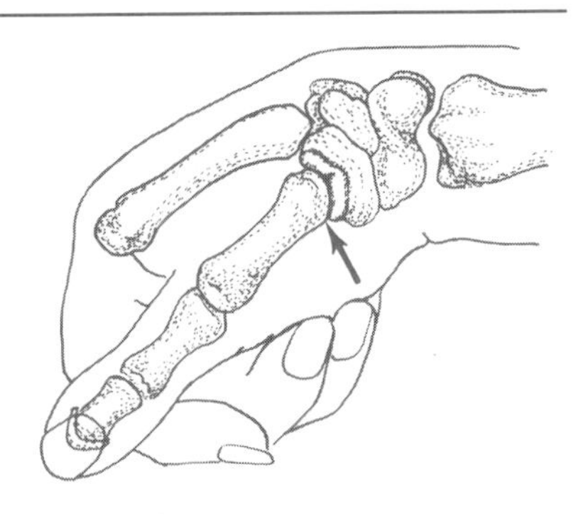

TABLE 12-1. Articulations of the Human Skeleton on the Basis of Structure [*Concluded*]

F. Ball-and-socket (enarthrosis). Ball-shaped head fits into a concave socket.
Movement: triaxial; widest range of all joints
Example: shoulder joint and hip joint

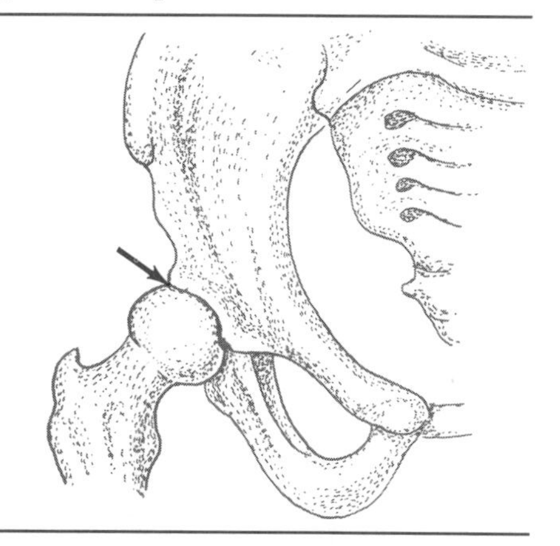

requires months to heal. A fractured femur, for example, may take six months to heal because sufficient calcium to strengthen and harden new bone is deposited very gradually. In addition, bone cells grow and reproduce slowly. Also the blood supply to bone is poor, which partially explains the difficulty in healing an infected bone.

The following steps occur in the repair of a fracture (Figure 12-8).

1. As a result of the fracture, blood vessels crossing the fracture line are broken. These vessels are found in the periosteum, Haversian systems, and marrow cavity. As blood pours from the torn ends of the vessels, it coagulates and forms a clot in and about the site of the fracture. This clot is called a *fracture hematoma*. It usually occurs six to eight hours after the injury (Figure 12-8A). Since the circulation of blood ceases when the fracture hematoma forms, bone cells and periosteal cells at the fracture line die.

2. A growth of new bony tissue develops in and around the fractured area (Figure 12-8B). This new tissue is called a *callus*. It forms a bridge between separated areas of bone so that they are united. The callus that forms around the outside of the fracture is called an *external callus*. The callus formed between the two ends of bone fragments and between the two marrow cavities is called the *internal callus*.

About forty-eight hours after a fracture occurs, the cells that ultimately repair the fracture become actively mitotic. These cells come from the osteogenic layer of the periosteum, the endosteum of the marrow cavity, and the bone marrow. As a result of their accelerated mitotic activity, the cells of the three regions grow toward the fracture. During the first week following the fracture, the cells of the endosteum and bone marrow form new trabeculae in the marrow cavity close to the line of fracture. This is the internal callus. Over the next few days, osteogenic cells of the periosteum form a collar around each bone fragment. The collar, or external callus, is replaced by trabeculae. The trabeculae of the calli are joined to living and dead portions of the original bone fragments.

3. The final phase of fracture repair is the *remodeling* of the calli. In the remodeling process, the dead portions of the original fragments are gradually absorbed. Compact bone replaces cancellous bone around the periphery of the fracture (Figure 12-8C). In some cases, the healing is so complete that the fracture line is undetectable, even by X-ray. A thickened area on the surface of the bone usually remains as evidence of the fracture site.

Locomotion

Movement is one of the most functional characteristics of animals. Such movement may involve the entire organism, or only individual body parts. In most animals, the contractions that bring about changes in

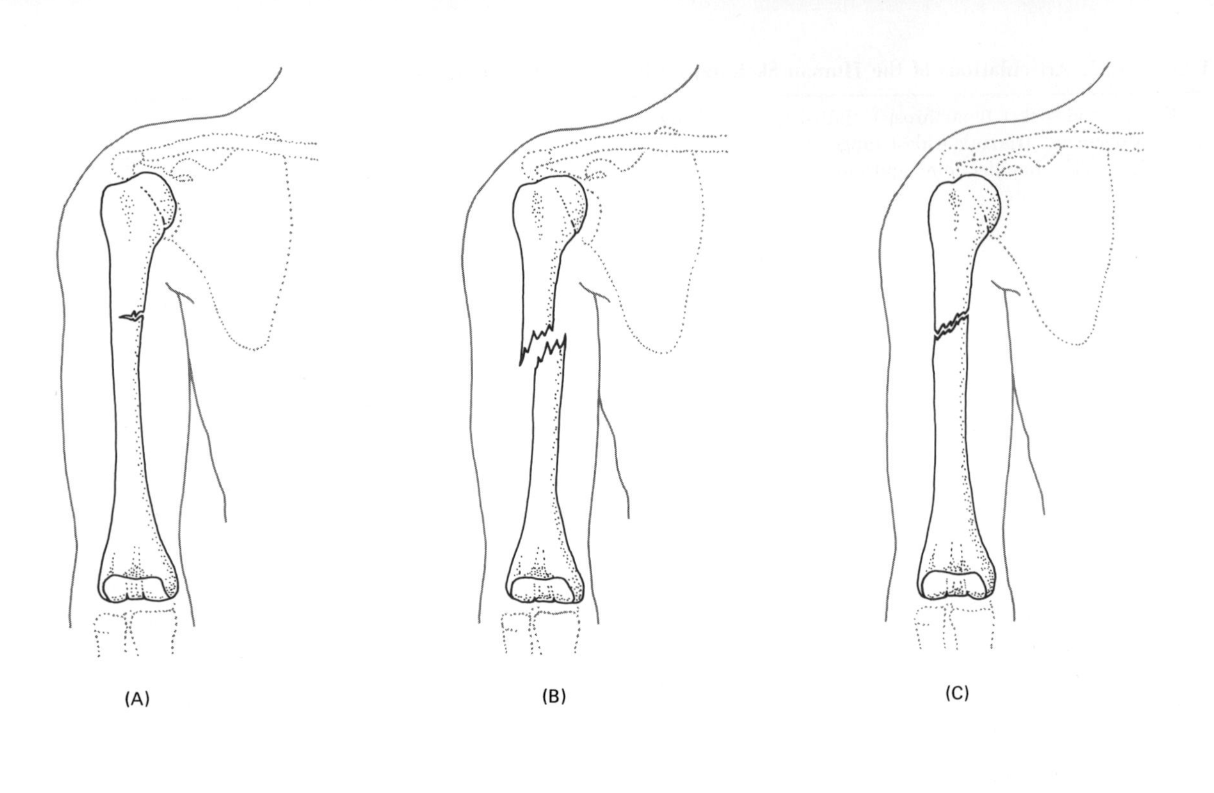

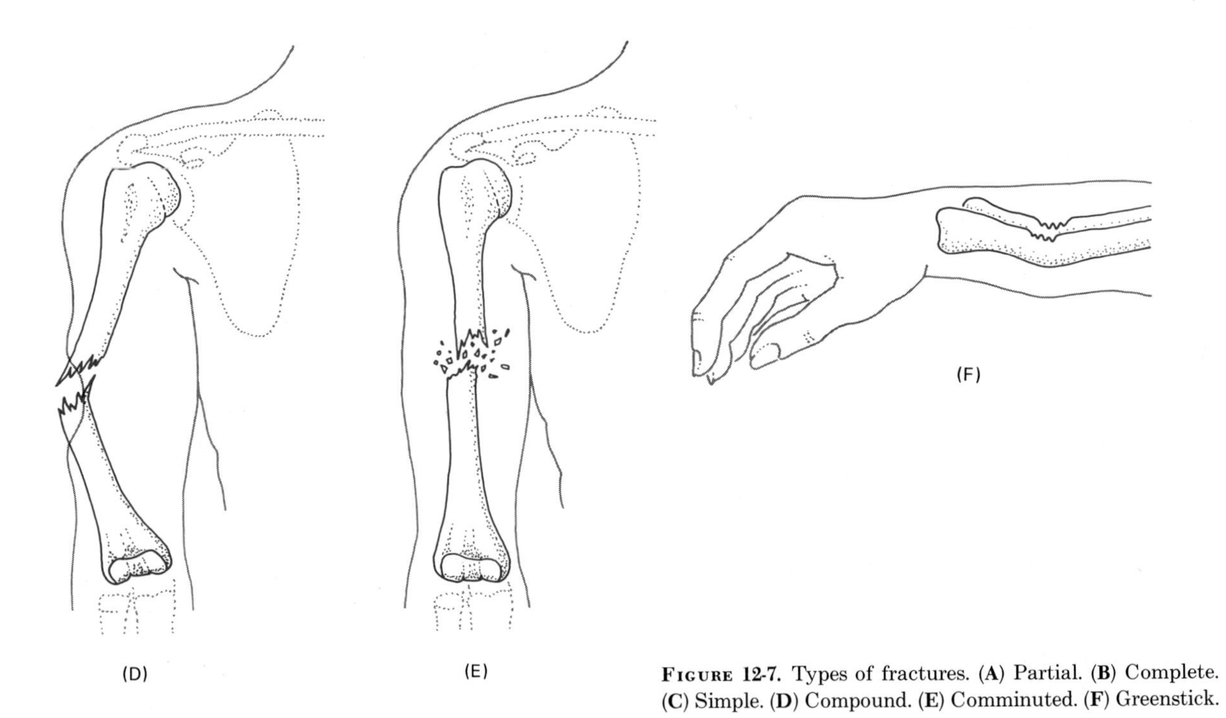

FIGURE 12-7. Types of fractures. (**A**) Partial. (**B**) Complete. (**C**) Simple. (**D**) Compound. (**E**) Comminuted. (**F**) Greenstick.

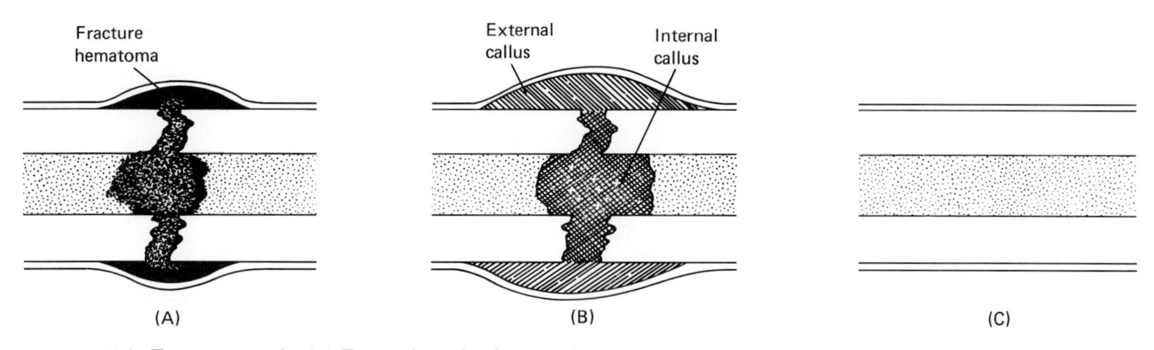

Figure 12-8. Fracture repair. (**A**) Formation of a fracture hematoma. (**B**) Formation of the external and internal calli. (**C**) Remodeling of the calli.

shape or form and locomotion are produced by special fibrils or muscular tissues. Even though plants show some movements, these are usually based upon growth or turgor and not on a specialization of contractile elements, as is the case with animals. Inasmuch as plant movements have been discussed in Chapter 10, they will not be treated here.

Protistan and Invertebrate Locomotion

Simple protozoans, such as *Amoeba,* move by using pseudopodia, a type of locomotion that results from a flowing movement of the cytoplasm alternating between solid and gel states. The movement of ciliated protozoans and flagellated organisms is generally attributed to specialized contractile elements within each cilium or flagellum. Other protozoans of more complex structure, such as *Vorticella,* have specialized contractile fibrils called *myonemes* (Figure 12-9A). The myoneme in the stalk of *Vorticella* superficially resembles a muscle fibril. The stalk which attaches the animal to the substrate consists of a central contractile myoneme within a flexible filament. When stimulated, the contraction of the myoneme bends the filament into a tight spiral and the animal rapidly shortens. Relaxation of the myoneme causes the stalk to straighten and the animal elongates.

All animals above subkingdom Eumetazoa, that is, organisms of the tissue and organ-system levels of organization, possess special cells that have contractile ability. The body wall of coelenterates contains T-shaped epitheliomuscular cells with contractile fibers in the basal part. These cells lie in opposed sets in the body wall so that the body can be reduced in either length or diameter (Figure 12-9B). One set of fibrils acts as longitudinal muscles that contract to shorten the body and tentacles; the other set acts as circular muscles to reduce the diameter and thereby extend the length of the body. Flatworms, such as planarians, usually have muscle fibers running in three directions—longitudinal, transverse, and dorsoventral (Figure 12-9C). The antagonistic action of the longitudinal and transverse muscles produces elongation or contraction of the body. Contraction of the dorsoventral fibers effects further alterations of body shape.

Earthworms have two layers of muscles in the body wall, an outer, circular layer and an inner, longitudinal layer (Figure 12-9D). Contraction of the outer layer causes the body to lengthen, and the action of the longitudinal muscles shortens it. On each body segment of an earthworm (except the first and the last) there are four pairs of minute bristles, the *setae,* which aid in locomotion. They dig into the soil and hold the front end of the worm steady while the tail end is pulled forward; the bristles of the rear then serve as anchors as the front end is extended.

Arthropods are the only invertebrates that do not possess muscle fibers arranged in layers. In these organisms, the muscles are separate and vary in size, arrangement, and attachments to move the body segments and the parts of the jointed legs and other appendages. These muscles are attached to the internal surfaces of the exoskeleton and act over hinge joints between adjacent parts. Because the appendages of arthropods are moved by muscles, they are capable of much more rapid

FIGURE 12-9. Protistan and invertebrate locomotion mechanisms. (**A**) *Vorticella* with myonemes. (**B**) *Hydra.* Sectional view (*left*) and enlarged section (*center*) of cells of body wall. Details of an epitheliomuscular cell (*right*). (**C**) Cross section of a planarian showing arrangement of muscle tissue. (**D**) Cross section of an earthworm showing muscular arrangement and setae.

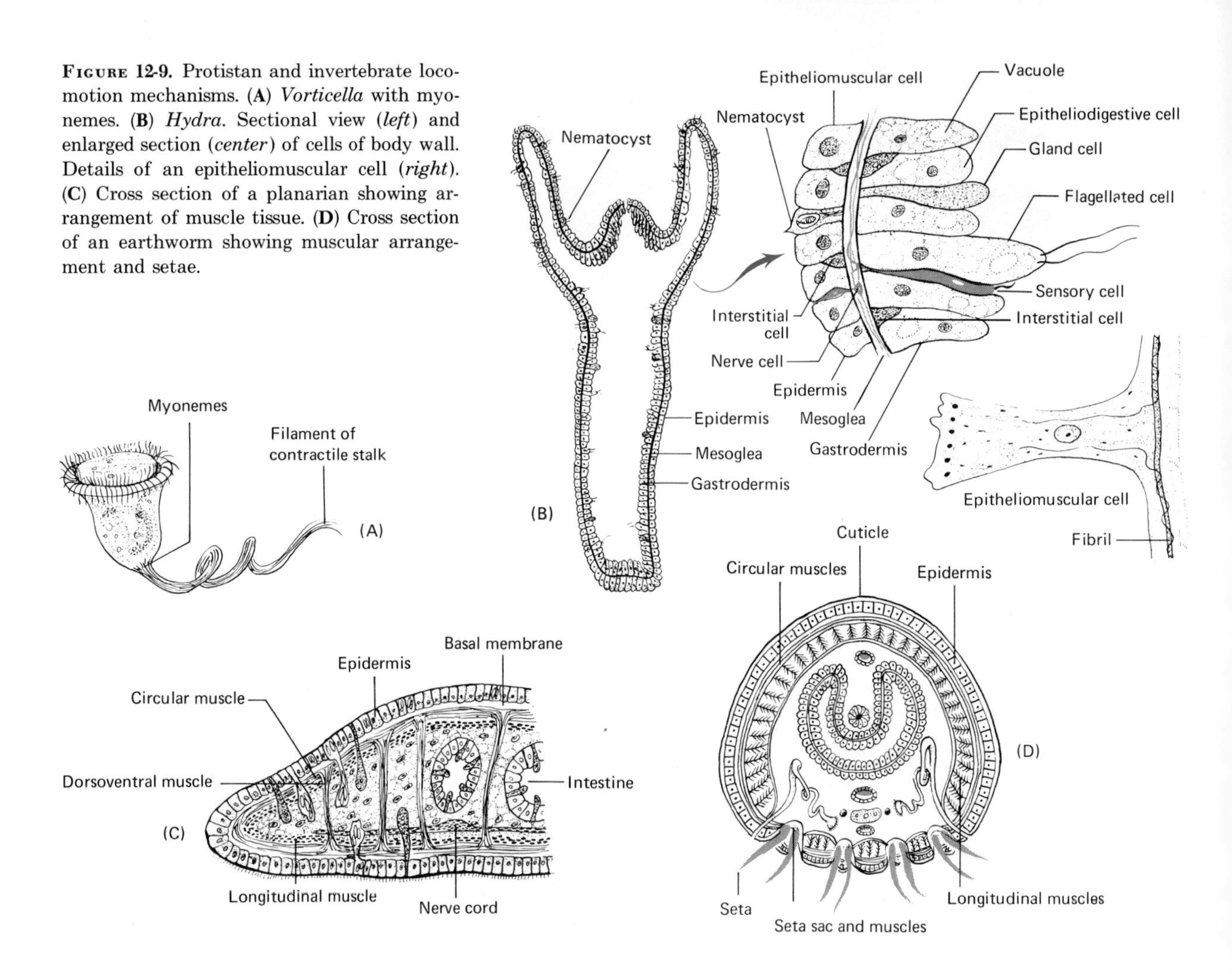

locomotion than are other invertebrates. Essentially, the appendages serve as levers and the muscles move the levers.

Vertebrate Locomotion

It was noted in Chapter 9 that vertebrates, including humans, contain three distinct types of muscle tissue—skeletal, smooth, and cardiac. In addition, the physiological properties of muscle tissue, their cellular structure, and functions were also discussed. In Chapter 7, the detailed structure of skeletal muscle and the physiology and biochemistry of muscle contraction were described. At this point, then, we will survey only the principal human skeletal muscles, their mode of operation, and the motions they produce.

The Human Muscular System

Properly speaking, the term muscular system means all the muscles of the body—those attached to the bones (skeletal), those located in the walls of numerous internal structures (smooth), and those that comprise the wall of the heart (cardiac). However, the term commonly refers to the skeletal muscles only and it is in this context that the term will be used here. Among the functions served by the muscular system are the maintenance of posture and heat production.

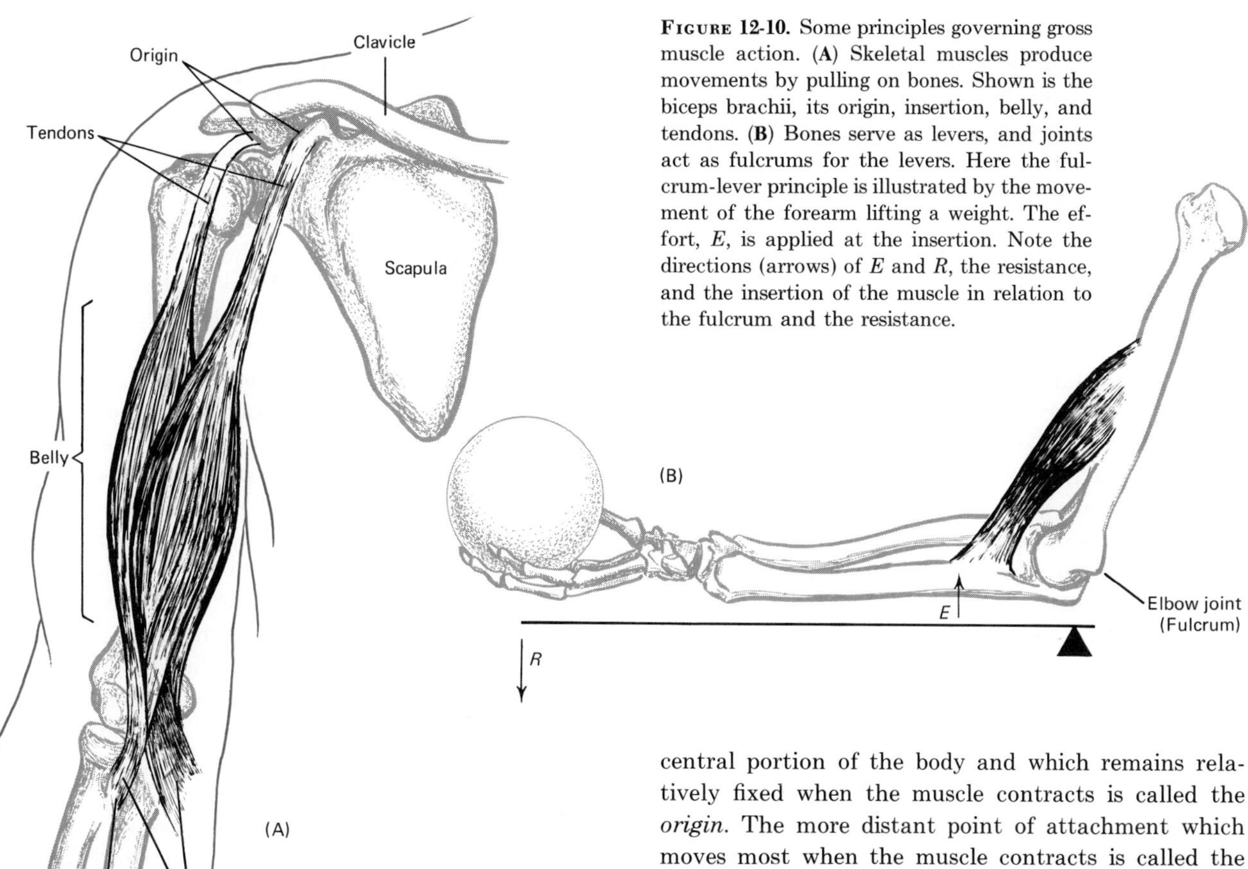

Figure 12-10. Some principles governing gross muscle action. (**A**) Skeletal muscles produce movements by pulling on bones. Shown is the biceps brachii, its origin, insertion, belly, and tendons. (**B**) Bones serve as levers, and joints act as fulcrums for the levers. Here the fulcrum-lever principle is illustrated by the movement of the forearm lifting a weight. The effort, *E*, is applied at the insertion. Note the directions (arrows) of *E* and *R*, the resistance, and the insertion of the muscle in relation to the fulcrum and the resistance.

Principles of Gross Muscle Action. Before discussing the skeletal system of the human, we will first examine a few basic principles involving muscular activity. First, skeletal muscles produce movements by pulling on bones or tendons. Most muscles span at least one joint and are attached to both articulating bones. Thus, when they contract, their shortening places a pull on both bones and this pull moves one of the bones at the joint, that is, draws it toward the other bone. Both bones do not have to move to equal degrees because one of them is usually stabilized by the contraction of other muscles or by its own less mobile structure. The end of the muscle attached nearer to the central portion of the body and which remains relatively fixed when the muscle contracts is called the *origin.* The more distant point of attachment which moves most when the muscle contracts is called the *insertion.* The fleshy portion of the muscle between the tendons of the origin and insertion is referred to as the *belly* (Figure 12-10A).

Second, bones serve as levers and joints act as fulcrums for the levers. A *lever* may be defined as a rigid rod that is free to move about some fixed point or support called the *fulcrum,* usually formed by a joint. A lever is acted upon at two different points by (1) the resistance (weight), which may be regarded as something to be overcome or balanced, and (2) the force (effort) which is exerted to overcome the resistance. In the body bones of varying shapes are levers, and the resistance may be a part of the body to be moved or some object to be lifted, or both. The muscular effort is applied to the bone at the insertion of the muscle and brings about the motion of work. As an example, consider the biceps muscle flexing the arm at the elbow (Figure 12-10B). In this case, when the forearm is raised, the elbow is the fulcrum, the weight of the forearm is the resistance, and the pull due to contraction of the biceps muscle is the effort.

A third basic principle involving the action of skeletal muscle is that muscles that move a part of the body usually do not lie over that part. For example, the biceps muscle that moves the lower arm does not lie over the lower arm (it is, however, attached to it by tendons). It lies in the upper arm (Figure 12-10A). Similarly, flexors of the wrist are located primarily in the lower arm and not the wrist.

Finally, skeletal muscles almost always act in groups rather than singly; most movements are produced by the coordinated activity of several muscles. Some of the muscles in the group contract while others relax. For example, several groups of muscles are involved in varying degrees in flexing the elbow. The *agonists,* or prime movers, give power for flexion; the opposing group, the *antagonists,* contribute to smooth movements by their power to maintain tone yet relax and give way to the movement of the flexor group. Other groups of muscles, called *fixation* muscles, act to hold the arm and shoulder in a suitable position for action. The *synergists* are muscles which assist the agonists by reducing undesired action or unnecessary movements.

Criteria for Naming Muscles. The almost 700 skeletal muscles of the human body are named according to several criteria. One of these involves the direction of the muscle fibers—rectus, transverse, and oblique. *Rectus* (meaning straight) *fibers* run parallel to the midline of the body; *transverse fibers* run perpendicular to the midline of the body; and *oblique fibers* are diagonal to the midline. Examples of muscles named on the basis of these criteria are the rectus abdominis, transversus abdominis, and obliquus abdominis internus (internal oblique), respectively. Another criterion employed is that of location. For example, the rectus femoris is so designated because of its proximity to the femur while the tibialis anterior is so named because it is located near the tibia.

Some muscles, such as the biceps (see Figure 12-10A), triceps, and quadriceps are named on the basis of the number of parts they contain. Other muscles such as the trapezius (trapezoid) and deltoid (triangular) are named on the basis of shape. Another criterion employed is the origin and insertion of a muscle. For example, consider the sternocleidomastoideus, a muscle that originates on the sternum and is inserted in the mastoid process of the temporal bone, and the stylohyoideus, which originates on the styloid process of the temporal bone and inserts on the hyoid bone.

A final criterion used in naming muscles is their action. *Flexors* (flexor carpi radialis) decrease the angle of a joint; *extensors* (extensor carpi ulnaris) return a part from flexion to a normal position by increasing the angle of a joint; *abductors* (abductor hallucis) move a bone away from the midline of the body; *adductors* (adductor longus) move a part toward the midline of the body; *levators* (levator scapulae) raise body parts; *depressors* (depressor labii infemoris) lower body parts; *sphincters* (sphincter and externus) decrease the size of an opening; *tensors* (tensor fasciae latae) fix a body part more rigidly; *supinators* (supinator) turn the hand palm upward, and *pronators* (pronator teres) turn the hand palm downward.

Table 12-2 contains a listing of a few of the principal skeletal muscles of the human body, and their actions. In each case, try to deduce the various criteria employed for their respective names.

Table 12-2. Selected Muscles of the Human Body, Derivations of Their Names, and Their Actions

Name of Muscle	Derivation of Name	Action
I. Muscles of facial expression and mastication		
Occipitofrontalis (actually two muscles—occipital and frontalis)	occipito = base of skull; front = forehead	Elevates eyebrows; wrinkles forehead horizontally.
Orbicularis oculi	orb = circular; ocul = eye	Closes eye; tightens skin of forehead.
Orbicularis oris	or = mouth	Brings lips together.
Buccinator	bucc = cheek	Compresses cheek and retracts angle of mouth.
Masseter	maseter = chewer	Closes jaw.
II. Muscles that control eye movement		
Superior rectus	superior = above; rectus = straight	Rolls eyeball upward.
Inferior rectus	inferior = below	Rolls eyeball downward.
Medial rectus	medial = closer to midline of body	Rolls eyeball medially.
Lateral rectus	lateral = farther from midline of body	Rolls eyeball laterally.
Superior oblique	oblique = slanting	Rotates eyeball on axis; directs cornea downward and laterally.
Inferior oblique		Rotates eyeball on its axis; directs cornea upward and laterally.
III. Muscles that move the head		
Sternocleidomastoid	sternum = breastbone; cleido = clavicle; mastoid = mastoid process of temporal bone	When both contract, head is extended (tipped backward); unilateral contraction turns head from side to side.
Semispinalis capitis	semi = half; spine = spinous process of a vertebra; caput = head	Extends head; bends it laterally.
Splenius capitis	splenion = bandage	Extends head; bends and rotates head.

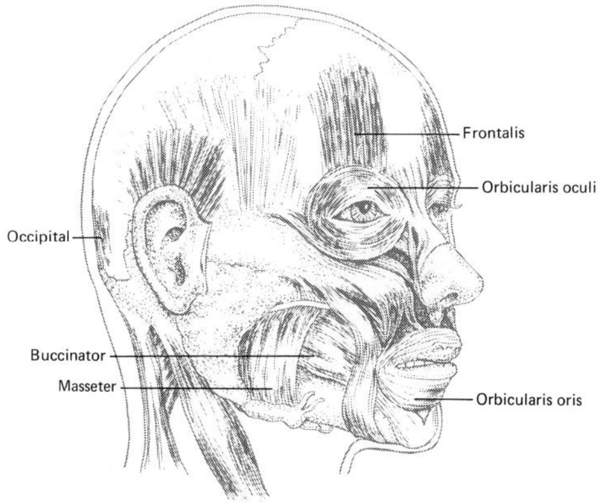

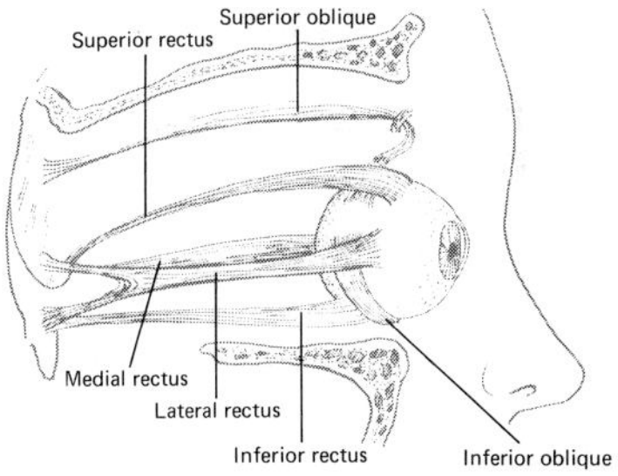

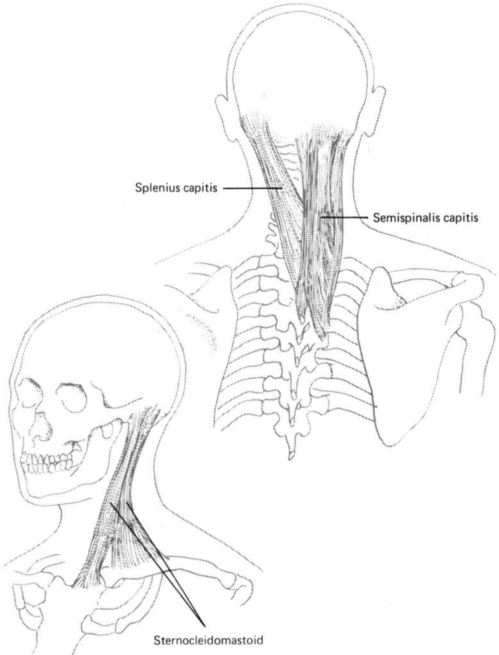

(*continued*)

TABLE 12-2. Selected Muscles of the Human Body, Derivations of Their Names, and Their Actions [*Continued*]

Name of Muscle	Derivation of Name	Action
	IV. Muscles that move the shoulder	
Trapezius	trapezoideus = trapezoid-shaped	Raises or lowers shoulders.
Pectoralis minor	pectus = breast, chest, thorax; minor = lesser	Moves shoulder down and forward.
Serratus anterior	serratus = serrated; anterior = front	Pulls shoulder forward; abducts and rotates it upward.
	V. Muscles that move the upper arm	
Pectoralis major	major = greater	Flexes and abducts arm.
Latissimus dorsi	dorsum = back	Extends and abducts arm.
Deltoid	delta = triangular-shaped	Adducts arm.
Supraspinatus*	supra = above; spinatus = spine of scapula	Abduction of arm.
Teres major	teres = long and round	Adducts, extends, rotates arm medially.
	VI. Muscles that move the lower arm	
Biceps brachii (biceps)	biceps = two heads of origin	Flexes forearm; supinates hand.
Brachialis	brachion = arm	Flexes forearm.
Triceps brachii (triceps)	triceps = three heads of origin	Extends lower arm.
	VII. Muscles that move the wrist	
Flexor carpi radialis	flexor = decreases angle at a joint; carpus = wrist; radialis = radius	Flexes, abducts wrist.
Flexor carpi ulnaris	ulnaris = ulna	Flexes, adducts wrist.
Extensor carpi radialis brevis	extensor = increases angle at a joint; brevis = short	Extends and abducts wrist.
Extensor carpi ulnaris		Extends and adducts wrist.

*Not shown in illustration.

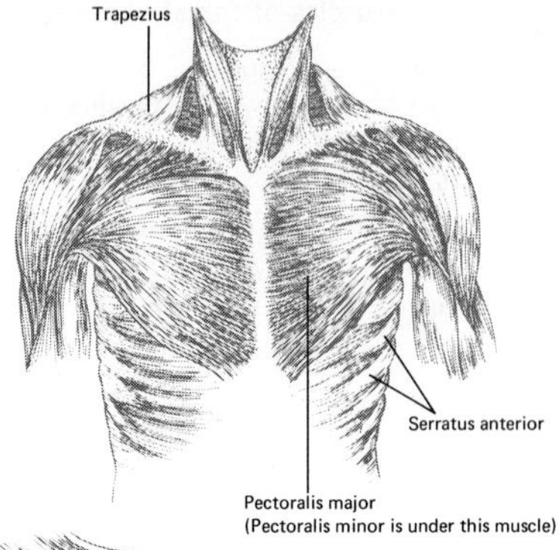

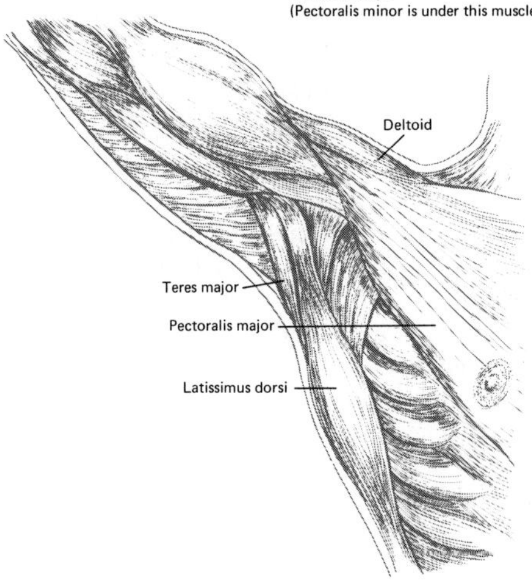

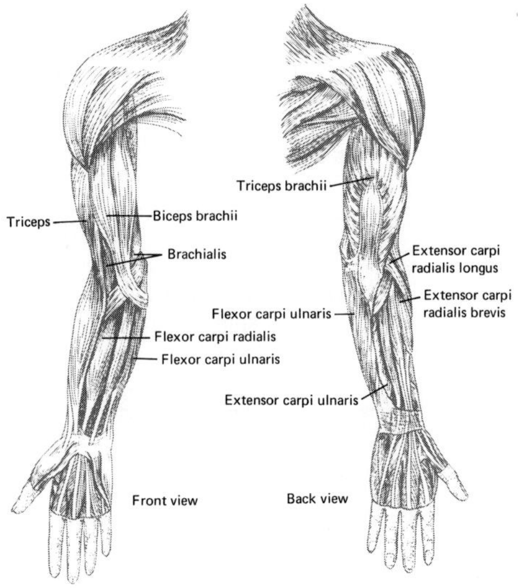

TABLE 12-2. Selected Muscles of the Human Body, Derivations of Their Names, and Their Actions [*Continued*]

Name of Muscle	Derivation of Name	Action
	VIII. Muscles that move the abdomen	
External oblique	external = closer to the surface	Compresses abdomen, depresses ribs.
Internal oblique	internal = farther from the surface	Compresses abdomen, depresses ribs.
Transversus abdominis	transversus = lying across; abdomina = belly	Compresses abdomen, and depresses sternum.
Rectus abdominis		Flexes vertebral column, compresses abdomen.
	IX. Muscles that move the femur	
Iliopsoas (iliacus and psoas major)	ilio = ilium; psoa = muscle of loin (Gk.)	Flexes thigh and trunk.
Rectus femoris	femoris = femur	Flexes thigh, extends lower leg.
Gluteus maximus	gloutos = buttock; maximus = largest	Extends, rotates thigh laterally.
Gluteus medius	media = middle	Abducts, rotates thigh medially.
Gluteus minimus*	minimus = small	Abducts, rotates thigh medially.
Tensor fasciae latae	tensor = makes tense; fascia = band; latus = broad, wide	Abducts thigh.
Adductor group (brevis,* longus, magnus)	adductor = moves a part closer to the midline of the body; longus = long; magnus = large	Adducts thigh.
	X. Muscles that move the lower leg	
Quadriceps femoris group		
1. Rectus femoris		Extends leg, flexes thigh.
2. Vastus lateralis	vastus = vast, large; lateralis = lateral	Extends leg, flexes thigh.
3. Vastus medialis	medialis = medial	Extends leg, flexes thigh.
4. Vastus intermedius*	intermedius = middle	Extends leg, flexes thigh.
Sartorius	sartor = tailor (refers to cross-legged position in which tailors sit)	Adducts and flexes leg.
Hamstring group		Flexes leg and extends thigh.
1. Biceps femoris		
2. Semitendinosus	tendo = tendon	Flexes leg and extends thigh.
3. Semimembranosus	membran = membrane	Flexes leg and extends thigh.

*Not shown in illustration.

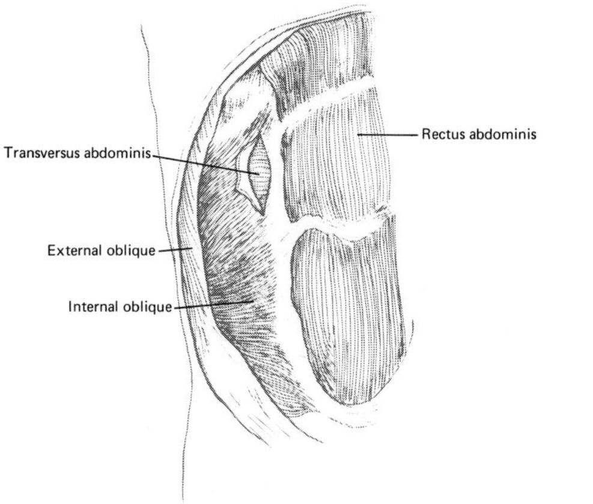

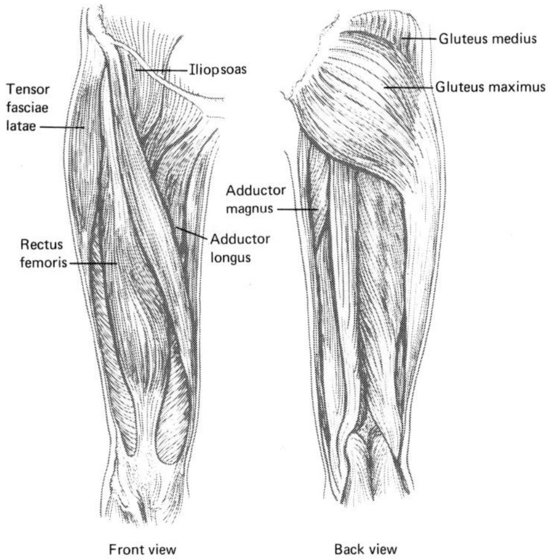

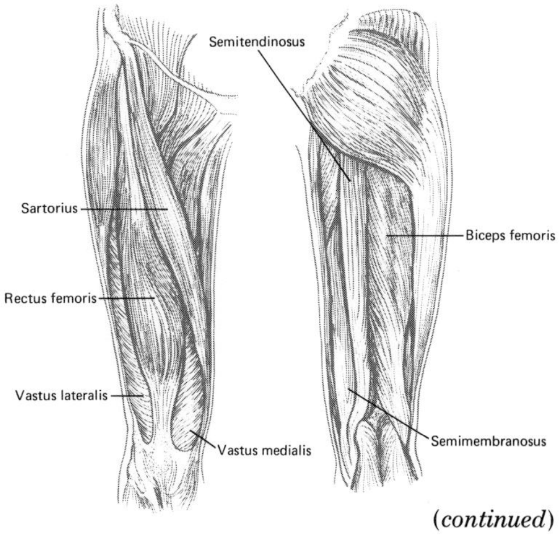

Front view

Back view

(*continued*)

Table 12-2. Selected Muscles of the Human Body, Derivations of Their Names, and Their Actions [*Concluded*]

Name of Muscle	Derivation of Name	Action
	XI. Muscles that move the foot	
Tibialis anterior		Flexes foot.
Gastrocnemius	gaster = belly; kneme = leg	Extends foot, flexes lower leg.
Soleus	soleus = sole of foot	Extends foot.

13 The Procurement and Digestion of Nutrients

GUIDE QUESTIONS

1 What activities are necessary prerequisites for the liberation of usable energy from nutrients? Into what groups are organisms divided with regard to methods of nutrition?
2 Define a nutrient. Explain in detail how nutrients are obtained by organisms.
3 Explain the role of absorption in plant nutrition. How do roots function in the absorption of nutrients?
4 How may heterotrophs be classified with regard to their nutritional patterns? Cite examples of each.
5 What is the major nutritional difference between an autotroph and a heterotroph?
6 Define digestion. What is meant by intracellular digestion? extracellular digestion?
7 Discuss nutrition and digestion in *Amoeba* and *Paramecium*. In what ways are nutrition and digestion in Porifera similar to that of Protista? In what ways are they dissimilar?
8 Why is the gastrovascular cavity of coelenterates referred to as a primitive digestive cavity? Compare digestion in *Hydra* with digestion in sponges. Which is more efficient?
9 How does the highly branched gut of the planaria benefit the organism? In what ways have parasitic animals become adapted with respect to their digestive tracts?
10 What is meant by a complete digestive tract? Discuss some of the anatomical and functional modifications that are possible with the complete digestive tract. Why are such modifications not found in a tract with only one opening?
11 Discuss in detail the various anatomical and functional structures in the human mouth. What variations in tooth structures are found in man? How does each function? How do teeth differ among carnivores and herbivores? Why?
12 What purpose(s) does mastication serve in the digestive process? How does swallowing take place? What involuntary movements occur during swallowing?
13 Discuss in detail the structure and various functions of the stomach in both mechanical breakdown and chemical digestion. What modifications in the stomach may be seen in other vertebrates?
14 Of what importance is the small intestine to the overall process of digestion? Discuss the structure of the small intestine. What are some modifications of the small intestine in other vertebrates? In what ways is the small intestine well adapted for its functions? Explain in detail.
15 Discuss the structure of the large intestine. What are its functions?
16 Outline in detail the various secretions of the digestive tract and associated glands. How are the secretions of the digestive glands controlled? Discuss the phases of gastric digestion.
17 Trace the following through the digestive tract of the human: roast beef, lettuce, mayonnaise on white bread, and a glass of milk. Indicate all changes (mechanical and chemical) that occur and in what regions they occur.
18 Define and describe the symptoms of the following digestive disorders: dental caries, peridontal diseases, peptic ulcers, and peritonitis.
19 Define an alcohol. How is an alcohol absorbed?
20 Describe how an alcohol is metabolized in the body.
21 Discuss the physiological effects of alcohol.
22 What is cirrhosis?
23 What are the six classes of nutrients?
24 Explain the importance of minerals to the functioning of the body. Select as many examples as you wish.
25 Define a vitamin. What are provitamins? Define avitaminosis.
26 Define a fat-soluble vitamin and a water-soluble vitamin. Give several examples of each.
27 For each of the vitamins described, discuss its source, role in the body, and deficiency symptoms.

IN CHAPTER 2 it was shown that a vast array of molecules function not only as the building blocks of organisms, but also as the functional units involved in the multiplicity of activities and chemical reactions of living systems. The concept that metabolism at the cellular level of organization is the basis for all activities that characterize a living form was stressed throughout Section Two. Among the principal features of cellular metabolism are energy procurement, energy transduction, and energy utilization. Such metabolic processes characterize organisms and communities as well as cells.

Basically, the liberation of usable chemical energy from nutrients implies three prerequisite activities. The first is the intake of already-synthesized high-energy compounds (a process called **ingestion** among heterotrophs) or the intake of raw materials from which high-energy compounds can be synthesized. Second, certain high-energy molecules must be enzymatically split into smaller ones that can be utilized by cells, a processing series of reactions called **digestion.** Third, processed food molecules must be distributed to all cells for metabolism. In this chapter, the procurement and processing of nutrients will be considered; in the next chapter, the distribution of nutrients will be discussed.

Organisms may be divided into two principal classes on the basis of their methods of nutrition—autotrophs and heterotrophs. *Autotrophs,* which include members of the Kingdom Plantae, some protistans, and some monerans, procure inorganic materials from the physical environment and synthesize them into organic food molecules. Inasmuch as these raw materials are small enough and soluble enough to pass through cell membranes, they require no digestion prior to intake. *Heterotrophs,* by contrast, including all animals, some protistans, and some monerans, cannot synthesize food molecules from inorganic precursors. Thus, they must procure previously synthesized molecules, either directly or indirectly from autotrophs, and since many of these prefabricated molecules are too large and not sufficiently soluble to be passed through cell membranes, they must first be digested.

Autotrophic Nutrition: Adaptations of the Green Land Plant

The green land plant will serve as a representative member of the Kingdom Plantae, and the adaptations of the green land plant will provide the basis for analyzing the procurement of nutrients for photosynthesis, the principal kind of autotrophic nutrition. The details of photosynthesis may be reviewed by referring to Chapter 6.

All phases of plant metabolism are dependent upon a constant supply of chemicals provided by the physical environment. The term **nutrient,** or **metabolite,** is used to designate any chemical substance that is utilized by plants as a raw material for metabolism. The environment not only supplies these raw materials, but is also the external medium in which plants carry on metabolic activities. Plants function in an environment characterized by daily and seasonal fluctuations of light, temperature, precipitation, and mineral availability. Consequently, the procurement, fate, and utilization of metabolites provided by the environment are also affected by these daily and seasonal variations.

It should not be thought that the movement of metabolites into the plant is in one direction only, and that the plant maintains itself at the expense of the environment. In fact, the interaction of the plant with its environment is cyclic in nature. While the plant is living, some raw materials are returned to the environment as waste products of various metabolic reactions. When the plant dies, decomposition is brought about through the action of microorganisms such as bacteria and fungi, and all nutrients are eventually returned to the environment.

The physical environment in which plants procure nutrients and perform metabolic processes is made up of three distinguishable, although interrelated, subdivisions. These are the atmosphere, the hydrosphere, and the lithosphere (geosphere). These components of the environment as well as biogeochemical cycles are discussed in Chapter 24.

In the performance of their metabolic activities, plants are continually taking in metabolites and allowing others to pass out. The process of intake through which metabolites of all kinds pass from the environment into plant cells is called **absorption.** The term environment in this context means both the external (physical) environment, in which materials enter the outermost plant cells, and the internal (biological) environment, in which the outermost cells pass metabolites to underlying cells. These nutrient materials ordinarily enter the plant through the root in water solution, although other parts of the plant may also serve as absorptive centers.

Before a plant can utilize metabolites provided by the environment, they must be absorbed and then distributed to sites of metabolic activities. The most important metabolites furnished by the environment consist of water, gases, and minerals in the form of ions of their salts. Among the processes involved in the absorption of materials are osmosis, passive absorption (simple diffusion), and active transport (Chapter 5).

Roots of higher plants are the principal organs of absorption. They are in direct contact with the subsoil, a medium which provides a means of anchorage and a reservoir of water and mineral nutrients. Extensive root–soil contact is accomplished not only by the numerous branchings of the root but also by the presence of root hairs. These epidermal projections enormously increase the area of root–soil contact, and there is much evidence that root hairs provide a major portion of the actual absorbing area of the root system.

Water and solutes move independently of each other through the walls of the root hairs. Water absorption is probably accomplished by osmosis. The total concentration of dissolved materials is normally less in the soil solution than in the cell sap (water and dissolved materials in the vacuoles), so water will pass from the soil through the plasma membrane of the root hairs and into the cell sap. This causes a dilution of the cell sap in the root hair. As a result, water passes from the root hair cell, by osmosis, and is absorbed by the cortex cell immediately adjacent to the root hair. This absorption process continues as water is absorbed by successively deeper layers of cells. After being absorbed by the layers of cortical cells, it is moved through the cells of the endodermis, the pericycle, and into the xylem vessels (Figure 13-1).

In the soil solution are dissolved a variety of mineral salts. If a particular mineral ion is at a greater concentration in the soil solution than in the sap solution, it will diffuse into the root hair if the selectively permeable membranes permit its passage. Certain ions can enter the cells of the root even though there is greater concentration of them inside the cells. This mechanism of ion accumulation against a concentration gradient is active transport, in which cellular energy is expended to transfer ions across plasma membranes. Once materials enter plants, they are conducted to various tissues and organs where they assume various metabolic roles. The mechanisms that are involved in the conduction of raw materials and nutrients through plants are discussed in Chapter 14.

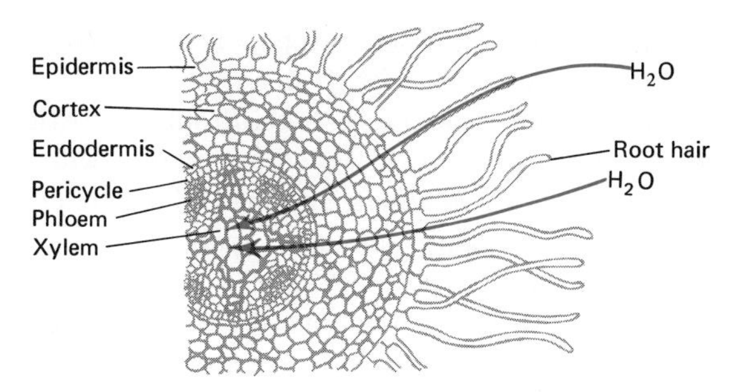

FIGURE 13-1. Pathway of the uptake of water and metabolites from the soil through a root. Most entering materials are absorbed by root hairs, although some may be taken in by epidermal cells. The processes of osmosis, diffusion, and active transport facilitate the transfer of materials from the soil solution into the xylem for distribution throughout the plant body.

Heterotrophic Nutrition

As stated earlier, heterotrophs are organisms that cannot synthesize food molecules from inorganic precursors; they must procure previously synthesized molecules, either directly or indirectly, from autotrophs. Inasmuch as the methods employed by heterotrophs in obtaining prefabricated high-energy organic nutrients are quite varied, these organisms may be further subdivided on the basis of the specific nutritional pattern employed. Accordingly, heterotrophs may be classified as *saprophytes,* organisms that live on dead organic matter, or *parasites,* organisms that live on or in other organisms. Moreover, animals may also be classified as *herbivores,* organisms that live exclusively on plants; *carnivores,* organisms that live primarily on animals that consume green plants; and *omnivores,* organisms that eat both plant and animal materials. Note that regardless of the type of heterotroph, all are ultimately dependent on autotrophs for their nutrition.

Before discussing examples of heterotrophic nutrition among other organisms, insectivorous plants will be considered first. These plants, although primarily photosynthetic (autotrophic), are also capable of trapping and digesting insects and other small animals in order to supplement their nutritional needs. In a nutritional sense, then, insectivorous plants are both autotrophic and heterotrophic. The most familiar of the insectivorous plants are the pitcher plant, the Venus's flytrap, and the sundew (Figure 13-2).

Insects crawl into the vase-like leaves of the pitcher plant and fall to the bottom where they are trapped and digested. The outer ends of the leaves of the Venus's flytrap resemble a spring trap. When insects alight on these leaves and touch the sensitive hairs, the two halves of the leaf close. The unique trapping device of the sundew consists of leaves covered with long, gland-tipped tentacles that curl about the insect and smear it with a mucilaginous, hydrolyzing secretion.

Heterotrophic organisms exhibit highly diverse specializations with respect to the procurement and preparation of nutrients. Yet, despite this diversity, certain commonalities do exist. For example, regardless of the gross structure and form of the nutrient procured, the chemical composition of all living forms is similar. Thus, the chemical activities that act upon nutrients to prepare them for cellular use are also basically similar throughout the living world.

The chemical process by which a complex nutrient is broken down into simple molecules that can pass through membranes and be utilized for cellular activities is termed **digestion.** Essentially, digestion is a series of hydrolytic reactions that proceed in the presence of enzymes. Throughout the living world, complex nutrients such as fats, proteins, and carbohydrates are split by hydrolytic enzymes. These enzymes, produced by living cells, are typically highly specific for the substrates upon which they act. The ultimate products of digestion—the molecules that are absorbed and utilized by cells—are amino acids from proteins, glycerol and fatty acids from fats, and monosaccharides from starches and sugars.

Among organisms, digestion may be viewed as occurring in two principal types, intracellular and extracellular. In *intracellular digestion,* the digestive process occurs within a living cell. In many protozoans and sponges, particles of organic matter or living organisms are taken in by ciliary ingestion or pseudopodial phagocytosis and once the nutrient is inside the cell, vacuoles form about the food, enzymes produced by the cell are introduced into the vacuole, and hydrolysis occurs there. The end products of digestion then diffuse through the vacuolar membrane into the cytoplasm for cellular use. Intracellular digestion also occurs in certain cells of coelenterates and turbellarian flatworms. In these organisms, however, some nondigestive cells receive nutrients from the digestive cells. Other organisms, such as some polychaetes, mollusks, echinoderms, and the horseshoe crab, carry on a degree of intracellular digestion. In many invertebrates phagocytic amoeboid cells assume a significant role in the ingestion and digestion of nutrients.

Organisms of more complex structural design (phyla above the flatworms) typically contain a tubular complex of tissues and organs (digestive system) that are adapted for the enzymatic hydrolysis of foodstuffs. This arrangement provides the basis for the second principal type of digestion, *extracellular digestion.* The enzymes involved in this extracellular process are produced by cells but are secreted into cavities where the nutrients are broken down. Fungi and bacteria also secrete extracellular enzymes into the environment prior to absorption by cells. It should be noted that intracellular and extracellular digestion are similar in that hydrolysis always precedes the actual absorption of complex nutrients across a membrane.

Although both the nutritional requirements and essentials of digestion are quite similar among organisms, the body plans and details of the process are frequently different. The digestive mechanisms of a variety of organisms will therefore be presented. Inasmuch as the evolution of patterns of nutrient procurement is more readily observable in certain groups of organisms, attention will be directed primarily to members of the Kingdom Protista and to invertebrates and vertebrates of the Kingdom Animalia.

Protistans

Protozoa possess varied mechanisms for the procurement and digestion of nutrients. The ameobae, for ex-

(A)

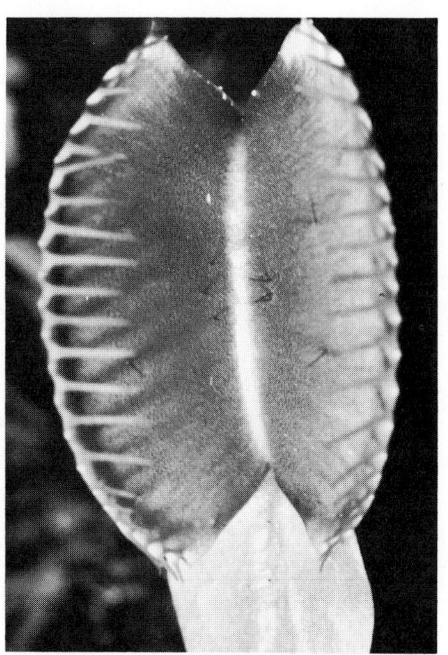

(B)

(B)

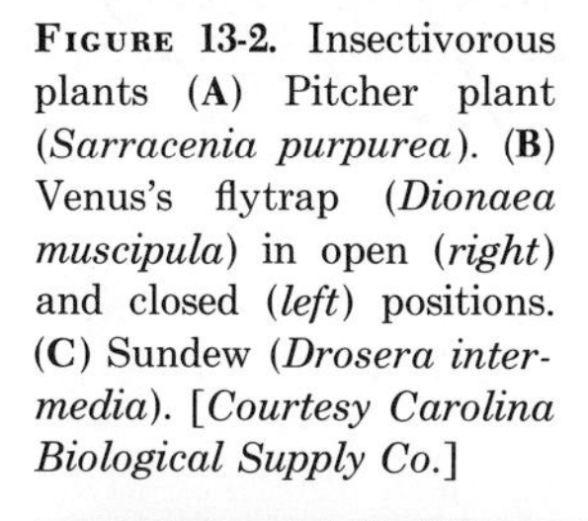

Figure 13-2. Insectivorous plants (**A**) Pitcher plant (*Sarracenia purpurea*). (**B**) Venus's flytrap (*Dionaea muscipula*) in open (*right*) and closed (*left*) positions. (**C**) Sundew (*Drosera intermedia*). [*Courtesy Carolina Biological Supply Co.*]

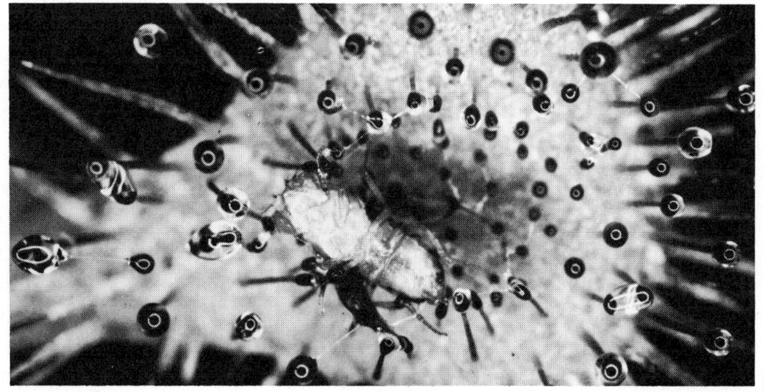

(C)

ample, send out cytoplasmic projections called **pseudopodia,** which engulf nearby food particles (Figure 13-3A). Once the food is completely surrounded by the pseudopodia, it is taken into the cytoplasm and enclosed in a *food vacuole* where it is digested. Digested nutrients are then absorbed from the vacuole into the cytoplasm of the cell where they are utilized. Although *Amoeba* is a typical example of a protozoan lacking specialized permanent digestive structures, the transitory food vacuoles it forms during phagocytosis are analogous to the digestive structures of more advanced organisms.

In other protozoans, such as *Paramecia* (Figure 13-3B), there is a considerable degree of digestive specialization within a unicellular framework. These ciliates possess an *oral groove,* a ciliated channel through

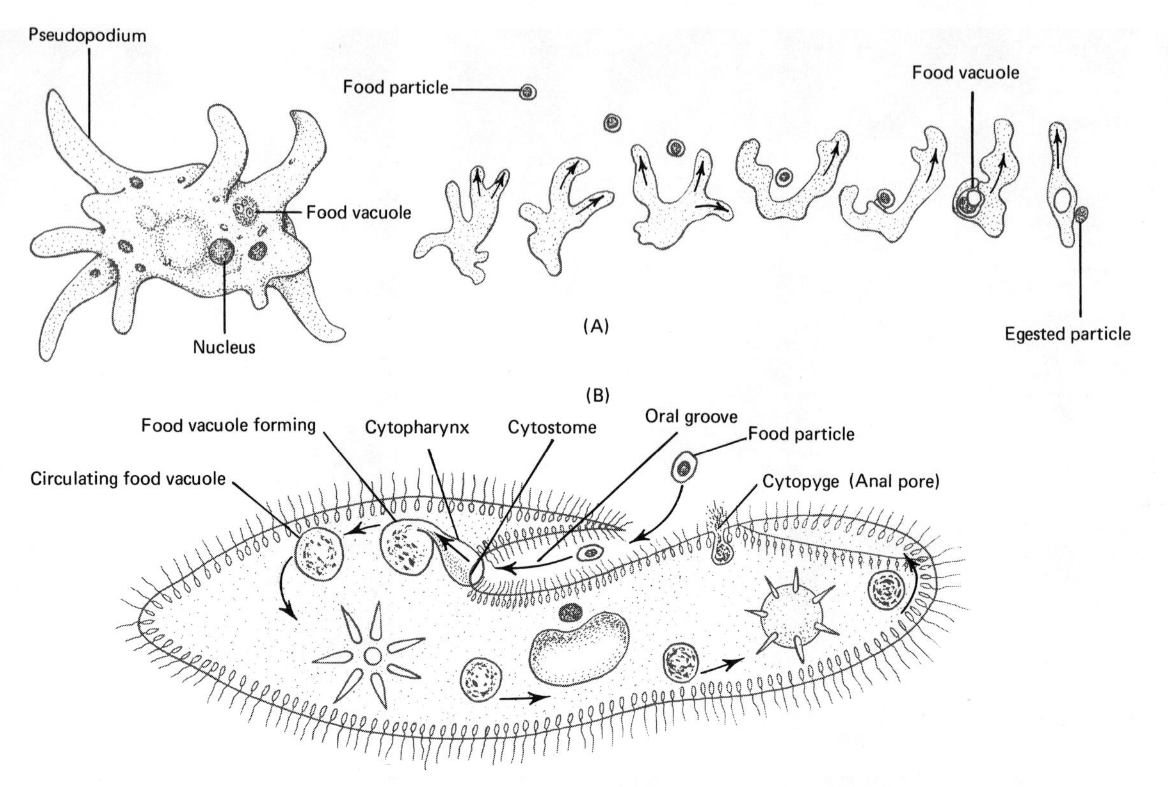

Figure 13-3. Ingestion in *Amoeba* and *Paramecium.* (**A**) *Amoeba. Left:* Generalized structure. *Right:* Sequence of events in the ingestion of food and the elimination of wastes (egestion). (**B**) *Paramecium.* Arrows indicate the path of food vacuole. Note elimination of wastes at the cytopyge.

which food particles are swept by water currents caused by the beating cilia. Nutrients carried down the oral groove pass into a *cytopharynx* and as food accumulates at the lower end of the cytopharynx, a *food vacuole* forms. Eventually the vacuole detaches and circulates throughout the cell. Digestive enzymes are secreted into the vacuole and the products of digestion are deposited within the cytoplasm. After the vacuole has circulated, it reaches a minute specialized portion of the cell surface, the *cytopyge,* or *anal pore.* At this point, the vacuole attaches to the surface, ruptures, and expels any undigested materials.

Invertebrates

Porifera. Sponges represent the simplest of the multicellular animals. Like most unicells, they do not possess specialized digestive structures. The microscopic food of sponges is captured by and digested in flagellated collar cells (*choanocytes*) that line certain interior canals of the animal (Figure 13-4). Digestion is thus intracellular as in protozoans.

Cnidaria. Representatives of the cnidaria possess a primitive digestive cavity called a *gastrovascular cavity* which is characterized by the presence of only one opening (mouth) through which both ingestion and egestion (elimination) occur. In *Hydra,* the single opening to the exterior is surrounded by movable tentacles in which are embedded numerous stinging structures termed **nematocysts** (Figure 13-5). When stimulated by prey, the nematocysts either penetrate the body of the prey (they possess a paralyzing poison) or entangle the prey. At this point the tentacles grasp the prey, draw it toward the mouth, and deposit it within the gastrovascular cavity. Once inside the cavity, extracel-

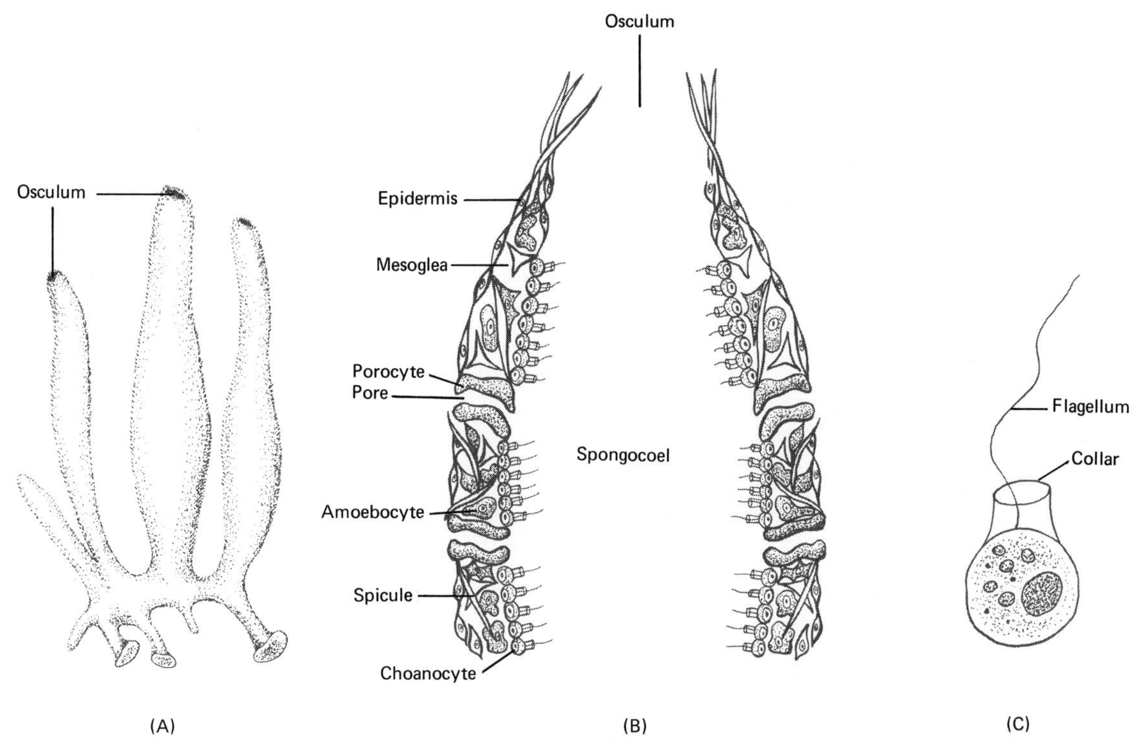

Figure 13-4. Ingestive and digestive apparatus of Porifera. (**A**) External aspect of a colony of *Leucosolenia,* a simple sponge. (**B**) Enlarged aspect of the top on *Leucosolenia* showing central cavity (spongocoel), opening of spongocoel (osculum), pores through which nutrients enter, spicules, and choanocytes, which function in food-getting and digestion. (**C**) Enlargement of a choanocyte (collar cell).

lular digestion begins. Such extracellular digestion in coelenterates is largely limited to proteins, and even these nutrients are not completely digested to amino acids. As soon as the food fragments are reduced to a certain critical size, they are phagocytosed by the cells lining the gastrovascular cavity. Within these cells intracellular digestion completes the digestive process. Any indigestible materials are expelled through the mouth. It should be noted that even though phagocytosis and intracellular digestion are required, extracellular digestion enables organisms to utilize considerably larger fragments of food.

Platyhelminthes. In comparison to the radially symmetric coelenterates, the bilaterally symmetric, free-living flatworms have a more complex gastrovascular cavity. The cavity (intestine), as exemplified by planaria, is a branched system of blind pockets which ramifies throughout the body and connects by channels to the mouth (Figure 13-6). The *mouth,* located on the ventral surface, opens into a muscular, tubular *pharynx,* which leads into the gastrovascular cavity. The highly branched cavity serves not only a digestive function but a circulatory function as well. In this regard, the cavity greatly increases the total absorptive surface area. It will be seen later that many organisms have evolved greatly subdivided absorptive surfaces so that a great deal of total surface area can be compacted into a comparatively small space. Although some extracellular digestion takes place in the gastrovascular cavity, the greater portion of the digestive process occurs intracellularly in the gastrodermal cells lining the cavity.

One class of flatworms, the tapeworms, has become so

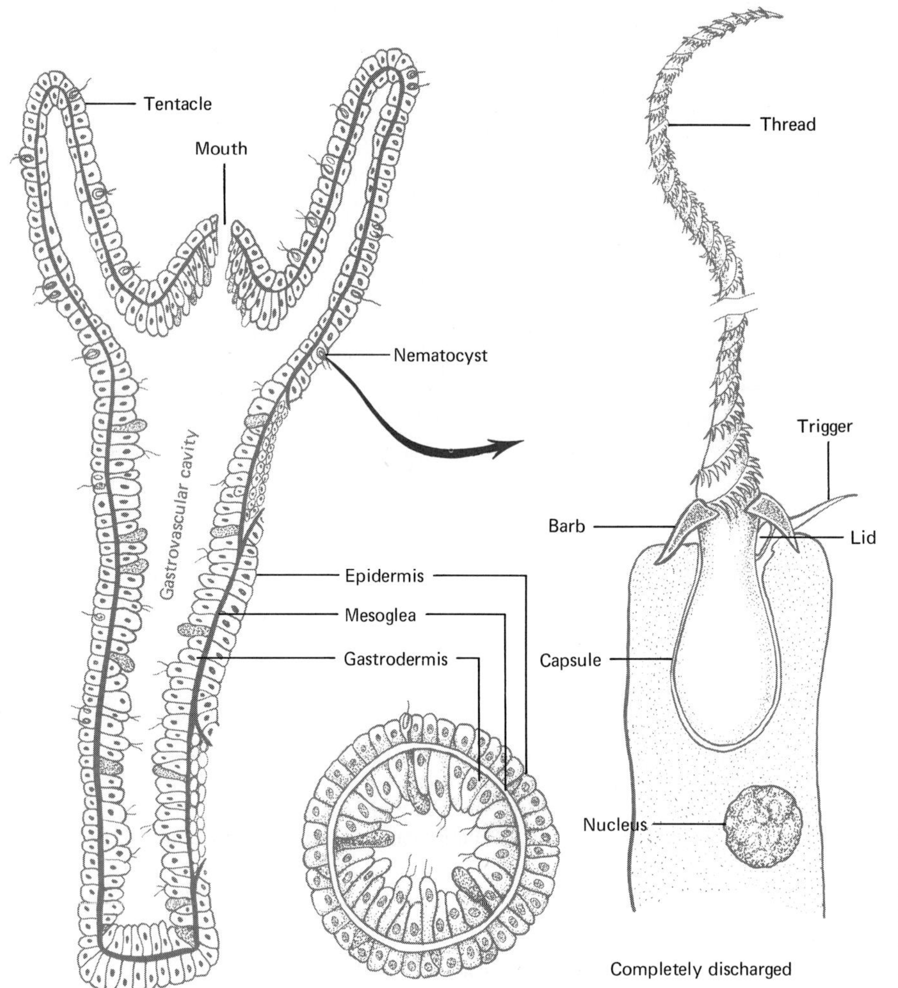

FIGURE 13-5. Ingestive and digestive apparatus of *Hydra*. Longitudinal section (*left*) and cross section (*center*) of *Hydra*, enlargement of a nematocyst (*right*). See text for details.

highly specialized as parasites that they have lost their digestive systems. These parasites inhabit the digestive tracts of other animals and since they are constantly bathed by the products of the hosts' digestion, they can absorb them without having to carry on any digestion themselves.

ANNELIDA. All other major animal phyla from the Annelida upward have a *complete tubular digestive tract,* that is, one with two openings, a mouth and an anus. Such an arrangement permits a one-way passage of food substances and allows for the important possibility of a succession of highly specialized anatomical and functional modifications of the digestive tube along its length. As the food passes through the tube, it is acted upon in a different way in each section. Thus, one section may be specialized for the mechanical breakdown of ingested foods, another for temporary storage, a different section for chemical (enzymatic) digestion, a section for the absorption of digested products, a section for water reabsorption, and still another for the storage and elimination of undigested food residues. The complete digestive system is not only more efficient, but also provides a potential for special evolutionary modifications that enable various organisms to adapt to various environments.

The digestive system of the earthworm exemplifies a

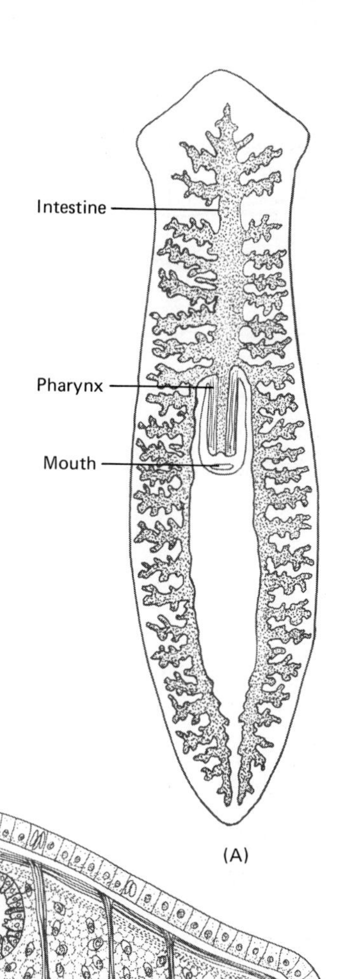

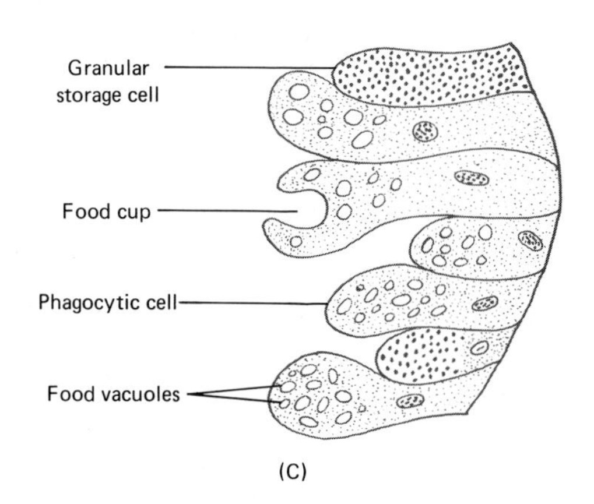

Figure 13-6. Ingestive and digestive apparatus of planaria. (**A**) Longitudinal section showing intestine, pharynx, and mouth. (**B**) Cross section showing intestine in relation to other structures. (**C**) Details of cells lining intestine (gastrodermis). Food particles are ingested and digested by the phagocytic cells and may be stored by the granular cells.

well-differentiated, complete digestive tract of a typical invertebrate (Figure 13-7A). Food, in the form of decaying organic matter mixed with soil, is drawn into the *mouth* by the sucking action of the muscular *pharynx.* The food then passes through the pharynx into the *esophagus,* which in turn carries it into the thin-walled *crop.* From the crop, which acts as a temporary storage chamber, the ingested material is channeled into a thick-walled, muscular *gizzard* where it is ground into small bits (small stones in the gizzard facilitate the grinding action). The ingested material, now in the form of a finely ground mass suspended in water, passes into the *intestines* where enzymatic digestion and absorption occur. In this case, digestion is extracellular since enzymes are secreted into the digestive tract by cells that line it. At the posterior portion of the intestines, some of the water utilized in digestion is reabsorbed, and the indigestible materials are eliminated via the anus. Note that the gizzard is an adaptation for eating sizable pieces of food and the crop (analogous to the stomach of man) enables the organism to ingest sizable quantities of food in a short time and then utilize this food over a long period. Moreover, the intestine of the earthworm is not a round tube internally. It possesses an infold of tissue, the *typhlosole,* which greatly increases the total absorptive surface area exposed to digested food (Figure 13-7B).

Vertebrate Digestive Systems

Among members of the phylum Vertebrata, the digestive tract reaches a high level of structural and functional complexity. Nonetheless, many fundamental features found in lower animals are clearly evident. The system is basically tubular in design from mouth to anus and is located in a coelom. In addition, regional specialization of the tract is exemplified by a complex of organs of varying structures with localized functions. In the following discussion of vertebrate digestive systems, the human digestive system will be emphasized, with limited reference to the digestive apparatus of other vertebrates.

The Mouth. The first portion of the vertebrate digestive system is the *mouth,* or *oral cavity.* In most vertebrates, it consists of an immovable upper jaw and a movable lower jaw with external folds of skin constituting the lips. The lips are replaced by a bill or horny

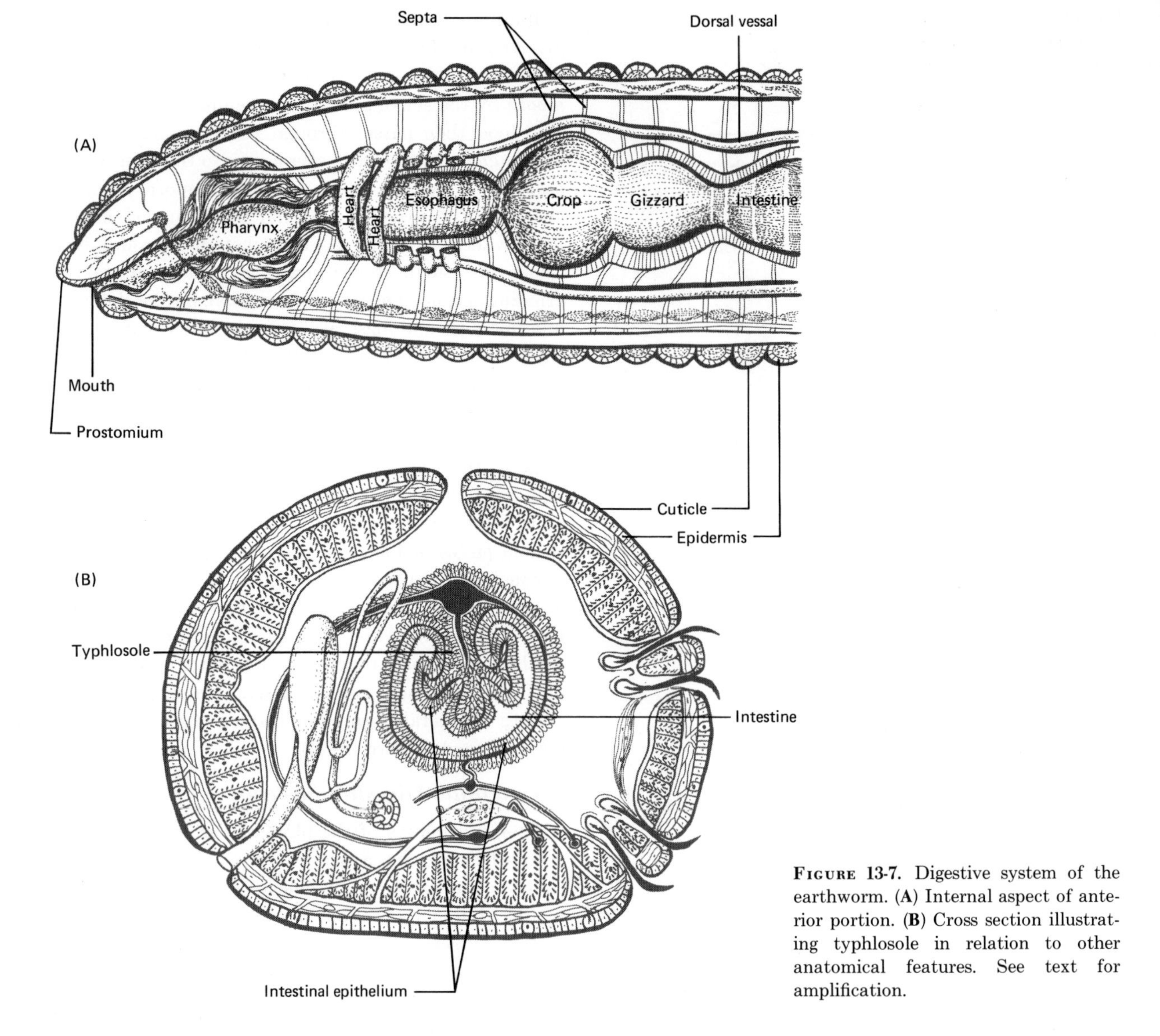

FIGURE 13-7. Digestive system of the earthworm. (**A**) Internal aspect of anterior portion. (**B**) Cross section illustrating typhlosole in relation to other anatomical features. See text for amplification.

beak in birds, turtles, and some primitive mammals. The entire oral cavity is bounded laterally by the cheeks and by the teeth embedded in the jawbones, on its floor by the tongue, and on its roof by the hard and soft palates. In addition, salivary glands pour their secretions into the oral cavity.

Tongue. The human *tongue* consists of skeletal muscle covered with mucous membrane. Several muscles that originate on skull bones insert into the tongue and because of the orientation of the fibers of these muscles, the tongue is able to move in many planes, all movement being under conscious control. The rough elevations of the tongue are referred to as *papillae* which may contain *taste buds.* The buds are richly supplied with sensory nerves (see Figure 11-14). The principal digestive functions of the human tongue are the manipulation of food for chewing (mastication) by the teeth and shaping the chewed food into a spherical mass, a *bolus,* prior to swallowing (*deglutition*). The tongue pushes the bolus into the pharynx (throat) and then into the esophagus.

In other vertebrates, the tongue may be quite elongated and greatly extensible, thus allowing the tongue to assume a role in food collection (e.g., trapping of insects by frogs). The human tongue is also important in speech. A fact of clinical importance in this regard is that if the *frenulum,* a fold of mucous membrane that anchors the tongue to the floor of the mouth, is too short for freedom of tongue movement, the individual is said to be "tongue-tied" and speech is faulty.

Teeth. The *teeth* of vertebrates function in the mechanical (physical) breakdown of food by their biting and chewing actions. Almost all vertebrate forms possess teeth, and the general evolutionary trend from a fish-like ancestor through amphibians and reptiles to mammals is a tendency toward reduction in the number of teeth present at a given time and in the extent of replacement of lost teeth. There is, however, an increased differentiation in types of teeth. In lower vertebrates such as fishes, several scores of generally similar teeth are present at any one time and are continually replaced as they are lost. Reptiles have fewer teeth than fishes and the number is further reduced in mammals. In humans, the number is reduced to thirty-two adult teeth and replacement only occurs in the teeth in front of the molars, and only once. The molars, which develop later in life, are never replaced once they are lost. The evolutionary trend toward a decreased number of teeth culminates in modern birds and certain reptiles in which teeth are totally absent.

With regard to the increased differentiation of teeth among vertebrates, a number of situations exist. In mammals, for example, four types of teeth occur variously among the different organisms. In front are the flattened *incisors* which are adapted for biting. Behind these are the pointed *canines* (bicuspids) which are adapted for tearing food. Next are the *premolars* and *molars* with flattened ridged surfaces. These teeth function in grinding, crushing, and pounding food. In each half of the upper and lower jaws man typically has two incisors, one canine, two premolars, and three molars. The number and arrangement of teeth may vary considerably among mammals. The teeth of carnivorous animals (dogs, cats) are more pointed than human teeth, the canines are longer, the premolars do not have flattened grinding surfaces, and the back molars are frequently lacking. Herbivorous animals, by contrast, have relatively large, flat premolars and irregularly surfaced molars. In general, the teeth of herbivores such as horses and cows are well adapted for grinding since the cellulose walls of the plants they eat must be broken up for adequate exposure to digestive enzymes. The teeth of carnivores are well adapted for piercing, cutting, and tearing flesh apart.

In mammals such as the human, the typical tooth is differentiated into three principal regions:

1. The *crown,* which protrudes above the gum line.
2. The *neck,* which is enveloped by the gum.
3. The *root,* which is anchored in a socket in the jawbone (Figure 13-8).

The external layer of the crown consists of an exceedingly hard substance, *enamel,* a complex of calcium compounds. The bulk of the tooth, composed of *dentin,* lies beneath the enamel and is similar in composition to bone. The interior of the tooth, called the *pulp,* is made up of connective tissue, nerves, and blood vessels.

Salivary Glands. The oral cavity also contains the openings of the ducts of three pairs of *salivary glands* which elaborate various secretions. These glands are referred to as the parotid, submaxillary, and sublingual glands. The *parotid glands,* located below and in front of the ear, secrete into a duct (*Stensen's duct*) that opens into the mouth on the inside of the cheek opposite the upper second molar. This pair of glands becomes quite inflammed and swollen if an individual contracts mumps. The *submaxillary glands* on the posterior portion of the floor of the mouth secrete into ducts (*Wharton's ducts*) opening on either side of the frenulum. The final pair, the *sublingual glands,* lie anterior to the submaxillary glands and contain several ducts (*ducts of Rivinus*) that open into the floor of the mouth.

The salivary glands and numerous minute glands of the mucosa of the mouth collectively secrete *saliva.* Saliva consists of both thin watery and thicker viscous mucous fluids that moisten and lubricate ingested food, thus ensuring its smooth passage through the pharynx and esophagus. In humans and most other mammals, saliva also contains a starch-digesting enzyme, *salivary amylase.* In humans, food remains in the mouth for such a short time that the principal end product of starch digestion is dextrin (95–97%); only 3–5% is re-

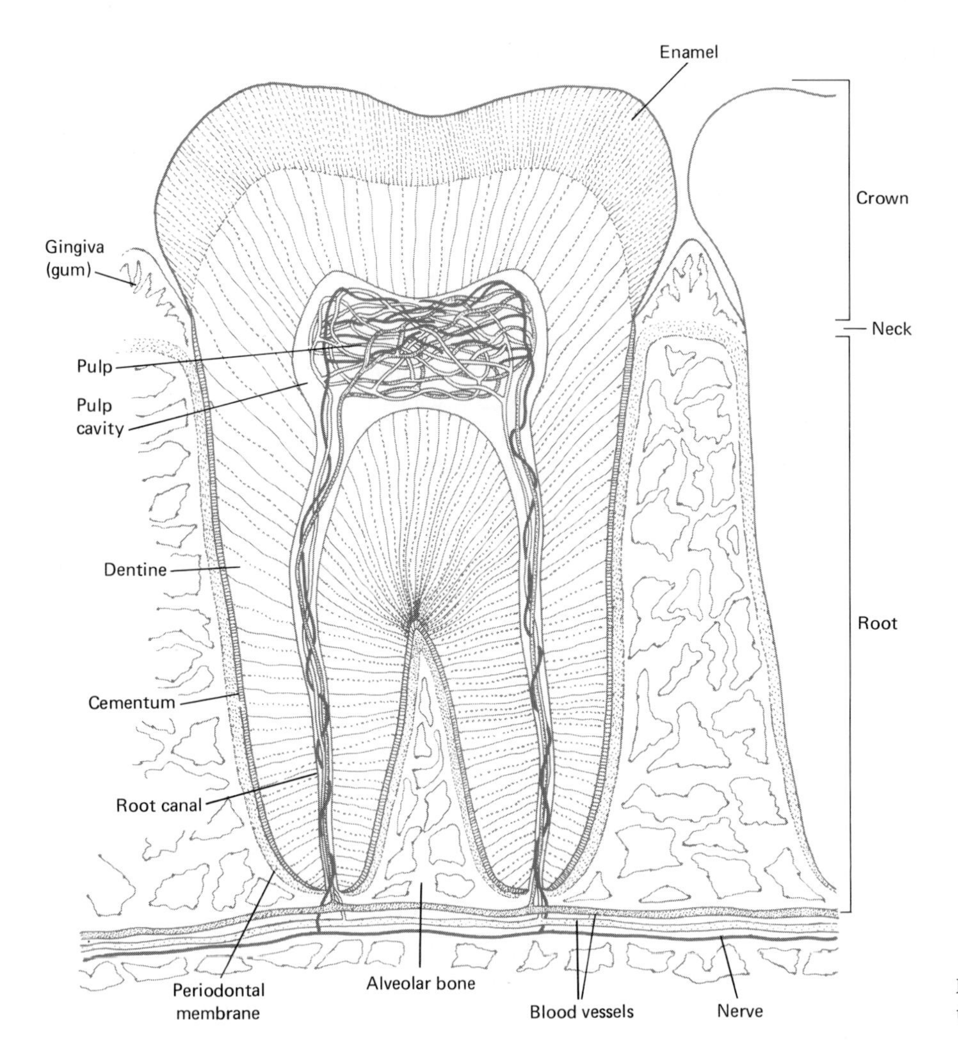

Figure 13-8. Principal parts of a typical tooth.

duced to the disaccharide maltose. The pH of human saliva ranges from 6.4 to 7.0 and approximately 1200–1500 ml are secreted in a 24-hour period.

Pharynx and Esophagus. Food that enters the oral cavity is reduced to a soft mass by the combined action of the lips, jaw bones, teeth, cheeks, and tongue preparatory to swallowing. During mastication, the food is mixed with saliva and shaped into a bolus. In the act of swallowing, the bolus is placed on the tongue and pushed backward into the pharynx. The *pharynx,* or throat, is a tube-like passageway about 5 inches long that extends from the base of the skull to the esophagus and lies just anterior to the cervical vertebrae (Figure 13-9). It not only serves as a passageway for swallowing food, but also functions in the transmission of air from the nose and mouth into the trachea. In the act of swallowing, the muscular walls of the pharynx contract, squeezing the bolus and forcing it into the esophagus. Simultaneous with swallowing, certain involuntary movements occur so that the bolus will not move into the nose, trachea (windpipe), or back again into the mouth. These include the elevation of the soft palate, which prevents food from entering the nose, the closing off of the trachea by the *epiglottis* so that food is channeled into the esophagus, and the movement of the base of the tongue and the muscular walls of the pharynx,

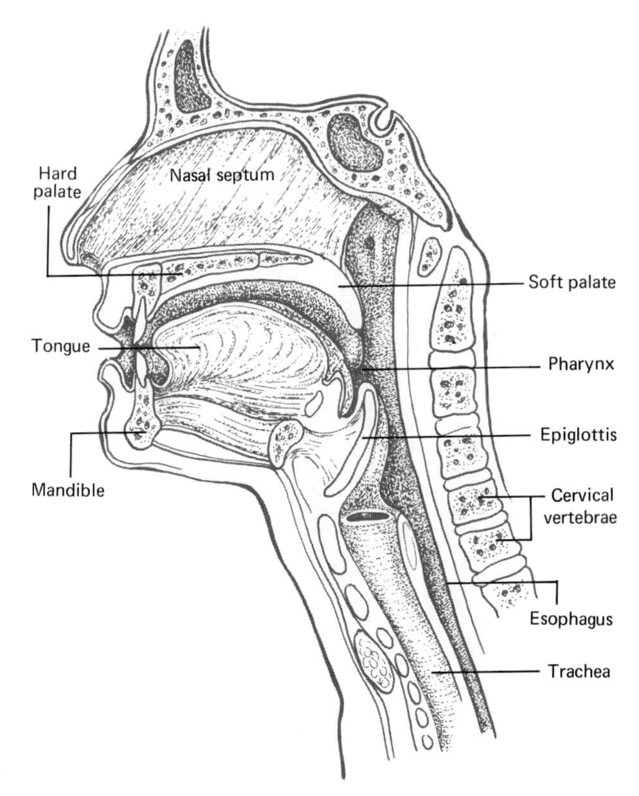

FIGURE 13-9. Pharynx and esophagus in relation to other anatomical structures.

thus preventing the reverse movement of the bolus into the oral cavity. Occasionally, these reflexes fail to occur and the food enters the larynx or trachea, causing the individual to choke or cough.

The *esophagus,* or foodpipe, is a collapsible muscular tube about 10 inches long that extends from the pharynx to the stomach, piercing the diaphragm in its descent. It is posterior to both the trachea and the heart. Solid or semisolid food is passed down the esophagus by a series of rhythmical automatic contractions of the muscular walls, a phenomenon called **peristalsis.** The passage of such foods to the stomach requires 4–8 seconds. Liquids and relatively soft foods move from one end of the esophagus to the other in about 0.1 second. The point at which the lower end of the esophagus joins the stomach is surrounded by smooth muscle, which is referred to as the *cardiac sphincter* (Figure 13-10A). Normally this sphincter muscle is closed so that the contents of the stomach are not regurgitated into the esophagus; it opens, however, when a wave of peristaltic action reaches it. The length of the esophagus varies from only a minor constriction in the gut of fish and amphibians to several feet in the giraffe.

STOMACH. Just below the diaphragm, the alimentary tube dilates into an elongated pouch-like structure, the *stomach,* which is connected at its upper end to the esophagus and at the lower end to the small intestine. It functions in the storage of food, the further breakdown of food through its powerful churning action, digestion through the action of digestive enzymes it secretes, and the acidification of food.

Anatomically, the human stomach is subdivided into three main regions (Figure 13-10A). The first of these is the *fundus,* which is the upper enlarged, dome-shaped portion lying nearest the diaphragm. The *body* is the central region and the *pyloric region* is the lower constricted, horizontal portion that connects to the small intestines. Between the pyloric region and the first part of the small intestines is a second muscular valve of smooth circular muscle, the *pyloric sphincter.*

The size of the cavity of the stomach depends upon the volume of its contents. Thus, when there is no food in it, it is contracted, but when food enters, it expands to accommodate the swallowed food. The musculature of the stomach is arranged in three layers. The smooth muscle fibers of the outer layer are arranged longitudinally, those of the middle layer are circularly oriented, while those of the inner layer run obliquely. Such an arrangement permits a wide variety of movements so that when food is present, it is swept by powerful waves of contraction which churn the food, mix it, and break it into smaller pieces. In addition to strong waves, weak, rippling peristaltic movements, called *mixing waves,* pass over the stomach about every 15–25 seconds. The end result of these stomach movements is the conversion of ingested food to a thin, liquid mass referred to as *chyme.*

The inner lining of the stomach consists of mucus-filled columnar epithelial cells and an underlying connective tissue layer in which are scattered *gastric glands.* The cells of these gastric glands are differentiated to produce three different secretions, the secretions being collectively referred to as *gastric juice.* One type of specialized cell produces mucus (*mucous cell*), an-

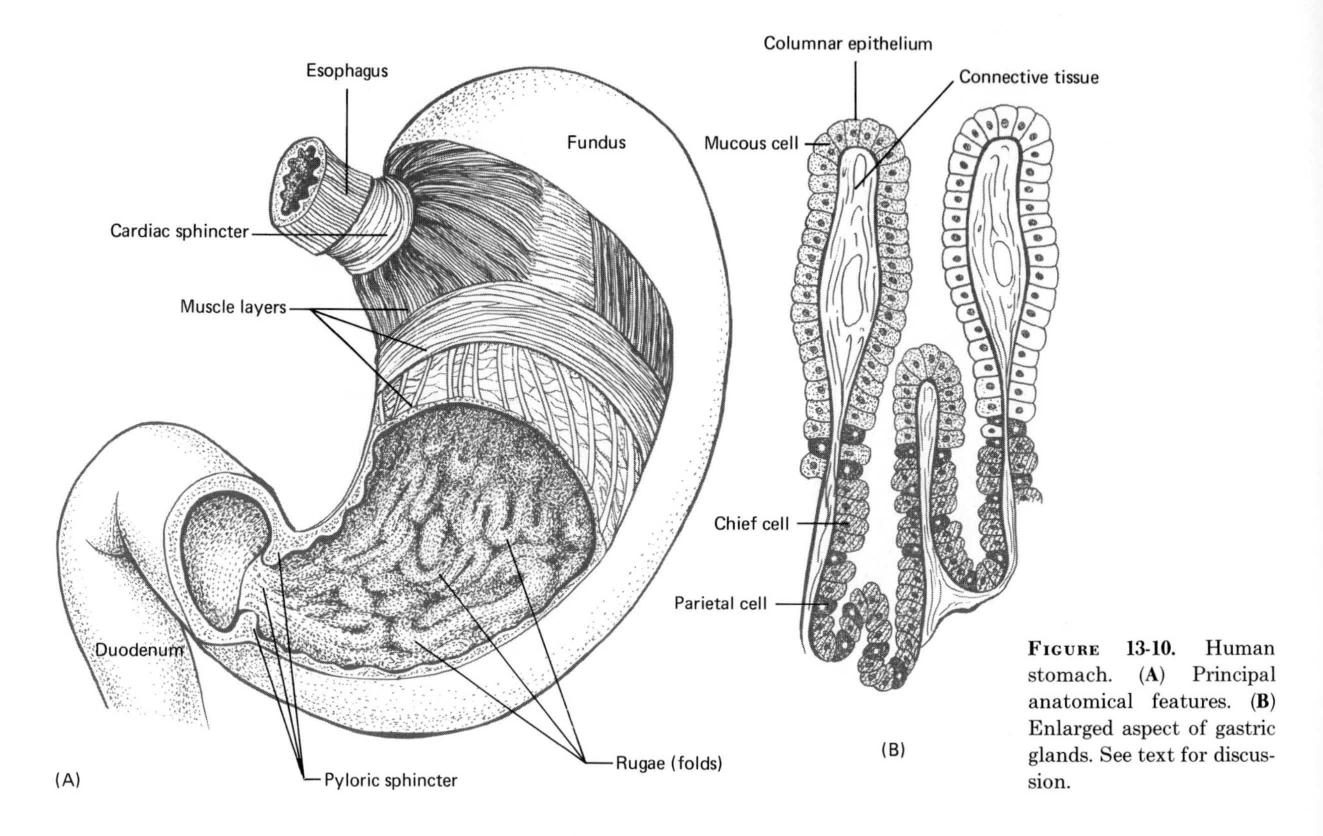

FIGURE 13-10. Human stomach. (**A**) Principal anatomical features. (**B**) Enlarged aspect of gastric glands. See text for discussion.

other the digestive enzymes (*chief cell*), and still another elaborates hydrochloric acid (*parietal cell*) (see Figure 13-10B).

Gastric juice in humans consists of water, the mucoprotein mucin (mucus), which is responsible for the viscosity of the juice, the protein-hydrolyzing enzyme pepsin, and hydrochloric acid (HCl). *Mucin* in the lining cells and in the juice assumes an important role in digestion in that it acts as a buffer against HCl and inhibits the enzymatic activity of pepsin. Thus, mucin prevents the digestion of proteins that compose the walls of the stomach. Moreover, the enzymes that are produced by the gland cells are isolated in an inactive form in zymogen granules which prevent the contact of these with cell contents. *Pepsin* is present in gastric cells in an inactive form called *pepsinogen*. Pepsinogen is activated to pepsin in the presence of HCl or previously activated pepsin. Pepsin is a proteolytic enzyme requiring an acid medium for its activity. It has the property of hydrolyzing proteins to yield polypeptides and peptides containing four to twelve amino acids. This action is preparatory to the more complex hydrolysis of proteins into amino acids in the intestine under the influence of other enzymes. *Hydrochloric acid,* which gives the stomach a final pH of about 2.5, in addition to furnishing suitable acid conditions both for the activation of pepsinogen and the optimal digestive activity of pepsin, also kills many microorganisms, thus preventing bacterial invasion and putrefaction of ingested foods.

The form of the stomach exhibits considerable variation among vertebrates. In some fishes the fundic portion is deeply folded, with blind pouches leading from the stomach. In other fishes and amphibians, as well as snakes and lizards, the stomach is typically straight. In birds, the anterior region of the stomach is modified as a crop where food is stored, and the posterior portion as the gizzard which is used for grinding food. Among certain mammals, such as cows, sheep, and deer (ruminants), the stomach usually consists of several distinct

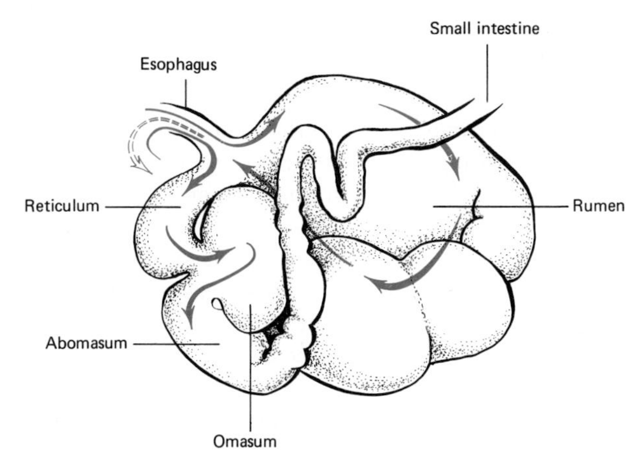

Figure 13-11. Structure of the "stomach" of a cow. Arrows indicate the route of ingested food.

chambers, each with a special function. The four-chambered structure of the stomach of the cow may serve as an example (Figure 13-11). The first three chambers are actually modifications of the esophagus, while the fourth is the true stomach. The chambers, however, are not connected in sequence to one another. Food from the esophagus proper enters the first chamber, the *rumen,* where it is stored and acted upon by bacteria and protozoans. Then it is gradually moved to the second chamber, the *reticulum,* where it is further softened by mixing with digestive enzymes. After being subjected to action by the rumen and reticulum, small masses of softened food are then passed back as cud to the mouth by way of the esophagus. In the mouth, it is thoroughly chewed and mixed with saliva. Upon being reswallowed, the food moves from the esophagus into the third chamber, the *omasum,* and finally into the true stomach (*abomasum*). In the true stomach, characteristic gastric digestion occurs.

In summary, then, the stomach serves as a place for temporary storage and facilitates both mechanical and chemical digestion. The mechanical aspects involve peristaltic waves in the stomach walls which churn and mix food with gastric juice. Chemically, protein digestion is initiated but not completed. Fluids pass through the stomach in about 20 minutes or less, whereas solids must first be converted to chyme and require from 3 to 4 hours before passage into the intestines. Ultimately, foods are discharged from the stomach into the small intestine in small amounts. This discharge is controlled by the pyloric sphincter.

Small Intestine. In humans, the *small intestine* is a tube measuring about 1 inch in diameter and about 12–15 feet in length in living subjects. It is contained in the central and lower portions of the abdominal cavity (Figure 13-12). The small intestine is arranged in this cavity in coils and loops and is attached to the dorsal wall of the cavity by a membranous mesentery which contains the blood vessels, lymphatics, and nerves that are distributed to the intestinal walls. The length of the intestine is correlated with feeding habits. In herbivorous organisms it is typically very long and highly coiled, considerably shorter in carnivores, and of intermediate length in omnivores such as humans. These differences are correlated with the difficulty of digesting the cellulose of plant material. An organism that illustrates this principle is the frog. The immature frog, called a tadpole, is herbivorous and the

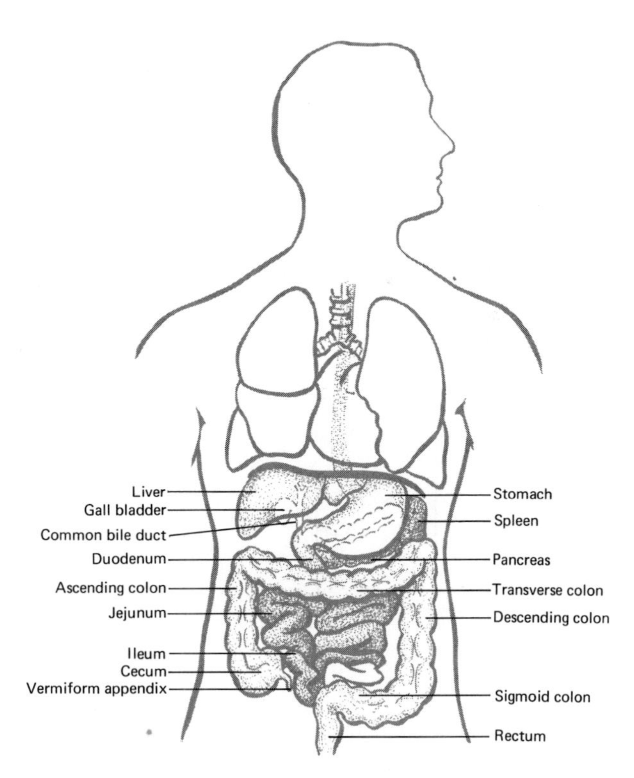

Figure 13-12. The human small intestine in relation to other organs of the abdominal cavity.

gut is exceedingly long and coiled. The adult, by contrast, is carnivorous, and has a considerably smaller and less coiled intestine.

For descriptive purposes, the small intestine is subdivided into three portions. The first of these is the *duodenum,* the shortest and broadest part of the small intestine. The duodenum is about 10 inches long and extends from the pyloric region of the stomach to the *jejunum,* the second subdivision. The jejunum, which measures about 5 feet in length, is also referred to as the empty intestine because it is always found empty after death. The final portion, the *ileum,* or twisted intestine, so called from its numerous coils, measures about 8 feet. The ileum extends from the jejunum to the large intestine which it joins at a right angle. The orifice is guarded by a sphincter muscle, the *ileocecal valve* (see Figure 13-15B).

Most chemical or enzymatic digestion and absorption occurs in the small intestine. The small intestine is well adapted structurally for its important roles (Figure 13-13). For example, the length of the intestine plays a role in increasing its absorptive surface area. Moreover, close examination of its internal surface reveals other structural adaptations that facilitate increased absorptive area. The mucosal lining is arranged in *plicae* (circular folds), which project into the *lumen* (space) of the tube (Figure 13-13A). Some of these folds extend all the way around the circumference of the intestine; others extend part of the way. Throughout the whole length of the intestine, the mucous membrane contains minute finger-like projections called *villi* (Figure 13-13B), which number between 4 and 5 million in humans. Each villus has a core of connective tissue, contains blood capillaries, a small lymph vessel called a *lacteal,* and has an

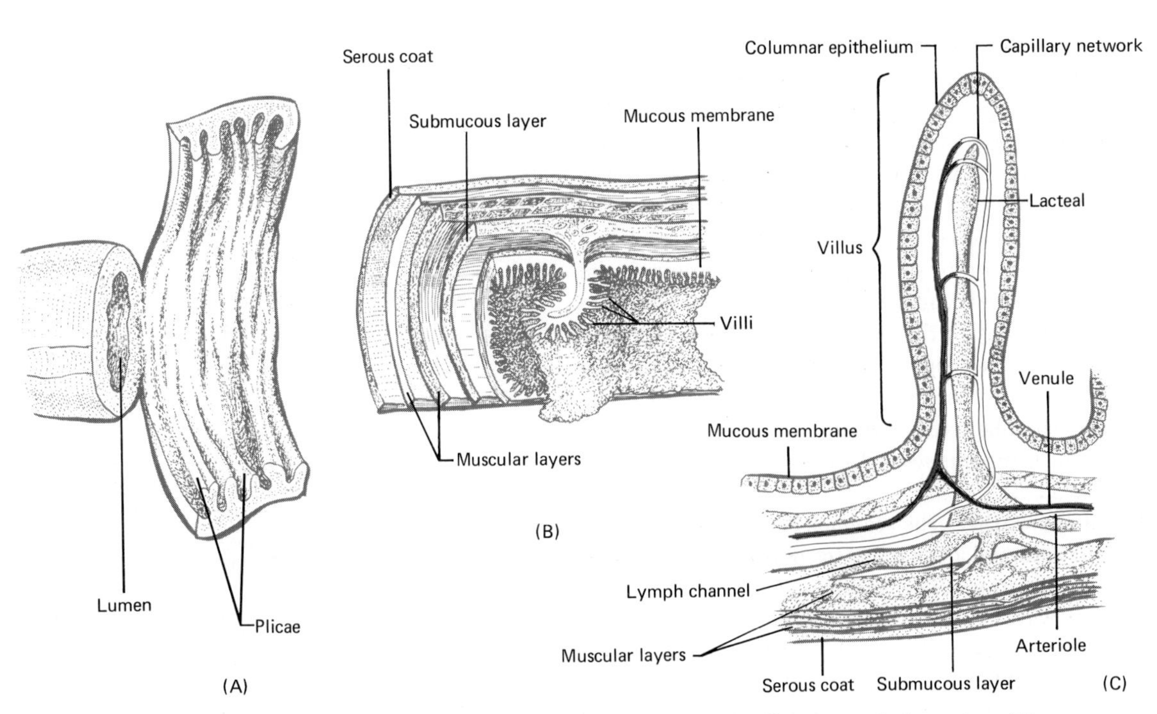

Figure 13-13. Structural adaptations of the human small intestine for digestion and absorption. (**A**) Section through small intestine showing plicae. (**B**) Villi in relation to the intestinal coats. (**C**) Enlarged aspect of a single villus.

outer covering of columnar epithelial cells (Figure 13-13C). After food has been digested it passes into the capillaries and lacteals of the villi. One other modification that increases absorptive surface area is the presence of *microvilli* (see Figure 5-4T) that border individual epithelial cells covering the folds and villi. Some vertebrates possess still other adaptations for increasing absorptive surface area. In many fish, blind sacs called *caecae* branch from the anterior end of the small intestine. Other vertebrates, such as primitive fish and sharks, contain spiral valves that are epithelial infoldings. These structures channel food to move along a spiral path so that a greater surface area is involved.

In addition to the intestinal glands present in the lining of the small intestine, the *pancreas* and the *liver* also assume a role in the digestive process that occurs in the small intestine. The pancreas, a characteristic organ of vertebrates, is roughly carrot-shaped and lies between the stomach and the duodenum. It is a highly glandular mass that has an endocrine as well as a digestive function. As an endocrine gland, its specialized *islets of Langerhans* secrete into the blood stream two different hormones (insulin and glucagon) essential for carbohydrate metabolism (see Chapter 10). As a digestive organ, the pancreas secretes a complex of enzymes collectively called *pancreatic juice.* This complex is delivered into the duodenum by way of the *pancreatic duct.*

Pancreatic juice contains several enzymes concerned with the hydrolysis of carbohydrates, fats, and proteins. The amount of pancreatic juice secreted each day varies between 600 and 800 ml and since the juice is alkaline (pH 7–8), it partially neutralizes the acid contents delivered by the stomach. The adjustment of the pH in the duodenum is necessary for the action of the enzymes produced by the pancreas. The starch-digesting enzyme of pancreatic juice, *pancreatic amylase* is similar in action to salivary amylase, catalyzing the hydrolysis of starch to disaccharides. A second enzyme called *lipase* causes the hydrolysis of fat to form fatty acids, glycerol, and other products. Several protein-splitting enzymes are also elaborated by the pancreas in inactive precursor states which are activated in the intestine. These include the inactive *trypsinogen* and *chymotrypsinogen,* which are converted to the protein-digesting enzymes *trypsin* and *chymotrypsin,* respectively. Through hydrolysis of peptide bonds, the pancreatic proteases break down proteins and polypeptides into dipeptides and some amino acids.

The *liver,* as the largest gland of the body, in addition to many other functions to be discussed subsequently, contributes to the digestive process through its synthesis and secretion of bile (Figure 13-14). *Bile* is a yellow, brownish yellow, or olive green complex of slightly alkaline reaction. The principal constituents of bile are water, bile pigments (*biliverdin* and *bilirubin,* breakdown products of hemoglobin after globin and iron have been removed), bile acids, bile salts, cholesterol, lecithin, and mucin. Bile is produced by the cells of the liver, largely from blood constituents, and ultimately emerges from the liver via the *hepatic ducts.* The hepatic ducts form a common hepatic duct which merges with the *cystic duct* of the gall bladder to form the *common bile duct* which enters the duodenum at a common opening with the pancreatic duct. The continuously secreted bile is stored in the *gall bladder* until it is needed.

Bile salts aid in the digestion of fats by lowering surface tension, which aids in the emulsification of fats. Once fats are *emulsified* (divided into smaller globules) the greater surface area provided enables lipase and other enzymes to act more effectively. In addition, bile enhances the absorption by the villi of fat digestion products and other lipid-soluble compounds, including the fat-soluble vitamins. This is particularly true of vitamin K, a necessary substance for the production of prothrombin, which is required for normal blood clot-

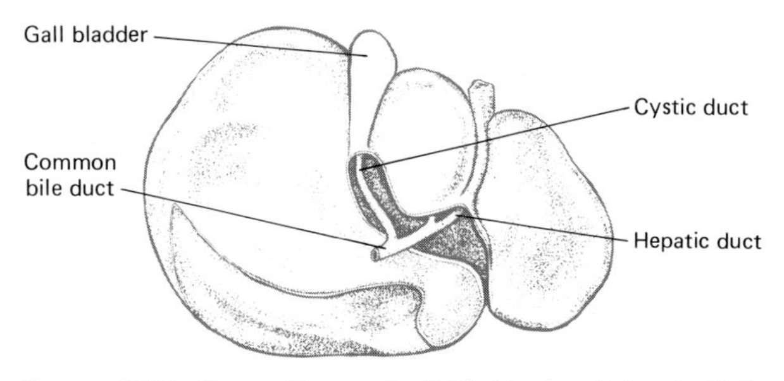

Figure 13-14. Human liver and gall bladder in relation to their system of ducts.

ting. Bile also stimulates intestinal motility and neutralizes the acid chyme, creating a suitable pH medium for pancreatic and intestinal enzyme activity. Finally, bile is an excretory medium for toxins, metals, and cholesterol.

Intestinal juice (succus entericus) is a collective designation for the complex of enzymes and fluids produced by the numerous digestive glands embedded in the wall of the small intestine. This secretion is a clear, yellowish fluid that is slightly alkaline in reaction. The daily volume secreted by the average adult is approximately 3000 ml. One of the enzymes present in intestinal juice is *enterokinase,* a catalyst that activates *trypsinogen* (a pancreatic secretion). *Aminopeptidase, carboxypeptidase,* and *dipeptidase* are intestinal enzymes (peptidases) that hydrolyze dipeptides to amino acids, thus completing the work begun by pepsin and trypsin. Also present are several enzymes that catalyze the breakdown of disaccharides to the monosaccharide level. These include *maltase* (maltose $\longrightarrow$ 2 glucose), *sucrase* (sucrose $\longrightarrow$ glucose and fructose), and *lactase* (lactose $\longrightarrow$ glucose and galactose). Finally, intestinal juice contains a *lipase,* as well as a number of enzymes that catalyze the hydrolysis of nucleic acids, nucleotides, and nucleosides.

Table 13-1 contains a listing of the principal digestive enzymes, their site of production, the substrates upon which they act, and the end products of hydrolysis.

The chyme entering the duodenum after an ordinary meal is normally free from coarse particles of food and is acid in reaction. The proteins are partially digested, some of the starch has been hydrolyzed, and fats are liquefied and mixed with other foods, but probably have not been hydrolyzed. It is in the small intestine that this mixture undergoes the greatest digestive changes. As with other parts of the digestive tract, these changes are both physical and chemical. The chemical changes, which have been described, are the result of the combined action of pancreatic juice, bile, and intestinal juice. The physical changes are related to several types of movements that occur in the small intestine as a result of its muscular activity. The movements of the small intestine are described as peristaltic, rhythmical, and pendular.

Peristalsis of the small intestine is essentially similar to the phenomenon that occurs in the esophagus and stomach. The *rhythmical movements* consist of a series of local constrictions of the intestinal wall, which occur at points where masses of food lie. These constrictions divide food into segments. Within a few seconds each of these segments is halved, and the corresponding halves of adjoining segments unite. Again constrictions occur,

TABLE 13-1. Summary of Chemical Digestion in Humans

Digestive Juices and Enzymes	Site of Production	Substrate	End Products
Saliva	Salivary glands	Starch	Dextrin (95–97%)
Amylase			Maltose (3–5%)
Gastric juice	Stomach		
Protease (pepsin) plus HCL		Proteins, including casein	Polypeptides and peptides containing 4–12 amino acids
Lipase (of little importance)		Emulsified fats (butter, cream, etc.)	Fatty acids and glycerol
Bile (contains no enzymes)	Liver	Large fat droplets (unemulsified fats)	Small fat droplets or emulsified fats
Pancreatic juice	Pancreas		
Proteases		Proteins and polypeptides	Dipeptides and some amino acids
Lipase		Bile-emulsified fats	Fatty acids and glycerol
Amylase		Starch	Maltose
Intestinal juice	Small intestine		
Peptidases		Dipeptides	Amino acids
Sucrase		Sucrose	Glucose and fructose
Lactase		Lactose	Glucose and galactose
Maltase		Maltose	Glucose

and these newly formed segments are divided, and the halves re-form. In this manner, every particle of food is brought into close contact with the intestinal lining and is thoroughly mixed with the digestive juices. *Pendular movements* are constrictions that move forward or backward for short distances, gradually moving the chyme onward and backward. Although pendular movements have been demonstrated in mammals such as rabbits, their presence in humans is doubted. The varied muscular movements of the small intestine increase the blood supply, bringing materials for secretion and removing absorbed materials faster. They assist the minute glands in emptying their secretions, mix the digestive fluids and foods thoroughly, and constantly bring absorbable materials to the mucosa, thus increasing the rate of absorption. The passage of chyme through the small intestine to the large intestine usually takes about 8 hours.

Large Intestine. The lower portion of the alimentary canal is referred to as the *large intestine* (Figure 13-15). It is about 5 or 6 feet in length and has an average diameter of $2\frac{1}{2}$ inches, decreasing somewhat toward the lower end of the tube. The large intestine is divided into the cecum, colon, and rectum.

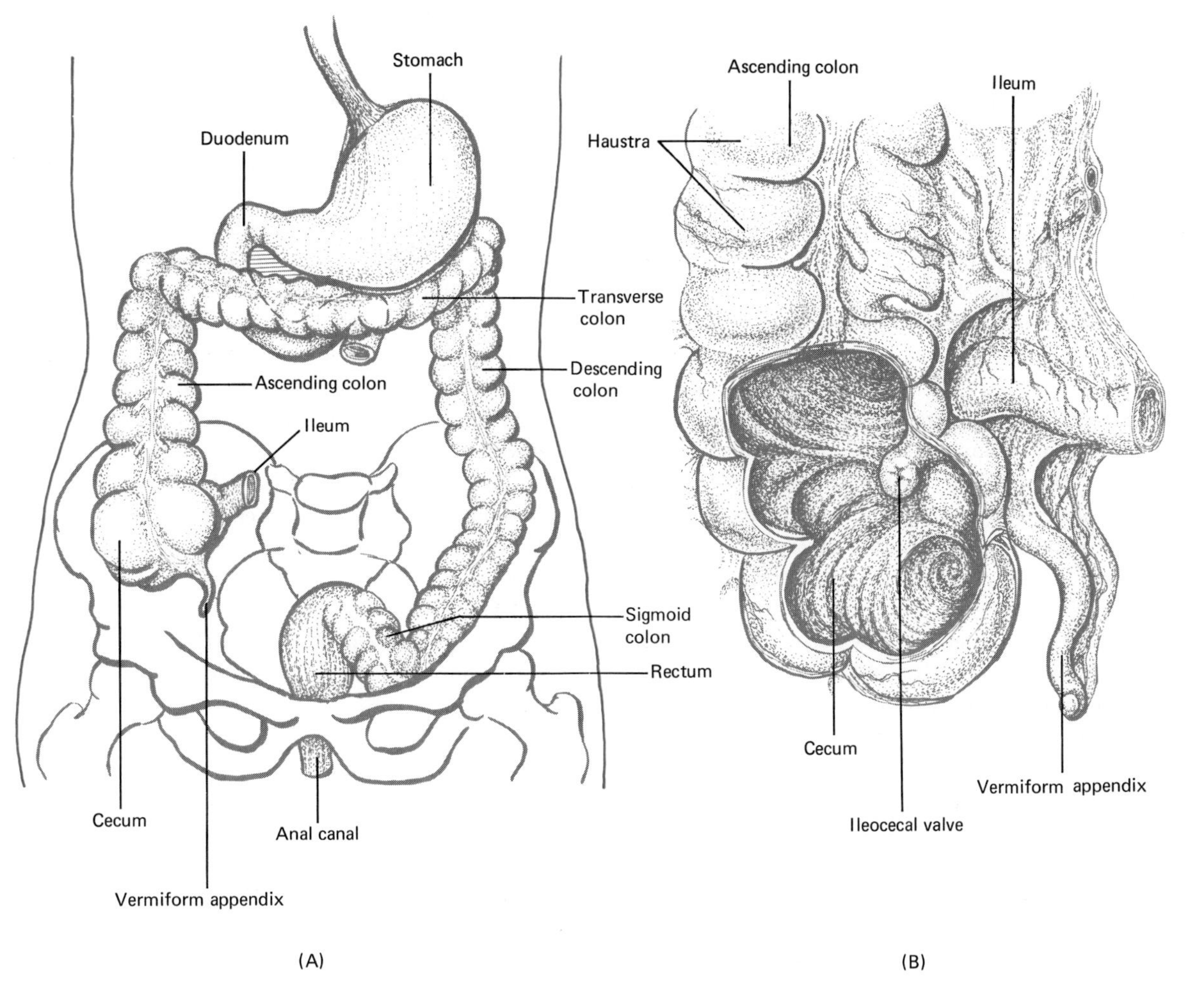

Figure 13-15. Human large intestine. (**A**) Position in the abdominal and pelvic cavities and principal subdivisions of the large intestine. (**B**) Enlarged aspect of the ileocecal region.

The small intestine opens into the side wall of the large intestine (the junction is at the ileocecal valve, which prevents the backup of materials into the small intestine) about $2\frac{1}{2}$ inches above the beginning of the large intestine. This $2\frac{1}{2}$ inches of large intestine forms a blind pouch called the *cecum.* This structure is especially well developed in herbivorous animals and serves as an active site of cellulose digestion. Attached to the cecum is a blind tube, the *appendix.* Although the appendix is similar in structure to the rest of the intestine, it is frequently subject to inflammatory and gangrenous conditions (appendicitis) in young adults. The appendix serves no known function in the human.

The *colon,* although a continuous tube, is subdivided into ascending, transverse, descending, and sigmoidal (pelvic) portions. The *ascending colon* lies in the vertical position on the right side of the abdominal cavity until it reaches the underside of the liver. At this point, the colon turns abruptly to the left and is continued across the abdomen as the *transverse colon.* On the left side of the abdomen, the colon curves beneath the lower end of the spleen and passes downward as the *descending colon.* The *sigmoid colon* is that part of the large intestine that makes a curve like a letter S (thus the name sigmoid) on a level with the margin of the crest of the ileum.

The *rectum* is about 5 inches long and is continuous with the sigmoid colon and *anal canal,* the terminal inch of the rectum. The opening of the canal to the exterior is guarded by two sphincter muscles; an internal one of smooth muscle and an external one of striated muscle. The opening itself is referred to as the *anus.*

The large intestine performs two principal functions. The first of these is the absorption of water and salts from the chyme that enters from the small intestine. The absorbed water includes not only that originally present in ingested food but also the large quantity contributed by the secretions of the salivary glands, stomach, intestinal glands, pancreas, and liver. If the passage of chyme is too rapid, water absorption is reduced and *diarrhea* results. This condition may be caused by excessive muscular activity of the walls of the large intestine as a result of physical irritation, drug action, emotional upset, and infection. The reverse situation, an unusually slow passage of chyme through the large intestine, leads to excessive water reabsorption and *constipation.* Inasmuch as the small intestine is highly adapted for digestion and absorption, little absorbable nutrient, other than water and salts, reaches the large intestine.

The second principal function of the large intestine is related to the extremely large and diverse bacterial flora present. These microorganisms synthesize certain vitamins (B complex, K), amino acids, and other substances which are absorbed into the walls of the large intestine and are useful in metabolism. Prolonged use of antibiotics may result in the destruction of some of these bacteria, which may lead to vitamin deficiency symptoms. Other organisms that constitute the normal flora not only completely hydrolyze undigested proteins but decompose amino acids into simpler groups. Among these products are hydrogen sulfide, skatole, indole, and phenol, which give characteristic odors to *feces* (undigested food residue).

Chyme passes into the large intestine several hours after eating. As the chyme moves through the ileocecal valve, the cecum becomes filled, and gradually the accumulation reaches higher and higher levels in the ascending colon. The contents of the ascending colon are soft and semisolid, but by the time they pass through the transverse and descending colons, they attain the consistency of feces. One type of movement characteristic of the large intestine is referred to as *haustral churning.* The pouches (*haustra*) that compose the large intestine (see Figure 13-15B) become distended and periodically contract and empty themselves. A second type of movement, called *mass peristalsis,* consists of the vigorous contraction of the entire ascending colon, which moves its contents to the transverse colon. Such movements take place only three or four times a day, last only a short time, and are usually related to eating.

Once the mass of ingested materials reaches the sigmoid colon and rectum, it is designated as *feces.* The feces consists of some water, undigested and indigestible parts of food, pigments of undigested foods and bile pigments, microorganisms of various types, products of bacterial decomposition, products of secretion, mucous and epithelial cells from the walls of the digestive tract, cholesterol, some purine bases, and inorganic salts. Normally the rectum is empty until just before *defecation* (elimination of feces). Various stimuli produce peristaltic action of the colon so that a small quantity of feces enters the rectum. This irritates sensory nerve endings and precipitates defecation.

Table 13-2 lists the various physical processes that occur in different regions of the alimentary canal.

CONTROL OF DIGESTIVE GLAND SECRETION. Digestive glands secrete enzymes when food is present in the alimentary canal or when it is seen, smelled or imagined. Complex reflexes and hormonal mechanisms control the flow of digestive juices so that they are produced in sufficient quantities when and for as long as they are needed.

Secretion of Saliva. The secretion of saliva appears to be entirely under the control of the autonomic nervous system and is governed by nerve impulses arising from both simple and conditioned reflexes. Chemical, mechanical, olfactory, and visual stimuli initiate afferent impulses to centers in the brain stem that send out afferent impulses to salivary glands, thereby stimulating them.

Gastric Secretion. Unlike salivary secretion, the control of gastric secretion is related to both neural and hormonal factors and occurs in three phases. Because stimuli that activate these mechanisms arise in the head, stomach, and intestine, the three phases are referred to as the cephalic, gastric, and intestinal phases, respectively. The *cephalic phase* of gastric secretion refers to reflex stimulation of the gastric glands through the autonomic nervous system by the sight, smell, or taste of food. The cephalic phase is partly reflex and partly psychic because a reflex mechanism controls gastric juice secretion at this time and psychic factors activate the mechanism.

The *gastric phase* refers to all the activities of the gastric glands which are brought about by conditions within the stomach itself. Certain substances such as meat extractions and products of protein digestion that have reached the pyloric portion of the stomach stimulate its mucosa to release a hormone called *gastrin.* Gastrin, like any other hormone, is released into the blood stream and, upon reaching the stomach, induces the gastric glands to elaborate gastric juice. It is also believed that the mechanical distention of the stomach, together with the presence of polypeptides released in the partial digestion of proteins by pepsin, further enhances the flow of gastrin.

The *intestinal phase* of gastric juice secretion is less clearly understood than the other two. It is, however, believed to be due to the presence of food in the small intestine which releases a hormone. The entry of lipids into the duodenum apparently causes the liberation by the intestinal wall of a hormone called *enterogastrone,* which inhibits gastric secretion.

Pancreatic Secretion. The secretion of pancreatic

TABLE 13-2. Summary of Physical (Mechanical) Digestion in Humans

Organ	Physical Process	Description
Mouth	Mastication	Chewing movements—reduce size of food particles and mix them with saliva.
	Deglutition	Swallowing—movement of food from mouth toward stomach.
Pharynx	Deglutition	See above.
Esophagus	Deglutition	See above.
	Peristalsis	Rhythmic muscular movements that move food into the stomach.
Stomach	Mixing waves	Forward and backward movements of gastric contents; peristalsis propels it forward, closed pyloric sphincter deflects it backward.
	Peristalsis	Moves material through stomach and at intervals into duodenum.
Small intestine	Rhythmic movements	Forward and backward movement within segment of intestine; purpose, to mix food and digestive juices thoroughly and to bring all digested food in contact with intestinal mucosa to facilitate absorption.
	Peristalsis	Purpose of peristalsis is to propel intestinal contents along digestive tract.
Large intestine		
Entire colon	Haustral churning	Churning movements within haustral sacs.
Ascending colon	Mass peristalsis	Entire contents of ascending colon moved into transverse colon; usually occurs after a meal.
Rectum	Defecation	Emptying of rectum (bowel movement).

juice is also controlled by hormonal and nervous mechanisms. The discharge of acid chyme into the duodenum elicits the release of a polypeptide hormone called *secretin* from the cells of the duodenal wall into the blood stream. When secretin reaches the pancreas, it stimulates pancreatic cells to secrete at a faster rate. But, the juice they produce has a low enzyme content. Another intestinal hormone, *pancreozymin-cholecystokinin,* stimulates them to produce enzymes. The presence of fats in the intestine serves as the stimulus for pancreozymin-cholecystokinin release by the intestinal mucosa. Nervous control is provided via the vagus nerve, which stimulates enzyme secretion.

Bile Secretion. Chemical mechanisms dominate the control of bile secretion. *Secretin,* the same hormone that stimulates pancreatic activity, also stimulates the liver to produce bile. The ejection of bile, however, from the gall bladder into the duodenum is largely controlled by *pancreozymin-cholecystokinin.* The release of fatty foods into the duodenum promotes its discharge by the mucosa of the duodenum into the general circulation. When the hormone reaches the gall bladder, the storage organ contracts and releases bile into the cystic and common bile ducts, and ultimately into the duodenum.

Intestinal Secretion. Knowledge concerning the control of secretion of intestinal juice is still somewhat obscure. It is thought that the presence of chyme in the small intestine causes the intestinal mucosa to release a hormone into the blood. This hormone, called *enterocrinin,* brings about increased intestinal juice secretion.

Disorders of the Human Digestive System

Now that we have discussed the structure and physiology of the digestive system, we shall look at some disorders that are related to it.

Dental Caries

Dental caries, or tooth decay, involves a gradual disintegration of the enamel and dentin. If this condition remains untreated, various microorganisms may invade the pulp cavity, causing infection and inflammation of the living tissue. If the nerves of the pulp are destroyed, the tooth is pronounced "dead."

No individual microbe is responsible for dental caries, but oral bacteria that create a pH of 5.5 or lower start the process. Acids can come directly from foods, such as the ascorbic acid of citrus fruits, or they may be breakdown products of carbohydrates. Microbes that digest carbohydrates include the bacterium *Lactobacillus acidophilus* and several of the streptococcal bacteria as well as some yeasts. Research suggests that the streptococci break down carbohydrates into *dental plaque,* a polysaccharide that adheres to the tooth surface. When other bacteria digest the plaque, acid is produced. Saliva cannot reach the tooth surface to buffer the acid because the plaque covers the teeth.

Certain measures can be taken to prevent dental caries. First, the diet of the mother during pregnancy is very important in forestalling tooth decay of the newborn. Simple, balanced meals are the best diet during pregnancy. Supplementation with multivitamins, with emphasis on vitamin D and the minerals calcium and phosphorus is customary because they are responsible for normal bone and teeth development.

Other preventive measures have centered around fluoride treatment because teeth are less susceptible to acids when they are permeated with fluoride. Fluoride may be incorporated in the drinking water or applied topically to erupted teeth. Maximum benefit often occurs when fluoride is used in drinking water during the calcification of teeth. Excessive fluoride may cause a light brown to brownish black discoloration of the enamel of the permanent teeth called mottling.

Brushing the teeth immediately after eating removes the plaque from flat surfaces before the bacteria have a chance to go to work. Dentists also suggest that the plaque between the teeth be removed every twenty-four hours with dental floss.

Periodontal Diseases

Periodontal disease is a collective term for a variety of conditions characterized by inflammation and/or degeneration of the gingivae, alveolar bone, periodontal membrane, and cementum (see Figure 13-8). The initial symptoms are swelling and inflammation of the soft tissue. Without treatment, the soft tissue may deteriorate and the alveolar bone may be resorbed, causing loosening of the teeth and receding of the gums.

Periodontal diseases are frequently caused by local

irritants, such as bacteria, impacted food, cigarette smoke, or by a poor "bite." The latter may put a strain on the tissues supporting the teeth. Methods of prevention and treatment include good oral hygiene to remove plaque and other sources of irritation. Periodontal diseases may also be caused by allergies, vitamin deficiencies, and a number of systemic disorders, especially those that affect bone, connective tissue, or circulation. In these cases, the systemic disorder must be treated as well.

Peptic Ulcers

An **ulcer** is a craterlike lesion in a membrane. Ulcers that develop in areas of the alimentary canal exposed to acidic gastric juice are called peptic ulcers. Peptic ulcers occasionally develop in the lower end of the esophagus. However, most of them occur on the lesser curvature of the stomach, in which case they are called gastric ulcers, or in the first part of the duodenum, where they are called *duodenal ulcers.*

The cause of ulcers is obscure. However, hypersecretion of acidic gastric juice seems to be the immediate cause in the production of duodenal ulcers and in the reactivation of healed ulcers. Hypersecretion of acidic gastric juice is not implicated as much in gastric ulcer patients because the stomach walls are highly adapted to resist gastric juice through their secretion of mucus. A possible cause of gastric ulcers is hyposecretion of mucus. Hypersecretion of pepsin may also contribute to ulcer formation.

Among the factors believed to stimulate an increase in acid secretion are emotions, certain foods or medications, such as alcohol, coffee, or aspirin, and overstimulation of the vagus nerve. Normally, the mucous membrane lining the stomach and duodenal walls resists the secretions of hydrochloric acid and pepsin, and no ulcer develops. In some people, however, this resistance breaks down, and an ulcer develops.

The danger inherent in ulcers is the erosion of the muscular portion of the wall of the stomach or duodenum. This could damage blood vessels and possibly produce a fatal hemorrhage. If an ulcer erodes all the way through the wall, the condition is called *perforation.* Perforation allows bacteria and partially digested food to pass into the peritoneal cavity, producing peritonitis.

Peritonitis

Peritonitis is an acute inflammation of the serous membrane lining the abdominal cavity and covering the abdominal viscera. One possible cause is contamination of the peritoneum by pathologic bacteria from the external environment. This contamination could result from accidental or surgical wounds in the abdominal wall or from perforation or rupture of organs exposed to the outside environment (appendix and stomach). Another possible cause is perforation of the walls of organs that contain bacteria or chemicals which are normally beneficial to the organ but are toxic to the peritoneum. For example, the large intestine contains colonies of bacteria that live on undigested nutrients and break them down so they can be eliminated more easily. But if the bacteria enter the peritoneal cavity, they attack the cells of the peritoneum for food and produce acute infection. As another example, the normal bacteria of the female reproductive tract protect the tract by giving off acid wastes that produce an acid environment unfavorable to many yeasts, protozoa, and bacteria which might otherwise attack the tract. However, these acid-producing bacteria are harmful to the peritoneum. A third cause may be chemical irritation. The peritoneum does not have any natural barriers that keep it from being irritated or digested by chemical substances such as bile and digestive enzymes. However, it does contain a great deal of lymphatic tissue and can fight infection fantastically well. The danger stems from the fact that the peritoneum is in contact with most of the abdominal organs. If the infection gets out of hand, it may destroy vital organs and bring on death. For these reasons perforation of the alimentary canal from an ulcer or perforation of the uterus from an incompetent abortion are considered serious. If a surgeon plans to do extensive surgery on the colon he may give the patient high doses of antibiotics for several days preceding surgery to kill intestinal bacteria and reduce the risk of peritoneal contamination.

Alcohol and the Digestive System

Chemically, alcohol is an alkyl group with a hydroxy group attached (see Table 2-2). The particular alcohol that is contained in all commonly ingested beverages is called *ethyl alcohol,* or simply *ethanol.* It is a clear, colorless liquid that has a specific gravity of about 0.8. Since ethanol has the ability to depress the action of the central nervous system, it may appropriately be classified as a drug. In fact, it can even be categorized as a mind-altering drug.

Absorption

Upon ingestion, alcohol is ready for immediate absorption into the blood. A small amount of absorption, about 20%, occurs in the stomach; the remainder is absorbed in the small intestine. Absorption of alcohol is very rapid because of its low molecular weight and high solubility in water. Although the rate of absorption depends on several factors, in general, alcohol is absorbed from the gastrointestinal tract into the blood in from 45 minutes to 2 hours.

Metabolism

At least 90% of all absorbed alcohol is metabolized and destroyed by the body and the processes occur primarily in the liver. The remainder of the alcohol is eliminated by excretory routes such as urine, perspiration, and exhalation.

The metabolism of alcohol begins through the action of enzymes in the liver. In the initial oxidative step, ethanol loses two hydrogen atoms to NAD^+. The reaction is catalyzed by alcohol dehydrogenase (ADH). As a result, ethanol is converted to acetaldehyde. Acetaldehyde is further oxidized to acetic acid in the presence of the enzyme aldehyde dehydrogenase. The hydrogen acceptor for this reaction is also NAD^+. Acetic acid then is combined into acetyl coenzyme A, enters the Krebs cycle, and is ultimately oxidized to CO_2 and H_2O (see Figure 6-15). The CO_2 is eliminated via the lungs. The oxidation of 1 gram of ethanol results in the production of 56,000 calories (5.6 kilocalories), which the body may use for its energy requirements.

In general, the amount of ethanol that can be converted to acetaldehyde is 0.5–1 ounce per hour. This approximates the alcoholic content of a 12 ounce bottle of beer or one mixed drink. The rate of conversion of ethanol to acetaldehyde in the liver is fairly uniform. Thus, the process of "sobering up" is dependent on the liver and not on drinking black coffee. Essentially, the caffeine contained in coffee is a nervous system stimulant; it does not accelerate liver metabolism.

The conversion of acetaldehyde to acetic acid takes place not only in the liver, but in many cells of the body including the cells of the central nervous system. The rate of this oxidation is rather important since accumulations of acetaldehyde adversely affect normal cell functions. For example, after ingesting large amounts of alcohol, acetaldehyde accumulations generally result in headaches, gastritis, nausea, dizziness, and other symptoms collectively called a "hangover."

Physiological Effects

If the proof of one beverage is higher than another, the one with the higher proof will be more intoxicating if the two are consumed in equal quantities. For example, 2 ounces of gin is more intoxicating than 2 ounces of beer. The proof of a distilled beverage is its alcoholic content and is about twice the given alcohol percentage. Thus, 100 proof equals about 50% alcohol.

Blood alcohol levels are measured in milligrams (mg) per 100 ml of blood and converted to percent. Thus a person with 50 mg of alcohol in 100 ml of blood has a blood alcohol content of 0.05%. The consumption of 0.5–1.5 ounces of alcohol results in a blood concentration of 0.03–0.05%. This range of concentration is likely to produce a slight feeling of muscle relaxation, a sensation of warmth due to cutaneous vasodilation, an increase in the flow of gastric juice, a slight decrease in reaction time, and a slight mood elevation.

The consumption of 1.5–3.5 ounces of alcohol raises blood concentration to between 0.05 and 0.15%. As the concentration rises, the previously described effects are intensified and motor impairment is initiated. Balance, speech, vision, and hearing are affected. Coordination becomes difficult and mental impairment increases. The individual also becomes much more responsive to his emotional environment. He may become happy, belligerent, or morose. Since alcohol interferes with the reabsorption of water by the kidneys, diuresis (increased urine flow) occurs. A blood concentration of 0.15% ethanol represents legal intoxication in all states, although values range downward to as low as 0.08% in Utah.

Alcoholic concentrations of 0.20–0.30% result in severe intoxication. The individual has minimum conscious control and must have assistance to move about. Concentrations between 0.40 and 0.50% result in unconsciousness and deep coma. At a blood concentration of 0.60%, death results from respiratory failure. The immediate cause of death from excess alcoholic consumption is depression of the respiratory center in the medulla oblongata.

The chronic abuse of alcohol, coupled with dietary deficiencies, is associated with a high incidence of fatty liver, or cirrhosis. *Cirrhosis* is a chronic disorder in which there is a progressive spread of connective tissue into areas of the liver where there was once functional tissue. The condition is believed to develop by a process of fat accumulation followed by dysfunction and, finally, fibrosis. Dietary deficiencies, especially of proteins, result in a low level of the lipotropic substances (choline, folic acid, vitamin B_{12}) that are necessary for the normal removal of fat from the liver. The excessive fat destroys liver cells by interfering with their metabolism. Once the cells die, they are replaced by fibrotic tissue.

Nutrients

Nutrients are chemical substances in food that provide energy, act as building blocks in forming new body components, or assist body processes. There are six major classes of nutrients: carbohydrates, lipids, proteins, minerals, vitamins, and water. Carbohydrates, proteins, and lipids are the raw materials for reactions occurring inside cells. The cells either break them down to release energy or use them to build new structures and new regulatory substances, such as hormones and enzymes. The chemical structure, physical and chemical properties, and functions of these three classes of molecules have been surveyed in Chapter 2. Some minerals and many vitamins are used by enzyme systems that catalyze the reactions of carbohydrates, proteins, and lipids. Water has four major functions. It acts as a reactant in hydrolysis reactions, as a solvent and suspending medium, as a lubricant, and as a coolant (see Chapter 2).

In its broadest sense, the word **metabolism** refers to all the chemical activities of the body. Since chemical reactions either release or require energy, the body's metabolism may be thought of as an energy-balancing act. Accordingly, metabolism has two phases, catabolism (energy-releasing reactions) and anabolism (energy-storing and -requiring reactions). Please refer to Chapter 6 to review the metabolism of carbohydrates, lipids, and amino acids and Chapter 7 for the biosynthesis of these molecules. At this point, we will discuss only the minerals and vitamins.

Minerals

Minerals are inorganic substances. They may appear in combination with each other or in combination with organic compounds. Minerals constitute about 4% of the total body weight, and they are concentrated most heavily in the skeleton. Minerals known to perform functions essential to life include calcium, phosphorus, sodium, chlorine, potassium, magnesium, iron, sulfur, iodine, manganese, cobalt, copper, and zinc. Other minerals, such as aluminum, silicon, arsenic, and nickel are also present in the body, but their functions have not yet been determined.

Calcium and phosphorus form part of the structure of bone. But since minerals do not form long-chain compounds, they are otherwise poor building materials. Their chief role is to help regulate body processes. Iron, magnesium, and manganese are constituents of some coenzymes. Magnesium also serves as catalyst for the conversion of ADP to ATP. Without these minerals, metabolism would come to a halt, and the organism would die. Minerals such as sodium and phosphorus work in buffer systems. Sodium regulates the osmosis of water and, along with other ions, is involved in the generation of nerve impulses. Table 13-3 describes the functions of some minerals vital to the body. Note that the body generally uses the ions of the minerals rather than the elemental form. Some minerals, such as chlorine, are toxic or even fatal to the body if they are ingested in nonionized form.

Vitamins

Organic nutrients usually required in minute amounts to maintain growth and normal metabolism are called **vitamins.** Unlike carbohydrates, fats, or proteins, vitamins do not provide energy or serve as building materials. The essential function of vitamins is

TABLE 13-3. Listing of Selected Minerals, Comments, and Importance in the Body

Mineral	Comments	Importance
Calcium (Ca)	Most abundant cation in the body; appears in combination with phosphorus in the ratio of 2:1.5; about 99% is stored in bone and teeth and the remainder is stored in muscle and other soft tissues and the blood; recommended daily intake for children is 1.2–1.4 grams* and 800 mg* for adults; good sources are milk, egg yolk, shellfish, and green leafy vegetables.	Structural component of skeleton and teeth; essential for all cellular activities; normal functioning of nerve tissue; normal permeability of cell membranes; required for excitability of muscle, blood clotting, and normal heart action.
Phosphorus (P)	About 80% is found in bones and teeth, the remainder is distributed among body cells; has more functions than any other mineral; recommended daily intake is 1.2–1.4 grams for children and 1200 mg for adults; good sources are dairy products, meat, fish, poultry, and nuts.	Important in muscle metabolism; constitutes a buffer system of the blood; essential component of every cell; constituent of nervous tissue; component of many enzymes; involved in the transfer and storage of energy (ATP).
Iron (Fe)	Most (66%) is found in the hemoglobin of blood, remainder is distributed among skeletal muscles, liver, spleen, and enzymes; recommended daily intake for children is 7–12 mg and for adults 10–15 mg; possible sources are meat, liver, shellfish, egg yolk, beans, legumes, dried fruits, nuts, and cereals.	As a component of hemoglobin it carries O_2 to body cells; essential for the functioning of enzymes involved in cellular respiration.
Iodine (I)	Found as an essential component of the thyroid hormone; thyroxin; estimated daily requirement is 0.15–0.30 mg; adequate sources are iodized salt; seafoods, cod-liver oil, and vegetables grown in iodine-rich soils.	Required by the thyroid gland to synthesize thyroxin; the hormone regulates the rate of metabolism.
Copper (Cu)	Normally stored in the liver and spleen; daily requirement is about 2 mg; good sources include eggs, whole wheat flour, beans, beets, liver, fish, spinach, and asparagus.	Required with hemoglobin for iron formation; necessary for melanin pigment formation.
Sodium (Na)	Most is found in extracellular fluids, some in bone; recommended daily intake of NaCl is 5 grams.	Assumes an important role in maintenance of pH of body fluids and normal distribution of water.
Potassium (K)	Principal cation in intracellular fluid; normal food intake supplies required amounts.	Functions with various enzymes; influences transmission of nervous impulses, skeletal muscle contraction, and cardiac action.
Chlorine (Cl)	Found in extracellular and intracellular fluids; principal anion of extracellular fluid; normal intake of NaCl supplies required amounts.	Assumes a role in acid-base balance of blood; secreted by gastric mucosa as HCl; maintains normal cardiac action.
Magnesium (Mg)	Component of soft tissues and bone; suggested minimum daily intake is 250–350 mg; widespread in various foods.	Required for normal functioning of heart, skeletal muscle, and nervous tissue; maintains normal structure of growing tissue; participates in bone formation.
Sulfur (S)	Constituent of many proteins (e.g., insulin) and some vitamins (thiamine and biotin); good sources include beef, liver, lamb, fish, poultry, eggs, cheese, and beans.	As a component of hormones and vitamins, it regulates various body activities.
Zinc (Zn)	Important component of certain enzymes; widespread in many foods.	Necessary for normal growth.
Fluorine (F)	Component of bones, teeth, and other tissues.	Appears to improve the structure of teeth.
Manganese (Mn)	Distribution throughout the body similar to that of copper; human daily requirement is unknown.	Activates several enzymes; needed for hemoglobin synthesis; required for growth, reproduction, and lactation.
Cobalt (Co)	Constituent of vitamin B_{12}	Associated with red blood cell formation.

* 1 gram = 1000 milligrams (mg); 1000 grams = 2.2 pounds; 1 ounce = 28.35 grams.

the regulation of physiologic processes. Some vitamins act with enzymes, and some serve as coenzymes.

Vitamins cannot be synthesized by the body from its own resources and must be obtained from a variety of sources. One source of vitamins is ingested foods—for example, vitamin C in citrus fruits. Another source is vitamin pills that contain several or all vitamins. Other vitamins, such as vitamin K, are produced by bacteria in the gastrointestinal tract. The body can assemble some vitamins if the raw materials are provided. Such raw materials are called *provitamins.* For example, vitamin A is produced by the body from the provitamin carotene, a chemical present in spinach, carrots, liver, and milk.

You will soon discover that vitamins are found in varying quantities in different foods and that no single food contains all required vitamins. This is probably one of the best reasons for eating a varied diet. The term *avitaminosis* refers to a condition in which there is a deficiency of any vitamin in the diet.

On the basis of solubility, vitamins are divided into two principal groups: fat soluble and water soluble. *Fat-soluble vitamins* are absorbed along with digested dietary fats by the lacteals of the villi of the small intestine. In fact, they cannot be absorbed unless they are ingested with some fat. Fat-soluble vitamins are generally stored in cells, particularly the cells of the liver, so reserves can be built up. Examples of fat-soluble vitamins are vitamins A, D, E, and K. *Water-soluble vitamins,* by contrast, are absorbed along with water in the gastrointestinal tract and dissolve in the body fluids. As the blood is filtered by the kidneys, excess quantities of the vitamins are excreted in the urine. Thus, the body does not store water-soluble vitamins well. However, it has the dubious advantage that very few cases of water-soluble vitamin overdoses have been observed. Examples of water-soluble vitamins are the B vitamins and vitamin C. Table 13-4 lists the principal vitamins, their sources, functions, and related disorders.

TABLE 13-4. Vitamins: Sources, Functions, and Deficiency Symptoms

Vitamin	Source and Comments	Function	Deficiency Symptoms
FAT-SOLUBLE VITAMINS			
A	Yellow and green vegetables, fish, liver oils, milk, and butter, stored in liver.	Maintenance of normal health and growth of epithelial cells, general health and vigor, assists in chemical reactions associated with vision. Excessive vitamin A (hypervitaminosis) causes drying and peeling of skin, loss of hair, enlargement of liver and spleen, and excessive bone fragility.	Inability to gain weight; dry skin and hair; increased incidence of respiratory disorders; skin, ear, and sinus conditions; infections of the urinary and alimentary tracts; nervous disorders; *xerophthalmia* (drying of cornea and conjunctiva with ulceration); and *night blindness* (decreased ability for dark adaptation).
D	Milk, butter, egg yolk, and fish liver oils. No special storage area. In the presence of sunlight, provitamin S of skin gland secretions is converted into vitamin D.	Increases absorption of calcium and phosphorus from the intestinal tract; utilization of same minerals in development of bones and teeth; related to the hormone of the parathyroid gland in calcium metabolism.	*Rickets;* general retarded growth, lack of vigor; possible loss of muscle tone.
E (*tocopherols*)	Meats, milk, eggs, fish, cereals, leafy vegetables, and wheat germ.	A factor in normal reproduction in male and female rats.	*Sterility* in rats. In humans it may serve as an antioxidant; helps hydrogen acceptors to perform their functions properly.
K	Spinach, cabbage, cauliflower, and liver; also known as the antihemorrhagic vitamin.	Essential for synthesis of prothrombin by liver and thus for normal blood clotting.	Delayed clotting time.

(*continued*)

TABLE 13-4. Vitamins: Sources, Functions, and Deficiency Symptoms [*Continued*]

Vitamin	Source and Comments	Function	Deficiency Symptoms
WATER-SOLUBLE VITAMINS			
B_1 (thiamine)	Good sources are whole-grain cereals, eggs, legumes, and pork.	Needed for normal carbohydrate metabolism; maintenance of normal appetite, digestion, and absorption.	*Beriberi* (gastrointestinal disturbances, paralysis, atrophy of limbs); *polyneuritis* (reflexes related to sense of position are impaired; stunted growth in children; poor appetite; decreased intestinal motility).
B_2	Meat, milk, eggs, fruit, liver, and green vegetables; also known as *riboflavin;* small amounts are supplied by bacteria of the digestive tract.	Component in all enzyme systems that regulate carbohydrate metabolism; normal growth, nutrition, and vitality.	*Dermatitis;* lesions of the eyes.
Niacin	Meats, liver, fish, poultry, whole-grain cereals, peas, beans, and nuts.	Utilized in metabolic reactions that release energy.	*Pellagra* (dermatitis, diarrhea, stomatitis, and dementia).
B_6 (pyridoxine)	Whole-grain cereals, milk, eggs, pork, liver, yeast, fish, and legumes.	Participates in metabolism of fats and amino acids.	Dermatitis of the eyes, nose, and mouth.
B_{12} (cyanocobalamin)	Liver, kidney, milk, egg, cheese, and meat; its absorption from the intestinal tract is dependent on HCl and the intrinsic factor secreted by the gastric mucosa; the only vitamin containing cobalt.	One factor in red blood cell formation.	*Pernicious anemia.*
Pantothenic acid	Egg yolk, kidney, liver, yeast, peanuts, heart, and meat; also synthesized by bacteria in the digestive tract.	Constituent of coenzyme A; utilized in fat metabolism, cholesterol synthesis, and antibody formation.	Practically unknown; possibly some neurological defects.
Folic acid	Green leafy vegetables and liver, also synthesized by bacteria in the digestive tract.	Factor in the synthesis of purines and pyrimidines in nucleoproteins; normal development of normal red blood cells.	*Macrocytic anemia* (Anemia marked by abnormally large red blood cells).
C (ascorbic acid)	Citrus fruits, tomatoes, and green vegetables.	Utilized in red blood cell formation; normal functioning of capillaries; maintenance of healthy gums and teeth; healing of wounds; normal metabolism; protection against infection.	*Scurvy;* poor wound healing, susceptibility to infection; retardation of growth; tender, swollen gums; pyorrhea.

14 Transportation Mechanisms

GUIDE QUESTIONS

1 Define cyclosis and diffusion. In what ways are they related to the transportation and distribution of nutrients in vascular plants? What other mechanisms are involved?
2 Of what importance to a plant is transpiration? How is it related to the transportation and distribution of nutrients?
3 How are nutrients transported in unicellular organisms? Compare the transportation and distribution of nutrients in lower Animalia with that of unicells. Indicate all similarities and differences.
4 Considering the structure of higher Animalia, explain why some type of vascular system is necessary for the distribution of nutrients. Discuss the nature of an open circulatory system. How does it show advancement over the distribution of nutrients in unicells and lower Animalia?
5 What role do respiratory pigments play in circulatory systems? How do organisms which lack such pigments function? In what ways is a closed circulatory system more efficient than an open system?
6 Discuss the various functions of the human circulatory system. Outline in detail the various components of blood and list their functions.
7 How are the blood groups determined? Of what significance is the Rh factor?
8 Explain in detail the clotting mechanism. How may clotting be hastened? hindered? Explain.
9 Trace the path of an erythrocyte through the heart from its entrance into the right atrium to its passage through the aorta.
10 Discuss the changes in heart structure of vertebrates from fish through birds. Indicate modifications correlated with the changes in respiration from gills to lungs.
11 Compare arteries, capillaries, and veins with regard to structure and function. What purposes do arterioles and venules serve?
12 Outline the systemic circulation, indicating the major arteries, regions, and organs to which they are distributed, and the major veins. Discuss the nature of the portal circulation. Of what significance is the portal circulation?
13 How does pulmonary circulation differ from the systemic circulation?
14 In what ways does fetal circulation differ from postnatal circulation?
15 Discuss the various factors involved in the maintenance of blood flow. How is blood pressure measured? Of what importance are cardiac output and peripheral resistance?
16 What is the lymphatic system? How is it related to the bloodstream?
17 What factors are involved in the transportation of digested nutrients? Relate the functions of the small intestine and liver to those of the bloodstream.
18 Describe the conduction system of the heart. What is an electrocardiogram? Explain the meaning of the waves in an ECG.
19 Discuss the major symptoms associated with arteriosclerosis and hypertension. What are some of the dangers of these disorders?
20 Describe the five risk factors that are implicated in heart attacks.

HAVING surveyed the principal mechanisms utilized by various organisms for the procurement and digestion of nutrients, we will now proceed to examine the transportation of digested nutrients. Once nutrients are digested, the organism is confronted with certain problems. The nutrients must be transported from their site of digestion to sites of utilization, and any by-products of metabolism must be eliminated by the organism.

All organisms take in nutrients and give off metabolic

wastes. In such an exchange, these substances must pass through cell membranes, which, because of their differential permeability, allow only certain kinds of materials to pass through them. Since these materials are either entering the cytoplasm or leaving it, they (except for lipid-soluble substances) must be soluble in water in order to pass through cellular membranes. Refer to Chapter 5 for a discussion of the factors related to membrane permeability.

Digestion and Transportation: Correlation

Ingested nutrients are digested within either a vacuole (intracellular digestion) or a body cavity lined with specialized digestive epithelium (extracellular digestion). Once digestion occurs, the end products are absorbed into either the surrounding cytoplasm (as in the case of unicellular organisms) or adjacent cells (as in the case of multicellular organisms). These absorbed nutrients must then be transported throughout the organism to the sites of metabolic reactions. This movement of materials requires a circulation of materials within a fluid medium. In unicellular organisms, such as *Amoeba* (Kingdom Protista), this is accomplished by cytoplasmic streaming. Multicellular organisms, such as sponges and coelenterates, rely on amoebocytes within an acellular jelly-like layer between their body walls. Higher acoelomate animals, such as flatworms and roundworms, depend upon body fluids within a *pseudocoel* (unlined coelom) to carry dissolved nutrients to body parts. Coelomate animals, such as annelids, mollusks, arthropods, and chordates, have developed specialized vascular systems to transport dissolved substances. The structural complexity of these animals is such that neither diffusion within body fluids in a cavity nor the activities of undifferentiated cells in a body wall would be sufficient to supply the nutritional needs of the entire organism. Such animals contain body fluids that are located within vessels either partly (open circulatory system) or entirely (closed circulatory system). The vessels are of various diameters and are branched extensively to reach all body cells. The dissolved nutrients that they carry are absorbed by the cells and then distributed to the metabolic sites within the cells.

Plants also must distribute nutrients from sites of synthesis or storage to metabolic reaction sites. Unicellular plants rely on cytoplasmic streaming within the protoplast. Nonvascular multicellular plants such as mosses and liverworts depend upon cell-to-cell diffusion. Such plants live in moist environments and so are in direct and constant contact with dissolved materials. Vascular plants, such as ferns, gymnosperms, and angiosperms, have differentiated tissues and greater nutritional needs. Dissolved nutrients (minerals, water) are carried by specialized conducting cells (xylem tracheids and vessels) from the roots, where they are absorbed, upward to the leaves where they are utilized in synthetic processes (for example, photosynthesis). Organic nutrients (carbohydrates, for example) are then translocated throughout the plant by the phloem tissue (sieve tube cells) to all plant cells where they are absorbed and distributed to the sites of metabolic activity within each cell.

Survey of Circulatory Mechanisms

Transportation in Vascular Plants

Conduction of Nutrients. Once water and mineral substances are absorbed by cells, internal distribution is accomplished mainly by *diffusion* and *cyclosis.* Both processes enable every part of the cell to receive adequate supplies of all metabolites that the cell absorbs at specific points on its surface. Diffusion not only transports materials within cells but may also occur between plasmodesmata of adjacent cells or through the permeable cell walls. Inasmuch as diffusion is a relatively slow transport process, the movement of materials may be appreciably increased by *protoplasmic streaming* (cyclosis). The streaming of protoplasm has been measured at rates of a few to several centimeters per hour. In considering a single cell, this rate of movement is fairly rapid, because a substance may travel the length of a cell in seconds. Lateral conduction of materials across roots, or other plant organs, probably involves the combined mechanisms of cyclosis and diffusion through vascular rays or other cell-to-cell transfers. The upward conduction of sap absorbed by roots, however, must take into account processes other than diffusion and cyclosis because, in considering the entire plant, both are relatively slow processes.

It has been known for many years that raw materials absorbed by the roots move upward in the plant through the tracheids and vessels of the xylem. The mechanisms involved, however, are not completely understood. Most of the water absorbed by the roots is returned to the atmosphere through the leaves of the plant. The distances separating leaves and roots are quite commonly of considerable magnitude. Many trees of large forests are over 200 feet high, and the tops of the tallest trees may exceed 400 feet. Therefore, any explanation of the rise of sap in xylem tissue must take into account the forces required to move materials along such vertical distances.

Two principal explanations offered to identify the forces capable of raising and conducting water in the xylem are (1) the root pressure hypothesis, in which a pushing force is involved, and (2) the cohesion tension theory, in which a pulling force is involved.

Root Pressure. The absorption of water by root cells creates a hydrostatic pressure in the root system known as **root pressure.** The manifestations of this pressure can be readily observed. It can be seen in some plants whose stems have been cut off a short distance above ground level, such as in the stump of a recently cut tree or freshly cut herbaceous plant. Sap will continue to flow, or *bleed,* from the surface of the cut for some time. If a glass tube is attached to the cut stump, the sap can be seen to rise in it against the force of gravity (Figure 14-1). Root pressure can also be observed when conditions are excellent for water absorption by roots, but the humidity is so high that minimal amounts of water are lost to the atmosphere by evaporation. Given these conditions, root pressure may force water out of the veins of leaves, forming droplets along the leaf margins. This process is called **guttation.** Both processes, bleeding and guttation, demonstrate the existence of root pressure.

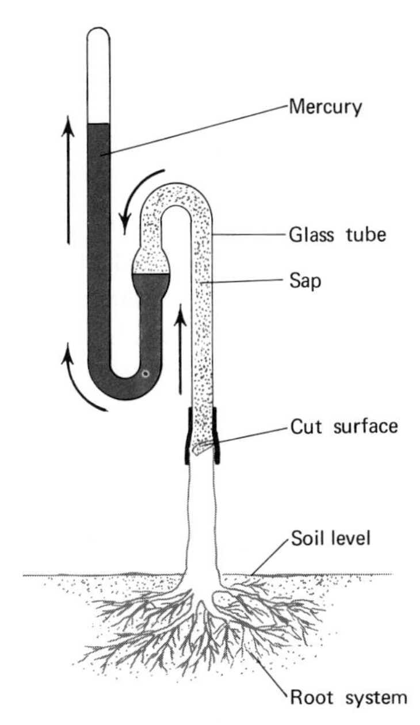

FIGURE 14-1. Measurement of root pressure. Sap exudes from the cut surface of a stem because of root pressure. The force of this pressure may be measured by placing a curved glass tube over the surface of the cut. Root pressure causes the sap to rise in the glass tube, which in turn causes the level of the mercury column to become elevated. Root pressure is determined from the distance through which the column of mercury is elevated. [*From M. S. Gardiner and S. C. Flemister,* The Principles of General Biology, *New York: Macmillan Publishing Co., Inc., Copyright © 1967.*]

Root pressure has been suggested as an important pushing force in the ascent of sap through xylem, but a number of serious objections make this explanation inadequate. First, not all plants bleed when they are detopped, and bleeding is negligible in summer when the evaporation is highest. Second, the root pressures that force the sap upward have been measured, and only in rare instances have these pressures exceeded 1–2 atmospheres (1 atmosphere at sea level is equal to a force of 14.7 pounds per square inch). These low values suggest that root pressure is not the principal force involved in the ascent of sap because a pressure considerably greater than 2 atmospheres is needed to lift sap from the roots to the tops of trees of ordinary height. Third, the rates at which exudation (bleeding) occurs are generally much slower than normal evaporation (transpiration) rates. Moreover, water movement through stems continues even though the roots have been removed. This can be demonstrated by the failure of cut stems or flowers to wilt when placed in adequate amounts of water.

Cohesion Tension Theory. This theory offers the most adequate explanation for the ascent of sap in plants. As contrasted to the root pressure hypothesis, the cohesion tension theory is based upon a pulling force rather than a pushing force. This theory assumes that water lost by transpiration through the leaves is pulled upward and that the pull is maintained by the attraction of water molecules for each other and to the walls of the xylem vessels.

Transpiration is the loss of water vapor from the internal tissues of living plants. Although such water loss may occur from any exposed part of the plant, most of it actually takes place through the leaves. Less than 10% of the total amount of transpiration from leaves occurs directly through the cuticle of the epidermis (cuticular transpiration). The greatest loss of water takes place through the stomata (stomatal transpiration). The quantity of water transpired by plants is very great. Of the total quantity absorbed by roots, as much as 98% of it is lost by transpiration.

Under natural conditions, the air of the atmosphere usually contains water vapor in a concentration of 1–3% (Figure 14-2). This concentration is considerably lower than the water concentration of living cells in the leaf. The parenchyma cells of the spongy mesophyll in leaves, as well as the air spaces that permeate them, are in

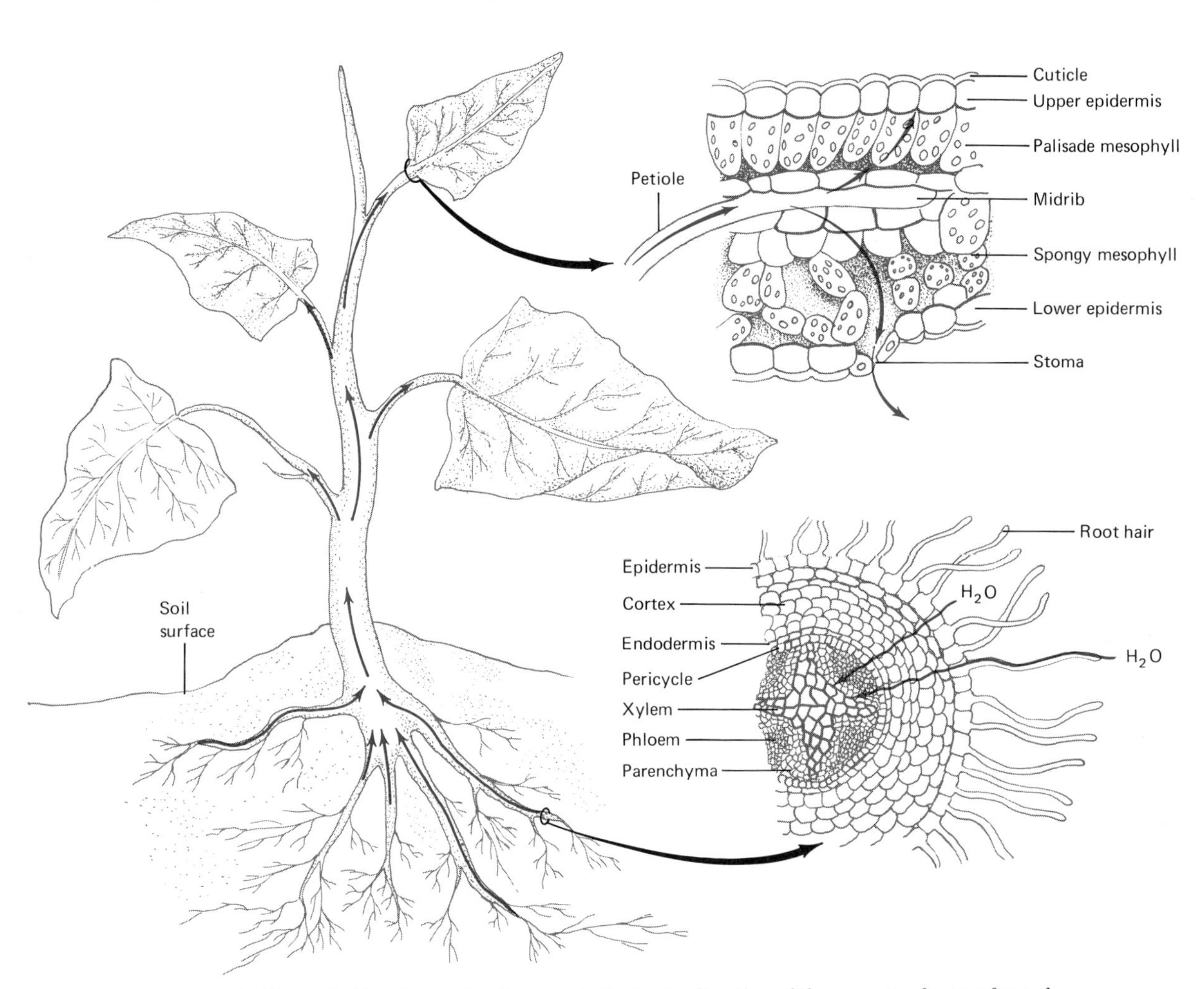

Figure 14-2. Transpiration. The colored arrows indicate the direction of the passage of water from the root system through the xylem of the stem to the leaves, where it exits via transpiration. The enlargements to the right show the paths of water through the root (*lower*) and through the leaf (*upper*). Consult text for details (also see Figure 13-1).

contact with the atmosphere through the stomata. Water which has been imbibed by the cell wall of these cells evaporates and saturates the air in the air spaces. The diffusion (vapor) pressure of water is then higher than in the outside atmosphere, even if the relative humidity is high. Water vapor molecules diffuse from the air spaces through the stomata into the atmosphere. Air currents, a dry breeze, or wind may facilitate the movement of this water vapor from the immediate vicinity of the plant. This water loss results in a slight drying of the air in the intercellular spaces. Such a water deficit is made up by the diffusion of water molecules from the wet cell walls into the air spaces. As water leaves the walls, this deficit is compensated for by the diffusion of water from the cell sap in the vacuoles and the cytoplasm. Water enters the water-deficient cells of the mesophyll by osmosis from neighboring cells. Ultimately, the water losses resulting from transpiration are made up by absorption from the xylem. Essentially, a diffusion gradient exists from the air spaces, through the mesophyll cells, to the water-filled xylem elements. In this manner, all the mesophyll cells are supplied with water from the xylem as it is lost through transpiration.

The removal of water from the top of the plant pulls a column of water upward. This column represents a continuous path between the water on the evaporating surfaces of the leaf mesophyll cells, the water in the xylem of the leaf, and the xylem running from the top of the plant to the root. If this continuity of water from the leaf cells to the root is broken in any way, such as by the entrance of air bubbles, that particular xylem system could not facilitate the ascent of sap. Water molecules being lost by transpiration must pull other water molecules behind them in such a way that there is no break in the water column. In this regard, it is assumed that there is some force in operation that holds water molecules together and that the tensile strength of the water is sufficient to form an unbroken column. The intermolecular force between water molecules is called *cohesion*. It is this force that affords the water column its tensile strength. If this force is greater than the force developed by transpirational pull, the water column will rise. Measurements of the tensile strength of sap in xylem tubes have shown that in some cases this exceeds 150 atmospheres. This is more than enough to withstand the osmotic pull developed by transpiration. It has been estimated that the tensile strength of water is not only sufficient to form an unbroken column against osmotic pull, but it is also adequate to overcome the frictional and gravitational forces encountered in its vertical rise in the plant. Furthermore, the attraction of water molecules for the walls of the xylem (*adhesion*) is very great.

Phloem Transport of Nutrients. Water and dissolved minerals absorbed by roots and transported by xylem represent the essential substances needed by plants for growth and development. These inorganic substances are the materials from which plant cells synthesize organic molecules for the maintenance of metabolic activities. The principal cells which assemble metabolites into nutrients are the photosynthetic cells of leaves. Anatomically, the distances separating photosynthetic cells from other living cells are relatively large. Considering this, the need for a rapid and efficient transportation system becomes apparent. The movement of dissolved nutrients is accomplished by the specialized cells of the phloem called *sieve tube elements*. These cells, like those of xylem, form a continuous conducting network that extends to every part of the plant.

A number of methods have been employed to ascertain that the phloem is the principal channel for the conduction of organic nutrients. Chemical analyses of phloem cell contents reveal that they consist largely of dissolved carbohydrates, usually sucrose, proteins, and other nitrogen-containing compounds. If the bark (which includes the phloem) is removed, a process called **ringing** or **girdling,** and the wood (xylem) is left intact, it can be shown that after several weeks nutrients will accumulate above the ring. In addition, the stem will bulge after the damaged tissues have regenerated. Girdling prevents the movement of nutrients to all parts of the plant below the ringed section, proving that phloem facilitates the downward movement of nutrients. The structure of sieve tubes also suggests a conducting function.

Although there is considerable evidence that organic substances are transported in the phloem, some evidence also suggests the bidirectional conduction of inorganic substances, such as salts. Studies using radioactive tracers have demonstrated the downward movement of salts in the phloem tissue. Radioactive phosphorus (^{32}P), for example, has been shown to move in such a manner. It would appear, then, that there is a bidirectional movement of both organic solutes and certain mineral salts in the phloem.

The term **translocation** is generally used to describe the movement of dissolved organic nutrients through phloem tissue. The directions of translocation are not only upward and downward but also laterally in vascular rays. The problem of translocation is a difficult one to resolve because of the interplay of a number of factors. For example, simple diffusion alone cannot account for the rapid movement of translocated materials. Furthermore, different materials move with different translocation rates through sieve cells that are filled with cytoplasm and contain sieve plates perforated by extremely small pores. In addition, translocation rates are affected by temperature, light, metabolic inhibitors, concentration gradients, hormones, and mineral deficiencies, among other factors. Considering all the aspects of phloem translocation, no single theory has gained general acceptance in explaining the mechanism involved. Two hypotheses, however, have suggested possible explanations.

Mass (Pressure) Flow Hypothesis. According to this hypothesis, there is a mass flow of water and solutes through sieve tubes owing to differences in turgor pressure. Cells of the leaf contain high sugar concentrations because of their photosynthetic activity. As a result of a high solute concentration, water from the xylem diffuses into them, thereby increasing their turgor pressure. Other cells of the plant, such as those that are actively engaged in respiration, growth, or storage processes, have a lower solute concentration because sugars are being consumed. This loss of sugar causes these cells to lose water, thereby decreasing their turgor pressure. At one end of the phloem, then, are cells with high turgor pressure, and at the other end are cells with low turgor pressure. The result of these varying turgor pressures at opposite ends of the phloem is a mass flow of the contents of the sieve cells from regions under high pressure to regions under lower pressure (Figure 14-3).

Protoplasmic Streaming Hypothesis. This theory assumes that solute particles diffusing into sieve cells are caught up in the circulating cytoplasm and are carried from one end of the cell to another. The particles are then passed on to the next sieve cell through plasmodesmata and are again picked up by the streaming cytoplasm.

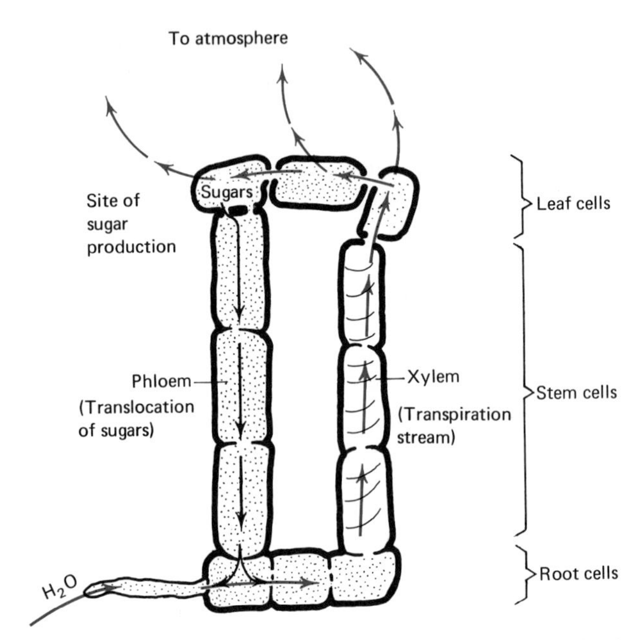

Figure 14-3. Diagram of the mass (pressure) flow hypothesis of phloem transport. Refer to text for details. [*Modified from A. S. Crafts,* Plant Physiology, *vol. 6, p. 1, 1931.*]

Circulation in Unicells

Unicellular organisms, such as *Paramecium* (Kingdom Protista), have no specialized system for the circulation of fluids. Such organisms depend upon cytoplasmic streaming for the distribution of absorbed and digested nutrients. A food vacuole within a *Paramecium* follows a path from the gullet where it was formed, posteriorly, then toward the anterior region where it turns and is moved by the cytoplasm toward the posterior end. As it moves with the cell, the vacuole distributes its digested nutrients to the various cell organelles, where they will be utilized in metabolic reactions (see Figure 13-3B).

Simple Animal Systems

Inasmuch as lower Animalia (sponges, hydras, and planaria) possess a relatively simple body design, they have no special need for an elaborate vascular system. Each body cell is close enough to the environment so that simple diffusion is sufficient for the transportation of materials throughout the organism. This diffusion of nutrients occurs between the internal digestive cells and the fluid material found separating these cells from the epidermal cells.

Sponges take in food through the *choanocytes* (collar cells), which line the *spongocoel* (internal cavity) (see Figure 13-4). These cells are specialized for the digestion of food particles, which are carried by water currents within the spongocoel. After digestion, the nutrients diffuse into the *mesoglea,* a jelly-like material in which amoebocytes are located. The *amoebocytes* may also receive dissolved nutrients and, by their movements throughout the mesoglea (mesenchyme) distribute the nutrients.

Hydra, a coelenterate, is also composed of two cell layers connected by a mesoglea. However, hydroid mesoglea is acellular and the layer is quite thin. Food is digested within a *gastrovascular cavity* (gastrocoel) lined with various types of epithelia (gastrodermis). Digested nutrients are absorbed by the gastrodermis and then diffuse through the thin mesoglea to the epidermal cells (see Figure 13-5).

Planaria is a flatworm composed of three cell layers: an ectoderm, a mesoderm, and an endoderm. The endoderm forms the cells lining the intestine which functions in extracellular and intracellular digestion. The mesoderm gives rise to a mass of cells known as parenchyma, muscle fibers, and internal organs. The ectoderm produces the epidermis. Nutrients are ingested by the contractions of the pharynx and are digested within the blind intestine. Epithelial cells lining the intestine absorb the digested materials, which then diffuse into the parenchyma cells. Fluids surrounding the parenchyma serve to distribute the products of digestion to all body parts (see Figure 13-6).

Open Circulatory Systems

Higher Animalia, such as the grasshopper, are structurally more complex and thus require a more complex mechanism for the distribution of nutrients. Instead of body fluids being within a body wall (for example, the mesoglea of sponges and hydras) or surrounding internal cells (as in the fluid around the parenchyma in planaria), they are contained within vessels that open into the body cavity. Such open vessels constitute an *open circulatory system.*

An open circulatory system is composed of a dorsal heart, a dorsal aorta, and various cavities or sinuses (Figure 14-4). The heart, located dorsally in the abdomen, is typically tubular in shape and is enclosed by a *pericardial cavity,* which is formed by a transverse diaphragm. A single *dorsal aorta* leads from the heart toward the anterior region of the body where it terminates by opening into a sinus or cavity (open cavity). In each segment through which the heart passes, there is an enlargement of the heart into a chamber that is perforated on each side by an opening (ostium). These ostia allow blood to enter the heart. Blood is carried forward via the aorta and flows out into spaces among the body tissues where it eventually moves into the thorax and abdomen. As the blood flows over the abdominal organs, nutrients are absorbed from the stomach and intestine. Blood returns to the heart via the ostia and is again pumped into the aorta and over the body tissues in order to distribute the dissolved nutrients. The blood of a grasshopper is typical of most arthropods in that it is a clear fluid containing colorless cells that act as phagocytes in the removal of foreign organisms. The chief function of the blood, however, is to act as a liquid transport medium for the distribution of nutrients and the collection of metabolic wastes.

As the internal structure of animals becomes more complex, there is a need for a more complex circulatory system. The composition of the blood also changes with the oxygen requirements of the organisms involved. In some invertebrates, the blood is colorless and gases are

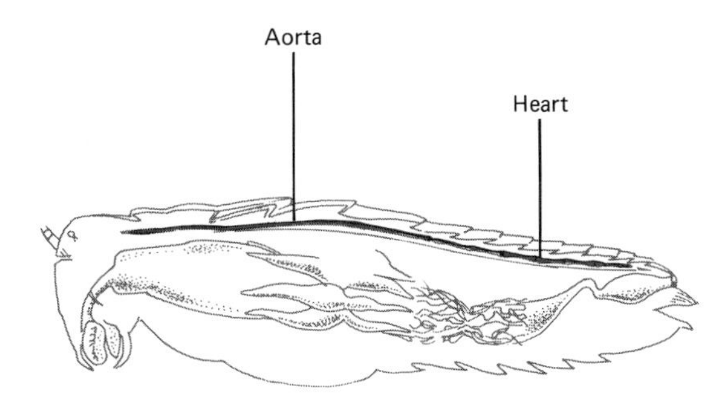

Figure 14-4. Open circulatory system of the grasshopper as seen from the left side with the body wall removed.

dissolved within the fluid (as are nutrients and wastes) with oxygen being delivered to the cells by diffusion from either the fluid or the environment (epidermal respiration). In other invertebrates, such as mollusks and certain annelids, the blood contains dissolved respiratory pigments that carry oxygen to the cells. Earthworms contain a blood pigment called *erythrocruorin,* an iron compound that is similar to hemoglobin (the respiratory pigment of vertebrates); the "blue blood" of many mollusks and crustaceans contains the pigment *hemocyanin,* a copper compound that turns blue in combination with oxygen. Invertebrate blood typically contains limited numbers of colorless *amoebocytes* which aid in removal of foreign particles.

Closed Circulatory Systems

Certain mollusks (squid), annelids (earthworms), and vertebrates (humans) have developed a circulatory system that is completely vascular in nature. Such a system, in which all the blood typically remains within vessels, is termed a *closed circulatory system.* A closed circulatory system allows for greater efficiency in rate of blood flow, economy of blood volume, and maintenance of blood pressure.

Annelidan circulatory systems typically have a contractile dorsal vessel that has been modified into "hearts" as is characteristic of the earthworm (Figure 14-5). In earthworms, these "hearts" (*aortic arches*) surround the esophagus and connect the *dorsal blood vessel* with the *ventral blood vessel.* The dorsal blood vessel is a muscular tube which, by its peristaltic action, moves the blood forward from the posterior end of the body to the "hearts." From there, blood is pumped into the ventral blood vessel and throughout the body. In each segment of the earthworm, there are paired branches of the ventral blood vessel called *segmentals* that form smaller branches leading to the body wall, skin, and nephridia and to the *lateral vessels,* which lie longitudinally next to the nerve cord. Blood is drained from these regions by the *parietal vessels,* which return the blood to the dorsal blood vessel. The nerve cord is drained ventrally by the *subneural blood vessel,* which empties into the parietal vessel in each segment. There are short vessels extending to and from the dorsal blood vessel to the intestine. Blood from the ventral blood vessel also enters the intestinal circulation.

The blood of an earthworm contains a fluid *plasma* in which are suspended many phagocytic leukocytes. Dissolved in the plasma is erythrocruorin, a respiratory pigment. The internal organs and the body wall are

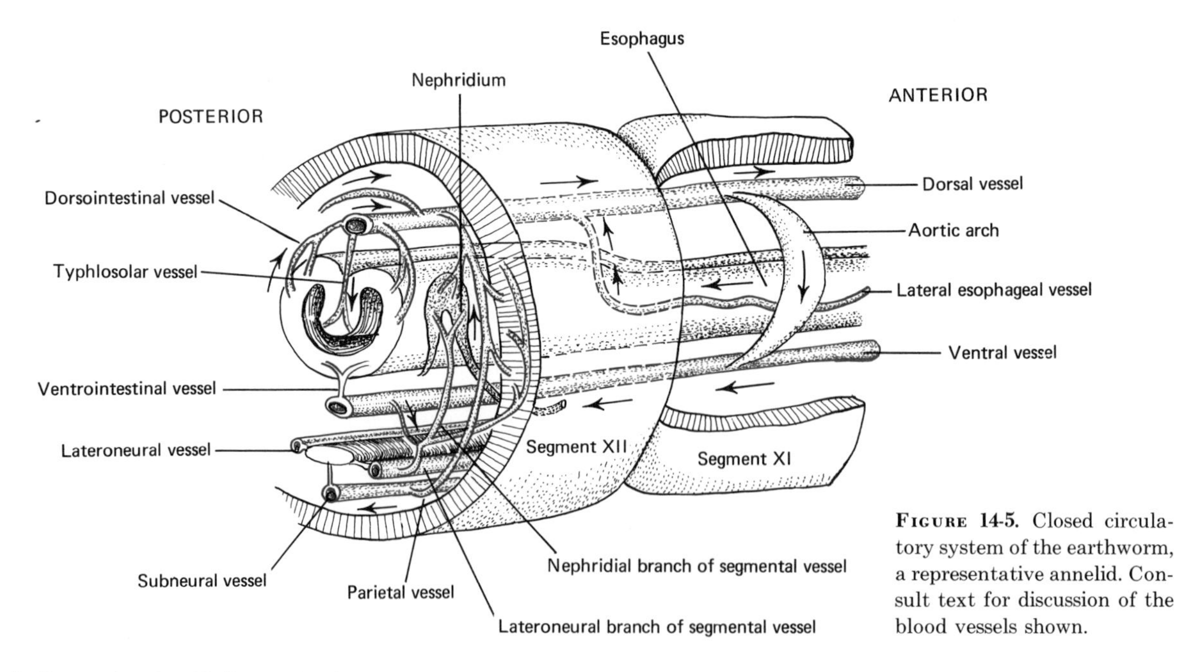

FIGURE 14-5. Closed circulatory system of the earthworm, a representative annelid. Consult text for discussion of the blood vessels shown.

richly supplied with capillaries that connect afferent and efferent vessels. These allow dissolved nutrients to diffuse easily into body tissues and rapid gas exchange to occur through the moist cuticle covering the body.

Circulation in Vertebrates

Functions of the System

A circulatory system, whether open or closed, serves several functions with respect to the organism and its metabolic activities. It has been stated that transportation of dissolved nutrients is a primary function of the circulatory system. Another function is the maintenance of homeostasis through the transportation of water and electrolytes, the control of fluid volume, pH, and regulation of body temperature. Hormones which are required for the control and coordination of body functions such as digestion and secretion are also carried by the bloodstream. In addition, the blood is responsible for the transportation of gases and of antibodies that are produced by various cells of the body; thus it contributes to the body's defense against foreign organisms or their products.

Components

Blood. Blood, a vascular tissue flowing within the circulatory system, is composed of cells, plasma, and dissolved or suspended solutes. Inasmuch as the structure of the three types of blood cells (erythrocytes, leukocytes, and platelets) has already been discussed in Chapter 9, only their functions will be noted here (see Table 9-3).

Erythrocytes contain the respiratory pigment *hemoglobin,* which is able to unite with four molecules of oxygen to form *oxyhemoglobin.* Carbon dioxide can also combine with hemoglobin to form *carbaminohemoglobin.* In this reaction, the globin part of the molecule rather than the heme part participates in the combining action. Both of these reactions (the addition of oxygen to from oxyhemoglobin and the addition of carbon dioxide to form carbaminohemoglobin) are reversible and will be treated in more detail in Chapter 16. Essentially then, erythrocytes function in the transportation of oxygen and carbon dioxide to and from the body cells.

In addition to functioning in oxygen and carbon dioxide transport, erythrocytes are also responsible for the properties of various blood types or groups that have been identified. The type of blood group is dependent upon the kinds of *antigen* (a substance that is capable of stimulating the production of antibodies) present on or in the membranes of erythrocytes. Most common are the A, B, and Rh factors. Others have been identified, but their details are beyond the scope of this textbook. Blood types are named according to the antigens present, the chief types being the following:

Type A	Factor A (antigen) present on membranes
Type B	Factor B (antigen) present on membranes
Type AB	Factors A and B (antigens) present on membranes
Type O	Neither factor A (antigen) nor B (antigen) present on membranes
Rh positive	Rh factor (antigen) present on membranes independent of presence of factors (antigens) A and/or B
Rh negative	Rh factor lacking

Blood plasma (the fluid portion of the blood) may or may not have antibodies that can react with the A, B, or Rh antigens of the erythrocytes. An important principle concerning this is that plasma will never have antibodies against its own erythrocyte antigens. If this were to occur, the erythrocytes would be destroyed. Equally important is the fact that the plasma will contain antibodies against antigens not present on its own cells. For example, a person with Type A blood (factor A present on erythrocyte membranes) would not have plasma antibodies against factor A but rather against factor B antigens not present on its erythrocytes. Likewise, Type B blood would contain antibodies against factor A antigens and not against factor B. Type AB blood possesses both factors A and B and therefore lacks the antibodies against both factors. Type O blood, on the other hand, lacks both factors A and B but contains the plasma antibodies against both factors (Table 14-1).

If the Rh factor is present on erythrocyte membranes, the blood group is termed as RH positive; if it is not present, the blood group is referred to as Rh negative. Normally, the blood plasma of Rh-negative blood does not have anti-Rh. However, if Rh-positive blood is introduced, the Rh factor acts as a foreign protein and stimulates the formation of anti-Rh antibodies.

Table 14-1. Blood Groups Related to Erythrocyte Antigens and Plasma Antibodies

Blood Group	Antigen on Membranes	Antibodies in Plasma	Reaction[a] of Antigens with A	B	AB	O
A	A	anti-B (b)	−	+	+	−
B	B	anti-A (a)	+	−	+	−
AB	A, B	none	−	−	−	−
O	none	anti-A, anti-B	+	+	+	−

[a] + = agglutination; − = absence of agglutination.

Leukocytes, or white blood cells, differ from erythrocytes in color, structure, number, and function.

Functionally, leukocytes are responsible for the defense of the body against pathogenic organisms and the promotion of tissue repair and regeneration. Some have been shown to become the phagocytic cells of loose connective tissue, the liver, and spleen. Leukocytes may ingest bacteria by phagocytosis or may be involved in the production of antibodies. They are able to synthesize growth-promoting substances directly from raw materials present in the blood and transfer these substances to the cells of epithelial and connective tissues for repair and regeneration. In addition, among the formed elements in blood, leukocytes are unique in having the ability to leave the vessels and to migrate toward microorganisms or other particles that may have invaded the tissues. This migration is primarily by *amoeboid movement,* which enables the leukocytes to move between the cells of the capillary walls and through tissue spaces in order to reach an infected area. Associated with the leukocyte migration is the loss of some blood plasma and some erythrocytes that are forced through the capillary walls. Upon infection by microorganisms, inflammation occurs; the symptoms of redness, heat, swelling, and pain are caused by irritation resulting from the toxins of the bacteria, the increased blood supply, and the collection of fluid in the tissues. When such conditions are present, there is a struggle between leukocytes and bacteria. If the leukocytes are successful, the bacteria are killed and all debris of the battle is removed. However, if the bacteria are successful, large numbers of leukocytes and other tissue cells will be destroyed, resulting in the formation of pus. *Pus* consists of dead and living bacteria, leukocytes, dead tissue cells, and material that exuded from the blood vessels, such as plasma and erythrocytes.

Platelets are small fragments of larger cells (*megakaryocytes*) that are produced in the red bone marrow. The platelets may stick to foreign bodies within blood vessels or to ragged tissue at the site of a wound. Adherence of platelets is believed to be the first step in the formation of blood clots.

Blood clotting is essential in stopping hemorrhage, thereby preventing excessive loss of body fluids. There are three basic steps in clot formation.

1. Formation of thromboplastin.
2. Conversion of prothrombin to thrombin.
3. Conversion of soluble fibrinogen to insoluble fibrin (Figure 14-6).

Thromboplastin may be produced by two mechanisms. In the *extrinsic scheme,* tissue thromboplastin is released by damaged cells. In the *intrinsic scheme,* platelets exposed to air or rough surfaces, such as the conditions accompanying a wound, disintegrate. The

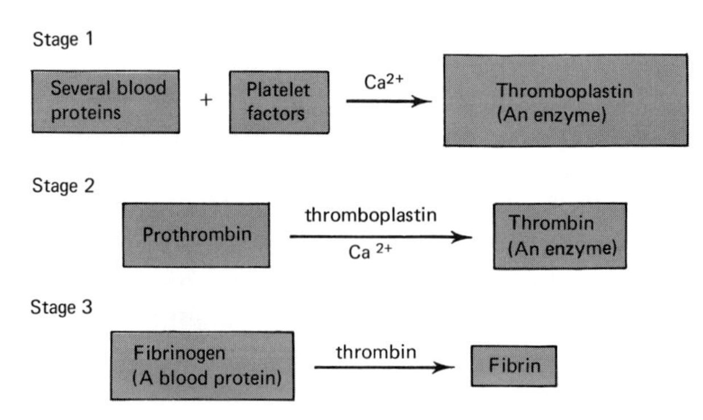

Figure 14-6. Sequence of events involved in blood clotting. Each of the stages is discussed in the text.

disintegration of the platelets releases granules which react with several blood proteins and calcium ions to form thromboplastin. Thromboplastin combines with *prothrombin* (a complex globulin protein in blood), calcium ions, and other blood proteins to form *thrombin.* Formation of prothrombin takes place in liver cells in the presence of vitamin K. *Vitamin K* is usually obtained in a balanced diet and in addition, certain bacteria in the intestine such as *Escherichia coli* are able to synthesize it. Vitamin K is a fat-soluble vitamin and bile is required for its absorption. If the bile ducts are obstructed or the liver is severely diseased and cannot produce sufficient bile, a patient cannot absorb enough vitamin K for normal prothrombin synthesis. Such a person would have a bleeding tendency and would require a vitamin K preparation as a preoperative safeguard.

Fibrinogen is another important blood protein that is soluble in the plasma. However, in the presence of thrombin, fibrinogen is converted to insoluble *fibrin* fibers. These fibers act as a mesh in which erythrocytes become entangled, forming a blood clot. The pale yellow fluid that is left after a clot is formed is *blood serum.*

Clotting may be hastened by injury to blood vessel walls, contact with a rough surface such as gauze placed on an open wound, venom of certain snakes, and rest. This final factor tends to prevent dislodgment of clots forming at open ends of blood vessels. Clotting may be hindered by calcium deficiency, deficiency or abnormality of platelets, vitamin K deficiency, fibrinogen deficiency, and introduction of anticoagulants such as *heparin,* which retards the conversion of prothrombin to thrombin, or of citrates or oxalates, which react with and bind calcium salts.

Plasma is the liquid portion of blood in which the various blood cells are suspended. Approximately 90% of plasma is water. This percentage is maintained by water intake and output by the kidneys, and exchanges of fluid between the blood, intercellular tissue fluid, and the cells. Various blood proteins are found in plasma, such as serum albumin, serum globulins, and fibrinogen. These proteins function in giving the blood colloidal characteristics, regulating blood volume, and serving as a source of nutrition for the body tissues. Moreover, they contribute to the transport of lipids, fat-soluble vitamins, bile salts, and hormones, and are related to osmotic pressure, which is essential for pulling water from tissue fluids into the blood vessels. In addition, prothrombin and heparin, previously discussed in blood clotting, are found dissolved in plasma.

Various end products of digestion, such as amino acids, glucose, and neutral fats, are present in plasma. An important lipid that is also present in plasma is *cholesterol.* Cholesterol functions as a precursor in the formation of hormones (for example, progesterone, testosterone, and adrenal cortical hormones), bile salts, and the corneum of the skin. It is also used in conjunction with other lipids in the skin to prevent evaporation of water.

Dissolved gases, such as oxygen, nitrogen, and carbon dioxide, are also present in the plasma. As noted earlier, oxygen and carbon dioxide may be carried by hemoglobin as oxyhemoglobin and carbaminohemoglobin, respectively. In addition, a small amount of carbon dioxide may be dissolved in the plasma. However, about two-thirds of the carbon dioxide carried by the plasma is in the form of bicarbonate ions. Some carbon dioxide diffuses into erythrocytes and combines with water to form carbonic acid. The bicarbonate ions diffuse into the bloodstream where they function as part of a buffer system that maintains the pH of the blood (see Chapter 2).

Finally, the plasma contains several substances that react with foreign organisms and their products and are known as antibodies. *Antibodies* are gamma globulins or a type of plasma protein that is formed by plasma cells in the lymph nodes, liver, spleen, bone marrow, and the lymphoid tissue of the gastrointestinal tract.

Heart. The *heart* is a muscular, four-chambered organ located in the thorax between the lungs and above the central depression of the diaphragm. It is somewhat cone-shaped with the broader end (base) upward and the blunt point (apex) lying on the diaphragm oriented toward the left (Figure 14-7A).

Covering the heart is a loose sac called the pericardium (Figure 14-7C). The *pericardium* is composed of two layers: a fibrous portion and a serous portion. The fibrous portion is composed of white fibrous tissue and is lined with a moist serous membrane. Between the serous portions of the pericardial cavity is a small quantity of pericardial fluid that prevents friction as the surfaces continually slide over each other with the

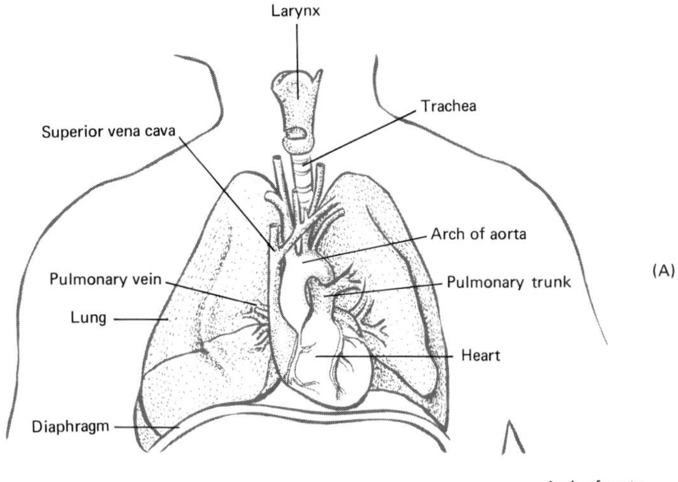

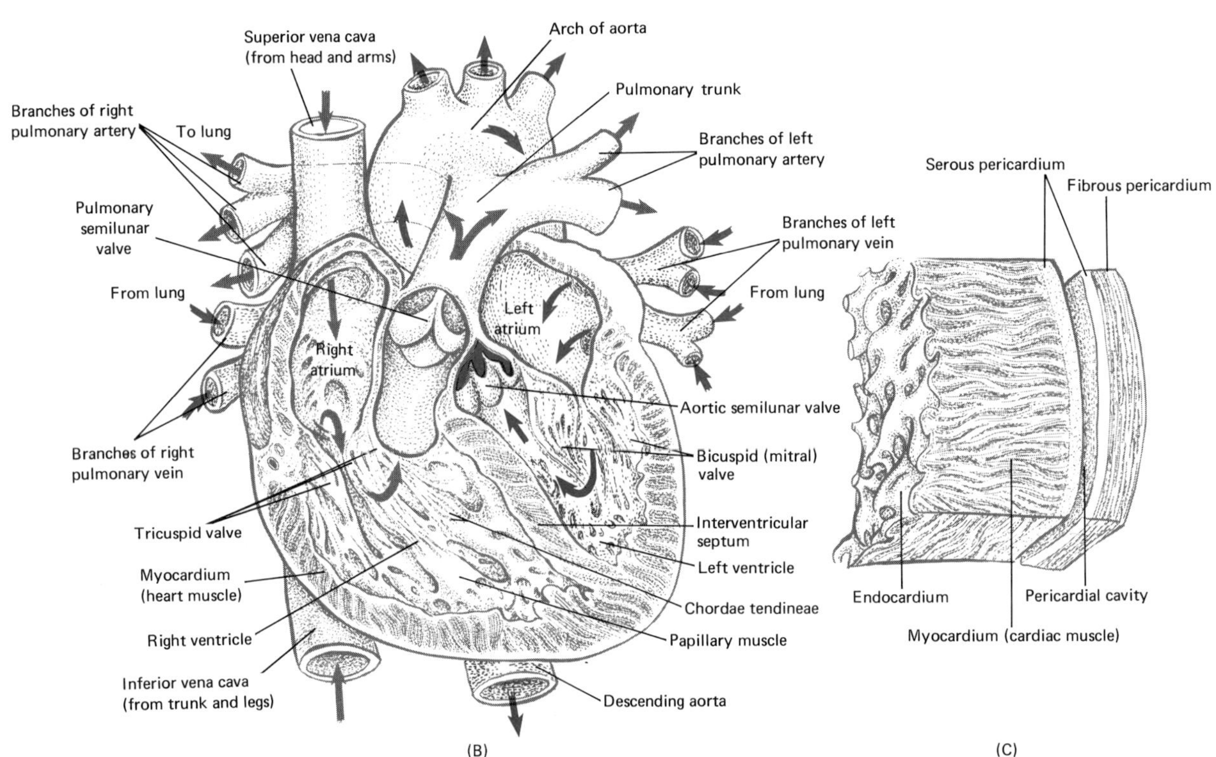

Figure 14-7. The human heart. (**A**) The position of the heart in relation to other thoracic organs. (**B**) Internal anatomy of the heart. Colored arrows indicate the path of oxygenated blood; gray arrows indicate the path of deoxygenated blood. (**C**) Enlarged aspect of a portion of the heart showing the structure of the pericardium, myocardium, and endocardium.

constant beating of the heart. The outer surface of the heart is covered by a serous membrane like that lining the pericardium.

The major portion of the heart wall, the *myocardium,* consists of *cardiac muscle*. This tissue forms the muscle bundles of the four chambers (two atria and two ventricles). Because the amount of work done by the ventricles in pumping blood is greater than that of the atria, the ventricular walls are thicker than those of the atria. In addition, the blood pumped into the systemic circulation by the left ventricle is under higher pressure than the blood pumped into the circulatory system by the right ventricle. Therefore, the left ventricle is more muscular than the right ventricle.

Lining the myocardium is a thin membrane composed of endothelial and connective tissue called the *endocardium.* This membrane covers the valves, surrounds the chordae tendinae, and is continuous with the linings of the large blood vessels. The endocardium contains many blood vessels, some bundles of smooth muscle, and parts of the conducting system of the heart.

The heart is divided into two sides (right and left) by two muscular partitions. The *interventricular septum* extends from the base of the ventricles to the apex of the heart (see Figure 14-7B), and the thinner *interatrial septum* is between the atria. Each side, in turn, is subdivided into two cavities: an upper atrium and a lower ventricle. The *atria* are thin-walled sac-like structures. They receive blood from the systemic and pulmonary circulations and pass it to the ventricles as soon as the ventricles relax after a contraction. On the other hand, the *ventricles* are thick-walled chambers that function in maintaining blood flow within the vascular network of the systemic and pulmonary circulations.

Valves that close automatically to prevent the backward flow of blood ensure one-way flow of blood through the heart. Between the right atrium and right ventricle is a valve with three flaps or cusps called the *tricuspid valve.* The cusps are anchored to the *papillary muscles* of the ventricle by the *chordae tendineae.* Similarly, the left atrium and left ventricle are separated by a cuspid valve with two flaps, the *bicuspid* (*mitral*) *valve.* Both cuspid valves are constructed so that blood may flow only from the atria into the ventricles and not from the ventricles into the atria. The cuspid valves remain open as long as the blood pressure is higher in the atria than in the ventricles. During ventricular contraction, blood pressure within the ventricles rises, the valves close and are prevented from everting into the atria by the chordae tendineae.

Separating the right ventricle and the pulmonary trunk is the *pulmonary valve,* a *semilunar valve* consisting of three half-moon-shaped cusps attached by their convex margins to the inside of the pulmonary artery. The left ventricle is separated from the aorta by a similar valve, the *aortic valve.* Both valves allow blood to pass freely from the ventricles into the arteries since their free borders project into the arteries. However, they normally form a complete barrier to the passage of blood in the opposite direction.

All of the phenomena associated with the contractions of the atria and ventricles, together with subsequent relaxation of the chambers, comprise the *cardiac cycle.* Both atria contract nearly simultaneously; then, as they relax, the ventricles contract and relax. During part of the ventricular contraction, the atria remain relaxed and then start the cycle over again.

Atrial contraction pumps blood from the atria into the ventricles. During this contraction, the cuspid valves are open, the ventricles are relaxed, and the semilunar valves are closed.

During part of ventricular contraction, the atria are relaxed and are receiving blood from the venae cavae and the pulmonary veins. The cuspid valves are closed, and the contraction forces blood from the ventricles through the open semilunar valves into the pulmonary artery and the aorta.

Evolutionary Development. The circulatory systems of all vertebrates from fish through humans are basically similar. In the evolution of the higher vertebrates from lower, fish-like forms, the major changes in the circulatory system occurred in the heart and are correlated with the changes in respiration from gills to lungs.

The heart of a fish (Figure 14-8A) consists of two chambers and associated arteries and veins (*sinus venosus, atrium, ventricle,* and *conus arteriosus*) arranged in a linear sequence. Deoxygenated blood from the veins drains into the sinus venosus. From here it is carried into the atrium, then to the ventricle and pumped out the conus arteriosus and into a large artery, the *ventral aorta.* The deoxygenated blood is then passed to the gills through five or six aortic arches. When the CO_2 is removed and the O_2 is added, blood leaves the gills and passes into the *dorsal aorta,* which distributes the oxygenated blood through its various branches to all parts of the body. Deoxygenated blood is eventually returned to the heart via the veins. Blood passes through the fish heart only once each time it makes a circuit through the

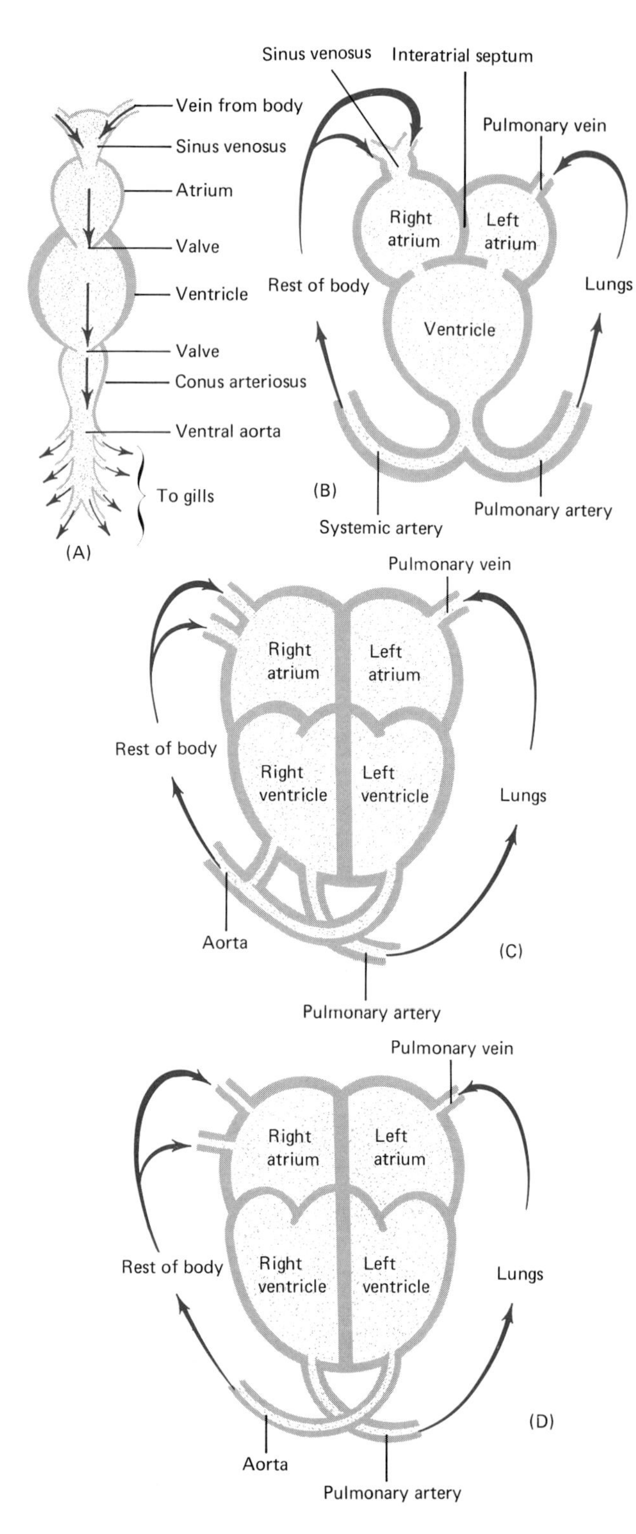

body. As contrasted with the double circulation (systemic and pulmonary) of higher vertebrates, the single circulation in fishes is less efficient.

The evolution of vertebrates to the land was accompanied by a change in the site of external respiration from gills to lungs. The heart of a frog, a representative amphibian, illustrates the partial conversion from single circulation to the beginning of separate systemic and pulmonary circulations (Figure 14-8B). In the frog heart, there are two *atria,* separated by a thin *interatrial septum,* which are anterior and dorsal to the single *ventricle.* The sinus venosus has shifted its position and connection so that it empties into the right atrium. Blood pumped by the ventricle is passed into the main artery which branches into the *pulmonary arteries* leading to the two lungs and other arteries which carry blood to other parts of the body (systemic circulation). Blood delivered via the pulmonary arteries is oxygenated in the lungs and is returned to the left atrium by the *pulmonary veins.* Deoxygenated blood from the remainder of the body is eventually carried into the right atrium. Since deoxygenated blood from the right atrium and oxygenated blood from the left atrium enter the common ventricle, there is some mixing of blood so that some deoxygenated blood may get into the aorta (instead of the pulmonary artery) and some oxygenated blood may be pumped into the pulmonary artery. The degree of mixing, however, is minimal since, at ventricular contraction, the deoxygenated blood is forced out first and tends to flow into the pulmonary arteries. The oxygenated blood leaves the heart last and tends to flow into the systemic arteries because the pulmonary arteries are already filled with deoxygenated blood.

In the evolution of reptiles from ancestral amphibians, a partition developed down the center of the ventricle. In lower reptiles (lizards), the ventricular partition is incomplete so that there is still some mixing of oxygenated and deoxygenated blood, although less than in the frog. In more advanced reptiles (alligators and

Figure 14-8. Comparative structural features of representative vertebrate hearts illustrating evolutionary changes that have occurred. (**A**) Fish heart. (**B**) Amphibian (frog) heart. (**C**) Advanced reptilian (crocodile) heart. (**D**) Bird (or mammalian) heart.

crocodiles), the probable descendants of a primitive ancestor that gave rise to birds, the ventricle is completely partitioned so that a four-chambered heart is formed (Figure 14-8C). In addition, the sinus venosus is reduced, foreshadowing its disappearance in the mammalian heart. Because of the anatomy of the three arteries leaving the heart of advanced reptiles, there is still some mixture of oxygenated and deoxygenated blood in the dorsal aorta.

The hearts of birds and mammals (Figure 14-8D) show the complete separation of right and left sides. The conus has split and become the base of the aorta and the pulmonary artery and the sinus venosus has disappeared as a separate chamber. The total separation of the right and left hearts forces the blood to pass through the heart twice each time it makes a circuit through the body. As a result, blood in the aorta of birds and mammals contains more oxygen than in the aorta of lower vertebrates. Since the tissues of the body receive more oxygen, a higher metabolic rate can be maintained, and the warm-blooded condition is possible. The warm-blooded condition would be impossible in other vertebrates (fish, frogs, and reptiles) because their blood cannot deliver enough oxygen to the tissues in order to maintain the high rate of metabolism necessary to sustain a high body temperature in cold environments.

Blood Vessels. There are three types of blood vessels: arteries, capillaries, and veins. *Arteries* are vessels that carry blood away from the heart to various body organs where the blood enters the capillaries. *Capillaries* are microscopic vessels that ultimately unite arteries to veins and that serve as the sites where the blood does all of its work for the body. Obviously, the large size of arteries and the small size of capillaries does not permit a direct union between them. Rather, arteries branch considerably into progressively smaller vessels, the *arterioles,* and these in turn unite with the capillaries. Arterioles function in regulating the flow and pressure of blood and redirecting blood to active organs. Similarly, the relative sizes of capillaries and veins make direct union impossible. Smaller branches of veins which join to capillaries are termed *venules. Veins* are vessels that collect blood from the capillaries and return it to the heart.

Three layers compose the wall of an artery: an inner coat consisting of two or three layers, a middle coat, and an outer coat. The inner coat (*tunica intima*) is composed of a layer of endothelial cells that lines the artery, a layer of connective tissue (found only in large vessels), and a layer of elastic fibers. Smooth muscle fibers or elastic connective tissue make up the bulk of the middle coat (*tunica media*). The outer coat (*tunica adventitia*) is constructed of loose connective tissue in which scattered smooth muscle cells or bundles of cells are arranged longitudinally. Considering the structure of the middle coat, the arteries are both extensile and elastic which contributes to the proper functioning of the arteries. Extensibility of the arteries permits them to receive the additional amount of blood which is forced into them with each contraction of the ventricles. Elasticity aids as a buffer with respect to the large volume of blood forced into the arteries with each heartbeat, thus helping in the regulation of systolic blood pressure. Functionally, arteries distribute blood to all body parts from the heart, eventually emptying into the capillaries.

Capillary walls consist of a single layer of endothelial cells which is continuous with the endothelium lining the arteries, veins, and heart. Functionally, the capillaries are the most important vessels in that the actual delivery and collection of the materials transported by the blood (O_2, CO_2, food, hormones, etc.) occur within these microscopic vessels.

Veins function in collecting blood from the capillaries and returning it to the heart. Structurally, veins are similar to arteries in that there are three coats: an endothelial lining, a middle layer of muscular or connective tissue, and an outer layer of areolar connective tissue. Veins differ from arteries in the development of a thinner middle coat, less elasticity, the presence of valves in many veins, and the tendency to collapse when not filled with blood due to the thinner walls. The valves are semilunar folds of the endothelium and characteristically consist of two flaps. Their function is to maintain the flow of blood toward the heart and to prevent reflux (back flow).

Circulatory Routes. Blood returning to the heart enters the right atrium through the superior and inferior venae cavae. The *superior vena cava* drains blood from the upper extremities and the head, whereas the *inferior vena cava* returns blood from the lower extremities and the trunk. This blood is poorly oxygenated since it has passed through all parts of the body and has brought oxygen to the body cells, picking up carbon dioxide in return. The right atrium passes this blood into the right ventricle from which point the blood will be forced into the *pulmonary artery,* thereby beginning the *pulmonary circulation* (see Figure 14-7B).

As the pulmonary artery leaves the right ventricle, it passes upward, backward, and toward the left where it divides into the *right* and *left pulmonary arteries* leading to the right and left lungs, respectively. Before entering the lungs, each of these arteries divides into branches to accommodate the several lobes of the lung. (The left lung has two lobes, whereas the right lung has three.) These branches, in turn, divide, finally subdividing into capillaries, which form a network within the walls of the air sacs (alveoli) where the exchange of O_2 and CO_2 occurs. The capillaries unite, forming larger vessels, the venules, which merge into the veins. These veins connect to form the *pulmonary veins* leading to the left atrium of the heart.

Upon its return to the heart from the lungs, the blood is passed from the left atrium to the left ventricle, which, upon contraction, forces the blood into the *aorta.* This is the beginning of the systemic circulation.

The *systemic circulation* (Figure 14-9) consists of

1. The aorta and its branches (all arteries and arterioles)
2. Capillaries connecting arterioles and venules.
3. All venules and veins that empty into the right atrium as well as the coronary veins that empty directly into the heart.

As the aorta leaves the left ventricle, it is known as the ascending aorta and is contained within the pericardium. There are two small branches of the ascending aorta, the *right* and *left coronary arteries,* which bring blood to the heart tissues. The aorta continues as the *aortic arch.* Branches of the arch supply the head and upper extremities. The arch is continuous with the descending aorta, which extends downward to the lumbar region of the abdomen. There are two regions of the descending aorta, the thoracic aorta and the abdominal aorta. Branches of the thoracic aorta supply the body wall of the chest cavity and the organs contained in the cavity such as the esophagus, lungs, and pericardium. Branches of the abdominal aorta supply the diaphragm, stomach, pancreas, spleen, liver, small intestine, large intestine, kidneys, gonads, and muscles and skin of the back.

At its lower end, the descending aorta divides into two branches, the *common iliac arteries.* These in turn divide into the *internal* and *external iliac arteries.* Branches of the internal iliac arteries supply blood to the pelvic walls, viscera, external genitalia, buttocks, medial side of the thigh, and, in the female, the uterus. The external iliacs enter the thighs and become the *femoral arteries.* There are small branches of the external iliacs to lymph nodes, abdominal muscles, and peritoneum. Branches from the femoral artery extend to abdominal muscles, muscles of the thigh, and external genitalia. A continuation of the femoral artery is the *popliteal artery* located at the back of the knee where it sends branches to posterior thigh muscles, the knee joint, and the skin and muscles of the lower leg. Below the knee, it divides into the *posterior* and *anterior tibial arteries* and *peroneal arteries* supplying the calf of the leg, the tibia and fibula, and the foot.

Systemic veins are divided into three major groups: (1) those that empty into the heart, (2) those that empty into the superior vena cava, and (3) those that empty into the inferior vena cava. Blood from the heart tissue returns directly to the heart via the *coronary veins* and *coronary sinus.* Blood returning from the head and face flows into two principal veins on each side of the neck, the *external* and *internal jugular veins.* These veins join the *right* and *left subclavian veins* that bring blood from the upper extremities and transfer the blood to the superior vena cava. Blood from the lower extremities returns via the *great saphenous* and *femoral veins,* which continue to form the *external iliac veins.* Joining with the *internal iliac veins,* the external iliacs form the *common iliacs,* which, in turn, unite to form the *inferior vena cava.* As the inferior vena cava ascends to the right atrium of the heart, it receives veins from the lumbar region, kidneys, diaphragm, liver, and gonads.

Veins returning from the spleen, stomach, pancreas, and intestines are included in the *portal circulation* (Figure 14-10A). Veins from these organs unite to form the *portal vein,* which passes upward to the liver where it divides into two branches before entering the right and left lobes of the liver. Within the liver, the branches of the portal vein form the capillary network of the liver

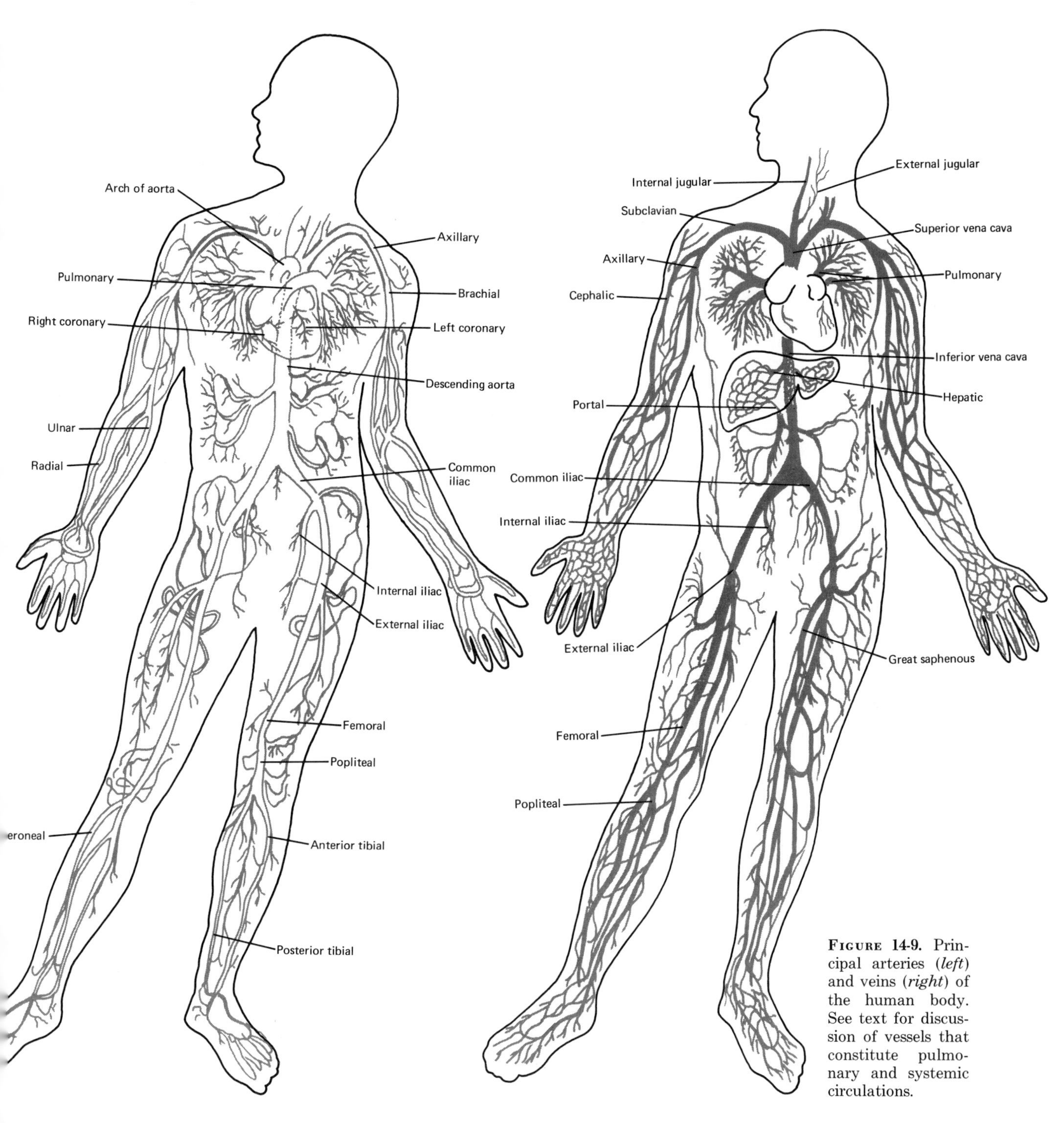

FIGURE 14-9. Principal arteries (*left*) and veins (*right*) of the human body. See text for discussion of vessels that constitute pulmonary and systemic circulations.

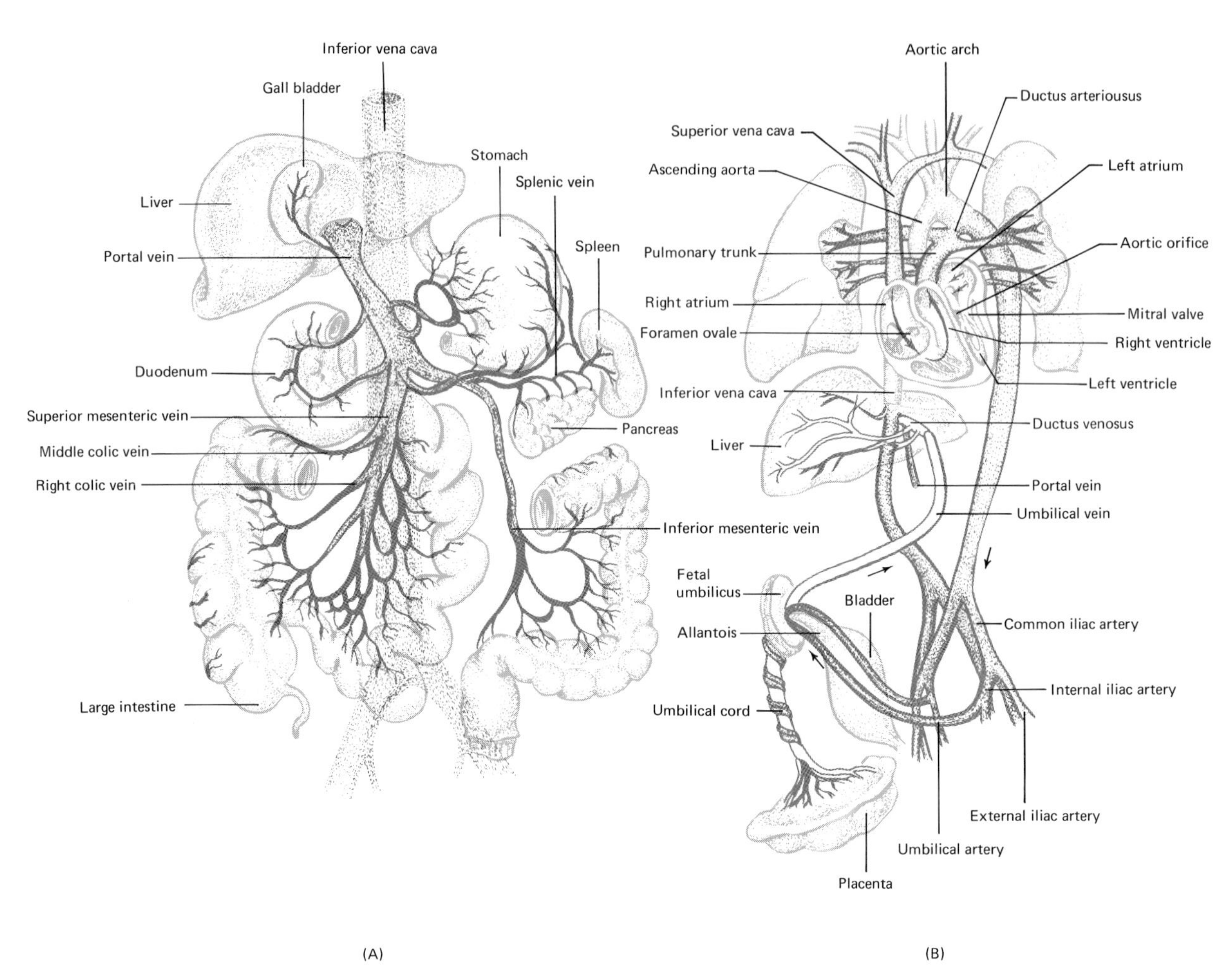

FIGURE 14-10. Portal (**A**) and fetal (**B**) circulation. Consult text for discussion.

referred to as *sinusoids*. Eventually, the capillaries unite and form the *hepatic vein,* which empties into the inferior vena cava. Thus, the liver receives blood from the hepatic artery and from the portal vein. The hepatic artery brings oxygenated blood whereas the portal vein carries blood with dissolved nutrients that have been absorbed from the digestive tract.

Prior to birth, the body also requires circulation. However, *prenatal* (*fetal*) *circulation* differs from circulation after birth for one reason—fetal blood receives oxygen and dissolved nutrients from maternal blood and not from its own lungs or digestive organs (Figure 14-10B). To accommodate this difference, there are additional blood vessels in the fetus to bring fetal blood close to maternal blood in order to effect exchange of materials. These structures are the two *umbilical arteries,* the *umbilical vein,* and the *ductus venosus.* In addition, there is a structure which functions as the site of exchange between the fetal and maternal blood known as the *placenta.* There are three structures within the fetal body which also play an important role in fetal circulation. One, the ductus venosus previously mentioned, provides a detour bypassing the liver as blood returns from the placenta. The other two, the *foramen ovale* and *ductus arteriosus,* serve as detours bypassing the lungs.

Essentially, the *umbilical arteries* are extensions of the internal iliac arteries and function in bringing fetal blood to the placenta. The *placenta* is a temporary structure formed during pregnancy which contains capillary nets of both fetal and maternal circulations and is attached to the uterine wall. It is within the placenta that exchange of oxygen and other substances between maternal and fetal blood occurs. The *umbilical vein* returns oxygenated blood to the fetus via the umbilicus. It sends some branches to the liver and then continues as the ductus venosus. The umbilical arteries and the umbilical vein constitute the *umbilical cord,* which is shed at birth along with the placenta.

There is an opening in the septum between the right and left atria called the *foramen ovale.* Most of the blood entering the right atrium is directed through the foramen ovale into the left atrium, thus bypassing the lungs. The small amount of blood that enters the right ventricle and the pulmonary artery is detoured around the lungs by the *ductus arteriosus,* which connects the pulmonary artery with the descending aorta.

Physiology of Circulation. Essentially, circulation refers to a blood flow through vessels arranged in such a manner as to describe a circuit or circular pattern. Maintenance of this blood flow is due to pressure gradients within the vascular system, which obey Newton's first and second laws of motion. Briefly, these laws state that

1. A fluid does not flow when the pressure is the same in all parts of it.
2. Fluids only flow when a higher pressure exists in one area than in another, that is, flow is from the area of higher pressure to the area of lower pressure.

A blood pressure gradient is simply a difference in blood pressure in two areas. For example, if blood pressure in an artery averages 100 mm Hg and pressure in a vein is 5 mm Hg, then a blood pressure gradient of 95 mm Hg exists between the two areas. (Pressure is expressed as mm Hg.)

Blood pressure in the arteries is determined primarily by the force and rate of heart contraction (cardiac output) and resistance to flow in the vessels. An increase in arterial blood volume would also increase the arterial blood pressure. Many factors contribute to determining arterial blood pressure, but we will concentrate on cardiac output and peripheral resistance.

Cardiac output refers to the volume of blood pumped by one ventricle into the arteries per minute. This volume is dependent upon the number of contractions per minute and the volume of blood ejected per contraction. Contraction of the heart is called *systole;* therefore, the volume of blood pumped by one contraction is referred to as *systolic discharge,* or *stroke volume.* Essentially, stroke volume is a relative measure of the force of ventricular contraction. In considering these relationships, then, cardiac output (CO) may be represented as follows:

$$\underset{(\text{ml/min})}{\text{CO}} = \underset{(\text{ml/beat})}{\text{stroke volume}} \times \underset{(\text{beats/min})}{\text{heart rate}}$$

Note from the equation that in general any factor that makes the heart beat stronger or faster may cause an increase in cardiac output and, therefore, an increase in blood pressure. Conversely, any decrease in heart force or rate may bring about a decrease in cardiac output and, therefore, a decrease in blood pressure.

Peripheral resistance, a second factor that determines blood pressure, is defined as ease of flow through small vessels branching from the aorta and as resistance to the flow of blood as a result of friction between layers of blood and the vessels through which it flows. Such a friction develops because of the viscosity (thickness) between the particles of blood and because of the volume of blood that flows from arteries into smaller arterioles and capillaries.

Blood pressure may be defined as the pressure exerted by the blood against the vessel walls which contain it. This pressure is greatest in the arteries during ventricular contraction or *systole,* and is referred to as *systolic pressure.* During ventricular relaxation or *diastole,* blood pressure falls, reaching a minimum pressure just prior to the next systole. The minimum pressure is termed *diastolic pressure.* Measurement of blood pressure is made with an apparatus called a *sphygmomanometer* which measures the amount of pressure equal to the blood pressure in an artery in terms of the number of millimeters the pressure raises a column of mercury in a glass tube.

A sphygmomanometer consists of an air bag within a rubber cuff attached by rubber tubing to a glass tube containing mercury and to a small hand pump. The cuff is wrapped snugly around the arm just above the elbow over the brachial artery and air is pumped into the bag. Thus, pressure is exerted against the outside of the artery. A stethoscope placed over the artery at the bend of the elbow makes it possible for the pulse to be heard. More air is added until all sounds cease. At this point, the air pressure exceeds the arterial pressure and compresses the artery. The bag is slowly deflated until the pulse can just be heard. The manometer reading at this time indicates *systolic pressure.* Deflation of the bag is continued, and the manometer reading just before the last sound of the disappearing pulse indicates *diastolic pressure,* the lowest arterial pressure caused by ventricular diastole.

Average arterial pressure during systole is about 120 mm Hg in an adult, while the average diastolic pressure is about 80 mm Hg. This is usually expressed as a blood pressure of $^{120}/_{80}$, the upper number representing systolic pressure and the lower number diastolic pressure. Considering these figures, it is obvious that there is a fluctuation in blood pressure during each heartbeat. The difference between these two pressures, 40 mm Hg in this example, is termed *pulse pressure.*

The shock wave transmitted through the column of blood in an artery and which causes alternate expansion and contraction of an artery constitutes the *pulse.* A finger placed over an artery that approaches the surface of the body and is located over a bone senses resistance which appears to be increased at intervals corresponding to the heartbeat. The pulse represents the pressure change brought about by the ejection of blood from the heart into the already full aorta and propagated as a wave through the blood column and arterial wall to the periphery of the body. There are two factors which contribute to the maintenance of a pulse: (1) intermittent ejections of blood from the aorta causing an alternately higher and lower pressure within the vessel, and (2) the elasticity of the arterial walls which enables them to expand with each injection of blood and then recoil. As the pulse is an indication of the frequency of the heartbeat, it follows that factors influencing heartbeat will also influence the pulse.

Lymphatic System. As blood passes through the capillaries of the circulatory system, a certain amount of plasma filters from the blood through the capillary walls into the surrounding tissue where it becomes known as *interstitial fluid* (Figure 14-11A). Most of this fluid is returned to the blood. However, some of it continues to flow over the body cells, eventually passing into blind-end vessels called *lymphatic capillaries.* Within these capillaries the fluid is termed *lymph* and is composed of interstitial fluid containing a variable number of leukocytes, carbon dioxide, lipids, glucose, salts, proteins, enzymes, and antibodies; there is very little oxygen and no platelets and erythrocytes are present. In considering these constituents, lymph may be compared to the plasma contained within the vessels of the circulatory system, with differences in cellular contents and protein concentration (lower in lymph).

Lymph capillaries function in maintaining volume and pressure conditions in the interstitial spaces so that fluids may be removed and the blood proteins dissolved in them returned to the bloodstream. Just as blood capillaries join to form veins, the lymph capillaries unite to form larger vessels that finally converge to form two main channels, the thoracic duct and the right lymphatic duct.

The *thoracic duct* drains the lower body, upper left trunk, left upper extremity, and left side of the head and neck. It enters the left brachiocephalic vein at the junction of the left internal jugular and left subclavian veins (Figure 14-11B). Lymphatics from the right side of the head, neck, right arm, and right upper part of the trunk enter the *right lymphatic duct,* which empties into the right brachiocephalic vein at the junction of the right internal jugular and right subclavian veins. Both ducts are guarded at their terminal ends by valves to prevent the passage of venous blood into the ducts.

Lymph vessels resemble veins in structure. Small vessels consist of a single layer of endothelial cells, whereas larger vessels have three coats similar to those of veins except that they are thinner and more transparent. Lymphatics differ from veins in that they contain more valves and contain *nodes* located at intervals along their course.

Lymph nodes are ovoid structures with an outer covering of connective tissue that sends fibrous partitions or bands into the substance of the node, thus dividing it into irregular, interconnected spaces. Within these spaces are masses of lymphoid tissue with narrow spaces (sinuses) between them and the fibers of the covering. These spaces or channels allow the passage of lymph

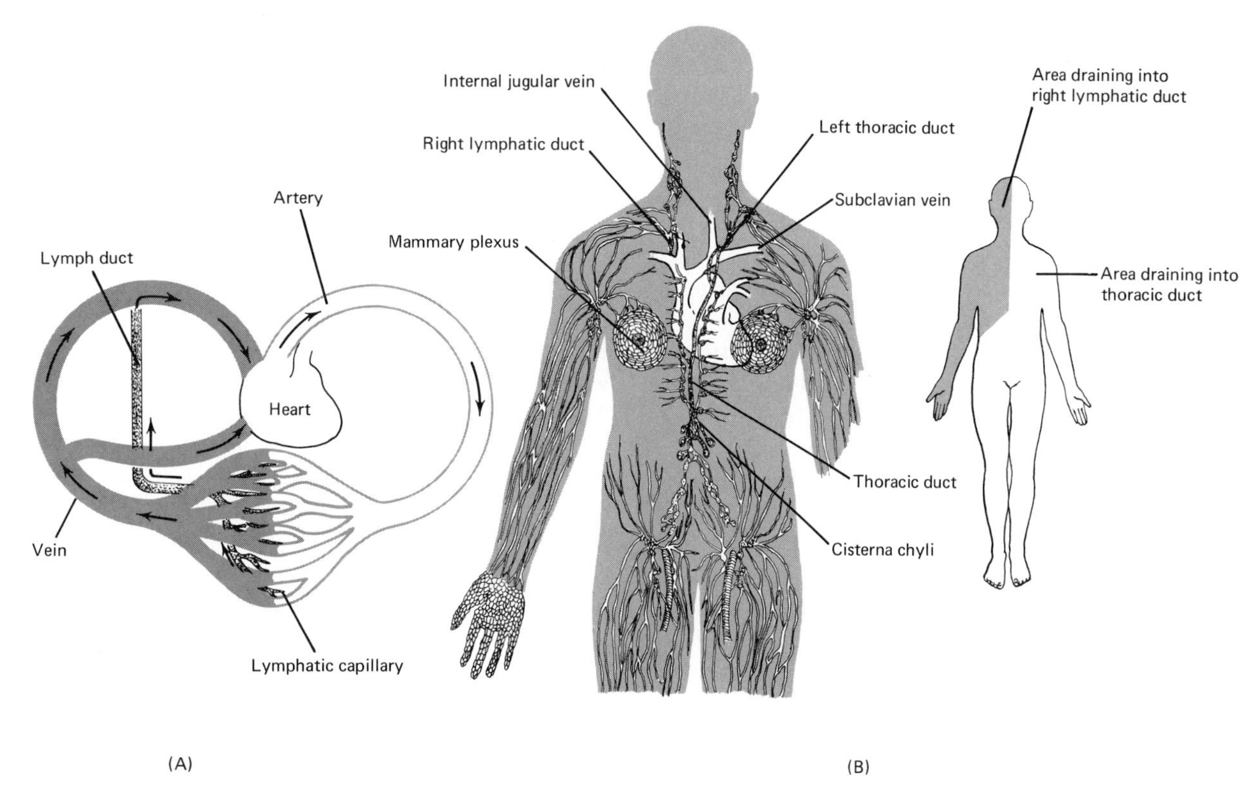

FIGURE 14-11. Human lymphatic system. (A) Relationship of the lymphatic system to the circulatory system. (B) *Left:* Location of right lymphatic duct, thoracic duct, and lymph nodes. *Right:* Regions from which lymph flows into the right lymphatic duct and the thoracic duct.

through the node. Lymph nodes are usually located in groups along the sides of great blood vessels. Other nodes are found on the back of the face and neck, at the underside of the rami of the mandible, around the sternocleidomastoid muscle, in the upper extremities at the elbow and axillary spaces, in the groin and abdomen, and in the lower extremities, such as the back of the knee (Figure 14-11B). Cells of the nodes include a population of fixed cells (macrophages) that can phagocytose microorganisms and cellular debris, develop into monocytes and lymphocytes, and develop directly into plasma cells that release antibodies. Lymph nodes may also trap cancer cells that have spread (metastasized) from an original site of malignancy.

Absorption of Digested Nutrients

The secretory and motor activities of the gastrointestinal tract are directed toward changing ingested food into substances suitable for absorption from the alimentary canal into the bloodstream. This absorption consists of the transfer of materials across cell membranes and involves such processes as diffusion, osmosis, hydrostatic pressure, and active transport. Particle size and concentration of the materials are important physical factors that influence these processes.

Among the conditions determining the amount of absorption that takes place within the alimentary canal are the absorptive surface area, length of time food remains in contact with the absorbing surface, concentration of fully digested material present, and rapidity with which absorbed nutrients are carried away by the blood.

These conditions are most favorable for absorption in the small intestine; therefore, the greatest amount of absorption takes place within this area. It has already been noted in Chapter 13 that the absorptive surface area of the small intestine is greatly increased by circular folds and villi (see Figure 13-13). Food remains within the small intestine for several hours during which time the most complete digestive changes occur.

Capillaries from the superior mesenteric artery bring blood to the wall of the small intestine. This blood is separated from the digested nutrients by the capillary walls and the intestinal mucosa. The relative concentrations of sugars, glycerol, fatty acids, and amino acids are high within the intestine and readily pass into the bloodstream by active and passive processes. Constant digestion of foods, muscular actions of the intestinal walls, and the activities of the villi maintain the movement of materials within the small intestine, thereby keeping material in contact with the absorbing surface. This continuous activity also increases the circulation within the capillaries in the villi, thereby aiding in the transportation of already absorbed materials and in keeping the concentration of nutrients in the blood relatively low. Absorption, both active and passive, facilitates the passage of substances through the cell membranes until all digested materials have been absorbed.

Within each villus of the small intestine, there are blood capillaries surrounding a central lymph channel called a *lacteal.* Fats are absorbed primarily into the lacteals and are carried to the larger lymphatics, and eventually to the bloodstream via the thoracic duct. Products of carbohydrate and protein digestion, and probably some of the glycerol and fatty acids, are absorbed by the blood capillaries, carried to the portal vein and then on to the liver for processing.

After entering the liver, the portal vein divides into a number of branches leading to the lobes of the liver. These branches are composed of small vessels whose walls are incomplete so that blood is brought into direct contact with each liver cell. The sinusoids, which have already been mentioned in the portal circulation, eventually converge to empty the blood into veins that unite to form the hepatic vein that leads to the inferior vena cava.

We will now consider the meaning of an electrocardiogram and a few common disorders of the circulatory system.

The Electrocardiogram

The heart receives fibers from the autonomic nervous system, but the autonomic neurons only increase or decrease the time it takes to complete a cardiac cycle. The chambers can go on contracting and relaxing without any direct stimulus from the nervous system. This is because the heart has a type of built-in private nervous system called the **conduction system.** The conduction system is composed of specialized nodal tissues that generate and distribute the electrical impulses which stimulate the cardiac muscle fibers to contract. These tissues are the sinoatrial node, the atrioventricular node, the atrioventricular bundle, the bundle branches, and the Purkinje fibers. The cells of the conduction system develop during embryological life from certain cardiac muscle cells. These cells lose most of their ability to contract and become specialists in impulse formation and transmission.

A *node* is a compact mass of conducting cells. The *sinoatrial node,* known as the SA node or pacemaker, is located in the right atrium beneath the opening of the superior vena cava (Figure 14-12A). The SA node initiates each cardiac cycle and thereby sets the basic pace for the heart rate. This is why it is commonly called the pacemaker. However, the rate set by the SA node may be altered by nervous impulses from the autonomic nervous system or by certain chemicals such as thyroid hormone, epinephrine, or acetylcholine. Once an electrical impulse is initiated by the SA node, the impulse spreads out over both atria and causes them to contract. From here, the impulse passes to the *atrioventricular (AV) node,* located toward the bottom of the interatrial septum. From the AV node, a tract of conducting fibers called a *bundle* runs to the top of the interventricular septum and then down both sides of the septum. This is called the *atrioventricular bundle,* or *bundle of His.* The bundle of His distributes the charge over the medial surfaces of the ventricles by way of the *bundle branches.* Actual contraction of the ventricles is stimulated by the *Purkinje fibers.* The Purkinje fibers are individual conducting cells that emerge from the bundle branches and pass into the cells of the myocardium.

Impulse transmission through the conduction system

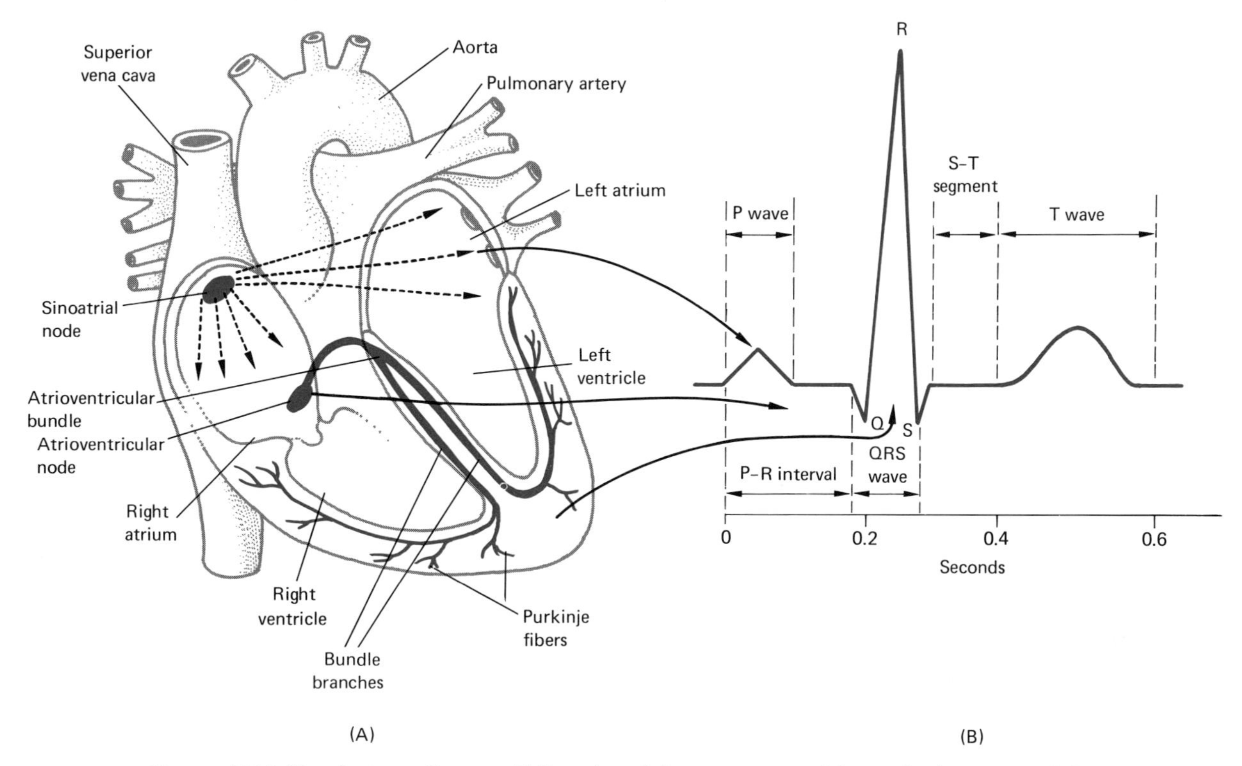

Figure 14-12. The electrocardiogram. (**A**) Location of the components of the conduction system of the heart. (**B**) Normal electrocardiogram.

generates electrical currents that may be detected on the surface of the body. A recording of the electrical changes that accompany the cardiac cycle is called an **electrocardiogram (ECG).** The instrument used to record the changes is an *electrocardiograph.*

Each portion of the cardiac cycle produces a different electrical impulse. These are transmitted from the electrodes to a recording needle that graphs the impulses as a series of up-and-down waves called *deflection waves.* In a typical record (Figure 14-12B), three clearly recognizable waves accompany each cardiac cycle. The first wave, called the *P wave,* is a small upward wave. It indicates the spread of an impulse from the SA node through the muscle of the two atria. A fraction of a second after the P wave begins, the atria contract. Following this, there is a complex called the *QRS wave.* It begins as a downward deflection, continues as a large, upright, triangular wave, and ends as a downward wave at its base. This deflection represents the spread of the electrical impulse through the ventricles. The third recognizable deflection is a dome-shaped wave called the *T wave.* This wave indicates ventricular repolarization and relaxation.

The ECG is invaluable in diagnosing abnormal cardiac rhythms and conduction patterns, detecting the presence of fetal life, determining multiple pregnancies, and following the course of recovery from a heart attack.

Human Circulatory Disorders

Diseases of the heart and blood vessels are the biggest single killers in the developed world. These diseases account for approximately 53% of all deaths. A recent comparison indicates that cardiovascular disease kills more people than cancer, accidents, pneumonia, influenza, and diabetes combined. Some of the cardiovascular problems involved are arteriosclerosis, hypertension, and various heart disorders.

Arteriosclerosis

A hardening of the arteries is described by the term **arteriosclerosis.** One type of arteriosclerosis, called *atherosclerosis,* is responsible for the most important and prevalent of all clinical complications. In this type, the inner layer of the artery becomes thickened with soft fatty deposits, called *plaques.* The plaque looks like a pearly gray or yellow mound of tissue on the inside of the blood vessel wall. It usually consists of a core of lipid (mainly cholesterol) covered by a cap of fibrous (scar) tissue. As the plaques increase in size, they not only calcify, but they may also impede or cut off blood flow in affected arteries. This causes damage to the tissues supplied by these arteries. An additional danger is that the lipid core of the plaques may be washed into the bloodstream. There, it could become an embolus and obstruct small arteries and capillaries quite a distance away from the original site of formation. A third possibility is that the plaque will provide a roughened surface for clot formation.

Hypertension

Hypertension, or high blood pressure, is the commonest of the diseases affecting the heart and blood vessels. Statistics from a recent National Health Survey indicate that hypertension afflicts at least 17 million American adults and perhaps as many as 22 million.

Primary hypertension, or *essential hypertension,* is a persistently elevated blood pressure that cannot be attributed to any particular organic cause. Specifically, the diastolic pressure continually exceeds 95 mm Hg. Approximately 85% of all hypertension cases fit this definition. The other 15% is called *secondary hypertension.* Secondary hypertension is caused by disorders such as arteriosclerosis and kidney disease. Arteriosclerosis increases blood pressure by reducing the elasticity of the arterial walls and by narrowing the space through which the blood can flow. Both kidney diseases and obstruction of blood flow to the kidney may cause the kidney to release renin into the blood. This enzyme catalyzes the formation of angiotensin from a plasma protein. Angiotensin is a powerful blood-vessel constrictor. It is one of the most potent agents known for raising blood pressure.

Heart Disorders

It is estimated that one in every five persons who reaches age 60 will have a **heart attack.** And it is also estimated that one in every four persons between 30 and 60 has the potential to be striken. Heart disease is epidemic in this country, despite the fact that some of the causes can be foreseen and prevented.

The Framingham (Massachusetts) Heart Study, which began in 1950 and continues still, is the longest and most famous study ever made of the susceptibility of a community to heart disease. Approximately 13,000 people in the town have participated in the investigation by receiving examinations every two years since the study began. The results of this research indicate that people who develop combinations of certain risk factors eventually have heart attacks. These factors are high cholesterol blood level, high blood pressure, cigarette smoking, overweight, lack of exercise, and diabetes.

People demonstrating three, four, or more risk factors form an especially high-risk group. The incidence of serious heart attacks in this high-risk group is far greater than in groups that have no risk factor or only one. The people who are most apt to develop arteriosclerosis—who run the highest risk of all—have these three risk factors: high cholesterol, hypertension, and cigarette smoking. Other researchers list emotional stress as an important risk factor, but there is still controversy among medical people as to the relative importance of this factor. The risk factors make a person more susceptible to heart trouble because they strain the heart or increase the likelihood that its oxygen supply will be shut off at some time.

15 Microbes, Disease, and Immunity

GUIDE QUESTIONS

1 What is meant by the term microbiology? What are the six principal groups of microbes? How are these groups related to humans?
2 Distinguish between sterilization and disinfection. Enumerate some of the physical methods of sterilization, chemical disinfection procedures, and treatments used in antibiotic therapy.
3 Define coagulation and denaturation. What is meant by a chemical analogue? Give an example.
4 Categorize disinfectants and antibiotics by type and use. What is an antibiotic? What are some of the criteria employed in selecting an antibiotic?
5 Define an infection. How may infections be spread? What are four factors that determine whether a pathogen can establish infection? Discuss each factor.
6 Outline and describe the nonspecific mechanisms of defense of the human body.
7 Describe the mechanism and importance of the inflammatory response.
8 Define antigen and antibody. How do complete antigens and haptens differ? Describe the various constituents of microbes that are antigenic.
9 Describe the kinds of antibodies and the modes of action of each.
10 Compare the template and clonal-selection theories of antibody formation.
11 What is immunity? Distinguish between active and passive immunity.
12 Discuss the allergic reaction and its relationship to immunity.
13 Define transplantation. How are the body's immunological mechanisms related to rejection of transplants? Explain the role of the thymus gland in antibody production.
14 Discuss some of the studies related to the use of radiation and the use of drugs to overcome rejection.
15 What criteria are employed in classifying transplants?
16 Define and give examples for each of the following: isograft, allograft, and xenograft. Discuss implantation and indicate the reasons for the relative success of implanted structures.
17 What are some of the implications of transplantation? How are these problems being resolved?

AT THIS point we will examine the types of microbes and their relationship to disease, body defenses against disease, the nature of the antigen–antibody response as a mechanism of immunity, and the relationship between the immune response and transplantation. As you will see, many of these processes are closely related to the activities of the circulatory and lymphatic systems.

The World of Microbes

The term *microbiology* is typically employed to describe the study of living forms of microscopic or submicroscopic size. From an evolutionary viewpoint, **microorganisms** or **microbes** are primitive and are mainly but not entirely unicellular. The individual cells are very rarely large enough to be seen by the naked eye, even though masses of them may be macroscopic. Normally, it is necessary to employ special microbiological techniques when growing and studying microbes under laboratory conditions.

Types of Microbes

Although various schemes may be utilized in classification of microbes, for purposes of this textbook the following selected groupings will be recognized:

1. Protozoans.
2. Fungi.
3. Bacteria.
4. Pleuropneumonia-like organisms (PPLO).
5. Rickettsiae.
6. Viruses.

Inasmuch as there is considerable controversy about the placement of some of these groups into various kingdoms, discussion of them will be restricted to groupings only. Moreover, because of space limitations, not all groups will be discussed and some will be treated more superficially than others. Table 15-1 contains a listing of some of the more important characteristics and relationships of these groups.

The Control of Microbial Growth

In a very practical sense, it is extremely important in many situations to destroy microorganisms, inhibit their growth, or to remove them. Varieties of microbes differ considerably in their susceptibility to various antimicrobial agents. Much depends on whether the organisms are on the skin, within the body fluid, or in the air, water, food, sewage, or room dust. Thus, the practical effect of applying antimicrobial agents is greatly influenced by circumstances. It is possible, however, to outline some of the basic principles underlying commonly used methods for controlling and destroying microbes.

An important factor is the degree to which various kinds of microbes are inhibited or destroyed. **Sterilization** is a process by which an object or material is made free of living organisms of any kind. There are no degrees of sterilization (practically sterile, nearly sterile, etc.); either the object is sterile or it is not. **Disinfection,** or **asepsis,** is the process of treating an object, usually with a chemical, to remove or kill pathogenic microorganisms. Other degrees of inhibition or destruction will be cited as they are discussed. Basically, microbial growth may be inhibited or stopped by three principal means. These are physical methods of sterilization, disinfection by use of chemical agents, and antibiotic therapy.

In general, antimicrobial agents have different modes of action in terms of killing or inhibiting the growth of microorganisms. Although the exact mechanisms are not completely understood, microbial cells are affected in more than one way. For example, heat and many chemical disinfectants cause the cell protoplasms to *coagulate* (e.g., albumin of a hard-boiled egg). Since many enzymes (proteins) are suspended in the colloidal cytoplasm, coagulation inactivates or destroys these vital cellular catalysts. Alcohols, phenol, and formalin have

TABLE 15-1. **Some Defining Characteristics and Relations of Microbe Groups**

Group	Characteristics	Relations to Man
1. Protozoans	Unicellular; free-living and parasitic, definite organelles; widely distributed; may form cysts.	Cause amoebic dysentery, malaria, African sleeping sickness, various inflammations of genitals.
2. Fungi	Unicellular and multicellular; nonchlorophyllous; saprophytic; reproduce by spores.	Fermentation; cause various superficial and systemic mycoses; sources of some antibiotics.
3. Bacteria	Unicellular; autotrophic and heterotrophic; universally distributed; reproduce primarily by fission; procaryotic nuclei; form highly resistant spores.	Commercial and industrial processes; implicated in a variety of human diseases and important in agriculture and in the cycles of the elements.
4. PPLO	Smallest free-living forms; varied forms; similar to bacteria in chemistry, metabolism, and structure.	May be implicated in some human diseases.
5. Rickettsiae	Resemble rod-shaped or spherical bacteria; smaller than most bacteria; closely associated with arthropods; typically require living cells for growth and multiplication.	Cause typhus, spotted fever, Q fever, trench fever.
6. Viruses	Ultramicroscopic; proliferate only in a living host; vary in shape; consist of a protein coat and nucleic acid core.	Cause chicken pox, warts, smallpox, measles, mumps, hepatitis, influenza, polio, rabies, common colds, and possibly some cancers.

strong coagulant action. Other chemical substances, such as chlorine, iodine, and strong acids and alkalies, **denature** (bring about structural changes in proteins with a subsequent loss of biological properties) cellular components.

Another group of chemical agents is known to cause injury to cell membranes and cell walls of microbes. Some substances, such as ammonium compounds, which reduce surface tension, tend to affect selective permeability and enzyme functions. Such disruptions may cause useful metabolic compounds to exit, while toxic products gain entrance. Penicillin, an antibiotic, interferes with the synthesis of the cell wall.

Many enzymes and coenzymes in microorganisms have side chains terminating in sulfhydryl groups (—SH) and do not function unless these remain free and reduced. Some compounds, such as hydrogen peroxide and salts of heavy metals, have the ability to oxidize sulfhydryl groups, and thus cause a loss of enzyme activity. Finally, some chemical agents interfere with the reaction between an enzyme and its normal substrate. Such compounds, collectively referred to as chemical analogues, are similar in structure to the enzyme involved and are capable of combining with a site on the enzyme molecule, preventing its union with the substrate. The bacteriostatic action of the sulfonamide drugs is an example of chemical analogue action (Figure 15-1).

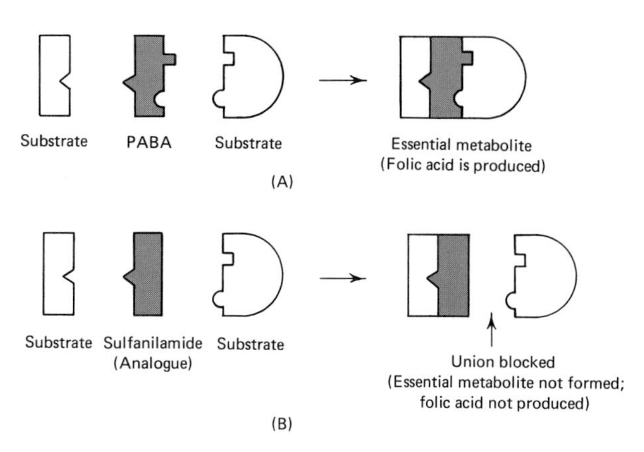

Figure 15-1. The sulfonamide drug sulfaniliamide is an analogue of PABA (*p*-aminobenzoic acid). (**A**) PABA combines with both substrates and forms the essential metabolite folic acid. (**B**) Sulfaniliamide combines with one substrate and blocks the formation of folic acid.

Sterilization by Physical Means. Among the practical methods of sterilization are flaming or incineration, radiation, dry heat, boiling, steam under pressure, and filtration.

Burning is a safe and economical means of ridding objects of microbes. Heating metal objects to red heat will render them sterile; other objects that may contain pathogens (swabs, sputum caps, dressings, etc.) are usually incinerated. Direct sunlight has a powerful germicidal action, a result of ultraviolet rays. Germicidal (ultraviolet) lamps are used extensively in operating rooms, bakeries, food processing and packing plants, hospital wards, and bacteriology laboratories. Since germicidal rays do not penetrate extensively, their effectiveness is confined to air or surface sterilization.

Any dry object that will not be damaged by high temperatures may be sterilized by heating in an oven. Dry heat, however, requires a relatively long period of time to bring about loss of viability. The usual laboratory practice is to heat objects at 150–180°C for 1–3 hours. Boiling, although effective in killing nonspore-forming bacteria in a short time, is relatively ineffective against bacterial spores. Some spores may resist the effectiveness of boiling (100°C) for as long as an hour or more.

Sterilization by steam under pressure is frequently carried out in hospitals and laboratories in an instrument called an *autoclave*. In the autoclave, sterilization is accomplished in an atmosphere of water vapor (steam) heated above the boiling point of water and under pressure. As the pressure increases, the boiling point of water also increases. In general, articles to be autoclaved are usually exposed to a pressure of 15 pounds (this gives a temperature of about 121°C) for 20–30 minutes.

Liquids or gases may be sterilized by permitting them to filter through an appropriate material. This material, which may be unglazed porcelain, asbestos, cellulose fibers, collodion, or diatomaceous earth, has pores small enough to hold back bacteria while permitting all or most of the other materials to pass through.

Disinfection by the Use of Chemical Agents. Many of the physical methods of destroying microbes are obviously unsuitable for sterilizing or sanitizing living tissues, either externally or internally. The chemical agents employed as disinfectants may appropriately be grouped into two principal categories: (1) compounds used freely to destroy microbes in bodily excretions and in the physical environment and (2) germicides (agents that kill all microbes regardless of their pathogenicity) used for direct local application to human tissues to prevent or treat infection.

The action of various chemical germicides and bacteriostatic agents is not completely understood; moreover, there is no ideal disinfectant since many factors influence effectiveness. Among these are the concentration at which the disinfectant is applied, how long it is applied, and the type of material acted upon. Table 15-2 contains a listing of some of the commonly used disinfectants. Some disinfectants are protoplasmic poisons, frequently toxic to living tissue and not especially selective in their action against microbes. Others block or poison enzyme systems or produce some other effect that may be reversible, at least within certain time limits. The net effect of these agents is primarily to inhibit growth. Actions of an irreversible nature are produced by a variety of chemical agents by denaturing cell proteins.

Antibiotic Therapy. In addition to the inhibition or destruction by physical methods of sterilization and chemical disinfectants, a third method is employed. This method involves the use of antibiotics. An **antibiotic** may be defined as a chemical substance derived from a living source that, in dilute solutions, has the capacity to inhibit the growth of or to destroy microorganisms. Although some antibiotics have been derived from plant and animal tissues, the vast majority are products of other microbes. Enormous numbers of antibiotics have been discovered in the last two decades, but only a relatively few have proved to be safe and useful in chemotherapy. At present there are only about twenty antibiotics in common use; others are less frequently employed.

The criteria for the selection of an antibiotic as a chemotherapeutic agent are

1. It must have low toxicity for host cells and tissues in doses effective against pathogens.

2. It should be effective against more than one specific organism.

Table 15-2. Types and Uses of Some Common Disinfectants

Type of Disinfectant	Uses
I. Chlorine–iodine group	
A. Chlorine (gas or liquid)	Disinfection of drinking water, swimming pools, sewage.
B. Chlorinated lime (calcium hypochlorite)	Water supplies, dairies, toilets, slaughterhouses, bedpans.
C. Tincture of iodine	Skin disinfection.
II. Carbolic acid group	
A. Carbolic acid (phenol)	Control for other disinfectants; (not used directly, has many undesirable properties).
B. Cresol (phenol derivative)	Hospital, laboratory, home disinfectants.
C. Biphenols	In soaps as skin antiseptics.
III. Mercury–silver group	
A. Bichloride of mercury	Disinfection of glassware, rubber articles, thermometers.
B. Mercurochrome	Skin disinfectant.
C. Merthiolate	Household disinfectant and skin disinfectant.
D. Silver nitrate	Prevent eye infections at birth.
IV. Alcohols and aldehydes	
A. Ethyl alcohol	Skin disinfectant; used as final step in "scrubbing up" before an operation.
B. Isopropyl (rubbing) alcohol	Same general uses as ethyl alcohol.
C. Formaldehyde	Disinfection of sputum, rubber gloves, excreta.
V. Dyes	
A. Gentian violet	Infected wounds, acute inflammation caused by certain fungus diseases.
B. Acridine dyes	Local skin irritations.

3. It should not produce side reactions, such as allergy, dizziness, neuritis, and nausea.

Not all of these properties are displayed by every useful antibiotic, but the best possible combinations are sought.

Despite the many positive effects of antibiotics, their use can lead to many complications. For example, certain antibiotics have considerable toxicity when administered over long periods of time. Streptomycin may cause damage to the auditory nerve, resulting in deafness. Other antibiotics only suppress the clinical symptoms of a disease without eliminating the infection. Moreover, antibiotic therapy can result in the development of resistant (mutant) strains of bacteria, can cause treated individuals to become hypersensitive to their use, and can suppress the normal flora (microbes naturally present) to the extent that other infections may result. A brief description of some of the more commonly employed antibiotics is given in Table 15-3.

TABLE 15-3. Some Commonly Used Antibiotics

Antibiotic	Comments
Penicillins	Obtained from molds (*Penicillium notatum* and *Penicillium chrysegenum*); most frequently used antibiotic; effective in the treatment of streptococcus, gonococcus, pneumococcus infections, syphilis, tetanus, anthrax; antibacterial action is manifested only during active division of the organisms by inhibiting synthesis of cell walls; many patients show allergic reactions, some severe.
Streptomycin	Obtained from the mold-like bacterium *Streptomyces griseus;* useful in treatment of tuberculosis, influenza, typhoid fever, some streptococcus and staphylococcus infections; may exhibit antimicrobial activity by altering permeability of the plasma membrane or protein synthesis; may affect the eighth cranial nerves.
Tetracyclines	A group of antibiotics, all derived from *Streptomyces;* broad-spectrum antibiotics; effective against many organisms, including larger viruses and rickettsiae; probably acts by altering protein synthesis; may suppress the normal flora of the large intestine.
Myostatin	Obtained from *Streptomyces noursei;* inhibits the growth of pathogenic fungi of skin, mouth, vagina, intestines.

Host-Microbe Relationships

Having considered some of the characteristics of microbes, as well as some of the methods used to control them, we turn now to the relationships of pathogens to man. This discussion will treat some basic factors relating to infection and the reactions of the host, the human organisms, to infection.

Sources of Infection

When any microbe enters and multiplies within the tissues of the body, it produces an *infection* and any disease that results directly from an infection is called an *infectious disease.* Many diseases of man are spread by direct contact or are communicated from person to person directly by coughing, sneezing, dirty hands, and other types of bodily contact. Such diseases are referred to as *communicable* or *contagious diseases* and include measles, mumps, whooping cough, chicken pox, influenza, and pneumonia. The great majority of infectious diseases are communicable. Other diseases, such as yellow fever, malaria, African sleeping sickness, typhus fever, and various types of encephalitis may be spread by carrier agents such as arthropods (ticks, mites, mosquitoes, lice). Such arthropods, which habitually suck the blood of humans or other animals, may act as *vectors* (agents that transmit the disease directly). Other insects, by contrast, serve as agents for the mechanical transfer of pathogenic microbes. Flies, cockroaches, and other crawling insects that have access to sewage or feces may carry dangerous microbes mechanically to food or body surfaces. Rodents also frequently cause disease in humans indirectly by contaminating food with their body discharges, which may contain pathogens.

Other indirect methods of the spread of disease are by contamination of drinking water and food. Also, inanimate objects called *fomites* may transmit disease. Common examples of fomites are towels, tableware, handkerchiefs, toilet articles, drinking glasses, and toys.

Pathogenicity and Virulence

Pathogenic microbes, that is, organisms that are capable of developing within the tissues of the body, are generally classified into two principal kinds: opportunists and true pathogens. Although such a distinction is useful, it is not absolute. *Opportunists* are those organisms capable of producing disease only when a special circumstance affords them entry to the body tissues (injury to skin or mucous membranes) or when natural resistance to infection is very low. The causative agent of tetanus is such an example. The organism has no power to invade the body itself, but it may be carried by dirt into an accidental wound. *True pathogens,* on the other hand, are able to invade the tissues of healthy individuals through some inherent power of their own. The common communicable diseases are caused by true pathogens, a relatively small number of species.

There are several important factors that determine whether or not a microbe can establish infection in any particular kind of situation. Among these are

1. The virulence (disease-producing power) of the invading organism.
2. The number of organisms that invade the host.
3. The specific path of entrance to the body of the host.
4. The degree of resistance of the individual host.

The factors that determine the virulence of a microbe are not completely understood. However, some contributing factors have been identified. For example, toxins produced by pathogenic bacteria are important in relation to virulence. **Toxins** are poisonous substances produced by certain microbes as part of their growth activities. Toxins may be *exotoxins,* substances which diffuse out of the intact bacterial cell into the surrounding tissue, or *endotoxins* which are contained within the bacteria and are released only when the cells are broken. Some exotoxins may produce localized killing of tissues, destroy red blood cells (*hemolysins*), affect nevous tissue (*neurotoxins*), destroy white blood cells (*leucocidins*) or at least inhibit their phagocytic activity, and destroy the cementing substance within tissues (*hyaluronidases*), thus accelerating the spread of microbes through tissues. Still other exotoxins dissolve clotted blood plasma (*fibrolysins*), whereas others cause blood plasma to clot (*coagulases*). Diseases associated with the production of powerful exotoxins are diphtheria, tetanus, botulism, gangrene, and anthrax. Endotoxins are not specific for each kind of organism; they often damage blood vessels and may cause fever.

A second major factor related to virulence is the capsules of certain bacteria. These capsules may have virulent properties and appear to protect the bacteria by resisting phagocytosis in the infected host.

Body Defenses Against Infection

In order to maintain its position and balance in an environment in which there are many antagonistic forces, the human organism has developed many defensive mechanisms. In all probability, these mechanisms have evolved continuously during the course of human history. Because of the virulence of pathogens, the human has had to develop certain protective mechanisms in order to ensure survival. For convenience, the body defenses against disease will be treated as two separate categories, nonspecific and specific.

Nonspecific Defensive Mechanisms. In a very broad context, **immunity** refers to the ability of a host to prevent or overcome invasion by virulent, pathogenic organisms. In other words, immunity refers to relative resistance toward some particular pathogen. Immunity depends on the interaction of many defensive factors, some nonspecific, that are effective against invading microbes of various kinds; other which are specific, act upon a particular organism or toxin. In the latter case the immunity is mediated through protective antibodies formed in direct response to the presence of microbes or other foreign matter in the body.

The most important, and also the best defined, host factors involved in nonspecific resistance are associated with activities of the skin and mucous membranes, cellular and fluid elements of the blood, the cells of the reticuloendothelial system (RES), the lymphatic system, and the inflammatory response.

The Skin and Mucous Membranes. Both of these structures, when intact, offer the first barrier to microorganisms that come in contact with the body. Mechanically, the design of the skin and mucous membranes is such that a continuous sheath of closely packed cells covers all exposed surfaces, supported beneath by dense connective tissue. In addition, the outermost surface of the skin is covered with tough keratinized cells; keratin is a tough, dense, waterproofing protein. The skin surface is covered with an acidic film derived from fats and proteins. At the opening of the respiratory tract, nasal hairs and secretions of the mucous membrane remove many particles that might otherwise enter. In the mouth, salivary gland secretions provide a mechanical flushing action. The physiological defenses of the skin include secretions of the sweat glands (sweat) and sebaceous glands (fatty acids) which have an antimicrobial effect. Another important secretion, *lysozyme,* present in nasal mucous, tears, and to a lesser extent in saliva and skin secretions, can break down the cell walls of certain bacteria. In the stomach, the acidity of gastric juice kills many microbes that are swallowed.

Blood Cells and Plasma. As noted earlier, blood consists of both formed elements (cells) and plasma. One nonspecific constituent of plasma is *complement,* so called because it contributes to the effectiveness of antibodies in some of their reactions. Complement acting alone has no destructive effect on foreign cells. However, in combination with antibodies, as well as calcium and magnesium ions present in plasma, complement brings about the destruction of foreign cells. Another plasma protein, *properdin,* has also been described as having antimicrobial properties when complement and magnesium ions are present.

The formed elements in blood that are of importance in nonspecific resistance to infection are the white blood cells that have phagocytic activity (Chapter 5). The most important of these are the neutrophils, lymphocytes, and monocytes. The presence of a foreign substance provides a chemical stimulus that attracts phagocytes to the infected area. Phagocytosis is greatly enhanced by the nonspecific components of plasma, together with the specific antibodies it may contain for particular microorganisms.

Reticuloendothelial System (RES). In addition to the blood cells, there is another system of relatively large tissue cells composed of plasma cells and phagocytic *macrophages.* Some of these are motile and able to migrate through the tissues. Others are fixed in a variety of tissues in the body such as lymphoid tissue, spleen, bone marrow, lungs, central nervous system, connective tissue, and the lining of blood vessels. Together the sessile and wandering cells constitute the reticuloendothelial system. Whatever their location, the principal function of the RES cells is the engulfment and removal of foreign or useless particles, living or dead, and the production of antibodies.

The Lymphatic System. This system of tissues also plays a significant role in nonspecific resistance to infection. Although lymphoid tissue is scattered throughout the body, nodes of the upper respiratory tract and cervical regions, the arms and legs, the gastrointestinal system, the genitourinary tract, and the inguinal area are strategically located to pick up microbes that first get past the main entry barriers. Here, most microorganisms are filtered out by phagocytes that are inside the nodes. The lymphatic tissues of the body are also important in the production of antibodies.

Inflammatory Response. When cells are damaged, the injury sets off an *inflammatory response.* The injury could result from mechanical means, such as a clean knife wound during surgery. Bacteria that give off poisonous chemicals could enter through the nose, pores in skin, or by way of a splinter or nail. Injury to cells can also occur if the blood supply is cut off, which causes the cells to "starve."

Inflammation is usually characterized by four fundamental symptoms: *redness, pain, heat,* and *swelling.* A fifth symptom can be the loss of function of the injured area. Whether loss of function occurs depends on the site and extent of the injury. In addition to these effects on the body, the inflammatory response, almost as a contradiction, serves a protective and defensive purpose. It attempts to neutralize and destroy toxic or poisonous agents at the site of injury and to prevent their spread to other organs. In other words, the inflammatory response is an attempt to restore tissue homeostasis.

The immediate inflammatory response to tissue injury consists of a complicated sequence of physiological and anatomical adjustments involving various parts of the body. These include the blood vessels, intercellular fluid mixed with parts of injured cells (called the exudate), the cellular components of the blood, and the surrounding epithelial and connective tissues. Other factors that affect the inflammatory response are the individual's age and general state of health. Healing processes of all types exert a great demand on the body's stores of all nutrients. Thus, nutrition, in addi-

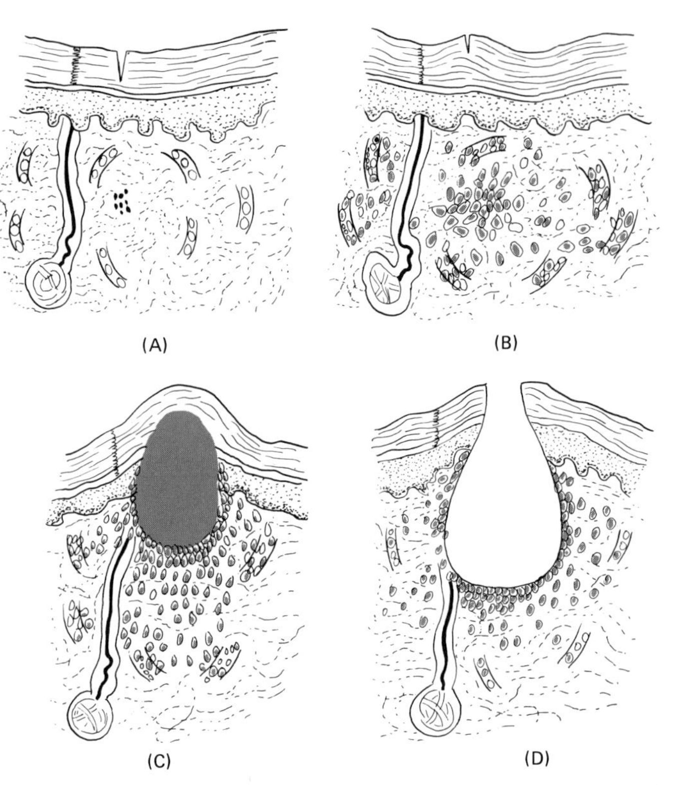

Figure 15-2. "History of a boil." **(A)** A cut section of the normal skin just after it has been pierced by a dirty pin that carried in staphylococci. **(B)** The early stages of inflammation; migration of leukocytes from swollen capillaries toward the infected area. The skin is becoming red, hot, swollen, and painful. **(C)** The accumulation of leukocytes has continued, and now there is a mass of purulent exudate about the germs. The boil has "come to a head," and the semiliquid pus can be seen through the thin cuticle over it. **(D)** The boil has finally ruptured and the pus has escaped, carrying with it the staphylococci. The migration of leukocytes has ceased and the capillaries are returning to their normal size. New tissue will grow in. [*Modified from C. P. Emerson, Jr., and J. S. Bragdon,* Essentials of Medicine, *17th ed. Philadelphia: J. B. Lippincott Co. Copyright © 1955.*]

tion to adequate circulation and tissue drainage, also plays an essential role in healing.

The following discussion outlines how the inflammatory response operates. In the example presented in Figure 15-2, the skin is broken by a dirty pin carrying staphylococci. The same response occurs, however, when bacterial invasion gives you a sore throat.

In all but very mild inflammations, a substance called pus is discharged. **Pus** is a thick fluid that contains living as well as nonliving white blood cells plus debris from other dead tissue. If the pus cannot drain to the outside of the body, an **abscess** occurs, which is simply an excessive accumulation of pus in a confined space. When inflamed tissue is shed many times, it produces an open sore, called an **ulcer,** on the surface of an organ or tissue. Ulcers may result from a prolonged inflammatory response to a continuously injured tissue. For instance, overproduction of digestive acids in the stomach may cause a steady erosion of parts of the epithelial tissue lining the stomach. Elderly people with poor circulation are susceptible to ulcers in the tissues of their legs. The ulcers develop when the tissues are continuously damaged by a shortage of oxygen and nutrients.

As we have seen, white blood cells are an important line of defense against penetrating injury. The blood supply drains off and blocks the spread of harmful materials. This removal cannot restore the tissues to normality, however, especially if the cells have been damaged. The body has some other method to replace the injured areas with new tissue. And the body also has to restore cells that are worn out through normal usage. The process by which tissues replace dead or damaged cells is called repair.

The injury stimulates tissue in the damaged area to release a chemical called histamine. *Histamine* is believed to increase the diameter of the blood vessels, or *dilate* them. It also increases the permeability of the plasma membranes of the blood vessel cells. Dilatation increases the amount of blood that can enter the area. The increase in permeability allows defensive substances in the blood to pass through the vessel walls and into the injured tissue. Such defensive substances include white blood cells, antibodies, oxygen, and scab-forming chemicals. The increased blood supply also carries off poisonous products and dead cells to excretory organs, thus preventing them from complicating the injury. These poisonous, or toxic, substances include waste products given off by invading microorganisms.

The body may also respond by increasing its metabolic rate and by quickening the heartbeat so that more blood per minute circulates to the injured area. Within minutes after the injury, the quickened metabolism and circulation and especially the dilatation and increased permeability of capillaries produce heat, redness, and swelling. The heat comes from the large amount of warm blood that accumulates in the area and, to some extent, from the heat energy given off by the metabolic reactions. The large amounts of blood in the area are also responsible for the redness. The increased permeability of the capillary walls allows quantities of fluid to move out of the blood and into the intercellular spaces in the tissue. Because the fluid moves into the intercellular spaces faster than it can be drained off, it accumulates in the tissue, causing it to swell or puff up. The swelling is called **edema.**

Pain, either immediate or delayed, is a cardinal symptom of inflammation. It can result from an injury to nerve fibers or from an irritation caused by the release of poisonous chemicals from microorganisms. Pain can also result from increased pressure from an accumulation of extracellular fluid within the tissues.

Very soon after the injury occurs, white blood cells are mobilized and rushed to the site from all over the body. The white blood cell, or *leucocyte,* is the body's first line of defense against the effects of the injury. The first white blood cells to arrive at the site of injury are neutrophils. These are actively phagocytic cells that can engulf foreign particles equal to their own size. When neutrophils die, their lysosomes release valuable enzymes that cause the decomposition of injured cells and bacteria in the affected area. Additional neutrophils then ingest the resulting bacterial debris and other refuse that the body cannot use. Other white blood cells, called *lymphocytes,* are believed to be involved either in the formation or release of *antibodies,* which are proteins that render invading bacteria and their chemicals harmless.

Nutrients stored in the body are used to support the defensive cells. They are also used in the increased metabolic reactions of the cells under attack.

The blood brings a soluble protein called fibrinogen to the site of injury. The increased permeability of capillaries produced by histamine causes leakage of fibrinogen to tissues. Fibrinogen is then converted to an insoluble, thick network called fibrin, which localizes and traps the invaders, thereby preventing their spread. This network eventually forms a blood clot, which prevents hemorrhage and walls off the infected area.

Specific Resistance to Infection: Antigens and Antibodies. When an individual recovers from most infectious diseases, there are accompanying physiological changes which result in a resistance of the individual to the specific microbe causing the illness. During the course of the disease certain reactions occur between the causative organism and the tissues of the infected host, and the individual, initially susceptible, becomes resistant toward the particular infecting agent. Such immunity may persist for several months or several years. The development of this kind of resistance is always associated with the appearance, first in the infected tissues, then in the blood, of antibodies capable of reacting specifically witht the particular organism (or toxin) which caused their formation and in some way provides this immunity.

Antigens. The terms *antigen* and *antibody* are so closely interdependent that a discussion of one must, of necessity, include a discussion of the other. Any substance that, when introduced into tissues, will stimulate the formation of an antibody is referred to as an **antigen.** The chemical nature or structure of the antigen must be foreign to the host or the antibody-producing mechanism since the body does not generally form antibodies against constituents of its own cells or tissues. Antigens may be proteins, polysaccharides, complex lipids, and certain other substances that combine with proteins. For example, practically all proteins are classified as *complete antigens,* that is, they have specific combining parts as well as components capable of stimulating antibody production. Bacterial cells, viruses, and other microbes, toxins, foreign proteins, and blood cells are complete antigens, each capable of stimulating antibodies specific for itself. Typically, complete antigens have a molecular weight of more than 10,000 and some are as large as 6 million; in general, the greater the molecular weight, the more powerful is the stimulus to antibody production. Some carbohydrates, such as those found in the capsules of certain bacteria, may also act as complete antigens.

Many other carbohydrates or lipids, however, are not able to stimulate antibody production by themselves. When linked with a protein molecule, these substances may become antigenic. In fact, when combined with protein, these substances often determine the specific character of the antibody that the whole compound produces. Such incomplete antigens are categorized as *haptens.* The haptens have much lower molecular weights than the complete antigens from which they are derived.

Many of the constituents of bacteria and other microbes are either proteins or substances linked to proteins, and thus are antigenic. Accordingly, a given microorganism may contain or produce many antigens that differ from one another and can induce the response of many different antibodies. The antigens of the bacterial cell itself, for example, are called *somatic* or *O antigens.* In some bacteria they are the endotoxins of the cell. Other cellular antigens occur only on the surface of the cell and are termed *surface antigens.* In species that form a capsule, *capsular antigens* are present; motile species contain *flagellar* or *H' antigens;* and some forms secrete exotoxins and other protein substances that act as antigens.

Antibodies. An **antibody,** defined in relation to an antigen, is a modified protein (globulin) substance whose surface has a shape complementary to the determinant group of the antigen and which is produced in response to the presence of a foreign antigen. The specialized structure of antibodies enables them to combine specifically with particular antigens. All antibodies are products of living cells of the host and are usually present in the blood. The molecular weight of most antibodies is about 160,000.

Antigen-Antibody Reactions. Antibodies react against specific microorganisms, their toxins, and other compounds. They can be used not only in the treatment of infection, but also to prevent infection and disease caused by these microbes. Antibodies are designated by names that describe their reaction *in vitro* (test tubes) or *in vivo* (the body) when they are allowed to act on certain types of antigens. Laboratory investigations utilize blood plasma, or serum, from humans and other animals as a source of antibodies, and antigens from varied sources, including microorganisms. The combination of the antigen and antibody results in one of several observable reactions. A few of these reactions, which determine the type of antibody involved, are described below.

Antitoxins are antibodies that are induced by exotoxins and combine with the exotoxins, neutralize them, and thus render them harmless. Although the exact mechanism of neutralization is not known, it has been suggested that the antitoxin simply covers the poisonous portion of the toxin so that it cannot reach the body tissues (Figure 15-3A). A second kind of antibody, an *agglutinin,* causes a visible agglutination (clumping) of antigens (Figure 15-3B). In attempting to understand the mechanism of agglutination, it should be realized that bacteria in solution behave like particles of a true colloid, that is, they remain separated and in suspension because of the repelling effect of the electrical charges that they carry. If an antibody is added, it coats the cells, causing a reduction of repellent electrical charges and probably makes the cells sticky so that they clump together. As larger and larger clumps are formed, they settle out of solution.

Another kind of antibody, called a *precipitin,* causes the precipitation of extracts of bacterial cells or other soluble antigens (Figure 15-3C). Such a reaction occurs in the same manner as an agglutinin reaction except that, in this case, the antigen must be in free solution, unassociated with any cell or particle. *Lysins* are antibodies that cause dissolution of bacterial or other cells that are specifically sensitive to their action (Figure 15-3D). Some of these lysins, termed *bacteriolysins,* destroy bacterial cells; others, called *hemolysins,* destroy red blood cells. It has been found that lysis depends on the interaction of two elements. In bacteriolysis, for example, the reaction is dependent upon the specific antibody (bacteriolysin) and complement. Complement, it will be remembered, is a normal constituent of blood plasma, and is the active element in causing bacteriolysis. However, lysis can occur only after the bacteriolysin has combined with the bacterial cell and prepared it for the action of complement.

Opsonins are antibodies that combine with the surface antigens of microorganisms or other cells and render the cells more susceptible to phagocytosis (Figure 15-3E). This process is also facilitated by the presence of complement. The importance of these antibodies in defense is obvious when one considers tissue invasion by virulent microbes. The opsonins assist phagocytes in capturing and removing harmful microorganisms.

It was once believed that separate and distinct antibodies were present in the serum of immunized individuals and that these antibodies gave rise to different

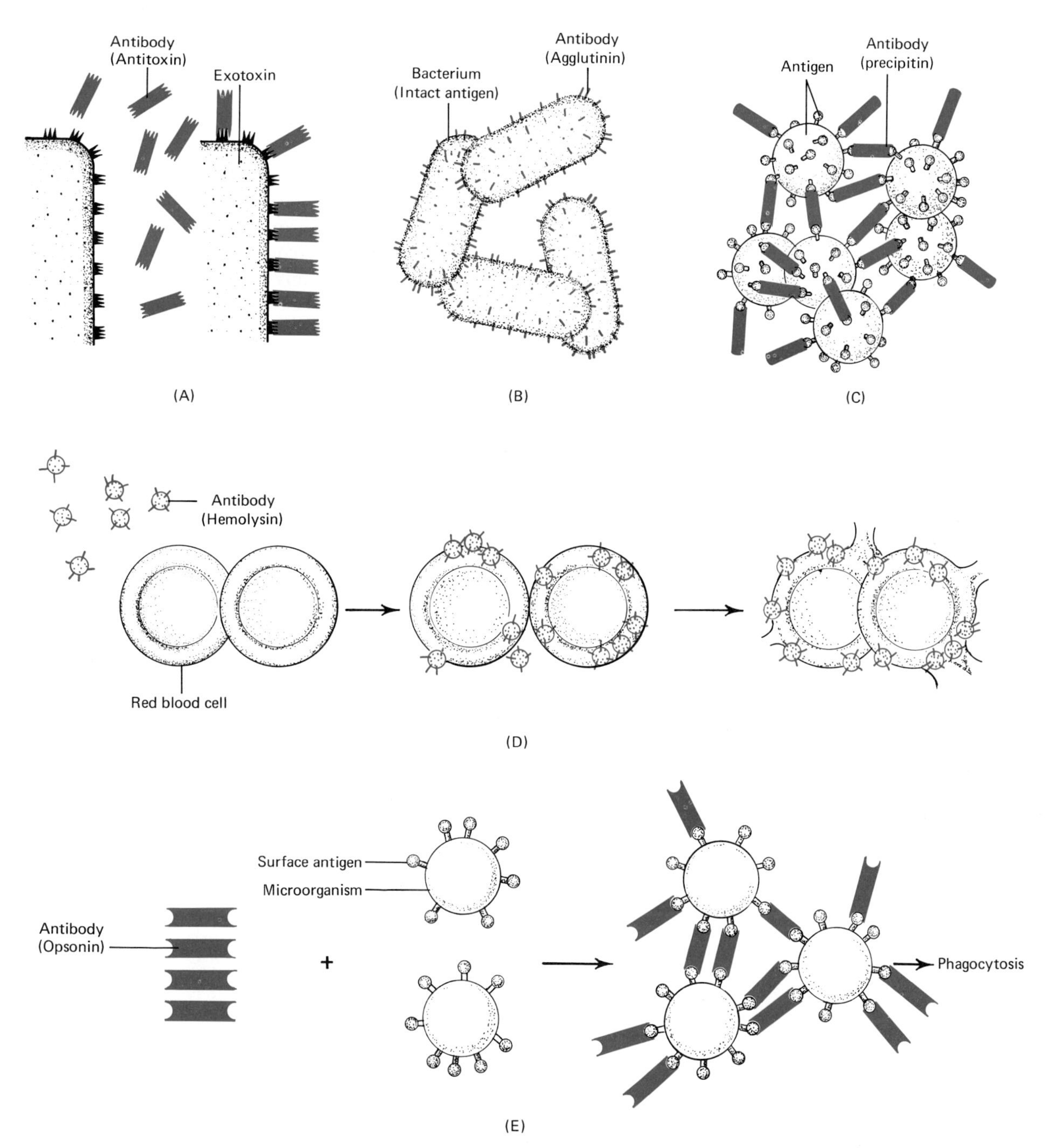

Figure 15-3. Reactions of antibodies on various antigens. In these highly diagrammatic representations, the proposed mechanisms of antibody activity are shown. (**A**) Antitoxin–toxin reaction. (**B**) Agglutination reaction. (**C**) Precipitation reaction. (**D**) Lysis reaction (hemolysis). (**E**) Opsonic reaction. See text for discussion of each type of reaction.

kinds of observed reactions (agglutination, lysis, precipitation, etc.). A more recently conceived hypothesis purports that for a given microorganism (antigen) there is only *one* antibody, but that this antibody may be demonstrated in several different ways, depending on the existing conditions. Thus, when an antibody is in solution with intact bacterial cells, it acts as an agglutinin; if the antibody is placed in a solution in which the cells are broken up and the cellular components are released into the solution, it acts as a precipitin; when phagocytes are present, the antibody acts as an opsonin; and if complement is present, the antibody behaves as a lysin.

Theories of Antibody Formation. The mechanisms by which antibodies are formed in the body are not completely understood. Several explanations have been given, some more plausible than others. Two of these will be discussed briefly. Various experimental data clearly indicate that antibodies are formed by plasma cells, morphological variations of lymphocytes. Recall that macrophages, both sessile and wandering, are components of the reticuloendothelial system.

INSTRUCTIVE (TEMPLATE) THEORY. According to the *instructive,* or *template, theory,* the antigen serves as a model (template) for the formation of the antibody complementary in configuration to the antigen (Figure 15-4A). Supposedly, the antigen becomes a direct template from which an antibody takes its final complementary form. A modification of this template theory states that the structural difference that distinguishes one antibody from another is due to a specific modification by an antigen of the RNA protein-synthesizing mechanism. Another modification of the theory suggests that an antigen may enter the nucleus of an antibody-producing cell and, by combining with the DNA, alter the information sent to the cytoplasmic templates for protein synthesis. The reason that the template theories are frequently called instructive theories is that the antigen is assumed to provide information or directions for the cell to follow in making highly specific antibodies.

Although the template theories explain most of the phenomena observed when bacteria or other cells act as antigens, they do not appear to cover satisfactorily certain features of antibody reactions involving soluble antigens. A second theory, the selective, or clonal-selection theory, has also been proposed.

SELECTIVE (CLONAL-SELECTION) THEORY. According to the *selective theory,* antibodies are synthesized in a manner similar to the synthesis of other proteins (Figure 15-4B). Instructions for the production of antibodies are provided by genetic elements in the nucleus of the plasma cell rather than from the antigen. It is postulated that the cells of the host are genetically competent and are able to synthesize antibodies specific for each of the thousands of foreign antigens that might possibly enter the cell. The role of the antigen, therefore, is to select the proper cell or cells and stimu-

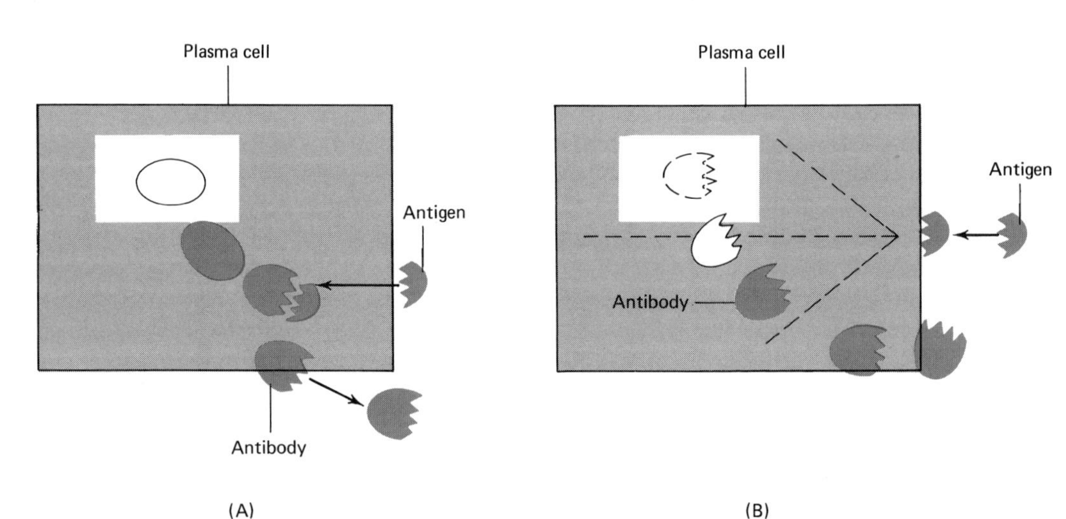

FIGURE 15-4. Template (**A**) and selective (**B**) theories on antibody formation. See text for discussion.

late their proliferation, thereby increasing antibody production.

A more recent explanation of the selective hypothesis assumes that the body contains pre-existing *clones* (a clone is a group of individuals descended from the same individual or cell) of immunologically competent cells that can react directly with an antigen and give rise to plasma cells that can synthesize a particular antibody. There is a separate clone for each antigen. According to this concept, then, when an antigen comes into contact with the appropriate cells, it causes them to form a clone that reacts with the specific antigen to produce an antibody. The clonal cells continue to proliferate until the antigenic simulus is eliminated.

Neither the template theory nor the selective theory has been completely validated. Neither appears to explain fully the mechanism of antibody formation and other aspects of the immune reactions. Investigators are still exploring the problem and many modifications of the leading theories have been proposed.

It was noted earlier that plasma cells, derived from lymphocytes, produce antibodies against foreign substances entering the body. Such lymphocytes are called *B-cells* and originate from the bone marrow and lymphoid tissue of the spleen, lymph nodes, and lymph nodules of the intestine. It is also believed that a second type of lymphocyte is present in the body. These are called *T-cells.* They originate in bone marrow and the thymus gland and require a chemical from the thymus to make them capable of reacting with antigens. Some of the T-cells are involved in the rejection of tissue transplants; others are capable of responding to a second exposure of a given antigen.

The Concept of Immunity

The nonspecific and specific defenses of the body provide **immunity,** that is, the ability to overcome the disease-producing effects of certain microbes. The immunity provided by antibodies consists of two types—active and passive.

Active Immunity

Whether or not a microbe makes you sick often depends on whether you have been exposed to it before. The first time a cell is exposed to an antigen, its antibody response is a little slow to start. This is because the cell needs time to make the proper adjustments in its protein manufacturing assembly lines. During this time, the microbes are free to multiply and produce toxins and other symptoms of the disease. If the same kind of microbe invades your body again in the future, your cells may still be geared to producing the antibody. In this case, the antigen–antibody response may occur before the microbes have a chance to bring on the symptoms of the disease. Such protection against future sickness is called *active immunity.* Active immunity may also be acquired through vaccinations with dead pathogens or with very low doses of their toxins. The proteins in the dead pathogens are capable of stimulating antibody production, but the dead microbes cannot hurt you. Toxins are given in doses just high enough to stimulate antibody-producing cells. The doses are too low, however, to cause disease.

Passive Immunity

Passive immunity is acquired through injection of antibodies from an animal or from another person who has previously been exposed to the disease. Examples are the globulin antibodies, which are effective against hepatitis and measles. Passive immunity gives you only temporary protection. It is generally used when a person has been exposed to a disease to which he is not immune and when he does not have enough time to manufacture his own antibodies.

Many cells of the body produce antibodies. Both the lymphocytes and the reticuloendothelial cells are capable of forming plasma cells, which are good antibody producers. These cells and their antibodies circulate freely through the blood and lymph and can get at an antigen quickly. However, the life span of these cells is too short to provide immunity against future attacks. It is suspected that long-term immunity is provided by long-lived cells, such as muscle cells, which are also capable of producing antibodies.

Immunity: A Mixed Blessing

The antigen–antibody response is essential to survival. However, under certain circumstances, it may create problems. Two such problems are related to allergy and transplantation.

Allergy

An individual who becomes overreactive to an antigen is said to be *allergic* or *hypersensitive.* The person with an allergy reacts to antigens differently from the way he did at some earlier time or differently from most other people. In addition, whenever an allergic reaction occurs, there is tissue injury.

The antigens that induce an allergic reaction are called *allergens.* Examples of allergens include certain foods, many antibiotics (e.g., penicillin), cosmetics, chemicals in plants (poison ivy, pollens), and even microbes. In the first stage of an allergic reaction, the cells of the body become sensitized by antibodies. That is, they become subject to injury if the allergen enters the body again at a later date. Sensitivity occurs when a person receives an initial dose of the allergen. When the allergen enters the body again in the second stage of an allergic reaction, tissue damage results.

In the allergic reaction, antibodies formed against the first dose of allergens remain attached to the cells that produced them. At this point, the cells become sensitized. When the antibodies react again with allergens introduced a second time, the antigen–antibody reaction destroys the cells as well as the allergens. Destruction of the cells triggers several physiologic responses, for example:

1. Injured cells release histamine. Large quantities of histamine cause tissue inflammation and contraction of smooth muscle fibers. This occurs especially in the breathing tubes and blood vessels, causing them to constrict. Histamine also increases the permeability of blood vessels so that fluid moves from the vessels into the interstitial spaces, causing edema.

2. In a severe allergic reaction, *anaphylactic shock* may result. This is caused by the prolonged effects of histamine. The respiratory tubes are continuously constricted, and the occurrence of edema is accelerated. This lowers the blood volume even more. If anaphylactic shock is not counteracted, death may result. The effects of anaphylaxis may be reversed by administering epinephrine or antihistamines, drugs that inactivate histamine.

Transplantation

The replacement of injured or diseased tissues and organs with natural and artificial ones, called *transplantation,* is emerging as one of the major biological challenges. The possibility of replacing tissues and organs is not a new idea. However, today this challenge is being approached using the full resources of the biological sciences and medical technology.

It was stated previously that the human organism possesses a number of defensive mechanisms or adaptations designed to inhibit, alter, or destroy foreign organisms or other products. Clearly, these defenses, both natural and acquired, are essential for the continuity of life. Immunity, and the ability of a living organism to acquire this immunity, is an adaptation that provides protection. It appears that immunity has had a significant effect on survival in the past and has been molded under specific evolutionary pressures; it is not something that has emerged accidentally in association with a distantly related function or evolutionary requirement.

Yet, in spite of this, immunity must be regarded as a mixed blessing because the immune reaction also provides the basis for rejection of foreign tissues by the body. The human body still accepts only what is genetically identical to the part replaced. Apparently the body is able to recognize some substances as "self" and others as "not self." This becomes quite clear when one considers the general nature of transplantation. For example, it has been known for years that if a piece of skin is grafted from one individual to another, the recipient may soon reject the graft. Similar reactions occur when organs such as lungs, hearts, and kidneys are transplanted from one individual to another. If however, a tissue or organ is transplanted from one identical twin to another, the transplant is accepted by the recipient; there is no rejection.

In a physiological context, transplants may be viewed as antigens since they elicit immune reactions similar to those precipitated by bacteria, viruses, toxins, etc. These antigens have a unique relationship to antibodies in that most antigens also consist of protein and in some unknown manner seem to be able to trigger their host into producing a highly specific antibody. It seems logical, then, that the way immunological capabilities develop and the way the body distinguishes between "self" and "not self" are basic to an understanding of the phenomenon of rejection.

The mechanism involved in the production of an

antibody remains as one of the great challenges of modern biology. Recall that plasma cells, the cells that produce antibodies, develop from lymphocytes. Many scientists believe that the thymus gland plays an important role in the development of immunologic capabilities by producing specialized lymphocytes during the early stages of life. Once manufactured in the thymus, the lymphocytes are distributed into the lymphatic system as freely wandering cells. Some become established in the spleen and bone marrow where they function in phagocytosis and antibody production. Lymphocytes produced by the thymus have the greatest ratio of nucleus to surrounding cell matter of any cells in the body. Some investigators believe that it is the high DNA content of lymphocytes that enables them to distinguish between "self" and "not self" and also what orders them to produce antibodies.

The crucial role of the thymus and lymphocytes in antibody production was demonstrated in an experiment on newborn mice. At birth, the thymus glands of an experimental group of mice were removed. Two to four weeks after the removal of the thymus the mice received skin grafts from different strains of mice. Normal (control) mice will violently reject such grafts within 8–10 days. But the thymus-less mice retained the foreign skin as if it had originally grown on their bodies. This experiment, as well as others, has shown that removal of the thymus from newborn mice severely inhibited their ability to produce antibodies. Such mice also showed no resistance to common pathogens. The hagfish, which has never evolved a thymus and has no immunological abilities, offers some supporting evidence.

Most of the data seem to indicate that early embryos lack immunologic mechanisms and that once they develop, all substances present in the body at the time are recognized as "self." It is believed that immunologic capabilities do not develop all at once, but rather gradually during late embryonic development and the first several weeks after birth. This suggests a mechanism involving different elements, each responsible for immunologic capabilities against different antigens. Accordingly, each such element matures normally if its antigen is absent from the system. If, however, its antigen is present at the time it starts to mature, the element fails to mature. Such reasoning could imply that substances can be identified as "self" and "not self" by the adult organism because the "self" substances were present at the critical time during development and destroyed or inhibited the immunologic elements that started to develop against them. Similarly, such reasoning could imply that "not self" substances were not present at the critical moment of development and, as a result, immunologic elements against them were free to mature.

The principal objective of current research dealing with rejection is to reach some sort of compromise with the barrier, but not to destroy it. In other words, scientists are attempting to discover immunosuppressive techniques to control rejection. They have pursued two major lines of investigation to overcome rejection, the use of radiation and the use of drugs. Since 1952, literally thousands of animals have been subjected to varying degrees of radiation and then have received organ transplants. The results showed that X-rays could indeed reduce an animal's immune response for short periods of time. They also demonstrated that living cells could withstand low doses of radiation fairly well, but that as the energy of the rays and duration increased, the cells died. Moreover, certain animal and human cells—those of lymph, blood, and blood-forming organs—are extremely susceptible to even low doses of radiation. Considering the fact that these cells are closely associated with the production of antibodies, radiation seems to offer a means of subverting the system and inducing the body to accept a foreign tissue transplant. It was soon found that in destroying the capabilities of cells that are believed to play a part in rejecting transplants, it was impossible to avoid destruction of other cell functions. Finally, the experiments revealed that only almost lethal doses of radiation suppressed the rejection response—and then only temporarily. Radiation alone was a partial answer at best—difficult to control, dangerous to the patient, and effective for only short periods. It also added another complication. Even when used effectively, radiation altered the entire immunity system and left the body open to invasion by all pathogens. Even relatively harmless organisms can bring about devastating effects in a patient whose internal defenses have been paralyzed by the effects of radiation. It was quite obvious that a different kind of immunosuppressive therapy was needed.

Investigators soon directed their interests toward more controlled therapy with chemotherapeutic agents which had been developed for the treatment of cancer. It was known, for example, that some new anticancer drugs that were temporarily effective against leukemia had a depressing effect upon the body's immunological defenses. This observation suggested a new and exciting possibility for transplantation. One of these drugs, *6-mercaptopurine* (6-MP), was used and was able to prolong acceptance of kidney transplants in dogs. Despite the many flaws that were uncovered in the use of 6-MP, the few successes in using it were indisputable. Shortly after this time, *Imuran* was developed and this drug is currently used as a basic immunosuppressive agent in all organ transplantation. Imuran acts by blocking the synthesis of DNA, which is necessary for the replication of cells. The most rapidly dividing cells when a graft is transplanted are the lymphoid cells that are reacting to the antigen (graft); these eventually attack the graft. Imuran blocks the synthesis of DNA and hence the multiplication of lymphoid cells, thus blunting the immunologic response and ensuring a degree of acceptance of the graft. However, the drug must be continued throughout the life of the individual. At approximately the same time it was shown that a corticosteroid, *prednisone,* could reverse a rejection reaction once it occurred.

More recently another agent, *antilymphatic serum,* has become available. This serum, produced in horses, acts against the lymphocytes of another species. Those patients that have been treated thus far with antilymphatic serum in conjunction with Imuran and prednisone have shown an improvement of survival to over 90% in kidney transplants lasting more than one year.

In addition to the development of immunosuppressive drugs in recent years, much progress is being made in the recognition of tissue typing (Figure 15-5). The objective of this field of research is to find the closest possible tissue match between donor and recipient. Tissue typing involves the identification of antigens in the leukocytes; the presence or absence of these antigens in the donor and recipient results in the determination of the donor–recipient compatibility. At the present time there are seventeen identifiable antigens in or on white blood cells.

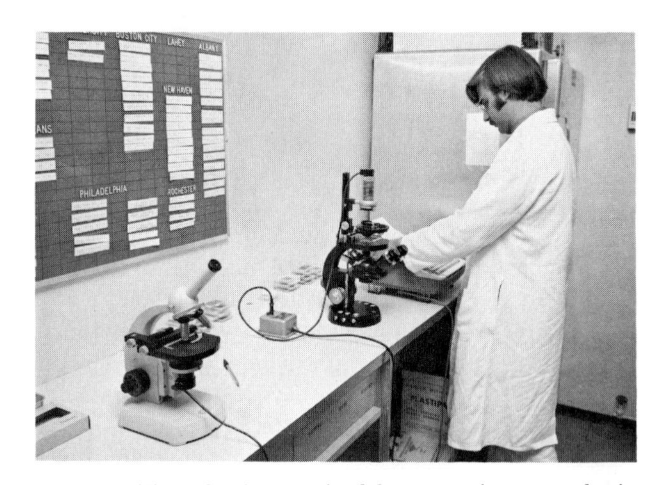

FIGURE 15-5. In the tissue-typing laboratory, tissues are classified according to antigen composition. The purpose of this procedure is to match the tissues of donor and recipient as closely as possible in order to reduce the rejection mechanisms. [*Courtesy of Peter Bent Brigham Hospital, Boston.*]

TYPES OF TRANSPLANTATIONS. The entire science of transplantation has successively moved from an impractical basis, to the theoretical realm, to a reality. Part of this evolution may be attributed to the development of new instrumentation, technological advances, improved and refined surgical procedures, and the realization that transplantation may some day become a routine method for the treatment of damaged or diseased body parts. Much of the impetus for this evolution has been provided by relatively successful transplants in experimental animals and by favorable results obtained in human subjects.

Transplantations may be classified in several ways.

1. On the basis of the relationship of the donor and recipient.

2. In terms of the nature of the transplant, that is, whether it is a tissue, a portion of an organ, or a complete organ.

3. On the basis of whether the transplant involves the use of an artificial device, in which case it might better be called *implantation.* However, the use of artificial devices and organs is so closely related to transplantation that this procedure might properly be considered a branch of this science.

The term **isograft** refers to a transplantation in

which the donor and recipient have an identical genetic background. This would include transplants between identical twins when working with humans or with closely inbred experimental animals. In addition, it would also include the transplantation of tissue from one part of the body to another. Since the genetic makeup of the transplants is identical, or very closely related, there is no rejection mechanism. This type of transplant has been most successful.

The term **allograft** refers to a transplantation between individuals of the same species but with different genetic backgrounds (Figure 15-6). The closer the relationship of donor and recipient, the more successful the transplant. The success of this type of transplant has been moderate. Frequently, it is used as a temporary measure where the transplant functions until the original tissue is able to repair or regenerate itself. Skin transplants and blood transfusions might properly be considered of this type.

The term **xenograft** refers to transplantation between animals of different species. This type of transplantation has had brief success and, in most instances, has been conducted with laboratory animals. It is a type of transplant limited primarily to research and experimentation at the present time.

Over the years transplants have been performed with tissues such as skin or portions of bone. The success of these types of transplants when small portions are used has been quite high and has led to the establishment of tissue banks. This type of transplantation has been used where there has been a serious burn of the skin area, where some type of corrective surgery is needed, or where bones have been broken or destroyed by disease. In some instances this type of transplantation is standard procedure since death of the individual generally would not result from the eventual rejection of the transplant. Organ transplantation, other than with experimental animals, is a decisive procedure used only in those cases where survival of the individual is in extreme jeopardy. The life expectancy of the individual is extremely short and transplantation represents the last resort.

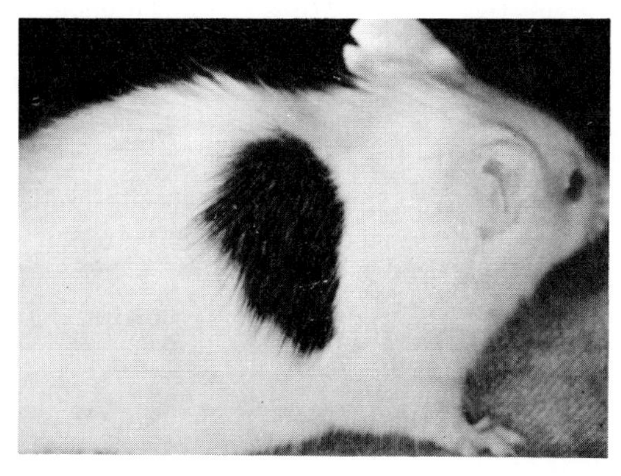

FIGURE 15-6. Photograph of an allograft from a species with dark-colored fur to one with light-colored fur. [*Courtesy of Anthony P. Monaco M.D.*]

The use of artificial replacements has been quite successful in many instances. The major problem has been to develop devices and materials that can duplicate the complicated physiological activities of the anatomical parts. Plastics have been quite successful and are widely used in replacing vessels, bones, and valves. The advantages of using plastics are that they are inert and do not trigger the body's rejection mechanisms. Artificial replacements are generally limited to relatively small replacement parts.

Considerable research has occurred in the development of artificial organs. The major problem, however, has been to produce a device that can duplicate the physiological function and still be small enough to be implanted. The artificial kidney has been most successful in duplicating the activities of this organ; however, it is too large for use in the body. It is hoped that some day the machine will be miniaturized to a point where it can be implanted.

One of the most outstanding accomplishments in the development of a device which can perform a physiological function is the *pacemaker.* It duplicates the function of a small network of fibers in the wall of the atrium from which the heart beat originates. The pacemaker sends electrical impulses to the heart, stimulating the heart to beat. These pacemakers are implanted within the body and are powered by batteries which last up to five years. At the end of that time the batteries can be replaced by a relatively simple operation. The recent utilization of atomic powered batteries will no doubt make such periodic replacement unnecessary.

At the present time research is proceeding in the

development of such artificial organs as the liver and heart. The research on artificial transplants and artificial organs is most encouraging and has stimulated much work in this area.

Organ Transplants. With the aid of new techniques such as radiation and the use of immunosuppressive drugs to increase the time before rejection of transplants, there has been an increase in the number of human organ transplants. In addition, tissue typing techniques which are similar to blood typing are making it possible to match more closely the immunological makeup of the recipient in transplants.

The evaluation of the success of organ transplants becomes a matter of degree. Except for certain kidney transplants, the lives of the recipients have been prolonged from a period of a few days to slightly more than a year. If by means of a transplant, a recipient is able to live even a brief period longer, this represents some degree of success. In addition, each transplant provides knowledge and leads to the development of techniques that will enable more successful transplants in the future.

Many of the technological and physiological problems have been solved in organ transplantation. The surgical techniques of making the necessary vascular and neuron connections have developed to a point where they no longer pose serious problems. The transplanted organ, if transferred immediately from donor to recipient, is capable of performing adequately and in most instances where there has been damage to the organ there is evidence of regeneration. It should be noted again that the major barrier to be overcome is that of the protective adaptation of the donor to reject the organ. The various means used to repress rejection have been temporary and their continued administration leaves the recipient without a major defense against disease and infection.

Implications of Transplantation. The transplantation of human organs inevitably arouses widespread public interest, both scientific and emotional, and it seems likely that as more progress is made, this interest will intensify. As a result of the widespread publicity given to transplantation, the legal profession, in conjunction with physicians, has established certain ethical guidelines. For example, in 1968 the Judicial Council of the American Medical Association developed the following guidelines.

1. No physician may assume responsibility in organ transplantation unless the rights of both donor and recipient are equally protected.
2. The physician should provide a prospective donor with the same care usually given others being treated for a similar injury or disease.
3. The death of a donor should be ascertained by at least one physician other than the recipient's physician.
4. The physician must have a full discussion of the proposed procedure with the donor and recipient or their responsible relatives or representatives.
5. Transplantation should be undertaken only by physicians who possess special medical knowledge and technical competence.

In addition to the above-mentioned moral and ethical aspects of transplantation, the legal implications are far reaching. These include the selection of donors, selection of recipients, ascertaining the exact time of death, fixing criminal responsibility, and determining the origin of the disease.

The selection of a donor and a recipient appear to be purely medical problems and their legal implications are obscure. The actual removal of organs, however, is a legal problem and according to the proposed Uniform Anatomical Gift Act, ". . . any individual of sound mind and eighteen years of age or more may give all or part of his body for any purposes deemed desirable to him." Moreover, ". . . the gift may be made by will and becomes effective upon death and can be made to any hospital, surgeon, or physician for medical or dental education, research, advancement of medical or dental science, therapy or transplantation, or to any accredited medical or dental school, college, or university, to any bank storage facility or to any specified individual for therapy or transplantation needed by him." This act, if accepted and made into law, would eliminate many of the legal barriers and problems that now exist.

Another section of the act states that ". . . the time of death shall be determined by a physician who attends a donor at his death, or if none, the physician who certifies death." In addition, ". . . this physician shall not participate in either the removal or transplantation of a part." This section of the act is very significant to the medical profession because it leaves the decision as to death in the hands of the medical profession.

To be sure, it may be years before these and related problems are settled, but unquestionably certain actions are being taken by the medical and legal professions so that future decisions will have some precedents.

The science of transplantation is a new one and truly represents one of the major frontiers. The advances that have been made up to the present time give some indication of what is to come in this area in the near future. Research and experimental transplants are being made and attempted with almost all parts of the human organism. The eventual results of this work will be commonplace successful transplants, and a greater knowledge of the functions of the human body.

16 Breathing and Waste Disposal

GUIDE QUESTIONS

1 Explain the relationships between O_2 and CO_2 and the bloodstream. Of what importance are respiratory pigments? In what way is respiration (breathing) related to digestion and circulation?
2 Discuss gaseous exchange in plants. How are photosynthesis and cell respiration related to each other in plants? What purpose does soil air serve? What is the importance of pH?
3 Compare external and internal respiration. How is each accomplished in annelids? humans?
4 Outline in detail the principal methods by which protists and animals obtain oxygen. State specific examples.
5 Explain skin breathing in annelids. Compare skin breathing and breathing in mollusks and insects.
6 In what ways do lungs show advances over gills? How do tracheal tubes compare in efficiency to lungs and gills?
7 Discuss gaseous exchange in fish and amphibians.
8 Outline the structure of the human respiratory system and give the function(s) of each part. How does breathing occur in humans? How are gas pressure gradients established?
9 How are gases transported in humans? Discuss the controls of respiration.
10 Describe the procedure for performing exhaled air ventilation.
11 Explain how you would perform external cardiac massage.
12 Discuss the effects of smoking on the body as reported in the Surgeon General's Report.
13 Describe the clinical symptoms and effects on the body associated with emphysema.
14 Explain the sequence of events involved in the development of a bronchogenic carcenoma.
15 Discuss the functions of excretory systems in organisms. In what ways are excretory systems related to circulatory systems? Explain. What types of wastes must be removed? How are wastes defined?
16 What processes are involved in waste elimination? Discuss waste elimination in plants. What modifications are found in protists for waste elimination?
17 Outline in detail the various excretory systems in invertebrates. How does each operate?
18 Discuss excretion in humans. Outline the structures, indicate what wastes are excreted, and how each structure functions.
19 Discuss the liver and its role in homeostasis. In what ways is the liver related to digestion, circulation, breathing, and excretion?
20 Describe how an artificial kidney operates.

LIVING organisms require nutrients for their metabolic activities, and, as already shown, these nutrients are obtained by various methods, digested, and finally transported to the sites of metabolic activity by the circulatory systems. Having been delivered to the various metabolic sites, the nutrients are processed by specific enzyme reactions to release their stored chemical energy for use by the cells. These reactions involve an oxidation of the nutrients in a controlled sequence of steps known as cellular respiration (see Chapter 7). As a result of cellular respiration, carbon dioxide is produced as an end product along with water and, of course, ATP. Considering these relationships, then, the processes of bringing oxygen to the cells of an organism and carrying away carbon dioxide are as essential as the digestion and absorption of food.

Exchange of Gases

In all aerobic organisms, oxygen is obtained from the environment by diffusion through cellular membranes. In some organisms, such as *Amoeba,* the diffusion is direct from the environment into the cell cytoplasm; in most higher animals, there are specialized structures for allowing diffusion of oxygen from the environment into the circulatory system for distribution to body cells. The exchange of oxygen and carbon dioxide between active cells and the environment is usually called *external respiration.* Organs used for this exchange are termed *respiratory organs* and are grouped into *respiratory systems.*

In general, oxygen will remain as O_2 whether as a gas or in solution and carbon dioxide will react with water to form carbonic acid, which, in turn, dissociates to form H^+ and HCO_3^- (bicarbonate) ions. In addition, if a respiratory pigment is present, such as hemoglobin or hemocyanin, both oxygen and carbon dioxide will combine with the pigment to form complexes of various types.

Gaseous Exchange in Plantae

During the day, two metabolic activities, photosynthesis and respiration (cellular respiration), occur in leaves and in green stems. Photosynthesis, the synthesis of carbohydrates, utilizes carbon dioxide and water as raw materials, whereas respiration utilizes oxygen in oxidative reactions. In addition, photosynthesis releases oxygen as a by-product, whereas respiration yields carbon dioxide. Both photosynthesis and respiration occur during the day, but, so far as gaseous exchange is concerned, photosynthesis masks respiration because it proceeds more rapidly. At night, photosynthesis ceases and respiration continues. Thus, at night, photosynthetic plants give off carbon dioxide and take in oxygen, a condition nearly the reverse of that during the day. These two metabolic activities, photosynthesis and respiration, are integral components of the oxygen cycle (see Figure 24-2), the carbon cycle (see Figure 24-2), and the water cycle (see Figure 24-4). In addition to the oxygen obtained from photosynthesis and the carbon dioxide absorbed from the atmosphere, land plants also receive these gases from *soil air.* This air, found in pore spaces in the soil, contains the same gases as the air present above the soil. Some of the oxygen and carbon dioxide in the soil air may be dissolved in the soil water that is found in the pore spaces and that forms a film around soil particles. Just as this water may diffuse into the roots, so may the gases dissolved in it. In this way, oxygen and carbon dioxide are obtained already dissolved in water which then enters the conducting tissues of the xylem for distribution through the stem to the leaves. As the water passes through the plant, it acts as a medium for the transportation of not only dissolved gases but also dissolved minerals, nutrients, and products of cell metabolism. Water not retained by the plant for assimilation or for photosynthesis is eventually returned to the atmosphere by the process of transpiration (see Figure 14-2).

A study of angiosperm leaf anatomy reveals a structure that is ideally adapted to regulate both gaseous exchange with the atmosphere and transpiration (see Figure 9-10). The **guard cells** play an important role in gaseous exchange in plants and their regulation of the stomata is affected by carbon dioxide concentration and osmotic pressure within the cells (Figure 16-1).

During the day, sugars such as sucrose are formed through photosynthetic activity, thereby increasing the osmotic concentration of the cells. As a consequence of this, the guard cells take up additional water (from the xylem conducting tissues and surrounding cells) and become more turgid, resulting in open stomata. Lowering of carbon dioxide concentration also causes the stomata to open. It is hypothesized that, during the day, photosynthetic consumption of carbon dioxide lowers the carbon dioxide concentration in the guard cells, and this results in the opening of the stomata. At

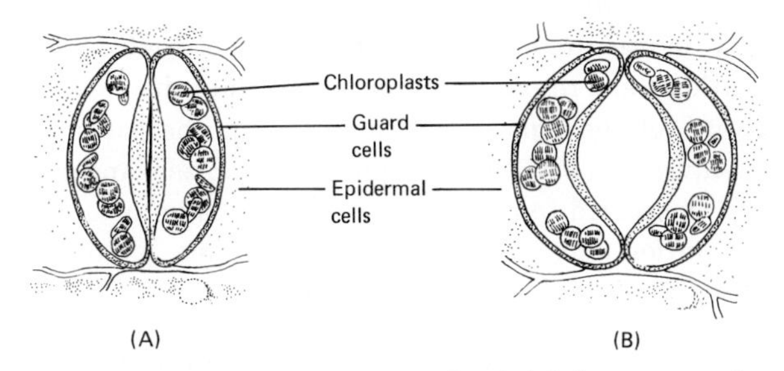

FIGURE 16-1. Stomata. (A) Stoma closed. (B) Stoma open. See text for discussion of mechanisms involved in stomatal closing and opening.

the biochemical level, it has been proposed that carbon dioxide has its effect on a pH-sensitive interconversion of starch and sugar. Carbon dioxide does this by dissolving in water to form carbonic acid which, in turn, dissociates, forming hydrogen ions and bicarbonate ions.

By using up the carbon dioxide in photosynthesis, the guard cell becomes less acidic. This change in pH is supposed to cause a shift in the equilibrium between starch and sugar in favor of conversion of starch to sugar. The hydrolysis of a starch molecule to many sugar molecules results in a greatly increased osmotic concentration (osmotic pressure being directly dependent on its molecular concentration) in the cells and, in turn, turgor pressure, thereby causing the stomata to open.

When the stomata are open during the day, gaseous exchange between the atmosphere and the air in the intercellular spaces of the spongy mesophyll takes place by diffusion. The oxygen produced by photosynthetic reactions within the mesophyll cells, which is in excess of the amount required for respiration, diffuses into the intercellular spaces and passes out of the leaf through the open stomata. Carbon dioxide in the atmosphere enters the leaf through the stomata and is utilized by the mesophyll cells in photosynthesis along with the carbon dioxide produced by respiration.

At night, photosynthetic activities cease, resulting in an increase of carbon dioxide concentration and a decrease in turgor pressure, which, in turn, closes the stomata. Respiration continues with an increase in carbon dioxide concentration in the leaf air spaces and cells. Upon resumption of photosynthesis, the entire process is reversed again.

Gases within the intercellular spaces of the mesophyll may diffuse into either mesophyll cells for their photosynthetic and respiratory functions, or into the vascular tissues of the phloem. In this manner, the gases are dissolved in the water of the phloem cytoplasm and may be transported to all cells of the plant body for use in respiration and, in the case of green stems, photosynthesis.

Gaseous Exchange in Protista and Animalia

In many protistans and invertebrates, the gas exchange is direct from air or water through membranes to the cells; but it is more complex in larger species and in those with dry or nonpermeable exteriors, or those more active. Respiration in these cases may be considered to occur in two stages: *external respiration,* an exchange between the environment and the respiratory organs; and *internal respiration,* an exchange between the body fluids and the tissue cells. Cellular respiration, the utilization of oxygen within the cells, is a part of metabolism.

Protistans and animals may obtain oxygen by one or another of five principal methods.

1. Simple diffusion from water or air through a moist surface into the body.
2. Diffusion from water or air through thin tissues to blood vessels.
3. Diffusion from air or water to a system of air ducts (tracheae).
4. Diffusion from water through gill surfaces to blood vessels.
5. Diffusion from air through moist lung surfaces to blood vessels.

It should be noted that the actual gaseous exchange occurs in a liquid medium; the breathing surface is exposed to water if the animal is aquatic or is kept moist by the animal (through secretions) if the environment is air. Gases must then dissolve in a film of liquid covering the breathing surface before they can diffuse in or out. Thus, a dry breathing surface is inert in terms of gaseous exchange.

Protista. There is no specialized mechanism for gaseous exchange in protistan organisms such as *Amoeba.* The oxygen dissolved in the water environment passes through the cell membrane by simple diffusion. Once inside the membrane, the oxygen dissolves in the cytoplasm, and, by means of cyclosis, is distributed to all cell organelles for their metabolic activities. Likewise, carbon dioxide, formed in cellular respiration, diffuses out of the *Amoeba* into the surrounding water.

Acoelomate Invertebrates. Animals such as sponges, coelenterates, and flatworms have relatively simple body structures. Body cells are either in direct contact with the aquatic environment or are close enough to allow for cell-to-cell transfer of gases. As in the unicells, diffusion of oxygen from the environment takes place through cellular membranes. In sponges and coelenterates, epidermal and gastrodermal cells are in direct contact with a water environment. Epidermal cells and choanocytes in the sponge take in oxygen directly and transfer it to the cells within the mesoglea. In *Hydra,* the mesoglea is very thin, acellular, and functions more as an internal support than as a medium for the transfer of gases. Oxygen diffuses directly into the epidermal cells from the water outside the body, and into the gastrodermal cells from the water within the gastrovascular cavity. Likewise, carbon dioxide diffuses out of the cells into the water.

Planaria also obtain oxygen from the environment by diffusion through the epidermis of the body exterior. The fluid in the parenchyma around the internal organs distributes the oxygen to all body cells. Similarly, carbon dioxide is picked up from the body tissues by the fluid and is carried to the epidermis where it diffuses into the environment.

Coelomate Invertebrates. Coelomate animals possess breathing surfaces within specialized breathing systems. In all other animals, and indeed, even in some of the coelomates (for example, the annelids), gaseous exchange takes the form of skin breathing; that is, CO_2 and O_2 diffuse across all body surfaces exposed to the environment. Internal gas distribution in such cases is adequately accomplished by direct diffusion between adjacent cells and by transport via lymph or blood. Even in animals with specialized breathing systems, the skin may serve as an accessory breathing organ (for example, centipedes and millipedes).

Mollusks, such as clams, aquatic snails, and squids, possess exchange surfaces known as *gills* (ctenidia) (Figure 16-2A). Most gills have finely dissected surfaces which expose a great deal of total surface area to the water environment. Thus, although the gas-exchange surface takes up a limited part of the animal, leaving the rest of the body to evolve relatively impermeable protective coverings, the surface-to-volume ratio of the exchange surface remains high.

Another characteristic of most gills is that they contain a rich supply of blood vessels. The blood in these vessels is often separated from the external water by only one or two cell layers; the single cell layer of the blood vessel alone and/or the single layer of the gill surface. Oxygen diffuses from the water across the cells and into the blood where it is usually picked up by a respiratory pigment. The blood then transports the oxygen throughout the body to all cells. Carbon dioxide,

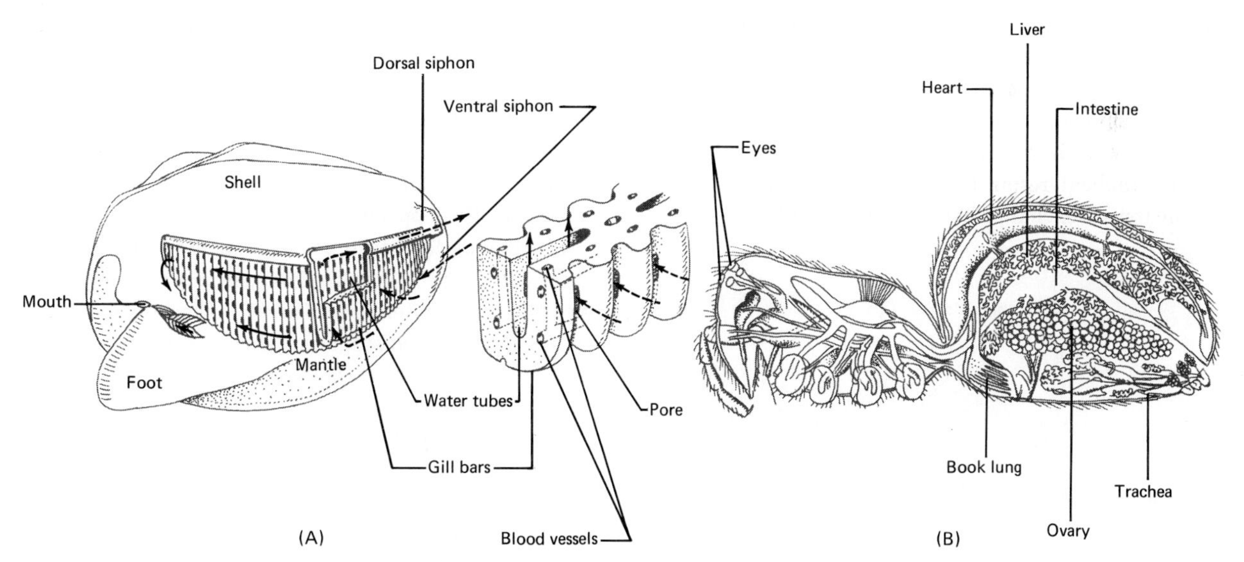

Figure 16-2. Respiratory apparatus of the clam and spider. (**A**) Outer left gill of a freshwater clam showing internal structure (*left*) and portion of a gill enlarged (*right*). (**B**) Structure of a spider with the left side removed, showing position of the book lung.

produced by cellular metabolism, diffuses from the cells into the blood, is transported to the gills, and is discharged into the surrounding water.

Land snails have adapted to terrestrial life and have evolved a modified breathing apparatus. Instead of gills being present, the mantle covering is well supplied with blood vessels. Mucous secretions of the mantle maintain a moist surface for gas diffusion. Thus, even on land, gas exchange occurs across a moist membrane.

A few land animals (spiders) possess modified gill-type structures (book lungs) that function in air exchange (Figure 16-2B). With such structures, however, there is considerable hazard in desiccation of the surfaces. Most terrestrial animals, therefore, have evolved invaginated exchange surfaces known as *tracheal tubes* which are characteristic of land arthropods, such as the grasshopper (Figure 16-3).

Along the lower sides of the thorax and abdomen of the grasshopper are ten pairs of *spiracles,* the openings of the respiratory system. These connect with air-filled cavities from which extend small branching tubes called *tracheae.* The branching is repeated over and over again until the terminal tubes, called *tracheoles,* are often less than 1 μ in diameter. Both tracheae and tracheoles ramify to every part of the body, and, except at their very ends, are filled with air. Oxygen and carbon dioxide move along the tracheal tubes by diffusion. However, the larger cavities near the spiracles are ventilated by the movements of surrounding muscles or exoskeleton. Foreign particles are prevented from entering the tracheae by small hairs around the edges of the spiracles.

This tracheal respiratory system provides a direct avenue for gas exchange between the outside air and the cells. The blood does not function in the gas exchange with the environment, but rather in transportation, distribution, and collection of O_2 and CO_2 in the vicinity of the tissues.

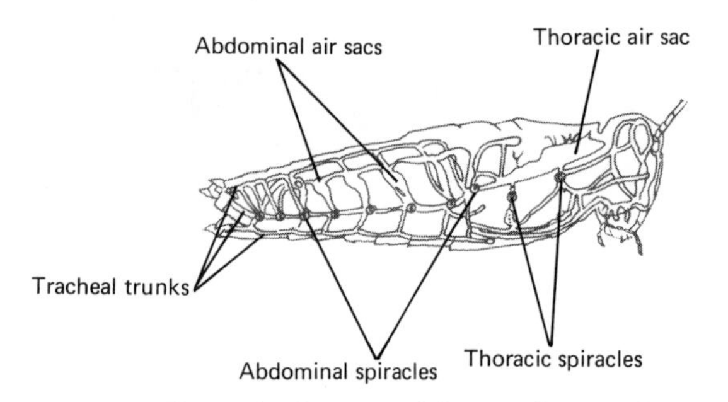

Figure 16-3. The tracheal system of the grasshopper. See text for amplification.

Vertebrates. Aquatic vertebrates such as fish have gills as breathing surfaces. Terrestrial vertebrates have evolved *lungs,* invaginated gas-exchange organs which are limited to a particular body region and dependent upon the blood transport system. Certain vertebrates, however, such as amphibians, may depend on external gills and dermal respiration during part of their life (larval development).

Aquatic respiratory systems are most highly developed in fishes. Gills are located on the edges of a series of lateral openings from the pharynx, just behind the mouth (Figure 16-4A). There are four or five gill slits on each side of the pharynx, opening into an external chamber. These chambers are actually outside the body of the fish, but are covered by a bony plate, the *operculum* (Figure 16-4B), which has a restricted opening at the posterior edge. Between adjacent gill slits is a skeletal *gill arch* that functions in the support of the *gill filaments* and in supplying them with blood vessels. Gill filaments bear minute transverse plates covered with thin epithelium and containing capillaries (Figure 16-4C).

The *gill lamellae,* located on the gill filaments, are composed of thin pleated coverings of epithelium overlying the branchial capillaries (Figure 16-4D). Oxygen is obtained by diffusion from the water through the cells of the epithelium and of the capillaries into the bloodstream. Carbon dioxide is transported to the gills and diffuses in the opposite direction.

In order to facilitate gas exchange by the gills, the water carrying the dissolved oxygen must not only pass over the gill filaments, but must be kept in motion. If the water remained stagnant, the oxygen supply in the vicinity of the gill surfaces would soon be depleted. Thus, water must pass continuously over the gills. This is accomplished in fish by the taking in of water through the mouth and allowing it to pass into the pharynx where it is forced through the gill slits and over the gills. The water leaves the area of the gills through the opening at the posterior end of the operculum.

Amphibians exhibit more diverse methods of respiration than any other group of animals. In the larval

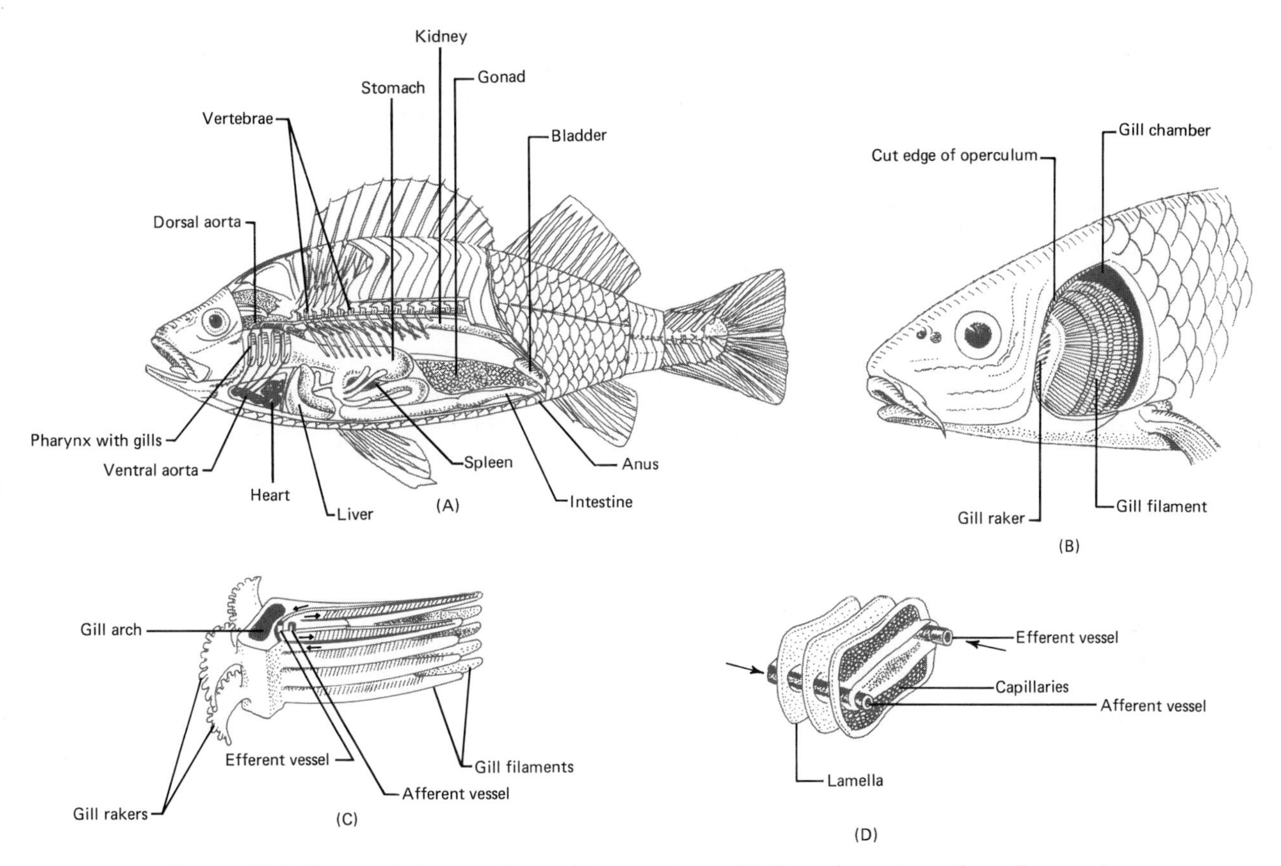

FIGURE 16-4. Representative aquatic respiratory systems. (**A**) General structure of a yellow perch showing the relationship of gills to other structures. (**B**) Gills of a carp in the gill chamber with the operculum removed. (**C**) Enlarged aspect of a carp gill showing arch, rakers, filaments, and blood vessels. (**D**) Enlargement of a portion of a filament of a carp showing lamellae and blood vessels.

stage, for example, the adults of strictly aquatic salamanders such as *Necturus* and *Siren* retain the external gills throughout their lives. Most amphibians respire to some degree through their skin and some respire also through the oral epithelium (for example, the plethodontid salamanders). In dermal breathing, the skin must be kept moist. This is not difficult in aquatic species. However, in terrestrial organisms, this must be accomplished by numerous mucous glands distributed over the body surface. With the exception of lungfishes, amphibians are the lowest vertebrates possessing lungs as sites of gas exchange. During metamorphosis, most amphibians lose their gills and a pair of lungs develops.

Air enters the respiratory tract of the frog (Figure 16-5A) through the *nostrils,* or *external nares,* located at the tip of the nose. The nostrils lead into the *nasal cavities* which, in turn connect with the *mouth cavity* by two small openings, the *internal nares.* Posterior to the tongue, in the region of the pharynx, is a slit, the *glottis,* leading to the *larynx* (Figure 16-5B). The larynx is a cartilaginous box with a membranous floor. Two posterior openings in the larynx lead directly to the ovoid, thin-walled lungs. The lungs are lined with a single layer of epithelium and are divided by septa into numerous small chambers or *air sacs,* which greatly increase the surface area, and are richly supplied with capillaries. Covering the lungs is a membranous *peritoneum,* separating them from other internal organs. Air is pumped into the lungs by a swallowing process. In addition to the development of lungs, amphibians are

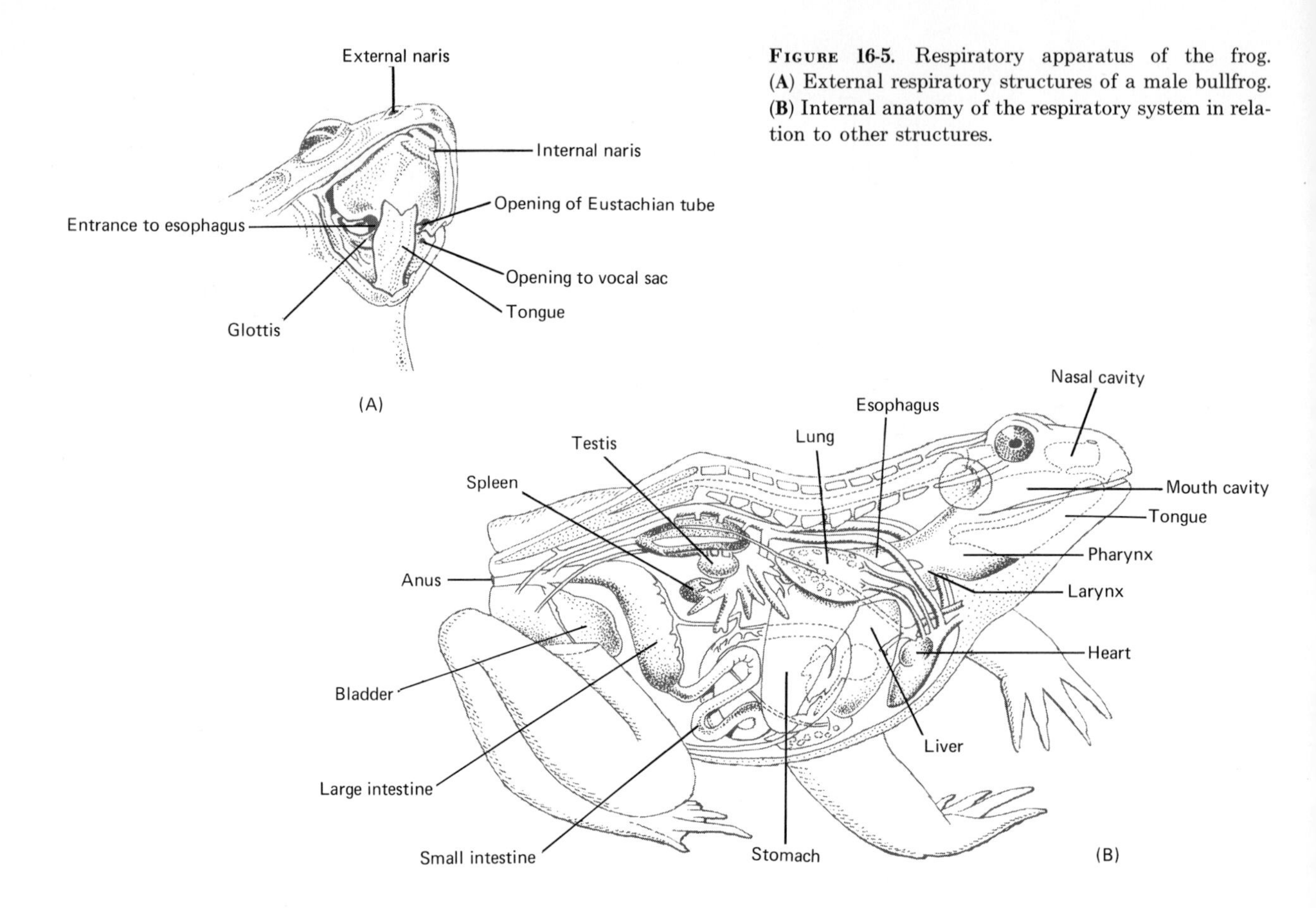

Figure 16-5. Respiratory apparatus of the frog. (**A**) External respiratory structures of a male bullfrog. (**B**) Internal anatomy of the respiratory system in relation to other structures.

the first vertebrates to have a voice (with the exceptions of a few fish). The sound-producing organs are a pair of elastic bands, the *vocal cords,* which are stretched longitudinally across the larynx. As air leaves the lungs, it passes through the larynx and across the vocal cords, causing them to vibrate. Muscles attached to the cords may vary the tension on them, resulting in different sounds. Males of many species of frogs possess *vocal sacs* connected to the sides of the pharynx which,when distended, serve as resonators.

Respiration in Humans

Breathing is the mechanism by which gases are moved into and out of the lungs. Lungs act as sites where oxygen from the air is passed on to some type of transportation system by which it can reach every cell. Methods of breathing are highly varied and have become extremely complex in the larger animals whose oxygen demands are great and whose body structure necessitates breathing systems that can meet the demands efficiently.

Several steps can be recognized in the overall process of respiration: (1) breathing, the movement of air into and out of the lungs, (2) external respiration, the exchange of gases (O_2 and CO_2) between the blood and the air, (3) transportation, the carrying of O_2 and CO_2 by the blood to and from the body cells, and (4) internal respiration, the gas exchange between the blood and the body cells. These steps are necessary and are accessory to the basic process of cellular respiration.

Anatomy of the System. In considering the aforementioned steps in the overall process of respiration, the respiratory system functions in the first and second steps, breathing and external respiration, whereas the circulatory system functions in the last two, transportation and internal respiration. The respiratory system consists of those organs that make it possible for gas exchange between blood and air. Such organs include the nose, pharynx, larynx, trachea, bronchi, and lungs.

Air normally enters the human system through the *nostrils* (anterior nares), the external openings of the *nose* (Figure 16-6A). The nostrils lead into the *nasal cavities* which are separated from the mouth below by the *palate*. The surface area of the nasal cavities is increased by bony folds called *turbinates* (conchae) and is lined with a mucus-covered ciliated epithelium. Air entering the cavities is warmed, cleansed, and moistened by secretions of nasal glands. Cleansing occurs by means of the mucus and by long cilia present in the anterior parts of the cavities; warming is accomplished by heat radiated from the many capillaries present beneath the epithelial lining.

Cleansed, warmed air passes from the nasal cavities into the *pharynx,* or throat (Figure 16-6B). In some instances, air enters the respiratory system through the mouth and passes into the pharynx without being cleansed or warmed. In either case, the pharynx serves as a passageway for the respiratory and digestive tracts, since both food and air must pass through the structure in order to reach the appropriate tubes, the esophagus and the *trachea,* respectively. In addition, there are two openings in the pharynx that lead to the Eustachian tubes which connect to the middle ears (see Figure 11-18A). In this manner, air pressure on the tympanic membrane is regulated.

Air passes from the pharynx through the glottis and into the *larynx,* or "Adam's apple." The opening of the larynx, the *glottis,* is always open except in swallowing. When swallowing occurs, a flap-like structure, the *epiglottis,* covers the glottis, thereby preventing saliva or food from entering the larynx. Below the glottis are fibrous and elastic ligaments which strengthen the edges of the glottis and provide elasticity. These ligaments, covered with mucous membrane, are attached at both ends to the cartilage of the larynx and are called the *vocal folds,* or *vocal cords,* because they function in the production of the voice.

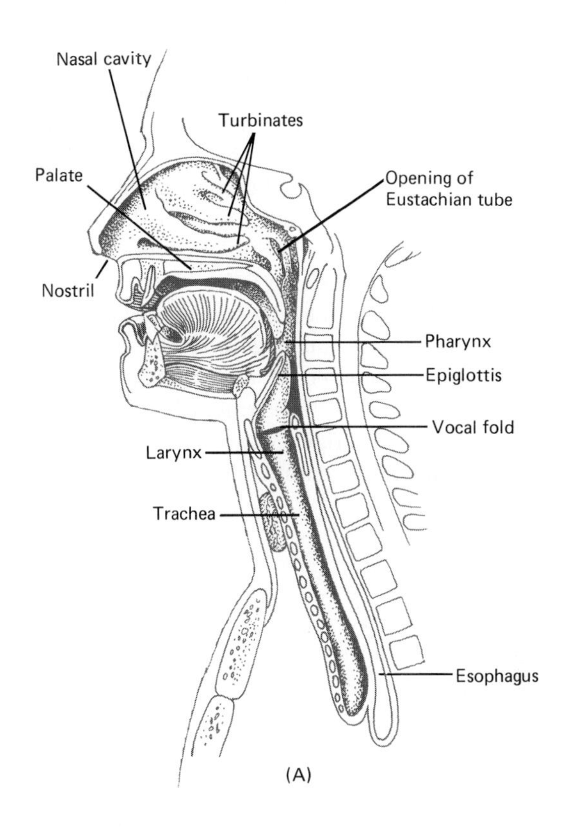

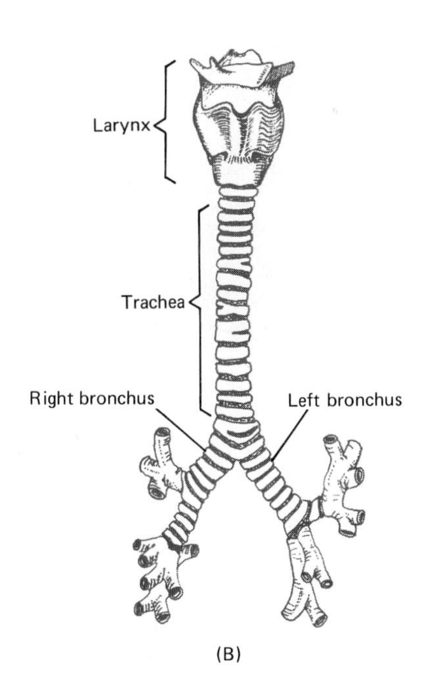

Figure 16-6. Human respiratory system. (**A**) Sagittal section of head, neck, and upper thoracic region showing respiratory passages down to the trachea. (**B**) Anterior view of larynx, trachea, and bronchi.

Air passing out of the lungs and up through the larynx causes the folds to vibrate. These vibrations cause the column of air over them to vibrate and thus give rise to the sound we call the *voice*. The pharynx, mouth and tongue, and nasal cavities above the glottis act as resonators which, together with the size of the larynx and length of the vocal cords, account for the characteristics of the human voice.

A long cylindrical tube, the *trachea,* leads from the larynx to the chest region (Figure 16-6B). Structurally, the trachea is composed of an epithelial lining and smooth muscle with C-shaped rings of cartilage at regular intervals that give rigidity to the wall and prevent it from collapsing. In the chest region, the trachea divides into the *right* and *left bronchi,* leading to the *right* and *left lungs,* respectively.

In the lungs, each bronchus branches repeatedly, forming progressively smaller tubes, the *bronchioles* (Figure 16-7A). At the ends of the smaller bronchioles are clusters of cup-shaped cavities, the **alveoli** (Figure 16-7B). Each alveolus is lined with a layer of flat epithelium (continuous with that of the bronchioles and bronchi) and is surrounded by networks of blood capillaries. In mammalian lungs, such as those of the human, there is a layer of lipoprotein about 50 Å thick lining the alveoli; this functions in lowering surface tension and enables the alveoli to remain open.

The lungs are porous and spongy, owing to the pres-

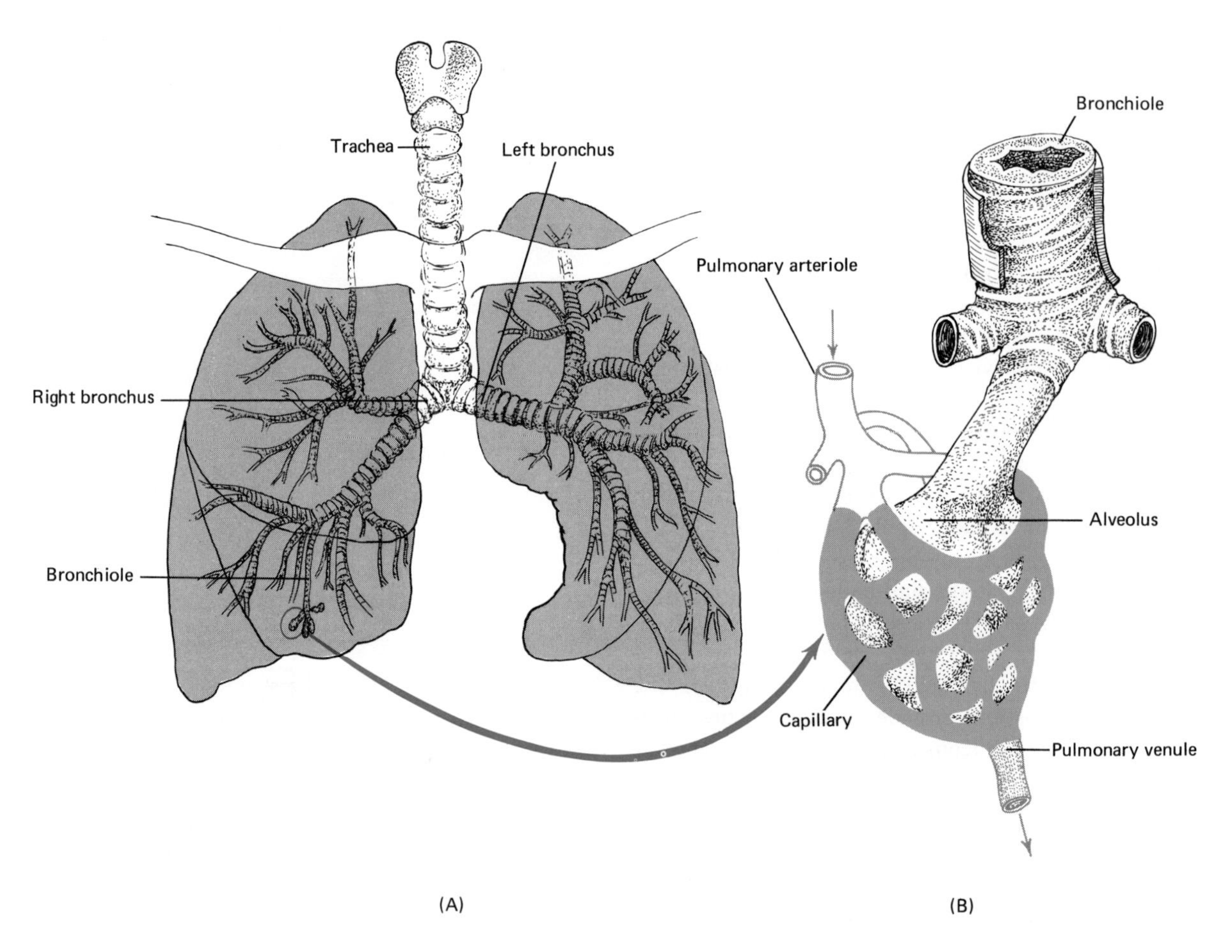

Figure 16-7. Human lungs. (**A**) Relationship of bronchi and bronchioles to lungs. (**B**) Enlarged aspect of an alveolus and a bronchiole.

ence of air within the bronchioles and alveoli and to the abundance of connective tissue that binds the various parts of the lung together. Externally, each lung is covered by a part of the membrane (pleura) lining the chest cavity, referred to as the *visceral pleura.* The pleura continues, from the point of attachment of each lung, as the lining of each pleural cavity and is called the *parietal pleura.* The two layers of the pleura are normally moistened with a small amount of fluid to prevent friction during breathing movements.

PHYSIOLOGY. Air moves in and out of the lungs for the same basic reason the blood flows in vessels—because of a pressure gradient. In this regard, a pressure gradient exists when atmospheric pressure is greater than pressure within the lung, resulting in air flow into the lungs. In other words, *inspiration,* or breathing in, takes place. When the pressure in the lungs exceeds that of the atmosphere, another pressure gradient results, and air flows down the gradient. However, this time it moves in the opposite direction, out of the lungs. In other words, *expiration,* or breathing out, occurs.

During inspiration (Figure 16-8A), these pressure gradients are established by changes in the size of the thoracic cavity by the contraction and relaxation of respiratory muscles, such as the diaphragm and intercostals. Upon contraction, the diaphragm descends, thereby enlarging the vertical dimension of the thorax. The intercostals may contract at the same time, elevating the sternum and ribs, thereby enlarging the thorax laterally and from front to back. Thus, on inspiration, the thoracic cavity enlarges and the pressure in the pleural space decreases to about 751 mm Hg. The lungs enlarge and a lower than atmospheric pressure is created in them. Since the atmospheric pressure is 760 mm Hg, air rushes into the lungs. This movement of air is an application of *Boyle's law* which states that "volume varies inversely with the pressure at constant temperature." Therefore, when the thoracic cavity increases in volume, the lungs expand to fill the additional space. This expansion results in an increase in lung volume, but a decrease in air pressure within the lung, thus establishing a pressure gradient between the lungs and the atmosphere.

In expiration (Figure 16-8B) the diaphragm relaxes and rises, thereby decreasing the volume of the thoracic cavity. This decrease in volume is associated with an

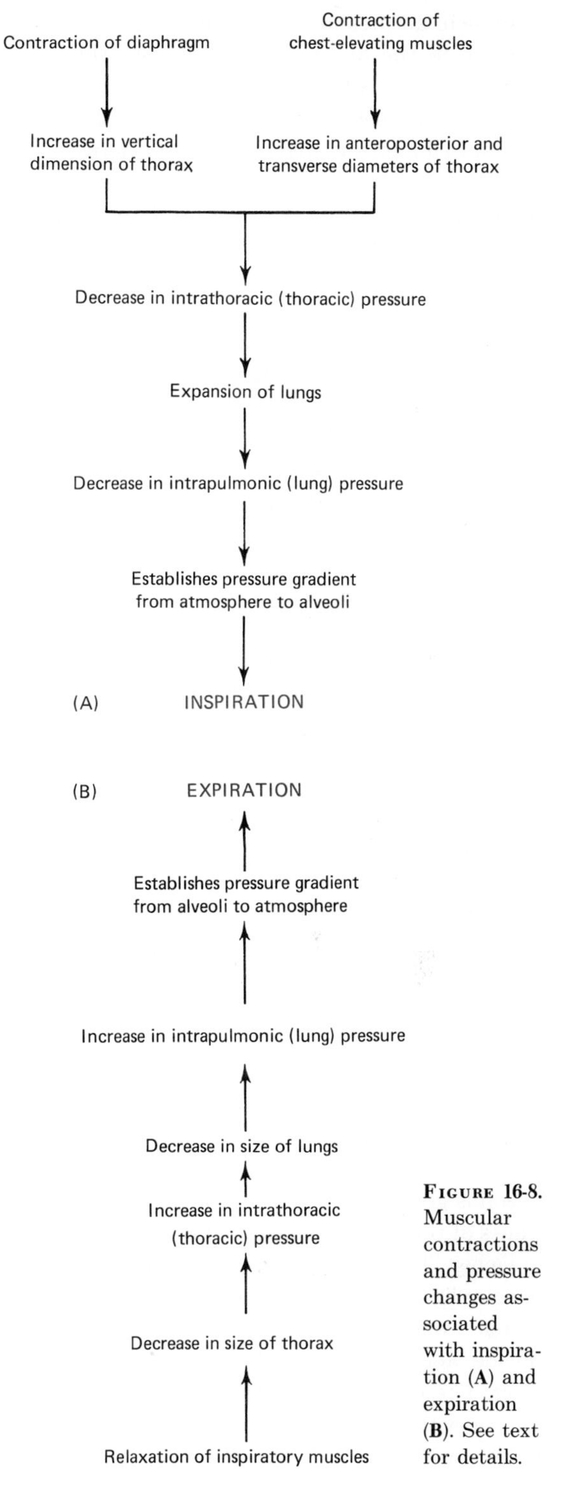

FIGURE 16-8. Muscular contractions and pressure changes associated with inspiration (**A**) and expiration (**B**). See text for details.

increase in air pressure within the lungs. The increased pressure in the lungs is above 760 mm Hg and, as a result, air is forced out of the lungs.

There is a direct relationship between the volume of air inspired and the pressure gradient between the atmosphere and the lung alveoli. In other words, the deeper the inspiration, the lower the pressure within the lungs, the greater the pressure gradient from atmosphere to alveoli, and the larger the volume of air inspired.

In normal adult breathing, about 500 cm^3 of air are taken in and out of the lungs with each inspiration and expiration. This is known as *tidal air*. If inspiration is forced, an additional 2500 cm^3 of air, known as *reserve inspiratory air,* may be taken in. After a normal expiration, an additional 1000 cm^3 of air, called *reserve expiratory air,* may be expelled. The tidal air, reserve inspiratory air, and reserve expiratory air (about 4000 cm^3) are collectively designated as the *vital capacity*. After a forced expiration, there remains in the lungs about 1200 cm^3 of air that cannot be removed. This is termed *residual air*. Thus, the *total lung capacity,* that is, the residual air plus the vital capacity, is about 5200 cm^3. Because of residual air it is possible to determine whether a dead infant has been stillborn or born alive. A lung of a stillborn child will sink, whereas that of a child who was born alive, and who had taken at least one breath, will float.

Transportation of Gases. Oxygen taken in during inspiration diffuses from the alveoli into the blood capillaries where it combines with hemoglobin to form oxyhemoglobin. Likewise, carbon dioxide from the blood (from carbaminohemoglobin and HCO_3^-) diffuses into the alveoli to be given off during expiration. This exchange operates strictly according to the principle of diffusion, aided by the enzymatic conversion of carbonic acid to CO_2 and H_2O. Oxygen concentration in the alveoli is greater than that of the blood, and the membranes of the alveoli and capillaries are both permeable and moist. Thus, oxygen goes into solution on the moist alveolar lining and diffuses from an area of higher to an area of lower concentration. The same is true of carbon dioxide, and both gases behave independently of each other. The combining reactions of oxygen and carbon dioxide with hemoglobin and, in the case of carbon dioxide, with water in the blood, have already been considered in Chapter 14. Oxyhemoglobin is relatively unstable, and when the blood reaches the capillaries in the body tissues where the oxygen content is low, the compound dissociates into hemoglobin and oxygen. The released oxygen does not diffuse directly into the cells, but rather dissolves in the interstitial fluid and, then, as this fluid passes over the cells, the oxygen diffuses into the cells. This same path is taken by carbon dioxide, only in the opposite direction. Carbon dioxide dissolves in the interstitial fluid, some of which returns to the blood where it then either combines with hemoglobin to form carbaminohemoglobin or with water to eventually form the bicarbonate ion. As activity increases in any structure, its cells necessarily utilize oxygen more rapidly. This decreases intracellular and interstitial oxygen concentration which, in turn, tends to increase the oxygen pressure gradient between blood and tissues and to accelerate oxygen diffusion out of the tissue blood capillaries. In this way, the rate of oxygen utilization by cells automatically tends to regulate the rate of oxygen delivery to the cells. As dissolved oxygen diffuses out of arterial blood, blood oxygen concentration decreases, thereby accelerating oxyhemoglobin dissociation to release more oxygen into the plasma for diffusion out of cells. Because of oxygen release to tissues from tissue capillary blood, oxygen pressure, oxygen saturation, and total oxygen content are all less in venous blood than in arterial blood. Similarly, carbon dioxide release from the cells increases the carbon dioxide concentration in tissue capillaries from its arterial level of about 40 mm Hg to its venous level of about 46 mm Hg, and favors the formation of carbaminohemoglobin and the bicarbonate ions.

Control of Respiration. The concentrations of CO_2 and O_2 and the pH of arterial blood are among the primary factors influencing respiration. Carbon dioxide concentration influences respiratory control centers in the medulla directly and, via its action, peripheral chemoreceptors. Normally, the range for arterial pressure of carbon dioxide is 38–40 mm Hg. If it is increased even slightly above 40 mm, certain neurons in the medulla, the *inspiratory centers,* are stimulated, thus producing faster breathing with a greater volume of air per minute moving in and out of the lungs. Conversely, decreased carbon dioxide produces the opposite effects, inhibiting respiratory centers and slowing respirations. In addition, decreases in blood pH (increase in acidity, or H^+) increase respirations.

The role of arterial blood oxygen is not entirely clear, however. Neurons of the respiratory centers, like all other body cells, require adequate amounts of oxygen in order to function properly. If these neurons become hypoxic, they become depressed and send fewer impulses to respiratory muscles, thereby decreasing the number of respirations, or resulting in failure of respiration to occur. A decrease in arterial oxygen to a point above the critical level stimulates chemoreceptors in the carotid and aortic bodies and causes reflex stimulation of the respiratory centers.

Among other factors influencing respiration are blood temperature (increases respiration), sudden pain stimulation (decreases respiration), stimulation of the pharynx and/or larynx (decreases respiration), sudden cold stimuli (decreases respiration), and exercise (increases respiration).

Before examining some of the mechanisms involved in waste disposal among organisms, we will first consider heart/lung resuscitation and two prominent disorders of the respiratory system—emphysema and lung cancer.

Heart and Lung Resuscitation

A serious decrease in respiration or heart rate presents an urgent crisis because the body's cells cannot survive long if they are starved of oxygenated blood. In fact, if oxygen is withheld from the cells of the brain for 4–6 minutes, brain damage or death will result. Heart/lung resuscitation is the artificial re-establishment of normal or near normal respiration and circulation. The two simplest techniques for heart/lung resuscitation are exhaled air ventilation (mouth-to-mouth resuscitation) and external cardiac compression (massage). Both techniques can be administered by a layman at the site of the emergency, and both are highly successful. They can be used for any sort of heart or respiratory failure, whether the cause is drowning, strangulation, carbon monoxide or insecticide poisoning, overdose of a drug or anesthesia, electrocution, or heart attack. However, the success of resuscitation is directly related to the speed and efficiency with which it is applied. Delay may be fatal.

EXHALED AIR VENTILATION. A technique for re-establishing respiration is *exhaled air ventilation*. The first and most important step is immediate opening of the airway. This is accomplished easily and quickly by tilting the victim's head backward as far as it will go without being forced. The tilted position opens the upper air passageways to their maximum size. If the patient does not resume spontaneous breathing after the head has been tilted backward, immediately begin artificial ventilation by either the mouth-to-mouth or the mouth-to-nose method. In the more commonly used mouth-to-mouth method, the nostrils are pinched together with the thumb and index finger of the hand. The rescuer then opens his mouth widely, takes a deep breath, makes a tight seal with his mouth around the patient's mouth, and blows in about twice the amount the patient normally breathes. He then removes his mouth and allows the patient to exhale passively. This cycle is repeated approximately twelve times per minute for adults. Atmospheric air contains about 21% O_2 (oxygen) and a trace of CO_2 (carbon dioxide). Exhaled air still contains about 16% O_2 and 5% CO_2. This is more than adequate to maintain a victim's blood O_2 pressure (p_{O_2}) and CO_2 pressure (p_{CO_2}) at normal levels if air is given at the prescribed rate and amount.

If the rescuer observes the following three signs, he knows that adequate ventilation is occurring.

1. The chest rises and falls with every breath.
2. He feels the resistance of the lungs as they expand.
3. He hears the air escape during exhalation.

The three most common errors in mouth-to-mouth resuscitation are inadequate extension of the victim's head, inadequate opening of the rescuer's mouth, and an inadequate seal around the patient's mouth or nose. If the rescuer is sure he has not made these errors and he is still unable to inflate the lungs, a foreign object is probably lodged in the respiratory passages. The rescuer's fingers should be swept through the patient's mouth to remove such material. An adult victim with this problem should next be rolled quickly onto his side. Firm blows should be delivered over his spine between the shoulder blades in an attempt to dislodge the obstruction. Then exhaled-air ventilation should be resumed quickly. A small child with an obstructive foreign object should be picked up quickly and inverted over the rescuer's forearm while firm blows are delivered over the spine between the shoulder blades. Then ventilation can be resumed. A more recently devised technique for expelling objects from the air passageways is known as the *Heimlich maneuver.* The principle employed in this antichoke first aid method is the application of force to compress the air in the lungs to expel foreign objects. If the patient is upright, the rescuer stands behind him and places both arms around his waist just above the belt line and grasps the right wrist with the left hand. The patient's abdomen is then rapidly and strongly compressed. If the patient is recumbent, the rescuer may rapidly and strongly force the patient's diaphragm upright by applying pressure. Then ventilation can be resumed.

If the patient is an infant or small child, the rescuer should make the following adjustments in his technique. The neck of an infant is so pliable that forceful backward tilting of the head may obstruct breathing passages, so the tilted position should not be exaggerated. The rescuer should also remember that the lungs of a small child do not have a large capacity. To avoid over-inflating the child's lungs, he should cover both the victim's mouth and nose with his mouth and blow gently, using less, volume, at a rate of 20–30 times a minute.

External Cardiac Compression (Massage). *External cardiac compression,* or closed-chest cardiac compression (CCCC), consists of the application of rhythmic pressure over the sternum. The rescuer places the heels of his hands on the lower half of the sternum and presses down firmly and smoothly at least 60 times a minute. This action compresses the heart and produces an artificial circulation because the heart lies almost in the middle of the chest between the lower portion of the sternum and the spine. When properly done, external cardiac compression can produce systolic blood pressure peaks of over 100 mm Hg. It can also bring carotid arterial blood flow up to 35% of normal.

Complications that can occur from the use of cardiac compression include fracture of the ribs and sternum, laceration of the liver, and the formation of fat emboli. They can be minimized by adhering to the following precautions:

1. Never compress over the xiphoid process at the tip of the sternum. This bony prominence extends down over the abdomen, and pressure on it may cause laceration of the liver, which can be fatal.
2. Never let your fingers touch the patient's ribs when you compress. Keep your fingers off the patient, and place the heel of your hand in the middle of the patient's chest over the lower half of his sternum.
3. Never compress the abdomen and chest simultaneously since this action traps the liver and may rupture it.
4. Never use sudden or jerking movements to compress the chest. Compression should be smooth, regular, and uninterrupted, and should comprise only 50% of the cycle; relaxation comprises the other half of the cycle.

Compression of the sternum produces some artificial ventilation but not enough for adequate oxygenation of the blood. Therefore exhaled air ventilation must always be used with it. This combination constitutes *heart/lung resuscitation.* When there are two rescuers, the most physiologically sound and practical technique is to have one rescuer apply cardiac compression at a rate of at least 60 compressions per minute. The other rescuer should exhale into the patient's mouth between every fifth and sixth compression. The sequential steps in emergency heart/lung resuscitation must be continued uniformly and without interruption until the patient recovers or is pronounced dead.

Smoking and the Respiratory System

Tobacco (*Nicotiana tabacum*) is an herb that has been smoked by European man for well over 300 years. The most comprehensive investigation of the relationship between cigarette smoking and health is the report to the United States Surgeon General dealing with smoking and health, probably best known by the conclusion it evoked.

Warning: The Surgeon General has determined that cigarette smoking is dangerous to your health.

Essentially, the investigating committee on smoking and health evaluated three principal kinds of scientific evidence:

1. Animal experiments.
2. Clinical and autopsy studies.
3. Population studies.

Effects of Smoking. As reported by the Advisory Committee on Smoking and Health:

(1) Cigarette smoking is causally related to lung cancer in men; the magnitude of the effect of cigarette smoking far outweighs all other factors. The data for women, although less extensive, point in the same direction. In comparison with non-smokers, average male smokers of cigarettes have approximately a 9- to 10-fold risk of developing lung cancer and heavy smokers at least a 20-fold risk.

(2) Cigarette smoking is the most important of the causes of chronic bronchitis in the United States, and increases the risk of dying from chronic bronchitis and emphysema.

(3) Male cigarette smokers have a higher death rate from coronary artery diseases than non-smoking males.

(4) Although a causal relationship has not been established, higher mortality of cigarette smokers is associated with many other cardiovascular diseases, including miscellaneous circulatory diseases, other heart diseases, hypertensive heart disease, and general arteriosclerosis.

(5) Pipe smoking appears to be causally related to lip cancer.

(6) Cigarette smoking is a significant factor in the causation of cancer of the larynx.

(7) There is an association between cigarette smoking and peptic ulcers.

(8) Women who smoke cigarettes during pregnancy tend to have babies of lower birth weight.

Emphysema and Lung Cancer. As noted previously, cigarette smoking increases the risk of dying from emphysema and is causally related to lung cancer. We will now describe some of the clinical symptoms and effects of emphysema and lung cancer.

One lung disease that starts with the deterioration of some of the alveoli is **emphysema.** The alveolar walls lose their elasticity and remain filled with air during expiration. The name of the disease means "blown up" or "full of air." Reduced forced expiratory volume is the first symptom. Later, alveoli in other areas of the lungs are damaged. The lungs become permanently inflated, that is, they cannot collapse because of the loss of elasticity. The patient has to work to exhale. Oxygen diffusion does not occur as easily across the damaged alveoli, blood p_{O_2} is somewhat lowered, and any mild exercise that raises the oxygen requirements of the cells leaves the patient breathless. Carbon dioxide diffuses much more easily across the alveoli than oxygen, so the p_{CO_2} is not affected initially. But as the disease progresses, the alveoli degenerate and are replaced with thick fibrous connective tissue. Even carbon dioxide does not diffuse easily through this fibrous tissue. If the blood cannot buffer all the carbonic acid that accumulates, the blood pH drops. Or, unusually high amounts of carbon dioxide may dissolve in the plasma. High carbon dioxide levels are toxic to the brain cells. Consequently, the inspiratory center becomes less active and the respiration rate slows down, further aggravating the problem. The capillaries that lie around the deteriorating alveoli are compressed and damaged and may no longer be able to receive blood. As a result, pressure increases in the pulmonary artery, and the right atrium overworks as it attempts to force blood through the remaining capillaries.

Emphysema is generally caused by any of a number of long-term irritations. Air pollution, occupational exposure to industrial dusts, and cigarette smoke are the most common irritants. Chronic bronchial asthma may also produce alveolar damage. Cases of emphysema are becoming more and more frequent in the United States, which is ironic because the disease can be prevented and the progressive deterioration can be stopped simply by eliminating the harmful stimuli.

As part of ordinary breathing, many irritating substances are inhaled. Almost all pollutants, including inhaled cigarette smoke, have an irritating effect on the bronchial tubes and lungs.

Close examination of the epithelium of a bronchial tube reveals that it consists of three kinds of cells (Figure 16-9A). The uppermost cells are columnar cells that contain the cilia on their surfaces. At intervals between the ciliated columnar cells are the mucus-secreting goblet cells. The bottom of the epithelium normally contains two rows of basal cells above the basement membrane. Researchers have learned that one of the most common types of lung cancer, **bronchogenic carcinoma,** starts in the walls of the bronchi.

The constant irritation of inhaled cigarette smoke and pollutants causes an enlargement of the goblet cells of the bronchial epithelium (Figure 16-9B). They respond by secreting excessive amounts of mucus. The basal cells respond to the stress by undergoing cell division so fast that the basal cells push into the area occupied by the goblet and columnar cells. As many as twenty rows of basal cells may be produced. Many researchers believe that if the stress is removed at this point, the epithelium can return to normal.

If the stress persists, more and more mucus is secreted

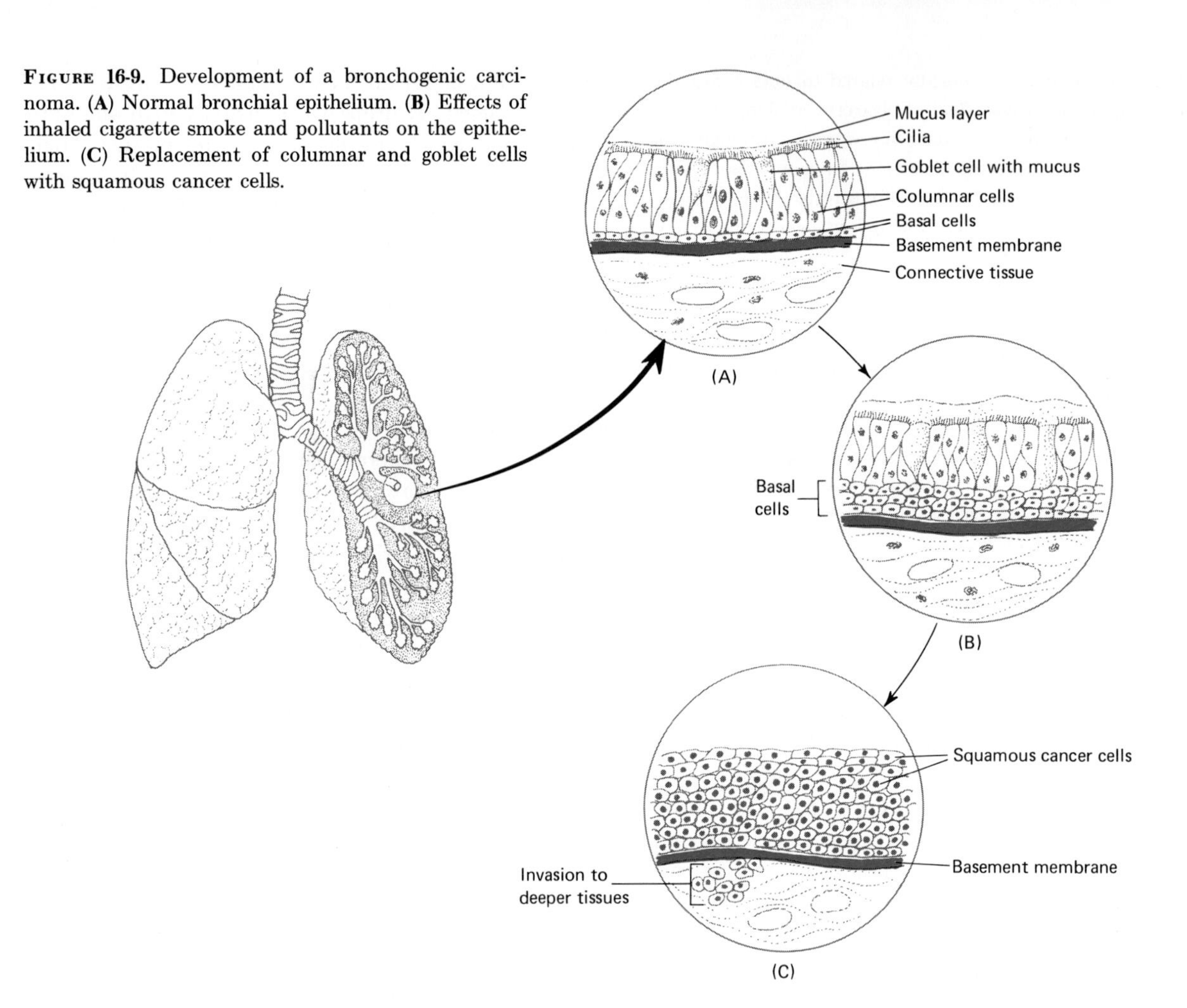

FIGURE 16-9. Development of a bronchogenic carcinoma. (**A**) Normal bronchial epithelium. (**B**) Effects of inhaled cigarette smoke and pollutants on the epithelium. (**C**) Replacement of columnar and goblet cells with squamous cancer cells.

and the cilia becomes less effective. As a result, mucus is not carried toward the throat; instead, it remains trapped in the bronchial tubes. The individual then develops a "smoker's cough." Moreover, the constant irritation from the pollutant slowly destroys the alveoli, which are replaced with thick, inelastic connective tissue. Mucus that has accumulated becomes trapped in the air sacs. Millions of the sacs rupture. This results in a loss of diffusion surface for the exchange of oxygen and carbon dioxide. The individual has now developed emphysema. If the stress is removed at this point, there is little chance for improvement because alveolar tissue that has been destroyed cannot be repaired, but further destruction of lung tissue can be avoided.

Assuming that the stress continues, the emphysema gets progressively worse and the basal cells of the bronchial tubes continue to divide and break through the basement membrane. At this point the stage is set for bronchogenic carcinoma. Columnar and goblet cells disappear and may be replaced with squamous cancer cells (Figure 16-9C). If this happens, the malignant growth spreads throughout the lung and may block a bronchial tube. If the obstruction occurs in a large bronchial tube, very little oxygen enters the lung, and disease-producing bacteria thrive on the mucoid secretions. In the end, the patient may develop emphysema, carcinoma, and a host of infectious diseases. Treatment involves surgical removal of the diseased lung. However, metastasis of the growth through the lymphatic or blood system may result in new growths in other parts of the body, including the brain and liver.

The processes that we have just described have been observed in some heavy smokers. One should realize, though, that there are other causes of emphysema, such as chronic asthma. There are also other factors that may be associated with lung cancer. For instance, breast, stomach, and prostate malignancies can metastasize to the lungs. People who have apparently not been exposed to pollutants do occasionally develop bronchogenic carcinoma. However, the occurrence of bronchogenic carcinoma is probably more than twenty times higher in heavy cigarette smokers than it is in nonsmokers.

Waste Disposal

Waste disposal, or **excretion,** is the process of eliminating metabolic wastes from the circulatory fluids or cells of an organism. If not removed, these wastes can upset the delicate physiochemical balance of the cellular protoplasm and fluids of an organism. Thus, the *excretory system* of an organism is responsible not only for eliminating wastes but also for maintaining the balance of the chemical environment.

Types of Wastes Removed

Metabolic by-products may be of various kinds. For example, digestion of carbohydrates may yield glucose molecules whereas oxidation of the glucose in cellular respiration yields carbon dioxide, water, and energy in the form of ATP. While glucose can be utilized by the cell, carbon dioxide cannot (with the exception of photosynthetic cells); therefore, carbon dioxide is a waste substance that must be eliminated. Nitrogen compounds such as ammonia, produced by the hydrolysis of amino acids, are additional waste products. Like carbon dioxide, ammonia also diffuses out dissolved in water. In simple aquatic organisms this poses no difficulty, since the elimination is fairly rapid. However, in terrestrial organisms, ammonia often must be converted to some other form because of its toxicity when accumulated. Most vertebrates convert ammonia to *urea* (humans) or to uric acid (birds and reptiles).

In addition to gaseous and nitrogenous wastes, excess water, salts, and organic materials must also be eliminated. The excretory system of any organism must be sensitive to concentrations inasmuch as water and salts are both wastes and nonwastes, the difference being one of amount, not kind. Therefore, one mechanism of excretion involves *selective filtration,* whether through the membrane of a unicell or of an excretory unit. In multicellular organisms, where all or most cells are not exposed directly to the environment, excretory systems must then operate in close association with conducting fluids and act as screening devices. In this regard, they must retain valuable substances within the fluids and collect only wastes for removal to the exterior. Thus, excretion involves three basic processes: *filtration, reabsorption,* and *secretion.*

Disposal in Protista

Waste disposal in protistans is a relatively simple process, since these organisms rely on diffusion across the cell membrane. *Amoeba* supplements the excretory diffusion by *phagocytic excretion;* that is, vacuoles containing undigested food particles and other wastes are moved to the cell membrane where the particles are "left behind" as the organism moves away. This is essentially the reverse procedure of that involved in phagocytic nutrition. *Paramecium* utilizes an *anal pore* (cytopyge) for the elimination of solid wastes from the remains of the food vacuole.

While gaseous wastes are eliminated by diffusion, excess water, which may contain dissolved urea, is removed by a specialized structure, the *contractile vacuole.* Each contractile vacuole goes through a cycle in which there are two stages: one in which it fills with liquid and becomes larger and larger, and one in which the vacuole contracts, thereby ejecting its contents from the cell. The contractile vacuoles (full and empty) of a *Paramecium* are shown in Figure 16-10.

Disposal in Invertebrates

Sponges and coelenterates lack specialized structures for waste elimination; they rely solely on diffusion into the surrounding water. The beginnings of an excretory system appear in flatworms such as planaria (Figure 16-11). Typically, flatworm excretory systems consist of

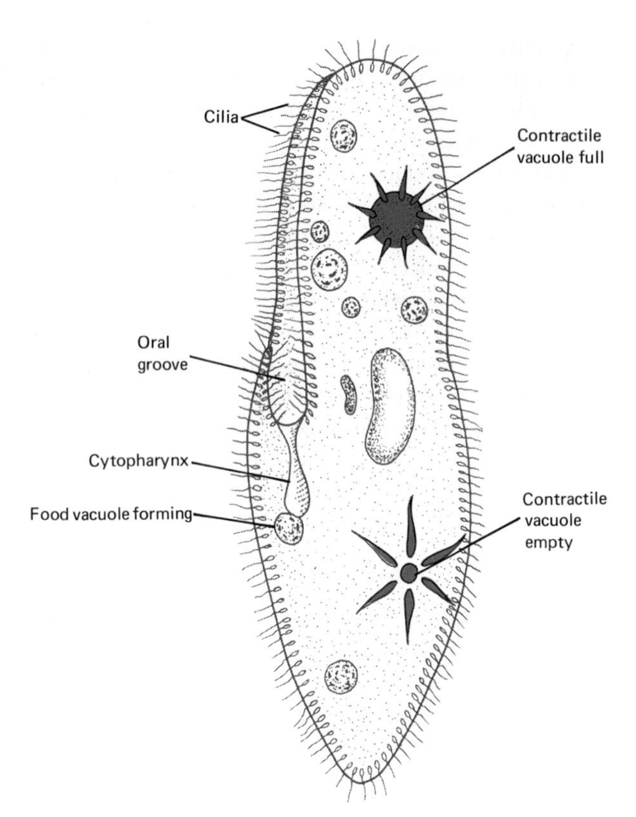

Figure 16-10. Contractile vacuoles of *Paramecium.* The upper vacuole is full and the lower one is empty.

two or more longitudinal branching tubules which may open to the surface through tiny pores or which may unite to form a bladder with a single exterior opening (as in flukes). The functional units of the systems are the many small bulb-like structures at the ends of the side branches. Each bulb consists of a hollow center into which a tuft of long cilia projects. The hollow centers are continuous with the tubules so that wastes move from the tissues into the bulbs and finally into the tubules for excretion through the pores. Currents are maintained by flickering movements of the cilia which give rise to the naming of these bulbs as *flame cells.*

Such "flame cell" systems appear to function chiefly in the regulation of water balance; most metabolic wastes are excreted from the tissues into the environment by direct diffusion or by ejection from the mouth.

Animals possessing circulatory systems utilize the blood as a fluid transport and collection medium for waste removal. For this reason, the blood vessels and the blood must be closely associated with both the body tissues and the excretory organs.

Earthworms, for example, illustrate a close association of the two systems (Figure 16-12). Within each segment is found a pair of excretory organs, the **nephridia,** which open to the outside and are independent of each other. Essentially, a nephridium consists of an open *funnel* with cilia to maintain currents, a coiled

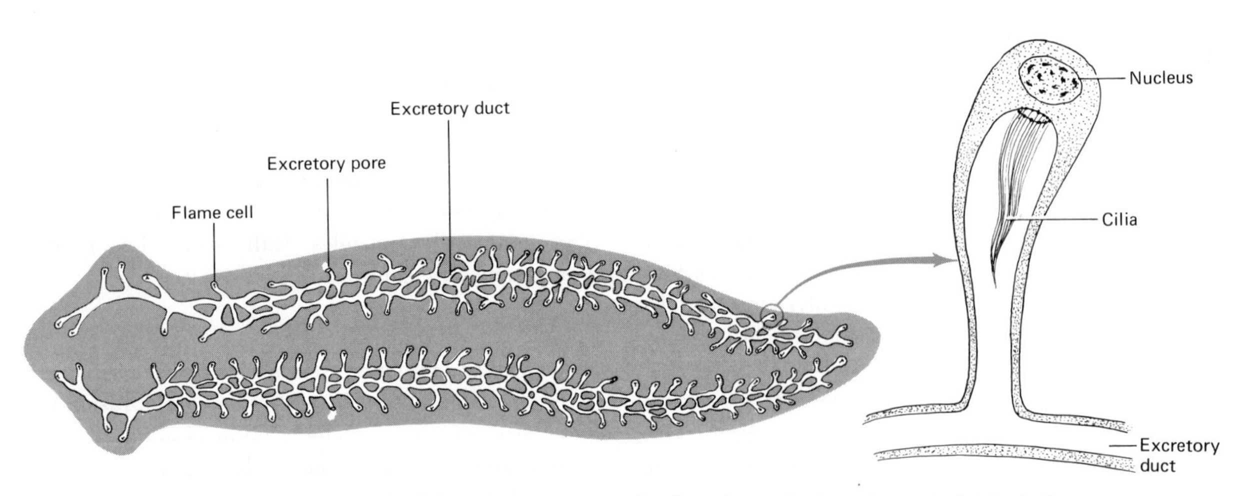

Figure 16-11. Components of the excretory system of a planarian and enlarged aspect of a single flame cell.

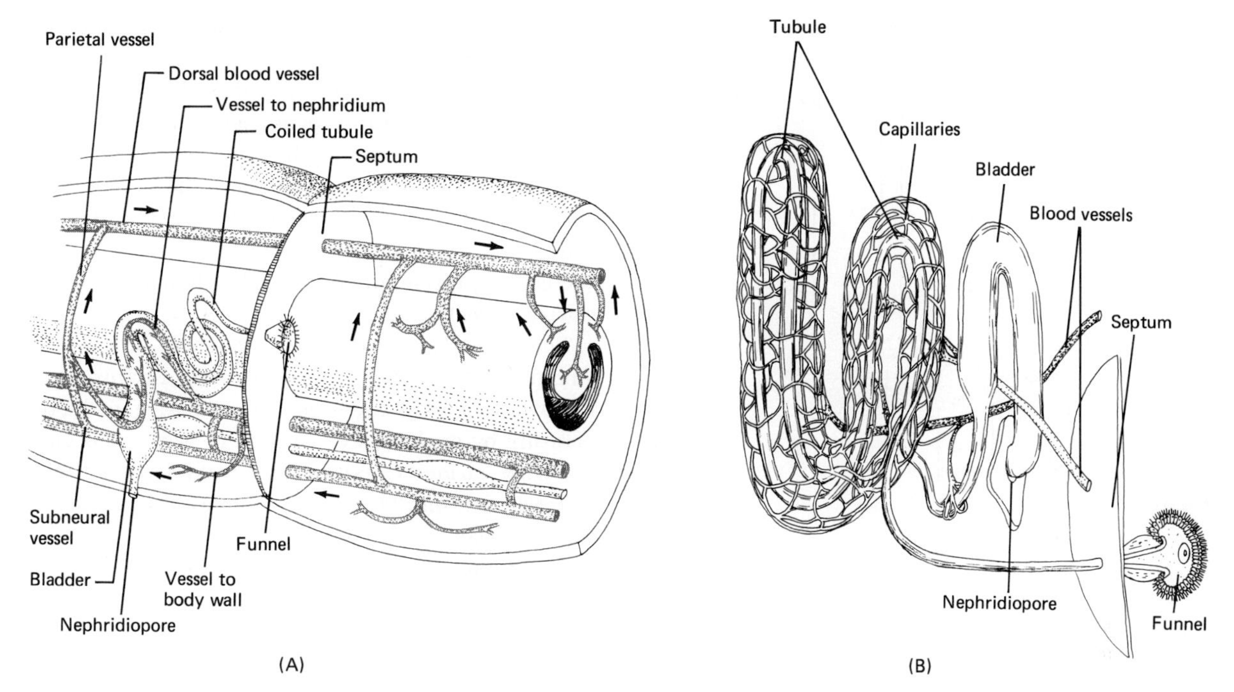

FIGURE 16-12. Excretory apparatus of the earthworm. (**A**) Relationship of a nephridium to the circulatory system. The nephridium has been omitted from the somite on the right for clarity. (**B**) Enlarged aspect of a single nephridium.

tubule connecting the funnel to a *bladder,* and a *nephridiopore* through which the materials are expelled from the bladder. Surrounding the tubule is a network of capillaries. Wastes in the coelomic fluid are picked up by the funnel and then are transported through the tubule to the bladder. Wastes within the bloodstream may diffuse directly from the capillaries into the tubule. In addition, there may also be some limited reabsorption of materials (water, salts) from the tubule into the capillaries. Collected wastes in the bladder are then expelled into the environment. Solid wastes, such as undigested food or inedible materials from the soil, pass from the intestine to the outside through the anus (see Figure 13-7A).

Insects have evolved a different type of excretory system that does not rely on capillary networks since they have open circulation (such a system lacks capillaries). Instead, the insects have structures called *Malpighian tubules,* diverticula of the digestive tract which are bathed directly by the blood in the open sinuses of the body (Figure 16-13). Structurally, these tubules are blind sacs, variable in number, and are located at the junction of the midgut (stomach) and hindgut (intestine). Fluid from the blood enters the ends of the tubules and as it moves toward the gut region, the nitrogenous wastes are precipitated as uric acid and much of the water and salts are reabsorbed by the blood. The concentrated urine passes into the hindgut and rectum where most of the remaining water is reabsorbed, leaving a very dry material for excretion. Thus, both nitrogenous wastes and undigested materials pass from the anus. The elimination of gaseous wastes (carbon dioxide) has already been considered earlier in this chapter.

Human Excretion

The human excretory system is basically similar to that of earthworms in that there is a close association between the closed circulatory system and the excretory system. Carbon dioxide removal involves capil-

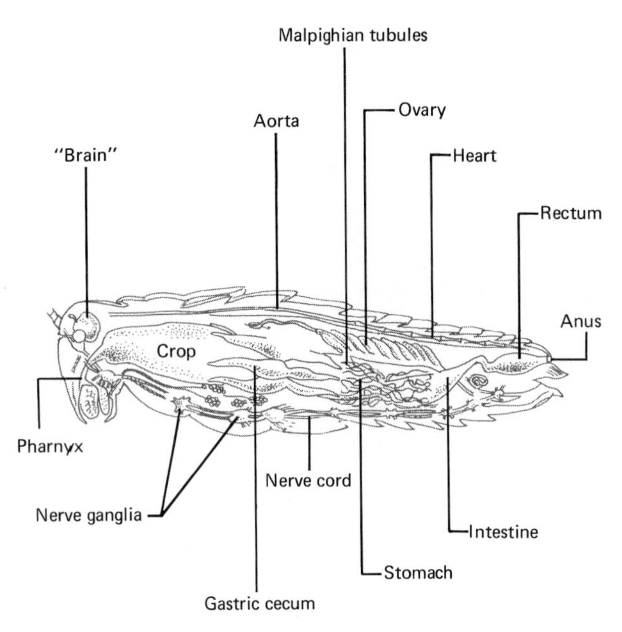

FIGURE 16-13. Internal structure of the left side of a grasshopper showing location of Malpighian tubules.

laries in the tissues of the body as well as capillaries of the alveoli in the lungs; intervening blood vessels serve as a pathway from tissues to lungs. Blood capillaries function in the absorption of nutrients from the small intestine, leaving only indigestible materials in the alimentary canal. These materials pass into the large intestine where much of the water and salts is reabsorbed into the bloodstream; the rest is passed to the rectum. Thus, the large intestine functions in the elimination of solid wastes (feces).

The skin also serves as an excretory organ in the elimination of water, salts, and heat. In this regard, however, the skin functions in regulating body temperature by enabling evaporation of water to occur at the body surface (perspiration). Salts dissolved in the water are left on the skin as a residue and may easily be washed away.

ANATOMY OF THE KIDNEY. Most of the elimination of excess water and soluble salts, urea, creatinine, uric acid, and sulfates, occurs by means of the *kidney* (Figure 16-14A). Human kidneys, paired bean-shaped structures located in the back of the abdominal cavity,

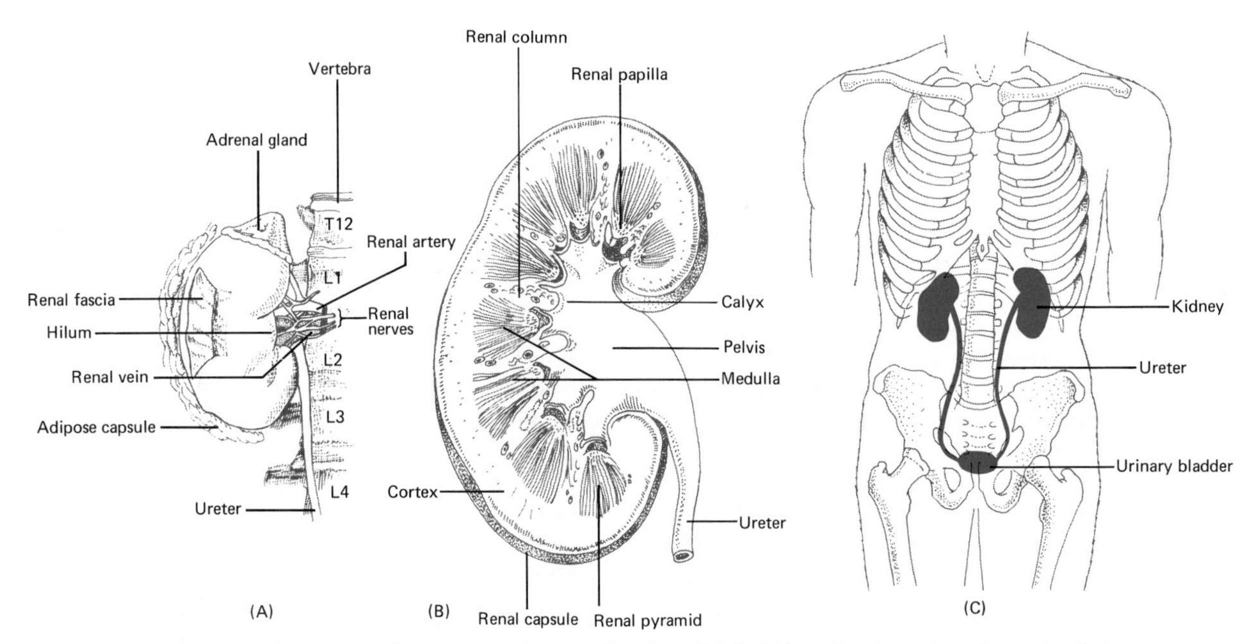

FIGURE 16-14. Human urinary system. (**A**) Anterior view of right kidney showing external aspects of a portion of the renal fascia capsule elevated. (**B**) Longitudinal section through the right kidney showing internal macroscopic anatomy. (**C**) Position of the urinary apparatus in the body.

are compact, discrete organs consisting of massed functional units called **nephrons.** Each kidney is covered by a fatty tissue, the *adipose capsule,* and this, in turn, is surrounded by fibrous connective tissue called the *renal fascia.*

Near the center of the concave border is an indentation referred to as the *hilum* (Figure 16-14A) which functions as a route of entry and exit for the blood vessels, lymph vessels, nerves, and ureters of the kidneys. If a kidney is cut longitudinally (Figure 16-14B), three internal regions may be noted: an outer portion, the *cortex;* an inner portion, the *medulla;* and the expanded cavity of the upper end of the ureter, the *pelvis.* The cortex and medulla are separated by arterial and venous arches. Structurally, the medulla is divided into several triangular wedges, the *renal pyramids.* The bases of the pyramids are toward the cortex, and their apices, or *renal papillae,* are oriented toward the center of the kidney. The pyramids have a striated appearance as compared with the smooth texture of the cortex.

Renal capsules, tubules, and blood vessels of the nephrons are contained within the cortex. Much of the bulk of the kidney (both cortex and medulla) consists of tubules with their associated blood capillaries and lymphatics surrounded by only enough connective tissue to maintain orientation.

Each nephron (approximately 1.5 million in each kidney) consists of a closed bulb, *Bowman's capsule* (or glomerular capsule), and a fairly long coiled tube (Figure 16-15). Tubules of the various nephrons empty into *collecting tubules* which, in turn, empty into the pelvis of the kidney. Ducts leading into the pelvis, the *ureters,* empty into the *urinary bladder,* which functions as a storage organ for urine (Figure 16-14C). Another duct, the *urethra,* drains the bladder and carries its contents to the outside.

Blood reaches each kidney via a *renal artery,* a branch of the descending aorta. As it enters the kidney,

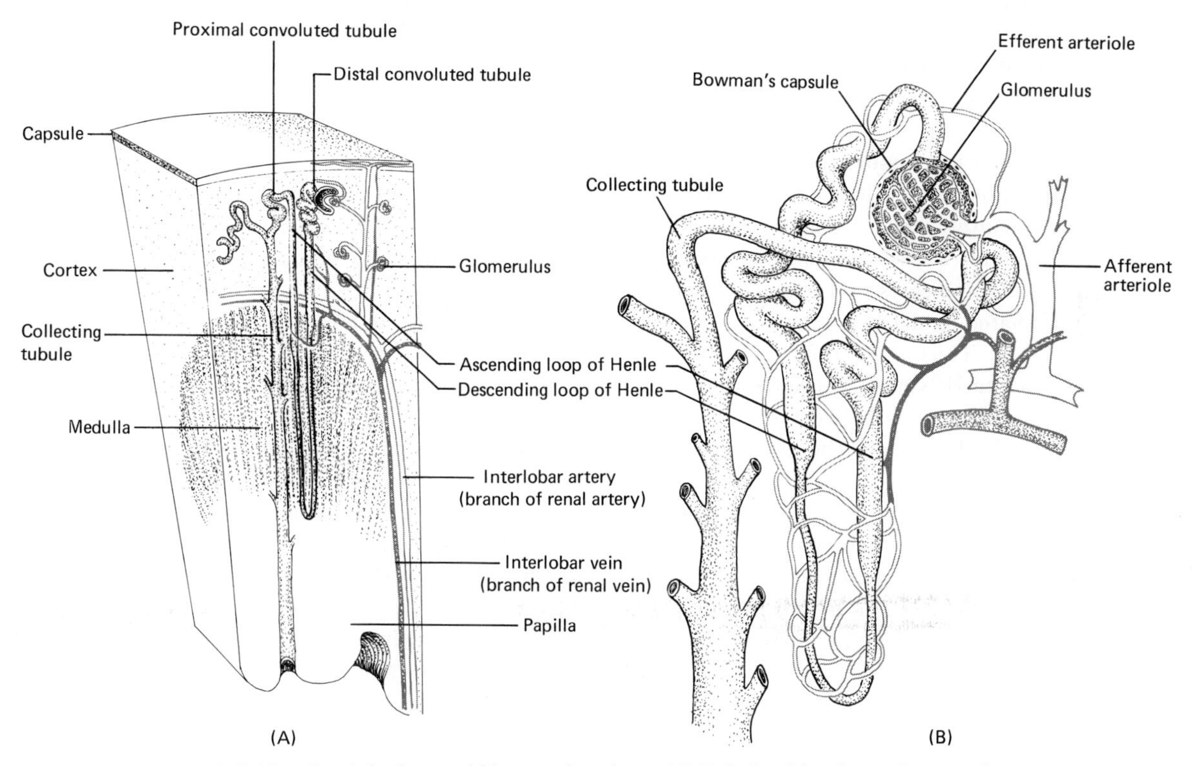

Figure 16-15. Details of the human kidney and nephron. (**A**) Relationship of a nephron to the cortex and medulla. (**B**) Details of a single nephron.

the renal artery divides into many branches that run through the medulla into the cortex, where each arteriole penetrates into a cup-like depression of the Bowman's capsule. Within each capsule, the arteriole subdivides into a tuft of capillaries, the *glomerulus*. Blood leaving the glomerulus enters an arteriole that then divides into a second capillary network around the tubule of the nephron. Finally, these capillaries unite to form a venule which unites with other venules to form the *renal vein* leading from the kidney to join the inferior vena cava.

Physiology. The functions of the kidney are to excrete urine and to regulate the water, electrolyte, and solute balances of the bloodstream. There are three basic processes involved: glomerular filtration, tubular reabsorption, and tubular secretion.

Glomerular filtration involves the movement, by pressure, of water and solutes from the glomerulus into the Bowman's capsule. This filtrate has essentially the same concentration of dissolved substances (glucose, urea, salts, amino acids, glycerol, etc.) as blood plasma. Lacking are the formed elements and most of the plasma proteins, all of which are too large to filter through the membranes of the glomerulus and capsule to any appreciable extent. It has been demonstrated that if the hydrostatic pressure in the glomerular capillaries is increased, the volume of the filtrate is increased proportionately, and if the pressure is decreased, the filtrate volume decreases proportionately. Additional evidence also indicates that changes in filtrate volume are not related to changes in oxygen consumption by the kidney. Thus, it appears that the cells of the glomerulus and Bowman's capsule do not carry out active transport but that the work is done by the heart as it forces the blood into the glomeruli under high hydrostatic pressure.

If the glomerular filtrate were expelled from the body without modification, many valuable substances (water, glucose, amino acids, salts) would be lost. *Selective reabsorption* of most of the water and many of the dissolved materials is one of the functions of the tubule, leaving a concentrated fluid called *urine* to be eliminated. Much of the reabsorption of salts occurs in the *loop of Henle,* where sodium is actively transported from the filtrate into the medullary interstitial fluid, thereby raising its osmotic pressure. This causes more water to be reabsorbed from the collecting tubule, producing a concentrated urine. In addition to the reabsorption of water, all of the glucose, almost all of the amino acids, and most of the salts are returned to the blood.

Tubule cells also function in *secretory* activities. For example, potassium and hydrogen ions are secreted into the urine, as is ammonia. This supplements glomerular filtration and increases the efficiency of the overall regulation of blood composition.

Role in Homeostasis. Regulation of osmotic pressure of extracellular fluids is accomplished by the kidneys. If large amounts of water are ingested, more water will be eliminated and the ratio between water and electrolytes will change.

The concentration of each electrolyte is also regulated by the kidneys. This is accomplished by tubular reabsorption and tubular secretion under the influence of hormones produced in the pituitary and adrenal glands (antidiuretic hormone and mineralocorticoids).

In addition to the elimination of metabolic wastes, particularly those of protein metabolisms (urea, uric acid, creatinine, and ammonia), the kidneys help to keep the pH of the plasma within normal limits. This is accomplished by the regulation of the excretion of hydrogen ions and electrolytes in tubular reabsorption and secretion.

The Liver and Homeostasis. The liver functions in a variety of ways to maintain and regulate homeostasis of body fluids and control of body processes.

After a meal, the blood in intestinal capillaries contains high concentrations of simple sugars and amino acids—concentrations much greater than those normally found in the circulatory system. As this blood enters the liver via the portal vein, the liver removes most of the excess glucose, converting it to glycogen. It also removes the fructose and galactose, converts them to glucose, and then stores the glucose as glycogen. If the liver has stored its capacity of glycogen, it converts glucose into fat, which can then be stored in various regions of the body as adipose tissue. This entire process is reversed if no food is ingested and no sugar is being absorbed from the intestine.

Enough glycogen is stored within the liver to supply glucose to the blood for a period of about 24 hours. If no new glucose enters the liver, the liver begins converting fats and amino acids into glucose, thereby maintaining the blood sugar level.

The liver also plays a role in secreting substances for intestinal digestion such as bile salts, bile pigments, and cholesterol. In addition to digestive substances, the liver forms and stores certain vitamins such as A, B_{12}, and D. Vitamin K is utilized in the formation of the blood-clotting protein prothrombin. Fibrinogen and heparin are also formed by the liver. Since the liver is an expandable organ, it has the ability to store large quantities of blood in its vessels and thus aids in the regulation of blood volume.

Finally, the liver is responsible for the detoxification of certain end products of digestion, for example, phenol, skatole, and indole. Certain drugs and heavy metals such as morphine, strychnine, and mercury are eliminated from the bloodstream, stored, and freed slowly so that their toxicity is diminished by dilution.

As a result of its many chemical activities, the liver provides a great deal of heat for the body, thereby maintaining optimum temperatures for other body organs. In summary, then, the liver functions as a manufacturing organ (formation of chemical compounds), a storage organ (storage of glycogen, vitamins), an excretory organ (excretes bile, urea, detoxification products), and as a heat source (resulting from the chemical reactions).

We will conclude the discussion of waste disposal by examining the principle of the artificial kidney.

Hemodialysis: The Artificial Kidney

If the kidneys are impaired so severely by disease or injury that they are unable to excrete nitrogenous wastes and regulate pH and electrolyte concentration of the plasma, then the blood must be filtered by an artificial device. Such filtering of the blood is called *hemodialysis*. *Dialysis* means using a semipermeable membrane to separate large nondiffusible particles from

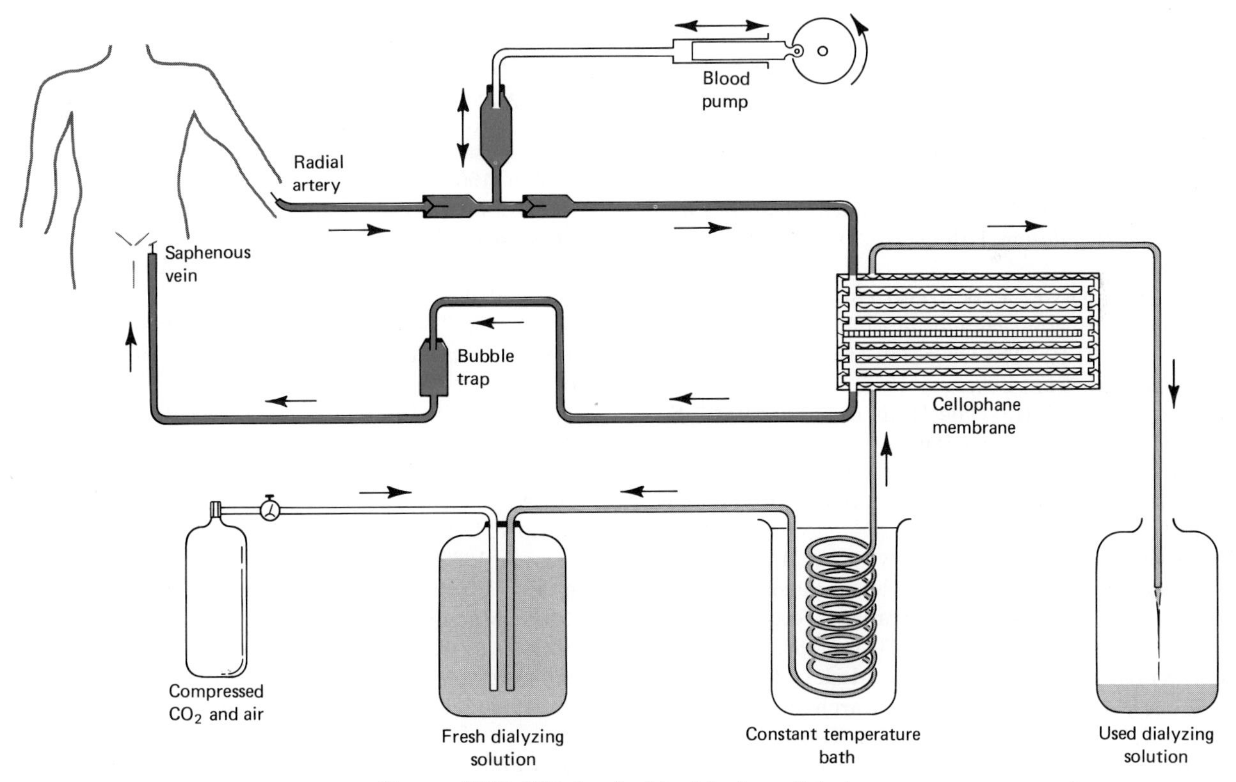

Figure 16-16. Filtering the blood by hemodialysis.

smaller diffusible ones. One of the most well-known devices for accomplishing dialysis is the kidney machine (Figure 16-16). When the machine is in operation, a tube connects it with a much smaller tube implanted in the patient's radial artery. The blood is pumped from the artery and through the tubes to one side of a selectively permeable cellophane membrane. The other side of the membrane is continually washed with an artificial solution called the dialyzing solution.

All substances (including wastes) in the blood except protein molecules and erythrocytes can diffuse back and forth across the semipermeable membrane. The electrolyte level of the blood is controlled by keeping the dialyzing solution electrolytes at the same concentration as that found in normal plasma. Any excess blood electrolytes move down the concentration gradient and into the dialyzing solution. If the blood electrolyte level is normal, it is in equilibrium with the dialyzing solution, and no electrolytes are gained or lost. Since the dialyzing solution contains no wastes, substances like urea move down the concentration gradient and into the dialyzing solution. Thus, wastes are removed, and normal electrolyte balance is maintained.

One additional advantage of the kidney machine is that an individual's nutritional status can be bolstered by placing large quantities of glucose in the dialyzing solution. While the blood gives up its wastes, the glucose diffuses into the blood. Thus, the kidney machine beautifully accomplishes the function of the nephron.

Obvious drawbacks to this artificial kidney system include the fact that the blood must be anticoagulated while dialysis is occurring, and a very large amount of the patient's blood must flow through this apparatus to make it work. The blood that has passed through the artificial kidney is treated with a substance (e.g., protamine) that is an antiheparin substance. Also, only about 500 ml of the patient's blood is in the machine at a time, a volume that is easily compensated for by vasoconstriction and increased cardiac output.

SUGGESTED SUPPLEMENTARY READINGS FOR SECTION THREE ORGANISMIC BIOLOGY: MAINTENANCE

ADOLPH, E. F., "The Heart's Pacemaker," *Scientific American,* March 1967.

BARRINGTON, E. J. W., *The Chemical Basis of Physiological Regulation.* Palo Alto, Cal.: Scott Foresman, 1968.

BARRINGTON, E. J. W., *Invertebrate Structure and Function.* Boston: Houghton Mifflin, 1967.

BENDITT, E. P., "The Origin of Atherosclerosis," *Scientific American,* February 1977.

BENZINGER, T. H., "The Human Thermostat," *Scientific American,* June 1961.

BIDDULPH, O., and S. BIDDULPH, "The Circulatory System of Plants," *Scientific American,* February 1959.

BLOOM, W. D., and D. W. FAWCETT, *A Textbook of Histology,* 9th ed. Philadelphia: Saunders, 1968.

BODIAN, DAVID, "The Generalized Vertebrate Neuron," *Science,* vol. 137, pp. 323–36, 1962.

CAPALDI, R. A., "A Dynamic Model of Cell Membranes," *Scientific American,* March 1974.

CAPRA, J. D., and A. B. EDMUNDSON, "The Antibody Combining Site," *Scientific American,* January 1977.

CASE, JAMES, *Sensory Mechanisms.* New York: Macmillan, 1966.

CLEMENTE, C. D., *Anatomy: A Regional Atlas of the Human Body,* Philadelphia: Lea & Febiger, 1975.

COMROE, J. J., "The Lung," *Scientific American,* February 1966.

COHEN, C., "The Protein Switch of Muscle Contraction," *Scientific American,* November 1975.

COOPER, M. D., and A. R. LAWTON III, "The Development of the Immune System," *Scientific American,* November 1974.

DAVIDSON, E. H., "Hormones and Genes," *Scientific American,* June 1965.

DILLON, L. S., *Animal Variety.* Dubuque, Ia.: Wm. C. Brown, 1967.

FRYE, B., *Hormonal Control in Vertebrates.* New York: Macmillan, 1967.

GARDNER, E., D. J. GRAY, and R. O'RAHILLY, *Anatomy: A Regional Study of Human Structure,* 4th ed. Philadelphia: Saunders, 1975.

GRIFFIN, D. R., and A. NOVICK, *Animal Structure and Function,* 2nd ed. New York: Holt, Rinehart and Winston, 1970.

GROLLMAN, S., *The Human Body,* 2nd ed. New York: Macmillan 1974.

GUILLEMIN, R., and R. BURGUS, "The Hormones of the Hypothalamus," *Scientific American,* November 1972.

GUYTON, A. C., *Textbooks of Medical Physiology,* 5th ed. Philadelphia: Saunders, 1976.

HEIMER, L., "Pathways in the Brain," *Scientific American,* July 1971.

HILLMAN, W. (ed.), *Papers in Plant Physiology.* New York: Holt, Rinehart and Winston, 1970.

HOLTER, H., "How Things Get Into Cells," *Scientific American,* September 1961.

HOYLE, G., "How Is Muscle Turned On and Off?" *Scientific American,* April 1970.

HUGHES, G. M., *Comparative Physiology of Vertebrate Respiration.* Cambridge: Harvard University Press, 1965.

JANICK, J., et al., "The Cycles of Plant and Animal Nutrition," *Scientific American,* September 1976.

KIMBER, D., et al., *Anatomy and Physiology,* 15th ed. New York: Macmillan, 1972.

LANGLEY, L. L., *Homeostasis.* New York: Van Nostrand Reinhold, 1965.

LESTER, H. A., "The Response to Acetylcholine," *Scientific American,* February 1977.

LIEVER, C. S., "The Metabolism of Alchohol," *Scientific American,* March 1976.

LLINAS, R. R., "The Cortex of the Cerebellum," *Scientific American,* January 1975.

LURIA, A. R., "The Functional Organization of the Brain," *Scientific American,* March 1970.

MARGARIA, R., "The Sources of Muscular Energy," *Scientific American,* March 1972.

MERRILL, J. P., "The Artificial Kidney," *Scientific American,* July 1961.

MONTAGNA, W., "The Skin," *Scientific American,* February 1965.

PATT, D. I., and G. R. PATT, *Comparative Vertebrate Histology.* New York: Harper & Row, 1969.

PROSSER, C. L., and F. A. BROWN, Jr., *Comparative Animal Physiology.* Philadelphia: Saunders, 1973.

SCRIMSHAW, N. J., and V. R. YOUNG, "The Requirements of Human Nutrition," *Scientific American,* September 1976.

STENT, G. S., "Cellular Communication," *Scientific American,* September 1972.

THOMAS, R. C., "Electrogenic Sodium Pump in Nerve and Muscle Cells," *Physiology Review,* vol. 52, p. 563, 1972.

TORTORA, GERARD J., D. CICERO, and H. PARISH, *Plant Form and Function.* New York: Macmillan, 1970.

TORTORA, G. J., *Principles of Human Anatomy.* San Francisco: Canfield Press, 1977.

VANDER, A. J., *Renal Physiology.* New York: McGraw-Hill, 1975.

WURTMAN, R. J., "The Effects of Light on the Human Body," *Scientific American,* July 1975.

FOUR
Organismic Biology: Continuity

17 Patterns of Reproduction

18 Developmental Processes

19 Principles of Inheritance

20 Human Heredity

21 The Evolution of Life

17 Patterns of Reproduction

GUIDE QUESTIONS

1 Define reproduction. How does asexual reproduction differ from sexual reproduction?
2 Define and describe each of the following: fission, budding, sporulation, vegetative propagation, and regeneration. Cite specific examples of each.
3 What is meiosis? List the stages and the principal events of each stage. What are the significant results of meiosis?
4 Discuss the timing of meiosis in the life cycles of organisms.
5 How is sexual reproduction accomplished in Monera?
6 Differentiate between isogamy and heterogamy. What is an alternation of generations?
7 Compare the life cycles of *Ulva,* a moss, and a fern. What are the essential similarities and differences?
8 What events are associated with microspore development and megagametophyte development in pine? How does pollination occur in pine?
9 Define a flower. List the principal parts of a flower and give the function(s) of each.
10 How does pollen develop? ovules? Describe pollination in a flowering plant. How does fertilization occur in a flowering plant?
11 What is a seed? What are some of the means by which seeds may be dispersed? Define a fruit. What is parthenocarpy? Relate exocarp, mesocarp, and endocarp to fruit structure.
12 Describe conjugation in *Paramecium.* What is hermaphrodism? How does sexual reproduction occur in the earthworm?
13 What is metamorphosis? parthenogenesis?
14 Describe and give the function(s) of the following structures of the human male reproductive system: testes, epididymis, vas deferens, seminal vesicles, ejaculatory ducts, prostate gland, Cowper's glands, and penis.
15 How are sperm formed in the human male? Describe the major parts of a sperm. What is seminal fluid?
16 Describe and the give the function(s) of the following structures of the human female reproductive system: ovaries, uterine tubes, uterus, vagina, vulva, and mammary glands.
17 Discuss the events associated with the development of a mature egg. What is ovulation?
18 How are the female sexual cycles hormonally controlled?
19 Explain the roles of the male and female in sexual intercourse.

IN A VERY broad context, *reproduction* may be viewed as a self-perpetuating process that is characteristic of all living forms and that results in the continuity of life from one generation to another. Reproduction occurs from the molecular and cellular levels through the organismic level, and the formation of new living units accounts for replacement and addition at every level of organization. Reproduction is not only a means of species continuity, that is, the extension of individual organisms from one generation to the next; it is also a process concerned with species maintenance. For example, as cells age, they tend to wear out. Through various synthetic processes parts of cells are replaced or, in some cases, entire living cells are replaced in response to wear and tear and death due to disease, accidents, or injuries. As noted earlier (Chapter 5), growth of new tissues and organs is dependent upon the continual addition of cells (cell division).

Irrespective of the variety of ways in which reproduction may occur, the essential feature of the process is the separation from the parent organism of a complement of DNA containing the genetic information for directing the growth and development of the newly produced unit. Thus, it is through the various types of reproduction that the specific characteristics of one generation are transmitted to a new one. The newly produced generation may exhibit these traits either exactly as the parent generation or with modifications due to changes in the genetic material as it is transmitted from one generation to the next.

Organisms exhibit considerable diversity with regard to reproductive patterns. The various processes, however, may be categorized into two principal types, asexual reproduction and sexual reproduction. Basically, *asexual reproduction* involves a single parent which gives rise to new offspring, all of which have hereditary characteristics identical to the parent. Inasmuch as special sex cells are usually not formed, there is little chance for new gene combinations in the offspring. *Sexual reproduction,* by contrast, involves the fusion of nuclei of specialized sex cells called **gametes** (sperm and egg) leading to the production of a fertilized egg referred to as a **zygote.** From this single cell, an offspring eventually develops which typically shows hereditary characteristics of both parents. Unlike asexual reproduction, sexual reproduction involves an elaborate preparatory process called *meiosis.* Through this process, the mechanisms of hereditary materials may occur. The importance of this phenomenon in terms of natural selection and evolution will be discussed in Chapter 21.

Asexual Reproduction

Asexual reproduction involves the transmission of genetic characteristics from a single organism to one or more daughter organisms. Since only one organism is undergoing the process, there is little chance for new gene combinations within the offspring. This situation generally results in stable characteristics within a species from one generation to another.

Fission in Monera and Protista

Fission, as it occurs in bacteria and protozoans, constitutes the simplest method of asexual reproduction. It is a process by which an organism splits into daughter cells. Bacteria (Kingdom Monera) lack an organized nucleus (procaryotic) and the entire cell functions as the reproductive unit. There is no true mitotic division since there are no actual chromosomes present. Most bacteria divide transversely (Figure 17-1A), although some may undergo longitudinal fission. Certain bacteria will also form endospores when environmental conditions become unfavorable. Such endospores, which form within the bacterial cell wall, are resistant to temperature extremes, chemicals, and desiccation.

Protozoans (Kingdom Protista) reproduce primarily by *binary fission* whereby the organism undergoes mitotic division and produces two daughter cells. Ciliates, such as *Paramecium,* may divide transversely (Figure 17-1B); flagellates, such as *Euglena,* divide longitudinally (Figure 17-1C); and sporozoans undergo multiple fission, a process by which a unicellular organism divides into several daughter cells.

Budding in Animalia and Fungi

In some multicellular organisms (*Hydra*) and certain unicellular fungi (yeast), **budding** may take place. In this process, a small protuberance develops and may be separated from the parent organism. Within the body wall of a *Hydra,* cells of both gastrodermis and epidermis undergo mitotic divisions, producing a projection which first appears as a small lump on the outside of the body wall and which further develops into an adult *Hydra* (Figure 17-2A). The bud may remain attached or may become detached and live independently of the parent *Hydra.* Unicellular yeasts develop a protuberance on the cell wall which continues to grow until a wall forms to delimit the bud from the parent cell. The bud may cohere or become independent (Figure 17-2B).

Spore Production in Plantae and Fungi

Sporulation involves the formation of a specialized cell called a *spore* within a resistant capsule. Such cells are produced by certain plants and fungi and may remain viable for very long periods of time under extreme environmental conditions. Whereas bacteria may produce endospores when conditions become unfavorable, other organisms produce spores as a means of reproduction and the spores are dispersed either by air or water.

In lower plants, spores are produced in cells called *sporangia.* The spores are flagellated and termed *zoospores* because of their motility. Such motility is an advantage since it permits wider distribution of the

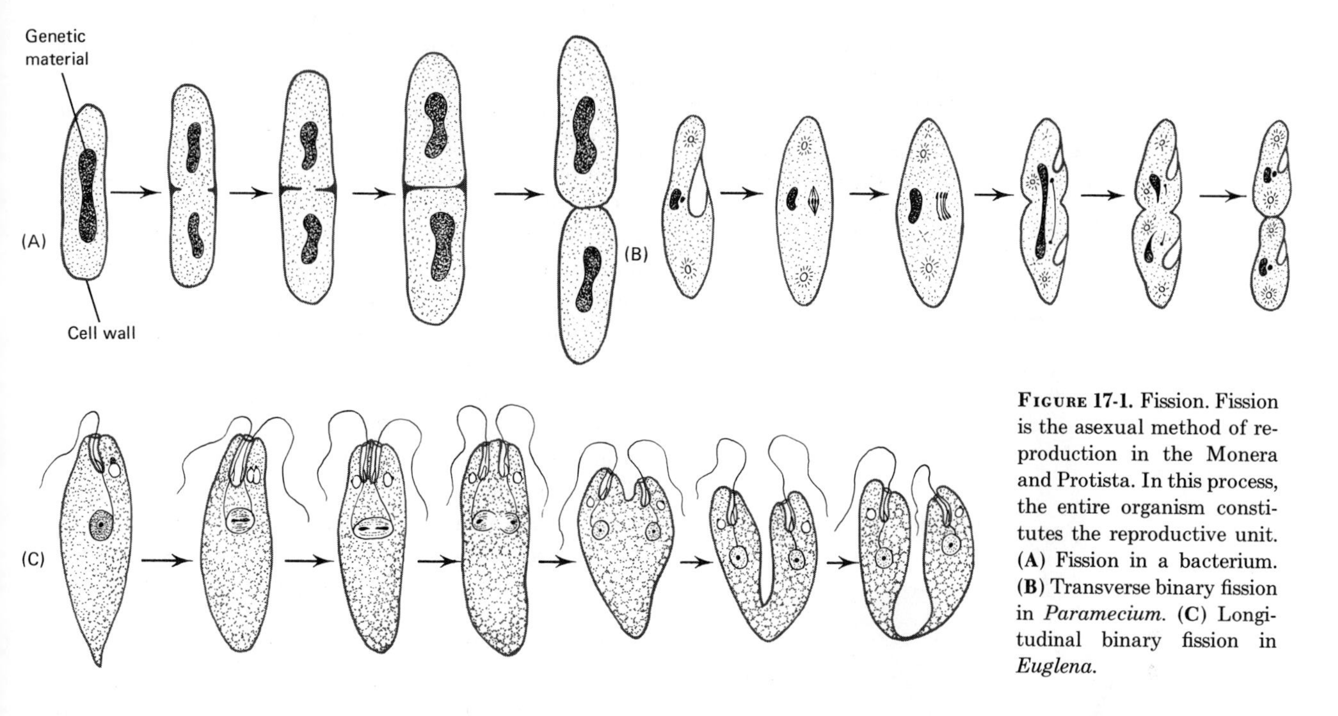

FIGURE 17-1. Fission. Fission is the asexual method of reproduction in the Monera and Protista. In this process, the entire organism constitutes the reproductive unit. **(A)** Fission in a bacterium. **(B)** Transverse binary fission in *Paramecium.* **(C)** Longitudinal binary fission in *Euglena.*

organism, thereby increasing chances for survival. *Ulothrix,* one of the Chlorophyta, is an example of a sessile organism that produces zoospores (Figure 17-3A). This particular form is a filamentous alga that grows in length by mitotic cell division. As the cells mature, differentiation occurs with the cell protoplast undergoing division without enlarging. Four zoospores, each with four flagella, are produced in the parent cell which, at this stage, is termed a *zoosporangium.* Upon release from the zoosporangium, each zoospore spends a short but active swimming period before descending to the bottom and fastening itself to the substrate. Each zoospore then undergoes cell division to become a new filament.

Fungus organisms are also spore-producing organisms. Like the lower plants, the chambers that give rise to the spores are called *sporangia,* although the name of the particular structure is determined by its general shape in certain groups. *Rhizopus stolonifer,* the common bread mold, is a representative fungus that may be used to illustrate spore production (Figure 17-3B). Sporangia are produced on ascending filaments called *sporangiophores.* The spores produced (sporangiospores) are released into the air and may be carried hundreds of miles before being deposited on a suitable substrate. A germinating spore develops into a hypha which then branches and becomes differentiated into *rhizoids* for anchorage and absorption, and sporangiophores. Lateral hyphae may also be formed which, when in contact with the substrate, begin the development of new hyphal units. These lateral branches are called *stolons.*

Vegetative Reproduction Among Plantae

Many higher plants are capable of reproducing by means other than seeds, a process called *vegetative reproduction* (*propagation*). Such reproduction occurs widely in nature and may also be carried out artificially by humans. Many cultivated plants, such as those that produce bananas, citrus fruits, pineapples, apples, pears, and potatoes, are propagated vegetatively.

Stems (aerial and underground), roots, and leaves serve as plant organs in vegetative propagation. *Cuttings* are small stems, usually with attached leaves, which have been removed from parent plants and placed in water or moist sand until adventitious roots develop. Cacti, ivy, geraniums, begonias, roses, and pineapples may be propagated in this manner. In the process called *layering,* the end of a branch of a plant is

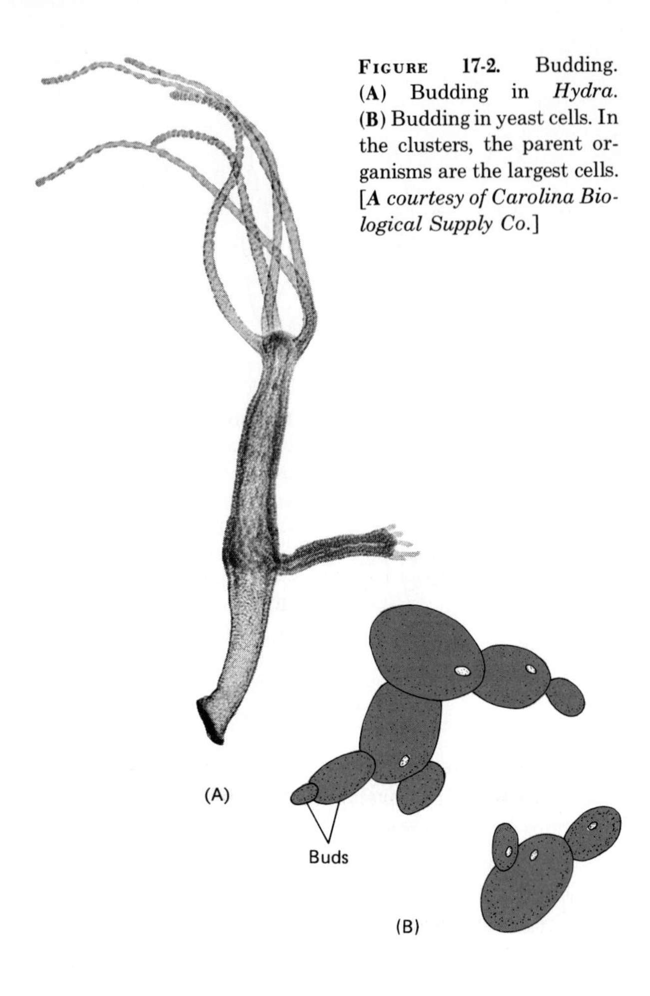

Figure 17-2. Budding. **(A)** Budding in *Hydra.* **(B)** Budding in yeast cells. In the clusters, the parent organisms are the largest cells. [**A** *courtesy of Carolina Biological Supply Co.*]

bent so that it makes contact with the soil (Figure 17-4A). After the stem produces adventitious roots, it is cut from the parent plant, removed from the soil, and transplanted. Layering is horticulturally employed for such plants as raspberry, blackberry, rhododendron, and other ornamental or food plants.

An example of vegetative propagation by an underground stem is the white (Irish) potato. The potato is a *tuber,* an enlarged portion of an underground branch (stolon). Prior to planting, the tuber (potato) is cut into pieces and each piece must bear at least one eye. Each eye develops into a shoot which produces adventitious roots from the underground portion of the stem. Subsequently, stolons develop from lateral buds, and the stolons expand at the tips into new tubers (Figure 17-4B). Another example of vegetative propagation by an underground stem is the crocus (Figure 17-4C). Some plants, such as crocus and gladiolus, contain a short, stout underground stem called a *corm.* When a corm is planted, several new corms may be produced at the bases of the new shoots that arise from the old corm.

One of the most important plants propagated by roots is the sweet potato (not related to the white potato). In commercial propagation, sweet potatoes are placed in moist sand or soil. In time they produce adventitious roots. The sprouts are then removed and planted in the field.

Kalanchoë is an example of a plant that reproduces vegetatively by leaves (Figure 17-4D). The leaves are fleshy and notched along their margins. Meristematic tissue is produced in the notches and this tissue gives rise to small plants (plantlets). The tiny plants, containing leaves, stems, and sometimes roots, fall from the parent leaf and continue their growth in the soil.

Regeneration Among Animalia

Starfish are probably best noted for their ability to regenerate missing rays (Figure 17-5A). *Regeneration* is the replacement of a body part or even an entire organism. Starfish are also capable of developing entire organisms from a single piece of ray and adjacent central disc. Crustaceans are able to regenerate lost chelipeds and antennae. *Hydra* (Figure 17-5B) and sponges are other examples of animals possessing the ability to regenerate body parts or entire organisms from a chunk of tissue. Planaria (Figure 17-5C) are able to develop new organisms from portions of the original body. Salamanders may regenerate a missing tail, but a salamander tail will not develop the rest of the animal. Man is able to regenerate tissue (skin) on a relatively small scale in the formation of scar tissue and healing of wounds.

In all cases of asexual reproduction, the extent of development of new organisms or parts varies with the size and complexity (position in the phylogenetic scale) of the reproductive unit or organism.

Sexual Reproduction

Sexual reproduction, the fusion of gametes resulting in the formation of a **zygote,** is typical of the overwhelming majority of living forms. Of the principal

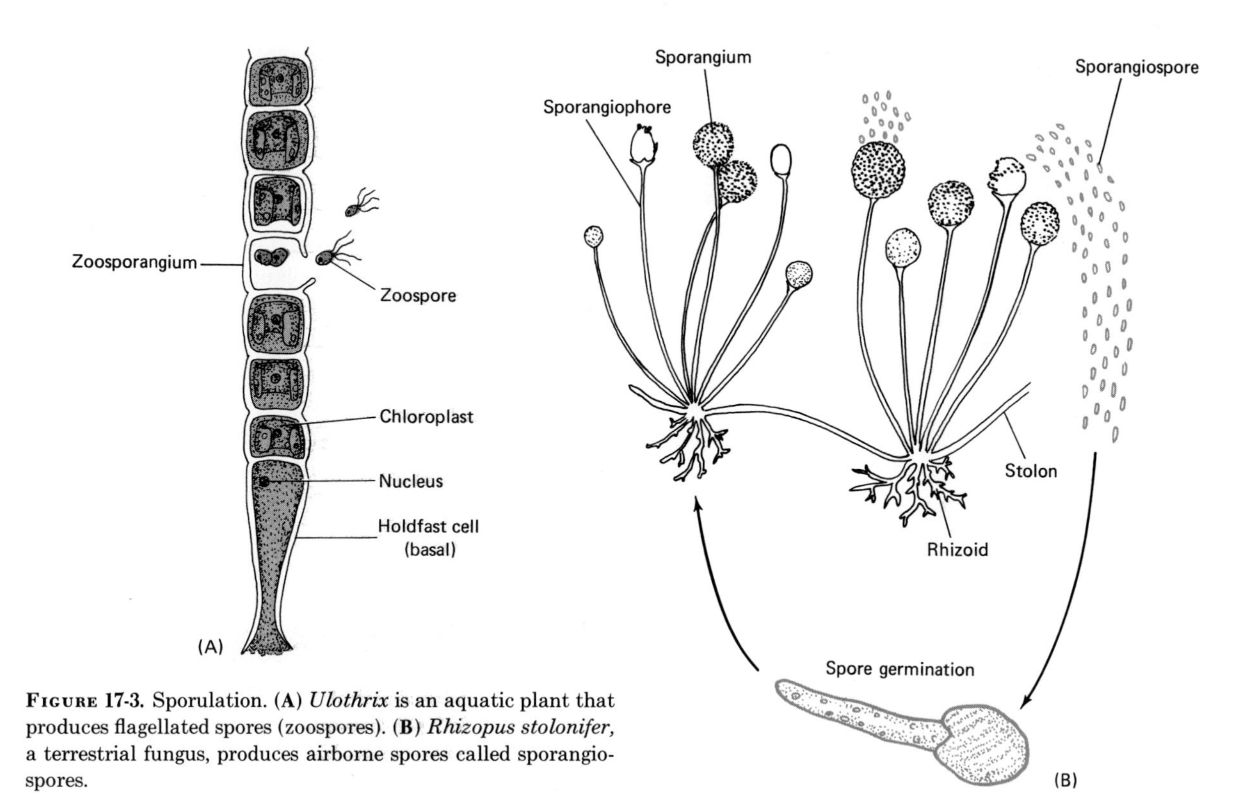

Figure 17-3. Sporulation. **(A)** *Ulothrix* is an aquatic plant that produces flagellated spores (zoospores). **(B)** *Rhizopus stolonifer,* a terrestrial fungus, produces airborne spores called sporangiospores.

groups of organisms, only the blue-green algae and a group of fungi (Deuteromycetes) have yet to be shown to undergo sexual reproduction.

All eucaryotic cells contain thread-like bodies, the chromosomes, in their nuclei. The somatic cells of most eucaryotic organisms are **diploid;** that is, each cell contains two sets (a pair of each kind) of chromosomes. The cells of some eucaryotic organisms are **haploid;** that is, there are no paired chromosomes, and there is only one of each type (one set) of chromosome present. The number of chromosomes within diploid (and haploid) nuclei is usually constant for each species. Consider the following chromosome numbers of diploid organisms: the nuclei of meristematic cells of corn plants have 20, pine meristematic cells have 24, somatic cells of a housefly have 12, and human somatic cells have 46 chromosomes. The chromosome number of the haploid organism *Chlamydomonas* is 8.

Inasmuch as the nuclei of two sex cells fuse during sexual reproduction and the chromosome number remains constant for all generations, a paradox is apparent concerning a numerically constant chromosome number. It would appear that the chromosome number would double with each succeeding generation. However, it does not double because of a special type of nuclear division called **meiosis.** This nuclear division occurs at some stage in the life cycle prior to the formation of sex cells, so each of the gametes contains only one-half as much hereditary material (chromosomes) as the parent cells from which they were formed. Thus, when two sex cells unite, each contributes one-half the number of chromosomes, so that the resulting zygote has the normal number of chromosomes for that species. At some time prior to a repetition of gamete fusion in the next generation, meiosis must occur. As will be shown later in this chapter, the time of meiosis is quite variable among organisms, but is carefully regulated within a species.

To understand the numerical distribution and constancy of chromosomes during sexual reproduction, consider the normal chromosome number (20) for cells of the shoot apex of corn. This full complement of chromosomes, the diploid number, is commonly designated as $2n$. In cells resulting from meiosis, the chromosome number is reduced from $2n$ to n, the haploid number, which is one-half (10, in the case of corn) the full complement of chromosomes. Therefore, when meiosis occurs in the cells of corn plants, each cell produced contains 10 chromosomes. Logically, when two such cells fuse to form a zygote during fertilization, the diploid number is restored. All organisms that reproduce sexually have both haploid and diploid phases, but in higher forms diploid cells are more numerous. Only the sex cells of higher organisms are haploid as a result of meiosis, and these are few in number relative to the diploid cells.

Meiosis

Complete meiosis involves two successive division sequences that result in four new cells, each of which is haploid. The first divisional sequence accomplishes the reduction in the number of chromosomes, whereas the second divisional sequence is primarily a mitotic one. In the following discussion of the meiotic process, a starting diploid cell containing only two pairs of chromosomes will be considered for purposes of simplicity.

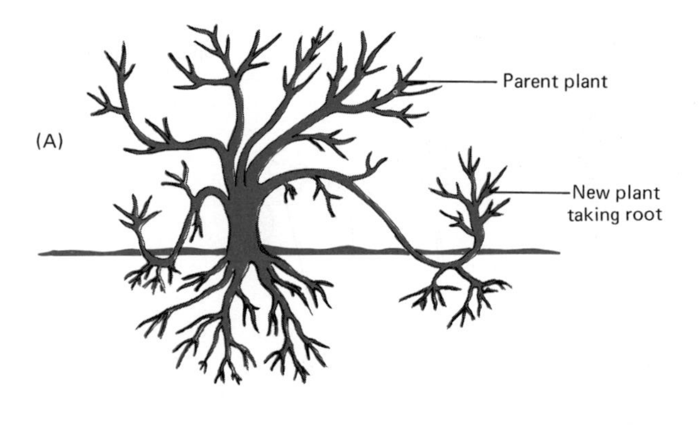

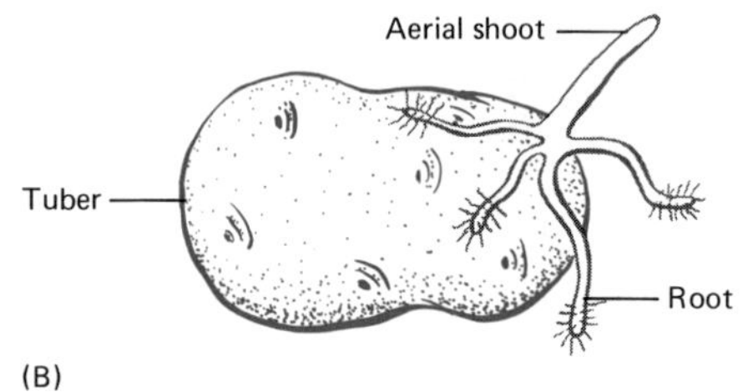

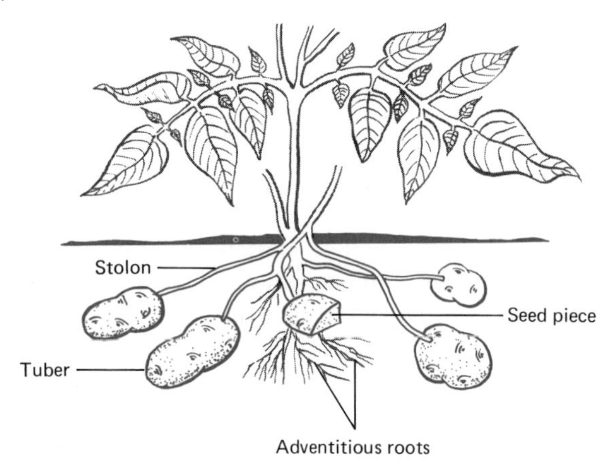

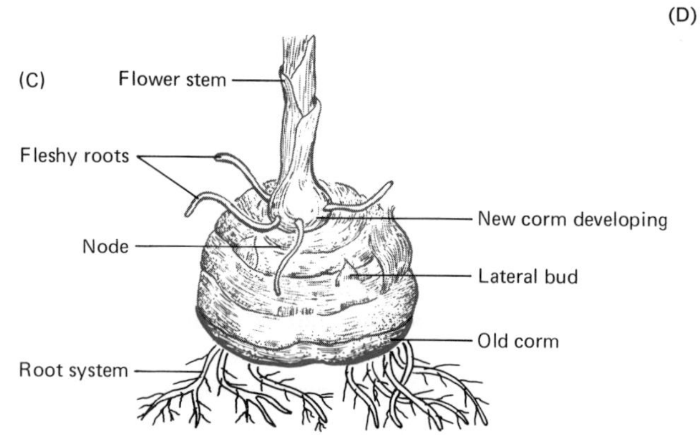

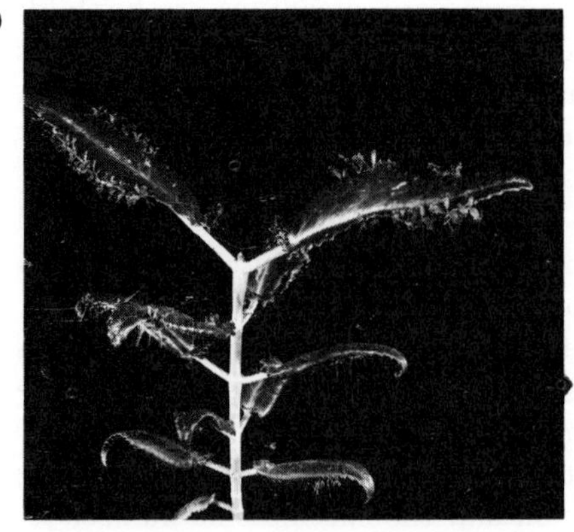

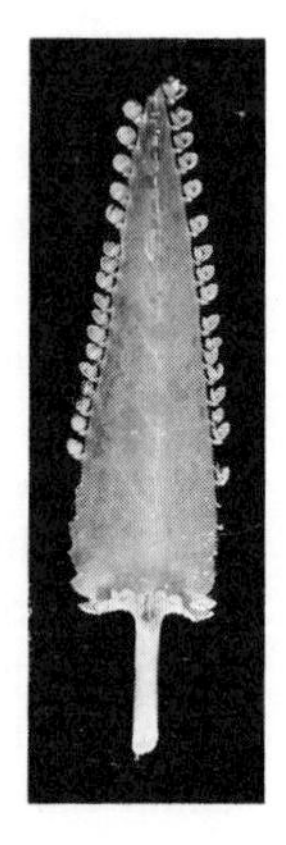

Figure 17-4. Vegetative reproduction in plants. (**A**) Layering. (**B**) Potato tuber and developing potato plant. (**C**) Corm of gladiolus. (**D**) *Kalanchoë*. Portion of a plant with leaves bearing plantlets (left) and a single leaf with plantlets around its margin (right). [***D** courtesy of Boyce Thompson Institute for Plant Research.*]

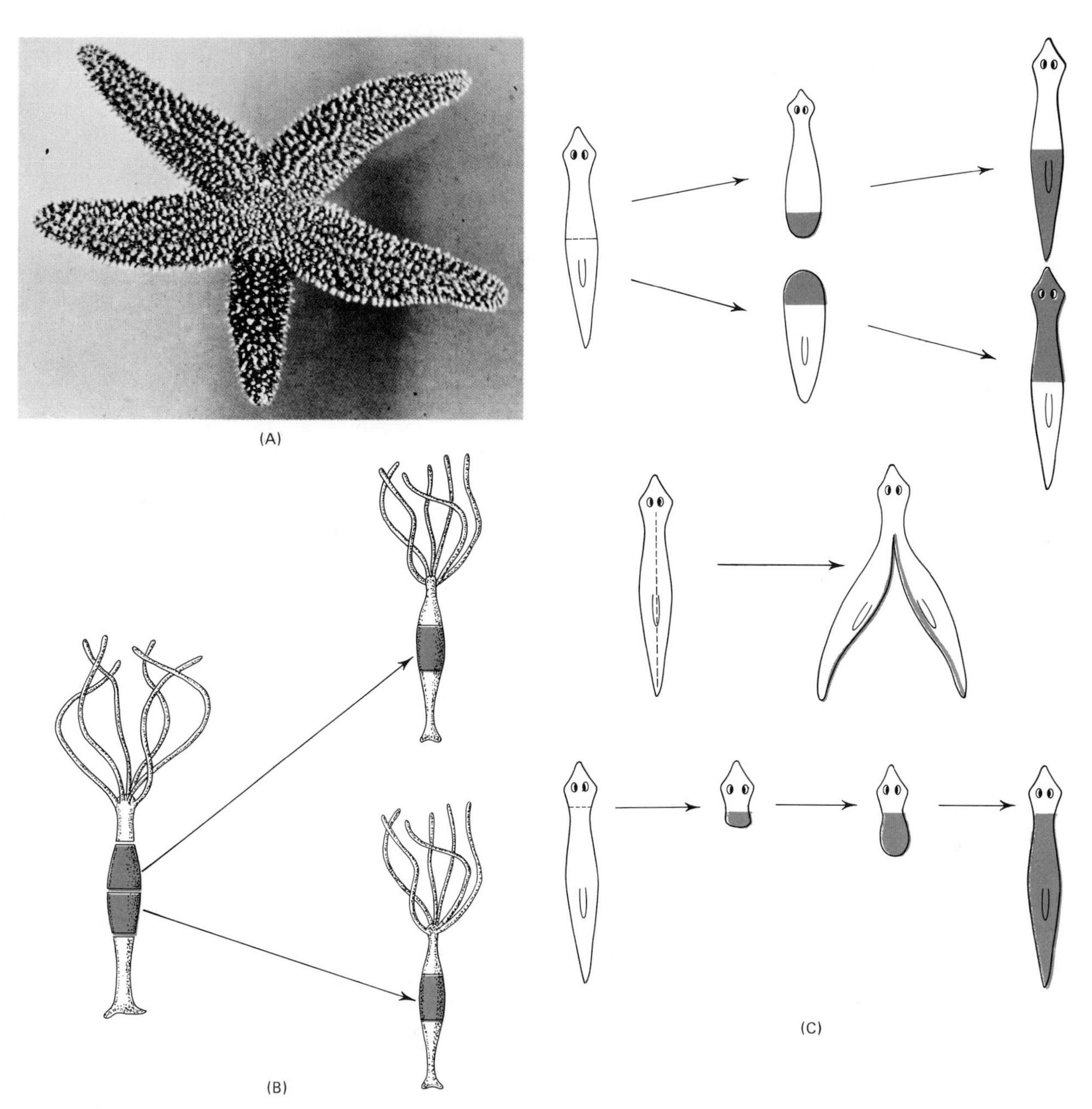

Figure 17-5. Regeneration among Animalia. (**A**) Partly regenerated arm of a starfish. (**B**) Regeneration in *Hydra*. If a *Hydra* is cut into pieces, a new organism will form from each piece. (**C**) Regeneration in planarians. The colored areas represent portions formed as new tissue during the early stages of regeneration. [**A** *courtesy of Carolina Biological Supply Co.*]

(A) Interphase I

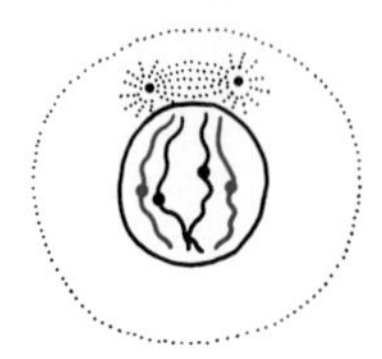

(B) Early Prophase I

Chromosomes become visible; centrioles divide and start to move toward opposite poles.

(C) Middle Prophase I

Homologous chromosomes synapse, shorten, and thicken.

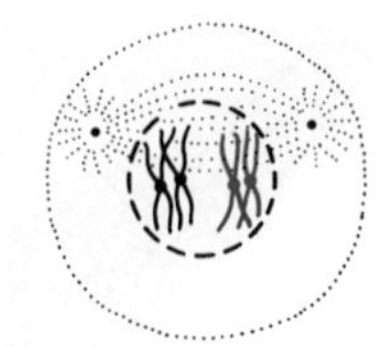

(D) Late Prophase I

Each chromosome separates into two chromatids that are held together at the centromere; nuclear membrane disappears.

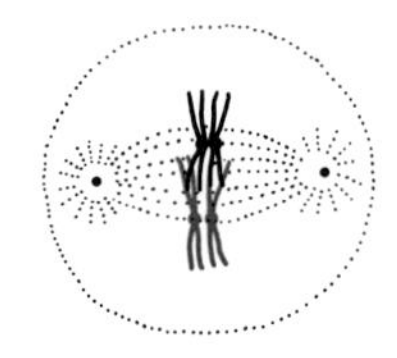

(E) Metaphase I

Each synaptic pair attaches to a fibril.

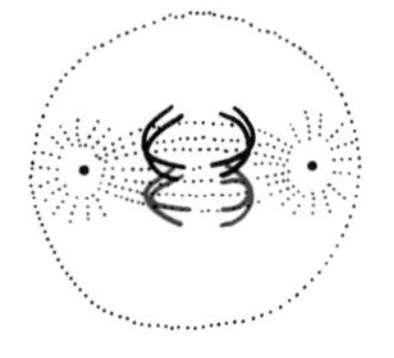

(F) Anaphase I

Double-stranded chromosomes migrate to opposite poles.

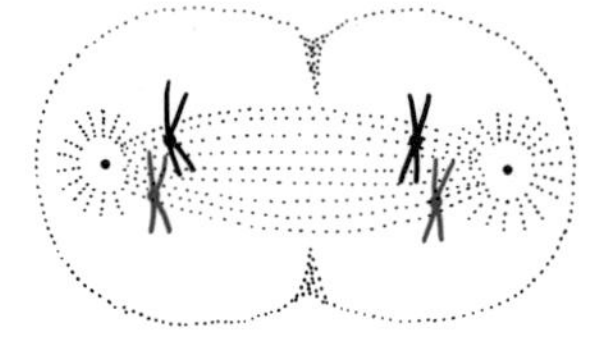

(G) Telophase I

Cytokinesis begins; nuclei re-form.

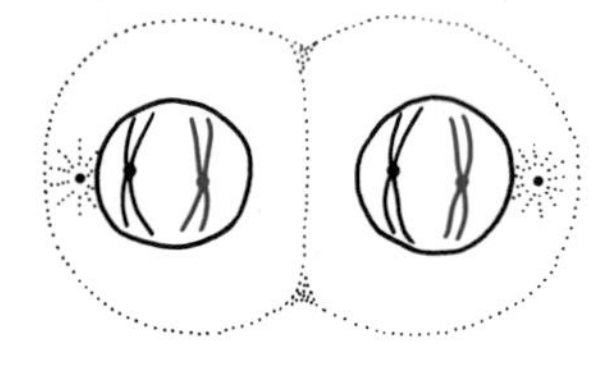

(H) Interphase II

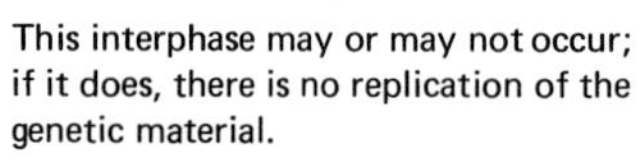

This interphase may or may not occur; if it does, there is no replication of the genetic material.

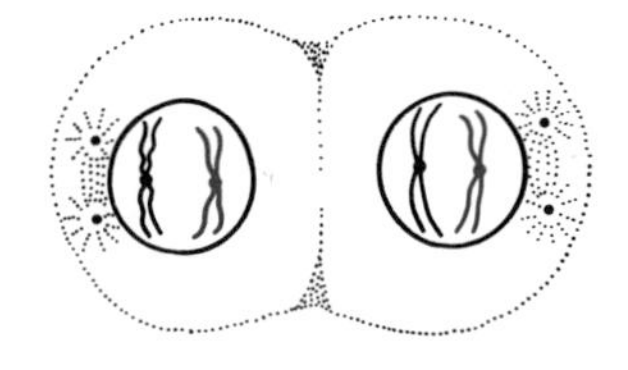

(I) Prophase II

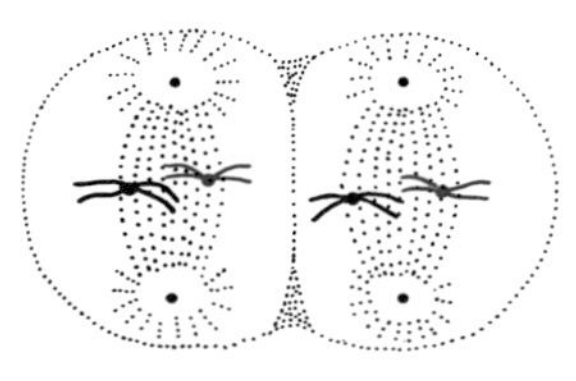

(J) Metaphase II

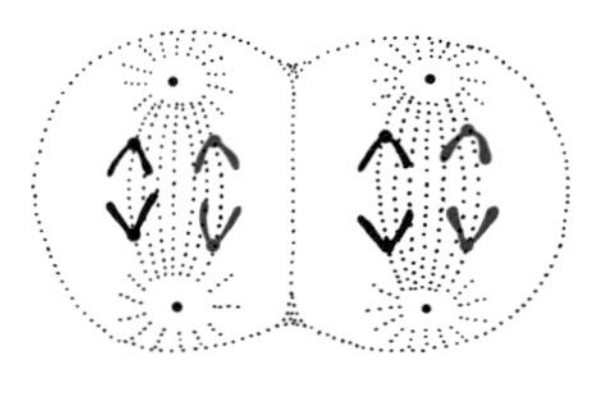

(K) Anaphase II

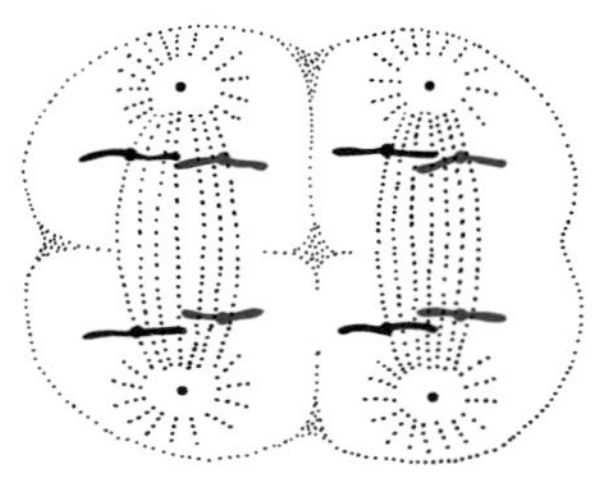

(L) Telophase II

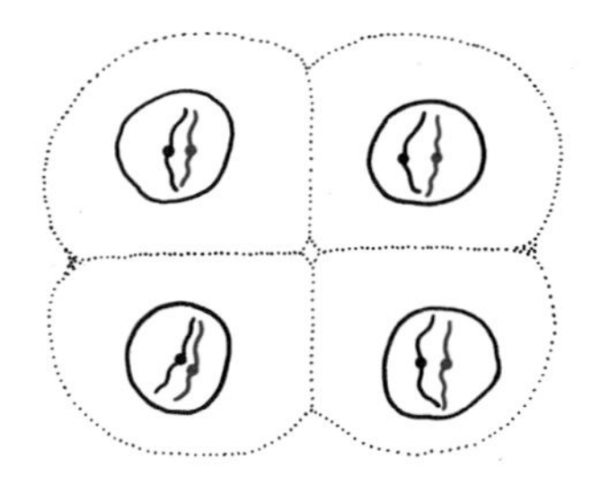

(M) Meiosis completed.

FIGURE 17-6. Meiosis in an animal cell. **(A–G)** Stage I of meiosis. **(H–M)** Stage II of meiosis. See text for amplification.

Meiosis is typically characterized by two stages (I and II), the second generally following immediately after the first. The complete scheme of the meiotic process is illustrated in Figure 17-6. The individual stages of the process will now be discussed.

INTERPHASE I. The *interphase* that comes before the first division of meiosis is similar to the interphase preceding mitosis. This stage involves the replication of the chromosomes (Figure 17-6A).

PROPHASE I. Some of the events that occur during *prophase I* (Figure 17-6B–D) are not unlike those associated with the prophase of mitosis. The chromosomes become visible and then thicken and shorten, the nucleoli and nuclear membrane disappear, and the spindle apparatus is organized. The replication of the genetic material takes place during the interphase prior to prophase I. Replication also occurs during interphase preceding the prophase of mitosis. A principal difference between the meiotic prophase I and the mitotic prophase is that in the former **homologous chromo-**

somes (members of a pair) attract each other and pair so closely that they appear as a single chromosome. They do not actually fuse, however, but do frequently intertwine. This pairing, called **synapsis,** occurs throughout the length of the homologous chromosomes.

During synapsis, a significant difference between chromosome behavior during meiosis and mitosis is visually evident. Frequently, in all but a few organisms, portions of one chromatid may be exchanged with corresponding portions of another one in the same *tetrad* (group of four chromatids). This process is called **crossing-over** and greatly increases the number and variety of recombinations of genes. Soon the chromosomes comprising a tetrad pull apart and there is visual evidence that crossing-over has occurred. There is a tendency for crossed-over chromatids to cling together, and they form temporary X-shaped configurations referred to as *chiasmata* (Figure 17-7). The migration of the synapsed homologous chromosome pairs to the equator of the spindle terminates prophase I and initiates the next phase.

METAPHASE I. In *metaphase I* (Figure 17-6E), the chromosome pairs, also termed *bivalents,* orient themselves on the equator of the spindle in such a way that the homologues of each one can eventually move to opposite poles. In this process, each bivalent acts independently of the others; this is a significant event since it results in random shuffling of the chromosomes into the daughter cells. This phenomenon, referred to as independent assortment, is discussed in Chapter 19.

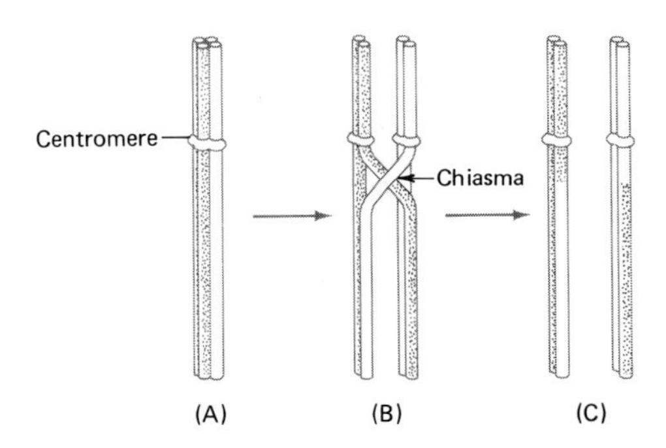

FIGURE 17-7. Synapsis, crossing-over, and chiasma formation. (**A**) Four chromatids (tetrad) in synapsis. (**B**) Formation of chiasma as chromosomes separate following crossing-over. (**C**) Separated chromosomes following crossing-over and the disappearance of the chiasma.

ANAPHASE I. During *anaphase I,* the separation of synapsed homologues takes place (Figure 17-6F). One longitudinally double chromosome of each pair moves to each pole. Recall that in mitotic anaphase, one of each of the chromosomes present of the entire chromosomal complement migrates to each pole so that each new nucleus has just as many chromosomes as the parent nucleus. In meiosis, however, entire chromosomes of each homologous pair, many having been modified by crossing-over, separate so that each pole receives either a paternal or maternal, longitudinally double chromosome of each pair. This occurs because in mitosis, division of the centromere takes place in anaphase, while in meiosis, centromere division is delayed until anaphase II.

TELOPHASE I. *Telephase I* is basically similar to the telophase of mitosis and is marked by the arrival of the chromosomes at the poles of the spindle (Figure 17-6G). Whereas in mitosis the two new nuclei formed have the same number of chromosomes as the parental nucleus, in meiosis each of the new nuclei has one-half the chromosomes that were present in the parental nucleus. Moreover, the chromosomes of telophase I are double-stranded. In some organisms, the sequence of events progresses directly from anaphase I to prophase II; in other organisms, there may be a short or long interphase (*interphase II*) between the first and second meiotic divisions. Interphase II of meiosis is similar to the interphase of mitosis except that in the former there is no replication of genetic material and, therefore, no new chromatids are formed (Figure 17-6H).

PROPHASE II. *Prophase II* is generally short and superficially resembles mitotic prophase except that the chromatids of each chromosome are often widely divergent; that is, there is no relational coiling (Figure 17-6I).

METAPHASE II. In *metaphase II,* also a brief stage, each of the double-stranded chromosomes attaches to a single fibril of the spindle (Figure 17-6J). Following the completion of these events the chromosomes become arranged in the equatorial plane.

ANAPHASE II. At the termination of metaphase II, the centromeres divide. The single-stranded chromosomes migrate toward opposite poles of the spindle during *anaphase II* (Figure 17-6K). Their arrival at the poles signals the close of this phase.

Telophase II. In *telophase II,* following the arrival of the haploid chromosomes at the poles, the chromosomes return to their long, reticulate configuration, nuclear membranes reappear, nucleoli re-form, and cytokinesis generally separates each nucleus from the others (Figure 17-6L).

Note, in summary, that in the first meiotic division two haploid cells are produced, each containing double-stranded chromosomes. In the second meiotic division, each of the haploid cells divides. Thus, a total of four haploid nuclei are produced from a single diploid cell.

There are two rather significant results of meiosis. One is obviously the reduction of chromosome number from the $2n$ to the n condition. The second result is that the cells produced in meiosis may be genetically unlike with respect to each other and with respect to the original cell that produced them. This is called *genetic recombination.* To illustrate genetic recombination, assume that the starting cell consists of two pairs of chromosomes and that a particular locus (place occupied by an allele) on each pair of chromosomes governs the same genetic characteristic but to varying degrees. In other words, assume that one pair of homologous chromosomes is responsible for height. One member of the pair contains allele A for tallness, and the other member of the pair contains allele a for shortness. An allele is one member of a pair of genes. Similarly, assume that a gene on the other pair of homologous chromosomes governs flower color. In this case, one member of the pair contains allele B for red flowers and the other member of the pair contains allele b for white flowers (Figure 17-8A).

During metaphase I of meiosis, these chromosomes and their respective genes may orient themselves in one of two ways. Either AB and ab or Ab and aB may result (Figure 17-8B). This chromosomal orientation is determined purely by chance, and the resulting cells produced after metaphase II of meiosis are also determined by this chance arrangement (Figure 17-8C). Note that the cells may contain genes AB, ab, Ab, or aB, that is, four different sets of genetic characteristics. Genetically, the first cell contains the genes for tall plants with red flowers, the second possesses a genetic expression for short plants with white flowers, the third for tall plants with white flowers, and the fourth for short plants with red flowers. Logically, the genetic combination achieved by nuclear fusion during fertilization will determine the genetic traits for height and flower color of the plant.

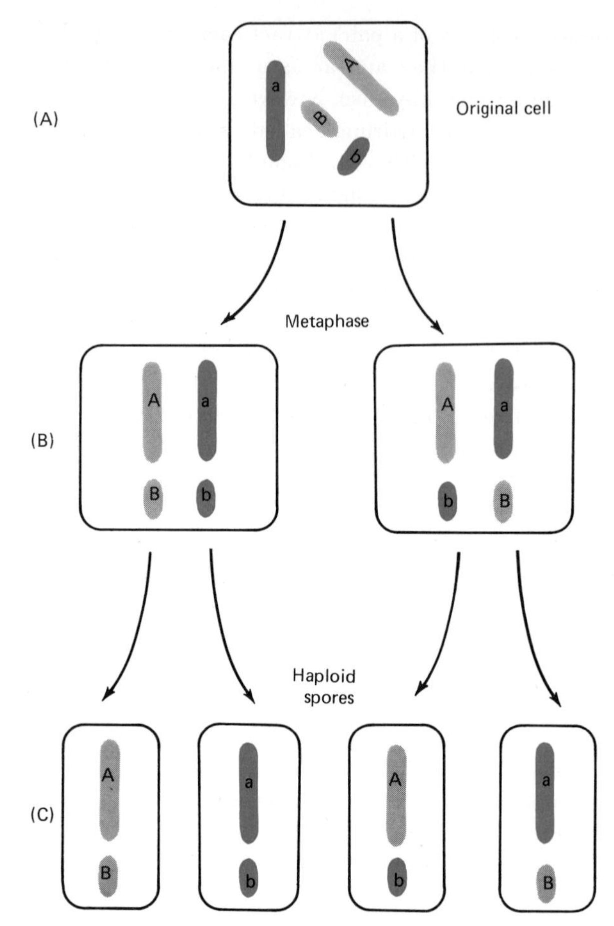

Figure 17-8. Genetic recombination during meiosis. Note the difference in genetic makeup between the original diploid cell (**A**) and the haploid spores (**C**) that may result from meiosis. Events associated with metaphase I and metaphase II are shown in **B.**

Comparison of Sexual and Asexual Reproduction

In reviewing the asexual and sexual processes, it may be noted that those organisms that undergo asexual methods tend to exhibit stable characteristics from one generation to another since there is no recombination of genetic material. Recall that recombination may occur as a result of crossing-over or the random distribution of chromosomes at meiosis. Organisms that undergo sexual reproduction have a greater chance of exhibiting a wider range of traits in succeeding generations owing to the fusion of the gamete nuclei and the resulting genetic recombination. Asexual reproduction is of advantage to an organism in that it may be carried out by one organism without dependence upon others and allows for rapid increases in population size. All that is required is favorable environmental conditions. However, should the environmental conditions change so as to become unsuitable for the organism, it is possible for the organism to become extinct owing to its inability to adapt to the change. Those organisms that undergo the sexual process have an advantage in that they may exhibit traits that better adapt them for environmental changes, thereby allowing them to survive conditions that might otherwise cause their extinction.

Meiosis and Life Cycles Among Organisms

Organisms differ with regard to when and where meiosis occurs in their life cycle. In the haploid organism *Chlamydomonas,* for example, meiosis takes place right after fertilization as a first step in the subsequent development of the zygote (see Figure 17-10). The process of fertilization gives rise to a diploid zygote and meiosis restores the haploid state immediately. Specifically, haploid gametes fuse and give rise to a diploid zygote, which develops into a zygospore. A *zygospore* is a thick-walled resting cell resulting from the fusion of gametes and is capable of surviving unfavorable environmental conditions. Prior to germination, the zygospore undergoes two successive divisions that transform the original cell into four cells. Prior to or in conjunction with these divisions, the chromosomes of the zygospore duplicate once ($2n \longrightarrow 4n$). The zygospore then undergoes meiosis and forms four haploid zoospores, each of which may develop into a new *Chlamydomonas.*

Such a life cycle characterized by haploid adults and zygotic meiosis is called a *haplontic life cycle* (Figure 17-9A). This kind of life cycle occurs in all Monera in which a sexual cycle has been demonstrated, some protistans, a few plants, and many fungi. Many haplontic organisms are also capable of producing spores between one fertilization and the next. Spore production occurs mitotically and, like the adults that produce them, the spores are haploid. Such spores are called *mitospores.*

In other organisms, meiosis occurs during the formation of gametes and the adult organism is diploid (Figure 17-9B). Such a life cycle is called *diplontic.* The only haploid stage in it is the mature gamete itself, and two haploid gametes must fuse in order to produce the adult diploid organism. Diplontic cycles occur in humans and all other animals, various plants, some fungi, and most protozoans. In diplontic organisms that produce spores, the spores are formed mitotically by the diploid adult and thus the mature spores are also diploid.

Note that in the haplontic cycle, meiosis occurs in the zygote. In the diplontic cycle, meiosis takes place during gamete formation. The third possible time of meiosis is the spore-production stage. In this instance, fertilization produces a diploid zygote and the developing adult remains diploid. Eventually, the adult develops a spore-producing structure in which diploid cells give rise to haploid spores. Since meiosis occurs at spore forma-

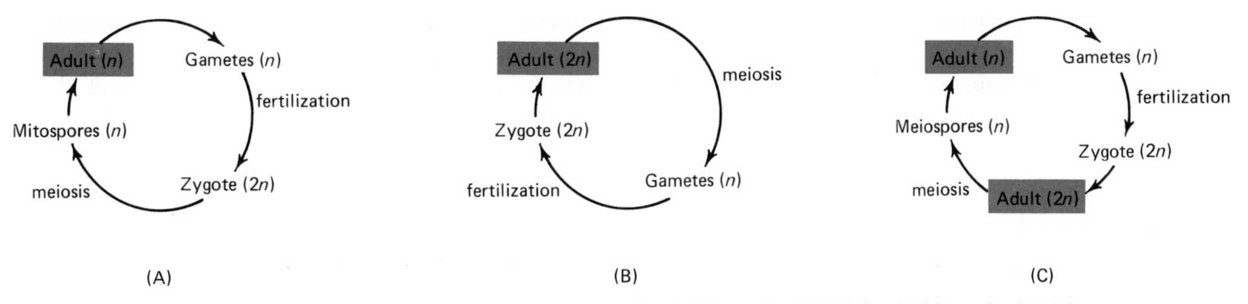

Figure 17-9. Life cycles among organisms. (**A**) Haplontic life cycle. (**B**) Diplontic life cycle. (**C**) Diplohaplontic life cycle.

tion, the resultant spores are termed *meiospores.* A haploid meiospore eventually develops into a new haploid adult, which later produces haploid gametes. The gametes fuse and the resulting diploid zygote initiates a new life cycle.

Such a life cycle consists of two generations as a result of meiosis occurring during spore formation, and each generation is represented by a separate adult. The diploid zygote gives rise to a diploid adult called the **sporophyte generation** (it later produces meiospores). The haploid meiospore then gives rise to a haploid adult termed the **gametophyte generation,** which later produces gametes.

This kind of life cycle characterized by meiosis during spore formation and by an **alternation of generations** (the sporophyte alternating with the gametophyte) is known as the *diplohaplontic life cycle* (Figure 17-9C). It is found among many fungi and most plants.

Sexual Reproduction in Monera

Sexual reproduction occurs in its most primitive forms in the Monera. In bacteria, for example, the transfer of genetic material takes place without the formation of gametes. Owing to their simplicity in protoplast structure, bacteria are able to acquire and incorporate new genetic material by simply absorbing DNA fragments through the cell wall and plasma membrane. This process is called **transformation** and involves DNA that has been released into the environment by dead donor cells. Traits received by the recipient cells appear in subsequent generations. A second type of recombination is **conjugation,** in which a donor cell directly transfers a portion of its genetic material to a recipient cell. Descendents of the recipient cell often possess genetic material that differs from that of the original preconjugation parental type. A third type of recombination, termed **transduction,** involves bacterial viruses called **bacteriophages** (see Figure 8-2).

Sexual Reproduction in Plantae

Sexual reproduction in more advanced organisms involves the formation and union of gametes and the establishment of conspicuous and persistent life cycles. If the gametes produced are alike in size and appearance, they are termed *isogametes* and their fusion is called *isogamy;* if the gametes are unequal in size, they are termed *heterogametes* and their fusion is referred to as *heterogamy;* if one heterogamete is small and motile while the other is large and nonmotile, their fusion is termed *oogamy.*

Isogamy

Isogamous sexuality occurs in the Chlorophyta and may be exemplified by *Chlamydomonas* (Figure 17-10A). Under certain environmental conditions, specifically those which may alter nitrogen balance, the haploid daughter cells of *Chlamydomonas* produced by mitosis may be stimulated to act as gametes. Two strains, plus and minus, have been identified as being necessary for sexual reproduction to occur. When these cells mix, clumping occurs followed by the emergence of free pairs. At the point of contact, the cell walls dissolve and fusion occurs, resulting in the formation of a diploid zygote. A heavy wall develops around the zygote, forming a *zygospore,* which is resistant to unfavorable environmental conditions. Once favorable conditions return, the spore germinates, giving rise to four zoospores (two of each strain) each of which becomes a new *Chlamydomonas.* Meiosis occurs during the germination of the zygospore so that the only diploid stage is the zygote.

Heterogamy

Heterogamous organisms also occur in the Chlorophyta. In *Oedogonium,* two kinds of gametes are produced (Figure 17-10B). The female gamete, or *egg,* formed within an enlarged vegetative cell called an *oogonium,* is nonmotile. Male gametes, or *sperm,* are produced within vegetative cells called *antheridia.* Sperm are motile and are released into the water where they swim to the oogonium. A pore in the wall of the oogonium permits a single sperm to enter and fertilize the egg. A *zygospore* develops and passes through a dormant period. Meiosis occurs with the formation of four *zoospores,* each of which develops into a new plant.

Alternation of Generations

As noted earlier, an alternation of generations whereby a sexual stage alternates with an asexual stage

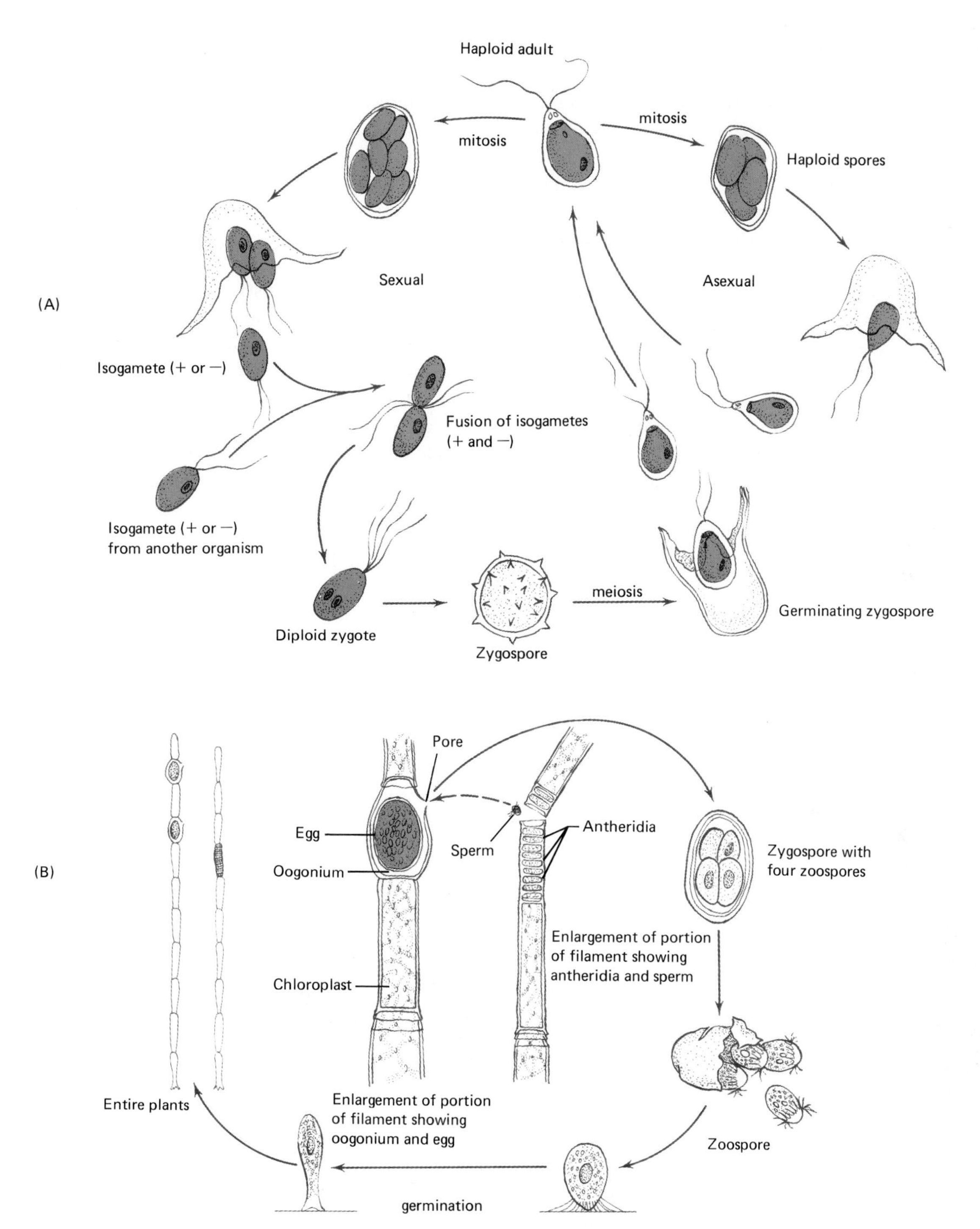

Figure 17-10. Sexual reproduction in Plantae. (**A**) Isogamy in *Chlamydomonas*. (**B**) Heterogamy in *Oedogonium*. See text for amplification.

occurs in all plants. In many, the gametophyte generation is larger, with the sporophyte represented by a single diploid cell, the zygote. In others, the sporophyte is dominant, so that the relative development of these two generations varies greatly. In still other plants, both generations may be nearly the same size, either similar or different in appearance.

ULVA. *Ulva* undergoes the alternation with little differentiation between the forms of the gametophyte and sporophyte generations (Figure 17-11). The gametophyte consists of either male or female plants that are haploid. Gametes produced by the gametophyte plants fuse in fertilization to form diploid zygotes. After many cell divisions, each zygote produces a diploid plant which represents the sporophyte generation. The sporangia of the diploid plant form cells (spore mother cells) that undergo meiosis and produce haploid *meiospores.* Upon germination, each meiospore, through successive mitotic divisions, develops into a haploid gemetophyte plant. Thus, the alternating cycle is completed.

BRYOPHYTES. Bryophytes (mosses and allies) exhibit a very definite alternation of generations which involves the alternation of a multicellular gametophyte with a multicellular sporophyte. At maturity, the gametophytes are usually larger and nutritionally independent, whereas the sporophytes are smaller and dependent upon the gametophyte. *Polytrichum* begins its life cycle as a haploid spore (Figure 17-12). Upon germination, the spore forms a multicellular filament, a *protonema,* containing chloroplasts. This filament forms branches, some of which lose the chloroplasts when they penetrate the substrate, becoming rhizoids which function in absorption and anchorage. Buds appear on the remaining branches and by cell division of an apical cell, form upright shoots with leaf-like structures. These shoots represent the gametophyte generation. Antheridia or archegonia are produced at the tips of the shoots. Haploid sperm are formed by cell division and bear two flagella which are used in swimming to the egg through the water in the environment. Once the sperm are on the female plant, they are attracted to the archegonia by chemicals. Fertilization takes place within the archegonium and the resultant diploid zygote is retained. The zygote undergoes several cell divisions to form an embryo which is protected by the gametophyte. Growth

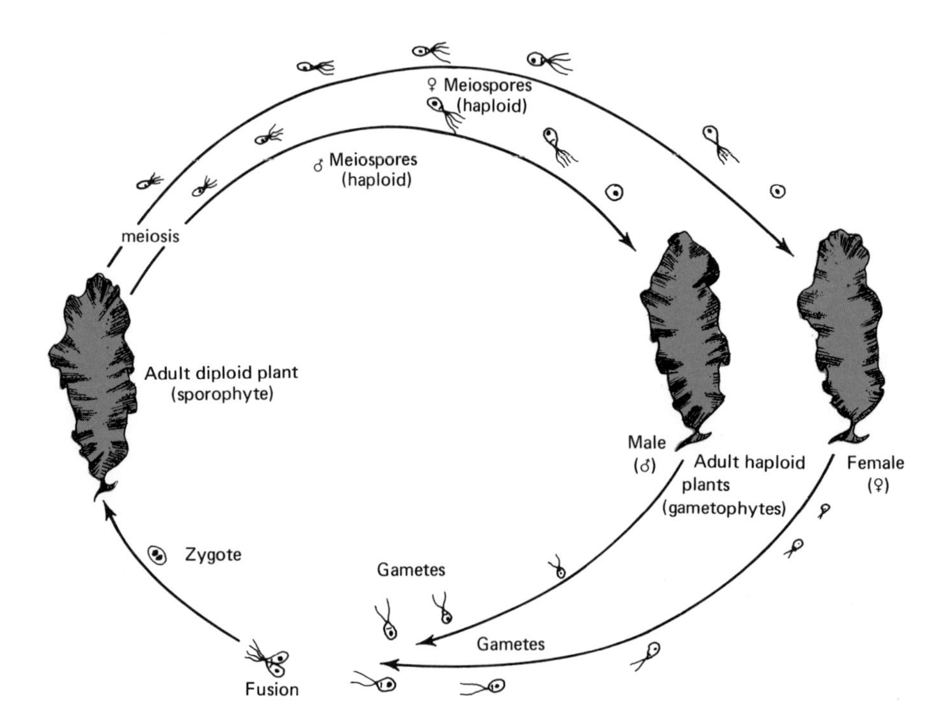

FIGURE 17-11. Alternation of generations in *Ulva* (sea lettuce). In this particular organism, there is little morphological difference between the gametophyte and sporophyte plants.

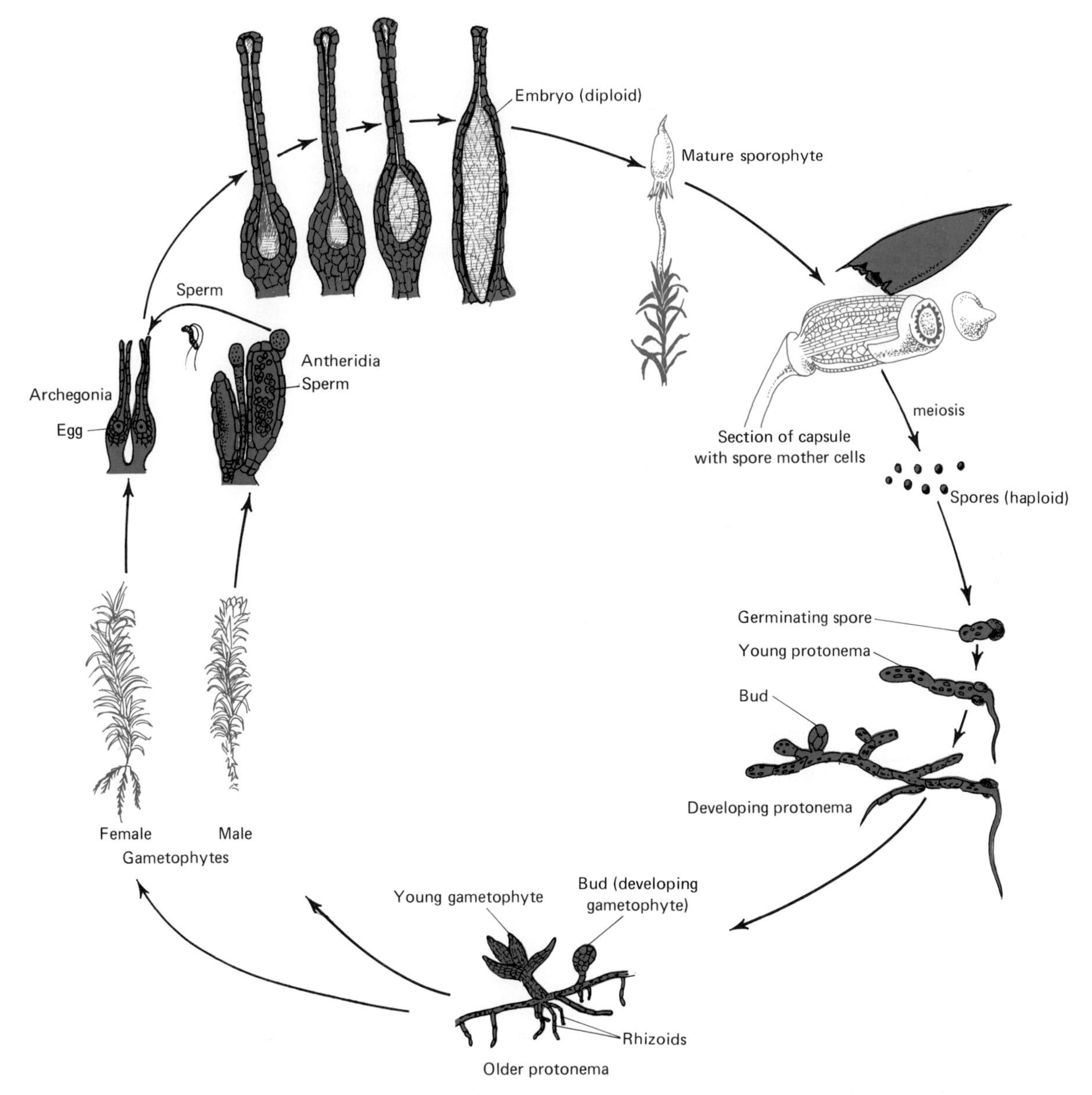

FIGURE 17-12. Alternation of generations in *Polytrichum,* a moss plant. In this life cycle, the gametophyte is the conspicuous plant.

of the embryo produces a vertical stalk which has its basal end embedded in the leafy shoot of the gametophyte. At the distal end, a sporangium or capsule is formed. Spore mother cells contained within the capsule undergo meiosis, producing four haploid spores, thus completing the cycle.

FERNS. Tracheophytes, or vascular plants, exhibit alternation of generations with a dominant, independent sporophyte generation. Fern sporophytes are characterized by large leaves, or fronds, and an underground stem with adventitious roots (Figure 17-13). On the underside of the fronds there are many small dots called

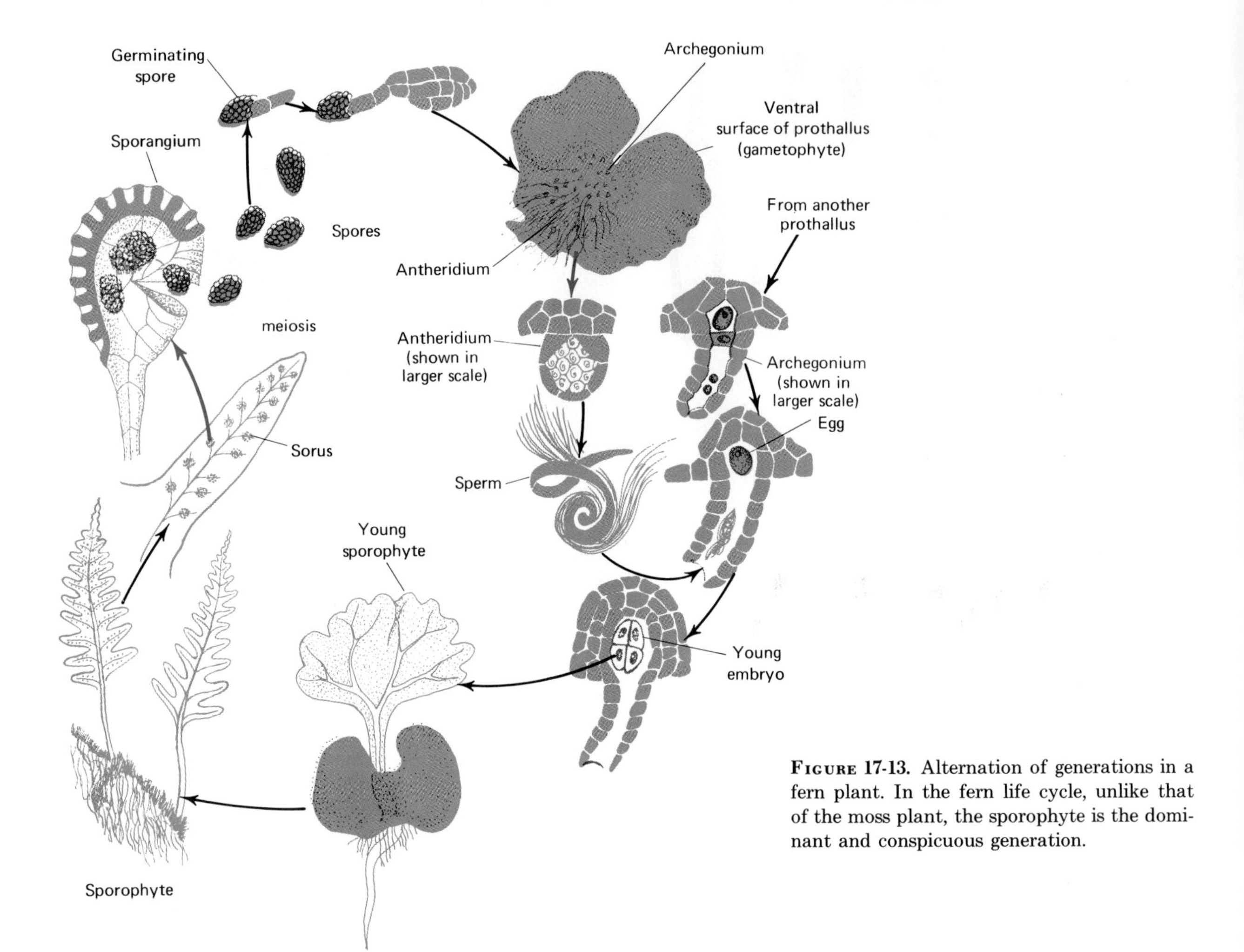

Figure 17-13. Alternation of generations in a fern plant. In the fern life cycle, unlike that of the moss plant, the sporophyte is the dominant and conspicuous generation.

sori. Each sorus is a multiple sporangium usually covered by a membrane known as an *indusium.* Immature sporangia contain spore mother cells that undergo meiosis, thereby producing haploid spores. A spore germinates on the ground and develops into a small, green, heart-shaped structure called the *prothallus,* which represents the gametophyte. The underside of the prothallus bears numerous rhizoids among which are found spherical antheridia. Archegonia are located near the notch of the prothallus. Sperm swim to the egg, usually on another prothallus since antheridia and archegonia on the same prothallus mature at different times. Fertilization occurs within the archegonium and the resultant diploid zygote develops into an embryo, which is nourished and protected for a short period by the gametophyte prothallus. Cell division produces the sporophyte plant, which becomes independent, and the prothallus disintegrates.

Angiosperms. Angiosperms, or flowering plants, are the most dominant organisms among the Plantae. They are characterized by the production of flowers, double fertilization, enclosed ovules and seeds, and a variety of pollination and seed dispersal methods. In the following discussion of angiosperms, special attention will be directed toward production of gametes (eggs and pollen), pollination, fertilization, and fruit and seed production.

Floral Structure. A **flower** may be defined as a specialized, branched stem bearing lateral appendages. It consists primarily of structures directly involved in reproduction and others which are only indirectly involved in reproductive activities. A cursory survey of a limited number of flowers would indicate that there is an enormous diversity in floral structure. Yet the similarities are greater than the differences, because all flowers are based upon the same structural design (Figure 17-14). The flower stalk is a stem which supports the flower and is known as the *pedicel.* The terminal portion of the pedicel consists of an enlarged structure, the *receptacle,* to which the floral parts are attached. The most familiar types of flowers have four kinds of floral organs attached to the receptacle. These organs, always arranged in the same order, are attached at successively higher levels. The first of these organs are the *sepals.* These appendages, attached at the base of the flower, are the outermost and lowest parts and usually green in color, closely resembling leaves. Collectively, the sepals are referred to as the *calyx.* The next appendages proceeding toward the center are the *petals,* collectively known as the *corolla.* The petals are usually variously colored, or white, leaf-like structures, which frequently extend beyond the sepals. The *perianth* is a collective designation for the calyx and corolla. The third group of appendages consists of the **stamens,** collectively termed the *androecium,* which are attached just above the base of the petals. Each stamen, or male reproductive structure, consists of a slender elongated stalk, the *filament,* and a somewhat expanded lobed structure at its distal end, the *anther.* The anthers contain pollen sacs in which large numbers of minute pollen grains are formed.

Attached to the uppermost site in the center of the flower is the female reproductive structure, the *gynoecium.* The basic structural unit of the gynoecium is the **carpel,** which is generally flask shaped and is composed of a number of prominent regions. The swollen basal portion, the *ovary,* contains *ovules* which develop into seeds. The ovary is connected to a stalk-like *style,* which terminates in an expanded area, the *stigma. Pistil* is another term referring to the megasporangial part of the flower and although the abandonment of the term has been advocated, it continues to be useful. The tissue of the ovary to which the ovules are attached is called the *placenta* and the arrangement of ovules, that is, the position of the placentae, is termed *placentation.*

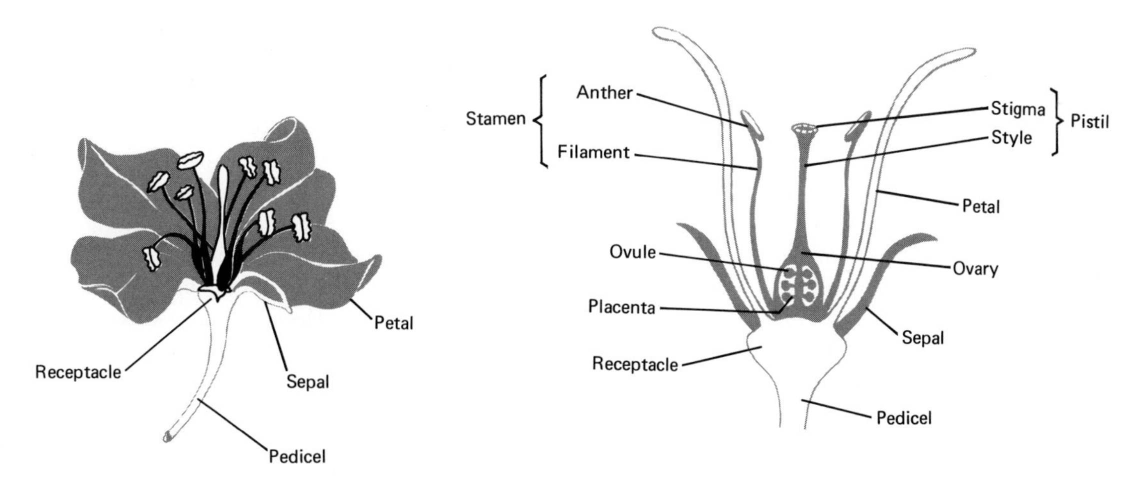

Figure 17-14. Generalized structure of a flower illustrating principal parts. *Left:* Entire flower. The petals collectively constitute the calyx; sepals collectively, the corolla. Together these form the perianth. *Right:* Enlargement of a longitudinal section through the center of a flower. Stamens collectively are the androecium; pistils collectively, the gynoecium.

Sepals and petals are referred to as *accessory parts* of a flower because they do not play a direct role in reproduction. Stamens and pistils, by contrast, are the *essential parts* of a flower because they are directly related to the reproductive process.

Most flowers show a definite numerical arrangement of parts. In the monocots (Monocotyledonae) the flowers generally have their parts in three or multiples of three. In the dicots (Dicotyledonae) flower parts are usually in fours and fives. Monocots and dicots are also differentiated on the basis of seed structure, leaf anatomy, and stem structure (see Figure 9-1).

Development of Pollen. A cross section of a young anther reveals that it is composed of four pollen sacs, or *microsporangia* (Figure 17-15A). Within each microsporangium there is a cluster of cells, the *microsporocytes.* These cells are characterized by their large size, abundant cytoplasm, and prominent nuclei (Figure 17-15B). Simultaneously with the growth of the anther, the microsporocytes (spore mother cells) undergo a series of divisions (meiosis) in which the chromosome number of each microsporocyte is reduced by half. In addition to a reduction in chromosome number, each microsporocyte undergoes divisions and gives rise to four cells, the *microspores* (Figure 17-15C). Essentially, then, each microsporophyte, which contains the full chromosome number (diploid) of the species of plant involved, produces four microspores containing only one-half the chromosome number (haploid). Each microspore eventually develops into a pollen grain.

The microspores undergo further mitotic division into two daughter nuclei (Figure 17-15D). There are no walls to separate these nuclei, so they are retained within the wall of the microspore. One of these nuclei, the *tube nucleus,* is larger; the other nucleus, the *generative nucleus,* is smaller; later it divides once again to form two sperm. Once the tube nucleus and generative nucleus are formed, the entire structure is referred to as a *pollen grain,* or young male gametophyte (microgametophyte). At about this stage of development, the wall between the microsporangia disintegrates, resulting in the formation of two pollen sacs (Figure 17-15E). With maturation of the flower, the anther splits along its long axis and the pollen grains are liberated.

Development of Ovules. The activities leading to the development of the ovules are not unlike those leading to the production of pollen grains. The ovule, one of many which may be produced in an ovary, first appears as a protuberance on the ovary wall. This protuberance consists of a *megasporangium* (*nucellus*) inside of which a single *megasporocyte* (spore mother cell) is developed (Figure 17-15F). As growth and development proceed, the megasporangium is raised on a stalk-like structure. One or two protective layers, the *integuments,* develop from the base of the megasporangium and completely surround it except for a small opening, the *micropyle,* that leads to the nucellus.

Accompanying these external changes, the diploid megasporocyte forms a linear tetrad of megaspores by meiosis (Figure 17-15G). Three of these, the ones nearest the micropylar end, disintegrate, leaving a single haploid megaspore (Figure 17-15H). The surviving megaspore undergoes successive mitotic divisions which result in eight haploid nuclei (Figure 17-15I). Three of the nuclei migrate toward the micropylar end, three move in the opposite direction, and two remain near the central region. Membranes develop around all nuclei except for those that are centrally located. This structure, consisting of six cells and two free nuclei, is now referred to as the mature *megagametophyte* or *embryo sac.*

Of the three cells at the micropylar end, one is the *egg;* the other two are referred to as *synergids.* The three cells at the opposite end of the embryo sac are termed the *antipodals* and the two nuclei in the center are called *polar nuclei* (Figure 17-15J). The fully developed embryo sac, composed of the polar nuclei, antipodals, synergids, and egg, together with the enveloping nucellus, integuments, and stalk, constitutes the mature *ovule.*

Pollination. **Pollination** refers to the deposition of pollen grains on the stigma of the female reproductive organ. The surface of the stigma is covered with a sticky secretion so that pollen grains will adhere more readily to its surface and therefore increase the chance that pollination will occur. Among the agents that facilitate pollination are wind currents, gravity, insects, birds, and other animals. Many flowers are *cross-pollinated;* that is, pollen is transferred from the anther of one plant to the stigma of the flower of another. Other angiosperms are *self-pollinated,* a process in which pollen is transferred from the anther to the stigma of the

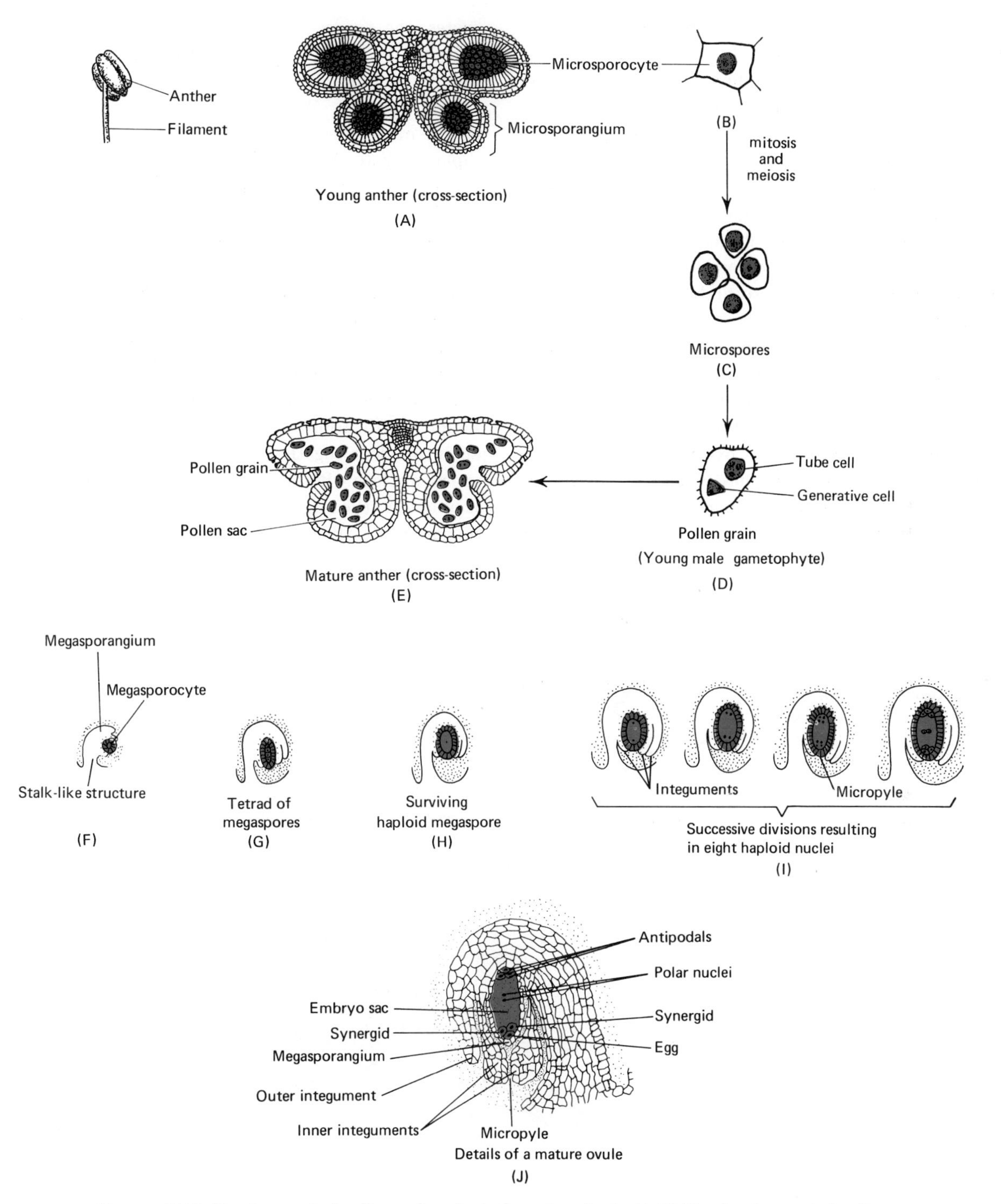

Figure 17-15. Development of pollen and ovules in flowering plants. (**A–E**) The development of pollen grains in the anther. (**F–J**) The development of the ovule.

same flower or another flower on the same plant. Cross-pollination occurs mainly by wind and insects. The most common insect pollinators are moths, bees, and butterflies. Flowers that are pollinated by insects generally possess large and conspicuous petals, secrete aromatic substances or nectar, and frequently have relatively small stigmas that lack the characteristic hairs and bristles of most wind-pollinated flowers. In addition, wind-pollinated flowers produce copious amounts of pollen grains to compensate for the large number of grains that fail to find receptive stigmatic surfaces.

Once a pollen grain is deposited on the stigma, it germinates (Figure 17-16). In this process, the pollen grain produces a cytoplasmic tube, the *pollen tube,* which grows (digests) through the tissues of the stigma, style, and ovary toward the ovule. The pollen tube produces extracellular enzymes which hydrolyze the tissues of the style. In this manner, energy for growth is provided. At this stage of development, the generative nucleus divides mitotically and produces two male gametes, the *sperm.* The sperm move down the pollen tube, the tip of the tube enters the embryo sac, bursts, and its contents are discharged.

Fertilization. Within the embryo sac, one of the sperm nuclei fuses with the egg nucleus, forming a diploid zygote (Figure 17-16). At about the same time, the other sperm nucleus fuses with the two polar nuclei (triple fusion) and the result is a primary endosperm nucleus. This nucleus also undergoes a series of mitotic divisions and gives rise to a multicellular food storage tissue, the **endosperm.** Subsequently, the zygote develops by many mitotic divisions into a multicellular *embryo,* or rudimentary plant (Figure 17-17). The antipodals and synergids disintegrate after fertilization. In addition, the tube nucleus also disappears. The term double fertilization is applied to the union of egg and sperm nucleus together with the fusion of the second sperm nucleus with the polar nuclei. This phenomenon is unique to the angiosperms. The entire life cycle of an angiosperm is illustrated in Figure 17-17.

Seeds. The success of the flowering plants has been

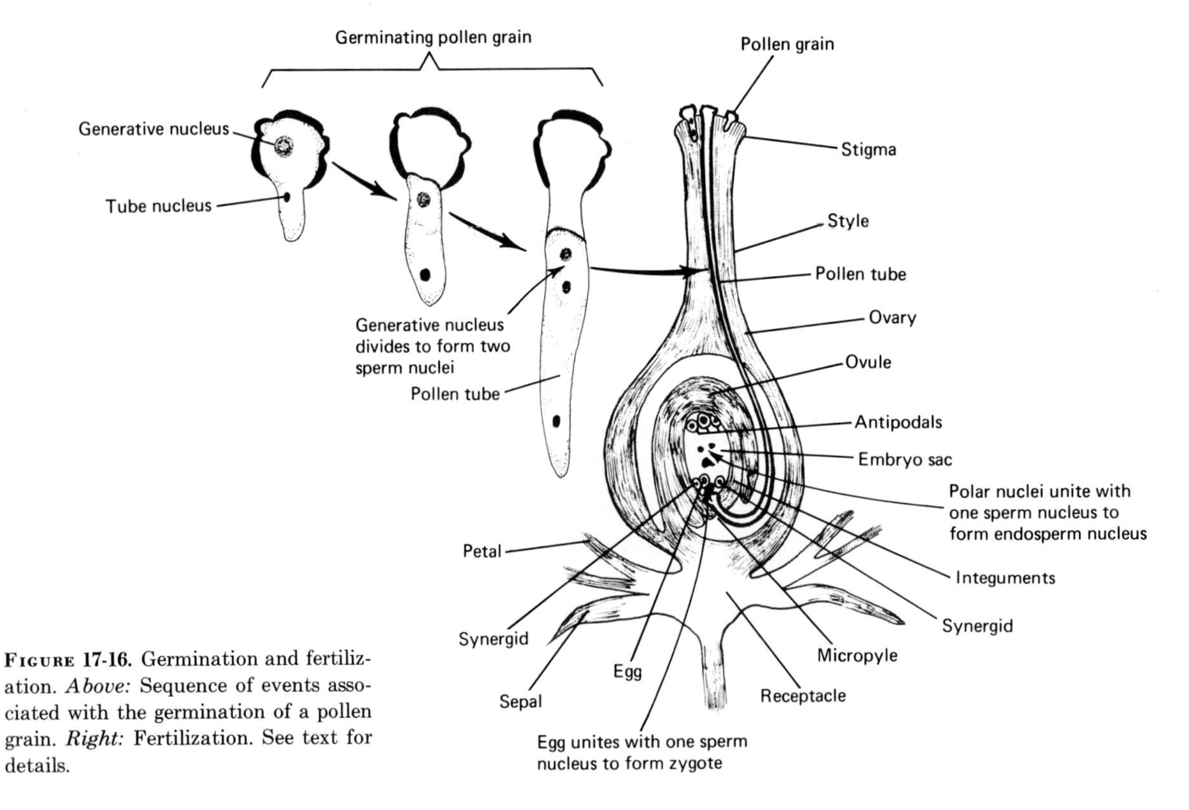

FIGURE 17-16. Germination and fertilization. *Above:* Sequence of events associated with the germination of a pollen grain. *Right:* Fertilization. See text for details.

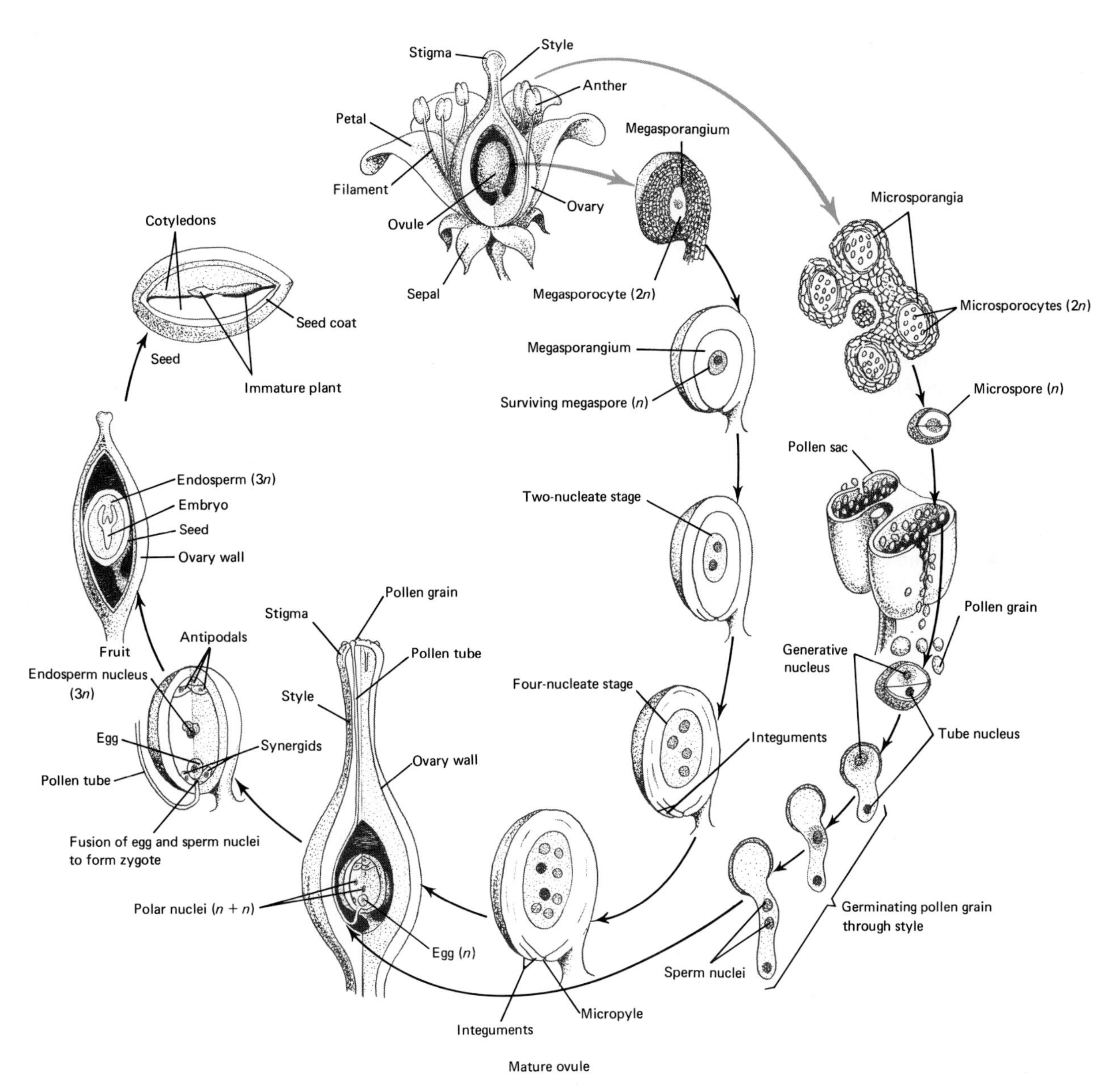

Figure 17-17. Life cycle of an angiosperm. See text for details.

due largely to the evolution of a mechanism for protecting the new generation within the old. This mechanism is accomplished by seed production within ovaries. A **seed** consists of an immature plant surrounded by a given quantity of stored food available for its early nourishment and a protective coat derived from the hardened surface layers of the integument. More simply, seeds are ripened ovules. The processes involved in the subsequent development of the seed into a mature plant are discussed in Chapter 18.

The dissemination of seeds serves as the mechanism for distribution of plants over a relatively wide geographic region. As with pollen, the principal agencies of seed dispersal are wind, animals, and water. A number of seeds are sufficiently light to be carried great distances by wind currents. Wind-dispersed seeds usually have devices such as wings, tufts, or hairs which make air travel easier. Some seeds and fruits float on water and frequently drift many miles before they again become land based. In some cases, the fruit may be eaten by animals and the enclosed seeds may survive the digestive process. Frequently, seeds possess spines, hooks, or a sticky secretion which enables them to adhere to the fur of mammals or the feathers of birds. Of course, man himself is a great disseminator of seeds, especially those which he finds economically worthwhile.

The length of time a seed may remain viable varies from a few weeks (silver maple) to more than 1000 years (Indian lotus), depending upon environmental conditions and the particular species. Seed **germination,** the resumption of the growth of the embryo, after a period of much-reduced metabolic activity, is discussed in Chapter 18.

Fruits. The botanist describes a **fruit** as a structure composed of one or more ripened ovaries with or without seeds, together with any other accessory flower parts which may be associated with ovaries. Such a definition includes grains of wheat and corn, pea pods, tomatoes, cucumbers, chestnuts, and many other structures, which, in familiar usage, are not regarded as fruits. It is commonly thought that a structure must be edible to be considered a fruit, but the ovaries of a large number of flowers develop into fruits that cannot be eaten. The question of whether or not a plant food should be called a fruit or a vegetable may be decided on the basis of the structural origin. If the structure develops from a floral ovary, it is a fruit; if it represents some other plant structure, it is usually a vegetable.

In most species pollination is necessary, not only for fertilization and the development of the embryo plant and seeds, but also for the development of fruits. If the stigma is not pollinated and fertilization does not take place, the flower usually withers and drops from the plant. Failure to set fruit may also result from unfavorable environmental conditions such as late frosts in the spring. Pollination normally prevents the abscission of the flower and initiates the development of the tissues of the fruit itself. Evidence indicates that hormones are present in pollen grains and initiate the development of the ovary. Without pollination, the egg cannot be fertilized. The development of fruits without pollination and fertilization is termed **parthenocarpy,** and such fruits are usually seedless. In some plants fertilization may occur, but the ovules fail to develop into mature seeds even though fruit develops normally. Natural parthenocarpy is found in certain kinds of citrus fruits, pineapples, bananas, cucumbers, and seedless grapes. Artificial parthenocarpy has been induced in some plants by treating them with dilute solutions of growth substances. Among the plants that have been induced to form seedless fruits by artificial parthenocarpy are watermelon, summer squash, cucumber, tomato, and holly.

The main structure of a fruit wall is the *pericarp,* which was formerly the wall of the immature ovary. The pericarp is differentiated into three morphologically distinct layers: the exocarp, mesocarp, and endocarp. The *exocarp* constitutes the outermost layer of cells and is sometimes characterized by epidermal hairs and stomata. The *mesocarp* is the middle layer of tissue, which varies in thickness and is often composed of parenchyma and vascular tissue. Of the three tissues, the internal *endocarp* is the most variable in structure, texture, and thickness (Figure 17-18). For example, in such fruits as the cherry and peach, the endocarp is the stony pit that encloses the seed.

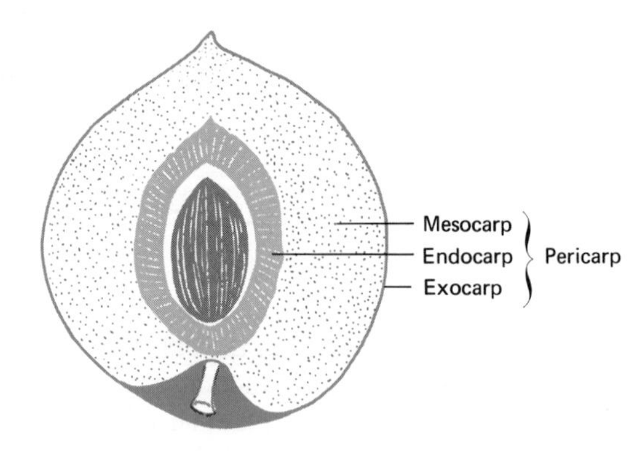

FIGURE 17-18. Diagrammatic representation of the pericarp showing its three subdivisions as they appear in a peach.

Sexual Reproduction in Protista and Animalia

Sexuality in Protista is fundamentally simple isogamy. In Animalia, heterogametes are produced and fertilization (heterogamy and oogamy) may be either internal or external.

Conjugation

Paramecia are unicellular animals which undergo sexual reproduction in the form of conjugation (Figure 17-19). Whereas bacterial conjugation involves a donor cell and a recipient cell, paramecia undergo a mutual exchange of nuclear material. Two individuals come into contact with each other at the region of the oral grooves. While they remain attached, a series of nuclear changes takes place in each organism. First, the micronucleus undergoes two divisions, forming four equal nuclei. Of these four, three disintegrate and the one remaining divides unequally. The products of this division are called *pronuclei.* The two smaller pronuclei in the conjugants are exchanged and fuse with the larger pronuclei (forming a zygote nucleus) which remains in each individual. While these changes have been taking

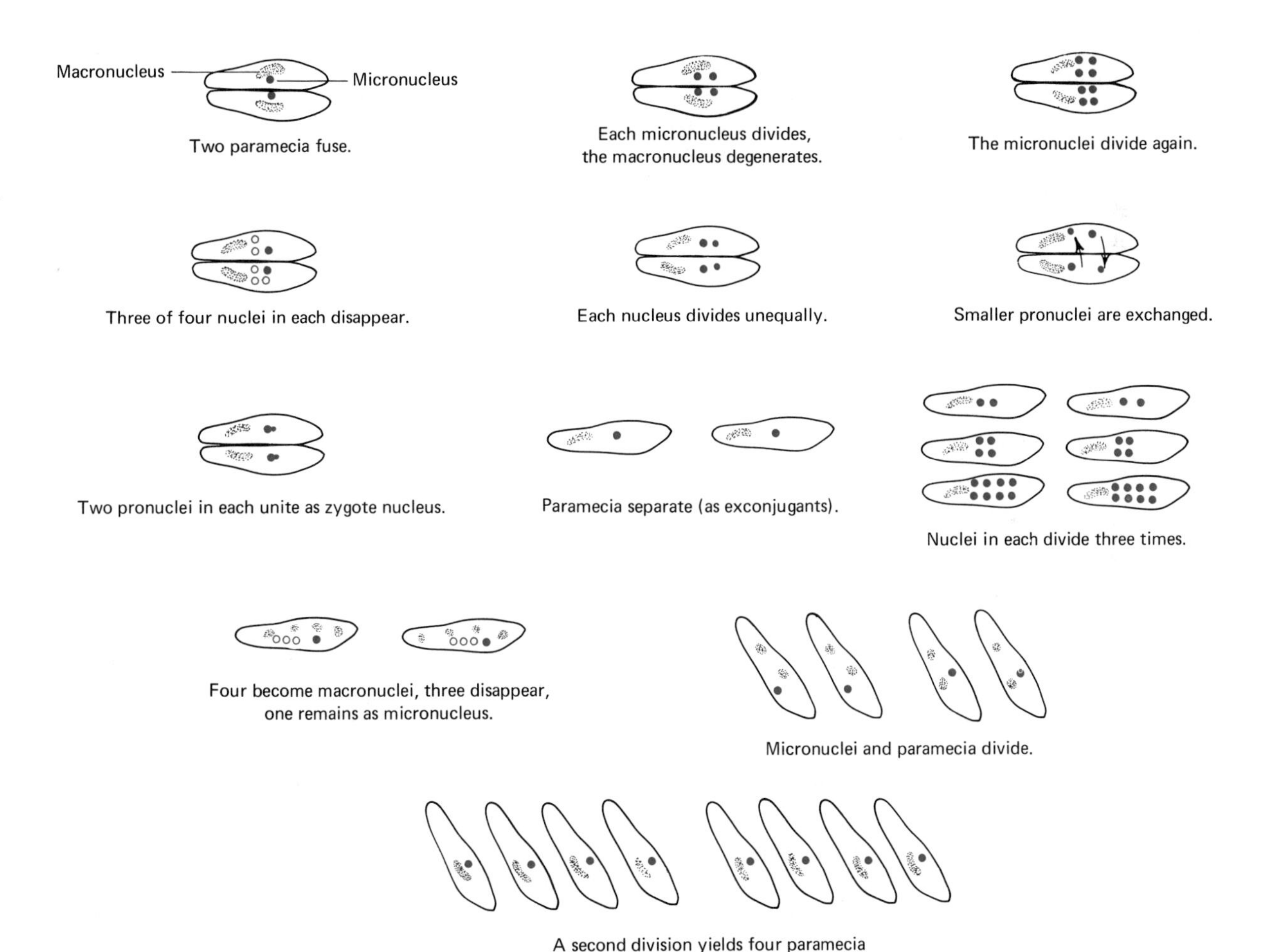

Figure 17-19. Conjugation in *Paramecium.* In this sequence of events, there is a mutual exchange of micronuclear materials. See text for description.

place, the macronucleus in each conjugant has been breaking down and has disappeared within the cytoplasm.

After fertilization is completed, the individuals separate. In each individual, the single micronucleus, the product of fertilization, undergoes successive divisions, forming eight micronuclei. Of these, four enlarge and become macronuclei, three disintegrate, and one remains as the micronucleus. Then, each individual undergoes two binary fissions, producing four new individuals each with one macronucleus and one micronucleus. The process of conjugation is completed with the formation of eight new paramecia from the original two individuals.

Hermaphroditism

In animals, the sexes are typically separate, although *hermaphroditism* (both sexes in one individual) is common, especially among sessile and sluggish animals.

Many invertebrates exhibit hermaphroditism. Although such organisms contain both male and female sex organs, self-fertilization rarely occurs. Sponges represent sessile hermaphrodites. Sperm are produced by ameboid cells within the body wall and are released into the water. Eggs are also produced within the body wall and are fertilized in place. A flagellated larva develops, escapes from the parent sponge, and, after a short period of free swimming, settles onto the substrate and develops into a new sponge. Earthworms are motile hermaphrodites; their reproductive organs are organized into systems complete with gamete-producing structures, ducts, and storage sacs.

Parthenogenesis

There are some organisms, for example, bees and ants, which normally produce new individuals from unfertilized gametes. This phenomenon is known as **parthenogenesis.** In isogamous organisms, either gamete type may develop by parthenogenesis. In oogamous animals, only the eggs are known in some cases to develop parthenogenetically. Parthenogenesis may occur either naturally or artificially. Artificial parthenogenesis may be initiated by mechanical methods, such as pricking the surface of an egg with a needle, chemical action, and radiation. A parthenogenetic gamate is functionally similar to a spore in that both develop into adult organisms directly.

Types of Fertilization

Fertilization is a complex process which involves the fusion of an egg and a sperm resulting in the activation of the egg and the establishment of the $2n$ condition. Normally the sperm is responsible for both results although it has been shown that the egg can become activated without requiring the $2n$ condition (parthenogenesis).

When sperm and egg are relatively close to each other, there are certain events which involve the egg surface and the sperm head. Most sperm have a cap, the acrosome, which is involved in establishing direct contact with the egg and in aiding the penetration of the egg by the sperm. Once contact is made, the acrosome ruptures and a tube forms, penetrating the egg to contact the plasma membrane. The sperm and egg membranes fuse and some of the egg cytoplasm bulges into the canal that has been formed. The sperm nucleus enters this cytoplasm which then withdraws, bringing with it the sperm nucleus and its associated parts. Following these events, there is an elevation of a new membrane (fertilization membrane) from the egg surface. This newly formed membrane may now serve as a protective barrier against the entrance of additional sperm cells.

After the sperm nucleus is incorporated into the egg cytoplasm, it is called a *pronucleus.* Another pronucleus is produced by the egg and the two pronuclei fuse to form the zygote nucleus. This fusion re-establishes the diploid condition of the species.

Fertilization may occur either externally or internally. Certain animals that lack copulatory organs will undergo external fertilization involving heterogametes. Earthworms, for example, utilize a slime ring to bring sperm and eggs together so that fertilization may occur. Copulatory animals have anatomical structures designed for the deposit and reception of sperm. Such animals have adaptations for either internal or external development of the embryo. Birds undergo internal fertilization but the embryo develops outside the mother's body. In the human, both fertilization and embryonic development are internal.

Care of the Young in Vertebrates

Parental care of the young is almost entirely restricted to vertebrates. Some insects lay their eggs in such a manner as to ensure a food supply for the young, but the parent does not remain with them. Only social insects, such as bees, ants, and wasps, exhibit care for the young by the adults. Even in vertebrates, parental care is not universal. Many species of fish, reptiles, and amphibians abandon their eggs after laying them. If the young develops within the mother's body, there is, of course, parental care. However, after birth, the young may be on their own, as exemplified by the shark and garter snake.

Many vertebrates carefully guard the eggs at least until hatching. With few exceptions, parental care after hatching or birth is limited to birds and mammals. In birds, this care extends to feeding, protecting, and training the young. Mammals follow a similar pattern, although the care may extend for a much longer period of time.

There is a relationship between the number of eggs and sperm produced by the parents, the number of young that develop and survive, and the amount of care given the young by the parents. A female Pacific salmon may lay about one million eggs in a spawning period, after which she dies. Of these, perhaps only ten embryos will survive to develop into adults. A human generally produces only one egg at a time to be fertilized. That one embryo has a much better chance for development into an adult than do the salmon embryos.

Vertebrate Reproductive Systems

Most vertebrates reproduce sexually. The organs of vertebrate reproductive systems include gonads (testes and ovaries) specialized for the production of gametes, ducts that transfer the gametes, and variously modified accessory organs such as glands, and, in many, copulatory structures. Although external genitalia are essential only for land life, copulatory organs did evolve among fishes. The claspers of the shark, which are modified pelvic fins in the male, are utilized for carrying sperm to the cloaca of the female. Many bony fishes also have fins that have been converted to copulatory organs. Among the strictly land vertebrates (most reptiles, birds, and mammals) a special copulatory organ, the penis, has evolved. In reptiles the penis is a pair of elongated masses of erectile tissue with a groove between. In mammals, the groove is closed to form a tube which is a direct continuation of the urethra. The terminal portion of the female tract has followed a parallel evolution by becoming modified into a tube-like receptacle, the vagina. Along with these changes has come the introduction of a large complex of hormones and certain nervous modifications that are needed to bring about the fusion of gametes.

Numerous vertebrates exhibit conspicuous secondary sex characteristics, which have the principal function of serving as visual, olfactory, mechanical, or auditory stimuli to attract the opposite sex and to elicit a mating response. Among some of the prominent secondary sexual characteristics of vertebrates are voice, features of the skeleton and musculature, and pigment patterns or specialized skin structures such as feathers or hairs, combs, and horns. Others include the release of chemical substances into the water by certain female fish, enlarged claws of some male turtles, mammary glands of female mammals, and the profusion and distribution of hair on the two sexes.

In the following discussion of vertebrate reproductive systems, emphasis will be placed on the human with only limited reference to the reproductive apparatus of other vertebrates.

Male Reproductive System

Testes. Vertebrate testes tend to be discrete compact organs of definite shape. In general, the **testes** are somewhat elongate in fish and typically oval in higher vertebrates. The principal functions of the testes are the production of sperm and the synthesis of male sex hormones.

The paired testes of the human male (Figure 17-20) are suspended in a pouch-like, skin-covered structure, the *scrotum.* A capsule of white fibrous connective tissue encases each testis and send partitions through its interior, dividing it into lobules. Each *lobule* contains one to four tiny coiled *seminiferous tubules* and numerous microscopic *interstitial cells* (Leydig cells) located between the tubules. The process of spermatogenesis,

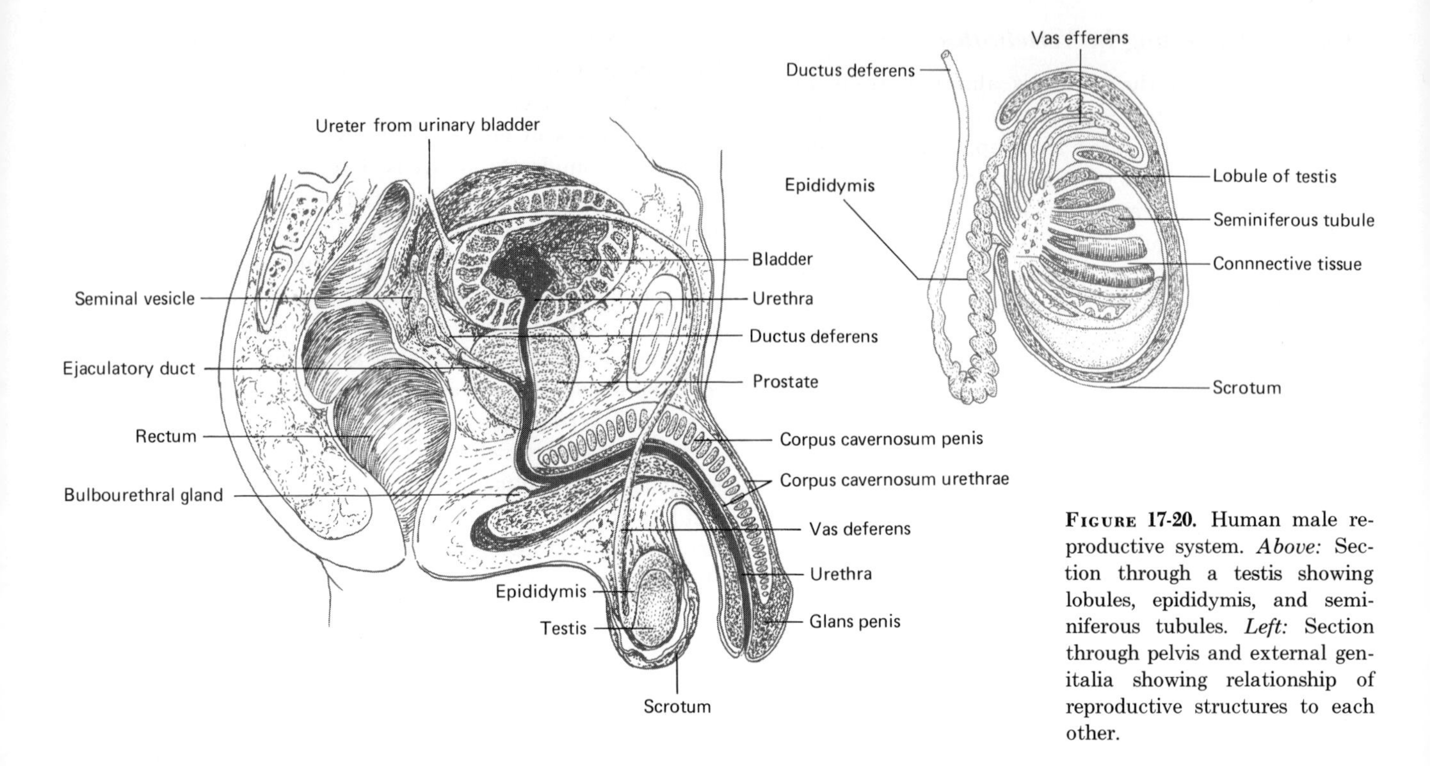

Figure 17-20. Human male reproductive system. *Above:* Section through a testis showing lobules, epididymis, and seminiferous tubules. *Left:* Section through pelvis and external genitalia showing relationship of reproductive structures to each other.

the production of sperm by meiosis, occurs within the seminiferous tubules. The secretion of hormones, chiefly *testosterone,* occurs by the interstitial cells. Among the principal functions of testosterone are the development and maintenance of male secondary sex characteristics, the stimulation of protein anabolism, and the inhibition of anterior pituitary secretions of gonadotropins.

At a given age, according to the vertebrate species under consideration, the testes begin to produce sperm and this continues throughout the reproductive life of the organism. In humans, sperm production begins at puberty. In a great number of mammals, the testes move from the body cavity in which they are formed into the scrotum during breeding season and are afterward withdrawn (bats, rodents). In other mammals, (most primates and carnivorous mammals), the testes remain in the scrotum permanently. Temperature is an important factor in sperm production for many mammals. The position of the testes within the scrotum provides a lower temperature than would be found within the abdominal cavity. Changes in environmental temperature cause a change in the degree of separation of the testes from the body. If the external temperature is warm, the testes are suspended lower than if the external temperature were cold. That this regulation of the position of the testes is important may be borne out by the fact that should the testes not descend but remain within the body cavity, no functional sperm are produced. This condition is known as *cryptorchidism* and results in sterile individuals. Where the condition has been surgically corrected, the individuals often become fertile. On the other hand, elephants and whales have testes permanently and functionally located in the abdominal cavity. In these organisms, temperature is apparently not a critical factor.

Spermatozoa. Male reproductive cells, or *spermatozoa,* vary in form among vertebrates, although all possess long flagella-like tails which serve as locomotive structures (Figure 17-21). Typically, sperm cells are composed of an anterior *head,* an intermediate *middle piece,* and a posterior *tail.* The head contains the nucleus with a haploid chromosome number and the acrosome, which, as noted earlier, effects penetration of the sperm cell into the egg. The middle piece contains numerous mitochondria which carry on respiration, an energy-releasing process that provides energy for locomotion. The tail, which consists of an internal arrangement of fibrils of a typical flagellum, facilitates locomotion of the entire cell. The length of vertebrate sperm cells ranges from 0.02 mm to over 2 mm; human sperm cells average about 0.05 mm. The number of sperm cells produced in the testes is relatively high. In the human, for example, a single ejaculation (approximately 4 ml) contains some 300 million cells; daily production gives rise to about 50 million cells. Such a large number is necessary in higher vertebrates since a human male who produces less than 60–80 million sperm cells per ejaculation is not likely to induce pregnancy.

Ducts and Glands. Once sperm are produced by the testes, they are transported to the exterior by a series of variously differentiated ducts. From the numerous seminiferous tubules, mature sperm cells are collected in a highly coiled tube, the *epididymis* (Figure 17-20). This tube, located on the surface of each testis, has a very small diameter but measures about 20 feet in length. The epididymis not only serves as a duct for the passage of sperm from the testis to the exterior, but also stores sperm prior to ejaculation and secretes a small portion of the seminal fluid (semen).

As it emerges from the testes, the epididymis becomes straight and somewhat enlarged and becomes differentiated as the *ductus deferens* (*vas deferens*) (Figure

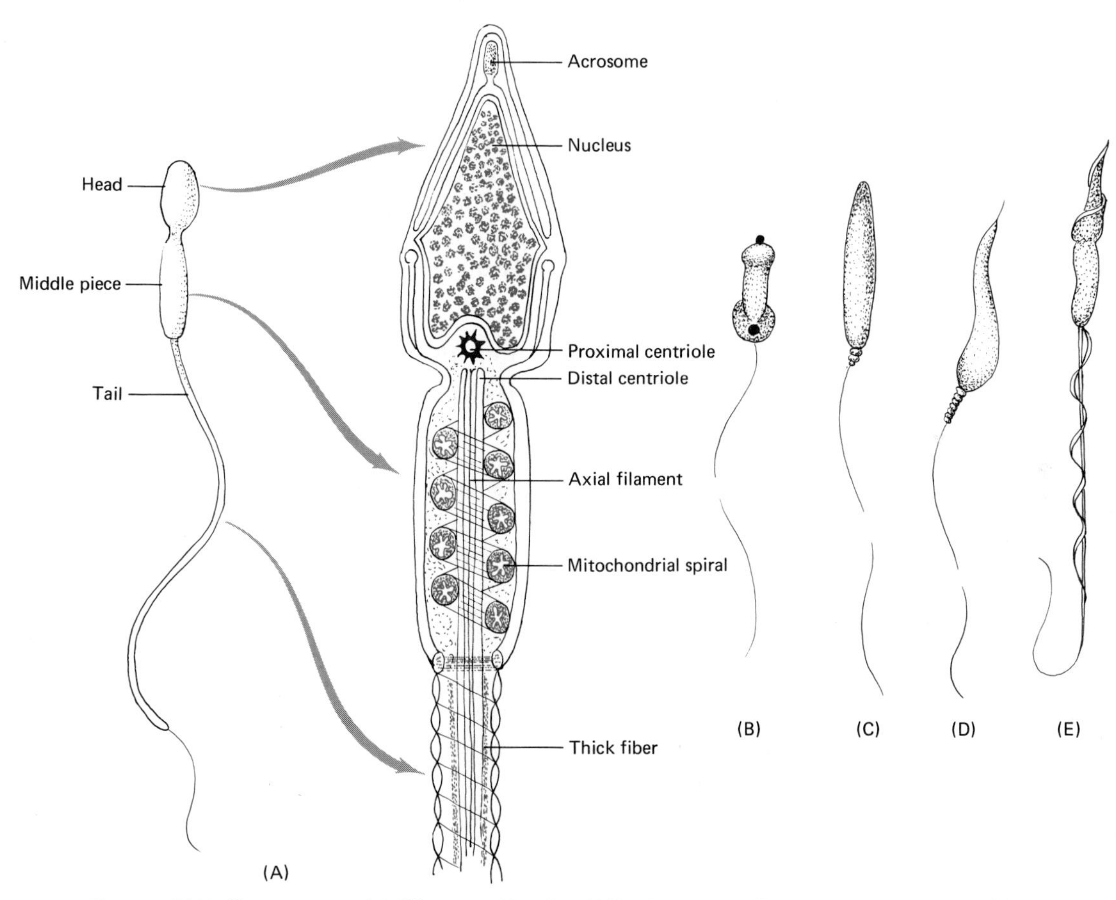

Figure 17-21. Spermatozoa. (**A**) Diagram of head, middle piece, and tail of a mammalian sperm (*left*) and enlarged aspect of the three principal regions (*right*). (**B–E**) Diagrams of various vertebrate spermatozoa. (**B**) Fish (sturgeon). (**C**) Amphibian (frog). (**D**) Reptile (turtle). (**E**) Bird (*Turdus*).

17-20). In mammals this duct passes into the inguinal canal and into the abdominal cavity, extending over the top and down the posterior surface of the urinary bladder. Behind the bladder in most mammals each ductus deferens receives a duct from one of a pair of glands, the *seminal vesicles* (Figure 17-20). Secretions from these glands contribute to the viscous portion of the semen. Each of the two *ejaculatory ducts* is formed by the union of a ductus deferens and the duct from a seminal vesicle. The ejaculatory ducts have thick muscular walls and serve in the emission of semen. In mammals the two ejaculatory ducts converge and join with the urethra from the urinary bladder so that a single tube transports both semen and urine. In most other vertebrates, however, the urinary and genital ducts open separately into a common chamber, the cloaca, which in turn expels substances to the outside.

Surrounding the first portion of the urethra is the *prostate gland* (Figure 17-20). This gland is a doughnut-shaped compound tubuloalveolar gland that encircles the urethra. This anatomical feature is of clinical significance since enlargement of the gland squeezes the urethra, frequently closing it and inhibiting urination. The prostate secretes an alkaline substance that constitutes the largest part of the seminal fluid. Its alkalinity helps to protect the sperm from the acid present in the male urethra and female vagina and thereby increases sperm motility.

Just below the prostate is a pair of pea-sized glands, the *bulbourethral glands* (Cowper's glands), which are connected by ducts to the membranous portion of the urethra (Figure 17-20). Like the prostate, these glands also secrete an alkaline fluid. The fluid serves as a lubricant during copulation.

Penis. Among most aquatic vertebrates, egg and sperm are simply discharged into the environment and fertilization occurs externally. In the vast majority of terrestrial forms, however, fertilization is internal; the moisture in the female tract in conjunction with the semen of the male provides the moist environment necessary for the union of sperm and egg. Male land vertebrates typically transfer sperm to females by two means. One involves cloacal apposition, a method common among most birds. Other birds, reptiles, and mammals transfer sperm by means of a penis during copulation.

The *penis* of male humans consists of three masses of erectile tissue enclosed in separate fibrous coverings and surrounded by skin (Figure 17-20). The two uppermost cylinders are referred to as the *corpora cavernosa penis,* whereas the smaller inferior one, the *corpus cavernosum urethrae,* contains the urethra. Under the influence of sexual emotion, the sinuses of the erectile tissue become filled with blood, causing the organ to become erect. At the distal end of the penis there is a slightly bulging structure, the *glans penis,* which, prior to circumcision, is surrounded by a fold of skin, the *prepuce,* or *foreskin.*

Seminal Fluid. *Seminal fluid,* or *semen,* is a collective designation for all of the secretions of the male reproductive system together with the sperm. Of the total volume produced, the testes and epididymis secrete about 5% of the semen, the seminal vesicles about 60%, the prostate gland approximately 30%, and the Cowper's glands about 5%. The testes, as noted earlier, also add hundreds of millions of sperm. In traveling the route from their place of origin to the exterior, the sperm pass from the testes through the epididymis, seminal duct, ejaculatory duct, and urethra. Ejaculation of the seminal fluid occurs by peristaltic movements as a result of sexual excitation.

Female Reproductive System

Ovaries. The **ovaries,** like the testes, produce hormones as well as gametes. The functions of these hormones will be discussed shortly. In a few fish and typically in amphibians, snakes, and lizards, the ovaries are sac-like and hollow, whereas in turtles, crocodilians, and mammals they are characteristically compact, ovoid organs. The ovaries of the human female (Figure 17-22A) resemble large almonds in both shape and size and are located below and behind the uterine tubes. Each ovary is supported by a series of ligaments.

The outer layer of each ovary consists of a layer of germinal epithelial cells, whereas the inner portion contains numerous nerve fibers, blood vessels, connective tissue, and muscle tissue (Figure 17-23). Immediately under the germinal epithelium is a layer of connective tissue beneath which is a thick layer of spherical groups of cells, or *follicles,* each enclosing an egg. At birth, hundreds of thousands of follicles are present in the ovary, but the number decreases steadily throughout

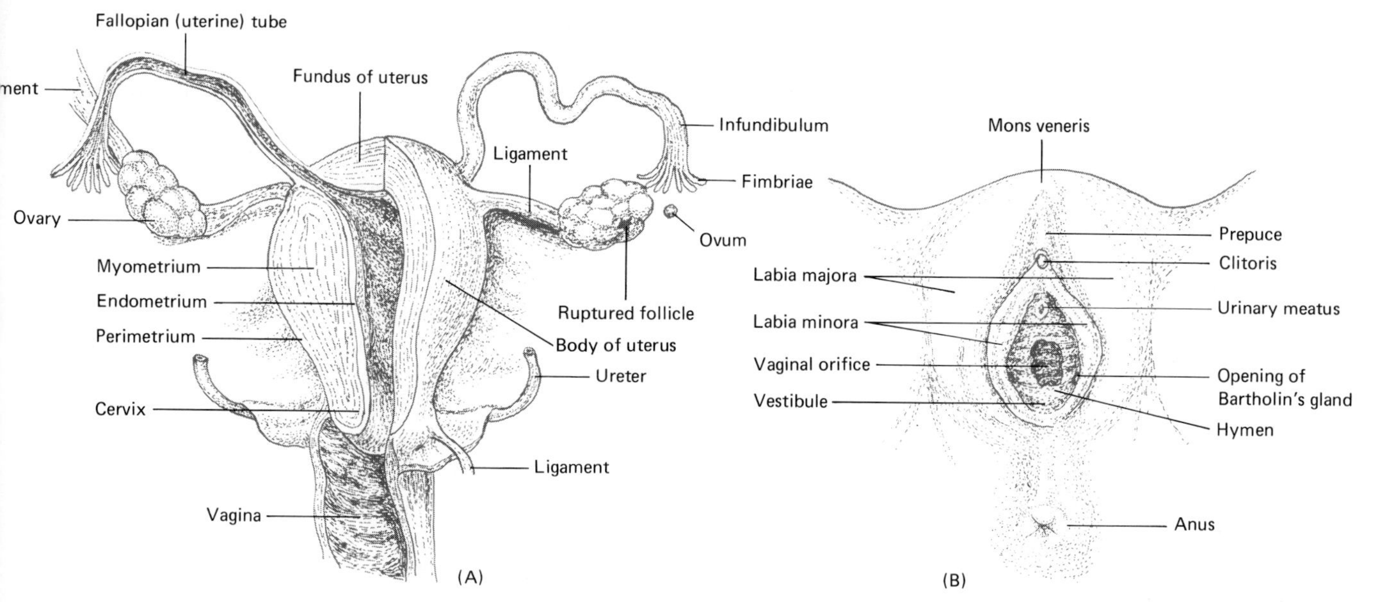

Figure 17-22. Human female reproductive system. (**A**) Internal aspects of the system. (**B**) External genitalia.

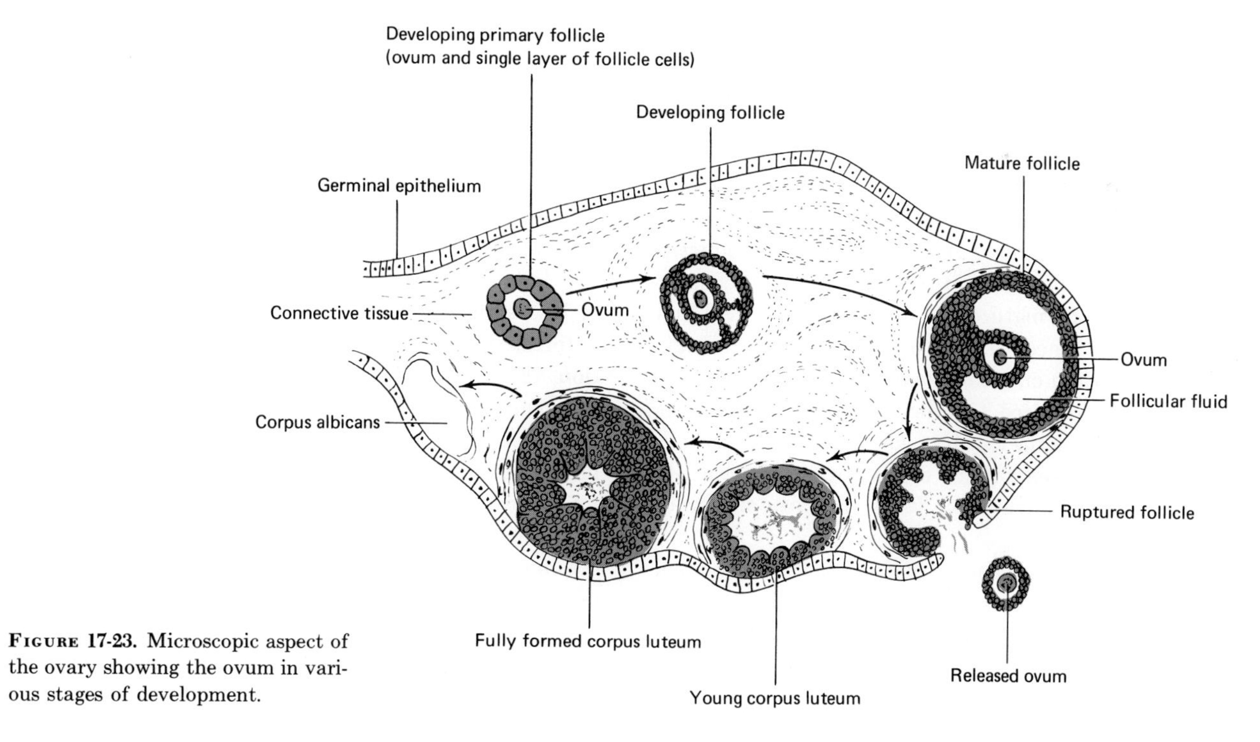

Figure 17-23. Microscopic aspect of the ovary showing the ovum in various stages of development.

life. During the reproductive years of the average female, only 300–400 follicles reach maturity and liberate their ova (eggs). This great limitation results because the maturation of ova and follicles does not begin until puberty and then usually only one follicle and ovum reaches maturity each month.

As part of the maturation process, the follicular cells proliferate by mitosis and fluid begins to accumulate. One of the female hormones (estrogen) is secreted by these developing follicular cells. The immature ovum is also growing and undergoing meiosis in which the chromosome number is reduced by half. As the follicle reaches maturity, it becomes greatly distended with fluid and bulges on the surface of the ovary. The entire sequence of events related to the maturation of the follicles and the secretion of estrogen is controlled by the follicle-stimulating hormone (FSH) secreted by the anterior pituitary gland.

Ovulation. Upon reaching a state of maturity, the follicular wall breaks and releases the egg from the ovary, a process called **ovulation** (Figure 17-23). Vertebrate ova are usually set free into the coelom where they are swept into an egg duct, the oviduct, for delivery to the exterior or to the site of internal implantation. In female frogs, for example, the body cavity becomes quite swollen with ripened eggs because the opening of the oviduct is detached from the ovary. In mammals, however, the opening of the oviduct is relatively close to the ovary so that eggs are seldom discharged into the body cavity.

Ovulation generally occurs every 21–30 days from puberty to *menopause,* the end of the reproductive period of life. An exception to this is during pregnancy when there is no maturation of follicles and ovulation does not take place. Usually ova are discharged singly, but, if more than one reach maturity at the same time, they may be released together. Once an ovum is discharged, it enters the egg duct where it may be fertilized if sperm are present; if the egg is not fertilized within 18–24 hours after ovulation, it disintegrates and disappears.

At the site of ovulation, the wall of the follicle collapses, and there is a small amount of bleeding into the follicular cavity. The follicular cells enlarge and a yellowish substance, *lutein,* accumulates in their cytoplasm. A follicle entirely filled with lutein is designated as the *corpus luteum* ("yellow body"). The lutein cells elaborate the female hormone *progesterone.* If the ovum is not fertilized, the corpus luteum continues to grow for about two weeks and then regresses, becoming a small, white ovarian scar. If, on the other hand, the ovum is fertilized, the corpus luteum enlarges considerably, forming the corpus luteum of pregnancy. It continues to function for about 4–5 months and then slowly degenerates, leaving a white scar on the ovary. Formation of the corpus luteum and the production of progesterone are under control of hormones secreted by the pituitary gland.

The Uterine Tubes. In lampreys and hagfishes, gametes of both sexes are discharged into the coelom and are carried to the exterior by a pair of pores in the body wall. In all other vertebrates, sperm ducts carry male gemetes, but eggs are characteristically discharged into the body cavity; a pair of tubes, the oviducts, transport the eggs out of the body cavity. The *uterine tubes,* or *fallopian tubes,* of the human female are attached to the uterus at its upper outer angles (Figure 17-22A). They extend upward and outward toward the sides of the pelvic cavity and then curve downward and backward. At the distal end, each tube expands into a funnel-like structure, the *infundibulum.* The open outer margin of the infundibulum resembles a fringe in its irregular form; the fringe-like projections are termed *fimbriae.*

The inner coat of the uterine tube is a mucous membrane arranged in a series of longitudinal folds. Some of its epithelial cells are primarily secretory, whereas others are ciliated. The middle coat is composed of circular and longitudinal layers of smooth muscle tissue that facilitate the peristaltic contractions of the tube. The outer coat is a serous membrane. Normally, fertilization, the union of a sperm and an egg, occurs in the fallopian tube. In some cases, however, the ovum does not enter the fallopian tube but remains in the abdominal cavity and is fertilized there. Such a condition is referred to as an abdominal fertilization. Also, the ovum may remain attached to the ovary and develop, a condition called an ovarian pregnancy. The word *ectopic* is used to describe implantation and development at any site other than the uterus.

The Uterus. In reptiles and birds, the lower portion of the oviduct, termed the uterus, is thickened and muscular and receives glandular secretions which contribute the shell to eggs. In placental mammals, the uterus serves as a site for the implantation of the fertilized egg and the subsequent development of the embryo.

The human *uterus* (Figure 17-22A) is a pear-shaped organ that measures about 3 inches in length, 2 inches in width at its widest part, and 1 inch in depth. Anatomical subdivisions of the uterus include the dome-shaped portion above the uterine tubes, called the *fundus,* the major central tapering portion, called the *body,* and the interior narrow portion opening into the vagina, called the *cervix.* Three coats compose the walls of the uterus: endometrium, myometrium, and perimetrium. The inner *endometrium,* which consists of a surface layer of columnar epithelium, a middle layer of loose connective tissue, and a basal layer of dense connective tissue, is sloughed off during menstruation and following delivery of a baby. The thick middle coat of the uterus, the *myometrium,* consists of layers of smooth muscle fibers. This coat is highly adapted to expel the fetus at birth. The outermost coat, the *perimetrium,* consists of a serous membrane, the parietal peritoneum, and a thin connective tissue layer. The uterus plays a role in three highly important functions: menstruation, pregnancy, and labor. *Menstruation* is a sloughing away of most of the compact and all of the spongy layers of the endometrium, accompanied by bleeding from the torn vessels. In pregnancy, the embryo implants itself in the endometrium where it undergoes extensive growth and development prior to birth. Labor consists of powerful, rhythmic contractions of the muscular uterine wall that results in expulsion of the fetus.

THE VAGINA. The *vagina* (Figure 17-22A) is a collapsible musculo-membranous tube, about 3½ inches long, lying between the bladder and the urethra anteriorly, and the rectum posteriorly. It extends upward and backward from its external opening and is capable of great distention. The upper end of the vagina surrounds and is attached to the cervix of the uterus while at the lower end there is usually a fold of mucous membrane in the virginal state, the *hymen* (Figure 17-22B), which partially closes the orifice. The vagina constitutes an essential part of the reproductive tract since it is the organ that receives the seminal fluid from the penis during copulation, serves as the lower portion of the birth canal, and acts as an excretory duct for uterine secretions and the menstrual flow.

THE VULVA. The *vulva* (Figure 17-22B) is a collective designation for the female external genitalia. Among these are the mons veneris, labia majora, labia minora, clitoris, urinary meatus, vaginal orifice, and Bartholin's glands.

The *mons veneris* (pubis) is a skin-covered pad of fat over the symphysis pubis. Coarse hairs appear in this region at puberty and persist throughout life. The *labia majora* ("large lips") are two prominent longitudinal folds of skin and underlying fat that extend backward from the mons veneris to the anus. The skin of the labia majora contains numerous hair follicles, sweat glands, and sebaceous glands. The *labia minora* ("small lips") are two small folds of mucous membrane situated between the labia majora. Anteriorly, the folds meet to form the *prepuce,* a covering over the clitoris. There are no hair follicles or sweat glands in the labia minora. The *clitoris* is a small organ composed of erectile tissue and is located just behind the junction of the labia minora. It corresponds in structure and origin to the penis and just as the foreskin covers the glans penis, the prepuce covers the clitoris. The *urinary meatus* is the small opening of the urethra situated between the clitoris and vaginal orifice. The *vaginal orifice* is the external opening of the vagina. The *greater vestibular* (*Bartholin's*) *glands* are two bean-shaped glands, one on either side of the vaginal orifice. Each gland opens via a single duct into the space between the hymen and the labia minora. They are homologous to the bulbourethral glands of the male and secrete a lubricating fluid.

The *perineum* is generally defined as the entire region of the pelvic outlet and contains all the structures found between the pubic symphysis and the coccyx. The clinician, however, defines the perineum as the region between the vagina and the anus. This area has great clinical importance because of the danger of its being torn during childbirth. If the tear is deep, it may extend all the way through the perineum and through the anal sphincter. To avoid this situation, an incision called an *episiotomy* is usually made in the perineum.

MAMMARY GLANDS. The *mammary glands,* or breasts, are functionally related to the reproductive system since they secrete milk for the nourishment of the young, but, structurally they are related to the skin.

The breasts lie over the pectoral muscles and are attached to them by a layer of connective tissue. Estrogens and progesterone, both ovarian hormones, control their development during puberty. Estrogens stimulate growth of the ducts of the mammary glands, whereas progesterone stimulates development of the alveoli, the actual secretory units.

A number of mechanisms operate in controlling lactation, that is, the secretion of milk. As noted earlier, the ovarian hormones (estrogens and progesterone) act on the breasts to make them structurally ready to secrete milk. A high concentration of estrogens, such as during pregnancy, inhibits anterior pituitary secretions of lactogenic hormones. Shedding of the placenta after delivery causes a decrease in estrogen concentration in the blood. The resulting drop stimulates anterior pituitary secretion of lactogenic hormones. In addition, the suckling movements of a nursing baby act in some way to stimulate both anterior pituitary secretion of lactogenic hormone and posterior secretion of oxytocin. Lactogenic hormone stimulates lactation. Milk secretion starts about the third or fourth day after delivery and may continue for 6–9 months or even longer. Oxytocin stimulates the alveoli of the breasts to eject milk into the ducts, thus enabling the infant to remove the milk by suckling.

Female Reproductive Cycles

During the years between *menarche* (the onset of the menses) and menopause, many changes recur periodically in the human female. Among these are cycles related to the changes in the endometrium, breasts, ovaries, vagina, hormone secretions, and even emotional attitudes. In this final section, some of the details known or hypothesized about two of the cycles, as well as hormonal interactions, will be examined.

In most mammalian species, the female experiences rhythmic variations in the intensity of sex urge that are coordinated with the time of ovulation. Such a period when the sex urge is greatest (when the animal is said to be "in heat") is referred to as *estrus*. The cyclic changes in the development, release, and maturation of the ovum, together with structural and physiological changes in the reproductive system periodically prepare the female for mating, and are collectively termed the *estrus cycle*. Some wild animals, such as deer and sheep, show only one estrus a year and are referred to as *monestrus*. Among *polyestrus* animals, such as cats and dogs, there are usually two estrus periods a year, whereas mice may have an estrus period as frequently as every 3 or 4 days.

The estrus itself is that part of the cycle during which the female of the species is both psychologically and physiologically prepared for mating. During this period, sexual desire and reproductive activity are at a maximum level. Prior to the estrus, maturation and growth of an ovarian follicle represent some of the most important changes. The onset of estrus is correlated with the rupture of the follicle and the release of the ovum. In this way the ovum is ready for fertilization at the same time the female is at a peak of readiness for accepting the male. As soon as ovulation occurs, the ruptured follicle collapses and is replaced by the corpus luteum, which secretes progesterone. The endometrium also undergoes changes during estrus. At the beginning of estrus, the endometrium progressively thickens and its vascular supply increases as follicular maturation proceeds. The release of the egg and the appearance of the corpus luteum occur while the endometrium builds up rapidly. In the event that fertilization occurs, the egg implants itself in the endometrium and the corpus luteum is maintained throughout most of the pregnancy. By contrast, if the egg is not fertilized, the ovum degenerates, the corpus luteum begins to disappear, and the endometrium returns to its original, thin, less vascular state.

The human female and other higher primates do not experience any distinct period of estrus. Among these organisms, receptivity of the male by the female is not confined to certain intervals; it is more or less continuous. In addition, regression of the endometrium is a more extreme process. The reproductive cycle of the human female is marked by menstruation, which occurs about every 28 days and lasts for 4–5 days. Essentially, menstruation is a phenomenon associated with the rapid deterioration of the uterine lining, resulting in the rupture of its blood vessels, the flow of blood and other fluids, and the sloughing away of portions of the endometrium.

The *menstrual cycle* is generally divided into four phases according to the major events occurring in each phase. These phases, characteristic of 28-day cycle (Figure 17-24), are the menstrual (destructive) phase, the proliferative (follicular) phase, the ovulatory phase, and the secretory (luteal) phase.

The *menstrual phase,* also called the *menses,* is characterized by the periodic shedding of the uterine lining and accompanying blood flow and normally occurs when fertilization does not take place. The endometrial lining is destroyed. When the ovum is not fertilized, the corpus luteum regresses and the subsequent deficiency of ovarian hormones causes the endometrium (unable to maintain its nutrition) to regress and disintegrate. In a typical menstrual cycle, the menses takes place on cycle days 1–5. However, there is some individual variation.

The *proliferative phase* occurs between the end of the menses and ovulation. This phase usually includes cycle days 6 to 13 or 14 in a 28-day cycle. During this phase, the endometrium, under the stimulation of estrogens, undergoes a process of growth (proliferation). The endometrium thickens as estrogen secretion rises.

During the *ovulatory phase* the concentration of estrin in the blood is high and reaches its peak at the time of ovulation. Ovulation is the rupture of the mature follicle with the release of the ovum. It occurs nearly always between the twelfth and sixteenth days, usually on the fourteenth day after the onset of the previous menstruation. However, ovulation occurs on different days in different length cycles, depending on the length of the proliferative phase. Inasmuch as the day of ovulation cannot be predicted with certainty,

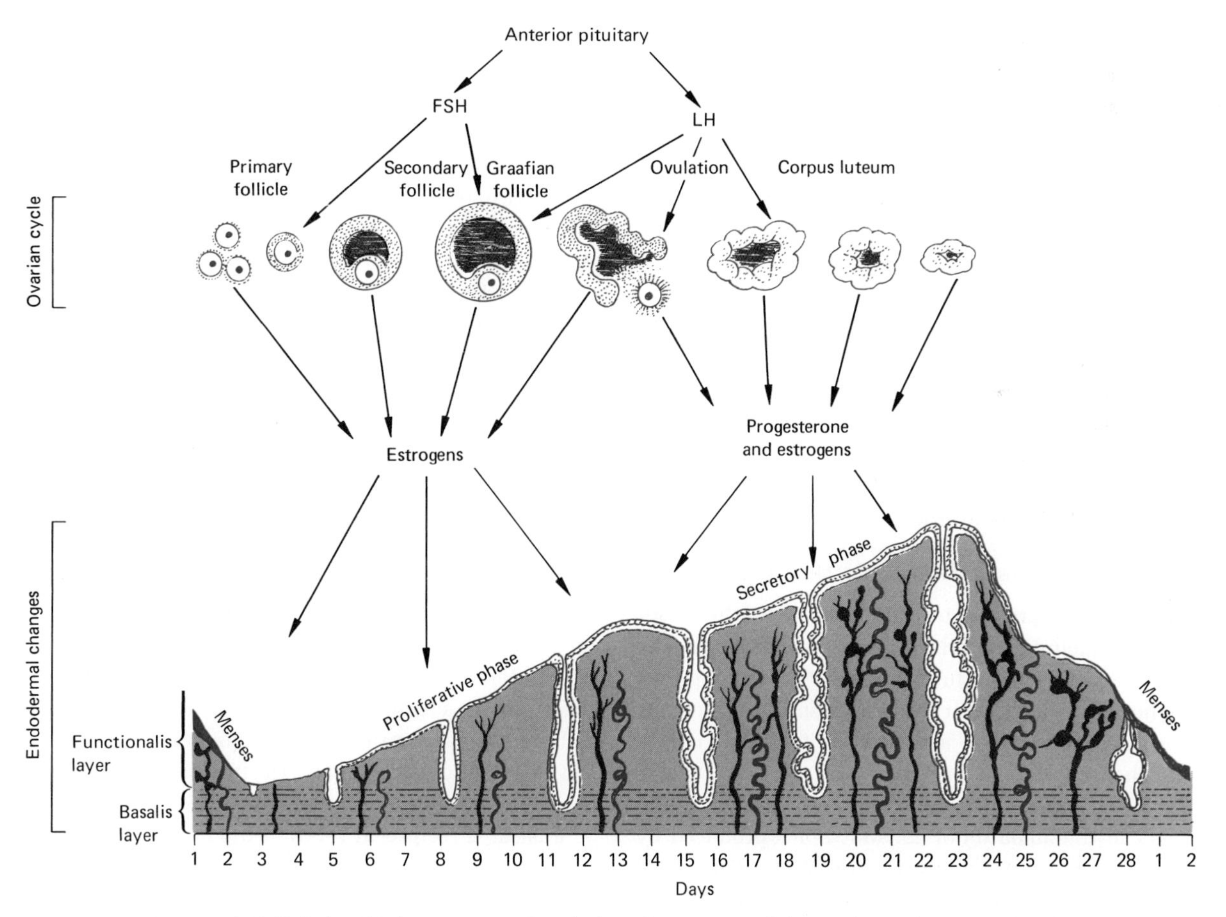

Figure 17-24. Relationship between anterior pituitary hormones and the ovarian and menstrual cycles.

this physiological fact probably accounts for most of the unreliability of the rhythm method of contraception. During the 24–36 hour period after ovulation there is very little appreciable change in the endometrium. However, as the ruptured follicle is transformed into the corpus luteum, the endometrium continues to proliferate under the influence of the remaining estrogen and the stimulatory effect of the newly formed progesterone. The presence of progesterone accounts for a temperature rise at ovulation and the temperature remains high until the onset of the next menstrual period.

During the *secretory phase,* which occurs between ovulation and the onset of the menses, the endometrium is mainly under the influence of progesterone and the remaining concentration of estrogens. As a result of the combined hormonal action, the endometrium continues to increase in size, and its glands become quite enlarged and tortuous and undergo their maximal secretory activities. This phase of the cycle lasts about 14 days, that is, cycle days 15–28. If the ovum is not fertilized, the corpus luteum disintegrates, and the concentration of progesterone falls sharply. The decrease in concentration of both progesterone and estrogen causes a cessation of secretion of the endometrial glands, the blood vessels of the endometrium involute, and the tissues of the uterine wall begin to die. These activities constitute a premenstrual phase. If the ovum is fertilized, it becomes embedded in the highly developed endometrium which then continues to develop into the maternal placental structures. These structures, as well as the developing embryo, are able to produce sufficient quantities of progesterone to permit them to continue to develop within the uterus.

Control of Human Female Sexual Cycle

Physiologists agree that hormones assume the major role in producing the cyclic changes that occur in the female during the reproductive years. Considering that these hormones have their most profound effects on the ovaries and the uterus, attention will be directed to cyclic changes in these organs.

Cyclic changes in the ovaries result from cyclic changes in the amounts of gonadotropins produced by the anterior pituitary gland (Figure 17-24). This gland secretes three hormones that activate and influence ovarian function. These hormones are the *follicle-stimulating hormone* (*FSH*), the *luteinizing hormone* (*LH*), and the *lactogenic hormone* (*LTH*). It is now known that these hormones are essential at the time of puberty and throughout the reproductive life of the ovaries for continued follicle development, ovulation, and the cyclic secretions of ovarian hormones.

FSH is carried to the ovaries via the bloodstream where it has two physiological effects. First, it stimulates the ovarian follicle to mature and, second, it stimulates the follicle to produce estrogen. The anterior pituitary is thought to produce estrogen for about 10 days after the menses. Estrogen not only influences the development of the secondary sex characteristics, but also stimulates the development of the endometrium. Several days before ovulation, the anterior pituitary starts to release increasing amounts of LH into the blood. LH brings about four ovarian changes:

1. Completion of the growth of the follicle and the ovum.
2. Increased secretion of estrogen by the follicle during the preovulatory phase.
3. Ovulation.
4. Formation of the corpus luteum.

At the time of ovulation, the level of estrogen, which has reached its peak, suppresses the secretion of FSH. Apparently the output of FSH is controlled by the concentration of estrogen in the blood and, conversely, the level of estrogen is regulated by the concentration of FSH. As the level of FSH falls, its stimulating effect on the follicle is decreased, and the secretion of estrogen is also decreased. As the concentration of estrogen falls, more FSH is secreted, resulting in the development of another follicle and more estrogen secretion.

LTH seems to be responsible for the secretory activity of the corpus luteum in the production of progesterone. Inasmuch as progesterone is necessary for the further growth and development of the endometrium, the corpus luteum must be sustained until the embryo has been able to sustain itself. If, for any reason, the corpus luteum is not maintained, the endometrium is shed, and if the individual is pregnant, the implanted fertilized egg is carried away in the menstrual flow.

Just as a reciprocal relationship exists between FSH and estrogen, one also operates between the levels of LH and LTH and progesterone. As the blood level concentration of progesterone increases, the secretion of

LH and LTH is suppressed. As a consequence, the corpus luteum begins to fade away, the concentration of progesterone decreases, and the endometrium sloughs off. The decreased level of progesterone stimulates the pituitary to secrete more LH and LTH. These hormones again exert their influence on the newly developing follicle. Most of the available data indicate that the reciprocal relationships between gonadotropins and ovarian hormones regulate the cyclic responses of the human female sex cycle.

Sexual Intercourse

Natural fertilization of an ovum is accomplished through *sexual intercourse,* in which spermatozoa are deposited in the vagina. The male role in the sexual act starts with erection, the enlargement and stiffening of the penis. An erection may be initiated in the cerebrum by stimuli such as anticipation, memory, and visual sensation, or it may be a reflex brought on by stimulation of the touch receptors in the penis, especially in the glans. In either case, parasympathetic impulses that pass from the sacral portion of the spinal cord to the penis cause dilation of the arteries of the penis, allowing blood to fill the cavernous spaces. These impulses also cause the Cowper's glands to secrete mucus which affords lubrication for intercourse. However, the major portion of lubricating fluid is produced by the female.

Tactile stimulation of the penis brings about emission and ejaculation. When sexual stimulation becomes extremely intense, rhythmic sympathetic impulses leave the spinal cord at the levels of the first and second lumbar vertebrae and pass to the genital organs. These impulses cause peristaltic contractions of the ducts in the testes, the epididymis, and the ductus deferens that propel spermatozoa into the urethra—a process called *emission.* Simultaneously, peristaltic contractions of the seminal vesicles and prostate expel semen and prostatic fluid along with the spermatozoa. All these mix with the mucus of the Cowper's glands, resulting in the fluid called semen. Other rhythmic impulses sent from the spinal cord at the levels of the first and second sacral vertebrae reach the skeletal muscles at the base of the penis, and the penis expels the semen from the urethra to the exterior. The propulsion of semen from the urethra to the exterior constitutes an *ejaculation.* A number of sensory and motor activities accompany ejaculation, including a rapid heart rate, an increase in blood pressure, an increase in respiration, and pleasurable sensations. These activities, together with the muscular events involved in ejaculation, are referred to as an *orgasm.*

The female role in the sex act also involves erection, lubrication, and orgasm. Stimulation of the female, as in the male, depends on both psychic and tactile responses. Under appropriate conditions, stimulation of the female genitalia, especially the clitoris, results in *erection* and widespread sexual arousal. This response is controlled by parasympathetic impulses sent from the spinal cord to the external genitalia. These impulses also pass to the Bartholin's glands and vaginal mucosa, which secrete most of the *lubrication* during sexual intercourse. When tactile stimulation of the genitalia reaches maximum intensity, reflexes are initiated that cause the female *orgasm* or *climax.* Female orgasm is analagous to male ejaculation, except that there is no expulsion of fluid, although there may be an increased secretion of cervical mucus. As part of the female orgasm, the perineal muscles contract rhythmically from spinal reflexes similar to those that occur in the male ejaculation. There is speculation that these same impulses also cause peristaltic movements of the uterus and uterine tubes, thus helping to transport the spermatozoa toward the ovum.

Birth Control

Methods of birth control include removal of the gonads and uterus, sterilization, contraception, and abstinence. Castration, or removal of the testes in the male; hysterectomy, or removal of the uterus, and oophorectomy, or removal of the ovaries in the female, are all absolute preventive methods. Once performed, these operations cannot be reversed, and it is impossible for the individuals treated to produce offspring. However, removal of the testes or ovaries has adverse effects on individuals because of the importance of these organs in the endocrine system, and these operations generally are performed only if the organs are diseased. Castration before puberty prevents the development of secondary sex characteristics. (For a broader discussion of birth control, see Chapter 26.)

18 Developmental Processes

GUIDE QUESTIONS

1 What three interrelated processes are involved in development? Give examples of each.
2 Discuss the principal steps in the development of a seed. What is double fertilization? How does the germination of a corn seed differ from that of a bean seed?
3 Compare root and shoot development. How are they alike? How do they differ? How is plant growth distributed?
4 What are the major events occurring in oogenesis and spermatogenesis? Describe the fertilization process.
5 Define cleavage, morula, blastula, and gastrula. Give an example of each.
6 How do determined and indetermined eggs differ? How is the quantity of yolk related to cleavage? How are eggs classified on the basis of the quantity of yolk present?
7 How does gastrulation occur? List some of the principal derivatives of the ectoderm, mesoderm, and endoderm. How are these derivatives formed?
8 Describe the events associated with cleavage and gastrulation of the human embryo. What is the placenta? How is it formed?
9 Outline the principal morphological features of a human embryo at various stages of development.
10 Discuss the three stages of labor. How is it controlled?
11 What is the role of the nucleus in development? Cite some researches to substantiate your answer.
12 What role does the cytoplasm play in development? Give several examples.
13 How do neighboring cells affect the developmental processes? What is induction?
14 Relate the effects of the external environment to development. List several examples.
15 What is growth? How does it occur? What is a growth curve? How does the pattern of growth differ between organisms with an endoskeleton and organisms with an exoskeleton?
16 Define allometric growth. How is growth terminated? What is aging? Describe some current theories of aging.

UNDER normal circumstances, from the moment an egg is fertilized by the sperm, there is a development of the zygote toward the adult form. For the biologist and the nonbiologist alike, the striking changes that take place between a single fertilized cell and a mature multicellular organism are both remarkable and mystifying. Although Monera and Protista exhibit some growth, development, and maturation, generally the new individuals closely resemble their parents almost immediately. In sexually reproducing representatives of the kingdoms Plantae and Animalia, the single fertilized cell undergoes a division, each resulting cell divides, and so on. From this process emerge all the tissues, organs, and characteristics of the parent species. This complex development is somehow determined from the start by the hereditary material passed on from the parents. The mature organisms are capable of reproducing offspring that, in turn, grow, develop, mature, and reproduce. Thus, the chain of life continues from generation to generation.

Processes of Development

In the development from the single fertilized egg to the mature adult, three processes take place.

1. **Cell division** and **growth,** an increase in number and size of cells.

2. **Morphogenesis,** the development of a definite complex structural pattern and form.

3. **Differentiation,** or cell specialization.

These three processes are not successive but are interrelated and proceed together, although at various times one process may predominate. Almost as soon as the egg is fertilized it begins to divide, and most of the resulting cells continue to divide throughout the life of the organism. This cell division and subsequent cell enlargement result in the natural growth of the organism. While there may be certain periods when growth is rapid, it usually decreases when a certain size is reached. However, there are certain parts of an organism in which growth continues throughout its lifetime.

Cell division and growth alone would result in large masses of cells, all resembling each other. It is obvious that the complex structures of multicellular organisms are composed of many different types of cells. Early in the development of the embryo, the morphogenesis of the cells gives rise to a generalized structural pattern in the form of primary germ layers. These layers give rise to cells which differentiate and develop into specialized tissues. The characteristic structure of the mature adult results from complex and intricate interrelationships among the developing specialized tissues. In higher animals, once the cells are well differentiated, they cannot revert to an undifferentiated form. By the time of birth, the anatomical pattern of the organism is fully established and, while there is some development toward maturation, subsequent changes are noted essentially in shape and size. This type of development is called *closed development.* In plants, differentiation of mature cells is also irreversible. However, the mature plant has extensive meristematic and embryonic tissues. Throughout the life of the plant, wood, bark, leaves, and flowers continue to develop. This type of development is called *open development.*

The **development** of an organism involves a complex series of events consisting of cell division and growth, morphogenesis, and differentiation. Through intensive study over the years, biologists have been able to describe what happens in the development of many different organisms. The real challenge is to determine how it happens, and it is in this area of development that much research is being conducted. The complexity of the pattern of development varies with the complexity of the mature organism, yet there are certain similarities among all organisms. Several representative patterns in multicellular plants and animals will be examined and the important events in their development described.

Development of a Flowering Plant

Development of the Seed

In Chapter 17, the events leading to the formation of a pollen grain and an ovule were discussed.

The **zygote** in the ovule of flowering plants is heterotrophic and is entirely dependent for its existence and development on the tissues that surround it. After fertilization, the zygote remains inactive; the endosperm nucleus, however, becomes active and commences rapid mitotic divisions. These free nuclear divisions form a multinucleate fluid in the embryo sac. Thus, the liquid endosperm develops first and absorbs the nucellar tissues within the ovule. Once extensive material has been developed around the zygote, cell walls are laid down in the endosperm and it becomes the nutritive material.

The development of the embryo (Figure 18-1) from the zygote is slow at first. Mitosis occurs in the embryo only after the endosperm is well developed; then embryonic development proceeds rapidly. After several divisions, the zygote produces a row of cells that becomes the **embryo** and a *suspensor,* which forces the embryonic mass of cells into the endosperm tissue. In time, the suspensor disintegrates. Further growth and development form the mature embryo, which consists of a short axis with one or two cotyledons. *Cotyledons* are food storage organs that also absorb food from the endosperm and that function as leaves after germination. In addition to cotyledons, the mature embryo also contains the *plumule,* which gives rise to the shoot; the *radicle,* which becomes the root system; and the *hypocotyl,* a region between the radicle and the cotyledons.

The **seed** is surrounded with the *testa,* or seed coat, which develops from the integuments of the ovule. Although integumentary growth practically closes the *micropyle,* its location may still be visible as a minute

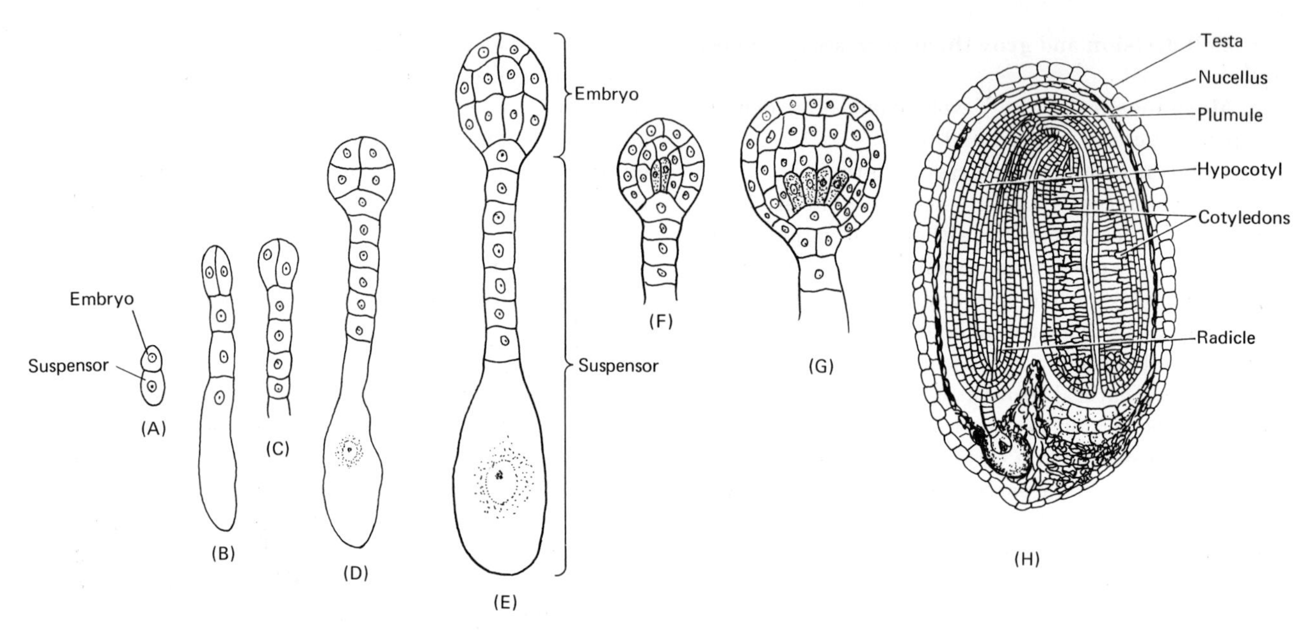

Figure 18-1. Representative stages in the development of the embryo of a flowering plant. Shown here is the dicot *Capsella bursa-pastoris*. [**A–G** *from R. M. Holman and W. W. Robbins,* A Textbook of General Botany, *4th ed. New York: John Wiley & Sons, Inc., 1939.*]

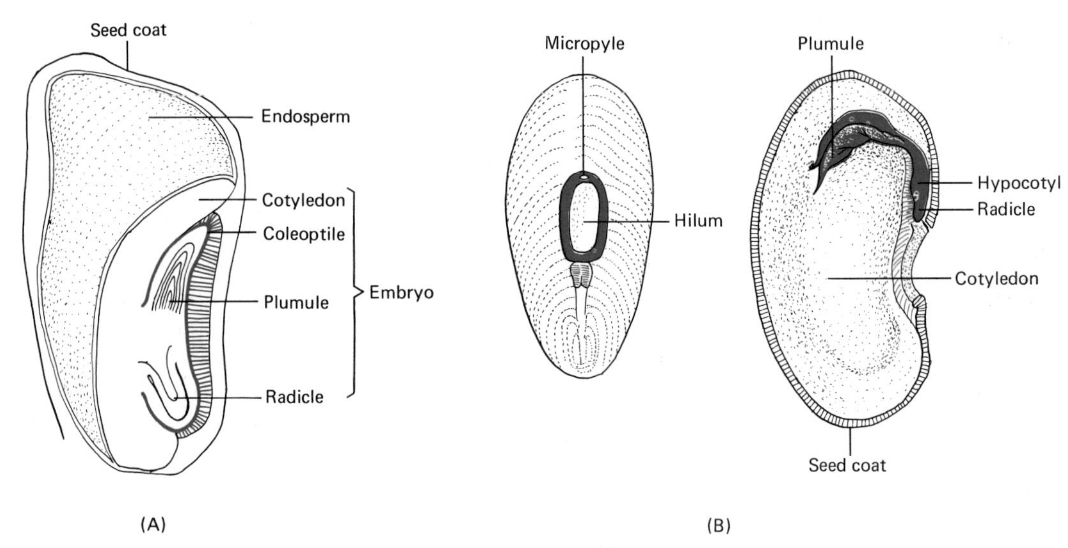

Figure 18-2. Monocot and dicot seed structure. (**A**) Corn grain in longitudinal section showing parts of the embryo and seed. (**B**) Garden bean. *Left:* External view. *Right:* Longitudinal section showing parts of the embryo and seed.

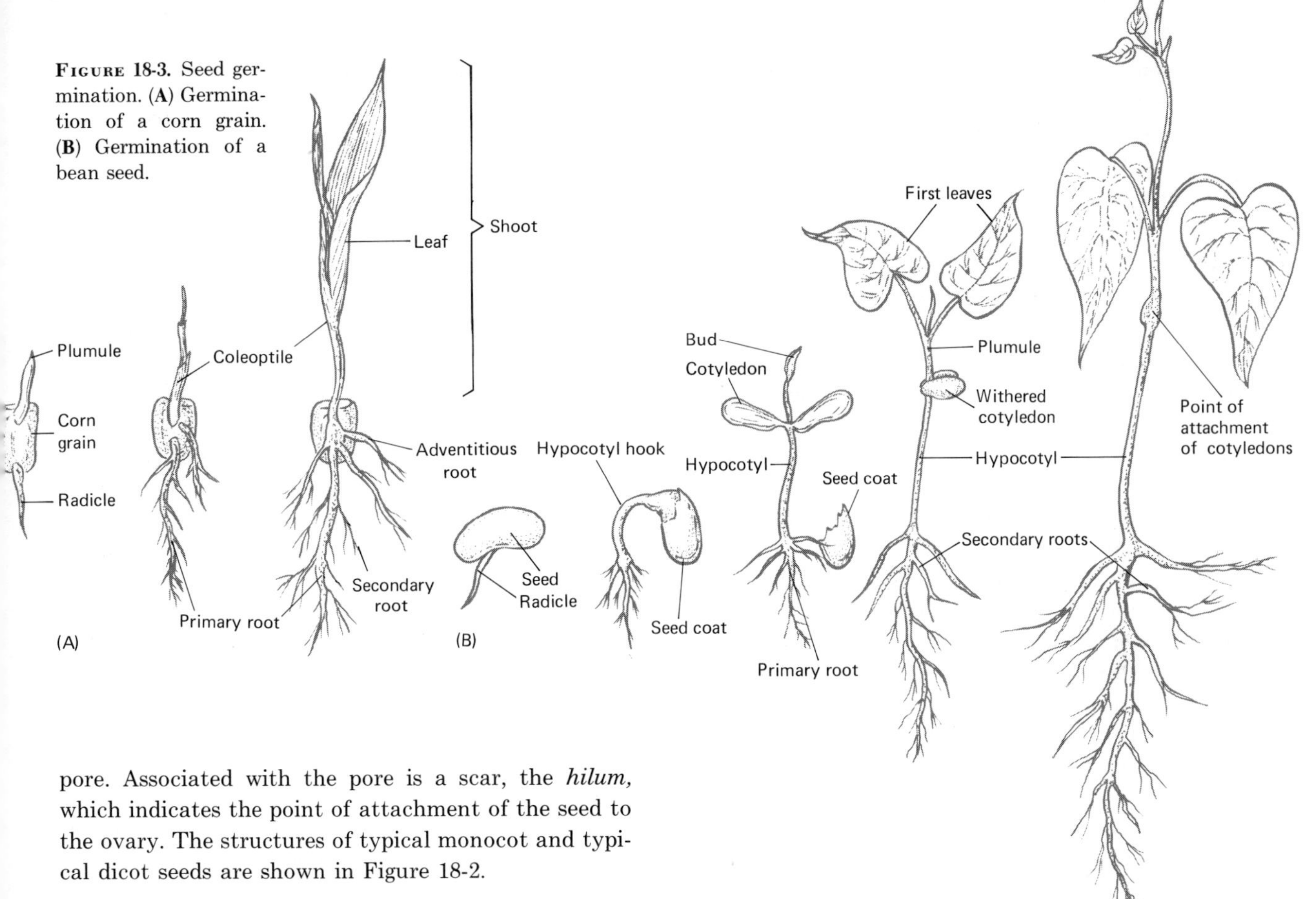

Figure 18-3. Seed germination. (**A**) Germination of a corn grain. (**B**) Germination of a bean seed.

pore. Associated with the pore is a scar, the *hilum,* which indicates the point of attachment of the seed to the ovary. The structures of typical monocot and typical dicot seeds are shown in Figure 18-2.

Germination of Seeds

Once the seed is distributed, if conditions are favorable, the seed germinates. Seed *germination* is the resumption of the growth of the embryo after a period of much-reduced metabolic activity. The most characteristic physiological effect of seed germination is the increased rate of respiration.

There is a dramatic increase in the rate of cell division in the meristematic region of the plumule and radicle just after germination takes place. From the time that the radicle emerges from the seed to the time that the plant can carry on an existence independent of the stored food reserves in the seed, the plant is referred to as a *seedling.*

Corn (a monocot) and lima bean (a dicot) represent the two typical patterns of seedling development. In the germination of corn (Figure 18-3A), the cotyledon absorbs digested food from the endosperm and transfers it to the growing region. The outside appearance of the radicle from the corn seed is quickly followed by the emergence of the *plumule sheath* or *coleoptile.* The hypocotyl does not elongate but remains with the cotyledon in the soil. Actually, the radicle becomes the primary root and persists for a short period until the adventitious roots develop. The plumule undergoes elongation and passes through the coleoptile. It then develops into a vegetative shoot, which becomes part of the independent plant.

By contrast, in lima bean germination (Figure 18-3B), the radicle emerges first and the hypocotyl ultimately grows above the soil. The early growth of the cells of the bean hypocotyl is unequal on the two sides of the axis, resulting in the formation of a *hypocotyl hook.* This hook soon straightens and so forces the cotyledons out of the seed coat and above the soil. At this time, the plumule undergoes rapid cell division and develops into the leafy shoot. When the food in the cotyledons has been used up, they die and fall off. The lowermost portion of the stem is the hypocotyl.

Development of the Root

As the root system continues to grow, there is considerable branching, penetration into the soil, and horizontal spreading of root branches. All roots terminate in delicate white tips that are usually less than $\frac{1}{16}$ inch in diameter. As has been described previously, these tips reveal four more or less distinct regions: the root cap, meristematic region, enlargement region, and differentiation region (see Figure 9-5).

As cells divide, enlarge, and differentiate, the root cap and meristem are forced farther and farther into the soil. A given cell starts as part of the *apical meristem.* As it ages, if it does not become part of the root cap, it then exists as part of the enlargement region and finally as a cell of the differentiation region. In effect, the cell has not changed position; it has merely changed in structure and function. Figure 18-4 represents these various stages of development; this example shows how a cell develops into a tracheid. For the sake of simplicity, other cells have not been shown. Bear in mind, however, that all cells which eventually become part of the tissues of the root undergo similar changes.

Development of the Shoot

The development of the shoot is very similar to that of the root. Growth and development of the stem are due to cell divisions in the apical meristem which comprises the meristematic zone. In effect, the meristem is constantly being carried upward by the growth of new cells it produces. The region just behind the meristem, characterized by rapid cell enlargement, is called the enlargement region. In this region, three conspicuous embryonic primary tissues are produced. These are the *protoderm, ground meristem,* and *procambium.* As development of these tissues occurs in the differentiation zone, each produces specific primary tissues (see Chapter 9). The protoderm gives rise to the epidermis; the ground meristem produces the cortex and pith; and the procambium differentiates into the vascular cambium, primary xylem, and primary phloem. It is also within this region that embryonic leaves, stem, and other organs differentiate.

Stages of Plant Growth

Fundamentally, all plant growth is cellular growth because tissue, organ, and organismic growth are all products of cell growth. The growth of organs of higher plants involves not only the enlargement and differentiation of cells already present but also the formation of new cells from previously existing ones. The formation of new cells by division of those already present and their subsequent enlargement and differentiation into

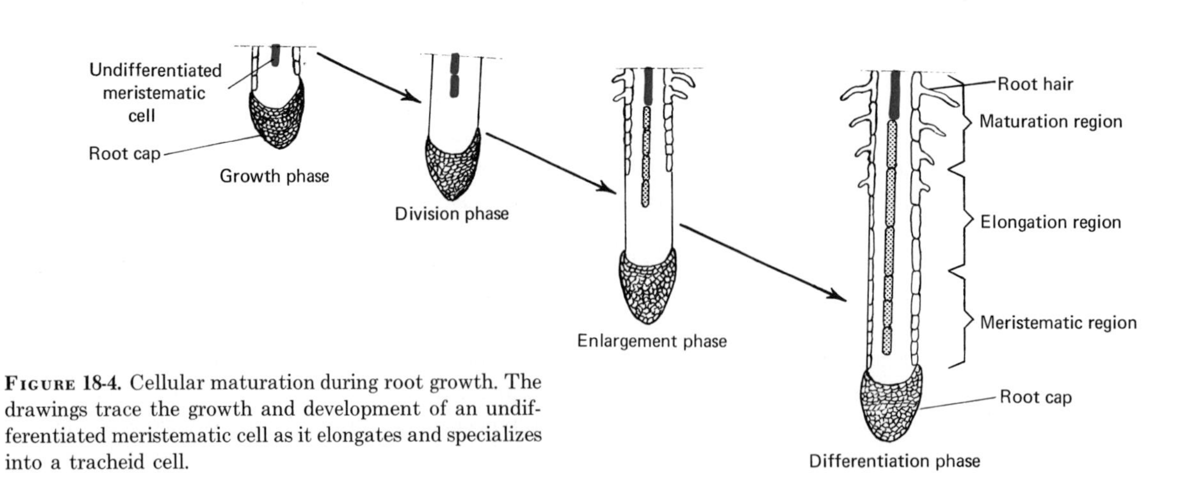

FIGURE 18-4. Cellular maturation during root growth. The drawings trace the growth and development of an undifferentiated meristematic cell as it elongates and specializes into a tracheid cell.

permanent structures provide the principal means of growth of any multicellular organism. In essence, then, the stages or phases through which cells pass to maturity are division, enlargement, and differentiation.

Division. In general, the process of growth is initiated in regions (meristems) of higher plants in which cell division takes place. *Apical meristems,* characteristic of root and shoot tips, engage in primary growth and the product of this activity is primary tissue. By contrast, the *vascular cambium* and *cork cambium* are involved with secondary growth, and the product of their activity is secondary tissue. Although the bulk of plant growth is vegetative growth, a small but extremely significant part of the growth is reproductive growth in which reproductive structures reach maturity.

Enlargement. After a cell has divided, the resulting two daughter cells together are at first only as large as the parent cell. Before further growth can take place, the daughter cells undergo much enlargement, the dominant and most obvious phase in the growth of the plant. *Enlargement* is essentially an increase in volume caused primarily by the absorption of water. As a cell elongates, the thickness of the wall is maintained. In addition, the vacuoles grow in size owing to the inward diffusion of water. The protoplasm appears as a narrow layer between the cell wall and the vacuole. Even though an elongating cell is also synthesizing protoplasm, the protoplasmic constituents appear to be sparse, owing to the increase in size.

Differentiation. As cells enlarge, they gradually assume their permanent shapes and forms. Both cell division and cell enlargement adequately account for an increase in size as the plant grows, but not for the specialization of cells. The *differentiation* of cells, involving a multiplicity of structural and functional specializations, terminates in a series of modifications of the original cell. As part of these modifications, cells may differentiate and become specialized with respect to size, shape, nature, and extent of secondary walls, or protoplasmic contents. Structurally, these modifications may be represented as changes in shape (tracheids), loss of end walls (vessel segments), perforation of end walls (sieve cells), impregnation with lignin (stone cells), and deposition of suberin (cork cells) or cutin (epidermal cells). Some functional specializations brought about by chemical changes involve the decomposition and death of protoplasm (tracheal elements), the formation of specialized organelles (chloroplasts), or the storage of certain types of foods or ergastic substances. Differentiation results in the transformation of apparently identical cells of the meristem into a number of highly specialized ones.

Distribution of Plant Growth

Plant growth is not uniformly distributed but is concentrated in specific growth regions. The distribution of plant growth was first demonstrated by Julius von Sachs in the middle of the last century. He marked young roots, stems, and leaves with inked lines and then observed the subsequent growth pattern for each organ. After the particular organ had grown for a period of time, it could be noted that the marks were no longer equidistant. Using this procedure, the growth rate for part of the organ can be determined from the distance between each pair of marks.

In roots, elongation occurs most rapidly in a region just behind the root tip. Similarly, stem elongation occurs in a region just below the shoot tip, but the length of the stem is determined not only by this elongation but also by growth of the internodes below. The expanding leaves of such broad-leaved plants as the tobacco have their own pattern of growth. In this case, expansion occurs in a relatively uniform manner throughout the entire surface of the leaf. Elongating leaves of narrow-leaved plants such as grasses grow primarily at the base, so the mature leaf tissue is being continuously expanded and pushed up by the growing region.

Development of an Animal Embryo: General Principles

Gamete Development

Before discussing the principal features of the development of an animal embryo, the processes of gamete development and fertilization will be examined.

The **gametes** of animals are the eggs and sperm, also called **germ cells.** These gametes originate within the **gonads** (ovaries and testes). *Primordial germ cells* are specialized gonadal cells that give rise to *gonial cells,*

cells capable of undergoing meiosis to form gametes. The female gonial cells are called *oogonia* and the male gonial cells are called *spermatogonia.*

Gonial cells are frequently larger than other cells and possess a large, ovoid nucleus with prominent chromatin. The multiplication of gonial cells is accomplished by mitosis. In the ovary, the meiosis of an oogonium to form haploid gametes (eggs) is called **oogenesis.** In the testes, the meiosis of a spermatogonium to form haploid gametes (sperm) is referred to as **spermatogenesis.**

In spermatogenesis (Figure 18-5A), spermatogonia, after ceasing mitotic division, go through a relatively long period during which they grow slightly larger before beginning the early phases of meiotic prophase I. At this stage, the cells are termed *primary spermatocytes.* These cells undergo a sequence of events for the meiotic division thereby forming two diploid daughter cells, the *secondary spermatocytes.* These, in turn, divide in the second meiotic division to produce haploid daughter cells termed *spermatids.* Four haploid spermatids are formed by the meiotic divisions of the original diploid primary spermatocyte. Each spermatid then undergoes transformation into a mature *spermatozoan,* or sperm cell. Mature sperm cells consist of a nucleus containing genetic material, a tail for locomotion, and an acrosome that aids in penetrating the egg cytoplasm.

Ooogenesis, or egg development, also begins after oogonial mitotic divisions cease (Figure 18-5B). In this regard, oogenesis is similar to spermatogenesis. The oogonial cell at this stage is referred to as a *primary oocyte;* it then undergoes a relatively prolonged period of growth prior to the first meiotic divisions. During this time, the size of the primary oocyte may increase many times. In addition, yolk is accumulated in the cytoplasm. **Yolk** is a mixture of proteins, phospholipids, and neutral fats and, in many animals, serves as a source of nourishment for the embryo (chicken) and young organisms (tadpoles). The first meiotic divisions of the primary oocyte are similar to those of the primary spermatocyte in that two diploid daughter cells are

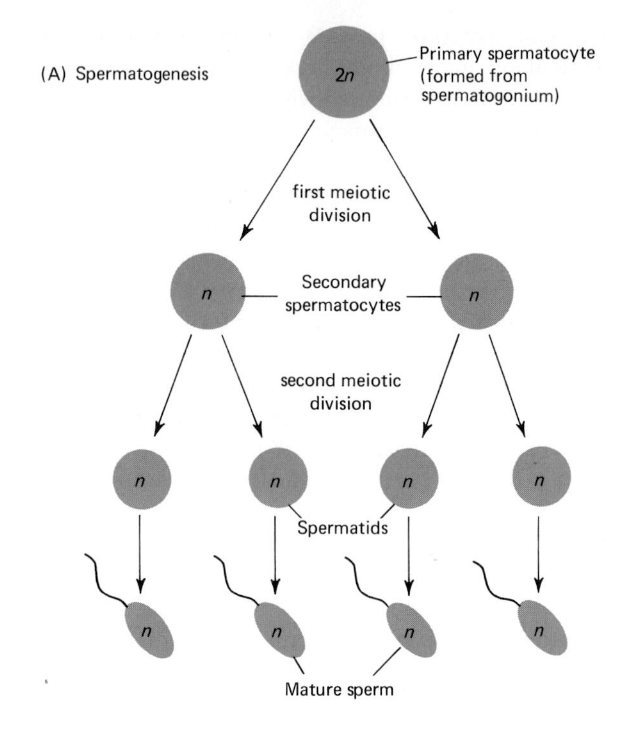

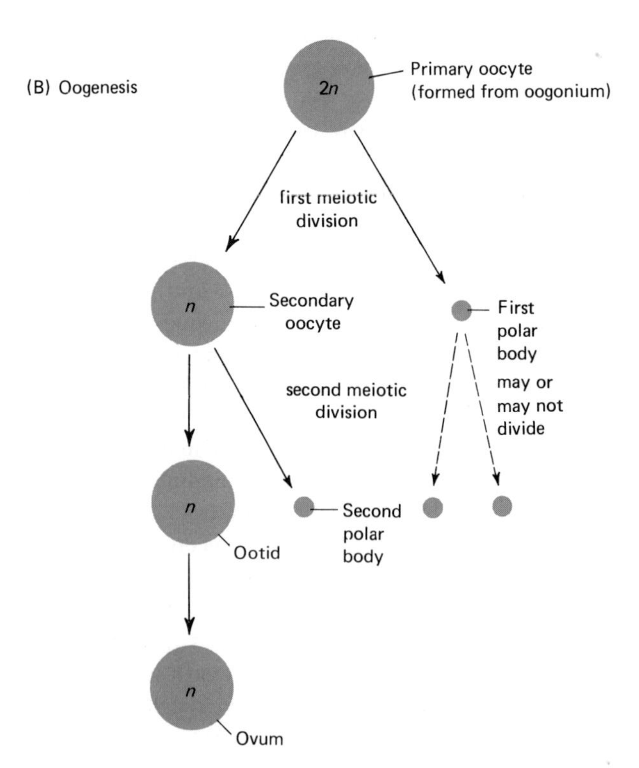

FIGURE 18-5. Comparison between spermatogenesis and oogenesis. (**A**) In spermatogenesis a single, diploid spermatogonium eventually produces four mature haploid sperm. (**B**) In oogenesis, by comparison, a diploid oogonium gives rise to a single, mature haploid ovum and several polar bodies, all of which disintegrate.

produced. However, these daughter cells are unequal in size. The cytoplasmic division in telophase I produces a large *secondary oocyte* and a small structure consisting almost entirely of nuclear material, the *first polar body*. In vertebrates, the egg normally stops developing in prophase of the first meiotic division. If fertilization occurs, or if the egg is artificially stimulated, the egg continues its development through the second meiotic division. The secondary oocyte undergoes the second meiotic divisions and also divides unequally in telophase II, producing a large *ootid* and a small *second polar body*. The first polar body may or may not divide to form two additional second polar bodies, but, in any event, the polar bodies eventually disintegrate. In time, the ootid matures into an *ovum* (egg). Thus, only one mature egg cell is produced from the original primary oocyte.

Fertilization

Fertilization is accomplished by the fusion of the gamete nuclei and is preceded by the activation of the egg. Upon contact of a sperm acrosome with the egg membrane (*vitelline membrane*) of the same species, there is an almost instantaneous response of the egg that is referred to as activation. *Activation* involves the formation of a *fertilization cone,* a raised mass of cytoplasm that bulges outward into the connection formed when the sperm and egg plasma membranes fuse. The sperm nucleus moves into this cone, which then recedes, carrying with it the bare sperm nucleus (without membranes), the middle piece, and the centrioles.

After the completion of these events, cortical granules beneath the plasma membrane disintegrate, releasing their contents which lift the vitelline membrane from the surface of the egg. This elevated vitelline membrane becomes the *fertilization membrane* which, in most species, persists while further development proceeds, and which acts as a barrier by preventing other sperm from penetrating the egg.

Once the sperm nucleus has entered the cytoplasm, it becomes vesicular and is called a *pronucleus.* A similar pronucleus is formed by the egg nucleus. The two pronuclei fuse to form the *zygote nucleus,* thereby re-establishing the diploid condition.

It has been found in lower animals that artificial stimulation such as pricking the egg with a fine needle will induce embryonic development. In fact, changing the salt content of the surrounding fluid, electric shock, agitation, or similar stimulation may trigger embryonic development. Haploid frogs, salamanders, and even rabbits have been produced by these means. Experiments in which the egg is deprived of nuclear control soon after fertilization show that some development still occurs, and this indicates that the cytoplasm carries some information for development. Evidently, the organization of the cytoplasm in the maturation of the ovum and penetration of the sperm prior to fertilization play an important part in early development.

Cleavage

Immediately upon fertilization (Figure 18-6A, B), a rapid cell division of the zygote takes place. This early division of the zygote is called **cleavage** and results in progressively smaller cells called *blastomeres* (Figure 18-6C–G). The division proceeds so rapidly that the egg subdivides with little cytoplasmic growth. These early cleavages are unique in that the blastomeres divide simultaneously. The result of this is a solid cluster of cells called a *morula* (Figure 18-6G), a mass not much larger than the original egg. With further division, there is a difference in the size of the resulting cells and they form a hollow sphere called a *blastula* (Figure 18-6H, I). The blastula cavity **(blastocoel)** is only a temporary structure, but it makes room for the first of a series of foldings that occur as the embryo develops. Much experimentation has been performed on embryos in the various stages of cleavage. It appears that the purpose of cleavage is not only to form a random mass of small cells, but also to establish organ-forming zones. On the basis of how these zones become established, eggs may be classified as either *determinate* or *indeterminate* and it is possible to separate the two cells resulting from the first cleavage to demonstrate this. In certain species, each of the resulting cells will not develop into a whole organism. In this type of egg, the *determinate* egg, the future of the different parts of the egg is fully established at the time of fertilization. This established pattern in the zygote is called *polarization* in which the head–tail, dorsal–ventral, and right–left axes are firmly established. On the other hand, if the cells of the first cleavage are separated and develop into complete embryos, then the egg is referred to as *indeterminate.* In

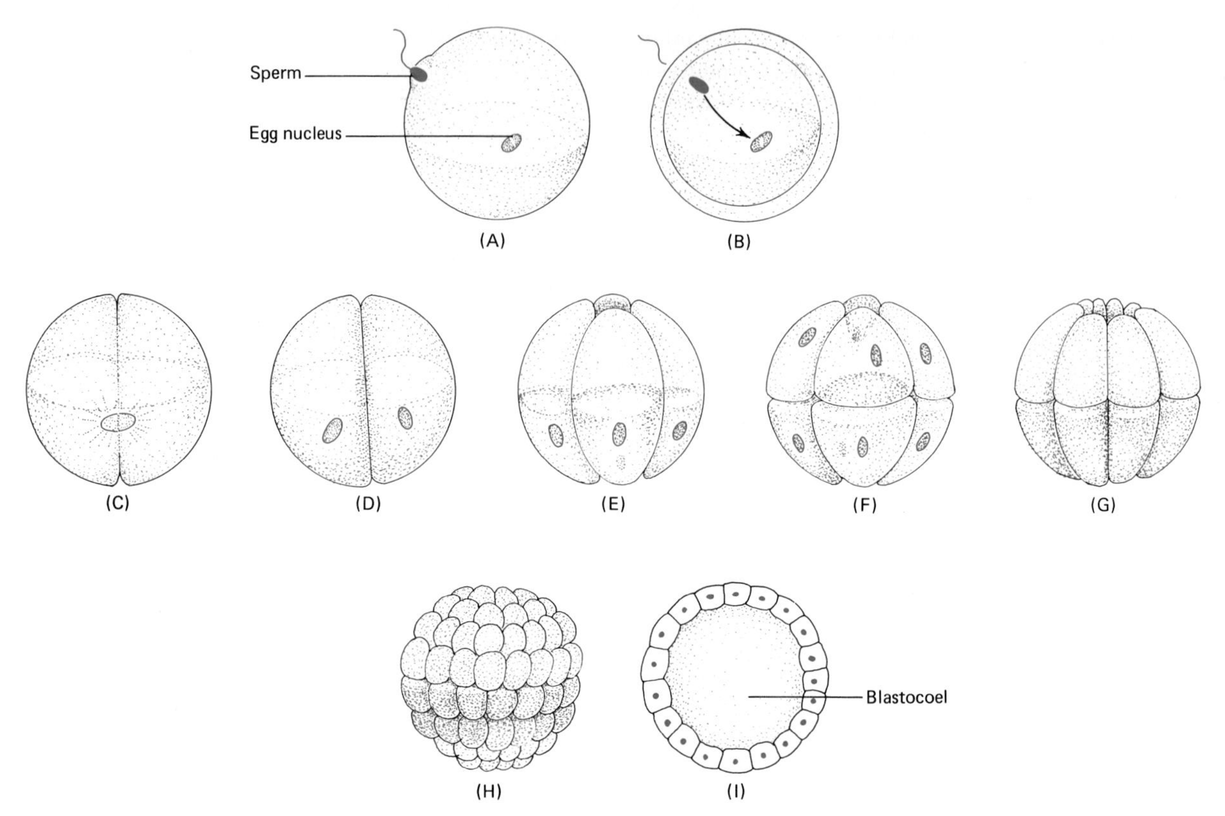

Figure 18-6. Fertilization, cleavage, and blastula formation in the development of the sea urchin. (**A–B**) Fertilization of the egg. (**C–G**) Cleavage stages. (**H**) Blastula seen externally. (**I**) Section through blastula showing blastocoel (primary body cavity).

this type of egg, the fixing of the various regions of differentiation occurs in later stages of cleavage. It is possible by separating the cells at various stages of cleavage to determine when differentiation occurs. Generally, the limit is reached at the eight-cell stage; any separation of blastomeres after this stage results in failure to develop into normal embryos.

The explanation for the differences in the two types of eggs lies in the direction of cleavage. Within the zygote, polarization occurs when one area becomes differentiated into the *animal zone* which contains little yolk, and the other area becomes differentiated into the *vegetal zone* which contains most of the yolk. If the first cleavage occurs along the axis of differentiation, the resulting cells, if separated, will develop into normal embryos. By contrast, if the first cleavage occurs across the axis, it separates the vegetal zone from the animal zone and the separated cells fail to develop into normal embryos.

Whether an organism produces determined or indetermined eggs depends on the plane of cleavage and is a genetically determined trait of the species. For example, worms and snails produce determined eggs while most vertebrates, including man, produce indetermined eggs. Occasionally, the separation of the blastomeres occurs in nature, resulting in multiple births or identical twins, triplets, etc. The reason for the natural separation of blastomeres is not known. *Identical* siblings develop from a single egg and have identical hereditary factors. When multiple births occur from several eggs being fertilized at the same time, the relationship is called *fraternal.* The offspring in this case do not have identical hereditary factors and need not be of the same sex or even resemble each other.

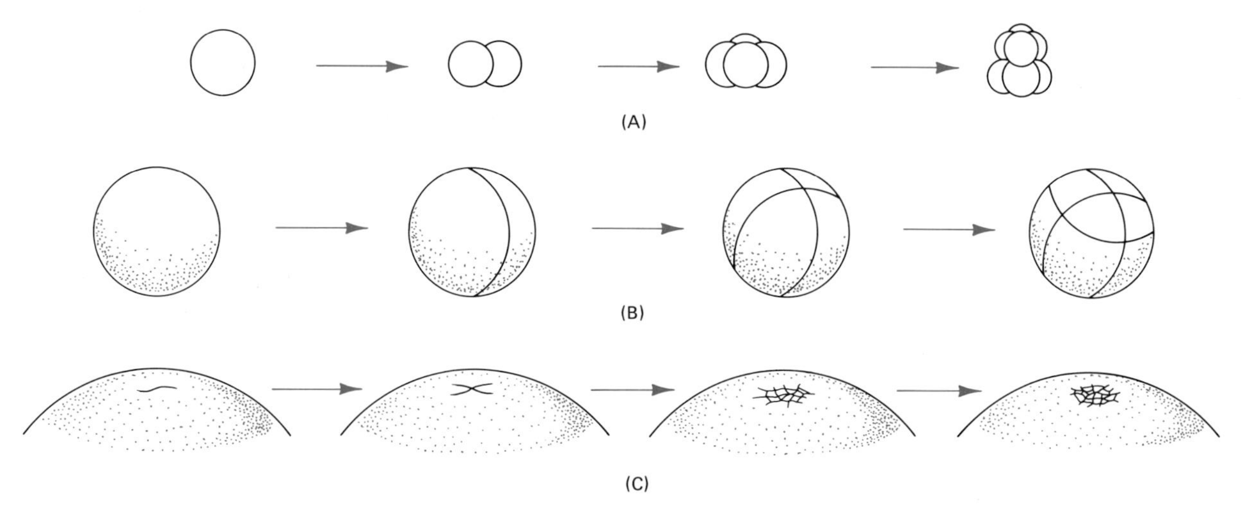

Figure 18-7. Relationship of yolk to cleavage. (**A**) *Amphioxus* egg contains very little yolk and cleavage is complete. (**B**) The frog's egg has substantially more yolk (colored area) and the transverse cleavage plane is displaced above it. (**C**) The egg of a hen is almost entirely yolk (colored area) and cleavage is restricted to a relatively small area.

Cleavage, as just described, is rather general and there is considerable variation among different species (Figure 18-7). Cleavage is often modified by the amount of yolk present. In the chicken egg, the large yolk confines cleavage to a limited plate-like area on the surface of the yolk. The developing embryo eventually consumes the yolk and occupies all the space present in the egg. In the frog egg, there is a large amount of yolk on one side. The whole egg cleaves, but, as more blastomeres are formed, those in the yolk region are considerably larger. The egg of *Amphioxus* has very little yolk and cleavage results in blastomeres of about equal size.

Gastrulation

Once the blastula reaches a size of about several hundred cells, cell division occurs with an accompanying cell growth. Although there are differences among the various species, it is possible to generalize the next major developmental event, gastrulation (Figure 18-8). **Gastrulation** is the development of cell layers; the resulting structure is called a **gastrula.** In most organisms, the three *primary germ layers* from which further differentiation and eventual organ development occur are the **ectoderm** (outer layer), **mesoderm** (middle layer), and **endoderm** (inner layer). These

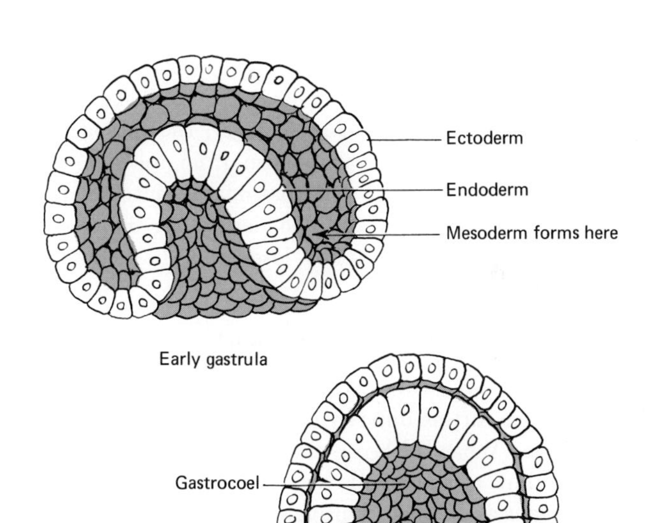

Figure 18-8. Gastrulation of an *Amphioxus* egg. See text for amplification.

three layers are already determined on the surface of the blastula. The simplest form of gastrulation is for the ball-like structure of the blastula to *invaginate,* that is, to fold in, in much the way that one would push in the side of a soft rubber ball. The result of this invagination is a cup-shaped gastrula with two layers: an outer layer, the ectoderm, and an inner layer, the endoderm. Regardless of modifications in gastrulation among the various species, there is typically an overgrowth of the ectoderm around the endoderm layer.

In later stages of gastrulation, the overgrowth of the ectoderm layer results in a two-layered, ball-shaped structure with an opening called a *blastopore.* In some species this becomes a single alimentary opening in the adult. In other species, it may become an anus or mouth when a second opening breaks through the opposite end of the developing embryo. The inner cavity of the gastrula, called the *gastrocoel,* will develop into the alimentary canal with its associated organs and also the respiratory organs, excluding the nose and pharynx. Gastrulation is essentially a morphogenic phase of development whereby the major outlines of the mature adult are set forth without specific cell type differentiation.

As soon as the endoderm layer is established, the development of the third germ layer, the mesoderm, takes place. This layer may form from folds in the endoderm or from single cells that migrate into an area between the ectoderm and endoderm. The mesoderm layer is complex and rapidly loses any characteristic layer-like shape.

Later Embryonic Development

The three layers of cells represent the primary tissues from which further differentiation and organ formation take place. When the layers first form they do not differ greatly but, as more cells are formed, distinct tissues and organs are formed. Through the use of special dyes, it is possible to trace the morphological movements of gastrulation in the development, shape, and structure of the principal tissues of the adult organism.

With the completion of gastrulation, the embryo has accomplished a major step toward becoming a functionally independent organism. Its tissues are arranged in a manner generally corresponding to their ultimate relationships and, with appropriate shaping and molding of the germ layers, the characteristic form of the organism will emerge. Morphogenesis in later embryonic development involves many processes, such as invaginations and evaginations of the cellular layers; fusion, detachment, or separation of cells, groups of cells, and cell layers; local aggregations and condensations of cells; cellular multiplication and local proliferation; migration of cells from the sites of their origins; differential growth; and destruction of cells. In order to illustrate how these processes contribute to the ultimate form of a developing embryo, some morphogenic changes in each of the primary germ layers of representative organisms will be described.

The ectoderm is the first germ layer to establish a major morphogenic change following gastrulation (Figure 18-9). In the amphibian embryo, the cells of the ectoderm on the future dorsal side become condensed into a compact, pear-shaped plate (*neural plate*) with raised edges (*neural folds*). The neural folds rise and fuse over the center of the original plate and the neural plate is transformed into the *neural tube.* This structure, which will later form the brain and spinal cord, is detached from the overlying ectoderm; groups of cells that comprise the neural crest are also detached from the fusing corners of the folds. The cells of the neural crest migrate extensively, moving to all parts of the body. Some form pigment cells which give color to the skin and its derivatives (hair, feathers, scales); others form cartilages of the jaw, spinal and sympathetic neurons, and neurons of the cranial ganglia.

After the neural tube is formed, the future forebrain region evaginates on both sides and forms the *optic vesicles* (Figure 18-10). These vesicles push through the loose mesoderm of the head and make contact with the head ectoderm. The ectodermal cells then elongate at the zone of contact, forming a *placode* (thickening) on each side, the future lens of the eye. The placodes then invaginate and detach from the ectoderm. As detachment proceeds, the optic vesicle reverses its outward bulge and invaginates, forming the *optic cup* (eyecup), which accommodates the invaginating lens. The lining of the eyecup ultimately forms the *retina* of the eye.

Two thickenings also arise on each side of the head anteriorly and invaginate to form the *nasal grooves,* from which develop the sensory epithelia of the olfactory organ. Two others form posteriorly to form the *otic*

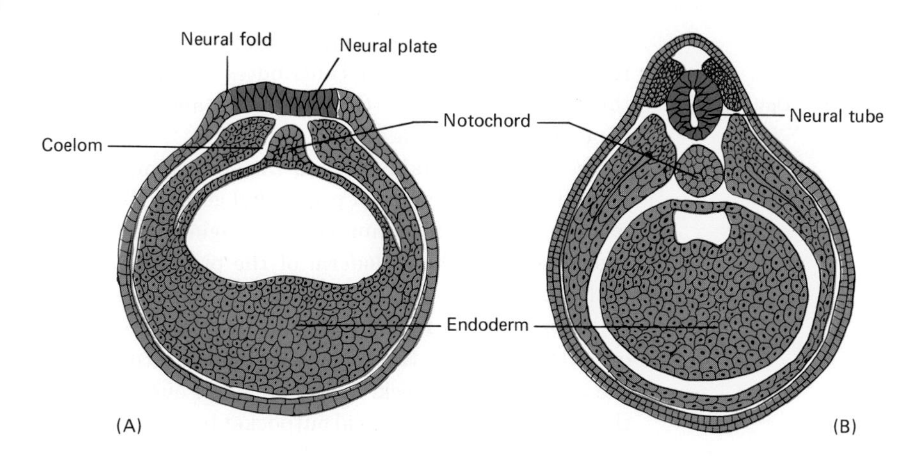

Figure 18-9. Neurulation in an amphibian embryo. (**A**) Neural plate stage. (**B**) Closed neural tube.

vesicles, which ultimately contribute to the inner ear. Other ectodermal thickenings appear in the head region and contribute to ganglia of cranial nerves.

Mechanisms similar to those that form the nervous system and sense organs also occur in structuring other derivatives of the ectodermal layer, particularly the integument and its derivatives. Refer to Table 18-1 for a more complete listing of contributions of the ectoderm.

Once formed, the mesoderm constitutes a massive middle layer that eventually contributes the great bulk of the body, thus determining its form to a major extent. Upon completion of gastrulation in the frog, as well as higher vertebrates, the mesoderm in the dorsal midline of the neurula detaches from the lateral mesoderm to form a rod-like **notochord** that runs almost the full length of the embryo (Figures 18-9 and 18-11). This structure is beneath the neural tube and provides rigid support along the embryonic axis. In higher vertebrates it contributes to the formation of intervertebral discs.

Lateral to the notochord on each side, the mesodermal cells of the segmental plate condense into a row of

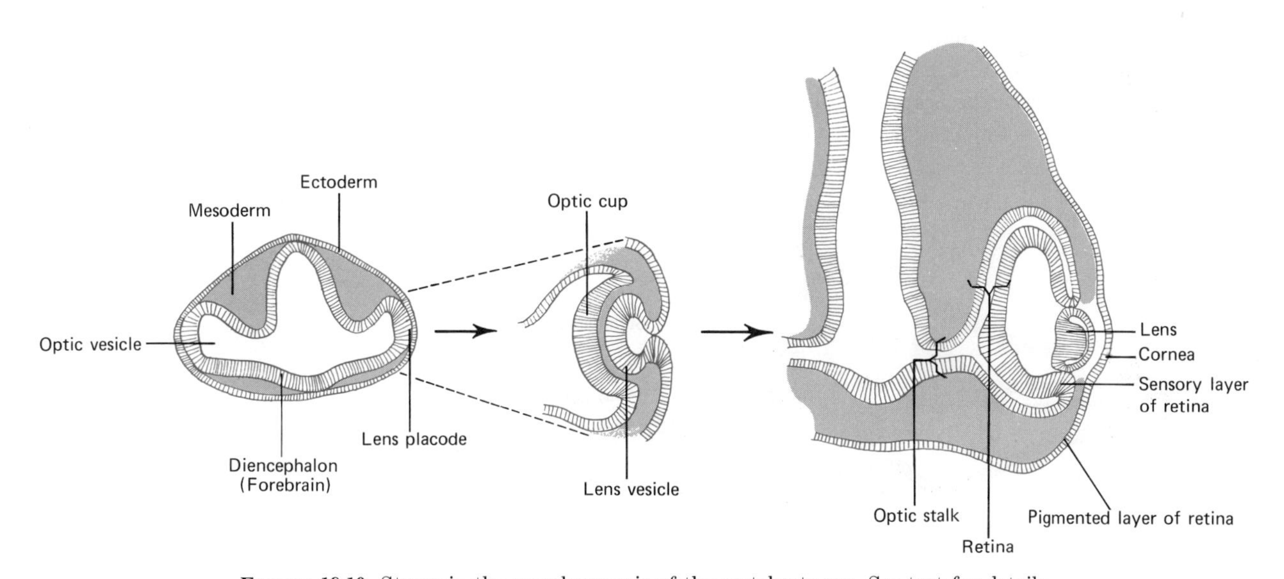

Figure 18-10. Stages in the morphogenesis of the vertebrate eye. See text for details.

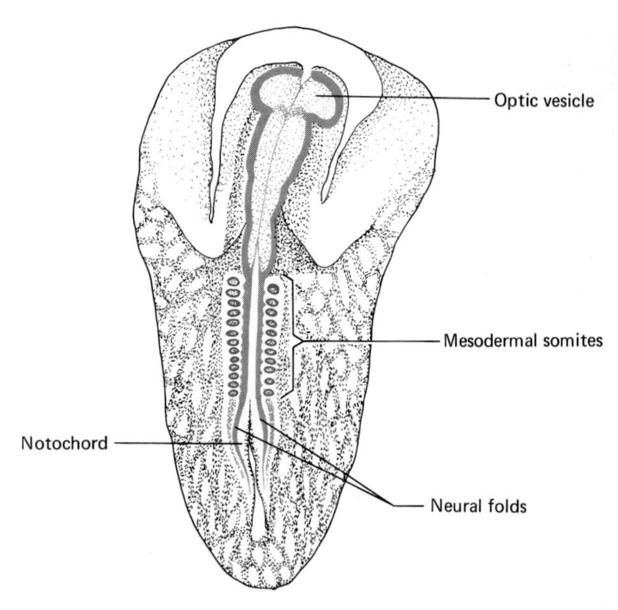

Figure 18-11. Mesodermal somites in a chick embryo, which lie on top of the massive ball of yolk.

tissue blocks called **somites** (Figure 18-11). These ultimately form the vertebral column and the musculature and dermis of the dorsal side and, depending on the organism, may contribute to the ribs. Lateral to the somites, there is a narrow intermediate zone of mesoderm. Beyond this zone, the mesoderm separates into two layers: an outer somatic layer and an inner splanchnic layer (Figure 18-12). The intermediate mass is the principal source of the urogenital organs and the adrenal cortex. The splanchnic layer contributes the heart, invests the digestive tract, and forms the mesenteries. The somatic layer contributes the peritoneum, the muscular layers of the body wall, and the paired appendages (see Table 18-1).

The developmental history of the endoderm includes long-distance migration of primordial germ cells, prominent foldings, and a number of evaginations of the digestive tube. The endoderm of the primitive gastrocoel becomes the inner lining of the digestive tract. Anteriorly, at the future pharynx, three outpocketings in both sides of the tract meet three inpocketings from the outside of the neck. These break through to form the *gill slits.* A single ventral outpocket behind the pharynx forms the liver bud, which develops into the liver and bile duct. An inpocketing of the ectoderm forms ventrally at the head region and similar one forms at the posterior end. In later embryonic life these break through to join the endoderm of the digestive tract. The ventral inpocketing becomes the mouth cavity; the posterior inpocketing becomes the anal canal, and both are lined with ectoderm. Other endodermal derivatives are shown in Figure 18-13 and in Table 18-1.

Human Embryology

Once spermatozoa and ova are developed through meiosis and the spermatozoa are deposited in the vagina, pregnancy can occur. **Pregnancy** is a sequence of events including fertilization, implantation, embryonic growth, and fetal growth that normally terminates in birth.

The term *fertilization* is applied to the union of the

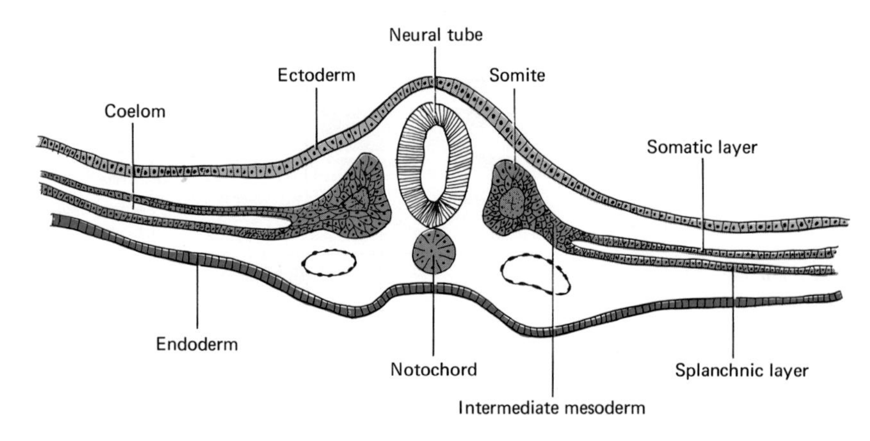

Figure 18-12. Section of an early chick embryo showing the relationship of germ layers.

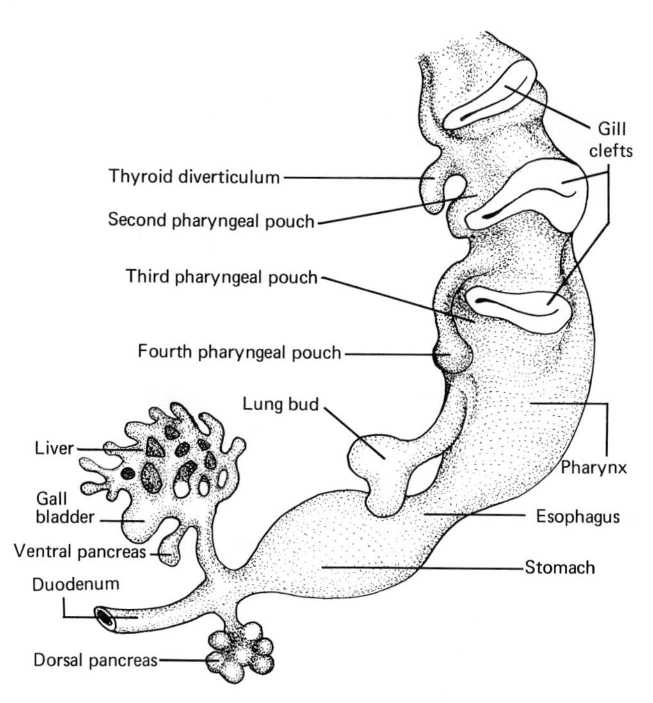

FIGURE 18-13. Some endodermal derivatives in the digestive tract of a mammal.

sperm nucleus and the nucleus of the ovum. It normally occurs in the uterine tube when the ovum is about one-third of the way down the tube, usually within 24 hours after ovulation (Figure 18-14). Peristaltic contractions and the action of cilia transport the ovum through the uterine tube. The mechanism by which sperm reach the uterine tube is still unclear. Some believe that sperm swim up the female tract by means of the whip-like movements of their flagella; others believe sperm are transported by muscular contractions of the uterus.

Sperm must remain in the female genital tract for 4–6 hours before they are capable of fertilizing an ovum. During this time, the enzyme hyaluronidase is activated and secreted by the acrosomes of the spermatozoans. Hyaluronidase apparently dissolves parts of the membrane covering the ovum (Figure 18-14A). Normally, only one spermatozoan fertilizes an ovum because once penetration is achieved, the ovum develops a fertilization membrane that is impermeable to the entrance of other spermatozoans. When the spermatozoan has entered the ovum, the tail is shed and the nucleus in the head develops into a structure called the *male pronucleus.* The nucleus of the ovum also develops into a *female pronucleus.* After the pronuclei are formed, they fuse to produce a *segmentation nucleus*—a process termed fertilization. The segmentation nucleus contains 23 chromosomes from the male pronucleus and 23 chromosomes from the female pronucleus. Thus, the fusion of the haploid pronuclei restores the diploid number. The fertilized ovum, consisting of a segmentation nucleus, cytoplasm, and enveloping membrane, is referred to as a *zygote.*

Immediately after fertilization, rapid cell division of the zygote takes place (Figure 18-14B). This early division of the zygote is called *cleavage.* The progressively smaller cells produced are called *blastomeres.* Successive cleavages produce a solid mass of cells, the *morula* (Figure 18-14C), which is only slightly larger than the original zygote.

Implantation

As the morula descends through the uterine tube, it continues to divide and eventually forms a hollow ball of cells. At this stage of development the mass is referred to as a *blastocyst* (Figure 18-14D). The blastocyst is differentiated into an outer covering of cells called the *trophectoderm* and an *inner cell mass,* and the internal cavity is referred to as the *blastocoel.* Whereas the trophectoderm ultimately will form the membranes composing the fetal portion of the placenta, the inner cell mass will develop eventually into the embryo. About the fifth day after fertilization, the blastocyst enters the uterine cavity.

The attachment of the blastocyst to the endometrium occurs 7–8 days following fertilization and is called **implantation** (Figure 18-14E). At this time the endometrium is in its postovulatory phase. During implantation, the cells of the trophectoderm secrete an enzyme that enables the blastocyst to literally "eat a hole" in the uterine lining and become buried in the endometrium, usually on the posterior wall of the fundus of the uterus. The portion of the endometrium to which the blastocyst adheres and in which it becomes implanted is the basalis layer. Implantation enables the blastocyst to absorb nutrients from the glands and blood vessels of the endometrium for its subsequent growth and development.

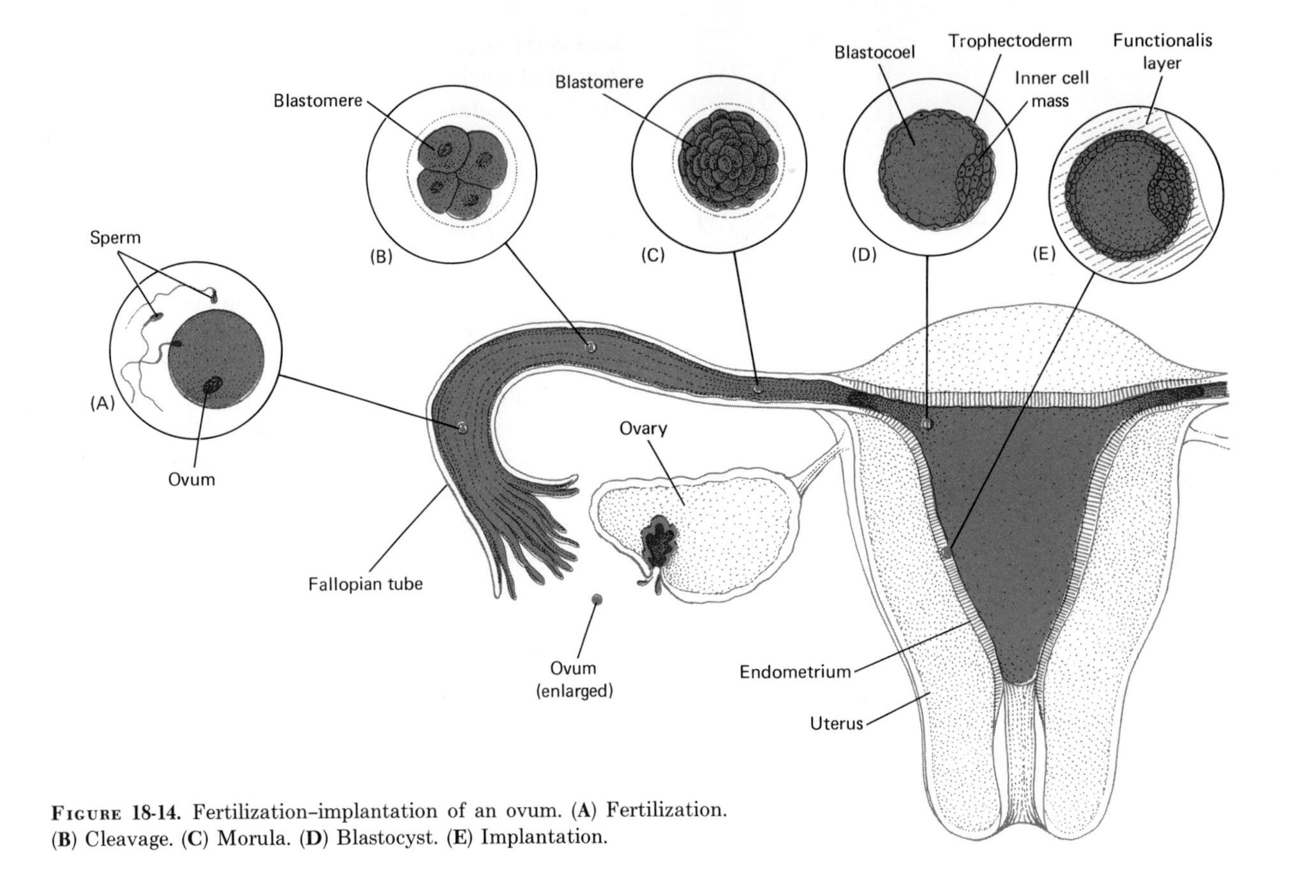

Figure 18-14. Fertilization–implantation of an ovum. (**A**) Fertilization. (**B**) Cleavage. (**C**) Morula. (**D**) Blastocyst. (**E**) Implantation.

Embryonic Period

The first two months of development are considered the **embryonic period.** During this period the developing human is called an *embryo.* After the second month it will be called a *fetus.* By the end of the embryonic period the rudiments of all the principal adult organs are present, the embryonic membranes are developed, and the placenta is functioning. Let us now examine these events in more detail.

Beginnings of Organ Systems. Following implantation, the inner cell mass of the blastocyst begins to differentiate into the three primary germ layers: the ectoderm, endoderm, and mesoderm. The **primary germ layers** are the embryonic tissues from which all tissues and organs of the body will develop. The fetal membranes, structures that lie outside the embryo and protect and nourish it, also develop from these three germ layers.

In the human being, the formation of the germ layers happens so quickly that it is difficult to determine the exact sequence of events. Before implantation, a layer of *ectoderm* (the trophectoderm) already has formed around the blastocoel (Figure 18-15A). The trophectoderm will become part of the chorion—one of the fetal membranes. Within 8 days after implantation, the inner cell mass moves downward so a space called the amnionic cavity lies between the inner cell mass and the trophectoderm (Figure 18-15B). The bottom layer of the inner cell mass develops into an *endodermal* germ layer.

About the twelfth day after fertilization the striking changes shown in Figure 18-15C appear. A layer of cells from the inner cell mass has grown around the top of the amnionic cavity. These cells will become the amnion, another fetal membrane. The cells below the cavity are called the *embryonic disc;* these cells will form the embryo. The embryonic disc contains scattered ectodermal, mesodermal, and endodermal cells in addition to the endodermal layer observed in Figure 18-15B. Notice in Figure 18-15C that the cells of the endodermal layer have been dividing rapidly, so groups of them now extend downward in a circle. This circle is the yolk sac, another fetal membrane. The *mesodermal* cells also have been dividing, and many have left the area of the embryonic disc and can be seen around the structures that are becoming fetal membranes.

About the fourteenth day, the scattered cells in the embryonic disc separate into three distinct layers: the upper ectoderm, the middle mesoderm, and the lower endoderm (Figure 18-5D). At this time the two ends of the embryonic disc draw together, squeezing off the yolk sac. The resulting cavity inside the disc is the endoderm-lined *primitive gut.* The mesoderm within the disc soon splits into two layers, and the space between the layers becomes the coelom, or body cavity.

As the embryo develops, the endoderm becomes the epithelium lining the digestive tract and a number of other organs. The mesoderm forms the peritoneum, muscle, bone, and other connective tissue, and the ectoderm develops into the skin and nervous system. Table 18-1 provides more details about the fates of these primary germ layers.

Embryonic Membranes. During the embryonic period, the *embryonic membranes* form. These

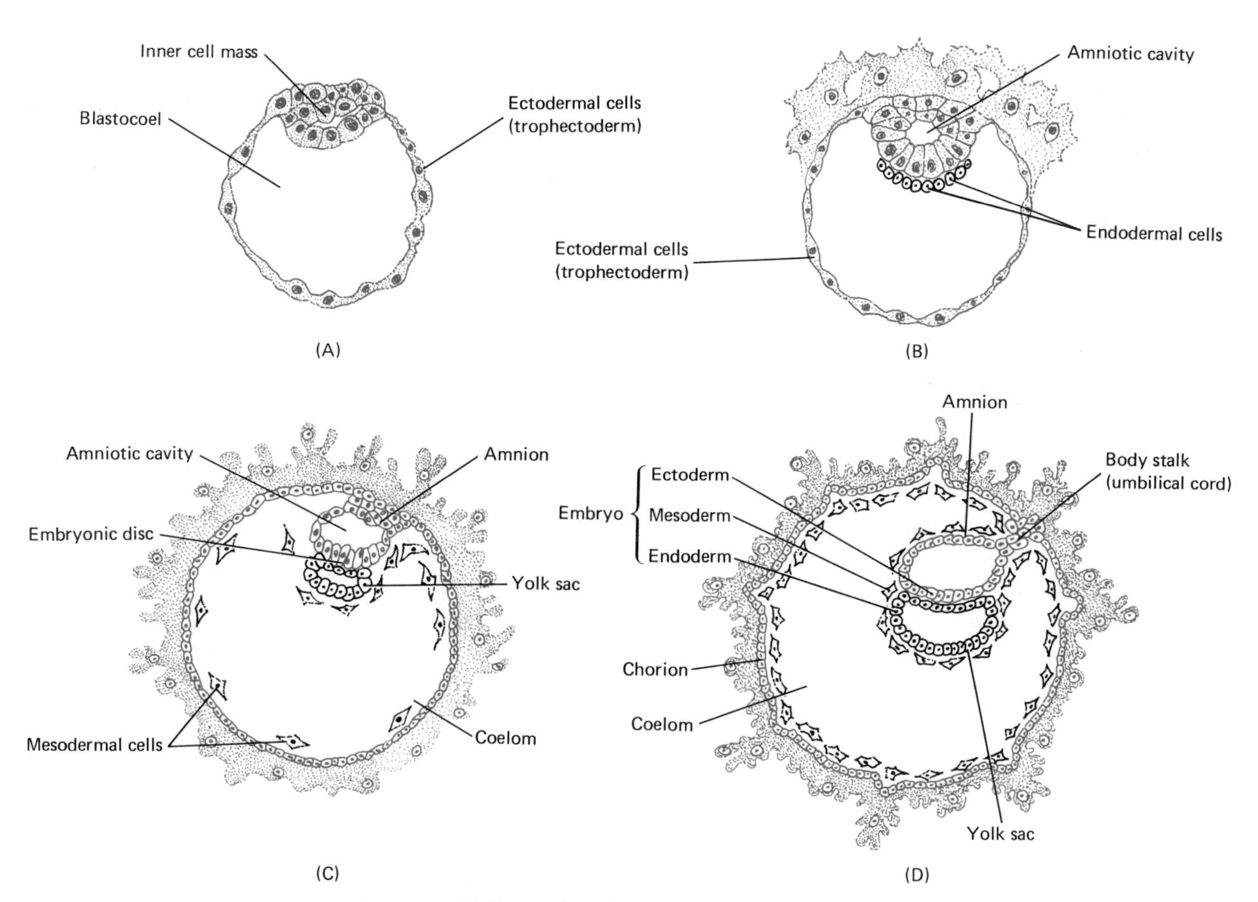

Figure 18-15. Formation of the three primary germ layers.

TABLE 18-1. Structures Produced by the Three Primary Germ Layers

Endoderm	Mesoderm	Ectoderm
Epithelium of digestive tract and its glands	Skeletal, smooth, and cardiac muscle	Epidermis of skin
Epithelium of urinary bladder and gall bladder	Cartilage, bone, and other connective tissues	Hair, nails, and skin glands
Epithelium of pharynx, auditory tube, tonsils, larynx, trachea, bronchi, and lungs	Blood, bone marrow, and lymphoid tissue	Lens of eye
Epithelium of thyroid, parathyroid, and thymus glands	Epithelium of blood vessels and lymphatics	Receptor cells of sense organs
Epithelium of vagina, vestibule, urethra, and associated glands	Epithelium of coelomic and joint cavities	Epithelium of mouth, nostrils, sinuses, oral glands, and anal canal
Adenohypophysis	Epithelium of kidneys and ureters	Enamel of teeth
	Epithelium of gonads and associated ducts	Entire nervous tissue, except adenohypophysis
	Adrenal cortex	

membranes lie outside the embryo and will protect and nourish the fetus. The membranes are the yolk sac, the amnion, the chorion, and the allantois (Figure 18-16).

The human *yolk sac* is an endoderm-lined membrane that encloses the yolk. In many species the yolk provides the primary or exclusive nutrient for the embryo, and consequently, the ova of these animals contain a great deal of yolk. However, the human embryo receives its nourishment from the endometrium. The human yolk sac is small, and during an early stage of development it becomes a nonfunctional part of the umbilical cord.

The *amnion* is a thin, protective membrane that initially overlies the embryonic disc. As the embryo grows, the amnion entirely surrounds the embryo and becomes filled with a fluid called *amniotic fluid.* Amniotic fluid serves as a shock absorber for the fetus. The amnion usually ruptures just before birth; it and its fluid constitute the so-called "bag of waters."

The *chorion* derives from the trophectoderm of the blastocyst and its associated mesoderm. It surrounds the embryo and, later, the fetus. Eventually the chorion becomes the principal part of the placenta, the structure through which materials are exchanged between the mother and fetus. The amnion also surrounds the fetus and eventually fuses to the inner layer of the chorion.

The *allantois* is a small vascularized membrane. Later its blood vessels serve as connections in the placenta between the mother and fetus.

THE UMBILICUS AND THE PLACENTA. Development of a functioning placenta is the third major event of the embryonic period. This is accomplished by the third month of pregnancy. The **placenta** is formed by the chorion of the embryo and the basalis layer of the endometrium of the mother (Figure 18-17). It provides an exchange of nutrients and wastes between the

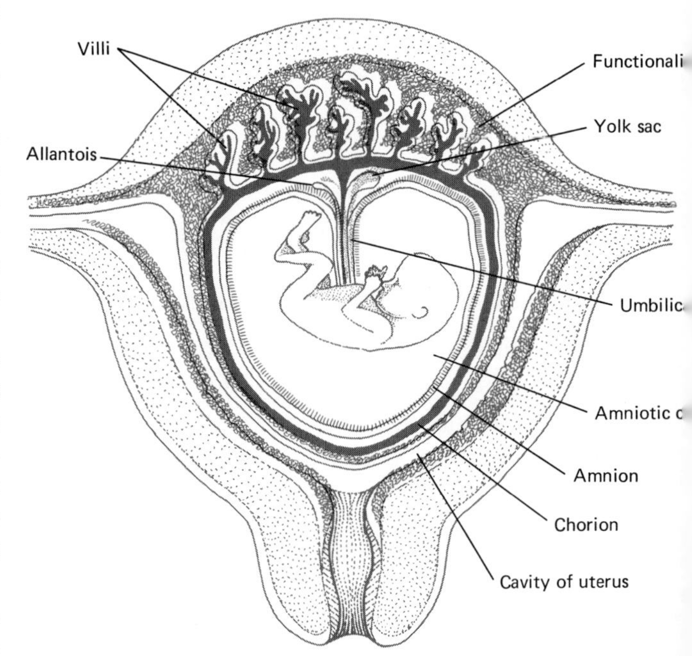

FIGURE 18-16. Embryonic membranes.

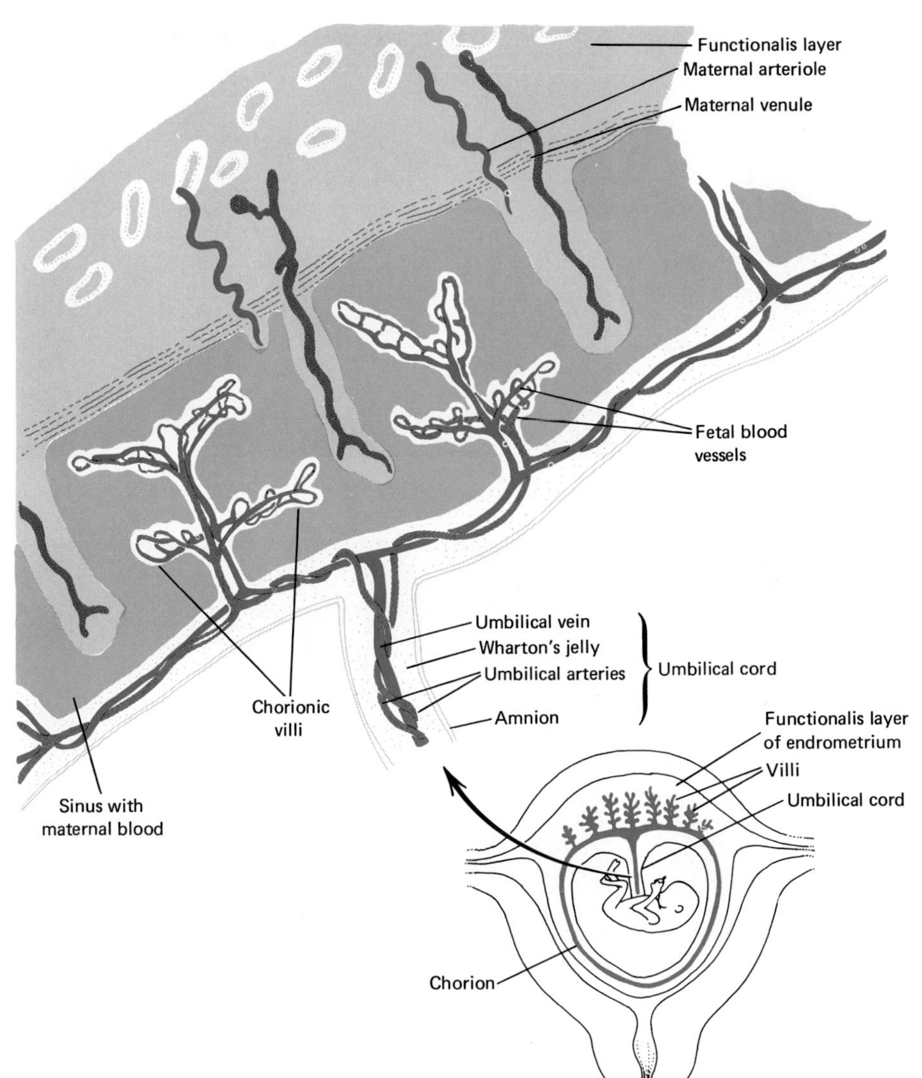

FIGURE 18-17. Structure of the placenta and umbilicus.

fetus and mother and secretes the hormones necessary to maintain pregnancy.

During embryonic life, finger-like projections of the chorion, called *chorionic villi,* grow into the basalis layer of the endometrium. These villi contain fetal blood vessels, and they continue growing until they are bathed in the maternal blood in the sinuses of the basalis. Thus, maternal and fetal blood vessels are brought into close proximity. It should be noted, however, that maternal and fetal blood do not mix. Oxygen and nutrients from the mother's blood diffuse across the walls and into the capillaries of the villi. From the capillaries the nutrients circulate into the umbilical vein. Wastes leave the fetus through the umbilical arteries, pass into the capillaries of the villi, and diffuse into the maternal blood. The **umbilical cord** consists of an outer layer of amnion containing the umbilical arteries and umbilical vein, supported internally by mucous connective tissue called Wharton's jelly. At delivery, the placenta becomes detached from the

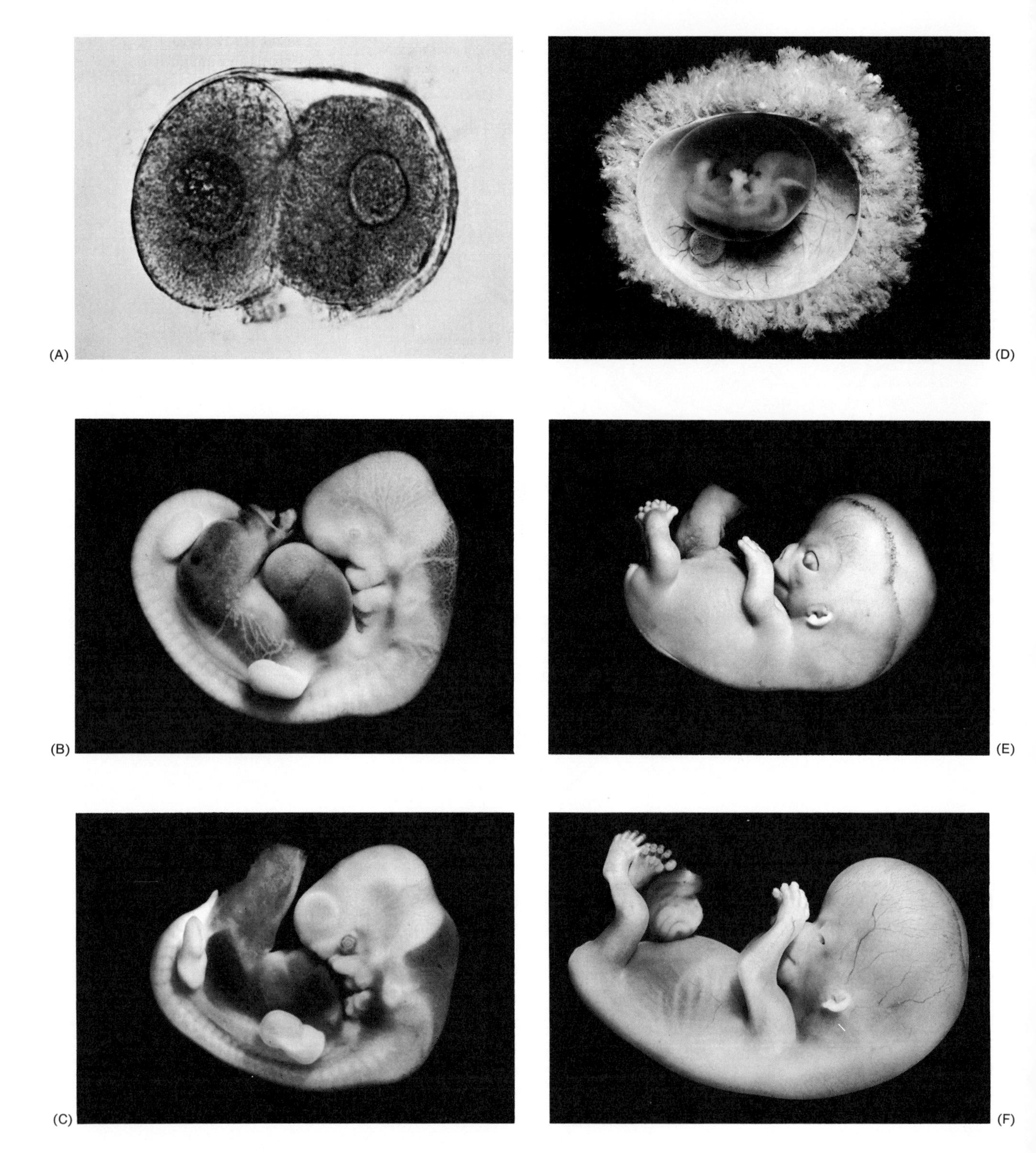

(A)
(B)
(C)
(D)
(E)
(F)

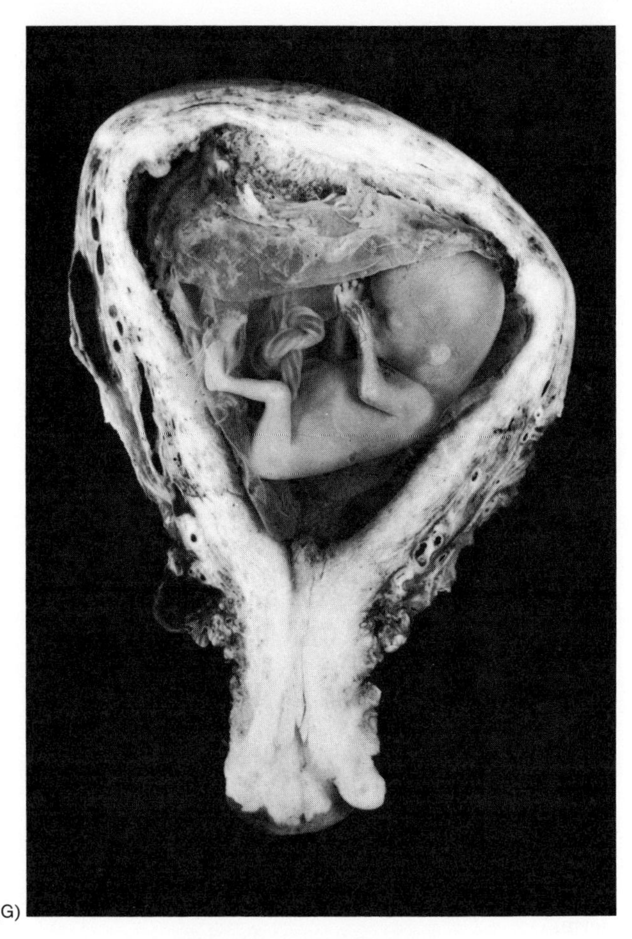

(G)

Approximate sizes and weights:

Third month: 3 in., 1 oz
Fourth month: 6 in., $\frac{1}{4}$ lb
Fifth month: 10 in., $\frac{1}{2}$ lb
Sixth month: 12 in., $1\frac{1}{2}$ lb
Seventh month: 15 in., $2\frac{1}{2}$ lb
Eighth month: $16\frac{1}{2}$ in., 4 lb
Ninth month: 20 in., 6–7 lb or more

Figure 18-18. Representative stages of gestation. The entire intrauterine period is conveniently divided into three principal stages. The first of these, the period of the ovum, extends from the time of fertilization to the time of implantation (8–10 days). Shown as an example of this stage is a two-cell human ovum at approximately 60 hours (**A**). During the period ovum segmentation occurs, the developing zygote becomes differentiated into embryonic and extraembryonic portions, and the primary germ layers become established.

The second period of development, the embryonic period, lasts to about the second month of development. During this interval the shape and appearance of the embryo are greatly altered and many of the important external features may be recognized. Each germ layer starts a course of differentiation into specific tissues, organs, and systems, and by the end of the embryonic period, all the major body systems are formed. (**B**) Thirty-one-day embryo. The first body organ to become functional is the heart (dark bean-shaped structure in the chest region). (**C**) Thirty-four-day embryo with remains of gill clefts, fore- and hindlimb buds, heart, and developing eye. (**D**) Forty-day embryo with intact amnion and one side of the chorion cut away. (**E**) Forty-four-day embryo. Note the continued development of the eye, ear, and limb buds.

The third stage of development, the fetal period, is the interval of time from about the beginning of the third month to the termination of intrauterine life. During this period some differentiation of tissues does continue, but the major changes are brought about by rapid growth of the body. In the third month, the head becomes relatively smaller, limbs become longer, nails appear, the eyelids fuse, external genitalia differentiate, and centers of bone formation become numerous. (**F**) Fifty-six-day embryo showing some of these features. (**G**) Seventy-day fetus shown in an open uterus. In the fourth month hair appears on the head and body, sense organs assume typical forms, and bones become distinct. In the fifth month a hair coat is present on the body and blood formation begins. In the sixth and seventh months the body form becomes better proportioned and the body has a wrinkled appearance. In the eighth month the fetus is capable of living if born prematurely. The testes descend into the scrotum, subcutaneous fat is deposited, and the body appears rounded and less wrinkled. In the ninth month growth continues, additional fat accumulates, the hair coat is shed, limbs become fuller, and nails extend to the tips of the digits.

Note also the approximate sizes and weights of the developing organism at various stages of development. [*Courtesy of Chester F. Reather, RBP, FBPA, Carnegie Institution of Washington.*]

uterus and is referred to as the "afterbirth." At this time, the umbilicus is severed, leaving the baby on its own.

Birth

The period of time that the embryo is carried in the uterus is called **gestation.** Some of the principal events of human gestation are summarized in Figure 18-18.

Assuming that the total gestation period is 280 days from the beginning of the last menstrual period, the date of delivery is estimated by adding 1 year and 7 days to the date of the last menstrual period and subtracting 3 months. It is customary to expect *parturition* (birth) at about 280 days from the last menstruation.

The actual mechanisms involved in the onset of labor

and birth are quite complex and poorly understood. Hormones produced by the placenta and ovaries assume a role in determining the onset of labor. Progesterone exerts a pregnancy-stabilizing effect and labor cannot take place until its effects have diminished. Estrogens promote rhythmic contractions of the uterus. Oxytocin from the pituitary also effects uterine contractions. The hormone relaxin serves to soften the ligamentous structures of the pelvis. Without proper hormonal balance and timing, labor would not occur, or the fetus and/or mother would be injured.

Labor is conveniently divided into three stages. The first, the *stage of dilatation,* is the period of time from the onset of labor to the complete dilatation of the cervix. The second stage, or *stage of expulsion,* is the period of time from complete cervical dilatation to birth. The third period is the *placental stage* from birth to the expulsion of the placenta and membranes. These stages are shown and described in Figure 18-19. The human infant unlike most mammals, is dependent upon the parent for survival for a long period of time.

Control of Development

As noted earlier, patterns of morphological changes and of differentiation have been followed for many species. However, while the changes are observable, the means by which these changes are brought about is not well known. Clearly, in the developmental differentiation of all species, there are two controlling factors, genetic and nongenetic. Understanding of the nature of the interaction and degree of control of these factors is the major task of developmental biologists and represents an area of active research.

Nuclear Determinants

It has been emphasized in previous chapters that, in cell division, a complete set of paired chromosomes is passed on to the somatic daughter cells. In reduction division (meiosis), one chromosome of each pair is present in the sex cells or gametes. Upon fertilization, the union of gamete nuclei restores the complete set of paired chromosomes present in the cells of the parents. As the zygote divides mitotically, each resulting daughter cell receives a complete set of chromosomes identical with those of the parent cell. It follows, then, that all the somatic or body cells of an organism are genotypically the same. With only a few rare exceptions, there is no indication that there is any change in chromosome number in normal mitotic cell division. Therefore, each body cell is endowed with all the hereditary characteristics of the mature organism. If all cells contain the same genetic information, it is the task of the developmental biologist to find those factors which are determined by the genes and those factors outside the hereditary material that control differentiation.

In 1952, R. Briggs and T. J. King of the Institute for Cancer Research in Philadelphia developed a technique whereby the nuclei of some cells could be removed and transplanted to other cells. They took eggs of the leopard frog (*Rana pipiens*), pricked them with a fine needle to stimulate development, and then removed the nuclei. They then took nuclei from embryos in the blastula stage of this frog and transplanted them to the enucleated eggs. It was found that the eggs with the transplanted nuclei from the blastula stage developed into normal embryos. When the experiment was repeated using nuclei from the gastrula and later embryonic stages, except for a small percentage, most of the developing embryos were abnormal and development was arrested at some stage. The abnormalities were specific. For example, if the nuclei were taken from the endoderm, the embryos resulting from the transplantation had well-developed endoderm but an abnormal ectoderm.

In 1962, J. B. Gurdon of the University of Oxford, England, performed similar experiments of transplanted nuclei using the African clawed frog (*Xenopus laevis*). His results were similar in that the later the stage of embryonic development of the transplanted nuclei to enucleated eggs, the greater the percentage of abnormalities. However, he was able to obtain some normal fertile adult frogs from nuclei taken from intestinal cells of tadpoles.

Two years later, F. C. Steward and his co-workers at Cornell University succeeded in growing whole carrot plants from freely suspended cells of embryo origin (Figure 18-20). They removed embryos from the flowers of the wild carrot and placed them in a nutrient medium. Soon the embryos proliferated and a suspension of cells from this proliferation was spread over an agar medium in a petri plate. The resulting growth consisted

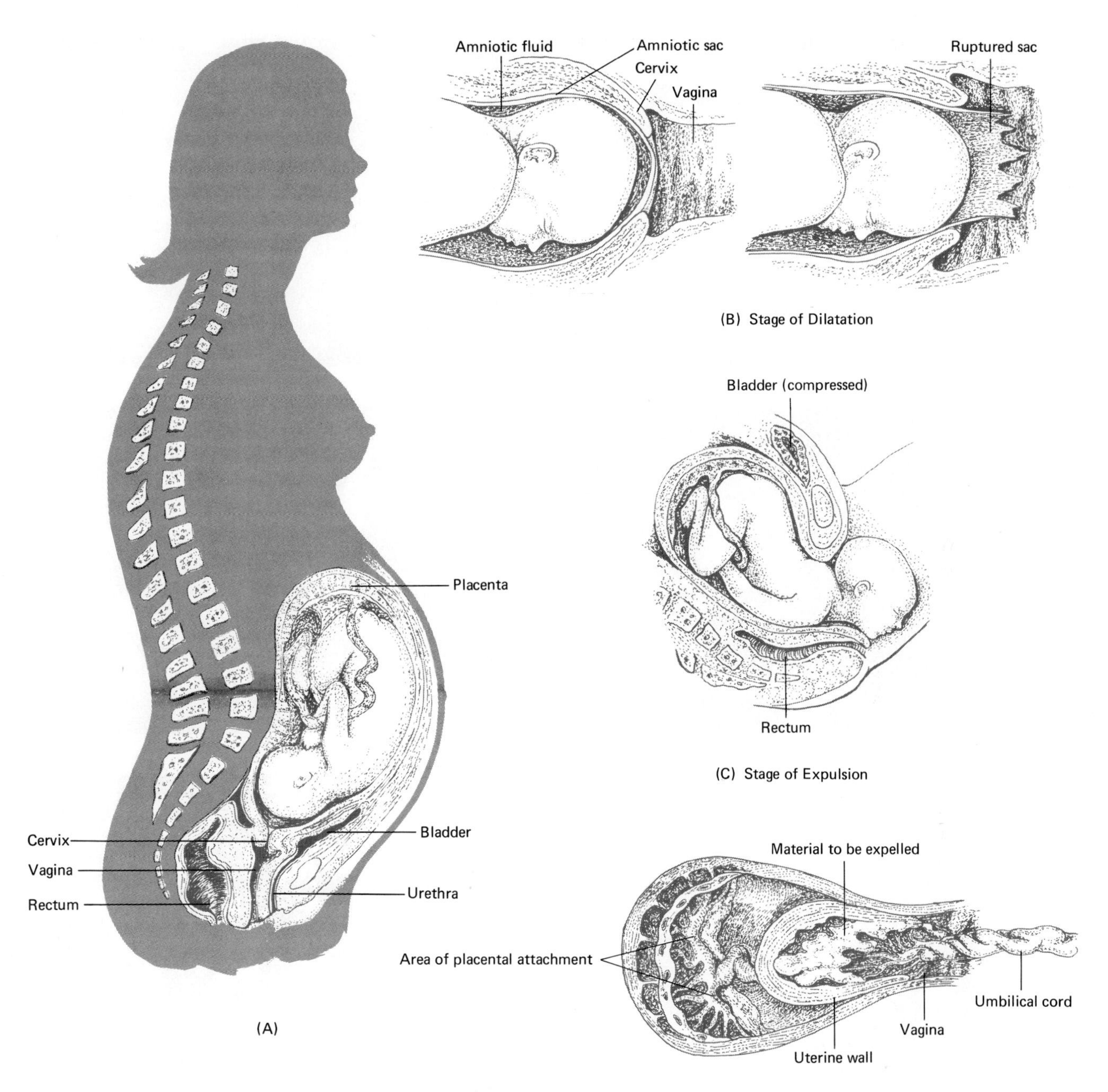

FIGURE 18-19. Birth. (**A**) Fetal position prior to birth. (**B–D**) Stages of labor. (**B**) Dilatation: protrusion of amniotic sac through partly dilatated cervix (*left*) and amniotic sac ruptured and cervix completely dilated (*right*). (**C**) Start of second stage (expulsion). (**D**) Placental stage showing area of placental attachment prior to birth and uterine wall contracting to expel placenta.

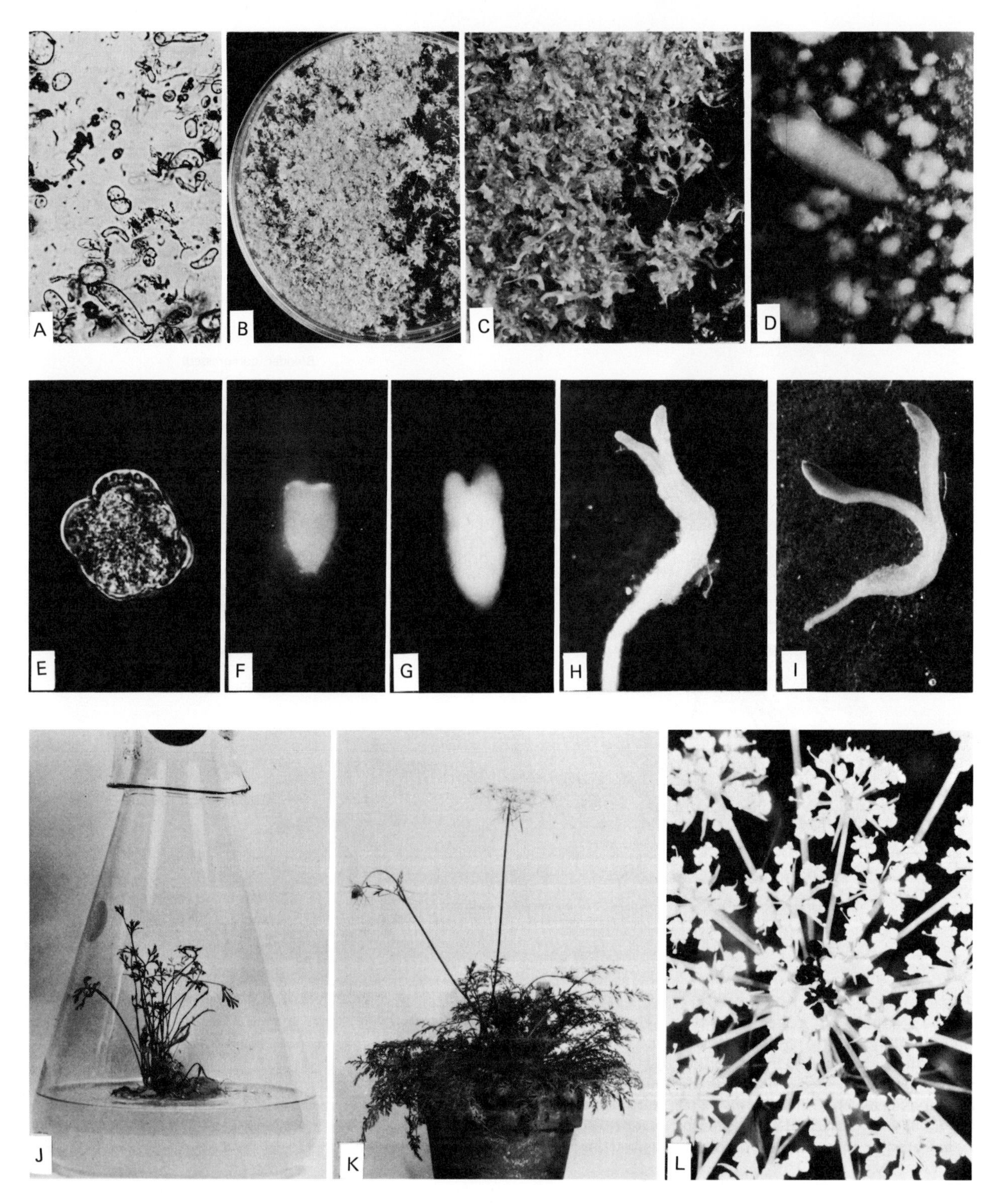
A
B
C
D
E
F
G
H
I
J
K
L

of undifferentiated cells that were well vacuolated and large, and a considerable number of embryo-like forms called embryoids. The embryoids went through the embryological stages of normal plant development. Eventually, they developed into viable plants.

Steward and his associates also developed normal carrot plants from freely suspended secondary phloem cells. Under the conditions in which living diploid cells exist in the intact plant, their totipotency (ability to develop into a complete organism) is restricted. It is quite possible that cells of immature embryos express their totipotency more readily than phloem cells because embryonic cells are not so completely differentiated. Additional studies by others with different plants have yielded similar results. This type of reproduction, whereby a comparatively undifferentiated cell gives rise to a complete organism, has yet to be duplicated with animals.

The results of research seem to indicate that the more differentiated a cell becomes, the more limited is the ability of the nucleus to control the complete potentialities of the cell. Yet, as has been shown, it is possible for a differentiated nucleus to dedifferentiate and control the normal development of an embryo when transplanted to an enucleated egg. The variability in the results of experimentation might be interpreted as a possible difference in species or specific problems in experimental techniques. Later studies, by J. B. Gurdon and his colleagues at Oxford, seem to establish the conclusion that genes are not necessarily lost or permanently inactivated during cell differentiation. Therefore, if the cell possesses all the potential genes of the organism, there must be additional variable factors which control specific genes. These variable factors must come from the environments in which the nucleus exists. This may be the surrounding cytoplasm, adjacent cells, or the environment external to the total organism.

Cytoplasmic Determinants

It has been mentioned earlier that the frog's egg is essentially polarized into two regions, a vegetal zone and an animal zone. The apex of each region is termed the *vegetal pole* or the *animal pole,* respectively. If cleavage occurred so that each resulting cell received a divided nucleus and cytoplasm from both regions, the two cells, when separated, should have resulted in the formation of two complete embryos. However, upon separation abnormal embryos were frequently formed. After investigation, it was discovered that between the animal zone and the vegetal zone was an area designated as the **grey crescent.** Only in those cleavages in which each cell received a portion of the grey crescent was there development of normal embryos. Similar results were found with the eggs of other species. This investigation showed that, although the nuclei were equivalent in the separated cells, differences in the cytoplasm determined the normal development in the early embryo.

The green alga *Acetabularia* is a single-celled marine organism. It is composed of a cap, stalk, and rhizoidal base containing a single nucleus. In the mature *Acetabularia,* the nucleus divides and the resulting nuclei are carried to the cap by cytoplasmic streaming where they form gametes. It has been shown that, if the cap is removed from the mature alga, the nucleus will not divide until a new cap is regenerated. On the other hand, if a mature cap is grafted onto an immature alga, premature nuclear division takes place. The cytoplasm of the cap appears to exert a definite control over the nucleus (see Figure 5-16).

In 1933, Emil Heitz and Hans Baur of the University of Hamburg first recognized giant chromosomes in the larvae of certain species of *Drosophila.* Since that time, these chromosomes have been a most important research tool of geneticists. When specially stained, these

Figure 18-20 [**opposite**]. Development of mature carrot plants from freely suspended cells of embryo origin. (**A**) Embryo cells in a liquid culture medium. (**B**) Embryoids on a petri plate developed from embryo cells in the liquid medium. (**C**) Higher magnification of (**B**). (**D**) Enlarged aspect of an embryoid. (**E–I**) Stages of embryology of normal carrot development. (**E**) Globular stage. (**F**) Heart-shaped stage. (**G**) Torpedo stage. (**H, I**) Two cotyledonary stages. (**J–L**) Further stages of development. (**J**) The plant on agar. (**K**) The plant bearing flowers. (**L**) The floral structure at high magnification. [*From F. C. Steward,* Science, *vol. 143, 1964, pp. 20–27. Copyright © 1964 by the American Association for the Advancement of Science.*]

chromosomes show patterns of light and dark bands called *Balbiani rings*. These bands have been considered as the equivalents of genes and have been used to identify specific chromosomes and genes. They have also provided important proof that the same chromosomes are present in all the cells of an organism. Wolfgang Beerman and Ulrich Clever of the Max Planck Institute for Biology in Tübingen, Germany, have concluded that there is no variation in the sequence of Balbiani rings along the chromosomes in different tissues, but they also found that there was a quantitative difference. By careful observation, it was found that there was a difference in size in certain of the rings described as *chromosomal puffs*. These puffs had been noticed before, but their significance had not been realized. Identical tissues had puffs in the same location on the chromosome, while different tissues had puffs in different locations. The specific cell types had a characteristic "puffing pattern" which changed as the cells developed. Analysis of the puffs showed that they were sites of active RNA synthesis. Those portions of the chromosomes (genes) which did not have puffs were relatively inactive and did not have RNA at those sites. Through an alteration of the cytoplasm by the addition of a hormone, it has been possible to change the puffing pattern. For example, in experimentation with insects, at the time the larva changes into a pupa, it has been possible to hasten the process by injecting the larva with a pupation hormone (ecdysone). The puffing pattern of the chromosomes rapidly changes to the pattern characteristic of the chromosomes in cells undergoing pupation. Further studies have shown that when chromosomes are transplanted from one tissue to another, they change their puffing pattern to conform to the new cellular environment. These studies emphasize the importance of the extranuclear environment in gene activity, resulting in specific differentiation.

Further differences in the cytoplasmic organization may also influence gene activity. As has been mentioned previously, the distribution of yolk and grey crescent are important in normal embryonic development. Other studies indicate that other areas of the egg are also important. When the region just below the surface of the egg, known as the *cortex*, has been altered, it has been possible to induce characteristic changes in development. Distribution of mitochondria differs among various cells. It has been found that cells possessing large numbers of mitochondria are capable of undergoing greater differentiation.

Regional distribution of the various cytoplasmic components within the cell is important. If the pattern of distribution of the cytoplasmic components is changed by centrifuging prior to cleavage, unusually abnormal embryos develop, with great differences in the amount and distribution of tissues.

The conclusion of these few studies mentioned indicates that the morphological and developmental changes that take place within a cell represent an interaction between the genetic material and certain cytoplasmic components. The sequential development involves a feedback mechanism acting upon the chromosomes which causes certain genes to be activated. This activation results in a switch mechanism which determines the particular direction of differentiation a cell will take.

Neighboring-Cell Determinants

Up to this point, the discussion of developmental control has been confined to the single cell. It is obvious that, in the development of the total organism, there must be coordination and communication between the various tissues and organs. Therefore, the environment external to the cell, that is, the environment directly surrounding the cell, as well as cells and tissues at a distance within the organism, must also exert an influence on development.

The environment directly surrounding the cell within the same tissue is referred to as the *microenvironment*. No two cells have exactly the same microenvironment. Some cells lie deeply within tissues while others are on the surface. The influence of chemical secretions of surrounding cells, differences in the concentration of available oxygen and carbon dioxide, and physical confinement, probably all greatly influence cell differentiation and development. Techniques have not yet been developed to measure directly the microenvironment. Yet, indirect experiments have indicated that the microenvironment probably has profound effects. For example, in the alternation of generations of the fern, the sporophytic ($2n$) plant, with its complex structure of rhizoids, stem, and frond, develops from the zygote within the female sex organ of the gametophytic (n) plant. The gametophytic plant is a small thallus-like structure with much less differentiation than in the sporophyte. It has been found that if the zygote ($2n$) is removed from the gametophyte and placed in the proper nutrient medium, it does not develop the embryonic parts characteristic of the sporophytic plant, but looks much like an embryonic gametophytic plant. Evidently, the microenvironment of physical confinement and differences in chemical gradients of the sex organs of the gametophytic plant control the early development of the fern plant.

Beyond the microenvironment, cells are surrounded by other tissues. The importance of surrounding tissue in the differentiation of other tissues has been known for many years. One of the earliest studies of the influence of surrounding tissues was done by Warren H. Lewis of Johns Hopkins University in 1905. The normal development of the eye of frogs occurs from outpockets (optic cups) in the brain. When these outpockets come into contact with the epidermis tissue, they induce the epidermis to produce an eye lens. Lewis transplanted one of these outpockets, called an optic vesicle, to the posterior part of a developing frog embryo. Although the optic vesicle no longer had connection with the brain, when it came into contact with the epidermis tissue where it had been transplanted, an eye lens developed. No eye developed in that portion of the embryo from which the optic vesicle has been removed. The vesicle had induced the epidermis to differentiate into an eye lens. Refer to Figure 18-10 for the location of the structures just discussed in conjunction with Lewis' experiment.

In 1924 Hans Spemann and Hilde Mangold of the University of Freiburg, Germany, performed a series of experiments which explained development by a phenomenon termed embryonic induction. **Embryonic induction,** or simply **induction,** is a process in which a developing structure (inductor) stimulates (induces) another structure to differentiate. In the early gastrula stage, the *dorsal lip* of the blastopore, a region consisting of a portion of the grey crescent in the egg, appears to exert an influence on the ectodermal cells stimulating differentiation into a brain and spinal cord. When Spemann transplanted the dorsal lip of the blastopore from a light-colored embryo to the belly region of a dark-colored embryo, two fused embryos developed, one from the dorsal lip of the recipient embryo, and one from the site of the transplanted lip. Most of the tissues of both embryos were dark, indicating that the transplanted lip had exerted control over some of the cells of the recipient embryo. Spemann and Mangold termed the dorsal lip an **organizer.** Since Spemann's original experiments, many other major and minor organizers have been found. In fact, tissue induction is a generalized occurrence in development.

Various chemical substances have been found to cause induction. However, in the living organism no specific compound has yet been identified. There is some evidence that these inducers are an RNA–protein complex of some kind. While the exact way in which these chemical substances exert their influence is not known, Hans Spemann and Oscar E. Schotte of the University of Freiburg in 1932 performed an experiment that suggested that inducers act upon the genetic material of the developing cells. Spemann and Schotte transplanted a small piece of ectoderm from the flank of a frog embryo to the mouth regions of a salamander embryo. The transplanted ectoderm developed into mouth tissue characteristic of the frog. Evidently, the endoderm of the salamander was able to induce the genes of the frog tissue to develop mouth parts but could not alter the hereditary frog determinants of the genes. Similar experiments have produced similar results. It has also been concluded that, in this type of differentiation, there is a feedback mechanism by which regulators induce specific genes to react; these in turn produce chemical substances that determine the specific cell types.

It has been known for many years that certain hormones affect growth as well as most other physiological functions. There are hormones that affect the growth and activity of hormone-producing tissues as well as hormones that act directly without the mediation of other hormones. For example, *somatotropin* regulates the size of tissue structure and is responsible for dwarfism or giantism, while another, *thyroxin,* stimulates differentiation and maturation of the same tissue.

It was believed for many years that hormones produced their effects by direct interactions in modifying the characteristic effects of enzymes. Evidence now indicates that hormones probably effect their action through control of the gene. The specific function of the hormone would be determined by the particular gene it activates, leading to the production of a specific enzyme.

Effects of the External Environment

In the normal development of an organism it is quite obvious that external factors such as temperature, light, humidity, gravity, and pressure play an important part in dvelopment. For example, in the Himalayan rabbit, temperature plays an important role in the development of pigmentation (see Chapter 19). There is also a mutant species of fruit fly that has vestigial wings. It has been found that if these genetically vestigial flies are grown at high temperatures they develop normal wings. Again, temperature is a determining factor.

D. M. Whitaker at Stanford University has studied the external factors which affect the development of the brown alga *Fucus*. In this alga, shortly after the egg is fertilized the zygote develops a slight protuberance. The first cleavage always occurs so that one of the daughter cells contains the protuberance and develops into a holdfast, a root-like structure. The other daughter cell forms the erect portion of the plant. Whitaker varied specific environmental factors to determine their effects on the development of the holdfast. He found that the protuberance forms on the side opposite the source of illumination. When illumination was equal on all sides, the protuberance developed on the warmer side when a temperature difference was maintained. When a similar difference in the pH of the growing medium was established, the protuberance grew on the side of lower pH. If all three environmental conditions were equal, the protuberance grew on the side nearest the other zygotes present in the medium. Finally, when zygotes were placed in a centrifuge, the protuberance developed on the side of the zygote farthest from the center of the centrifuge. The development of the protuberance represents a response to those environmental conditions for which the holdfast is adapted in its functioning of anchoring the plant.

Another example of the effect of external environmental conditions upon development is shown in the growth of a seedling. If a seedling is grown in the dark, chlorophyll is not synthesized. The direction in the growth of roots, stems, and leaves is a response to external conditions of light, gravity, water, and nutrients. How these external factors exert their control upon the nucleus and cytoplasm of cells, resulting in a specific differentiation, is not yet known.

Some Developmental Changes After Birth

The degree of development reached at the time of birth varies greatly among different living forms. While most of the basic organs and shape of the embryo are well established at the time of birth, there are still great changes brought about by growth and other types of development. Certain species undergo metamorphosis, while others are capable of various degrees of regeneration of body structures. Growth proceeds at varying rates and development continues throughout aging and ceases only at death.

Growth

Among most organisms, once morphogenesis and differentiation of embryonic tissue have taken place, the major developmental process is growth. The rate at which maximum growth is attained varies greatly among living forms. Some birds and mammals reach their maximum size within several months, while many woody plants continue to grow for hundreds of years. In the human embryo, most morphological changes and differentiation have occurred by the end of 12 weeks. Growth then continues for approximately 25 years.

For growth to take place, the complex molecules that make up the organism must be synthesized and assimilated at a rate greater than their breakdown. The raw materials for these molecules are obtained from the physical environment. Some of these molecules are synthesized and used for the actual increase in size of the organism, while others provide the energy to carry on the life processes. Growth, therefore, consists of gaining more material than is given off as wastes. However, growth is not just a mere accumulation of material. Materials from the environment are reassembled within the organism to form the unique characteristic molecules of the given species. Autotrophs take simple inorganic molecules and build them up into complex biochemical compounds. The heterotrophs, in consuming organic matter, break down the complex biochemicals into intermediate-sized molecules and reassemble them into the characteristic biochemicals of the species. Growth by the constant breakdown and buildup of molecules is one of the most characteristic properties of life.

Growth may occur by two means: by an increase in the number of cells (hyperplasia) and by an increase in cell size (hypertrophy). In vertebrates, difference in size is generally due to the adult having many more cells than the infant. Of course, immediately upon cell division the daughter cells undergo some enlargement, but most growth results from an increase in the number of cells. In certain of the smaller invertebrates, growth occurs until a certain number of cells is reached; further growth occurs through the enlargement of the existing cells. In monocots there may be considerable growth in the thickness of stems. This type of growth occurs through the enlargement of the cells of the stem.

Growth patterns are usually measured by calculating the increase in body weight, which is roughly equal to an increase of size during a given period of time. These patterns may be modified by environmental conditions or the hereditary characteristics of the species. Generally, the pattern of growth may be plotted as an S-shaped curve (Figure 18-21). Growth is slow at first, but soon increases rapidly. Then the rate begins to decelerate and tapers off as adult size is reached.

The shape of the curve may have modifications depending upon whether rapid growth occurs for a shorter or longer period of time. There may also be other modifications. For example, in mammals, growth begins relatively slowly and then increases rapidly until the time of weaning, then the rate slows down (Figure 18-22). After a time of slow growth, there is another period of rapid growth until puberty, when the rate slows down again. After another period of slow growth, the rate increases and then finally decelerates, ceasing when adult size is reached.

In mammals, growth is regulated to some degree by the way the bones elongate. During the period of growth, the ends of the bones are separated by a layer or plate of cartilage tissue. Growth occurs by the addition of bone tissue at these plates. In the adult, the cartilage layer disappears and the separate sections of the bone fuse, with further growth no longer possible. This type of bone growth is not found in all vertebrates, for example, the bony fishes and birds.

A different growth pattern is found among those animals with an exoskeleton such as that found among the arthropods (Figure 18-23). Since the rigid exoskeleton that encases the body greatly limits growth, the organism undergoes a series of molts in which the exoskeleton is shed. During this time there is a burst of

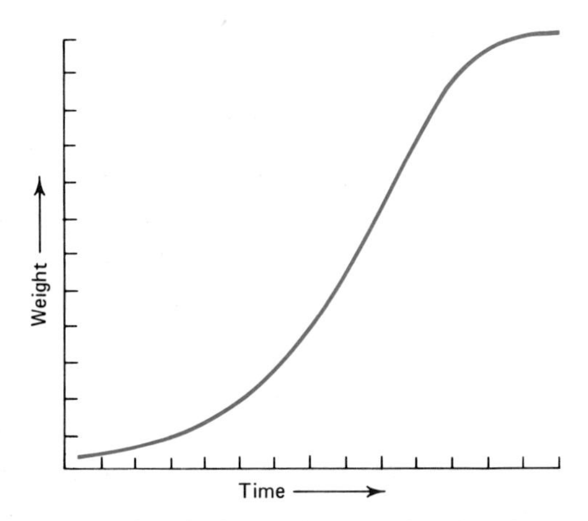

Figure 18-21. Standard growth curve for the whole body. Growth begins slowly and then after a period of time it increases rapidly. Following this period of maximum growth, the rate gradually decreases and tapers off. In many organisms height increases with weight. The growth curve indicated is also known as a sigmoid (S) curve.

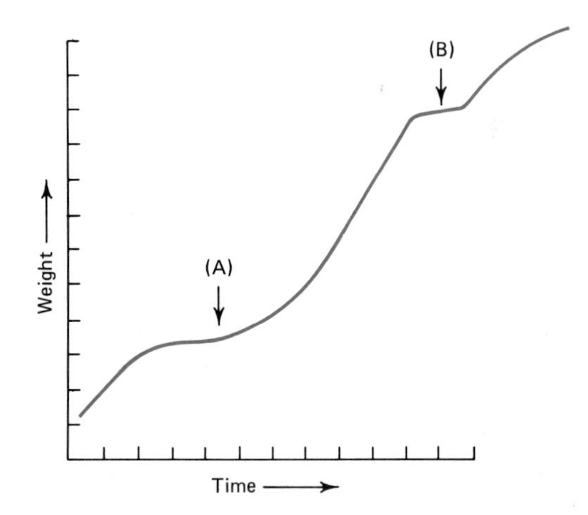

Figure 18-22. Growth curve of mammals. Growth of a mammal proceeds rapidly until the time of weaning. Adjustment of the organism to the new food causes growth to slow down (**A**). Growth then resumes at rapid rate until the time of puberty (**B**) when it slows again because of physiological changes. The rate of growth then increases again until maturity is reached.

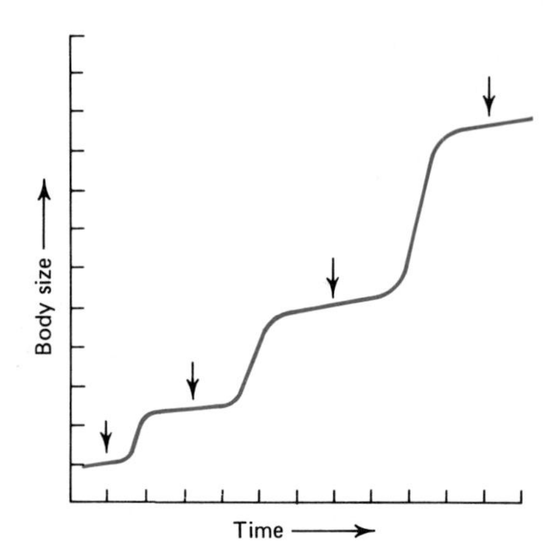

Figure 18-23. Growth curve of arthropods. The step-like pattern of growth in arthropods results from periods of rapid growth following each molt, when the old exoskeleton is shed, until the new one has fully hardened. The hardened exoskeleton prevents any appreciable growth (↓).

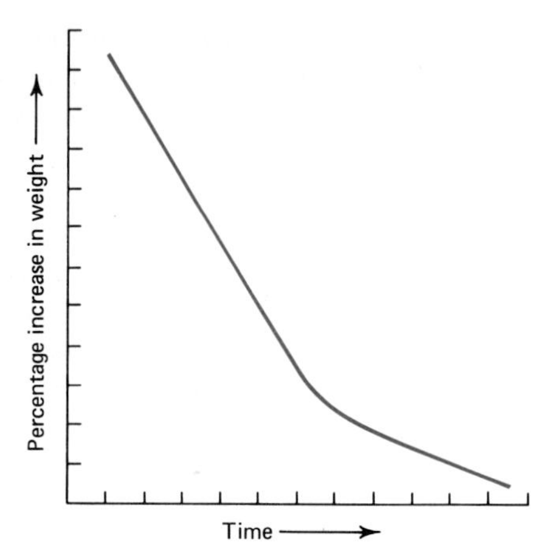

Figure 18-24. Growth curve of mammals based upon percentage increase in weight. The percentage increase is calculated by dividing the increase in weight over a given period of time by the total weight at the beginning of the period and multiplying by 100. It can be seen that the percentage increase during the time period rapidly decreases and tapers off at maturity.

very rapid growth before the new exoskeleton hardens. A graph of such growth takes a characteristic step-like pattern.

One of the characteristics of growth is that it is dependent upon the amount of living material already present in the organism. Therefore, a more accurate description of growth can be made by considering the percentage increase in weight rather than the absolute increase. For example, a baby weighing 20 pounds, by increasing his weight 4 pounds, is growing more rapidly than an adolescent weighing 100 pounds and increasing his weight 10 pounds in the same time period. In contrast to the typical S-shaped growth curve, the curve obtained by plotting percentage increase does not show periods of maximum growth, but rather shows that growth is more rapid in the earlier period of life and tends to slow down throughout life (Figure 18-24).

The consideration of growth thus far has implied that the growth pattern is the same throughout all parts of the organism. However, when comparing an infant with an adult human, or a baby chick with an adult chicken, the differences are not just a matter of size. There are also changes in shape. These changes in shape are the result of differential growth rates in various parts of an organism. This difference in the rate of growth of different parts is quite apparent in plants where the apical meristem is the area of rapid growth. In humans, the head of the infant is much larger in proportion to other parts of the body. By contrast, the legs of the infant are proportionally smaller than the other parts of the body. There is continued growth of the legs during the entire period of growth.

This difference in the rate of growth of different parts of the body was first recognized by Julian Huxley who referred to the process as **allometric growth.** Allometric growth has been shown to be related to evolutionary changes. Frequently, two different species appear to be quite different from one another, yet, through disproportionate growth of different parts, the difference is superficial. For example, although the porcupine fish and ocean sunfish look quite different, in reality the difference occurs only because the tail region of the sunfish grows more rapidly in a dorsal–ventral direction than that of the porcupine fish (Figure 18-25). In many plants, differences in the leaves as, for example, between the pin oak and white oak, are due only to various rates of growth in different portions of the leaf.

Termination of Growth

The complex interactions resulting in a particular growth pattern of a given species are poorly understood.

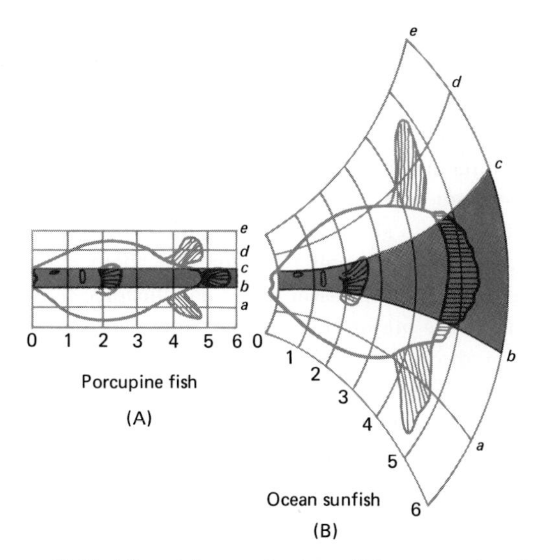

FIGURE 18-25. Allometric growth of two different species of fish. The porcupine fish (*Diodan*) (**A**) and the ocean sunfish (*Orthagoriscus*) (**B**) are quite different superficially. However, if a grid of coordinates is drawn over both fish, it can be seen (colored areas) that the difference between them is based upon different amounts of growth in corresponding regions. Comparison of other grids also illustrates different amounts of growth in other corresponding regions. [*From D. W. Thompson,* On Growth and Form, *abridged ed. New York: Cambridge University Press. Copyright © 1961.*]

Although the characteristic adult size of a given organism is determined by heredity, the environment can modify this. If conditions are poor, maximum growth is not attained. Sometimes the interaction between the hereditary material and the internal environment within the organism changes and there is continued growth when it should normally cease. This condition has been an area of continued study because it generally results in cancer.

Food factors also play an important part in normal growth. Aside from the need for sufficient materials, there are specific compounds, such as essential vitamins and essential amino acids, that the organism must obtain from its environment for normal growth. Lack of sufficient quantities results in stunted growth or death. Hormones also play an important part in normal growth. For example, the pituitary hormones control metabolic rates. Deficiency of these hormones results in stunted, abnormal growth, while excessive amounts result in giantism and other abnormalities of rapid growth.

Metamorphosis

While growth is the general pattern of development after birth, there are some organisms, for example, those that undergo metamorphosis, in which considerable differentiation occurs after birth. Development requires a source of energy and materials, and, if sufficient material is present in the egg or is supplied by a parent long enough for the organism to develop to a stage of independent existence, the species does not undergo metamorphosis. For example, bird and reptile eggs have sufficient yolk to enable the embryo to develop to a point where, upon hatching, they can seek food for themselves or be fed by the parent. Mammalian eggs have almost no yolk but the materials for development are provided directly by the mother. Development among these organisms proceeds rather directly from infant to adult. In contrast, a large number of invertebrates begin postembryonic life in a larval form and then undergo changes (metamorphose) resulting in an assumption of adult form.

These larval stages play an important adaptive function in the life histories of the species. Many aquatic animals that are sessile, or permanently attached, such as barnacles, corals, mussels, and sponges, hatch as tiny larval forms. These larvae swim or are carried by currents, thus aiding in the dispersal of the species to new locations. At the end of the larval stage, they attach permanently to a substrate and metamorphose to the adult form. Other aquatic animals such as frogs, crabs, and lobsters hatch in the larval form and feed upon microscopic plankton. After a period of larval development and growth, they metamorphose into the adult form. The process of metamorphosis for these organisms is an adaptation to alternative food sources rather than dispersal of the species. The adult forms of these organisms are carnivorous and consume large prey.

Another adaptive developmental pattern based upon special feeding phases occurs in the metamorphosis of many insects. Most feeding occurs in the larval stage, some adult insects not feeding at all. Where both the larval and adult forms feed, the food they eat is usually different.

In the development of insects that undergo *complete*

metamorphosis, the larval forms undergo a series of molts (see Figure 10-15). After larval development the insects enter the *pupa* stage. In this stage, the larvae enclose themselves in a case, or *cocoon.* The larval tissue is gradually destroyed, some of it being used to provide energy for the differentiation and development of the adult form. Within the larva are small plates or embryonic cells which do not undergo much development. During the pupal stage, the adult insect differentiates and develops from these plates of cells. The resulting adult is a totally different organism from the larva. Most of its cells are newly produced.

Some insects, such as grasshoppers (Figure 18-26), cockroaches, and lice, do not undergo complete metamorphosis. The young of these insects somewhat resemble the adult form but with certain body parts, such as the wings and reproductive organs, poorly developed. In the immature stages, the insects are called *nymphs.* Through a series of molts, a *gradual metamorphosis* takes place, resulting in the adult form. There is no larval or pupal stage.

The metamorphosis of organisms represents a specialized developmental pattern. Considerable study of the metamorphosis of various species has been made. Hormones play an exceedingly important part in this type of postembryonic development. Hormonal control in metamorphosis has been discussed in Chapter 10.

Aging

With the passage of time, mature organisms continue to undergo developmental changes. These changes, resulting in a gradual deterioration of the mature organism and eventual death, are collectively called *aging.* Until relatively recently, the study of aging had been neglected. With the advances of medical science and technology, the human life expectancy has been greatly extended. Because more people are living for a much longer period of time, many scientists have been studying the problems of aging. These studies not only involve humans, but other organisms as well. The results of the studies so far show that the process is complex and the effects show great variability among different organisms. Much is yet to be learned about aging, but there are some well-established conclusions.

One of the major theories receiving considerable attention during the past decade is the Free Radical Theory. Free radicals are a group of bonded atoms with one or more unpaired electrons and are generally quite unstable. These radicals have been found to be produced in many biochemical reactions throughout the body. Since they are unstable, they immediately combine with available molecules, particularly those with double bonds, in oxidation-reduction reactions, causing changes in the normal biochemical pathways. It has been found that free radicals such as —OH and —COOH are produced in the mitochondria in the normal reactions of oxygen combining with hydrogen in the biological oxidation scheme. These radicals readily combine with unsaturated fatty acids and other lipids to produce lipofuscin, the so-called age pigment or "old age spots." These radicals and other free radicals produced throughout the body may cause cumulative oxidative alterations in cell and organelle membranes, collagen, elastin, and chromosomal material.

Among the characteristics of aging to which free radicals may contribute are loss of elasticity in the skin, atherosclerosis from irritated blood vessel walls, senility resulting from free radical reactions in the brain, alterations in chromosomal material, and improper mechanical manipulation as a result of connective tissue be-

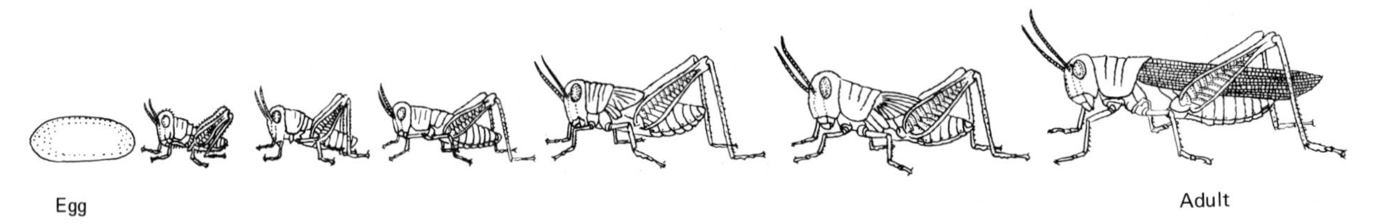

Figure 18-26. Incomplete metamorphosis of a grasshopper. Note that at each succeeding nymphal stage, the insect becomes more like an adult.

coming rigid. While the extent to which free radicals cause the aging process is not fully known, they definitely are a contributing factor.

Some biologists believe that aging is caused by changes in the environment. Particular attention has been placed upon the effects of sublethal doses of radiation (X-rays and cosmic rays). It is known that radiation can cause mutations in sex cells, and there is no reason to believe that the body cells are not affected in a like manner. Other biologists place emphasis upon the stresses of life in general upon the body. The ceaseless wear and tear upon the body tends to alter the body's ability to maintain a physiological steady state. Some support has been found for this theory. Organisms that lead a more active life, such as rats, and mice, which mature soon after birth, age more rapidly than humans who mature more slowly.

Many biologists believe that aging is genetically determined. Support for this theory is found in the characteristic average ages of various species of organisms. While the environment may affect the aging process by slowing it down or speeding it up, it can do so only within the limits set by the hereditary material. Even under the most ideal environmental conditions, organisms age.

A final theory of aging that should be mentioned is one in which the aging process is related to antibodies. This theory, often referred to as the *immunological theory,* purports that with age, individuals become allergic to substances produced by their own cells. The human organism protects itself from antigens (bacteria, viruses, toxins, etc.) by producing antibodies which inactivate or destroy the foreign proteins. Under normal circumstances, antibodies are synthesized only in response to foreign proteins and not to native ones. However, according to the immunological theory, it is believed that this protection deteriorates as the organism ages. There seems to be an increased incidence of many degenerative changes that are related to the production of antibodies against native proteins. The mechanisms operative in such a reaction are poorly understood and their elucidation must await further investigation.

19 Principles of Inheritance

GUIDE QUESTIONS

1 Differentiate between heredity and variation.
2 What is a monohybrid cross? Describe Mendel's monohybrid cross utilizing tall and short pea plants.
3 Define dominant, recessive, homozygous, heterozygous, genotype, phenotype, allele.
4 What is the law of unit characters? the law of segregation? the law of dominance?
5 What is the basis for predicting genetic ratios? Give an example. Define a test cross. When might it be employed?
6 What is a dihybrid cross? What is the law of independent assortment?
7 Define and give an example of each of the following: incomplete dominance, complementary genes, epistasis, collaboration, modifying genes, multiple genes, gene penetrance, expressivity, multiple alleles.
8 How did Sutton provide one of the first clues to the relationship between genes and chromosomes?
9 What was Morgan's contribution to the chromosome theory of heredity? What is a sex chromosome? an autosome? What is meant by the XY method of sex determination? Discuss several other patterns of sex determination.
10 How does the phenomenon of nondisjunction provide proof for the chromosome theory?
11 Distinguish between a sex-influenced and a sex-limited trait. Give an example of each.
12 Compare linkage and crossing-over. What is the significance of each?
13 How are chromosomes mapped? What is a map unit?
14 Define a chromosomal aberration. What is ploidy? euploidy? aneuploidy? Cite several examples.
15 Define and illustrate deletion, inversion, translocation, duplication.
16 Differentiate among cistron, muton, and recon.
17 What is a gene mutation? What is a tautomer? How are tautomers important in gene mutations? Describe the significance of deamination in gene mutation.
18 Define a base analogue. How many analogues contribute to gene mutation?
19 What is the difference between a lethal mutation and a visible mutation? How frequently do genes mutate? List several mutagenic agents.
20 Describe some experiments that indicate that genetic determinants may be located elsewhere than in chromosomes.
21 How does the Himalayan rabbit illustrate the influence of the environment on heredity? How does sunlight influence the expression of genes in plants?

ALTHOUGH the methods and patterns of reproduction, whether sexual or asexual, vary among different types of organisms, the salient outcome is that a new generation of living organisms is produced. Generally, newly produced generations undergo periods of development in which growth and differentiation are the dominant vegetative activities. After this period of development, these organisms are able to reproduce and give rise to future generations.

The Basic Plan of Inheritance

Among the many unique qualities of protoplasm is its ability to produce more protoplasm of the same kind (self-propagation). The external manifestation of this protoplasmic property is the continual tendency of offspring to resemble their parents in all defining features of form and function. The phenomenon by which offspring resemble their parents is called *heredity.* Heredity implies a basic blueprint of inheritance, distinctive for each type of organism, in which future generations will resemble both parents and past generations.

All organisms inherit certain characteristics from their parents, so it is not very difficult to distinguish between fern plants and pinetrees, or birds and mice, for example. At the same time, subtle but recognizable differences exist between even parents and their offspring. These differences are collectively termed *variation.*

The specialized branch of biology concerned with heredity and variation is called *genetics.* Today, most people casually accept the fact that offspring inherit certain characteristics from their parents. However, it was not until 1866 that any data were systematically recorded and published concerning the basic concepts of inheritance. This important contribution to biology, formulated by Gregor Mendel, went unrecognized until 1900. In his document, *Experiments in Plant Hybridization,* the fundamentals of genetics were established.

The Work of Mendel and Its Modern Interpretation

Gregor Johann Mendel (1822–84), an Austrian monk, devised a series of experiments which ultimately uncovered the fundamental principles of inheritance. In his experiments, the garden pea plant (*Pisum sativum*) was used. This proved to be an excellent plant for experimentation for a number of reasons. First, pea plants exhibit a number of distinctive contrasting traits. This fact made it easy for Mendel to study the inheritance of one trait at a time and to maintain careful records of his findings. Moreover, pea plants, although normally self-pollinating, can be cross-pollinated with relative ease. Another reason for the selection of pea plants was that they produce relatively large numbers of seeds.

Monohybrid Cross

One of the first steps in Mendel's experimental procedure was to study the inheritance of a single trait. As an example, height was considered. He crossed (pollinated) different pea plants from a variety averaging about 6 feet tall among themselves for many generations, and only tall plants were produced. On this basis, he concluded that these plants contained some factor for tallness. In a similar manner, he crossed pea plants averaging 1 foot in height and once again concluded that these plants contained a factor for shortness. Each of these two groups of plants was considered to be pure for their respective traits. That is, tall plants contained no factors for shortness, and short plants contained no factors for tallness.

Having developed pure lines of tall and short plants, Mendel next crossed these with each other. This cross, consisting of parents differing in a single trait, is called a *monohybrid cross.* To keep accurate records of the experimental crosses, Mendel designated the generations of peas by specific names. For example, the cross made in the starting generation was called the *P* (*parent*) *generation.* The first generation of offspring to be produced from the P generation was called the F_1, or *first filial* (*daughter*), *generation;* the second generation was referred to as the F_2, or *second filial, generation;* and so on.

In the P generation Mendel crossed pure tall with pure short plants. He found that the F_1 generation contained all tall plants; none were either short or intermediate in height. Next he pollinated the tall pea plants of the F_1 generation and observed that approximately three-fourths of the F_2 generation were tall and one-fourth were short (Figure 19-1).

Based upon these observations, Mendel drew certain conclusions. It was obvious that some factors were present that determined the inheritance of height and that these factors occurred in pairs because mating within the F_1 generation produced both tall and short plants, and, therefore, the F_1 plants must have contained both short and tall factors. Mendel concluded further that one factor (referred to as **dominant**) may mask the expression of the other because when tall and short pea plants of the P generation were crossed, all the F_1 plants appeared tall, but both tall and short plants appeared in the F_2 generation. Thus, the F_1 plants must have carried the factor for shortness, but it was obscured by (was **recessive** to) the factor for tallness. A final conclusion reached by Mendel was that when gametes are formed, the factors separate from each other. Each of these conclusions later became known as Mendel's laws, but, before stating these laws, the results of Mendel's experiments will be analyzed in terms of modern principles of genetics.

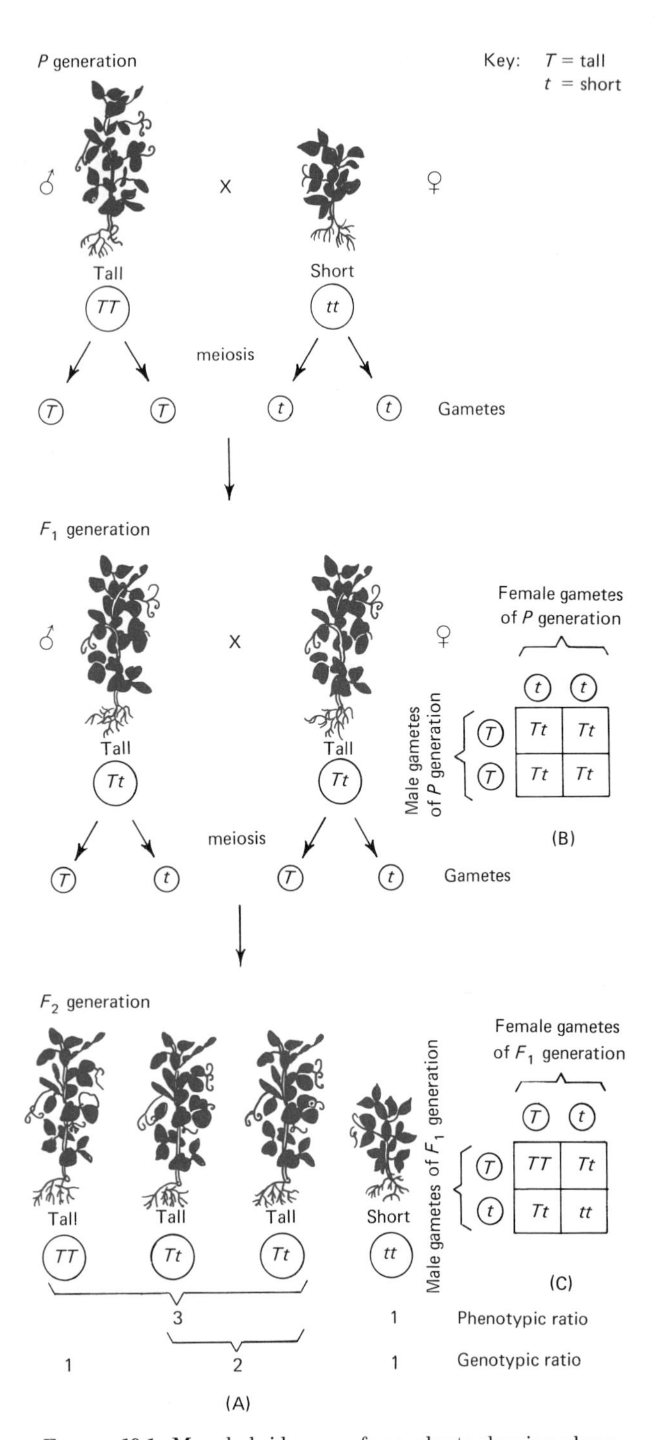

Figure 19-1. Monohybrid cross of pea plants showing phenotypes, products of meiosis, and procedure for employing Punnett square technique. See text for amplification.

In modern genetics, the factors to which Mendel made reference are called **genes** and the genes are usually symbolized by letters. Capital letters are generally used for dominant traits and the same lowercase letters are usually used to designate the recessive traits. Capital letter *T* is the symbol that represents the gene for tallness, and lowercase *t* represents the gene for shortness. To interpret Mendel's results, the tall pea plants for the P generation would be represented by *TT*, because genes come in pairs on homologous chromosomes, and because the tall P generation pea plants are pure or **homozygous** for tallness; there are no genes for shortness. Similar reasoning indicates that the short pea plants of the P generation should be represented by *tt* (Figure 19-1A). Mendel's first law of genetics, called the *law of unit characters,* states that various hereditary characteristics are controlled by genes and that in diploid organisms, these genes occur in pairs.

When homozygous tall pea plants are crossed with homozygous short pea plants in the P generation, the F_1 generation consists entirely of tall pea plants. Remember that the genetic makeup, or **genotype,** for the tall pea plants of the P generation is *TT* and that this represents the diploid state. Similarly, the diploid genotype for the short pea plants of the P generation would be *tt*. For sexual reproduction to occur, both *TT* and *tt* must undergo meiosis, so that haploid gametes will be produced (Figure 19-1A). This separation of genes during meiosis is the basis for Mendel's second law of genetics, the *law of segregation,* which states that a pair of genes located on homologous chromosomes are separated from each other during the formation of haploid products (gametes) of meiosis. The term **allele** is used to describe individual members of a gene pair. In other words, a particular gamete produced subsequent to meiosis contains only one gene of a pair, the other having been passed to another gamete. If this principle is applied to the cross under consideration, the diploid cell *TT* would eventually produce, after meiosis, haploid gametes containing only *T* and the diploid cell *tt* would give rise to haploid gametes containing only *t*.

Having analyzed the genotypes for the P generation plants as well as the constitution of the gametes they produce just prior to fertilization, the next obvious step is to determine how the gametes combine with each other to produce fertilized eggs (zygotes) that will eventually develop into F_1 generation plants. Special charts (*Punnett squares*) resembling checkerboards are used to predict possible gamete combinations (Figure 19-1A, B). Assume that the homozygous tall pea plants of the P generation are male plants (pollen-producing) and the homozygous dwarf plants of the P generation are female (egg producing). Typically, the male gametes are placed on the side of the chart and the female gametes are placed on the top. The four spaces in the chart represent the various possible combinations of male and female gametes; these possible combinations are the genotypes of fertilized eggs in the next generation—in this case, the F_1 generation. The possible gamete combinations are determined by "dropping" the female gamete on the left into the two lower boxes and by "dropping" the female gamete on the right into the two lower spaces; the upper male gamete is then moved across to the two spaces in line with it, and the lower male gamete is moved across to the two spaces in line with it (Figure 19-1A, B).

It should be noted that all F_1 plants are tall, similar in appearance to the tall plants of the P generation. This observed characteristic, that is, the external appearance, is called the **phenotype.** In this regard, then, the phenotypic characteristic of tallness is common to the pea plants of the P generation used as pollen plants and to all the plants in the F_1 generation. Note also that the genotype for the P generation tall plants is homozygous (*TT*), whereas that of the F_1 plants is **heterozygous** (hybrid). In other words, a heterozygous organism contains contrasting genes, or alleles (*Tt*), for the same trait.

Based upon the results of the F_1 generation, it is clear that even though the gene for shortness is inherited, none of the plants express the short trait. This observation is the basis for Mendel's third law, the *law of dominance,* which states that one gene of a pair may mask or inhibit the expression of the other. In this particular instance, the gene for tallness is dominant and that for shortness is recessive. This explains the appearance of all tall pea plants in the F_1 generation (Figure 19-1B).

When these tall plants of the F_1 generation are crossed with each other, the F_2 generation consists of a ratio of three tall to one short plant. This 3:1 ratio is a phenotypic ratio (Figure 19-1C). It does not mean that two plants were crossed in the F_1 generation and three tall and one short plant appeared in the F_2. It implies that many F_1 plants were crossed and they produced, in F_2, a ratio of 3:1. Actually, after many trials, Mendel recorded 787 tall pea plants and 277 short ones in the F_2 generation. These numbers closely approximate a 3:1 ratio. It should also be noted that in the F_2 plants a genotypic ratio also exists (Figure 19-1A). This ratio consists of a 1:2:1 distribution of homozygous tall to heterozygous tall to homozygous short plants. Both the phenotypic and genotypic ratios may be deduced by tracing meiosis in the F_1 plants (Figure 19-1A) and determining the possible gamete combinations by using the checkerboard (Figure 19-1C).

An example of the inheritance of a single trait may also be noted in an animal. The inheritance of coat color in guinea pigs represents such an example. If a homozygous black (*BB*) male guinea pig is mated with a homozygous white (*bb*) female guinea pig, it will be found that all F_1 offspring are heterozygous black (*Bb*). Thus, in guinea pigs, the black trait is dominant to the white trait (Figure 19-2). It will also be found that one-half of the eggs formed by the F_1 generation organisms will carry the gene for black coat (*B*), and one-half will carry the gene for white coat color (*b*). The same situation is true of the sperms. Fusion of the F_1 gametes will produce F_2 organisms in the ratio of one-quarter homozygous black (*BB*), one-half heterozygous black (*Bb*), and one-quarter homozygous white (*bb*), the same ratio Mendel obtained in his monohybrid cross with pea plants. Mendel's principles apply in animals as well as in pea plants.

Probability and Genetics

Calculating or predicting genetic ratios is not unlike calculating the expected results of tossing two coins. Nothing is really changed except that instead of using gametic characters, coins are used. If two coins are flipped simultaneously and the head is considered dominant (of course, it isn't), one of three possible combinations may result. These are head-head (*HH*), or the homozygous dominant trait; head-tail (*HT*), or the heterozygous trait; or tail-tail (*TT*), the homozygous recessive trait. The *HT* combination will occur with twice the frequency of either *HH* or *TT* combinations. Assume that the two coins are flipped simultaneously

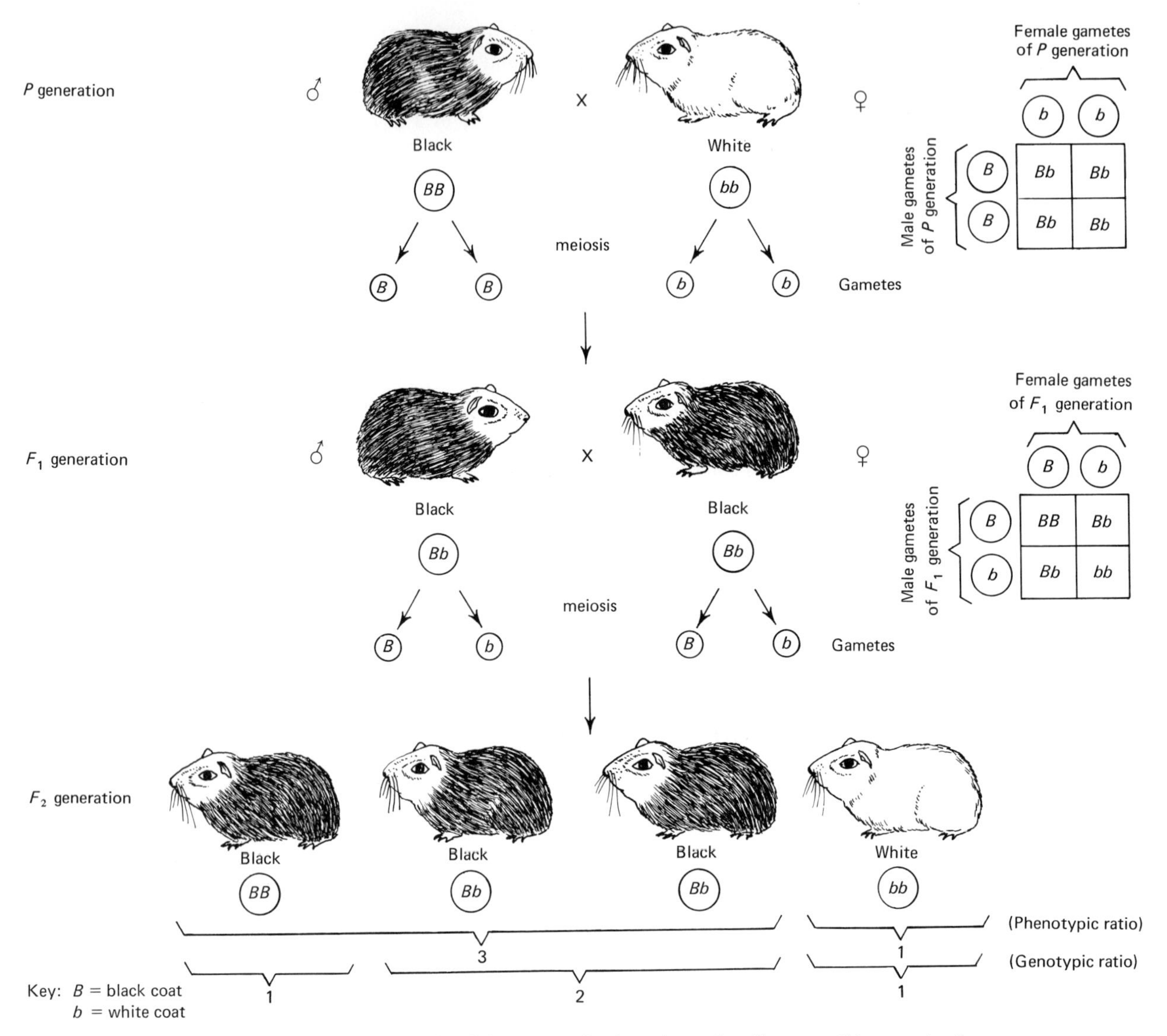

FIGURE 19-2. Monohybrid cross involving coat color in guinea pigs. Compare this cross to the one shown in Figure 19-1.

only four times. Although it is possible to get a 1:2:1 ratio, it is not probable based on only four trials. If the coins are flipped 40 times, however, the probability of getting a 1:2:1 ratio increases. Similarly, 400 or 4000 trials would further increase the probability. The point to be made here is that either actual or expected genetic ratios are most accurate when based on large numbers of trials.

Test Cross

Returning to the cross among pea plants, it will be noted that, phenotypically, the tall plants of the P generation are similar to those of the F_1 generation. Genotypically, however, the P generation plants are homozygous and the F_1 tall plants are heterozygous. These are known facts because the heredity of height in pea plants has just been traced. Suppose that a geneticist or a florist wanted to determine the genotype of a tall plant of unknown parentage. To do this, a **test cross** is employed. Such a cross consists of mating this plant of unknown genotype with a plant exhibiting the homozygous recessive trait. In this case, tall plants of unknown genotype (*TT* or *Tt*) are crossed with short plants (*tt*). Note that the genotype of the recessive plant is always known.

If the test-cross plants produce a ratio of about half tall plants and half short plants, the genotype of the tall plants is *Tt*, since a tall plant with the genotype *TT* would produce all tall plants (Figure 19-3A). Even if only one or a few test-cross offspring plants are short, it can be concluded that the tall parent is heterozygous. If, on the other hand, all the test-cross offspring are tall, after a dozen or so offspring are produced it can be concluded that the parental genotype is *TT* (Figure 19-3B).

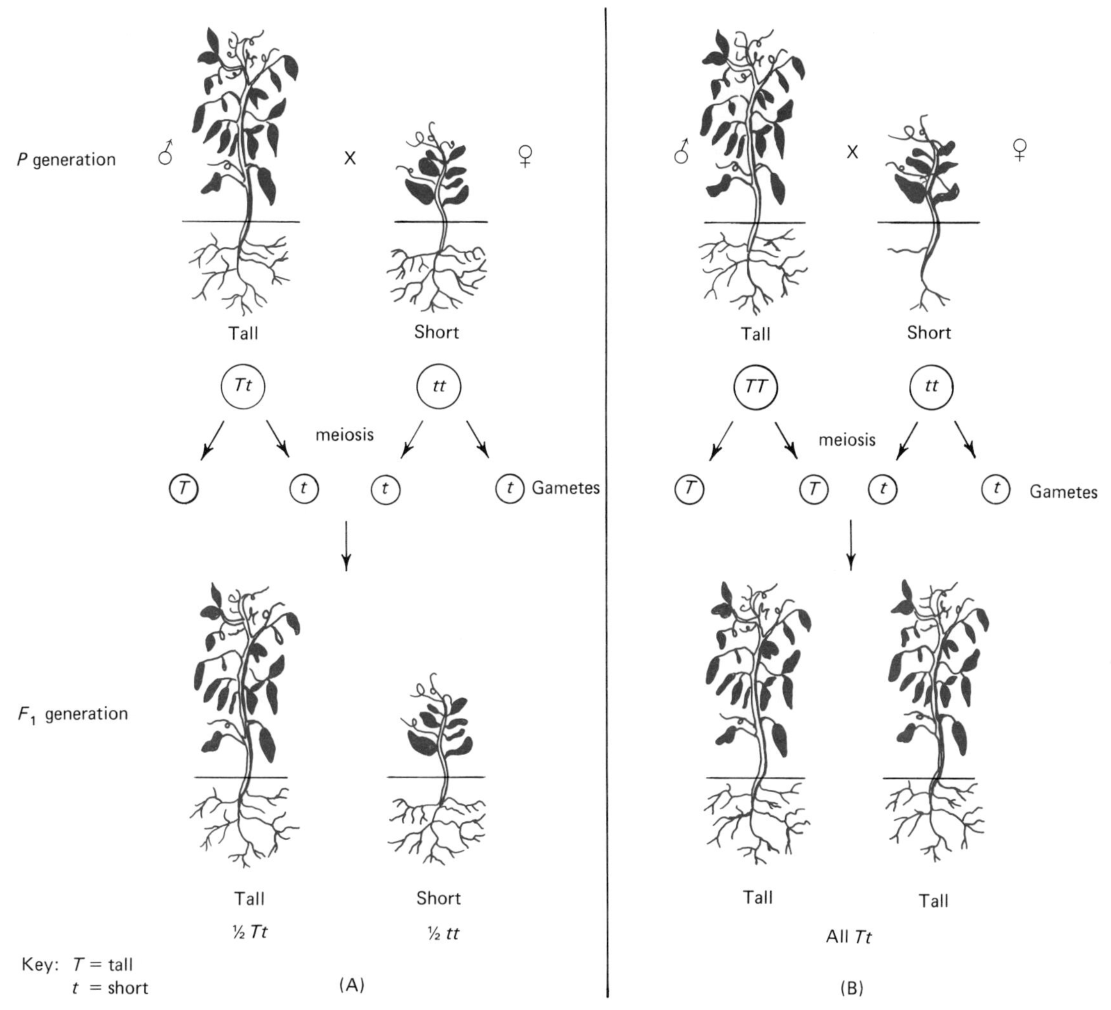

Figure 19-3. Test cross. The purpose of a test cross is to determine the genotype of an organism showing a dominant trait by crossing it with another organism showing the recessive. In pea plants, where tall dominates short, if the unknown genotype is heterozygous (**A**), about 50% of the F_1 generation plants will show the recessive trait. If, however, the unknown genotype is homozygous (**B**), no recessive traits will appear in the F_1 generation plants.

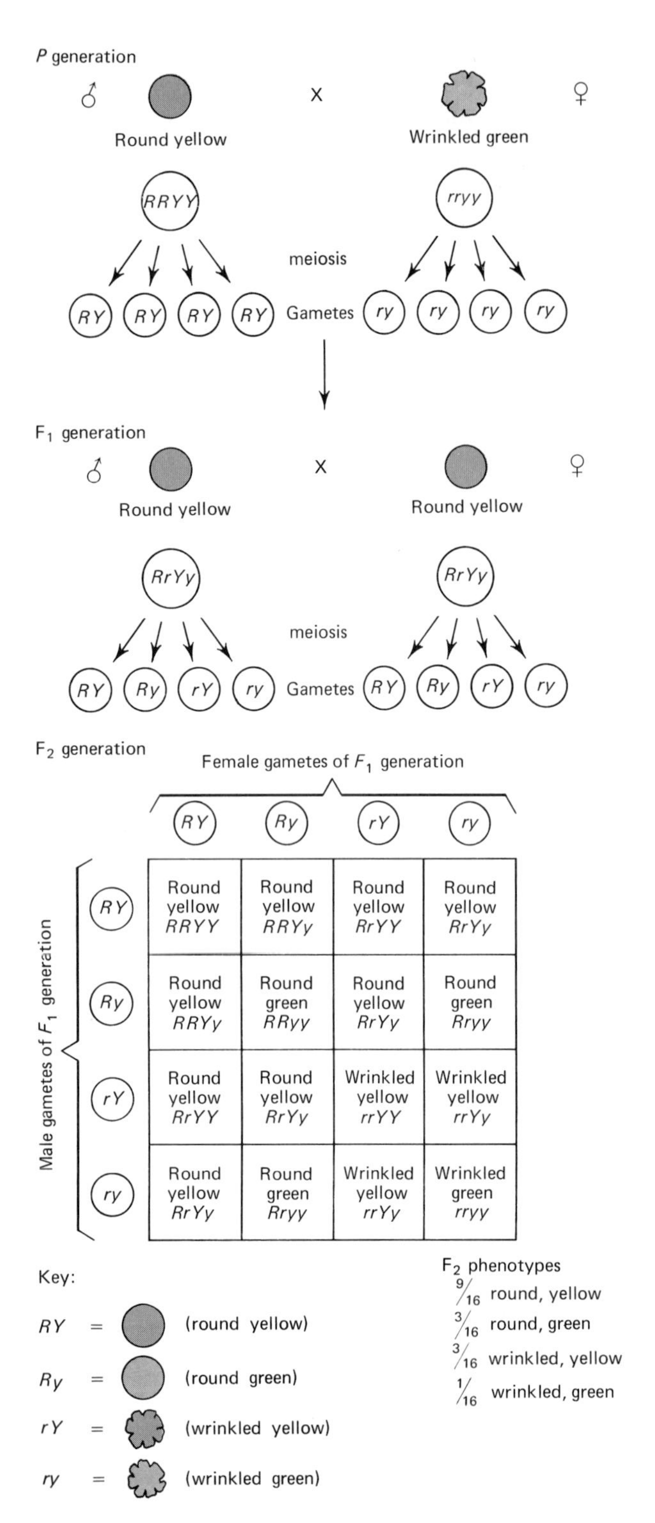

Dihybrid Cross

In addition to studying the inheritance of single traits, Mendel also performed a series of experiments involving the inheritance of two or more characteristics. A cross between individuals differing in two distinct characteristics is called a *dihybrid cross.* One of Mendel's dihybrid crosses consisted of mating plants having round yellow seeds with plants having wrinkled green seeds (Figure 19-4). These parents produced F_1 offspring all of which contained round yellow seeds. As in the monohybrid cross, the F_1 plants are heterozygous for the two dominant traits. When these plants were crossed among themselves, the F_2 generation consisted of four different phenotypes: 315 had round yellow seeds, 101 had wrinkled yellow seeds, 108 had round green seeds, and 32 had wrinkled green seeds. In simple terms, the four different phenotypes occurred in an approximate 9:3:3:1 ratio.

Mendel's dihybrid experiments provided important data not available from monohybrid crosses. For one thing, the results indicated that a dihybrid cross could produce two new types of plants phenotypically unlike either the P or F_1 generation plants. The new types developed either wrinkled yellow or round green seeds. The implication to be drawn from the appearance of new phenotypes is that the genes for seed color and the genes for seed texture do not necessarily stay together in the combinations in which they were present in the P generation. By further implication it can be reasoned that the genes for seed color are located on homologous chromosomes and that the genes for seed texture are on a different chromosome pair. Thus, genes for these two traits segregate independently during meiosis. These observations provide the basis for Mendel's fourth law of heredity, the *law of independent assortment,* which states that the separation (segregation) of allelic genes of gene pairs on one pair of homologous chromosomes during meiosis is entirely independent of the segregation of allelic genes on other pairs of homologous chromosomes.

Figure 19-4 represents the mechanism of dihybrid

FIGURE 19-4. Dihybrid cross. Pea plants with round yellow seeds are pollinated from plants having wrinkled green seeds. All F_1 plants produce round yellow seeds. Upon pollination of the F_1 plants, a 9:3:3:1 ratio results in the F_2 generation.

cross. R represents the allele for round seeds, r for wrinkled seeds, Y for yellow seeds, and y for green seeds. The round yellow parent can produce only gametes RY. The wrinkled green parent can produce only gametes ry. When RY gametes of one parent combine with ry gametes of the other, all the F_1 plants are heterozygous for both dominant traits ($RrYy$) and show the round yellow phenotype of the dominant parent. Each of these F_1 plants can produce four different types of gametes: RY, Ry, rY, and ry. When these four possible gametes for one F_1 plant are combined with the same four possible gametes from another F_1 plant, sixteen zygotic combinations are possible. These possible combinations include nine different genotypes, which determine four different phenotypes in the ratio 9:3:3:1.

Genetics Since Mendel

Mendel's observations on inheritance in the garden pea were subsequently confirmed by numerous investigators and his general conclusions were found to apply in other plants and in animals. However, subsequent studies also showed that certain modifications of Mendel's laws were necessary.

Incomplete Dominance (Intermediate Inheritance)

Genetic studies made after Mendel's work have clearly demonstrated that while dominance and recessiveness occur in many cases, they are rarely, if ever, absolute. Two common instances among plants in which the F_1 offspring exhibit an intermediate characteristic as compared with parental types are found in flower color in four-o'clock and snapdragon. This type of inheritance in which offspring are phenotypically intermediate between parental phenotypes is called *incomplete dominance.* When homozygous red-flowered (RR) four-o'clocks are crossed with homozygous white (rr) four-o'clocks, all the F_1 flowers are pink (Rr) (Figure 19-5). The pink phenotype is intermediate between red and white and indicates a lack of dominance for both red and white alleles. Subsequent crossing of the pink-flowered four-o'clocks produces the typical monohybrid genotypic ratio of 1 homozygous red to 2 heterozygous pink to 1 homozygous white.

In addition to the example of intermediate inheri-

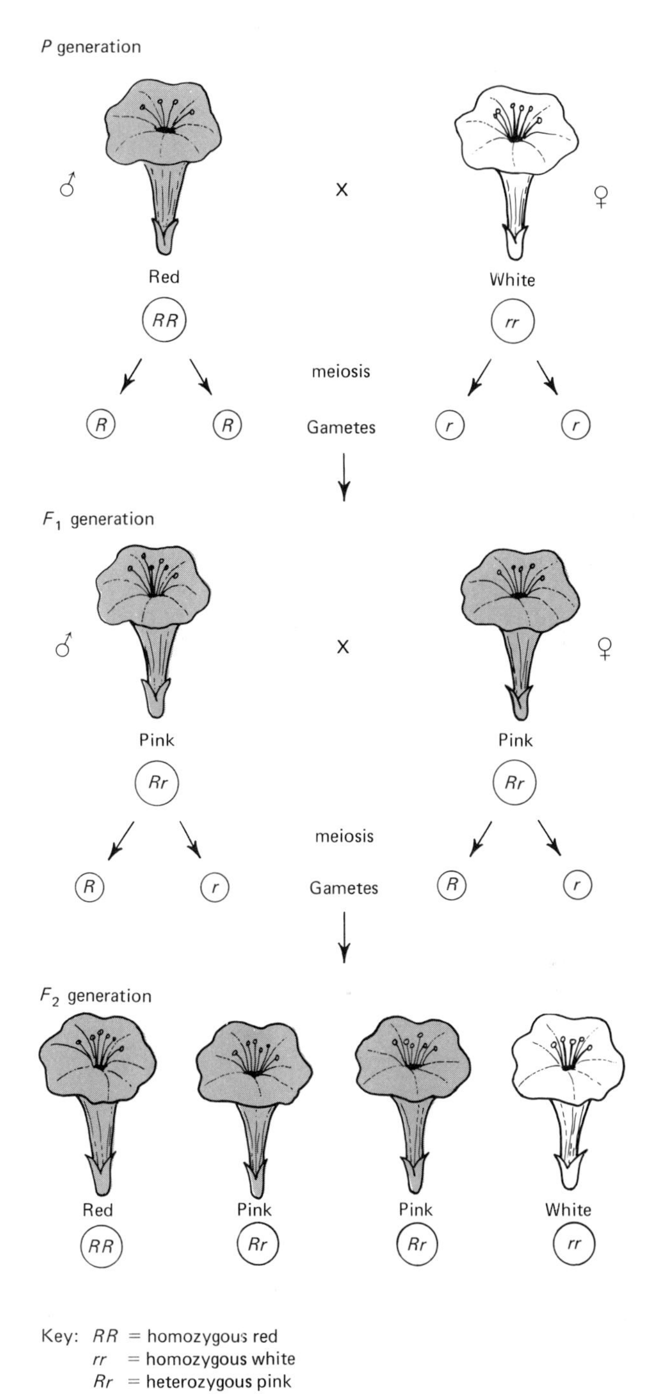

Figure 19-5. Incomplete dominance (blending inheritance) in four-o'clock flowers.

tance just noted in which the heterozygous phenotype is obviously an intermediate between the homozygous phenotypes, other cases are not quite so obvious. In other words, the heterozygous expression of two different genes is somewhat different from either parent. For example, if a black chicken of a certain strain is crossed with a splashed white chicken (white with black splashes), the offspring are "blue." The "blue" chickens in this case form a special breed of domestic fowl, the Andalusian chickens. When crossed together, these "blue" chickens yield black, "blue," and splashed white in the ratio of 1:2:1. Thus, the two parental alleles (black and splashed white) interact to give rise to a phenotype ("blue") somewhat different from the gray that might be expected.

In patterns of inheritance in which incomplete dominance is operative, it can be noted that the F_1 offspring of a monohybrid cross between parents, each of which is homozygous for a different allele, have a phenotype somewhat different from both parents. Moreover, note that the F_2 phenotypic ratio resulting from hereditary patterns involving intermediate inheritance is 1:2:1 and not 3:1.

Interactions of Genes

It is now recognized by geneticists that many inherited traits are controlled by more than one gene. Such an inheritance is called a *plistropic interaction.* Indeed, it is reasonable to assume that many genes of an organism exert some influence on the development of every hereditary trait, with one, two, or several genes having a predominant effect. In this regard, these minor genes may assist the major genes to express a trait through various catalytic or inhibitory biochemical reactions. Generally, when several genes operate together to determine a trait, the trait will not appear if any one of the genes is missing. If several gene pairs are involved in the expression of a single characteristic, and if no dominance exists between the alleles, their effect will be cumulative. When two or more different genes interact with each other in the expression of a single trait, the calculated ratios are sometimes different from the basic Mendelian ones. A few examples of gene interaction will now be considered.

Complementary Genes. One of the commonest types of gene interactions involves two or more

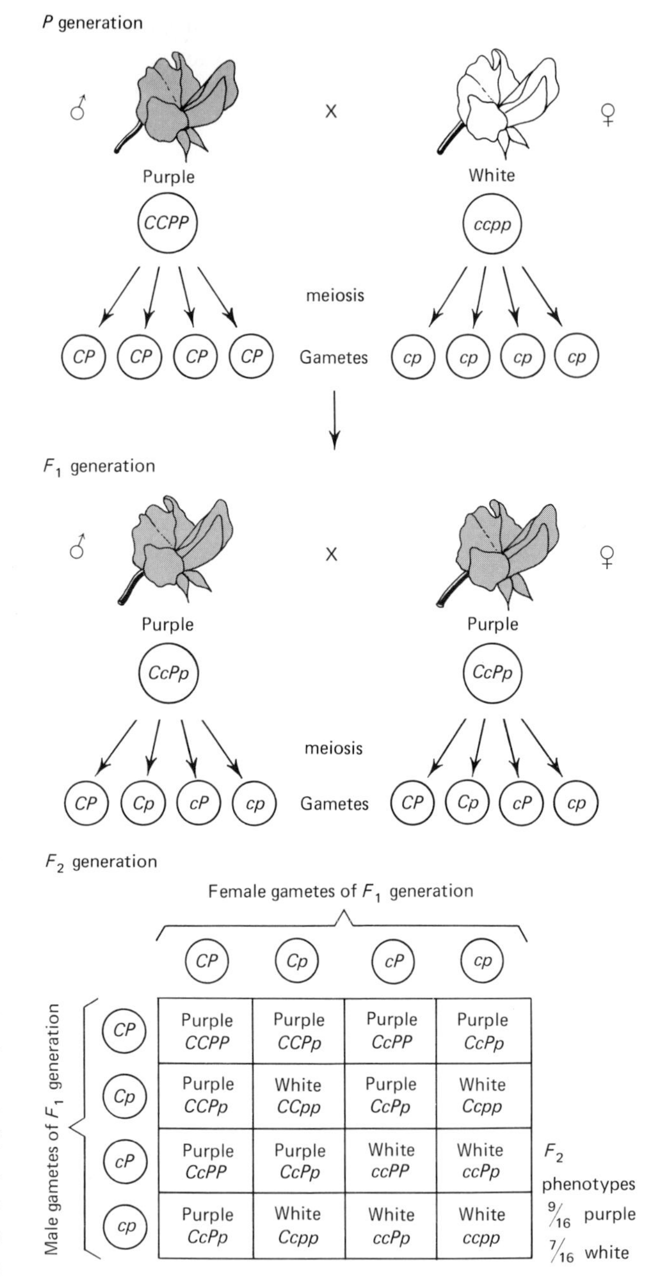

Figure 19-6. The operation of complementary genes in flower color of sweet peas. Note that the two complementary genes, *C* and *P*, must be present in order for the dominant condition (purple flowers) to be expressed. In the absence of either of the two dominant alleles, the recessive condition (white flowers) is expressed. Compare this cross to the one shown in Figure 19-3.

independent pairs of genes called **complementary genes.** These may be defined as two or more dominant genes that interact to produce a phenotype although neither gene produces a phenotypic effect itself. In order for a single given dominant trait to be expressed, all complementary genes must be represented by at least one of the dominant alleles. The alternate phenotypic character is produced by the homozgous recessive condition for any or all of the complementary genes.

An example of the operation of complementary genes may be seen in flower color in sweet peas (Figure 19-6). In these flowers, the two complementary genes, *C* and *P*, must be present in the dominant condition for pigment (purple) formation to occur. In the absence of either dominant allele, the flowers are white. Thus, one white variety of sweet peas may have the genotype *CCpp* (it is white because *P* is lacking), while another white variety may have the genotype *ccPP* because the *C* is lacking. The flowers containing the recessive genotype *ccpp* would also be white.

In the results shown in Figure 19-6 note that the four groups of genotypes in the F_2 generation are the ones that in a normal dihybrid cross would produce four different phenotypes in the ratio of 9:3:3:1. But in this case, they have produced only two phenotypes in the ratio of 9:7. In sweet peas, gene interaction, in which neither dominant allele can exert its effect unless the other dominant is present, has been explained experimentally. The pigment in colored flowers is anthocyanin. This complex compound is synthesized in a chain reaction, each link of the chain being catalyzed by a specific enzyme. If any of the enzymes is inactive, anthocyanin is not produced. Experimental data indicate that *C* controls the production of an enzyme that catalyzes the formation of a necessary raw material for the anthocyanin pigment, and that *P* controls the production of an enzyme that catalyzes the transformation of this raw material into anthocyanin. In this regard, it is not sufficient for the plant to possess the *C* allele and produce the raw material if it has no *P* allele to convert the raw material into purple pigment. Similarly, it is not sufficient for the plant to contain the *P* allele if it has no *C* and thus produces no raw material. In both cases, lack of one dominant allele results in the production of white flowers since no anthocyanin is synthesized.

Epistasis. Whenever a gene at one locus (the

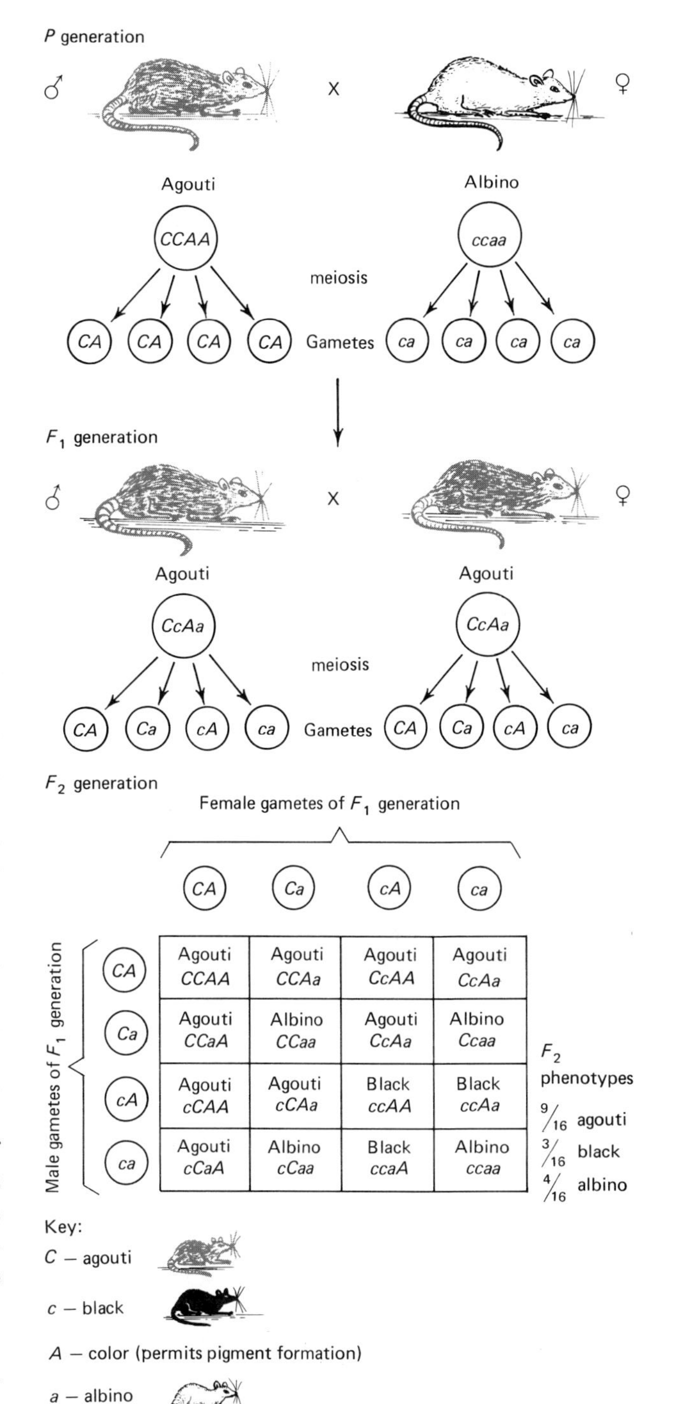

	CA	Ca	cA	ca
CA	Agouti CCAA	Agouti CCAa	Agouti CcAA	Agouti CcAa
Ca	Agouti CCaA	Albino CCaa	Agouti CcAa	Albino Ccaa
cA	Agouti cCAA	Agouti cCAa	Black ccAA	Black ccAa
ca	Agouti cCaA	Albino cCaa	Black ccaA	Albino ccaa

Figure 19-7. Epistasis in coat color in mice. Consult text for amplification.

position on a chromosome occupied by a gene or any of its alleles) of a chromosome influences the phenotypic expression of a gene at another locus, the first gene is said to be **epistatic** to the second. Epistasis may be illustrated by analyzing coat color in mice (Figure 19-7). The typical wild mouse has a coat color called *agouti,* a peculiar shade of grey. Some mice, however, have black coats when homozygous for a recessive gene which causes the hair pigment to be black rather than agouti. Mice, like most other animals, also carry a few genes for albinism, which result in a white coat when the genes are homozygous. Albino is not an allele for agouti and black. Rather, when homozygous, this gene for albinism prevents the formation of a pigment of any kind, regardless of the genes for agouti or black which the mouse may carry. This is because the dominant allele of the albino gene seems to produce an enzyme which allows the formation of a pigment, and, in the absence of this enzyme, no pigments are synthesized. If a homozygous agouti male carrying the two dominant alleles of the gene for albinism, *CCAA,* is crossed with an albino female carrying two genes for black, *ccaa,* all offspring are agouti (*CcAa*). When these are crossed among themselves, however, agouti, black, and albino mice are obtained in the phenotypic ratio of 9:3:4 in the F_2 generation. This unusual ratio results from the fact that all mice homozygous for the gene for albinism are albinos, and the mice which carry the agouti genes and the black genes are heterozygous for the albino genes and cannot be distinguished. It should be made clear that although dominance and epistasis are superficially alike, dominance is the phenotypic expression of one member of a pair of alleles at the expense of the other, whereas epistasis is the masking by one gene of the phenotypic effect of another entirely different gene. In this particular example, gene *a* (which is recessive to its own allele, *A*) actually masked the effect of either *C*– or *cc,* so that any *––aa* individual is albino. Essentially, dominance results from the interaction between alleles and epistasis results from an interaction between nonallelic genes.

Collaboration. A third type of gene interaction is referred to as **collaboration,** a phenomenon by which two different genes influencing the same trait interact so as to produce single trait phenotypes that neither gene by itself could produce. Such an example is

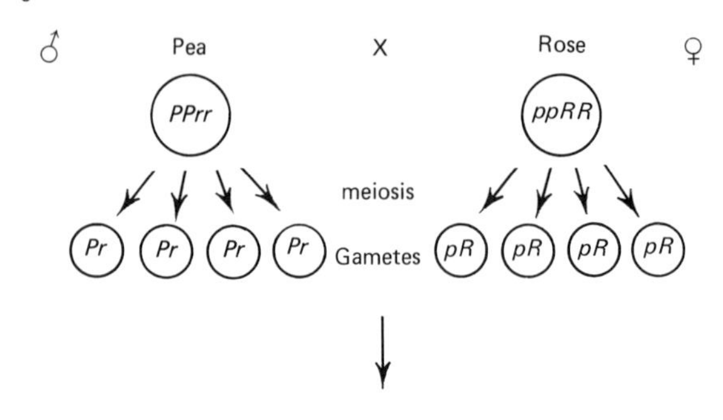

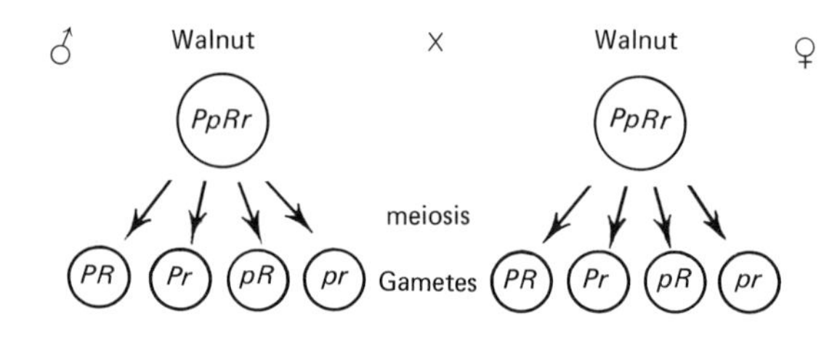

F_2 generation

Female gametes of F_1 generation

Male gametes of F_1 generation	*PR*	*Pr*	*pR*	*pr*
PR	Walnut *PPRR*	Walnut *PPRr*	Walnut *PpRR*	Walnut *PpRr*
Pr	Walnut *PPRr*	Pea *PPrr*	Walnut *PpRr*	Pea *Pprr*
pR	Walnut *PpRr*	Walnut *PpRr*	Rose *ppRR*	Rose *ppRr*
pr	Walnut *PpRr*	Pea *Pprr*	Rose *ppRr*	Single *pprr*

F_2 phenotypes
- 9/16 Walnut
- 3/16 Pea
- 3/16 Rose
- 1/16 Single

Key:

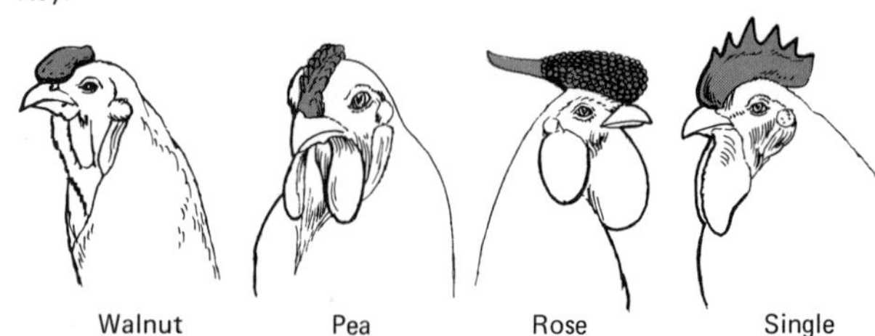

Figure 19-8. Collaboration illustrated by the inheritance of combs in poultry. See text for details.

illustrated by the inheritance of combs in poultry (Figure 19-8). The gene for *rose* comb, *R*, is dominant to that for *single comb, r*. Another gene governs the inheritance of *pea comb, P*, while its recessive allele, *p*, also produces the single comb. A single-combed fowl must have the genotype *pprr*, a pea-combed fowl is either *PPrr* or *Pprr*, and a rose-combed fowl is either *ppRR* or *ppRr*. When a homozygous pea-combed fowl (*PPrr*) is mated to a homozygous rose-combed one (*ppRR*), the F_1 offspring have neither pea nor rose comb but a completely different type called a *walnut* (*PpRr*) *comb*. The phenotype of walnut comb is produced whenever a fowl has one or two *R* genes plus one or two *P* genes. The mating of walnut-combed fowls gives rise to F_2 offspring in the ratio of 9 walnut (*R–P–*), 3 rose (*R–pp*), 3 pea (*rrP–*), and 1 single-combed fowl (*rrpp*).

Multiple Genes. Up to this point, consideration has been given to hereditary traits that depend mainly on the action of single genes, although, as has been noted, modifying genes may affect the phenotype produced by the main gene. We will now consider traits determined by the cumulative or additive action of several genes. The term *polygenic inheritance* or *multiple factor inheritance* is employed when two or more independent pairs of genes affect the same characteristic in the same way and in an additive fashion. Genes having such action are referred to as *polygenes*, or *multiple genes*, and the traits they determine are of a quantitative nature, that is, they control characteristics such as height, weight, breadth, degree of pigmentation, and so on.

As an example of multiple gene inheritance, consider two varieties of a certain hypothetical plant, one tall and one dwarf (Figure 19-9). The tall variety has an

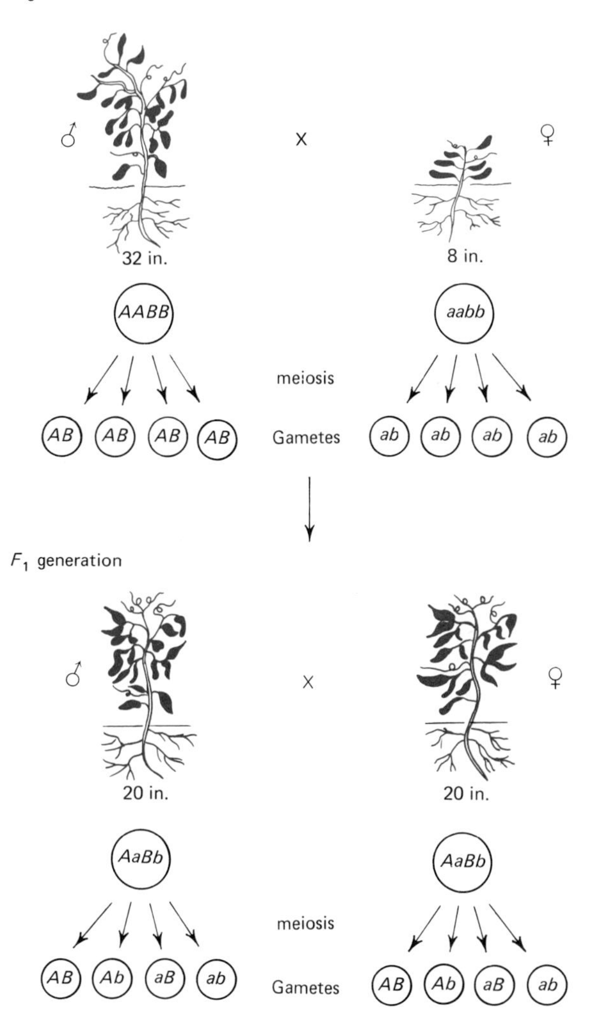

F_2 generation

Female gametes of F_1 generation

Male gametes of F_1 generation	*AB*	*Ab*	*aB*	*ab*
AB	32 in. *AABB*	26 in. *AABb*	26 in. *AaBB*	20 in. *AaBb*
Ab	26 in. *AABb*	20 in. *AAbb*	20 in. *AaBb*	14 in. *Aabb*
aB	26 in. *AaBB*	20 in. *AaBb*	20 in. *aaBB*	14 in. *aaBb*
ab	20 in. *AaBb*	14 in. *Aabb*	14 in. *aaBb*	8 in. *aabb*

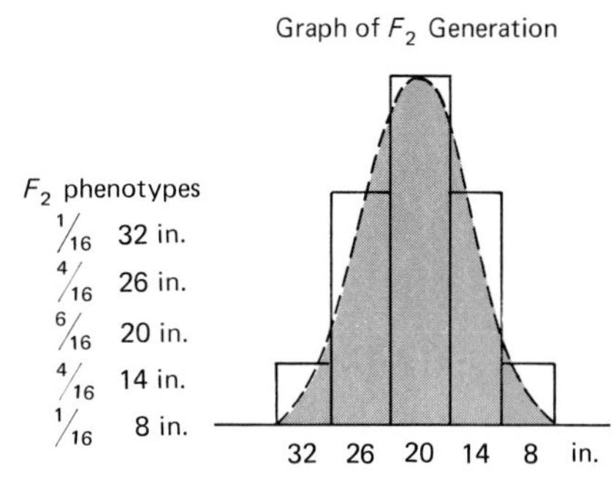

Figure 19-9. Multiple gene inheritance in two varieties of a hypothetical plant. At lower right is a graphic representation of the relative number of individuals having each of the heights indicated at the bottom of the graph. See text for discussion.

average height of 32 inches, whereas the dwarf variety averages 8 inches in height. Upon crossing these two varieties, it is found that the F_1 hybrids have an average height of 20 inches. Subsequent crossing of the F_1 plants gives rise to offspring showing great variability, ranging in height all the way from 8 inches to 32 inches, but with an average of 20 inches. In fact, the heights of the F_2 offspring show an approximation to a normal distribution; graphically, such an approximation would be represented by a bell-shaped normal frequency curve.

The genetic basis of phenotypes such as these can be explained by assuming that two or more separate genes affect the same character in the same way in an additive fashion. In other words, it can be assumed that the quantitative difference (height) is dependent upon two or more pairs of genes having cumulative effects. If the tall variety has the genotype *AABB* and the dwarf plants are genotypically represented as *aabb,* then the *aabb* genotype produces 8 inches of height and any dominant alleles present add to this. Consider that gene *A* and gene *B* have the same effect in increasing the height of the plants above the basic 8 inches. The difference in height between the two varieties is 24 inches. There are four uppercase genes in the genotype of the tall variety. Thus, each allele produces a height increase of 6 inches. A genotype of *aabb* gives a height of 8 inches and substituting an allele for the corresponding lowercase gene increases the height by 6 inches. Thus, a plant with the genotype *Aabb* is 14 (8 + 6) inches high, a plant with the genotype *AAbb* is 20 (8 + 12) inches high, and so on, to the limiting genotype, *AABB,* which produces a plant 32 (8 + 24) inches high. Although for the sake of simplicity, the contributions of the genes have been indicated in absolute units (inches), frequently the contributions may be a percentage increase, with each allele causing the height to increase by a certain percentage. The results of such a cross among hypothetical plants are shown in Figure 19-9. Note the approximation to a normal frequency curve of the graph of F_2 phenotypes in the figure.

Thus, this type of inheritance may be readily explained by assuming the existence of pairs of genes having an additive effect. Moreover, more than two pairs of genes may be involved in a quantitative difference. Three pairs might be involved in the case just cited. In such a case, the 32-inch-high variety would have the genotype *AABBCC,* whereas the 8-inch-high variety would have the genotype *aabbcc*. This means that each allele would produce an increase of 4 inches ($\frac{1}{6} \times 24$) over the basic eight. The F_1 plants would have the genotype *AaBbCc* and would give rise to eight kinds of gametes. Calculation of the possible combinations would show that the F_2 offspring occur as follows: $\frac{1}{64}$, 32 inches; $\frac{6}{64}$, 28 inches; $\frac{15}{64}$, 24 inches; $\frac{20}{64}$, 20 inches; $\frac{15}{64}$, 16 inches; $\frac{6}{64}$, 12 inches; and $\frac{1}{64}$, 8 inches. Note, in this case, that there are seven phenotypes instead of the five present when only two pairs of genes are involved. These F_2 results even more closely approximate a normal frequency curve.

GENE PENETRANCE AND EXPRESSIVITY. According to the principles of inheritance as illustrated by Mendelian genetics, it was shown that when an organism has a dominant gene, the phenotypic effect of that gene is expressed, and when an organism has a recessive gene, the phenotype associated with it is expressed only when the homozygous condition exists. In other words, Mendelian genetics assumes that genes have complete penetrance, that is, genes always produce their expected phenotype when present in the proper condition. It was shown also that when two or more genes interact and the phenomenon of epistasis and complementarity are operative, the expression of one gene is dependent on some other one. As such, it is possible for an organism to carry a dominant gene and yet not express it phenotypically. Moreover, it is possible for modifying genes to suppress the expression of a given dominant gene, and the expression of such a gene may have many degrees of intensity. In considering these facts, then, gene action may be spoken of as having partial penetrance and variable expressivity.

Penetrance is defined as a condition in which a dominant gene produces the expected phenotype in every organism possessing the gene or a recessive gene produces the expected phenotype in every individual homozygous for that gene. The degree of penetrance may be expressed as the percentage of individuals that actually expresses the phenotype as determined by a given genotype. **Expressivity,** by contrast, is related to the manner in which the phenotype is expressed. It is a situation in which individuals with the same genotype show different phenotypes.

These two phenomena, penetrance and expressivity,

may be illustrated by considering the gene in humans which is responsible for *blue sclera* (the whites of the eyes appear blue). The sclerotic coat of the human eye is normally white and the part of it that is seen at the front is sometimes referred to as the white of the eye. A certain dominant gene can cause the coat to be blue. This gene usually behaves as a simple dominant and those individuals who possess the gene (either homozygous or heterozygous) should express the blue sclera phenotype. Experimental data, however, indicate that about 10% of the people who carry this gene do not express the blue sclera but have a normal white sclera. Thus, the penetrance of this gene is about 90%. These individuals can transmit the gene to their children and the offspring may express this dominant gene. The gene for blue sclera also illustrates variation in expressivity. Among the 90% of those who carry the gene and express it, there is a considerable variation in the shade of blue of the sclera. In some cases, it is such a pale blue that it is difficult to distinguish from white, but in others, the blue may be so dark that the sclera appears almost black. Logically, both penetrance and expressivity are aspects of the same phenomenon. Lack of penetrance of the gene produces white color, which may be regarded as one extreme of the expressivity gradient from white through pale to dark blue.

MULTIPLE ALLELES. Our discussion of the principles of genetics thus far has considered only those genes that exist in pairs. These genes may exhibit a dominant–recessive relationship or an intermediate one. At many gene loci more than two alleles occur. **Multiple alleles** are defined as three or more alternative conditions of a single locus which produce different phenotypes. Among the members of a population, any given individual may possess any two of the genes but never more than two. In a similar context, any gamete can have only one. But in the population as a whole, there may be distributed three or more alleles.

One of the well-known examples of multiple allelic inheritance is that of human blood types. Human blood types, designated as A, B, AB, and O, are under genetic control and it is known that there are three main alleles controlling the inheritance. The type of blood an individual has depends on the presence or absence of specific substances (antigens) on the red blood cells. These antigens are designated as A and B. A person with antigen A is considered to be type A; one with antigen B, type B. An individual with both antigens A and B has type AB blood, and a person with neither A nor B has type O blood (see Chapter 14).

The idea of multiple alleles is that only one gene location (locus) in only one pair of chromosomes is involved. This locus may be occupied by any one of several allelic forms of the same gene. These multiple forms of the same gene are said to constitute a series of multiple alleles. In considering the inheritance of blood groups, there is a series of three alleles. These alleles are designated as I^{A}, I^{B}, and i. Although the three alleles may be paired in any combination, only two of them can be present in any one individual. Both I^{A} and I^{B} are fully expressed in the presence of each other; that is, they produce antigens A and B, respectively. By contrast, both I^{A} and I^{B} are dominant to i, and it is only when both I^{A} and I^{B} are absent (genotype ii) that an individual has type O blood. The relationship of the antigens to phenotypes and genotypes can be listed as follows:

Phenotype	Genotype	Antigen on Cells
Type A blood	$I^{A}I^{A}$ or $I^{A}i$	A
Type B blood	$I^{B}I^{B}$ or $I^{B}i$	B
Type AB blood	$I^{B}I^{B}$	A and B
Type O blood	ii	None

The Chromosome Theory of Heredity

Genes and Chromosomes: Sutton's Hypothesis

Mendel's experiments and the accurate results he derived from them were quite remarkable considering the status of biological knowledge back in 1866. In 1876, the first clear demonstration was made of the fusion of the egg and the sperm nucleus during fertilization. Shortly afterward, it was realized that the union of gamete nuclei accounts for the equal contribution of both parents in the transmission of inherited characters to the offspring. The beginnings of knowledge concerning chromosomes, mitosis, and meiosis took place during the 1880s. It was not until the early part of the twentieth century, however, that evidence finally accumulated to indicate conclusively that the hereditary substances of cells is located on the chromosomes.

Walter S. Sutton, a graduate student at Columbia

University, provided one of the first clues to the solution of the problem dealing with the relationship between genes and chromosomes in 1902. Initially, he reasoned that the biological link between generations of multicellular organisms is their gametes and that genes must be located somewhere within the sperm and egg cells. Next, he reasoned that the egg and sperm make precisely the same genetic contribution to the organism. Evidence of this was supplied by Mendel by making reciprocal crosses of pea plants in which first the male and then the female plant carried the dominant trait he was studying. These crosses produced exactly the same kind of offspring; it made no difference which parent expressed the dominant trait.

Assuming that the genetic contribution of both sperm and egg is essentially similar, some structure common to both gametes must be responsible for inheritance. Considering that the nucleus of the egg is very similar to the nucleus of the sperm and that the cytoplasm of the egg is very different from that of the sperm, Sutton concluded that the nuclei of the gametes probably contain the genes. Experiments indicating that the nucleus carries within it the information for the development of a new individual have been discussed in Chapter 18. Within the nucleus lie the chromosomes; various data indicated that chromosomes behave much like Mendel's "factors" of heredity. For example, Sutton observed that a gamete has only one of each kind of chromosome (one member of each homologous pair). This corresponds to the Mendelian requirement that only one of each pair of genetic factors be present in each gamete. Second, the union of sperm and egg, each with a single set of chromosomes, re-establishes the diploid chromosome number of the organism. This is similar to the requirement that genes be contributed equally by each parent. Third, each pair of homologous chromosomes separates independently of all other pairs during meiosis. The four possible combinations of two pairs of chromosomes correspond to what Mendel had found to hold true for different pairs of traits when they were followed together (yellow and green seed color, round and wrinkled seed texture).

Careful consideration of these parallelisms between genes (Mendel's factors) and chromosomes led Sutton to propose the hypothesis that genes are located on chromosomes. In other words, one allele of a pair of genes is located on each member of a homologous pair of chromosomes.

Discovery of Sex Chromosomes and Sex Linkage

More evidence for the chromosome theory of heredity was provided by experiments using the fruit fly, *Drosophila melanogaster.* Thomas Hunt Morgan, in about 1910, raised thousands of fruit flies and found that one fly had white eyes instead of the normally red eyes. The white-eyed fly was a male and it was mated with a homozygous red-eyed female. The results of this cross showed that the F_1 generation contained all red-eyed organisms (Figure 19-10). According to Mendelian genetics, this meant that the allele for white eyes is recessive and the allele for red eyes is dominant. The F_1 flies were crossed and a ratio of 3:1 (red to white eyes) was obtained in the F_2 generation. These data also agreed with Mendel's F_2 results for single trait inheritance until it was noticed that all the white-eyed flies were males. None of the female flies had white eyes; all had red eyes as did about half of the males. Note that although there is a 3:1 Mendelian ratio of red to white eyes, only the males had white eyes.

Careful analysis of the chromosomes in the cells of *Drosophila* indicated that there is a difference between those of males and females (Figure 19-11). Of the four pairs of chromosomes in each cell, three are identical; one pair is different. The chromosomes in which members of the pair are similar are referred to as **autosomes.** The chromosomes in the pair with unlike membranes in the two sexes are called **sex chromosomes.** The straight rod-shaped chromosomes of this pair in the female are designated as X chromosomes (Figure 19-11A). Only one of these occurs in the male, along with a hook-shaped chromosome (found in the male only), referred to as the Y chromosome (Figure 19-11B). Thus, the male of *Drosophila* has three pairs of autosomes and one X sex chromosome and one Y sex chromosome. The female, by contrast, has three pairs of autosomes and two X sex chromosomes.

Morgan's observations of the chromosomes of *Drosophila,* correlated with the mechanism of meiosis, led him to conclude that males produce two different kinds of sperm. Half would carry an X sex chromosome and one chromosome from each pair of autosomes. The

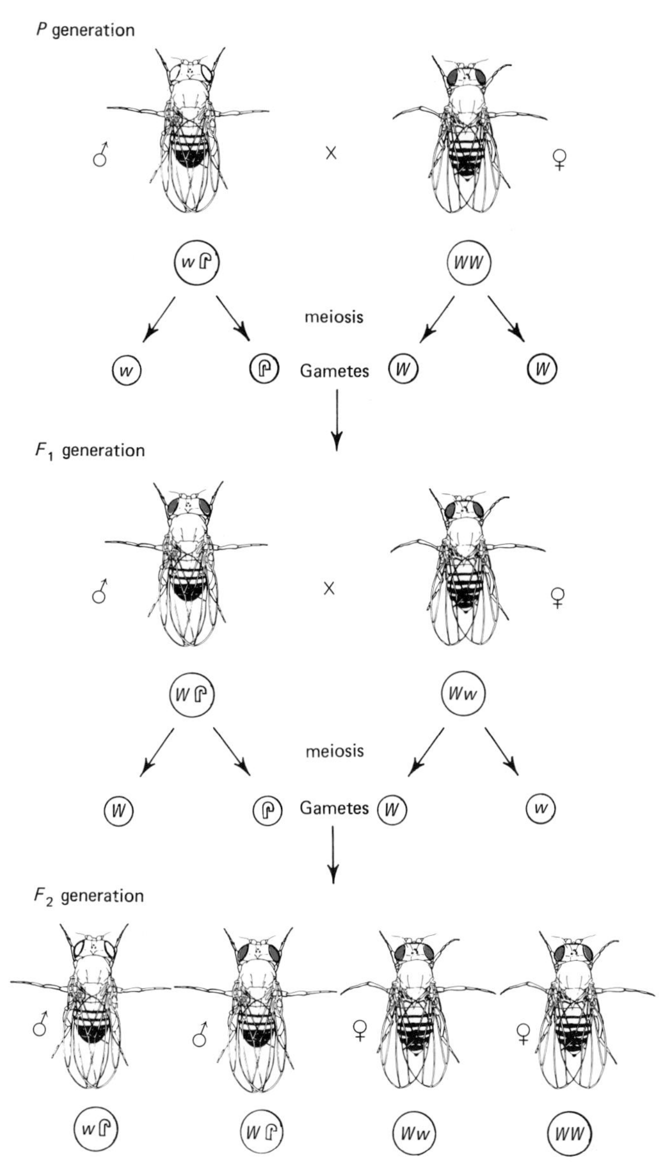

Figure 19-10. Inheritance of white eyes in *Drosophila.* The symbol ℙ designates the Y chromosome of the male; it is used in place of a letter since it makes no genetic contribution to the characteristic under consideration.

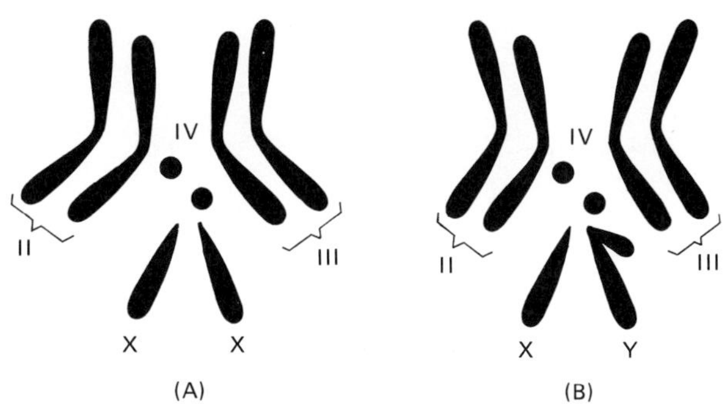

Figure 19-11. Autosomes and sex chromosomes of a diploid cell of *Drosophila.* (**A**) Female. Autosomes (II, III, IV) and sex chromosomes (XX). (**B**) Male. Autosomes (II, III, IV) and sex chromosomes (XY).

other half would carry the Y sex chromosome and the remaining chromosomes from each pair of autosomes. He also deduced that females produce only one kind of egg. Every egg would carry one X sex chromosome plus one chromosome from each of the pairs of autosomes.

The discovery of sex chromosomes in *Drosophila* suggested how the sex of a new individual is determined. Figure 19-12 shows that the inheritance of X and Y chromosomes determines the sex of an offspring. Since all the egg cells contain one X chromosome, if a sperm with an X chromosome enters an egg, the result is XX, or a female. If a sperm with a Y chromosome enters an egg, the result is XY, or a male. In recent years, it has been possible to study the chromosomes of cells of many animals and the same method of sex determination was found to hold true in many other organisms. In human cells, for example, there are 23 pairs of chromosomes in all. Of these, the male has 22 pairs of autosomes and one X and one Y chromosome, and the female has 22 pairs of autosomes and two X chromosomes.

Discovery of the Y chromosome in the male fruit fly cast some doubt as to whether it had an allele to match each gene on the X chromosome. Basically, Morgan explained the inheritance of white eyes in *Drosophila* by assuming that the gene for white eyes was carried on the X chromosome and that there was no corresponding allele on the Y chromosome; the Y chromosome assumed no role in the inheritance of white eyes. He also assumed that some of the genes on the X chromosome had no corresponding alleles on the Y chromosome, so that any trait carried by the X chromosome (including a recessive one) would appear in male flies unless there was a dominating allele on the Y chromosome.

In order to test these hypotheses, Morgan mated a red-eyed male with a white-eyed female. As shown in

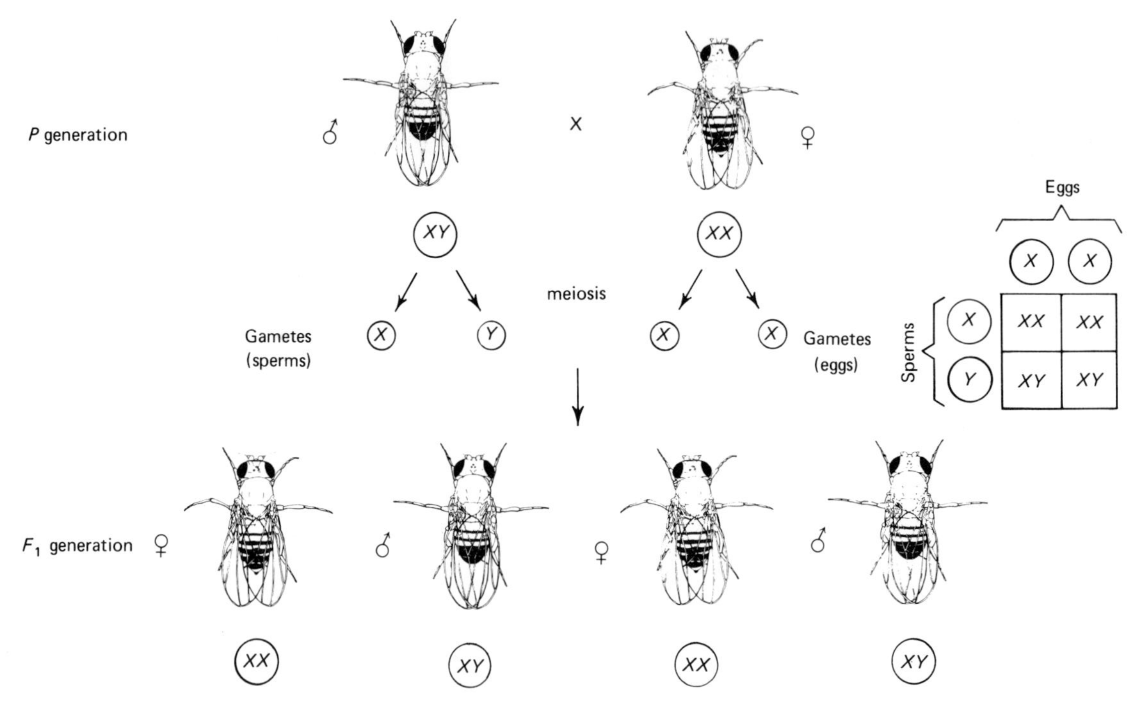

Figure 19-12. The relation of chromosomes to sex determination in *Drosophila.* The Punnett square at the right illustrates the procedure for determining egg and sperm combinations in sex determination.

Figure 19-13, in the F_1 generation all the males had white eyes and all the females had red eyes. In the F_2 generation, in both males and females, half had red eyes and half had white eyes. In summary, Morgan's work gave strong support to Sutton's chromosome theory since his experiments demonstrated that there was a direct relationship between a particular hereditary trait (white eyes) and a particular chromosome (X chromosome).

Nondisjunction: Further Proof for the Chromosome Theory

Calvin B. Bridges, one of Morgan's graduate students, worked with a gene for vermilion eye color in *Drosophila.* This gene, located only on the X chromosome, is recessive and determines the bright red eye color of certain *Drosophila* mutants. A cross involving this eye type is shown in Figure 19-14. Notice that the normal separation of X and Y chromosomes during meiosis results not only in the determination of sex in the offspring, but also in the distribution of the gene for vermilion eyes. This observation agrees with the theory that the vermilion gene is physically located on the X chromosome. The Y chromosome does not have a gene for vermilion eyes. Even though the gene for vermilion eyes is recessive, it appears in males since there is no allele on the Y chromosome. Thus, in the cross outlined, all the F_1 females (but none of the males) would have normal red eyes, whereas all the F_1 males (but none of the females) would have vermilion eyes.

Bridges made similar crosses many times and noted that occasionally (about 1 in every 2000 flies) a female with vermilion eyes and a male with normal eyes were produced. These observations apparently contradicted the chromosome theory of heredity since, according to that theory, the production of vermilion-eyed females implied that two genes for vermilion eyes were present. This would have been so because the vermilion trait is recessive to normal red eyes. Since one of the two genes required for vermilion eyes was contributed by the female, the origin of the other was obscure.

Careful analysis of Figure 19-14 reveals that there are

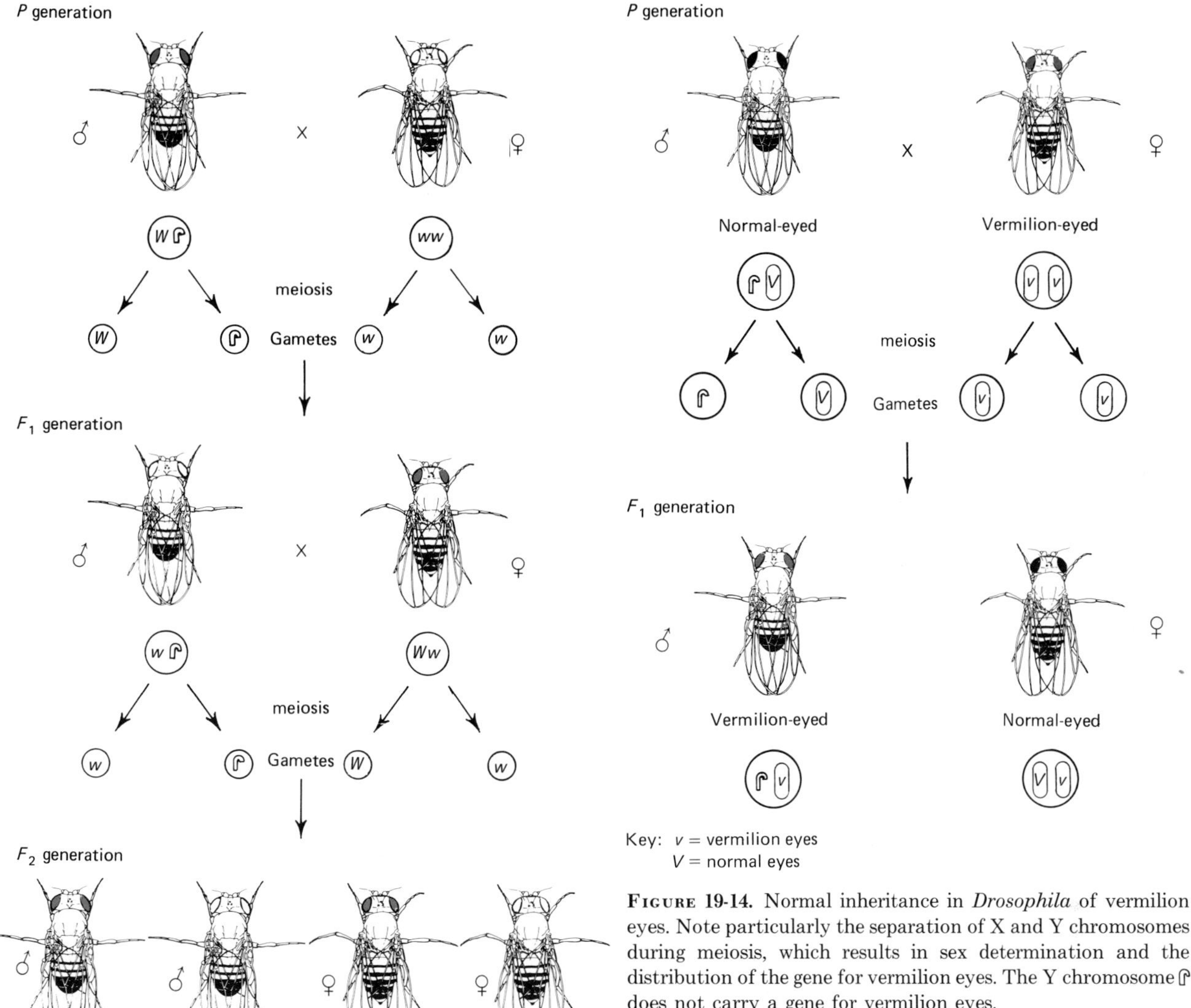

Figure 19-13. Morgan's experiment demonstrating the relationship between a hereditary characteristic and a particular chromosome. See text for amplification.

Figure 19-14. Normal inheritance in *Drosophila* of vermilion eyes. Note particularly the separation of X and Y chromosomes during meiosis, which results in sex determination and the distribution of the gene for vermilion eyes. The Y chromosome does not carry a gene for vermilion eyes.

no vermilion genes in the male parent. The obvious conclusion, therefore, is that the second vermilion gene must also have originated in the female parent. Bridges reasoned that this could be explained by the hypothesis that infrequently there was a failure of the two X chromosomes to separate (disjoin) during meiosis. This failure of the two X chromosomes to separate during the anaphase of the first or second meiotic division is referred to as **nondisjunction.** If the two X chromosomes do not separate from each other they pass into the same egg. In the cross representing nondisjunction (Figure 19-15) note that when the two X chromosomes do not separate, two kinds of eggs are produced. One kind contains unseparated X chromosomes, and therefore two vermilion genes, and the other kind has no X chromosomes at all. When both kinds of eggs resulting from nondisjunction are fertilized by normal sperms, the fertilized eggs contain an abnormal chromosome number. Thus, when the eggs with two X chromosomes are fertilized by normal sperms, half will be XXY and half will be XXX. When the eggs without X chromosomes are fertilized, the genetic makeup of half will be only X and the chromosome complement of the other half will be only Y. Inasmuch as the fertilized eggs with either three X chromosomes or no X chromosomes are exceedingly abnormal, they do not survive.

The experimental results shown in Figure 19-15 enabled Bridges to assume that the exceptional flies were due to the presence or absence of certain X chromosomes. The final and convincing proof of Bridges' hypothesis was to ascertain whether the exceptional flies actually carried these particular chromosomes in their cells. Examination of the cells revealed that his hypothesis of nondisjunction was supported by observations between the inheritance of vermilion eyes, the sex of the fly, and the chromosomes of the fly. For example, the exceptional vermilion-eyed females, which were thought to have the genotype *vvv,* did in fact have two X chromosomes and one Y chromosome. The *vv* is used here to designate the Y chromosome as having no allele for vermilion eyes; each *v* comes from an X chromosome. In substance, Bridges' experiments proved beyond any reasonable doubt that the gene for vermilion eyes is carried on the X chromosome, and in a broader context, his results demonstrated that genes are located on chromosomes and that sex-linked genes were part of the X chromosome.

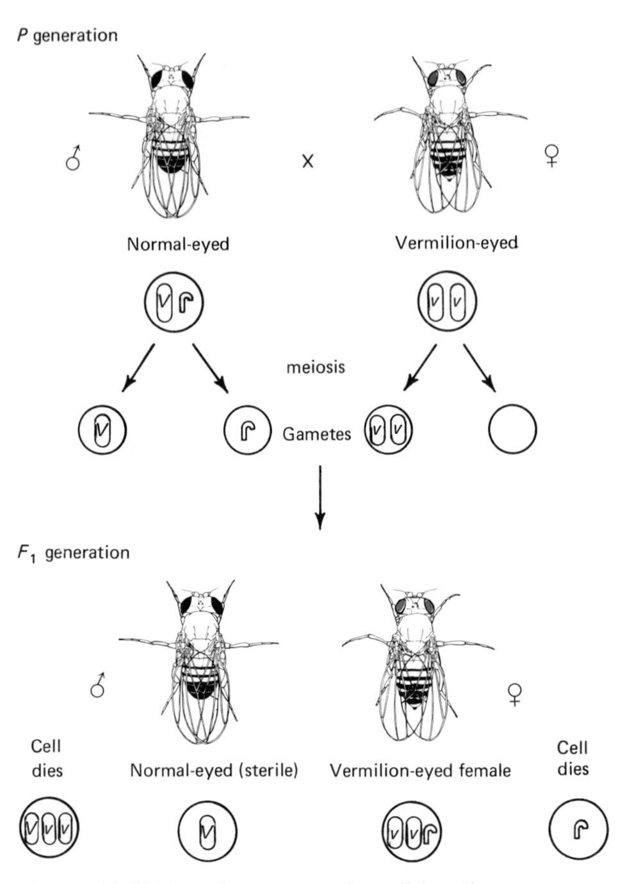

Figure 19-15. The phenomenon of nondisjunction.

Sex-Influenced and Sex-Limited Traits

In addition to **sex-linked traits**—traits that are expressions of genes borne on the sex chromosomes (in all mammals on the X chromosome)—some genes produce different effects in the two sexes but are not always found on the sex chromosomes. In other words, several genes are known that are not located on the sex chromosomes but whose phenotypic effects are influenced by the sex of the individual. Autosomal genes that are expressed differently in the male and female are referred to as **sex-influenced genes.** An excellent example of this type of gene is found in the domestic sheep. Some breeds of sheep, such as the Dorset, have horns in both sexes; other breeds, such as the Suffolk, have horns in neither sex. Inasmuch as the Dorset sheep are homozygous for the horned gene and the Suffolk are homozygous for the hornless gene, there is no distinction between the sexes in these pure breeds. When pure-bred horned sheep are crossed with pure-bred hornless sheep, however, all horned males and hornless females are produced. Although the F_1 generation sheep are all heterozygous, the gene for horns acts as a dominant in the males and as a recessive in the females. When two such heterozygous individuals are crossed, offspring in the ratio of three horned to one hornless occurs among the males, but one horned to three hornless among the females. This is the expected ratio on the basis of a sex-influenced gene. In several of the instances studied, the action of sex-influenced genes appears to be determined somehow by sex hormones.

In other cases, the effects of some autosomal genes are limited to one of the sexes. Such genes are called **sex-limited genes** and are responsible for the development of the primary and secondary sex characteristics. A number of examples of sex-limited traits in birds produce a sexual dimorphism in many species which brings about remarkable differences between the sexes. The characteristic plumage of the male pheasant is highly accentuated in contrast to the drab pattern which typifies the female. In the domestic fowl, however, the male may exhibit the female feather pattern if he fails to receive the gene combination necessary to produce the male plumage. In a few breeds, males as well as females, may show the hen-feathered condition. It has been found that this condition in the male is due to a dominant gene, *H*. Any male that receives at least one *H* will be hen-feathered, but those males that are homozygous for the recessive, *hh,* will exhibit the cock-feathered condition. Females are all hen-feathered regardless of genotype. The inheritance of this sex-limited trait is shown in Figure 19-16.

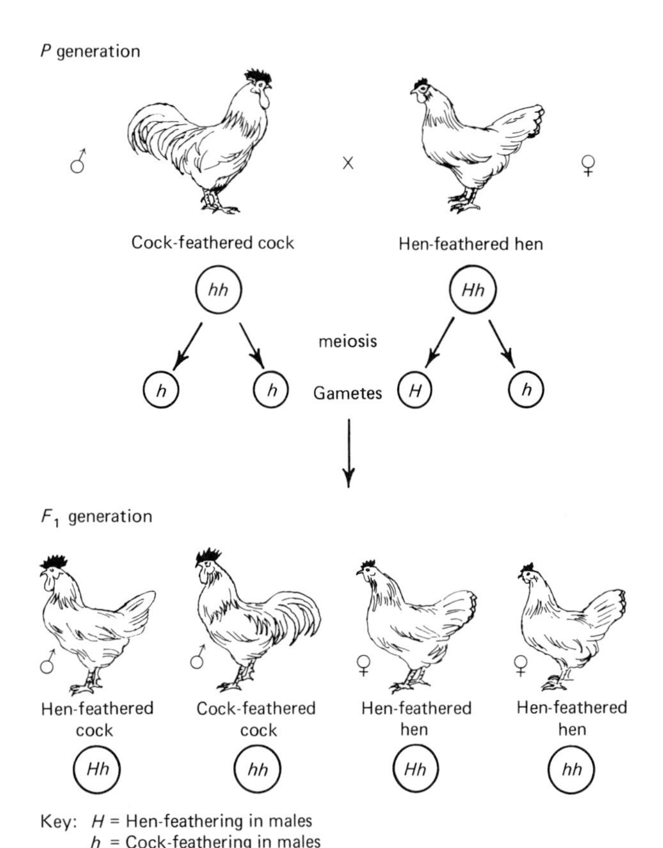

Figure 19-16. Inheritance of a sex-limited trait (hen-feathering) in the domestic fowl.

Linkage and Crossing-Over

Recall that Mendel found that each of the genetic differences he followed in his experiments with pea plants were inherited independently. His heterozygous mating involving two different traits produced a 9:3:3:1 phenotypic ratio in the F_2 generation. This ratio was a consequence of independent assortment of the two pairs of alleles. Specifically, the 9:3:3:1 ratio is obtained only if the two pairs of alleles are on different pairs of homologous chromosomes. Thus, if genes are located on sex chromosomes or if the two pairs of alleles are on the same chromosome, Mendel's rules will not apply.

It should be noted that Mendel had no knowledge of chromosomes and that his principles were derived exclusively from the data obtained from his crosses. Moreover, even though Sutton's work appears to substantiate the work of Mendel, his theory that genes are located on chromosomes is incompatible with the unmodified form of Mendel's law of independent assortment. Consider that *Drosophila* has 4 pairs of chromosomes, pea plants have 7 pairs, and humans have 23 pairs and that each species has thousands of different genes. By implication, then each chromosome must contain many different genes. In addition, since entire chromosomes segregate independently during meiosis, only genes that are located on different chromosomes can segregate independently of each other. Genes that are found on the same chromosome cannot separate and thus must move as a unit during meiosis. Such genes are said to be **linked** and they do not separate independently but are inherited together.

As an example, consider the mating of two organisms, one having the genotype *AABB*, and the other having the genotype *aabb* (Figure 19-17). If independent assortment does occur, the F_1 offspring will have the genotype *AaBb* and the gametes formed will be *AB, Ab, aB*, and *ab* (Figure 19-17A). In addition, nine genotypes expressing the phenotypic ratio of 9:3:3:1 in the F_2 will result. By contrast, if both dominant genes are located

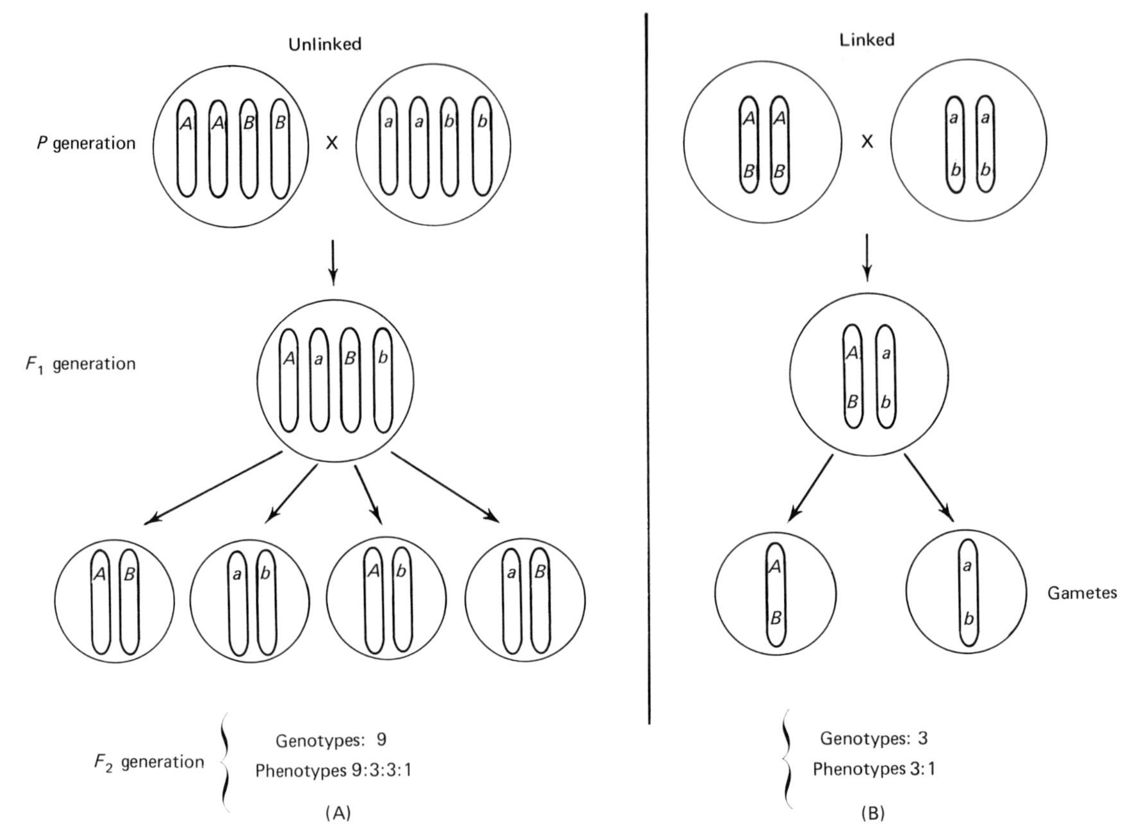

Figure 19-17. Comparison between independent assortment (**A**) and gene linkage (**B**).

on the same chromosome and linkage is complete, only two types of gametes will be formed: *AB* and *ab* (Figure 19-17B). As a result, instead of a 9:3:3:1 phenotypic ratio in the F_2 generation, a 3:1 ratio will result, because the two genes are inherited as though they were a single gene, similar to monohybrid inheritance.

For this reason, gene linkage tends to modify phenotypic ratios as originally stated by Mendel. In summary, then, gene linkage represents an exception to complete independent assortment because of the relationship of linked genes to their chromosomes and the segregation of entire chromosomes during meiosis.

Although complete linkage of two or more genes appears to occur in certain hereditary patterns, it is comparatively uncommon for every gene on a chromosome to be passed down to the next generation as a linkage group. It has been found that some genes originally linked on the same chromosome do not always remain linked. Careful examination of meiosis shows that genes on one chromosome may be exchanged for corresponding genes of the homologous chromosome by a process referred to as **crossing-over.**

To understand the mechanism of crossing-over during meiosis, consider a cell consisting of only one pair of homologous chromosomes. During the initial stages of meiosis, the chromosomes pair (Figure 19-18A). Each chromosome is present as two chromatids (Figure 19-18B). During this pairing, or synapsis, the chromatids of homologous chromosomes become intertwined (Figure 19-18C). As a result of breakage and rejoining of chromatids, there is an exchange of parts between chromatids (crossing-over) and the broken ends then become joined in new sequences (Figure 19-18D). The net effect of this crossing-over is that there is an exchange of hereditary

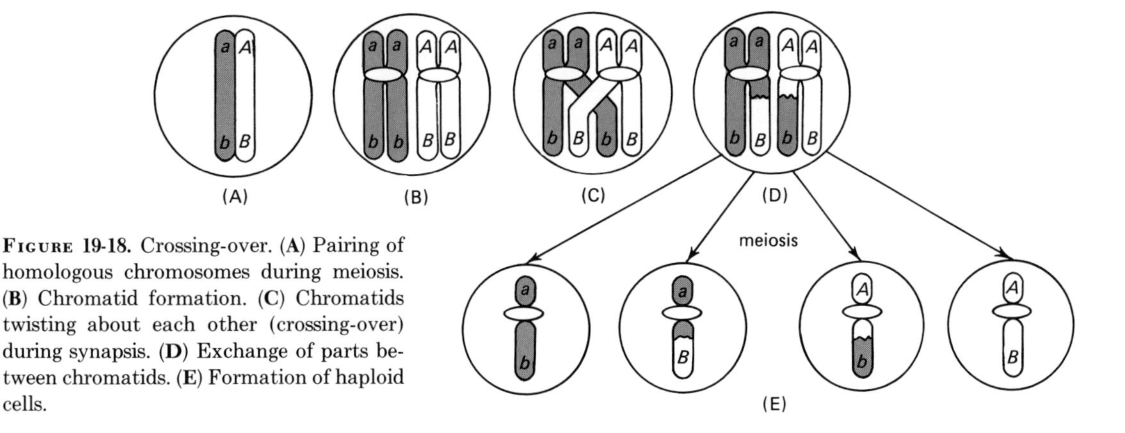

Figure 19-18. Crossing-over. (**A**) Pairing of homologous chromosomes during meiosis. (**B**) Chromatid formation. (**C**) Chromatids twisting about each other (crossing-over) during synapsis. (**D**) Exchange of parts between chromatids. (**E**) Formation of haploid cells.

material between chromatids of homologous chromosomes and the resulting haploid gametes contain four different types of chromosomes (Figure 19-18E). If crossing-over did not occur, the gametes would be of only two types: *AB* and *ab;* however, with crossing-over, four types of gametes are produced: *AB, Ab, aB,* and *ab.*

Mapping Chromosomes

With the discovery of crossing-over, it became clear that not only are the genes carried on the chromosomes, but they must also be arranged in a definite, fixed, linear order along the length of the chromosome. Also, all the genes must be located at corresponding loci on different chromosomes. If this were not true, exchange of sections of chromosomes would not produce an exact exchange of alleles.

As other genetic traits were studied, it became clear that the percentage of crossing-over between any two genes was different from the percentage of crossing-over between two other genes. In addition, these percentages were fixed and predictable. A. H. Sturtevant, in 1913, assumed that the percentage of crossovers was related to the distance between the genes. He postulated that genes are arranged on a linear sequence on chromosomes and he tested his hypothesis by crossing-over studies of the chromosomes of *Drosophila.* Since Sturtevant's studies, extensive chromosome maps have been worked out for quite a number of organisms.

Morgan assumed that the breakage of a chromosome is about equally possible at any point along its length. It follows, then, that the farther apart two linked genes are on the chromosome, the higher the frequency of breakage between them since there are more points between them at which a break may take place. Stated another way, the frequency of crossing-over between any two linked genes will be proportionate to the distance between them. Although the percentage of crossing-over provides no information about the absolute distances between genes, it does provide relative distances. As a result, geneticists use an arbitrary unit of measurement, the map unit, to describe distances between linked genes. A *map unit* is equal to 1% of crossing-over; that is, it represents the linear distance within which 1% crossing-over takes place.

As an example of the principle of *chromosome mapping,* assume that the percentage of crossovers between genes *G* and *T* on the same chromosome is 40% and that the percentage of crossovers between genes *T* and *M* is 20%. According to the data obtained from crossing-over studies, the distance between gene *G* and gene *T* is twice as great as the distance of gene *M* from gene *T.* In considering this, two distinct possibilities exist for the relative positions of the three genes on the chromosome. The sequence could be

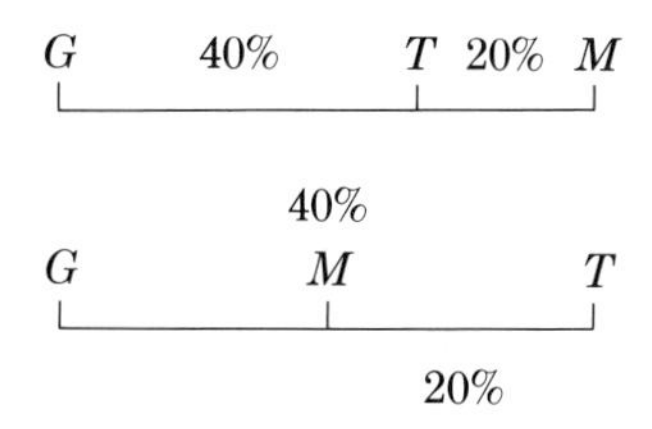

In order to establish the correct sequence, it is necessary to determine the frequency of crossovers between genes *G* and *M*. If the percentage of crossovers between these two genes is 60%, then the sequence *GTM* is correct. If, by contrast, the percentage of crossovers between genes *G* and *M* is 20%, then the sequence *GMT* would be correct.

By employing this procedure, chromosome mapping has been carried out in a variety of organisms including viruses, bacteria, molds, and various plants and animals. The chromosome map of *Drosophila,* the most commonly studied species, shows that more than 500 different genes have been located in relationship to one another on the four chromosome pairs (Figure 19-19). Perhaps the most significant conclusion of all mapping studies is that they established further proof that genes occur along the chromosome in an ordered, linear arrangement.

Essentially, the construction of chromosome maps based upon crossing-over percentages provided theoretical proof for the linear orientation of genes on chromosomes. In time, geneticists were able to construct less theoretical maps by correlating observations of chromosomes with the results of breeding experiments. It was discovered in *Drosophila,* as well as other insects, that in certain cells the nuclear material continues to replicate many times without separation, resulting in thick interphase chromosomes (*giant chromosomes*).

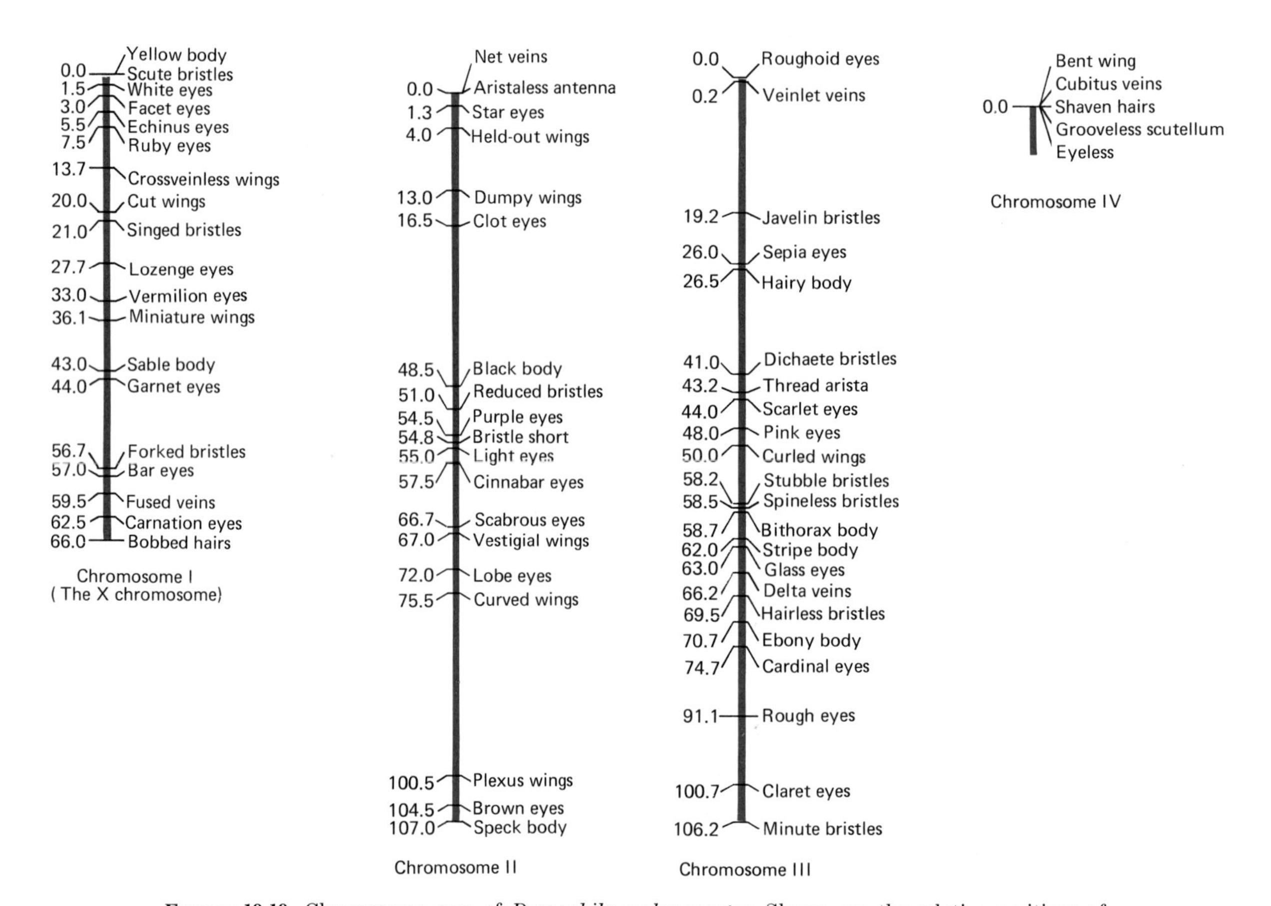

Figure 19-19. Chromosome map of *Drosophila melanogaster.* Shown are the relative positions of some of the principal genes. The numbers at the left of each figure refer to distances from the upper end of the chromosome as determined from the percentages of recombination observed in linkage experiments.

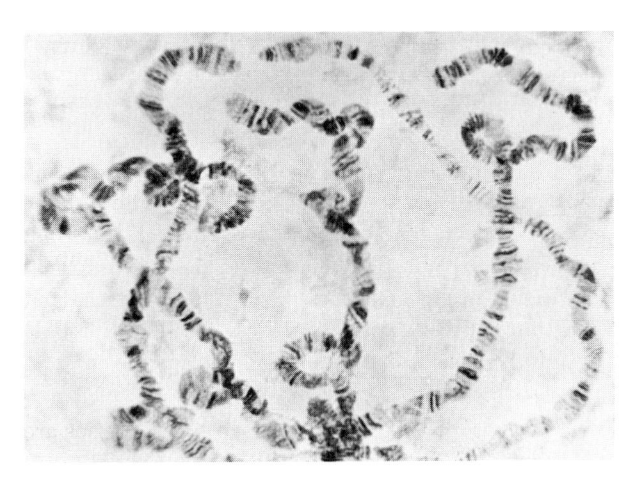

Figure 19-20. Giant chromosome of *Drosophila* showing crossbands (deeply staining) and internodes (light areas). [*Courtesy of Carolina Biological Supply Co.*]

The *Drosophila* salivary chromosomes may be about 100 times larger than chromosomes of ordinary cells. Microscopically, they exhibit a distinctive pattern of deeply staining crossbands and light internodes throughout their lengths (Figure 19-20). These crossbands differ from one another in shape, size, and distinctness and they make it possible to recognize various regions of the chromosome. The dark bands are composed chiefly of nucleic acid (DNA), and it has been estimated tentatively that the number of genes contained by a haploid chromosome set in *Drosophila* ranges from 5000 to 15,000. As a result of studies involving deletions, translocations, and inversions of chromosomes (see page 513), it has been possible to associate many genes with distinct bands. In other words, it has been possible to construct chromosome maps in which bands represent the position of genes, and these maps show the same gene order as those made from crossover data. Thus, giant chromosomes represented a powerful cytological means for checking inferences that were being drawn solely from genetic data.

Chromosomal Aberrations

In the normal course of meiosis, chromosomes synapse, segregate in an orderly manner, and gametes are regularly produced with the haploid chromosome number. Gene loci also typically retain the same sequence they have possessed for many past generations. However, irregularities sometimes occur in nuclear divisions so that cells or entire organisms with aberrant **genomes** (specific chromosome complement) result. These irregularities, in the form of abnormalities in chromosome number, structure, or arrangement, may produce phenotypic changes, alterations of expected genetic ratios, or changes in the linkage relationships of certain genes.

In general, chromosomal aberrations may be categorized into two principal kinds. The first of these, *ploidy,* involves changes in the number of whole chromosomes. Ploidy may involve either entire extra sets of chromosomes, a condition called *euploidy,* or loss or addition of a single chromosome, a phenomenon referred to as *aneuploidy.* The second type of chromosomal aberration involves structural alterations.

Alterations of Chromosome Number

Aneuploids. One of the first examples of aneuploid chromosome distribution was discovered by Bridges during his work on *Drosophila.* This work, involving the inheritance of red and white eyes in males and females, has already been considered so that it will suffice at this point to note that the phenomenon of nondisjunction is operative in the production of aneuploids. Recall that the failure of the two X chromosomes to separate during meiosis produces some eggs having two X chromosomes and others having none. Fertilization of these eggs produces offspring with genotypes XXX, XXY, X, or Y. Those with the genotype XXX are referred to as "super females" and are completely sterile, have a low viability, and seldom live long enough to be seen. Offspring with the genotype XXY are normal females; the extra chromosome, the Y, seems to have no effect on the appearance or fertility of such females. *Drosophila* with only a Y chromosome die almost at once since they lack important groups of genes on the X chromosome. Finally, offspring with the genotype X are sterile males since the missing Y chromosome carries genes for male fertility. Phenotypically, males lacking a Y chromosome are indistinguishable from normal males. Thus, it can be seen that nondisjunction may result in a number of aneuploid conditions, some of which affect reproductive ability and the potential to live.

Nondisjunction is known to occur in other species as well as *Drosophila*. One of the more definitive studies of this occurrence has been made in the Jimson weed by A. F. Blakeslee and J. Belling (1924). This organism shows a considerable amount of morphological variation in many traits, particularly in fruit structure. The normal chromosome number for this plant is 24. When nondisjunction of one of these twelve pairs occurs, a plant is produced with either 25 or 23 chromosomes. When one of the twelve kinds of chromosomes was present in triplicate (eleven pairs and three homologous chromosomes), there were certain recognizable phenotypic effects. In the normal diploid, the genes are in a balance and give rise to normal phenotypes, but with an extra chromosome present, the balance is upset. Accordingly, a different phenotype would be expected for each chromosome present in triplicate. This was found to be true; Blakeslee and Belling recognized twelve different phenotypes that deviate from the normal, each of which was due to the presence of a different chromosome present in triplicate. Thus, *trisomy* has different effects according to which chromosome has the trisomic condition. Corresponding monosomic patterns (23 chromosomes) were not found because this condition is usually lethal in Jimson, as it is in other organisms.

EUPLOIDY. This condition, unlike aneuploidy, involves entire sets of chromosomes. *Euploids* are characterized by possession of entire sets of chromosomes in which monoploids (haploids) carry one set, or genome (n), diploids have two genomes ($2n$), and polyploids possess three or more genomes. *Monoploidy* is rare in adult animal forms, although the male honeybee is an outstanding exception. In plants, the condition is more common. Recall that in all bryophytes, the monoploid represents the dominant part of the life cycle and is the commonly recognized plant. In the vascular plants, the monoploid is short-lived and inconspicuous; the diploids are the dominant forms in the life cycles.

Polyploidy is a condition in which organisms have three or more haploid sets of chromosomes. Thus, if the normal chromosome number of a given species is 8 ($2n = 8$), individuals with a chromosome number of 12 ($2n + n$) are triploids, those with a chromosome number of 16 ($2n + n + n$) are tetraploids, and so on. Essentially, all but the diploids may be referred to as polyploids. As with the monoploidy, the polyploid condition is relatively rare among animal species, probably because of an imbalance of sex-determining mechanisms. By contrast, polyploidy is quite common among plants.

Many examples of tetraploids exist in nature among plants, and in some animals as well. This condition may arise through an abnormal mitosis or through abnormal meiosis. In the mitotic abnormality, there is a normal duplication of chromosomes up to the metaphase and then the process ceases. This leaves a cell with double the normal chromosome number (tetraploid) when it enters the next cyclic interphase. Such a cell may then undergo normal cell division and produce a group of tetraploid cells. If this occurs in the growing tip of a plant, the resulting plant tissues will contain cells that are exclusively tetraploid. Vegetative propagation of this tetraploid plant part by cuttings or grafting would result in the continuity of the condition through many successive generations. Considering that tetraploids are frequently hardier, more vigorous in growth, and have larger flowers and fruits, such a procedure is commercially valuable. Common tetraploids include varieties of Portland and Fredonia grapes, McIntosh apples, and certain varieties of cherries, pears, blueberries, cranberries, and blackberries. In all cases, these tetraploids are considerably larger than their diploid counterparts. A series of photos illustrating a comparison between diploid and colchicine-induced tetraploid azalea plants is shown in Figure 19-21.

In addition to the occurrence of polyploidy in somatic cells (abnormal mitosis), polyploidy may also arise during meiosis. In the first meiotic division, the chromosomes may synapse and prepare for a normal reduction division, but fail to separate, so that one gamete receives the entire diploid complement and one receives no chromosomes at all. Although the latter may die, the diploid cell may go through a second meiotic division and produce diploid gametes. The union of such a diploid gamete with a normal haploid gamete gives rise to a zygote with three sets of chromosomes ($2n + n$) or a *triploid*. In addition to exhibiting certain atypical phenotypic effects, triploids are also sterile. In plants, however, triploids may be maintained by grafting, cuttings, or through propagation of bulbs and roots. Seedless grapes, Baldwin apples, and Keizer's tulips are examples.

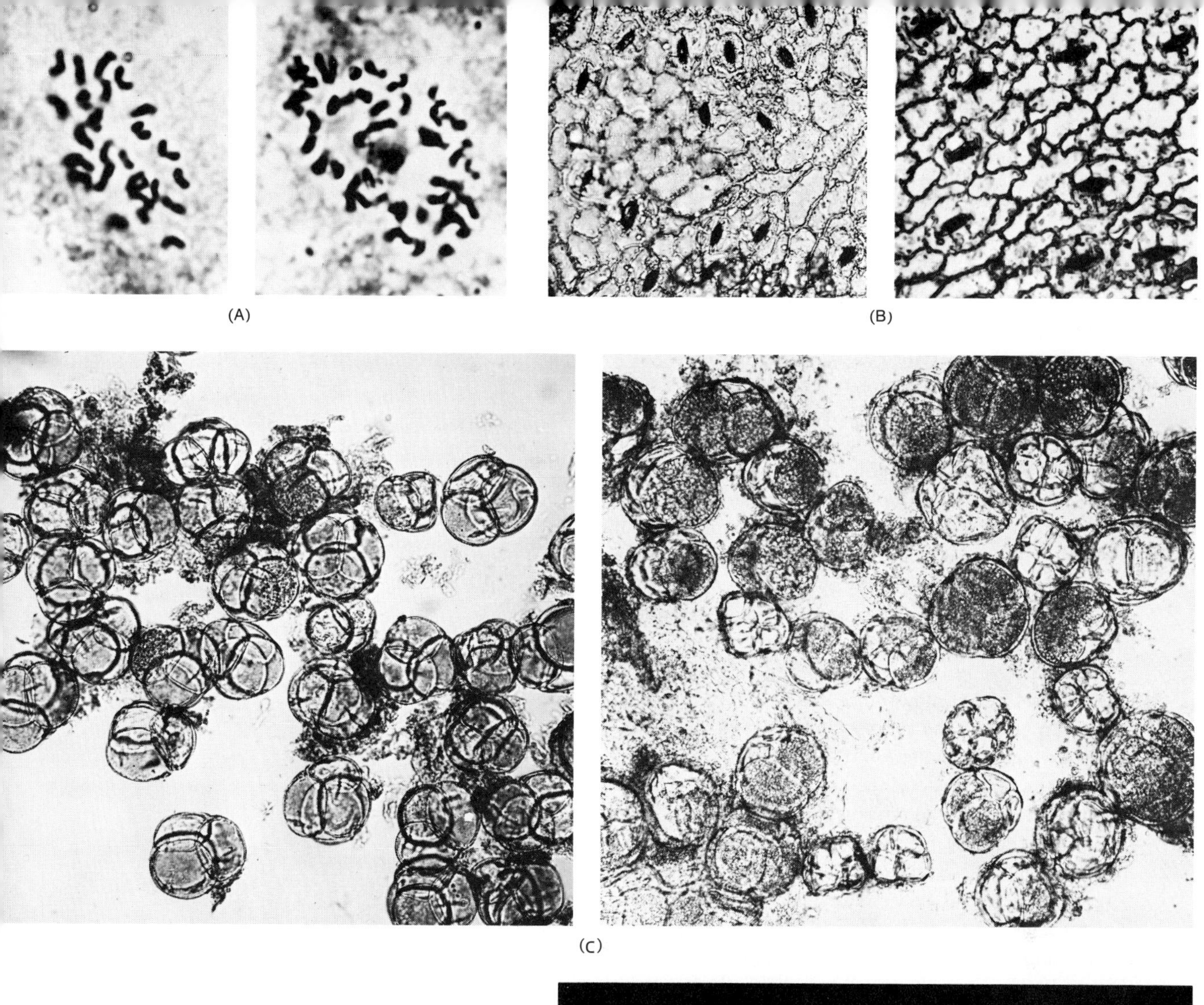

Figure 19-21. Comparison between diploid (*left*) and tetraploid (*right*) azaleas. (**A**) Chromosomes. 750×. (**B**) Stomata. 75×. (**C**) Pollen grain tetrads. 150×. (**D**) Flowers. [*USDA photos.*]

Polyploids may arise naturally or be artificially induced. In nature, for example, among plants, polyploids may arise from diploids as a result of interference with cytokinesis once replication of the chromosomes has taken place. Some data indicate that chilling may accomplish this in natural populations. The most efficient and widely used method of artificially induced polyploidy is achieved by the application of a poisonous chemical called *colchicine*. This alkaloid substance, derived from the autumn crocus, is applied either as a liquid or in a lanolin paste to certain plant tissues. Although colchicine does not interfere with chromosome replication, it prevents normal spindle formation and the double number of chromosomes becomes incorporated within a single nuclear membrane. Subsequent nuclear divisions are normal so that the polyploid line of cells, once initiated, may be maintained. Polyploidy may also be induced artificially by other chemicals such as vertrine and acenaphthene or by exposure to cold or heat.

The term *autotetraploid* is applied to an organism all of whose sets of chromosomes are those of the same species. In general, the gametes produced by autotetraploids are highly infertile. On the other hand, polyploids may develop and be developed by hybrids between different species. These organisms are referred to as *allopolyploids,* the most commonly encountered type having two genomes from each of the ancestral species.

In 1928, the Russian cytologist G. D. Karpechenko synthesized a new genus from crosses between vegetables of two different genera, the radish and the cabbage (Figure 19-22). Even though these plants show considerable differences in their morphology, they each have a chromosome number of 18 and can be crossed to yield a hybrid. The genes in the two, however, are so different that the hybrids are usually sterile. Karpechenko did find that a few of the hybrids were fertile; these were tetraploids and each had thirty-six chromosomes. The allotetraploid was named *Raphanobrassica* (*Raphanus,* the genus name for the radish, and *Brassica,* the

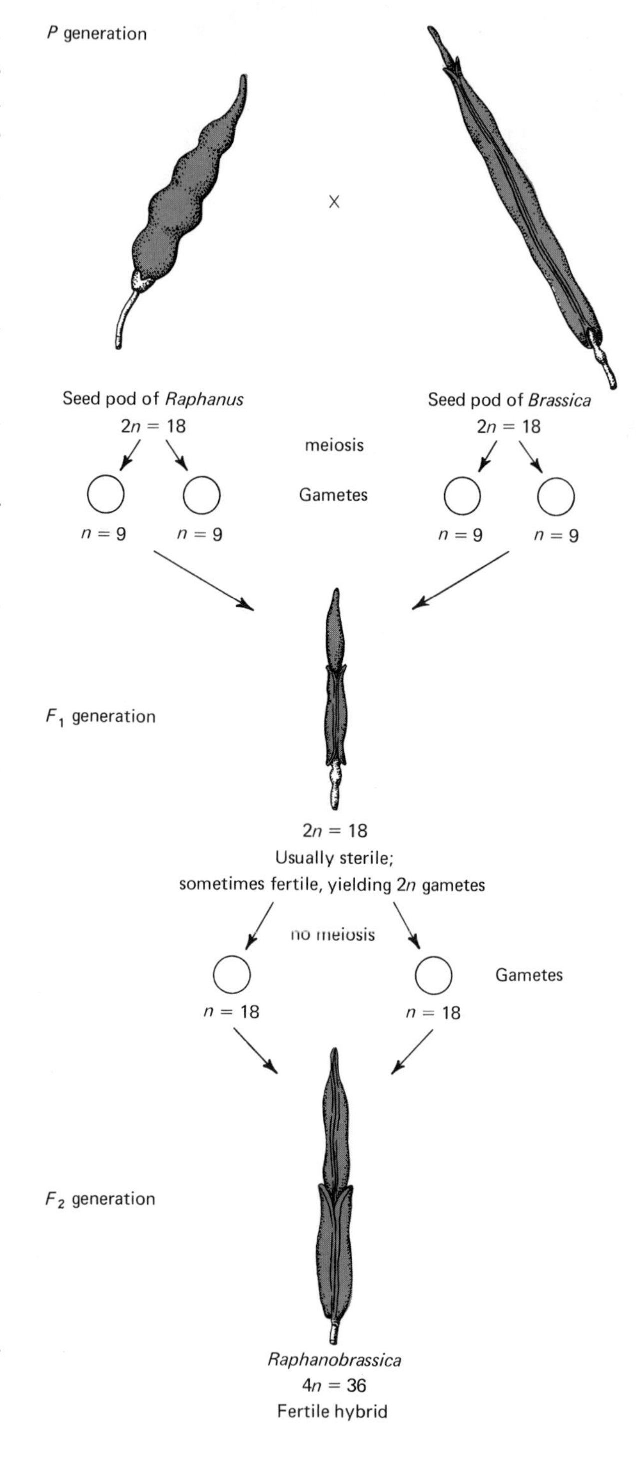

FIGURE 19-22. Diagrammatic representation of the origin of *Raphanobrassica* showing seed pods and chromosome numbers. [*After G. D. Karpechenko,* Z. Indukt. Abst. Vererb., *vol. 48, p. 27, 1928.*]

genus name for the cabbage). This new form bred true and when it was crossed with either of its progenitors, highly infertile triploids resulted. Even though *Raphanobrassica* is of no direct economic importance (root of a cabbage and leaves of a radish), it does represent a method by which interspecific hybrids may be produced. Interspecific hybridization combined with polyploidy offers some evidence by which new species may arise suddenly in natural populations.

Structural Alterations of Chromosomes

It has been shown that chromosomes normally undergo breakage and reattachment during the prophase of the first meiotic division in the process known as crossing-over. Inasmuch as this is a very common process resulting in no abnormal chromosomes or gene arrangements, there are no negative effects on the organism. It is possible, however, for breakage to occur without reattachment to the homologous chromosomes, and when this occurs, various kinds of abnormal chromosomes may result.

Recent evidence indicates that chromosome mutations may produce changes in important regulator or regulatory genes. As discussed in Chapter 8, the regulator genes affect the transcription of structural genes. Thus changes in the regulatory genes may be far more important for adaptive purposes than those in structural genes. Some biologists suggest that for similar species the genetic material is essentially the same, but differences arise from slight rearrangements of the genetic material.

The simplest result of breakage is the loss of a part of a chromosome, a phenomenon referred to as **deficiency,** or **deletion** (Figure 19-23B). If a portion of a chromosome becomes detached from the remaining portion of it bearing the centromere, it will usually be left behind in the cytoplasm as the chromosomes follow their centromeres to the poles and the nuclear membranes are formed. Such a detached portion, excluded from the normal surroundings in the nucleus, soon disintegrates. If a germ cell carrying a chromosome deficient in the deleted genes unites with another gamete carrying a normal chromosome, a zygote is produced that carries the deleted genes in a single dose.

In another type of chromosomal alteration called an **inversion,** a chromosome may twist about itself and where the ends cross, break and exchange positions (Figure 19-23C). This is most likely to occur when the chromosome orients itself into a loop-like configuration, since such twisting makes it easy for the broken ends of the deleted portion to become reattached in a new position. Inversion, which typically takes place during meiosis, produces a gamete carrying a chromosome with genes in the reverse order on the inverted portion.

A third kind of chromosomal alteration, referred to as **translocation,** is a phenomenon in which a portion of one chromosome may attach to another nonhomologous chromosome (Figure 19-23D). A cell heterozygous for a translocation produces interesting synaptic figures since the genes tend to pair their alleles even though they are located in a portion which has become translocated to a nonhomologous chromosome.

Duplication of a gene locus constitutes a fourth type of chromosomal alteration. In the process of duplication, a portion of a chromosome is represented more than twice in a normally diploid cell (Figure 19-23E). The extra segment may be attached to the chromosome whose loci are repeated, to a different linkage group, or even be present as a separate fragment. Such alterations may cause an imbalance of gene activity which will reduce the viability of the organism.

The Nature of Genes

It will be recalled from Chapter 8 that a series of convincing experiments provided many data supporting the idea that DNA is the actual material of which genes are made. The Watson–Crick model of DNA provided an explanation of how the hereditary material duplicates and various researches showed that genes exert their influence through specific enzymes which they produce. In other words, genes exert their control over cellular activities by controlling the enzymes that govern the biochemical reactions of a cell. Once this relationship was established, researchers began to investigate the genetic code, that is, how the information carried in genes determines the structure of proteins. Once the genetic code was deciphered, other investigations showed how gene activity is controlled (operon model).

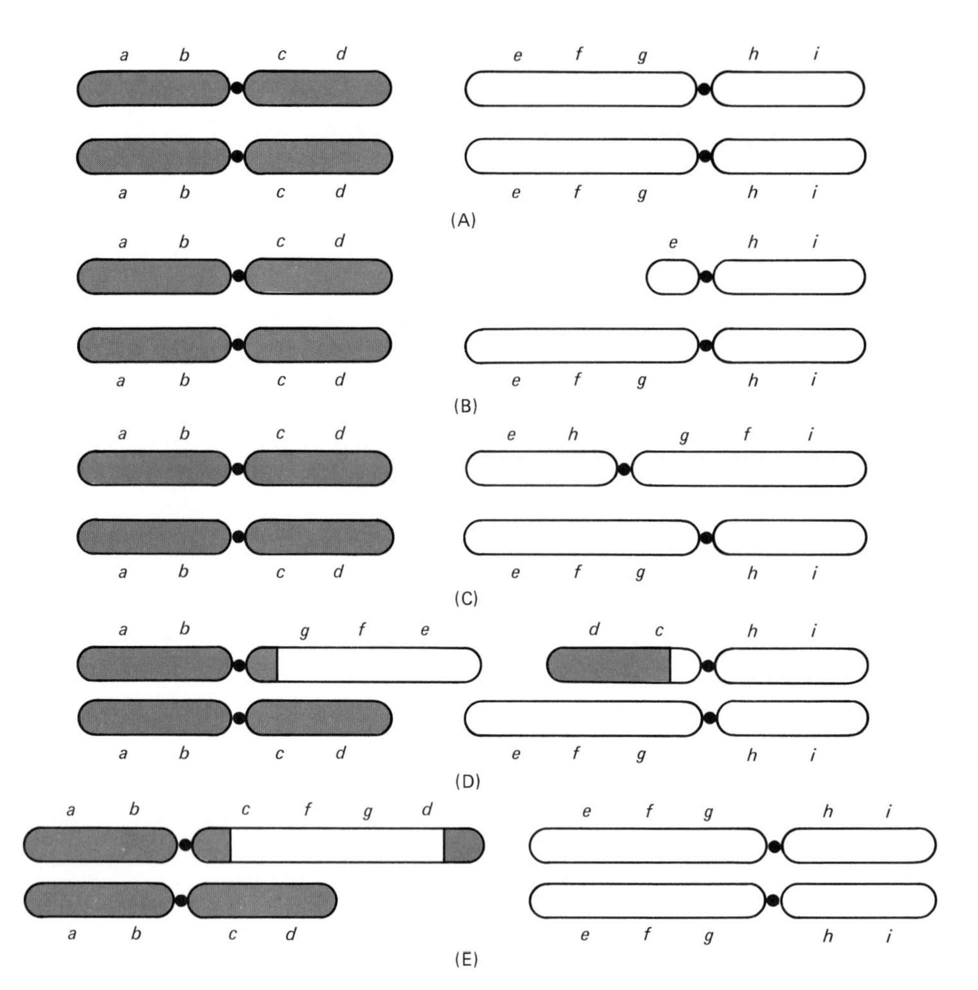

Figure 19-23. Kinds of chromosomal aberrations. (**A**) Normal chromosomes prior to any structural modifications. (**B**) Deletion. (**C**) Inversion. (**D**) Translocation. (**E**) Duplication.

The Fine Structure of Genes

Mendel referred to genes as "factors" and from the early part of the twentieth century onward, it was agreed that the factors of inheritance are intimately associated with chromosomes. Morgan's work with *Drosophila* led to the concept that genes are the smallest units of recombination. Physically, genes were regarded as discrete particles arranged in linear sequence on a chromosome. Recall that recombination of genes was explained on the basis of crossing-over. Other experiments by geneticists also led to the concept that the gene is the smallest unit of mutation and the smallest unit of function. Thus, for many years, the gene was regarded as the unit of recombination, mutation, and function.

In recent years, it has become increasingly clear that the units of recombination, mutation, and function are probably not identical. In this regard, Seymour Benzor has applied different terminology to the various units. For example, the term **cistron** has been applied to that segment of DNA which specifies one polypeptide chain. In other words, the cistron is regarded as the unit of function. Considering that an average polypeptide chain has approximately 300–500 amino acids in its sequence, then it may be assumed that the average cistron is about 900–1500 nucleotides long (refer to the codon concept in Chapter 8).

Within the cistrons, there appear to exist a large number of mutable sites. The gene as a unit of mutation is smaller than a cistron, consisting of fewer nucleotides. This unit, called a **muton,** is the smallest length of DNA capable of mutational change. Such an area may be as small as a single nucleotide pair in DNA. Finally, the term **recon** is applied to the smallest unit of DNA capable of recombination. Like the muton, the recon is very small and, in all probability, likewise may consist of as little as one nucleotide pair. It is also probable that mutons and recons are structurally indistinguishable from each other.

Genic Mutations

The term **mutation,** broadly defined, is any change in the genetic material that gives rise to an alteration in the genotype. In practice, mutation may designate a change in genes or a change in chromosomes. Inasmuch as chromosomal aberrations have already been considered, only gene mutations will be discussed at this point.

Gene mutations or *point mutations* are changes of the base sequence within the DNA molecule. Such mutations result in different informational content of DNA which is transmitted through replication to cell progeny or by transcription to *m*RNA.

A gene mutation may have only a minor effect in a cell. In fact, the effect may be so small that it cannot be easily detected. At the other extreme, however, if a mutation affects the production of a necessary enzyme, the change may result in the death of the cell. Since the cell or organism cannot survive if such a mutation occurs, it is referred to as a *lethal mutation.* Most mutations that have been investigated lie somewhere between these two extremes, that is, they precipitate observable effects but are not serious enough to cause death.

In *Drosophila,* white eyes result from an inability of the organism to synthesize the red and brown pigments that characterize the wild-type (red), or normal, eyes. This mutational event, the change of eye color from red to white, results in a change in the genotype such that the mutagenic allele (white, *w*) lacks the genetic information required for the production of an enzyme involved in the synthesis of these pigments. Thus, there is the loss of a specific genetic function. In addition to white eyes, there are many other mutations in *Drosophila* involving eye color. Among these are brown eyes, scarlet eyes, sepia eyes, vermilion eyes (bright red), and white-apricot eyes (yellowish pink). Such eye color mutants belong to a much larger class of mutations called *visible mutations* in which the phenotypic effects are seen as some alteration of the morphology of the organism. Among the visible mutations in other organisms are albinism in humans, lack of tails in mice and Manx cats, variegated patterns on leaves, and white flower color in sweet peas. Many visible mutations represent the loss of an essential structure or function.

Various studies on genes indicate that there is a wide range of variation in the rate of gene mutations. Some genes mutate rather frequently (one in every 2000 gametes), whereas others mutate far less frequently (one in several billions of cell division). Most genes, however, mutate about once in a hundred thousand to a million gametes each generation. In some bacteria, mutation rates may be as low as one in ten thousand million cells. Even though these mutation rates seem fairly low, they do provide a great deal of genetic variety. For example, it has been estimated by some investigators that a human has 10,000 genes. Assuming that this is true, and that each gene mutates at about the same rate, then more than one in every 100 gametes would have a mutation of at least one gene. Considering the large number of individuals in most species, it is reasonable to assume that the number of new mutations is very large.

The causes of most mutations are not known. They may occur at any time in the life of an organism in either somatic or reproductive cells. A mutation that occurs in a body cell may not be detected at all. Only those mutations that take place in gametes are of major importance in the genetics and evolution of sexually reproductive organisms. Techniques have been devised to increase the frequency of mutations by artificial means. The most direct way to increase the frequency is to treat the reproductive cells with external agents of various kinds. Among these agents are temperature (increase), chemicals (mustard gas, formaldehyde, nitrous acid, and peroxide), and radiation (X-rays, beta and gamma rays, neutrons, and ultraviolet radiation).

In addition to the aforementioned characteristics of mutations, it should also be noticed that most are recessive and some are capable of reversal. In other words, it is possible for some mutations to revert to their normal allelic condition.

Heredity and Environment

The form and function of any organism are dependent upon the interaction of two simultaneously operating factors: heredity and environment. Whereas the potential for growth and development is determined by heredity, the actual degree to which the potential is realized is governed by the environment. In other words, genes are inherited but are not expressed characteristics themselves. The expression of a gene or a group of genes is a function of the hereditary environment. This environment includes both the genetic environment (other genes present) and the physical environment. Among the principal factors in the physical environment that might affect the heredity of an organism are temperature, sunlight, and diet.

One of the best examples of the effect of the environment on phenotypes is the effect of temperature on the fur pattern known as Himalayan in domestic rabbits. This pattern is caused by a simple recessive gene and rabbits homozygous for this gene have white fur over most of their body but black fur on their tails, ears, and tips of their legs and noses (Figure 19-24). This pattern is produced through the differences in temperature in the different regions of the body. The skin temperature is higher on the back and abdomen than on the extremities since there is a greater loss of heat by radiation at the extremities. Thus, the Himalayan gene in rabbits does not produce the characteristic coat pattern directly, but merely causes black pigment to be produced in the fur on all areas of the skin which remain below a certain critical skin temperature (92°F). Variations in temperature above this point limit pigment production and the hair is white. This phenomenon may be proved by the following experiment. If white hair is plucked from the back of a rabbit of the Himalayan variety and if the plucked area is covered by an ice pack to reduce the temperature, the newly replaced hairs will develop black pigment. If the rabbit is kept at normal temperatures, the black spot will gradually disappear, for these hairs will be shed and replaced with new white ones. The Siamese coat patterns of cats are due to a similar temperature relationship.

Among certain flowers, a similar condition exists. In one variety of the Chinese primrose, for example, there are homozygous genes which produce white flowers at temperatures above 86°F. The same genotype gives rise to red flowers at a temperature of 68°F. The final floral color of these plants depends upon the temperature existing during a certain critical period in the early formation of the flower. It is thus possible, by changing a plant back and forth to hot and cold rooms, to get both red and white flowers on a single plant.

Numerous other examples could be cited and in each case it can be seen that only genes are inherited and the characteristics are expressed through environmental action on the gene-controlled processes of development. In other words, the penetrance and expressivity of genes may be affected by environmental conditions.

Sunlight also assumes an important role in the expression of many genes. It has been shown in Chapter 10 that light duration has significant effects on various

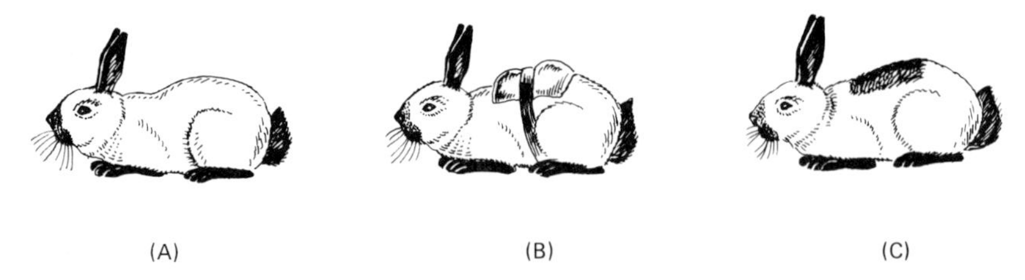

FIGURE 19-24. Effect of environment on phenotype. In this example, temperature affects the expression of a gene for coat color in the Himalayan rabbit. (**A**) Normal appearance of the rabbit. (**B**) Ice pack applied to an area from which fur has been plucked. (**C**) Regrown fur appears black. [*From A. M. Winchester,* Genetics, *3rd ed. Boston: Houghton Mifflin Co. Copyright © 1966.*]

plant processes, especially flowering (photoperiodism). In addition to light duration, light intensity and quality also affect the phenotypes of certain plants.

Prior to any known physiological cause-and-effect relationship between light and growth, certain peculiar developmental patterns of plants grown in darkness had been observed. The most striking influence of light on plant growth can readily be seen by comparing plants grown in complete darkness with ones grown under ordinary conditions of day and night (Figure 19-25A). The complete absence of light affects not only the rate of growth but also certain anatomical features of plants. In the dark, leaves tend to be yellow because of their inability to synthesize chlorophyll; stems elongate abnormally and become spindly, with poorly developed vascular tissue; and leaves fail to enlarge and remain folded as though still embryonic. A plant exhibiting these traits is said to be **etiolated** and the phenomenon is called *etiolation.*

Plants grown in shade (not darkness) also exhibit some characteristics of etiolation such as excessive stem elongation and, in addition, they show marked variation in leaf anatomy. For example, leaves grown in light have more sugar and less water than leaves grown in shade and, as a result, tend to be small in area, thick, and compact with a heavier cuticle and more conducting tissues and support tissues (Figure 19-25B, left). Note that the structure of sun leaves is such as to prevent excessive transpiration and to facilitate translocation of photosynthetic products. Shade leaves, by contrast, have a loose palisades mesophyll consisting of a single layer of cells and a very loose spongy mesophyll filled with air spaces. In addition, the epidermal cells lack heavy cutinization (Figure 19-25B, right).

All plants do not respond similarly to variations in light intensity. Some require more light than others and different species of plants grow best at different levels of light intensity. Some plants, such as sunflowers, tomatoes, and grasses, require direct bright light for maximum growth and development. Other plants, such as violets and ferns, have a relatively low light requirement and develop normally in shade areas.

Light varies in quality in different regions of the spectrum. The effects of different wavelengths of the visible spectrum have been studied by growing plants

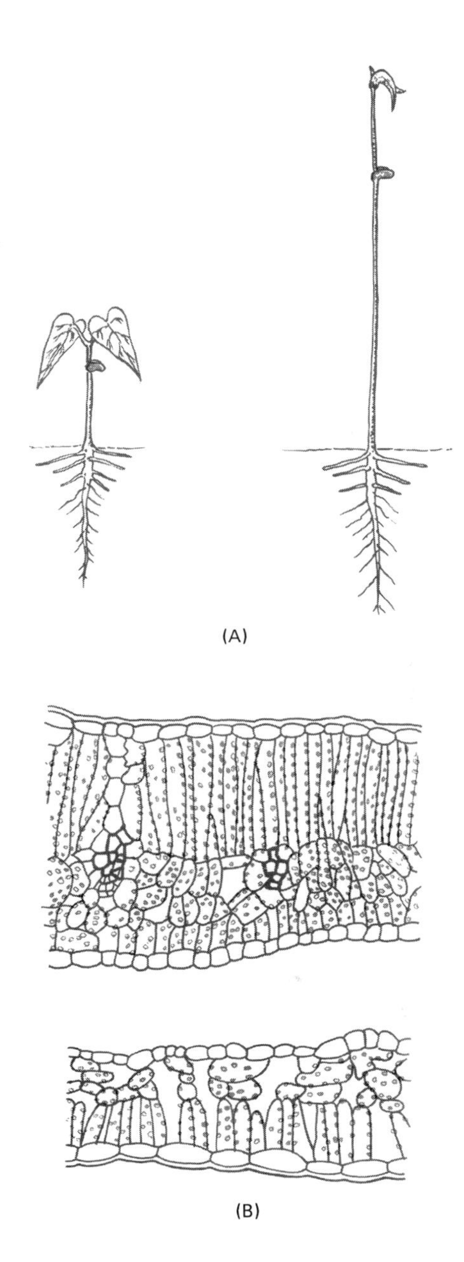

FIGURE 19-25. Light and plant growth. (**A**) The effect of darkness on the growth of bean plants. *Left:* In the light, stem growth is inhibited, while the growth of leaves is promoted. *Right:* In the darkness, bean plants develop long stems, smaller leaves, and other etiolated characteristics. (**B**) Sun and shade leaves of sugar maple. *Left:* Sun leaf has well-defined internal tissues. *Right:* Shade leaf exhibits a loose mesophyll, has less vascular tissue, and lacks heavy cutinization.

under glasses that transmit limited regions of the spectrum. At any given light intensity, various wavelengths of light influence plant growth and development in different ways. For example, it has been found that plants grown under the red-orange part of the spectrum exhibit certain aspects of etiolation. Plants grown under the blue-violet end of the spectrum exhibit growth patterns similar to those grown under the full visible spectrum.

Many striking results are noted when plants are grown in light from which the blue-violet portion of the spectrum has been eliminated. Plants grown under this condition elongate very rapidly during the first few weeks of growth. Some plants, such as *Coleus,* tomato, and four-o'clock, attain greater final height under these conditions, whereas others, such as sunflower and Sudan grass, do not. In either case, regardless of the final height, plants grown in the absence of blue-violet light have thinner stems, less well-developed vascular tissue, poorly developed flowers and fruits, and a higher degree of hydration than plants grown in the full light spectrum.

In some plants, such as corn, there are a number of genes which result in the expression of the albino trait. If a seedling plant is homzygous for albinism, it will grow several inches using the stored food in the seed, but will soon die because it fails to develop chlorophyll. In this instance, the genes for albinism are lethal since chlorophyll is necessary for the manufacture of food. Inasmuch as this condition may be carried by heterozygous (normal) plants, albinism continues to appear in many corn plants. Normal plants have genes that control the production of chlorophyll in the presence of sunlight; the chlorophyll itself is not inherited. This can be shown by raising corn seedlings in the dark where chlorophyll never develops because of the absence of sunlight. According to the laws of probability, if seeds from heterozygous parents are germinated, the ratio of chlorophyllous to albino plants should be 3:1. It is found, however, that all are albino if germinated in the dark. If these plants are then laced in sunlight, the green color of chlorophyll will begin to appear in those capable of producing it, giving the 3:1 ratio.

The nutrition of an organism is another environmental factor that affects the expression of genes. In *Drosophila,* for example, there is a gene that controls the development of giant body size. If fruit flies containing this gene are raised under crowded conditions in which the food supply is low, then a very few flies will express the gene for giant body size even though all may be homozygous for the trait. If the food supply is exceedingly sparse, none will show the trait. By contrast, if the food supply is abundant, then the majority of *Drosophila* having the genes will be giants.

20 Human Heredity

G U I D E   Q U E S T I O N S

1 Discuss the ways in which human genetics differs from the genetics of other organisms.
2 Explain multiple gene inheritance. How does this kind of inheritance differ from multiple allelic inheritance? Cite specific examples of each to illustrate differences.
3 Discuss in detail some human diseases that arise from gene mutations. How do gene mutations differ from chromosomal abnormalities?
4 Explain Down's syndrome. What is meant by "$^{15}/_{21}$"? Of what importance is nondisjunction in Down's syndrome?
5 Explain the nature of Turner's syndrome and Klinefelter's syndrome. What role is played by nondisjunction in each case?
6 What is amniocentesis? Why is the procedure important?
7 Of what significance are twin studies to human genetics? In what ways do fraternal and identical twins differ? How may the zygosity of twins be determined?
8 Discuss the roles of heredity and environment with regard to human genetics.

THE human organism, like all other species, exhibits wide variations in phenotypes. A cursory survey of a limited number of human beings would reveal considerable variety with respect to height, weight, eye color, skin color, mental ability, facial configurations, and so on. This is just what would be expected, since the basic human genetic mechanisms are the same as those of other organisms. This chapter will therefore be devoted to an analysis of some of our principal hereditary patterns.

In a very practical context, the study of human genetics necessarily differs from the genetics of other organisms, especially lower ones. For example, the geneticist is not able to arrange the matings of the human individuals being studied. In *Drosophila,* pea plants, or mice control is quite easy; with people, neither closely controlled mating nor standardized environment is feasible. In addition, the long human life cycle and the relatively small number of offspring produced are unfavorable factors in terms of certain standard research techniques. These, as well as other problems, make the study of human genetics difficult, but certainly not impossible.

Hereditary Patterns in Humans

Multiple Gene Inheritance

It was noted in Chapter 19 that certain hereditary traits are determined by the cumulative or additive action of several genes. Genes having such action are called multiple genes, or polygenes, and the phenotypes which they determine are quantitative in nature, that is, height, weight, breadth, and degree of pigmentation. A few examples of multiple gene inheritance will now be considered.

SKIN COLOR. One of the classic investigations of polygenic inheritance in humans was conducted by Dr. Charles Davenport on the heredity of skin color in black/white crosses in 1931. His research was conducted in Jamaica and Bermuda, and it suggested that the difference in skin color between whites and blacks was due to the presence on different chromosomes of two pairs of genes exhibiting incomplete dominance. One allele of each pair was responsible for the production in the skin of certain quantities of pigment (melanin) in an additive manner. Each gene is responsible for the ap-

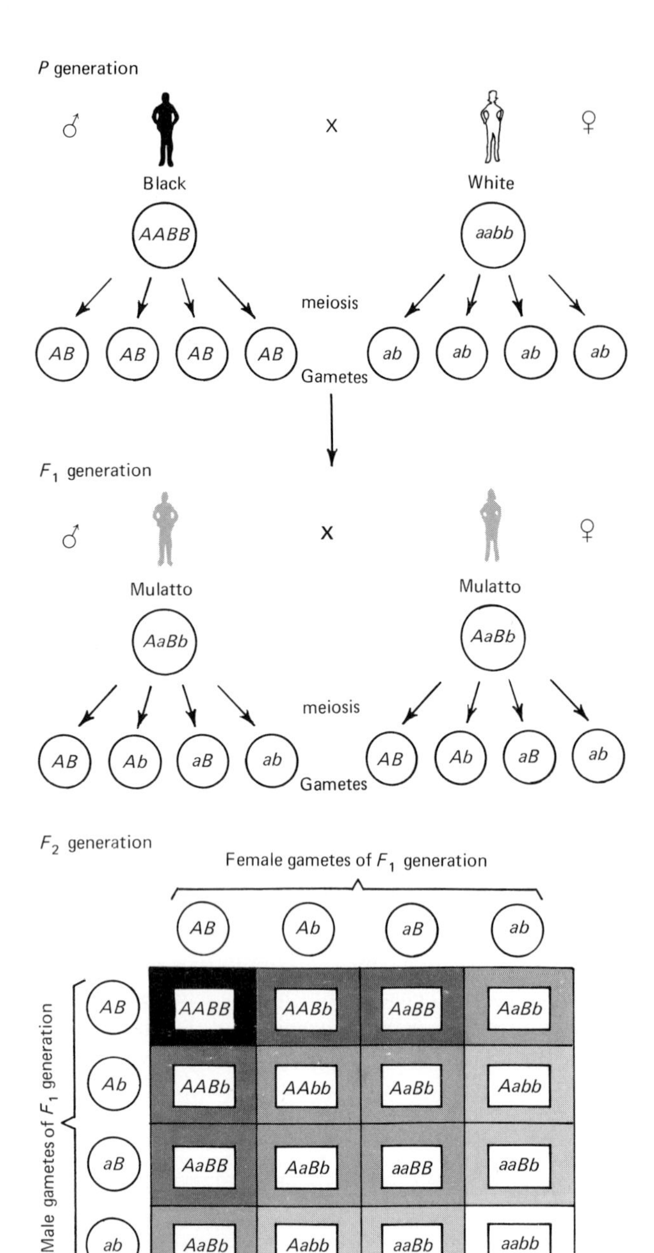

Figure 20-1. Multiple gene inheritance. Shown here are the offspring resulting from matings between mulattos. Since Davenport's classic investigation of human skin color, there has been some doubt as to the actual number of pairs of genes involved in the difference between white and black pigmentations.

F_2 phenotypes
$\frac{1}{16}$ 100% black
$\frac{4}{16}$ 75% black
$\frac{6}{16}$ 50% black
$\frac{4}{16}$ 25% black
$\frac{1}{16}$ 100% white

pearance of a given amount of pigment whether the other genes for pigmentation were present or not. Davenport used the symbols A and B to represent genes responsible for the production in the skin of a certain quantity of pigment in a cumulative fashion. The corresponding alleles, a and b, designated a lack of pigmentation. Thus, a full black would have the genotype $AABB$ (all four genes for pigmentation), and a full white person would have the genotype $aabb$.

If a full black person, $AABB$, is mated with a full white person, $aabb$, all F_1 offspring are mulatto (intermediate skin color) with the genotype $AaBb$. The offspring resulting from matings between mulattos, however, exhibit a wide range of skin colors from white to full black (Figure 20-1). This is possible because a mulatto produces four different kinds of gametes (AB, Ab, aB, and ab) and the random union of these gives rise to several genotypes that result in five different possible phenotypes. Note in Figure 20-1 that phenotypically there is statistical possibility of having $\frac{1}{16}$ as dark as a full black grandparent ($AABB$), $\frac{1}{16}$ as light as the full white grandparent ($aabb$), $\frac{6}{16}$ the same as the mulatto parents ($\frac{4}{16}$ $AaBb$, $\frac{1}{16}$ $AAbb$, $\frac{1}{16}$ $aaBB$), $\frac{4}{16}$ slightly lighter than the mulatto parents ($\frac{2}{16}$ $Aabb$, $\frac{2}{16}$ $aaBb$), and the remaining $\frac{4}{16}$ slightly darker than the mulatto parents ($\frac{2}{16}$ $AABb$, $\frac{2}{16}$ $AaBB$).

The pioneer work of Davenport has clearly demonstrated that skin color inheritance depends on the effects of cumulative multiple genes. There is, however, some disagreement as to the actual number of pairs of genes involved in the difference between white and black pigmentations. Research conducted by R. Gates (1949) in the United States suggests that three pairs of genes are involved in black and white pigmentation. Still other investigations (C. Stern in 1953 and 1954) are purported to show that the number of operative gene pairs is four, five, or six.

Eye Color. Traditionally, the inheritance of human eye color has been regarded as an excellent example of Mendelian inheritance based on a single gene pair. Most people think that since brown eyes dominate blue eyes, a simple dominant-recessive relationship is operative. Some careful thought, however, would reveal that some individuals cannot readily be classified as either brown-eyed or blue-eyed; there are all sorts of shades that are more or less intermediate between these—green, hazel, gray, light brown, dark brown, and black. Examination of the eye shows that color depends on the presence in the iris of the pigment melanin. If no observable amount of melanin is present, the eye appears blue. The quantity of melanin varies all the way from complete absence to such a large amount that the eye appears black. Such a continuous variation in a quantitative characteristic is usually found to depend on cumulative multiple genes. Although many investigations suggest that eye color depends on multiple genes, the number of pairs of genes involved has yet to be elucidated.

Other Polygenic Traits. In addition to skin color and eye color, the great range of continuous variations in hair color suggests that this trait is also influenced by multiple genes. In all of these, differences in phenotype are related to the amount of pigment produced. Other characteristics that vary quantitatively in a relatively continuous manner and do not involve the deposition of pigment include stature, weight, and intelligence. In any given population, stature varies continuously from shortest to tallest, with an average height near the midpoint between the extremes. Bell-shaped, normal frequency distribution curves are commonly approximated whenever sufficient data on stature are graphed and this suggests that inheritance is based on multiple genes acting continuously. Considering that the total lengths of body parts (head, neck, legs, and trunk) do no always vary correspondingly, and that the genes for stature are influenced by hormones and diet, the number and mode of action of genes involved in the determination of stature are uncertain. Weight is also a quantitative characteristic that varies continuously between extremes. Once again, even though variations in weight suggest polygenic inheritance, differences in diet have such effects that analysis of the exact genetic mechanism is quite difficult. At present, there is no accurate means for measuring inherent intellectual ability because of inadequate measuring devices and numerous variable factors. However, correlation of grades or the results of intelligence tests (approximate bell-shaped curves) with twin studies provides some evidence as to the inheritance of intelligence.

Multiple Alleles

The human ABO blood group was discussed in Chapter 19 as an example of multiple allelic inheritance. Another series of multiple alleles related to blood antigens constitutes what is referred to as Rh factors. Discovery of the red blood cell antigen related to Rh factor was reported by K. Landsteiner and A.S. Wiener in 1940 when blood from a rhesus monkey was injected into guinea pigs. The guinea pigs produced antibodies that will agglutinate (clump) the red cells of all rhesus monkeys. This observation suggests that the red blood cells of all rhesus monkeys contain a certain specific antigen. This antigen was designated as the *Rh factor.* It was also noted that when human blood was tested by the guinea pig serum, the cells of some persons clumped whereas the blood of others was not affected. On the basis of these findings, it was concluded that some individuals have the same Rh antigen as that found in the blood of rhesus monkeys, while others do not have it. Individuals containing the antigen are classified as *Rh-positive,* and those lacking it are designated as *Rh-negative* (13% of Caucasians); among the Mongoloid and Negroid races the Rh-negative percentage is considerably lower.

Although no individuals are known who contain natural anti-Rh antibodies, Rh-negative individuals can develop these antibodies upon exposure to the Rh antigen and this fact is medically important. Assume, for example, that an Rh-negative person is in need of an immediate transfusion and that a suitable donor with the correct ABO blood group is found. Assume also that the donor is Rh-positive. If the recipient has never had a positive transfusion before, there will be no reaction since there are no natural antibodies to Rh antigens. However, the presence of the Rh antigen on the cells received will stimulate the recipient to produce Rh antibodies. Now suppose that the same recipient requires another transfusion at a later date and the same donor is selected. In this instance the cells of the donor will clump as they react with the antibodies now present and the recipient may die. As a rule, no Rh-negative individual should ever be given a transfusion from an Rh-positive individual.

The Rh factor is also of clinical significance in childbirth. If an Rh-negative woman marries an Rh-positive man, some of their children may develop a condition called *erythroblastosis fetalis* (Figure 20-2). This abnormality is characterized by anemia due to hemolysis (breakdown) of red blood cells and jaundice. The disease may be so severe as to cause death before birth or within several days after birth. Frequently, such a marriage causes no problems since there is no direct connection between maternal blood and fetal blood and thus the bloods are not mixed. In some cases, however, in the late stages of pregnancy, a defect in the placenta may allow some mixing of blood between the two systems. If such a defect occurs in an Rh-negative mother bearing an Rh-positive fetus, the mixing of fetal and maternal blood causes the mother to begin synthesizing Rh antibodies. This mixing has the same effect as giving the mother an Rh-positive transfusion. Ordinarily, the antibodies produced by the mother do not reach appreciable quantities before the birth of the child. Subsequent children may, however, develop erythroblastosis fetalis if seepage across the placenta again develops. It should be noted that it is not a foregone conclusion that the second positive child from a negative mother will develop erythroblastosis. Physicians can test a mother's blood for antibody level and, if necessary, a blood transfusion can be given to a newborn infant. Moreover, if the husband is heterozygous for the Rh-positive factor, there is a fifty-fifty chance that any given child will have Rh-negative blood like the mother. Finally, it is possible that no placental seepage will occur.

The Rh factor represents one of the most complex of all the allele systems known in humans. The Rh series includes at least nine different antigens, and some investigators hold that the inheritance of these antigens is best explained by multiple allelic inheritance. In other words, this theory assumes that there are multiple versions of one gene. A second theory is that Rh inheritance is related to three or more separate genes that lie close together on the same chromosome and have very similar functions. Even though the Rh-positive phenotype can be produced by any one of several slightly different alleles, the genetics of Rh blood types may be treated like a simple monohybrid cross.

Sex and Inheritance

It has been noted that certain hereditary traits are related to sex chromosomes. Figure 20-3 shows a complete set of human chromosomes and the main regions in the X and Y chromosomes. Since these two pair together in synapsis during spermatogenesis, it is inferred that a certain portion of each is homologous to a corresponding portion of the other. In man, the X and Y chromosomes synapse end to end instead of side by side, and, therefore, homologous portions are probably small. Considering that evidence from many sources indicates that genetic alleles pair with each other, it would seem that at least a few pairs of genes are common to the X and Y chromosomes. Such genes are called *incompletely,* or *partially, sex-linked genes.* The greater portion of the X chromosome is not homologous with the Y chromosome, however. Genes in this portion of the X

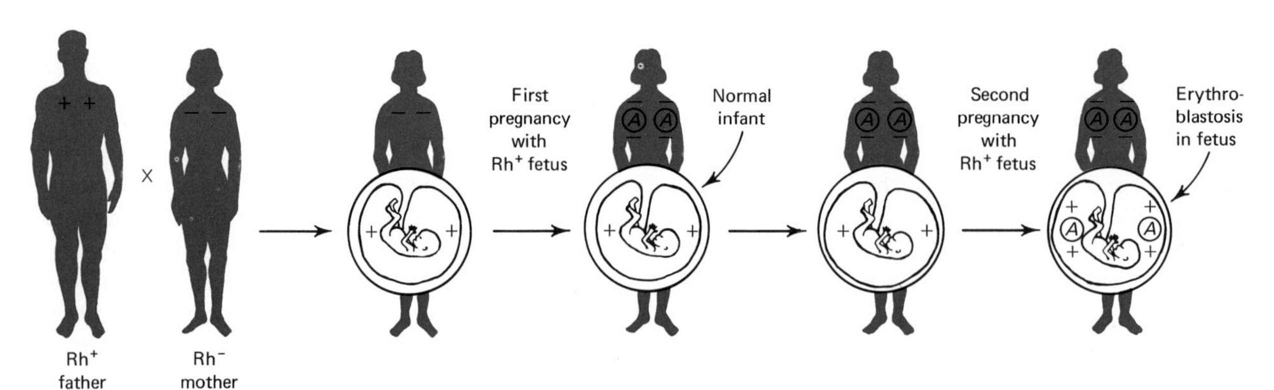

FIGURE 20-2. Sequence of events associated with erythroblastosis foetalis. The symbol (A) represents Rh antibodies. See text for details.

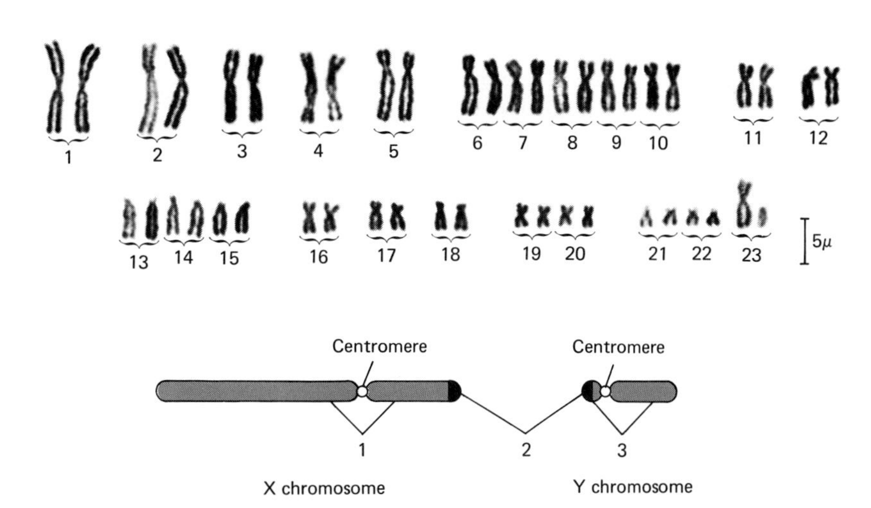

FIGURE 20-3. Human chromosomes and regions of the X and Y chromosomes. *Above:* Photo of the 23 pairs of chromosomes found in somatic cells of humans. The last pair (number 23) are the sex chromosomes. *Below:* Regions of the X and Y chromosomes: (1) segment of X chromosome not homologous to the Y chromosome, (2) segments of the X and Y chromosomes homologous to each other (synaptic sites), (3) segment of the Y chromosome not homologous to the X chromosome. [*Photo courtesy of Carolina Biological Supply Co.*]

chromosome, which have no alleles in the Y chromosome, are referred to as **X-linked (sex-linked) genes.** Similarly, a portion of the Y chromosome is not homologous to the X chromosome. Genes present in this portion of the Y chromosome are referred to as **Y-linked,** or **holandric, genes.** Still other genes are involved in sex-limited traits (characteristics limited to one sex) and sex-influenced traits (characteristics in which the dominance of an allele depends on the sex of an individual).

SEX LINKAGE (X LINKAGE). More than fifty sex-linked human traits have been described, most of which appear to be due to recessive genes. Of the completely X-linked human genes, the best known and most commonly encountered is the gene for *Daltonism,* or *red-green colorblindness.* Some data suggest that approximately two out of every 25 white males are red-green colorblind, but fewer than one in 200 white females are so affected. Such colorblind individuals have difficulty in distinguishing red from green. Since the gene for red-green colorblindness is recessive, the following possibilities exist if the symbols *C* (normal allele) and *c* (allele for colorblindness) are employed:

X^CX^C	Normal female
X^CX^c	Normal female
X^cX^c	Colorblind female
X^CY	Normal male
X^cY	Colorblind male

Hemophilia is a condition in which the blood fails to clot after a surface or internal injury, or clots very slowly. It is a much more serious defect than colorblindness; persons with extreme cases can bleed to death from even a small cut. Because clotting time in different hemophiliacs varies somewhat, it has been suggested that the condition is affected by varying modifying genes, perhaps even a series of multiple alleles. Male hemophiliacs (X^hY) occur with a frequency of about one in 10,000 and heterozygous females (X^hX^H) occur at about the same frequency. Hemophilic females (X^hX^h) occur in one of every 100 million births. The frequencies are lower in adults because some males die before reaching reproductive age and hemophilic females almost certainly die with the onset of menstruation.

Among a few other sex-linked traits in humans are nonfunctional sweat glands, forms of diabetes, types of deafness, uncontrollable rolling of the eyeballs, absence of central incisors, night blindness, a form of cataracts, white forelocks, juvenile glaucoma, and juvenile muscular dystrophy.

HOLANDRIC (Y-LINKED) INHERITANCE. Any genes that may be found in the nonhomologous portion of the Y chromosome will be passed directly from a father to all his sons but to none of his daughters. Direct inheritance through males but never through females characterizes Y-linked inheritance. Analysis of human traits that have been suggested as examples of

Y-linked inheritance are difficult because of the relative genetic inertness of the Y chromosome and because there is some difficulty in distinguishing holandric from sex-limited genes. Although no unequivocal cases have been defined in man, the most probable example is that of *hypertrichosis* (hairiness) of the pinna of the ear. The trait is variable with respect to the amount of hair developed; some males develop only three or four hairs on the pinna and in some the hair does not develop until the males are between 20 and 30 years of age.

SEX-LIMITED TRAITS. Sex-limited genes are those whose phenotypic expression is determined by the presence or absence of one of the sex hormones. Thus, their phenotypic effects are limited to one particular sex and they are typically responsible for secondary sex characteristics. The development of a beard in human males is one illustration of a sex-limited character. Since there is no significant difference between males and females in the number of hairs per unit area of skin surface, but rather only in their development, sex hormones appear to assume a role in limiting beards to men.

SEX-INFLUENCED GENES. An example of a sex-influenced trait in man is *pattern baldness.* Although this baldness may arise from several causes, including excessive exposure to high-energy radiation, diseases such as syphilis and seborrhea, and thyroid defects, pattern baldness is genetic in nature. It is more prevalent in males than in females and appears to be determined by a single pair of autosomal genes. The trait is dominant in men and recessive in women. A man is bald if he has only one gene for baldness, whereas a female must receive two such genes to be bald. It appears that a single gene can operate only in the presence of the male hormone.

Inheritable Defects in Metabolism

To date, almost one hundred human diseases have been described as metabolic and inheritable, the result of aberrations arising from gene mutations. These so-called inherited diseases develop as a result of an inherited molecular abnormality. All known inherited defects of metabolism have been ascribed either to a defective protein, usually an enzyme, or to a defect in the mechanism for regulating protein synthesis.

Alkaptonuria

Alkaptonuria, a disease of relatively rare occurrence, is characterized by the darkening of the urine upon standing. The urine turns black upon exposure to air because it contains a substance, *alkapton* (homogentisic acid), as a result of a genetic metabolic defect. Alkapton accumulates and is excreted because the individual is deficient in an enzyme that catalyzes its oxidation to carbon dioxide and water. Many persons with alkaptonuria suffer no apparent ill effects, but in later life, some suffer from degenerative arthritis. This appears to be caused by crystallization of some of the acid in the cartilages of the body with advancing age. There may also be a progressive darkening of the cartilages of the body, most noticeable in the nose and ears. In some cases, the darkening spreads to the fibrous tissue of the skin and the sclera of the eye. The gene defect responsible for alkaptonuria is inherited as a simple recessive trait.

Phenylketonuria

Phenylalanine is an essential amino acid that cannot be synthesized by humans and must therefore be supplied in the diet. Once ingested, phenylalanine may be (1) incorporated into cellular proteins, (2) converted to phenylpyruvic acid, or (3) converted to tyrosine, another amino acid. Tyrosine is a precursor for melanin pigments, thyroxine, and epinephrine and is also a protein constituent. The conversion of phenylalanine to tyrosine is catalyzed by the enzyme *phenylalanine hydroxylase* (parahydroxylase). Individuals with the genotype *pp* fail to produce the enzyme so that there is an accumulation of phenylalanine (phenylketones) in the blood. Instead of a normal blood phenylalanine level of 1 mg/100 ml, phenylketonurics may have a level of 50 mg/100 ml. Such individuals are said to have phenylketonuria (PKU), a condition that results in serious mental and physical retardation. Since most phenylketonurics have an IQ below 20, they are classified as idiots. Early diagnosis coupled with low phenylalanine diets for the first five years results in more nearly normal brain development.

Sickle Cell Anemia

Most individuals have normal hemoglobin and red blood cells that remain circular in outline even when deprived of oxygen. The erythrocytes of certain individuals, however, become quite distorted when deprived of oxygen, the distortion being referred to as "sickling." When sickling occurs, the sickled cells clog blood vessels where they are subsequently destroyed. Individuals with this condition, called sickle cell anemia, have a relatively short life expectancy. It has been determined that sickle cell anemia results from abnormal hemoglobin. Investigations of the polypeptide sequence of the abnormal hemoglobin S reveal that the only difference from the normal hemoglobin A polypeptide sequence is the replacement of a single amino acid of the nearly 300 comprising part of the molecule. In sickle cell hemoglobin, the amino acid valine replaces glutamic acid at a specific site in the protein structure, the rest of the molecule remaining unchanged. Using the symbols Hb^A and Hb^S to represent the genes for normal and sickling hemoglobins, we may designate the following genotypes:

$Hb^A Hb^A$	Normal individual
$Hb^A Hb^S$	Individual who carries the trait without showing the symptoms
$Hb^S Hb^S$	Individual having sickle cell anemia

Note that in order for an individual to have sickle cell anemia, both genes for the abnormal trait must be present (homozygous condition). An individual with one gene for normal hemoglobin and one for sickling hemoglobin (heterozygous condition) carries a sickling tendency but does not show symptoms of the disease.

Chromosomal Abnormalities

Down's Syndrome

It has recently been found that a very important instance of trisomy in man involves a condition referred to as *mongoloid idiocy,* or *Down's syndrome* (Figure 20-4A). This disorder, formerly known as mongolism, is characterized by mental retardation, retarded physical development (short stature, stubby fingers), distinctive facial structures (large tongue, broad skull, slanting eyes, and round head), and malformations of the heart, ears, and feet. Sexual maturity is rarely attained.

There have been many theories as to the cause of mongolism. In 1959 it was discovered that individuals with the disorder usually have 47 chromosomes instead of the normal 46. Further chromosome studies have shown that the syndrome is associated with the extra chromosome. All chromosomes are in pairs except chromosome 21, which is present in triplicate (Figure 20-4B). This trisomic condition is believed to arise from a nondisjunction of the chromosome 21 pair during meiosis in the mother, and possibly (rarely) in the father.

Although mongoloids may be born to mothers of any age, there is a tendency for them to be born to older mothers. For example, the chances of a woman of 20 bearing a child with Down's syndrome are about one in 3000 and by age 45 the probability rises to one in 40. The age of the father does not appear to be significant; the percentage of Down's syndrome in children from 20-year-old fathers is about the same as that from 60-year-old fathers. Some experimental evidence suggests that the tendency to nondisjunction during oogenesis increases with the age of the mother. For example, studies of *Drosophila* were performed in which virgin females were aged for varying periods of time before mating was conducted. The females were then treated with X-rays to increase the frequency of nondisjunction. The aged females showed a much higher percentage of nondisjunction in their offspring than did younger females. In mammals, all of the oocytes a female will ever have are present at birth or shortly thereafter. In the human female, the aging of the oocyte plus the accumulating effects of environmental stimuli could account for the increase in nondisjunction with age. In the human male, by contrast, aging of primary sex cells is of no consequence because sperm production is a continuous process during all of the reproductive years.

Some mongoloids are also found in which 46 rather than 47 chromosomes are present. In such instances, one of the chromosomes is found to be exceptionally long because of a translocation in which the extra chromosome 21 becomes attached to one of the other autosomes (usually number 15). The type of translocation is frequently referred to as "15/21." A mongoloid with a 15/21 chromosome also has two regular 21 chromosomes. Although the observed number of chromosomes is not increased over the normal 46, the substance

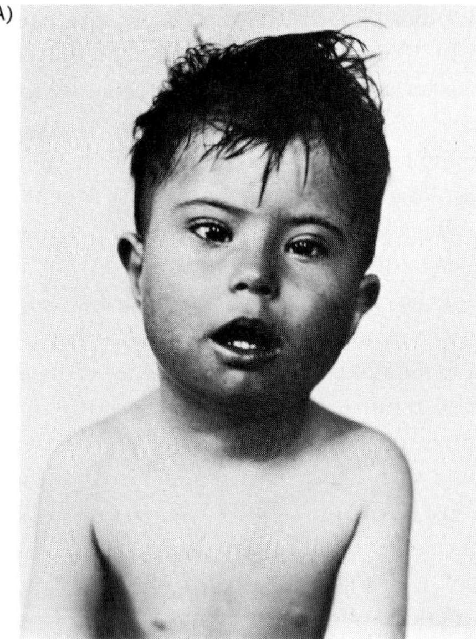

Figure 20-4. Down's syndrome. (**A**) Photo of a child illustrating some physical abnormalities associated with Down's syndrome. (**B**) Karyotype (size and appearance of metaphase chromosomes) of a trisomic mongoloid idiot having 47 chromosomes. The triploid condition at chromosome 21 is circled. [*From George W. Burns,* The Science of Genetics, *3rd ed. New York: Macmillan Publishing Co., Inc. Copyright © 1976 by George W. Burns.*]

(genes) of three number 21 chromosomes is still present. Thus, whereas in triplo-21 mongoloids the extra chromosome 21 is present as a separate entity, it is attached to another chromosome in translocation mongolism. Examination of parents of translocation mongoloids usually reveals that one of them has only 45 chromosomes including one 21, one 15, and one fused 15/21. Since the genetic material is present in the proper amount, such individuals are phenotypically normal.

Turner's Syndrome

In other instances involving nondisjunction of the sex chromosomes, individuals exhibit various abnormalities in the growth and development of sex organs. If an egg is produced in which only autosomes are present and if the egg is fertilized by a sperm containing an X chromosome (and its complement of autosomes), the resulting zygote will contain, in addition to 44 autosomes, only one X chromosome (Figure 20-5). Human X0 individuals are characterized by a combination of traits known as *Turner's syndrome.* About one out of every 3000 female births results in a child with this abnormality. Although the individuals are phenotypically females, adolescent and adult traits do not develop normally and they never reach functional maturity. Persons afflicted with Turner's syndrome are dwarfed physically, averaging only about 4 feet, 10 inches when adult, and often are mentally retarded. There are other characteristics, too, such as "webbing" of the skin on the side of the neck, and wide-spaced nipples of the immature mammary glands. Autopsies performed on such individuals reveal that the ovaries consist primarily of connective tissue, so that there is little or no secretion of female hormones (estrogens).

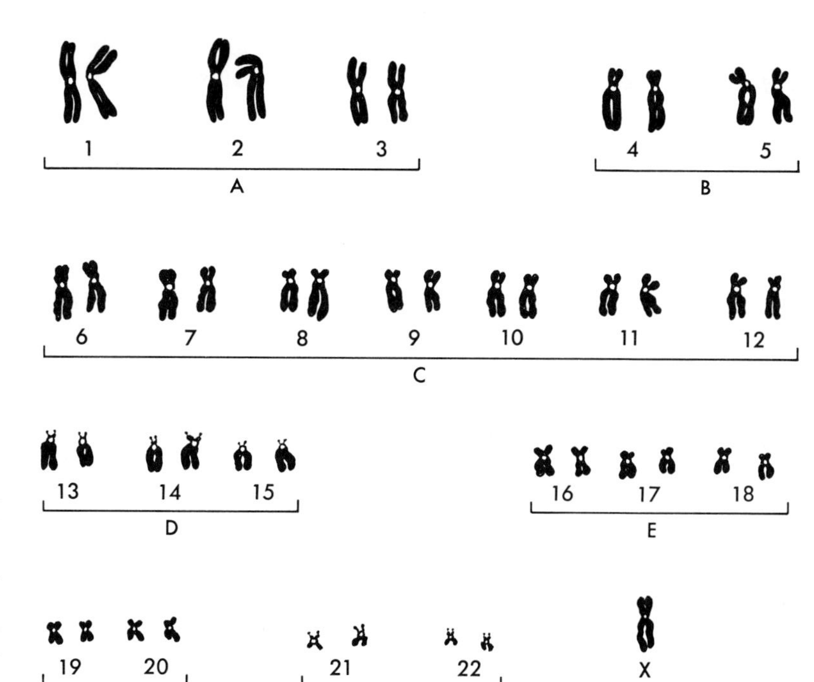

FIGURE 20-5. Karyotype of an individual with Turner's syndrome. Note that such persons are X0 with only 45 chromosomes; only one X chromosome is present. [*Reprinted with permission from Dr. Victor A. McKusick,* Journal of Chronic Diseases, *vol. 12, pp. 1–202, 1960. Copyright © 1960 by Pergamon Press.*]

Klinefelter's Syndrome

A second major sex anomaly is represented by *Klinefelter's syndrome.* Individuals with this disorder are phenotypically male, though genitalia are underdeveloped and body hair is sparse. Most cases have some degree of breast development and tend to have longer than normal arms and legs. Intelligence is also generally subnormal. All Klinefelter's individuals are sterile. Cytological studies of persons with Klinefelter's syndrome show that genotypically they are XXY, having 47 chromosomes instead of the usual 46 (Figure 20-6).

Prenatal Detection: Amniocentesis

The amniotic fluid that bathes the fetus has many living cells floating in it. These cells are derived mostly from the skin and respiratory tract of the fetus. **Amniocentesis** is a technique of withdrawing some of this amniotic fluid by hypodermic needle puncture of the uterus, usually 16–20 weeks after conception. A small amount of amniotic fluid is removed, and its cells are examined for biochemical defects and/or for abnormalities in chromosome number or structure. All disorders of chromosome number usually can be detected by this procedure early in pregnancy, and approximately forty different diseases involving enzyme deficiencies usually can be detected as well.

Heredity and Environment: Twin Studies

The human geneticist is confronted with a multiplicity of problems because his experimental organism is man. As noted earlier, humans have comparatively few offspring, they do not lend themselves well either to genotype standardization or to maintenance in a uniform environment, and planned matings are impossible. It is for this reason, as well as others, that twin studies are so important to geneticists. The experimental design in humans that most closely approaches the requirements for the geneticist is a comparison of identical and fraternal twins. Identical twins represent the only human example of individuals with exactly the same genotype and permit studies of the effects of the environment on similar genotypes, whereas fraternal twins permit studies of the effects of similar environments on different genotypes.

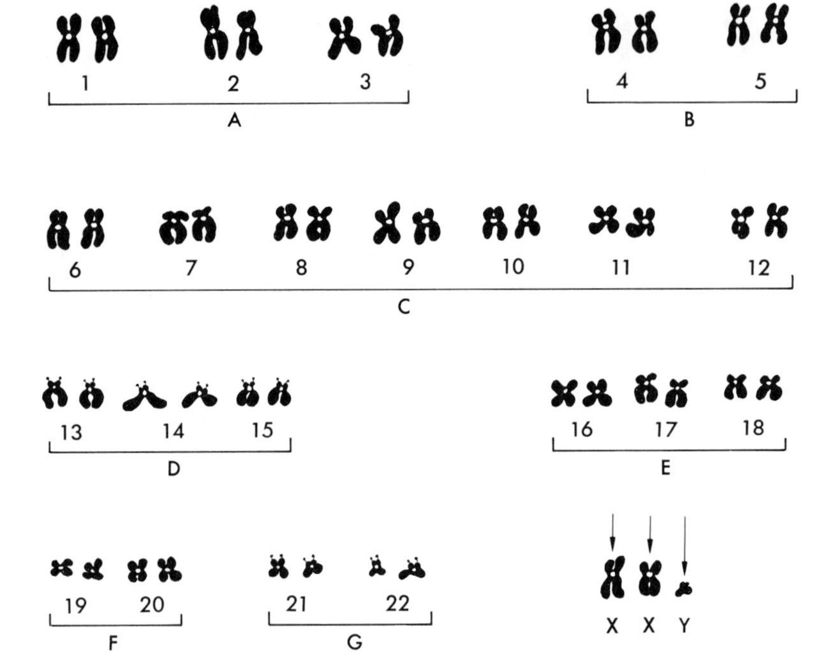

Figure 20-6. Karyotype of an individual with Klinefelter's syndrome. Note that such individuals are XXY and have 47 chromosomes. [*Reprinted with permission from Dr. Victor A. McKusick,* Journal of Chronic Diseases *vol. 12, pp. 1–202, 1960. Copyright © 1960 by Pergamon Press.*]

The two kinds of twins just referred to differ with respect to origin and it is for this reason that they also differ genetically. *Fraternal,* or *dizygotic, twins* result from the fertilization of two different eggs by two separate sperm and are no more similar genetically than other siblings born at different times. *Identical,* or *monozygotic, twins* arise from a single fertilized egg that begins development and subsequently splits into two embryos. The individuals thus produced are genetically identical and are always of the same sex.

Quantitative traits such as stature, weight, and intelligence are the ones most commonly studied in twins in order to estimate the relative importance of heredity and environment. Typically, researchers assume that if genetic components are involved in the expression of quantitative traits, the average difference between pairs of monozygotic twins should be less than the difference between pairs of dizygotic twins. Results of a study of stature do in fact reveal that there are significant differences between the two types of twins. For example, it was found that differences in stature are much greater between dizygotic twins of the same sex than between monozygotic pairs of the same sex. Such results indicate that there is a strong genetic component in stature. With regard to weight, variations are much greater than for stature. The authors of the study dealing with weight concluded that, unlike stature, weight appears to be predominantly under environmental control.

The underlying theory in applying the twin method to the study of intelligence is the same as that applied to stature and weight; that is, variance in monozygotic twins should be less than it is in dizygotic twins. The evidence seems conclusive that there is a genetic as well as an environmental component in determining an individual's performance on intelligence tests. The mode of inheritance, however, although probably consisting of multiple genes, is not yet known.

In any experimental design, the researcher attempts to introduce only one variable factor at a time while all other factors are kept constant. Thus, in order to ascertain the roles of heredity and environment, two parallel experiments should be designed: (1) one in which the heredity is constant but the environment is varied and (2) one in which the environment is constant but the heredity is varied. Although these conditions cannot be duplicated exactly in humans, an approximation of the first type of experiment is possible by studying monozygotic twins separated early in life and raised in separate environments. Such studies conducted by various investigators reveal that although the monozygotic twins were reared apart, often without knowledge of the co-twin's existence, the twins continued to be remarkably alike in many psychological traits. On the other hand, an approximation of the second experimental situation consists of studying dizygotic twins reared together in the same house or investigations of unrelated children reared together in foster homes or institutions. Children reared together in an orphanage (constant environment) show just as wide variability in intelligence as children reared separately in their own homes. Even when children are adopted early in infancy, there is a greater correlation between the intelligence of the child and its true parents than between the child and its foster parents. Such data seem to indicate that although the upper limits of an individual's intelligence are genetically determined, the degree to which the potential is developed is determined by environmental influences, training, and experience.

21 The Evolution of Life

GUIDE QUESTIONS

1 List three fundamental aspects of life. How do inorganic and organic evolution differ? What is implied by organic evolution?
2 Discuss how the following areas of investigation support the concept of evolution: comparative morphology, comparative embryology, comparative physiology and biochemistry, genetics, geographical distribution, taxonomy, and paleontology and geology.
3 What are homologous organs? What is recapitulation? Define a fossil. How are fossils dated?
4 What are the tenets of Lamarck's evolutionary theory? Discuss the components of Darwinian evolution.
5 What is the basic difference between Darwin's theory of evolution and the mutation theory of evolution? What is meant by the synthetic theory of evolution?
6 What is the importance of genetic recombination to the concept of evolution?
7 In what respects do mutations provide the raw materials for evolution? Distinguish between gene mutations and chromosomal mutations.
8 How has the concept of natural selection been modified since Darwin's time? What is genetic equilibrium?
9 Explain the role of natural selection in the evolution of the horse and the peppered moth.
10 Why may the genotypes of a population be represented by many alleles while that of an individual may be represented by a single pair of alleles for a given gene?
11 What is a gene pool? How may it be characterized? Of what use is a Punnett square in representing gene frequencies?
12 What is the Hardy–Weinberg law? Discuss some of its practical applications. Under what conditions is this principle said to be operative? What actually occurs in nature that modifies this principle?
13 Discuss what is meant by indeterminant evolution or genetic drift.
14 Explain why changes in populations occur mainly because of selection acting upon new gene combinations. Why do mutations play only a minor role in population change? What is mutation pressure?
15 What relationship exists between environmental conditions and genetic variability? What is population isolation? How does migration affect gene pools?
16 What is meant by reproductive isolation? Define speciation. Distinguish and discuss spatial and genetic isolation. Outline the major steps involved in speciation.
17 Define adaptation. How are adaptations related to evolution?

ALL forms of life exhibit both diversity and unity. It is through the concept of evolution that these two aspects are understood to complement rather than oppose each other.

The Meaning of Evolution

Any attempt to convey the meaning of evolution must first consider certain fundamental aspects of life. One aspect of living things is that they all consist of a unique combination of nonliving chemicals that are assembled into a particular structural organization. From the molecular level of organization through the ecosystem, the properties of life are related to the ways in which the components are organized into orderly patterns. Thus, life is characterized by organizational states. A second basic aspect of life is its capacity for reproduction. Through this mechanism, each kind of living form is capable of producing new individuals. Essentially, the offspring resemble the parents in many ways, although certain variations may exist. A third principal characteristic of all living forms is their ability to adjust to an ever-changing world—this phenomenon is called **adaptation.** Every living form adjusts constantly to both sudden and gradual changes in its surrounding environment. Since the origin of the first living form, the adaptive mechanism has produced the enormous diversity that characterizes life; this diversity provides strong support for the theory of evolutionary change.

What, then, is meant by evolution? **Evolution,** in its simplest and broadest sense, means a gradual and orderly change from one condition to another. For nonliving systems the term is *inorganic evolution.* For example, geologists study the evolution of land masses, astrophysicists are concerned with the evolution of chemical elements, and so on. We, however, are interested in evolution of living organisms—strictly speaking, in *organic evolution.* For us, then, evolution is a concept of biological change in organisms according to an orderly sequence of events over a long period of time. Evolution also means that species have been altered, some have become extinct, while still others have arisen. Organisms resemble each other because they have acquired hereditary traits from a common ancestral form, and related groups of organisms differ from each other because of changes in these hereditary traits that have accumulated since they became separated in their lines of descent.

Since evolution means continuous change, certain implications necessarily follow. It is generally believed that life originated only once, some 3–4 billion years ago, and that, from this point on, organisms developed through evolutionary change. Second, all life comes from life through inheritance. Third, all life is related in an unbroken chain from the time of its origin to the present.

Most biologists agree that the first living organisms were very simple forms, simpler even than the most elementary forms of life known today. Moreover, the concept of the origin and early development of life as outlined in Chapter 3, although speculative and tentative, is generally accepted. From these primitive beginnings, other organisms of greater complexity and diversification have evolved through the incorporation of numerous adaptations into individuals and groups of individuals. Although processes leading to simplifications have occurred in certain groups of organisms, in general, the trend has been to ever-increasing structural complexity and specialization.

Evidence for Evolution

The supportive data for evolution, derived from various lines of research, demonstrate that evolution is not limited to the past but is still going on today. Among the areas from which supportive data have come are comparative morphology, comparative embryology, comparative physiology and biochemistry, genetics, geographical distribution, taxonomy, paleontology, and the geologic record.

Comparative Morphological Data

The science of structure, or **morphology,** provides some important and indeed some of the original evidence for evolution. In fact, the principle of classification, that is, orgaizing all living forms into related groups, is based primarily on that of similarities and differences in structure. (Some taxonomists also employ biochemical, behavioral, and ecological criteria for establishing relationships.) Usually the greater the structural similarities, the more closely related the organisms. In general, structural similarities indicate common ancestry, and common ancestry implies close relationship. Conversely, pronounced structural differences suggest a more distant relationship with a common ancestor ages ago. As one example, consider the angiosperms (flowering plants). All of these plants are structurally similar in that they produce flowers. Many subdivisions of flowering plants exist with regard to the particular type of flower, leaf pattern, and type and arrangement of internal tissues, and, based upon these differences, subgroups are easily defined. As a whole, however, angiosperms are more closely related to each other than they are to gymnosperms (plants that usually bear cones rather than flowers). Other structural features used to classify plants are wood anatomy, internal tissue development, and seed structure.

The morphological resemblances among animals that indicate genetic relationships are perhaps more conspicuous than those found in plants. Organs that have the same basic structure, the relationship to other organs, and the same type of embryonic development are referred to as **homologous organs.** An outstanding example of such homologous organs is found in various vertebrate animals. The great similarity in skeletal structure of the arm of a human, the fin of a whale, the wing of a bird, and the forelimb of a crocodile indicates a homologous relationship (Figure 21-1). It would be quite difficult to account for these likenesses on any basis other than that of common origin. Accordingly, the more recently two groups of organisms have shared an ancestor, the greater the number of homologous organs they have in common.

Comparative Embryological Data

Embryological characteristics provide an extremely valuable means of relating organisms in still other ways. The developing organism often passes through a series of changes, some of which seem to repeat, at least partially, the adult traits of lower and presumably more ancient forms, even though the mature individuals are quite different. These changes are interpreted as a repetition, in very condensed form, of the stages through which the particular species has passed in its evolution from ancestral forms. This idea that embryonic development repeats that of ancestral organisms is called **recapitulation.** As one example, consider the embryonic development of vertebrate embryos at comparable stages (Figure 21-2). Some structures appear during the development of the most advanced vertebrates just as they do in the more primitive species, but they either disappear or become greatly modified in the later stages of development in advanced vertebrates. All vertebrates, as noted earlier, develop pharyngeal pouches during early embryonic stages. In fish, these pouches become functional gills; in amphibians, they represent functional gills only during the aquatic larval stage but become modified and nonfunctional in the adult; in reptiles, birds, and mammals, the pouches appear only during embryonic stages and either disappear or become modified into parts of the ear, pharynx, tongue, and

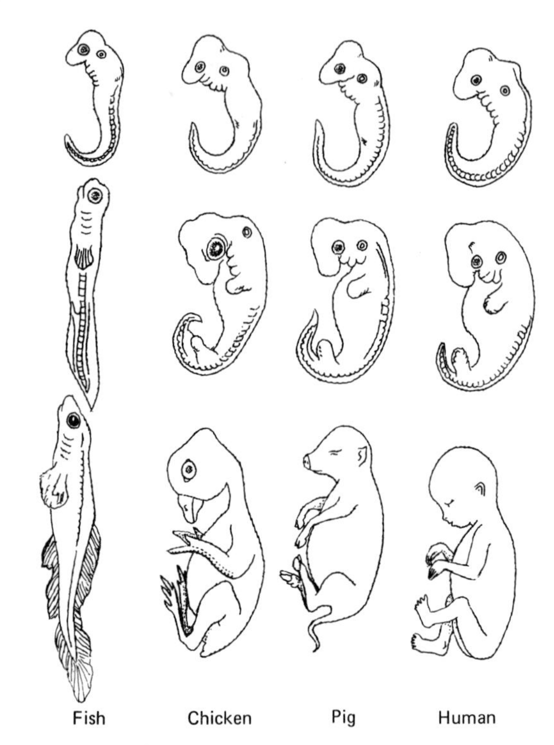

FIGURE 21-2. Representative stages in the development of the embryos of four vertebrates. Note the high degree of similarity among the earlier stages and the progressive structural differences among later stages.

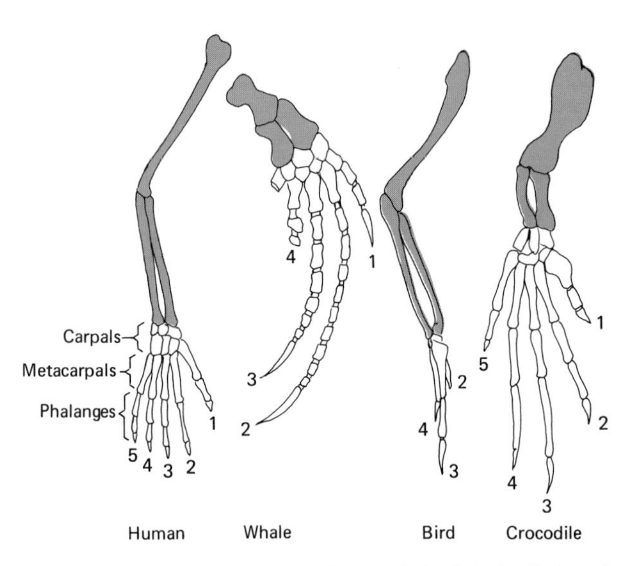

FIGURE 21-1. Homology in the bones of the left forelimbs of representative vertebrates. Note that differences in length, shape, and numbers of digits adapt each kind of forelimb for specialized functions such as handling, swimming, flying, and crawling, respectively.

certain glands in the adult. The temporary possession of a tail and a two-chambered heart also illustrates recapitulation in the human embryo.

Among plants, recapitulation is also observed. Germinating spores of mosses produce a short filament of green cells (protonema) that resembles a filamentous green alga (see Figure 17-12). After a period of growth, the protonemata develop into mature moss plants. For a brief period of time, however, these plants pass through a stage reminiscent of the algae from which they probably evolved.

Comparative Physiological and Biochemical Data

Just as morphological and embryological evidence show relationships through homologies, comparative physiological and biochemical data provide support for evolution through similar functional activities. In other words, functional as well as structural recapitulation occurs. All organisms, from microorganisms to humans, possess a basic biochemical and physiological resemblance, as expressed in their chemical composition, chemical reactions, and physiological activities. At the molecular level, all living organisms consist of the same basic chemicals—nucleic acids, proteins, carbohydrates, lipids, water, and minerals. Moreover, all organisms possess, with some variation, essentially similar or related types of chemical reactions. For example, studies of mammalian hemoglobins reveal close structural similarities, especially among species that are believed to be closely related. DNA and/or RNA is found in virtually every living organism, and it is assumed that these nucleic acids contain the same hereditary controlling mechanism in all forms.

In vertebrate animals, the introduction of any protein into the bloodstream of an animal in which the protein is not normally found causes a specific reaction to occur. In such a reaction, the foreign protein is called an *antigen* and the counteracting substance produced in response to its presence is referred to as an *antibody.* Various serological (blood and body fluid) techniques have shown that antigens of a given animal react very strongly with the body fluids of other animals. Moreover, the reaction is strongest in those animals that have close morphological resemblances to the animals whose antigens are involved, and reactions are weaker in those animals in which morphological similarities are more distant. Such antigen–antibody reactions support the already established morphological evidence that humans are more closely related to the primates than to any other groups of mammals. The uniformity of biochemical organization among different groups of organisms supports the conclusion that these groups descended from a common ancestor; similarities would be less likely to exist if there had been more than one ancestral form.

Similarities in physiological processes at higher levels of biological organization also provide support for the evolutionary resemblances and closeness of ancestry among organisms. For example, in vertebrates, the physical mechanisms of organ function of the digestive, excretory, respiratory, nervous, endocrine, and reproductive systems all operate in basically the same manner. Among green plants, their universal capacity to carry on photosynthesis suggests a common ancestry.

Genetic Data

Resemblances among organisms are directly dependent upon the similarities and differences of their genotypes, that is, their gene makeup, since genes control morphological and physiological traits. Thus, variations and similarities among living forms are essentially genetic ones. One of the most convincing proofs of the concept of evolution is the observation of the origin of new forms and an analysis of the mechanisms that brought these new forms into existence. Such evidence has been provided by demonstrating that organisms have a potential for change. Simple experiments in hybridization (combining parents with the most desirable traits) using plants may result in the production of a new species. One of the best-known examples is *Raphanobrassica,* derived from the cultivated radish and the cultivated cabbage (see Figure 19-22). Genetic evidence of evolution is also found in mutation, a genetic change, that may give rise to new species.

For thousands of years, selective breeding has been applied to produce more desirable varieties of domesticated animals and cultivated plants, and forms that vary greatly from the ancestral stock have been produced. The ability to bring about evolutionary change is readily apparent in the enormous diversity of domesticated dogs; the many breeds of cows, horses, sheep, chickens, and goats; and the many recently developed fruits such as nectarines and seedless oranges. Thus, from genetic studies, through cultivation and domestication, and from such specific examples in nature as the emergence of races of insects that are resistant to DDT and other insecticides and the resistance of certain bacteria to antibiotics, there is ample evidence of the potentiality of living organisms for change—a potential that is a major theme of evolutionary theory.

Data from Geographical Distribution

A brief examination of even a limited geographical region would reveal that organisms are not evenly distributed. Notable examples of discontinuous distribution are the geographic habitats of camels (Africa and Asia), llamas (South America), magnolias (China, Japan, and the eastern United States), and alligators (southeastern United States and the Yangtze River in China). To a large extent, such a distribution of plants and animals may be explained on the basis of an environment conducive to growth and continuity. Still unexplained, however, is the existence of habitats throughout the world that do not contain organisms that would presumably be well adapted there. The Hawaiian Islands, for example, provide a suitable environment for frogs and salamanders, yet neither of these forms is found there. Similarly, numerous species of cacti that are abundant in the desert areas of the southwestern United States are not found in the African deserts.

In attempting to explain the occurrence of discontinuous distribution, it can be assumed that all organisms are the product of evolution and that each species evolved at a given place on the face of the earth. It can further be assumed that any subsequent dispersal of the species would depend upon the efficiency of its dispersal mechanism (locomotion structures in animals; spores and seeds in plants) and the presence or absence of barriers to migration (mountains and bodies of water). Moreover, correlation of the present-day distribution of plants and animals with the fossil record indicates that geographically isolated organisms were at one time continuously distributed over a vast area. As a result of climatic and geological changes, however, many groups of organisms could no longer survive and thus became extinct. Essentially, the distribution of various organisms demonstrates that species are changeable, another fact that lends strong support to the theory of evolution.

Taxonomic Data

Various systems of taxonomy attempt to arrange living forms into a natural sequence to show evolutionary relationships among various groups. Regardless of the size of a particular group, all of its members resemble each other with regard to structural, embryological, biochemical, or physiological similarities. As stated previously, structural similarity indicates common ancestry and common ancestry implies close relationship. Conversely, pronounced structural differences usually suggest more distant relationships with a common ancestor. For example, organisms as diverse as humans and fish are classified in the same large group because at some stage of development they have certain characteristics in common, all of which are inherited from some ancient form that was ancestral to both. Inasmuch as organisms can be arranged into distinct groups from simple to complex forms, such a classification at least suggests that evolution must have taken place.

Data from Paleontology and Geology

The final evidences to be presented that evolution occurs are paleontological and geological data. The study of fossils (paleontology) has provided by far the most direct evidence for evolution, and, based upon fossil records, scientists have attempted to construct a time sequence for the appearance of organisms living today.

A **fossil** may be defined as the remains of a once-living organism or an impression of that organism. Inasmuch as the fossilization process occurs only under ideal conditions, most ancient organisms decomposed completely and did not fossilize. Therefore, fossils constitute a very small proportion of the total number of organisms that have evolved during various periods of the earth's history, and the fossil record is incomplete. Two examples of fossils are shown in Figure 21-3.

The fact that only certain kinds of organisms are suitable for fossil preservation and the fact that there is no orderly process of fossil formation eliminate any hope that the history of the evolution of all organisms can be obtained. Moreover, there are many gaps in the fossil record, so that there is never an unbroken chain of evolutionary history. Yet, in spite of these inadequacies, paleontological investigations do substantiate the concept of evolution. For example, study shows that different kinds of fossils occur in various layers of rock, and from this observation it can be concluded that different types of organisms have inhabited the earth at different periods during its history. In other words, the organisms that have inhabited the earth have been changing constantly. Furthermore, the kinds of fossils found in each rock layer are generally morphologically similar to those found in adjacent strata. Such similarities indicate close relationships among organisms and suggest that the organisms of various periods have developed from those of earlier times. In addition, the fossil record indicates that the structure of organisms has become more complex and specialized as the earth has aged, and that many forms have arisen, lived for a time, and then become extinct.

Most logically, if fossil specimens are to be of any value, their age must be determined. Geologists have deduced that different geological strata are characterized by unique assemblages of fossils, and that particular strata could be identified by their fossil contents. The various sedimentary rock strata have been deposited in the order of the time of their formation. Thus, the oldest rock strata are those found at the bottom of any given sequence of layers and the most recently

(A)

FIGURE 21-3. Fossils. (**A**) Leaf fossil. (**B**) Fossil of a dinosaur foot. [*Courtesy of Field Museum of Natural History, Chicago.*]

(B)

formed ones occur above these in a definite sequence. Geological dating procedures make it possible to establish the age of various fossils and strata within fairly accurate limits.

One such dating procedure involves the use of radioactive decomposition measurements of isotopes of elements found in rocks. Radioactive elements undergo transformations into more stable forms by the emission of subatomic particles, such as helium nuclei or electrons. This transformation, or decay, continues until all the atoms of a radioactive element have been converted to a stable form. Atoms of ^{238}U (uranium), for example, decay to the stable form of ^{206}Pb (lead). The essential feature of such decay is that it proceeds at a constant, measurable rate which differs for isotopes of different elements. Thus, the *half-life,* or period of time required for one-half of the atoms of the element to decay, for ^{238}U is 4.5 billion years, and that for carbon (^{14}C to ^{12}C) is about 5760 years. By comparing the ratio of ^{238}U to ^{206}Pb or the ratio of ^{14}C to ^{12}C in a given rock, its age can be determined.

Inasmuch as uranium is relatively rare as a component of rocks and fossils, its value as a measuring device is somewhat limited. The uranium–lead method, however, has proved to be highly valuable, particularly in determining the age of the earth. For example, the oldest rocks known are estimated by this method to be 3.6 billion years old. The age of the earth is estimated to be between 4 and 5 billion years, probably 4.5 billion. Coincidentally, a rock sample from the moon returned to earth by Apollo 12 has been found to be 4.5 billion years old. A more satisfactory method of determining the age of fossils, especially recent ones, is *radiocarbon dating.* This procedure depends on the well-established theory that a small but constant proportion of the carbon utilized by living organisms is ^{14}C. This means that a material, such as wood, which was at one time part of a living plant, can be analyzed for the amount of radioactive carbon present, thus determining its age. The accuracy of the radiocarbon method is limited to materials that are no more than about 50,000 years old. Table 21-1 contains a listing of isotopes, half-lives, and types of materials tested.

By a variety of dating methods, geologists have devised a time scale that categorizes the geologic history of the earth. The scale, called a *geologic time scale,* is shown in Table 21-2. It can be seen that the geologic history of the earth is divided into five successive main divisions of time called *eras.* The three most recent eras are further subdivided into a number of successive *periods.* The beginning and terminal dates of the eras and periods are designated in terms of major geologic events known to have occurred at those times. The transitions between eras, for example, were characterized by great upheavals consisting of mountain-building processes and severely fluctuating climates. Periods were terminated by minor geologic revolutions.

Table 21-1. Representative Isotopes, Half-lives, and Types of Materials Tested

Radioactive Isotope	Half-life (years)	Types of Materials Tested
Carbon-14	5,760	Wood, charcoal, shells
Protactinium-231	34,300	Deep-sea sediments
Thorium-230	80,000	Corals, shells, deep-sea sediments
Uranium-234	248,000	Corals
Beryllium-10	2,700,000	Deep-sea sediments
Potassium-40	approx. 1.3 billion	Volcanic ash, lava

The various eras and periods of the history of the earth are determined not only by certain physical differences among rock formations but also by the types of fossils found in each layer. Accordingly, geologists postulate that rock formation processes occur at about the same rate in the present as they did in the past and that the thicker the layer of rock, the more time required for its formation. Moreover, it is assumed that deposition of rocks has occurred in an orderly process, the older layer in a series being the bottom, and that the fossils found in a given stratum represent life of the period during which deposition of the rock occurred. Based upon these data, it is believed that the oldest rocks contain fossils of only the most primitive organisms while the younger formations contain fossils of more complex and more recently evolved organisms.

Considering the results of all methods of estimating geological time, there is little doubt that evolution has occurred over a period of many millions of years. Moreover, paleontologists are reasonably certain of the periods of time that various groups of organisms first appeared and when some of them became extinct. In all

Table 21-2. Geologic Time Scale

Eras and Duration (millions of years)	Periods (millions of years from present)	Major Geological Events	Advances by Organisms as Indicated by the Fossil Record
Cenozoic (65)	Quaternary (0–2.5)	Periodic glaciation	Rise of civilization; first true humans; increasing dominance of herbs; extinction of many trees.
	Tertiary (2.5–65)	Warm climate initially, but gradually cooled; formation of Alps and Himalayas.	Arthropods and mollusks abundant; appearance of modern invertebrates; modern mammals evolved; dwindling of forests; rise of herbs; worldwide distribution of modern forests; modernization of flowering plants.
Mesozoic (160)	Cretaceous (65–136)	Swamps in early part; Rocky Mountains and Andes formed	Extinction of ammonites and dinosaurs; spread of insects and birds; rise of primitive mammals; angiosperms dominant; gymnosperms dwindling.
	Jurassic (136–190)	Great continental seas in U.S. and portions of Europe	Ammonites and insects abundant; dominance of dinosaurs; first birds; early mammals; first known angiosperms; cycads and conifers dominant.
	Triassic (190–225)	Warm climate; great desert areas	Marine invertebrates decline; first dinosaurs; mammal-like reptiles; increase of cycads, ginkgo, and conifers.
Paleozoic (345)	Permian (225–280)	Glaciation and aridity; Appalachians and Urals formed	Last of trilobites; expansion of ammonites and reptiles; dwindling of club mosses and horsetails; first cycads and conifers.
	Carboniferous (280–345)	Great coal swamps; warm climate; mountain building; shallow seas	First insect fossils; first reptiles; spread of sharks; rise of amphibians; extensive coal formation in swamp forests; primitive gymnosperms.
	Devonian (345–395)	Emergence of land; few arid regions	Brachiopods flourishing; decline of trilobites; first amphibians; rise of fishes; early land plants; primitive lycopods, horsetails, ferns, seed plants, and gymnosperms.
	Silurian (395–430)	Climate mild; great inland seas; Taconic mountains formed	Corals and brachiopods; first land invertebrates; rise of primitive fishes; first known land plants; marine plants dominant.
	Ordovician (430–500)	Submergence of land; mild climate in Arctic areas	Climax of trilobites; cephalopods, starfishes, corals; first vertebrates (armored fishes); rise of land plants(?); marine plants dominant.
	Cambrian (500–570)	Mild climate; lowlands and inland seas	Trilobites dominant; mollusks, sponges, brachiopods; some lower plants established.
Proterozoic (930)	Precambrian (570–1500)	Rocks mostly sedimentary; glaciation; Grand Canyon formed	Few fossils of sponges and protozoa; most invertebrate phyla evolved; bacteria and blue-green algae; some simple plants.
Archeozoic (3000)	(1500–5400)	Rocks mostly igneous or metamorphosed	No fossils found; indirect evidence of life from organic material in rocks; all organisms probably unicellular or very simple.

probability, scientists will never be able to trace the evolution of all living organisms through the fossil record of their ancestors. Nonetheless, the presence and distribution of fossils already discovered and analyzed provides some of the most direct evidence for the theory of evolution.

The Evolutionary Process: Theories

No biologist has ever witnessed the origin by evolution of a major group of organisms; the evolutionary processes which gave rise to major groups of organisms took place in the remote past. Nevertheless, from what is known about these origins, there is a great deal of evidence to indicate that the mechanisms which brought them about were quite similar to those found in modern groups of organisms. In the preceding section, some of the accepted proofs for the occurrence of evolution were presented. In this section, a brief historical development of evolutionary thought will be outlined.

Lamarck's Theory

The French biologist Jean Baptiste Lamarck was the only pre-Darwinian biologist to offer a well-developed account of how living forms might have evolved. His hypothesis was predicated on the fact that an environmental change of any magnitude would result in the need for a corresponding alteration in a species living in that environment.

In order to illustrate his hypothesis, Lamarck used many examples. He reasoned that at one time ancestors of snakes probably possessed legs and short bodies like lizards. In response to crawling through narrow openings which required less and less use of their legs over long periods of time, the unused legs disappeared and their bodies became longer. Similarly, he reasoned that webbed feet of ducks were a product of their use in swimming; horns and antlers developed in response to contact in battle; cacti produced needle-like leaves in response to a need to conserve water; and the giraffe developed a long neck over many generations by stretching for leaves on trees.

It might appear, at least superficially, that this explanation of evolution by use and disuse of a part and its inheritance by future generations is plausible. Experimental evidence, however, has been amassed to refute Lamarckian evolution. While it is true that parts of plants and animals do change as a result of use and disuse, it has been shown that such acquired traits are not necessarily passed on to the next generation. In one particular experiment, mice whose tails had been removed were mated. The offspring of the next generation possessed tails, and their tails were also removed. This procedure was employed for twenty generations and the mice of the twenty-first generation had tails as long as those of the first generation. As a whole, the theory of Lamarck is not substantiated by experimentation and receives little support from the scientific community.

Darwinian Evolution

As a result of many years of careful study and observation, Charles Darwin, in 1859, proposed his theory of evolution. Essentially, he hypothesized that new species came about by a process called **natural selection.** In formulating this hypothesis, Darwin made a number of basic assumptions. First, he assumed that organisms tend to increase in geometric ratio from generation to generation; that is, in each generation of any organism the offspring produced outnumber the parents so that each generation should be larger than the one before (2, 4, 8, 16, 32, 64, and so on). Despite this geometric progression within groups of organisms, Darwin could see no evidence of an enormous population increase. His second assumption, therefore, was that except for minor fluctuations, the population of any given species remains fairly constant. Based upon these assumptions, Darwin concluded that a struggle for survival exists among living organisms; a competition for space, food, water, light, and other factors takes place and many organisms do not survive the competition.

Next, Darwin showed that variations occur in every species he studied. In other words, no two individuals are exactly alike. This is readily apparent if one considers the innumerable variations that exist among individuals of the human species. On the basis of this assumption and the conclusion that all organisms continually struggle for existence, Darwin generalized that under a given set of environmental circumstances, those individuals whose variations best fit that environment will be most likely to survive. Conversely, organisms with less favorable variations would die. Inasmuch as favorable variations can be passed on to offspring, these would be accumulated over a period of time and, as a result, organisms would become so different from the original species that a new species ultimately would evolve.

Natural selection, therefore, requires first, reproduction and, second, hereditary variation of such a kind as to influence the success of reproduction under existing circumstances. It offers not only a mechanism of how evolution might have taken place, but also an explanation of how a new species could be produced from an old one.

The Problem of Variation

Despite the apparent completeness of Darwin's evolutionary theory, it had one serious flaw. He knew nothing about the causes of hereditary variation and for years after the principles of evolution had been formulated, the problem remained. Why and how do variations in living organisms occur, and how are they transmitted from one generation to the next? It was at last clear that evolution functioned through the selective preservation of inherited differences among individuals and yet it was not known how such differences came about in the first place.

The early part of the twentieth century was a period of rapid development in biology, especially in the area of genetics. But the geneticists of the early 1900s did more to obscure the process of evolution than to clarify it. Specifically, conflict arose when the phenomenon of mutation was discovered. Geneticists attempted to explain all of evolution on the basis of mutation. They held that once a mutation occurred, the mutant organism was far superior to other individuals of the same species. Moreover, based upon certain observations, it appeared that some mutations created new species in great and sudden jumps. Essentially, proponents of the *mutation theory* maintained that each new mutant form was an incipient species and that only through mutation could sudden and drastic changes (variations) be produced in an organism; natural selection was merely a negative factor that eliminated the unfit. The Darwinian concept of evolution was that the evolutionary process resulted from the full effects of many slight variations accumulated over exceedingly large numbers of generations.

The Synthetic Theory of Evolution

During the 1930s the major factors of evolution began to fall into place. Not only had mutation and natural selection been subjected to careful experimentation, but other factors had also been implicated as part of the evolutionary process. Mutation no longer was accepted as the prime agent of evolution. Natural selection was considered as the major force, with mutation assuming a supplementary role. But if mutation alone could no longer be given credit for the countless adaptations of the natural world and therefore for evolution, various researches showed that mutation at least supplied the raw material for these changes. Without the new opportunities produced by mutations, organisms would be unprepared to cope with constant changes in the physical environment or with variations of more advantageously adapted organisms.

The conflict between geneticists and naturalists was eventually resolved by the formulation of a modern theory of the evolutionary process. This view, called the *synthetic theory,* incorporates and unifies several factors that are known to be operative in effecting variations among groups of organisms. Included are natural selection and mutation, which were each considered at one time to be only explanations of the evolutionary process. The synthetic theory of evolution recognizes four principal, interrelated processes. Two of these are genetic recombination and mutation, processes that provide the genetic variability without which change cannot take place. The other two, natural selection and reproductive isolation, act upon the sources of variability to guide populations of organisms into adaptive channels.

The Evolutionary Process: Mechanisms

Genetic Recombination

In meiosis, one set of chromosomes is passed on to each gamete. In sexual reproduction, the union of any two specific gametes is a chance occurrence. This means that through meiosis two alleles are separated. Homologous alleles are then reunited during sexual reproduction. This process of genetic mixing is known as *genetic recombination.*

In any sexually reproducing diploid organism there

will be at least two alleles for each gene, each producing a different phenotypic effect. Since there are many genes on each chromosome, the number of possible gene combinations is so large that no two individuals are likely to be genotypically identical. For the same reason, the potential for genetic variability within an interbreeding population through recombination is far greater than that which is expressed phenotypically. Thus, through genetic recombination there is a constant source of genetic variability resulting in various new combinations of existing genes. Genetic recombination is one of the most important processes in providing variability.

Mutation

Recall that a mutation may be regarded as any change in the hereditary material. Mutations are considered to be the source of new and different genetic material of a given population of organisms. Biologists recognize two principal kinds of mutations, gene mutations and chromosomal aberrations (Chapter 19).

As has been stated previously, no two sexually reproducing organisms are exactly alike and all individuals exhibit a certain degree of variation. Such differences are the result of either hereditary factors or environmental conditions. Some environmental conditions which may cause variation are diet, disease, climate, chemicals, and radiation. Generally, the resulting variation does not affect the sex cells and is not passed on to the next generation. By contrast, gene mutation, changes in chromosomes, and genetic recombination within sex cells will be passed on directly to the offspring. While the occurrence of a given mutation in a specific individual is based upon chance, mutations are not truly random. The structure and composition of DNA determine and limit which kinds of changes are possible, which changes are not possible, and which changes are likely to occur more readily. There are limitations upon the types of mutations that can occur.

If a mutation occurs that is neutral or adaptive, and the organism survives, the mutation becomes an addition to the gene pool, the totality of genes possessed by a given group of like organisms. The specific characteristic expressed by the gene mutation may then appear in future generations to be acted on by natural selection. If the expressed variation is adaptive, it will survive within the population. For example, suppose a mutant black-fur rabbit appears in a natural population of brown-fur rabbits. If the mutant rabbit survives and reproduces, it then passes the mutant gene for black fur on to its offspring. Assuming that the black-fur offspring are not placed at a great competitive disadvantage, the mutant gene will be carried to future generations.

In the fruit fly, *Drosophila,* one of the best known and, genetically, most extensively studied organisms, mutations have been observed since 1909. One of the first observed mutations in *Drosophila* was the appearance of a single white-eyed male among a group of normal red-eyed individuals. Since then, approximately 5000 spontaneous mutants have been identified. For example, mutations are known which affect color and shape of the eyes, color and size of the body, size and shape of the wings, and number and distribution of bristles on the body. In various higher plants, mutations are known which affect the shape and color of flowers, resistance to disease, size, shape, lobing of leaves, and the branching pattern of the entire plant. In fact, gene mutations are known for every kind of organism that has been subjected to genetic analysis.

Since various types of changes can occur, it can be concluded that genes are not completely stable. In *Drosophila,* for example, about one gene mutates for every twenty gametes produced; each gamete contains about 20,000 genes. Genes can also mutate at a characteristic rate. In corn (*Zea*), the gene for seed color mutates 492 times per 1 million gametes, whereas the gene for shrunken corn seeds mutates at a rate of 1.2 times per 1 million gametes. Moreover, more than one kind of mutation is possible for any particular gene and many genes are known to mutate from a mutant form back to a normal type.

Although gene mutations are rare and often harmful to the organism, they still provide enough new variability for evolution to occur. This occurs because there are thousands of genes in each gamete. There may be thousands or perhaps millions of individuals producing gametes each generation; and there are numerous generations of individuals over the span of evolutionary time in which some opportunities for mutation occur. Thus, even though a particular gene mutates only rarely, the many genes, numerous individuals, and millions of years of evolutionary time collectively enable mutations to provide ample opportunity for genetic variability. Gene mutations are the ultimate source of hereditary variation but in themselves do not cause evolution.

Although gene mutations provide the raw materials for the evolutionary process, by themselves their effect upon variations in a population is usually slight. Essentially, any given individual is the result of interactions involving the entire genotype (all genes), and every gene assumes a role in the expression of hereditary traits. In this regard, the interaction of a mutated gene with other existing gene (if not all) is frequently of greater significance in the development of new variations than the direct effect of the mutated gene alone. For example, a mutation of a gene needed to provide the information for the production of cartilage may affect bone structure, muscular movements, and many other activities.

Since each hereditary characteristic is so closely interrelated with every other characteristic, mutations are generally harmful. However, since there are two determinants for each characteristic, the deleterious effects of a mutant gene may be masked by the normal gene. In this way, a mutant gene may be carried and multiplied within a population and lost only when two sex cells, each bearing the mutant gene, come together because the deleterious effect is then expressed. Through environmental changes, the effect of the expression of the mutant gene in the diploid state may place an organism in a more advantageous position, both structurally and reproductively. The mutant gene may then become firmly established and a part of the normally existing genetic material of the gene pool.

In addition to gene mutations, chromosomal aberrations also occur. If chromosomes undergo spontaneous modifications, the effect is typically more pronounced than if only a single gene mutates. This is probably because gene mutations, if not lethal (causing death), are generally masked by the already existing allele. One type of observed chromosomal mutation is a change in the chromosome number such as a loss or gain of part of a set of chromosomes, a loss of an entire set of chromosomes (a lethal condition in diploids), and the addition of one or more chromosomal sets. Another form of chromosomal mutation involves a change in structure. In this case, there can be a change in the number of genes by loss or addition, or a change in the arrangement of genes through an exchange of chromosome segments. In both instances, chromosomal alterations contribute to variability by rearrangement of existing genes. Such rearrangements (Chapter 19) may markedly affect both genotypic and phenotypic variability since the expression of a phenotype is largely a function of the interaction of many genes.

Population Genetics

The genotype of an individual does not change during its lifetime and therefore an individual cannot evolve. A unit which does change with time, however, is the **population,** a group of individuals capable of interbreeding. Such a unit is characterized by a changing genotypic composition throughout an existence which may span many millions of years. This change and consequent modifications in phenotypes constitute evolution. In order to understand evolution, we must move from the inheritance of individuals to that of populations.

Ultimately, all new alleles in a population arise by mutation and once a variety of alleles is in existence, recombination provides thousands of genotypic variations in the population. Both mutation and genetic recombination provide the genetic variability upon which natural selection can act to produce evolutionary change. The guiding force of evolution is natural selection, which has two basic requirements. First, there must be reproduction and, second, there must be hereditary variation of a kind that influences the success of reproduction under existing circumstances.

Natural Selection

Genetic variability, while appearing by chance within a population, does not occur completely at random. Since hereditary material is chemical and obeys both chemical and physical laws, there is a limit to the types of changes that can take place. A **population** may be defined as a group of individuals capable of interbreeding and therefore sharing the same gene pool. Genetic variability is repetitive; the same type occurs over and over again in individuals. A mutant gene can mutate back or a type of recombination can reverse itself. In the absence of outside factors, the gene pool tends to reach an equilibrium; that is, the appearance of genes

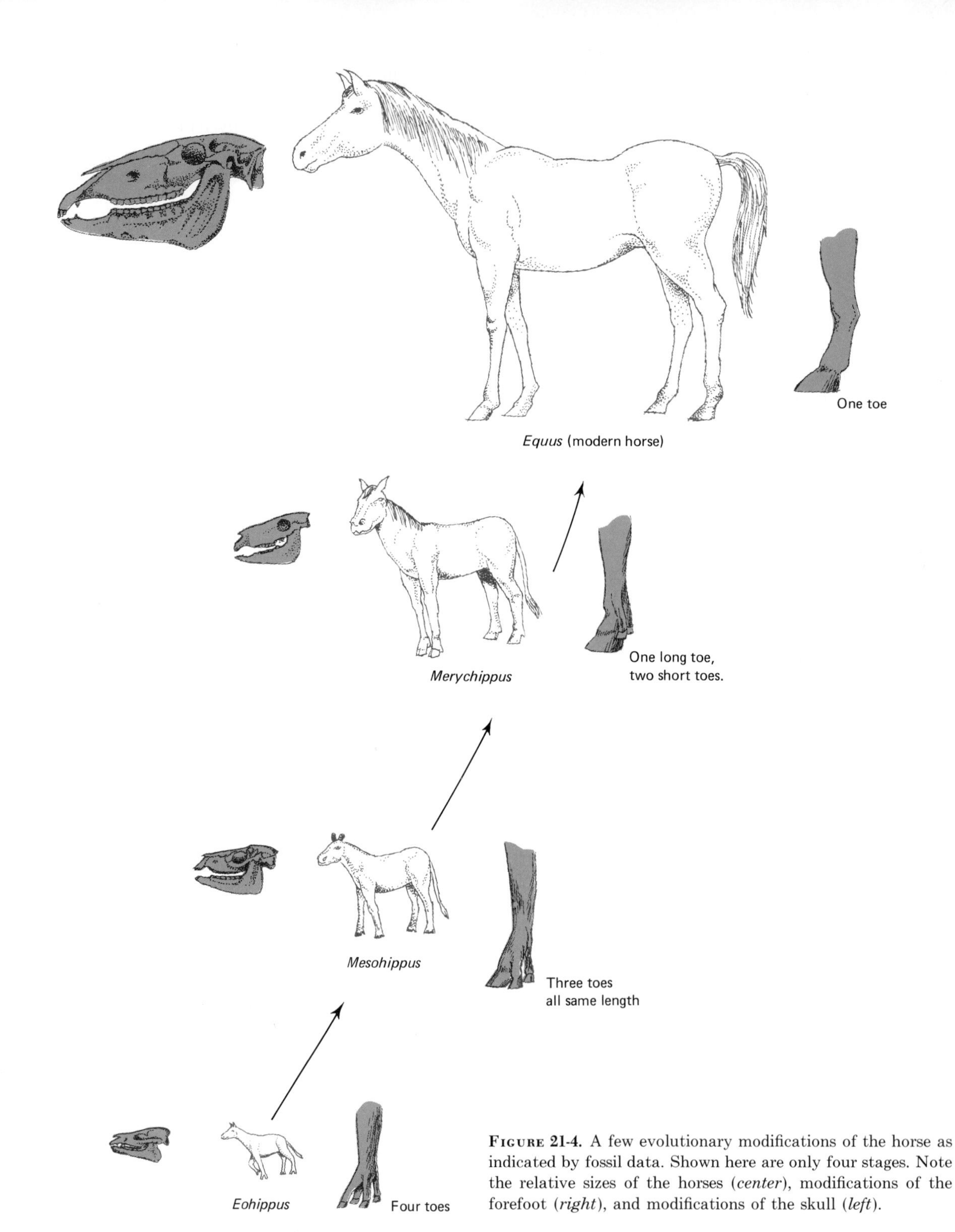

FIGURE 21-4. A few evolutionary modifications of the horse as indicated by fossil data. Shown here are only four stages. Note the relative sizes of the horses (*center*), modifications of the forefoot (*right*), and modifications of the skull (*left*).

and genotypes in a population theoretically would become constant generation after generation. This genetic stability is known as **genetic equilibrium.** This does not mean that all individuals will be genetically alike but that the same genetic variations will reappear. Some of these recurring variations are expressed phenotypically because of adaptive gene combinations, others are expressed genotypically and remain a potential source of evolution. For example, the horse as shown in paintings made during the Middle Ages is easily recognized and with slight variation (thanks to selective breeding practices) is very similar to the modern horse. Yet, fossil evidence shows that the horse has undergone considerable evolutionary change over a period of 60 million years (Figure 21-4). The phenotypic expression of genetic variability within the generations of horses over the past hundreds of years has proven to be adaptive to the existing environmental conditions. The gene combinations occurring within the population of modern horses are adaptive. The key to understanding the evolution of the horse—of any organism, for that matter—lies in determining how genetic equilibrium is modified so that the composition of a population changes. However, it must be emphasized that natural selection can act upon genetic variation only when such variation is expressed phenotypically. Moreover, as noted earlier natural selection may be more important at the regulatory gene level than at the structural gene level.

Among modern horses, through genetic variability, there will be genotypic differences which will be expressed (phenotypic) so that there will be some variation. If the expression of a variation places a horse at a disadvantage within the environment, the chances of survival and reproduction are greatly reduced in comparison with other members of the population. The particular gene combination can then be said to be less adaptive; selection will favor those horses without the particular expressed variation. Although all horses have the potential to express the less advantageous variation, those that continue to survive and reproduce do so because it has not been expressed. By contrast, suppose that the given expressed variation enables a horse to survive better, live longer, and produce more offspring. It can then be said that the particular gene combination is adaptive. Within the population, the genetic combinations favoring adaptation to the given environment tend to be perpetuated. It is now clear that some may be neutral and behave as though they are evolving without selection (i.e., by drift). This process of selection is called *natural selection.* It must be emphasized that natural selection promotes a direction and control of evolution; it is not a predestined or preplanned phenomenon.

Natural selection promotes adaptive relations between the population and its environment by favoring certain genetic combinations, rejecting others, and constantly modifying the gene pool. If environmental conditions were to remain constant, natural selection would result in the perpetuation of the given living form that was most adaptive and the population would eventually exhibit little change. However, environmental changes have taken place and have been extremely numerous and complex. They have involved alterations of climate, shifts and redistribution of land and sea forms, and often (and most importantly) changes and shifts in plant and animal communities. As a result of these environmental changes, populations have either undergone changes in their previously adaptive gene combinations in accordance with these changes or have become extinct.

In the evolution of the horse from an ancient ancestral form, changes in genetic variation that have been favored by changes in the environment have resulted in the natural selection of the modified populations. Of course, the genetic potential of an ancestral form of the horse is still within the genetic material of the existing population. However, the changes in genetic combinations have been so great over the sixty million years of evolution that reversal within a short period of time would be highly improbable. A reversal to the same primitive genotype would be an infinitely small probability, even in a billion years. Moreover, even though reversal to a horse with the phenotype of the original primitive horse is possible, the genotype would be quite different.

The evolution of the horse has come about through a process that is nonrandom with respect to adaptation. That is, the phenotypic expression of genetic variability has tended to be specifically and directionally toward the adaptiveness of the evolving populations of horses. The changes that have taken place in the evolution of the horse or any organism have been through the slow modifications in existing populations. The changes came about by the random appearance of new forms through mutations and the action of natural selection upon the ones that were best adapted. The gradual evolution of all organisms has come about essentially through a process of nonrandom reproduction. That is, there is a differential in the reproduction rates among the various genotypes within the population. That segment of the population possessing genotypes that better suit the evolving existence of a population in a given environment, reproduces more rapidly. In modern usage, nonrandom reproduction is synonymous with natural selection because evolution takes place through effective reproduction of populations.

One of the classic examples of natural selection is the case of the peppered moth, *Biston betularia*. More than one hundred years ago, insect collectors noted that certain species of the moths appeared in two forms—light-colored and dark-colored. Up until 1845 all known specimens were light in color, but in that year a single black moth was found in Manchester, England, a growing industrial center. It was estimated at this time that only 1% of the total population was black. Several decades later, however, more and more black moths were found, and by 1895 the black moths comprised 99% of the Manchester population.

An explanation for the change from light- to dark-colored moths is found in natural selection. During the daytime, moths frequently rest on trunks of trees and, prior to the development of industry in England, the moths were afforded protection from predators such as birds since they blended in well with lichens that covered the tree barks. With the spread of industry, fumes and soot killed the lichens and the trees became blackened with industrial residues and pollutants. Once this occurred, the light moths were at a disadvantage because their coloration no longer blended in with the tree trunks. Dark forms were selected and the balance for survival swung in their favor. This phenomenon, called *industrial melanism,* has also been observd in other European countries as well as the United States; black forms, originally rare, increased in numbers in conjunction with industrial development (Figure 21-5).

Proof that natural selection did operate under such conditions was provided by an English biologist, H. B. D. Kettlewell, in the 1950s. In his study, he released known numbers of light and dark moths into two different areas. One of these, Birmingham, an industrial area, had a local population consisting of 90% black moths. After releasing the moths, he observed birds feeding on the released moths. After capturing the survivors, he found that 40% of the black moths and only 19% of the light moths had been recovered. At the other site, Dorsetshire, an area free of industry, he followed a similar procedure. In this case, the results were the opposite; whereas 12.5% of the light moths survived, only 6% of the dark moths were recaptured.

Differences between dark- and light-colored moths are controlled by genes. Since the original forms were light-colored, only mutation and genetic recombination could have produced the dark moths. With a change in the environment precipitated by industry, natural selection enabled the dark forms gradually to outnumber the light ones. The rapidity of change from light to dark forms within natural populations in industrial areas provides strong support for the role of natural selection as a directive evolutionary force. It should be noted that even though selection interacting with variation does account for a great deal of evolutionary change, other factors are apparently implicated.

Faced with the constant changes in the environment, genetic variability, and natural selection, related populations have often become modified in different ways in response to similar environmental change. In addition, many populations have been able, through natural selection, to move into previously uninhabited areas of the environment. For example, suppose there is a limited supply of a given type of food, such as berries, to which a given population of birds is adapted. A limited food supply will limit the size of the population. Now further assume that through a slight modification in the mouth parts of some of the birds a segment of this population is able to supplement its diet by eating seeds. Gradually, through natural selection, there may be an evolution of a population of birds adapted to survive on a total diet of seeds. Through evolution, the seed-eating population of birds has been able to exploit a part of the environment previously uninhabited. Darwin's finches provide such an example (see Figure 23-8).

Figure 21-5. Industrial melanism. (**A**) *Biston betularia,* the peppered moth, and its black form, *B. carbonaira,* at rest on a lichen-covered tree trunk in an unpolluted countryside, Dorset, England. (**B**) *Biston betulaira* and its black form at rest on a soot-covered oak trunk near Birmingham, England. [*From the experiments of Dr. H. B. D. Kettlewell, University of Oxford.*]

The Gene Pool

While the genotype of a diploid individual can be represented by only two alleles for a given gene, such is not the case with populations. In a population, there may be many alleles for a given gene. For example, assuming that one gene determines flower color in a given species, a single diploid individual will contain only two alleles for flower color (possibly one for white color and one for red color). By contrast, there may be other alleles for such colors as blue and purple among other individuals within the population. All four alleles represent possible flower color within the gene pool of the population. All the possible alleles for all the genes of a given population represent the total **gene pool.** In other words, the gene pool constitutes the total genetic makeup of an entire population. It should be noted that because of genetic interaction and integration, a single gene may have wide-ranging effects on the expression of many other genes. A change in the selective advantage of one gene can therefore affect the fitness of many other genes.

A gene pool can be characterized by the frequency of given alleles, generally expressed as a ratio or percentage within a population. For example, suppose there are two alleles determining flower color, one for red (R) and one for white (r). Further assume that red is dominant. To simplify calculations the frequency will be calculated per 100 individuals of the natural population. Therefore, the numbers involved in the decimal expression of the ratio will be the same as the percentage. When the flowers are counted, it is found that 4 plants are white and 96 are red. Since the union of gametes occurs randomly at the time of fertilization, the phenotypic ratio of white flowers to red flowers (4:96) is the result of the genotypic frequency of white alleles in sperm and eggs. In other words, 4% are white and 96% are red. Inasmuch as the white alleles are recessive, the 4 white flowering plants (rr) can result only if the frequency of r (white) alleles within the population is 20% or 0.2 because $0.2r \times 0.2r = 0.04rr$. The frequency of R (red) alleles must then be 80% or 0.8 ($100\% - 20\% = 80\% = 0.8$). By placing these frequency data in a Punnett square, it is possible to determine all the genotypes of the population (Figure 21-6). From Figure 21-6 it can be seen that

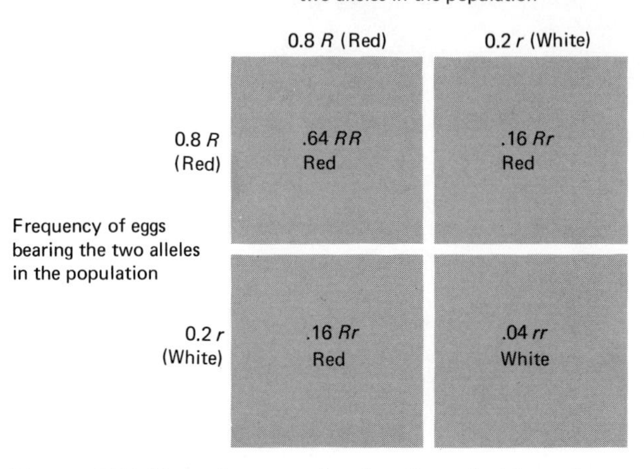

Figure 21-6. Punnett square showing the ratio of genotypes based upon the ratios of alleles within the population. Consult text for explanation.

the ratios of genotypes per 100 individuals are $0.64RR$ (homozygous red flowers):$0.32Rr$ (heterozygous red flowers):$0.04rr$ (homozygous white flowers). In other words, 96% (64% RR + 16% Rr + 16% Rr), or 96 flowers, are red, and 4% (4% rr), or 4 flowers, are white.

Let us now determine what happens when this same population produces offspring, assuming that there are no forces preventing random fertilization. When fertilization occurs, there are four possible combinations of the alleles of eggs and sperm.

1. An R egg may unite with an R sperm.
2. An R egg may unite with an r sperm.
3. An r egg may unite with an R sperm.
4. An r egg may unite with an r sperm.

The ratios of these possible combinations are determined by the ratios of the alleles in the population. Table 21-3 shows the calculations of genotypes based upon the ratios of alleles in the population. The conclusions that can be drawn from these data are that gene frequencies have not changed in the offspring and that the numbers of phenotypes have remained the same in the theoretical population. The same results would be obtained regardless of how many generations were included and regardless of the number and types of gene pairs considered.

Table 21-3. The Frequency of Genotypes in Offspring Based Upon the Frequency of Alleles in the Parent Generation

Alleles of Parent Generation: Egg	Alleles of Parent Generation: Sperm		Frequency of Alleles in Ratios		Ratio of Genotypes in Offspring	Number of Phenotypes per 100 Plants	
R +	R	⟶	8×8	⟶	64 RR	64 red	total: 96 red
R +	r	⟶	8×2	⟶	16 Rr	16 red	
r +	R	⟶	2×8	⟶	16 Rr	16 red	
r +	r	⟶	2×2	⟶	4 rr	4 white	

The Hardy-Weinberg Law

The same genotypic frequencies can be obtained mathematically by expanding the binomial expression $(p + q)^2$, where p represents the ratio of the allele R and q represents the ratio for allele r:

$$(p + q) \times (p + q) = p^2 + 2pq + q^2 \quad (\text{or } R^2 + 2Rr + r^2)$$

By placing the ratios for the two alleles in the algebraic expression, the same results are obtained as with the Punnett square:

Binomial expansion	$R^2 \quad + \quad 2Rr \quad + \quad r^2$
Substituted allele ratios	$(8)(8) + 2(8)(2) + (2)(2)$
Ratios of genotypes	$64\ Rr \ :\ 32\ Rr \ :\ 4\ rr$

The algebraic expression can also be expanded to include more than two alleles. For example, if there are three alleles (p, q, and r), the algebraic expression would be $(p + q + r)^2$. However, as the number of alleles is increased, the calculations become more complex.

The mathematical expression for calculating allele frequencies is referred to as the *Hardy-Weinberg law,* a basic principle of population genetics. Essentially, the principle states that the proportion of genes and genotypes in any large population of diploid, cross-fertilizing individuals will remain relatively constant generation after generation. This law operates provided the population is large enough not to be affected by random changes in gene frequencies, no mutations occur (or mutation in the two directions is equal), no migration takes place, reproduction is completely random, and there are no environmental changes. Given these contingencies, the population maintains genetic equilibrium, that is, it maintains its proportion of homozygous and heterozygous individuals.

There are many practical applications of the Hardy-Weinberg principle to breeding practices. For example, it was noted earlier that rose comb is dominant over single comb in domestic poultry (Chapter 19). Since certain breeds are required to have rose comb, it is advantageous for a poultry breeder to know what proportion of the rose-combed birds carry the gene for single comb. This information may be of value if the poultryman wishes to eliminate the recessive gene from his flock.

It is also important to know how often certain genes occur in a population, especially those genes that are related to health. Diabetes, hemophilia, mental illness, muscular dystrophy, and many other diseases appear to have a hereditary component. The more that is known about the distribution of these diseases in the population, the more success there will be in treating and controlling them. In other instances, application of the Hardy-Weinberg principle provides valuable information on the frequency of blood groups in the population. These include the ABO groups already discussed as well as the MN blood types. Thus, the Hardy-Weinberg principle has value in studies of alleles that show quantitative inheritance. Quite obviously, one of today's major problems is to ascertain the genetic effects of high-energy radiation. There is a need for an accurate means of determining the frequency of harmful genes in the population. In this regard, it is important to know if the total numbers of harmful mutations are being increased because of radiation associated with fallout, or because of industrial uses of nuclear energy, and medical and dental practices.

Factors Affecting Genetic Equilibrum

Recall that evolution has been defined as a change in the genetic composition of a population. If the frequencies of all genes in a population always remained constant, evolution obviously could not occur. We will now briefly consider some of the conditions that may affect **genetic equilibrium** and therefore the evolutionary process.

Population Size. In deriving the Hardy–Weinberg law, it was assumed that the population is infinitely large for chance to be completely ruled out as a causative factor capable of changing gene frequencies. In practice, no population is infinitely large; however, most natural populations are large enough so that the chance genetic variability that occurs among different members probably does not greatly affect the overall gene pool. Within large populations, all degrees of variability are represented so that any chance variability that might occur in a small segment of the population does not greatly affect gene frequencies. However, if a population is small and isolated from any gene flow from other populations, there is a tendency for the gene pool to be susceptible to chance fluctuations in gene frequency. This means that genes may be eliminated or fixed in a population by chance alone. As a result, there may be fluctuations from year to year that are not the result of selection.

Since this type of change is based upon chance, it is called *indeterminant evolution*. The genetic equilibrium of a small population may result as a chance occurrence in which the gene pool is as likely to move (drift) toward the loss of one allele of a gene as toward the loss of another allele of the same gene. This indeterminant evolution is also known as **genetic drift.** Random phenomena such as found in genetic drift can only be of evolutionary significance when a certain number of important genes is completely eliminated.

Mutation. Mutation is a second factor modifying the genetic characteristics of a population. The Hardy–Weinberg law assumes that genes reproduce themselves accurately, that is, either no mutations occur in a population or there is a mutational equilibrium in which the number of forward mutations per unit time is equal to the number of back mutations. It is an established fact that mutations are always taking place, even though the rate of mutation for different genes varies. If, for example, allele A_1 changes by mutation to allele A_2, the population will be losing some A_1 and gaining A_2. This is called *mutation pressure*. Moreover, very rarely does the rate of forward mutation equal the rate of backward mutation.

$$A_1 \underset{\text{backward}}{\overset{\text{forward}}{\rightleftharpoons}} A_2$$

Mutation pressure, if not in equilibrium, tends to cause a slow shift in gene frequencies of a population. Accordingly, whereas more stable alleles tend to increase in frequency, more mutable alleles tend to decrease in frequency (excluding other factors that might alter mutation pressure). It should be noted, however, that mutation pressure is probably seldom a major factor in altering gene frequencies in populations since mutation is a relatively slow and random process. Thus, although mutations ultimately provide the raw materials for evolution, they occur at random and rarely determine the direction of evolutionary change.

Migration. The third prerequisite condition for the maintenance of genetic equilibrium according to the Hardy–Weinberg law is that there must be no immigration or emigration. In nature, very few populations are completely isolated from other populations of the same species. In fact, a high percentage is subject to at least a small amount of gene migration in which immigrants from other populations add new genes to the gene pool and emigrants to other populations remove their genes from the gene pool. Migrations of individuals with slightly different genotypes thus cause changes in the gene pools of populations. Despite the fact that some populations do not migrate and other populations experience only negligible migrations, it is possible that migration does affect at least some populations in nature.

Random Reproduction. A fourth assumption of the Hardy–Weinberg law is that in order for genetic equilibrium to occur, reproduction must be completely random. In genetic terms, reproduction refers to a large number of factors, all of which contribute to the population's potential for reproductive continuity and not simply the mating of individuals. For example, reproductive continuity involves certain factors in mate selection, performing the mating process efficiently and with periodicity, fertility among mating individuals, the number of zygotes formed, the percentage of zygotes that undergo complete and successful embryonic development, the number of actual births, the survival of the young until reproductive age is reached, and so forth. Random reproduction means that all of these factors, as well as others, must be independent of the genotypes of individuals in the population. Since these factors are probably always related to the genotype of an individual, it is reasonable to assume that complete random reproduction does not occur in natural populations. Conversely, nonrandom reproduction (selection) is always operative in populations and this factor must be regarded as an integral component of evolutionary change.

Genetic variability, drift, and natural selection are constantly disturbing the equilibrium of the gene pool. Recall that in the original example of the flower population, the genetic equilibrium contained the following ratios or genotypes, 64 RR:32 Rr:4 rr, with a phenotypic expression of 96 red flowers and 4 white flowers per 100 flowers. Now suppose a species of insect migrates into the population and that the insects respond more favorably to white flowers than to red flowers. The chances of insect pollination between white flowers would increase. The union of the gametes is no longer random since the white flowers would be visited more frequently than previously. Selection increases in favor of the white flowers. This means that more gametes bearing the r (white) allele will participate in fertilization. Through our previous means of calculation of allele ratios by the number of white flowers in the population, the ratio of r alleles may increase to 3. Correspondingly, the R (red) allele will then have been reduced to 7. By substituting these values into the binomial expression previously described, the following results are obtained:

Binomial expansion	$R^2 + 2Rr + r^2$
Substituted allele ratios	$(7)(7) + 2(7)(3) + (3)(3)$
Ratios of genotypes	$49\,RR : 42\,Rr : 9\,rr$

It can be seen that the ratio of the rr genotype is increased by 5 (9 − 4) and that the frequency of white flowers per 100 is increased to 9. By deduction, it can also be seen that the sum of ratios of the Rr and RR genotypes is 91 (49 + 42) and that the frequency of red flowers per 100 is now decreased to 91.

Changes of the genotypes in following generations will continue as long as selection by the insects favors the white flowers. A point may be reached when random fertilization will again take place with the frequency of the previously dominant allele R (red) being much lower. However, it can be seen that if selection continues to favor one allele, the other may eventually be lost from the gene pool.

Environmental Conditions. In reality, most of the variations within a population are the result of the interaction of multiple genes which are influenced by environmental conditions. For example, if we were to take and weigh each individual in a population of a given species of squirrels, we would find that, in plotting their weights, the distribution would follow a bell-shaped curve (Figure 21-7, curve A). Since variability is the result of both genetic and environmental factors, theoretically, two similar curves could be drawn. One curve (Figure 21-8A) could be drawn where the environment was constant and variation caused by genetic factors (different genotypes), and another (Figure 21-8B) where the genotype was constant and variability was caused by environmental conditions.

Suppose that environmental conditions were to

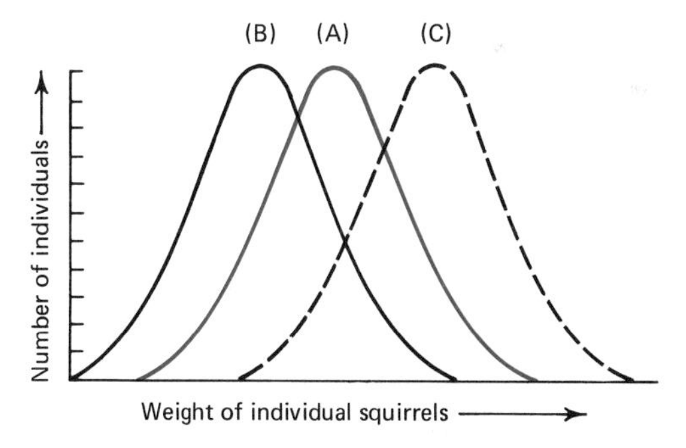

Figure 21-7. Normal distribution curves of weights of squirrels in three hypothetical populations. The weights of the squirrels in a natural population are distributed over the normal curve of probability (curve A). If selection favors squirrels with lighter weights, then the curve moves to the left (curve B). If, on the other hand, selection favors squirrels with heavier weights, then the curve moves to the right (curve C). Once the selection favors one or the other weights, a new distribution of weights is established following the normal curve of probability.

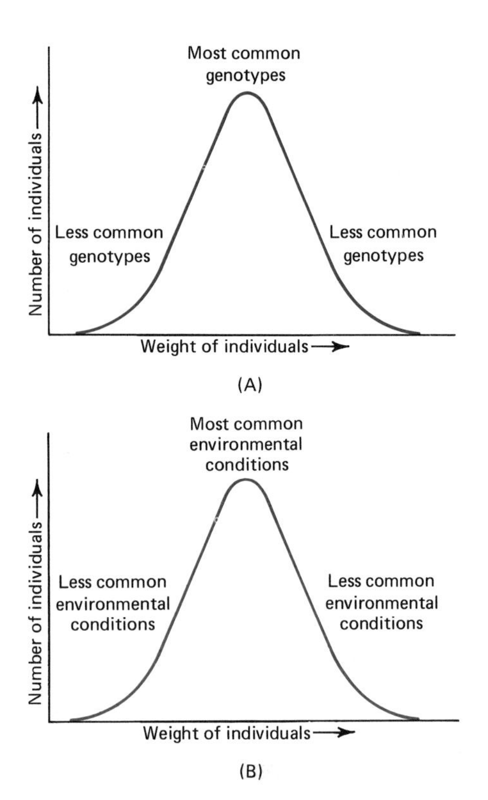

FIGURE 21-8. Bell-shaped curves of distribution of variability in a theoretical population. Variability in a population is the result of the interaction of genetic and environmental factors. Theoretically, the distribution of individuals will follow the normal curve of probability whether determined only by genotypes (**A**) or only by environmental conditions (**B**).

change, causing a reduction in available food for squirrels. Further assume that the combination of alleles among the lighter squirrels enables them to survive to reproductive age on a smaller food supply. We would then find in successive generations a shifting of the bell-shaped curve in the direction (left) of the number of squirrels with lighter weights (Figure 21-7, curve B). If environmental conditions favored a heavier squirrel, the curve would shift in the direction (right) of the number of squirrels with the greater weight (Figure 21-7, curve C). Similar shifts in the distribution of phenotypes may occur if interactions of new combinations of alleles result in either an increase or decrease in reproductive capacity of squirrels in the given environment.

Reproductive Isolation and Speciation

Up to this point the mechanisms of evolution that have been discussed have been related to changes that take place within a specific population over long periods of time. There is another major process in evolution by which a single population will gradually give rise to two or more new descendent populations. This process is called *speciation.*

All environments are diverse and complex in space and time and are constantly changing. Therefore, the interrelationship between a population in various locations and its environment will be different. This has the effect of dividing a population up into smaller local units called *demes.* The individuals within a deme tend to be more similar to each other than to individuals in another deme of the same population for two reasons.

1. They are more apt to interbreed within the deme and are therefore more closely related genetically.

2. They are exposed to more similar environmental conditions and are therefore subjected to nearly the same forces of natural selection.

For example, a grove of a given type of pine tree in a forest, or all the squirrels living in a woodland lot, or all the trout in a stream represent demes. Each is more likely to interbreed within the group than with members of the same population in other locations. However, demes are not clear-cut permanent units of population. While interactions are more likely to occur within the deme, individuals may migrate to other locations and mate, or pollen may be carried great distances and fertilize another member of the same population. A population inhabiting a large contiguous area may be broken up into demes with no definite boundaries and much overlap. For example, a population of squirrels distributed throughout a large forest will tend to form local groups (demes); individuals will interact more frequently within the group, but they may interact with members of other groups. Through interbreeding any adaptive gene combination will then be intensified within the deme. This means that the genetic combinations may be expressed in one deme and not in another.

In studying a total population, it will be seen that there is variability so that generalized, although not clearly defined, subdivisions may be made. The first of these will be slight differences within groups as represented by the demes. Demes in turn may be grouped in larger subdivisions called races in which the organisms within each race, although capable of interbreeding, will differ from each other in various ways that are quite recognizable. If races become geographically isolated from each other, they are called subspecies. Thus, races and subspecies are genetically similar individuals that have an intimate temporal and spatial relationship to one another and may be viewed as open genetic systems that may be affected by gene flow from adjacent races or subspecies. If one subspecies becomes reproductively isolated from another, then it is called a *species.* The two are no longer capable of exchanging genes. As contrasted with races and subspecies, species are closed genetic systems protected against gene flow from other species.

For example, the song sparrow is recognized by its size, color pattern, and song. Throughout North America, the species is divided into various subspecies which may be distinguished by differences in song, color patterns, and size (Figure 21-9). These subspecies are located in given geographical areas. Interbreeding of the subspecies takes place only in those areas where there is overlap. The gene combinations of a subspecies tend to

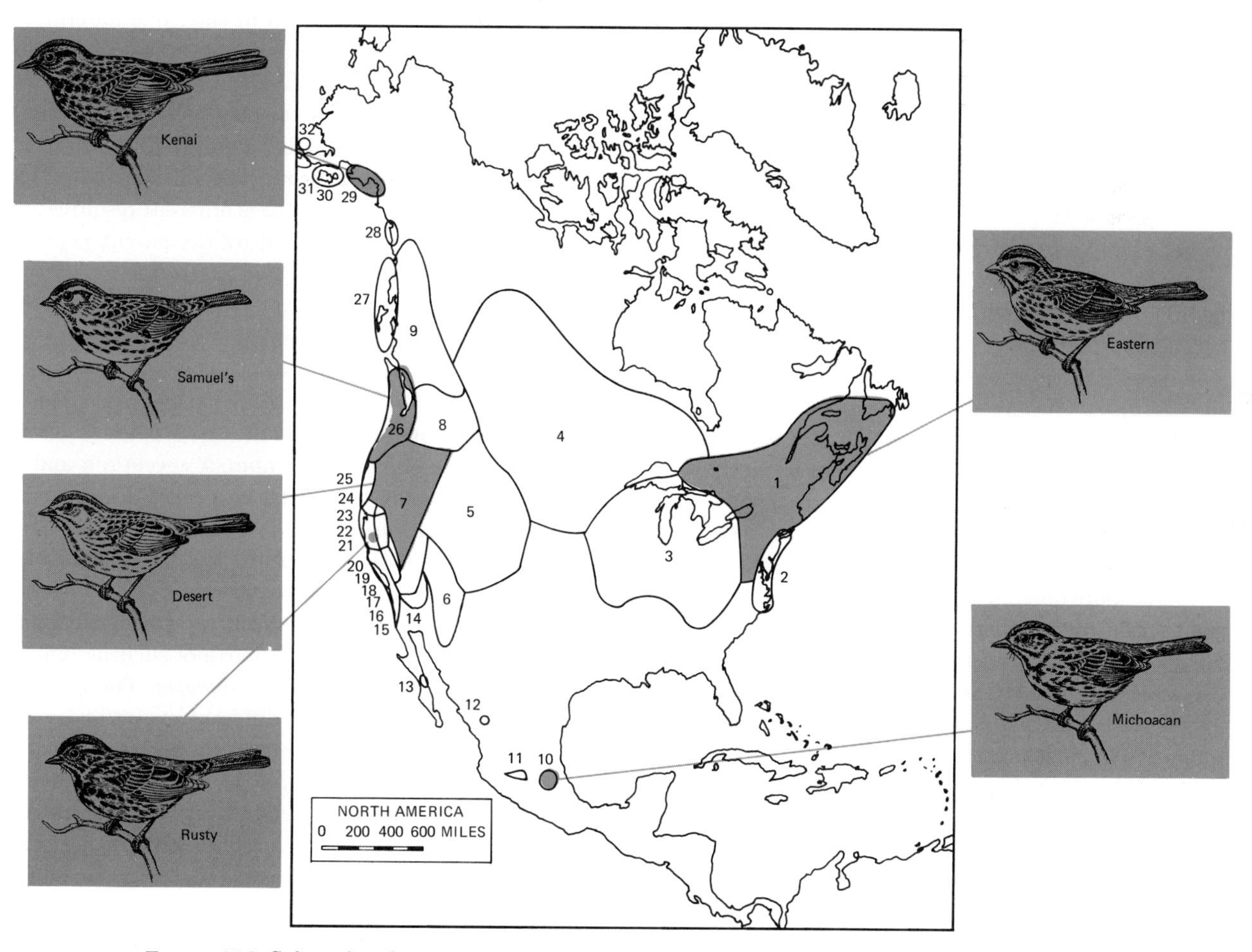

FIGURE 21-9. Subspecies of song sparrows and geographic areas occupied by each. Of the 32 subspecies, only 6 are shown; numbers in the map indicate the areas occupied by the remaining 26 subspecies. [*Modified from G. L. Stebbins,* Processes of Organic Evolution, *pp. 88–89. Englewood Cliffs, N.J.: Prentice-Hall, Inc. Copyright © 1966. By permission.*]

be quite similar within each geographical area and thus the sparrows possess similarities which enable an observer to distinguish one subspecies from another. Within each geographical area further division of the subspecies of the song sparrow into races may be found. For instance, in the San Francisco Bay area, those sparrows inhabiting the hill slopes are recognizably different from those inhabiting the marshes. The sparrows located in overlapping areas tend to resemble each other more closely so that the differences are represented by a gradual change. It should be made clear that in the field it may be quite difficult to recognize races, subspecies, and species.

Through genetic variability, natural selection, and reproductive isolation various subdivisions of a population may eventually develop such great differences in their gene pool that they are no longer capable of interbreeding. They have evolved into a new species by a process known as *speciation.*

The phenomenon of speciation is generally regarded as an adaptive process in which a single species gives rise to two or more new species. Although a variety of factors are involved within a population that gives rise to a new species, one of the most important is isolation. The term **isolation** refers to any mechanism or condition that prevents any two groups within a species from interbreeding. It has been shown that mutation, genetic recombination, and natural selection in a changing environment may result in a change in a population. However, in order that these changes may become established to bring about an evolutionary development of two or more different populations, a certain degree of isolation in the existing population is necessary. There must be an effective means for preventing access to the common gene pool by the evolving population (species). Free interbreeding without isolation tends to propagate a relative homogeneity among individuals of a population.

The most important means of isolation is through reproductive isolating mechanisms. Reproductive isolating mechanisms may be classified into two groups.

1. *Prezygotic,* or those in which interbreeding is prevented when both species are reproductively active or the union of gametes is prevented after mating or cross-pollination has occurred.

2. *Postzygotic,* or those in which hybrids fail to develop fully or hybrid fertility is reduced.

Table 21-4 summarizes prezygotic and postzygotic isolating mechanisms.

In order to establish reproductive isolation there must be a change in the genetic systems of the emerging species to the extent that related populations are no longer able to interbreed. The emergence of new species may occur in several ways.

1. A population becomes divided geographically, a process also known as *spatial* or *geographical isolation.* This means of evolving new species is known as *allopatric speciation.*

2. A population may remain in the same habitat or environment, but genetic recombination, alterations in the genes at the regulatory level, and natural selection may produce changes in premating mechanisms (anatomical, physiological, and behavioral) that result in reproductive isolation of a group within the population. This new emerging species utilizes different resources in the habitat from those utilized by the parent population. This is known as *sympatric speciation.*

3. As in sympatric speciation, the population is not geographically isolated, but changes in the genetic system resulting in pre- and postmating reproductive isolating mechanisms occur simultaneously, as the emerging species begins to utilize different resources of the habitat. This is known as *parapatric speciation* and is considered by some biologists as a special variety of sympatric speciation.

In allopatric speciation spatial or geographical isolation refers to geographical barriers that tend to segregate organisms. Large populations are frequently split into smaller ones by physical barriers such as mountains, deserts, glaciers, or bodies of water. The isolated populations may advance in different evolutionary directions. The various segments of the isolated population are referred to as *allopatric,* that is, occupying different ranges or areas. Even though geographically isolated populations originally contain the same gene pool, mutation and genetic recombination occur differently in each group, and natural selection acts, in each case, on the subsequent variations. For example, Darwin observed that the finches on the various Galápagos Islands, although quite similar, nevertheless differed in

TABLE 21-4. Classification of Reproductive Isolating Mechanisms

Mechanism	Description
I. Prezygotic	Interbreeding is prevented when both species are reproductively active or the union of gametes is prevented after mating or cross-pollination has occurred.
A. Ecogeographical[a]	A population becomes allopatric because of a geographical barrier and the population remains this way even after the barrier is removed because one or both species can no longer occupy the intermediate area. Example: Two groups of plants become specialized as a result of a change in environmental conditions (e.g., climate). If, in time, the original climate returns, either or both groups can no longer tolerate it.
B. Seasonal	Populations that breed during different seasons of the year. Example: Similar species of plants that are capable of interbreeding but are prevented from doing so because one completes flowering prior to the other.
C. Behavioral or ethological (animals only)	Populations with different or incompatible behavior prior to mating. Example: Similar species of animals with different mating calls or different visual displays.
D. Habitat	Populations occupying different habitats in a range. Example: Similar species of aquatic organisms, one occupying the lower depths, the other the shallow areas of a pool or one occupying the rapid moving water of a stream while the other occupies quiet pools.
E. Mechanical	Populations with structural differences in reproductive organs. Example: Similar species of plants in which the pollen tube is prevented from penetrating the ovary or similar species of animals in which there is a difference in the shape or size of the genitalia. (It should be noted that recent observations indicate that this mechanism is relatively ineffective in preventing cross-mating in animals.)
F. Gametic isolation	Populations in which mating may occur but union of gametes is prevented. Example: The genital tract of the female provides an unsuitable environment for the sperm which dies before union with the egg.
II. Postzygotic	The failure of hybrids to develop fully or the reduction of hybrid fertility.
A. Hybrid inviability	Populations in which fertilization occurs and the embryo dies at some stage of development. Example: The sperm fertilizes the egg but there is an abnormal development of the embryo.
B. Hybrid sterility	Populations in which sterile hybrid offspring are produced. Example: The gonads or other reproductive organs of the offspring are abnormal or fail to develop. The mule, a cross between a donkey and a horse, is an example of hybrid sterility.
C. Hybrid breakdown	Populations cross-breeding but producing weak hybrids or hybrids with low reproductive rates. Example: The offspring of a cross between two species of plants is more susceptible to a local paraiste or disease.

[a] All of the populations for each kind of reproductive isolating mechanism are sympatric or parapatric except for the ecogeographical, which are allopatric.

striking ways. Since finches will not readily fly across wide stretches of water, it is supposed that the initial colony was established by a species of finches that were perhaps blown by a high wind from the mainland of South America to one of the islands. As time passed, members of this colony were blown to or accidentally flew to another island and so on until other colonies were established on the various islands. In time each of the colonies evolved by the various processes previously discussed. Another often-cited example of geographic isolation is that of the Kaibab squirrel, which inhabits the north side of the Grand Canyon, and the Abert squirrel, which inhabits the south side. The temperatures on the canyon rims are typical of 7000-foot plateaus, while at the base of the canyon, temperatures reach 120°F, typical for a desert environment. Although the rims face each other, they are separated by an inhospitable desert and river. It is supposed that the Kaibab and Abert squirrels evolved from the same ancestral species, but over a long period of geographical isolation slowly became different from each other (Figure 21-10).

The speciation of geographically separated populations is initiated when the gene pools of the separated segments have become so altered that interbreeding is no longer possible. In some instances, if the isolated environments are similar, there may be no actual reproductive isolation because genetic variation may proceed in essentially the same direction. However, in the vast majority of cases, variations acted upon by selection create such differences in populations that they will not normally interbreed if they are brought back in close contact and occupy the same range or area. Thus, there is a good chance that isolated populations, even though identical at one time, will evolve different genotypic and phenotypic variations over a span of years.

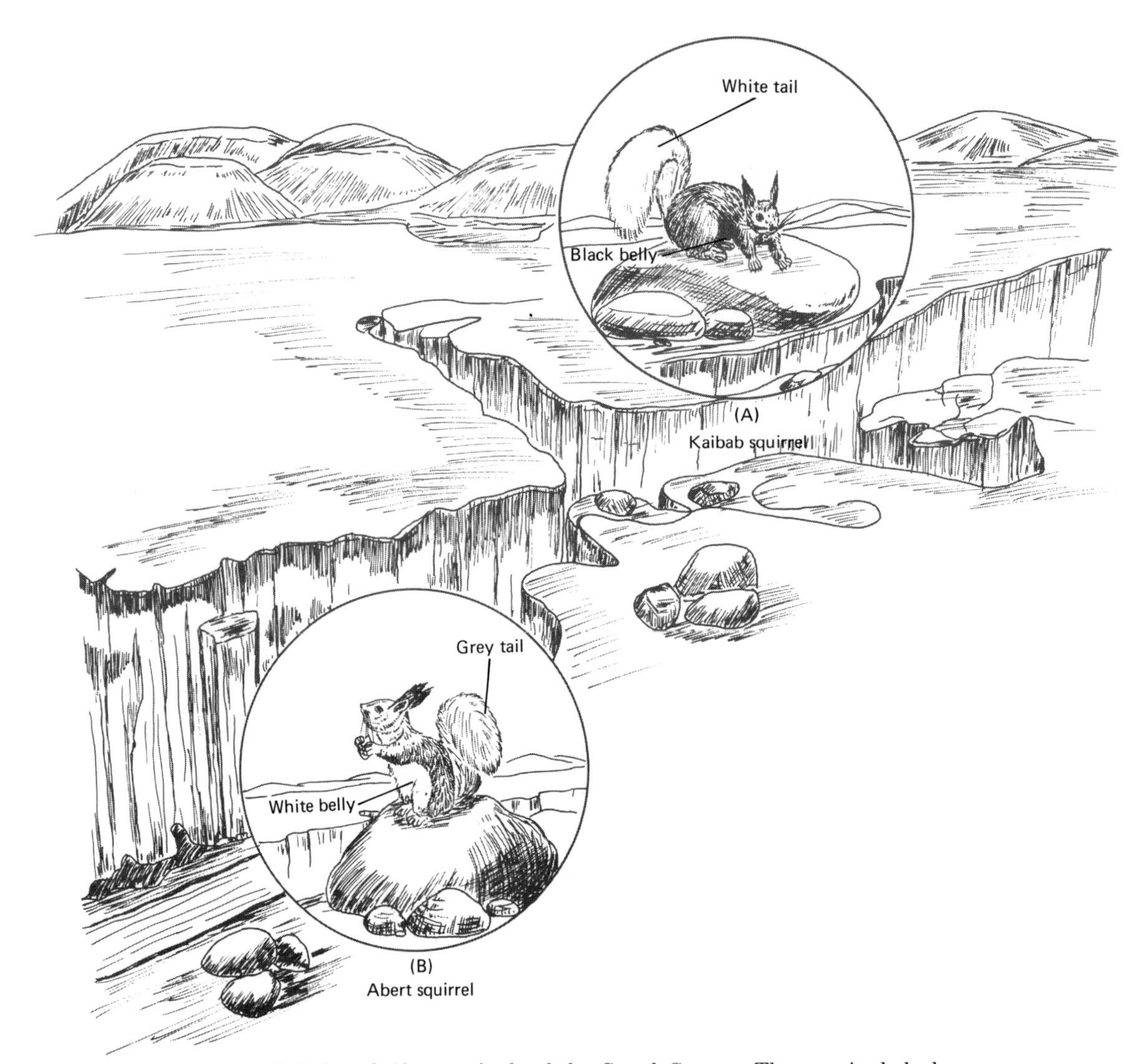

Figure 21-10. Kaibab and Abert squirrels of the Grand Canyon. These squirrels had a common ancestral population that became geographically isolated. (**A**) The Kaibab squirrels inhabit the north rim of the Canyon and have black bellies and white tails. (**B**) The Abert squirrels inhabit the south rim of the Canyon and have white underparts and grey tails.

Isolating mechanisms that prevent interbreeding (separate genetically) of related populations occupying the same range are not as easily determined as those found in allopatric speciation. For example, similar species of maple, the sugar maple (*Acer saccharum*) and the red maple (*Acer rubrum*) of the eastern United States, or similar species of pine, the ponderosa pine (*Pinus ponderosa*) and the digger pine (*Pinus sabiniana*) of California, occupy the same areas but remain distinct from one another. Apparently they are incapable of exchanging genes under natural conditions. Recent laboratory and field research shows that parapatric and sympatric speciation may also play an important role in the evolution of new species. Studies show that the action of the regulatory gene in controlling the transcription of other genes may play a major role in evolution. A single gene substitution or chromosome rearrangement at the level of the regulatory gene can initiate speciation if it can affect the gene flow between divergent populations through the development of reproductive isolating mechanisms.

As was previously stated, parapatric speciation occurs when reproductive isolating mechanisms arise while a new population is becoming adapted to the utilization of new aspects of its existing environment. For example, in areas where the soil has become contaminated with zinc, copper, or lead by mining operations the evolution of heavy-metal-tolerant races of plants has occurred within a relatively short period of time. Even though these new races exist within several feet of the parent population, strong reproductive isolating mechanisms are developing such as differences in flowering time and more effective pollination among members of the emerging races.

Sympatric speciation occurs where reproductive isolating mechanisms arise and the new population utilizes a different aspect of the environment in which it lives. For example, the race of hawthorne fly (*Rhagoletis pomonella*) that is now widespread on apple trees was derived from the hawthorne-infesting race of this species. This new race most likely originated in the Hudson River Valley where it was first reported and where both apple and hawthorne trees grow.

In summary, speciation is an adaptive process that involves the establishment of barriers to gene flow between closely related populations by the development of reproductive isolating mechanisms. These mechanisms evolve through the genetic variability that normally occurs in all populations.

Patterns of Evolution and Adaptation

It has been shown in the preceding discussion that genetic variability in organisms is caused by genetic mutation, chromosomal mutation, genetic recombination, drift, and gene flow. Through natural selection and reproductive isolation acting upon variability, populations may evolve that are able to function more efficiently in a given environment. In other words, evolutionary change results in the emergence of organisms that are better adapted to their environment than any of their predecessors, which may not have survived. Evolution leads to adaptation and the myriad adaptations that have occurred over the long span of evolutionary history have led to the establishment of distinct groups of organisms.

Any characteristic that is advantageous to a given organism or population is referred to as an **adaptation.** Stated another way, an adaptation is any genetically controlled trait that aids an individual or a species to survive and produce offspring in a particular environment. Every organism consists of many adaptations all of which contribute to success in growth, development, and reproduction. Adaptations may be under the influence of only a few genes or they may be controlled by a large number of genes. In terms of various levels of structural organization, adaptations may involve entire systems or organs, or they may involve specific cells or various cellular organelles. Moreover, adaptations may be highly specific in that they are useful only under limited circumstances or they may be general in that they may be advantageous under many and varied situations. Despite the implications that adaptations are of several types, it should be noted that their separation into categories is quite different since one is typically reflected in another. There are circumstances, however, in which arbitrary subdivisions of adaptations can be made based upon the predominance (not independence) of one type over another. Such subdivisions are called structural, physiological, and behavioral adaptations (Chapters 27–28). Evolution implies that the surviving species are more adaptive than those that do not survive.

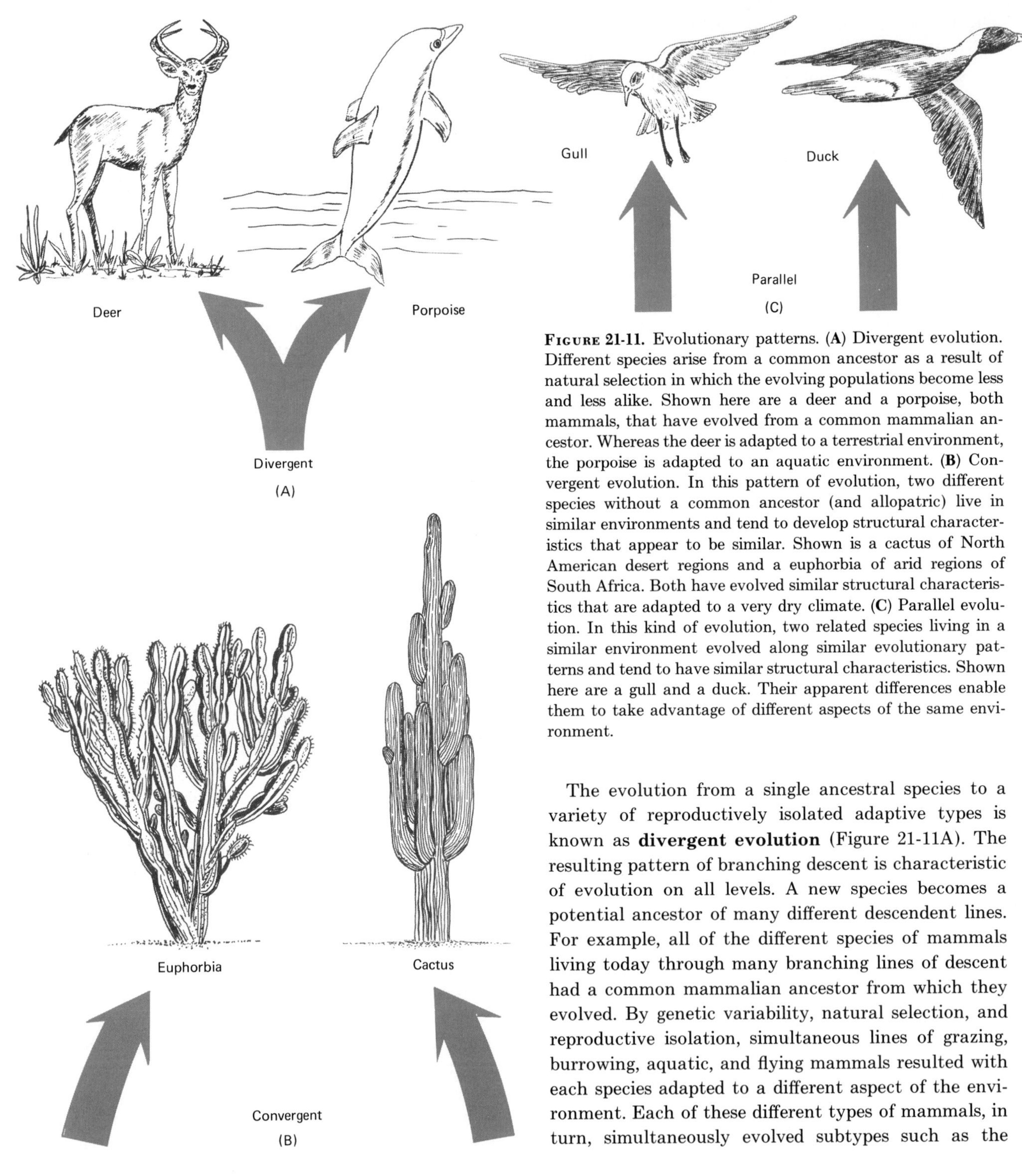

Figure 21-11. Evolutionary patterns. (**A**) Divergent evolution. Different species arise from a common ancestor as a result of natural selection in which the evolving populations become less and less alike. Shown here are a deer and a porpoise, both mammals, that have evolved from a common mammalian ancestor. Whereas the deer is adapted to a terrestrial environment, the porpoise is adapted to an aquatic environment. (**B**) Convergent evolution. In this pattern of evolution, two different species without a common ancestor (and allopatric) live in similar environments and tend to develop structural characteristics that appear to be similar. Shown is a cactus of North American desert regions and a euphorbia of arid regions of South Africa. Both have evolved similar structural characteristics that are adapted to a very dry climate. (**C**) Parallel evolution. In this kind of evolution, two related species living in a similar environment evolved along similar evolutionary patterns and tend to have similar structural characteristics. Shown here are a gull and a duck. Their apparent differences enable them to take advantage of different aspects of the same environment.

The evolution from a single ancestral species to a variety of reproductively isolated adaptive types is known as **divergent evolution** (Figure 21-11A). The resulting pattern of branching descent is characteristic of evolution on all levels. A new species becomes a potential ancestor of many different descendent lines. For example, all of the different species of mammals living today through many branching lines of descent had a common mammalian ancestor from which they evolved. By genetic variability, natural selection, and reproductive isolation, simultaneous lines of grazing, burrowing, aquatic, and flying mammals resulted with each species adapted to a different aspect of the environment. Each of these different types of mammals, in turn, simultaneously evolved subtypes such as the many different herbivorous and carnivorous mammals.

A term often associated with divergent evolution is adaptive radiation and frequently the differences are a matter of degree rather than kind. **Adaptive radiation,** like divergent evolution, is a process by which various representative species lead away from a common ancestral type. However, the species involved in adaptive radiation generally show greater similarity to each other than they do in divergent evolution. Adaptive radiation usually refers to divergent evolution on a much smaller scale. For example, the large varieties of herbivores show divergent evolution while all the various species of antelopes (herbivores) show divergence; the evolution is often represented as adaptive radiation.

The second pattern of evolution is observed when two or more unrelated species are adapted to similar environmental conditions. Although their common ancestors were less alike, the existing species resemble each other. This is known as **convergent evolution** and is represented by such examples as the resemblance between dolphins (mammals) and fishes, cacti and euphorbias (Figure 21-11C), and the wings of insects and birds. It is also represented by species mimicry in which two unrelated species may appear almost alike, or camouflage, in which several species will develop coloration similar to their environment. These adaptations have survival value because they offer protection.

Convergent evolution comes about through natural selection among different species subjected to similar conditions in the environment. This results in structural and functional adaptations that are basically alike. Generally, the adaptive structures of the different species are not homologous and the genetic variability that gives rise to the changes in existing structures is different. For example, the fin of the dolphin evolved from the forelimb of some terrestrial ancestor, while the fin of the fish did not. However, their similarity in appearance and function show that this type of structure is best adapted for rapid movement through the aquatic environment.

The third general pattern of evolution is referred to as **parallel evolution** (Figure 21-11B); it occurs where there is similarity in the adaptation among closely related organisms. In convergent evolution, the species are so different that the convergent character of the adaptation is obvious. By contrast, in parallel evolution, the convergence is less obvious and appears to follow more or less the same course as a result of similar genetic variability. The structural origin of the adaptation among the different species is essentially the same. Parallel evolution occurs as a result of natural selection where there are limited morphological mechanisms. For example, the structure of the rigid endoskeleton must conform to specific architectural principles if the organism is to be supported and at the same time capable of locomotion. Although there are some skeletal modifications among the various species, the similarity is great. The same is true with regard to eyes. While there are differences between the various types of eyes among different species, they must conform to certain optical principles. Where there are a limited number of solutions based upon physical principles and materials available, natural selection has been forced to work within certain limits. In the branching descents from a common ancestor, the similarity in adaptation represents a parallel evolution.

SUGGESTED SUPPLEMENTARY READINGS FOR SECTION FOUR ORGANISMIC BIOLOGY: CONTINUITY

AUSTIN, COLLIN, *Ultrastructure of Fertilization.* New York: Holt, Rinehart and Winston, 1970.

———, *Germ Cells and Fertilization.* New York: Cambridge University Press, 1972.

BEARN, A. G., and J. S. GERMAN, "Chromosomes and Disease," *Scientific American,* November 1961.

BODEMER, C., *Modern Embryology.* New York: Holt, Rinehart and Winston, 1968.

BONNER, D. M., and S. E. MILLS, *Heredity,* 2nd ed. Englewood Cliffs, N.J.: Prentice-Hall, 1964.

BURNS, GEORGE W., *The Science of Genetics,* 3rd ed. New York: Macmillan, 1976.

COHEN, S. N., "The Manipulation of Genes," *Scientific American,* July 1975.

DOBZHANSKY, T., "The Present Evolution of Man," *Scientific American,* September 1960.

———, *Genetics and the Origin of the Species,* 3rd ed. New York: Columbia University Press, 1964.

Etkin, William, "How a Tadpole Becomes a Frog," *Scientific American,* May 1966.

Folkman, J., "The Vascularization of Tumors," *Scientific American,* May 1976.

Friedmann, T., "Prenatal Diagnosis of Genetic Disease," *Scientific American,* November 1971.

Goldsby, R. A., Race and Races, 2nd ed. New York: Macmillan, 1977.

Grant, V., "The Fertilization of Flowers," *Scientific American,* June 1951.

Jinks, J. S., *Extrachromosomal Inheritance.* Englewood Cliffs, N.J.: Prentice-Hall, 1964.

Kerr, N., *Principles of Development.* Dubuque, Ia.: Wm. C. Brown, 1967.

Lane, C., "Rabbit Hemoglobin from Frog Eggs," *Scientific American,* August 1976.

MacArthur, R., and J. Connell, *The Biology of Populations.* New York: Wiley, 1966.

McKusick, V., *Human Genetics,* 2nd ed. Englewood Cliffs, N.J.: Prentice-Hall, 1969.

Mettler, L., and T. Gregg, *Population Genetics and Evolution.* Englewood Cliffs, N.J.: Prentice-Hall, 1969.

Moody, P. A., *Introduction to Evolution.* New York: Harper & Row, 1970.

Raff, M. C., "Cell-Surface Immunology," *Scientific American,* May 1976.

Saunders, J., *Animal Morphogenesis.* New York: Macmillan, 1968.

———, *Patterns and Principles of Animal Development.* New York: Macmillan, 1970.

Segal, S. J., "The Physiology of Human Reproduction," *Scientific American,* September 1974.

Stahl, F. W., *The Mechanics of Inheritance.* Englewood Cliffs, N.J.: Prentice-Hall, 1965.

Stebbins, G. L., *Flowering Plants: Evolution Above the Species Level.* Cambridge, Mass.: Harvard University Press, 1974.

Steward, F. C., "The Control of Growth in Plant Cells," *Scientific American,* October 1963.

Sturtevant, A. H., *A History of Genetics.* New York: Harper & Row, 1965.

Sussman, M., *Developmental Biology.* Englewood Cliffs, N.J.: Prentice-Hall, 1973.

Tietze, C., and S. Lewit, "Abortion," *Scientific American,* January 1969.

Torry, J. G., *Development in Flowering Plants.* New York: Macmillan, 1967.

Waddington, C. H., *Principles and Problems of Development and Differentiation.* New York: Macmillan, 1966.

Wallace, B., *Chromosomes, Giant Molecules, and Evolution.* New York: Norton, 1966.

Weaver, R. G., "The Cancer Puzzle," *National Geographic,* September 1976.

FIVE
Ecosystems

22 The Abiotic Environment

23 Populations Within Ecosystems

24 Dynamics of Ecosystems

25 Types of Ecosystems

26 Human Populations and the Biosphere

22 The Abiotic Environment

GUIDE QUESTIONS

1 Distinguish between the abiotic and biotic environments. Why is the earth regarded as an open system with regard to energy? What are some of the factors causing cyclic change within the abiotic environment?
2 What are the physical characteristics of soil? What are the major constituents of soil and how may soil be classified? How is soil formed?
3 Describe and give examples of each of the four ways water may be related to soil.
4 Give three properties of water and tell how they may affect the biotic environment.
5 Into what two major divisions are inland waters divided? Cite specific examples.
6 What is meant by spring and fall overturns? What is their importance?
7 What are the various zones of the ocean and where are they located? How are ocean currents produced?
8 What are the major components of the atmosphere? What important roles does the atmosphere play in the abiotic environment?
9 How are wind currents formed? What are the major air currents of the earth?
10 What factors determine climate? What are the major climatic regions of the earth?

FROM the smallest microbe to the largest redwood tree, all living forms are inseparably related to the environment in which they grow and reproduce. The kinds and numbers of different organisms in a given space depend on available materials and nutrients and the conditions under which these are supplied. The tropical forest with its warm temperatures, abundant water, and necessary minerals can support large and diverse plant and animal populations. Life in the treeless tundra of the Arctic Circle is limited by long periods of extreme cold that freeze the water and limit the availability of necessary minerals. The desert regions may have warm temperatures and sufficient amounts of minerals, but here the lack of water is the limiting factor that determines the existing flora and fauna.

While the environment affects the number and types of living organisms, they, in turn, can exert profound influences upon the environment. The evolutionary development of adaptation within organisms has been a response to external environmental forces, whereas many of the characteristics of land and sea areas and the composition of the atmosphere are the result, in part, of the activities of living organisms. For example, plants are totally adapted to their available water supply, while in the atmosphere the high concentration of oxygen is a result of the photosynthetic activities of green plants. The total living world and all aspects of its environment constitute what is called the **biosphere.**

For convenience, the biosphere may be subdivided into two principal components: (1) the **abiotic environment,** or nonliving aspects, and (2) the **biotic environment,** or living aspects. The abiotic environment includes such factors as solar radiation, temperature, atmospheric pressure, topography, wind, precipitation, and inorganic and organic materials. The biotic environment, by contrast, includes all living organisms. Although the abiotic and biotic environments may be separated arbitrarily for discussion, they are inseparably interrelated. The total biotic components of Monera, Protista, Fungi, plants, and animals interact in a fundamentally energy-dependent relationship within the abiotic or physicochemical environment. The kinds of organisms and their way of life in any given area are ultimately determined by the raw materials provided by the abiotic environment and the conditions under which they are utilized.

The study of the interrelationships of organisms within the physicochemical environment is called **ecology.** The associations of organisms may be studied at various levels. For example, study of the interrelationships among the various individuals of a population or of the interactions and dynamics of a population as a complete entity is **population biology.** Study of the relationships among populations within a given area is **community ecology.** If the circulation, transformation, accumulation, and loss of materials and the transductions of energy within such a community are considered, the subject of study is an **ecosystem.**

The Importance of the Abiotic Environment

All aspects of the physical world are constantly changing. These changes are brought about by unceasing variations in weather and climate, and by geophysical phenomena, various geochemical activities, and life. Many of the constant changes in the environment are due mainly to changes in the balance between energy received at the earth's surface in the form of light and other radiant energy and energy leaving the earth's surface in the form of reflected light and invisible heat radiation. Such a free exchange of energy results in what is called an *open system.* A *closed system* is one in which energy neither enters nor escapes. The variations in energy input are produced primarily by the earth's rotation on its axis and its revolution around the sun, which result in daily cycles and yearly climatic changes, respectively.

Another force that initiates cyclic changes of great significance is the gravitational attraction of the moon and its effects upon the oceans. These cycles within the abiotic environment in turn create less obvious cycles, all of which act upon the biotic environment.

The abiotic environment may be divided into three distinguishable, although interrelated, subdivisions. The *lithosphere* comprises the solid components of the abiotic environment, including the soil and parent material from which the soil is derived. The *hydrosphere* includes all bodies of water such as rivers, streams, ponds, lakes, and oceans. The *atmosphere* is the layer of gases that surrounds the earth.

The Lithosphere—Land

The **lithosphere,** or solid component of the abiotic environment, consists of the various land masses of the earth. It plays two important roles in the economy of living forms: it provides the substrate required by most plants and many animals, and it is the source of minerals essential for growth and maintenance, and therefore continuity, of all life.

Soil

Soil is the most important and interesting aspect of the lithosphere, for it is the result of the interaction between the *parent material* (*bedrock*) and other factors of the abiotic and biotic environments. Soils are relatively complex mixtures of rocks, minerals, water, air, living organisms and products of their decay in which numerous physical and chemical changes are constantly occurring. A *soil profile,* or vertical cross section of soil, shows three general layers called soil horizons (Figure 22-1). The uppermost layer, called the *topsoil,* is the most important as far as living organisms are concerned because of their direct contact with this layer. It is also the principal source of essential materials needed for growth. In the *middle layer,* the soil material consists principally of large rock particles formed from the lowest layer, the *parent material.*

Topsoils, as well as soil profiles in general, exhibit considerable variations in physical texture, chemical constituents, origin, depth, and fertility. Despite these differences, the distinct components of the soil mixture may be separated into the following categories.

1. Rock particles and mineral matter.
2. Soil water.
3. Soil air.
4. Organic matter.
5. Organisms.

Rock Particles and Mineral Matter. The largest percentage of soil consists of a mixture of

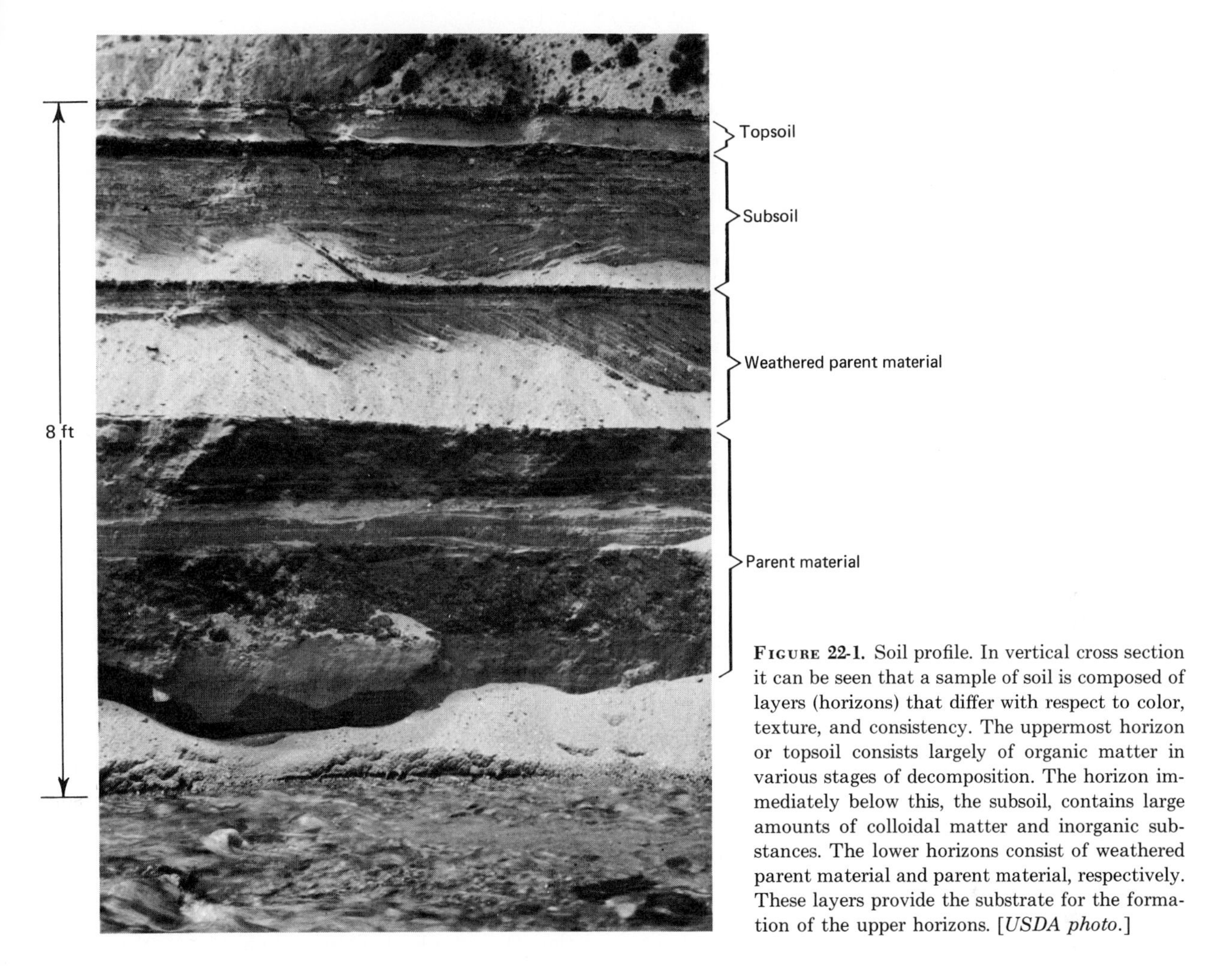

FIGURE 22-1. Soil profile. In vertical cross section it can be seen that a sample of soil is composed of layers (horizons) that differ with respect to color, texture, and consistency. The uppermost horizon or topsoil consists largely of organic matter in various stages of decomposition. The horizon immediately below this, the subsoil, contains large amounts of colloidal matter and inorganic substances. The lower horizons consist of weathered parent material and parent material, respectively. These layers provide the substrate for the formation of the upper horizons. [*USDA photo.*]

rock fragments and particles that have been formed primarily by the weathering of pre-existing rock. Various weather phenomena such as precipitation, temperature fluctuations, and wind are the principal agents of mechanical breakdown of rock. Chief among the rock materials that are subjected to weathering to form soil particles are shale, feldspar, sandstone, limestone, and granite. The nature and size of the mineral particles thus formed are directly related to the degree of weathering, the chemical composition of the parent rock, and the resistance of the rock to breakdown. The sizes of the rock particles in a soil provide a basis for distinguishing various soil textures. The *texture* of a soil is a relative measure of the size of the most abundant particles in a given soil mixture. Particle sizes are classed as coarse sand, fine sand, silt, and clay (the finest particles).

In general, soils do not consist exclusively of sand, silt, or clay but contain proportions of each of these constituents in varying amounts. They are named on the basis of the percentage of each they contain. A *sandy soil* is a mixture in which the silt and clay particles comprise less than 20% of the material by weight. A *clay soil* is one which contains at least 30% clay particles. A *loam soil* contains about 50% sand and 50% silt. The pores, which make up between 30 and 60% of the soil volume, contain air and water. Larger particles have larger pore spaces between them. Accordingly, sandy soils have large pore spaces which permit excellent penetration of water and air. Such penetration in clay soils is somewhat more difficult because of the extremely small pore spaces that exist.

The ability of soil particles to retain water is called the *water-holding capacity*. In sandy soils, aeration is good and water is absorbed rapidly, but retention is poor and water drains away quickly. Clay soils, by contrast, have greater water-holding capacity than sandy soils, even though aeration and water absorption are less adequate because most clays are of colloidal size and this results in a tremendously large surface-to-volume ratio. Once water has penetrated, a clay soil may hold three to six times as much water as a sandy soil. The addition of organic matter increases the water-holding capacity of soils because it contains relatively small pores and because it, like clay, is colloidal in nature.

Soil Water. Water may be related to soil in any of four ways. *Gravitational water* percolates down through the soil and is lost, owing to gravitational flow. It is available to plants and animals for only a short time following rain or irrigation. *Capillary water* fills the tiny pores between the soil particles and is retained by capillary forces after gravitational water has percolated down to the water table. Capillary water is the soil water that is most readily available to plants. *Hygroscopic water* adheres closely to soil particles, forming a layer only one or two molecules thick. This water is almost always present, even in desert soils, but is not readily available to organisms. *Bound water* is bound chemically in soil minerals and is even less readily available to organisms than hygroscopic water (Figure 22-2).

The importance of soil water, particularly for plants and other soil organisms, is that it may be utilized as a metabolite and is the medium in which all dissolved substances enter soil organisms. Various inorganic salts such as nitrates, phosphates, and sulfates and other water-soluble substances are the principal materials found dissolved in soil water.

Just as water is held to the surface of soil particles to form a film, so also are mineral ions held to these surfaces. This condition is possible because soil particles act as negatively charged ions and are able to attract and hold positively charged mineral ions such as K^+, Ca^{2+}, and Mg^{2+}. Clay soils and organic matter, because of their colloidal nature and large surface areas, are especially effective in the adsorption of ions. Because of this property of soils, fine soil particles act as a reservoir of mineral nutrients and prevent the leaching of minerals out of the topsoil through water drainage. Sandy soils, in constant need of irrigation, require a continual application of fertilizers because of the leaching of essential minerals. Negatively charged ions such as nitrate (NO_3^-) and sulfate (SO_4^{2-}) are found as part of the soil solution. They are not adsorbed to the surfaces of soil particles. While the total amount of minerals present in the lithosphere far exceeds the amount needed to sustain life, the quantities available in any given area may vary greatly. The least plentiful essential mineral circulating in any given ecosystem will determine the kind and quantity of life existing within the area. For example, the microscopic forms of life present in areas of the oceans are determined by the amount of available nitrates in these areas.

Most mineral nutrients initially become available through chemical reactions in the parent material of the lithosphere which convert inorganic matter into soluble forms. Once dissolved, the minerals may be transported from one area to another. The minerals may also enter the hydrosphere through drainage into streams, rivers, and eventually into the oceans. Minerals dissolved in the ground water may be brought to the surface of the soil by evaporation and by capillary action. Once made available to living forms, the minerals are circulated in the biosphere by passing back and forth through life and decay. The use of mineral nutrients by living organisms and their return to the lithosphere for re-use represent a major cycle in the biosphere (see Chapter 24).

Soil Air. For plants and animals living in the soil an adequate supply of *soil air* is essential. This air, found in pore spaces, contains the same gases as the air present above the soil. In general, however, the relative proportions of soil air gases may differ from those of the atmosphere. For example, the average percentage by volume of atmospheric carbon dioxide is lower than that of the soil air. This disproportionate condition exists because roots of higher plants, together with other plant and animal life of the soil, produce carbon

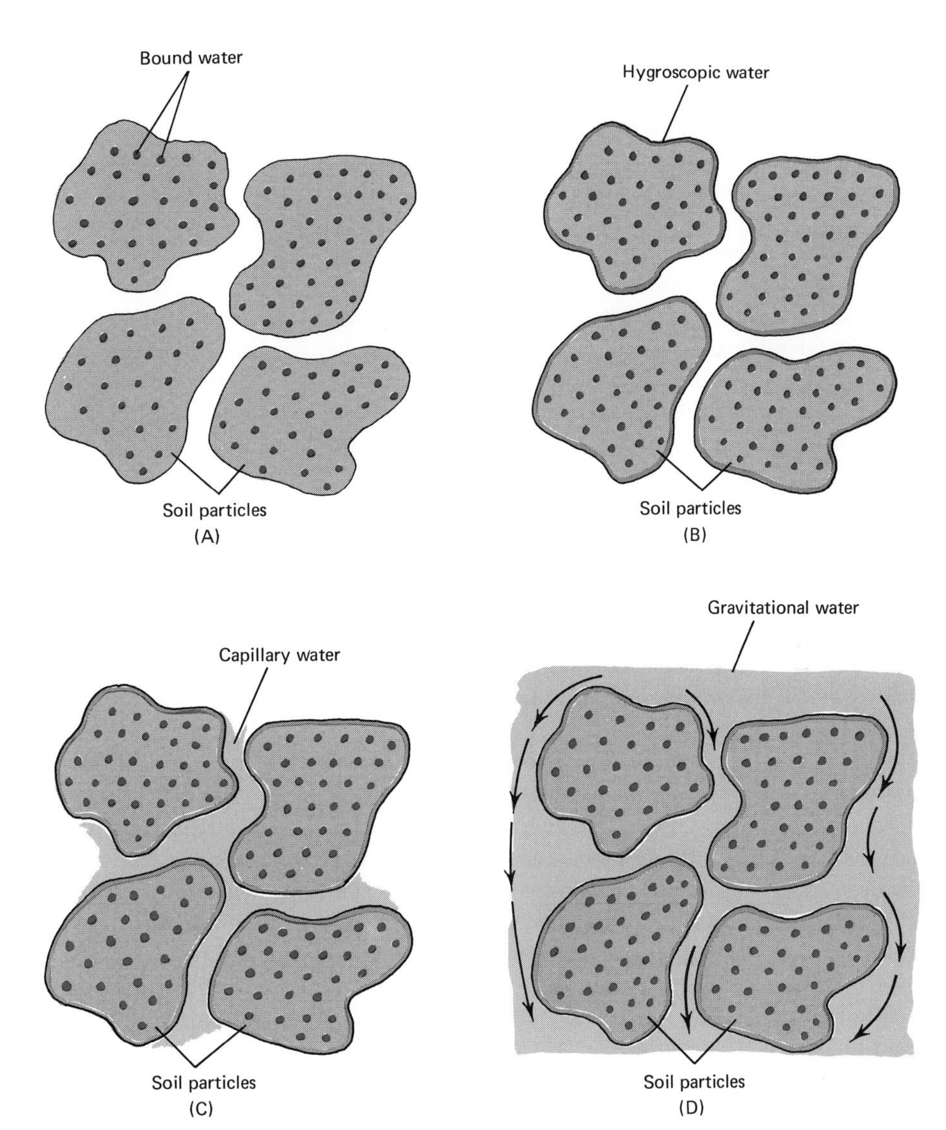

FIGURE 22-2. Relationships of the types of water with soil particles. (**A**) Bound water is chemically bound. (**B**) Hygroscopic water adheres to soil particles. (**C**) Capillary water is retained by capillarity. (**D**) Gravitational water percolates through soil. See text for amplification.

dioxide during respiration. Inasmuch as roots and soil organisms require oxygen for respiration, the oxygen concentration of the soil decreases while the carbon dioxide content increases. The soil air, however, is never saturated with carbon dioxide or depleted of oxygen, because carbon dioxide diffuses out of the soil and oxygen diffuses into it. For most plants and animals, the amount and movement of soil air depends not only upon adequate pore spaces but also upon the water content of the soil.

ORGANIC MATTER. All organic matter in the soil is derived from the decayed remains of plants and animals, the waste products of living organisms, and the activities of various fungi and microorganisms such as bacteria. The organic matter in the soil consisting of carbohydrates, proteins, lipids, lignins, mucins, and other materials is ultimately reduced to simpler inorganic compounds such as ammonia, water, carbon dioxide, and various nitrate, sulfur, phosphate, and calcium compounds. The greater portion of the organic matter of soils is of plant origin, principally from dead roots, decaying wood and bark, and fallen leaves. The breakdown of these organic substances through the action of soil organisms produces and maintains a continuous supply of inorganic substances which plants and other organisms require for growth. In most agriculturally important soils, organic matter comprises about 2–10%. Swamps and bogs, by contrast, contain a higher content of organic matter, up to 95% in some peat bog soils.

A considerable part of the organic matter in soils occurs as *humus,* a colloidal complex. This dark-colored material is composed chiefly of organic materials that are resistant to decay, such as cellulose and lignin. In this regard, humus is partially decomposed organic matter.

The addition of organic matter, either completely or partially decomposed, is essential to continued soil fertility. In addition, because of their spongy nature, organic materials loosen the soil to prevent the formation of heavy crusts and increase the proportion of pore spaces in the soil, which increases aeration and water retention.

Soil Organisms. The soil contains not only innumerable bacteria and fungi, which are important in decay, but also many animals, ranging from microscopic forms to insects, millipedes, centipedes, spiders, slugs, snails, earthworms, roundworms, mice, moles, gophers, and reptiles. Most of these are beneficial in some way or another in that they promote some mechanical movement of soil, thus keeping the soil loose and open. In addition, all soil organisms contribute to the organic matters of soils as a result of their waste products and decomposition.

Without the presence of microorganisms, especially bacteria, the soil would soon become unfit to support life. Bacteria influence soils in many ways. Some decompose organic matter into simple products. In these reactions, nutrients are made available for re-use. Other soil bacteria are associated with transformations of nitrogen and its compounds, so that an available supply of nitrogen is continually maintained.

The Hydrosphere—Water

The second major subdivision of the abiotic environment is the **hydrosphere,** the water areas of the earth. Water is by far the most abundant substance; it covers almost four-fifths of the earth's surface.

The waters of the hydrosphere may be broadly classified as fresh water (aquatic environments), in which the mineral content is relatively low, and marine water (marine environments), in which the mineral content is high (specifically with reference to sodium chloride). The mineral content is a major factor in determining the type of animal and plant life found in the hydrosphere. Aquatic and marine organisms are generally confined to their specific environments because of their osmotic regulation capabilities. Most marine organisms cannot live in fresh water and most fresh water organisms cannot live in salt water.

The Physical Properties of Water and the Abiotic Environment

The physical properties of water greatly influence the abiotic environment, helping to regulate the activities within this environment. The extremely high heat capacity of water resists rapid changes in temperature and has a moderating effect on the temperature and climate of adjacent land masses (heat capacity about 0.2 calorie per gram). This accounts for the relatively late spring in land areas near great bodies of water. Accordingly, these same land areas have moderating temperatures which extend well into the fall as the surrounding waters give up their heat to the land. By contrast, land masses will show a marked temperature increase compared with water areas of similar size under the same conditions of solar energy intensity and duration. Climatic conditions of land areas not located near large bodies of water are subjected to rapid and greatly contrasting changes in temperature and climate.

A unique property of water is its density at various temperatures (Figure 22-3). Most substances contract continually as the temperature drops and they cool. As their volume decreases, their density (weight per unit volume) increases. If the substance is a fluid or liquid, the cooler portions will drop to the bottom and the warmer portions will circulate toward the top. Therefore, most substances will freeze from the bottom upward. Water continually contracts, as do other liquids, down to a temperature of 4°C; then it behaves in an unusual manner. As the temperature goes below 4°C water expands; hence, the coldest layers (either water

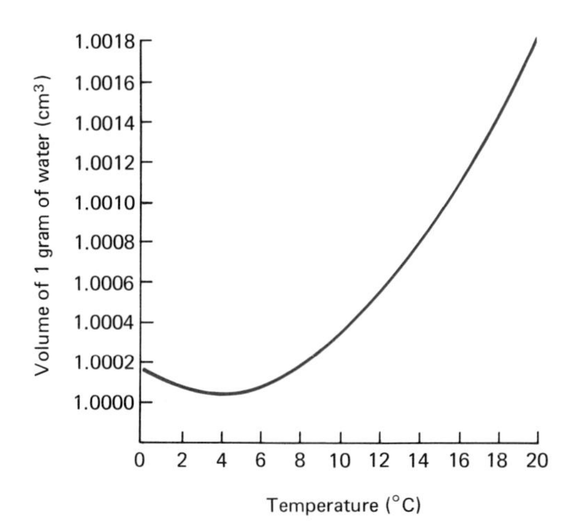

FIGURE 22-3. Graphic representation of the expansion curve of water. The curve shows the expansion (volume) of water with temperature variation. For example, at a temperature of 14°C the volume of 1 gram of water is approximately 1.0008 cm^3, whereas at 18°C it is about 1.0014 cm^3. Note that the density of water (1.0000) is greatest at 4°C.

or ice), because they are less dense, are located at the surface.

In addition to the physical properties of water, both the terrain of the land and the transfer of solar energy will determine the specific characteristics of a given area of the hydrosphere. The terrain of the land controls the area and the depth of the body of water. Moreover, large open bodies of water are subjected to the action of the wind. The constant churning of water by wind action establishes currents which distribute oxygen to lower depths and transfer heat from the upper to lower levels. Shallow bodies of water tend to have the same temperature throughout while deeper waters have specific layers of different temperatures.

Solar energy, in the form of light and heat, is probably the most important controlling factor acting upon any body of water, as it is on any land surface. All the energy received by the surface of a body of water is not absorbed; a portion is reflected. The amount reflected is determined largely by the angle at which the rays from the sun strike the surface. The greater the angle of incidence, the greater the amount of energy reflected. In the winter the angle of the sun's rays striking the earth's surface is greater than at any other season; therefore, much of the energy is reflected to the atmosphere.

As flowing water passes over the varying land types, it picks up and carries dissolved minerals and suspended particles. When this water empties into standing bodies of water it contributes to the mineral content and increases their turbidity. Pure water is highly transparent to light. However, natural water contains dissolved substances, suspended materials, and organisms which inhibit the transmission of the light and increase the turbidity. The depth to which light will penetrate into a given body of water will vary with its turbidity. It is for this reason that plants are limited to the upper layers of water.

The amount of heat energy in a body of water determines its temperature. This thermal property is the controlling factor in many chemical, physical, and biological reactions. The solubility of minerals in water is determined by temperature. The higher the temperature, the greater the amount of mineral matter that will dissolve; therefore, warmer waters will have a higher mineral content than cooler waters. On the other hand, gases are less soluble in warm water. By contrast, the cooler waters of the hydrosphere will have a higher oxygen content. It is the balance between mineral and oxygen content that will determine the amount of aquatic life that can be supported in a given region of the hydrosphere.

Inland Waters

The inland water areas of the hydrosphere may be subdivided into two types: standing water or *lentic* habitats and running water or *lotic* habitats. Standing water includes bodies such as lakes, ponds, swamps, or bogs, while running water includes springs, brooks, creeks, or rivers. There are no sharp differences between these two types, but rather a gradual change or gradient between them.

STANDING WATERS. Standing waters vary greatly in their characteristics. They range in area from the smallest woodland ponds to large bodies of water such as the Great Lakes. They also vary in chemical content from fresh to salt water and in turbidity from clear to muddy. Each has its own characteristic temperature, oxygen and carbon dioxide content, depth, and currents.

Lakes. A lake is a large body of standing, open water occupying an area known as a *basin.* The basins of lakes are formed as a result of depressions in the earth caused by movements of the earth's crust, volcanic activity, glacial action, wind action, stream activity, and landslides. Some lakes have been formed through the construction of dams.

The source of water in lakes is precipitation, either directly in the form of snow or rain, or indirectly in the form of underground water accumulation. Land surface drainage accounts for most of this underground water. Lakes are not only the source for many rivers and streams but they are also fed by rivers and streams. Many lakes represent a temporary widening of the river bed and are termed *open lakes.* The basin of the lake serves as a settling basin for sediment brought in by flowing water. This is evidenced by the observation that entering rivers and streams are turbid while those leaving the lakes are generally clear. In those lakes that are *closed,* that is, lakes in which there is no outlet, there is a great accumulation of minerals; the most abundant of these is sodium chloride. The Great Salt Lake and Caspian Sea are two examples of closed lakes.

One of the most significant phenomena with regard to the biotic environment of the lakes in colder climates is the purely physical interaction between water and air temperature during the various seasons (Figure 22-4). In winter, the colder, less dense, ice covers the surface of the water; below this, the temperature of the water is just above the freezing point (1–4°C) and is relatively uniform all the way to the bottom. In the spring, as the ice melts, the surface temperature of the water increases from about 1.5 to 4°C, and the water begins to circulate as convection currents are created. The denser water at 4°C sinks, bringing any cooler water to the surface. These currents, aided by the wind, serve to mix lake water until a uniform temperature of 4°C is reached and the water is at its maximum density. This process mixes oxygen and minerals throughout the lake and is called the *spring overturn* or *spring circulation period.*

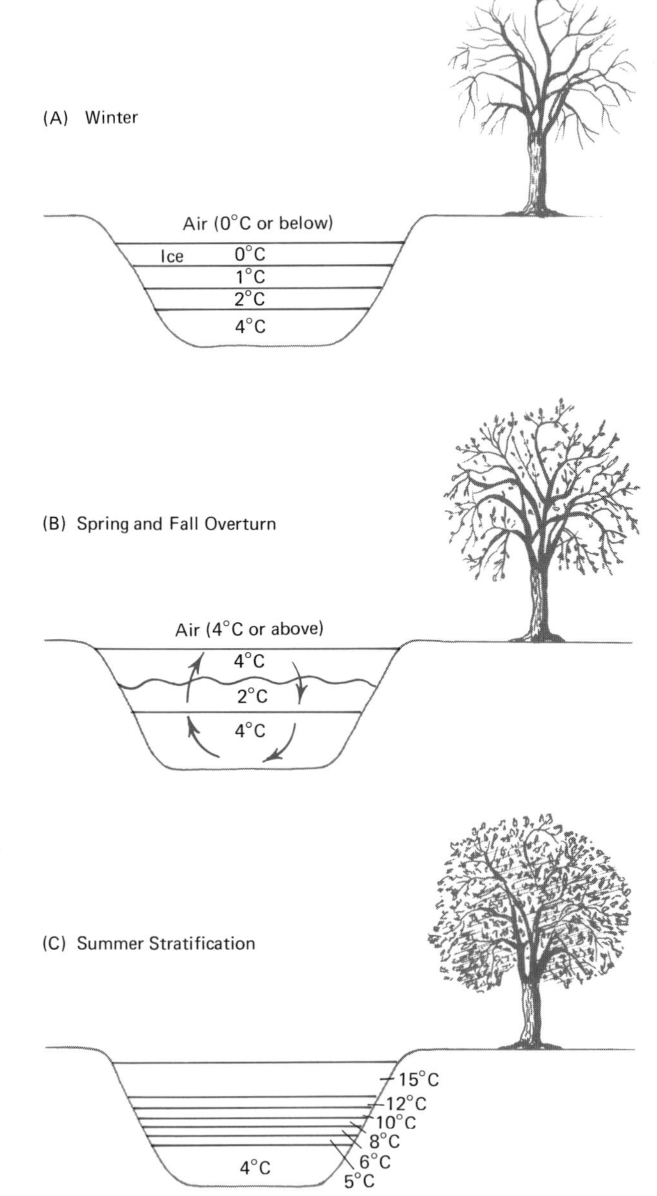

Figure 22-4. The influence of climate on lake stratification. (**A**) In winter the denser water at 4°C is at the lower level, while the colder water and ice are at the upper levels. (**B**) In spring, as the surface water approaches 4°C, the denser water sinks, setting up currents. A similar condition also occurs in the fall. These are known as the spring and fall overturns. (**C**) In summer the warm layer is at the top, the cool water at the low level. In the middle zone there is a sharp thermal change in the water. In summer there is relatively little mixing.

As summer approaches, the water at the surface warms up rapidly and expands, becoming lighter than the lower layers. This lighter layer of water resists mixing with the lower more dense layers so that by midsummer there are three clearly defined layers: the warm upper layer with little temperature variation (the *epilimnion*), a middle layer where there is a rapid drop in temperature through the layer (the *thermocline*), and a uniformly cooler lower layer (the *hypolimnion*). The thickness of each respective layer is determined by the climate and the size and depth of the lake. In addition to temperature differences each of the layers differs with regard to the amount of dissolved oxygen and minerals.

In autumn, as the surface water cools and becomes more dense, convection currents are again set up. These currents keep circulating the water by moving the cooler water to the lower levels until a uniform temperature of 4°C is reached throughout the lake. This mixing of water is called the *fall overturn* and results again in mixing oxygen and minerals throughout the lake. As the surface temperature decreases, the coldest water and ice, because they are the least dense, remain at the surface.

This constant change plays a most important role in the ecosystems of lakes. Such continuous fluctuations circulate oxygen and minerals so that a great variety of aquatic life can be supported at the different levels. By contrast, lakes in warm climates do not have this constant change. The warmer, less dense water remains on the surface, since the air temperature always remains well above 4°C. The net result of this stability is that life-supporting oxygen does not reach the lower levels. Therefore, aquatic organisms are limited to the upper layer of such lakes, where light can penetrate to support the photosynthetic release of oxygen.

Ponds. *Ponds* are small bodies of relatively shallow water. They may be permanent (contain water throughout the year) or temporary (dry up for a part of the year). Natural ponds are plentiful where there is sufficient rainfall; however, beavers dam up streams and human beings create ponds even in the more arid regions.

In some ways ponds are similar to lakes, but in many ways they differ. Owing to the shallow depth of ponds, convection currents are unimportant and thermal layers never form as they do in lakes. Consequently, distribution of minerals, gases, and temperature is quite uniform at the various depths of a pond, in contrast to the lake. In addition, sunlight penetrates to the bottom of ponds, so that photosynthetic activity is carried out even in the deepest water. Because of photosynthetic activity and the ready availability of gases and minerals, ponds support life in abundance.

Bogs. *Bogs* are poorly drained areas and do not support large numbers of different species. The water level is at ground level or slightly above, and the soil in a bog is waterlogged. The constant leaching of the soil by water, the low oxygen content, and reduced rate of decay result in a highly acidic condition and accumulation of organic matter. Many bogs are former shallow lakes and ponds that have become filled by organic matter. They will continue to fill until they become a terrestrial environment (see Figure 24-10). Bogs are most common in cool temperature regions. *Swamps* are similar to bogs except that they are capable of supporting large trees and shrubs and are most common in warmer climates.

FLOWING WATER. Flowing waters, the second major subdivision of inland waters, differ from standing water in their volume, currents, turbidity, temperature, and gas content. Of the water that falls on the earth as precipitation, some evaporates, some penetrates the soil to form ground water, which may later emerge as springs, and some flows over the land surface as *ground runoff*. This runoff is the source of most flowing waters.

Because of the great differences in the nature of the rock and soil formations of the earth and because land areas are generally sloping, runoff accumulates in small gullies that form rivulets. Rivulets coalesce into brooks, which form streams and, eventually, rivers. The direction of flowing water is from areas of higher elevation to those of lower elevation, typically emptying into the ocean.

Generally, flowing water can be divided into three areas, or horizontal zones. The *upper course* or rapid zone originates from underground springs and runoff water and consists of brooks and small, swiftly moving streams. The *middle course* or pool zone includes the foothill areas where the streams have lost some of their velocity but are still swiftly moving. In the *lower course* streams have joined to form a slow-moving river that meanders across a plain and eventually empties into a larger river or the ocean.

Springs and Brooks. Wherever the ground water level (*water table*) reaches the surface of the land the water will seep out in the form of a *spring*. Springs, alone and in combination with draining runoff water, are the source of many *brooks*. Small brooks are sometimes affected drastically by climatic conditions, especially if formed mainly from runoff. Brooks formed mainly from springs are affected little by climatic conditions. Below a depth of about 30–40 feet in the earth variations in weather have very little effect. The rock mantle stays at a uniform temperature that is the same as the yearly average temperature for that locality. In the United States this temperature is somewhere between 40 and 60°F. The ground water is practically the same temperature as the rock mantle, so springs and spring-fed brooks are relatively cool and constant in temperature.

As a brook flows down a steep slope, there is considerable aeration of the water. Since cool water can hold relatively large quantities of gases, brooks contain a good supply of oxygen. Because of the rapidly flowing waters, drifting flora and fauna cannot remain in one area, and the nature of the lotic ecosystem differs from that of the lentic ecosystem at least in part as a result of the rapid movement of the water. Brooks generally carry their water to streams.

Streams and Rivers. *Streams* and *rivers* are larger bodies of water than brooks and flow in the course of least resistance toward a lower elevation and eventually into the sea. Typically, the water moves rapidly as a stream at first and picks up large amounts of suspended particles. The rapidly moving water and resulting aeration allow the air temperature to moderate the temperature of the water. In addition, there is considerable gaseous exchange between the air and water. Light is able to penetrate the shallow depths, thereby permitting photosynthetic activity. Minerals and organic matter are added to the water from the bordering land areas. As streams flow into rivers, the beds are wider and the volume of water greatly increases. The land begins to level and the water moves more slowly. The larger suspended particles are deposited and there is an accumulation of organic matter and silt on the bottom. The water remains rather muddy owing to the suspended particles of soil colloids. The turbid water prevents penetration of sunlight and thereby limits photosynthetic organisms to the surface areas. The temperature of the larger rivers tends to be more stable than brooks and streams because the volume of water is larger and there is less aeration.

The nature of the course made by a flowing body of water is determined by the incline, the types of soil and rock over which it flows, and the fluctuation in the volume of water it carries. For example, a river flowing through a plain, which has little variation in topography, meanders, making many wide bends and turns called *oxbows*. A river greatly changes the character of the land areas it traverses and may be a dominant factor in adjoining terrestrial ecosystems as well.

Oceans

The *oceans* make up the largest portion of the hydrosphere; in fact, they cover about 70% of the earth's surface. There are areas such as the Mariana Trench in the northern Pacific where the ocean reaches a depth of 35,600 feet. However, the average depth of the oceans has been estimated to be about 12,500 feet.

While the oceans have remained essentially the same over millions of years, some changes have taken place. Changes were probably caused by diastropic movements—that is, the uplifting or upbuckling of large areas that raised them above sea level. Because of geophysical changes not yet clearly understood, there have been alternating cycles of warming and building of the polar ice caps. (It has been estimated that when the present polar ice caps melt, the sea level will be raised about 60 feet. Of course, this melting, if it takes place, will do so over tens of thousands of years.) In the past, the sea level has varied by approximately 300 feet between the periods of maximum and minimum glaciation. This volume of water is relatively small in comparison to the total volume of the oceans.

The Ocean Basin. The oceans occupy areas called *basins*. These basins have a gentle slope extending from the shoreline of a continent for about 100 miles out to sea on the average (Figure 22-5A). This slope is called the *continental shelf*. It terminates in a sharp dropoff called the *continental slope*. This slope is marked by deep gorges, canyons, and caverns. After dropping for several thousand feet, the continental slope levels off to the ocean floor, which is called the *abyssal plain*. There are great variations in this plain. At some points mountains arise to such a height that they break the surface of the ocean, forming islands, for example, the Azores, which extend far above the abyssal plain (Figure 22-5B). In other areas there are deep

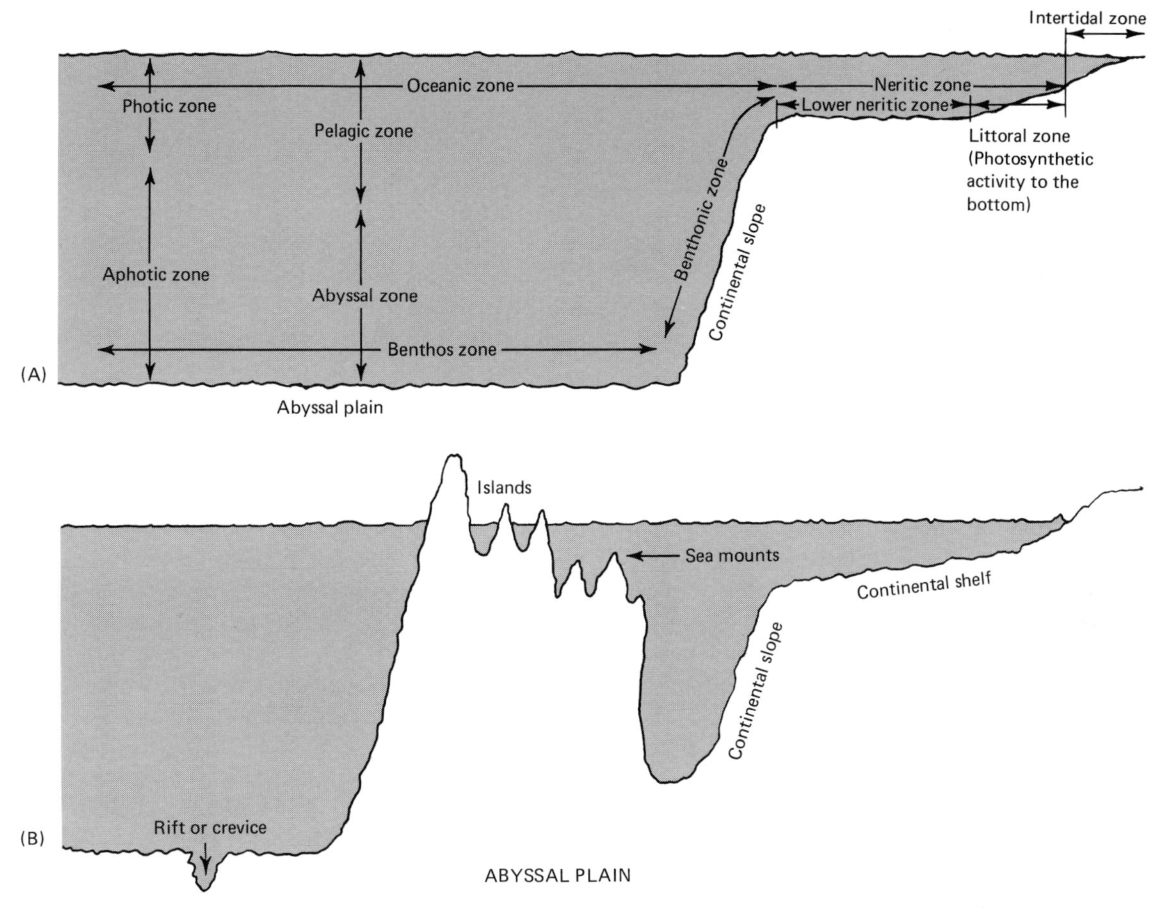

Figure 22-5. Diagrammatic representation of the ocean. (**A**) The various zones of the ocean and its basin. (**B**) Topographical variations in the abyssal plain.

gorges and canyons (rifts or crevices) extending thousands of feet below the ocean.

The ocean may be divided into several zones for study. The part of the ocean between high and low tides is called the *intertidal zone.* This area is generally rich in both flora and fauna. The part of the ocean lying over the continental shelf is known as the *neritic zone.* A major subdivision of the neritic zone is the *littoral zone,* an area in which photosynthetic activity can take place to the ocean floor. Beyond the continental shelf is the *oceanic zone,* which is divided vertically into the *pelagic zone* and, at great depths, the *abyssal zone.* The pelagic zone is in turn subdivided into the *photic zone,* the depth to which sunlight will penetrate, and the *aphotic zone,* a region of complete darkness. The depth to which sunlight will penetrate varies from a few feet to several hundred feet, depending upon the mineral and organic content of the water and the angle at which the sun's rays strike the water (Figure 22-5).

That portion of the ocean bottom lying over the continental shelf and slope is called the *benthonic zone* while that portion of the ocean bottom over the abyssal plain is called the *benthos zone.*

Sea Water. The most outstanding characteristic of sea water is its salinity or high mineral content. While the total salinity will vary from area to area, the proportion of each of the different types of salts is practically constant. More than 75% of the total min-

TABLE 22-1. Composition of Ocean Water

Name of Salt and Chemical Formula	Grams of Salt per 1000 Grams of Water
Sodium chloride (NaCl)	23
Magnesium chloride ($MgCl_2$)	5
Sodium sulfate (Na_2SO_4)	4
Calcium chloride ($CaCl_2$)	1
Potassium chloride (KCl)	0.7
Other (minor) ingredients	1.3
Total	35.0

eral content is sodium chloride. Magnesium, calcium, and potassium salts are the next most prevalent (see Table 22-1). The salinity of the open ocean at a depth of 1000 feet is used as a standard. Here concentration is 3.5% or 35 parts of salt by weight per 1000 parts of water. The surface water contains a slightly higher salt concentration. Total salinity varies from area to area because of differences in temperature, precipitation, and runoff water. Where there is high temperature and low precipitation there is extensive evaporation, thus increasing the concentration of salt (the Red Sea has a salt concentration of over 4.0%). In the cooler latitudes, fresh water from the melting ice caps and reduced evaporation results in lower salinities. Lower salinities also occur in areas where large rivers empty into the ocean (the Baltic Sea has a concentration of about 0.7% because of the cooler temperatures as well as the large number of rivers emptying into it).

The concentration of dissolved nutrients such as nitrates and phosphates is generally quite low. It is a limiting factor in determining the flora and fauna of the various regions of the oceans. However, low concentration of dissolved nutrients at any one time is not necessarily indicative of a scarcity of living forms since these substances may be utilized from the environment as rapidly as they are made available. The concentration of dissolved nutrients varies with use and with the seasons and zones of the ocean. During the warm summer months, there is temperature layering in many ocean areas much the same as that found in lakes. The less dense warmer water layers float on the more dense cooler layers. The boundary between the two layers is called the *thermocline* and prevents vertical mixing. The organisms in the upper layers soon exhaust the nutrients present and the populations decrease. During the spring and fall the surface and deeper waters approach similar temperatures and vertical mixing occurs, bringing nutrients from the lower to the upper levels. This same effect occurs in the extreme parts of the northern and southern hemispheres throughout much of the year. The net result is that during certain seasons of the year and at certain latitudes in the ocean, larger than normal populations can be supported. In areas of the ocean where the water is relatively warm most of the year, there is very little vertical mixing, resulting in the lack of replenishment of nutrients in the upper layers, leaving these areas void of many living forms (Figure 22-6).

CURRENTS AND TIDES. The ocean is in continuous circulation. Differences in air temperature between the poles and the equator, the rotation of the earth on its axis, major wind currents, and the specific variations in the density of water with changes in temperature result in massive movements of water called *currents*. These major currents move in a clockwise direction in the northern hemisphere and a counterclockwise direction in the southern hemisphere. Some of the most notable currents are the California Current of the northern Pacific, the Humboldt Current of the southern Pacific, and the Gulf Stream of the northern Atlantic (Figure 22-7).

In certain regions of the oceans there is an upwelling of the water which brings rich nutrients to the surface from the lower depths. This occurs along the continental shelf and in the extreme northern and southern areas of the world's oceans. These areas are very rich in both plant and animal life. These same upwellings affect the climate over both surrounding land and water areas.

Tides dominate the sea, especially the shore areas. Tides change approximately every 6 hours, with low tides and high tides occurring twice daily. This periodicity greatly affects intertidal and littoral ecosystems. Every two weeks, both the moon and sun are directly exerting their gravitational forces on the same side of the earth, thereby causing much higher and much lower tides. Conversely, at the middle of such a two-week period they are acting in opposite directions and there is

Figure 22-6. Temperature layering in the ocean. The layering of ocean water follows a pattern similar to that of lakes. (**A**) During the summer a thermocline forms between the warmer upper layer and the cooler layer. The thermocline represents a barrier to mixing; therefore, there is little vertical circulation of water. (**B**) In the spring and fall the surface layers are the same temperature as the lower layers; the thermocline disappears and there is vertical mixing. The vertical mixing brings up nutrients from the lower levels and ocean bottom, which brings about a "bloom" in marine life.

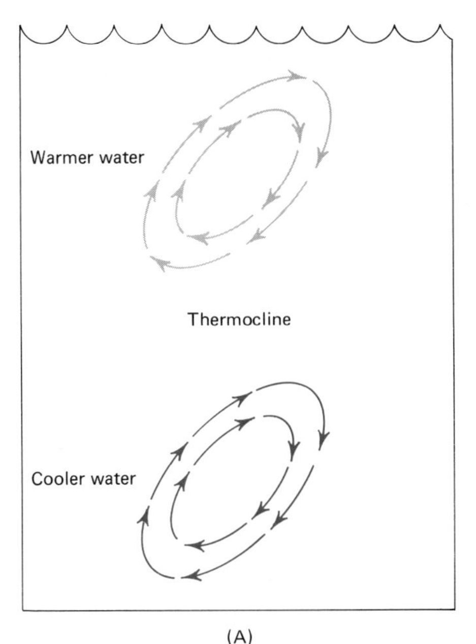

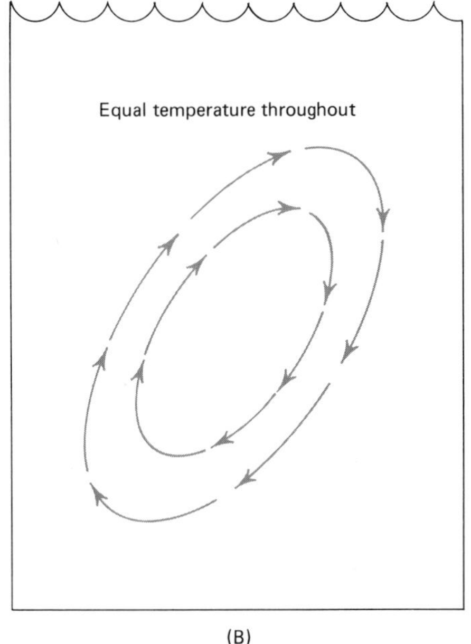

less of a change in the tides (Figure 22-8). Tides vary from less than 1 foot in open seas to as much as 50 feet in certain bays, such as the Bay of Fundy.

Inland Marine Waters. Those water areas that are extensions of the oceans such as bays, estuaries, tidal marshes, swamps, and creeks are subjected to great variations in temperature, salinity, and nutrients. Salinity will vary as streams, rivers, and creeks carry runoff water to marine waters. In addition, there will be variation due to differences in evaporation rates. The shallow waters of tidal marshes, swamps, and bays will also vary with changes in seasons or weather and by the changes of tides. Generally, in summer the water is warmer in these inland marine waters than that of the open ocean and will have a lower dissolved gas content. This condition also occurs by the reduced aeration through lack of wave action in these sheltered bodies of marine water. The close proximity to land areas results in increased concentrations of mineral and organic matter which results in a rich flora and fauna in these water areas.

The total marine environment varies as greatly as the terrestrial environment. Therefore, there will be areas that will be able to support large populations while other areas will be completely barren of any living forms. With the exception of adequate moisture, the hydrosphere is controlled by essentially the same factors as the lithosphere.

The Atmosphere

The third major subdivision of the abiotic environment is the **atmosphere,** the layer of gases that surrounds the earth. The atmosphere or air is a mixture of gases held to the earth by gravitational attraction. Its density is greatest at sea level and decreases rapidly as elevation increases. About 97% of the atmosphere lies within 18 miles of the surface of the earth.

Composition of the Atmosphere

Pure dry air is colorless and odorless and contains approximately 78% nitrogen and approximately 21% oxygen. The remainder consists of several other gases including carbon dioxide, which makes up 0.03% of the atmosphere. Carbon dioxide is important because it is one of the principal reactants in the photosynthetic process and it also absorbs heat radiation from the sun and earth.

The atmosphere also contains water vapor (humidity), an important factor in determining the weather.

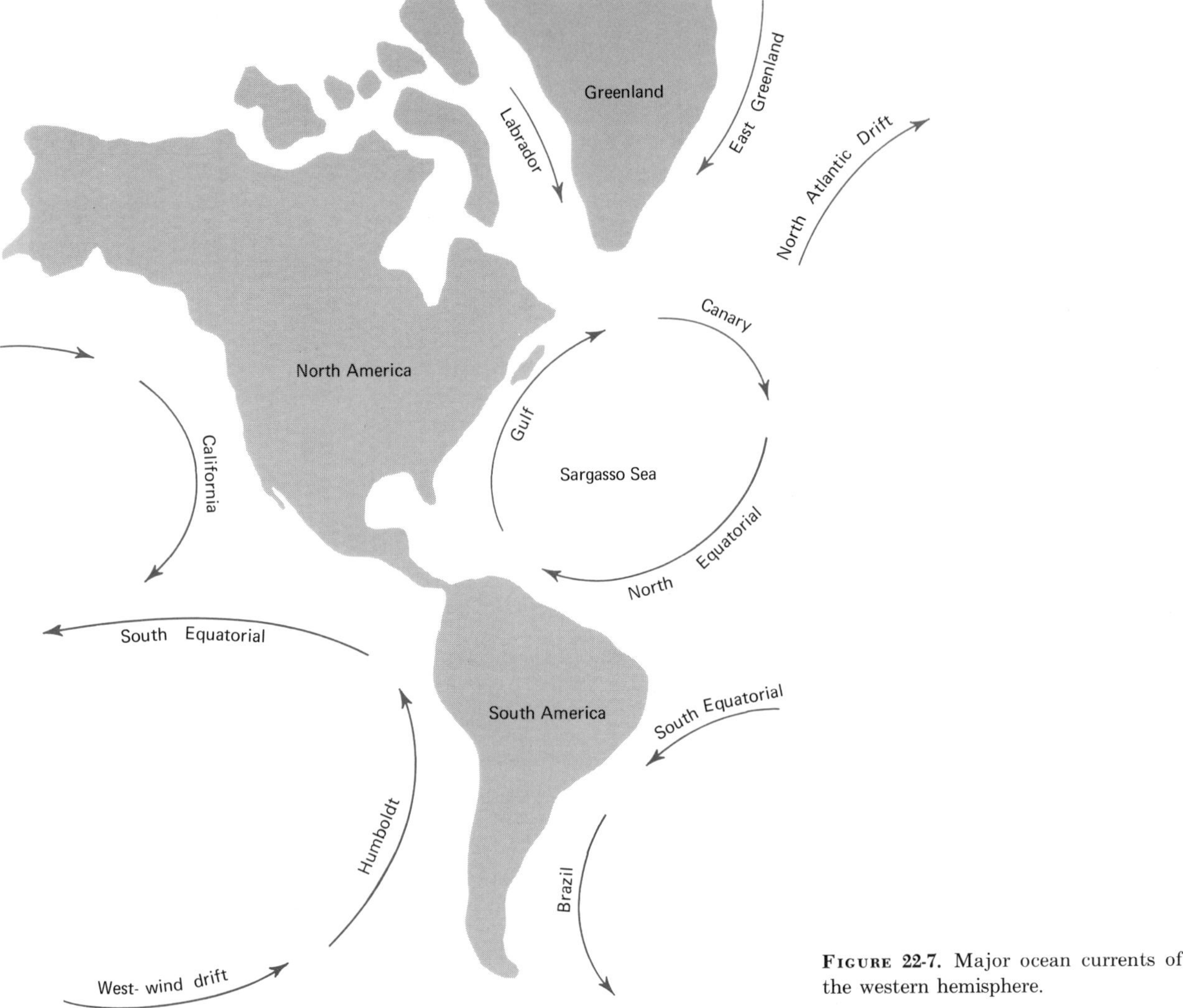

Figure 22-7. Major ocean currents of the western hemisphere.

When water vapor condenses it forms clouds and fog. If there is excessive condensation, rain, hail, or snow may result. The amount of water vapor in the air will vary from extremely small amounts in hot desert areas to very large amounts in hot tropical rain forests.

An important component of the air is particulate matter which enters the air as dust from the dry earth, pollen, soot, fumes, and even minute crystals of salt from ocean spray that has evaporated. All of these particles are carried by air currents. Particulate matter in the atmosphere helps to produce the brilliant colors of the sunrise and sunset, and more importantly, provides nuclei, or centers, around which water vapor condenses to produce clouds.

The heat of both the atmosphere and the surface of the earth is obtained directly from the sun in the form of solar energy. The amount of solar energy received at a particular place on the earth, *insolation,* is dependent upon the angle at which the rays strike the earth and the length of time the area is exposed to the rays. These factors are controlled by seasonal changes, weather patterns, and latitude. Water vapor, like carbon dioxide,

(A)

(B)

Figure 22-8. High and low tides. (**A**) High tide. (**B**) Low tide. The change of tides in the waterways around Nova Scotia is dramatic and extreme. In most areas of the earth, the difference between high and low tides is not as pronounced. Aquatic life is adapted to rapid currents, periods of submersion, and exposure to air as a result of the changes in the tide. [*Courtesy of Nova Scotia Information Service.*]

is capable of absorbing heat from the sun and the earth. As a result, it acts much like an insulating blanket that permits the radiation from the sun to penetrate, but retards the rapid escape of heat from the earth. This phenomenon of trapping energy is frequently called the "greenhouse effect." Carbon dioxide and water vapor also reflect much of the radiation from the sun back into space as it hits the atmosphere. Some of the radiation that reaches the earth is reflected back into space, but there is considerable shielding as a result of water vapor and carbon dioxide in the atmosphere which tend to moderate temperatures.

Winds

Winds are the result of many complex forces. Air flows from an area of high pressure to an area of low pressure and continues to flow until the pressure differences disappear (Figure 22-9). The greater the pressure differences, the higher the wind velocity. Large- and small-scale differences in pressure at various areas of the earth are determined by differences in thermal energy from the sun and earth causing differences in atmospheric pressure. As air heats it expands, rises, and exerts less pressure. As air cools it becomes more dense and exerts greater pressure. In addition to temperature, moist air affects air pressure since moist air is less dense and therefore exerts less pressure than dry air.

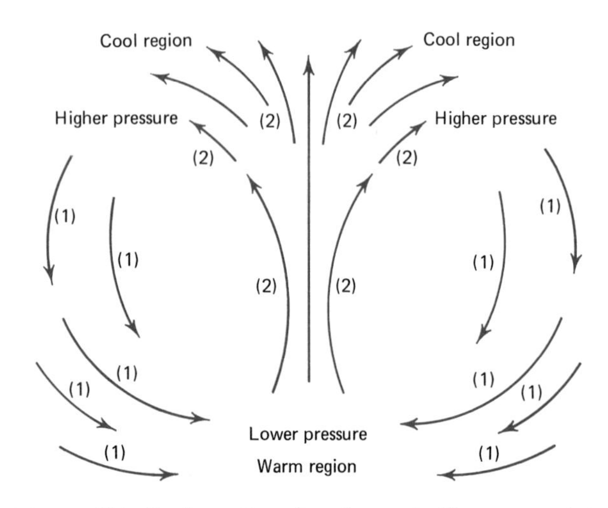

Figure 22-9. The formation of wind currents. Since warm air is less dense than cool air, it exerts less pressure. (1) The surrounding cold air with greater density and pressure pushes in on the warm air because of the unequal pressures. (2) This forces the warm air to rise and the cold air takes its place. These variations in pressure and temperature result in a circulation of air.

In general there are three major categories of winds. The first category is the *breezes* or *local winds* of a given area, which vary from day to day. An example of local winds are the "sea breezes." Land areas near large bodies of water warm up more rapidly during the day than the air over the water. The warm air rises over the land and the cool air from over the water travels toward the land. At night the converse is true. Land cools off more rapidly than the water. The warmer air over the water rises and the cooler land air flows in to take its place (Figure 22-10).

Mountain–valley winds arise in much the same way as land–sea breezes. During the day the winds blow from the valley upward along the mountain slopes and at night down from the mountain to the valley because of the unequal heating of the areas. During the day the mountain slopes warm up more rapidly than the valley and cool off more quickly during the evening.

Local winds in other areas are caused by the unequal rates at which the land heats up and cools off. Among the factors that cause differences in heating are the terrain, the amount of vegetation, and the natural color of the land areas.

The second major category of winds is represented by the movement of large currents of air which, in contrast to local winds and breezes, extend over large areas of the earth. These currents of air form wind belts around the earth. There are series of these belts extending north and south of the equator (Figure 22-11).

While the normal direction of wind currents would be expected to flow in a straight direction from the north or the south, from areas of high pressure to areas of low pressure, this does not occur. The rotation of the earth on its axis and the curvature of the earth deflect the air currents just as they do ocean currents. This deflection, caused primarily by the earth spinning on its axis, is called the *coreolis effect.*

There are three major wind belts in each hemisphere: the *northeasterly* and *southeasterly trades,* the *prevailing southwesterlies* and *prevailing northwesterlies,* and the two *polar easterlies* in the northern and southern hemispheres, respectively. Between each of the wind

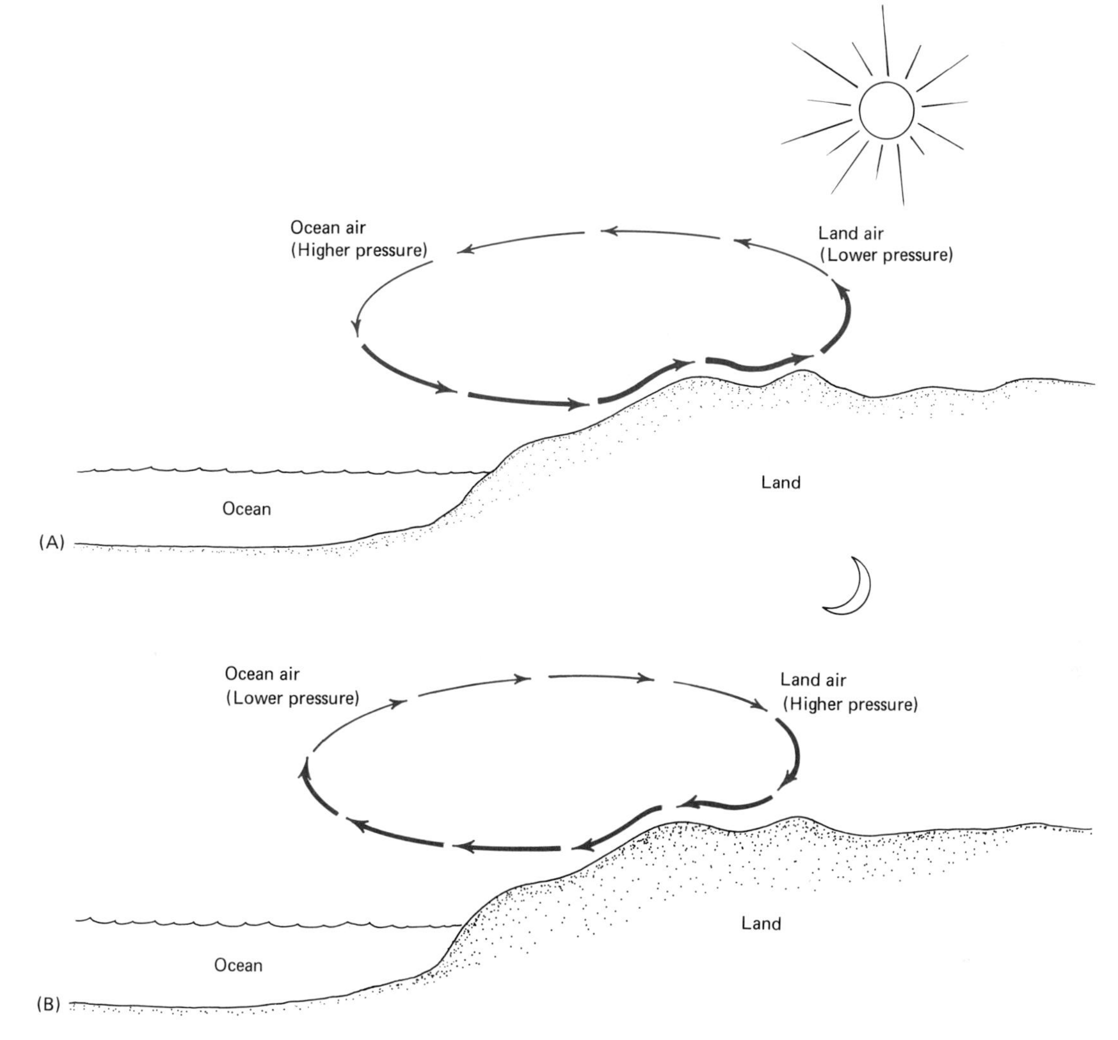

Figure 22-10. Land-sea breeze. (**A**) During the day the land is rapidly warmed by the sun. The cooler air (high pressure) from the ocean pushes in, and the warm land air (low pressure) rises. (**B**) At night, the land cools more rapidly than the ocean and the wind direction is reversed.

belts there are areas where the direction of the currents of air is generally vertical to the surface of the earth. The *doldrums* at the equator are areas of low pressure where warm rising air is replaced by cooler air moving in from areas of higher latitude. Between the trades and prevailing westerlies is the so-called *horse latitude belt* of high pressure, where the cooler air descends toward the earth. Generally, in both these areas there is great calm. Between the westerlies and the polar easterlies are the *subpolar lows,* areas of low pressure where warmer air is rising. The differences in temperature between the polar easterlies and prevailing westerlies is so great that these areas are relatively stormy regions.

Since these wind belts are dependent on pressure belts which in turn are dependent on temperatures of the earth's surface, they are constantly shifting as the seasons change. These shifting wind belts with their different temperatures, land-to-sea directions, and moisture content, greatly determine the weather of the various regions of the earth that they cross.

FIGURE 22-11. The major air currents of the earth. Consult text for further explanation.

Above the various belts of air surrounding the earth is the third and largest of the atmospheric circulations, consisting of the major currents between the equator and the poles. The solar energy entering the atmosphere of the earth at the lower equatorial latitudes is consistently greater than at the higher latitudes near the poles. This results in a large-scale movement of warm air in both northerly and southerly directions from the equator toward the poles. This movement of air adds to the complexity of the wind belts that surround the earth (Figure 22-11).

The winds of the earth greatly control the major ocean currents. In addition, they determine the characteristic climate and weather of the different regions. As such, they exert a controlling effect on the abiotic and biotic environments in any given region.

Climate

Within the abiotic environment, temperature and precipitation are the two major variables that determine the extent and kind of life that can be supported in the biosphere. Of course, for aquatic life, temperature and oxygen content are the major variables. Together, temperature and precipitation are also the major factors of climate and are the basis for climate classifications.

Using temperature as a basis of climate classification, three clearly recognized climate groups can be defined. The first is the *equatorial-tropical* group in which the temperatures are warm and there is no winter season. The lowest mean temperature for any given month in this group does not drop below 64°F. The second group is the *middle latitude* in which there are alternating winter and summer seasons. This group formerly was considered the temperate zone; however, the extremes in temperature and storminess in the southern part of the zone were not suggestive of temperate conditions. The warmest months of the middle latitude groups have an average temperature above 50°F. The third group is the *polar-arctic* group in which there is no true summer. The monthly average temperature for the warmest month is not above 50°F (Figure 22-12). In any analysis of climate, temperature extremes and yearly distribution of temperature are important considerations.

Temperature is not satisfactory as the sole basis for climatic classification because it does not distinguish between desert and humid regions. Therefore, climate is further subclassified according to the amount of precipitation. There are essentially five major climate groups based upon the mean annual rainfall. These groups are arid, semiarid, subhumid, humid, and very wet (Table 22-2). Basing the classification solely on the mean an-

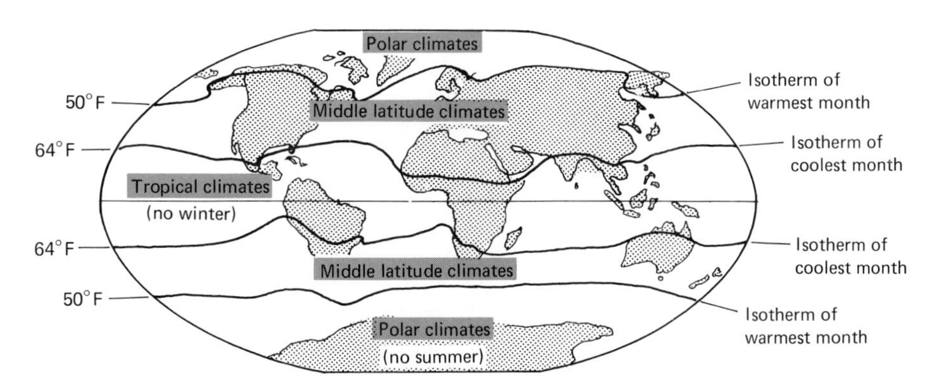

FIGURE 22-12. The major climatic zones of the earth based upon temperature range. These zones are not uniform in their distribution, owing to differences in the specific heat of land and water and the effects of prevailing winds. The term *isotherm* refers to a line joining the points on the surface of the earth having the same temperature at a given time (or the same mean temperature for a given period). [*Modified from Arthur N. Strahler,* Physical Geography, *2nd ed. New York: John Wiley & Sons, Inc. Copyright © 1960.*]

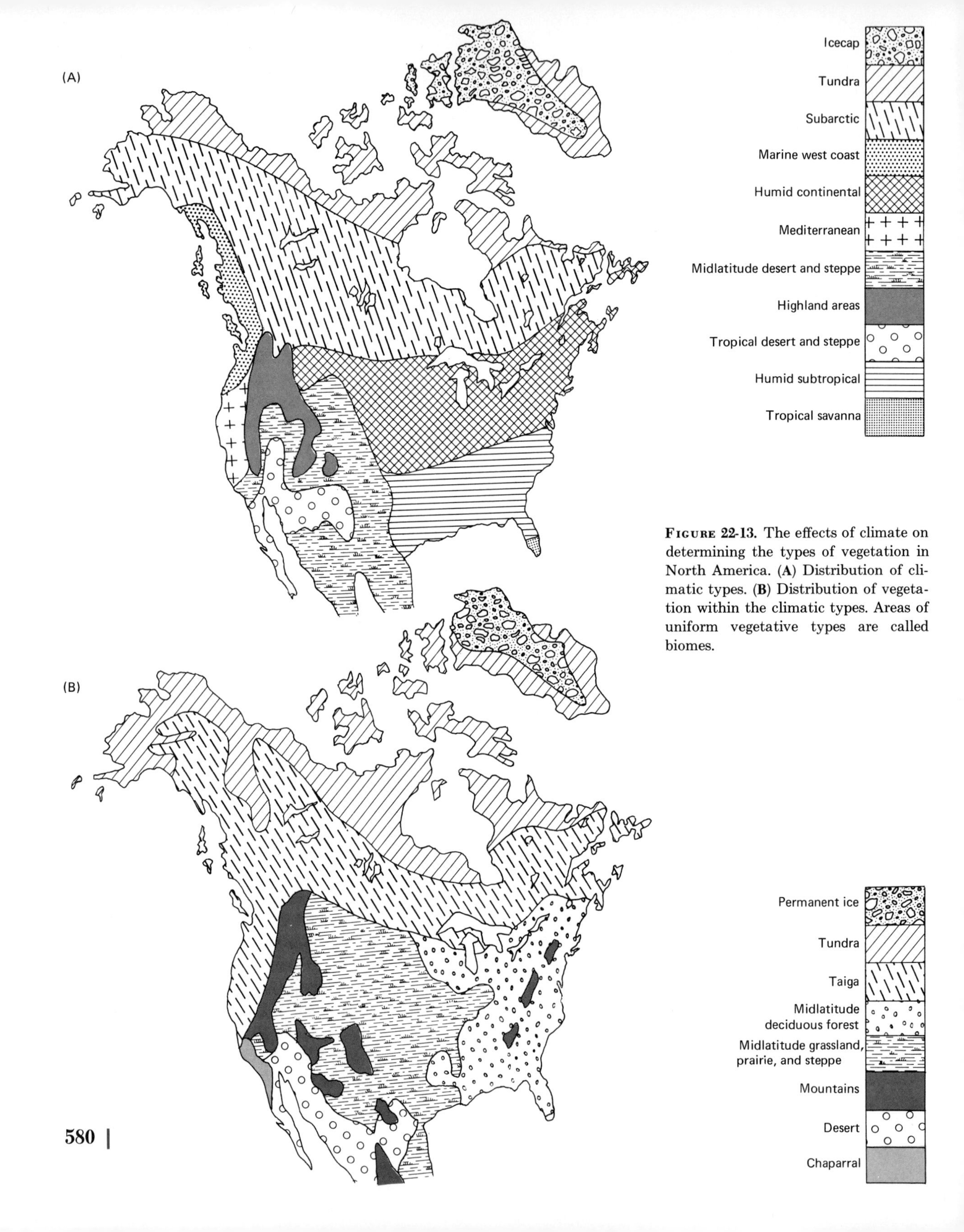

Figure 22-13. The effects of climate on determining the types of vegetation in North America. (**A**) Distribution of climatic types. (**B**) Distribution of vegetation within the climatic types. Areas of uniform vegetative types are called biomes.

TABLE 22-2. Climatic Regions of the Earth Based upon Precipitation

Climatic Type	Annual Rainfall (inches)
Arid	0–10
Semiarid	10–20
Subhumid	20–40
Humid	40–80
Very wet	80 or more

nual rainfall does not, however, give an indication of seasonal distribution, which is considered in any detailed analysis of climate (Figure 22-13A).

Any study of climate types based upon temperature and precipitation will show that there is tremendous variation from place to place. In spite of this variation, patterns will emerge; there are large areas over which there is at least a degree of uniformity with respect to temperature and precipitation. This uniformity will be reflected in the types of flora and fauna that can be supported in the area. For example, the deserts of southwestern United States and the rain forests of tropical South America are large areas with a degree of uniformity in climate that dictates a degree of uniformity in both flora and fauna within them.

These large areas having uniformity of climate, and therefore uniformity in vegetation (Figure 22-13B), are called biomes. They are usually classified and named according to the type of vegetation and will be discussed in Chapter 25.

23 Populations Within Ecosystems

GUIDE QUESTIONS

1 Explain the relationship among biotic potential, environmental resistance, and carrying capacity and how they are related to population growth.
2 Explain how population growth is regulated.
3 Give three generalized patterns of the distribution of populations.
4 Distinguish between intraspecific and interspecific interactions.
5 Give several examples of intraspecific competition, social organization, and territoriality.
6 Describe the social organization of bees.
7 Give several examples of interspecific competition.
8 Distinguish between habitat and ecological niche. Give several examples of each.
9 Distinguish between a community and ecosystem.
10 Distinguish among and give specific examples of commensalism, mutualism, parasitism, and predation.
11 Distinguish between obligatory mutualism and facultative mutualism.
12 Explain how competition, predation, territoriality, and social organization establish a balance in nature.

THE abiotic environment to a great degree determines the various forms of life as represented by the different species populations, more commonly referred to as populations. A **population** is a group of organisms of the same species occupying a particular space at a particular time. The different populations make up the **biotic environment.** Descriptions of populations are representative of the group and not necessarily of any one individual in the group. For example, a population of the moss *polytrichium* occupying a given meadow is described by such characteristics as height, capsule size, and color. These represent the predominant characteristics of the group and not those of an individual organism. Among individuals of the population there will be degrees of variability, within limitations.

Each population possesses statistical and specific characteristics that are unique to populations, including population growth or decline, the ability to exchange genetic information, and intraspecific and interspecific interactions. These characteristics, as well as hereditary traits, distinguish one species from another and, to a large extent, represent specific adaptations to the abiotic and biotic environment and account for variability among the different populations. Examples of variability can be seen in predominant types of vegetation in one area as compared with another. On a large scale one area may be dominated by grasses (grasslands), while another may be dominated by beech and maple trees (deciduous forest).

Where the changes in the abiotic environment are gradual, different populations will show gradual changes in abundance. For example, as one travels south in the eastern part of the United States, there is a gradual change in dominance from deciduous forests to pine forests. Farther south, tropical vegetation begins to appear, most obviously evidenced by various species of palms interspersed with the pines. Continuing south to the tip of Florida, one sees that tropical vegetation becomes dominant. While this description covers large geographical areas, the same phenomena occur in all areas, large or small. Where there are sharp changes in the abiotic environment, there will be sharp changes in the populations. This can be observed in the rapid changes in mountain vegetation as a result of the great variation in climatic conditions that accompanies increasing altitude. In addition, biotic changes in the environment brought about by human beings and other living forms will also cause changes and shifts in populations. As examples, an increase in a predator population will cause a decrease in the prey population, while an increase in a prey population may result in an increase in the predator population; or industrial wastes dumped in waterways may alter or destroy existing aquatic populations, and open strip mining will have similar effects upon terrestrial ecosystems. Understanding the effects of the rapid increase in human population upon existing populations, upon ecosystems, and, finally, upon the total biosphere is one of the most important aspects of the study of ecology.

While this chapter will be concerned with individual populations, it must be emphasized that every geographical area contains assemblages of populations that together make up the ecological community. Within the ecological community each population is directly or indirectly affected by other populations within as well as outside the area. The utilization and cycling of available nutrients and the flow of energy within the ecological community, as well as the interactions among and between populations in association with their abiotic environment, comprise an ecosystem. The balance between input and output of available resources of the abiotic and biotic environments as they are cycled back and forth is known as the *balance of nature*. This balance is the result of interactions between the abiotic environment and the populations that make up the biotic environment.

In order to understand the dynamics of ecosystem balance, it is essential to have an understanding of some of the characteristics of populations, including population size or density and intra- and interspecific interactions within communities and ecosystems.

Population Size

Population size is affected by several factors: natality (births), mortality (deaths), immigration, and emigration. From these factors certain other attributes of a population can be derived. These include age distribution, dispersion (distribution of individuals in space), and genetic composition.

Growth Rate

The ability of a population to grow and develop is dependent upon its birth rate and death rate. In its simplest form this relationship is expressed by the equation

$$\text{Growth rate} = \text{Natality} - \text{Mortality}$$

However, when considering growth rates of populations, one must keep in mind the fact that the birth and death rates of populations are the result of several forces. Among these forces is the inherent ability of each species to reproduce at a theoretical maximum rate called **reproductive potential.** Opposing this is the inherent death rate or life span; that is, disregarding other forces, there is a point in time when an organism simply dies of old age. For a population to survive, the reproductive capacity must exceed the inherent death rate. This would then mean that a population should continue to grow indefinitely and without limit. However, this does not occur over long periods of time. Opposition to unlimited growth comes from many forces of the physical and biological environment in which the population exists, forces that tend to limit the achievement of reproductive potential and increase mortality. These forces are known as **environmental resistance.** This may be formulated as

$$\begin{matrix}\text{Growth} \\ \text{rate}\end{matrix} = \begin{matrix}\text{Reproductive} \\ \text{potential}\end{matrix} - \begin{matrix}\text{Environmental} \\ \text{resistance}\end{matrix}$$

Since reproductive capacity and aging are genetically determined, it is possible by studying a given population to predict an average number of offspring per individual of a given age as well as an average natural life span. For a population the accumulation of these averages is often called its **biotic potential.** The determination of the biotic potential takes into account reproductive potential and environmental resistance. However, biotic potential in turn is controlled by relationships within the community known as intraspecific and interspecific interactions such as competition, symbiosis, parasitism, and predation. This, too, may be expressed as a formula (in this case the second factor may have a positive or a negative effect).

$$\text{Growth rate} = \text{Biotic potential} \pm \text{Species interactions}$$

While population size may be determined or estimated at any given moment, over time it is affected by many different factors and forces. Within any ecosystem there may be observable changes in the number of individuals of a given population. Sometimes these changes are represented by orderly growth or decline in numbers. At other times the changes may be drastic or erratic.

Birth Rate (Natality)

The **birth rate** is determined by the number of individuals born to a given population during a given period of time. It always represents an addition of new or young individuals but does not necessarily mean a net increase in population size because of death, immigration, or emigration. The maximum natality represents the maximum reproductive capacity of a population; that is, the number of births is limited only by the inherent physiological ability of a species to reproduce and is a constant for a given population.

Comparison of the actual birth rate with the maximum natality may indicate whether favorable or unfavorable forces are acting upon a given population as well as upon the ecosystem. Natality represents only one component of the dynamics of population size.

Death Rate (Mortality)

The **death rate** represents a decrease in a given population during a given period of time. It, too, may not represent a net decrease in population size because of births, immigration, and emigration. Generally, the deaths of individuals in a population are expressed by specific mortality, the mortality for given age groups. In the consideration of a population the number of deaths is not as important as the number of individuals that survive (survivorship). For example, although death rates may be relatively high, the birth rate and survival to reproductive age may be much higher and the population therefore increases. It is the living individuals that determine the characteristics of a population.

Specific mortality is frequently expressed in logarithmic survival curves. When expressed this way, the characteristic survival and mortality of a set population during a given period of time can be theoretically viewed without the complications of birth rate and migrations. For example, if all individuals of a given population had the same capacity to survive, death would occur at approximately the same time for all individuals. Figure 23-1 shows that a given laboratory population of starved fruit flies will survive for approximately the same period of time and then die at approximately the same age. It is interesting that the human survival curve in well-developed countries approximates the curve of laboratory fruit flies (Figure 23-1). The advances of medical science have reduced the number of deaths of infants, allowing a larger number of the human population to reach the maximum physiological age.

If the mortality rates are about equal at each age level, a straight-line relationship is obtained (Figure 23-1). The hydra in the laboratory exhibits this type of survival curve; the numbers dying at each age are about equal. In other populations, although large numbers of individuals are produced, infant mortality is exceed-

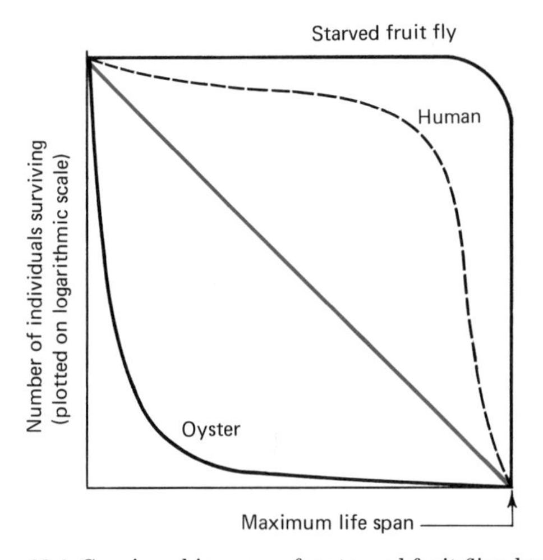

FIGURE 23-1. Survivorship curves for starved fruit flies, humans, and oysters. The colored graph line is a representation of what would result if the death rate was constant throughout the life span.

ingly high, with a resulting rapid reduction in the population of the newborn. However, once a given age is reached, the existing members generally reach the physiological age limits (Figure 23-1). Many aquatic organisms, such as the oyster, produce large numbers of larvae with an exceedingly low survival value, but individuals that reach adulthood generally survive to maximum physiological age.

Age Distribution

An important characteristic of a population is the number of individuals at each given age or age distribution. Age distribution to a considerable degree affects birth and death rates. Logically, if a population is composed of a relatively large number of members of reproductive age, its reproductive capacity is high, whereas if it is composed of a large number of members beyond the age of reproduction, its reproductive capacity is low. The various age levels of individuals in a population can be classified into three major groups: juveniles (below the age of reproductive ability), reproductives (capable of reproducing, and matured adults (beyond the reproductive age). When a population is analyzed and the age distribution graphed, usually one of three different types of age pyramids will result (Figure 23-2).

If a population is composed of a greater number of juveniles than reproductives and mature adults, barring unusual outside forces, it is safe to assume that the population is growing and will continue to increase in size as the juveniles reach the reproductive age (Figure 23-2A). The numbers of individuals at the various lower age levels also give an indication of how rapidly a population will increase in size. In populations where the numbers of individuals in the three major groups are about equal, the population size is stationary (Figure 23-2B). Finally, if a population is composed of a greater number of mature adults than juveniles and reproductives, the population size is declining (Figure 23-2C). From such an analysis it is possible to predict the future status of a population and its effect on the ecosystem.

In recent years considerable detailed analysis has been made of the age distribution of human populations in various areas of the world. These analyses have proved to be most useful in predicting changes in human population size and the problems encountered by a changing population.

Immigration and Emigration

Immigration results in an increase in population size and emigration results in a decrease. In most population studies immigration and emigration are usually considered to have a negligible effect or to balance each other. In studies of large populations it is difficult to measure the effects of these on population size. By a field technique known as capture-recapture where individuals of a given population are tagged or otherwise marked it is possible to determine the number of individuals in a population entering or leaving a given area. If a population is properly sampled, it is possible to separate births from immigration, this process yields what is known as the *dilution rate* (births plus immigration); the separation of deaths from emigration is known as the *loss rate* (deaths plus emigration). Generally dilution and loss rates do not play a significant role in population size. However, gross changes in either of these rates within a specific population may indicate a change in the character of the ecosystem as a result of external or internal forces and provide a useful tool for the ecologist.

Environmental Resistance to Population Growth

The **biotic potential** of a population is defined as its ability to grow. Under optimal environmental conditions the biotic potential may be represented graphi-

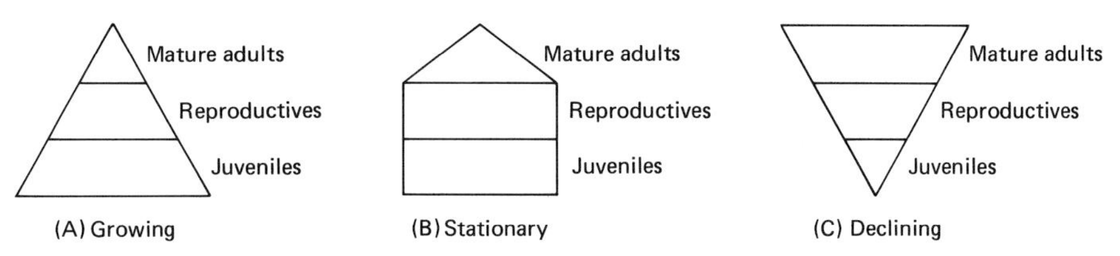

FIGURE 23-2. Three types of age pyramids.

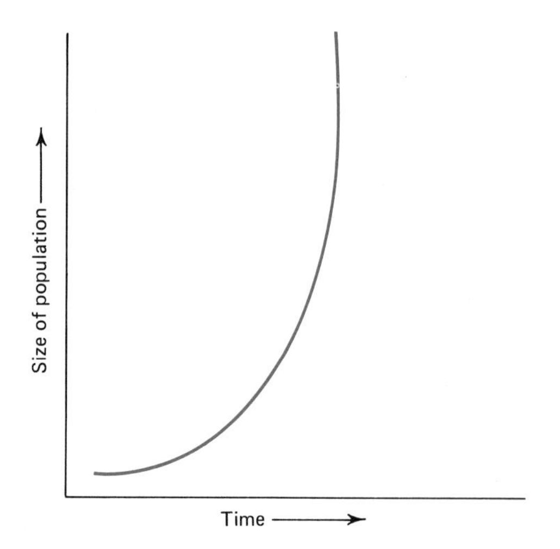

Figure 23-3. The unimpeded growth curve of a population under optimum environmental conditions. See text for discussion.

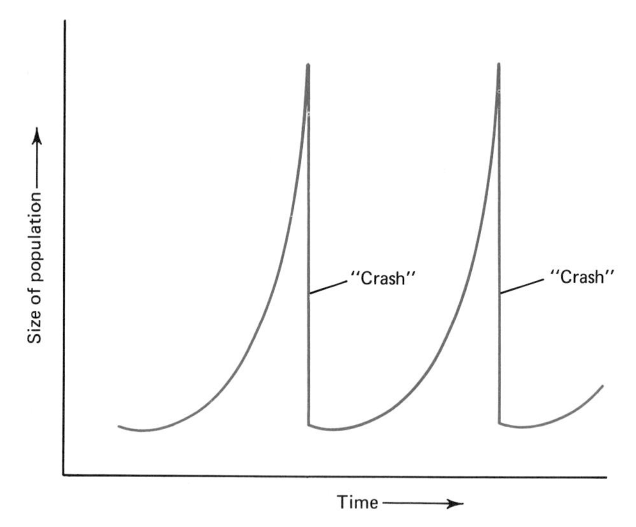

Figure 23-4. The theoretical growth curve of a population going beyond the carrying capacity of the ecosystem, resulting in the characteristic "crash."

cally by an unlimited growth curve as shown in Figure 23-3. Under optimal environmental conditions a given population increases slowly at first and then rapidly increases to an unlimited extent. Of course, in nature this may occur for short periods of time, but for all natural populations a point is reached where natural limitations come into play and slow or even stop this unlimited growth. In other words, under natural conditions the biotic potential is limited. The factors that contribute to this limitation are called **environmental resistance.** In most populations this resistance may be lack of food, accumulation of wastes, disease, behavior patterns, or various other interactions of the biotic and abiotic environments. Even if these factors were not present, unlimited population increase would result in a limitation of space due to sheer numbers. The net result of the interaction between biotic potential and environmental resistance is an alteration of the unlimited growth curve. As environmental resistance increases with population size, the growth rate will level off, giving an S-shaped curve, or will, in some species drop drastically, giving a J-shaped growth curve (Figure 23-4).

The maximum population size that an environment can support is known as the *upper carrying capacity* of the environment for the population under consideration. Occasionally, when the growth of a species population goes beyond the carrying capacity of the ecosystem, certain resources become exhausted, resulting in a sudden decrease in the size of the population called a *crash.* This type of population growth occurs among certain species of plants and insects that have high reproductive potential; it may be represented by an oscillating J-shaped curve (Figure 23-4). There is a rapid increase in the population, overshooting the carrying capacity, with the resulting characteristic crash. The population then increases again, only to follow the same pattern.

The more unusual growth curve displayed by most larger plants and animals with somewhat limited reproductive potential and/or well-developed species interactions is the S-shaped curve (Figure 23-5). It can be seen that environmental resistance causes the population growth rate to slow up and population size to stabilize at or near the upper carrying capacity of the environment.

These graphic representations are largely theoretical; however, in nature, the actual growth rates of populations usually approximate somewhere between the ideal J curve or the ideal S curve.

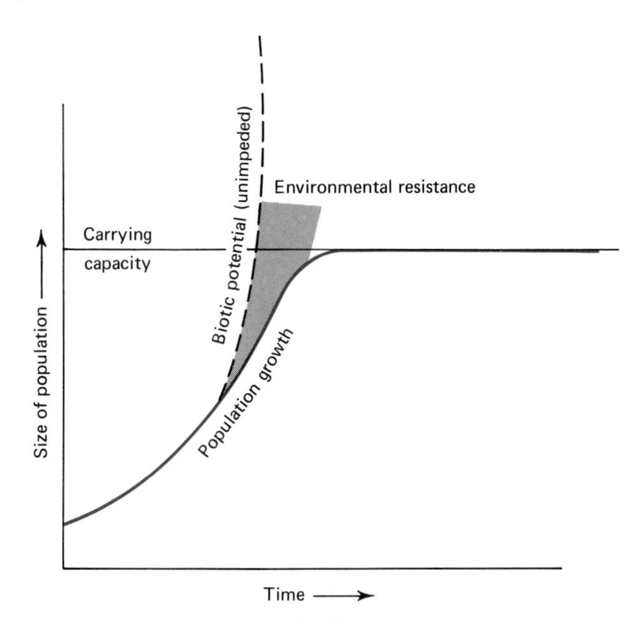

FIGURE 23-5. Environmental resistance exerting an effect upon the biotic potential, resulting in an S-shaped population growth curve.

Distribution of Populations

The nature of the interactions that exist will determine the distribution of individuals in a population and the populations within the community. Distribution occurs in three generalized patterns:

1. Random.
2. Uniform.
3. Clumped.

Random distribution is rare and occurs where there is a relatively uniform environment. For example, in a grassland where there is a large stable herd of grazing animals, the location of a given carnivore cannot be predicted. Of course, the predators will be located near the herd, but their distribution will be random.

Uniform distribution occurs where there are a few major factors controlling the environment and competition is very great. An example of this type of distribution is desert vegetation which is equally spaced because of the competition for the limited moisture content of the soil. Each individual occupies a given amount of area.

Clumped distribution is the most common pattern. Among animals of the same species, individuals tend to form groups of certain sizes for purposes of reproduction and social behavior. Other reasons might be that the mineral content and water distribution is not uniform in a given area, or, as in the case with some grazing animals, herds are formed for protection and because the grass is limited to specific areas. Clumping may also occur among some organisms because the species is *sessile,* attached to the substrate, or *sedentary,* remaining in one place most of its life. Seeds of many plants fall near the sessile parent plant, resulting in the aggregation of the species. Oysters and other marine organisms that are free-swimming in larval form aggregate into large colonies as sedentary adults because suitable substrates are limited.

The populations making up a community within an ecosystem may show all three types of distribution, depending upon the adaptations and specific needs of the individual populations. For example, within the desert community, as previously stated, vegetation may be uniform, whereas desert rodents such as gophers and mice will clump together for protection, reproduction, and social organization. Predators such as hawks and owls will be randomly distributed over a wide area to reduce the competition for the limited amount of prey.

Population Regulation

The theoretical graphs of population growth allow comparison with actual species-population growth patterns in an ecosystem and may indicate where the affecting factors may be sought. The factors that affect population growth may be classified into two major categories of environmental resistance: density-independent and density-dependent. The *density-independent* factors are the controlling factors that are not dependent upon numbers of individuals making up the population, but on some aspect of the physical environment. For example, with many insects there is ample food supply for a continued increase in population, but seasonal changes restrict growth. The *density-dependent* factors are those which depend upon numbers of individuals within the population (density) to exert their effect, usually becoming more forceful with increasing density. For example, the grazing animals of the grasslands of the tropical zone have an ideal physical environment. However, if not checked by other factors they would increase in such numbers that they would soon destroy their food supply. Another example of the density-dependent category is the predator–prey relationship. As was shown previously, the sizes of the predator population and prey population show a positive correlation.

While it is apparent that food supply is one of the major determinants for most populations, rarely are animals found starving in the wild. This had led to various proposals as to how they keep their population numbers low enough to prevent starvation. Two factors that may contribute to this are territorial behavior and pecking orders (dominance) which allow the more vigorous members to feed and reproduce and prevent the weaker members from reproducing, thus controlling population size.

Various other theories of population regulation have been proposed and supported by laboratory experimentation. However, it appears that population regulation is not the result of any one mechanism but an interaction of many complex factors.

Population Growth and the Dynamic Ecosystem

Logically, if a population is increasing, decreasing, or stable, it is reflecting the response of all its members to the various forces that operate within the ecosystem. The continued growth, continued existence, or eventual extinction of a population represents the degree of success of its adaptations to a given environment and to the changes that may occur within that environment. Some changes in population size may be the result of immediate sharp changes in the environment such as fire, drought, adverse climate, or disease. Other changes in population size, such as those that occur as an ecosystem passes through the various natural changes, take place more slowly. Still other changes take place over very long periods of times; these are evolutionary changes (see Chapter 21).

Having considered the statistical factors of populations we now will deal with intraspecific interactions, habitat and niche, communities, and interspecific interactions. It should be noted that interactions among organisms include behavior, which is the subject of Section Six. In this chapter we are concerned with interactions from the point of view of population control and regulation.

Intraspecific Interactions

Interactions within a population are frequently termed intraspecific interactions, that is, interactions between individuals of the same species. Among the intraspecific interactions are competition, social organization, and territorial segregation.

Intraspecific Competition. There is intraspecific competition for food, minerals, water, sunlight, territory, and mates. As population increases, competition among the individuals of the same species also increases. Additional increase results in mechanisms coming into play which tend to limit any further increase in the population. For example, a population of trees in a forest is controlled as much by competition for light and water from members of the same species as it is by such factors as weather and climate. Extensive root systems of mature trees reduce the amount of soil, water, and nutrients available for young plants. This also reduces reproduction and kills off all but the most vigorous seedlings. There is evidence that certain species of plants secrete substances from their roots that inhibit the growth of other plants. Intraspecific competition among animals may be illustrated by the Alaskan seal. The males come to shore in advance of the rest of the herd to stake out territories for breeding purposes (Figure 23-6). Each male fights for a harem of from five to twenty females and defends his own territory vigorously during the breeding season. Only the strongest males of the species are able to maintain their territories and contribute to reproduction.

In addition to regulating population size, intraspecific competition exerts a selective influence within the population. Those individuals that are able to survive and produce offspring pass on their specific characteristics to the next generation. If a particular characteristic is beneficial, genetically controlled, and not possessed by all individuals, gradual change in the species results (see Chapter 21). This is one of the major bases of evolu-

Figure 23-6. Territoriality of Alaskan seals. During the breeding season the male Alaskan seals come ashore before the females and stake out territories. Each male seal defends his territory and harem of five to twenty females from other male seals. [*Courtesy of Dr. Marie Kuhnen, Montclair State College.*]

tionary changes that take place in a population over long periods of time.

Social Organization. Although all plants and most animals typically associate in a clumped distribution, in actuality they live as independent individuals within the population except during reproductive periods. The intraspecific interactions are largely indirect. For example, competition really occurs without any conscious reasoning power on the part of the individual member of the population. Plants do not purposely try to overpower other members of their species, nor is there a direct cooperative effort to overcome the limiting factors of the environment by most species. Rather, the survival of the species within the ecosystem is a result of adaptations that ensure continued individual existence. There are, however, a few exceptions to the indirect interactions of most living organisms and these are found in the animal kingdom. Some animals form a social organization in which there is direct interaction, usually cooperation, among the members of the same species. In this chapter we will discuss the social organization of these species in relation to the interactions within the ecosystem; the behavioral aspects will be discussed in Chapter 28.

Social organization occurs among certain species of insects such as bees, ants, and wasps and among certain vertebrates such as wolves, birds, and deer. The social organization among insects is based upon structural specialization; that is, certain members are structurally and physiologically different from other members of the species. This is known as **polymorphism.** The direct interdependence of each member upon other members of the population results in communal living in hills, hives, or nests. As a result of polymorphism the members within a population are divided into structurally distinct castes. For example, among honeybees there are three different ranks: one queen, a relatively few drones, and thousands of workers (Figure 23-7). The main purpose of the queen is reproduction. The workers build the hive, gather food, care for the young, and ward off enemies. Reproductively, they are sterile.

Similar types of social organization with polymorphism and castes are found among ants and termites. There is a species of ants that grows fungus gardens, while another species keeps aphids (small sap-sucking insects), which they care for much like dairy cattle and from which they receive a sweet secretion.

Communal living has considerable survival value for

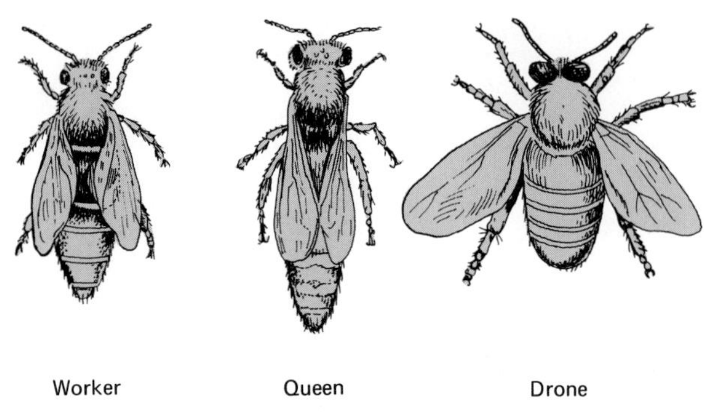

FIGURE 23-7. Polymorphism of honeybees. The principal function of the queen and drones is reproduction, while the infertile workers carry on the various other duties of the hive. Each caste is structurally designed to carry on a specific function or functions.

the colony. For example, a large population can ward off the attack of a predator. Moreover, division of labor makes for efficient utilization of resources by the population. The location of sources of food can be communicated by some members of the population (see Figures 28-3 and 29-1), while maintenance of the hive or nest and care of the infant members can be delegated to other members of the colony. If the population becomes too large for available resources, a colony may divide; that is, a portion of the population may leave and establish a new colony elsewhere. In such a case, the group that leaves the original colony consists of sufficient numbers of individuals for the new colony to succeed. Among other species of communal organisms the whole population may move to a more favorable location. Although such communal organization of the various activities of the colony may result in the sacrifice of individual members, it may be presumed to be beneficial for the survival of the population as a whole.

The high degree of specialization shown in polymorphic insects may represent a disadvantage. Death of the queen may end the hive or colony. Loss of specialized food sources may starve the population. However, these disadvantages of the societal organization are offset by the high rate of reproduction and the advantages of the resulting large population.

Social organization among vertebrates, in contrast to insects, is based upon a functional specialization. Structurally, except for the usual male and female differences called *dimorphism,* all members are more or less alike. Specialization is based upon physical strength, skills, and mental acuity. The schools of fish, packs of wolves, herds of deer, and flocks of birds mainly represent a protective interaction (Figure 23-8). An individual in a group is more readily warned of danger and also more effectively protected from attack. Other groups, such as beavers, interact cooperatively in the building of dams and nests. This not only provides a shelter for the group but affords protection as well.

Among many animal species a hierarchy of social dominance is established. Through a series of encounters of combat or other displays of force one member of a group becomes socially dominant over the others and the status of the other members is represented in descending order of dominance. Evidence indicates that individuals in such a group feed better and are more successful at rearing young than in groups where there is constant combat among the members. Such social organization has the effect of population control. In crowded populations the lower members of the social hierarchy may be cast out; this probably results in their death if they are unable to establish a new colony or find their place in another population of small size. This social intolerance is probably quite effective in limiting population size among species that have a high reproductive capacity, such as rodents.

TERRITORIALITY. Closely associated with the social organization of some vertebrates is the phenomenon of *territoriality.* Within a given population each family or mating unit has its own physical territory. Usually these territories are defended against invasion by other members of the same species. Among seals the male stakes off a given mating territory and drives off other males that may try to enter the area. Females are welcomed. Among certain species of birds, the male will stake out a territory, try to attract a female, and defend the territory against other males.

Generally, the territory will provide sufficient room for mating and nesting and an adequate food supply to rear the young. During the time when territories are established there is intense intraspecific competition for suitable areas. Those that do not establish territories are surplus members and become territorial outcasts. This competition serves as a mechanism of natural

FIGURE 23-8. Intraspecific interactions among Rocky Mountain mule deer. Among vertebrates, social organization has a functional basis. The photograph shows a family of deer consisting of a buck, does, and fawns. The family structure provides protection. The territory of each family, which is closely guarded against competitors, ensures a supply of food for the group. [*U.S. Fish and Wildlife Service photo.*]

selection, permitting only those who are successful in selecting and defending a territory to breed. Survival value for those possessing a territory is high while for territorial outcasts survival is low. Territoriality has the effect of population control. Since there is a limit to the number of suitable territories within a geographical area, the population is spaced. This spacing, while limiting the number of breeding members, also reduces the spread of epidemic diseases and lessens the effectiveness of predators. Spacing also tends to keep the population size within the limits of the carrying capacity of the ecosystem by dividing the needed resources prior to mating.

Habitat and Niche

The physical location where an organism lives and interacts with the abiotic and biotic environments is called the **habitat.** Within the habitat an organism has a distinct functional relationship to everything else in the environment and occupies a specific position and status called an **ecological niche.** The ecological niche is an abstract concept and does not have clearly defined parameters. All the activities that a living organism participates in, as well as its adaptation to the particular habitat, are considered part of its ecological niche. While many different populations can have the same habitat, only rarely, if ever, can two different populations completely occupy the same ecological niche. Habitat and niche should not be considered synonomous. Upon superficial observation it may appear as if two or more populations are occupying the same niche; however, careful observation and investigation will show otherwise. When one observes various waterfowl foraging in the shallow waters of a pond, it would appear that they are all seeking the same type of food. Further investigation will show that one species of bird is seeking small fish; another, crustaceans; and still others, algae. When Darwin studied the finches of the Galápagos Islands, he found many different species, each adapted to its ecological niche. One species had a bill adapted for catching insects on the ground, another for catching insects in the trees, and still others for eating seeds and fruits (Figure 23-9). Within the same pond there may be water bugs similar in appearance yet belonging to two different species. One species may prey upon other insects, while the other may feed upon decaying plants.

The ecological niche represents an optimum utiliza-

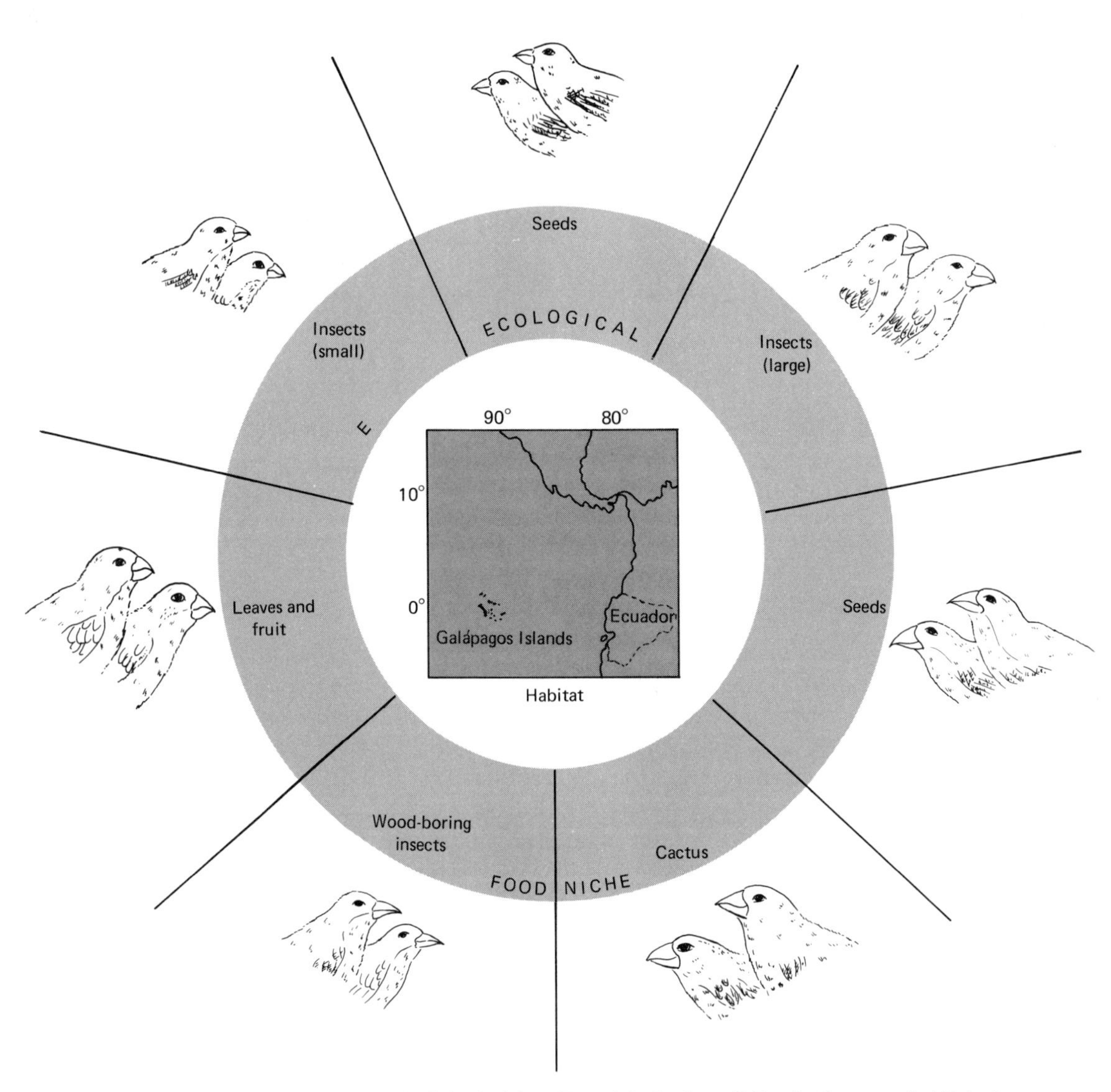

Figure 23-9. Habitat and ecological food niches. Darwin's finches all live in the same habitat, the Galápagos Islands. Within the habitat, each species of finch occupies its own ecological niche. The beaks of the various species are adapted for obtaining a specific type of food. Thus, each species also occupies a different ecological niche with regard to food.

tion of nutrients and energy resources in a particular habitat for which a species is adapted. The more similar two niches are for two different species, the greater the competition between them. If two species seek the same niche, or if the resources of the habitat are severely limited, competition may have three possible consequences.

1. An adaptive characteristic of one species may be superior to the other; this results in the extinction of the less adaptive species.

2. A specific adaptation in one of the two species may differ so slightly that one species may tolerate a slightly different nutritional requirement, energy resource, or climatic environment and the inferior species might literally be pushed to a different niche or region of the habitat. This frequently occurs in intertidal zones where one species is better able to withstand desiccation at low tide than its competitor.

3. In rapidly reproducing species the intense pressure of competition may result in the natural selection of individuals with characteristics that minimize the competition, resulting in divergent species with slightly different needs, and therefore different niches.

Communities

The ecological difference between communities and ecosystems is a subtle one. Frequently, when one begins discussing or studying a community it ends up as a discussion or study of an ecosystem. The problem is that a community is an assemblage of interacting differing populations. The various populations are related to an ecosystem in much the same way as an individual is related to the population and a population to a community. There are no discrete entities sharply delimited one from the other.

A community can be viewed as an assemblage of populations which are the biotic part of a single major ecosystem or the biotic part of interacting ecosystems. Therefore, study of a population cannot be limited to *intraspecific* interactions, that is, *within* a population, but must also include *interspecific* interactions, *between* populations. The following discussion will consider some of the interspecific actions exerted upon populations by other populations within the community.

Interspecific Interactions

Interactions between different populations in the community are termed *interspecific interactions.* Within the ecosystem each population occupies its own ecological niche. However, since many different species may be occupying the same habitat, there tends to be overlapping. This overlapping results in various interactions. Among these are interspecific competition, symbiotic interactions, and predation.

Interspecific Competition. Competition between species is a continuous process tending to maintain differences in populations. It is one of the main factors in the development of ecological niches. Almost anything may become the object of competition within the environment. In plants it may be sunlight, water, or available mineral nutrients. Among animals it may be territory, water, or nesting places. However, among animals the most frequent object of competition is food.

If two species compete directly, that is, occupy the same niche, the typical outcomes are the elimination of one by extinction, the driving out of one species to a different, less competitive location, or the hastening of natural selection to increase differences between them so that they do not attempt to compete by occupying the same ecological niche. For example, the marsupials that once existed in the Neotropical region became extinct when the land bridge to the Nearctic region was re-formed. The more successful placentals moved into the Neotropical region and competed with the native marsupials. Eventually, the placentals took over almost all the ecological niches occupied by the marsupials and the marsupials became extinct (except for three species.)

J. H. Connell found in his study of a community of barnacles on the coast of Scotland that there was one species, *Chthamalus,* which lived at the upper level of the intertidal zone while another species, *Balanus,* lived at a lower level. The species found at the upper level was able to live at both levels, but the lower level species could not live at the upper level because it could not tolerate periods of desiccation. Although the *Chthamalus* could live in both zones, it was kept at the higher level because the lower level species reproduced so rapidly it crowded out any upper level barnacles that attempted to establish themselves in the lower level. *Chthamalus* was driven to the upper level where there was less competition (Figure 23-10).

An example of natural selection eliminating competition was found by Robert MacArthur while at the University of Pennsylvania. He studied several species of warblers all of which appeared to occupy the same ecological niche in spruce trees. However, after careful analysis, he found variations in their feeding habits. Each species sought food in a different area of the tree. In effect, as a result of natural selection, each had its own ecological niche and was specialized to occupy that niche.

Interspecific competition has the effect of limiting the number of species within an ecosystem and, like intraspecific competition, the number of individuals within a population. Communities that contain many different species tend to be more stable because they have a greater number of competitive relationships and there will be fewer population fluctuations.

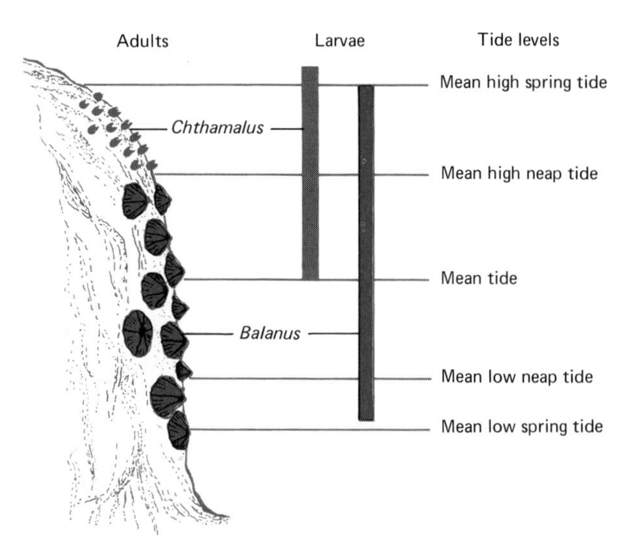

FIGURE 23-10. Competition among two species of barnacles. *Chthamalus* occupies the upper level of the intertidal zone and is capable of existing at the lower level. *Balanus* occupies the lower level and cannot exist at the upper level. The barnacles at the lower level dominate the region by reproducing so rapidly that they physically push the other species to the upper level. The upper level barnacles remain in their region because competition is not as great. [*Modified from Joseph H. Connell,* Ecology *vol. 42, p. 22, 1961. Copyright © 1961 by Duke University Press.*]

SYMBIOTIC INTERACTIONS. The term **symbiosis** means "living together" and generally, refers to the interactions between two or more different species. These interactions show all degrees of variability. If two populations, or individuals of two populations, show no interaction or have no effect at all on each other, they exhibit *neutralism.* When individuals may live and benefit from contact with other species but also may live without that contact, they exhibit *proto-cooperation.* On the other hand, if two species are dependent upon each other, the interaction is known as mutualism. Other interactions between species may be classified into several categories, the most common of which are commensalism, parasitism, and predation.

Mutualism. **Mutualism** is an interaction of two species in which both derive benefit. The degree of association may be a loose sort of relationship such as that which exists between the tickbird and the rhinoceros. The bird receives a supply of food by eating parasitic insects while the rhinoceros is relieved of unwanted guests. Both may exist without each other, so the association is *facultative.* On the other hand, a relationship might be so interdependent that neither species can exist without the other; in this case the association is *obligatory.* An example of obligatory mutualism is shown by termites and certain protozoans. Termites are unable to break down the cellulose of wood which they consume. Protozoa living within the digestive tract of the termites secrete enzymes which break down the wood cellulose in sufficient quantity to supply their own nutritional needs as well as those of the termites. The termites would starve without the protozoans while the protozoans could not exist without the shelter and supply of cellulose provided by the termite. Another example of obligatory mutualism is shown by the lichens (see Figure 24-11A). The lichens, which are very common in the tundra regions and are found on rocks and the bark of trees elsewhere, consist of a single-celled alga and a filamentous fungus. The alga produces food sufficient for itself and the fungus. The fungus provides moisture, nitrogenous wastes, and anchorage for the alga. Because of the close association, lichens can exist in areas of great climatic variation. Within the extremes of the example given are various degrees of interdependence on the part of one or the other of the interacting species.

Commensalism. The interactions of **commensalism** occur when two different species live together and one derives benefit while the other is neither helped nor harmed. The host neither resists nor fosters the interaction. Examples of this type of interaction are the orchid and Spanish moss (Figure 23-11). The commensal orchid and commensal Spanish moss obtain water and minerals that collect in the various cracks and pockets of the host tree. The host tree is neither harmed nor benefited. Another example is seen in certain algae which attach themselves to the bodies of turtles, whales, and crustaceans. The commensal algae benefit by being carried to different environments but do not harm the host organism. Among animals, the pilot fish swims closely by the shark, eating scraps of food as the shark tears its prey. Many bivalves, such as oysters, clams, and mussels, have small crabs living in the shelter of their shells.

Commensalism is essentially a facultative association. The commensal, while deriving benefits from the association, is capable of living apart from the host. The host is not at all dependent upon the commensal. If the association represents greater dependence it is considered mutualism or parasitism.

Parasitism. Of the various symbiotic relationships, parasitism is the most complex and unique. **Parasitism** is essentially a nutritional relationship in which one organism lives in or on its host for most or part of its life cycle. The parasite is harmful to the host to varying degrees. Since it is a food-getting interaction, parasitism, with the exception of a few green plants, occurs among the heterotrophs (viruses, bacteria, fungi, and animals). Parasites are of two types, internal and external. *Internal parasites* are found any place within the body of their hosts; however, a given species is usually localized. The *external parasites* attach to the outer skin or hair.

The internal parasites are more specialized than external parasites, both in their life cycles and in their structures. Because of the unique internal environment in which they live, many internal parasites lack certain organs and organ systems. The tapeworm has no digestive system because it lives in the digestive tract where it can directly absorb the digested material of the host. By contrast, certain other body features are highly developed. The body walls of parasites living in the digestive tract have a cuticle to resist *hydrolysis* (digestion). Various specialized hooks and suckers are present around the head of many parasites of the digestive system to prevent expulsion from the host.

FIGURE 23-11. Spanish moss. An epiphyte that grows on trees and other vegetation, Spanish moss lives in a commensal association. The trees provide support and the moisture and minerals that collect in the cracks and pockets of the trees provide nutrients for the moss. [*Courtesy of Mississippi Department of Agriculture and Commerce.*]

The life cycles of the internal parasites are particularly unusual. Many intricate relationships have evolved in the perpetuation of the species. The Chinese liver fluke serves as an excellent example (Figure 23-12). The liver fluke that parasitizes humans begins its life cycle as an egg laid inside a human that is eliminated in the feces. Through various means of disposal, the feces reach freshwater ponds where the eggs hatch into *miracidium larvae.* The larvae swim about and enter snails. Within a snail they form a sporocyst. The sporocyst develops into many individuals of another form called the *rediae,* which enter the liver of the snail and continue to reproduce. The rediae form eventually gives rise to a third, free-swimming form, the *cercariae.* This free-swimming organism enters fish and encysts in the muscles. After the fish is eaten, the encysted form develops in an adult liver fluke, and the cycle is completed.

Most internal parasites produce exceedingly large numbers of eggs in order to ensure completion of their life cycle. When a parasite is dependent upon specific hosts, as in the life cycle of the Chinese liver fluke, it is *host-specific.* By contrast, ticks which parasitize a large variety of hosts are called *host-nonspecific parasites.* The interaction of the host and the parasite must reach a balance if it is to be a successful interaction. It is no real benefit to a parasite to kill its host. If there is wide-scale extinction of the host, the parasite also be-

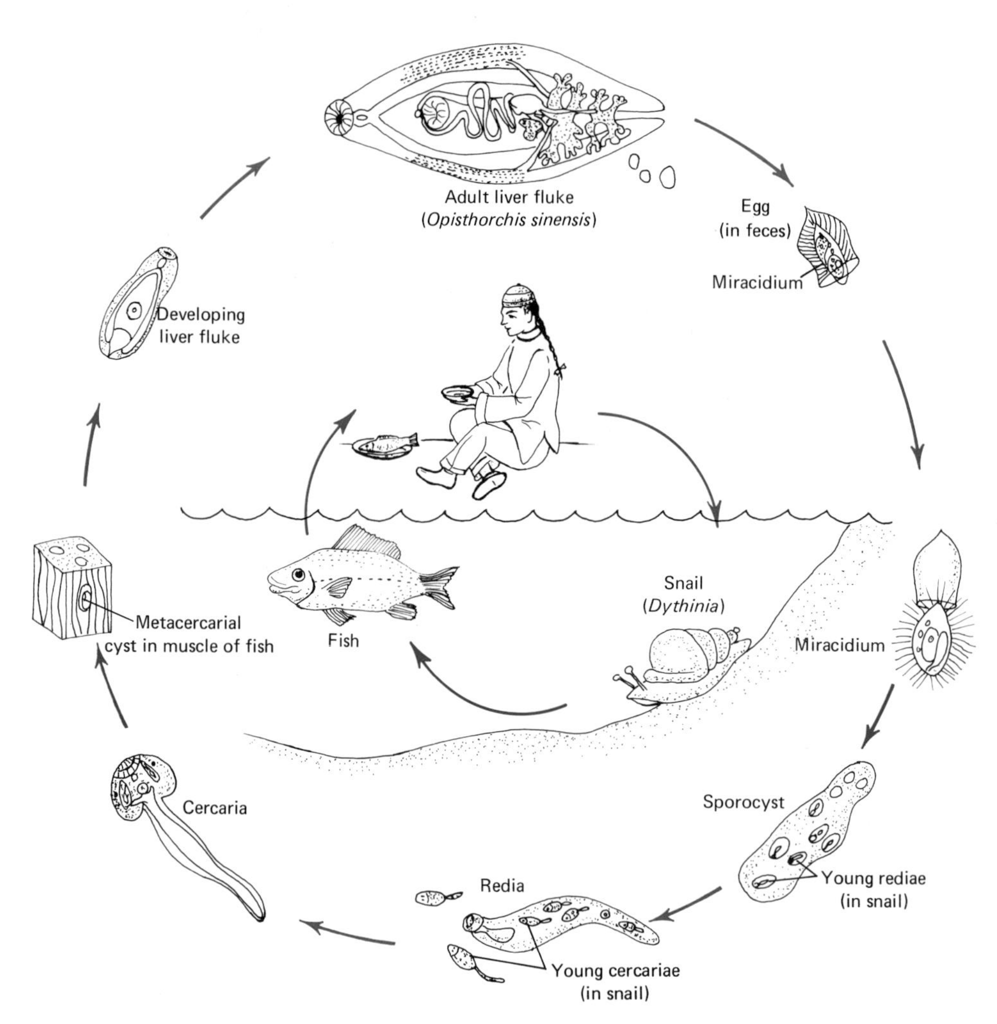

FIGURE 23-12. The life cycle of the Chinese liver fluke, a host-specific parasite. The Chinese liver fluke moves from human to snail to fish and back to human. Each step of its life cycle is dependent upon the previous one. This life cycle shows the degree of dependence and adaptation that a parasite undergoes in its association with its host. Consult text for details.

comes extinct. Through the evolution of parasites and their hosts there tends to be adaptation toward adjustment on the part of both to ensure less serious disturbances. Where there is great destruction of the host by the parasite, it can be assumed that the interaction is a relatively new one. Over long periods of time both host and parasite evolve toward a more balanced relationship and over even longer periods closer types of relationships evolve.

Predation. The difference between parasitism and predation is one of degree. The body louse attaches itself to the human body and sucks blood. It is considered a parasite. The mosquito lights on a human, sucks blood, and flies away. It may be considered a predator. The distinguishing difference is that the louse spends part of its life attached to the host while the mosquito is relatively free living.

Most predators are animals, although there are a few insectivorous plants. Within the meaning of the term *predator* are included the herbivores as well as the carnivores. The herbivores generally do not kill their prey, plants. This is also true of many carnivorous insects. Generally, carnivores among the higher animals must kill their prey in order to obtain food. Unlike the parasite, the predator will usually attack a large number of different species of prey; predators are less specific in their needs.

Predator and prey populations affect each other, so that fluctuations in one population are soon reflected as fluctuations in the other. On rare occasions, particularly if some drastic change has taken place in the ecosystem, or if the ecosystem is simple in species composition, predators may have a very severe effect upon the prey population, at times reducing it to near extinction level. The most common result of the predator–prey relationship in well-established ecosystems is, however, stabilization of population sizes. The degree of stability achieved is dependent upon the ability of the predator to utilize other prey and the positions of the predator and prey within the ecosystem.

24 Dynamics of Ecosystems

GUIDE QUESTIONS

1 What are the major components of an ecosystem? Give an example of each.
2 How are the major components of an ecosystem interrelated? Give examples.
3 What are the functions of each of the following: producers, consumers, and decomposers? How are they adapted for the roles they play?
4 Explain how energy flows through an ecosystem. Explain what is meant by trophic levels.
5 What is meant by a food chain? Give specific examples of each of the following types of food chains: predator chain, parasitic chain, and saprophytic chain.
6 What is the relationship of energy to the food chain?
7 Give examples of ecological pyramids. What information do they convey?
8 How do tolerance levels and limiting factors affect an ecosystem. Give examples.
9 Give examples of natural and artificial changes that may occur in an ecosystem and tell how they affect an ecosystem.
10 Give the several steps of shallow lake succession and terrestrial succession.
11 Distinguish between primary succession and secondary succession.
12 Give several examples of how ecosystems tend to establish equilibrium.

ANY area in which the abiotic and biotic portions of the environment interact and in which there is a flow of energy as well as a cycling of nutrients is called an **ecosystem.** For convenience, the ecosystem can be divided into two basic components: the abiotic and the biotic. The abiotic component, discussed in Chapter 22, can for present purposes be divided into three broad parts. The first contains the inorganic substances—nitrogen, water, carbon dioxide, and various other inorganic elements and molecules. These substances are initially supplied to the biotic environment from the lithosphere, hydrosphere, and atmosphere. They provide the basic raw materials to produce high-energy compounds and to maintain a stable internal environment within the body of organisms to carry on life processes. While some of these inorganic substances are continuously entering ecosystems, many of them are being cycled for re-use by the biogeochemical cycles (see page 602).

The second part of the abiotic component contains the organic substances that remain after an organism dies or organic substances excreted as waste products of living organisms. Some of these substances may be further broken down to simpler substances by organisms such as bacteria and fungi, while others are broken down by physical and chemical forces within the abiotic environment.

The third part consists of climate and other physical factors, such as the type of substrate upon which terrestrial organisms exist or the type of water in which aquatic organisms live.

Biotic Components of Ecosystems

Functionally, the biotic components of an ecosystem may be divided into two principal categories on the basis of energy source.

1. The **autotrophs,** primarily photosynthetic organisms, take inorganic materials from the abiotic environment and, with energy from the sun, build them into complex organic substances.

2. The **heterotrophs,** which cannot carry on photosynthesis, utilize either directly or indirectly the complex substances produced by the autotrophic component.

It should be noted that chemosynthetic bacteria are frequently considered as producers; however, within the ecosystem they are more properly seen as an intermediary between the autotrophic and heterotrophic components. These bacteria obtain their energy by oxidizing simple inorganic compounds and radicals such as ammonia, nitrites, and sulfides.

The biotic components of ecosystems may also be classified according to nutritional relationships. On this basis we have three groups: producers, consumers, and decomposers.

The autotrophs comprise the **producer** component and are capable of producing their own food as well as providing nutrients for other components. The green plants, which carry on photosynthesis, and certain Monera and Protista, which carry on either photosynthesis or chemosynthesis, belong to this group.

The organisms of the heterotrophic type make up the other two biotic components, the consumers and decomposers. Animals and parasitic plants, together, make up the **consumer** component. As has been previously stated, consumers are unable to produce their own organic compounds for nutritive purposes and must obtain them from the producers. Among the consumers are the **herbivores** or **primary consumers,** which obtain their organic nutrients by directly consuming the producers. **Secondary consumers** are omnivores (consume both plants and animals) and *carnivores,* which depend partially or wholly on other consumers for their food. **Tertiary** and **quaternary consumers** are predators that depend upon secondary or tertiary consumers as their source of organic nutrient.

The **decomposer** component is the second group of the heterotrophic type. The fungi and bacteria are the major members of this component. These organisms are **saprophytic;** that is, they obtain their organic nutrient from nonliving organic matter originating from excrement or dead organisms. The decomposers secrete enzymes that break down the complex dead or decaying matter into simpler organic compounds. These simpler organic compounds are then absorbed by the decomposers to carry on their life processes. In addition, some of the simpler organic compounds produced by the decomposers are released back into the abiotic environment where they can be re-used by green plants. These compounds are a vital part of the nutritional patterns of autotrophs. As such, the decomposers provide an essential part or link in the cycle of life. Were it not for these organisms, essential nutrients would be permanently locked up in insoluble complex organic molecules and could not be re-utilized by producers and, indirectly, by consumers (Figure 24-1).

Almost all ecosystems can be analyzed in terms of the abiotic and biotic components. With few exceptions all three are present in an ecosystem. However, there do exist some incomplete ecosystems where all three of the biotic components are not present. For example, in the abyssal depths of the oceans no light penetrates, and therefore no photosynthetic organisms (producers) exist. (If chemosynthetic bacteria are present, their role is so slight that they make no real contribution to maintaining the ecosystem.) The abyssal ecosystem is, therefore, made up only of consumers and decomposers. They receive their nutrients from organic matter that settles from the photosynthetic zone or by predation upon other members of the ecosystem. Under certain conditions, incomplete ecosystems may be composed only of producers and decomposers. This has occurred where there have been massive growths of toxic algal populations. The toxins produced by the algae exclude the existence of any consumers. This type of ecosystem can be found when occasionally, in warm coastal waters, the flagellate *Gymnodinium brevis* undergoes a massive and rapid increase in population and produces what is known as the *red tide.* The flagellate secretes a chemical so toxic it kills all consumers in the area. Such an ecosystem is extremely unstable, and changes take place rapidly to restore a more complete ecosystem.

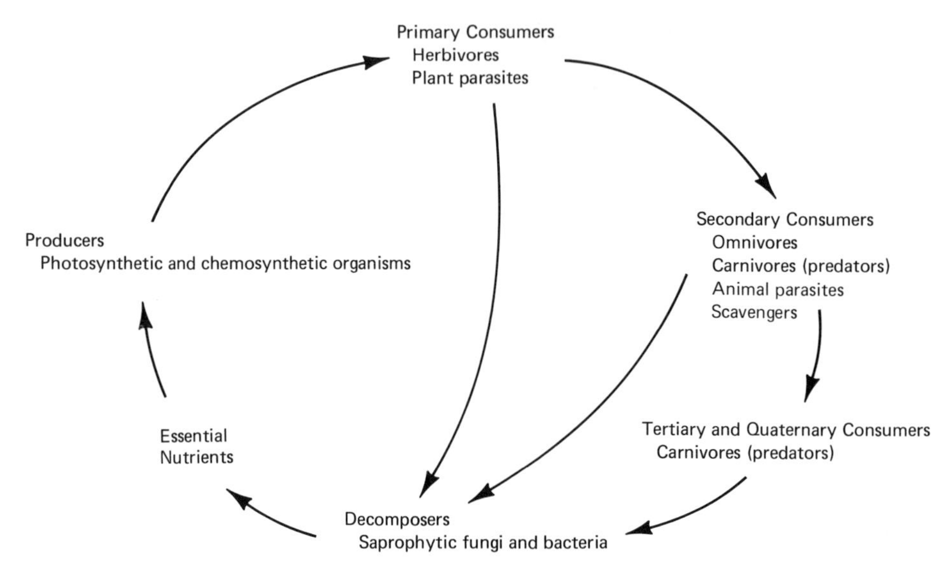

FIGURE 24-1. Ecological cycle of life made up of the biotic components.

Functions of the Biotic Components of Ecosystems

Producers

The basic production in an ecosystem is the photosynthetic process. The biochemical aspects of photosynthesis have been discussed in Chapter 6. From an ecological point of view photosynthesis plays three important roles.

1. It is essentially the only means by which energy from the sun can be captured and utilized by the ecosystem.

2. It is the means by which complex organic compounds are synthesized directly or indirectly.

3. It plays an important role in the releasing of molecular oxygen into the atmosphere, which is essential for life.

Photosynthetic organisms, in addition to producing glucose and other organic molecules in the photosynthetic process, can synthesize more complex organic compounds by using the energy obtained by the breakdown of the glucose they produced. Among the more complex compounds are polysaccharides such as starch and cellulose, lipids such as fats and phospholipids, pigments such as chlorophyll and xanthophyll, and proteins and vitamins. While it is possible for consumers and decomposers to synthesize complex organic compounds by using the energy from the breakdown of glucose, certain essential compounds can be obtained only from the producers. These include some of the vitamins and the essential amino acids because only photosynthetic organisms have the ability to take the nitrates produced by nitrogen-fixing bacteria (see page 605) and produce amino acids, the subunits of proteins.

Because photosynthetic activities are the ultimate source of all energy, organic molecules, and free oxygen for living organisms, the total photosynthetic activity or gross productivity in an ecosystem becomes extremely important in the analysis of any ecosystem. Gross productivity may be measured in several ways, most usually by oxygen evolution or carbon uptake in photosynthesis. That portion of gross production (from 10 to 50%) not used in plant respiration is used for plant growth and storage. The plant tissue that results is then available for consumers and decomposers.

Another important method of analysis of production in an ecosystem makes use of **biomass,** the weight of the various biotic components. A measurement of the increase in the biomass of the producer component represents the *net production* (productivity) of the given ecosystem. This increase in biomass or net production is the portion of total or gross production that is not used by the producers themselves in respiration and becomes available to the other components of the ecosystem. Net production is most often measured in increase of grams of dry organic matter per square meter per day. This measurement may be converted to calories or other units of energy to represent available energy. The producer component of an ecosystem will have an increase in biomass from 0.5 to 20 grams/m^2/day. However, this production is subject to considerable variability based upon length of day, intensity of light, availability of nutrients and water, and the inherent ability of the producer to synthesize, among other factors. For example, Lake Erie in the winter has shown a daily biomass productivity of 1.0 grams/m^2/day, whereas in the summer it has shown a biomass productivity of 9.0 grams/m^2/day. Certain coral reefs in the Pacific produce 18.2 grams/m^2/day, while in Silver Springs, Florida, production has been as high as 35.0 grams/m^2/day. This gross production, however, does not mean that the biomass produced is available to the consumer. Some of the matter produced is metabolized by the producer to carry on its life activities and some of the matter is in a form that cannot be used by the consumer.

Consumers

The existence of consumers is totally dependent upon the overproduction of nutrients by the producers. The structure of the producer's body is composed mostly of cellulose and provides food for the consumer either directly or indirectly. Most animals cannot directly metabolize cellulose. Within their digestive system exist populations of bacteria that are capable of breaking down cellulose and converting it into a form that can be metabolized. Essentially, the bacteria predigest the cellulose for the consumer by converting it to glucose, which can then be absorbed and metabolized. Herbivores are dependent upon their association with these bacteria for their nutrients.

Herbivores provide an essential link between producers and other types of consumers. Among the most obvious of the terrestrial herbivores are the large grazing animals of the grasslands, rodents, and larger plant-eating animals of the forest. The less obvious primary consumers are those found in the aquatic environment. While there are identifiable fish, crustacea, and birds which are grazing on the algae of large lakes and seas, the aquatic environments are teaming with microscopic organisms called plankton. Plankton is composed of microscopic photosynthetic phytoplankton (producers) and grazing microscopic zooplankton (consumers). The microscopic zooplankton provide an essential link in providing food for larger secondary consumers and still larger tertiary and quaternary consumers.

In most ecosystems there are consumers that do not directly consume producers but other consumers. These are the carnivores. At any given time the level of a consumer is determined by the level of the consumer upon which it preys. For example, a consumer that devours a secondary consumer is a tertiary consumer, and so on. Also, the level of a consumer may vary from time to time since many consumers are not too specific in their choice of prey. For example, at one time a hawk may prey upon a small rabbit (primary consumer) and be classified as a secondary consumer, while at another time it may prey upon a snake (secondary consumer) and be classified as a tertiary consumer. Generally, the highest level is a quaternary consumer, although there are exceptions. The number of levels of consumers is limited because of the loss of energy transfers from consumer level to consumer level.

A special type of consumer is the *scavenger*. At times it is difficult to classify the scavenger as to a particular level consumer. These consumers feed upon dead or decaying plant and animal matter. They are distinguished from predators inasmuch as they feed on organisms that have died. They are considered consumers however, since they reconstitute the organic matter that they consume, using some for energy and some for growth and development. In function they are somewhat related to the decomposers for they do take dead organic matter and recycle it. The distinguishing difference between scavengers and decomposers is that the scavengers actually consume the dead organic matter and pass it through their digestive systems. Decomposers essentially digest the organic matter outside their bodies, breaking it down into simpler molecules and then absorbing them.

The *parasites* are another group of consumers to

which it is difficult to subscribe a specific level of consumption. Obviously, animals and plants that parasitize other plants are primary consumers. However, those parasites that attack various high-level consumers are difficult to assign to a level of consumption.

Decomposers

All organisms in their metabolism excrete organic wastes and, therefore, accomplish some decomposition. However, the true decomposers are those whose primary source of energy for growth and development is obtained through decomposition. As previously stated the decomposition process is external to their bodies. The decomposers perform three vital roles in ecosystems. First, they break down complex organic molecules into simpler forms, thereby contributing to the recycling of nutrients. Second, they also provide food for other organisms. No single type of decomposer completely degrades organic substances. One group of decomposers breaks down (degrades) a substance to one point, and the process is continued by another group of decomposers. For example, one group may degrade glucose to ethanol; another group will then take over and convert this alcohol to acetic acid; and still another group will degrade the acid to carbon dioxide and water. A third function of decomposers is to produce substances that exert a regulatory control in the ecosystem. These substances are called **ectocrines** or environmental hormones. Substances secreted by decomposers will limit the growth of other populations. The best known of these is penicillin. The *Penicillium* mold releases substances that inhibit the growth of bacteria. In fact, many of the antibiotics that have proved so beneficial are ectocrines secreted by the decomposers. Secretion of ectocrines is not limited to the decomposers. It is known that higher plants and animals may secrete substances that inhibit the growth of various populations within their particular ecological niche. However, the decomposers are the most important source of ectocrines.

Decomposers act upon all organic molecules. Many carbohydrates and most lipids and proteins are rapidly degraded. Some materials, such as cellulose, lignin, hair, and bones, are more resistant to decomposition. Of these, cellulose is the most important. It strongly resists bacterial decomposition. As a result, if conditions are not optimum for its decomposition, it may accumulate as debris within the habitat and inhibit the growth of major populations within an ecosystem. It is for this reason that natural ground fires are essential to maintain certain ecosystems. The burning of the debris returns the nutrients to the soil and allows the normal ecological processes to continue. The role of fires will be discussed later in this chapter.

Decomposers are extremely abundant in an ecosystem. However, because of their size, they are the least obvious members of an ecosystem; except for a few of the larger fungi, most are microscopic fungi and bacteria. Large populations of decomposers are essential for any ecosystem. If their numbers are greatly reduced either by natural or artificial causes, their loss can grossly alter the ecosystem or result in the complete destruction of all life in a given area.

Energy and Biogeochemical Cycles

In the analysis of ecological relationships there are essentially two basic processes that are universal. The first is the transduction and flow of energy and the second is the cycling of materials.

Radiant energy from the sun is essentially the only significant source of energy for any ecosystem. Photosynthetic organisms in the presence of sunlight are capable of taking carbon dioxide and water from the environment and forming high-energy compounds (see page 48). In the process a small amount of radiant energy from the sun is transduced to chemical energy. The energy is captured within high-energy compounds and serves as the energy to carry on the life processes of living organisms. The nonphotosynthetic organisms obtain their energy by consuming directly or indirectly those organisms capable of carrying on the photosynthetic processes. As one organism consumes another organism, a certain amount of energy is lost at each sequential step, according to the principle known as the Second Law of Thermodynamics (see page 6). The passage of energy through an ecosystem is unidirectional and noncyclic. This is one of the most important concepts governing ecosystems. There is an intimate relationship between the flow of energy and the recycling of inorganic nutrients because energy is required in their cycling or regeneration.

In the process of building up protoplasm, organisms also directly or indirectly incorporate certain inorganic compounds and elements, inorganic nutrients, within their bodies. Among these are carbon dioxide, water, ammonia, nitrogen, phosphorus, sulfur, magnesium, calcium, and at least fifteen other elements. These substances are obtained directly, by absorption from the environment, or indirectly, by consumption of other organisms. When an organism dies, through decomposition nutrients are returned to the environment and may then be re-used. Nutrients are also returned to the environment through excretion products. The cyclic movements of inorganic compounds and elements through the ecosystems are called **biogeochemical cycles.** Generally, these nutrients, unlike energy, may be recycled and are not irretrievably lost. However, there are a number of ways in which these substances may be removed from cycling for long periods of time and lost to currently existing ecosystems. Since many of these substances are in limited supply and may be removed from recycling, factors controlling their cycling constitute another important concept governing ecosystems.

Biogeochemical cycles may be generally classified into two types, the **gaseous cycles** and **sedimentary cycles.** In addition, each type of cycle may be further distinguished into two parts. In one part large amounts of the substances are removed from rapid recycling and slowly pass through the abiotic environment, which acts as a future source or reservoir. The second part is an active portion, in which a smaller amount of the nutrients are rapidly recycled between the abiotic and biotic environments.

The gaseous cycles are represented by substances that at some point in their cycling are in a gaseous form. The reservoir for these substances is the atmosphere and hydrosphere. The sedimentary cycles are represented by those substances that are solids, both soluble and insoluble. Correctly or incorrectly these substances are referred to as "minerals." The major reservoir for these substances is the lithosphere.

Gaseous Cycles

Carbon Cycle. All organic compounds contain carbon. The source of most of the inorganic carbon that is used to synthesize organic compounds is the carbon dioxide of the atmosphere and that dissolved in water (Figure 24-2). The first step in the utilization of carbon dioxide occurs in photosynthesis by chlorophyll-containing organisms, the most dominant of which are the photosynthetic protists and the green plants. The carbon dioxide becomes a part of simple carbohydrates. These organisms, by synthesizing and resynthesizing these simple carbohydrates, ultimately form polysaccharides, lipids, proteins, and other complex organic molecules. Animals eat the photosynthetic organisms and the organic molecules are digested and resynthesized. These animals are eaten by other animals and the organic materials are again digested and resynthesized. Thus, the original carbon atoms as they are transferred from organism to organism are shifted from organic molecule to organic molecule. In the process, some of

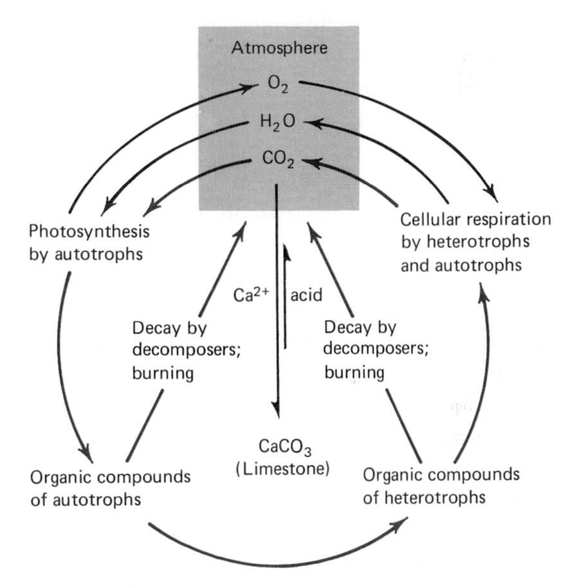

Figure 24-2. The carbon and oxygen cycles. In the carbon cycle (colored arrows), carbon, obtained from the atmosphere as carbon dioxide, is used by photosynthetic organisms for the synthesis of organic compounds. These organic compounds follow various pathways throughout the biosphere. Eventually, by activities such as respiration, burning, and slow oxidation (decay), most of the carbon is returned to the atmosphere. Some may be permanently bound in limestone ($CaCO_3$). In the oxygen cycle, oxygen is produced during photosynthesis and is used in respiration by living organisms. It is returned to the atmosphere as carbon dioxide and water, only to be released again during the photosynthetic process. The carbon and oxygen cycles are closely interrelated.

the high-energy compounds are broken down by respiration.

Since the amount of available carbon dioxide from the atmosphere is relatively small (0.03%), the supply would soon be exhausted if it were not replenished. The return of carbon dioxide back to the atmosphere for re-use completes the *carbon cycle.*

Some carbon dioxide is directly released to the environment as an end product of respiration of all living organisms. It then becomes immediately available to start the cycle over again. However, much of the carbon remains in the tissues of organisms until they die or is eliminated in complex organic waste. If this carbon were not converted to carbon dioxide, life would cease to exist. It is in this aspect of the ecosystem where decomposition is critical. Certain organisms consume and digest the organic remains and reduce them to simpler and simpler compounds with the result that most of the carbon of organic compounds is converted back to carbon dioxide.

Carbon dioxide may be returned to the atmosphere by other means, the most familiar of which is through fire and slow oxidation. When organic compounds are burned or slowly oxidized, carbon dioxide is given off. However, these processes are not continuous or widespread enough to provide a sufficient supply to maintain the life of photosynthetic organisms.

Some carbon may be removed from the cycle for long periods of time, and some of it may never be returned. For example, the dead bodies of some organisms may not be decomposed. The accumulation of this undecayed matter may be converted into coal, petroleum, and natural gas and buried deep below the surface of the earth. For a long time little of the carbon in these materials was available. Much carbon is also locked up in the form of natural limestone. Animals and plants, such as certain algae and coral, incorporate carbon in their structures as calcium carbonate, limestone. Though some of the limestone is broken down by natural acids and weathering, much of it is buried deep in the earth and probably will never be available.

By far the most important part of the carbon cycle is represented by the activities of living organisms. The rate at which plants remove carbon dioxide from the air is far greater than the rate at which it can be released by decomposition of dead organisms alone. Therefore, green plants depend, to a great extent, on animal respiration for their carbon dioxide. The cycle emphasizes the interdependence of all organisms within the ecosystem. The existence and activities of one group are dependent upon the existence and activities of all others.

Interrelated with the carbon cycle is the *oxygen cycle* (Figure 24-2). During the photosynthetic process, in addition to the production of simple carbohydrates, the photosynthetic organisms give off oxygen. This oxygen is essential to the existence of all aerobic organisms. In respiration, the aerobic organisms give off carbon dioxide and water as end products.

Nitrogen Cycle. Nitrogen is needed by all organisms for the synthesis of proteins as well as other nitrogen-containing compounds. Atmospheric nitrogen (N_2), although found in great abundance, cannot be used by most organisms. It must be combined with other elements such as oxygen and hydrogen to form nitrates (NO_3^-) or ammonium salts (NH_4^+) before it becomes available for synthesis. The most common usable nitrogen source is the nitrate ion (NO_3^-). The chemical and physical forces operating in the soil and water, together with the activities of certain bacteria, blue-green algae, and fungi, are important factors in converting nitrogen to usable forms.

Photosynthetic organisms absorb nitrates from the soil or water. In synthesizing proteins, these organisms utilize the nitrogen from nitrates and add it to the carbon, hydrogen, and oxygen which have been organized during photosynthesis. Eventually, the organic nitrogen compounds are returned to the soil or water by either plants or animals. For example, plants may die after a season or two, they may shed their leaves, or they may be eaten by animals. The animals, in turn, use plant protein to synthesize their cytoplasm, and either during excretion or at death, they also return the nitrogen to the soil. Plants, in association with humans and other animals, compose a cycle in which nitrogen is continually being converted from one form to another. These conversions and close associations constitute the basis for the *nitrogen cycle* (Figure 24-3).

When an organism dies, the process of decay or decomposition results in the breakdown of its fairly com-

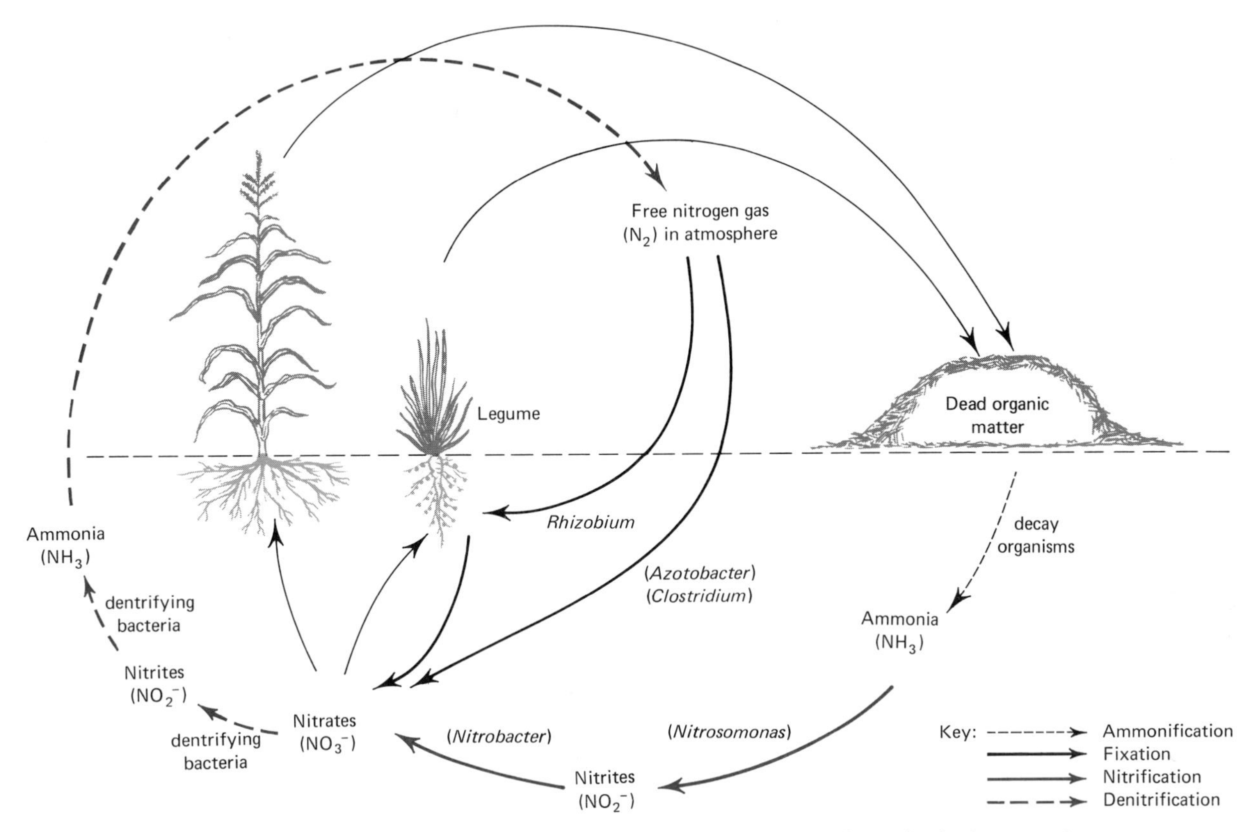

FIGURE 24-3. The nitrogen cycle. Through continual circulation and transformation in the atmosphere and the soil by microorganisms, nitrogen is converted into a form that plants can utilize for the synthesis of proteins and other nitrogen-containing compounds. These nitrogen-containing compounds are passed on to the heterotrophs when they consume plants. See text for details of the cycle.

plex components into a number of simple compounds, which are returned to the environment. The agents of decay are certain bacteria and fungi that are associated with the various stages in the complex process of decomposition. The end result of decay is the formation of ammonia from organic nitrogenous compounds, a process called *ammonification.*

$$\text{(1)}\qquad \text{Dead or waste products} \xrightarrow{\text{decay}} \underset{\text{ammonia}}{NH_3}$$

Living in the soil are two genera of bacteria, *Nitrosomonas* and *Nitrobacter,* that are capable of oxidizing ammonia. In the oxidation process, the energy liberated is used by these bacteria to synthesize various compounds just as the photosynthetic autotrophs utilize the energy from sunlight. The oxidation of ammonia (*nitrification*) involves two steps.

1. Ammonia is converted to nitrites (NO_2^-) by the *Nitrosomonas.*

2. Nitrites are changed to nitrates (NO_3^-) by the *Nitrobacter.*

At various points in the cycle, atmospheric nitrogen is either added or removed. The loss of nitrogen from the cycle is called *dentrification,* while the addition of nitrogen is termed *nitrogen fixation.* There are many species of *denitrifying bacteria* that are capable of converting nitrate into atmospheric nitrogen; the sequence of reactions is as follows:

$$(2)\ \underset{\text{nitrate}}{NO_3^-} \longrightarrow \underset{\text{nitrite}}{NO_2^-} \longrightarrow \underset{\text{ammonia}}{NH_3} \longrightarrow \underset{\text{atmospheric nitrogen}}{N_2}$$

Thus, denitrifying bacteria deliver free or molecular nitrogen to the atmosphere and remove nitrates from the soil. From the standpoint of soil fertility, this is an unfavorable process. However, denitrifying bacteria are anaerobic, that is, they do not function well in the presence of oxygen, therefore denitrification is most active in soils that are poorly aerated (oxygenated).

Nitrogen fixation involves the change of free atmospheric nitrogen into nitrates.

$$(3)\quad \underset{\text{atmospheric nitrogen}}{N_2} \longrightarrow \underset{\text{nitrates}}{NO_3^-}$$

Certain types of bacteria and blue-green algae are unique in their ability to use N_2 as a source of nitrogen for the synthesis of organic compounds. Essentially, there are two groups of *nitrogen-fixing bacteria* distinguishable on the basis of where they live. Some live in the roots of legumes (beans, peas, clovers) and are called *mutualistic bacteria,* while others live free in the soil with no direct plant association and are termed *nonmutualistic bacteria.*

The most common mutualistic genus is *Rhizobium.* These bacteria live in swellings (*nodules*) on the roots of legumes. Mutualism is a nutritional relationship in which two organisms live together and both derive benefit. In this case, the bacteria supply nitrogen in the form of NH_3 to the leguminous plants while the plants supply the bacteria with an adequate supply of carbohydrates as a source of energy to produce NH_3. The free-living, nonmutualistic bacteria, mainly *Azotobacter* and *Clostridium,* also supply the soil with considerable amounts of NH_3, which is converted by other bacteria to NO_3^-.

Water Cycle. All life is dependent upon water and the movement of water through the ecosystems is cyclic. This movement constitutes the basis for the *water cycle* (Figure 24-4). A direct cycle occurs in the abiotic environment when water leaves the atmosphere, usually in the form of rain or snow. It may fall directly into the ocean or on land areas where it may form brooks, streams, rivers, ponds and lakes or seep into the soil and form underground channels. All these bodies of water eventually lead to the ocean. In all cases a certain portion of the water returns to the atmosphere through evaporation.

The cycle within the biotic environment occurs when organisms obtain their water directly by drinking or by absorbing it from the soil. Some of the water is used to synthesize complex organic substances within the organism and it is released when these substances are broken down. In addition, all the chemical reactions taking place within the bodies of organisms occur in an aqueous medium. Autotrophic organisms transpire large quantities of water during respiration and photosynthesis; and heterotrophic organisms release water from respiration by breathing, through evaporation from surface areas, and through discharge of wastes. In any event, the water taken in by organisms is returned to the atmosphere only to become reincorporated within the cycle.

Sedimentary Cycles

Gaseous substances are more rapidly cycled than sedimentary substances. Since the reservoir for the gaseous cycles is the atmosphere and hydrosphere, the various gases can be quickly moved from place to place by air or water currents. Although there can be local disruptions in the cycling of these gaseous substances, adjustments are generally quickly made, at least from the point of view of large geographical areas.

Sedimentary substances, on the other hand, since they are part of the lithosphere, are nonmobile and their circulation within the reservoir is dependent upon their solubility or a geological upheaval. In many instances these substances are removed from cycling between the abiotic and biotic environments for long periods of time. Certain substances, such as calcium and phosphorus, become incorporated into the relatively permanent structures of bones and shells of living organisms, and their availability for recycling is determined by the life span of the organism and the rate at which these structures can be decomposed. It is well known that bones and shells can exist over extremely long

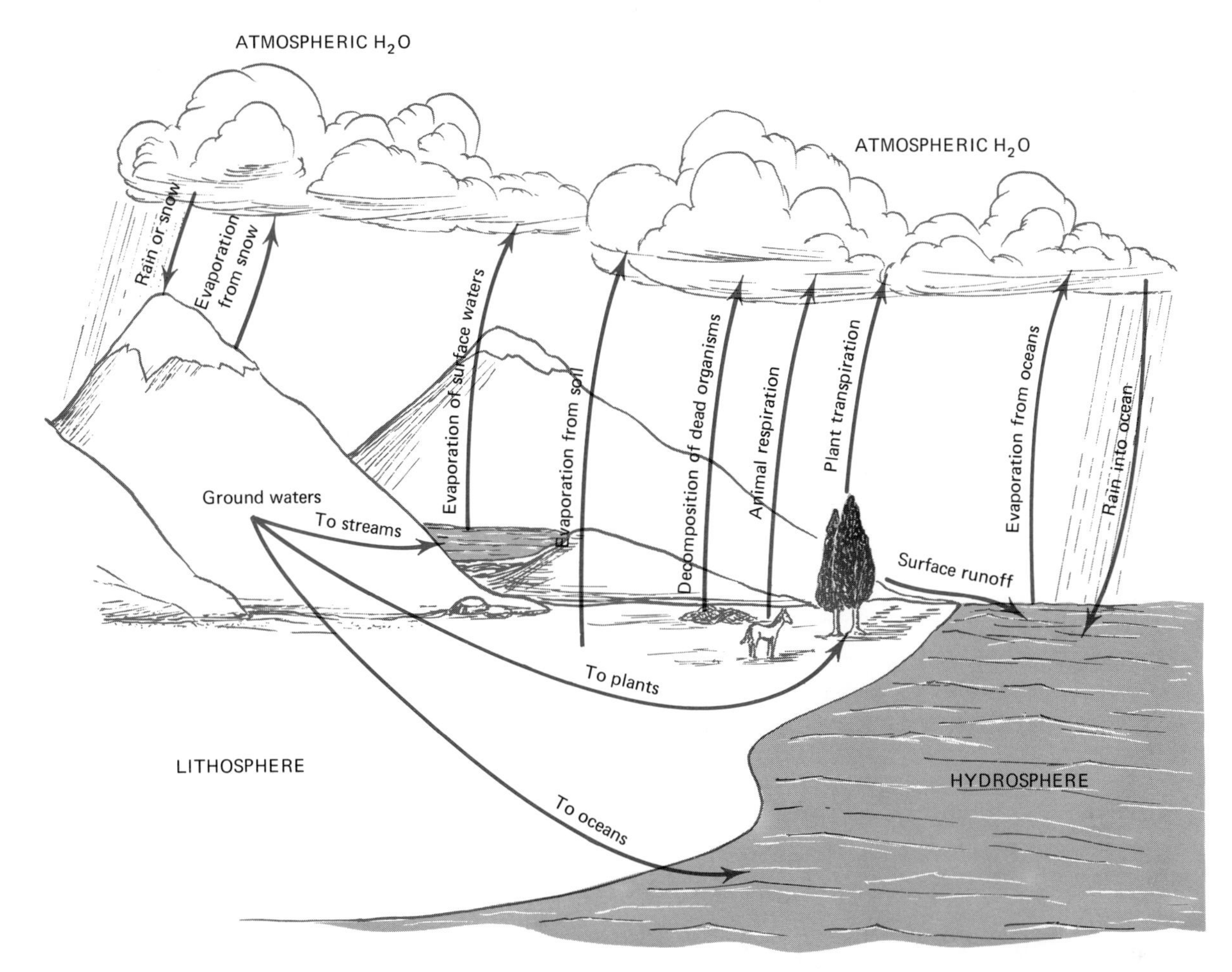

Figure 24-4. The water cycle. Water, the principal metabolite required by plants and animals, is returned directly to the atmosphere by evaporation from bodies of water and the soil as well as indirectly by transpiration of plants, by respiration of living organisms, and as a product of decomposition reactions. It is returned to the lithosphere and hydrosphere as various forms of precipitation.

periods of time with little or no decomposition. Provided there are no sharp changes in ecosystems there is a balance between the availability of sedimentary substances and their utilization. When substances are removed for long periods of time and the availability from native sources is limited, this removal represents a *leakage* from the cycle or ecosystem. The greater the homeostasis established between the abiotic and biotic environments, the lower the leakage. Leakage of sedimentary substances is to be expected in all ecosystems. However, ecosystems can and have been known to adjust to the removal of these substances from the cycles, provided there are no great alterations in the abiotic and biotic environments.

The sedimentary substances are the rocky crust of the lithosphere and the waters of the hydrosphere. As has been previously stated, these substances may be removed from cycling for long periods of time. This occurs because frequently the minerals become a part of the deep sediments of the ocean bottom and only through geological upheaval do they again become available to living organisms. Indications are that, throughout the biosphere, the availability of essential elements such as calcium, potassium, magnesium, iron, and phosphorus is decreasing. The supply of phosphorus appears to be the most limited, and eventually it may become critical. Inasmuch as the various mineral cycles are quite similar, only the *calcium cycle* will be described (Figure 24-5).

Calcium Cycle—An Example. The calcium compounds of the lithosphere are found in mineral deposits such as limestone (calcium carbonate), gypsum (hydrous calcium sulfate), and dolomite (calcium magnesium carbonate). These minerals are found, as are most minerals of the lithosphere, in an insoluble form. The action of weathering, natural acids, or secretions from living organisms changes some of these minerals to a more soluble form. As the terrestrial organisms consume or absorb the water, calcium in the form of soluble ions is taken into the living organism. These ions play an important role in maintaining electrolytic balance, functioning as activators for enzymes, and forming the actual skeletal structure of higher animals. These ions accumulate as the organism continues to grow and remain incorporated with the organism until it dies. Of course, any excess of an ion is given off by the organism

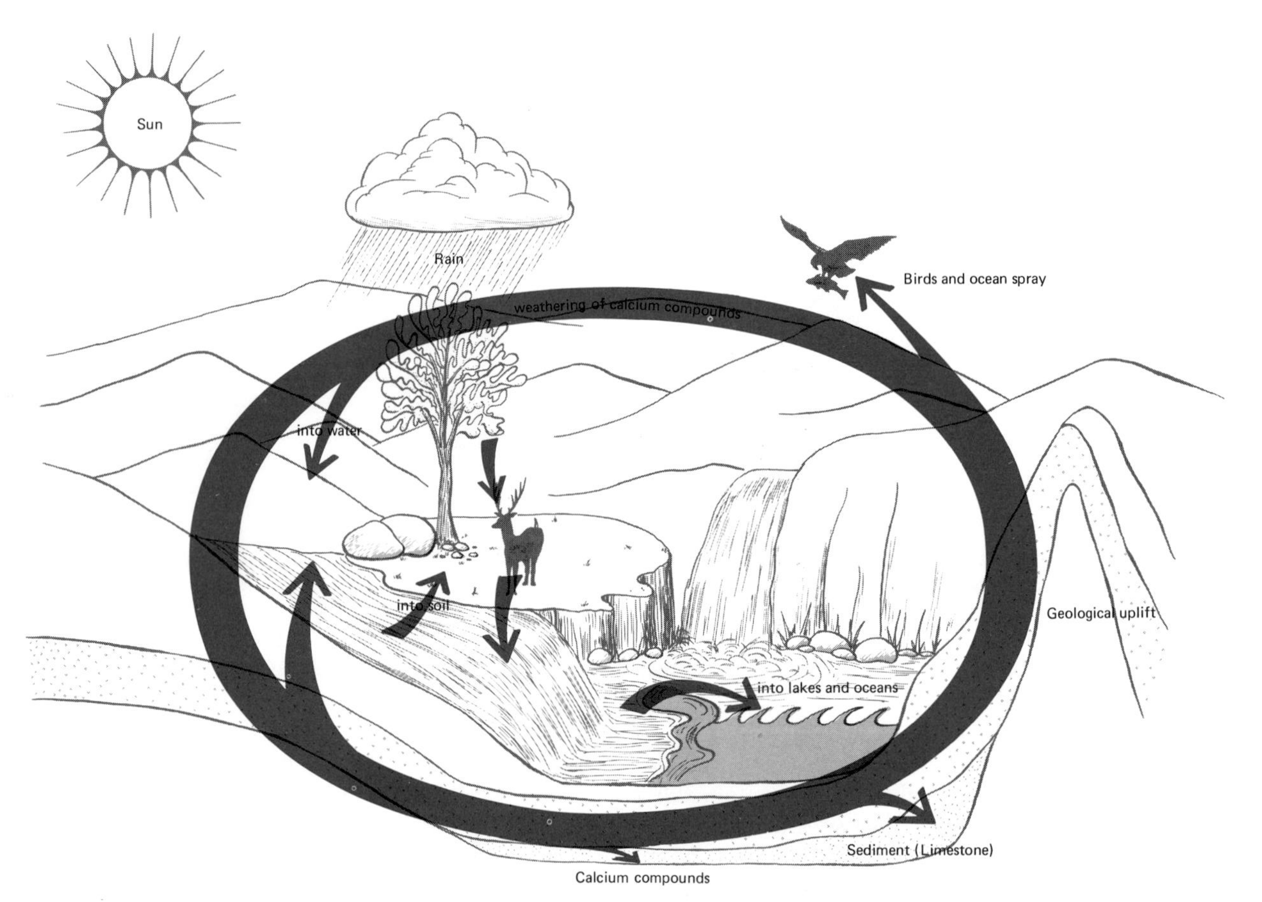

Figure 24-5. The calcium cycle. Like all sedimentary cycles, this cycle is called incomplete because much calcium is not readily recycled. It finds its way to the ocean, becomes incorporated in the body structure of marine organisms, and eventually forms part of the sediment (limestone). From there it can only be returned to the biosphere by geological upheaval. However, small amounts of calcium are returned to the biosphere for re-use when terrestrial organisms consume marine organisms or when ocean spray evaporates and forms minute dust crystals that may be carried to land by wind.

as waste. When the organism dies, the decomposers act upon the dead organic matter and free the calcium for re-use. However, one of the major problems with calcium, as with most minerals, is that it is washed away with runoff water or seeps deeply into the soil and becomes part of the ground water, both of which eventually find their way into the oceans. Here, the marine organisms absorb or consume the water in much the same manner as terrestrial organisms except that a large number of marine organisms such as coral, mollusks, and bivalves incorporate the calcium ions into a hard exoskeleton or shell of calcium carbonate. When these animals die, the shells become part of the sediment that eventually forms limestone. Some of the calcium from the limestone is released by the action of natural acids in the water, but it remains in the ocean. The only manner in which calcium is returned to the lithosphere to any great extent occurs when aquatic organisms are consumed by terrestrial organisms such as birds or when there is a geological upheaval. Trace amounts of calcium as well as other minerals are carried as ocean spray by air currents over coastal areas. When the ocean spray evaporates, minute crystals of these minerals are formed and may be carried considerable distances by wind currents.

The major consideration in the sedimentary cycles is that gaseous compounds are not formed and the substances may be removed from the cycle. Within a well-balanced ecosystem these substances are available for recycling through decomposition of excrement and dead organisms. Maintaining conditions within ecosystems to enhance the recycling of these substances and reduce leakage is a most important aspect of ecology.

Concentration of Minerals. Of all the naturally occurring elements, approximately fifteen are essential for the metabolic activities of living organisms. Since the amounts of required minerals will differ with various organisms, there will be selectivity in the quantity maintained. This results in a concentration of minerals at various points along the food chain. For example, phosphorus accounts for slightly more than 0.1% of the crust of the earth, and the percentage in plant tissues is about 0.2%. In humans, by contrast, phosphorus accounts for almost 1% of body weight, which is a ten-fold concentration over environmental levels.

Different species selectively concentrate different minerals for their specific use. It is essential that they have this ability in order to accumulate sufficient quantities of essential minerals for structural needs, for electrolytic balance, and for the synthesis of hormones and enzyme systems. The organism, in concentrating these minerals far above environmental levels, makes use of the process of active transport, by which ions are moved to the site of use through the expenditure of energy.

Materials such as calcium and phosphorus enter the food chain by being taken in and assimilated by the green plant. They are then converted and concentrated by subsequent organisms into insoluble compounds that make up parts of the skeletal system or a hard outer shell. Other elements, such as zinc, copper, magnesium, iron, and iodine, are also synthesized into large complex organic molecules. Some of these molecules are maintained within the body of the organism because they cannot pass through the cell membrane. Some smaller molecules and ions, which readily pass out of the cells, are reclaimed from the excretory system by various metabolic processes, the most important of which is active transport. The concentration of the various materials is of great importance within the ecosystem. In addition to certain essential elements that must be concentrated above environmental levels, there are products of modern society that although present in the environment in trace quantities, may also be absorbed and concentrated. Among these products, which may reach levels of toxicity within living matter, are radioactive wastes, pesticides, and other pollutants. When released into the environment, these substances may enter the various physical and biological cycles. Some of these toxic substances have chemical and physical properties similar to those found within many organisms. Thus, once absorbed, they may concentrate and reach levels that can greatly alter normal metabolic activities and eventually may result in death.

Energy Flow

The living world is characterized by dynamic change. Activities such as growth, maintenance, reproduction, and movement of individual organisms or the interactions within and between populations in the various ecosystems can only occur through the utilization of energy. In fact, ecosystems represent vast complexes that facilitate the transfer and transduction of energy (Figure 24-6).

The sun supplies an enormous amount of energy to the earth's surface each day. For example, in one day the energy striking the earth is equivalent to 684 billion tons of coal. However, most of this energy is reflected back into space or absorbed as heat. Of the light that strikes the leaf of a green plant, 98% is reflected and 2% is absorbed. Of the 2% absorbed, 1% or less is in the wavelengths utilized in the photosynthetic process; the rest is absorbed as heat. Thus, in terms of utilization of light from the sun the efficiency of green plants is 1% or less. An even larger proportion of sunlight is reflected from the surface of large bodies of water. Therefore the photosynthetic plankton are able to utilize even less than 1%.

Photosynthetic organisms produce far more high-energy compounds than are essential for their existence. Of the radiant energy captured by photosynthetic organisms, about one-sixth is used for their own maintenance, the other five-sixths being available for use by consumers. When an animal or other primary consumer eats plants, the potential energy in the plant body is transduced into thermal, mechanical, chemical, light, and even occasionally electrical energy to carry on life functions. Of the possible potential energy within the plant body, the primary consumer is able to utilize only about 50%. The other 50% is indigestible and passes through the body unused. In the utilization of the usable energy by the primary consumer, only about 5–20% is built up into protoplasm; the rest of the energy is used to carry on the life functions of the primary consumer or is lost to the environment as heat. The secondary consumer is able to utilize about 70% of the potential energy contained within the body of the primary consumer; the other 30% is passed off as waste. But, again, only 5–20% of the potential energy is converted into protoplasm that can be passed on to other consumers; the remainder is lost as heat. Through this chain the original radiant energy from the sun flows from organism to organism and in the process is converted into other usable forms of energy or is lost as heat. The transduction of energy is expressed in the First Law of Thermodynamics (energy may be converted from one form to another but can be neither created nor destroyed).

The high-energy organic compounds within living organisms provide a potential source of energy for life activities. As these compounds are converted to energy, there is less potential energy in reserve within the body of the organism and it must be replenished from an outside source. Heterotrophic organisms obtain new supplies by consuming other organisms. It would appear that the same amount of energy, although in different forms, could be passed continuously from organism to organism without loss. However, this does not occur. First of all, some of the energy is passed on to the nonliving world. This loss of free or available energy is simply stated by the second law of thermodynamics (every energy transformation results in a reduction of the usable or free energy).

Since usable energy in the sequence of transfer from organism to organism within the ecosystem is rapidly dissipated, the amount of energy available at each succeeding step along a food chain becomes progressively smaller. It is for this reason that food chains rarely contain more than four or five trophic levels or food niches and that numbers of organisms and biomass usually become smaller at each trophic level. For example, the number and weight of plants is usually far greater than the number and weight of herbivores. The number and weight of herbivores far exceeds the weight and number of secondary consumers, the carnivores. At each succeeding level there are fewer and fewer organisms, with the dominant predator being smallest in number and occupying the most precarious food niche.

Small reductions in any of the organisms at lower trophic levels will greatly reduce or even eliminate the number of organisms at the higher trophic levels. For example, in the tundra the exclusive prey of the Arctic fox is the lemming, a small rodent. The reproductive potential of the lemming is extremely high so that within three or four years the carrying capacity—the number of organisms or biomass a portion of the ecosystem can support—is exceeded. With the increase in the lemming population there is a correspondingly rapid increase in the number of Arctic foxes. However, when the carrying capacity of the ecosystem can no longer support the lemming population, large numbers suddenly die or migrate and in their migration throw themselves into the sea. This drop in the number of

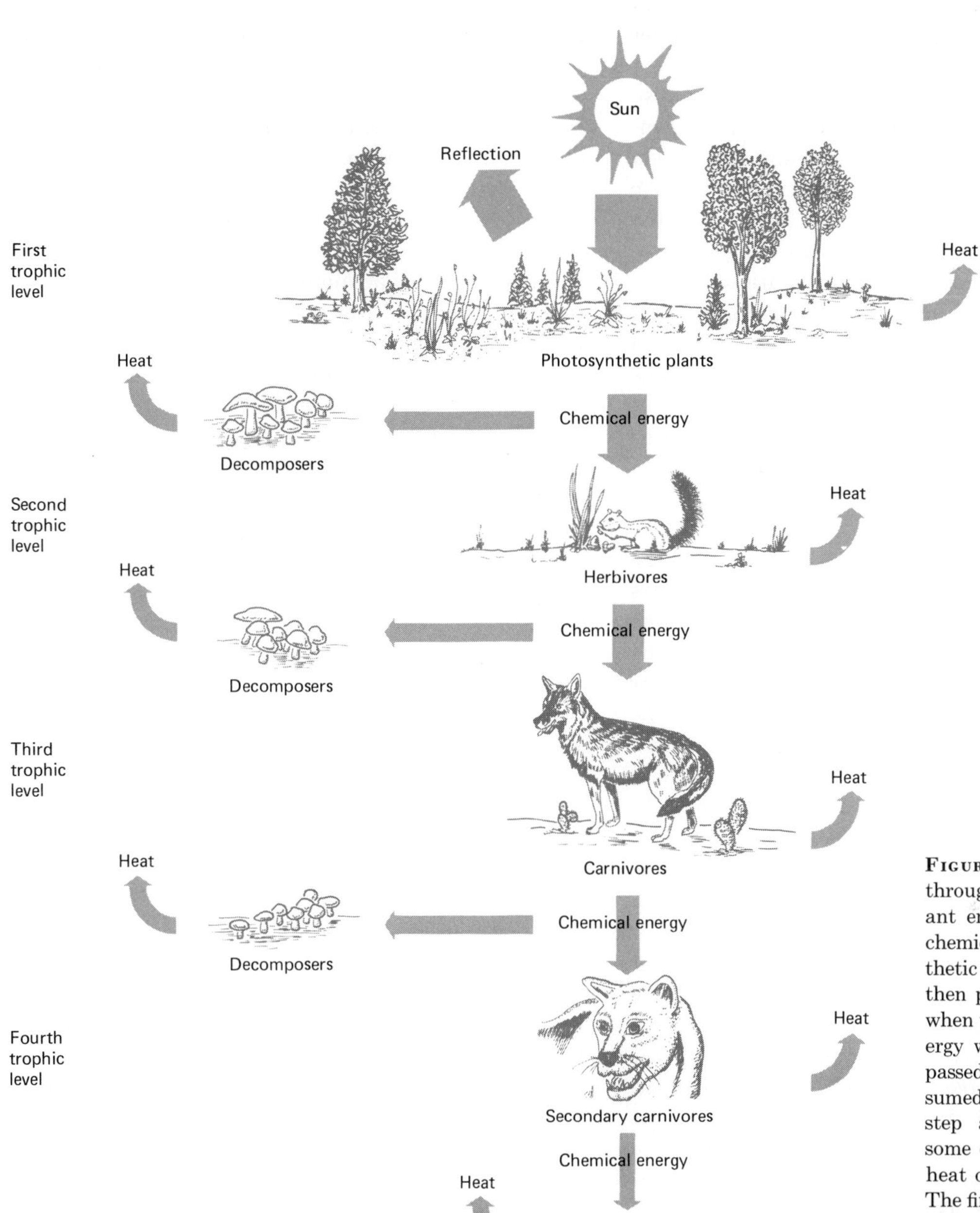

FIGURE 24-6. Energy flows through the ecosystem. Radiant energy is converted into chemical energy by photosynthetic plants. This energy is then passed on to herbivores when they eat plants. The energy within the herbivores is passed on when they are consumed by carnivores. At each step along the food chain some of the energy is lost as heat or used in maintenance. The final carnivores may only utilize a small fraction of the radiant energy that originally entered the system. The original radiant energy from the sun is able to support less and less of a biomass because of the loss of energy at each step. The decomposers represent the final steps in the chain.

lemmings is followed by a sharp decline in the number of Arctic foxes.

The fluctuations of the lemming population also affect other ecosystems. Another predator of the lemming is the snowy owl. As the lemming population decreases, the snowy owl moves farther south to the northern coniferous biome and there preys upon the snowshoe hare. The increase in predators in this ecosystem in turn greatly reduces the population of snowshoe hares. Thus, there are periodic increases and decreases in populations within several ecosystems as a result of a decrease in the population of one animal in a single ecosystem. These fluctuations are most apparent and prevalent in harsh environments where predators are dependent upon the few species present.

Food Chains and Trophic Levels

The movement of energy in the form of food from the producers to one consumer after another constitutes what is known as a **food chain.** At each step of the chain, as has been discussed, a large portion of the potential energy is lost. In fact, in most ecosystems about 80–90% of the energy of each step is lost. The consequence of this loss of energy is that rarely can a food chain be composed of more than four or five components. An example of a simple food chain would be grass ⟶ antelope ⟶ lion. Since at each step of a food chain large amounts of energy are lost, those chains that are the shortest provide the least loss of energy and the greatest production of biomass.

Food chains are of three general types. The *predator chain* begins with those animals that eat plants. These small animals are then eaten by other animals, which in turn may be eaten by still other animals. The *parasitic chain* begins with larger animals, usually herbivores, parasitized by smaller organisms, which in turn may be parasitized by other organisms. The *saprophytic chain* begins with dead plants and animals that are consumed by fungi and bacteria. The fungi and bacteria that die may be acted upon by other fungi and bacteria. This will usually continue until the organic compounds are finally converted back to inorganic compounds.

In reality, food chains are not separate and independent but are interlocked into what is called a **food web** (Figure 24-7). Within the web the consumer steps from organism to organism are referred to as **trophic levels.** For example, green plants, the primary producers, represent the first trophic level. Herbivores, animals that eat plants, are at the second trophic level; carnivores that prey upon herbivores are at the third trophic level; animals that prey upon the third trophic level animals are at the fourth trophic level; and so on.

The concept of trophic levels is relatively straightforward, and it has great implications within the food web of an ecosystem. The more complex the trophic structure, the more stable the ecosystem. That is, if there are a great many populations at the first trophic level (producers) and many at subsequent higher levels (consumers), the reduction or extinction of one population will have much less effect on the ecosystem. However, ecosystems with simple trophic structures are more vulnerable to disastrous ecological change. One of the simplest ecosystems, the tundra ecosystem of the Arctic, is extremely vulnerable to ecological change (Figure 24-8). If the populations of lichens, or any other component part of the system, were greatly reduced, the ecosystem would be destroyed. Since this ecosystem is so vast in size, its loss would have considerable effects upon other ecosystems more or less associated with it as well as the total biosphere. It is for this reason that conservation-oriented groups have viewed the development of petroleum reserves in tundra regions with such concern.

Ecological Pyramids

In the analysis of the trophic structure of ecosystems use is frequently made of graphical representations called **ecological pyramids.** There are three types of ecological pyramids. The first is the number pyramid. In this type of graph the number of organisms is calculated at each trophic level (Figure 24-9A, B). In a temperate forest there may be few producers, but they may be very large in size and capable of supporting a large number of consumers. Grasslands, on the other hand, are composed of a large number of small producers so that the number of producers in relationship to the number of consumers is exceedingly high (Figure 24-9A).

The second type of pyramid is that of biomass. This pyramid is based upon the weight of the total living material at each given trophic level, usually measured in grams per square meter (Figure 24-9C–E). Like

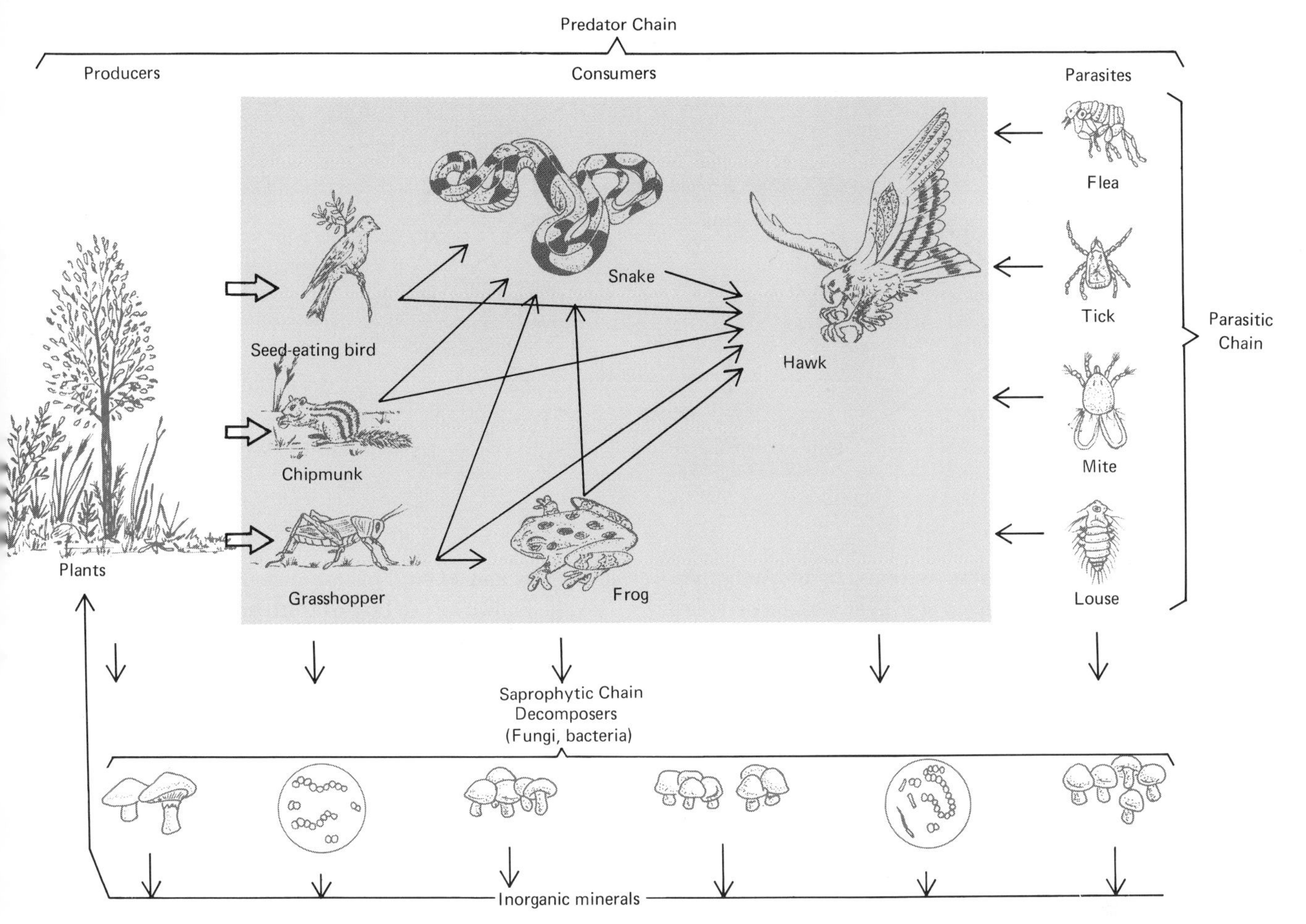

FIGURE 24-7. Diagram of a food web. The web is formed by an interaction of the predator chain, in which larger animals prey upon smaller animals, the parasitic chain, in which smaller animals prey upon larger animals, and the saprophytic chain, in which dead organic matter and waste products are decomposed. Each of the different chains is intimately woven into a web of interactions. In nature the webs are far more complex than this simple one. [*Drawn by R. Califre.*]

number pyramids, biomass pyramids can be analyzed to show important relationships among trophic levels. For example, a lower trophic level is able to maintain itself at a lower biomass than a higher trophic level (Figure 24-9C). If the biomass of this ecosystem is measured at any given moment a higher trophic level may have a higher biomass than a lower level. This indicates that the lower trophic level has a very high rate of reproduction and is able to replenish itself at a much higher rate.

The third type of pyramid is that of energy, which is usually measured and presented in kilocalories per square meter. This type of pyramid, unlike the number and biomass graphs, is always pyramidal in shape because there is always a decrease in available energy at each successively higher trophic level (Figure 24-9F, G).

Through the use of these three types of graphic representations much can be learned about the comparative efficiencies of various ecosystems and the values of biomass, numbers of organisms, and energy relationships. Ecological pyramids have a certain predictive

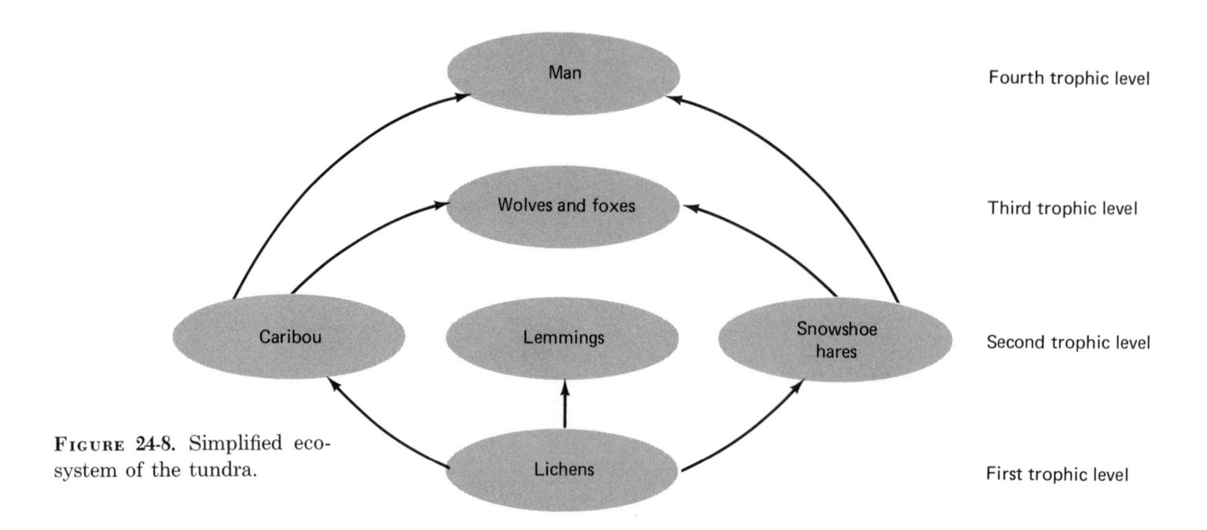

FIGURE 24-8. Simplified ecosystem of the tundra.

value for studies of new ecosystems. In addition, pyramid profiles taken of a given area over various periods of time can show the influence of both natural and artificial changes in the habitat.

Limiting Factors and Tolerance Levels

It is readily apparent that there is considerable variation among the species of flora and fauna in the different areas throughout the world. Less obvious are the differences among species in small geographical areas that are adjacent to one another, but differences do exist. In certain areas the number and kinds of species are relatively constant and have been for long periods of time. In other areas the numbers and kinds of species are undergoing changes. These conditions result from the differences in resources for life provided by the abiotic and biotic environments.

In 1840 Justus Liebig, while studying the effects of various nutrients upon plants, found that growth was often limited by some raw material required in minute quantities. He stated that the growth of a plant is limited by that nutrient which is found to be present in a minimum quantity. This has come to be known as *Liebig's law* of the minimum. Later, when they were applying Liebig's law to species found in their natural habitat, ecologists expanded it to include all factors necessary for life—moisture, temperature, nature of substrate, and so on. This expansion of Liebig's law has become known as the **law of limiting factors.** The law of limiting factors can be shown by experiments on the rate of photosynthetic activity by varying the amounts of CO_2, light intensity, and temperature. If light intensity is low, the rate of photosynthesis is low in spite of optimum temperature and sufficient quantities of CO_2. The rate is also low if CO_2 concentration is low, even if temperature and light intensity are optimal, or if temperature is low and there is high light intensity and CO_2 concentration. While in these experiments the low rate of photosynthetic activity was shown to be the result of any single limiting factor, in nature photosynthetic rate is dependent upon all three factors—CO_2 concentration, light intensity, and temperature—and any combination of limitations will affect the rate.

The specific requirements of any given population are determined by its inherited adaptations to the given environment in which it lives. There are many obvious explanations for the existence of certain species in a given habitat. Freshwater fish are physiologically adapted to an aquatic environment in which the mineral content is relatively low. By the same token saltwater fish are adapted to water with a high mineral content. Most species of marine fish do not have the physiological apparatus to withstand the osmotic pressure of fresh water, nor can freshwater fish survive in a marine environment.

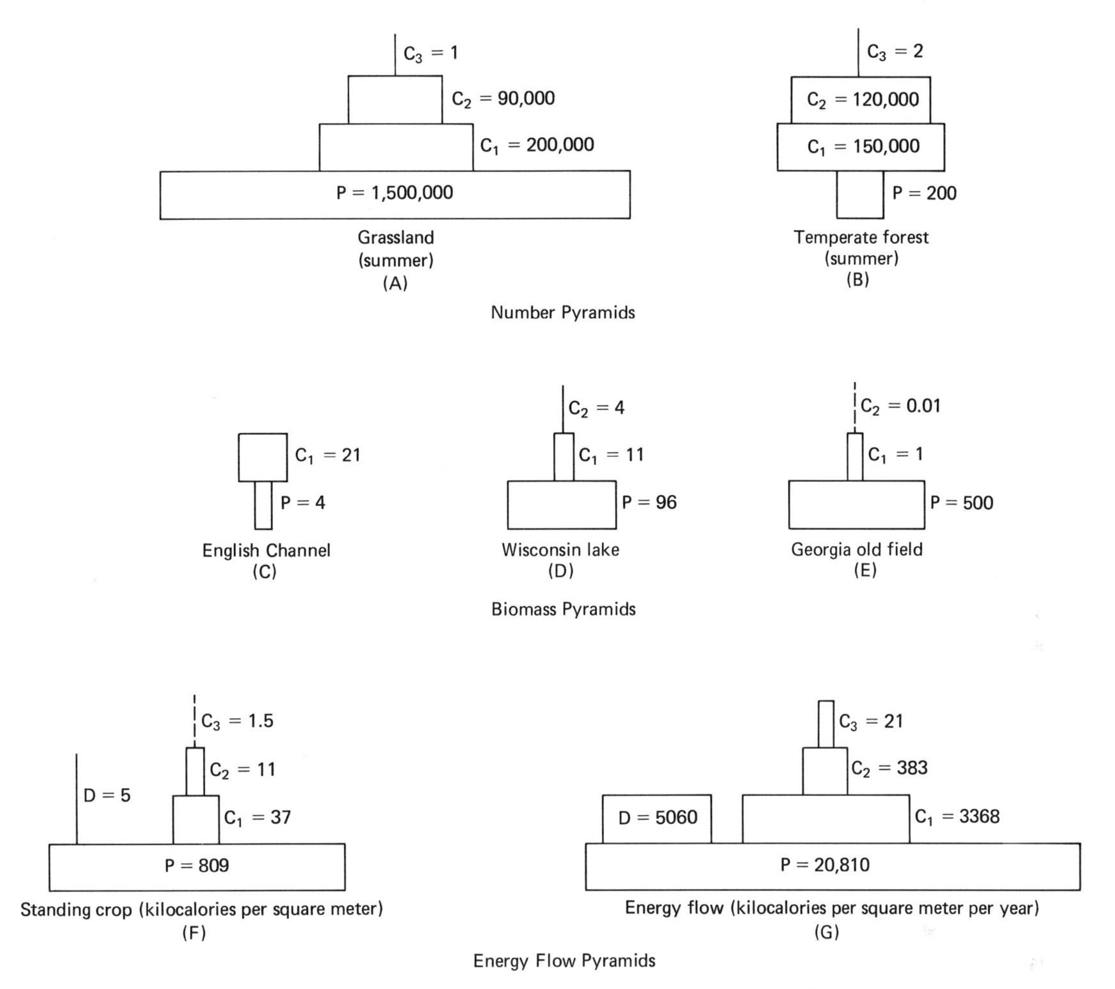

FIGURE 24-9. Ecological pyramids. (**A, B**) Number pyramids based upon individuals per 0.1 hectre exclusive of microorganisms and soil animals. (**C, E**) Biomass pyramids based upon grams of dry weight per square meter. (**F, G**) Energy flow pyramids based upon kilocalories per square meter per year. [*Modified from Eugene P. Odum,* Fundamentals of Ecology, *3rd ed. Philadelphia: W. B. Saunders Company. © 1971.*]

Given the limiting factor of a high mineral content of the marine environment, it has been found that certain species are able to tolerate great differences in salt concentration. Eels can readily move from an area of high mineral content to one of low mineral content and exist for long periods of time. Salmon spend most of their adult lives in the open ocean, but at spawning time seek out freshwater rivers in which to lay their eggs. Oysters, on the other hand, are extremely sensitive to slight changes in salinity. Changes in mineral content as a result of runoff of fresh water into a marine environment can considerably limit the oyster population. While species are limited by a specific factor in their environment, there is variability within species as to what constitutes an absolute minimum value that is tolerable.

About fifty years ago an American ecologist, Victor Shelford, set forth certain principles that today are known as Shelford's **law of tolerance.** Essentially, it states that the abundance of distribution of a species can be controlled by factors exceeding the maximum or minimum levels of tolerance for the organisms comprising the species. In other words, Shelford expanded Leibig's law to include both minimums and maximums of environmental factors. A population will tolerate set minimum and maximum values for any factor that the environment must provide. Tolerance limits to a given condition of the environment may control a population's existence throughout its total life span or for a given period during its life cycle. Tolerance levels are particularly narrow during the prereproductive period for certain species. The law of tolerance emphasizes the ecological requirements of climate, substrate, nutrients, and biological interactions within and among species. One of the principles inherent in the law are that organisms may have a wide range of tolerance for one factor in the environment and a narrow range for another. For example, in 1908 the Hungarian partridge was introduced into the United States from Europe as a game bird. The species became well established in Montana where the winter climate was much colder than the European habitat but did not become established in Missouri where the winter climate was about the same as in Europe. The Missouri summers were warmer than the European habitat and it was at first concluded that the birds could not tolerate the heat. However, it was later discovered that the rainfall in Missouri exceeded that of the European habitat during the nesting and fledgling periods. It was shown that the birds had a higher tolerance for variations in temperature but narrow tolerance for rainfall during the early part of their life.

Another principle derived from Shelford's law is that organisms with a wide range of tolerance for all factors will be widely distributed. It is for this reason that starlings, which were introduced into the United States in the 1800s, have spread throughout the whole North American continent.

Occasionally, when conditions of the environment are not optimum with respect to one factor, tolerance may decrease for another factor. It has been found that if nitrogen levels for certain species of grasses are low the water requirements are much higher to resist wilting. One the other hand, some species exist in habitats that provide a less-than-optimum level of one environmental factor because some other factor is limiting. For example, certain species of plants found growing in the filtered light of the forest floor will actually grow better in full sunlight, but they cannot tolerate the heating effect of full sunlight for extended periods of time. Understanding tolerance limits of the various species is becoming increasingly important, especially when assessing the changes that man has made in the environment.

Indicator Species

Indicator species are species that, because of their tolerance limits, can be used to evaluate an environment for certain environmental conditions. For example, the presence of saguaro, the giant tree cactus of Arizona and Northern Mexico, indicates that there are no prolonged periods below freezing. The presence of trees below the tundra indicates at least one month with a mean temperature of 50°F or higher.

Knowledge of the limiting factors of a large number of species makes it possible to determine the nature of a given environment from a survey of the species present. For example, if a given area, although currently quite dry, contains populations that are known to require a moist habitat, it is probably safe to assume that the area is generally marshy, at least for a portion of the year. Species that are used to determine particular characteristics of the environment are called **indicator species,** and they have become important in relation to environmental pollutants. Indicator species have proved to be most useful in determining the extent and source of pollution in waterways. Bacterial count of certain species along the intertidal zone of beaches has long been used to evaluate water quality for swimming as well as to identify sources of pollution. In addition, if the normal fish population of bass, trout, and other favorable species of a freshwater stream is decreasing and being replaced by carp, catfish, and other less favorable types, the water probably is becoming increasingly polluted. Extreme pollution is indicated by the absence of large species and the presence of various types of diatoms.

Both limiting factors and tolerance limits are important to all ecological concepts. Defining limiting factors and tolerance levels is essential to formulation of predictive theories in ecology, which are useful in assessing the impact of environmental alterations caused by humans. With the growth of human population and technology, understanding of the limiting factors of the various ecosystems and the tolerance levels of the various species will go a long way in providing solutions to man's effect on the biosphere.

Changes in Ecosystems

Ecosystems are dynamic phenomena of life. While they tend to maintain themselves in a steady state, they are subject to changes. These changes may be a result of natural phenomena such as long periods of drought, sharp changes in temperature patterns, natural catastrophes like fire and flood, and biological processes such as decomposition, deposition, and disease. Man-made changes also have had profound effects upon ecosystems, and these have extended to the whole biosphere, as discussed in Chapter 26.

Natural Changes

Succession. The series of changes in communities that occupy an area over a long period of time are called **succession.** The first organisms to occupy an area constitute a *pioneer community.* The types of flora and fauna of pioneer communities are determined by the abiotic environment. Once established, a community brings about changes in the environment such as addition of humus to soil, changes in pH, and increased water retention of the soil. Eventually, the total environment is altered to the point that another community can displace the pioneer community from the area. Community after community establishes itself and in turn is replaced until a climax community is established. The **climax community** is a community that will not be replaced by another unless there is a basic change in climate or land form.

The type of succession, the time span of its development, and the eventual climax community all vary with the abiotic and biotic conditions of the area. Although there are differences in successions of different areas, study has shown that they have basic underlying similarities.

The various phases that the ecosystem goes through are called *seral stages,* and the whole series of communities that develop under given environmental conditions is called a **sere.** When the ecosystem, through succession, has reached the stage where the dominant and subdominant species have culminated in a relatively steady state, the *ecological climax* has been reached. The climaxes within the various areas of the earth result in the biomes and biogeographical regions. Climax communities are perpetuated as long as local environmental conditions are not altered drastically. Naturally, if environmental conditions are seriously changed by climatic changes, geological changes, fire, flood, or human activities, the climax stage is destroyed.

Both terrestrial and aquatic ecosystems go through successional changes. Freshwater ponds and lakes with their shallow waters and their characteristic flora (Figure 24-10) show successional changes very clearly. As soil and silt are washed in and dead vegetation accumulates, the lake becomes shallower. The shallow water becomes a marshy region, with its characteristic community, and the movement of the shoreline progresses toward the center of the lake. As soil and humus build up, the marshy community is replaced by grasses and shrubs. These in turn may be replaced by trees. The succession ends with a climax of trees.

Lakes that are in an early or pioneer stage of succession are said to be successionally young and are called *oligotrophic lakes.* Those lakes that are in later seral stages are said to be aging and are called *eutrophic lakes.* Generally, oligotrophic lakes are clear, deep, and relatively infertile, while eutrophic lakes are murky, shallow, and very fertile. The process of the successional aging of lakes is called *eutrophication.*

Climate usually determines the basic type of succession in an area, but the modifications seen as succession unfolds are produced by the communities themselves. A community tends to alter the area that it inhabits in such a way as to make it less favorable for itself and more favorable for another type of community. For example, the accumulation of dead plant bodies in an area of vegetation fills in the shallow water zone, thus making it a less desirable environment for an aquatic community and more desirable for a terrestrial community.

Although the types of succession vary with local conditions, certain generalizations apply to all successions.

(1) Succession is a continuous process from pioneer community to climax. However, the earlier steps progress more rapidly than the later seral stages. The re-

(A)

Pioneer stage

Submerged vegetation sere

Emergent vegetation sere

Temporary pond and prairie sere

Climax forest stage

(B)

Figure 24-10. Lake succession. (**A**) The diagram shows the various seral stages from bare bottom to climax of deciduous trees. Each step depends on the preceding accumulation of soil matter. Eventually, there is little indication that a lake formerly existed in the area. (**B**) The various seral stages from open water to deciduous trees are shown in the photograph. Compare with the accompanying diagram. [*USDA photo by Freeman Neim.*]

placement of the marsh community occurs in a much shorter period of time than does the development of the forest sere.

(2) The number of species increases with each seral stage and then reaches a more or less stable number as the climax is approached. For example, in terrestrial succession the lichens are among the few species of organisms that can exist under the severe conditions of living on bare rocks (Figure 24-11A). The lichens, in their soil-building activities, provide sufficient substrate for mosses (Figure 24-11B). When these die, more soil accumulates in sufficient quantities for seed plants. The number of species increases greatly. However, as these communities are replaced by other communities, the number of species does not increase significantly and may, in fact, decrease (Figure 24-11C).

(3) There is a steady increase of nonliving organic matter and the biomass with the greatest amounts present in the climax stage (Figure 24-11D). Once the climax stage is reached, the numbers of species present and the synthesis and degradation of organic materials come into a near balance. If it were not for this relative equilibrium, the climax would bring about further changes and the ecosystem would go through additional seral stages.

(4) There is a steady increase in the complexity of the ecosystems toward a greater utilization of the available energy. The well-defined stratification of the tropical forest biome shows utilization of the greater amount of energy received in this region of the earth. The relatively simpler structure of the tundra biome is a result of the smaller amount of energy available in this region. However, both biomes represent the development of ecosystems that utilize the available energy to the maximum.

When succession begins in a sterile area such as bare rock or in an area not previously occupied by a similar community, or when a lake community is eventually replaced by a forest community, it is called *primary succession. Secondary succession* results when there are severe changes in climate or other factors such as fire, cultivation, and grazing that cause the ecosystem to revert to an earlier stage (Figure 24-12). Farm areas that have been cleared and then abandoned are examples of secondary succession. Secondary succession progresses more rapidly than primary succession because soils and physical conditions have been altered to a certain extent by previous communities and have not been completely eradicated. Hence, seral stages can develop more rapidly. The abandoned field quickly grows up in herbs and weeds. These are replaced by shrubs and tree seedlings. These pioneer trees grow well in unshaded areas, but will be replaced in later communities by shade-tolerant trees, with the most shade-tolerant species making up the climax stage. The seral stages have been repeated; thus, secondary succession has taken place.

The present biomes and biogeographical regions have existed for thousands of years. There is ample fossil evidence that there have been great changes in the world of living things over periods of time. During the ages, through evolution, various living forms have dominated for periods of time, only to be eventually replaced. Evolutionary changes have occurred on a far greater time scale than presently occurring successional changes. While evolutionary changes will affect all living things, the time span in which they occur is so great that it is impractical to consider them as successional changes. Only changes in the activities within the various ecosystems that affect the present and foreseeable future generations are considered in successional studies.

The fact that all ecosystems are not in the climax stage tends to show that changes can and do occur. Seral stages may be disturbed to the extent that they are completely destroyed, set back to an earlier stage, or merely retarded in developing into the next seral stage.

Climatic Changes. Generally, ecosystems are adapted to the normal climatic variations of a given area. It is only when there are drastic changes in the climatic pattern that the balance in the ecosystem is altered. A severe winter with excessive snowfall may prevent a large deer population from obtaining its normal food so that its members will strip bark from trees. The trees die, exposing the shaded ground cover to the summer sun and the eroding and leaching rains. The topsoil is lost, and the land can no longer support a large community. The whole ecosystem has been set back several seral stages. The ecosystem starts a secondary succession toward the climax stage.

Fire. Naturally occurring fires are caused by lightning and play a major part in the forest and grassland

(A)
(B)
(C)
(D)

Figure 24-12. Secondary succession. The forest was cleared to prepare the land for farming. After abandonment of the farm, the trees gradually returned. Eventually the trees will replace the cleared land. [*USDA photo.*]

biomes. Fires differ in their effects. For example, severe crown fires, that is, fires in the upper portion of the vegetation, generally completely destroy most of the organisms in a given area and the development of the ecosystem must begin over again (Figure 24-13). While succession occurs more rapidly, provided there are no other factors involved, it still takes many years to reestablish the forest biome. Light ground fires or fires in the litter and leaves are selective in their effect and exert a great influence upon the type of vegetation and their associative organisms. In grasslands, periodic burning prevents the development of more dominant species of shrubs which might eventually replace the grass and the grazing animals associated with the biome. The long-leaf pine forest of the southeastern United States has been maintained by fire. The terminal buds of the pine are fire resistant. Periodic ground fires destroy the young seedlings of deciduous trees which would eventually cause shade and prevent the development of the young pine seedlings. Thus, the selectivity of fire has maintained these forests in a subclimax stage of development and has provided a continuous supply of one of the world's best timber trees as well as the other organisms associated with the long-needle pine (Figure 24-14). The case of the long-needle pine does not apply to all ecosystems. In other areas, fire-resistant species may not be as desirable and may retard the development of a more productive ecosystem.

Deposition and Decomposition. The deposition of organic matter upon the surface of the earth in any given region is dependent upon the amount of vegetation and the rate at which the decomposers break it down. Generally, in regions of cold or where there are long periods of dry weather, decomposition does not proceed rapidly. In most warm regions decomposition

Figure 24-11 [opposite]. Terrestrial succession. (**A**) Lichens, the pioneer vegetation, fasten to the bare rock where they aid in the formation of soil. (**B**) Mosses, grasses, and herbs begin to replace the lichens. (**C**) As soil and organic matter continue to accumulate, the grasses and herbs are replaced by taller herbs and shrubs (foreground). (**D**) Finally, deciduous trees complete the succession. [**A** *USDA photo;* **B, C** *U.S. Forest Service photos;* **D** *Courtesy of D. C. Anderson, Canadian Forestry Service, Department of Fisheries and Forestry, Sault Ste. Marie, Ontario.*]

(A)

(B)

FIGURE 24-13. Results of a crown fire. (**A**) A crown fire. (**B**) The intense fire killed the trees and other vegetation and also destroyed much of the organic matter of the soil. It will take many years of succession before this ecosystem can again support climax forest vegetation. [*U.S. Forest Service photos.*]

proceeds rapidly and there is little accumulation of organic matter; the rate of deposition is about equal to the rate of decomposition in these areas.

The amount of undecayed organic matter can greatly affect the ecosystem in a given area. A deep accumulation of dry organic matter can prevent the development of seedlings. It also makes the area more susceptible to fire. The typical lake succession as described previously could not occur at all if large amounts of undecomposed organic matter did not accumulate. If through disease or export by runoff water or wind, sufficient organic materials are not returned to the soil, the area quickly becomes deficient in essential nutrients and cannot sustain the existing ecosystem.

Artificial Changes

The artificial changes that affect an ecosystem are essentially caused by man. These changes have affected practically all regions of the world and continue to do so on an ever-increasing scale. While these changes are generally made to benefit man in some way, they have had profound effects on the various aspects of the ecosystem. At times these changes have proved to be directly harmful to man rather than beneficial. It has been only relatively recently that man has begun to realize the effects of his changes upon the ecosystems. Within the limits of this text it is possible only to relate in a general way a few of the major changes that man has caused in existing ecosystems.

AGRICULTURE. The major purpose of agriculture is to provide man with his basic needs of food, shelter, and clothing. To achieve this goal, whether the enterprise is farming, grazing, or forestry, man in some way modifies and controls ecosystems in a manner which diverts the maximum biomass and energy to his immediate goals.

It is important to realize that natural ecosystems represent the development of an area to its fullest potential in the utilization of energy within the controlling physical factors of available minerals, water, and climate. The degree of long-term success of agriculture depends upon how closely it approximates the natural ecosystem with regard to the flora and fauna managed or raised. Most plants and animals are specifically adapted for the biome type in which they naturally occur, and are less productive or cannot even live outside their natural habitat. Domestic plants and animals have been derived from wild ancestors, and, like their ancestors, are most productive in the specific environment for which they are adapted. Large dairy cows cannot be maintained in the arid high altitude of the Andes where the llama and alpaca thrive, nor can oranges be grown in Canada or wheat in the everglades of Florida.

The long-term success of any agricultural venture depends upon the manner and degree to which the existing ecosystems are altered. History is full of examples of economic and social disasters in which the limitations presented by the biotic and abiotic environments or the simple dynamics of ecosystems were not recognized or heeded. For example, earlier in this century the settlers of the Southwest found the dry grasslands a fertile area for farming. On these vast treeless plains the accumulation of organic matter over the years provided a thick layer of topsoil. The grass was plowed under, and crops were planted and flourished. Then came the climatic cycle of drought, which was predictable, and the annual amount of rainfall diminished. Crops failed, and the bare soil, without the cover of drought-resistant grass, was exposed to the baking sun and eroding winds. This large area became known as the "dust bowl." Much of the area still remains barren; however, some has been reclaimed by modern methods of irrigation and cultivation (Figure 24-15).

The most successful use of such a grassland area is for grazing (Figure 24-16) because it most closely follows the typical ecosystem of the region. Grazing animals control the growth of larger shrubs and at the same time return some of the organic matter in the form of waste products. In other words, the cattle simply take over the niche vacated by the bison in the ecosystem.

In the farming and grazing of large land areas, many of the native plants and animals have been excluded from their natural habitat; in some instances they have been excluded to the point of, or near, extinction. As a result, the biogeographical regions represent a historical classification rather than an actual one. The bison no longer roam the Nearctic region, while the giraffe, zebra, and antelope are rapidly diminishing in the Ethiopian region and the kangaroo from the Australian region.

In recent years the widespread use of pesticides in agriculture has had a devastating effect on the local ecosystems as well as the total biosphere. Insects are an important link in the beginning of many food chains. A reduction in their number results in a reduction in the size of other populations of consumers that depend upon them for food. Most insecticides are not selective, and many desirable insects and other animals are affected by them indiscriminately. Those insects that are resistant or receive sublethal doses may retain within their bodies some of the insecticide. These are then eaten in great quantities by larger consumers. The amount of insecticide accumulated in these larger organisms may reach toxic levels. In this way, many desirable populations such as songbirds, fish, and even mammals are inadvertently harmed by insecticides.

Pesticides enter the air and water and are then circulated over long distances and wide areas. Therefore, the pesticides exert their effects not only upon the ecosystem in which they are applied but even on ecosystems far removed, where their accumulation might greatly affect the community composition of aquatic and terrestrial ecosystems.

One of the most-studied insecticides is DDT. DDT (dichlorodiphenyltrichloroethane) was developed during World War II. Its use saved millions of people from typhus carried by body lice and malaria carried by mosquitoes. Wide-scale programs of spraying and dusting proved to be so effective that the use of DDT was extended to control a large number of different insect pests. Not only has it been used extensively on food crops, but it has also been sprayed on large forest areas for the control of such insects as spruce budworm and the Gypsy moth and over large land and marshy areas to control mosquitoes.

As a result of extensive use, DDT residues are found in practically all areas of the earth, including areas that have never been treated with the insecticide. The reason for this lies in the properties of the DDT molecule. DDT is extremely stable and resists breakdown. When it does break down, many of the products are just as toxic as the pesticide. It has been found that DDT and its breakdown products remain in the environment for decades. The molecule has also proved to be quite mobile; it does not remain where applied, but is carried throughout the earth by air and water currents. While the molecule is only very slightly soluble in water, it is

(A) (B)

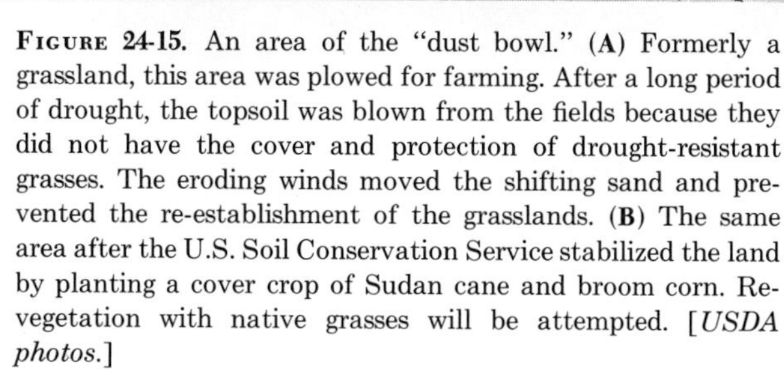

FIGURE 24-15. An area of the "dust bowl." (**A**) Formerly a grassland, this area was plowed for farming. After a long period of drought, the topsoil was blown from the fields because they did not have the cover and protection of drought-resistant grasses. The eroding winds moved the shifting sand and prevented the re-establishment of the grasslands. (**B**) The same area after the U.S. Soil Conservation Service stabilized the land by planting a cover crop of Sudan cane and broom corn. Revegetation with native grasses will be attempted. [*USDA photos.*]

soluble in lipids (fatty tissue of living organisms). As algae and other aquatic organisms absorb the substance, it concentrates in fatty tissue. This causes the insecticide to be accumulated in living organisms rather than being "lost" in the abiotic environment, and its removal from the water allows more to be dissolved.

DDT is toxic to higher animals because it is a hepatic enzyme inducer. This means that it causes the liver to overproduce enzymes. These enzymes modify the steroid sex hormones, thereby affecting reproductive processes.

Insects are quickly affected by small quantities of DDT because of their small body size and relatively short life cycles. Larger animals accumulate larger quantities, and the effects are more subtle because of the longer and more complex life cycles. Although initially present in the environment in small and apparently harmless quantities, as DDT passes through the food chain, concentrations increase to toxic levels. For example, algae may contain up to 2 parts per million by weight of DDT. Small herbivores such as fish, insects, and birds accumulate DDT in their fatty tissue as they eat such plants. These animals are in turn preyed upon by secondary consumers who further concentrate the insecticide within their bodies. The concentration increases progressively up the food chain to levels of 1000

FIGURE 24-14 [OPPOSITE]. A long-leaf pine forest in Georgia. This forest ecosystem has been maintained by ground fires. Periodic ground fires prevent the growth of shrubs and deciduous trees. The terminal buds of the long-leaf pine are fire resistant and are not injured by ground fires. The grass growing below the trees plays an important part in facilitating fire. [*U.S. Forest Service photo.*]

Figure 24-16. Grazing cattle of the grassland. This area was once the habitat of bison and other grassland animals. The native fauna have been replaced by domesticated cattle. Man prevents use of the area by any animals except those that suit his purpose. [*USDA photo.*]

or 2000 parts per million. Eventually toxic levels are reached. Most seriously affected are those consumers at the end of long food chains such as carnivorous fish and birds. One of the most apparent effects of the insecticide is found in the decrease in the populations of fish-eating birds such as the osprey, bald eagle, and brown pelican. In birds, estrogen, a steroid sex hormone, controls calcium metabolism. Female birds of prey concentrate DDT to levels where estrogen production is reduced and insufficient calcium is metabolized, resulting in thin-shelled eggs. Such eggs are either easily broken in the nest or permit water to evaporate so rapidly that the embryo becomes dehydrated and dies. Such bird populations are reduced by low reproductive successes even though the doses of DDT existing in adult individuals of these species are sublethal.

DDT is found in the bodies of penguins in the Antarctic and in seals of the Arctic, demonstrating the mobility of the substance. Airborne molecules or molecules carried off in runoff water reach the ocean where they are picked up by currents. Plankton absorb some of the DDT. The plankton in turn are consumed by large schools of fish, which travel over large distances and eventually reach the feeding grounds of seals and penguins. This evidence indicates that it is not possible to limit such long-lasting pollutants to any given area. At each level of the food chain the pollutant has adverse effects either directly or indirectly upon the ecosystem.

DDT has been a great help to mankind. It has provided an effective means for controlling insects and their adverse effects. There has been a considerable reduction of many insect-carried diseases. Food production has been greatly increased through the use of this insecticide. However, the effects of DDT upon the environment have pointed to the necessity of using all insecticides and herbicides with extreme care. At the present time the United States government has placed severe restrictions upon the use of DDT.

The ever-increasing output of agricultural products also creates an environmental problem in that it removes large quantities of minerals from the soil. While some of these may be replenished by fertilizers, much of the mineral matter is permanently lost in the incomplete mineral cycles. Thus, there is a decreasing fertility with a resultant decrease in productivity. Through the indiscriminate removal of large forests for lumber and the clearing of land for farming and grazing the protective cover of the soil has been removed, exposing it to the eroding effects of the wind and water (Figure 24-17). In recent years there has been an effort to control erosion, but the damage that has been done in the past has permanently altered many ecosystems.

Even the efforts to protect desirable wildlife have occasionally created havoc because of a lack of understanding. When natural predators are reduced or eliminated, the natural populations of desirable wildlife increase beyond the carrying capacity of the habitat. Mass starvation usually follows. It may then take years for the food supply and wildlife species to re-establish former levels.

Another way in which man has changed ecosystems is through the introduction of an alien species in a favorable environment where there are no natural enemies. In Australia rabbits were introduced to provide food and pelts. Within a relatively short time the rabbit population, without any predators, increased to such great numbers that the grasslands were threatened not only for domestic animals but for the native fauna as well. At the turn of the century the water hyacinth was brought to the United States from Japan to be used as an ornamental flower. It soon found its way into rivers and streams. Since it has no insect parasites or diseases in the United States, it grows to such an extent that it clogs rivers and streams (Figure 24-18). It is changing many aquatic ecosystems and making the waterways less navigable. At the present time the only way to control this plant is to remove it physically, a most difficult and expensive task at its present rate of growth.

Urbanization and Industrialization. As

Figure 24-17. The results of overgrazing. The fence has prevented cattle from grazing in the grassland to the right. To the left, overgrazing has caused deterioration of the grassland. It is reduced to a minimum cover of snakeweed and sagebrush. The area on the left is more subject to erosion. [*USDA photo.*]

Figure 24-18. Water hyacinth growing in Louisiana. The water hyacinth is an alien species that now thrives in the water areas of the southern United States. In the absence of natural enemies, the plant has grown and spread so rapidly that it blocks the waterways and alters the ecosystems of the regions. So far, various means of eradication or control have not been successful. The photo shows a stream completely overgrown with the plant. [*Courtesy of Louisiana Wild Life and Fisheries Commission.*]

the human population increases, greater demands are placed upon both the abiotic and biotic environments. Large areas of the earth are converted to the exclusive use of humans. With improved methods of agriculture, large farms can be maintained by fewer workers. This has resulted in a movement of the human population from rural to urban areas. These urban regions are ever expanding. Large land areas are cleared for new highways, housing, and factories to support the growing population. The waste products, both household and industrial, accumulate in such quantities that removal is a major problem. Mountains of garbage accumulate around cities, and the hydrosphere and atmosphere become polluted with the wastes. While much can be done to control the disposal of wastes, the damaging effects, even if the wastes are disposed of properly, cannot be entirely eliminated and the results are permanent changes in the existing ecosystems.

These changes are not limited to the ecosystem in which the disruption initially occurs but are transferred to other ecosystems that may be a great distance away. For example, major littoral and intertidal ecosystems of coastal areas have been destroyed by pollution of river waters that has its source many miles upstream. Liquid wastes dumped into rivers and streams soon find their way into estuaries and the oceans, resulting in pollution. Solid wastes used as fill can destroy the usefulness of marshland areas for spawning and nesting and for the production of certain materials needed by the biotic community (Figure 24-19). This results in large-scale loss of important fish and wildlife habitats. In a ten-year period (1955–64), over 45,000 acres of marshland were destroyed along the coast from Maine to Delaware. The loss of marshland areas affects not only the wildlife but people as well. With the loss of 240 square miles of marshland around San Francisco Bay, there was an equal loss in oyster bottom area in the bay, the soft-shell clams have vanished, and the amount of Bay shrimp has decreased from a harvest of 6½ million pounds in 1936 to 10,000 pounds in 1966, thereby greatly diminishing a human food source.

Increasing human population has caused the growth

Figure 24-19. Assorted solid wastes used as marshland fill. While these areas provide additional land for use by man, filling in destroys the marshland community. Feeding, nesting, and spawning sites for such organisms as fish, amphibians, and birds are destroyed. There is also the possible pollution of associated water areas through seepage. [*USDA photo.*]

of large sprawling urban centers in which thousands of square miles of land are removed from natural production merely to provide sites for homes, schools, stores, or factories. The ever-increasing mileage of paved sidewalks, streets, highways, parking lots, and airport runways has not only contributed to this loss but also opened up previously inaccessible areas with the inevitable effects upon the natural ecosystems.

Increased human population creates greater demands for agricultural and industrial products. This results in the destruction of more natural ecosystems to make way for the production of agricultural and industrial products. Modern technology has been admirably successful in developing new products and new processing methods, but industry has not kept pace in respect to controlling the emissions and effluents of industrial processes. These have caused tremendous and usually entirely unplanned and unwanted effects upon ecosystems. These effects range from removing the soil down to rock by strip mining to the destruction of life in a stream by the dumping of hot, though pure, water. Even climatic conditions in some areas have been changed as a result of smokestack emissions from distant factories. Whatever the cause, and whatever the effect, the wastes of industrialization and human population alter existing conditions and in so doing alter existing ecosystems. The flora and fauna that are not able to adjust to the changed conditions cease to exist.

Conservation

In becoming the dominant species in terms of effect upon ecosystems, human beings have put themselves in the unenviable position of not being able to avoid changing the ecosystems upon which they depend for life. Although many of these changes are harmful, not all human activities are detrimental. Production in arid and semiarid regions has been increased by irrigation, and in infertile areas through the use of fertilizers. Great effort has been made to preserve natural areas in their native condition. In many instances where human activities of necessity interfere with nature, attempts have been made to lessen these effects, as illustrated by improved farming, grazing, and forestry practices. Conservation programs are becoming more common practice. With the growing awareness of man's effect upon nature has come a widening search for better control of industrial waste products and for methods of land use that are more harmonious with nature.

Balance Within the Ecosystem

Just as there are control mechanisms at the molecular, cellular, and organismic levels of biological organization, so also are there controls at the population, community, and ecosystem levels. Thus, despite constant changes, there exists a set of checks and balance that serves to maintain a dynamic equilibrium at these levels, too.

Changes that occur in ecosystems over a geological time span are met by evolutionary changes resulting from natural selection. Those changes that occur naturally as a result of normal succession are met by changes in species makeup of the communities involved. The ecosystem always makes the most efficient use of the available energy within the limitations of the physical environment. Changes also occur over a shorter time span that cannot be considered a part of the normal successional changes. Whether naturally or artificially caused, they are met by adjustment of the species present or the creation and filling of new ecological niches by species that are new to the ecosystem. At the population and community levels of organization, controls result from the many intraspecific and interspecific relationships that exist. Basically, the ecosystem and its component units, the population, and the community represent a constant interaction between the biotic and abiotic factors, the sum total of which gives design and organization to the biosphere.

25 Types of Ecosystems

GUIDE QUESTIONS

1 What are the different types of aquatic ecosystems and terrestrial ecosystems? Give examples of producers, consumers, and decomposers in each.
2 How are the various producers, consumers, and decomposers of each of the various ecosystems interrelated? How are they adapted to the specific roles they play?
3 How are the various abiotic and biotic components of the various ecosystems interrelated?
4 What specific environmental conditions determine the fauna and flora of a region?
5 What are the six major biomes, what are their characteristics, and where are they located?
6 What are the six major biogeographical regions, what are their characteristics, and where are they located?
7 What are the advantages and disadvantages of dividing the biosphere into biomes and biogeographical regions?

THE purpose of this chapter is to set out a framework or foundation for the study of specific ecosystems and to provide a basis for understanding research and literature on ecology, but it is beyond the scope of this text to deal intensively with specific ecosystems. Previous chapters have presented some of the fundamentals of ecology; this chapter will analyze generally the various types of ecosystems and then present a description of the major floral and faunal regions of the earth.

Ecosystems are of two major types: aquatic and terrestrial. Within these generalized types there are further subdivisions, according to specific habitats. The basic principles of the generalized categories can be applied when studying the ecosystems of specific localities.

Aquatic Ecosystems and Their Components

The physical aspects of the aquatic environment were discussed in Chapter 22. The aquatic environment consists of fresh waters, which are categorized as lentic (standing water) or lotic (running water), and the marine environment, which is subdivided into the ocean and inland marine waters.

Standing Freshwater Ecosystems

The major differences between pond and lake ecosystems result from the differences in size and depth. Along the edge of a pond or lake is a shallow-water region called the littoral zone where light will penetrate to the bottom. Beyond this is the open water region, which is divided vertically into an upper limnetic zone, the depth to which there is effective light, and a lower profundal zone, the bottom and deep-water area into which light cannot penetrate. The last two zones may or may not be present in ponds, depending upon the depth and turbidity of the water (Figure 25-1).

LITTORAL PRODUCERS. The producers of the shallow-water zone are represented by many types of plant life. In the shallowest part of the zone is emergent vegetation—the roots and lower portion of the plants are submerged, while the leaves and stems are above the water—predominantly cattails, bullrushes, and other reed-like plants. These plants form an important link between the terrestrial and aquatic ecosystems. They provide food, shelter, and protection for animals such as water snakes, muskrats, and beavers. In addition, they provide ready access to the water for smaller animals and insects.

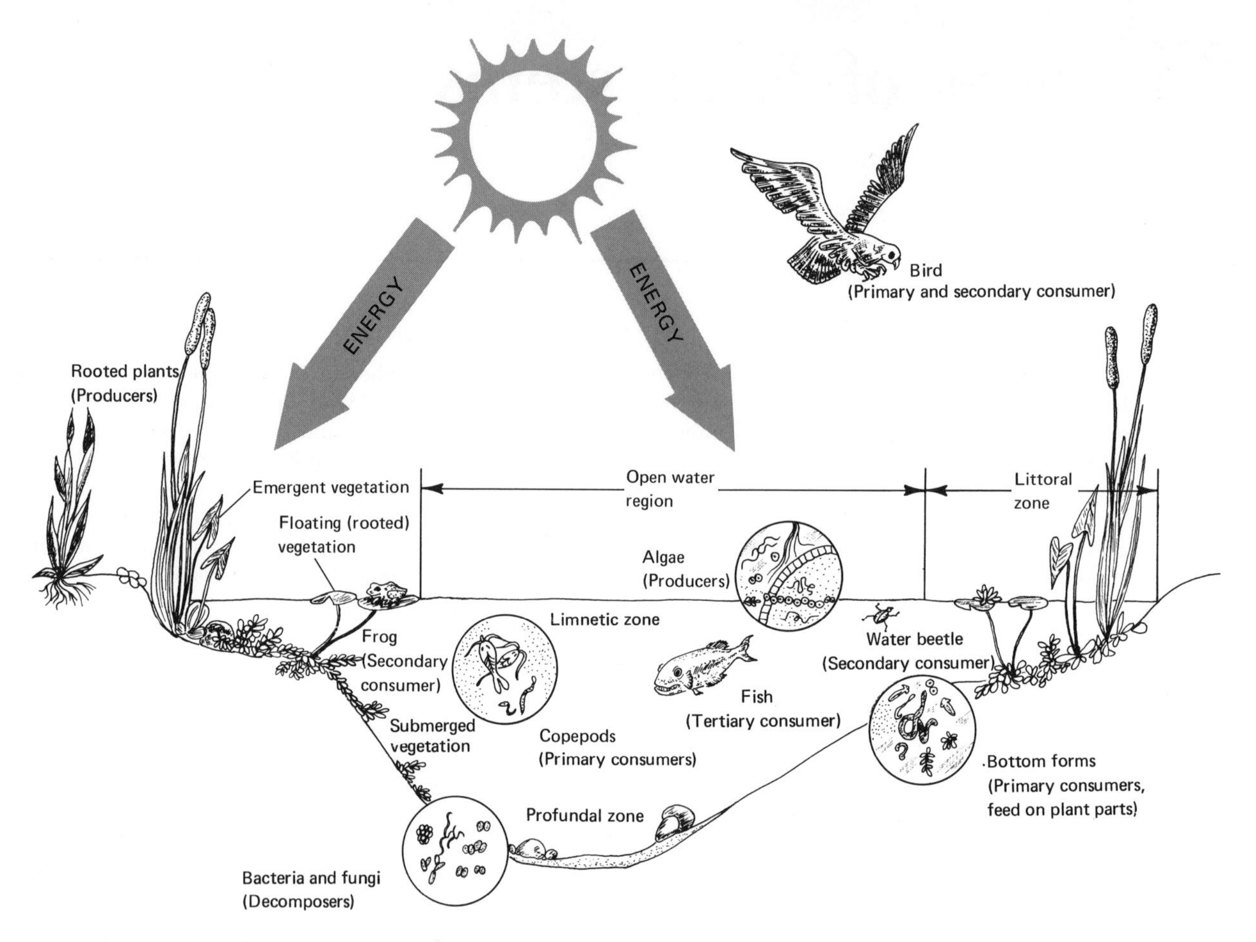

FIGURE 25-1. Standing freshwater ecosystem. The producers are phytoplankton (algae) and rooted plants. Plants may be emergent or submerged with floating leaves; algae, submerged or free-floating. Producers are found in the littoral and limnetic zones where radiant energy from the sun is available for photosynthesis. The consumers, which are found in the littoral, limnetic, and profundal zones, include such diverse organisms as protozoans, crustaceans, insects, fish, frogs, and birds. The decomposers on the bottom are bacteria and fungi. Within a stable ecosystem there is a balance among the three groups. [*Drawn by R. Califre.*]

Beyond the zone of emergent vegetation is the zone of rooted plants with floating leaves, of which pond lilies are the most common forms. These plants also provide food, shelter, and protection for smaller animals such as frogs, fish, and insects.

The zone of submerged vegetation is made up of plants completely submerged. The leaves of these plants tend to be thin to allow easy nutrient exchange with the water environment. This vegetation includes algae as well as the higher plants. Many animals have found their ecological niche within the submerged vegetation. It is an area in which many fish and amphibians such as frogs and salamanders spend their preadult lives.

The nonrooted producers of the littoral zone include the blue-green algae, diatoms, and green algae. The blue-green algae perform much the same function as the nitrogen-fixing bacteria of the soil. All of the producers contribute oxygen and organic nutrients to the other components of the freshwater ecosystem.

LITTORAL CONSUMERS. The littoral zone of

standing freshwater ecosystems contains more species than any of the other zones. This probably occurs because of the many types of subhabitats that are present. In shallow water a consumer might live on or around the vegetation or it might live on or under the bottom. In addition, free-swimming organisms find a suitable habitat here, as do those organisms that live on the surface of the water. Each of these subhabitats is populated with a distinctive fauna. Among the primary consumers located on the stems and leaves of plants are snails and various insect larvae that are found only here. In the same location are secondary consumers such as dragonfly and damsel fly larvae, which live by preying upon the herbivores. On and immediately under the bottom are both primary and secondary consumers such as snails, worms, insect larvae, and crayfish. The swimming organisms of this zone and those that live on the surface also include both primary and secondary consumers.

Amphibians are limited almost exclusively to the littoral zone, living as primary consumers during larval stages and moving to secondary or even tertiary consumers as adults. Water snakes, turtles, some species of waterfowl, and even mammals such as the muskrat can be considered to be a part of the consumer component of littoral zone ecosystems.

While some species of fish are permanent residents of this zone, practically all invade the littoral regions to spawn and feed. These larger fish, although invaders, represent the end of the food chain.

Zooplankton, small inconspicuous animals that float freely in the water, are found in this zone in large numbers. Among the species of zooplankton are found primary consumers, and in spite of their microscopic size, secondary consumers. The role of the zooplankton in a shallow-water ecosystem, while considerable, is not nearly as important here as in open ocean systems where the lower trophic levels are occupied almost exclusively by these minute animals.

Littoral Decomposers. The decomposers are generally the least obvious components of any ecosystem. As the dead organic matter of plants and animals settles to the bottom, various bacteria and fungi break it down into inorganic matter. This inorganic material can then be reused. Decomposition is generally slow in comparison to growth in water or moist habitats so that frequently there is great accumulation of materials. Peat bogs are the result of this type of accumulation.

Limnetic Producers. The limnetic zone is located in open water above the profundal zone. Since sunlight does not penetrate to the bottom, all production must be carried out by plants that float freely in the water. In other words, production in the limnetic zone is limited to phytoplankton. Among the phytoplankton found here are blue-green algae, various types of diatoms, and both unicellular and filamentous Protista and green algae. The phytoplankton goes through annual increases and decreases in population size. The increases are called *blooms* and usually correspond to the spring and fall overturns which were discussed in Chapter 22.

Limnetic Consumers. The consumers of the limnetic zone are essentially zooplankton and fish. There are relatively few species of these consumers, although numbers of zooplankton may be great at certain times of the year.

The number of freshwater fish feeding upon zooplankton is relatively small because of the seasonal fluctuation in phytoplankton levels. Therefore, the fish at the secondary and tertiary trophic levels are also limited. It is for this reason that the large schools of fish frequently found in the oceans are rarely present in lakes and ponds.

Profundal Zone Components. Since there is no light reaching the profundal zone, there are no producers; food must come from the littoral or limnetic zone. The constituents of the deep-water profundal zone are decomposers, saprophytes, and carnivores of the secondary or tertiary trophic level. The decomposers are various species of bacteria and fungi. The consumers are blood worms, annelids, and certain species of clams and animal larvae. Organisms existing in this zone are capable of withstanding low concentrations of oxygen and the absence of light.

Flowing Water Ecosystems

The major differences between standing and flowing water ecosystems are the currents or movement of water, the close association with surrounding land areas, the constant high oxygen content, and small thermal and chemical stratification. The rapidity with which the flowing water moves will determine to a considerable extent the existing ecological communities. The communities present will differ between areas where a stream moves rapidly and where a stream pools or is slow moving. The stones and hard bottom of a rapid-water zone offer surfaces upon which plants and animals may attach. In rapidly flowing water the plants are generally submerged and are chiefly made up of various species of algae or mosses. It is remarkable to see the dense growth of these plants covering the rocks in the pounding waters of rapids and waterfalls. The shifting sandy bottoms of slowly moving rivers and streams do not provide adequate anchorage for plants, and this area is generally limited to burrowing aquatic animals such as worms, clams, and insect larvae. Where the water forms pools and the silt settles to the bottom, there is less plant and animal life. Plankton may multiply but are not an important part of the ecosystem in the flowing water of a stream. While the number of producers in flowing water is small, the close association of land areas and the dense growth surrounding most streams provides a source of food. Thus, surrounding land areas are an important factor in the ecosystems of flowing water.

Horizontal zonation tends to occur along the length of rivers and streams, in contrast to the vertical zonation of standing water. This zonation starts at the headwaters as the rapid zone and progresses downstream into a series of pool zones. As the stream gets larger and slower moving, the size of the organisms within the ecosystems gets larger. This does not necessarily mean that there is greater biomass, that is, the total weight of all living organisms in an ecosystem. In fact, as the size of the organisms increases, their numbers decrease so that the biomass tends to remain almost the same.

Ocean Ecosystems

It has been only recently that the ecosystems of the ocean have been studied to any great extent. Most of these studies have been limited to coastal waters, and relatively little is known of the ocean at great depths. In addition, the great variability in environmental conditions results in a large and variable number of different ecosystems. Aside from the vast area, many regions are relatively inaccessible for study with present technology. Intensive study requires satisfactory underwater craft as well as underwater laboratories.

Within the scope of this text it is possible only to generalize about the various ecosystems of the oceans. The various marine ecosystems are classified according to the various zones of the ocean described in Chapter 22 (see page 570). The marine ecosystems include the intertidal zone, which is located between high and low tides; the littoral zone, which is the photosynthetic area above the continental shelf and is part of the neritic zone, which covers the total area above the continental shelf; the pelagic zone, which is the region above the continental slope; and the ocean zone and abyssal plain, which are areas of the open ocean beyond the continental slope. There is also vertical zonation in the open ocean area based upon the depth of penetration of sunlight. The upper light-penetrated zone is called the photic zone, and the zone into which light does not penetrate is the aphotic zone (Figure 25-2).

INTERTIDAL ECOSYSTEMS. Within the intertidal and subtidal zones there is considerable variation in abiotic conditions within short vertical distances. For example, certain areas are exposed to air for relatively long periods of time, while some areas are exposed rarely, if ever. Certain areas are subjected to strong wave movements; other areas are relatively calm. The salinity, oxygen content, temperature, and quality of light received will also differ over very short distances. This variation in environmental conditions has resulted in a clearly defined vertical stratification of different species of plants and animals that are adapted to the specific environmental conditions of the area. For example, the producers of the intertidal zone are certain green and brown algae that are capable of enduring exposure to air; in fact, exposure is essential in their life cycles. Associated with these algae are various types of consumers such as snails, barnacles, and crustaceans that are also capable of tolerating exposure to air. At the lower depths will be found red algae, which are adapted to the diffused light at this level.

NERITIC ECOSYSTEMS. ***Neritic and Littoral Producers.*** Although there are some plants attached to the bottom, the most important producers in the neritic zone, as in all parts of the open ocean, are plankton. The *phytoplankton* of the oceans is the largest group of photosynthesizers on earth. Directly or indi-

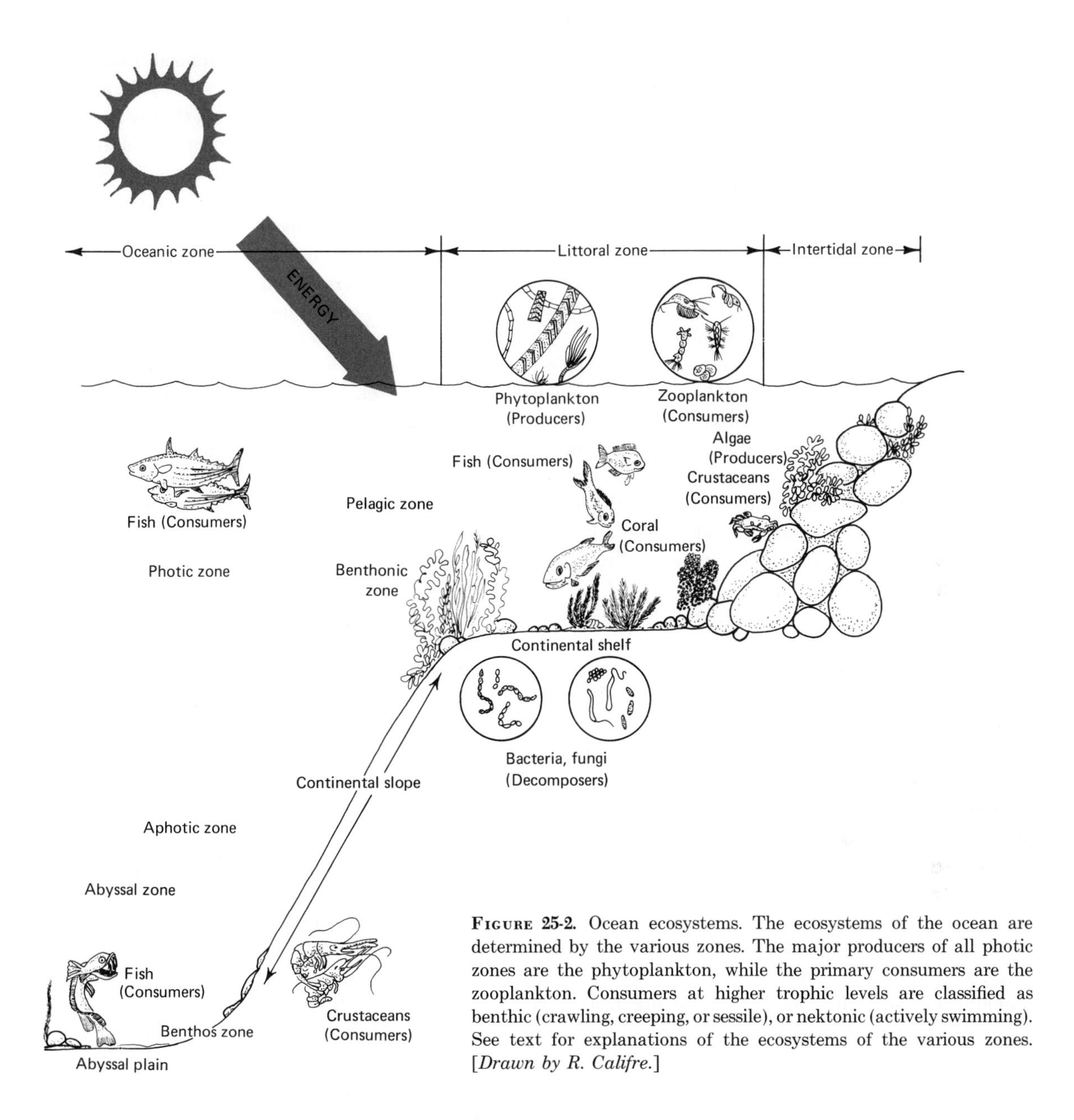

FIGURE 25-2. Ocean ecosystems. The ecosystems of the ocean are determined by the various zones. The major producers of all photic zones are the phytoplankton, while the primary consumers are the zooplankton. Consumers at higher trophic levels are classified as benthic (crawling, creeping, or sessile), or nektonic (actively swimming). See text for explanations of the ecosystems of the various zones. [*Drawn by R. Califre.*]

rectly, all ocean life is dependent upon the phytoplankton. The most abundant forms are the single-celled diatoms and dinoflagellates.

Seemingly, the surface waters provide an ideal habitat for phytoplankton. Bathed in sunlight and surrounded by minerals and water, these organisms are afforded a relatively stable environment; yet plankton are not evenly distributed throughout the oceans. The uneven distribution of phytoplankton is related to the abundance or scarcity of certain raw materials or nutrients. In cooler water where there is vertical mixing and in those areas where there are upwellings there is a constant replenishment of nutrients so that maximum populations can be supported. In warmer waters layering occurs with relatively little mixing; the net result is an exhaustion of the supply of nutrients and a decrease in the abundance of phytoplankton and the consumers dependent upon them.

In the more shallow parts of the neritic zone, and in the intertidal zone, the producers do not necessarily have to remain afloat in order to keep within the zone of light. In fact, the problem becomes one of remaining firmly attached to a solid substrate because of the wave movements and tides. Where there is a hard and rocky substrate, great masses of marine life abound.

Neritic and Littoral Consumers. Although they are not immediately apparent, among the major consumers of the oceans are the *zooplankton,* which are found floating among the phytoplankton. Many of the species of zooplankton found in marine environments are adults of certain species; many more are the larval forms of larger sea animals. In addition to floating zooplankton, consumers of the ocean are further classified as

1. *Benthos*—crawling, creeping, and sessile organisms.
2. *Nekton*—active swimming organisms, such as fish, which are capable of changing locations at will.

The benthos organisms of the neritic zone, like the algae, show marked stratification. In the upper limits of the neritic zone the consumers must be adapted to meet possible desiccation and sharp changes in temperature since they are only covered with water or ocean spray for brief periods of time. Differences in length of time of exposure to air and the type of substrate to which they attach demand variations in adaptations. As a result, even closely related species possess adaptational differences. In addition to being adapted to meet a rapidly changing physical environment, these organisms are adapted in many unique ways to obtain food. Many are sessile and rely on currents to bring the food to them; they extract their food from the water as it flows by. The characteristic consumers of the rocky areas are various species of barnacles, mollusks, and periwinkles, while the characteristic consumers of the sandy areas are various species of clams, crabs, and shrimp.

The nekton organisms include fish and such mollusks as squid and scallops. While these organisms have freedom to move from one area to another, they are generally quite limited in geographical range and ocean depth. The flounder, halibut, and sole are broad, flat fish adapted to swimming along the ocean floor. Near to the surface are large numbers of schooling fish, such as herring, sardine, anchovy, and menhaden, which feed mainly upon plankton. These are important groups for they form a basic link in the food chain as the prey of larger fish which are unable to consume plankton. The greatest numbers of fish are found in the cooler waters of the neritic zone where there is upwelling. The major fish of the colder waters are herring, cod, haddock, and halibut. In the warmer waters are found mullet, bass, and mackerel. In warm tropical waters the numbers of fish are much smaller than in cooler waters. However, the number of species is far greater. Some of the most exotic aquatic organisms are found in warmer waters.

In addition to the zooplankton and fish, many varieties of birds, turtles, and seals are tertiary consumers of the neritic zone. Close studies of these animals reveal that each has its characteristic ecological niche and is as integral a part of the ecosystem as the plankton and fish.

Neritic and Littoral Decomposers. Bacteria are the major decomposers of the ocean. They are most predominant in areas where phytoplankton is in greatest abundance. Dead organic matter is regenerated into inorganic compounds by these organisms. Depending upon the regional conditions, bacteria may exist in the water level just below the phytoplankton or may be found in the bottom sediment. The regenerated inorganic materials may rise to the upper levels in areas of upwelling, to be used again. The numbers of bacteria vary from one per milliliter of sea water in the open ocean to 10^8 per milliliter of sea water in the shallow areas of the littoral zone.

PELAGIC ECOSYSTEMS. The photic areas of the pelagic zone or open ocean do not support populations as large or as great in variety as those of the coastal regions. There are large areas in which the annual temperature does not vary greatly and therefore, with few exceptions, the nutrients are quickly exhausted. In certain of the cooler currents passing through warmer areas, plankton may exist in sufficient numbers to support a variety of sea life. In the cooler regions of the ocean the phytoplankton during certain times of the year support immense populations of shrimp-like crustaceans known as **krill.** During the warmer months in the extreme northern and southern seas large numbers of krill rise near to the surface of the ocean to feed on the phytoplankton, which are present in large numbers as a result of an increase of nutrient from overturn and longer periods of sunlight. Krill vary in length from about $\frac{1}{8}$ inch to 2 inches at maturity. Baleen whales follow long migratory routes from the Arctic in June to the Antarctic in December to feed upon the seasonal increase in populations of krill. Krill also supports large populations of seals and birds such as penguins.

As a result of the characteristic directions of the wind and ocean currents there are certain regions of great calm. Within these areas there may be accumulations of multicellular floating algae, which are considered to be phytoplankton. These algae are equipped with specialized air bladders to keep them afloat. The most outstanding of these areas is the Sargasso Sea in the mid-Atlantic. This extensive area of "seaweed" is relatively isolated, resulting in the evolution of distinct flora and fauna that are found nowhere else.

The microscopic zooplankton of the pelagic zone are generally the adult forms of many species. The secondary and tertiary consumers are few in number but rather large in size. The chief fish of the open ocean are herring, mackerel, bonito, and tuna. The whale, a mammal, is another secondary consumer of this zone.

Birds may be found far out into the open ocean. Albatross, frigatebirds, and certain species of terns spend all of their lives far from land except for breeding seasons. While it may appear difficult to place these birds in a specific ecosystem of the open oceans, careful examination shows that they have a definite ecological niche.

In the deep-water aphotic zone of the oceans there are no producers, and organisms of this ecosystem depend upon organic matter drifting down or entering this zone from the other zones. The bottom of the ocean is covered with fine sediments in warm areas and with bare clay in others. Among the organisms found at this depth are certain species of crustaceans, echinoderms, and mollusks. The fish of this zone are a curious group. They are adapted to external darkness, extreme pressures, and a carnivorous existence. Many of the fish have mouths large enough to swallow prey larger than themselves, an excellent adaptation for a region where meals may be few and far between. Some species of fish possess bioluminescent lures to attract prey or a possible mate. These organisms are another illustration of the diversity of living forms and the degree to which they have adapted to their environments.

Terrestrial Ecosystems

Of the various ecosystems, those of the terrestrial environment are the most variable. The broad general differences among the biomes and biogeographical regions and the many local variations within these regions afford thousands of habitats with innumerable ecological niches.

Several important factors control and determine the nature of all terrestrial ecosystems regardless of their locations.

(1) All living organisms require water. Thus, water becomes a major limiting factor in terrestrial life. The availability of an ample and continuous supply of water will determine the amount of freedom or conservation an organism will observe in its use. In arid regions, all organisms are physiologically adapted for the conservation of water. Generally, terrestrial organisms living in moist habitats freely consume and excrete large quantities of water.

(2) The variability in daily and annual climatic changes over land areas requires special adaptations. For example, animals with constant internal temperatures have various adaptive skin coverings while animals without a constant internal temperature may hibernate, have periods of dormancy, or have resistant stages to survive seasonal extremes. Like animals, vegetation is controlled by the annual changes in climate. Variability in climate results in fluctuations in populations, migrations, cyclic periods of increased and decreased activity, and competition for more ideal locations.

(3) All terrestrial organisms live upon or in the soil, which is complex and varies in nutrient content from one area to another. Given other favorable factors, the fertility and structure of the soil will determine the amount and type of vegetation that can be supported.

This, in turn, will determine the number and varieties of consumers within the ecosystem.

(4) Land has geographical barriers that prevent free movement and result in the development of ecosystems free from outside biotic forces.

The diversity of terrestrial living forms and the variety of possible habitats make any ecosystem classification an oversimplification. However, organisms living in terrestrial ecosystems can be easily classified according to food niche; producer, consumer, and decomposer may be applied to terrestrial as well as aquatic ecosystems (Figure 25-3).

Terrestrial Producers

The producers of the terrestrial ecosystems are the photosynthetic organisms. There are a few terrestrial producers of the Monera and Protista kingdoms, but their contribution is small and limited to specialized ecosystems. The dominant producers are the higher forms from the Plantae kingdom, most specifically seed-bearing plants.

Within the terrestrial ecosystem, the producers perform three major functions.

1. They provide the initial source of food in the food chains.
2. Large plant structures provide habitats directly and indirectly for other organisms.
3. They are prime agents in soil formation and in modifying the abiotic environment.

To a great extent the size and complexity of the ecosystem depend upon the degree and nature of the plants' adaptation to the given abiotic environment.

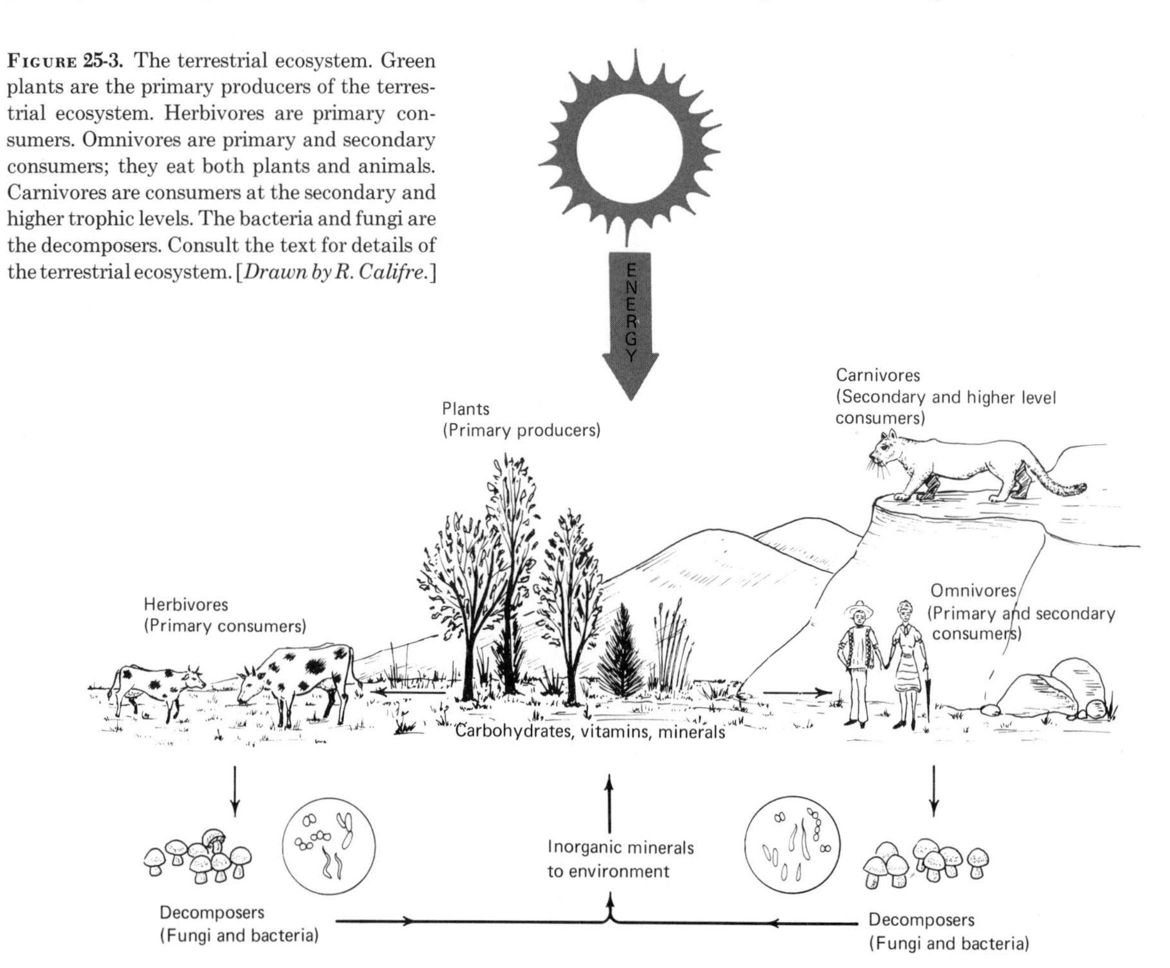

FIGURE 25-3. The terrestrial ecosystem. Green plants are the primary producers of the terrestrial ecosystem. Herbivores are primary consumers. Omnivores are primary and secondary consumers; they eat both plants and animals. Carnivores are consumers at the secondary and higher trophic levels. The bacteria and fungi are the decomposers. Consult the text for details of the terrestrial ecosystem. [*Drawn by R. Califre.*]

For example, the sparse vegetation of arid regions limits the size and number of organisms that can be supported. The plant bodies and seeds provide food and water for birds, insects, and a few small rodents. Within the large cacti, woodpeckers carve out nests. The abandoned nests provide shelter for an array of other birds, reptiles, and insects. Secondary consumers such as birds, snakes, insects, and lizards prey upon the primary consumers. Thus, the ecosystem is directly dependent upon the kinds and numbers of plants present.

Terrestrial Consumers

The number of species of primary consumers of the terrestrial ecosystems is far greater than the number of species of producers upon which they feed. The primary consumers have representatives from all the kingdoms of living things. From the Monera and Protista kingdoms species of bacteria and single-celled organisms infest plant bodies and cause disease. Various species of fungi and certain species of plants parasitize all members of the Plantae. Among them are parasitic rusts and smuts, which can devastate fields of grain, and the dodder and mistletoe, parasitic Plantae, which can cause widespread damage to trees and shrubs. Consumers from the Animalia kingdom include organisms varying in size from microscopic mesozoan parasites to the largest living land animal, the elephant. Man, the greatest consumer, not only uses plants for food but also uses them for shelter, clothing, and other aspects of modern living. In his consumption of plants, man has considerably modified and altered the terrestrial ecosystems.

The trophic levels above the primary consumers include the various carnivores. Depending upon their prey, they may be considered second, third, and infrequently, fourth or fifth level consumers. Rarely are there consumers beyond the fifth level. The activities and growth of consumer populations are directly related to the vegetation. Annual changes in plant growth and foliation, as well as unpredicted changes such as drought, flood, disease, and pollution, will increase or decrease the number of producers and primary consumers. The number and kinds of primary consumers will then determine the number and kinds of other consumers. The consumers play an important part in controlling the size and activities of populations within an ecosystem.

Terrestrial Decomposers

The decomposers of the terrestrial ecosystems, as with the other ecosystems, are the bacteria and fungi. These organisms obtain energy to carry on their life functions by breaking down the organic compounds of dead organic matter and organic wastes. In doing this, they prevent the accumulation of a mass of organic matter and also release essential minerals for re-use. The decomposers are found mostly in the soil and the heterotrophic bacteria are by far the most numerous. The rate of decomposition is dependent upon temperature and moisture. In regions where there are prolonged dry or cold periods, there may be large accumulations of undecomposed organic matter. By contrast, in moist, warm, tropical climates there are relatively small accumulations of organic matter. In fact, in tropical regions, when the land is cleared of the dense growth of trees and vines for farming, decomposition is so rapid that essential minerals are quickly leached from the soil. The decomposition process converts the essential minerals to a soluble form which is removed in runoff water from the heavy rain. In order to maintain the productivity of the soil, very specialized methods of cultivation are necessary. The climate of the middle-latitude zone with its alternating winters and summers allows the accumulation of a moderate amount of organic matter. During the warm moist weather, decomposition accelerates and at this time the greatest amount of plant growth takes place. Therefore, most of the essential minerals are re-used as they are released by the decomposers and are not lost through leaching and runoff water.

The terrestrial environment is subjected to far greater changes than the more stable aquatic environment. Aquatic life, because of the properties of water, is protected from rigorous climates and weather. The only constant stress placed upon aquatic systems is the relatively low oxygen availability, while in terrestrial ecosystems the availability of oxygen is never a problem. However, unpredictable climate and weather, water supply, and other natural forces such as flood and fire, and artificial forces such as industrialization, farming, and grazing all dramatically alter the terrestrial ecosystems.

Biomes and Biogeographical Regions

Biomes and biogeographic regions are closely related to the concept of climax. If there are no destructive forces acting upon a developing ecosystem, natural changes occur, resulting in stages of succession of communities (see page 617). Each stage is more stable than the preceding stage. When a stage in succession is reached in which, under the given environmental conditions, the ecosystem is utilizing the available resources of nutrients and energy to an optimum, it is considered to be a climax community. A climax community has a more complex organization and larger biomass and is a more stable stage than those that preceded it. It is able to withstand the normal natural changes in the environment so that it is self-perpetuating. Such climax communities may exist for thousands of years. It must be emphasized, though, that a climax community does change; barring drastic changes in the environment, however, it does so slowly.

One of the major problems in considering climax communities is that they do not have clearly defined boundaries. As one travels southwest through the grasslands of the plains states to the arid desert regions of the southwestern United States, there are no abrupt boundary lines between the two types of vegetation. Instead, there is a broad range of overlap where both types of vegetation coexist.

Because of variability in environmental conditions within a given area, it is difficult to generalize as to broad areas representing climax communities. Nevertheless, ecologists have found it convenient to recognize that there are certain areas of the earth that are dominated by characteristic types of vegetation. For example, much of the far north is devoid of trees and large types of vegetation. This area is characterized by low shrubs, grasses, and lichens and is called the tundra. Such a large region with uniform climate, vegetation, and animals is called a biome. It must be emphasized, however, that within a biome, because of local variability in environmental conditions, both abiotic and biotic, there may be communities that are vastly different from the more general type. For example, within the deciduous forest biome there may be areas of grassland or even areas that are characterized by desert vegetation.

When the ecologist studies life on the entire earth, he finds that the earth may also be divided into regions based upon characteristic fauna. The study of these biogeographic regions allows one to place the existence of characteristic animal species in a historical perspective. The earth and its organisms are constantly changing, and the present distribution of species is the result of great changes that have taken place in the past. For example, there is evidence that at one time all the continents of the earth were part of a single large land mass. Over millions of years large land masses shifted apart and became separated by large expanses of oceans. On these isolated land masses evolved characteristic species, some similar to those found elsewhere and some quite dissimilar.

Biomes

Biomes are large areas with a uniformity of climate and vegetation and its associated fauna. The vegetative types are determined mainly by climatic conditions. Even though seeds and spores show many adaptations for wide dispersal (many are readily carried by wind and water even over such geographic barriers as wide expanses of oceans, deserts, or mountains), plants and fungi are not evenly distributed over the earth. They are limited in distribution to specific areas by such factors as temperature, soil conditions, precipitation, and specific adaptations of the particular species. The tundra, taiga, deciduous forest, grassland, desert, and tropical rain forest are the six major types of biomes.

TUNDRA. Within the polar climate group is found the *tundra* biome (Figure 25-4). It is the most geographically continuous of the biomes and includes the northernmost land areas of North America, Europe, and Asia. The climate is cold and the subsoil is permanently frozen. The surface area is characterized by gently rolling plains with many lakes, ponds, and bogs. Trees, when present, are quite small and scrubby. There are numerous, small, perennial herbs, grasses, sedges, and dwarf woody plants. However, the dominant flora are low ground mosses such as sphagnum and lichens such as reindeer moss. The plants of this region are well adapted to the stresses of the cold climate, the continu-

Figure 25-4. Tundra. The characteristic floras of the tundra are lichens, mosses, small perennial herbs, and grasses. Trees and shrubs, if present, are dwarfed. The subsoil of this area is permanently frozen. The land is characterized by gently rolling hills. [*USDA photo.*]

ous night during the winter season, and the continuous daylight of low intensity during the summer season. During the short growing season great splashes of color cover large areas of the tundra where brilliantly colored flowers bloom simultaneously.

In spite of the extremely severe environment, the tundra teems with life. There are considerable numbers of certain mammals such as the caribou, musk ox, arctic fox, arctic hare, and lemming. In addition, vast numbers of migratory birds are found during the short summer. Conspicuous among animal life are swarms of flies and mosquitoes. Although the number of individuals is large, there are relatively few species, showing that while successful adaptation has occurred in only a few species, once it is achieved, large populations can thrive.

Taiga. South of the tundra, stretching in broad belts across the northern continents, are vast evergreen forests. These constitute the *taiga* biome (Figure 25-5). Here winters are long and severe, and the growing season is limited to a few months during the summer. However, the soil completely thaws for a short period of time, thereby making sufficient water available to support large evergreen forests. The dominant species are spruces, firs, hemlocks, and pines. Some deciduous trees such as the paper birch and aspen are also present. In those land areas where mountains extend down into the southern zones as they do in parts of North America, the taiga may be continuous with the evergreen forests of the mountains much farther south. The taiga, like the tundra, has many lakes, ponds, and bogs. The number of species of animals, however, is larger than in the tundra. Moose, black bear, wolf, lynx, squirrel, and small rodents abound. Large numbers of birds are present during the short summer. Like the tundra, the taiga is subject to pronounced seasonal periodicity of populations.

Deciduous Forest. South of the taiga biome the regions show wide variations in temperature and precipitation that result in great differences from one area to another. In those parts of the middle latitude climatic zones where there are warm summers, cold winters, and abundant rainfall, the *deciduous forest*

Figure 25-5. Taiga. The taiga is made up of dense evergreen forests of spruce, hemlock, fir, and pine. The winters are severe, but the summers are warm enough to allow the soil to thaw completely. The photograph shows a portion of the taiga in Canada. [*Courtesy of D. C. Anderson, Canadian Forest Service, Department of Fisheries and Forestry, Sault Ste. Marie, Ontario.*]

biomes are found (Figure 25-6). These are dominated by broadleaf hardwood trees such as oak, maple, ash, and hickory. The seasonal variations in temperature are great in this biome. The plants are adapted to the low temperatures of winter by ceasing to grow. The number of species is greater than in the taiga, but less than in the more humid biomes. The familiar animals of the deciduous forest biome are deer, bears, racoons, foxes, squirrels, and many varieties of birds and insects. The major deciduous forest biome is distributed over the eastern United States and Europe.

Grassland. The *grassland* biomes are found in both the middle latitude and equatorial tropical climatic zones (Figure 25-7). The major controlling factor is the rather sparse or erratic rainfall of about 10–40 inches annually. The grasses vary from short buffalo grass to tall elephant grass and bamboo.

There are various regional designations for the grassland biomes. The *savannas* are found in the tropical plains of Africa, India, and South America. The climate is characterized by a long dry winter season alternating with a wet summer season. The temperature remains relatively warm throughout the year. The *prairies* are widespread in central United States, eastern Europe,

FIGURE 25-6. Deciduous forest. The deciduous forest is characterized by broadleaf hardwood trees, such as oak, maple, ash, and hickory, which shed their leaves prior to the advent of winter. [*U.S. Forest Service photo.*]

and in the South American countries of Argentina and Uruguay. There are great annual variations in temperature because of their middle latitude location, and rainfall tends to be scanty. The grasses vary from tall to short species. The *steppe,* which is drier than the prairie, contains short grass species and is located in interior plains and plateaus that are semiarid. Within the United States the higher plains extending from the Canadian border to Texas represent this type of biome. Steppe grasslands are found in western North America but are most extensive in central and southwest Asia. Grasslands support a large number of species of animal life. Most notable is the great variety of grazing mammals and rodents.

DESERT. Major *desert* areas are located on each continent except Europe (Figure 25-8). The desert shrub and desert waste biomes are found wherever the average annual rainfall is less than the annual evaporation. Rain, when it falls, occurs in a few heavy cloudbursts and evaporates rapidly soon after it reaches the earth. Both plant and animal forms are uniquely adapted to this biome. Some plants flower and produce seeds during a brief period of time after a rainfall. Other plants, such as the cacti of the Western Hemisphere and the euphorbia of Africa, have thick stems that store water. Some desert shrubs have thick, succulent leaves for water storage, while others shed their leaves during prolonged dry periods to conserve water. Desert animals are small, and many burrow in the soil. Certain species of lizards and insects are dominant. Warm-blooded animals are generally comparatively rare because of the shortage of water and the difficulty in maintaining a constant body temperature where daily air and land temperatures vary from hot during the day to extremely cold during the night. The absence of large bodies of water and the rapidity with which the land

FIGURE 25-7. Grassland. The grasslands are characterized by grasses of different heights and low shrubs; trees are usually absent. This photograph shows the grasslands in the foreground with their characteristic fauna, the antelope, while in the background the rising mountains show mountain zonation with the characteristic regions of shrubs and trees. [*USDA photo.*]

FIGURE 25-8. Desert. The North American desert is characterized by sparse vegetation. The dominant species are cacti and various types of shrubs. Note how uniformly each member of the same species is distributed. [*Courtesy of Dr. Marie Kuhnen, Montclair State College.*]

Figure 25-9. Tropical rain forest. The vegetation of the rain forest is characterized by tall broadleaf trees and vines, with a lush, dense undergrowth. [*USDA photo.*]

warms up and cools off result in extreme changes in temperature on a daily basis, sometimes as much as 100°F.

Tropical Rain Forest. The *tropical rain forest* biomes are found where there is abundant rain (80–200 inches annually) with no dry season and fairly high temperatures (Figure 25-9). This biome is the most complex in terms of species makeup. There is a dense growth of tall broadleaf evergreen trees and heavy vines. The dominant trees form a tight canopy over the rest of the forest. Beneath this canopy are other well-developed layers: the subcanopy, overstory, understory, and ground layers. This vertical stratification has different floral species dominating at each level. Direct sunlight and rain do not reach the forest floor, but the shaded lower levels are continually bathed by water dripping from the trees. Near the forest floor the temperature remains constantly high and it is exceedingly humid.

The fauna of the tropical forest biome is also stratified. Certain species live high up in the canopy, others just below in the large branches and along the tree trunks, while others are limited to the forest floor. Much of the fauna is largely nocturnal. This biome covers large areas of central Africa, south and central Asia, Central America, and the Amazon Basin of South America.

While the biomes are represented as clearly defined areas of characteristic flora and fauna, there will be overlapping where one biome merges into another. In fact, the changes tend to be gradual rather than sharply defined. The overlapping areas may be classified as minor biomes such as chaparrals, tropical scrub forests, mediterranean scrub forests, and tropical semideciduous forests.

Mountain Zonation. Because of the differences in physical and climatic conditions with increasing altitude, the various biomes that one would see going from the equator to the North Pole are more or less repeated as one goes up a mountain (Figure 25-10). The number of biomes represented will depend upon the location and height of the mountain. Mountains in the wet tropics will have the tropical rain forest at the base followed by the deciduous forest, coniferous forest, low shrub, and finally, tundra at the snow line. Logically, mountains of the colder regions have fewer biomes. The various biomes found along the mountain slopes occur within a few thousand feet of elevation.

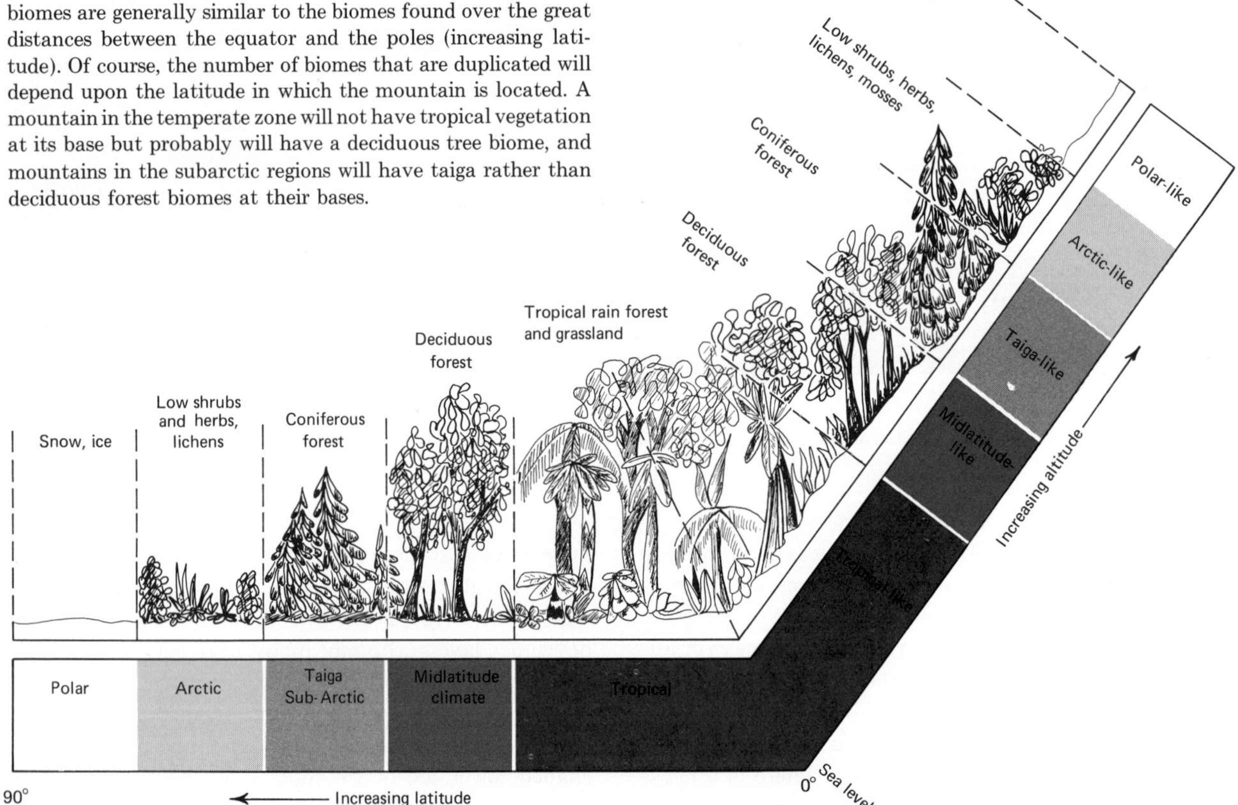

Figure 25-10. Mountain zonation. Within the short distances from sea level to mountain top (increasing altitude), the various biomes are generally similar to the biomes found over the great distances between the equator and the poles (increasing latitude). Of course, the number of biomes that are duplicated will depend upon the latitude in which the mountain is located. A mountain in the temperate zone will not have tropical vegetation at its base but probably will have a deciduous tree biome, and mountains in the subarctic regions will have taiga rather than deciduous forest biomes at their bases.

Biogeographical or Faunal Regions

Unlike the biomes, the **biogeographical regions** are greatly limited by geographical barriers. Wide expanses of oceans, deserts, mountain ranges, and plains may completely inhibit the movement of certain animals. A forest separating two grasslands may represent a barrier for grazing animals. Water separating two land masses may present a barrier for some species, while flying and swimming animals may cross it easily. The various barriers to dispersal and routes of dispersal help to provide an understanding of the distribution of organisms throughout the regions of the earth. Yet this only represents a partial explanation. The potential for organisms to survive physiologically in new regions, to compete with established species in new regions, and the means for physical dispersion into such regions, all determine the kinds of living organisms in an area. Additional factors that must be considered when studying the distribution of animals are geological and evolutionary developments as well as present environmental conditions.

Any area of the earth, small or large, may be analyzed for biogeographical distribution. However, there are six generally recognized major faunal regions on the earth: the Australian, the Neotropic, and the four regions of the "world continent," Nearctic, Palaearctic, Oriental, and Ethiopian (see Figure 25-11).

Australian Region. The *Australian region* is the most distinctive of all the regions. There are many species found nowhere else on earth, and many species that are widespread elsewhere are completely lacking. Geological and biological evidence indicates that Australia was separated from the Eurasian land mass millions of years ago (Figure 25-11). Since that time a water barrier has kept it in isolation. However, some minor migration may have taken place between the western islands of the East Indies and Australia.

The most distinctive faunal aspect of the Australian region is the marsupials. These are a group of mammals in which the young are born prematurely and then suckled in a pouch on the female (a familiar member of this group is the opossum, the only North American marsupial). In placental mammals, by contrast, the entire embryonic development takes place within the uterus. The only native placental animals of the Australian region are the bats, which could fly over the water barriers, and small rodents, which probably arrived from Asia by swimming or upon floating debris. The similarities among certain marsupial and placental animals are striking. There are, among others, marsupial mice, weasels, wolves, anteaters, rats, ground hogs, and bears. Since the marsupials were subjected to similar physical and biological stresses as their placental counterparts in other faunal regions, the similarities present strong evidence for the theory of parallel evolution (see Chapter 21). There are some marsupial mammals that bear no resemblance to placental mammals yet fulfill similar ecological functions. For example, the kangaroo is a grazing animal much the same as the bison and antelope of the North American grasslands. Such differing animals that play the same role in different environments are called *ecological equivalents.*

The marsupials were quite successful until human occupation destroyed many of their native habitats. The introduction of the more successful placental animals by man is another factor that led to the extinction of some members of this group of animals.

Neotropic Region. The *Neotropic region* includes the continent of South America. This region is in some respects like the Australian region. Almost completely surrounded by water, it is attached to the North American continent by the land bridge at the thin Isthmus of Panama (see Figure 25-11). Over many millions of years, it has been alternately separated from and joined to the North American continent. The fossil record shows that marsupials were once common in the Neotropic region. However, as the land bridge formed between the continents, placental animals were allowed to migrate from the north and, except for three species (opossums, marsupial mice, and marsupial rats), marsupials are now extinct in this region. Among the distinctive animals of this region are the camel-like llamas and alpacas, sloths, New World monkeys, rheas, anteaters, guinea pigs, and tapirs. It is interesting to note that the only other region where tapirs are found is in Southeast Asia. This may indicate that there was once a continuous land bridge between the Neotropic and Oriental regions and that the climate of the northern land mass was quite different from what it is today. While it appears that most migration took place in the direction from the Nearctic region toward the Neo-

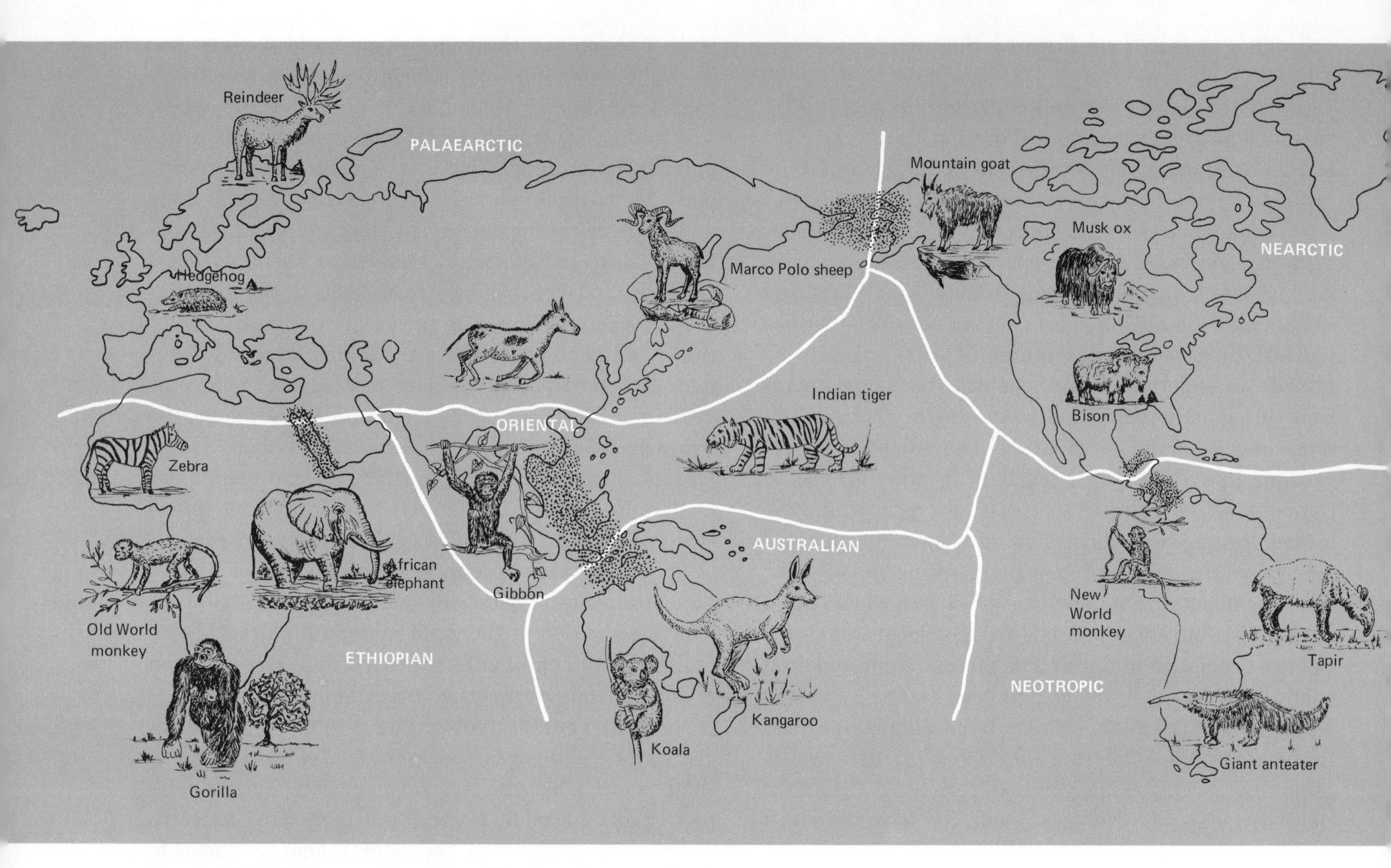

FIGURE 25-11. Biogeographic regions. The six regions are classified according to the distribution of characteristic fauna. The various species have been limited to the given areas because of major geographical barriers such as water, deserts, mountains, and temporary land bridges. The dotted areas represent former major land bridges between biogeographic regions. [*Designed by R. Califre.*]

tropic region, the North American porcupine, marsupial opossum, and armadillo probably originated on the South American continent.

THE WORLD CONTINENT. The practically continuous land mass between Africa, Europe, Asia, and North America is known as the "world continent." While the North American continent is separated from Asia by the Bering Straits, there is strong evidence that there was a land bridge until relatively recent geological times (Figure 25-11). The world continent is divided into four faunal regions: the *Ethiopian* region, which is the African continent south of the Sahara Desert; the *Palaearctic* region, which includes the area around the Mediterranean Sea, all of Europe, and northern Asia; the *Oriental* region of southern Asia; and the *Nearctic* region of North America. The continuity of this land area has resulted in similarity among species of the various regions. Those differences that exist are probably caused by major geographical and ecological barriers, which restricted migration, such as the Sahara Desert and Himalayan mountains.

There is a closer similarity between the Palaearctic region and the Nearctic region than any of the other regions. In fact, many biologists group the two together and call them the *Holarctic* region. The fauna of the Holarctic region include the timber wolf, hare, moose, bear, and elk. Certain species of deer are limited to the Nearctic region, while the wild boar (not to be mistaken for the peccary or javelina of the United States) is limited to the Palaearctic region.

The Oriental region is inhabited by such species as the tiger, gibbon ape, water buffalo, and Indian elephant. The Ethiopian region is characterized by such species as the giraffe, zebra, antelope, lion, aardvark, African elephant, gorilla, and Old World monkey (those lacking long, prehensile tails).

Dividing the land areas into major regions helps to explain the similarities and dissimilarities between the various flora and fauna of the areas as well as some of the mechanisms of evolution. Both biomes and biogeographical regions represent large general areas and partially explain how the limitations of climate and geographical barriers determine the existence of different species. Within each of these broad divisions the various fauna and flora have evolved relatively free from forces outside of their region. Within the regions there are considerable differences in the various species of animal and plant life as well as differences in distribution. This can be attributed to the results of living organisms interacting with the biotic and abiotic environments.

26 Human Populations and the Biosphere

GUIDE QUESTIONS

1 In what ways does an increasing population contribute to the pollution problems? How does urbanization contribute to the pollution problem?
2 Define sterilization. How is it accomplished in the male and female?
3 Describe the use of the condom and diaphragm as contraceptives.
4 What is an intrauterine device?
5 Explain how the pill prevents conception.
6 How do technological and industrial changes contribute to pollution?
7 Discuss the sources and effect upon living organisms and aquatic ecosystems of thermal pollution.
8 What are the causes of smog and how might it be controlled?
9 Discuss the three major types of pollution and their effects upon ecosystems.
10 What are the types and sources of water pollution?
11 What are the types and sources of air pollution?
12 What are the types and sources of land pollution?

PRECEDING chapters have discussed evidence that both external and internal forces regulate the size of populations among living forms. Within any stable ecosystem the size of the populations represents the optimum biomass that can be supported at the specific stage of succession. While there may be rapid increases in a given population over a short period of time, there are natural checks and balances over which organisms have little or no control that operate to decrease or stabilize the number of individuals. For example, for the larger population there may be an increase of predators, insufficient food supply, or the rapid spread of disease. These controls over population size represent one of the balances in nature.

Unlike all other living organisms, humans can influence the external and internal forces that act to limit the growth of his own population. As a result of our increasing control over the biotic and abiotic environments, the rate of growth of the world human population has become critical. The present population of more than 4 billion people will be about 8 billion by the end of the century. The effects of this increased population upon the total biosphere are, and will continue to be, profound. Even at the present time over 3 billion people are either underfed or undernourished. Many of these individuals, besides being undernourished, are underhoused and undereducated. Only 450 million people of the 4 billion plus live in abundance; the rest live at subsistence levels or less. There is an ever-widening gap between the "haves" and the "have nots," and it has many social, political, and biological implications in our modern world.

Population Theories

Malthusian Theory

The effect of an increasing human population has been studied for some time and various theories have been proposed over the years. However, these theories have been generally limited to economic and social implications. One of the earliest theories and probably the one that has had the greatest impact was proposed in 1789 by Thomas Malthus, an economist and historian. His theory was based upon two postulates.

1. Food is necessary to the existence of man.

2. The desire and capacity of man to reproduce remains about the same.

Malthus stated that, with unchecked growth, population increases geometrically while food production increases arithmetically. That is, population increase follows a sequence such as 1, 2, 4, 8, 16, 32, while increased food production follows a sequence such as 1, 2, 3, 4, 5. Therefore, the power of the population to increase its number is infinitely greater than the power of the earth to produce food. Malthus received his greatest criticism on the basis of this last statement. If the reproductive capacity of humans were realized every 25 years, within 150 years the world population would increase to 64 times its original size. Of course, while population has increased tremendously, this increase has not followed an exact geometric ratio. In addition, the increase in food production has been greater than an arithmetic ratio. Factors influencing population growth and food production are significantly different from those which Malthus could have foreseen or anticipated. In later years Malthus made some modification with regard to controls on population growth but the theory remained essentially the same. His theory essentially has been correct. The only error was in predicting the time factors.

Malthus was correct in his view that population would continue to increase rapidly in the absence of certain controls which he called positive and negative checks. Historically, his theory of population and his concept of positive and negative checks provided Charles Darwin with a basis for his theory of evolution involving natural selection.

As the nineteenth century and the industrial revolution progressed, greater optimism for the future developed and Malthus' theory lost much support. Among the more advanced countries, scientific and technological progress tended to weaken Malthus' theory.

Although Malthus has been severely criticized, his population theory was the first to be presented in a clear objective manner based upon facts available to him. In addition, he contributed greatly in bringing the population problem to public attention. Modifications of many of Malthus' ideas have found their way into modern theories of population. While Malthus' theory is essentially a naturalistic theory, succeeding theories of population growth may be broadly classified into natural law theories and social theories.

Natural Law Theories

The search for a natural law theory of population has been pursued along two different lines.

1. A search for "laws" of nature that apply to living organisms including humans, as well as to inanimate objects.

2. A search for means to improve man's lot.

While there are differences with regard to the causes, both types of theorists proposed that there is a natural tendency for populations to decrease or at least to remain relatively constant in size. Michael Sadler (1780–1835) proposed that the desire to reproduce decreases as the population increases and that the desire is innate. Thomas Doubleday (1790–1870) believed that man's increase in numbers was inversely related to food supply. As the food supply increased there was a slower increase in population size. There is no scientific basis for these beliefs. Herbert Spencer (1820–1903) stated that as individuals pursue personal advancement there is a decrease of interest in reproduction. While there is evidence that there is a decline in the rate of population growth among those countries that are most advanced technologically and have a higher standard of living, this decline is not due to a "natural law." Garrado Gini suggested that there is a cyclic rise and fall of population due to rapid changes in biological traits of entire populations among the various countries or regions of the world.

Essentially, all of these natural law theories place no responsibility upon humans for controlling increases in population. Nature is assumed to achieve this independent of man through a weakening of reproductive capacity and by an increase of personal, scientific, and economic development. While there is some evidence to support each of these theories, this evidence is only found in various segments of the world's population and does not apply to the total world. In addition, each theory proposes a decline in reproductive capacity but it is difficult to distinguish whether the proponents of a particular theory mean a physiological decrease in fertility or a psychological desire not to reproduce.

The natural law theories proposing an eventual control of population through natural laws are far less pessimistic than Malthus' theory. Yet, the present plight of most of the world's population makes it difficult to visualize when, how, and if the natural forces will begin to take world-wide effect.

Social Theories of Population Growth

Unlike the natural law theories, which seek a law in nature, the social theories place emphasis upon changes in the social system such as the family, community, and governmental levels. The plight of the poor of the world is attributed to faults inherent in the social structure regardless of whether the economic system of the government is based upon capitalism or upon socialism. The proponents of social theories do not assume some underlying natural law but look for social factors that determine population growth at a particular time. These theories propose that as the economic position of people improves, they will exercise some voluntary control over the rate of reproduction.

This text will not debate the relative merits of the natural law and the social theories of population growth. This brief introduction to some of the major theories is meant only to place the problem in its proper perspective.

Scope of the Population Problem

Geography and the World Population

The distribution of the world's population is not uniform. Some areas are teeming with humanity while other areas are practically uninhabited. Between these two extremes there is great variability in population density. At the present time about 50% of the world's population is living on 5% of the earth's surface, while 57% of the land supports only 5% of the people.

The distribution of human population has been centered in those areas that are the most ideal to support large numbers of all living organisms (Figure 26-1). These land areas are generally located on expansive plains and flood plains in the middle latitudes and subtropics, with a few areas in the tropics. These areas are located on portions of the five major continents. The least populated lands are essentially the non-food-producing areas. These include deserts, ice caps, and tundra which together comprise more than 50% of the earth's land surface. Technological advancements have enabled portions of the tundra and desert to become food producing, but these provide only a small contribution to the increasing needs of the ever-growing population.

Those land areas that support the largest populations have the highest birth rates and the lowest standards of living. As a result of recent advances in medical science and sanitation, there has been a sharp decline in the death rate and a comparable increase in the population has occurred in these areas. This has resulted in an uneven distribution of the world's population. The greatest population growth is concentrated in regions which have surpassed the capacity to absorb increases. With increasing populations these regions progressively become poorer. This growth in population represents an overwhelming obstacle as the people in these areas make efforts to improve their economic condition.

The problem of human population growth is two-fold. Populations are growing at an explosive rate and this growth is occurring most rapidly in those areas that can least afford such growth in the face of attempts to reduce poverty and raise standards of living.

Biological, Social, and Political Implications

All areas of the earth, whether they are highly developed technologically and industrially or relatively underdeveloped with low standards of living, are affected by the rapid increase in population. The interdependence of all human populations biologically, economically, politically, and morally makes the problem extremely complex. Highly industrialized areas place high demands upon the existing ecosystems. Statistically, the average American consumes as much nonreusable resources as approximately thirty-five average individuals in an underdeveloped country such as India. The natural flora and fauna are permanently removed or changed to satisfy the particular immediate needs of man. In addition, large quantities of wastes, the by-products of a highly industrialized and technological society, produce alterations in ecosystems over very expansive areas.

In the underdeveloped areas the basic minimum needs of food, shelter, and clothing for a rapidly increasing population have placed an ever-increasing burden upon the natural environment. The fertility of the soil is depleted through overuse. In order to raise graz-

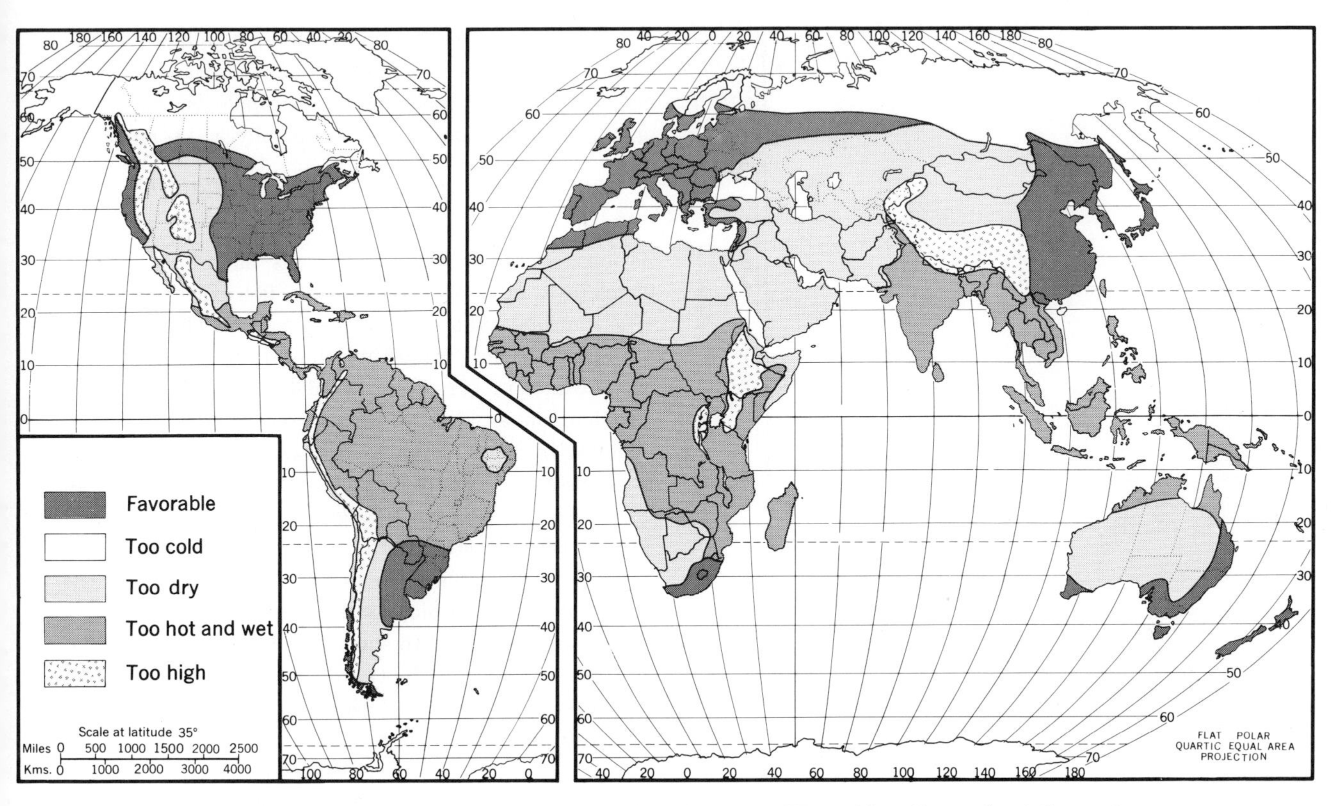

Figure 26-1. Distribution of land attractive and hostile to man. [*From Glenn Trewartha,* A Geography of Population: World Patterns. *New York: John Wiley & Sons Inc. © 1969.*]

ing animals and exportable crops, large forest areas have been cleared only to become barren because of the loss of a protective covering for the soil and a resultant reduction of the water table.

In attempts to raise the overall economic level of underdeveloped nations, many governments have established one-crop economies. For example, a nation may plant most of its cultivated land with coffee, sugar, tea, or rice. This use of the land limits the amount of basic necessities that a nation can supply its own population. In addition, it places the economic stability of the nation upon the whims of a world market as well as natural catastrophies such as flood, drought, and pestilence.

It has been estimated that the human population in the year 6000 B.C. was about 5 million people. Some 8000 years later, A.D. 1650, the population reached about 500 million (Figure 26-2). This means the world population doubled about every 1000 years. By 1850 the population reached 1 billion—this doubling took about 200 years. The 2 billion level was reached in 1930. At the present time there are more than 4 billion people and probably there will be 8 billion by the year 2000. Such an increase represents a doubling of the population every 35 years. If population growth continues at this unprecedented rate, a point will be reached where there will not be sufficient room for the population, let alone areas to provide for even the barest necessities. For example, at the present rate, in 900 years there would be 100 persons for every square yard of the surface of the earth, land and sea. However, such predictions tend to make the problem appear as conjecture and do not call attention to its immediacy.

To intensify the problem further, the demands by the increasing population in the developed countries upon the world's resources create further imbalance between

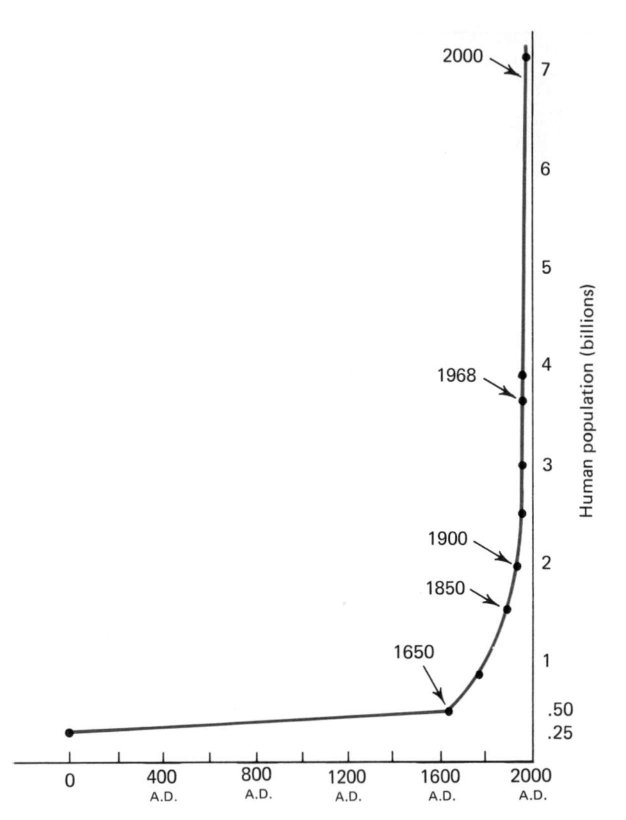

Figure 26-2. Growth of world population. At the beginning of the Christian era the world population has been estimated to have been between 133 and 300 million. By 1650 the population had doubled, and by 1850 there were 1 billion people. In the past 100 years the growth rate has greatly accelerated. It is estimated that by the year 2000 the earth will have approximately 8 billion inhabitants.

the "haves" and "have nots." For example, the United States has one-fifteenth of the world's population yet it uses over half of all the world's raw materials consumed each year. By the year 2000 the United States will have less than one-fifteenth of the world's population but may require up to 80% of the raw materials. If the present standard of living (rate of consumption) is maintained, it is difficult to see how any balance of supply and demand can be reached. Also, the gap between supply and demand will become even greater as the underdeveloped countries attempt to raise their present standards. Even without much further consideration it becomes obvious that the population problem is the most important problem facing man and its solutions require the contributions of the biologist as much as the economist, statesman, anthropologist, psychologist, religious leader, legislator, or physical scientist.

Composition of the World's Population

Demography is the statistical study of human populations, including such aspects as birth rates, death rates, rate of population growth, life expectancy, age distribution, and sex ratios (Table 26-1). This information helps to determine the composition of the population, which in turn helps to define the scope of the problem on a world-wide basis as well as for given regions. In most instances the term population, when humans are considered, applies to the total number of persons in a given area at a given time. The biologist is able to take the statistics of the demographer and view population as an active, living entity subject to the same physical and biological forces experienced by all living organisms and to show the effects of the population upon the total biotic and abiotic environments.

The accuracy of population statistics varies greatly. Those from North America and Europe are relatively accurate while those from other parts of the world are incomplete or unknown. Considerable efforts are being made by world organizations in an attempt to standardize procedures and obtain more accurate statistics and thus better understand various aspects of the population problem.

Age and Sex. Areas of the earth differ greatly in the relative distribution of the population with respect to age and sex. Since the sex and age groups of a population represent the foundation of a given society, these aspects of population are important from a biological and social point of view. That combined portion of the population under the age of 15 years or over the age of 60 years is known as the *dependency ratio*. The importance of this dependency ratio stems from the fact that every member of the population is either a consumer or a producer. The *producers* are generally considered to be individuals between 15 years and 60 years of age. The *consumers* are either under or over this age group. A country with a small dependency ratio is generally more productive economically. Consequently, the dependency ratio is an important economic factor in underdeveloped countries. Many underdeveloped countries have

Table 26-1. Estimated and Projected Population of the World's Major Regions[a]

Region	Population Estimates Mid-1975 (Millions)	Current Rate of Population Growth (Percent)	Approximate Number of Years to Double Population	Births per 1000	Life Expectancy from Birth (Years)	Population Under 15 Years (Percent)
Africa	401	2.6	30	47	43	43
Asia	2256	2.1	33	35	50	40
North America	237	0.9	90	17	71	31
Latin America	324	2.7	29	38	60	44
Europe	473	0.6	110	16	70	25
Oceania	21.3	2.0	35	23	71	30
USSR	255	1.0	70	18	70	36
World totals or averages	3967	1.9	42	32	53	36

[a] Sources: *United Nations: Demographic Yearbook 1975* (New York: United Nations, 1975), p. 139, and William Petersen, *Population* (New York: Macmillan, 1969), pp. 328–32.

experienced a rapid drop in death rates while their birth rates have remained high. This increases the relative number of young people who are dependent at a time when a country, economically, cannot produce enough to maintain its existing population.

One of the most important aspects of the present population crisis is that almost half of the population of the underdeveloped countries is under the age of 15 years. Within this decade this group will move into their reproductive years and it is difficult to foresee anything but an even more explosive growth of the world population. The problem is not only confined to the underdeveloped countries. For example, in the United States, although the rate of population growth has been reduced, those children born during the "baby boom" following World War II have now reached their reproductive years; even with a reduced birth rate, the population will continue to grow, at least into the next century.

Almost as important as age distribution in population statistics is sex ratio, that is, the ratio of males to females. Differences in this ratio in various areas of the world help to explain differences in births, deaths, marriages, and social customs, as well as other differences, for example, the consumption of specific products. These factors in turn determine to some extent the demands made upon the environment.

Growth Rates. The growth of world population is frequently termed an "explosion," which describes the sudden increase in population that is taking place primarily in the underdeveloped countries. This acceleration of the rate of population growth is a natural consequence of scientific developments that have taken place over several decades. These developments include curative methods of disease control, such as the use of sulfa drugs and antibiotics, and preventive methods of control, such as immunization and inoculation and the use of insecticides.

While the more developed countries have clearly benefited from the use of new drugs and immunization, the most striking drop in death rates has occurred in underdeveloped countries where infectious diseases formerly ran rampant. The more developed countries have established high standards of sanitary cleanliness and have applied scientific health knowledge over several decades. This has reduced the death rates from infectious disease to a point where such deaths are almost completely eliminated. Thus, the newer curative and preventive methods have had little effect in lowering the death rate in these countries. Even with the comparatively low levels of sanitary cleanliness present in underdeveloped countries, these new methods of disease control are effective. The use of new medicines is so simple that a single trained health worker can inoculate a whole village against smallpox and administer curative drugs for many other infections in a short period of time. The effects of these medicines are so readily apparent that there is relatively little resistance by the uneducated in accepting them.

The prevention of malaria or yellow fever by spraying to kill mosquitoes also meets with comparatively little cultural resistance. Most of the work can be performed by skilled teams spraying swampy areas which have little agricultural value and therefore this procedure does not greatly affect the cultural or working life of the local people. However, in the long run, spraying may adversely affect the existing ecosystems and eventually lower the productivity of the region.

The growth rate is a relationship between the number of births and the number of deaths. It is generally expressed per 1000 people. For example, if the birth rate is 35 per 1000 per year and the death rate is 15 per 1000 per year, the rate increase is 20 per 1000 per year ($35 - 15 = 20$). When expressed as a percentage this growth rate is 2% a year ($20 \div 1000 \times 100 = 2\%$). From this it would appear that it would take 50 years to add another 1000 to the population, but such is not the case. This would assume that none of the offspring would have children during the 50-year period. Logically, children born in the earlier part of the 50-year period would start reproducing during this time period. Based upon estimates and actual data, Table 26-2 shows the time it takes for the population to double.

In general, throughout the world and especially in the underdeveloped countries, the death rate has been reduced within a few years to a point where it is only half of what it was about 20 years ago, or less. On the other hand, there has been little or no reduction in the birth rates. This means that in many areas of the world the rate of population growth is between 2 and 3% each year (see Table 26-1) and in some areas even higher. At these rates the world population will double in from 35 to 24 years.

TABLE 26-2. The Time It Takes a Population to Double Based upon Annual Percent Increase in Growth Rate

Annual Percent Increase	Doubling Time (Years)
1.0	70
2.0	35
3.0	24
4.0	17

No society in modern times has succeeded in reducing its birth rate significantly until the standard of living has been well above subsistence level. In addition, there must be a desire on the part of parents to have a still higher standard of living for their children than that which they have had for themselves. It is very difficult to raise the standard of living if the number of children born is near the physiological maximum.

It must be realized that the present decrease in the death rate in underdeveloped countries has not been due to an increase in per capita intake of food but to improvement of health conditions. However, in the industrialized Western countries the increase in per capita consumption of food and other necessities has been the dominant factor in reducing death rate because of the previous high standards of health.

URBANIZATION. Urbanization is the process whereby an increasing proportion of the population becomes concentrated into large cities. This phenomenon associated with modern man is increasing in practically all regions of the world (Table 26-3). The causes of urbanization are many and complex and differ from one region to another. Urbanization of the Western countries began with the industrial revolution and proceeded as the nations moved from an agrarian to an industrial society. Initially, there was a migration of the population from rural areas to centers of manufacturing and trade. While this was a major factor in the initial increase in the population of cities, it is no longer an important factor. Since cities represent a concentration of the population, their population increase is due to their naturally greater reproductive capacity. As the population increases, cities tend to spread into adjacent outlying areas. Such growth has resulted in the development of large urban areas in which there is an overlapping between major cities. In the United States this growth is most evident in the Northeast where there is practically one continuous suburban area between Boston and Washington, D.C., and in the Far West in the area between San Diego and Los Angeles. The same phenomenon has occurred outside the United States in the areas surrounding London, Paris, and Rome.

The concentration of population in suburban areas within the developed countries has had the advantage of allowing the consolidation of rural farming areas into larger units. These larger units have enabled the mech-

TABLE 26-3. Total World Population and World Urban Population, 1800–1980[a]

Year	Total World Population (Millions)	Population Living in Localities of 20,000 Inhabitants or More		Population Living in Localities of 20,000 to 100,000 Inhabitants		Population Living in Localities of 100,000 Inhabitants or More	
		Millions	Percent of World Population	Millions	Percent of World Population	Millions	Percent of World Population
1800	906	21.7	2.4	6.1	0.7	15.6	1.7
1850	1171	50.4	4.3	22.9	2.0	27.5	2.3
1900	1608	147.9	9.2	59.3	3.7	88.6	5.5
1950	2400	502.2	20.9	188.5	7.8	313.7	13.1
1960	2998	753.0	25.1	—	—	—	—
1980	4330	1380.0	31.9	—	—	—	—

[a] From G. Trewartha, *A Geography of Population: World Patterns.* New York: John Wiley & Sons, Inc., © 1969.

anization and greater utilization of land areas with resulting maximum yields. For example, although the farm population in the United States declined from 32.5 million in 1916 to 20.5 million in 1960, agricultural productivity has continuously increased.

Urbanization in the developed countries has followed economic development, while in the underdeveloped countries urbanization is the result of sheer biological increase at an enormous rate. Both urban and rural areas of the underdeveloped countries are increasing in population. This presents a special problem with reference to rural areas since the land is continuously being subdivided, with the resulting inefficient utilization and overuse.

The concentration of large populations within relatively limited areas has profound effects upon biotic and abiotic environments. Pollution of air, water, and land areas is a major problem. Large land areas are consumed to provide physical space for expanding population and industry. The technological, scientific, and economic advantages of the developed countries provide a basis upon which to find solutions to the problems presented by their increasing populations. The underdeveloped countries do not have these advantages to help solve their population problem and the increased demands by the developed countries place the underdeveloped countries at a further disadvantage.

In addition to providing for the physical needs of an ever-increasing population, the massing of large groups of people presents new social and cultural problems. For this reason emphasis must be placed upon a better understanding of human behavior. An understanding of the interrelationships of all aspects of life, not just the areas of the natural sciences, is essential for logical solutions to the population problem.

The overall world problem raises two questions.

1. Is it possible to support an ever-increasing world population? The answer to this question involves a scientific evaluation of the potential of the abiotic and biotic environments to produce a sufficient food supply and the technological advances required to realize this potential.

2. Is it probable that such scientific and technological knowledge, if available, will be applied in sufficient time to avoid an increasing loss of human lives? The answer to this question involves a re-evaluation of present political, economic, social, and ethical ideas, essentially through world-wide education and cooperation on a scale heretofore unknown.

Food Supply

Although it is impossible to present a complete discussion of present and future sources of food, an attempt will be made to discuss briefly some of the most important aspects of this topic.

Crops

The constantly increasing crop yields in Europe, North America, and the Soviet Union tend to lend support to the misleading idea that there is no foreseeable limit to the size of the world population that can be supported. However, plants depend upon available water, minerals, and suitable climate. There are limitations set by these needs of plants. In addition, plants are limited in their ability to utilize carbon dioxide and solar energy as well as in their capability to absorb water and mineral substances through a root system. While new strains are and will be developed to increase crop yield, this increase occurs only at the expense of limited water and mineral reserves. These high-yielding crops also need large amounts of fertilizers to attain their full potential. Increased crop yield alone is not sufficient. Evidence indicates that in many instances additional crop production has resulted in an increase in the carbohydrate (starch and sugar) content of food with a comparable decrease in protein content. Living organisms require at least 12% of their total caloric intake in the form of protein. The net result is that humans, as well as livestock, must consume a greater quantity of the newer strains of plant crops to obtain needed protein or else supplement their diets with other sources of protein. For example, in high-yielding rice strains, the protein content is reduced to 5 or 7%. Where rice constitutes a staple food in the diet, two or three times as much rice must be consumed to prevent malnutrition. Present-day hybrid corn fed to hogs must be mixed with protein concentrate while in the past years hogs could be raised almost exclusively on this grain.

In many instances increased agricultural production does not represent an increase in yield per acre but is the result of the opening of new land areas. As was stated in Chapter 24, the biome represents a climax in which the flora and fauna of the area have evolved to a point where they are utilizing the available resources to a maximum. When land is cleared and replaced by food crops, the resulting farmland is created at the expense of the ecosystem and has far-reaching results. For example, large-scale removal of forests throughout various areas of the earth has had devastating effects. The forests have the ability to retain water and regulate its flow. Forest removal has resulted in a substantial lowering of the ground water table. In addition, forests retard mineral depletion and soil erosion by preventing the rapid flow of run-off water and protecting the soil from direct wind contact.

Even more immediate drastic consequences result when tropical forests are cleared for farming. When the trees are removed, the soil is exposed to the sun and the air. Complex chemical changes take place and the soil turns to a hard rock-like substance called *laterite*. The area becomes totally unproductive. Laterite is so durable it has been used as a building material. At the present time there are no modern agricultural methods for preventing laterization. The only method of prevention is the ancient method of farming a small clearing for a year or two and letting the jungle reclaim the area. Although the consequences are known, tropical forest areas are being cleared because of the food demands of a growing population. The problem of laterization is increasing.

The needs of an ever-expanding population over the past several decades have resulted in an ever-expanding area of land brought under cultivation. This has reached a point where agricultural experts in all parts of the world agree that the additional areas available for the growing of food crops are exceedingly limited. It is estimated that there are about 750 million acres yet available for cultivation. The rate at which the population is growing requires about 125 million additional acres each year. Unfortunately, these reserves are not equally distributed and those areas of the world which have the greatest need have the smallest reserves.

There are certain areas of the earth, such as deserts and dense tropical forests, which might provide additional land areas for growing food. However, technological advances have not proceeded to the point where utilization of this land on a large scale is economically feasible, at least in the near future. If and when these areas can be brought under control, due consideration must be given to the major problems of maintaining their productivity.

Through various modern agricultural practices such as multicropping, irrigation, and crop rotation, the productivity of a given land area can be greatly increased. However, each of these practices must be pursued with great care and with an understanding of the total ecosystem. Multicropping is the practice of planting two or more crops on the same plot of land during a single growing year. While this most certainly increases the agricultural productivity for a given area, it also requires considerably greater inputs. The soil must be replenished by the addition of commercial or natural fertilizers as well as by the addition of water.

As has been stated previously, the productivity of a given land area can also be increased through the use of hybrid, high-yielding strains of grain. Yet this high yield may represent a decrease in protein content. Through selection it is possible to develop high-yielding strains with a high protein content. The development of high-protein grains by the International Rice Research Institute of the Philippines, high-lysine corn in the United States, and high-protein wheat in Mexico has had encouraging results. Productivity of proteins in certain areas of Pakistan, India, the Philippines, and Mexico has been increased considerably as a result. However, such plant breeding takes time and careful control. Through a change of crops it is possible to switch to plants which have a high protein yield, for example, switching from rice or wheat to soybeans. Of course, such switching and planting high-protein hybrids does not represent a case of obtaining something for nothing. High-protein crops rapidly deplete the soil of nitrates, and fertilizers and an increased water supply are required to replenish the soil.

Clearly, it is possible to increase greatly the production of food crops. Such an increase could help to provide a large source of food. However, all aspects of crop production must be carefully considered. Greater cooperation in production and distribution must be pursued on a world-wide basis in order to prevent a further concentration of the food supply in limited areas of the earth.

Livestock

Much of what has been said about plant crops also applies to livestock and poultry because animals are completely dependent upon crops and pasture lands. While there has been considerable progress in developing high-yielding livestock and poultry, this has been to a great extent at the expense of utilizing other food reserves, where they exist, at a more rapid rate. Animals, as consumers, are further along on the food chain. Therefore, there is a greater expenditure of energy in the production of animals for food.

It is in livestock and poultry production that the greatest inequalities exist in the world-wide distribution and use of foodstuff. In the developed countries of North America and western Europe modern-day livestock and poultry production shows little relationship to the amount of land needed to produce the quantities of meat and meat products consumed by these nations. The chicken on the family dinner table has probably never seen the light of day, much less had any intimate contact with the land that supplied its food. Livestock production has become an industry in which there is an input of feed and an output of consumable food.

In the raising of animals, providing the minimum protein needs is essential to the production of sufficient quantities of meat or meat products. Therefore, most of the protein-rich plant products are retained within the developed countries and, more importantly, large quantities of feed and protein-rich fishmeal are imported from the underdeveloped countries. The two major fishmeal protein producing areas are located off the western coast of Africa and the western coast of South America.

Approximately two-thirds of the output of fishmeal and fish oil from the South American area is exported to western Europe and one-third to the United States—practically the entire output. Presently, the Peruvian and Chilean fisheries export more protein-rich fishmeal than all the protein obtained from the production of livestock in South America. The fishmeal exported from Africa each year could reduce the protein shortage in that region of the world by 50%. This inequity in protein distribution will increase as the population in both the developed and underdeveloped countries increases. At this time it is difficult to conceive of any clear-cut solution. In the developed countries meat and animal products form an important staple part of the diet. It is quite unlikely that diets of the developed countries will change drastically. At the same time the nutritional needs of the underdeveloped countries involved in fishmeal production, while at the critical stage presently, will contine to increase beyond this stage. To further complicate the situation, the exportation of protein-rich fishmeal from the underdeveloped countries is intimately involved with the economy and politics of these nations.

Parasites, Disease, and Waste

A considerable increase in the food supply of the world would result from better control of parasites, disease, and waste. Each year, millions of tons of food are lost because of this lack of control, yet, it is within our present scientific and technological capability to control this loss. Again, the underdeveloped areas suffer the greatest loss through parasitism and disease. However, even the developed countries are not free from this problem. A measure of control can be exercised if there is careful and intelligent use of pesticides. There can also be further development in breeding more disease-resistant strains of crops and livestock.

Probably the least justifiable loss is that of food which has been gathered, processed, and stored. Millions of tons of grain are lost to various types of vermin and fungi. Improper processing and storage result in large-scale spoilage of meat and other types of food. These losses not only represent an actual loss of food but also cause an increase in the cost of remaining food.

In the processing of food, especially within the developed nations, much waste results. Selection of only choice portions with the discarding of large amounts of edible portions which are nutritionally as high as or higher than those retained represents a substantial loss of a limited world food supply. Recently there has been greater utilization of these waste products as a source of animal food but there can still be substantial improvements in this area.

The Ocean

The ocean has been providing food for man ever since earliest recorded history. As the population has grown, so too has interest in the ocean as an increasing major resource in food production. The ocean is essentially the only foreseeable major recourse within the reach of present technology that is not yet being utilized fully. However, as has been true with many terrestrial areas of the earth, exploitation has already occurred in certain ocean areas and has drastically altered ecosystems and greatly reduced productivity.

Estimates indicate that the present production of approximately 60 million tons of fish a year, of which half is consumed directly by humans and the other half converted to fishmeal that is fed to livestock, could be tripled. However, these estimates must be viewed with caution. It is possible that heavy harvesting may drive many species into extinction and greatly reduce the productivity of the sea. The increase in fish production from the ocean represents one of the few instances in which an increase in foodstuff has continued to exceed the growth of the population. In order to maintain this resource in a constantly renewable condition, continuous ecological research and cooperation on an international scale are necessary. One of the major advantages of the ocean is that it provides so much protein-rich food. Moreover, its resources are readily available to many underdeveloped countries which lack large agricultural resources to great industrial potential.

In the utilization of the ocean over the years, there has been increased emphasis upon using the catch for fishmeal for animal feed rather than for direct consumption. This has several advantages.

1. It utilizes large schools of certain species of fish that are too small, too bony, too oily—or undesirable because of other characteristics—to be used as human food.

2. Utilization of these fish tends to maintain a better ecological balance by limiting the number of species caught.

3. Many of these less desirable species depend upon phytoplankton as their source of food, and as such they are closer to the beginning of the food chain and utilize the available energy with less loss.

Logically, there is a limit to the amount that can be removed from the natural supply of fish year after year. For example, the amount of cod and haddock taken from the North Atlantic has not increased but rather appears recently to have decreased. As a source of food the ocean must be viewed as potentially providing for large segments of the world population. The extent to which the oceans are able to continue to increase production depends upon the size and potential of unfished stocks such as those in the Indian Ocean. It is also dependent upon the potential of those areas that are presently being fished, and our ability to manage fishing on national and international levels in order to sustain maximum yields.

Already, overfishing and pollution have diminished supplies close to populated land areas. This is especially true of shellfish such as oysters, clams, and shrimp. Each year the number of factory ships increases in the major fishing areas of the extreme northern and southern regions of the oceans. These ships are able to process hundreds of tons of fish daily and remain at the fishing grounds for long periods of time. Large quantities of fish are being removed with little attention paid to the resulting consequences on the total ecosystem of the ocean.

Additional Sources of Food

Various additional sources of food have been suggested. These have limited application or are not practical from an economic or technological point of view at the present time. In any event, they represent a very limited supply of food.

In Japan, marine algae have been cultivated on a large scale. In addition, there has been extensive experimentation with single-celled freshwater algae. In most instances, algae have not proved to be wholly acceptable from a nutritional point of view. The cellulose and cellulose-like composition of plant structures cannot be digested and metabolized by humans. Conversion to a product that can be metabolized by humans is too expensive at the present time and the products are not entirely palatable.

Plankton from the sea has also been suggested as a source of food. While the total quantity in the ocean is enormous, the amount in any given volume of sea water is exceedingly small. Obtaining sufficient quantities would necessitate filtering millions of tons of sea water. At the present time, this is not economically feasible. In addition, certain species of plankton are harmful, and the necessary control and inspection would be difficult. There is also the most important consideration that these organisms are essential for the total ecology of the ocean and are also the largest group of photosynthetic organisms. Large-scale removal probably would have drastic effects upon the total biosphere.

Nonedible wastes from food processing are assuming greater importance. Waste products such as banana peels, orange peels, and coffee grounds contain valuable nutrients which can be processed into excellent animal feed. Greater utilization of this type of waste can provide an important source of food. Sewage also contains large quantities of valuable nutrients which, if processed, could be used to produce algae which in turn could be utilized as animal feed. The processed sewage could also be used to increase the algal content of lakes for raising more fish. This has been accomplished on a minor scale in Japan. However, the full possibilities for the utilization of wastes have yet to be realized.

The Food Shortage Problem

The discussion thus far points to the fact that our present food supplies are insufficient to feed the growing world population. While estimates vary, it is almost certain (barring some alternative holocaust such as a mass nuclear war or unknown plague) mass starvation on a tragic scale will take place within this century. The realization that this will occur within 25 years indicates the immediacy of the problem and makes it the most important and serious problem facing mankind.

The contrasts in the resources and standards of living between the developed and underdeveloped countries are striking. As the populations in the developed countries increase, so do the demands upon and within the underdeveloped countries. While the people of the underdeveloped countries are unschooled, they are not unaware of the huge differences in the standards of living in the different regions of the world. With few exceptions, people in even the most isolated areas of the earth have been introduced to the wonders of science in the production of food and goods and in promoting better health. In the urban areas, which include the greater mass of human population, motion pictures, television, posters, and transistor radios as well as items displayed in stores have increased the expectations of the peoples in these underdeveloped areas. If their expectations are not fulfilled at least to a reasonable degree, there will be increasing unrest, discontentment, and frustration, if not open conflict.

An immediate solution to the problem would appear to be an increase in the production of food. However, with few exceptions this has not proven to be successful. In Chapter 24 it was shown that within the ecosystem the natural populations tend to stabilize when they reach the carrying capacity of the environment, that is, the optimum utilization of raw materials and energy. Man, like all living organisms, is part of and dependent upon the natural environment. There are many who believe that man has reached the carrying capacity of the biosphere, if not exceeded it. There is a limit to the size of the human population that the resources of the earth can continue to support. Therefore, there must be some sort of balance established between the rate of reproduction and the rate of food production and other resources. This balance can only be attained through

1. Increasing the efficiency of our productive capacity and broad distribution of products and resources.
2. Population control through severe reductions in birth rates.

Greater efficiency of food production can be obtained by

1. Discovering and applying methods of recycling all resources.
2. Applying known and discovering new methods of control of parasites and disease, pollution, and sheer waste.
3. Applying known and discovering new methods in agriculture to increase crop and livestock production through a maximum utilization and understanding of existing ecosystems.

Population Control

The world's population will continue to increase as long as birth rates exceed death rates. To stabilize the population, birth rates must be decreased or death rates increased. Of the alternatives, increasing the death rate is the least desirable. In fact, the total effort of man in some way is related to extending the length of life. Therefore, the most acceptable solution is to reduce the birth rate to the level of the natural death rate or below. The enormity of the problem can be visualized when it is understood that almost 50% of the present world's population will reach their reproductive age within this decade.

Even if there is an increase in food production through the various methods suggested, there are clearly limitations in supplying food for an ever-increasing population. People, although subjected to the same physical and biological forces as all living organisms in the biosphere, are capable of using their intellect to enable them to survive under adverse conditions. This ability to survive under adverse conditions results in the increase in population. From a moral, ethical, and social point of view, population control must be directed toward limiting the number of human beings. This must be done through control of the birth rate.

Birth Control

Various methods of birth control have been known and practiced in varying degrees throughout recorded history. These methods may be broadly classified as preventive, contraceptive, or corrective.

Preventive Methods. Preventive methods include abstinence, castration, and sterilization. Among small groups, complete or partial abstinence has been practiced. Usually, this has been done for religious, cultural, or moral reasons. However, this procedure is not a practical means of birth control on a worldwide scale.

Removal of the testes in the male (*castration*) and removal of the uterus (*hysterectomy*) or ovaries (*oophorectomy*) in the female are absolute preventive methods. Once performed, the operation cannot be reversed, and it is impossible to produce offspring. These methods are normally not a means of birth control. Removal of the testes or ovaries has adverse effects on the individuals because of the importance of these organs in the endocrine system. These operations are generally performed only if the organs are diseased.

Sterilization in males is called a **vasectomy** and is a simple operation in which a portion of each vas deferens is removed (Figure 26-3A). An incision is made in the scrotum, each ductus deferens is located and tied in two places, and the portion between is cut out. The operation is 99% effective and once sterilization occurs it is a complete method of birth control. A vasectomy may be reversible. Statistics differ on the success of reversing the operation, varying from 30 to 80%, depending upon the sample considered. Rumors and superstitions concerning vasectomy have been widely spread, especially in underdeveloped countries. These include a loss of sexual potency and desire, ill health, and absolute irreversibility. Although none of these conditions can be attributed to a vasectomy, unfortunately, as a result of such misinformation, the procedure has not been readily accepted in most high population growth areas. At the present time there are appreciable sterilization programs only in India and Japan.

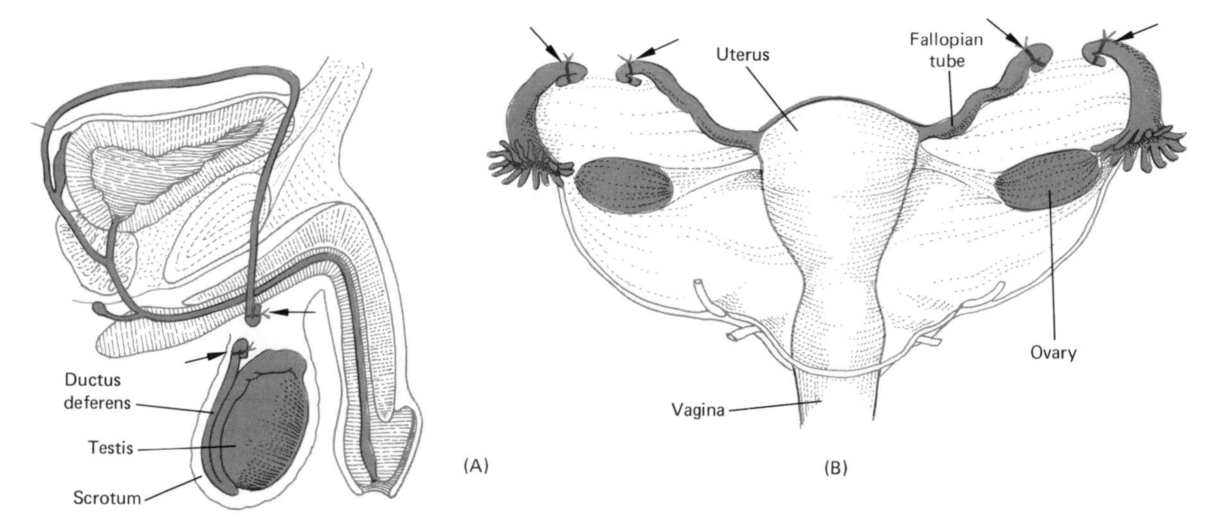

Figure 26-3. Sterilization. (**A**) Vasectomy. Through an incision in the scrotum, the ductus deferens is cut and tied. (**B**) Tubal ligation. Through an incision into the abdomen, the fallopian tube is cut and tied.

Sterilization in females is generally achieved by an operation on the fallopian tubes (Figure 26-3B). A special procedure, called *tubal ligation,* has been developed that is rapid and quite effective. The tubes are squeezed and a small loop called a *knuckle* is made. A suture of catgut is tied very tightly at the base of the knuckle and the knuckle is then cut off. After 4 or 5 days the suture is digested by body fluids and the two severed ends of the tubes separate. The ovum is thus prevented from passing to the uterus and the sperm cannot reach the ovum. This method of sterilization requires a short period of hospitalization and is not as simple as a vasectomy. Generally, the operation is performed immediately after childbirth since it has been found to be simpler at that time and adds little to the normal recovery period.

Of all the various methods of birth control, the preventive method of sterilization is the most effective. Once the operation has been performed successfully, there is no chance of conception. However, the probability of a universal application of this method of birth control within the foreseeable future is extremely remote.

Contraceptive Methods. Contraceptive methods are those in which sexual potency is maintained but fertilization of the ovum is prevented. They include "natural," mechanical, and chemical means. The "natural" method includes the "rhythm method" or safe period. This method takes advantage of the natural physiological process of the menstrual cycle of the female. Fertilizable ova are only present during a period of 3–5 days in each menstrual cycle. During this time couples must refrain from intercourse. One of the difficulties in this method is that few women have absolutely regular cycles and ovulation cannot be predicted far enough in advance to define the fertile period. An additional difficulty of the rhythm method is that while the probability of conception is low during certain periods, statistically there is never a period when conception is impossible.

Mechanical means of contraception include the condom used by the male and the diaphragm by the female. The *condom* is a nonporous, elastic covering placed over the penis that prevents deposition of sperm in the female reproductive tract. The *diaphragm* is a dome-shaped structure that fits over the cervix and is generally used in conjunction with some kind of sperm-killing chemical (Figure 26-4). The diaphragm stops the sperm from passing into the cervix, and the chemical kills the sperm cells.

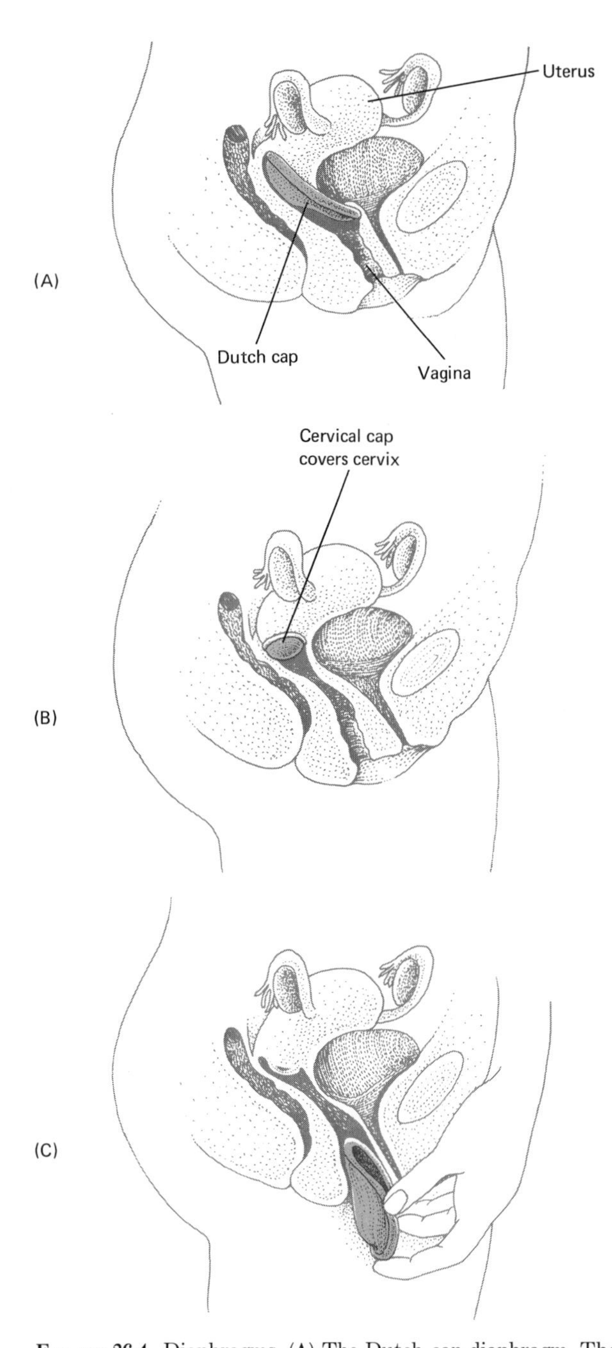

Figure 26-4. Diaphragms. (**A**) The Dutch cap diaphragm. The thin spring around the margin of the diaphragm opens out, presses against the wall of the vagina, and stretches across the entire top of the vagina. (**B**) The cervical cap diaphragm fits directly over the cervix of the uterus. (**C**) Procedure for inserting the Dutch cap.

Another mechanical method of contraception is an *intrauterine device* (*IUD*). Some examples of IUDs are shown in Figure 26-5. These are small objects such as loops, spirals, and rings made of plastic, copper, or stainless steel that are inserted into the cavity of the uterus. The operation of IUDs is not fully understood. Some investigators believe that they cause changes in the lining of the uterus resulting in the production of a substance that destroys either the sperm or the fertilized ovum. Some evidence indicates that IUDs cause tissue destruction and uterine perforation.

There have been modifications in these devices, and all are in various stages of development. Many users of the device spontaneously eject it, bleed, or suffer discomfort. Further research is taking place in an attempt to discover better shapes, better materials, easier methods of insertion, and the exact way the devices function.

Wide-scale programs throughout the world have introduced the IUD. The method provides a very economical means of birth control and, unlike sterilization, is easily reversible. Once the device, which is usually a simple looped tube, is removed, normal reproductive function can occur. International programs encouraging the use of the IUD in the underdeveloped countries and those areas where population increase is the greatest have been increasingly effective.

Chemical means of birth control include the use of various vaginal douches, suppositories, creams, and foams that make the vagina unfavorable for sperm survival. Of the newer chemical means, oral contraception ("the pill") has found rapid and widespread use, particularly in the more developed countries. This means of contraception utilizes various types of synthetic steroid drugs that inhibit follicle development and ovulation. When taken, these drugs prevent the release of pituitary hormones leading to ovulation. The disadvantage in the use of the "pill" on a world-wide scale is that this medication is relatively expensive and beyond the economic means of a large portion of the world's population. Also, there are indications of possible undesirable side effects. Interestingly enough, it has been found that a woman faces greater dangers from pregnancy and childbirth than from taking the pill and not becoming pregnant. The use and success of these drugs have stimulated research on physiological contraception. Undoubtedly, better and more effective

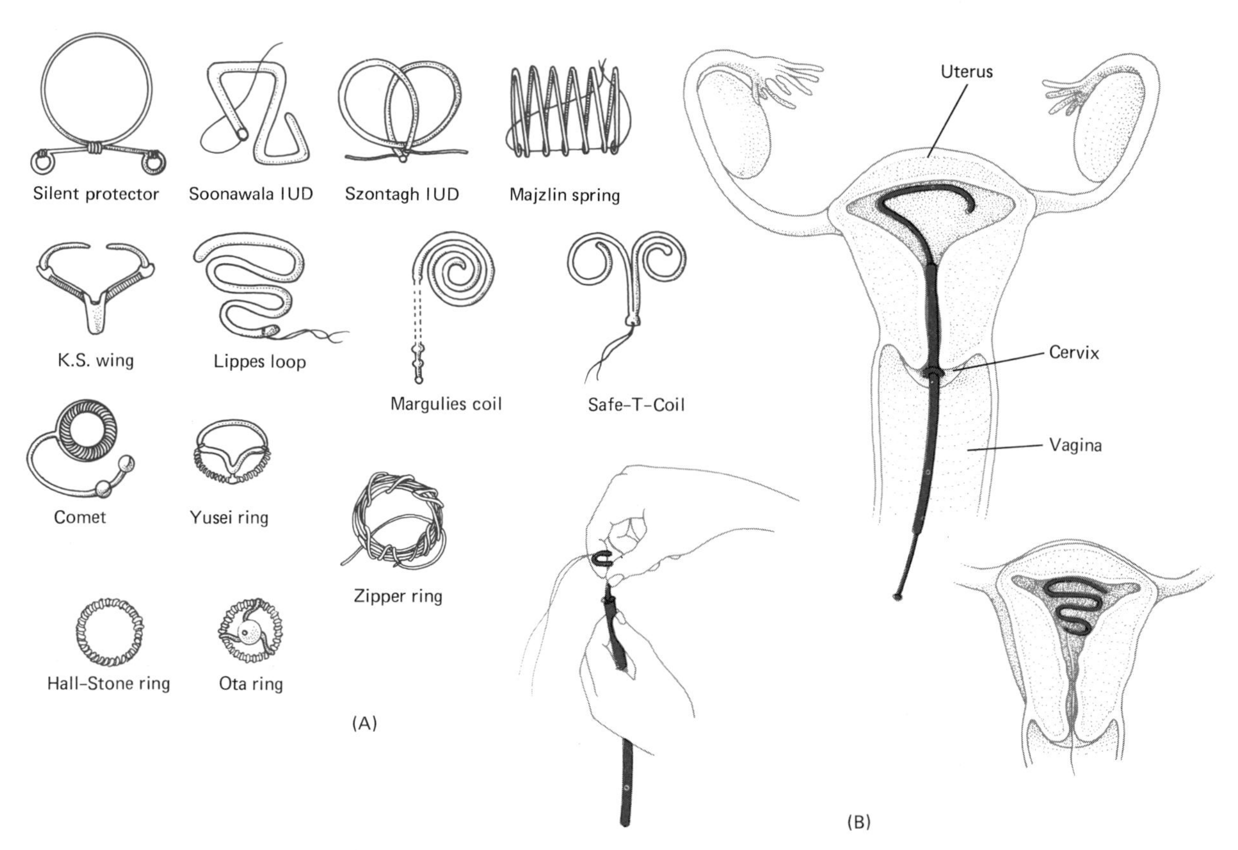

FIGURE 26-5. Intrauterine devices. (**A**) Representative types of intrauterine devices. (**B**) Procedure for insertion of intrauterine devices. IUD's are inserted into the cervix. The device is threaded into a long, narrow-bore tube that is passed through the cervix. Once in position in the uterus, it spreads out to its former shape. Most IUD's have a thread or chain projecting into the vagina. The thread or chain, which may be detected by a finger, indicates if the device is still in place; it also may be used for removing the device.

drugs will be developed in the near future. A summary of the contraceptive methods can be found in Table 26-4.

The use of contraceptive methods as a means of birth control requires some degree of education and self-control regardless of the particular methods or techniques used. The degree of effectiveness of these methods varies, unlike the preventive methods, which are absolutely effective.

CORRECTIVE METHODS. Corrective methods of birth control are either elimination of the fetus before birth (abortion) or the elimination of the child after birth (infanticide). While infanticide has been practiced from time to time throughout history, it has not found any acceptance among modern societies. On the other hand, induced abortion has been and continues to be practiced within all societies. In fact, it may be the most widely used single method of birth control in the world today. Although there is considerable controversy in the United States with regard to abortion, because of ethical, social, and legal problems involved, legalized abortion is a major phenomenon in many countries and in some states of the United States. The legalization of abortion in Japan rapidly cut the birth rate in half. Social standards of Western countries have placed emphasis in other directions than abortion for answers to an increasing birth rate; however, there are indications that this may change.

Table 26-4. Methods of Contraception[a]

	Description	Effectiveness and Acceptability
	A. Contraceptive Methods Requiring Consultation with Physician	
1. Oral contraceptive (the pill)	It is generally accepted that the synthetic hormones contained in oral contraceptives (estrogens and progestins) inhibit ovulation. There are two oral contraceptive pill methods. The most commonly used method is often referred to as the "combination" or "balanced progestin-estrogen" method. This method is considered to be slightly more certain in preventing unwanted pregnancy. Each pill contains a combination of both synthetic estrogen and progestin to assure inhibition of ovulation. When no egg is released from an ovary, a woman cannot become pregnant. The other, called the "sequential method," employs two different pills each month. When this method is used, a pill containing synthetic estrogen is taken daily for the first 15 or 16 days. This pill inhibits ovulation. A second pill containing a mixture of synthetic estrogen and progestin is then taken for five days to assure orderly bleeding within 3–5 days after the last pill is taken each cycle. The pills are usually taken for 20 or 21 consecutive days each cycle. A variation of the 21-pill schedule employs 7 additional inactive pills for continuous daily pill taking to simplify the schedule.	Except for total abstinence, or surgical sterilization, the combination pill is the most effective contraceptive known. True "method failures," exclusive of patient errors, are extremely rare. The sequential method appears to be slightly less effective than the combination method in preventing pregnancy, even with perfectly regular use. Approximately one-third of women will experience one or more of the commonly observed side effects during the first cycle or two of use. Experience indicates that these side effects (nausea or other gastrointestinal distress; spotting or "unexpected" breakthrough bleeding; breast tenderness, enlargement, or secretion; fluid retention and weight gain) are usually quite mild and transient, and tend to diminish sharply in incidence and severity or disappear completely after first cycles. Occasionally, other side effects may occur. Since some of these may be serious, patients should discuss them with their physicians. A wide variety of adverse experiences has been observed in association with the use of oral contraceptives, some of which have been attributed to the medication. The evaluation of these findings is extremely difficult because the conditions in question may occur among women who are not "on the pill" and may, therefore, be expected to occur from time to time among the millions of users of oral contraceptives. The one important condition for which a statistical association has been definitely established is thromboembolic disease, a disturbance of blood clotting, including its sometimes fatal outcome, pulmonary embolism. While this condition occurs several times as frequently among users of oral contraceptives as among nonusers, it is a rare side effect even among users, with about 1 woman in 2000 hospitalized during a given year and 1 death among 30,000 women resulting from the medication. These risks should be evaluated against a background of the known risk to life associated with pregnancy and childbirth which (in developed countries) is on the order of 1 death per 5000 pregnancies, excluding the risks of illegal abortion. The laboratory studies of the liver and various body functions in women using oral contraceptives have revealed, in some instances, deviations from normal values. The possible relationships are not fully understood.
2. Intrauterine devices	Small objects (loops, spirals, rings) made of plastic or stainless steel; inserted into uterus by physician. May be left in place indefinitely. How the device prevents pregnancy is not completely understood. They do not prevent the ovary from releasing eggs. They	Since the devices do not require continued attention on the part of user, this method is especially effective in situations where sustained motivation is lacking. Some women who try the devices cannot use them satisfactorily because of expulsion, bleeding, or discomfort. The devices are not recommended for women who have not had a child (except in special cases) because the uterus of such women is too small and the cervical canal too narrow. Infrequently, the

[a] Prepared by Christopher Tietze M.D., Associate Director, The Population Council.

TABLE 26-4. Methods of Contraception [*Continued*]

	Description	Effectiveness and Acceptability
	possibly cause changes in the lining of the uterus which, in turn, produce a substance that may destroy either the sperm or the fertilized egg when it arrives in the uterus.	insertion of the device may lead to a mild or severe inflammation of the pelvic organs; in a few instances, death has resulted from this condition. The level of protection offered is probably not greater than provided by traditional methods used with perfect regularity.
3. Diaphragm	Flexible hemispherical rubber dome used in combination with cream or jelly; woman inserts into vagina to cover cervix, providing barrier to sperm. Must be left in place at least 6 hours after intercourse and may be left in place as long as 24 hours. Must be fitted by physician; refitted every 2 years and after each pregnancy.	Offers a high level of protection although occasional method failures may be expected because of improper insertion or displacement of the diaphragm during sex relations. A rate of 2–3 pregnancies per 100 women per year seems to be a generous estimate for consistent users. If motivation is weak, much higher pregnancy rates must be expected.
4. Rhythm (While some couples have successfully worked out this system for themselves, most will require assistance from a physician or rhythm clinic.)	Depends on abstinence from intercourse during time of month when woman is fertile. Due to menstrual irregularity in many women and inability accurately to determine time of egg release, success with the method may require abstinence for as long as half the month.	Correctly taught, correctly understood, and correctly practiced, the effectiveness of the rhythm method in women with regular menstrual cycles may approach that of mechanical and chemical contraceptives. However, such successful use implies periods of abstinence longer than most couples find acceptable. Self-taught "rhythm," haphazardly practiced, is one of the most ineffective methods of family planning. The effectiveness of the rhythm method can be greatly increased by measuring and recording the body temperature each morning before getting up. A small rise in temperature will show that an egg has been released 1 or 2 days earlier and can no longer be fertilized.
	B. Contraceptive Methods Not Requiring Consultation with Physician	
1. Condom	Thin, strong sheath or cover, made of rubber or similar material worn by man to prevent sperm from entering vagina.	Offers high degree of protection if man will use it correctly and consistently. Some couples find the use of condoms objectionable. Failures are due to tearing of the sheath or its slipping off after climax. The condom rates in effectiveness with the diaphragm.
2. Chemical methods	Products inserted into vagina. Purpose is to coat vaginal surfaces and cervical opening, and to destroy sperm cells; may act as mechanical barrier as well. Provide protection for about 1 hour.	The effectiveness of chemical contraceptives used alone is believed to be lower than the effectiveness of the chemical preparations used in combination with a diaphragm or a condom. Nevertheless, significant reductions in pregnancy rates may be obtained by the use of these simple methods.
(a) Vaginal foams	Cream packed under pressure, inserted with applicator.	Among the various forms of chemical contraceptives, the vaginal foams appear to be most effective, followed by the jellies and creams. Foaming tablets and suppositories are the least effective of the chemical methods.
(b) Vaginal jellies and creams	Inserted into vagina with applicator.	Drainage of the chemical materials from the vagina is objectionable to some couples. Foaming tablets may cause temporary burning sensation.

[*continued*]

Table 26-4. Methods of Contraception [*Continued*]

	Description	Effectiveness and Acceptability
(c) Vaginal suppositories	Small cone-shaped objects that melt in vagina. Must be inserted in time to melt before the sex act.	
(d) Vaginal tablets	Moistened slightly and inserted into vagina; foam is produced. Must be inserted in time for tablet to disintegrate before the sex act.	
3. Coitus interruptus (withdrawal)	Man withdraws sex organ from woman's vagina before emission of semen. Requires that man practice great self-control. Even then, some sperm may escape before the climax.	Coitus interruptus has been responsible for many failures of family planning, probably because semen may be deposited without the man being aware of it. However, statistical studies have also shown coitus interruptus may be relatively effective. This is the principal method by which the decline of birth rate in Western Europe was achieved from the late eighteenth century onward. This method is unacceptable to a large number of couples because it may limit sexual gratification of the man or woman, or both.
4. Douche	Flushing the vagina immediately after intercourse to remove or destroy sperm.	Considered to be a poor method of contraception because sperm enter the cervical canal within 90 seconds after ejaculation. Statistical studies have confirmed this low level of effectiveness.

Education

From the previous discussion it can be seen that there are scientific means of achieving a reduction in birth rate. Objections to the use of contraceptive methods are based upon economic, religious, or cultural grounds, or upon the belief that the available means are too technical and sophisticated for mass utilization. In addition, probably the most significant factor preventing the use of contraceptive methods is the lack of proper motivation. To be acceptable, a method must conform to religious and cultural patterns, be economical, and be readily available. While no one method meets all these requirements, there would appear to be at least one method available and appropriate for each particular society. This does not mean that the acceptable method would be completely effective, but it could result in a substantial reduction in birth rates.

The major problem is getting large numbers of people to use some means of birth control. Ultimately the responsibility for limiting the birth rate rests with the individual family. The way to reach individual families is through education. Education has a two-fold result. First, couples can come to realize that it is no longer necessary to have large numbers of children to ensure survival of two or three and that large numbers in a family no longer ensure economic independence. Second, through education and development of skills the family can become more economically independent, thus raising the standard of living. However, raising the standard of living for a family is not sufficient in itself. Parents must want their children to attain a higher standard of living than they themselves have. In order to give their children advantages, they must limit the number of children. Evidence indicates that as the standard of living increases, the birth rate drops.

Increasing population is of concern to every human. Even if each family today limited itself to two children, the population would reach 10-12 billion people before stabilizing (births equaling deaths). With more than half the world's population inadequately nourished, it is difficult to believe that food production and distribution will be increased within this decade to feed the known increase in population. The various methods of birth control can reduce the number of births. However, their success has been so minor that they have had little effect in reducing the total growth of world population.

The Changing Environment

Modern man has made drastic changes in the biotic and abiotic environments. These have been made both consciously and unconsciously as man purposely changes the environment to suit his needs. Large land areas are cleared, arid land is irrigated, the courses of rivers are changed or are dammed to make large artificial lakes, mountains are moved, and the currents of inland marine waters are altered. Man has even been able to make intentional minor changes in the weather, and it is within the realm of possibility that he will be able to make even greater changes. The flora and fauna of the various biomes have been purposely changed, substituted, or destroyed.

As a result of these activities, many changes have occurred that were not expected. Ecosystems are self-maintaining and self-sustaining systems. As man changes the ecosystems, he removes important factors in their self-sustaining property. Their balance is destroyed, the natural cycles are interrupted, and the natural regenerative properties are lost. Since man is dependent upon these systems as resources, he must now face the dilemma that he is destroying a source of his basic needs to satisfy other needs.

Population Changes

Much of the pollution problem is the natural and direct result of a rapidly increasing population (Figure 26-6). However, this increase in itself does not account for the total extent of the problem. The concentration of large populations into urban and metropolitan areas produces human, household, and industrial wastes (especially in the highly industrialized nations) in quantities that overwhelm the natural decomposition processes of the environment (Figure 26-7). An additional factor contributing to the pollution problem is the nature and standard of living of a specific society. The underdeveloped countries are not yet faced with pollution problems involving industrial, transportation, or household wastes to the extent that the developed countries are.

Within the developed countries, urbanization has resulted in special problems of pollution control. This is especially true where a high standard of living exists in an affluent society. The quantities of fuel, water, and food consumed by urbanites are astronomical and continue to increase as the standard of living is raised and urbanization continues. The average city dweller in the United States directly or indirectly uses 150 gallons of water, consumes 4 pounds of food, and burns 19 pounds of fossil fuel each day. This results in 120 gallons of sewage, 4 pounds of food refuse including its containers, and 1.9 pounds of air pollutants. When one considers that 80% of the United States is urban, the daily wastes are measured in billions of gallons and millions of pounds. These wastes are the result of activities that an affluent society considers essential. There are additional tons of wastes resulting from items that are discarded because of changes in styles or fads or because the cost of repair exceeds the cost of new items.

One of the outstanding characteristics of the modern affluent society is its mobility. With increased "urban sprawl," the people are located at greater distances from their work and centers of commerce. Public and private means of transportation have exceeded that which seemed possible only a century ago. In the United States most families consider an automobile a necessity and the "two-car family" is commonplace. The millions of automobiles, trucks, and buses on the highways pour millions of pounds of waste products into the atmosphere daily.

Technological and Industrial Changes

Industrial and technological changes are interdependent within a growing affluent population. Early industry was located near lakes, rivers, or harbors in order to have an adequate water supply for power and ease of transportation. The water was also a convenient means for waste disposal. As the cities grew and a higher standard of living demanded increasingly better and greater amounts of consumer goods, still higher demands were made upon industry. This resulted in increased production and greater technological advancement and increasing yields of wastes. Not only are larger amounts of wastes dumped into the waterways, but much pollution also results from gaseous emissions into the atmosphere. In addition, newer types of pollutants result from industrial and technological changes. These large quantities of wastes build up and accumulate in the environment.

With advances in technology, the chemical industry

FIGURE 26-6. Large land areas cleared as a result of population growth. The increase in population has changed large areas of the biosphere. [*Courtesy of Chamber of Commerce, Fort Wayne, Ind.*]

has assumed a leading role as a source of pollution. Synthetic materials have relieved or removed the heavy demands upon natural products. In addition, improved means of processing and production involve greater uses of chemicals. Waste chemicals now accumulate in greater and greater quantities. Technological advancement also accounts for the accumulation of large amounts of scrap wastes. Machinery, household appliances, vehicles, etc., soon become outmoded, to be replaced by newer and technologically better models. The discarded items become waste. In most instances it is more profitable to produce directly from raw products than to reclaim scrap wastes.

Within recent years experimentation with and utili-

(A)

(B)

Figure 26-7. Mounds of refuse. The amount of waste has increased to a point where it overwhelms the ability of the environment to decompose it. (**A**) Vast expanse of dead algae along the shore of Lake Michigan. (**B**) Smoke pouring from burning mounds of refuse. [***A*** *Federal Water Quality Administration photo;* ***B*** *National Air Pollution Control Administration photo.*]

zation of nuclear weapons have produced in radioactive wastes that can have far-reaching effects. When radioactive fallout reaches the earth, it makes food unfit for consumption, concentrates at various trophic levels of the food webs, and results in mutation and even death. In the production of electricity by nuclear power, millions of gallons of water are needed each day to remove waste heat. The cool water enters the power plant and is emptied as warm water into streams, creeks, and rivers. This warm water has resulted in a unique type of pollution called *thermal pollution.*

In the increased effort to produce more food, more effective insecticides and herbicides have been developed. Also, new methods of cultivation require the spraying or dusting of large areas by means of mechanized sprayers on trucks or by airplanes. These insecticides and herbicides cover large areas and are carried into the atmosphere. Many of these chemicals retain their potency for long periods of time. That which falls to the ground is carried by runoff and ground water. In addition to agricultural uses, insecticides, herbicides, and pesticides are used to control disease-carrying insects, weeds, and vermin. This use also has far-reaching effects upon the ecosystems. One of the most serious problems was that of DDT. Intensive research on the distribution and effects of DDT has been one of the most effective means of drawing attention to the importance of pollution.

Research is being conducted to discover effective but far more selective chemical insecticides. Recent studies have indicated the possibility of using hormones, which, when applied at definite times, interrupt the normal life cycles of a given species of insect (Figure 26-8). Certain of these hormones have been isolated as a result of studies of insect physiology. The advantages of such types of insecticides arise from their selectivity and from the fact that the species is not able to develop an immunity to them because they are a part of the biochemistry of the insect and are needed at one point in the normal developmental cycle.

FIGURE 26-8. Insecticide research. In the search for ways to control insects without using insecticides, scientists of the U.S. Department of Agriculture have synthetically developed certain chemicals that keep insects from maturing completely, thereby making it impossible for them to reproduce. This photo shows the effects of a synthetic hormone. When sprayed in microscopic amounts on a yellow mealworm, the hormone caused an adult head and thorax to develop but prevented an adult abdomen from developing. It might be possible to control insect pests with such chemicals. [*USDA photo.*]

Climatic Effects

Studies on smog have indicated that it persists despite a considerable reduction in industrial and household gaseous emissions (Figure 26-9). This research has further established that in many cases smog is associated with a climatic condition known as *temperature inversion.*

With increasing elevation, the air temperature usually falls. Sometimes weather conditions are such that the temperature first rises with increasing height above the ground so that two distinct overlying air masses, a

FIGURE 26-9. Smog layering above the city of Denver, Colorado. [*National Air Pollution Control Administration photo by Charles E. Grover.*]

warm upper layer and a cool lower layer, exist. In the absence of winds these two layers remain stationary and may remain for days at a time. Pollutants gradually accumulate to high concentrations in the lower air mass. The action of the ultraviolet light from the sun upon some of these pollutants converts them into more toxic substances. As a result of the combination of thermal inversion and pollution, smog occurs over heavily populated areas such as Los Angeles, New York, and London. At times the concentration of pollutants reaches levels that endanger human life and cause damage to plants and property.

Pollution

Pollution may be broadly classified as water pollution, air pollution, and land pollution. In certain areas of the world one type may be more serious than another, depending upon the degree of development of the society. However, within the developed countries all three types present tremendous problems.

Water Pollution

As was shown in Chapter 22, the water areas of the hydrosphere are generally interconnected. Streams lead to rivers, rivers to lakes, lakes to rivers, and finally to bays and oceans. Generally, water falling as rain finds its way to streams or seeps into the ground and eventually finds its way to the sea. Pollutants can therefore be carried great distances and produce effects far from their sources.

The main uses of water in the United States, in order of decreasing quality required, are public water for drinking, bathing, and cooking, fish or shellfish propagation, industrial water, agricultural, hydroelectric power, navigation, and disposal of sewage. In order to maintain the quality of water for specific uses there must be control over pollution. The most important contributors to pollution of water are sewage, bacteria, and chemical wastes. These are the results of pollution from household, industrial, and agricultural wastes.

Types of Water Pollution. A very broad method of classifying water pollutants is to divide them into degradable and nondegradable pollutants. **Nondegradable pollutants** are those that are not altered by natural biological processes. These pollutants are mainly inorganic chemicals such as salts (especially chlorides), metallic oxides, and colored, toxic, and taste-producing materials.

Degradable pollutants are those that can be changed by biological, chemical, or physical processes found in natural water. The most common of this type is domestic sewage. Sewage is composed of organic substances that are readily converted into bicarbonates, nitrates, sulfates, and phosphates by bacteria and other organisms living in water. A small amount of this type of pollution in large streams acts as a fertilizer and may increase the number of producer organisms. If the water is not too heavily loaded, the natural water will "self-purify" aerobically (with oxygen) and will not produce offensive odors, tastes, and toxic waste products. However, if the amount of sewage exceeds a certain level, biological degradation occurs anaerobically (without oxygen) and produces toxic and noxious hydrogen sulfide, methane, and other gases.

Aerobic and anaerobic processes occur in all natural waters and are utilized in sewage treatment plants. Degradable pollutants are generally organic wastes. Domestic sewage is the most widespread; however, industry produces an equal amount in certain limited areas. For example, a sugar beet processing plant may produce wastes in one year equal to that produced by a city of half a million.

In evaluating the amount of pollution that can be tolerated in water, there are several methods or indicators which may be used. For nondegradable pollutants there is a relatively simple method of calculating the concentration (usually parts per million by weight) of the pollutant and control is enforced to prevent it from exceeding a given amount. Dilution is probably the only factor involved in the dispersion of nondegradable pollutants. With degradable pollutants the measurement of concentration is more complex and involves the *biochemical oxygen demand* (*BOD*) which indicates the rate at which dissolved oxygen is removed from the natural waters. This rate is dependent upon such factors as temperature, quantity, and kind of pollutant involved. Toxic substances, by killing the living organisms present in natural waters, may reduce the rate of BOD to a point where the water is bacteriologically "dead." A greatly reduced oxygen content limits the number of higher organisms that the water can support. The natural ecosystem is greatly altered and the normal organisms are generally replaced by less desirable forms. If wastes are properly processed and the ecosystem of the natural waterway understood, it is possible to dispose of a maximum amount of wastes without creating a pollution problem.

Domestic Water Pollution. Domestic pollution results from household wastes which pass through municipal sewerage systems. This pollution is composed mainly of human wastes; however, the amounts and types of other kinds of wastes from modern living are continuously increasing. Some of them pose particular processing problems for municipal sewerage systems. For example, some synthetic detergents, while only mildly toxic, produce large quantities of foam. They are difficult and expensive to remove because they require special processing in order to prevent the foam from inhibiting the passage of sewage through the system. Large quantities of food wastes, which were formerly disposed of as garbage, are now ground up by modern disposal units and are emptied directly into the sewerage system. New complex compounds such as cleaners and water-based paints also find their way into municipal sewerage systems. Constant research in processing sewage is required to keep up with the many new products entering the market and finding their way into disposal systems.

Human wastes in the water supply may result in infectious diseases such as typhoid, dysentery, and cholera. Most recently considerable attention has been placed on the presence of viruses in municipal drinking water. Some common varieties have been found to be resistant to disinfectants and are quite viable outside the human body.

Fortunately modern methods are able to process sewage; however, the major problem is the sheer quantity and economics of dealing with it. As the population increases and continues to urbanize, it continues to place an ever-increasing burden upon sewage systems.

Industrial and Agricultural Water Pollution. With constant scientific and technological advancement, the waste products of industry and agriculture not only increase in quantity but in kinds. Many of these wastes, mostly complex organic compounds, are very resistant to breakdown and in some instances destroy the natural degrading processes. Some of these wastes are emptied directly into the natural waterways, with disastrous consequences for the existing ecosystems (Figure 26-10). The emptying of industrial and agricultural wastes requires careful and constant control and surveillance. As commerce increases on the waterways and industry utilizes the offshore areas, wastes composed of petroleum products and trash continue to increase in quantity. The shoreline in many areas is becoming covered with layers of oil and mounting trash blocks the waterways.

With the increased utilization of atomic energy as a major source of power, the problem of thermal pollution is assuming greater importance. Billions of gallons of warm water are emptied into natural waterways, causing substantial temperature changes. Temperature is one of the primary controls of life. Most aquatic life such as fish, protozoans, and algae are unable to control their body temperature and are particularly sensitive to thermal changes in the environment. Each species is adapted to its particular environment. Their existence and life cycles are based upon a particular temperature

Figure 26-10. Sewage from an industrial plant emptying into a waterway. [*Federal Water Quality Administration photo.*]

of the environment and they are usually not able to adjust to abrupt changes. The use of water as an industrial coolant is so extensive in some localities that it poses a real threat to aquatic life.

From studies of the effect of temperature upon aquatic life, it has been found that most of the effects of higher temperature are upon the rate of metabolism. As the temperature rises, the rate of chemical reaction increases. This causes the rate of metabolism to increase and results in a greater demand upon the existing supply of oxygen. For example, many fish need four to five times more oxygen to live at 65°F than they do at 39°F. The different activities of fish in response to a rise in temperature are complex and variable, depending upon the species. For many fish, as the temperature rises swimming speed increases. However, for trout, swimming speed slows down to such a point that at 70°F these fish are unable to catch prey. Probably the most adverse effect of unusually high temperatures is on the reproduction of aquatic life. Above certain temperatures eggs do not develop, spawning is inhibited, or hatching does not take place. Aquatic ecosystems have within them delicate and complex series of food webs. If the temperature seriously affects the proliferation of one species, this factor can affect all links within the web. The solution to the problem of thermal pollution is the development of a technologically economic means of removing heat from waste water before this water is returned to its source.

Air Pollution

Of the major types of pollution, air pollution is the most complex and most difficult to control. Most air pollutants are gases, and these are more difficult to dispose of than liquid and solid wastes, which can be more easily transported from one area to another. When released into the atmosphere, air pollutants are rapidly dispersed over great distances and over wide areas. Atmospheric conditions can control and concentrate the distribution of air pollutants, and sunlight can alter their chemical characteristics. Sources of air pollution are much more generalized and widespread than those of other types of pollution.

Industrial and Agricultural Air Pollution. Within recent years considerable pressure has been placed upon industry to control air pollution (Figure 26-11). A major source of power has been coal and petroleum. The combustion of these fuels results in such pollutants as sulfur dioxide (SO_2), carbon monoxide (CO), hydrogen sulfide (H_2S), and hydrocarbons. As the chemical industry grew, hydrochloric acid (HCl), nitrogen dioxide (NO_2), and oxides of zinc, lead, and arsenic were added to the pollutants of air. While the effects of air pollution were known and studied for many years, it was not until about 1945 that the problem developed to such proportions that it became a major concern. Some controls were placed upon industry, but conditions did not improve. The study of radioactive fallout gave some indication of the scope of the problem. Today severe restrictions are being placed upon industry in order to control air pollution. However, constant and increasingly more control is required to reduce industrial air pollution to a harmless level.

Domestic Air Pollution. Because of a rap-

Figure 26-11. Industrial air pollution. Emission of smoke from a chemical plant covers a large area. [*National Air Pollution Control Administration photo by Robertson Studio.*]

idly increasing population with an increased standard of living in the United States, air pollution from domestic sources has become a major problem. The sources of domestic air pollution are far more generalized than those of industrial air pollution. Each household heating unit is a potential source. Of even greater significance is air pollution resulting from the automobile. The exhaust from automobiles emits hundreds of thousands of tons of hydrocarbons into the atmosphere daily. The solution for this pollution is not to be found in legislation forbidding the emission of wastes but through technical advances which will enable automobile engines to burn fuel completely. Household heating is the second greatest source of air pollution from domestic sources. The most widely used fuels in the United States are coal and oil, with the latter being the most common. While low sulfur content fuel oil is helping to reduce the sulfur dioxide from a given household unit, the increasing number of homes required to satisfy the growing population has almost canceled any gains.

Radioactive Pollution. Radioactive pollution of the atmosphere is any increase in the natural background radiation as a result of human activities. As the use of nuclear energy and the application of radioactive materials increases, the danger of radioactive pollution also increases. Disastrous experiences with radioactive fallout, a result of the exploding of atomic bombs in the atmosphere, have emphasized the importance of immediate control. Since it has been shown that radioactive fallout can occur thousands of miles from the source, this is an international problem as well as a national problem.

Radioactive contaminants may arise from several sources such as the production of nuclear fuels, the operation of nuclear reactors, medical uses, the application of radioactive substances in industry, agricultural and scientific research, and nuclear tests. While safe levels of concentration have been established for the atmosphere, radioactive contaminants may be concentrated indirectly by passing through a food chain. Radioactive strontium from nuclear testing that falls upon grass is passed on via cow's milk. The strontium then becomes part of bone. Since radioactive strontium is relatively long lasting, it may reach high levels of concentration. While there has been much research on the effects of radiation upon organisms (Figure 26-12), further studies are needed to determine the effects on organisms of long exposures to low levels of radiation.

Some of the known effects on humans of high-level exposure to radiation are glandular hyposecretion, fibrosis of connective tissue, anemia, cancers such as epithelioma and leukemia, shortened life span, premature aging, and genetic effects. Similar results are found with other organisms. However, the effects of radiation upon the total biosphere are not fully understood.

Land Pollution

As a result of population increases, large land areas have been and are being changed, with consequences for the total biosphere. These changes occur through the disposal of wastes, from obtaining raw materials for industry, and from the utilization of land areas for dwellings and transportation. The problem is greatest among the developed countries where modern living requires greater material demands. For example, in the United States, as a result of massive production over 9 million automobiles and trucks, millions of tons of paper, 28 million bottles and 48 million cans are scrapped each year. Most of this waste is not reprocessed. The transportation of such wastes and where to place them are major problems.

Industrial Land Pollution. In the United States each year industry discards over 165 million tons of solid wastes. That which is not emptied into the waterways or burned is dumped on the countryside. The major sources of industrial pollution are pulp and paper mills, oil refineries, smelters, chemical manufacturers, and power and heating plants (Figure 26-13). Most industrial furnaces generate "fly ash," a gray, powdery residue of unburned materials. Most industries have installed collectors to remove this waste from their smokestacks. The amount collected is over 20 million tons each year and results in mountains of this material covering the landscape. Recently it has been found that this waste is an excellent filler for cement, concrete blocks, bricks, and asphalt paving. However, large areas of the earth have already been excessively marred.

Surface mining, which has greatly expanded since World War II, has had harsh effects on the landscape. Individual miners going deep into mineshafts have given way to giant earth-removing machines moving billions of tons of topsoil, subsoil, and rock to expose the valua-

(A)

(B)

Figure 26-12. Experimentation on the effects of radioactive fallout upon plants. (**A**) View of a portion of the Brookhaven gamma field. The shadow at the lower right indicates the position of the cobalt-60 gamma-ray source. The plants in the foreground are gladioli. At less than 1000 roentgens per day (r/day) the plants are virtually unaffected. At more than 4000 r/day the plants do not survive. (**B**) Stunted tomato plants reveal the effects of different amounts of gamma radiation (r/day). The plant at the right was heavily irradiated, the one at the left received no radiation at all. [*Courtesy of the Brookhaven National Laboratory.*]

Figure 26-13. View of land destroyed by smelter fumes. [*U.S. Soil Conservation Service photo.*]

(A)

(B)

Figure 26-14. Effects of surface mining. (**A**) Strip mining in the mountains helped to create the flood conditions shown in the photo. (**B**) Large area showing the scars of strip mining. [*Courtesy of Billy Davis, Director of Photography, Courier-Journal and Louisville Times.*]

ble ore (Figure 26-14). This surface material covers many acres of the countryside. More than 3 million acres, an area larger than the state of Connecticut, has been disturbed by surface mining. Strip mining of coal has cut deep slits in the earth extending miles in length. Soil and rocks spill down slopes, exposing the land to erosion. Runoff water brings silt and sediment to nearby streams, eventually filling them. While efforts have been made to reclaim some of this land, the process is difficult and costly.

Tons of refuse from the chemical and petroleum industry may permanently destroy the landscape, making reclamation impossible. Occasional accidents, such as the emptying of large amounts of oil, escape of toxic chemicals, or uncontrollable mine fires permanently destroy the productivity of specific land areas.

Agricultural Pollution. Modern agricultural methods have contributed to soil pollution by the use of insecticides, fumigants, herbicides, and chemical fertilizers. Some of these chemicals are applied directly to the soil, but even if applied to the foliage, some of the residues find their way into the soil. Most of these chemicals are quite stable and remain in the soil for long periods of time. They not only kill living forms on the surface of the soil but, when mixed with the soil by cultivation, reach many more organisms.

As has been mentioned previously, irrigation may increase the salinity of the soil to such levels that it is no longer able to support food crops (Figure 26-15). For example, the Imperial Valley of California is probably the most fertile land area in the United States. Continuous irrigation has resulted in raising the water table. Salts are brought to the surface and concentrate through evaporation. Without control, the salinity of the soil will increase to a level where productivity of the area will be greatly reduced.

Intensive indiscriminate use of chemical fertilizers causes the soil to lose its ability to fix nitrogen. This necessitates the continued use of ever-increasing amounts. These in turn find their way into water supplies and not only endanger human health, but also disrupt aquatic ecosystems.

Domestic Pollution. There is a steady flow of gadgets, furniture, appliances, and groceries into the American home every day. Out of these homes go cans, cartons, bottles, broken gadgets, worn-out furniture and appliances. These solid wastes amount to almost a ton per person each year and the volume is increasing. In addition, the nature of these wastes is changing. Continued trends toward nonreturnable, nonreusable containers present special problems. Glass, aluminum, and plastics are durable and are not even completely consumed by incineration. Their litter can be expected to remain indefinitely.

Figure 26-15. Effects of overirrigation. The evaporation of irrigation water brings materials to the surface and increases their concentration. The view shows a good barley crop abruptly stopped by a "salt spot." In many instances the salt is more uniformly distributed than in the field shown. Usually there is no sharp demarcation between healthy plants and no plants. [*USDA photo.*]

The great bulk of solid wastes is placed upon unused land at the edge of the city or town, where it destroys the natural beauty of the landscape. Two other results of this practice are air pollution from burning refuse and water pollution from seepage. With increased urbanization there is decreasing room for dumps. For example, San Francisco no longer has any area of its own in which to deposit solid wastes. A neighboring town has arranged to accept the wastes in a marshy area. Once this area is filled in, there will be no area within a reasonable distance to accept the wastes.

Increased transportation has resulted in a major problem of automobile disposal. The economics of reclaiming scrapped vehicles is complex, and in many cases recycling these vehicles is not economically feasible.

One of the least justifiable types of land pollution is litter. This expensive habit of an affluent society costs over a half billion dollars a year. Refuse spread over the countryside is a fire hazard, creates public health problems, and destroys the natural beauty of the environment. Most litter is the result of carelessness or sheer lack of concern and can be prevented. With the rapidly diminishing free land areas close to cities and suburban areas, litter has become a major problem.

The Effects of Pollution

The biosphere, the area of the earth which is capable of supporting life, is a thin mantle that covers the globe. Throughout the millions of years that life has existed, a complex interrelationship has evolved between the biotic and abiotic environments. In this interrelationship, organisms are adapted to specific ecological niches. Raw materials are consumed and returned to the environment to be used over again. The result is a closed system composed of complex cycles and a wide diversity of organisms, resulting in a dynamic balance of life. Under natural laws, forces are exerted upon the biotic environment to maintain this balance. This has been shown in the prey-predator relationship or in the relationship between plants and soil nutrients. If an alien component is added to an ecosystem, a series of changes take place to restore a balance. When primitive man exhausted the resources of a given area he moved on, allowing the depleted area to replenish itself. Modern man no longer moves in this way. When the resources are exhausted in the area in which he lives he imports resources, disturbing the balance in far distant areas. The pressures of modern man upon his environment are ever increasing.

Air Pollution Effects

Air pollutants have immediate effects upon man as a health hazard. Some of the illnesses possibly attributable to air pollution are chronic cough, sore throat, smarting of the eyes, nausea, intestinal disturbances, and for people with already-existing respiratory and cardiac diseases, death. Air pollution has been shown to cause death in different species of animals. Little research outside of the laboratory has been done on the effects of air pollution upon native fauna, but it is likely that there are effects similar to those experienced by man.

Intensive studies of the effects of sulfur dioxide, fluorine compounds, and smog upon plants have been made (Figure 26-16). In low concentrations, these pollutants affect the photosynthetic process. In higher quantities they actually destroy foliage and in still higher quantities can be fatal. Much of the damage to plants is visible. In California alone, millions of dollars of damage is done to agricultural crops each year. While much research has been conducted regarding the direct and indirect affects upon agricultural crops, little has been done on natural flora.

Effects of Water Pollution

The direct effects of water pollution upon human and aquatic life are quite extensive (Figure 26-17). Polluted water may be unfit for human consumption because it contains pathogenic bacteria and viruses or noxious and toxic substances. In much the same way, this pollution affects aquatic life. Pollution results in a reduction in the number of species, and in many instances, the water becomes dominated by a few generally less desirable forms. Valuable food sources such as fish and shellfish can be destroyed. Dramatic water shortages can occur within the sight of ample water supplies. For example, New York City has experienced water shortages, yet the Hudson River flows next to the city. But Hudson River water is too polluted to use. While one can speculate upon the effects of agricultural and industrial wastes

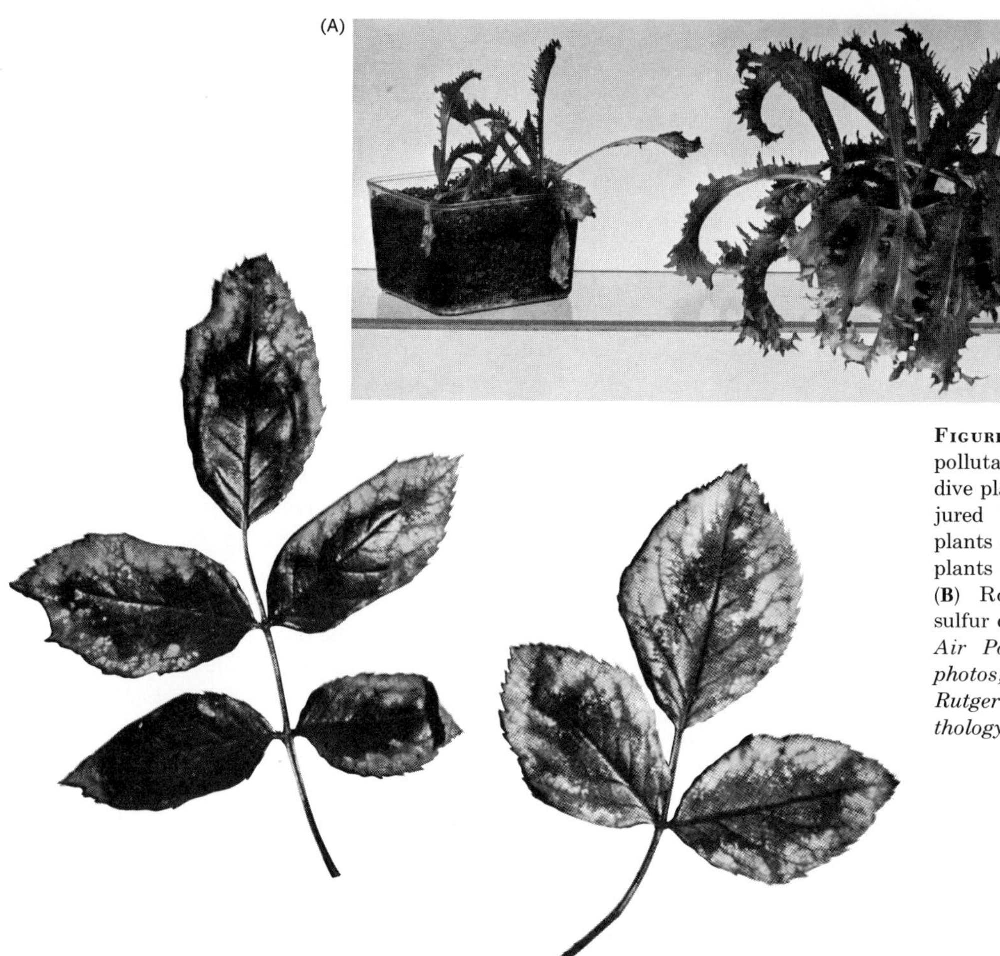

Figure 26-16. Effects of air pollutants upon flora. (**A**) Endive plants on the left were injured by air pollution. The plants on the right are healthy plants grown in filtered air. (**B**) Rose plants injured by sulfur dioxide gas. [**A** *National Air Pollution Administration photos;* **B** *by Dr. Ida Leone, Rutgers University Plant Pathology Department.*]

directly emptied into the waterways or indirectly through seepage, the full effects upon the biosphere are unknown.

Effects of Solid Waste Pollution

While solid wastes placed upon land areas do not have as wide-scale effects as air and water pollution, the rate at which solid wastes are accumulating has become a major problem. Solid wastes occupy millions of acres of land. In many instances such wastes destroy soil, drive fauna and flora from their natural habitat, and destroy the natural beauty of the landscape. In addition, waste seepage into the ground water and runoff water pollutes rivers, streams, lakes, and estuaries, and if burned pollutes the air. While homes, roads, and industrial complexes are not properly termed pollutants, their continued growth is consuming millions of acres of natural land area. The resultant effects upon existing ecosystems within these areas directly and indirectly affect associated ecosystems.

Total Effect of Pollution

As was pointed out, the world population will double by the year 2000 and is expected to quadruple in an-

Figure 26-17. Pollution of a waterway. [*Federal Water Quality Administration photo.*]

other 75 years. As a consequence, at that time there will be approximately 1000 people per square mile of arable land. If further pollution continues at present-day rates, the earth's ability to support large populations will be greatly diminished or even completely eliminated.

Reduction of Pollution

While pollution is a worldwide problem, its solution is primarily dependent upon the developed countries. The waste products of an industrial and technological society are the major causes of world pollution. As the underdeveloped countries advance technologically and industrially, this development must include pollution control.

Most ecologists are not optimistic about the prospects of dealing with the complex problem of pollution. Much irreversible damage has been done, and there is still much to be understood. However, ideas do exist on how to help balance the effect of man upon his environment. Many of the solutions are to be found through research, legislation, and funding. Solutions to pollution problems usually involve the expenditure of large sums of money. While the immediate cost may seem high at present, the ultimate cost of delaying such solutions because of economic considerations may be tenfold.

Biological and Chemical Research

Greater knowledge must be gained regarding the interrelationships of all living organisms. Study of these interrelationships will disclose the effects of each species upon its ecosystem and the total biosphere. Through the development of disease and insect-resistant livestock and plant crops, the use of pesticides could be reduced. The natural enemies of certain pest species might also be used without dire consequences to the whole community. Organic fertilizers might be used to replace chemical fertilizers. A complete understanding of the demands of an increasing population upon the total environment is needed. Research is needed to develop new chemicals for use in present-day products which will allow these products to be readily decomposed into nontoxic, non-noxious wastes, and which do not place a burden upon the environment. Research is also needed to find a means of economically converting into re-usable products those chemical substances for which there are no substitutes.

Technological Advancement

Technological advances continue to yield better and more efficient means of helping people meet their needs. However, these advances must be accomplished within the context of the total biosphere. Economically feasible ways must be found to convert the millions of tons of wastes into useful products. Greater control must be exercised over the escape of pollutants into the environment from industrial and domestic sources.

Legislative Control

Biological and chemical research and technology have provided and will continue to provide methods to control pollution. However, these methods are not and will not be applied universally unless there is legislative action and enforcement. Where there are technological methods available and as new ones are discovered for pollution control, such methods should become mandatory immediately. The developed countries should employ regional and national attacks upon the problems of pollution. In addition, since the effects of pollution are so widespread, the need for international regulation and cooperation is clear.

Fundamental to any solution of the problem must be a new way of thinking. Pollution is a problem of all mankind and the beginning of a start to a real solution is to realize that natural resources are finite.

SUGGESTED SUPPLEMENTARY READINGS FOR SECTION FIVE. ECOSYSTEMS

ANDREWARTHA, H. C., *Introduction to the Study of Animal Populations.* Chicago: University of Chicago Press, 1963.

ATWOOD, GENEVIEVE, "The Strip-Mining of Western Coal," *Scientific American,* December 1975.

BERNARDE, MELVINE A., *Our Precarious Habitat.* New York: Norton, 1970.

"The Biosphere," *Scientific American,* September 1970.

BOERMA, ADDEKE, "A World Agricultural Plan," *Scientific American,* August 1970.

BORGSTROM, GEORG, *Focal Points: A Global Food Strategy.* New York: Macmillan, 1973.

BOUGHEY, A. S., *Ecology of Populations.* New York: Macmillan, 1973.

BRILL, W. J., "Biological Nitrogen Fixation," *Scientific American,* March 1977.

BRUBAKER, S., *To Live on Earth: Man and His Environment in Perspective.* Baltimore: Johns Hopkins Press, 1972.

BROWER, P. B., "Ecological Chemistry," *Scientific American,* February 1960.

BUSH, GUY L., "Modes of Animal Speciation," *Review of Ecology and Systemics,* vol. 6, 1975.

CARSON, RACHEL, *Silent Spring.* Boston: Houghton Mifflin, 1962.

CHRISTIAN, J. J., and D. D. DAVIS, "Endocrines, Behavior, and Population," *Science,* vol. 146, pp. 1550–60, 1964.

CLARK, JOHN R., "Thermal Pollution and Aquatic Life," *Scientific American,* March 1969.

COLE, L. C., "The Ecosphere," *Scientific American,* April 1958.

EDWARDS, C. A., "Soil Pollutants and Soil Animals," *Scientific American,* April 1969.

EHRLICH, PAUL R., *End of Influence.* New York: Ballantine Books, 1974.

EHRLICH, PAUL R., *The Population Bomb.* New York: Ballantine Books, 1971.

"Food and Agriculture," *Scientific American,* September 1976.

EHRENFELD, DAVID, *Biological Conservation.* New York: Holt, Rinehart and Winston, 1970.

EMERY, K. O., "The Continental Shelves," *Scientific American,* September 1969.

GATES, D. J., *Man and His Environment: Climate.* New York: Harper & Row, 1972.

HANDLER, PHILIP, *Biology and the Future of Man.* New York: Oxford University Press, 1970.

HORN, H., "Forest Succession," *Scientific American,* May 1975.

"The Human Population," *Scientific American,* September 1974.

HUTCHINSON, G. E., "The Biosphere," *Scientific American,* September 1970.

KENNEDY, C. R., *Ecological Animal Parasitology.* New York: Wiley, 1975.

KINNE, OTTO (ed.), *Marine Ecology,* Vols. I and II. New York: Wiley, 1976.

KORMONDY, EDWARD J., *Concepts in Ecology.* Englewood Cliffs, N.J.: Prentice-Hall, 1976.

——— (ed.), *Readings in Ecology.* Englewood Cliffs, N.J.: Prentice-Hall, 1976.

KOZLOWSKI, T. T., and C. E. AHLGREN, *Fire and Ecosystems.* New York: Academic Press, 1974.

LEVINS, RICHARD, *Evolution in Changing Environments.* Princeton: Princeton University Press, 1968.

MACARTHUR, ROBERT, *Geographical Ecology.* New York: Harper & Row, 1972.

MACINTYRE, F., "The Top Millimeter of the Ocean," *Scientific American,* November 1973.

MARCHANT, R. A., *Where Animals Live.* New York: Macmillan, 1970.

ODUM, EUGENE, *Fundamentals of Ecology.* Philadelphia: Saunders, 1971.

RICHARDS, B. N., *Introduction to Soil Ecosystems.* New York: Longman, 1974.

SMITH, R. L., *Ecology and Field Biology.* New York: Harper & Row, 1974.

STEELE, JOHN H., *Structure of Marine Ecosystems.* Cambridge, Mass.: Harvard University Press, 1974.

"The Solar System," *Scientific American,* September 1975.

STRAHLER, A., *Principles of Earth Science.* New York: Harper & Row, 1976.

TEAL, J., and M. TEAL, *Life and Death of a Salt Marsh.* Boston: Little Brown, 1969.

TREWARTHA, GLENN T., *A Geography of Population: World Patterns.* New York: Wiley, 1969.

TURK, J., et al., *Ecosystems, Energy, Population.* Philadelphia: Saunders 1975.

YOUNG, G., "Dry Lands and Desalted Water," *Science,* vol. 167, pp. 339–43, 1970.

WAGNER, R. H., *Environment and Man.* New York: Norton, 1974.

WHITTAKER, R. H., *Communities and Ecosystems.* New York: Macmillan, 1970.

WHITTON, B. A. (ed.), *River Ecology.* Berkeley: University of California Press, 1975.

WILLIAMS, C. M., "Third-Generation Pesticides," *Scientific American,* July 1967.

SIX
Behavior

27 Behavior: Its Nature and Study
28 Behavior as an Individual Adaptation
29 Social Behavior
30 Human Behavior

27 Behavior: Its Nature and Study

GUIDE QUESTIONS

1 Distinguish between taxis, kinesis, and tropism. Give an example of each.
2 What are three major approaches to the study of behavior and what are the distinguishing differences among the three methods?
3 What are some of the difficulties encountered in the study of behavior.
4 Distinguish between reflex, conditioned reflex, and operant conditioning.
5 Why are the terms "instinctive" and "innate" presently unpopular among behaviorists?
6 What are the criteria and components of species-specific behavior? Give a specific example of species-specific behavior.
7 Give three examples and possible mechanisms of cyclic behavior.
8 Distinguish between hibernation and estivation and give examples of each.
9 Name the two major types of migration and give an example of each.
10 What are the major types of learned behavior? Define and give an example of each.
11 Distinguish between the two types of memory and the theoretical physical basis for the manner in which they function.
12 In Chapter 21 evidence was presented for evolution having occurred. Review the kinds of data. How do they apply to the evolution of behavior?
13 What major systems are involved in controlling behavior? How do they work? In what ways are they similar? different? Do they control the same kinds of behavior?
14 What are the advantages of rhythmic or cyclic behavior? What explanations can be given for some species showing it but others not?

THE term **behavior** includes everything the living organism does as it interacts with its abiotic and biotic environments. Moving out of the hot sun into the shade, building a nest, eating grass, movement of leaves or plant body toward the light, catching a field mouse, and simply sleeping are examples of behavior.

Behavior, like any other subject in science, may be studied for the sheer satisfaction of gaining knowledge of the world around us. However, there are even more important reasons for studying behavior. It should be obvious at this point that all organisms in the biosphere are more or less dependent upon each other. Understanding how organisms react will enable people to live in better harmony with nature, to the benefit of all living things. In addition, the study of behavior may provide general ideas and theories to help us understand and control our own behavior.

The Study of Behavior

The difficulty in studying behavior is that each response involves a series of complex interactions. In many instances, the mere fact that the organism is being studied may alter the type of response and thus preclude a true account of a behavioral pattern. A notable example of this was the discovery of the "Hawthorne effect." A group of psychologists attempted to determine the conditions under which employees in the General Electric plant in Hawthorne, New Jersey, would produce more and be happier in their work. They initiated a series of steps, each of which made the working conditions better. Each time a change to improve working conditions was made, the efficiency and production improved. Then, in order to show that the greater production and better morale were a result of the improved working conditions, they began to make changes that became less and less ideal. To the surprise of the researchers, it was found that production continued to increase and morale was still very high. It was learned through interviews with the employees that because they were the subject of an experiment they felt that they were special, and a part of something much more than just the manufacture of a product. This continued improve response, regardless of the conditions, became known as the "Hawthorne effect." The purpose in citing this example is to illustrate the difficulty and complexity encountered in the study of behavior, particularly human.

Approaches to the Study of Behavior

Man has constantly observed his behavior and that of other living forms around him. However, the study of behavior utilizing a truly experimental approach is a relatively new science. With the realization that behavior is complex and that the number of factors affecting it are far beyond those that are exhibited by outward appearances, researchers have learned to approach the study of behavior more critically and systematically.

The historical development of the study of behavior as a science helps to affirm the impact of Charles Darwin upon the whole field of biology. In one of his later works, *Expressions of the Emotions of Man and Animals,* he theorized that there was an evolutionary basis for behavior. He suggested that similar responses among different, but closely related, species might have evolved from a common ancestor. Unfortunately, in spite of the care which Darwin took in presenting his theory, others extended it beyond a firm scientific basis. This resulted in teleological explanations. A *teleological theory* postulates that behavior in an organism has a final purpose or design toward which it knows it is evolving. For example, a teleological explanation for the long neck of the giraffe, and one proposed by Lamarck (see Chapter 21) to explain evolution, was that the giraffe knew it had to grow a long neck if it was going to be able to eat the leaves from the tops of trees. Still others gave an anthropomorphic interpretation of behavior; that is, they assigned human behavioral characteristics to other organisms. Examples of this would be attributing the movements of leaves toward the light or roots toward water as a "desire," or the killing of prey by predators as "anger." While giving a specific complex behavior pattern a "name" was convenient, it did not provide a satisfactory explanation. As a result of these faulty extensions of Darwin's original ideas, the evolutionary basis of behavior lost favor.

A new approach to the study of behavior emerged that was independent of the teleological and anthropomorphic interpretations. This approach had its beginnings in Europe. C. Lloyd Morgan in 1896 wrote *Habit and Instinct,* a book on animal behavior, in which he stated that behavior should be interpreted in the simplest neural mechanisms possible. Through careful study that precluded all factors not susceptible to experimental control, he ruled out many of the teleological and anthropomorphic interpretations of behavior.

THE PSYCHOLOGICAL APPROACH TO BEHAVIOR. In the United States, the simpler forms of behavior were studied by Jacques Loeb who attempted to describe behavior in physical and chemical terms. Herbert Jennings studied the role of taxis and trial and error on invertebrates while John Watson began to study behavior in terms of simple response. Although contemporaries of Loeb, they realized that behavior was far more complex than a simple chemical or physical response. Gradually, the study of behavior through carefully controlled experimentation with the use of laboratory animals became widely accepted in the United States to the extent that there developed the American school of behavior. Among the many American behaviorists are B. Frederick Skinner of Harvard University and Harry F. Harlow of the University of Wisconsin. These scientists are more concerned with understanding the nature of behavior than the behavior of an organism within the ecosystem. From the results of his studies upon pigeons, rats, and other animals, Skinner has developed theories of learning that have had wide acceptance. The development of programmed learning and teaching machines has been an outcome of his principle of immediate reward or reinforcement. Harry F. Harlow has worked with the relationship of mother and child among rhesus monkeys. From his experiments he has shown the importance of bodily contact and the infant's need for the attention of a mother (Figure 27-1). The American behaviorists place their greatest emphasis upon *learning*. Their studies are essentially *experimental*, in which variables are controlled in order to study one aspect of behavior.

The Ethological Approach to Behavior. While the psychological approach to behavior has flourished in the United States, there has been a revival of the evolutionary approach to behavior through the work of several notable European behaviorists, and there is now a European school of behavior. Among the better-known scientists of this school are Konrad Lorenz and Niko Tinbergen, both Nobel Prize winners in 1974 for their studies of behavior.

These behaviorists make use of observation and detailed analysis, carefully avoiding the faults found in earlier evolutionary theories. Rather than studying the organisms in a laboratory situation, they are concerned to understand the behavior of a given species within the context of its natural environment. These behaviorists call their approach **ethology**, which is the biological basis of behavior. The emphasis of the ethological approach is on the study of **species-specific** or **species-typical behavior**—the behavior that characteristically develops in a species and is observable when the animal exists freely in its natural surroundings. The ethologist studies all aspects of this behavior, its development, its physiological basis, its adaptive or ecological significance, and its comparison with similar behavior in other species. Species-specific behavior previously was called *instinctive* or *innate* behavior, but these terms led to confusion because they mean different things to different people. Their usage suggests that more is known about the causation of the behavior than in fact is known. Thus, the terms "instinct" and "innate" are controversial at the present time, and it is best to avoid their usage until the controversy has been settled.

Much of the work in the development of ethological theories has been done with birds, such as ducks, geese, and seagulls; with fish such as sticklebacks, guppies, and

(A)

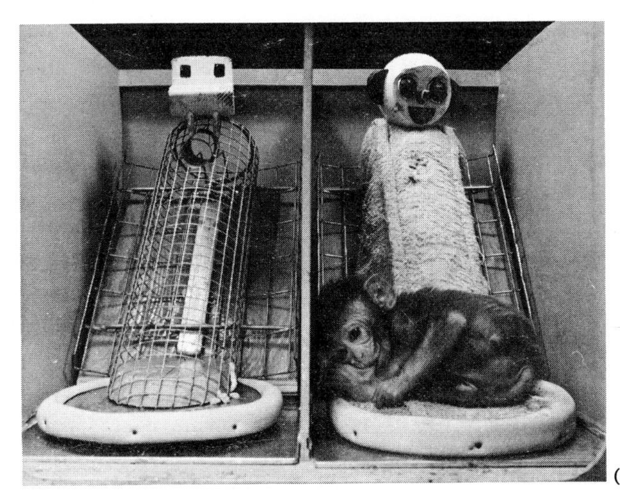
(B)

Figure 27-1. Reactions of an infant monkey to two "mother" models. In his studies with mother–infant relationships, Professor Harlow has attempted to determine the importance of bodily contact. The terrycloth nonmilk-producing "mother" provides a soft body with which the infant can feel close and secure. The wire milk-producing "mother" supplies the needed nourishment. (**A**) The hungry infant monkey continues to maintain contact with the terrycloth model while nursing at the wire model. (**B**) The terrycloth model provides greater security and comfort for the resting infant monkey. From the results of these experiments, the importance of bodily contact has been demonstrated. Among his other research findings, Professor Harlow has shown that the mother–child relationship is critical in the development of normal behavior patterns of maturing monkeys. [*Courtesy of Prof. Harry Harlow, University of Wisconsin.*]

cichlids; and with insects such as wasps, ants, and bees. Since the ethological approach is a naturalistic approach, it is more directly related to the biological sciences than the psychological approach. Ethology seeks an understanding of behavior within the environment.

The Neurophysiological Approach to Behavior. Within recent years, a considerable number of studies have been made concerning the physical and chemical basis for the behavior of organisms. These investigations involve the complex effects of hormones, external and internal stimuli, and experiences upon the nervous and motor systems of organisms. The researchers using this approach to behavior belong to the neurophysiological school. Their major interest is the determination of the mechanisms of an organism's behavior. Therefore, most of the research is confined to individual organisms and is conducted in a laboratory situation.

New approaches to the study of animal behavior are being developed all the time. For example, there is currently a great deal of interest in the ways genes control behavior. This field of study is called *behavioral genetics*. In addition, it is becoming apparent to biologists that no single approach to the study of behavior is capable of answering all questions. While one approach has techniques that are most valuable in determining one aspect of behavior, other approaches provide other answers. Through an integration of the various approaches, a more complete understanding of the "Why?" and "How?" of behavior of a given species is obtained. However, this requires a cooperative effort between specialists in each of the various approaches.

Problems of Behavior-Study Methods

In the study of behavior the three main approaches—the psychological, the ethological, and the neurophysiological—developed independently with little or no overlapping of methods and concepts. The net result was occasional disagreements as to the nature of behavior. The most notable has been the psychologists' emphasis upon learning and the ethologists' emphasis upon species-specific behavior. The problem essentially is one of definition of terms and the difficulties involved in deriving definitions for broad aspects of behavior.

The complexity of the investigations and the broad area that behavior includes have led to the development of a high degree of specialization among researchers. Some of the areas of research include the reactions of the total organism, while others are concerned with the behavior of a specific organ. Techniques have been devised by which minute electrodes can be inserted into the nervous tissue, trace amounts of chemicals can be analyzed in (or introduced into) living systems, or the activities of organisms can be followed by electronic equipment. There are researchers studying animals in the laboratory under artificial conditions while others are studying organisms in their natural environments. While there is great diversity in their approach to the study of behavior and the selected organisms, a common aim among all behaviorists is developing. The previous disagreements are beginning to be understood as a result of too broad or too narrow an interpretation of concepts. While the American and European schools still maintain their methods and techniques in the study of behavior, there is no longer the division that once existed. There are advocates of each school on both sides of the Atlantic, and the work done by all is contributing to understanding the behavior of all living things.

There are essentially two methods for behavioral study.

1. Observation of the organism or organisms in their natural environment.
2. Experimentation and observation under controlled conditions.

Users of both methods seek to learn what the organism does—how it behaves. Once the behavior is described, it is possible to analyze the environment or conditions that evoke the behavior, the modification of the behavior as a result of anatomical and physiological limitations, and the individual and social factors affecting the actions of the organism.

Observation of the behavior of organisms is best done under natural conditions. This approach is time consuming and requires well-structured procedures. However, through the use of observations over long periods of time, one can learn much about the behavior of individuals, as well as the daily and seasonal behavior of populations. For example, observations of a herd of seals for an hour a day, at different times of the day, over a period of several months will show daily routine behavior, seasonal cyclic behavior, and changes in the behavior of the younger members as they mature.

Ideally, the accumulation of facts through the ethological approach of observation may suggest theories that can then be tested by experimentation in the laboratory and in the field resulting in a greater accuracy of explanations and more refined theories. It is most important to emphasize that these two methods of studying behavior should be mutually dependent.

Stimulus Response

The study of behavior, regardless of the approach, elicits a basic and universal theory of behavior: *a stimulus causes a response.* Any change in the biotic or abiotic environments capable of eliciting some sort of reaction or response in a living organism is termed a **stimulus**. The organism may or may not be aware of the stimulus: in fact, most responses are the result of stimuli of which the organism is not conscious. For example, the act of breathing, the physical act of movement, or the act of seeing all occur through stimuli of which the organism is not aware. The survival of the organism is dependent on responses to stimuli. There are broad general types of stimuli to which most, if not all, living organisms respond, such as temperature, pressure, radiation, or gravity. There are also specific stimuli which affect only certain organisms and not others. As examples, the mating behavior of the family dog is not aroused by the courting antics of a robin on the lawn, while the dog is able to respond to sound frequencies that are not audible to the human.

Stimuli may originate from within the organism (internal) or may originate from the environment (external). The *internal stimuli* help the organism to respond to such needs as those for water, food, elimination of wastes or oxygen. Most internal stimuli are independent of any external stimuli, yet evoke definite responses such as seeking food or water. *External stimuli* are received as some form of energy: light energy, sound vibrations, radiant heat energy, or changes in pressure. These forms of stimuli are received by receptors sometimes localized in sense organs and may or may not evoke a response. The variations in the ability of different organisms to respond to different stimuli are adaptations which have come about in the evolution of behavior to enable the specific organisms to survive in their specific environments.

For an organism to respond to stimuli the reaction must be controlled and coordinated. There are two main types of control systems found among animals. They are the nervous system and the endocrine system. Neither of these systems is found in the Monera, Protista, the phylum Porifera (sponges), and Plantae, and the control mechanisms of these groups of organisms are not well known. The behavior of specific organisms in these groups has been studied and there is a certain amount of predictability, as well as explanation for the survival value of their responses. However, the present limited knowledge of the behavior of these groups contributes little to understanding of the evolution, development, and physiology of the behavior of higher organisms.

The nervous and endocrine systems of higher organisms act in coordination to produce specific behavior. Both systems rely on the production of certain chemical substances, in some instances identical, to function as information carriers. The nervous system has the added feature of electrical signals to convey information from one part of the body to another.

In many animals most of the endocrine glands are accumulations of nerve cells that are modified for secretion. The nervous system provides the means by which behavior manifests itself, while the endocrine system may both control and modify behavior. The endocrine system and nervous system were discussed in Chapters 10 and 11. It would be of some benefit to review these chapters while studying this section on behavior.

Simple Behavior

The most obvious behavior is movement of some sort. If an organism remains motionless, very little can be said about its behavior. Inactivity may represent a failure to perceive stimuli or an actual response to a change in the environment. Little can be learned from this type of response. The simplest response is a simple movement such as a change of position, the flick of an ear, or the closing of an eyelid. Although the distinction may be arbitrary, simple movements may be classified as tropisms, taxes (singular; taxis), kineses, or reflexes.

This classification of simple behavior is not fully agreed upon by all behaviorists. However, it is quite satisfactory for an introduction to the study of behavior.

Tropisms, Taxes, and Kineses

The simplest pattern of behavior is one in which a single specific stimulus evokes a single specific response. This simple behavior is represented by the tropisms of Plantae and Fungi and the taxes and kineses of Protista and Animalia.

A **tropism** is an involuntary movement or response to a stimulus. The tropism is determined by differences in the intensity of stimulation in different areas of the organism. The response is an inherited rigid behavior pattern that cannot be controlled or modified by the organism. Tropisms are characteristic of organisms that lack a nervous system and are therefore under the control of hormones. Tropisms are typically growth movements (see Chapter 10).

It is generally agreed that the responses of sessile organisms such as plants are called tropisms and the automatic simple orientation responses of freely moving organisms are called taxes. A **taxis** is generally a movement toward or away from a stimulus and has definite direction, in contrast to a **kinesis,** which is an undirected locomotory reaction. For example, the single-celled *Euglena* possesses a light-sensitive spot of pigment that detects the direction of light. Somehow the effect of a light stimulus is transmitted to the flagellum which moves the *Euglena* toward the light. The organism is positively *phototactic.* If the light intensity is excessively high, the unicell will swim away. Such a negative response is called an *avoidance reaction.* If, on the other hand, paramecia are placed in a culture dish and a small amount of weak acid is introduced to one side, they will congregate near the acid. This response is not the result of swimming toward the acid stimulus but comes about through random swimming of the paramecia. (Recent studies show that the response of paramecia results from stimulation of the cell membrane.) When the organisms reach the acid environment, they slow down and turn less frequently. Thus, they congregate in this area of the water. Such a response is called a kinesis. The kinesis has a definite survival value for the paramecia since the decaying bacteria upon which they feed are generally found in regions with a lower pH than the surrounding water (Figure 27-2).

Taxes among higher organisms have been the subject of much controversy. The original work of J. Loeb tended to interpret all behavior on the basis of taxis or tropism. However, the theory met with strong opposition. While much of the behavior of the Protista can be explained by taxes, the responses of higher organisms indicate a higher level of behavior. Taxes and tropisms represent responses which appear to be little more than the basic irritability of protoplasm or a specialized simple response that is chemically determined and con-

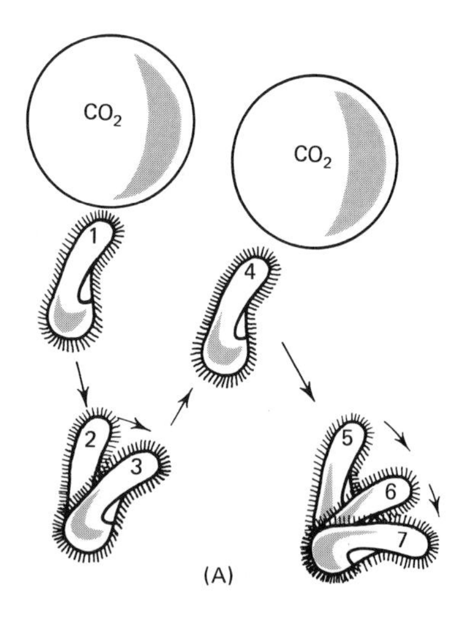

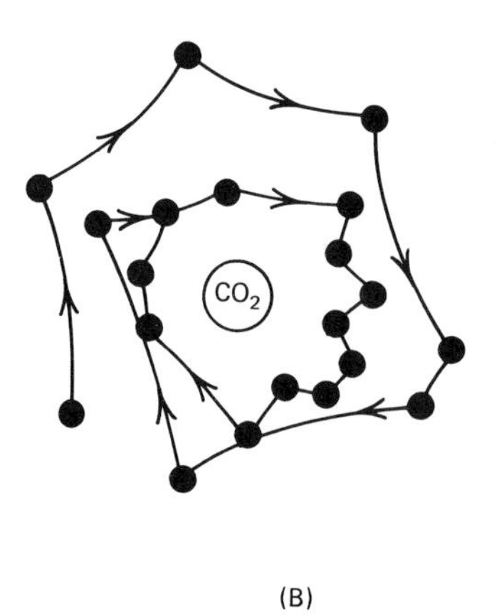

FIGURE 27-2. Paramecia responding to CO_2. Carbon dioxide in water makes the solution slightly acidic. (**A**) When a single paramecium comes in contact with the CO_2, it reverses direction, turns through an angle of about 30 degrees, and swims forward again. (**B**) As paramecia swim randomly and come in contact with the slightly acidic environment, their rate of movement is decreased. Eventually, many paramecia congregate in the slightly acidic area. They tend to remain at a precise distance from the CO_2, avoiding the stronger acidic conditions closer to it.

trolled. The transition to a higher level type of response is not a clear-cut step but shows a gradual increase in the complexity of behavior.

Recent research has shown that changes in direction of locomotion and the avoidance reaction of paramecia are due to stimuli altering the cell membrane permeability which in turn alters (reverses, slows, or speeds up) the beat of the cilia. The same stimulus affects the anterior and posterior ends differently, so that it might be said that the physiological basis of the behavior resides in the adaptive specialization of the cell membrane.

Simple Reflex

While there is evidence that numbers of neural connections may be affected by experiential factors (that is, established by experience), the nervous system develops within a distinct and definite organization determined by heredity and environment. Impulses evoked by stimuli must travel through the existing network of nerves. As the complexity of the nervous system increases among the different species of organisms, the responses, and therefore the behavior, become more complex. However, there are simple, unlearned, unmodifiable responses among organisms possessing a nervous system called **reflexes.** These reflexes are genetically determined and constitute much of the behavior of lower organisms; however, they also occur among the higher forms of life. Blinking your eye is an example.

Among some of the Cnidaria the classical reflex arc of receptor, sensory and motor neurons, and effector is not present (see page 260). For example, in the body wall of the Hydra contractile elements are found at the base of special epithelial cells. Scattered between the epithelial cells are sensory cells which have processes that connect with nerve cell processes or directly to the contractile cells. When the sensory cells (receptors) are stimulated they link directly to the contractile cells (effectors). Only two components are present, receptor and effector.

In organisms with a central nervous system the reflex arc involves three major components: receptor, sensory and motor neurons, and effector. In the simple reflex the sensory cells (receptor), when stimulated, transmit impulses via sensory nerves (afferent) to the spinal cord. Here there are direct synaptic connections with motor neurons, which in turn transmit the impulses either to effector cells, such as muscles, or to interneurons that carry the impulses within the central nervous system to other motor neurons, which cause the responses.

Although the reflex may be considered a simple response, it may in fact involve coordination in the contraction and inhibition of many muscles, especially among the higher organisms. Reflexes might best be thought of as immediate functional extensions of the nervous system. Normally, they develop as the embryo develops and, once functional, probably undergo little further modification.

Frequently, it is not possible to determine if a particular reaction is a simple reflex or a learned response. Newborn babies do not exhibit a blinking of the eyes when an object approaches their face; the response develops gradually. It might be assumed that the response is gradually learned as the hand or an object touches the eyeball. Since a reflex depends upon a pathway of nerves, a more plausible explanation is that there is maturation of the organization that initiates the blinking response.

The simple reflex helps to explain some of the behavior of the lower organisms and the simple unlearned responses of higher organisms. When the response involves many neurons, alternative modes of response, and the so-called higher centers of the brain, it becomes too complex to be considered a simple reflex.

Species-Specific Behavior

Species-specific behavior may be defined generally, as the behavior exhibited by an organism living freely in its environment. Lower animals such as insects show complex but highly rigid and stereotyped patterns of species-specific behavior. The high degree of rigidity and stereotype in this behavior is believed to be largely due to genetic control in the sense that it is preprogrammed in the central nervous system. In arthropods, for example, "command nerve fibers" are recognized. One such command fiber may control one complex behavior pattern while another command fiber may control another complex behavior pattern. Learning seems to play a very minor role. Insects can learn from experience, however.

As one goes up the phylogenetic scale, the learned

elements in the species-specific behavior become increasingly evident, and the behavior is really a mosaic of the more rigidly genetically controlled components and those that are learned.

Criteria for Species-Specific Behavior

There are several specific characteristics that distinguish species-specific behavior. The first is that the specific response is found among all members of the species. The building of a web by a spider is an example of the universal nature of species-specific behavior (Figure 27-3). Each species of spider builds a web with a characteristic pattern. A given species follows a complex series of responses in building a specific characteristic web even though it has not previously been exposed to the pattern. Some species of birds, if raised without hearing another bird, are capable of producing the same characteristic song as other members of the same species. The second characteristic is that species-specific behavior has a physiological basis. With improved research techniques in the field of neurophysiology, it appears that species-specific behavior is physiologically definable. Neurophysiological research is beginning to confirm this hypothesis. Through various surgical and chemical techniques, it is possible to affect various nerve centers evoking specific behavior patterns. For example, lesions in the median region of the hypothalmus can provoke continuous eating behavior in certain laboratory animals. If a gas that inhibits aggressiveness is injected into a cat's nervous system, the cat will be terrified at the sight of a mouse.

FIGURE 27-3. Web of the spider *Zygiella x-notata*. Each species of spider builds a web with a characteristic pattern. Any member of a given species of spider follows a particular complex series of responses in constructing its web, even though it has not previously been exposed to the pattern.

The species-specific mating behavior of canaries is started when the time of daylight lengthens in the springtime, stimulating the production of primary hormones. These hormones, in turn, stimulate a long chain of physiological and behavioral events which result in the rearing of young. If a spider is given a drug, the spider produces a different type of web. Clearly its species-specific behavior has been physiologically changed. The third characteristic of species-specific behavior is its *inheritability*. The inheritability of species-specific behavior has been demonstrated by Konrad Lorenz by crossing two different species of ducks. The offspring of these crosses carried out mating behavior that was a combination of the traits of both parents. William C. Dilger of Cornell University succeeded in crossing the African peach-faced lovebird, which carries nesting materials by tucking them in the feathers of the rump, with a closely related species, Fisher's lovebird, which carries nesting materials in its bill. The offspring of the cross started out by carrying the nesting materials tucked in their rump feathers (Figure 27-4). However, this was not as efficient as carrying the materials in their bills so they used their bills to carry the nesting materials but still went through the motions of trying to tuck the materials into their rump feathers first.

Each of these characteristics tends to emphasize the genetic basis for species-specific behavior. Those opponents of species-specific behavior base their criticism

Figure 27-4. A female *Agapornis roseicollis* tucking strips of paper in her rump feathers. Professor W. C. Dilger of Cornell University has experimented with the instinctive behavior of these birds, which tuck nesting materials in their rump feathers. He mated this species with a closely related species that carries nesting materials with their bills. The offspring at first tried to tuck nesting materials in their rump feathers and then carried the materials with their bills. The results of this experiment seem to indicate that the characteristic species-specific behavior has a genetic basis. [*Courtesy of Prof. W. C. Dilger* (*photo by David G. Allen*).]

upon the fact that it is impossible at the present time to state exactly what it is that is inherited in behavior. In other words, what is it that is specifically coded in the DNA molecule that is passed on to the offspring? This is yet to be discovered. While species-specific behavior is rigid, there is the possibility for slight change and variability. If this were not so, there would be no evolution of behavior and the behavior of the organism would become nonadaptive.

Components of Species-Specific Behavior

The ethologists set forth at least three components within the organism that are involved in species-specific behavior. The first is known as an *appetitive behavior* phase. This phase is essentially a buildup within the organism that results in a restlessness until the behavior is released. This restlessness is the result of such factors as hormones, influence of neural centers such as the cerebral cortex, or previous recent stimulation. The second component is the *releasing mechanism,* which is viewed as neural mechanisms that are selectively responsive to the stimuli called *releasing stimuli* or *sign stimuli.* There are a great many stimuli received by an organism within the environment. The releasing mechanism responds specifically to one or a combination of sign stimuli. There is evidence that different responses have different releasing mechanisms. The third component, the *consummatory act,* removes the stimuli which originally released the appetitive behavior. For example, the sensation of hunger causes an animal to begin to search for food. It searches in appropriate places. This is what is called appetitive behavior. This appetitive searching behavior brings the animal into a stimulus situation in which it senses food. The food acts as a sign stimulus on the releasing mechanism controlling predatory behavior. Thus, the eagle flies over an area where it knows there are field mice. The sight of a mouse elicits predatory behavior. Eating the prey is the terminal act in the sequence and it can be called the consummatory act. Having eaten the food, the animal is no longer hungry, so it no longer shows appetitive behavior for food, but now perhaps it begins to search for a mate, and so on.

Ethologists are constantly studying the behavior of different species in order to further substantiate and clarify their concepts. Among the various researches that have been done, there are several which have become classics. The three-spined stickleback is a small fish which in winter is found in deep cold fresh water or in the coastal sea areas. In the spring when the length of the day increases, the light activates certain glands to produce hormones which stimulate reproductive behavior. These hormones initiate the appetitive behavior phase of reproductive behavior. The sticklebacks migrate to shallow freshwater spawning grounds. The warmer temperature and the availability of nesting areas stimulate the male to swim away from the school and stake out a territory. At the same time the ventral side of the fish changes from a dull green to a brilliant red color. This coloration is the first of a series of sign stimuli which will direct the mating cycle. The selection of a nesting site puts an end to the first part of the reproductive behavior and is a consummatory act.

The red underside (sign stimuli) of the male stickleback evokes a response of aggressiveness in other males and an attraction in the females. Once the nesting site is secured from other males, the stickleback makes a shallow pit in the sand by carrying small bits of sand away in its mouth. It then collects filaments of algae with which it constructs a nest. Once the nest is built, the male responds to the females swollen with eggs (sign stimulus). Through a series of intricate courting movements the male and female fish respond to one another. The specific movement of the male acts as sign stimulus to the female and releases a specific movement, which in turn acts as a sign stimulus for the male. In this way, the male maneuvers the female to the nest where she subsequently lays her eggs (Figure 27-5). Several females are courted and attracted to the nest where the eggs are laid and then fertilized by the male. The consummatory act in this chain of mating behavior is not fertilization but the presence of eggs in the nest. This in turn initiates a new set of behavior in which the male rears the young.

In order to study the components of species-specific behavior, various experiments were performed. The female stickleback has been found to be attracted to crude models of the male stickleback with a red ventral side rather than to an exact model without the red color (Figure 27-6). The color is evidently the sign stimulus that activates the releasing mechanism in the female, after which the female is stimulated to lay her eggs in the nest through prodding or poking by the male. She will not lay the eggs unless this stimulation is received. If prodded by a stick or glass tube she will lay her eggs. The prodding is the stimulus that activates the releasing mechanism, resulting in the consummatory act of

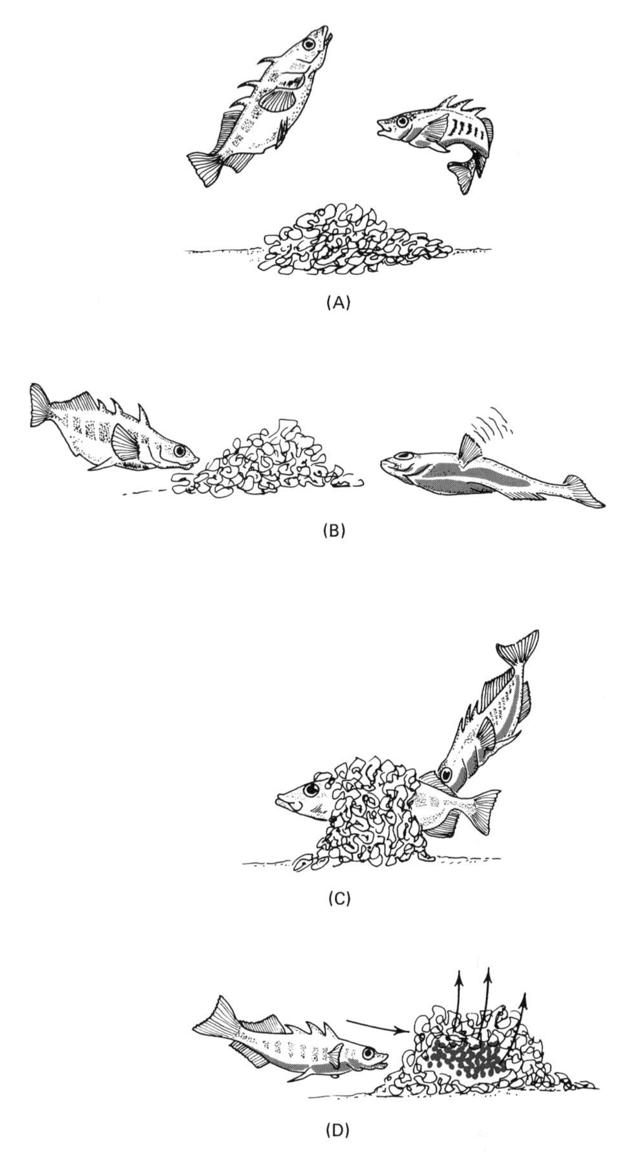

FIGURE 27-5. Mating behavior of the stickleback. (**A**) A male stickleback responds to the female swollen with eggs, while the female responds to the red-colored underside of the male. The responses are to these specific sign stimuli. (**B**) The male, through a series of intricate courting procedures, lures the female to the nest. (**C**) Once the female lays her eggs in the nest, the male fertilizes them. The mating behavior is composed of a series of movements, each of which is evoked by a specific sign stimulus, and terminated by the consummatory act of laying eggs in the nest. (**D**) The male then takes care of the eggs, fanning them with his pectoral fins.

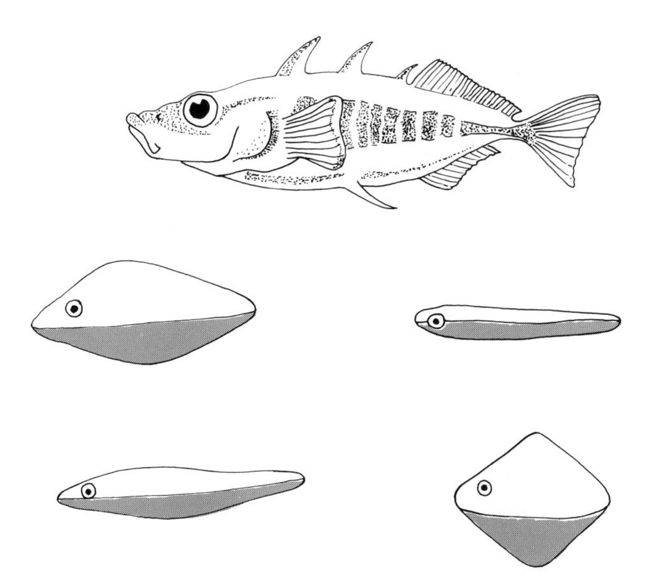

FIGURE 27-6. Models of male sticklebacks used by Niko Tinbergen. Tinbergen found that the important sign stimulus for evoking aggressiveness in the male for attracting the female is the red-colored underside. The variably shaped colored models evoked a greater number of responses than the facsimile (top), which did not have the red-colored underside.

egg laying. The male stickleback will continue courtship behavior if the eggs are removed from the nest.

Throughout the whole world of living things there are sign stimuli that will release similarly highly specific behavior patterns. The parent bird will respond by giving food only to young with specific markings. The orange- and red-colored mouths of gaping young birds stimulate their parent to give them food (Figure 27-7). If the young bird is too weak to open its mouth wide enough, it will not be fed. The adults of certain species of gulls have a red spot on their bills. This red spot is a sign stimulus for the young bird. When the parent bird comes near, the young bird will peck at the spot. The pecking is a sign stimulus for the parent bird to give food. If the young bird does not peck on the bill of the parent, no food will be offered.

There are cases in which there is a supernormal stimulus, as with the English cuckoo. This bird lays its eggs in the nests of other birds. The newly hatched cuckoo has a large head and wider gape than the other birds in the nest. As a result, the parent bird favors the cuckoo over its own fledglings.

The theory of ethologists has not been advanced without criticism. There are those who believe the terms are poorly defined and tend to mask the real underlying mechanisms. Others believe the behavior might in reality be learned or be the result of maturation. For example, the behavior of the parent bird might create stimuli which the young bird may learn while it is still in the egg. Although the answers to many of the criticisms have not been fully resolved, recently there is less and less oposition to the concepts. The ethological approach to behavior has stimulated much research and has resulted in the development of many new methods and techniques of study.

Cyclic or Rhythmic Behavior

Among the most interesting aspects of behavior are the cyclic or rhythmic responses of living organisms, that is, the adaptation of timing responses to the changing conditions of the environment. For many years biologists have observed and have been somewhat amazed at the cyclic responses of organisms which correspond to the change from daylight to darkness, seasonal change, tides, and phases of the moon. However, it has been only relatively recently that this aspect of behavior has been studied to any great extent. It has been found that cyclic behavior is equally important in both plants and animals. The daily opening and closing of flowers, the growth patterns in plants, daily periods of food seeking and rest in animals, seasonal activities such as reproduction, hibernation, estivation, migration, and foliation are all examples of responses which occur and reoccur in a cyclic or rhythmic pattern.

Circadian Rhythms

The cyclic behavior which is most universal and most familiar is that based upon the 24-hour cycle of day and night. Such responses are called **circadian rhythms.** Activity that occurs at night is called *nocturnal* and that which occurs during the daylight hours is called *diurnal.* The diurnal migration of plankton in the photic zone of the ocean is an example of a circadian rhythm. During the night the plankton are distributed

Figure 27-7. Nestlings with gaping mouths. The gaping mouths with colors and markings evoke behavior in the parent birds that results in the feeding of the young. The parent birds respond only to characteristic colors and patterns. If the young do not have the right markings, they are not fed. [*U.S. Forest Service photo.*]

throughout the surface layers of the ocean, but during the day they move to the lower part of the photic zone to avoid the intense sunlight (Figure 27-8). This is a typical example of a circadian rhythm. Organisms that feed upon plankton correspondingly move up and down in the water, coming to the surface to feed at night. These movements are called vertical movements.

The leaf movements of many plants reach a maximum at twilight when the leaves fold or close, and at dawn when the leaves unfold or open. An interesting aspect of this movement is that if a plant is subjected to continued darkness it will open and close at approximately the same time as it would under normal conditions. The swimming and food-seeking activities of certain marine animals brought into the laboratory for study correspond closely to the changes of the tides in

FIGURE 27-8. The diurnal movements of plankton in the photic zone. During the night (*left*) the plankton are distributed throughout the surface layers of the ocean. During the day (*right*) the plankton move to the lower part of the photic zone. Fish that feed upon plankton follow a similar pattern, following the source of food. This daily migration is a typical example of circadian rhythm.

their natural environment. Fiddler crabs show a color change based upon the tides. At low tide they are dark in color and lighter at high tide. Fiddler crabs from Woods Hole, Massachusetts, and from Martha's Vineyard were placed together in an aquarium and kept in total darkness for long periods of time. It was found that the fiddler crabs from Woods Hole consistently changed color 4 hours before the crabs from Martha's Vineyard. The low tide at Woods Hole is reached 4 hours earlier than at Martha's Vineyard. There are many examples of circadian rhythms among all living organisms, from the simplest to the most complex forms.

One of the major areas of research on circadian rhythms is the determination of whether they are externally or internally controlled. Many experiments have been performed upon different organisms in which environmental factors such as temperature and light have been controlled. Yet, the rhythmic response persists. One of the variations made in the rhythmic stimuli has been a change in the length of day. It has been found that certain organisms, such as deer mice and cockroaches, can adjust their daily cycle to the new change in stimulus within limits. If the new day is less than 16 hours (8 hours light and 8 hours dark) or more than 29 hours long, the organisms will revert to an approximate 24-hour cycle.

It appears from the various studies that external factors play a role in controlling circadian rhythms. Yet, there is ample evidence that the internal factors exert a degree of specificity as to those external rhythmic stimuli to which the organism will respond. Clearly, both external and internal factors are involved in circadian rhythms, but the exact mechanism of response is not known.

Seasonal Cycles

Organisms that live for more than a year exhibit behavior of a seasonal nature. Among the various seasonal activities are reproductive behavior and avoidance of adverse environmental conditions by hibernation, estivation, dormancy, or migration. The same factors, that is, external and internal stimuli, that affect circadian rhythms, affect seasonal behavior. However, there is evidence that internal factors play a more dominant role. It has been reasoned that since erratic changes in seasons are far less likely to occur than daily changes, there evolved an internal control that is less likely to be affected by daily changes, which can show extreme variations. The seasonal rhythms may be compared to a clock; the timing mechanism itself may be internal, but similar to a clock, it must be set by referring to external factors such as light–dark cycles, seasons, or tides.

Reproductive Behavior

The breeding seasons of most species occur annually. Breeding may occur at any time of the year but is synchronized so that the birth of offspring will occur when environmental conditions are most favorable. While reproductive capacity may last as long as three months among some species, it is very often extremely restricted. For example, the palolo worm, a marine organism of the Pacific and Atlantic oceans, reproduces only during October and November and when the tides are at their lowest point for the year. In contrast, the grunion fish of California reproduces only during the highest tides. They are washed up on the beach while riding the waves at high tides. As the female deposits her eggs in the sand, the male curls around her and releases his sperm. The eggs are fertilized in the sand (Figure 27-9). The adult fish are washed out with the next wave. Water does not reach the eggs again until the next highest tides, about two weeks later. In the meantime the eggs develop into fish in the warm, moist sand. They are then washed out into the open sea.

In an effort to develop an understanding of the cyclic nature of reproductive behavior, experiments have been performed on animals transported from the southern hemisphere to the northern hemisphere. The seasons are reversed in the two hemispheres; that is, when it is summer in North America, it is winter in South America, and vice versa. Animals that normally mate during the summer were found to become physiologically receptive for mating soon after being brought from winter in South America to the summer in North America. There is considerable evidence to show that the length of daylight has some influence. However, as with all cyclic behavior, the exact cause and all the mechanisms involved are as yet unknown.

Hibernation and Estivation

Hibernation occurs when an organism goes into a period of greatly reduced activity and remains relatively dormant in nests, burrows, or caves during the winter months. It is most characteristic of animals in cold regions. The inception and duration of hibernation vary greatly among different animals. Some go into hibernation long before the extreme cold weather. The stimuli which may trigger hibernation are cooler temperatures, lack of the proper food, length of daylight, or whether the animal has sufficient fat storage within its body. During hibernation, the animal undergoes many physiological changes. The metabolic rate and body temperature are greatly reduced, respiration decreases, and there is a lowering in the rate of circulation of body fluids. Among the animals that hibernate are insects, amphibians such as frogs, toads, and salamanders, reptiles such as snakes and lizards, and small mammals such as rodents and bats. Contrary to belief, bears do not hibernate; they go through carnivoral lethargy in which they spend long periods of time sleeping. Some of the factors that play a role in bringing about the change in metabolic activities are the carbon dioxide concen-

FIGURE 27-9. Grunion spawning. As the female deposits her eggs in the sand, the male curls around her and releases his sperm. See text for details. [*Courtesy of Dr. Richard A. Boolootian, Science Softwares Systems, Inc., West Los Angeles, Cal.*]

tration of the hibernation den and hormonal activity. The exact processes that bring on hibernation and the process of hibernation itself are not fully understood.

Estivation is dormancy during the hot dry periods of the summer in the warmer regions of the earth. It enables the organism to survive periods of extreme drought and excessive heat. The organism generally buries itself in the ground where temperatures are lower and a constant high humidity can be maintained. Many of the same type of physiological changes that occur in hibernation also occur during estivation. Alligators, crocodiles, and turtles estivate during periods of drought.

Migration and Homing

One of the most interesting types of cyclic behavior is the migration of different species of animals. Migration essentially represents long-distance orientation behavior, and there are two types: one-direction orientation and goal orientation. *One-direction orientation* occurs when the animal travels in a specific compass direction using some sort of environmental clue such as the sun or stars. For this orientation the animal needs only to know the changing position of the sun or stars at a given time. In *goal orientation* migration, the animal seeks a particular geographical location. In order to do this, the animal must know not only time but his present geographical position.

Many studies have been made to determine how the organism orients itself. A young salmon hatches in a shallow freshwater stream and then swims to the ocean where it remains for several years. No matter how far into the ocean the salmon travel, when they attain sexual maturity and are ready to spawn, they always return to the same stream where they were hatched. Through experimentation it was found that salmon have acutely sensitive chemoreceptors. Those salmon in which the olfactory tissues have been destroyed or the nostrils have been plugged cannot distinguish among odors and therefore cannot locate the original spawning streams. Not only must the salmon be capable of distinguishing minute differences in the odors of the various streams, but they must also remember the odor of the original stream over the years they remain in the

Figure 27-10. The geographic distribution and migratory patterns of the Atlantic and Pacific golden plover. (**A**) The Atlantic golden plover follows one route south in the fall to its winter habitat in South America and another route north in the spring. The Pacific golden plover flies over the open ocean from its breeding territory in Alaska to its winter habitat in the Hawaiian Islands, the Marquesas Islands, or the Taumotu Archipelago. (**B**) An Atlantic golden plover in its Arctic habitat. [***B** courtesy of Dr. Marie Kuhnen, Montclair State College.*]

ocean. This may involve some kind of olfactory imprinting process

The Pacific golden plover journeys over thousands of miles of open ocean, where there are no landmarks, in its migration from Alaska to the Hawaiian Islands (Figure 27-10). Since the older birds depart several weeks before the birds that have hatched during the season, the younger birds cannot "learn" the direction from the older birds. Therefore, the plover must be born not only with the ability to determine direction but also with a specific map of the journey somehow "engraved" in its nervous system.

Various experiments have been performed with night-traveling migratory birds in a planetarium. It has been found that the birds orient themselves in the direction of the stars as during the normal migration. Other experiments with the use of mirrors show that day-traveling birds orient themselves by the sun. While these experiments give some indication of the stimuli that evoke the response, little is known about the internal mechanism of the response itself. Various hypotheses have been set forth with regard to an internal receptor or an internal physiological clock, but none has been proved. Much research remains to be done in this field. The difficulty in researching migratory birds is that migration occurs only during certain times of the year and only with wild populations.

Closely related to migratory behavior is the homing behavior of birds. When an adult bird is removed from its nest and taken considerable distances, it will often return to the nest. In fact, birds have been known to return from distances as far as 3000 miles across the Atlantic Ocean.

Homing is even more remarkable than migration. Migratory birds generally travel in only a north–south direction and with other members of the species. A homing bird, with no prior knowledge of the direction in which it will be displaced, is capable of returning alone.

It has been known for a very long time that pigeons are excellent homing birds. Their homing is not limited to the breeding season, as is the migration of wild birds. In addition, pigeons can be domesticated and, therefore, are excellent experimental animals for study of the nature of homing, which may in turn give some insight into the migratory behavior of wild populations.

Various studies have been made with pigeons in which their vision has been impaired and olfactory passages been plugged. It has been found that while returning pigeons use the sun and landmarks and odors to return to their lofts more rapidly, both of these senses are not essential, as many of the impaired birds returned also. Other studies have been made in which small magnets were placed on the birds to destroy their possible response to the earth's magnetic field. While this did affect the pigeons accuracy to some extent, again, it did not completely inhibit the birds from returning to the lofts. Still other studies have attempted to determine if the birds had some sort of internal navigational sense. Birds were taken on a twisted and turning course to the release site. When released, the birds had no difficulty in returning directly. Evidently, this is not a significant factor in homing. From the various experiments it appears that landmarks, sun, magnetic fields, and odors may all play some part in homing but like migration much more research is needed to understand this behavior.

Learned Behavior

Learning is a process which brings about an adaptive change in the behavior of an individual through experience. The change in behavior may be more or less permanently acquired or modified. While it is based upon heredity, the ability to acquire the behavior can only come about through experience. Here again, as with species-specific behavior, learned behavior may not always be clearly distinguishable. A complicating factor is the problem of maturation; species-specific behavior may not exhibit itself until a particular point in the life cycle of the organism and then may be mistaken for learned behavior.

Types of Learned Behavior

Habituation. *Habituation* is a simple type of learning in which there is a decrease in the probability of a response to a stimulus upon repeated presentation of the stimulus. It essentially consists of becoming accustomed to a stimulus. Habituation is probably one of the most widely occurring types of learned behavior. For example, young chickens and pheasants as well as other young birds crouch or give alarm behavior when a flying object moves overhead. This is probably a species-specific defense response to predators. After a while, however, they no longer give the defense response to all overhead objects. Familiar objects are ignored. Habituation does not occur with unfamiliar flying objects, for if a new shape flies overhead or one which has come to be recognized as harmful, the young birds will again crouch.

Habituation has been exhibited by some of the lowest forms of life which has led to the speculation that these living forms exhibit at least some form of learning. For example, amoebae will respond to light. If a bright light is shone down upon an amoeba it will contract suddenly and remain motionless. However, after a short time it will ignore the light and resume its normal activities.

Habituation can perhaps best be thought of as the elimination of protective or escape responses to potentially harmful stimuli when these stimuli prove not to be harmful. A change in the strength of the stimulus usually results in the organism again showing the response.

Conditioned Response. Another of the simplest types of learned behavior is the *conditioned response.* This is behavior acquired through experience that results in a response to a stimulus different from the one which orginally evoked it. Most of the studies of this type of behavior have been done in laboratories by psychologists whose major interest has been to discover the nature of learning. As a result of these studies, two basic concepts of conditioned behavior have evolved, classical conditioning or conditioned reflex, and operant or instrumental conditioning.

Classical Conditioning or Conditioned Reflex. The essential mechanisms of the *conditioned reflex* were set forth by the Russian physiologist Ivan Pavlov in his classical dog conditioning experiments. Food (unconditioned stimulus) was placed in the mouth of a dog, whereupon the dog salivated (unconditioned response). This was probably due to a simple reflex arc made up of the taste buds (sensors), sensory neurons (carrying the stimulus to the brain), association neurons, and motor neurons, which caused the glands to salivate (response). Thereafter, each time before the dog was fed a bell was sounded. Eventually the dog salivated upon the sounding of the bell without being given the food. The dog learned to respond to a substituted stimulus. This behavior became known as the *conditioned response* or *conditioned reflex* and represents a form of associative learning, that is, behavior associated with a particular stimulus.

There have been numerous experiments performed upon many different animals which demonstrate conditioned reflexes to a large variety of stimuli. However, there is considerable debate as to the part conditioned reflex plays in the behavior of animals within their natural environment.

Operant or Instrumental Conditioning. *Operant* or *instrumental conditioning* has been most thoroughly studied by B. F. Skinner of Harvard University. In demonstrating this type of learned behavior, a single apparatus called the *Skinner box* is used (Figure 27-11). It is a box containing an object which the animal can manipulate, such as a lever or bar that a rat can press or a disc or key that a pigeon can peck. The animal is placed in the box, which is free from outside stimuli. In exploring the box, the animal will depress the bar or peck the disc, whereupon a pellet of food drops into the box. The number of responses increases (conditioning) very rapidly as a result of receiving the food. This system of reward is called *reinforcement.* The apparatus can be designed so that the food is dropped only after the lever is pressed a given number of times (fixed-ratio schedule of reinforcement), or after a given period of time (fixed interval schedule of reinforcement). Through the use of this technique animals have learned to perform a complex series of behavior patterns such as intricate movements, acrobatics, or manipulation of

(A)

(B)

FIGURE 27-11. Operant (instrumental) conditioning. (**A**) In the Skinner box shown here, a rat is applying pressure to a lever in order to obtain food. (**B**) In this Skinner box a pigeon is shown pecking at a disc in order to obtain food. See text for discussion. [*Courtesy of Dr. B. F. Skinner, Harvard University* (*photos by Will Rapport*).]

certain devices. Operant conditioning differs from reflex conditioning in that the animal actively participates in the development of the response.

Trial and Error. Trial-and-error behavior occurs when there is an unsatisfactory response to a stimulus. After responding in different ways or distinguishing between different stimuli (trials) a satisfactory response may be attained. For example, a young toad may strike at all flying insects, some of which may have a disagreeable taste. By trial and error the toad will learn to distinguish between the edible and inedible insects. Another example is a rat placed in a maze. By trying various passageways he eventually finds the one that leads to the food source. After a series of trials the rat will eventually learn to run the maze without error.

Trial and error is a fundamental type of learned behavior. While it is found in some lower forms of animals, it is much more effective and prevalent among the higher forms. An earthworm requires many more trials to perform a simple act than does a dog or a human.

Imprinting. The concept of imprinting was recognized early in this century by Charles Whitman at the University of Chicago and Oskar Heinroth at the Berlin Zoo. However, Konrad Lorenz studied the phenomenon in depth and saw the broader application of imprinting to the development of behavior in many animals. Lorenz found that newly hatched ducklings would follow the first moving object of reasonable size emitting a reasonable sound, much the same as they would follow a mother duck. He called this type of behavior imprinting. *Imprinting* is the process of learning to recognize and become attached to an object (Figure 27-12). It occurs early in the life of the organism during a period known as the *critical time*. This time may vary from several hours to several days after birth, depending upon the species. Imprinting among ducklings was most effective between 13 and 16 hours after hatching, while for chickens it varied between several hours and several days. In later years Lorenz modified his original concept of imprinting and came to recognize it as a special type of conditioning. It differs from the other types of conditioning in that there is no reward or association evoking the response. It is also limited to a very definite time of short duration; once imprinting occurs it is almost irreversible even in the laboratory. In nature, the process is essentially irreversible and beautifully serves the needs of the young animal.

Figure 27-12. Professor Konrad Lorenz with goslings imprinted to accept him as their "mother." Imprinting is the process of learning to recognize and become attached to an object, in this instance Professor Lorenz himself. His work on imprinting has contributed much information to the study of behavior. [*Courtesy of Prof. Konrad Lorenz (photo by L. Rubelt).*]

Other authorities, such as W. H. Thorpe in England, consider imprinting as a separate kind of learning. Indeed, imprinting has been referred to as one-trial learning by other behaviorists. Still others refer to imprinting as an inborn disposition to learn. From all this, it seems that imprinting can rightly be thought of as a special kind of learning.

While most of the research on imprinting was done with birds, recent studies show similar effects with other animals. There has been some speculation that the grouping or aggregating of animals may be a result of imprinting. There are studies in progress extending the concepts of imprinting into early childhood. It is becoming clear that experiences of infants during the first two or three months affect their behavior later in life.

Insight. One of the most complex and highest forms of learning is insight, although some behaviorists prefer to call it *reasoning*. *Insight* or reasoning is a kind of learning in which the organism gains information about the environment without necessarily making any overt response. So equipped, the organism is able to perceive the immediate situation, call upon past experi-

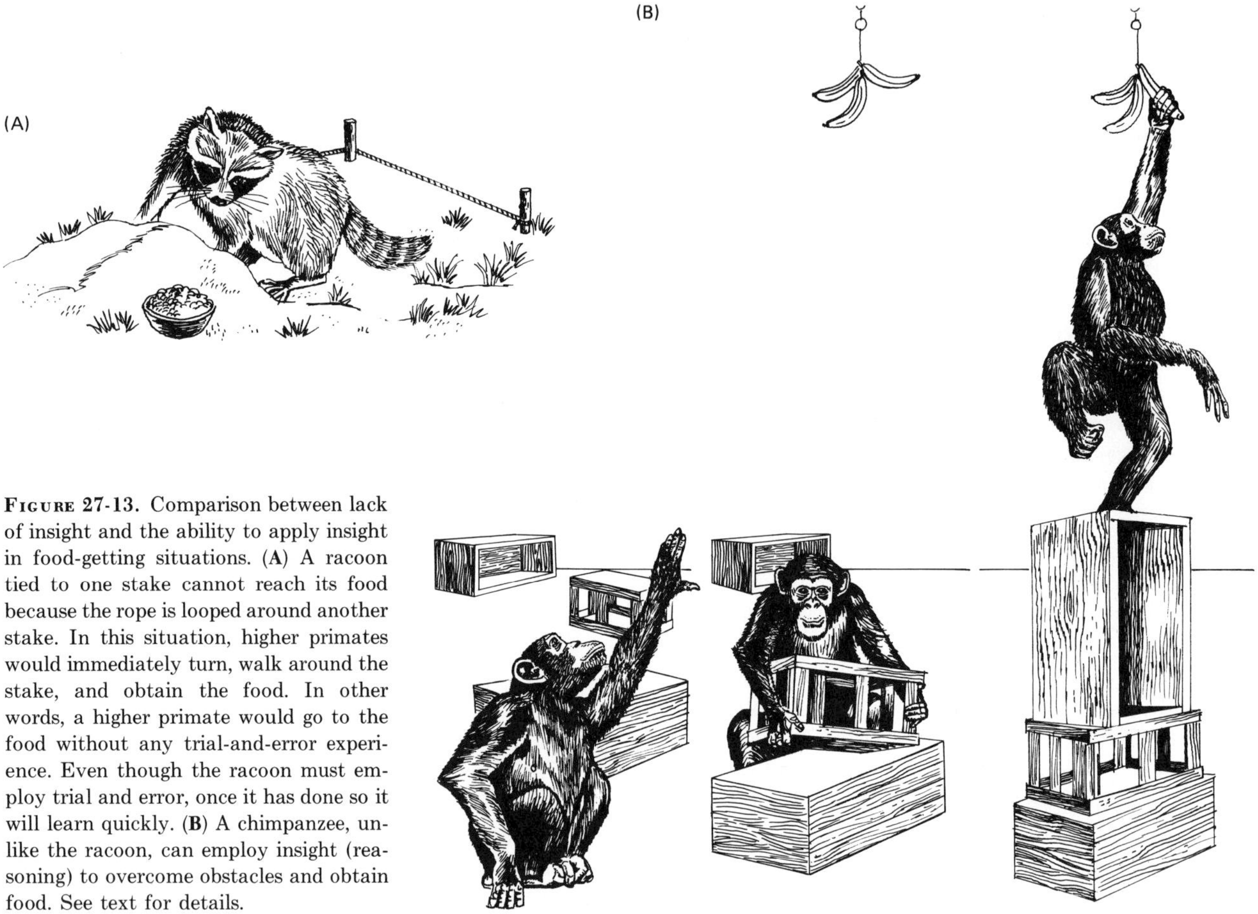

(A)

(B)

Figure 27-13. Comparison between lack of insight and the ability to apply insight in food-getting situations. (**A**) A racoon tied to one stake cannot reach its food because the rope is looped around another stake. In this situation, higher primates would immediately turn, walk around the stake, and obtain the food. In other words, a higher primate would go to the food without any trial-and-error experience. Even though the racoon must employ trial and error, once it has done so it will learn quickly. (**B**) A chimpanzee, unlike the racoon, can employ insight (reasoning) to overcome obstacles and obtain food. See text for details.

ence, deduce a logical solution, and make a response. It is essentially a problem-solving situation. An example is one in which a chimpanzee is put in a cage with boxes and a banana that cannot be reached by the animal using its arm only (Figure 27-13). The chimpanzee will pile one box upon another and will climb up to reach the banana. Insight is the ability to respond correctly without trial and error by applying previous learning to the new situation through a mental process. It is important to distinguish insight from *generalization*. Generalization is characteristic of all learning and occurs when the stimulus differs just slightly from a previously experienced stimulus. The same response is evoked by a generalized stimulus. Pavlov found that his dogs would salivate not only when the sound of the bell was 1200 cycles per second but also at 1000 cycles and 1400 cycles. This response to slightly different stimuli is not the result of insight. In insight, the stimulus must be a completely new one and the correct response the result of the ability to make an abstract (mental) decision. Insight is generally limited to the higher primates and humans. Although many animals can solve mazes and perform complex responses, they are essentially trained through trial and error or through the types of conditioning. They lack the mental ability to develop concepts.

Among the higher forms of learning involving insight are principle learning, symbolism, and language. All of these require abstract thought and have been recognized only among humans.

Memory

The basic foundation of all learning is memory. If an organism is to alter its response as a result of past experiences it must have a means of remembering. This means is called *memory.* Memory is essential if learning is to be retained.

The nature of memory has intrigued people over the years. Early in this century, investigators had compared the brain to a complex communication exchange with lines of communication between the sense organs bringing electrical impulses to the brain and carrying impulses to effector organs (muscles or glands). These lines of communication have been fairly well established and mapped out. Various regions of the brain control various bodily functions via nerve fibers leading to and from the brain to various parts of the body. Interest in the functions of the brain gave rise to the doctrine of separate localization of mental functions in specific established pathways called *memory traces* or *engrams.* It was found that electrical stimulation of the cortex of the brain of epileptic patients could cause recollection of past events. This seemed to support the theory that memory traces were stored in specific pathways in particular areas of the brain.

The late Karl Lashley of Harvard University attempted to substantiate this theory. He taught rats and other animals to solve various problems. He then removed portions of the brain and allowed the animals to solve the problem. His results showed that there was no localization within the brain of that portion which remembered the solution; there was only relative forgetting. For example, if 25% of the cortex was removed, the animal forgot 25% of what it had learned; if 50% was removed, the animal forgot half, and so on.

In later experiments he attempted to sever connections between various brain areas of cats in an effort to disrupt memory and thus find a guide to memory traces. However, his experiments were not successful. In a report, *"In Search of the Engram,"* he stated the following: "This series of experiments has yielded a good bit of information about what and where the memory trace is not. It has discovered nothing directly of the real nature of the engram." He further concluded that

1. There are no defined reflex paths in the brain.
2. It is not possible to demonstrate the isolated localization of a memory trace anywhere within the nervous system.
3. The so-called association areas are not storehouses for specific memories.
4. Since different regions of the brain are equal in their retention of memory, it points to multiple representation.

From his studies, Lashley indicated that possibly external stimuli somehow set up complex patterns of excitation in nerve cells throughout the brain.

D. O. Hebb, a Canadian psychologist and former student of Lashley, and Sir John Eccles, a Nobel Prize-winning physiologist from Australia, in their search for a physical basis of memory, speculated that learning was a matter of a physiological change at the synapses of the neurons in the brain. The concept of *reverbatory circulation* was formulated from their work. According to this concept, impulses must circulate many times around a pattern that is to be "remembered." Through the work of these men and many others such as J. Z. Young and Brian Boycott of Oxford University, working with the brains of octopuses and other animals, there emerged two concepts of memory; *short-term memory,* in which forgetting occurs after a short period of time, and *long-term memory,* in which there is little forgetting.

These two types of memory are given a physical basis in the establishment of a short-lived electrochemical process in the brain after every experience. This is the physiological mechanism that produces the short-term memory. Within a few seconds or minutes, however, this process decays and disappears. If continued or renewed stimulation occurs, the short-term electrochemical process triggers a second series of events, resulting in a second chemical process involving primarily the production of new proteins and other macromolecules, or the release of new RNA molecules. There has been considerable research supporting this explanation. Dr. Murray Jarvik of the Albert Einstein School of Medicine, working with rats, devised an experiment to interrupt the short-term electrochemical process before it had an opportunity to establish the long-term process. A rat was placed on a platform a few inches from the floor. The rat would step off the platform within a few seconds no matter how many times it was placed back on the platform. On one occasion when the rat stepped down he was subjected to a painful electrical shock. When the rat was later placed back on the platform, even 24 hours later, it remained there, demonstrating the ability for long-term memory. Jarvik then took another rat, trained it, and then subjected it to the electrical shock. However, immediately after the shock, a mild electrical current was passed across the rat's brain. It was found that when this rat was placed on the platform, it immediately stepped off, not remembering the previous electric shock. Evidently, the mild electric current through the brain activated the neurons and thus disrupted the short-term memory process and its association with the long-term memory process.

When the experiment was repeated and the electrical current passed through the brain 20 minutes after the electric shock on contact with the floor, the rat remembered and would not get off the platform. This indicated that there was a time factor in establishing the long-term memory process.

Bernard Agranoff of the University of Michigan took the approach of inhibiting the manufacture of proteins in the brain and determined its effect on memory. He trained goldfish to swim from one side of an aquarium to the other in order to avoid an electric shock. He established that goldfish are capable of long-term memory. Using another set of goldfish, he injected *puromycin* (an inhibitor of protein and RNA synthesis) into their brains immediately before and after training. The injected goldfish were not impaired in their learning of the shock avoidance task since presumably the short-term memory process that enables an organism to remember from one task to another does not involve the synthesis of proteins or RNA. When the goldfish were tested a day or two later they had no retention for the task, indicating the dependence of the long-term memory process upon continued protein and RNA synthesis.

The importance of RNA in the memory process was first set forth by Holger Hydén of the University of Göteborg in Sweden. He found that RNA concentration increased within the brains of rats as he trained them in complex tasks. On the basis of his work he proposed that memory was coded within the configuration of RNA.

Further studies by others such as W. C. Corning of Fordham University and E. R. John of New York Medical College indicated that while RNA is intimately involved in the storage of memory, it probably is only one small link in a complicated chain of cellular processes. James V. McConnell of the University of Michigan, on the basis of his own research and that of others, hypothesized that RNA may play two distinct and somewhat different roles. First, incoming sensory signals are first "coded" by physiochemical changes in RNA. The changed RNA would then produce different types of proteins, resulting in a different function of the neuron. Second, the RNA may occasionally serve to transport coded information from one part of the brain to another. This might account for the fact that Lashley was unable to find the engram.

In another series of interesting, although somewhat controversial experiments, McConnell conditioned planaria to perform specific tasks. He then allowed unconditioned planaria to eat the conditioned planaria that had been cut up into pieces. He found that the unconditioned planaria that ate their trained relatives learned more rapidly than a control group. McConnell hypothesized that it was a result of the transference of RNA.

David Krech and a group at the University of California have performed experiments similar to those of Hydén in which they have attempted to determine the chemical and morphological differences (weight, size, RNA, macromolecules, proteins) in the brains of rats which have had stimulating and active lives in contrast to rats that had been "intellectually" impoverished. They found, as did Hydén, that there is a difference between the chemical and morphological characteristics in the brains of rats that are "trained" and those of rats that are "untrained." From their results they have hypothesized some interesting and possibly far-reaching applications.

In originally performing their experiments they found that they did not know how to create a "psychologically enriched environment" for training rats, so they included a wide variety of activities. The cages were brightly lighted, there was a variety of sound, and the rats were given playmates, games to manipulate, maze problems to solve, and new areas to explore. The researchers then realized that all of these activities may not have had an equal effect on the brains of the trained rats. They then controlled the types of activities which they gave the rats. One of the tentative conclusions was that most of the activities provided did not have much of an effect on the chemical and morphological characteristics of the rat brain. A second conclusion was that only those tasks which had a survival value for the rats in their normal environment caused changes. The bases for these tentative conclusions were that

1. Handling, taming, petting, sheer exercise, and varied visual stimulation cause little or no changes in the brain.

2. The experiences of learning their way through tunnels and dark passages and climbing upon, burrowing under, and crawling through piles of objects in the cage did cause changes.

This has led Krech and colleagues at the University of California to hypothesize that for each species there exists a set of *species-specific experiences* (related to the way of life of the organism within its normal environment) that are maximally enriching and are therefore maximally efficient in developing its brain. A most interesting application of this hypothesis would be learning in humans.

The study of memory has made great progress in recent years and there are some who believe that many of the answers are in rough form. However, it is still an area in which there is continued and active research.

28 Behavior as an Individual Adaptation

GUIDE QUESTIONS

1 Discuss the adaptive characteristics of behavior.
2 How are the following types of behavior adapted to the specific way of life of plants: autonomic growth movements, paratonic growth movements, and turgor movements?
3 In what ways are the behavioral patterns of simple animals similar to plants? How are they dissimilar?
4 List several basic responses characteristic of all organisms.
5 Give specific examples of the adaptations of organisms to the following types of behavior: feeding, drinking, breathing, tasting, hearing, smelling, and seeing.
6 How does electrical communication work as a signal between animals? Is this similar to the way electrical impulses operate in the nervous system?
7 What types of insect behavior are regulated by hormones?
8 What are pheromones? Give specific examples.
9 What are the two types of sleep and what is the role of the reticular activating system center in sleep?
10 How does natural selection direct the evolution of behavior?

While there is some uncertainty and controversy about the exact nature of the various types of behavior, behaviorists are in agreement on one thing—the behavior of an organism is adapted to the environment in which it lives. Through natural selection, behavioral patterns have evolved that enable organisms to survive within their ecological niches. These patterns have evolved in much the same way as other aspects of the species such as anatomical and physiological features. Occasionally, the adaptability and the evolutionary character of behavior are not readily apparent because of their complexity.

The adaptive characteristics of behavior are much more than simple reactions to physical forces. In other words, an organism does not react to stimulation according to the simple laws of physics. For example, if a ball is tapped with a stick, it is possible to predict with reasonable certainty that it will move in the direction in which the force is applied and will travel at a speed and distance proportional to that force. On the other hand, if you tap a dog with a stick, there are many things that might occur. He might run, turn around and bite, run a distance and bark, roll over, and so on. His reaction may be far greater than the force that was applied or he might even ignore the force. This example tends to show that most behavior cannot be explained as simple responses to physical forces. Furthermore, behavior results from the interaction of both external stimuli, such as physical forces, which represent certain meaningful stimuli, and internal stimuli, such as the organism's physiological state and its motivation. The effectiveness of one in releasing or causing behavior to occur may be dependent on the strength or weakness of the other. To add to the complexity, there may be conflict as to which behavior to exhibit. For example, the impulse to stand and fight may conflict with an impulse to run away. All of these considerations present major difficulties in studying the mechanisms of behavior.

Individual Adaptation

The behavior patterns of individual organisms can be understood in their adaptive function much the same as anatomical or physiological characteristics. In fact, all aspects of the individual organism are interrelated. The individual behavior of an organism is greatly controlled by its anatomy and physiology.

The most important function of all living things in maintaining life is simply the ability to remain alive until the reproduction of the species is accomplished. Any behavioral pattern that maintains an organism up to and during its reproductive age makes the animal better fitted than others of the species that do not reach reproductive age. This specific behavior pattern is adaptive. While there are many aspects of an individual organism that aid in survival, such as sharp claws, fangs, physical strength, or poison glands, they are no more important than such behavior as response to light, sound, touch, or temperature.

Within the ecological niche of an individual organism, the nature of the interaction with the abiotic and biotic environments will determine the level and type of behavior of the species. Among sessile organisms such as certain plants and fungi, responses are controlled by hormones. Their behavior represents a reaction to an environment over which they have little or no control. In motile organisms lacking a nervous system, the cytoplasmic membrane probably acts as a transducer of external stimuli in chemotaxis.

It will be recalled that experimental research indicates that the membrane may play an important part in transducing other external stimuli into a response of these organisms. This transductance is electrochemical in character. While it is not possible to say that the type of simple behavioral response limits the size of the living form, this mechanism has proven to be of survival value.

In organisms possessing a nervous system, the integration of hormonal and neural control of behavior increases with organismic complexity as one moves up the phylogenetic scale. Behavior that is adapted will enable the organism to compete better, reproduce, and thereby provide for the perpetuation of its species. Since behavior develops through evolution, there are occasionally vestigial forms of behavior for which there appears to be no value. For example, consider the circling of the dog before lying down. Possibly, this action represents some type of bedding behavior of an ancestor to the dog. This and other similar behaviors have persisted in the face of change because they are not detrimental to the present living form.

Adaptive Behavior of Plants

Generally, in discussions of behavior the major focus tends to be placed upon animals rather than plants. This is probably because of the complexity of animal behavior and also because it is more readily observed. However, the protoplasmic composition of plants is basically the same as animals and must therefore solve the same problems. Essentially, the lower forms of plant and animal life respond in much the same way. The response to stimuli is a simple total response of the whole cell or microscopic organism. However, among the higher plants responses to both external and internal stimuli differ greatly from the higher animals. In plants it is primarily under hormonal control, yet the responses are adapted to the structural and functional characteristics of the plant living in a given ecological niche.

The autonomic growth movements of nutations and circumnutations resulting in nodding or spiral movements have no apparent adaptive function for the plant. These movements may be the result of the physical forces of the uneven distribution of growth in the characteristic shape of the various species of plants. On the other hand, the twining movement that occurs in various types of vines, such as morning glories and grapes, represents an adaptation that provides some support for a relatively fragile stem. The nastic movements that occur in organs that are flat in shape, such as bud scales, leaves, and petals, enable the various organs to protect the growth areas from unfavorable temperatures or place the structure in a more favorable position to carry on such functions as pollination or photosynthesis.

Paratonic growth movements are those that are the result of external stimuli causing localized growth in a specific part of the plant. Such stimuli as gravity cause roots to grow downward, anchoring the plant firmly in the soil. A response to light induces leaves to position themselves in order to obtain the maximum amount of light for photosynthesis. The growth of roots toward water helps to ensure an adequate supply. The stimulus of contact causes tendrils, which are modified leaves or stems, to wrap around objects with which they come in contact. These responses represent the major means by which plants have adapted through advantageous adjustment to environmental conditions.

Turgor movements, which occur through a change in water pressure, unlike growth movements, take place in fractions of a second or a few minutes at the most, and are reversible. Many turgor movements do not appear to have a physiological or adaptive significance. The presence of some of these responses apparently does not make the individual plant any better fitted to survive within its environment. Among those turgor movements that do not have an apparent adaptive value are rhythmic oscillations, sleep movements, and reactions to contact.

Paratonic turgor movements are among the most important adaptive responses of plants. These movements occur as the result of external stimuli, particularly light and temperature, effecting a change in water pressure within cells. The movement of the guard cells in controlling the opening and closing of the stomata is an adaptive response to water and gas regulation in photosynthesis. The movements of the leaves of the insectivorous Venus' fly-trap and sundew (see Figure 13-2B, C) result from the stimulation of contact by an insect. Since the habitat of these plants is a moist swampy area where nitrogen content is low, it is hypothesized that the capture of insects is of survival value. The protein of the insect is digested by plant enzymes and supplies usable nitrogen.

There are other reactions of plants, all of which are controlled by hormones. These include such activities as flowering, change of color of the plant body, and abscission or dropping of the leaves. Most of these simple behavioral responses represent adaptations that enable the plant to survive better in its specific habitat.

Adaptive Behavior of Simple Organisms

The behavior of organisms lacking a nervous system is similar to that of plants since it is under chemical although not hormonal control. Inasmuch as these organisms are relatively small, the responses can occur quite rapidly since the chemical agent need not travel great distances to evoke the response. The responses of these simple living forms are generally total, that is, rather than a specific area of the organism reacting, the whole organism responds. Occasionally, as a result of responding totally, the living form is placed in a less favorable position. However, the reaction of living forms must generally be a favorable adaptation or else the species would cease to exist.

Adaptive Chemical Responses

The irritability of protoplasm is the only means by which living forms without a nervous system interact with the biotic and abiotic environments. This irritability is essentially a chemical response. Living forms lacking a nervous system are relatively small, with the exception of some species of sponges. These living forms generally live in water or in a moist habitat. If there is a lack of water in the environment or a sharp change in temperature, these organisms rapidly respond by going into a dormant spore or encysted state. This adaptive response has survival value when environmental conditions are unfavorable.

Organisms with a nervous system also possess, among others, adaptive chemical responses. There are various ways in which these chemical responses occur. First, within the organism chemical stimulation evokes such responses as food seeking, cell division, and habituation. These responses enable the organism to survive from a purely physiological point of view. Second, chemicals from the environment activate chemicals within the living system, evoking various types of responses. Chemicals emitted by food evoke an attractive response. Also certain substances given off by other members of the species tend to attract, thus aiding an organism in finding a mate. Third, a change in energy from the environment will evoke a response through the chemicals within the living form. Sharp changes in temperature, light intensity, pressure, or contact may result in either a positive or negative response. All responses tend to direct the individual to that environment or factors within the environment that are best fitted for the organism's way of life. Thus, behavioral responses to chemical stimuli enable organisms to occupy successfully their ecological niche and thereby survive to reproduce.

Specific Individual Behavior

Many aspects of the behavior of living organisms involve interactions with other members of the species or community. However, most responses are an individual adaptation of the organism to suit its needs within the environment. While many of these responses are basic to all organisms, the manner in which they react may be quite specific.

Feeding, Drinking, and Breathing

Feeding, drinking, and breathing are important in maintaining the homeostatic condition of the organism and are essentially physiological. Homeostatic reactions are a result of internal feedback systems within the living organism which maintain a constant internal environment in spite of changes in the external environment. Thus, C. P. Richter found that when the adrenal cortex, which controls sodium chloride balance and retention, was removed from rats, they died. However, if provided with a sodium chloride solution, rats that had their adrenal cortex removed would compensate for the nonretention of sodium chloride by drinking large quantities of the solution and thus remain alive. This response did not involve any conscious drive by the rats but satisfied the stimulus of some inner homeostatic control.

Feeding Patterns. Feeding behavior shows great variability as well as complexity. There are living forms that feed continuously; an example is the marine organisms that extract food by filtering it from the water. Other forms, such as certain parasitic insects and certain carnivorous vertebrates, feed rarely. There are other organisms which at some point in their life cycles do not feed at all. For example, male seals when they come ashore to stake out their territory do not feed for long periods of time. During hibernation or estivation organisms do not eat.

Most organisms feed in a cyclic pattern varying from several minutes to several hours. While feeding behavior among the different species varies greatly, it does follow a general pattern. It usually begins with exploring the habitat where food is generally found. This is the appetitive phase of the behavior. The appetitive phase is then followed by selecting, preparing, and ingesting food. This is the consummatory act. These activities may be interspersed with other types of activities, such as sleeping, resting, and drinking.

Among vertebrates, many factors affect the internal stimuli which drive the animal to seek food. Higher centers in the brain responding to nerve impulses from the alimentary tract, temperature, body weight, and metabolites in the blood are other important internal factors in the drive for food.

External influences play an important role in food behavior patterns. All organisms are selective in choosing food. This is especially important in determining an organism's given ecological niche. No two species living in the same habitat have exactly the same staple food. Within nature, specific preferences for food may be based upon an anatomical adaptation such as the size of a bird's bill or a specific filtering apparatus. There are also underlying mechanisms in food selection which are not fully understood. Cattle show preference for certain types of grasses as well as grasses at a specific stage of growth.

The characteristics of the food such as texture, hardness, shape, odor, and color are important in food selection. Within nature the final choice of food by an organism is the result of internal and external stimuli. The amount and kind of food consumed by an organism depend upon previous feeding activities as well as experience. Organisms learn through experience that certain foods have a noxious taste, smell, texture, or other adverse property.

Within nature it has been found that frequently the size of the population of the prey will cause a shift in feeding habits. When a new prey appears within the environment it is ignored until it reaches a certain density; then it is sought. This may require new behavior patterns or new means of searching. It has been suggested that this shift in feeding habit is a type of conditioning. It is interesting that this type of feeding tends to maintain a certain control in the size of populations especially among certain bird-insect relationships.

Drinking Patterns of Behavior. The drinking and feeding habits of organisms are closely related; feeding has an effect upon drinking behavior and vice versa. Drinking has the same type of cyclic property as feeding. Previous deprivation of water disrupts the normal drinking patterns. Dryness of the mouth and throat plays a part in evoking drinking behavior. Although there are areas of the hypothalmus of the brain of higher animals which stimulate drinking, no inhibitory or satiety center has been found.

Individual organisms differ in the quantities of water they require and their means of metabolizing it for physiological use. For species requiring large quantities of water, deprivation for even a short period of time may be fatal. In regions where there are dry and wet periods, the onset of the dry period may evoke estivation behavior.

Breathing Behavior. For terrestrial animals, breathing patterns do not follow the cyclic type of response associated with feeding and drinking behavior. Even though there are variations in the rate of breathing based upon the amount of physical activity, as well as other factors, breathing occurs constantly. Among aquatic animals such as whales, porpoises, and seals, breathing behavior becomes more complex. While these organisms can spend long periods of time submerged, they must periodically return to the surface for air. Interesting studies have been made of the Weddell seals of Antarctica. These seals base most of their underwater activities around a series of air holes in the ice. The exact mechanism by which they seek and locate the air holes is not known.

Feeding, drinking, and breathing play the most dominant part in determining the total behavior of an organism, but as previously emphasized, they are ultimately related to successful reproduction. The organization of these activities determines the amount of time necessary for survival. Where the physiological makeup of an organism in relationship to its environment does not require large allotments of time to perform survival activities, such as breathing among terrestrial animals and drinking among aquatic organisms, it frees the animal to pursue other activities. Since each of these behaviors represents survival activity, there results an overall pattern in which one activity takes priority over another. This priority results in cyclic or rhythmic behavior.

Among the most remarkable characteristics of the drinking, feeding, and breathing behaviors are the various anatomical and physiological adaptations of the different organisms.

Chemoreception and Behavior (Taste and Smell)

The specific adaptations of taste and smell are important to feeding behavior and predator–prey relationships. These senses determine to a large degree the basic feeding and protective behavior of organisms. While tastes and smells appear to elicit simple chemical responses, the basic mechanism by which these stimuli are interpreted by the brain is still a mystery. There are a number of theories of how chemical energy at the receptor is transduced to the electrical energy of nerve impulses, but the mechanisms by which any of the sensory stimuli are transduced is not known.

Some arthropods have hair-like structures called *setae,* which are located on the antennae (Figure 28-1). These structures respond to minute amounts of chemicals in the air. Higher vertebrates have taste buds located on the tongue and olfactory receptors in the nose. These receptors transmit the stimulation to the brain. Taste comes about as a result of direct contact of the substance with the chemoreceptor. Taste is essentially the detection of the degree of those qualities called sweet, salty, sour, and bitter.

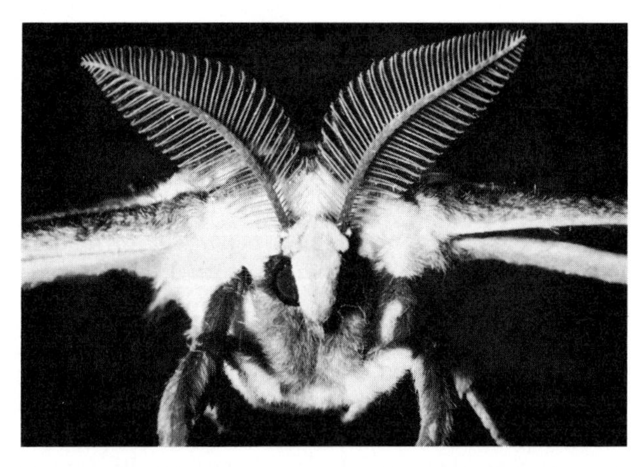

Figure 28-1. The plumed antenna of the male Cecropia moth is an adaptation that enables it to detect chemicals given off by the female moth. These evoke a mating response in the male and a search for the female. [*Courtesy of Dr. Ray R. Kriner.*]

Smell is the result of airborne substances coming into contact with olfactory receptors. The sense of smell can detect a far greater range of substances than taste. Human beings can distinguish at least 10,000 different odors. It has been found that many other organisms, such as bees, can detect almost as many different kinds of odors. In certain organisms, taste and odor detection may be more highly specialized but limited in range. For example, certain parasitic wasps are extremely sensitive to the odor of a specific prey. The prey can be located although it is completely hidden from view and buried deep in the ground or wood.

Chemoreception can play an important part in the selection of a habitat. Aquatic organisms can detect subtle changes of oxygen and carbon dioxide content in water or changes in pH. Terrestrial insects can detect water at great distances. Many parasites are directed to their host by means of chemoreceptors. For example, human body odor attracts the louse, while it has been found that carbon dioxide and other factors direct the female mosquito to warm-blooded animals.

While certain responses to odors and tastes are species-specific, much chemoreceptive behavior is learned. The association of a particular new odor with some factor of the environment continuously occurs throughout the lifetime of many organisms, resulting in behavioral differences between young and old.

Vision and Behavior

Among vertebrates and many invertebrates, vision, or the response to light, is the most important sense in determining behavior. Through the sensation of vision, rapidly moving organisms are able to avoid obstacles and to orient themselves. In addition, through sight they can respond to stimuli that are located at a considerable distance from the organism.

Among the invertebrates, phototaxis is frequently used in selecting a favorable environment. When fly larvae are newly hatched, they are positively phototactic, which aids in dispersal from the hatching area, but as they mature they become negatively phototactic, which helps in protecting them from predators. Light is important for the orientation of the body among many animals, both invertebrate and vertebrate, regardless of whether or not they have gravity receptors. For example, fish have otoliths in their inner ears for gravity perception and crustaceans have statocysts serving the same function. However, both shrimps and many species of fishes characteristically orient with their dorsal sides toward the light. This is known as the "dorsal light reaction." If the light is moved so as to shine from the side of the aquarium, the fish tips over to an angle of approximately 45° because its body position is a function of both light and gravity responses.

Visual stimuli are used by some birds and other vertebrates for habitat selection. Certain anatomical features such as body size, size and position of legs and feet, and size and shape of the bill will also affect the organism's choice of a habitat. However, experiments have been performed in which birds were given a choice of similar nesting areas, and it appeared that the only basis for their selection was visual. The degree to which vision is used for habitat selection is an area of recent investigation and definite conclusions have not, as yet, been drawn.

Among other vertebrates vision represents a direct response to a stimulus. It has been found that many predatory fish will automatically strike at any moving object of appropriate size. A frog cannot learn to respond to an object that is not moving. If an insect is motionless, a frog will not strike. In fact, from a behavioral point of view, the frog cannot even see it because it has been found that certain nerve fibers leaving the frog's eye do not carry impulses unless they are stimulated by a moving object.

The nature of responsiveness to light differs among various animals. Not all animals see the same things. There is a wide difference in the range of the spectrum to which different organisms will respond. For example, insects are able to respond to the ultraviolet spectrum, but cannot see red. The structure of the eye varies among different organisms. The octopus eye is the most advanced among the invertebrates. The image received by this eye is less precise than the vertebrate eye but it can still distinguish shapes. The insect eye is a compound eye consisting of small units, each of which is a small eye in itself. This type of image is probably quite complex, with one of the units receiving a complete image of the object on which the eye is focused. The surrounding units receive only a portion of the image.

The human eye is probably the most superior of the vertebrate eyes in reproducing images. The image has a great degree of fineness. Some lower organisms have light-sensitive structures, but no image-forming structures are present. These organisms can only detect light and dark in much the same way as warmth and cold are detected.

Visual acuity—that is, the ability to distinguish forms—determines many behavior patterns. It can enable an animal to see prey or a predator at a distance. It helps in establishing social behavior by enabling recognition within the group. Vision is important in the behavior of diurnal animals, that is, those animals whose activities are confined to the daylight hours. Among nocturnal animals, the highly developed sense of smell rather than vision provides the greatest amount of information about the environment; however, many animals, such as owls and cats, hunt at night and use vision. Their eyes are adapted for night vision in a number of interesting ways. For example, there are reflecting layers which serve to amplify the low level of light by sending it back across the retina; there are also more light-sensitive rods per ganglion which serve to multiply the low light levels.

The highly developed ability of humans to focus may represent an adaptation correlated with the development of the hand. The fine and detailed movements of the hand in manipulating objects can only be accomplished with the use of eyes.

Species-specific behavior of organisms is released by visual stimuli. As previously noted, the color changes in the male stickleback evoke an attraction in the female. The wide gaping of brilliantly colored mouths of young birds evokes feeding responses in the adults. Many of the ritualized mating procedures represent behavior evoked by visual stimuli.

The use of vision as well as the various types of visual sense organs represent an adaptation toward more complex behavior patterns. These behavior patterns enable the organism to occupy its particular ecological niche with greater success.

Sound and Behavior

Physical movement in air, water, and soil generates vibrations called sound which, if great enough, can be detected at some distance from the point of origin. These vibrations can indicate the presence of the moving object as well as the direction, distance, and nature of the movement. Organisms have developed various means of perceiving these vibrations and use them to determine the nature of the environment, to avoid danger, to locate food, and to communicate with other members of the population and community. The structures that have evolved for sensing sound vary according to the quality of the sound most important to the survival of the organisms. Some organisms respond more specifically to frequency (number of vibrations per unit of time) while others react more specifically to the direction of sound. Although there are great differences in sound receptors, there are essentially two basic mechanisms. One mechanism registers the actual movement of the air molecules causing the sound, while the other registers changes in pressure. The structure of hearing receptors is either membranous or made up of hair-like cells (Figure 28-2). The location of the sound receptors in the various organisms differs. In insects, they may be located on hairs of the antennae or in a membrane on the leg. In fish, the labyrinth of the inner ear, the lateral line receptors along the sides of the fish (which pick up low frequencies), and the swim bladder of some fish act as sound receptors. In other vertebrates, the sound receptors are located on the head.

Many animals can hear as well as emit sounds not audible to humans. It has been found that bats emit sounds of extremely high frequency. These sounds bounce off obstacles which are then perceived by the bat. It is by this means that bats can navigate in complete darkness as well as locate small insects for food. Some predatory insects locate prey in the bark of trees through sound while some animal behaviorists believe owls locate mice at night mostly through sound. A large number of organisms emit sound and the most widespread function of sound is social communication. Sound has the advantage of being emitted by one animal and rapidly perceived by another.

A great variety of mechanisms are used by animals to produce sound. Insects usually produce sound by rubbing two parts of the body together; crickets and grasshoppers rub their legs or wings against a specialized structure. Some fish generate sounds by using the air bladder as a drum or by passing gas from one section of the bladder to another. Vertebrates produce sounds by a

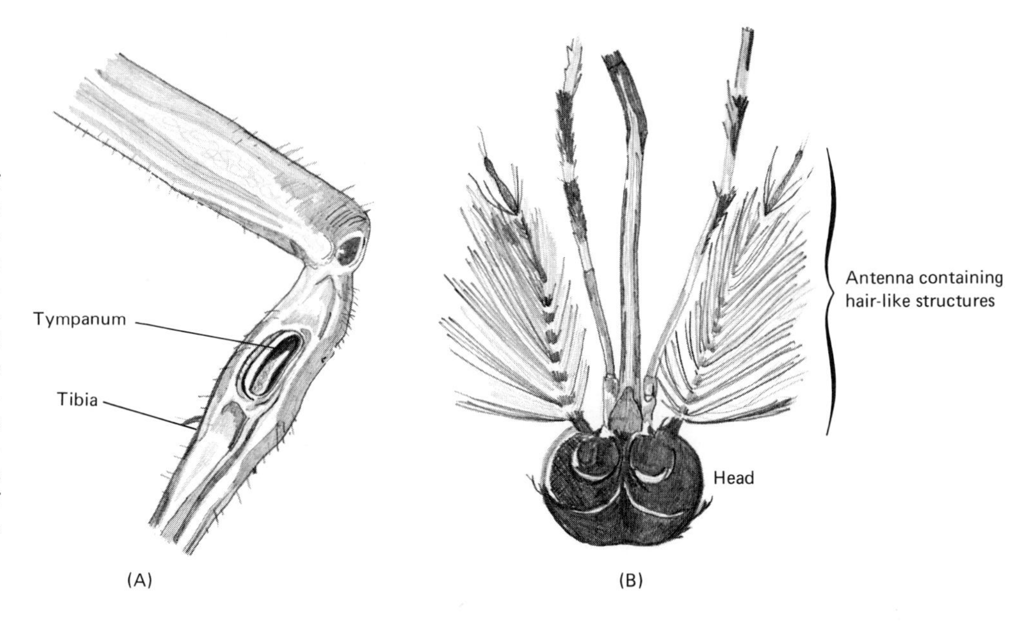

FIGURE 28-2. Types of hearing structures. In general, hearing structures are of two types: membranous and hair-like. (**A**) The membranous type is typified by the grasshopper tympanum on the tibia. (**B**) The hair-like modification is represented by the structures on the antennae of the mosquito. [*Drawn by R. Califre.*]

specialized structure in the throat. The structure is made up of thin membranes that vibrate as air passes between them.

The production of sounds by certain species may occur only when evoked by some external stimulus such as a predator or competitor. Among some species, sound may be confined to the reproductive seasons, while among others, especially birds, the sound changes during the reproductive seasons. It has been found that among crickets the male can produce a calling song, a courtship song, an aggressive song, and a recognition song. The songs of birds are as varied as the number and kinds of birds themselves. Considerable study has been made of the songs of birds and these studies indicate that there are probably many communication signals within the song.

Sound responses represent an adaptation that enables some organisms to survive better; it informs them of an unseen danger, the location of food, or the presence of a mate as well as familiarizing the organism with other aspects of the environment.

Electricity and Behavior

One of the most unique means of communication among animals is the use of electrical impulses or signals. Electrical impulses are transmitted into the immediate environment. These impulses may bounce off objects and be perceived by the sender to describe the environment in much the same way that bats use sound waves. However, it has a more important function of causing responses in other nearby organisms. The extent to which this type of stimulus is used among different species is not known but it appears to be limited to certain aquatic species.

Several divergent species of freshwater fishes of South America and Africa and some marine Elasmobranchs are capable of generating and responding to electrical impulses. Specialized organs in these fish, similar to muscle or nerve tissue, are capable of generating high electrical potentials across the membranes within such organs. The current generated is sufficient to shock both prey and predators. The current is also used to detect objects in the environment and to communicate among other organisms. The electrical signals are received by electroreceptors embedded in the fish's skin.

Studies have shown that the generated electrical signals show diversity in the shape of the electrical field, waveform, discharge frequency, and timing patterns. This diversity provides a means of communication between sender and receiver. The use of these signals in communication has been compared to the light signals emitted by fireflies.

Both field and laboratory studies show that these fish use the electrical signal for recognition of species, sex, age, and class, especially during the breeding season. It may also be important during other times to establish social groups and social behavior. The study of this unique means of communication is helping to provide answers to the possible relationships between the evolution of communication systems and social behavior among animals.

Hormones and Behavior

Relatively little is known about hormones among most of the animal phyla, but it is an area of active research. Hormones of the same chemical composition have different effects among different species. For example, the hormone that stimulates the production of milk by the mammary glands in mammals evokes a response to incubate eggs in hens.

Among insects, hormones regulate various general types of behavior. They affect cryptic behavior, enabling grasshoppers and caterpillars to change color and blend with the colors of the habitat. Hormones also regulate the rhythmic behavior seen in the circadian rhythms of feeding and resting. In addition, hormones also affect sexual activities, the molting of insects, and the social behavior of such insects as ants, bees, and termites.

Hormones released by one individual of a species that can exert remarkable effects upon the behavior of other individuals of the same species are called **pheromones**. Almost imperceptible amounts of these secretions may affect behavior. Pheromones are of two general types. The first are those that alter the endocrine and reproductive systems of the animals. These pheromones are generally passed on by mouth and are ingested. For example, the queen bee passes a pheromone on to developing members of the hive, resulting in the development of morphological females who are workers and do not produce eggs. Among termites, any given individual can develop into a worker, soldier, king, or queen, depending upon the pheromones passed among the members of the colony. The second general type may be airborne, laid down in a trail, or released into water. It will cause an immediate response in an individual. One such pheromone emitted by certain female moths attracts male moths from several miles away. It has been estimated that no more than a hundred molecules per cubic centimeter is sufficient to evoke a response. Of this second type of pheromone, those given off by ants are most interesting. One of these causes an alarm response that will unite other members of the colony against a common enemy. Another pheromone, laid down by an ant who has located food, will direct other members of the colony to the source (Figure 28-3). There is evidence that other pheromones among ants evoke such behavior as grooming, care of the young and queen, and seeking new nest sites. The second type of pheromone is not long lasting, for if it were, it would lead to confusion in communication.

There is evidence that certain aspects of vertebrate

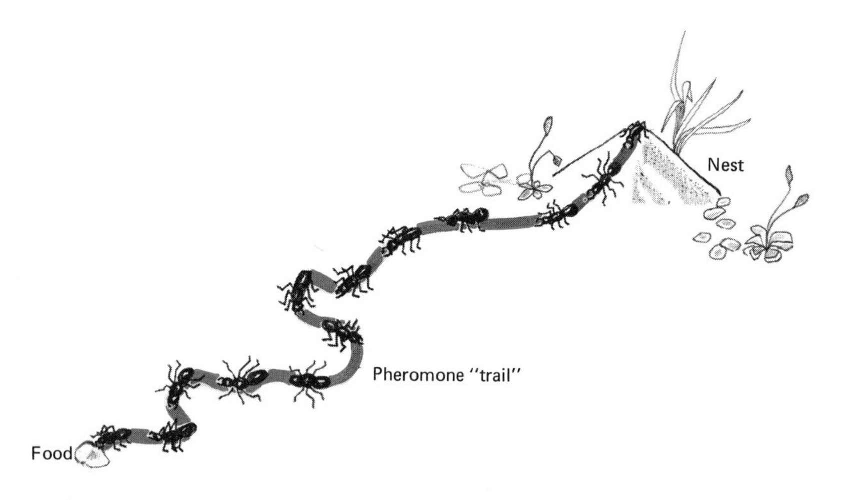

FIGURE 28-3. The effects of pheromones on an ant population. In this instance a pheromone laid down by an ant that has located a source of food will direct other members of the colony to the source over the same pathway.

behavior patterns such as attack, escape, and sexual responses are regulated by hormones secreted by the organism itself. The role of adrenalin in eliciting the emergency response and the role of the pituitary secretion of gonadotropin are excellent examples of hormones affecting behavior (see Chapter 10).

In many instances behavior will influence the secretion of hormones. In female cats and rabbits mating will cause ovulation while in certain birds the sight of a male building a nest will stimulate the secretion of pituitary hormones regulating ovulation in the female.

Recent studies of birds have shown the significance of internal interactions of hormones and neurons and external factors such as length of day, presence of a mate, and nesting site in evoking a sequential behavior pattern of sexual activity. If certain of the internal factors as well as external factors are not present, the reproductive cycle will not start, and if started will not continue.

Through the interaction of hormones the organism has an internal control mechanism which has enabled it to adapt better to its environment. The nature of the interaction of hormones and nervous system is complex.

Sleep

Among the least understood behavioral patterns of living organisms is sleep. Sleep is distinguished from other inactivity, loss of consciousness, estivation, and hibernation in that it is a natural daily rhythmic period of relative inactivity. During sleep there is a decrease in responsiveness. However, there are periods of movement which generally increase with the length of sleep. The general posture of sleep is a reclining position with the muscles greatly relaxed. Heart rate decreases and blood pressure is lower; body temperature may decrease as much as 1°F. All of these physiological reactions are evidently the result of reduced metabolic activity. Recent studies of electroencephalograms, which show the electrical activity of brain waves, indicate there is a decided change in the waves during sleep.

When organisms are deprived of sleep there is increased weariness, irritability, and mental disorganization. In addition to these effects, humans may suffer hallucinations and nightmares. Various theories have been proposed for the mechanism of sleep as well as its cause; yet none provides a satisfactory explanation. Biologists consider it a positive period of inactivity, in contrast to theories based upon a cessation of the active waking state.

It has been known for several years that within the brain there is a center for maintaining wakefulness. This is called the *reticular activating system* (RAS) center and is located in the hypothalamus. It has been found that when this center is stimulated the organism is aroused from sleep. When the center is destroyed the organism goes into a permanent coma. While at one time it was believed that sleep came about when the RAS center was not stimulated, recent studies indicate that there may be centers which evoke sleep by repressing the RAS center. Further evidence indicates that there is a center in the raphe of the brain which acts as a brake on the RAS. As portions of the raphe in the brain of cats were removed, the cats became more sleepless. Further research has shown that the raphe of the brain secretes large amounts of a substance called *serotonin*, which plays an important role in sleep behavior.

There are at least two distinct levels of sleep, light sleep and deep sleep. In light sleep, the level of brain activity decreases but the organism may be aroused relatively easily. Deep sleep sets in after a period of time. The brain waves increase rapidly and are distinct from those of light sleep. The organism is not aroused easily.

From an adaptive point of view, sleep appears to regenerate activities which the organism performs when awake. Many studies have been made on the amount of sleep required as well as specific patterns. The results of these studies indicate that the amount of sleep and sleep patterns are quite individualistic and variable. Most organisms can function normally with less sleep than is found in their normal behavior patterns, yet there is a certain minimum amount of sleep necessary for survival.

Evolution of Behavior

Behavior, like any other aspect of living forms, is adapted to provide for the survival of the organism within a given ecosystem. The great variety of behavioral patterns has evolved throughout long periods of change. The evolution of the structure of organisms has been traced through the study of fossils, but this method cannot be similarly applied to the evolution of specific behavior patterns. Such aspects of behavior as migration, circadian rhythms, and social behavior do not leave fossil records. However, in some cases a fossil record of behavior is available. Dr. Carl Berg has studied the holes found in fossil shells of marine snails. By comparing these with holes drilled in modern shells by living marine snail predators, he has been able to trace the evolution of this behavior.

Behavior, like all adaptations, has a genetic basis, and as such it is subject to variation. The selection pressures upon behavior patterns are the same as for any other aspect of adaptation. Since the mechanisms of evolution are fairly well defined, it is possible to infer the evolution of some aspects of behavior. Mutation and gene recombination produce variety in behavior patterns. If a new behavior pattern enables the organism to compete successfully and to survive better, the pattern will be passed on to the offspring. However, the number of advantageous deviations from the normal behavior patterns is small. In most instances, deviations in behavior place the organism at a selective disadvantage just as deviations in structure and function place the organism at a disadvantage in natural selection. The evolution of behavior, like other aspects of evolution, is a slow, selective process. Changes that occur must adapt the population to a total pattern within the ecosystem.

29 Social Behavior

GUIDE QUESTIONS

1 Give various examples of how animals communicate.
2 Can you think of similarities between the social behavior of wolves and humans? Do wolves also use "body language"?
3 Cite the different levels of social behavior and give a specific example of each.
4 What are the specific steps in mating behavior and what are the roles of the male and female?
5 What part do sign stimuli play in mating behavior? Give examples.
6 What are the advantages of family and group behavior?
7 Give three examples of sign stimuli and the role they play in family and group behavior.
8 What part does aggression play in group behavior? What are its advantages and disadvantages?
9 Give examples of releasing mechanisms and the part they play in group behavior.
10 What are the advantages of dominance and territorial behavior to the survival of the species?
11 Give examples of the following interspecific behaviors: plants and insects, predator and prey, deflection devices, concealing coloration, advertising mimicry, and Batesian mimicry.

WHILE organisms of the same species have similar predictable behavior patterns, their reactions are essentially independent of other members of the population and community. The tropisms of plants, the taxes of protists, and much species-specific behavior of lower forms of animal life take place without regard to other members of the species or community. This does not mean that the various species do not exert a profound effect upon each member of the community. The interdependence is so great that the existence of each organism is dependent to a considerable extent upon other organisms. The interdependence takes place without any knowledgeable or purposeful behavior. For example, plants do not purposefully produce oxygen and food for the heterotrophs, nor do worker ants gather food for the purpose of sustaining the egg-laying queen. Rather, the behavior of the individual organism has evolved to a point where it has adapted the individual to survive best within the ecosystem to ensure the survival of the species. This survival is essential for the reproducing population or deme. As the individual and its way of life become more complex, the type of behavior pattern becomes more intimately involved with and modified by other members of the population and other species within the community. This interaction between other members, that is, social behavior, shows an increase in complexity from the lower forms of life, reaching the greatest complexity in human social behavior.

Communication

The social behavior of animals is based upon communication. While verbal symbolic language is the main means of communication among humans, we also make use of expressions, postures, movements, and even odors. In the animal world sounds are an important means of communicating but equally important and more important in some species are the other means. All social interaction is the result of communication and all social behavior involves communication. Among some groups such as army ants and bees some communication occurs by physical touching. For most groups of animals, however, like fish, birds, and grazing animals communication occurs across space as the individual moves freely in the habitat.

Communication through sound, visual sign, or odor may serve as a sign or releasing stimulus to evoke appetitive behavior in other members of the group. The releasing of appetitive behavior will result in a consummatory act or specific type of behavioral activity. The "trigger features" or social releasers are those stimuli present in the behavior of one animal that release a response in another animal.

In the previous chapter the various means of communication—visual, oral, chemical, and electrical—were discussed in relation to individual behavior. We shall now consider some examples of social communication. On a given summer evening one may hear the sound of several different species of crickets singing at the same time. The songs are communications acting as social releasers that elicit male-seeking behavior by the reproductively ready females. Experiments have shown that, although the songs of the various species differ, it is the pulse rate of the vibrations that is the only determining factor, that is, the sign or releasing stimulus. The female is attracted to the male of the same species by the number of pulse beats per second. In communities where different species of crickets exist together each of the species will have songs of different pulse rate structure. This form of social communication specific to each species helps to ensure survival of that species.

The songs of birds have been intensively studied. They are elaborately patterned communications that do not appear to have any resemblance to the behavior they attempt to communicate. Through the concurrent evolution of vocal apparatus, sensory mechanisms, and behavioral patterns birds are able to recognize rapid sequences of tones. Through their song they are able to signal territoriality, courtship, alarm, feeding, and many other social relationships.

Attempts have been made to determine the extent of vocal communication among the higher primates. Laboratory experiments have shown that these animals are capable of learning to respond to a large number of sounds but in nature they have about thirty different vocal signals, about the same number as the most complex songs of birds.

Communication among wolves is an excellent example of communication by sounds and body movements. Wolves wander in packs in search of food. Within the pack there is a hierarchy of dominance. One wolf, the alpha wolf, leads the pack, has the pick of the females, and dominates the others until deposed by a younger wolf. Wolves rarely fight among themselves, probably because they are able to communicate so well. The way a wolf holds his tail indicates threat, surrender, or position in the pack. There are also appeasing gestures to those higher in dominance. The growl of a wolf is a warning, while his bark, if short, is an alarm; if continuous, a challenge. The howl of the wolf is quite intricate and characteristic of the pack. It may help a straying member to find his way back to the pack as well as indicate to other packs ownership of a given territory.

One of the most elaborate communication systems found in nature is that of the honeybee. Bees returning to the hive from a new source of food communicate by means of a "dance" the type of food, its abundance, how far away it is, and in which direction with respect to the hive. The odor and pollen of the flowers of the food source visited by the bee adhere to its body hairs. During the dance, fellow workers touch the bee and are able to determine the type of food. Dances generally last from several seconds to about three minutes. The duration of the dance gives some indication of the abundance of the supply. If the source is particularly rich the bee will dance longer and with greater vigor. If the distance from the hive is within an 85-meter radius, the dance is a round dance. If the distance is greater, the bee performs a tail-wagging dance. The direction is indicated by the direction of the tail-wagging dance. On a vertical surface the bee orients itself by gravity. A point directly overhead is a reference point for the position of the sun. The bee then proceeds to do the wagging dance in a straight line, forming an angle from the reference line of the sun indicating the angle between the food source and the sun. The number of times the dance is repeated over the same angle indicates the distance (Figure 29-1).

In summary, the degree of communication among members of the same species and with other species depends upon the complexity to which the social organization has evolved.

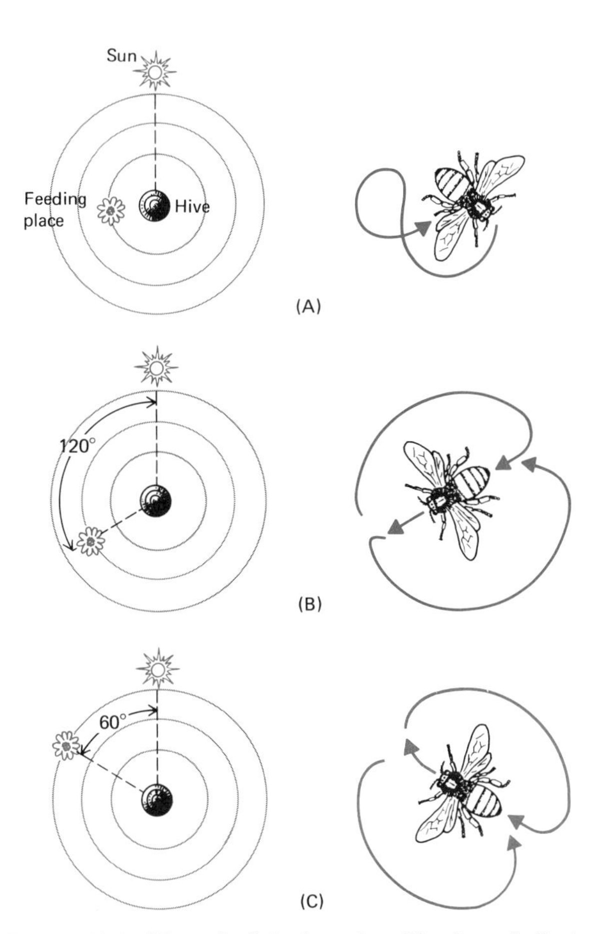

Figure 29-1. "Dance" of the honeybee. The dance indicates the location of a food source. See text for amplification.

The Nature of Social Behavior

Not all aggregations of organisms are social. Groups of organisms may come together independent of each other as a result of some external factor. For example, the congregation of hundreds of insects around a lighted lamp at night is the result of the phototactic response of each individual insect. They do not react to or with one another. On the other hand, aggregation of certain species of birds prior to fall migration is a direct result of interactions. Where there is some form of interdependence, the interaction is social behavior; the influence of coming together is not the result of mere attraction, but cooperation among individuals. Social behavior may be organized at several different levels. There is interaction between two members of a species such as that occurring during the mating season. This association may be temporary or of long duration. Family interactions involve several members of a species, the parents and their offspring (Figure 29-2). Families may interact, resulting in small groups or packs. Beyond this, a large population of a species may congregate and interact at the colony level (Figure 29-3). Within a given ecosystem various populations may interact, resulting in community behavior.

Figure 29-2. Female Traill's flycatcher feeding its young. In many species the family represents the basic unit of social behavior. Here, interaction between parents and young is essential for the perpetuation of the species. The family provides the basic needs of food, shelter, and protection. [*U.S. Forest Service photo.*]

FIGURE 29-3. A colony of black-billed gulls. A large population of a species may congregate and interact at the colony level. The colony provides for greater survival of the species than individuals on small groups. [*Courtesy of Dr. Colin Beer.*]

Mating Behavior

For many species, reproduction consists of the release of gametes into the environment and the union of the sex cells independent of any interaction on the part of the adults producing them. The only condition necessary for fertilization is the release of eggs and sperm at approximately the same time; fertilization occurs by chance. In this type of reproduction, the number of eggs and sperm, of necessity, is exceedingly great to ensure survival of the species. The male oyster, for example, may release hundreds of thousands of sperm cells which are carried by water to eggs released by a female. There they are fertilized and hundreds of thousands of larvae develop. Of these, only a relatively small number will survive. Each step of the reproductive process is independent of any cooperative behavior. Once again, the only prerequisite is that of timing. By contrast, among those species where there is cooperation in mating, the number of gametes is greatly reduced. This cooperation requires a highly adjusted and coordinated behavior. This is usually done by chemical, visual, or vocal signals which prepare the individuals for breeding and for making physical contact. Physical contact is a very important factor because most organisms avoid it. This avoidance of physical contact may originate as a defensive adaptation against predators. Since the female generally carries the eggs for a period of time and also takes care of feeding and protecting the young, she is most important in perpetuating the species. The male, by contrast, can often fertilize more than one female and is generally in a less defenseless position.

Courtship is a kind of social behavior that is adapted for the exchange of gametes. Courtship behavior breaks down the physical avoidance of one animal for another. Often the aggressiveness of the male must be sublimated; this is usually achieved by the female "answering" his aggressive behavior movements by submissive ones. Courtship often resembles fighting behavior in its early stages.

With all species there must be a synchronization or timing pattern to ensure the uniting of the sperm and egg. For those species that physically unite or copulate there must also be spatial coordination. This means that there must be a specific behavior pattern to bring the species together. During copulation the organisms must bring their genital organs into physical contact. This requires a certain orientation of their bodies. Of prime importance to the perpetuation of the species is the need for reproductive isolation. That is, only members of the same species mate. Hybridization is rare in nature. In many instances hybrids are sterile, but of even greater significance is the fact that a hybrid is generally unable to fit into the ecological niche of either parent species. Mating behavior may be summarized as:

1. Synchronization, whereby two animals come together and are physiologically ready to reproduce.
2. Courtship, whereby the two animals break down one another's tendency for physical avoidance.
3. Copulation, whereby sex cells, usually sperm, are transferred.

The *synchronization*, or time, of reproductive behavior generally occurs in the spring. Various birds, fish, and mammals migrate toward their respective breeding grounds. This behavior is usually brought on by the lengthening of daylight, which triggers some internal mechanism which in turn causes the secretion of hormones, resulting in the growth of the sex glands. The timing of reproductive behavior may not occur simultaneously in all members of the species. There may be a considerable variation between males and females. In certain animals, sign stimuli from the male are needed to induce reproductive behavior in the female, and in other cases the sign stimuli originates with the female. Often *courtship* plays an important part in readying an animal physiologically for reproduction and is essential in making certain that *copulation* occurs. Without courtship there would be little or no fertilization because in very few instances in nature can the female be induced to copulate against her will. As was previously shown in the mating of the stickleback, there was a complex interaction of sign stimuli, appetitive behavior patterns, and fixed motor patterns, and finally, the consummatory act which resulted in the laying and fertilization of the eggs. Each step in the pattern of behavior was essential for the succeeding step. With the initiation of mating behavior, the individuals may undergo considerable change. The male becomes more aggressive, may establish a territory, and may change color to become more conspicuous.

Although members of a species may be sexually receptive, mating may not occur immediately. For example, among sticklebacks, there is a certain courting procedure in order to get the female into the nest. Among gulls there are a series of head bobbings, soft callings, and mutual regurgitation prior to copulation (Figure 29-4). This courtship and appeasement is essential in competition among rival males for a female during the breeding season. If the series of sign stimuli in courting behavior were not followed, the female would be attacked. The courting behavior of the female is such that it does not evoke threat response behavior.

Individuals of many species only come together during the mating season. Therefore, during this time members of a species must be attracted or oriented to each other. This orientation is brought about by various means. Among birds, frogs, and certain insects there may be a specific song. In such widely different species

(A)

(B)

(C)

FIGURE 29-4. Head bobbings of the black-headed gull. Although members of a species may be sexually receptive, mating may not occur immediately. Among black-headed gulls there is a series of head bobbings (**A-C**) and soft calling before mating can occur. A female that does not respond with similar sign stimuli may be mistaken for another male and be attacked. [*Courtesy of Dr. Colin Beer.*]

as moths and dogs or cats there may be a specific odor to bring the individuals together. Some species, such as birds, use visual stimuli such as attractive feathers, specific posture, or change of color during the mating season. The proper sign stimulus is essential in each case if the organisms are to orient themselves for reproduction.

The mating behaviors of all the various species except man, and possibly some of the apes, are the result of responses to a series of well-defined internal and external stimuli. The various stages of synchronization, courtship, and copulation are pursued without any foresight or thought of the ultimate end. The adaptation of the mating behavior enables the survival and perpetuation of the species.

Family and Group Behavior

While mating behavior occurs between two members of the same species, cooperative behavior, in a family or group, is much more complicated. It involves interaction between parents, between parents and offspring, between offspring, and between different adults making up the population. The ultimate ends of this behavior vary. These may be to provide food, shelter, and protection against the environment and predators. The sign stimuli are numerous and complicated, involving such actions as preventing parents from eating their young, protecting the young until they are able to fend for themselves, indicating to the group a source of food, and warning the group of danger. An important aspect of these sign stimuli is that there must be reciprocal stimuli, that is, while one member evokes a positive beneficial response the other must supress a negative harmful response. For example, when the chicks of gulls hatch, a whole new series of parental behavior patterns emerges, involving such activities as feeding, protecting, and guiding the chicks. The shape of the body of the parents evokes a response in the chicks. The chicks in turn must peck at a red spot on the bill of the adult herring gull before it will give up food to the young. The feeding behavior follows a pattern of sign stimuli between the young and adults.

Family Behavior. The longer the period necessary for the care of the young, the more complicated the behavior patterns become. For example, the dependence of the young upon the parents among songbirds exists for a relatively short period of time, rarely exceeding 6 weeks. In contrast, the young among certain species of monkeys are dependent upon the parents for as long as a year. Most of the family behavior patterns among songbirds are limited to the basic necessities for physical survival including food, shelter, and protection. Among monkeys, in addition to the basic necessities, social behavior is more fully developed in such activities as mutual grooming, play among siblings, and physical contact between parents and young.

The establishment of certain stimuli in maintaining the parent–offspring relationship is exceedingly important. If the sequence of stimuli does not follow the established behavioral patterns, the parent–offspring relationship will cease to exist. Among many mammals such as cats and rats, licking behavior plays an important role in the care of the young. Licking the mammary glands and external genitalia during and after pregnancy fosters a longer and closer bond between mother and offspring. In the feeding of birds, as was stated previously, the gaping mouth and persistence of the young begging for food evoke the feeding response in the parents. Usually, the hungriest bird begs the most and therefore is fed first.

One of the most important factors in family life is the means by which parents are prevented from eating their young. In many species of fish the young are so small they look like prey to the parents. The female cichlid fish carries the young in her mouth and in doing so refrains from eating all prey and thus is prevented from accidently eating her own young. As young gulls mature they begin to look more and more like the adults. In order to prevent aggression on the part of the parent birds, the young assume a submissive attitude in their posture which is similar to that assumed by the female during courting behavior; this appeases the adults.

The animal family relationships studied by Harry Harlow were mentioned in Chapter 27. In his studies of rhesus monkeys, Harlow has shown that the family as well as the group are held together by what he calls "affectional systems." These represent several different categories of relationships—mother to child, child to mother, peer to peer, male to female, and adult male to adult male. From his research he found that when monkeys are reared alone, away from other members of the species, they show abnormalities in behavior. As adults they do not mate but react aggressively or in fear of other monkeys. There is little of the social grooming and mutual cleaning that is found among monkeys in nature. There is no socialization with other members. In most instances the isolated monkeys sat staring in space for long periods of time. Many of them mutilated parts of their body by tearing and pinching their skin. It is interesting to note that similar behavioral patterns are expressed by children deprived of their parents and by adolescents and adults in mental institutions.

As a result of further experimentation he found that the relationship of the mother to the child among monkeys, in providing body contact, food, protection, and other physical necessities, was essential in rearing normal offspring. The complexity of the mother–child relationship and the degree of interaction increases from simple to higher forms of life. However, only in humans is there behavior that is determined by clearly defined future goals on the part of the individual.

Group Behavior. As the young of a species mature, they become less attractive to the parent and new behavior patterns are established. The parents become more aggressive and provide less and less care. The young may become members of a group such as packs, herds, or flocks in which different types of interactions take place. Among the advantages of forming these groups is their effectiveness in providing a defense against predators. They may also allow cooperation in obtaining food; in some species, they simply provide social relationships. Among some vertebrates where grouping appears at first to be just an aggregation, studies have shown that the organisms living in a group are capable of living longer and establish better eating habits than when living solitarily. It has been found that goldfish secrete substances from their body which seems to "condition" the water for other goldfish by neutralizing poisons. The combined effect of a large group of fish is more effective than that of small groups. Among chickens, those that are in a group will eat more than those that are solitary.

In some populations a hierarchy of dominance such as a pecking order may prevail. Among chickens a definite pecking exists in which the highest member of the flock is not pecked at all while the lowest may be pecked by all (Figure 29-5). Other animal groups, such as monkeys, may be dominated by mature males. The establishment of dominance within a community is partially determined by the level of sex drives as well as the physical strength of the individual. Position in the pecking order or group hierarchy may determine priority of food, water, territory, and among males, choice of a mate.

A hierarchy of dominance in a population reduces the number of conflicts and increases the degree of intraspecific and individual recognition. This enables large flocks to specifically identify their young for feeding and protection.

It appears as though early imprinting plays an im-

Figure 29-5. Group behavior. In this illustration of white Leghorn hens, a number of dominance relationships are shown. (**A**) The hen on the left, in her first contact with the one on the right, submits. As she turns away, the dominant hen pulls a feather from her back. (**B**) The dominant hen in the flock may even peck at the largest hen in the flock. (**C**) While the dominant hen is asserting her authority, three others, intermediate in the pecking order, rush to the food.

portant role in the establishment of groups. Once the organism has been imprinted with other members of the group, the imprint remains. In order for organisms to establish a group there must be effective sign stimuli. Among birds these stimuli are usually visual, such as characteristically colored plumage, or among fish, a spot on the body which aids in identifying and keeping the school together. Specific sounds and call notes also help to establish and maintain the group. Social insects recognize members of their colony through characteristic odors.

Most relationships in a group depend to a great degree upon the movements of others within the group. When one bird in the flock becomes alarmed, the others also become alarmed. Feeding, flying, and sleeping are closely coordinated, with each individual watching the others. Communal attack among birds, called mobbing, is a type of a social cooperation. If a small bird comes upon a predator such as a roosting owl or large hawk, it will raise an alarm that attracts the other birds in the neighborhood who then join in driving off the predator.

Territoriality. While there are many favorable factors in group behavior within the ecosystem, there are limited amounts of resources such as food, space, and energy as well as mating partners. Therefore, there is competition for these resources. This competition is very clearly defined in the establishment of territories, but once the territory is established, it leads to reduced competition. An organism may establish a definite area within the habitat and defend it from all other members of the population. While this may represent a disadvantage to certain members of the population because it may exclude them from an ample supply of the resources, it has definite survival value for the species. Those individuals that establish and defend a territory are the more vigorous and better fitted members of the population. They are more assured of producing offspring and thus perpetuating their more favorable attributes within the species.

Territorial behavior occurs among a wide group of organisms such as insects, fish, amphibians, birds, and mammals. Of the various species, territorial behavior has been most thoroughly investigated among the birds; yet many aspects of the findings have been found to be common to other species. For example, the territorial impulse is a well-developed part of the reproductive behavior of marine mammals such as the elephant seals. The same is true for many species of fish. The size and shape of the territory depend upon the habitat and the species. Where there are no differences or only slight differences in the various sections of the habitat, the territorial areas are roughly circular because of individual defense of adjacent areas. Where there are differences in an area in terrain, food supply, and size, the shape of the territory will vary.

The length of time that a given organism will maintain a territory will depend upon its life cycle or its specific habits. Most species generally maintain a territory during breeding season. However, there are variations. Some birds, such as the starling, may start defending a territory in the fall before the breeding season and will progressively defend it more vigorously as the spring breeding season approaches. A few species that are migratory may return to the same areas while still others will select a new site each season. Generally, the male of the species defends the territory more vigorously than the female; the female, however, will defend the nest and area where the young are located with great vigor. Insects such as wasps will defend a given spot. Hermit crabs do not defend a geographic area; they vigorously defend their shells, which may be thought of as a sort of portable territory or microhabitat. Certain fish such as sunfish, bass, and trout will defend a nesting site. Although many mammals show territorial behavior, there has been difficulty in observing certain species which are nocturnal.

Territorial behavior serves various functions depending upon the species. In some species, only a small area around the nest site is defended and feeding occurs in other areas with other members of the population. Other species defend an area at a distance from the nest. However, the vigor of the defense decreases with distance from the site. Essentially, the first function of territorial behavior is to provide a nesting site. A second function is to provide a food supply. This applies especially in the wintertime or other times when food is in short supply. The English nutcracker will defend an area in which it has stored nuts in the winter. The male hummingbird will defend an area where flowers and insects are abundant. A third function of the territorial behavior is to provide an area in which to demonstrate courtship display. These are intricate and elaborate rituals by which the various species of males attract females. Once a female enters this area she is courted and then mating occurs. The female may nest within the area and the male will take part in raising the young, or the female may leave the area and raise the young alone in a different area.

Of the various types of territorial behavior, the most highly developed is evidenced among colonial animals. In contrast to a territory established and defended by an individual, a territory may be claimed and defended by a group. Among hummingbirds, a group of males may take over an area, excluding other males. As females enter the area they are inseminated and then they nest elsewhere. In other types of colonies the male may defend an area in which there are several nesting females. Among seals, the male will set up a territory within the colony in which he has a harem of up to twenty females. He usually defends this territory against other males. In colonial territoriality there are clearly defined relationships of dominance and submission. Various types of cooperative and social behavior such as mutual feeding and attack and mutual grooming and play take place. There are some instances in which certain species set up a territory against other species. However, defense of a territory against another species is relatively rare since different species occupy different ecological niches and essentially do not compete for the same resources.

While a species may be able to defend a territory against another species, it is usually an unnecessary expenditure of energy. However, when this occurs it is for adaptive reasons. For example, species of marine damsel fishes defend territories in which the eggs are laid by the female and cared for by the male. Most fish that enter the territory are chased away, especially those species which feed on the bottom and therefore would eat the eggs. Familiar but different species are usually ignored. It is obvious that recognition is an essential factor in this type of territorial defense.

In recognition, a series of sign stimuli are used. Other signs are used if a threat display is elicited. The various display mechanisms are usually sufficient to ward off intruders. Physical combat over a territory is rare, and mortal combat is even rarer. Often, the display or sheer persistance of one of the participants is sufficient to expel the other from the territory.

Aggression. The absence of physical combat over a territory does not mean that there is no aggressive behavior within the population. Aggressive behavior is most prominent during the breeding season and is called reproductive fighting. In addition, there is some aggression displayed when dominance relationships are being established in a hierarchy. The means of displaying aggression vary among the different species. Dogs, gulls, and some fish bite, while some hoofed animals kick. Animals with antlers or horns butt each other (Figure 29-6). Although there may be considerable aggression displayed, rarely are any of the individuals seriously injured. The display is usually sufficient. In addition to visual display, threats may be made by depositing scents or by making sounds. Reproductive fighting generally occurs among males and only occasionally among or with females. Fighting usually takes place in a specific area where one member is setting up a region in which mating and nesting is to take place. One of the most important results of aggression is that it provides individuals with a territory or a mate, which is essential for reproduction.

Some behaviorists use the term *agonistic* in place of aggressive, which indicates that in fighting behavior or display there is present a fear or flight motivation as well as an aggressive one. It is often difficult to say to what extent a behavioral pattern demonstrated in fighting is motivated by aggression and to what extent by fear.

It is interesting to note that overt fighting, that is, actual combat, rarely occurs between animals with structural adaptations, such as poison glands, sharp pincers, claws, or horns, that have the capacity to wound mortally, or between animals belonging to small island populations. In these species, fighting is ritualized, consisting mostly of species-specific threat movements such as the butting contests of mountain goats and sheep and the pushing contests of Galápagos giant lizards. Restricting fighting to species-specific threat movements is important if the members of the species are to survive.

While fighting is usually associated with territoriality, dominance, or mating, once a territory is established the organism will attack all trespassers but will flee from attack when outside its territory. The display of aggression is generally a result of internal factors,

FIGURE 29-6. Aggressive behavior exhibited by two bull elks. [*U.S. Fish and Wildlife Service photo.*]

such as hormones, or specific sign stimuli from the environment. These stimuli serve a dual function; they elicit a display of aggression when a stranger is present in the territory or cause fleeing from a stranger when outside the territory. David Lack relates in his book *The Life of the Robins* an experiment that he performed with sign stimuli. When a mounted robin with all brown feathers was placed in an occupied territory, it was not attacked. However, a crude model with a few red feathers evoked an aggressive response. The red feathers were the sign stimulus evoking aggressive behavior. Similar aggressive responses were also evoked by a phonograph recording of another male singing. Among lizards, special movements display certain colors that will evoke aggression in other lizards. The importance of the red belly of the stickleback as a sign stimulus was discussed in Chapter 27.

Dominance. Aggressive behavior in animals may occur for reasons other than possession of territory or a mate. There may be aggressive behavior over food or a roosting place. In these instances learning may reduce the amount of fighting. Once a hierarchy of dominance is established, those individuals subordinate to another learn that they will not be attacked if they recognize the dominance. This results in minimal fighting within a group. However, if some individuals are removed, or become weaker through illness or old age, or new individuals are introduced, fighting to establish a new order may result.

The means of demonstrating dominance within a given group vary among the species. For example, a mouse will assume a specific posture and vibrate its tail upon meeting another mouse. If the mouse does not retreat, it will engage in combat. Eventually, one of the mice will retreat and recognize the dominance of the other. Among cattle and other hoofed animals there will be a series of threats that may actually result in the physical combat of butting heads. In fact, the adaptation of antlers or horns is probably primarily for establishing social rank rather than for fighting a predator.

Studies have shown that social rank occurs in many invertebrates and vertebrates. The essential steps in establishing rank are advertisement, threat, and if necessary, actual fighting. Among the factors that determine social rank within the group are age and early experiences. Usually the younger individuals are low ranking, moving up only after there has been a development of physiological and anatomical capabilities. If an individual is placed in a subordinate position early in its life, it may remain there even after maturity.

The results of social dominance play an important part in the survival of the individual. Those at the top of the social order have mating priority and better access to food. This results in a certain selectivity within the species as the attributes of those at the top of the social rank are more often passed on.

Behavior Between Species

Up to this point the discussion of social behavior has dealt with interactions within the species or population. However, in the community there are interactions between different species. In the symbiotic relationship the behavior of the interacting species must be closely coordinated. The success of the organisms that benefit from the association is directly related to established behavior patterns. If the resulting behavior is such that one or the other is placed at an extreme disadvantage, as in certain types of parasitism, the relationship soon ceases because of the death of one of the organisms.

In contrast to the physiological and behavioral associations in symbiosis, there are relationships among species in which there is specific cooperation or interaction based upon similar sign stimuli. Certain species have evolved a set of mechanisms or anatomical features that release a response in an organism of another species. Other species have devices or anatomical features that serve to avoid evoking a response in an organism of another species.

Releasing Mechanisms

Among the prominent mechanisms by which organisms attempt to cause a response in another are the color and structural devices that ensure pollination in plants. Many flowers are adapted through brilliant coloration to evoke an attractive response in pollinating agents. One of the classical studies of this type of response was that done by Karl von Frisch who showed that honeybees could distinguish among colors. Clearly, the brilliant colors of the flowers are sign stimuli evoking an attractive response in insects and birds. Most insects are not sensitive to red color; however, those insects which respond to red flowers are responding to ultraviolet light. Red flowers and berries are most prominent in regions where pollinating birds such as hummingbirds live. Some flowers (e.g., iris and lily) have patterns of dots or stripes arranged in such a way as to direct the insect to the pollen (Figure 29-7). In other flowers, scent is the most important stimulus in evoking a response in insects. Various species of plants produce odors similar to that of decaying meat. This attracts flies and other scavenger insects. As the insects move from flower to flower, the pollen is transferred. In all of these flower–insect relationships the behavior of the insects is species-specific. The sign stimuli release the behavior pattern, which is consummated when the insect feeds on the nectar and at the same time inadvertently transfers pollen. Some insects, such as bees, learn to specialize by visiting only one species of flower for a time and then moving to another species.

In the flower–insect relationships discussed so far the behavior results in a benefit for both organisms; the flower is pollinated while the insect obtains food. There are other species interactions in which the stimulus evokes behavior that is of benefit to only one of the species. Among those animals of the benthos zone, the deepest ocean depths, there are various luminescent lures that evoke an attractive response in another species. The organism attracted to the lure is quickly snatched up for food.

In some types of orchids the shape of the flower is so similar to the female of certain species of wasps that pollination occurs when the male attempts to mate with the flower. Of course, this is of no benefit to the insect.

Another mechanism used in causing a response in the predator–prey relationship is that of conspicuous coloration. There are essentially two types within this category. In the first type the colors have no significance to the predator until it interacts with the prey. This type of coloration is found in the bright color pattern of the monarch butterfly. The colorful monarch butterfly is distasteful to birds. It flies slowly so that birds have the opportunity to recognize it. Once a bird has attempted to eat a monarch butterfly, it does not attack another. In the second type of defense the coloration is a "bluff" because the species is harmless, being neither distasteful nor obnoxious. For example, the polyphemus moth is

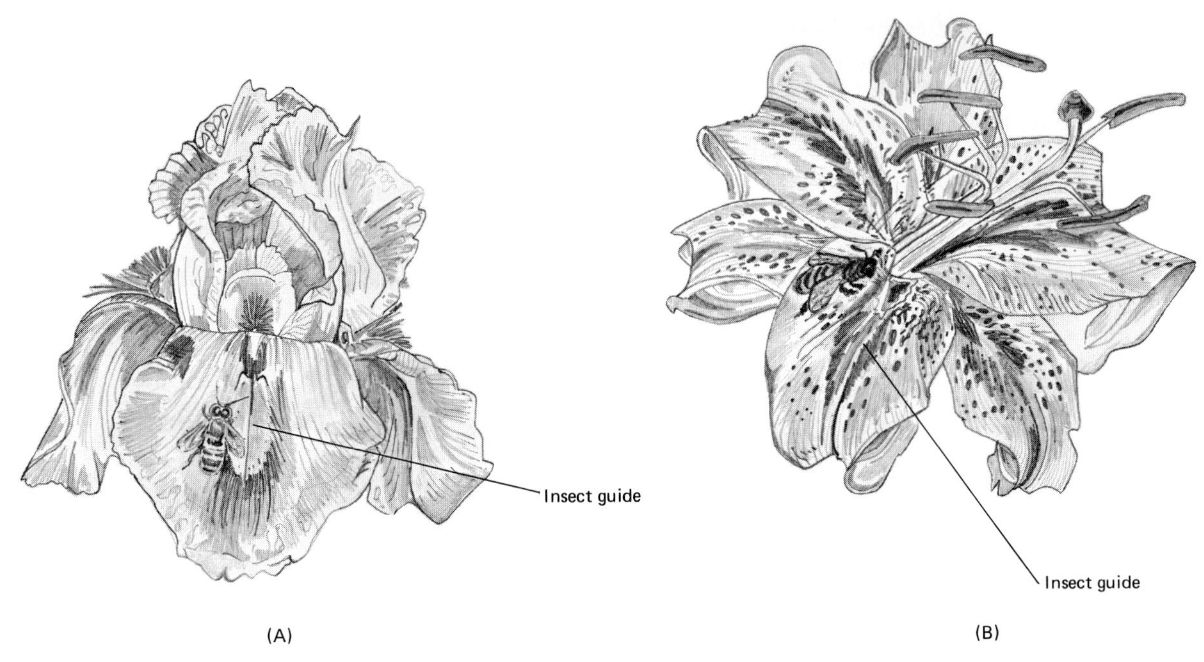

Figure 29-7. Insect guides to pollen in flowers. (**A**) The stripes of different colors offer a striking contrast in the iris. The insect is attracted to a stripe and follows it in to obtain nectar. At the same time pollination occurs. (**B**) In the lily there are both stripes and dots of contrasting colors. Again, pollination is aided. [*Drawn by R. Califre.*]

somewhat camouflaged when at rest. When a possible predator approaches, it quickly spreads its wings, exposing large spots that resemble eyes (Figure 29-8). Experiments have shown that birds are definitely frightened by this display. When the spot was covered on the insect it was readily attacked. Part of the effectiveness of this type of coloration is a result of the suddeness of its display.

Avoidance of Release Responses

Various species have specific devices by which they attempt through visual means to prevent an active response in another species. This adaptation is prevalent in predator–prey relationships. These devices are concealing coloration, advertising mimicry, and deception mimicry, and they represent some of the most unique and interesting adaptations of living forms.

Concealing Coloration or Camouflage. When the color of an organism blends in with its natural habitat, the adaptation is called *concealing coloration* or *camouflage.* Colors and/or patterns enable the organism to blend in with its natural surrounding provide greater survival value. The white winter coat of the snowshoe rabbit blends in with the snow. The light and dark spotted pattern of the fawn, as well as that of many other animals, blends with the light and shadow of the typical forest habitat (Figure 29-9).

Some animals have the ability to change their color and to match their background. This ability is found among lizards, frogs, fish, and crabs as well as other animal groups. The flounder can change not only color but also patterns to match its background. By remaining motionless the camouflaged organism blends into the background, escaping predators.

In addition to coloration, there are some species of organisms that are adapted to resemble objects within the environment. The organism is so well disguised by shape as well as color that it can be recognized only if it moves. There are insects susceptible to attack by birds which resemble twigs and leaves. There is a species of

Figure 29-8. Bluff defense coloration. (**A**) At rest, the Polyphemus is susceptible to attack by a predator. (**B**) When disturbed by prodding, the moth spreads its wings and flashes its "eyespots." This sudden exposure of "eyes" startles the predator. (**C**) Diagram of a head of an owl. Compare this with the "eyespots" in (**B**) to note the similarity. [*Drawn by R. Califre.*]

frog living in the tropics that, when crouched motionless upon a leaf or branch, resembles the excrement of birds. These adaptations are most effective for survival since birds are visual predators and simply do not recognize these prey.

Advertising Mimicry. There are some organisms with highly effective protective devices such as a sting, repelling odor, poison secretion, or bad taste. If these organisms were inconspicuous, it would take a predator a much longer time to learn to avoid attacking them. Thus, a greater number of the species would be destroyed. F. Muller first described the *advertising mimicry* phenomena, which have since been called *Mullerian mimics.* The monarch butterfly has been cited as an example. The skunk is another example. Presumably, the attractiveness of the coloration of these species evokes an initial response. However, once the predator has learned the unpleasantness of the experience, there is avoidance. There are a number of insects not belonging to the same species yet all having a similar obnoxious protective device and all resembling each other. For example, different species of bees, wasps, and hornets have a similar shape and outstanding yellow and black stripes. Different species of butterflies which are bad tasting have similar colors and patterns. This adaptation of advertising similarities aids the species; for once a predator learns to avoid any one of the species it tends to avoid the others. It also represents an excellent example of convergent evolution (see Chapter 5).

Deceptive Mimicry. Certain species that do not have an obnoxious defense resemble species that do. This type of mimicry was described by H. W. Bates in 1862 and is called *Batesian mimicry.* Some nonstinging flies resemble bees or hornets. Many species of butter-

Figure 29-9. Fawn of a white-tailed deer. The light spots on the back of the fawn contrast with the darker surrounding fur. This coloration aids the fawn in blending with its surrounding habitat by giving the illusion of a shadow effect. [*U.S. Forest Service photo.*]

Figure 29-10. Batesian mimicry. The viceroy butterfly on the right is quite palatable to predators. The monarch butterfly on the left is not palatable and discourages attack by predators. Because of the close resemblance between the two species, the viceroy is protected from attack.

flies resemble species which are bad tasting. One of the most outstanding examples is the viceroy butterfly, which very closely resembles the bad-tasting monarch butterfly (Figure 29-10). Deceptive mimicry is an advantage only for the mimics. The model or the species that the mimic is copying is at a disadvantage because it is attacked by predators that have not learned to avoid the bad-tasting species and also by predators that have eaten the mimic and have found it palatable. The mimic will benefit from the deception if it is relatively rare and if it appears within the environment after predators have learned to avoid the bad-tasting model. An additional advantage of the mimic is that it is generally smaller than the model. Predators will usually select the largest prey when given a choice.

Each of the various adaptations discussed have evolved to enable a particular species to survive better. While the adaptations are essentially anatomical or morphological, they are intimately related to behavior. Each of the adaptations considered results in behavior patterns that enable the survival of the species.

30 Human Behavior

GUIDE QUESTIONS

1 What are some of the major problems in studying human behavior?
2 Into what three major categories did Sheldon classify individuals?
3 Give examples of studies of the effects of biochemicals on behavior.
4 In what ways is the brain associated with human behavior?
5 Give two examples of glandular and metabolic influences upon human behavior.
6 Give two examples of social influences upon human behavior.
7 What are the levels of social behavior? Give an example of each.
8 What are two major forces in the social environment?
9 What are the various types of group roles? Give an example of each.
10 What is meant by social personality and social motivation?
11 What are social institutions? How did they come about?

MANY of the mechanisms and processes of behavior of all organisms can be found more or less in certain aspects of human behavior. However, because man has been able to exert a degree of control over the abiotic and biotic environments, he has developed the most complex behavior of all living organisms.

Psychobiological Approach to Human Behavior

While the study of human behavior has been oriented mainly within the field of psychology, studies in recent years have revealed that many aspects of behavior are closely related to various aspects of biology. It is for this reason that a new approach called the *psychobiological approach* to the study of behavior has emerged. Although it is beyond the scope and purpose of this text to give a full consideration of human behavior, there are certain aspects that are essential for a more complete understanding of the total world of living things.

Humans have many anatomical and physiological characteristics in common with other vertebrates and some invertebrates. In fact, the anatomical and physiological differences among species are, in many instances, a matter of degree rather than kind. For example, the structure of the eye and the image it is capable of perceiving vary among the species. There is considerable difference between the structure and image perceived by the insect in comparison to the human eye. However, there is slight difference between the human eye and the eye of a mammal such as an ape. Since structural and functional characteristics are so closely related to the behavior of an individual, it follows that many of the responses in humans are similar to, or the same as, those in other vertebrates. These similar responses, however, are limited in most instances to simple survival reactions. Like all animals, humans seek warmth when cold, food when hungry, and water when thirsty.

Humans, like many other animals, interact with other members of the species, and this interaction constitutes social behavior. While civilization has greatly modified human social behavior, it still retains many similarities to that of the primates and other vertebrates. Evidence of the similarity between humans and other animals has been shown with studies of the Australian aborigines. It was found that tribes were made up of about 500 persons defending a given territory. The size of the territory was directly related to the environmental resources. The similarity in number of persons and size of the territory among the various tribes indicated that these dimensions represented some sort of optimum. A smaller population could not adequately defend the territory and a larger population resulted in a splitting of the tribe because of lack of resources within the given territory. Within the tribe there was social rank, with a chief and subordinates. As has been shown in the previous chapter, this same type of population control, territoriality, and social rank is also found among many other species of animals.

Within recent years, scientists using modern instruments have been able to explore areas of behavior which previously were inaccessible. Physiologists have implanted electrodes in different areas of the brain of experimental animals to receive information regarding the various processes taking place. Endocrinologists, through the use of radioactive isotopes, are able to trace and discover the mechanisms of metabolism, glandular action, and chemical changes within the organism. The information gained from these researches is being applied to broaden the interpretation of human behavior. The morphological and physiological processes of heredity, brain function, glandular and metabolic mechanisms, and social influences on biological functioning all contribute to the behavior of man as well as other organisms. Behavior is complex and little is known of the interrelationships of the processes and mechanisms in establishing any given behavior patterns. However, research in this area is constantly adding information which is helping to supply some of the answers regarding the processes underlying human behavior.

Heredity

Since anatomical and physiological processes which are so intimately related to behavior are inherited, it is possible that predisposition to certain behavioral patterns is also inherited. Studies of the heredity of man have indicated that patterns of growth, certain skills and abilities, and physical appearance have a considerable genetic basis. Whether or not human intelligence has a genetic basis is a question of great current interest and controversy. The degree to which personality and the potential ability for adjustment or maladjustment is genetically controlled is not known. The learning and complexity of behavior patterns of man are so much greater than that of other organisms, that frequently it is difficult to distinguish that which is acquired through environmental conditions and that which is determined genetically. An added difficulty is the fact that studies of human behavior are based primarily upon observation.

Behavior and Body Build

Body types constitute one of the most persistent areas of research in the study of hereditary effects upon behavior and various theories have been developed on this subject. These theories attempt to show that a particular inherited body build has a characteristic temperament and behavior associated with it. For example, it has been theorized that there is a fat-jolly type, a thin-nervous type, and a muscular-athletic type. Various theories date back to the time of Hippocrates, who assumed there were four humors—black bile, yellow bile, phlegm, and blood—they were passed on from parent to child in varying amounts, resulting in different personalities and temperaments.

Since the time of Hippocrates various classifications have been made. Among those that have received the widest support was that set forth by W. H. Sheldon, in 1942. He classified individuals into three major categories. *Ectomorphs* are thin and tall and tend to be highly sensitive, introverted, and self-conscious. *Mesomorphs* are well-proportioned and muscular. They tend to be rugged, athletic, and extroverted. *Endomorphs* are relaxed, sociable, and easy-going fat types. One of the major criticisms of this categorization is that rarely is there a person who fits exactly all the characteristics of a given body type. In addition, changes in environment as well as natural changes that occur with age can cause changes in body build. Many studies have been made, some giving support, other criticizing the various theories. From observation there is some indication that there is a relationship between body type and behavior; however, this relationship may not be genetically determined. It is probably that the relationship of body type to personality is the result of an interaction between heredity and the physical and social environments.

Evidence from Heredity. The aspects of behavior that are inherited in man have not been conclusively determined. Conditions of the environment such as diet, habits, and living conditions are difficult if not impossible to control in order to get empirical data. Much of the evidence on inherited behavior of man has been obtained from studies of twins. In recent years biochemical analysis has been useful in determining certain hereditary aspects of behavior. As scientists unravel the mechanisms whereby genes determine the various linkages and patterns of chemical function, it is hoped that the manner and degree in which the biochemistry of the body determines behavioral patterns will also be uncovered.

Factors Modifying Behavior

While heredity provides the physiological and anatomical characteristics of an individual, which in turn may influence behavioral patterns, there are factors within the environment which also greatly affect behavior. Studies have indicated that behavior may be altered before and after birth by certain environmental factors.

Prenatal Influences. It has been found that the developing embryo may be affected by various chemicals, diseases, and physical factors. Deprivation of adequate oxygen because of delayed birth, circulatory irregularities of the mother, or incompatible Rh blood types may result in effects similar to physical brain injury. Oxygen deficiency in the later development of the fetus may lead to a predisposition to quarrelsomeness, hypermobility, lack of ability to concentrate, and a variety of other behavioral and learning defects in the later development of the child.

Certain chemicals and other materials in the blood system of the mother may permeate the placenta and affect the growth and development of the embryo. Among the external influences that may affect the embryo are narcotics, toxins, alcohol, drugs such as thalidomide, X-rays, and other forms of radiation.

Some prenatal conditions do not appear to affect the growth and development of the fetus but in some manner affect activity. Certain children within the first weeks of birth show great nervousness and are highly sensitive to stimulation or are "high strung." Others are quiet, unaggressive, and quite placid. While these appear to be inherited characteristics, there is the possibility that other factors may be involved. The mother may be overemotional, secreting larger quantities of hormones, such as epinephrine, which may reach the fetus. Other external stimuli may cause biochemical reactions in the mother which in turn affect the fetus. These external influences affect the internal environment in which the fetus is developing and result in behavior which cannot be ascribed to heredity alone.

Postnatal Influences. The environmental factors after birth may also greatly affect the development of specific behavior patterns. Injury to the brain or meninges during birth may affect areas of the brain that control behavior. It has been found that the cerebral tissue of males is more delicate and vulnerable to injury at birth than females. It has been postulated that this may account for the greater incidence of delinquency and criminal behavior in males. It is quite possible that such behavior as hyperactivity, short attention span, inability to concentrate, and erratic behavior may be the result of undetected brain damage during the birth process.

The effects of defective sensory organs as well as other hampering conditions in the early environment of the child will greatly shape behavioral development. Children born deaf, blind, or neurologically and anatomically impaired will respond differently to environmental stimuli than normal children. Where factors in the physical and social environment will produce satisfaction in a normal child, they may give rise to frustration in an impaired child. This results in restricted personality development.

Another important, although not quite so obvious, factor is the mother–child relationship early in life. The studies of H. Harlow with rhesus monkeys discussed in Chapter 29 have been applied to human behavior. If the relationship between mother and child is not allowed to develop, many of the expected behavioral patterns of normal children are lacking. There may be a social withdrawal and abnormal personality development. As the individual becomes older, other family relationships such as the presence of other siblings and other family members influence behavior development. As the child becomes more social, relationships with people outside the family as well as the environment in which the child develops takes place. This will be discussed later in this chapter.

The Brain and Behavior

Clearly, the origins of individual behavior are centered in the brain. The innumerable neurons in the brain operate in receiving and sending impulses which influence thought, judgment, insight, and other mental activities. With various modern instruments and techniques scientists are able to study the chemical and electrical influences upon brain activity which are important factors of human behavior.

Evidence indicates that a complex series of electrochemical events is involved in brain function. The reception, processing, and transmission of information and commands to the muscular system underlies behavior. The brain waves of healthy individuals are clearly distinguishable from those of indivuals with brain damage or disease. Most research on the electrical activity of the brain has been done upon animals. Prolonged electrical stimulation of the brain of monkeys has resulted in peptic ulcers similar to those found in extremely nervous humans. In other experiments in which electrodes were inserted within certain areas of the brain of cats, it was found that a quiet, docile animal could be changed into a snarling, savage cat when the electric current was turned on. The studies made of the electrical activity of the brain of humans have been done upon psychotic individuals. It has been possible to evoke feelings of fear, sadness, fright, and terror by electrically exciting specific areas within the brain. The exact relationship between the electrical activity of the brain and the development of specific behavior patterns is not known. Studies have shown that there is an alteration in chemistry of the stimulated areas.

The initial investigations of the influence of chemicals upon brain activity were begun when it was observed that there were striking behavioral changes in people who ate certain plants such as the peyote cactus or a species of Mexican mushrooms. When these plants were analyzed for the effective chemical agent, it was found that *psilocybin* from the mushroom and *mescaline* from the cactus caused the changes in behavior. Further work led to the discovery of synthetic chemicals similar to the natural chemicals which could produce essentially the same effects to a greater or lesser degree. Most notable among these synthetic chemicals is *lysergic acid diethylamide* (LSD). When LSD is injected into normal persons, symptoms of psychoses, compulsive thought and speech, visual imagery, hallucinations, feelings of unreality, and a rapid vacillation of emotions result. With increased use of LSD outside the laboratory and the resulting adverse effects, more intensive study has been made of this chemical and related compounds. It has been found that there may be permanent effects upon the behavior of the individual as well as an alteration of the sex cells, the carriers of hereditary material. *One of the major conclusions to be derived from these studies is that the use of drugs that alter or change behavior even temporarily may have far greater and more lasting effects on the individual than originally suspected.*

Recent studies of other compounds normally found within the human body may under certain circumstances affect behavior. For example, tryptamine, which occurs naturally within the human body, is detoxified and eliminated by the kidney. If the compound is not eliminated at a certain rate, it is converted to other intermediate metabolites which may cause disturbances of behavior. The various studies of the effects of other substances within the body indicate that unusual accumulations of various chemical compounds may be a major determinant in the variation of adaptive behavior of humans.

Additional studies of the brain involve the effects of abnormal stimulation. It appears that the normal functioning of the brain is dependent upon continued environmental stimulation. Experiments in which individuals were subjected to extreme deprivation of external stimulation, such as being confined in specially designed cubicals free from sound and visual forms, resulted in abnormal sensations. The exact mechanism evoking this abnormal behavior is unknown. One theory proposes that there is a depletion of RNA and essential proteins in the neurons. This results in the inability of the neurons to carry normal electrical impulses. However, regardless of the validity of this theory, there is definite evidence that the proper functioning of the brain is dependent upon the varied and continuous input of stimuli.

Glandular and Metabolic Influences on Human Behavior

While the foundation on which behavior is established is genetic and the administration of the mental motor activities occurs in the brain, the regulation of the growth and developmental processes is directly related to the endocrine glands and the hormones that they secrete. The effects of endocrine glands and their hormones have been discussed in Chapter 10. However, the specific mechanisms in determining behavior for most hormones are unknown. Present research of hormones is taking place in many areas. There are two clearly defined areas that appear to be important to human behavior. The first of these deals with the effects of hormones upon the activities of the nervous system and the second deals with hormonal effects on the metabolic processes that affect behavior.

It is well known that malfunctions in the endocrine system can cause profound changes in behavior. Hyposecretion of the thyroid gland can result in sluggishness, apathy, and lack of energy, while overfunction can cause irritability, excitability, and excessive energy. Underactivity of the adrenal glands can result in fatigue while overactivity can result in a highly tense physical state. The secretions of other endocrine glands have both direct and indirect effects upon the behavior patterns of man. The results of external stimuli affect the secretion of hormones. These in turn affect behavior. The mechanisms by which these stimuli are translated into a physiological functioning of the endocrine glands are not known.

Since the hormones are regulators of the chemical reactions within the organism, they play an important part in determining behavior. Studies show that an improper metabolism results in a change of behavior. For example, a deficiency of glucose or oxygen brings about a dulling of perception and judgment as well as depression, anxiety, and neurotic symptoms. Certain vitamin deficiencies can result in feelings of fear, apprehension, irritability, hostility, poor memory, and poor social adjustment.

Social Influence on Human Behavior

Man is one of the most social of all organisms and frequently, in the analysis of behavior, it is difficult to distinguish the degree to which psychobiological and social forces influence behavior. Often many behavioral disorders are psychosomatic, that is, social conditions manifest themselves in physiological disorders. For example, some people react to anxiety in a social situation by developing an upset stomach, headache, an attack of diarrhea, asthma, or some other physical disorder. The physiological activity somehow serves as a symbolic response to some external social experience.

It has been found that most stomach ulcers are a disease of cities. Among primitive tribes where there is a well-structured social order, ulcers are almost completely absent, while in the cities with loose social structures ulcers are common. In some way the social behavior affects the physiological function of the individual. Other observations show that the social environment in which an individual lives will determine susceptibility to emotional disorder as well as specific patterns of behavior.

All the factors that contribute to the integrated behavior of the individual represent an adaptation to the biotic and abiotic environments. Throughout the evolution of man these behavioral adaptations have enabled man to survive. Those behavioral attributes which developed within the species that enabled man to adapt better to the forces of environmental change have been the legacy of living man.

Human Social Behavior

The social character of the human has resulted in development of societies which represent both an environment and a type of behavior. Society, as an environment, is composed of two forces—the actions of other persons and the physical structures of civilization and culture. These forces are a part of the total biotic environment to which the individual or group is adapted in order to survive. Like other aspects of the environment, they are subjected to change.

As a type of behavior, society consists of the interactions which occur between individuals and groups. The behavior of society or social behavior is a result of the complex interactions of (1) the genetic heritage of individual humans, (2) the psychological, physicobiological, and physical forces which determine human behavior, and (3) the environment in which they live. Since society is a form of behavior, it is adapted to the biotic and abiotic environments. Social behavior has been subjected to change and has evolved in much the same manner as other aspects of the living world.

Levels of Social Behavior

As with all animals, there are levels of social behavior among humans. These levels are dependent upon the degree of social interaction among individuals. The first level is that of an aggregation. An *aggregation* is a group of individuals performing the same type of behavior but who do not interact in any systematic way. This may be a group of gulls feeding in the same area of the ocean or a herd of cattle grazing in a field. Among humans it may be represented by a group of young children playing in the same area but separately and independently. The next highest level of social interaction is *interpersonal behavior.* Here interaction is between two individuals in which both are affected by presence of the other as well as by the environment. This level of social behavior includes mother–child relationships, friendship, sexual behavior, and fighting. As was discussed previously, the mother–child relationship in monkeys and humans is an important factor in determining adult behavior. In human society, as in the animal communities, certain patterns of behavior between individual peers are necessary for acceptance. Whenever three or more people interact, the next level of social behavior, *group behavior,* takes place. Among humans, group behavior may be a free social interaction or a culturally determined interaction regulated by institutions, laws, and tradition. In group behavior the individual adjusts his activities according to the accepted norms and traditions of the society. Group behavior among humans may not represent a clear-cut survival adaptation as it does among most animals. While all behavior represents an adaptation for survival, many of the group activities of humans such as play, rituals, and cooperative creativity do not serve the purposes of direct physical survival. Play among lion or bear cubs prepares the individual for later physical survival, whereas play among children probably contributes to psychological and social survival in later life. With the higher powers of reasoning and man's application of technology against the forces of the environment, physical survival is no longer a major problem in human society. Therefore, psychological and social survival is of far greater importance. The highest level of social behavior is *culturally patterned activities.* These activities represent large broad patterns of behavior which are an intimate part of the civilization. Included in these activities are religion, government, and education. The individual within the society adapts to certain patterns of behavior such as attending or not attending college, practicing a specific religion, or participating in such activities as a particular type of social entertainment.

The adaptation of the individual's behavior and resulting personality is influenced and determined by social and cultural factors. Culture itself is a product of human personality which is adapted to the biotic and abiotic environments. In much the same manner as the beaver builds its home of sticks and the bird builds its nest of grass and leaves, according to its own makeup within its environment, so does man. The igloo of the Eskimo, the thatched hut of the grassland tribe, and the brick and cement apartment house of the city dweller represent a response to the physical environment in which man lives. Man's clothing, tools, and social interaction are to a great extent a response to the environment. On the other hand, society plays as much a part in defining an individual's behavior and personality as does his biological nature.

The Social Environment

The social environment is composed of two major forces: (1) the behavior of other individuals at the various levels of interaction and (2) the culturally determined organizations or social institutions which make up the society. Within the social environment the behavior of the individual determines his position or role as much as the role sustains the behavior of the individuals. The behavioral differences between individuals are the result of the effects of other members of the group, the structure of the institution, and the genetic makeup of the individual. The behavior of the individual in turn creates a social environment for himself and others. For example, the individual football player may have certain determined attributes which enable him to fit a specific position on the team. The quarterback is generally fast, agile, and of a slighter build than the heavier, slower moving, and larger defensive lineman. The individual activity or behavior of each player contributes to the total success or failure of the team. The success or failure of the team in turn affects the behavior and activity of each of the individual members.

Within society there are many different possible roles for the individual such as parent, worker, leader, teacher, etc. In addition, most individuals fill more than one role. Each individual is better fitted to fill certain roles rather than others. The personality of the individual reflects his life roles in society and the manner in which he fulfills them.

Interpersonal Roles. The most fundamental roles of the individual are the positive interpersonal relationships of individual to individual. These include husband to wife, mother to child, sibling to sibling, friend to friend, and so on. While these interpersonal roles are generally considered positive, there may be interpersonal roles that may be negative in character, such as between two fighters or political opponents. All interpersonal roles persist as a result of a certain amount of give and take. As has been shown previously, the interaction at the person-to-person level plays an important part in determining individual behavior or personality. Of course, there are all degrees of interaction on a person-to-person basis. Some of these may be extremely impersonal, such as that between members of a family and the postman or saleswoman. Others may be quite authoritative, such as that exhibited by a watchman, a guide, and so on.

Group Roles. Some individuals spend much of their lives on an interpersonal relationship level. That is, they interact with members of their family and fellow workers but with few other people. Others, in addition to interpersonal relationships, involve themselves to a greater or lesser degree in groups such as community clubs, religious groups, or large social groups. When an individual becomes involved in a group, he assumes a particular role for which he finds himself best adapted. Group roles vary in complexity, social function, and cultural significance.

Levels of Complexity in Group Membership. The social roles vary greatly from mere attendance by a member to that of leadership. As the role assumes greater importance, it becomes more complex. The difference between a group role and that of an interpersonal relationship is clearly defined. For example, two close friends in an interpersonal relationship modify their interactions when placed in a group. When alone, they will act and say things to each other that they would not say in a group.

The simplest type of group role is *membership*. Here the behavior of the individual is to a great extent determined by the specific group, whether it be a class, social club, or family group. The second level of group roles is that of the observer. In this role, the individual observes the group and reports and interprets what the group does. This role requires objectivity and an impersonal relationship. These individuals include such persons as authors, judges, police officers, and work supervisors. The third level of group roles is a *promoter*. These are individuals who promote activity among the members. While not assuming the role of leadership, it is a person who may be "the life of the party," "the trouble-maker" or one who through his own behavior stimulates activity of other members. At the highest level of group roles is *leadership*. The leader makes decisions and represents the group when it is not together. The behavior of the leader is a complex interaction of membership, social observation, human relations, and decision-making activities.

While the group roles of humans reach a higher level of complexity than among other organisms, there are many characteristics which they have in common. As has been noted previously, there is a social hierarchy among some species of animals, with the various members of the community assuming different roles. While the leadership role among animals is gained by physical strength, size, or "bluff," among humans it is essentially based upon personality and ability rather than physical attributes. As with animals, interactions between members of the family differ from those between other members of the population.

Function of Group Roles. Group roles differ in function as well as complexity. The function of the role is adapted to fulfill the needs of the group and its members. Among group functions are those of obtaining food, achieving communication, rearing children, and dealing with enemies and catastrophe. In addition there are group functions in social, family, and audience roles. The social activities of the family center around the rearing of children. The functions of providing food, shelter, clothing, and protection by the parents represent a primary means of survival. The audience roles are the means by which societies are organized. These roles include such activities as education, religion, government, and the development of artistic and literary culture. Central in promoting this type of role interaction are teachers, clergy, politicians, artists and authors. Work roles are primary in promoting economic, building, scientific, and distribution activities. Work roles can be simple or complex and are the prime factors in shaping adult personalities. In much the same manner as each species fills its ecological niche, so too the various individual and group roles of humans fulfill a niche in a society. The success of the individual within the society to a considerable extent depends upon how well he fills and is adapted to his particular role in much the same way as an ecological niche is filled by other organisms.

SOCIAL PERSONALITY. Group roles and their functions provide the social environment in which the individual adjusts his behavior. When individuals choose a particular role there are social pressures imposed upon them. These pressures may force the individual to change or to establish new behavioral patterns. Essentially, each individual's personality is organized and defined according to the relationship between ability, behavioral characteristics, and the role structure of the society.

Adjustment of the individual to the social environment in which he lives requires that he assume different roles in the family, at work, in recreational groups, in community groups, and so on. In each of these roles the individual modifies his behavior and displays different aspects of his personality.

There are many factors that determine role selection. Some of these factors, such as age, sex, and physical ability, are also found among other animals in establishing social rank. Others, such as intelligence, mechanical ability, artistic talents, and emotional stability, are more or less limited to humans. Important in determining a particular role that an individual will play within society is the size of the family, the occupation of the parents, and the family's religious and ethnic character. In addition, the nature of the community, whether it is rural, suburban, or urban, will help to direct the social role of the individual. The social environment places strong pressures upon individual behavior and personality.

Once an individual has identified with a group, there are strong social pressures from within the group placed upon him. The individual's patterns of behavior, motives, and opinions are influenced by the standards of the group. Some people conform to the standards more than others. The external pressure of the group may also have some effect on the internal determinants of the behavior of the individual. For example, acceptance or rejection of certain external pressures from the group may be translated into physiological reactions through the brain of an individual. Feelings of well-being or anxiety may result. The manner in which the individual perceives the external pressures may affect the secretion of specific hormones. These in turn may affect particular behavior patterns.

SOCIAL MOTIVATION. Unlike most animal behavior, man's behavior is largely goal directed; that is, there is *social motivation.* Man is capable of placing his actions in the context of the past, present, and future. Within this framework he sets teleological goals, educational goals, social goals, scientific advancement, technological advancement, success in business and public life, plus many other goals. Social motivation may result in achievement, frustration, or conflict. Individual behavior is guided by standards of performance set by social groups. These groups operate at every level of human activity. The level of aspiration set by individuals will vary greatly. Some function at a low level, others at a high level. Abilities determined by genetic and biological makeup limit certain aspirations yet, within the social environment there are opportunities for each individual to attain success. Obviously, there are persons who do not have the physical attributes to succeed in certain career goals. Furthermore, there are others who are limited in their physical dexterity, artistic ability, and intelligence, and are unable to succeed at goals which they set for themselves. Within the social environment there are attainable goals to which each individual is adapted to enable him to fulfill his aspirations. If there is considerable difference between the aspirations and attainment of goals, frustration results. The level of frustration and the degree of acceptance of frustration by an individual may contribute to more serious social problems such as conflict, prejudice, and an antisocial personality.

Although man is an animal, he differs from all other animals in that he is endowed with certain characteristics which enable him to develop a culture. Through the evolutionary process he has physically and mentally evolved to a point where his behavior is the highest and most complex form in the animal world. Physically, man has evolved the characteristic of standing on his hind limbs, thus freeing his forelimbs to manipulate objects. The forelimbs, with a ball-and-socket joint at the shoulders, a hinge at the elbow joint, a rotating joint at the wrist, and a set of digits plus a highly developed visual sense, enable man to lift and handle heavy objects and to grasp and carry out exceedingly fine manipulations. The senses of man are adapted for social communication. The combined perceptual abilities of vision, hearing, speaking, smelling, and touch and higher powers of reasoning enable man to communicate and socialize on a much higher level than his animal ancestors and all present living forms.

Memory, learning, insight, perception, and language have developed to such a degree that with his physical endowments man is capable of existing in almost any type of environment. While other organisms have some of the attributes of man, the differences between man and other organisms are great. Much of the behavior exhibited by living forms is largely genetically determined and is species-specific; however, this type of behavior is less important to the existence of man. It is possible that through the growth of learned behavior the function of species-specific behavior in man had less survival value. Those elements of species-specific behavior which man has, so far as is known, are more generalized and are probably better classified as drives or needs, although some patterns of behavior, such as the smiling response, seem to be universally present in all peoples.

Through processes of evolution, the behavior of man has been adapted to enable each individual to cope with the forces of the abiotic and biotic environments. Through his behavior man has gained control over all other living organisms as well as the ability to survive and exist in the most severe physical environments. However, this unique position has been the result of both conscious and unconscious interactions with other humans, particularly through social institutions.

Social Institutions

The development of social institutions probably was the result of two major adaptive shifts in the evolution of man. The first was the upright position, which freed the hands for manipulation; the second was the shift from a totally vegetarian diet to a plant and animal diet. This second shift placed importance upon the use of tools and weapons. It also placed strong selective value upon increased intelligence and complex social interactions. The hunting for large game required greater intelligence than picking available fruits and berries and required the cooperation of many individuals working toward a common end.

The upright posture, the long gestation period, and the lengthy rearing period (longer for the human than any other species) before an infant could care for itself placed a high selective value upon the evolution of family life. With the institution of the family, groups of families found greater survival value living together cooperatively. There was mutual effort in food gathering and protection from common enemies. With changes in environment and an adaptation to an omnivorous diet, the survival of a group which built its own dwellings and engaged in agriculture was more successful than that of wandering tribes.

With the development of an agricultural society, many institutions evolved. The domestication of animals had an important effect on the development of social organizations. Animal and agricultural products had value and became mediums of exchange. While certain members of the community farmed, others carried on other necessary activities within the community. As the community survived and grew, institutions evolved which made the society more complex.

Hierarchies of leadership evolved into government. In order to explain the unknown and provide mental security, religions were instituted. The complexity of social organizations increased with the development of technology. The evolution of different types of social interactions with their institutional counterparts has made the physical evolution of man of less survival value. Prior to the development of complex social institutions such as education, medicine, technology, government, and social welfare, man's destiny was to a considerable degree determined by adaptation to the physical environment. With the increasing complexity of human society, man's evolution and adaptation depend upon a greater emphasis of self-awareness. This self-awareness will enable our social institutions to be realistically based upon the physiological, psychological, and social aspects of man as an existing entity within a dynamic ecosphere. The effects of man and his society have had dramatic effects upon the total abiotic and biotic environments. Man's individual and social behavior will determine to a great degree the future of present living forms as well as man himself.

SUGGESTED SUPPLEMENTARY READINGS FOR SECTION SIX. BEHAVIOR

ABLE, E. L., *Drugs and Behavior.* New York: Wiley–Interscience, 1974.

AGRANOFF, BERNARD W., "Memory and Protein Synthesis," *Scientific American,* June 1967.

ALCOCK, JOHN, *Animal Behavior, An Evolutionary Approach.* Sunderland, Mass., Sinauer Associates, 1974.

BEECH, H. R., *Changing Man's Behavior.* Baltimore: Penguin Books, 1969.

BENTLY, D. and R. HOY, "Neurobiology of Cricket Song," *Scientific American,* August 1974.

BENZER, S., "Genetic Dissection of Behavior," *Scientific American,* December 1973.

BLEIBTREU, J. N., *The Parable of the Beast.* New York: Macmillan, 1968.

BROWN, F. A., J. HASTINGS, and J. PALMER, *The Biological Clock: Two Views.* New York: Academic Press, 1970.

BROWN, J., *The Evolution of Behavior.* New York: Norton, 1975.

CAMPBELL, B., *Human Evolution.* Chicago: Aldine, 1974.

DAVIS, D. E., *Integral Animal Behavior.* New York: Macmillan, 1966.

DILGER, W., "The Behavior of Lovebirds," *Scientific American,* January 1962.

EDMUNDS, M., *Defense in Animals.* New York: Longman, 1974.

EIBL-EIBESFELDT, I., *Ethology, The Biology of Behavior.* New York: Holt, Rinehart and Winston, 1975.

ETKIN, WILLIAM (ed.), *Social Behavior from Fish to Man.* Chicago: University of Chicago Press, 1967.

HASLER, A. D. *Underwater Guideposts: Homing of Salmon.* Madison: University of Wisconsin Press, 1966.

HEBB, D. O., *The Organization of Behavior.* New York: Wiley, 1961.

HESS, E., *Imprinting.* New York: Van Nostrand Reinhold, 1973.

HINDE, R. H., *Biological Basis of Human Social Behavior.* New York: McGraw-Hill, 1974.

HONIG, W., and P. H. R. JAMES, *Animal Memory.* New York: Academic Press, 1971.

JENNINGS, D. H., and D. L. LEE (eds.), *Symbiosis.* New York: Columbia University Press, 1975.

KEETON, W. T., "The Mystery of Pigeon Homing," *Scientific American,* December 1974.

KLOPFER, P. H., and J. P. HAILMAN, *An Introduction to Animal Behavior.* Englewood Cliffs, N.J.: Prentice-Hall, 1974.

LEVINE, S., *Hormones and Behavior.* New York: Academic Press, 1972.

LONGO, V. G., *Neuropharmacology and Behavior.* San Francisco: Freeman, 1972.

LORENZ, KONRAD, *Evolution and the Modification of Behavior.* Chicago: University of Chicago Press, 1965.

———, *Studies in Animal and Human Behavior.* Cambridge: Harvard University Press, 1970.

MANNING, A., *An Introduction to Animal Behavior.* Reading, Mass.: Addison-Wesley, 1972.

MCGAUGH, J. L., N. M. WEINBERGER, and R. E. WHALEN (eds.), *Psychobiology.* New York: Academic Press, 1971.

PALMER, J. D., "Biological Clocks of the Tidal Zone," *Scientific American,* February 1975.

PAPPENHEIMER, J. R., "The Sleep Factor," *Scientific American,* August 1976.

PENGELLEY, E. T., and SALLY J. ASMUNDSON, "Annual Biological Clocks," *Scientific American,* April 1971.

PETERSON, F. A., "Short-term Memory," *Scientific American,* July 1966.
SCHMITT, F. O., "The Physical Basis of Life and Learning," *Science,* vol. 149, pp. 931–43, 1963.
SCOTT, JOHN P., *Animal Behavior.* Chicago: University of Chicago Press, 1972.
SKINNER, B. F., *The Behavior of Organisms: An Experimental Analysis.* New York: Appleton-Century-Crofts, 1966.
———, *About Behaviorism.* New York: Knopf, 1974.
———, *Walden Two.* New York: Macmillan, 1976.
STOKES, A. W., *Territory.* Stroudsburg, Pa.: Daiden, Hutchinson and Ross, 1974.
STREET, PHILIP, *Animal Migration and Navigation.* New York: Scribner, 1976.
TINBERGEN, NIKO, "The Curious Behavior of the Stickleback," *Scientific American,* December 1952.
———, *The Herring Gull's World.* New York: Basic Books, 1963.
VON FRISCH, KARL, *Bees: Their Vision, Chemical Sense and Language.* Ithaca: Cornell University Press, 1971.
———, *Animal Architecture.* New York: Harcourt Brace Jovanovich, 1974.
WALCOTT, CHARLES, "The Homing of Pigeons," *American Scientist,* vol. 62, no. 5, 1974.
WHALEN, R. E. (ed.), *Hormones and Behavior.* New York: Van Nostrand Reinhold, 1967.
WICKLGREN, W. A., "Long and Short of Memory," *Psychology Bulletin,* vol. 80, no. 6, 1973.
WICKLER, W., *Mimicry in Plants and Animals.* New York: McGraw-Hill, 1968.

APPENDIX

The Five-Kingdom System of Classification According to R. H. Whittaker

Kingdom Monera

THE Monera, which go back furthest in evolutionary history, are believed to have been descended independently from the first living form. The Kingdom Monera includes all organisms in which the cells lack plastids, mitochondria, and complex flagella (organelles for locomotion). In terms of nuclear organization, monerans lack well-defined chromosomes and nuclear membranes and are therefore procaryotic. Most are solitary-unicellular; some are colonial-unicellular. The predominant mode of nutrition is absorption, although some monerans are autotrophic, deriving the energy for synthesis from solar radiation (photosynthetic) or inorganic chemicals (chemosynthetic). Reproduction in this group is primarily asexual by a process called binary fission in which the genetic material divides, a cross wall develops, and the newly formed cells separate. Locomotion in the group is either absent or by means of simple flagella or by gliding.

Living representatives of the Monera comprise two major groups of organisms: bacteria and blue-green algae. Although the exact evolutionary relationship of bacteria to blue-green algae is somewhat obscure, certain data suggest that the kingdom to which they belong is a naturally related category.

Bacteria

Bacteria (*Myxobacteriae, Eubacteriae, Actinomycota,* and *Spirochaetae*) are probably the simplest and smallest living organisms possessing cellular organization. All are visible only with the aid of a microscope and generally assume one of three morphological forms: (1) cocci (spherical), (2) bacilli (rod-shaped), and (3) spirilla (spiral). Some representative forms are shown in Figure A-1. Bacteria probably outnumber all other organisms on earth and are universally distributed. While some species are autotrophic (chemosynthetic or photosynthetic), most are heterotrophic (saprophytic or parasitic). Although most are known for diseases that they cause, many of them assume a significant role in food processing, organic decay, and soil fertility.

Blue-Green Algae

Blue-green algae, members of the *Phylum Cyanophyta,* are distributed in fresh and marine waters and wherever there is dampness: in soil, along stream banks, on tree trunks, near hot springs or glaciers (Figure A-2). They all contain a special blue-green pigment but, because the blue-green hue is commonly hidden by other pigments, some exhibit a variety of colors including black, purple, red, yellow, green, blue and intermediates of these. The Red Sea, for example, is said to have been named because of the abundance of a blue-green algal

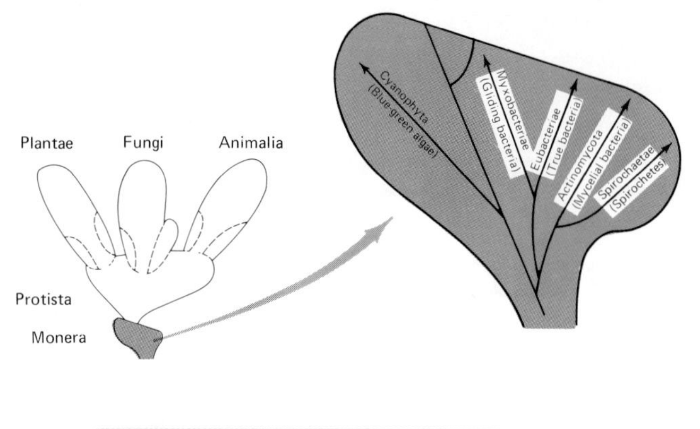

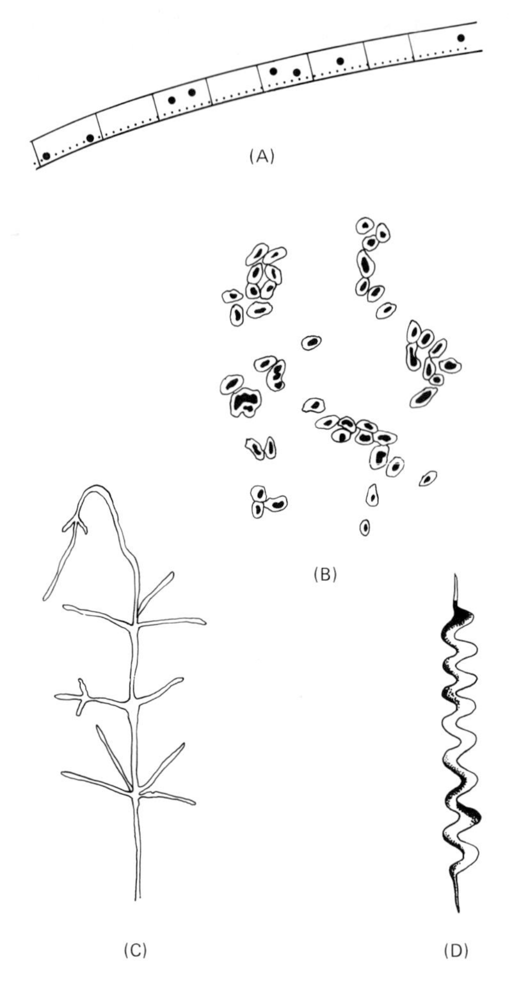

Outstanding Characteristics
1. Unicellular organization
2. Procaryotic
3. Lack plastids, mitochondria, and advanced flagella
4. Nutrition primarily by absorption; some photo- or chemosynthetic
5. Reproduction typically asexual

Figure A-1. Kingdom Monera: Myxobacteriae, Eubacteriae, Actinomycota, and Spirochaetae. (**A**) Phylum Myxobacteriae. These bacteria are closely related to the blue-green algae in that they do not possess flagella, and movement, if present, is by gliding. They consist of flexible cells that form filamentous colonies. Shown here is *Beggiatoa alba,* a bacterium found in fresh and polluted waters. (**B**) Phylum Eubacteriae. This phylum includes the most numerous and best-known kinds of bacteria. They are spherical or rod-shaped; motile species have flagella arising all over the cell; some rod-shaped species form spores. Illustrated is *Klebsiella pneumoniae,* the causative agent of pneumonia. (**C**) Phylum Actinomycota. This phylum, represented by *Streptomyces,* is characterized by branching, coenocytic filaments. (**D**) Phylum Spirochaetae. These bacteria are long, slender, helical, and motile. *Treponema,* the causative agent of syphilis, is shown.

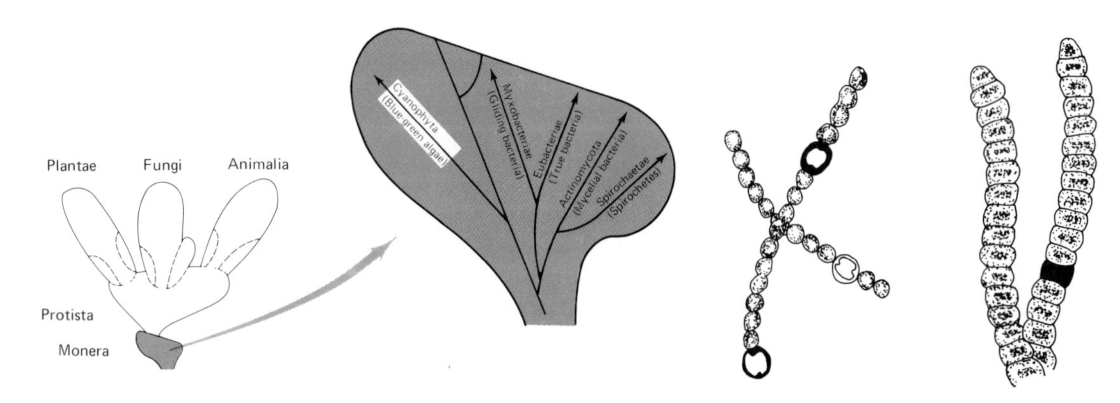

Figure A-2. Kingdom Monera: Cyanophyta. (**A**) *Nostoc.* (**B**) *Oscillatoria.*

form with heavy red pigmentation. Some move with a slow gliding motion and their nutrition is plant-like, involving photosynthesis. Because of this autotrophic characteristic, they are an important source of food for aquatic animals. Some have established a symbiotic relationship with members of the Kingdom Animalia. If allowed to grow unchecked, they become so abundant in reservoirs that they can clog the filtering system and give the water a fishy taste.

Kingdom Protista

The Kingdom Protista includes an enormous assemblage of organisms, solitary-unicellular or colonial-unicellular, which lack tissue formation and all of which are believed to be descendents of very early cells that evolved eucaryotic nuclei. In all cells of the Protista, the genetic material exists as distinct structures, the chromosomes, that are in turn delimited within a true nucleus bounded by a well-defined nuclear membrane. Nucleoli are also universally present.

In the course of evolutionary development, the presence of chromosomes made possible two new types of cell division: mitosis (Chapter 5) and meiosis (Chapter 17). Both of these processes, which originated with the Protista, are more complex cellular processes than the characteristic fission of the Monera. The development of meiosis also provided the potential for the production of genetically dissimilar offspring through sexual reproduction, a phenomenon not possible through fission. Reproduction in the Protista then, is both by asexual and sexual means.

The cytoplasm of protistans, capable of cyclosis, or streaming, contains rather prominent vacuoles that function in a storage, excretory, or water-balancing capacity. None of these features is found in the monerans. Photosynthetic protists (algae) also contain true plastids, particularly chloroplasts, which may have evolved through a symbiotic relation between pre-procaryotic and protistan ancestors. In addition to the photosynthetic pigment chlorophyll *a*, which is found in the Monera, autotrophic protistans contain one or more varieties of chlorophyll (*b*, *c*, *d*, and *e*) as well.

Protista have evolved many processes for nutrient procurement. Some are photosynthetic, others are heterotrophic, subsisting as parasites, saprophytes, or holotrophs. Some primitive modern Protista still retain a combination of these nutritive patterns; advanced types have concentrated on one of these and lost the others. Movement is by advanced flagella, or other means, or they are nonmotile.

Autotrophic Protistans

One of the autotrophic phyla included in the Kingdom Protista is *Euglenophyta*, or as they are commonly called, the euglenoids. These minute unicellular organisms resemble the green algae in that they typically have chlorophyll in chloroplasts and are thus photosynthetic autotrophs. *Euglena* is such an example (Figure A-3A). In some respects, however, they are more like protozoans than plants. They are spindle-shaped with one, two, or even three flagella that propel them through their freshwater environments; their cell walls are not rigid, but rather pliable and deformable; and some have gullets which are passageways for ingested foods.

Reproduction is generally asexual, the body dividing lengthwise. Unlike the major groups of the Plantae that store food as starch, the euglenoids store food as *paramylum*, a polysaccharide chemically related to starch. An example of a nonchlorophyllous (heterotrophic) euglenoid is *Peranema* (Figure A-3B).

A second autotrophic Protista phylum is *Chrysophyta*, or the chrysophytes. These organisms, either unicellular or colonial, inhabit damp soil, fresh water, and marine water. This phylum is enormously diversified and includes a great many structural types. Diatoms are the most abundant chrysophytes (Figure A-4A). In the ocean, when diatoms decay or are eaten by animals, their cell walls, impregnated with silicon, accumulate on the ocean bottom and form deposits known as *diatomaceous earth*. Ancient deposits, in areas that once were seas, are mined, and the diatomaceous earth is used in filters, for insulation, as an absorbent for liquid nitroglycerin in dynamite, and as a mild abrasive in toothpastes and metal polishes.

Diatoms are the principal component of *phytoplankton*, the photosynthetic organisms that float on the surface of waters and that are used as a primary source of food for many organisms that live in water. The diatoms contain chromoplasts (specialized organelles containing pigments, including chlorophyll) and gener-

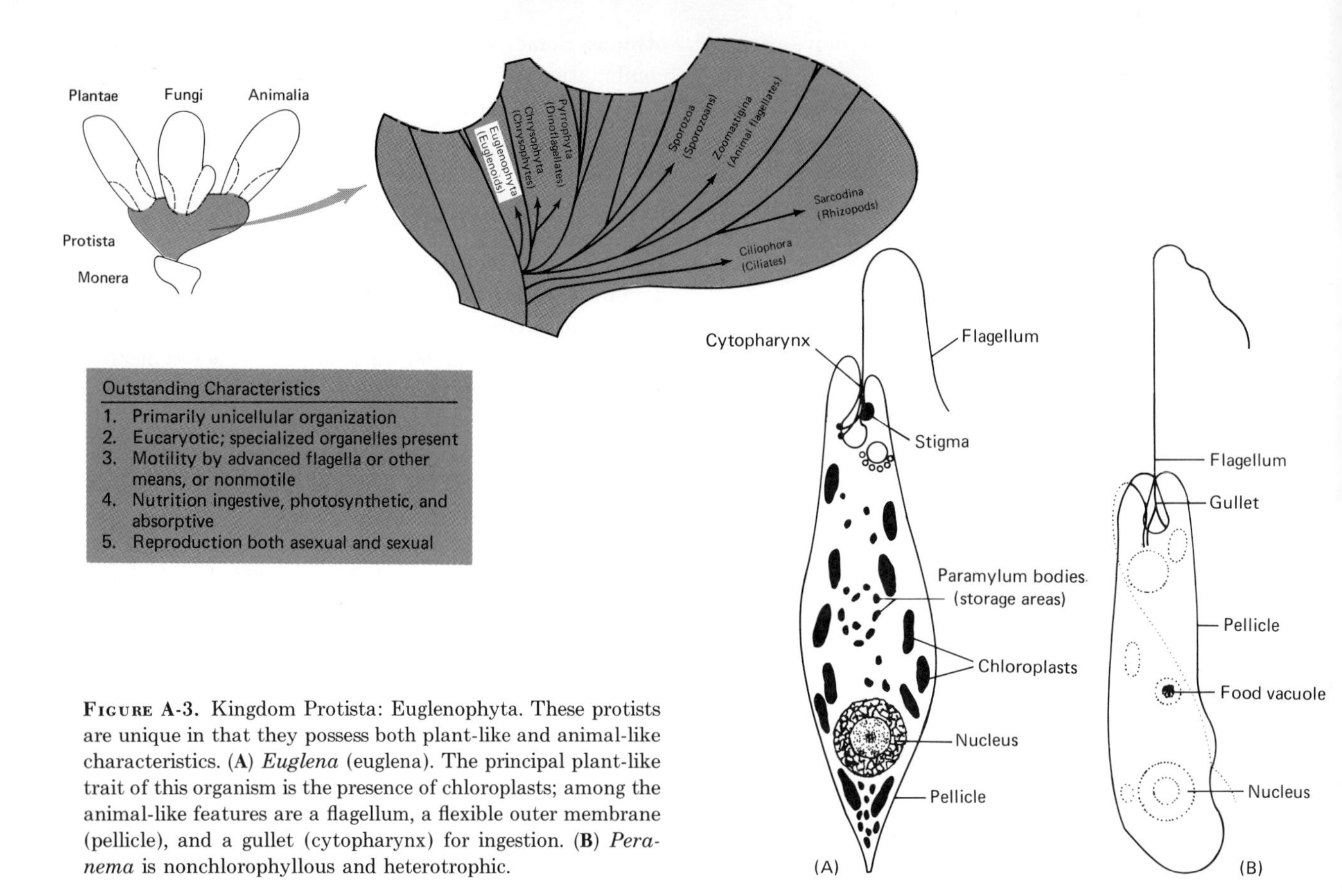

Figure A-3. Kingdom Protista: Euglenophyta. These protists are unique in that they possess both plant-like and animal-like characteristics. (**A**) *Euglena* (euglena). The principal plant-like trait of this organism is the presence of chloroplasts; among the animal-like features are a flagellum, a flexible outer membrane (pellicle), and a gullet (cytopharynx) for ingestion. (**B**) *Peranema* is nonchlorophyllous and heterotrophic.

ally store food in the form of oils. Less numerous than diatoms are other chrysophytes (golden algae) that are structurally more complex and live mainly in cold water (Figure A-4B).

The final autotrophic protistan phylum to be considered is *Pyrrophyta*, or fire algae. Some species of these structurally simple organisms have a reddish, fiery hue, thus providing the name for the phylum (Figure A-5). These are the organisms responsible for the so-called "red tides" in which thousands of fish die. Most pyrrophytes are unicellular, marine organisms that possess two flagella for locomotion. These organisms are also referred to as *dinoflagellates*, many of which are components of plankton. Some are photosynthetic, others are parasitic, and still others feed like animals. Many marine dinoflagellates are phosphorescent, glowing at night when the water is disturbed.

Heterotrophic Protistans

The phyla to be considered here are collectively referred to as protozoans. They do not contain chlorophyll, all are heterotrophs, and the unicellular condition is almost universal. The four phyla that comprise the protozoans are Sporozoa, Zoomastigina, Sarcodina, and Ciliophora.

Members of the phylum *Sporozoa*, or sporozoans, are internal parasites living in blood cells, muscles, or internal organs of their hosts. In most species, the life cycle includes a stage in which the cell is enclosed in a resistant wall forming a spore. The species is distributed during its spore stage, often being transported by mosquitoes, ticks, or flies. One of the most familiar sporozoans is *Plasmodium vivax*, the causative agent of malaria (Figure A-6A). This parasite is transmitted to humans

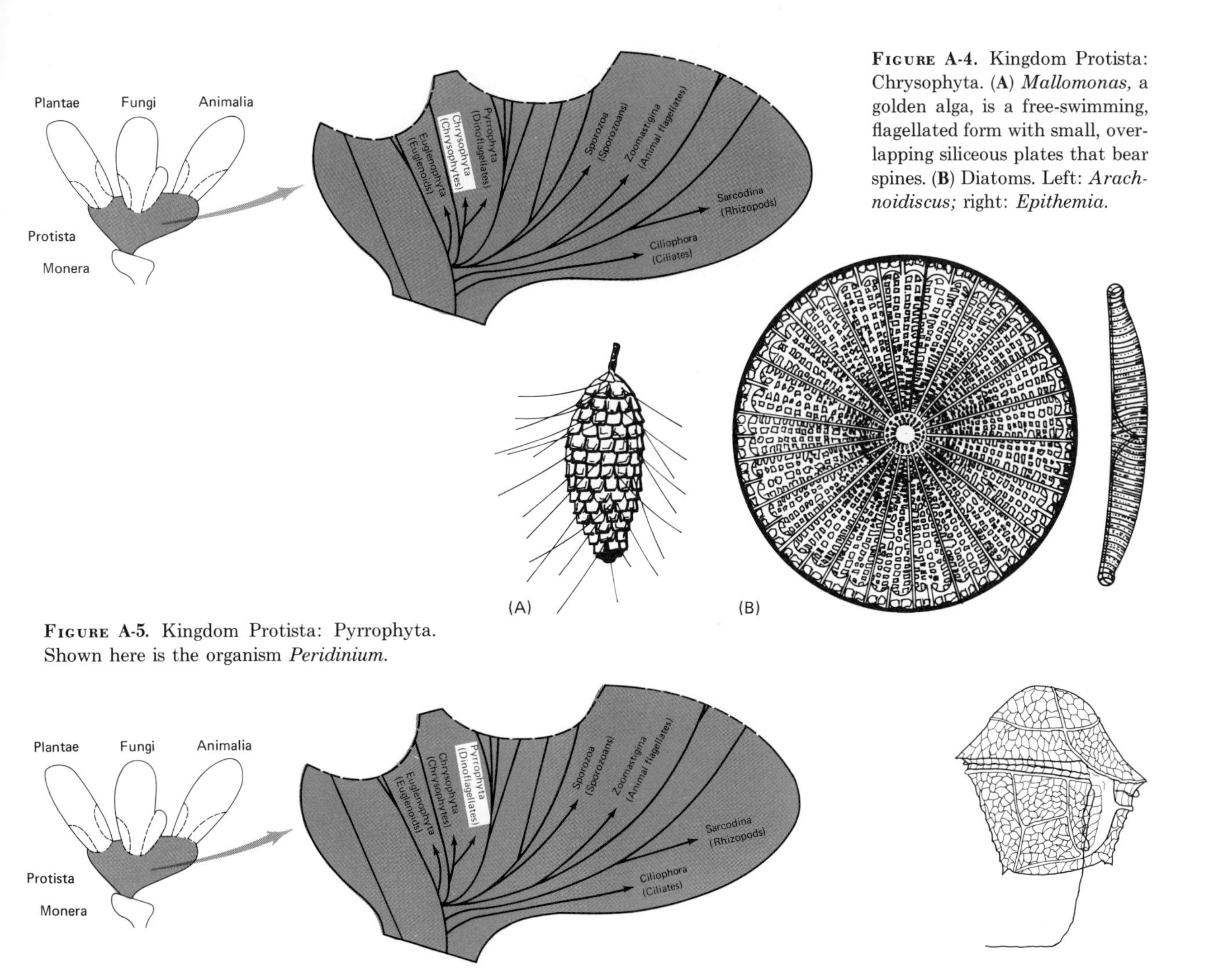

Figure A-4. Kingdom Protista: Chrysophyta. (**A**) *Mallomonas,* a golden alga, is a free-swimming, flagellated form with small, overlapping siliceous plates that bear spines. (**B**) Diatoms. Left: *Arachnoidiscus;* right: *Epithemia.*

Figure A-5. Kingdom Protista: Pyrrophyta. Shown here is the organism *Peridinium.*

by the *Anopheles* mosquito, which serves as the intermediate host.

Animal flagellates, which belong to the phylum *Zoomastigina*, are among the most primitive of the protozoans. They move by means of flagella and feed on microorganisms and debris. Some flagellates are free-living species; others live in a mutualistic (mutually beneficial) relationship with other animals. For example, *Trechinympha* (Figure A-6B) lives as a wood-digesting symbiont in the gut of termites where it helps them digest their food. Still others are parasitic in the blood of various vertebrates. One such species, *Trypanosoma gambiense,* causes sleeping sickness in humans (Figure A-6C).

Species of the *Sarcodina*, or rhizopods, are among the simplest of the protozoans, yet are found in various and changing forms. Some, like the *Amoeba* (Figure A-6D), constantly change their body shape by putting out protoplasmic extensions, called pseudopodia, which are used for locomotion and food getting. Other rhizopods have shell-like skeletons containing calcium with openings through which the pseudopodia protrude (Figure A-6E). Species with calcareous shells have created immense chalk deposits, which, when uplifted by the

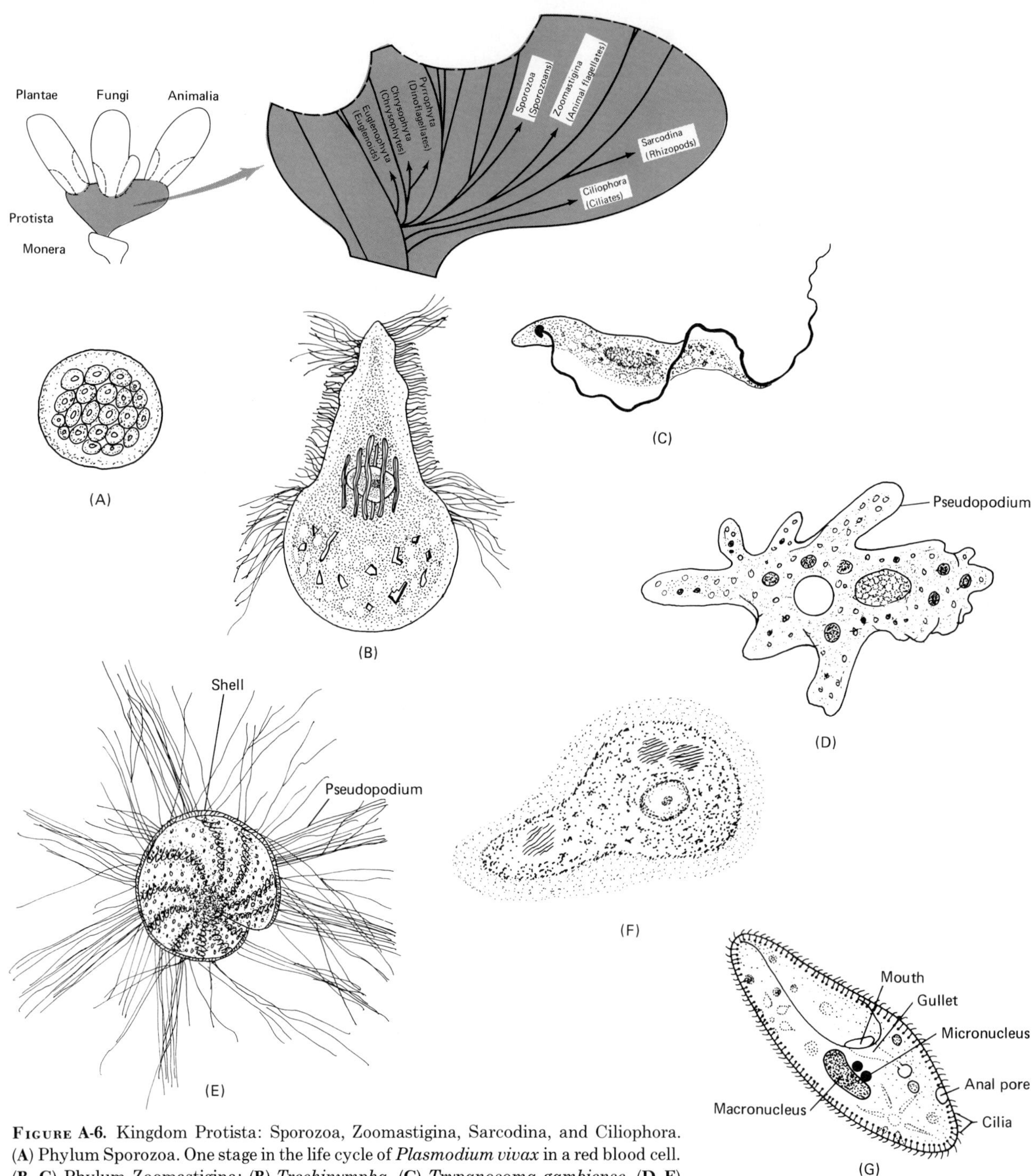

FIGURE A-6. Kingdom Protista: Sporozoa, Zoomastigina, Sarcodina, and Ciliophora. (**A**) Phylum Sporozoa. One stage in the life cycle of *Plasmodium vivax* in a red blood cell. (**B, C**) Phylum Zoomastigina: (**B**) *Trechinympha,* (**C**) *Trypanosoma gambiense.* (**D–F**) Phylum Sarcodina. (**D**) *Amoeba,* (**E**) *Elphidium,* (**F**) *Entamoeba histolytica.* (**G**) Phylum Ciliophora, represented by *Paramecium.*

earth's faulting, contribute to various land formations. A prominent example is England's White Cliffs of Dover. A parasitic rhizopod of humans, *Entamoeba histolytica,* is the causative agent of amoebic dysentery (Figure A-6F).

The final protozoan phylum, the *Ciliophora*, or ciliates, contains the most advanced and structurally the most complex representatives. These organisms possess cilia, tiny hair-like processes, that are utilized for locomotion and, in some species, as a means of obtaining food. Though single-celled, ciliates have very complex and specialized organelles. They have a mechanism to coordinate the movements of the cilia, a mouth, gullet, anus (anal pore), and two nuclei, one of which functions only in sexual reproduction (Figure A-6G).

Kingdom Plantae

Included in the Plantae are multicellular organisms, the cells of which are walled, vacuolate, and eucaryotic and contain the photosynthetic pigments in specialized organelles called plastids. The principal mode of nutrition is photosynthesis, although a few organisms, such as mistletoe and dodder, are parasitic and partially feed by absorption. Most live anchored to a substrate and are therefore nonmotile. Structural differentiation has progressed to the point where organs of photosynthesis (leaves), support (stems), and anchorage (roots) have evolved in higher forms. In these forms, there is a further specialization into photosynthetic, vascular (conducting), and covering tissues. Reproduction in plants is primarily sexual with cycles of alternating haploid (gametophyte) and diploid (sporophyte) generations, the former being progressively reduced in size among the higher members of the kingdom.

Red Algae

Species of the phylum *Rhodophyta*, or red algae, are primarily marine organisms. Like other algal forms, they possess chlorophylls (*a* and *d*), and in addition, they contain a dominant red pigment, phycoerythrin, which masks the green color of the chlorophylls. The red pigment makes it possible for rhodophytes to utilize the blue rays of the light spectrum for photosynthesis. Inasmuch as red algae live in deep waters, it appears that phycoerythrin is an adaptation to their deeper, dimmer environment. Red algae are almost all stationary and occur in a wide range of delicate, branched forms (Figure A-7).

Brown Algae

Commonly known as brown seaweeds or kelps, members of the phylum *Phaeophyta* live mainly in the shallow waters of the intertidal zones, particularly on the rocky coasts of cold waters (Figure A-8). Among the identifying pigments of these algae are chlorophylls *a* and *c* and the brown pigment fucoxanthin, the latter present in sufficient amounts to mask all others. Rubbery strands of giant kelps can reach a length of 300 feet, streaming with the tides and currents and anchored by holdfasts on the bottom. Among the useful products derived from brown algae are iodine, and algin, which is is used in the manufacture of ice cream and cosmetics. Certain species of brown algae are also used as a food in oriental countries.

Green Algae

The chlorophytes, or organisms of the phylum *Chlorophyta*, commonly live in fresh water, in damp spots in the soil, or on tree bark and stones. Although some green algae are solitary, others form large colonies which exhibit a high degree of coordination. Reproduction may be either asexual or sexual, and is often complex. Primitive green algae are represented by types such as *Chlamydomonas* (Figure A-9A); colonial types, such as *Volvox* (Figure A-9B), are slightly more complex and specialized; somewhat more complexly organized are the filamentous forms, of which *Spirogyra* (Figure A-9C) is an example; and the most complex green algae types, such as *Ulva* (Figure A-9D), consist of sheets of cells. In an evolutionary context, the Chlorophyta are quite important since they are believed to be the stock from which higher plants evolved.

Stoneworts

The stoneworts of the phylum *Charophyta* are the most complex of the algae. Stoneworts live in freshwater ponds and possess structures resembling the roots, branching stems, and leaves of more advanced types of plants (Figure A-10). As a group, stoneworts may represent an evolutionary step which advanced the basic

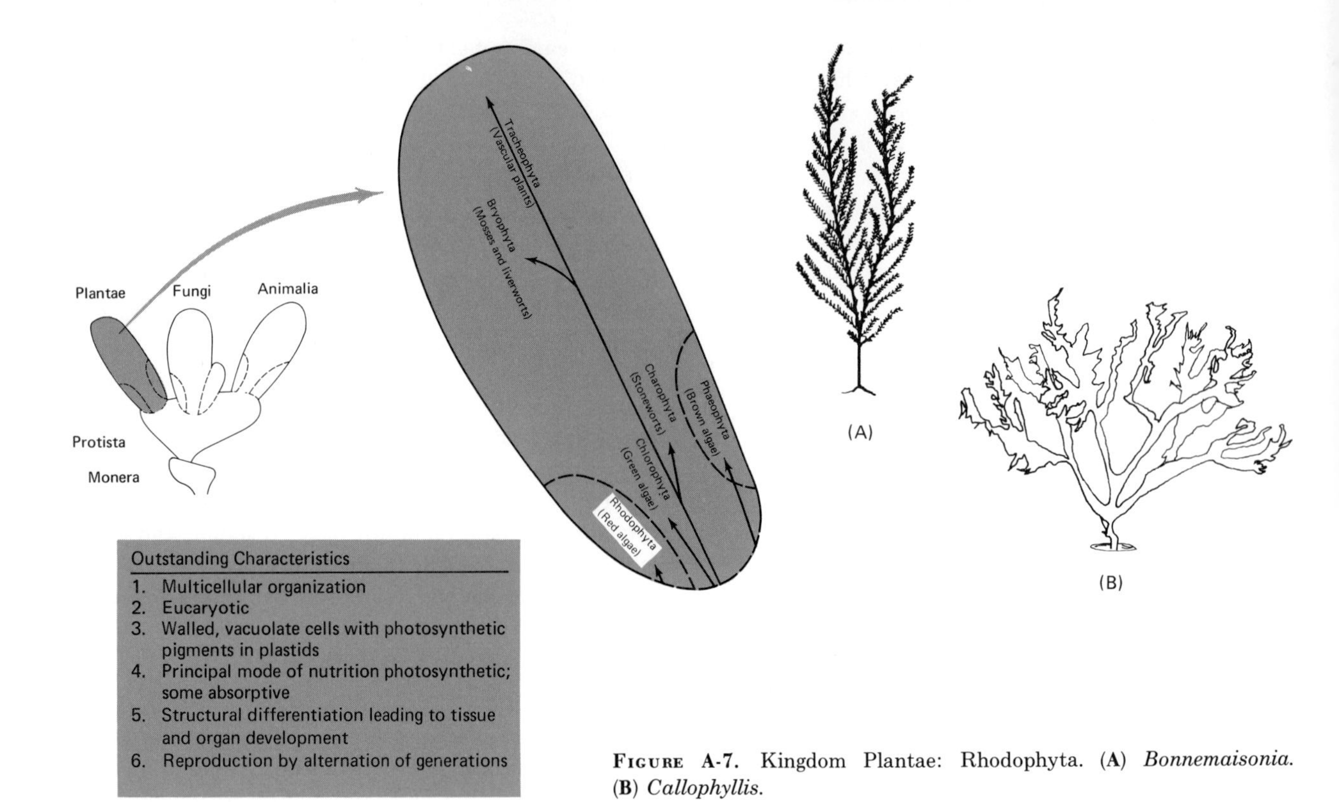

Figure A-7. Kingdom Plantae: Rhodophyta. (**A**) *Bonnemaisonia.* (**B**) *Callophyllis.*

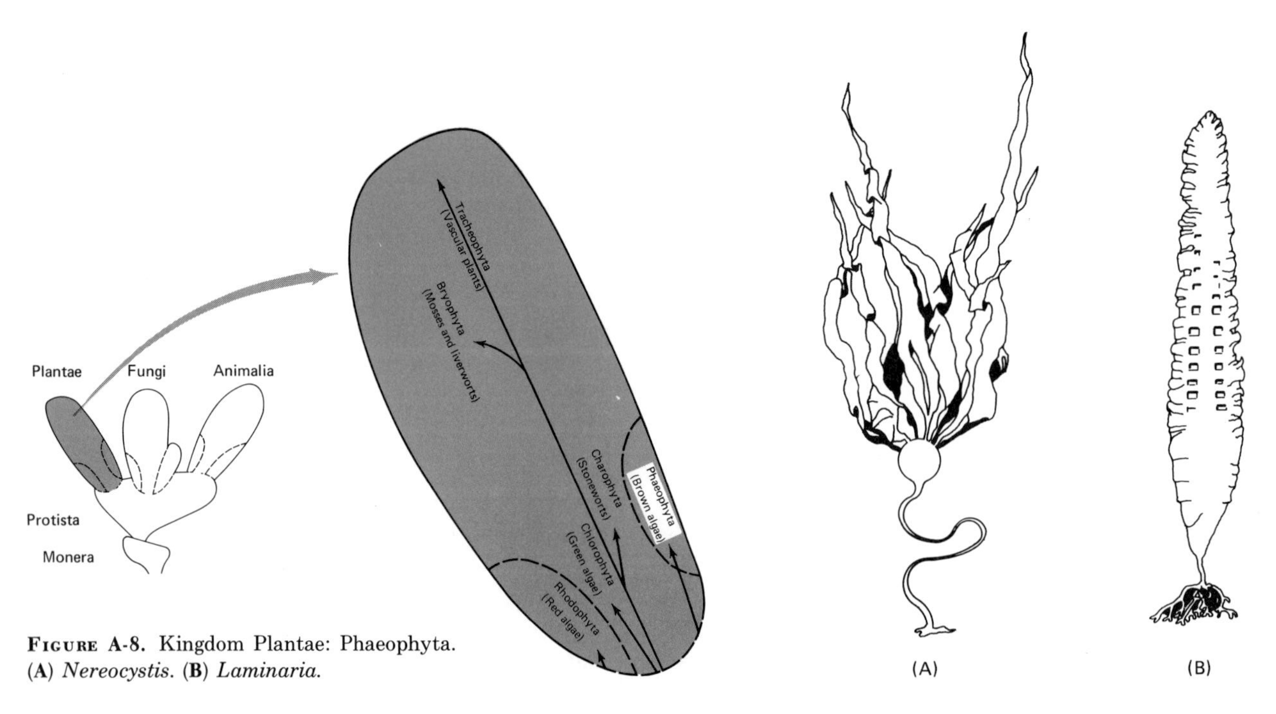

Figure A-8. Kingdom Plantae: Phaeophyta. (**A**) *Nereocystis.* (**B**) *Laminaria.*

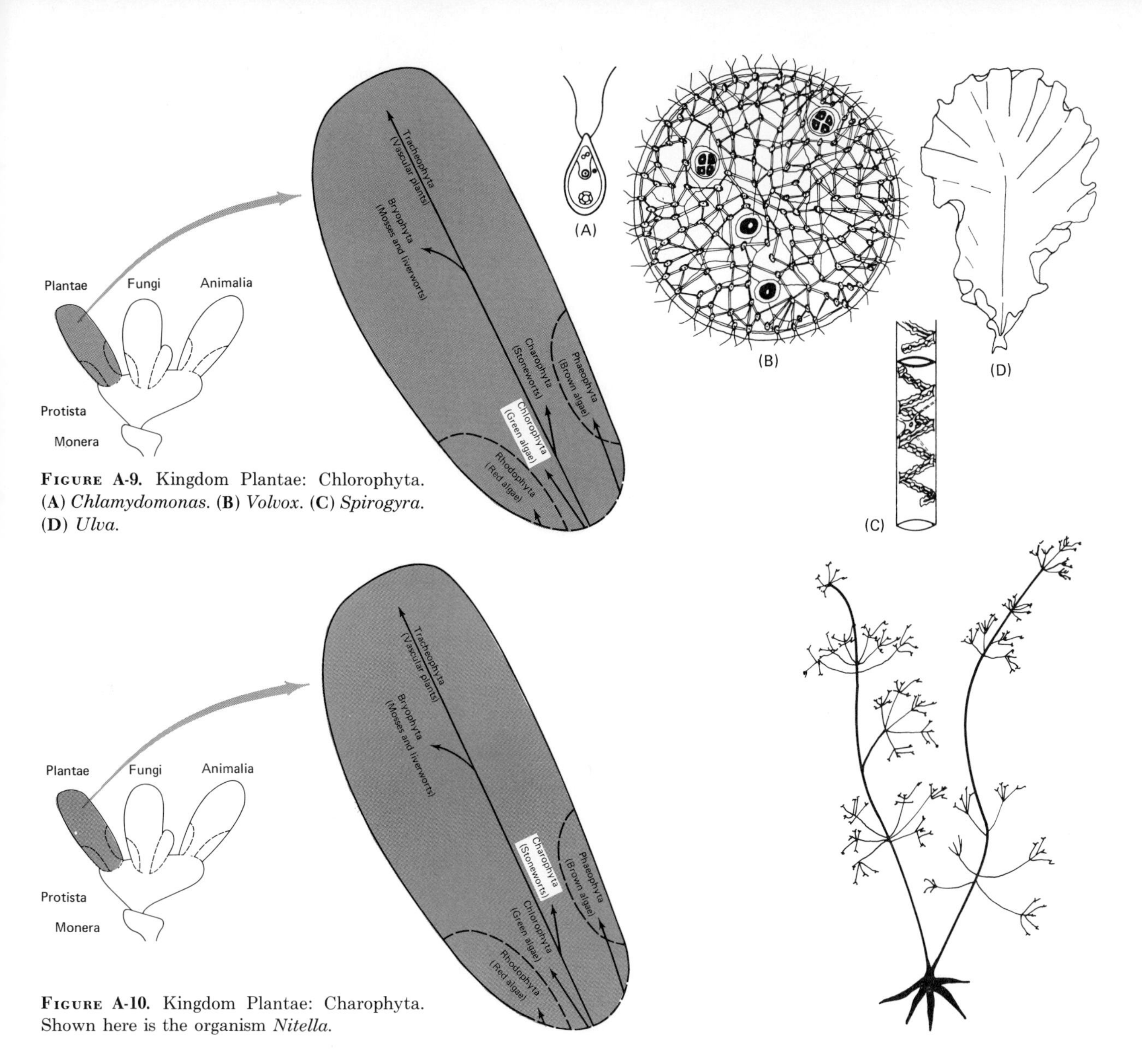

Figure A-9. Kingdom Plantae: Chlorophyta. (**A**) *Chlamydomonas.* (**B**) *Volvox.* (**C**) *Spirogyra.* (**D**) *Ulva.*

Figure A-10. Kingdom Plantae: Charophyta. Shown here is the organism *Nitella.*

algal organization in a direction actually taken later by land plants.

Bryophytes

The phylum *Bryophyta* consists of three universally distributed groups of plants: (1) mosses, (2) liverworts, and (3) hornworts. In general, they occur in more or less shady, moist places, where the danger of desiccation is minimized. Bryophytes have no true roots, stems, or leaves but have structures which superficially resemble them. None contains vascular tissue (xylem and phloem), that is, conducting tissue. Reproduction is by alternation of generations in which the gametophyte (gamete-producing plant) is dominant (conspicuous), and the sporophyte (spore-producing plant) is dependent upon the gametophyte for nutrition. Representative members of the three classes of bryophytes are shown in Figure A-11 along with a description of each.

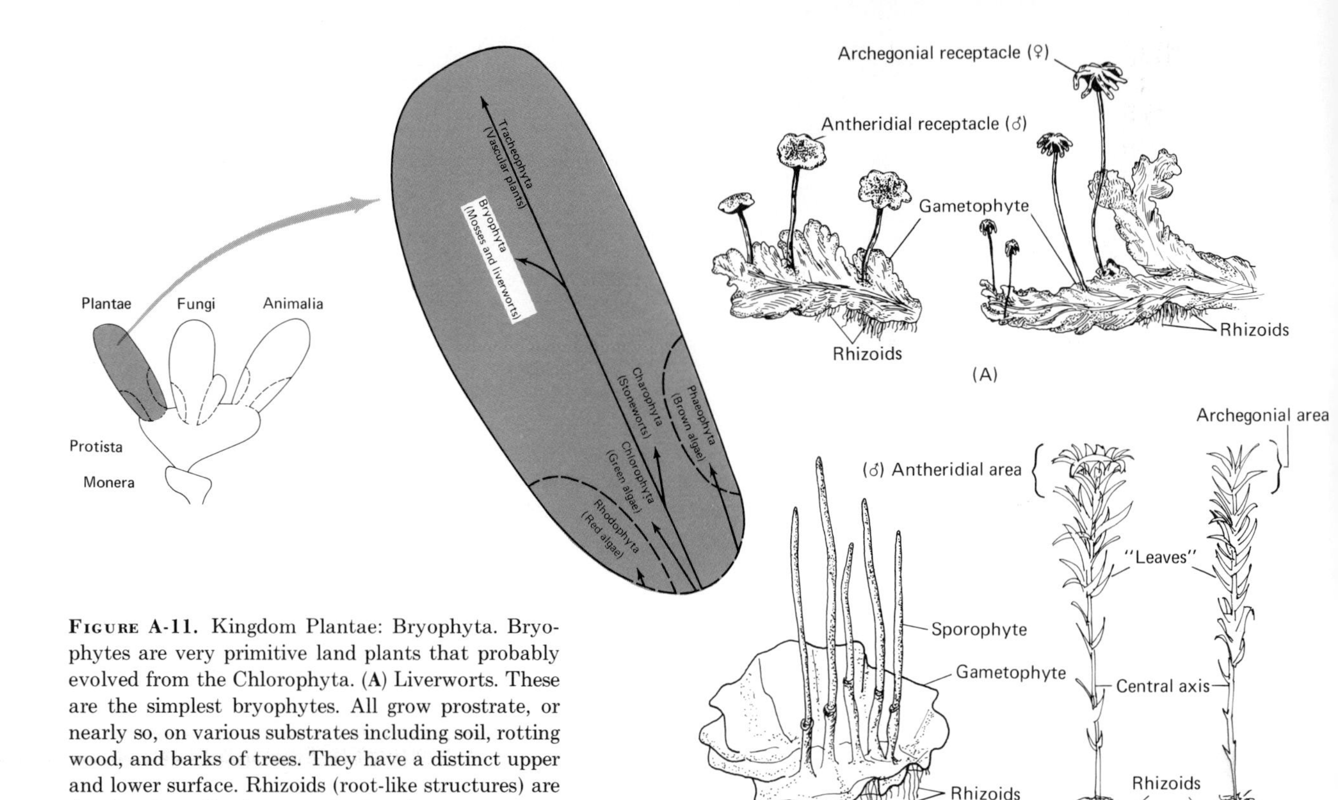

Figure A-11. Kingdom Plantae: Bryophyta. Bryophytes are very primitive land plants that probably evolved from the Chlorophyta. (**A**) Liverworts. These are the simplest bryophytes. All grow prostrate, or nearly so, on various substrates including soil, rotting wood, and barks of trees. They have a distinct upper and lower surface. Rhizoids (root-like structures) are developed on the lower surface, and sex organs are borne on the upper surface. Shown is a thallose liverwort, *Marchantia.* (**B**) Hornworts. These represent a small group of bryophytes commonly found on the edges of streams or lakes. Shown is *Anthoceros,* the best-known genus. (**C**) Mosses. These bryophytes are usually short and are common in habitants of wooded areas and moist environments. An individual moss plant consists of an erect "stem," bearing "leaves," that is anchored to the substrate by rhizoids. Sex organs are usually borne at the apex. In some moss plants, male and female sex organs are developed on separate individuals.

Tracheophytes

Vascular plants, or tracheophytes, members of the phylum *Tracheophyta*, represent the largest phylum with regard to the number of species and size of some members of the Kingdom Plantae. They are characterized by the presence of xylem (water-conducting tissue) and phloem (food-conducting tissue). Virtually all tracheophytes possess true roots, stems, and leaves. In direct contrast to the bryophytes, the sporophyte generation is independent and dominant and the gametophyte is greatly reduced. The four lines of tracheophyte descent are represented by the psilopsids, club mosses, horsetails, and large-leafed plants (ferns, coniferous seed plants, and flowering seed plants). An example of each of these lines is shown and described in Figure A-12.

Kingdom Fungi

Organisms of this kingdom are primarily multinucleate (more than one nucleus in each cell) with eucaryotic nuclei dispersed in a walled and often partitioned hypha (tubular filament which comprises the body). Plastids and photosynthetic pigments are lacking and nutrition is absorptive. However, the slime molds, which are placed in this kingdom for convenience, normally engulf or ingest their nutrients. Reproductive cycles include

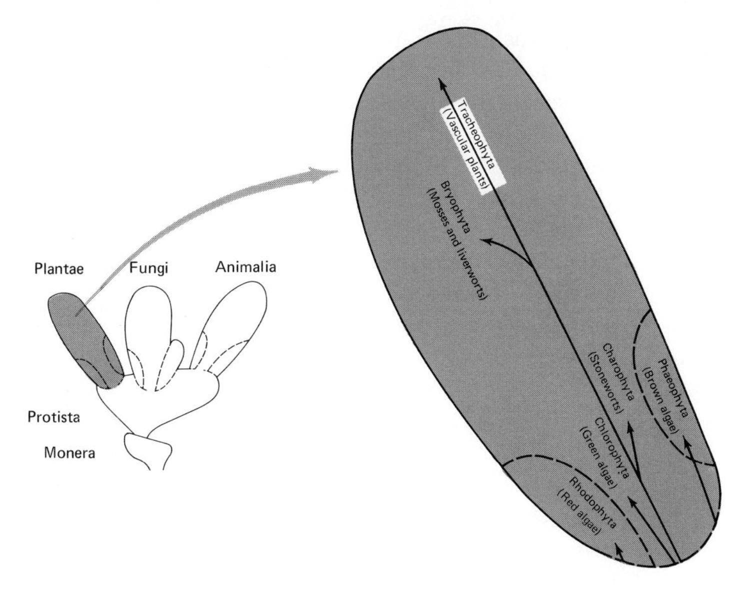

Figure A-12. Kingdom Plantae: Tracheophyta. Many morphological data indicate that tracheophytes probably evolved directly from the Chlorophyta. (**A**) Psilopsids. These are the oldest and most primitive of the vascular plants. Shown is *Rhynia,* a small rush-like plant consisting of an upright dichotomously branched stem anchored to the substrate by rhizoids. (**B**) Club mosses. These are trailing or creeping plants with small leaves. In the representative, *Lycopodium,* note the main stem, which branches freely and is prostrate, the upright stems growing from the horizontal stem, and the strobilus, a spore-containing structure. (**C**) Horsetails. The only living genus of this group is *Equisetum.* Many species inhabit cool, moist places and most are characterized by the presence of silica in the stems. The stem is hollow and jointed, leaves are small, and spores are borne in strobili. (**D**) Ferns. In these plants, leaves are usually large and divided into distinct segments (pinnae). The vascular system is well developed, and spores are produced on the underside of the leaves. (**E**) Gymnosperms. These plants are represented by such common trees as the pine. These are fairly large trees, mostly with needle-like leaves, typically evergreen, with seeds produced in cones. (**F**) Angiosperms. Flowering plants are the most highly developed, diversified, and widespread of all plant groups. Their most outstanding characteristic is the flower, which gives rise to a seed and a fruit. [**A** *from R. Kidston and W. H. Lang, Pts. 1–5,* Transactions of the Royal Society of Edinburgh, *1917–1921. By permission.*]

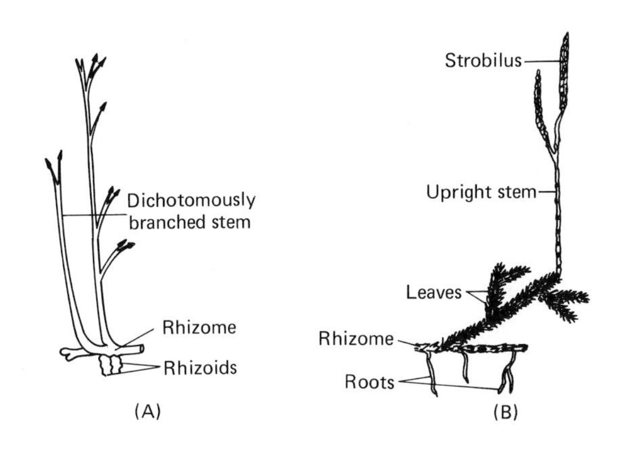

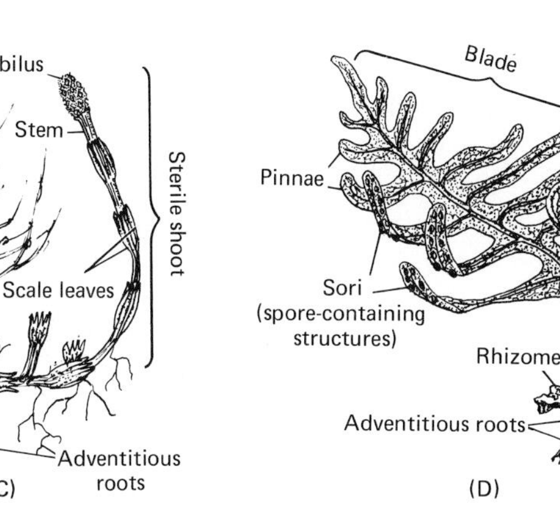

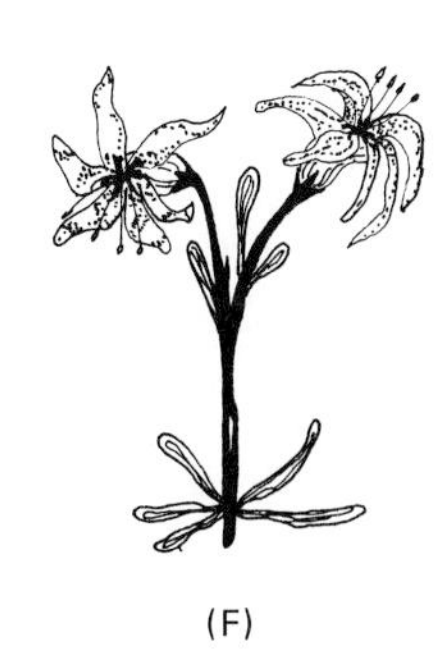

both asexual and sexual processes. Two major groups comprise the Kingdom Fungi. These are the slime molds, which are fungus-like organisms, and the true fungi.

Slime Molds

The term *slime mold* is applied to at least four distinct groups of fungus-like organisms which differ in structure and physiology and have characteristic life histories. One of these groups is probably best classified as true fungi. The other three are *Phylum Acrasiomycota* (cellular slime molds), *Phylum Myxomycota* (plasmodial slime molds), and *Phylum Labyrinthulomycota* (net slime molds). Slime molds have naked, amoeboid assimilative stages and are also characterized by a holozoic mode of nutrition. They also produce spores in characteristic fruiting bodies. Only one group, the cellular slime molds, will be discussed here.

Cellular slime molds are soil organisms. When their spores are dispersed and find a suitable substrate, they germinate and give rise to amoeba-like cells which feed on bacteria, grow, and divide (Figure A-13). After several divisions, a large amoebal population results. When the food supply is diminished, some of the cells secrete a hormone-like substance, *acrasin*, which attracts other cells, causing them to stream toward a center of aggregation. Several such streams of cells are established and all migrate toward a common point. The result of these aggregations is the formation of a *pseudoplasmodium* ("slug"). At a given stage of development, the cells become physiologically differentiated into *sorocarps* for sporulation (spore production).

The slug at this stage raises itself upward and forms a nipple-like structure at its free end. The amoebae at the free end begin to migrate downward through the center of the slug, secreting cellulose walls and forming a cylinder toward the base. When the cellulose cylinder reaches the base, the amoebae at the base begin to migrate up the stalk and eventually form a mass at the tip. There they are enveloped in thin, cellulose walls and become mature spores. Upon liberation, the spores germinate and a new cycle begins.

True Fungi

Most people are familiar with the fruiting bodies (sporophores) of many fungi (mushrooms, puffballs, stinkhorns). These are only the spore-bearing structures and represent but a small part of the fungal life cycle. The soma (thallus) of a fungus, the portion that grows and obtains food and eventually produces the sporophore, is generally small and inconspicuous. It consists of microscopic tubular filaments called **hyphae,** which branch and rebranch in the substrate from which the fungus obtains nourishment. The entire system of hyphae is referred to as a **mycelium.** In the life cycle of a fungus, the mycelium produces the spores and the spores produce the mycelium. Some unicellular fungi, such as the common yeast, do not have a mycelium.

Nutritionally, fungi are heterotrophic and must be supplied with organic molecules synthesized by autotrophs. As heterotrophs, they exhibit varying degrees of either saprophytism or parasitism. Most fungi are *obligate saprophytes*; that is, they must obtain nutrients from the dead bodies of other organisms. Many fungi are *facultative parasites* (or *facultative saprophytes*) in that they are capable of attacking living organisms, thus causing disease, but they are also able to grow on nonliving organic matter. Still other fungi are *obligate parasites.* These organisms live only on living protoplasm and attack certain species of organisms (hosts).

Reproduction among fungi is both asexual and sexual. Asexual reproduction in unicellular fungi is accomplished by budding (see Chapter 17). The most important method of asexual reproduction in fungi is the production of spores. Asexual spores may be borne in sac-like structures (*sporangia*), in which case they are called *sporangiospores*, or they may be borne directly on hyphae (*conidia*). Sexual reproduction occurs in three stages. The first of these, **plasmogamy,** is the mechanism by which two compatible nuclei in a single protoplast are brought together. In the second stage, **karyogamy,** the fusion of the two nuclei takes place. Karyogamy results in a zygote nucleus which is diploid. Finally, **meiosis,** the third stage, occurs and spores produced as a result of meiosis are termed *meiospores.* Many fungi produce both asexual and sexual spores.

Phylum Chytridiomycota. Organisms of the phylum *Chytridiomycota* are typically aquatic but some live in moist soil. Many parasitize algae and small water animals. One economically important chytrid is *Synchytrium endobioticum,* which causes a disease of potatoes called "black wart." One of the most interesting,

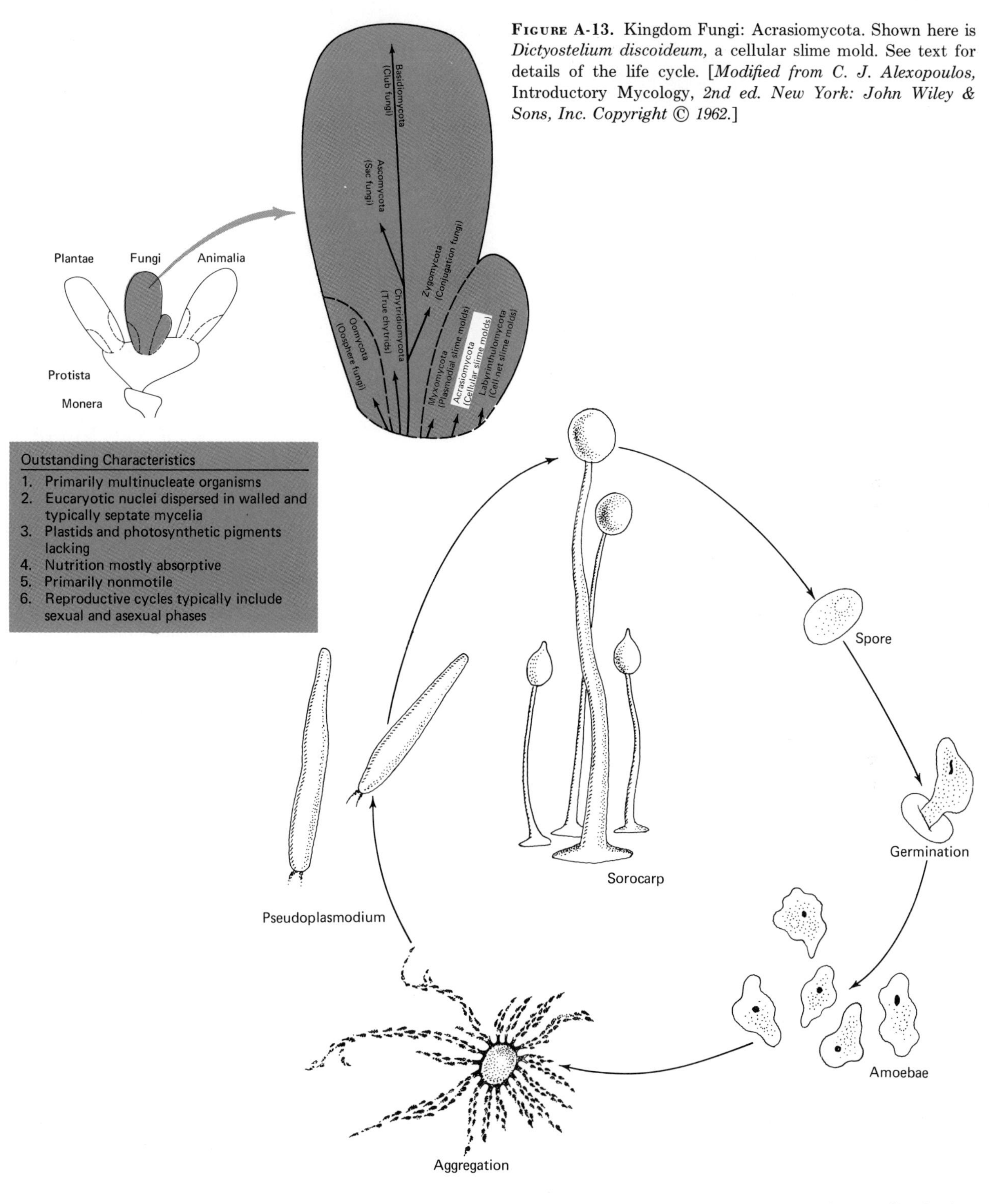

Figure A-13. Kingdom Fungi: Acrasiomycota. Shown here is *Dictyostelium discoideum,* a cellular slime mold. See text for details of the life cycle. [*Modified from C. J. Alexopoulos,* Introductory Mycology, *2nd ed. New York: John Wiley & Sons, Inc. Copyright © 1962.*]

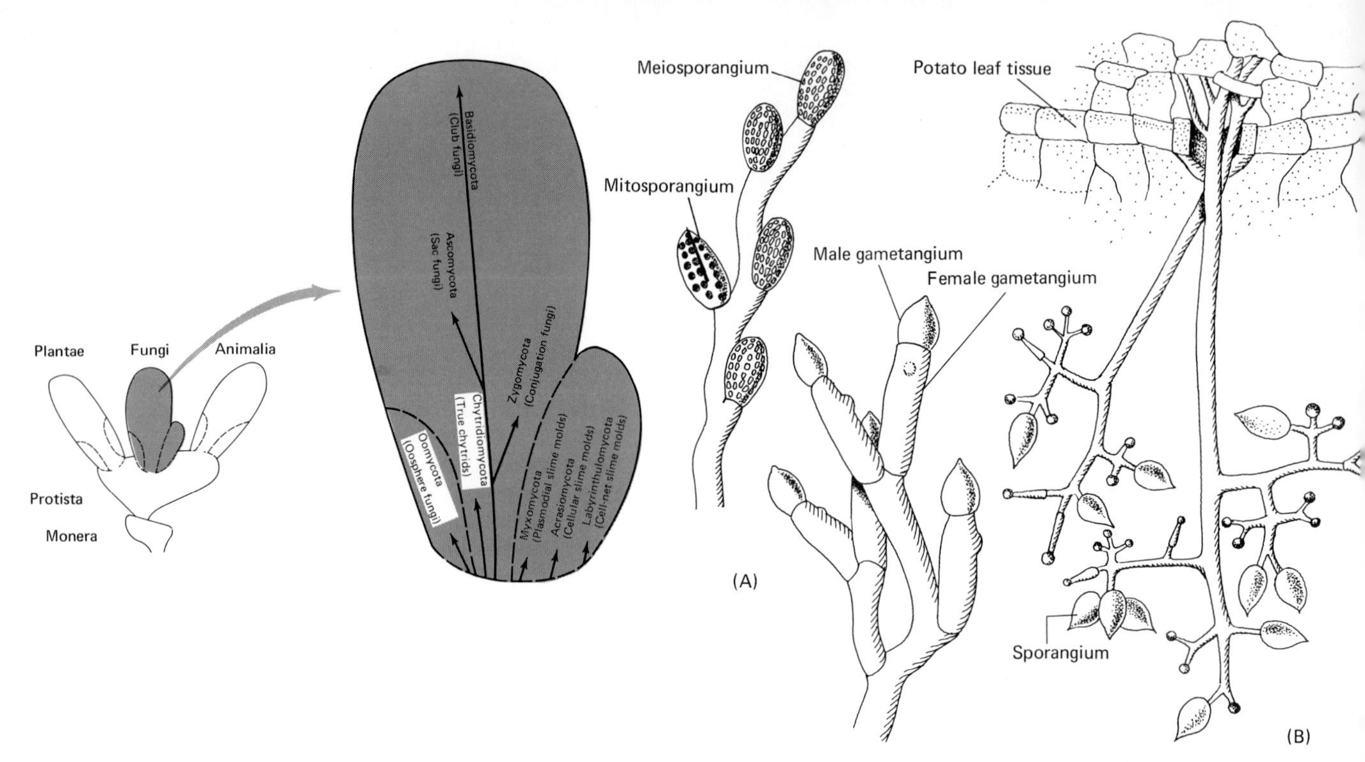

FIGURE A-14. Kingdom Fungi: Chytridiomycota and Oomycota. (**A**) Phylum Chytridiomycota. *Allomyces macrogynus* (water mold) is the only filamentous fungus known that exhibits a true alternation of generations; a diploid filamentous sporothallus (*left*) that produces spores alternates with a haploid filamentous gametothallus (*right*) that produces gametes. Spores (diploid) produced by the mitosporangium germinate and grow into new sporothalli. Spores produced by the meiosporangium undergo a period of rest before germinating. At the time of germination, their nuclei undergo meiosis and the haploid meiospores develop into new gametothalli. Male and female gametes produced by the gametangia fuse in pairs to form zygotes. Each zygote grows into a diploid sporothallus. (**B**) Phylum Oomycota. *Phytophthora infestans* (potato blight fungus) on the lower surface of a potato leaf.

although not the most typical, members of this phylum is the water mold, *Allomyces* (Figure A-14A).

PHYLUM OOMYCOTA. Members of the phylum *Oomycota* constitute a fairly large and very important group of fungi. Some live in water, others in soil, and still others in association with land plants. Economically important fungi of this group include *Saprolegnia parasitica*, which attacks fish and fish eggs; *Aphanomyces eutyches*, the cause of a serious disease of sugar beets and other plants; and various other organisms which cause downy mildews (*Plasmophora viticola*), white rusts, and blights (*Phytophthora infestans*) (Figure A-14B).

PHYLUM ZYGOMYCOTA. Fungi of the phylum *Zygomycota* are terrestrial organisms. They produce sexually by the fusion of two gametangia (gamete-producing structures) which results in the formation of a zygospore (thick-walled spore resulting from the fusion of gametangia). The most familiar of the conjugation fungi is probably *Rhizopus stolonifer,* the common bread mold (Figure A-15A).

PHYLUM ASCOMYCOTA. The phylum *Ascomycota* (sac fungi) constitutes an enormous and varied group of fungi many of which are extremely important in medicine, agriculture, and industry. Apple scab, brown rot of stone fruits, and various powdery mildews of grapes, roses, and peaches are caused by sac fungi. The fungus *Claviceps purpurea* is the original source of

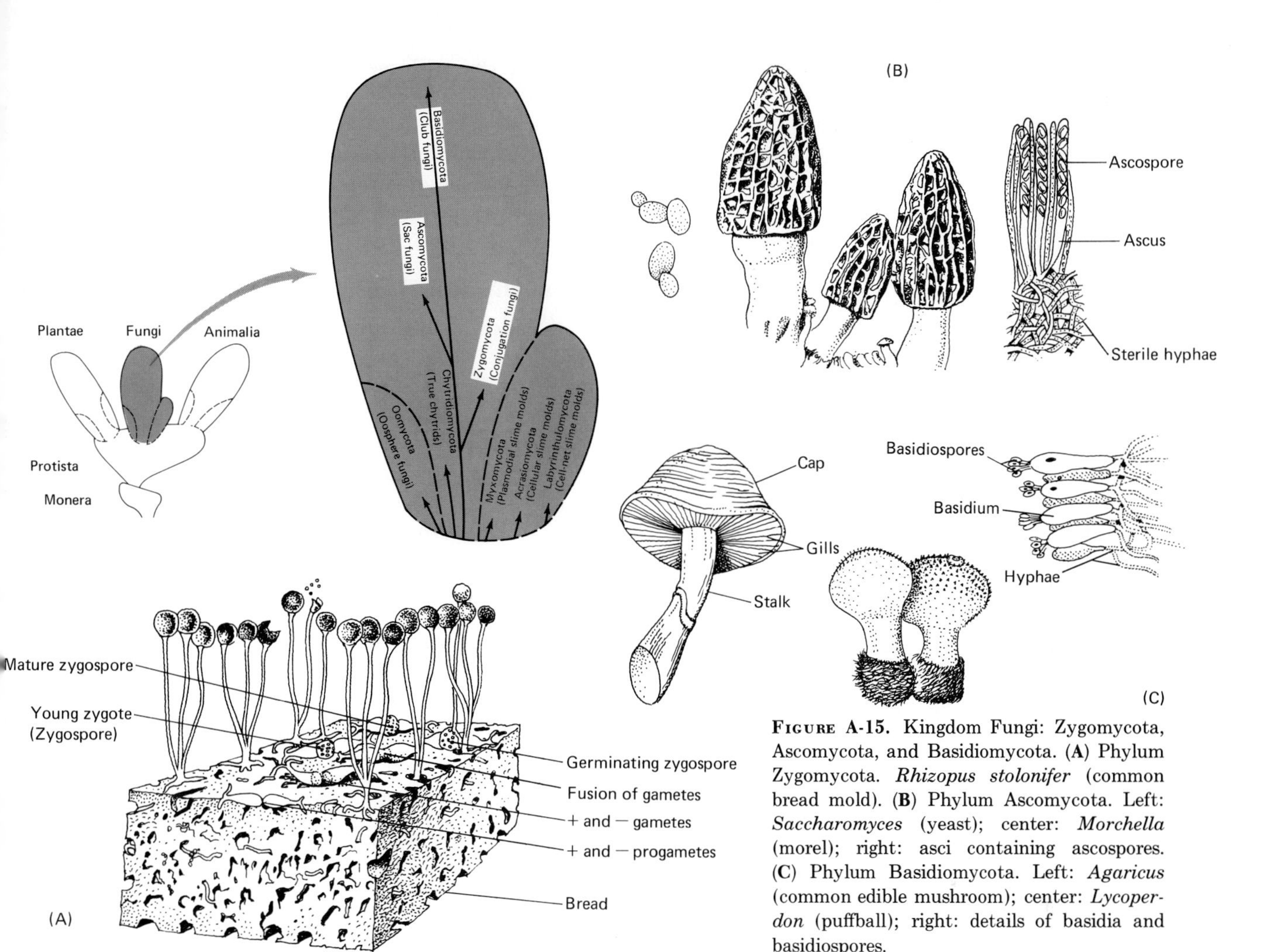

Figure A-15. Kingdom Fungi: Zygomycota, Ascomycota, and Basidiomycota. (**A**) Phylum Zygomycota. *Rhizopus stolonifer* (common bread mold). (**B**) Phylum Ascomycota. Left: *Saccharomyces* (yeast); center: *Morchella* (morel); right: asci containing ascospores. (**C**) Phylum Basidiomycota. Left: *Agaricus* (common edible mushroom); center: *Lycoperdon* (puffball); right: details of basidia and basidiospores.

LSD, the much-discussed hallucinogenic agent. Among the commercially important sac fungi are baker's or brewer's yeast and truffles. The one important characteristic that distinguishes sac fungi from all other fungi is the **ascus,** a sac-like structure in which ascospores are formed following karyogamy and meiosis (Figure A-15B).

Phylum Basidiomycota. The phylum *Basidiomycota* (club fungi) encompasses a great variety of fungi, including rusts, smuts, jelly fungi, mushrooms, puffballs, stinkhorns, bracket fungi, coral fungi, earthstars, and the bird's nest fungi (Figure A-15C). These grow in the soil, in or on logs and tree stumps, and in the living plant bodies which some club fungi parasitize. Just as the ascus distinguishes the sac fungi, the basidium characterizes the club fungi. A **basidium** is a structure on which basidiospores are produced following karyogamy and meiosis. Whereas sac fungi produce their ascospores inside an ascus, club fungi produce basidiospores on the surface of a basidium.

Kingdom Animalia

Members of the Kingdom Animalia possess eucaryotic cells that lack walls, plastids, and photosynthetic pigments. Nutrition is primarily ingestive, with digestion taking place in an internal cavity; some forms are absorptive and a number of groups lack an internal digestive cavity. The levels of organization and tissue differentiation in higher forms are considerably more well developed than in other kingdoms. All animals have evolved sensory-neuromotor systems and motility based on a contractile mechanism. Reproduction is typically sexual.

As with the discussion of the other kingdoms, no attempt will be made to treat all groups of animals; only the major ones will be mentioned. Before doing this, however, we will first note the principal taxonomic criteria that are used to place various animals into respective phyla. The entire animal kingdom is subdivided into three principal subkingdoms: (1) Agnotozoa, (2) Parazoa, and (3) Eumetazoa. Members of the *Agnotozoa* derive nutrition through absorption and ingestion by surface cells; internal digestive cavities and tissue differentiation are lacking; and movement is by cilia. In the *Parazoa*, nutrition is primarily ingestive by individual cells lining internal water canals; cell differentiation is present but tissue differentiation is either lacking or very limited; and although individual cells of the organism are motile, the organism as a whole is nonmotile. Organisms grouped under the *Eumetazoa* have an advanced multicellular organization with tissue differentiation. They range in complexity of structure from the jellyfish to the human.

Since the Eumetazoa contains the vast majority of animal phyla, it is further subdivided into smaller groups on the basis of other distinguishing characteristics. For example, *Branch Radiata* contains organisms with radial symmetry, while the remaining group, *Branch Bilateria*, includes animals with bilateral symmetry. Finally, the Bilateria are again subdivided into three major groups. In this case, the subdivisions are made on the basis of the nature of the coelom, a body cavity. Members of the *Grade Acoelomata* lack a body cavity; organisms of the *Grade Pseudocoelomata* have a false coelom, that is, one not lined by peritoneum; and animals classified under the *Grade Coelomata* possess a true body cavity.

Subkingdom Agnotozoa

The phylum *Mesozoa* is the only one represented in this subkingdom. It consists of minute, worm-like parasites that are less than one-quarter of an inch in length, with the simplest structure found among multicellular animals. One such mesozoan, *Dicynema*, is a common parasite of the kidney of octopuses and squids (Figure A-16). All are composed of a ciliated outer cell layer and an inner mass of reproductive cells. Some biologists believe that mesozoans arose directly from the protozoans; others maintain that they represent extremely degenerate flatworms.

Subkingdom Parazoa

Included in this subkingdom are two phyla, one of which is extinct. The extant phylum *Porifera* consists of the sponges, the most primitive of all multicellular animals. The body of these typically marine organisms contains a single cavity (spongocoel) into which water is drawn through numerous pores, and from which water exits through a single opening, the osculum (Figure A-17). The internal skeleton is comprised of calcareous and/or siliceous needles (spicules). Sponges are sessile, ranging in size from $\frac{1}{2}$ inch to 6 feet in height and varying in shape from flat, encrusting forms to balls, cups, fans, and vases.

Subkingdom Eumetazoa

As noted earlier, the subkingdom Eumetazoa consists of a vast assemblage of animals with advanced multicellular organization and tissue differentiation. Moreover, the organisms are further distinguished on the basis of symmetry. We will first consider those with radial symmetry.

RADIALLY SYMMETRIC EUMETAZOA. The principal phylum of this group of animals is *Cnidaria*, also known as the coelenterates. Their radially symmetric bodies consist of two layers of cells—an external epidermis and an inner gastrodermis; they are the most primitive animals with definite tissues. Among the other characteristics of this phylum are tentacles with stinging cells (nematocysts) surrounding the mouth; a saclike digestive (gastrovascular) cavity lacking an anus; primitive contractile fibers in the base of the epidermal tissue; a diffuse network of nerve cells; and reproduction commonly by alternation of generations. Representatives of this phylum include the jellyfish, corals, hydras, and sea anemones and the Portuguese man-of-war (Figure A-18).

BILATERALLY SYMMETRIC EUMETAZOA. In order to organize the multitude of bilaterally symmetric

Figure A-16. Kingdom Animalia: Mesozoa. *Dicynema* has the simplest structure of any multicellular animal. It consists of an outer layer of ciliated cells surrounding one or more long axial reproductive cells.

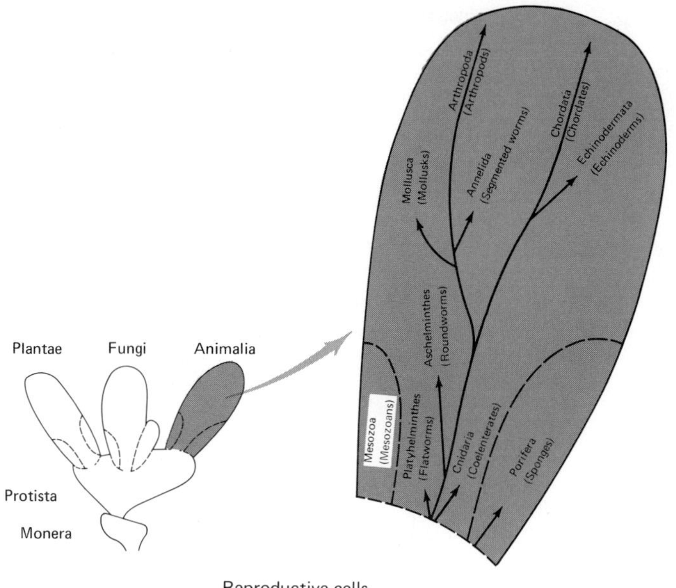

Outstanding Characteristics

1. Multicellular
2. Eucaryotic cells lacking plastids and photosynthetic pigments
3. Nutrition primarily ingestive with digestion in an internal cavity
4. High level of organization and tissue differentiation in higher forms
5. Evolution of sensory-neuromotor systems
6. Motility by contractile fibrils
7. Reproduction primarily sexual

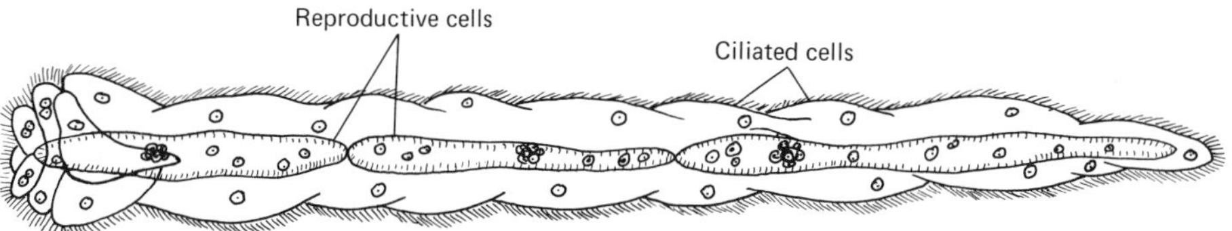

Subkingdom Agnotozoa

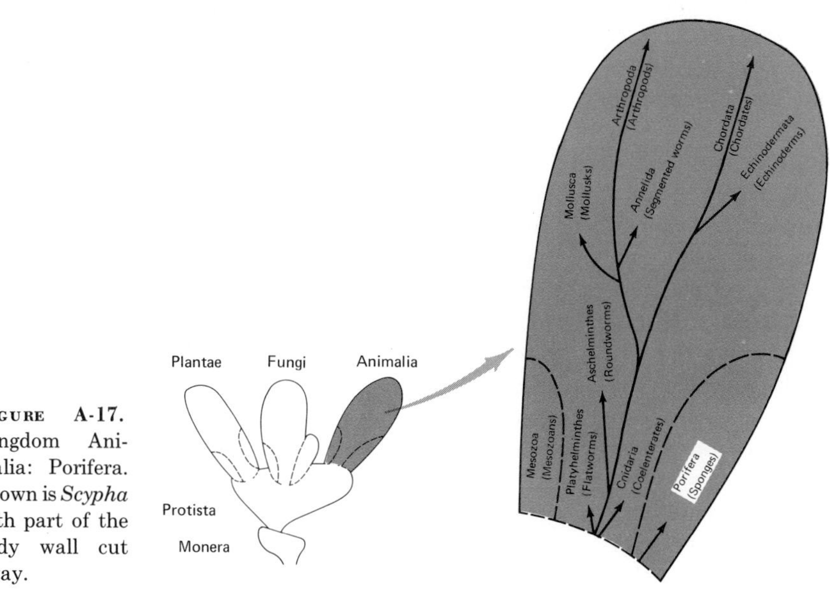

Figure A-17. Kingdom Animalia: Porifera. Shown is *Scypha* with part of the body wall cut away.

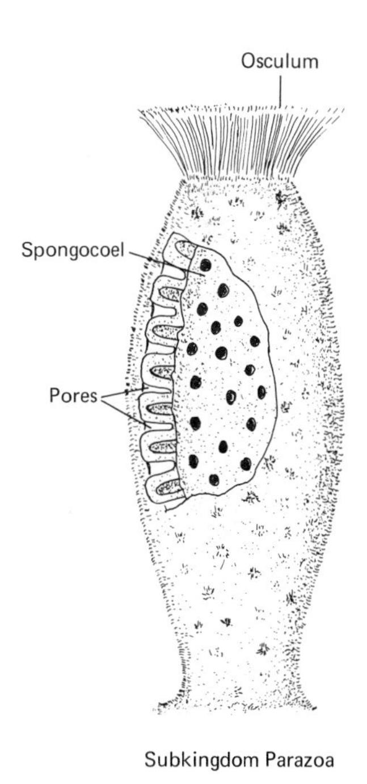

Subkingdom Parazoa

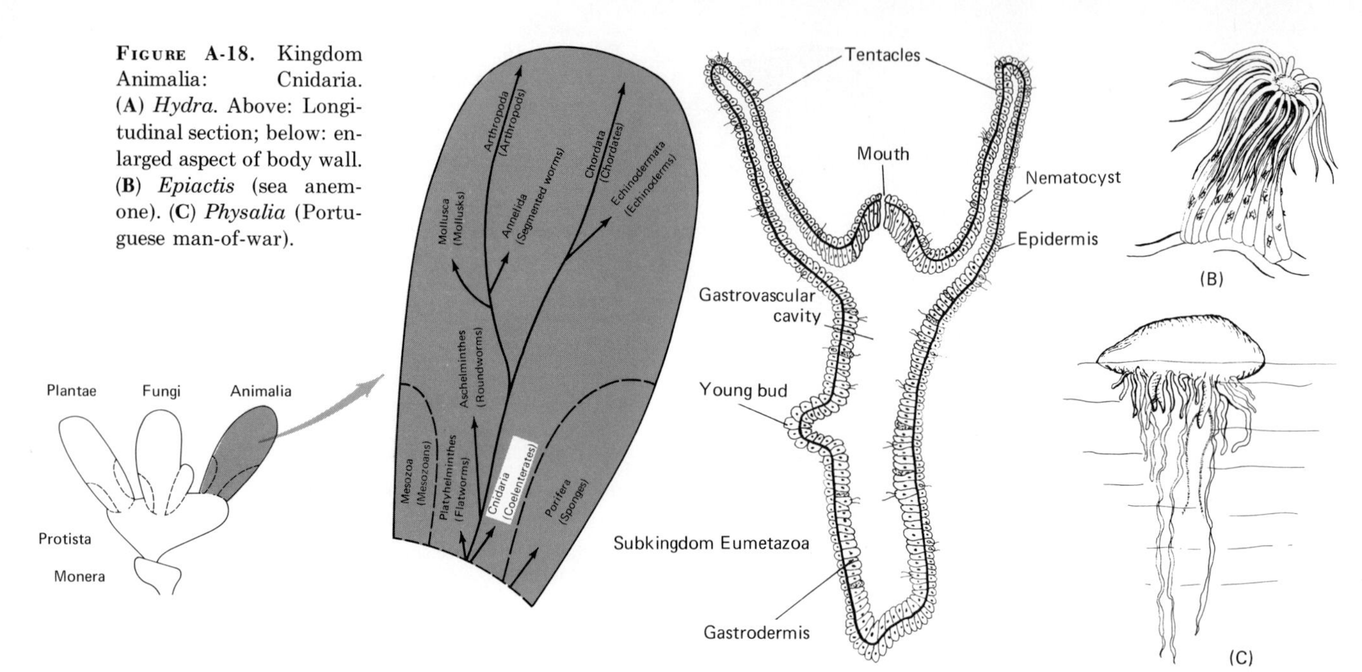

Figure A-18. Kingdom Animalia: Cnidaria. (**A**) *Hydra.* Above: Longitudinal section; below: enlarged aspect of body wall. (**B**) *Epiactis* (sea anemone). (**C**) *Physalia* (Portuguese man-of-war).

animals into convenient groups for study, they will be separated on the basis of the type of coelom present, if any. Accordingly, the Eumetazoa with bilateral symmetry are separated as acoelomata, pseudocoelomata, and coelomata.

Acoelomata. The phylum *Platyhelminthes* (flatworms) is one group of organisms lacking a coelom. They show many advances over the Porifera and Cnidaria. Among these are (1) bilateral symmetry, with anterior-posterior and dorsal-ventral aspects; (2) a primitive nervous system with nerve cords extending along the body; (3) the addition of a third germ layer, the mesoderm; (4) layers and bundles of muscles for various movements; and (5) internal gonads (ovaries and testes) with permanent reproductive ducts (oviducts and vas deferens) and copulatory organs (vagina and penis). Included here are some of the most common parasitic worms as well as free-living species (Figure A-19).

Pseudocoelomata. Of the three phyla included in the Pseudocoelomata only one, the phylum *Aschelminthes*, will be discussed. This phylum contains several groups of often worm-like animals that are mostly of small to microscopic size and slender in form. They inhabit marine, freshwater, and terrestrial habitats. The worm *Ascaris* (Figure A-20), a representative member of this

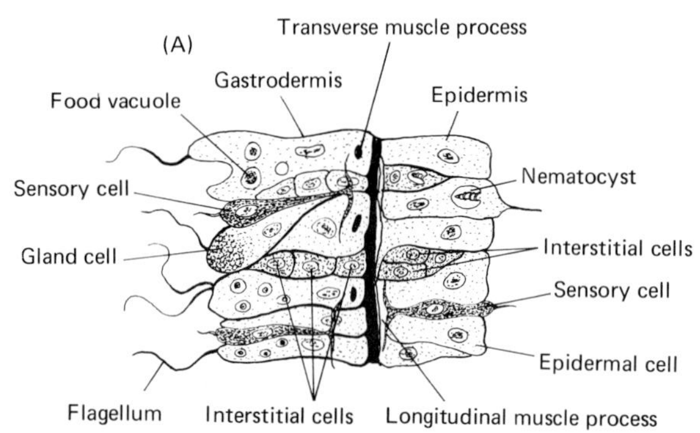

phylum, possesses a very well-developed cuticle (thin, tough, noncellular covering external to the epidermis) that is periodically shed and affords excellent protection. Other characteristics of the phylum include a complete alimentary canal, a pseudocoel, longitudinal muscles for locomotion, and glands in the tail region which secrete a sticky substance that helps to hold the animal in place.

Coelomata. All phyla within this large grouping contain a true coelom. The origin of the coelom, however, provides a further basis for subdividing the Coelomata into two distinct categories: (1) subgrade Schizocoela, in which the coelom originates in the mesoderm,

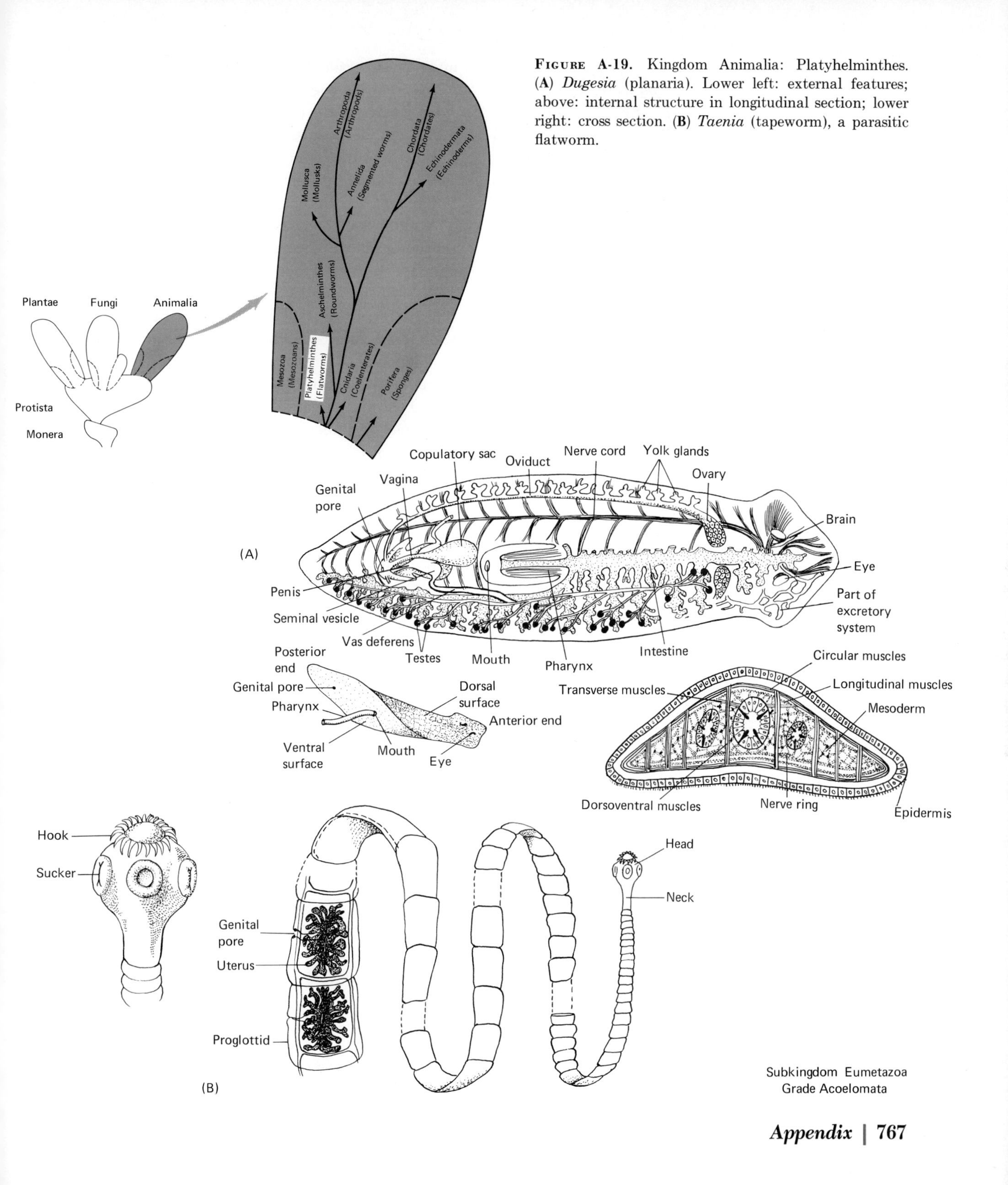

Figure A-19. Kingdom Animalia: Platyhelminthes. (**A**) *Dugesia* (planaria). Lower left: external features; above: internal structure in longitudinal section; lower right: cross section. (**B**) *Taenia* (tapeworm), a parasitic flatworm.

and (2) subgrade Enterocoela, in which the coelom forms from pouches from the embryonic gut.

Only three of the eight phyla of the subgrade Schizocoela will be discussed. The first of these, phylum *Mollusca*, or mollusks, includes such diverse animals as snails, oysters, and octopuses and is one of the most familiar of all the large animal groups (Figure A-21). The molluskan body consists of four principal parts: (1) the head, which contains the "brain," sense organs, and mouth; (2) the foot, which is a muscular locomotor organ; (3) the mantle, which is a fold of the body wall that secretes the shell and protects the body organs; and (4) the visceral mass, which lies above the foot and is composed of digestive, excretory, circulatory, and reproductive organs.

The second phylum of the Schizocoela, the phylum *Annelida*, includes earthworms, leeches, and many other kinds of segmented worms (Figure A-22). In contrast to the worms of other phyla, those of the Annelida have bodies composed of many essentially similar ringlike segments called somites; each segment is bounded internally by partitions, the septa. Annelids have a well-developed cuticle, as well as both circular and longitudinal sets of muscles. An alimentary canal consisting of a mouth, anus, and several highly specialized regions is also typical of the phylum. Well-differentiated excretory, circulatory, nervous, and reproductive systems are present. An important new feature of this phylum is the appearance of a closed circulatory system, that is, one in which blood is distributed via a series of closed and interconnected vessels.

The Phylum *Arthropoda* is the largest of all phyla and actually includes most of the world's known animals. It includes crabs, shrimps, barnacles, crayfish, insects, spiders, ticks, centipedes, and millipedes (Figure A-23). The arthropod body is conspicuously divided into head, thorax, and abdomen, each composed of several segments which may be fused in various ways. The appendages are jointed and are differentiated in form and function to serve special purposes. All exterior surfaces are covered by an exoskeleton containing chitin. The nervous system, eyes, and other sense organs are proportionately large and well developed. In direct contrast to the Annelida, the circulatory system of arthropods is closed to the environment, that is, blood is pumped through arteries to organs and tissues

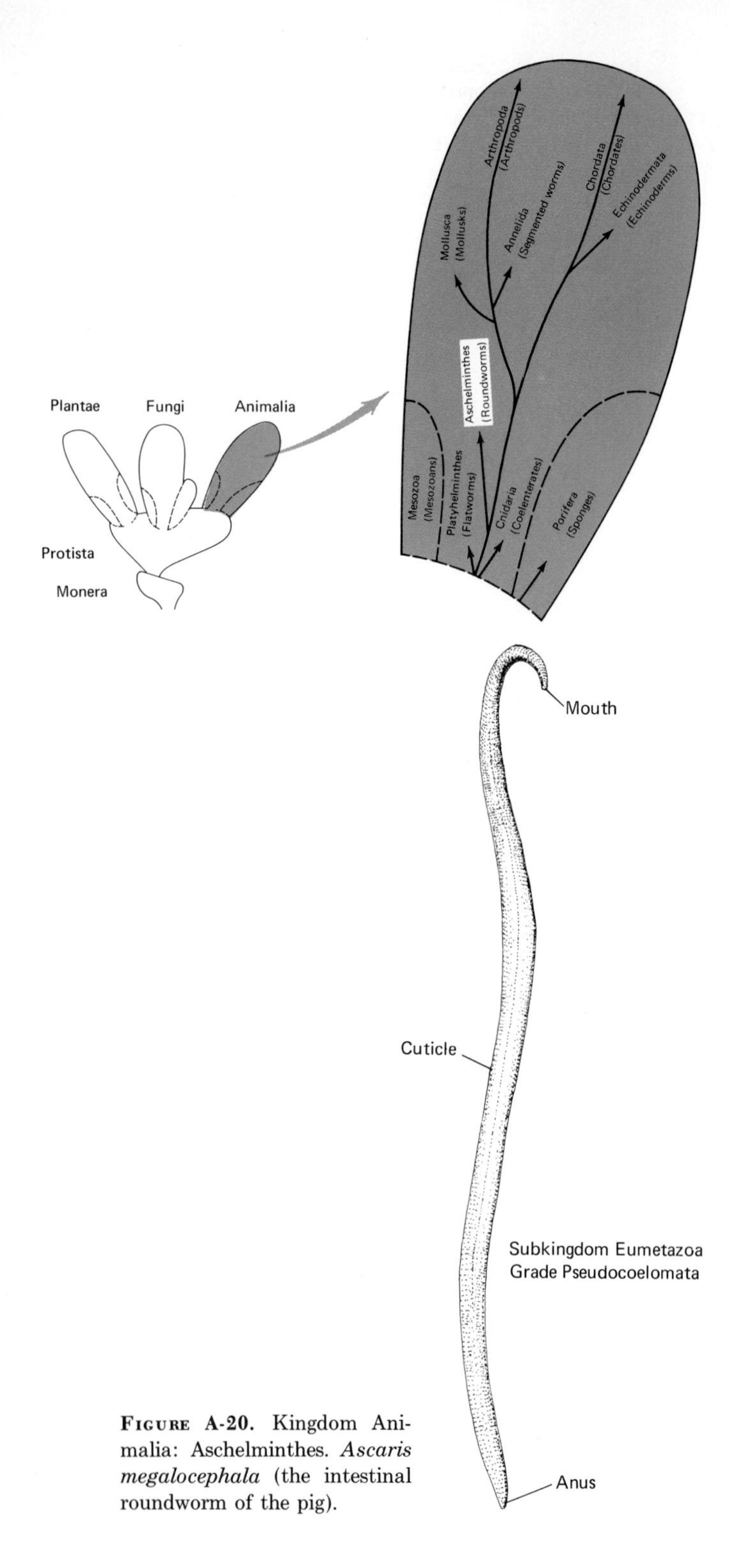

Figure A-20. Kingdom Animalia: Aschelminthes. *Ascaris megalocephala* (the intestinal roundworm of the pig).

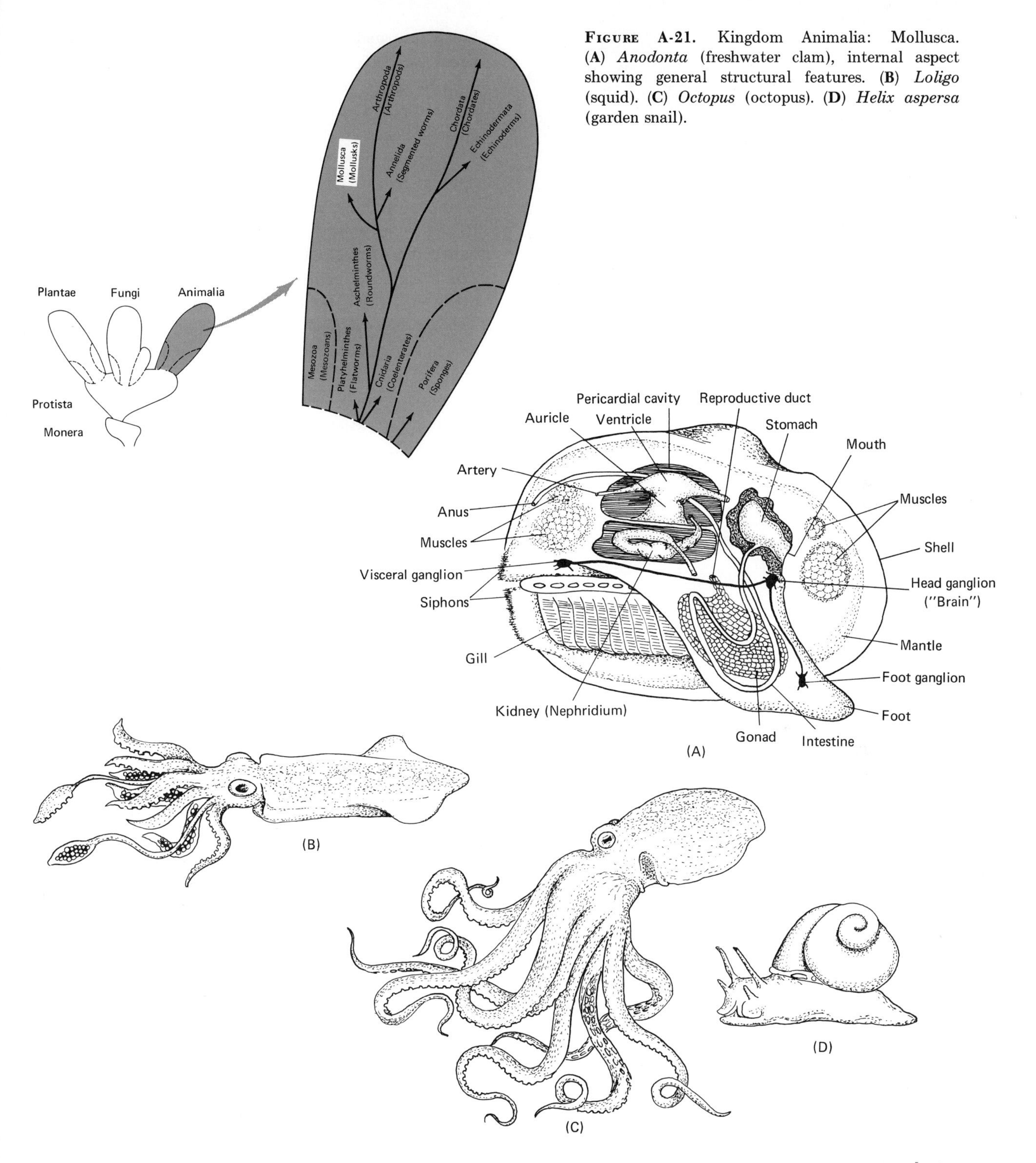

FIGURE A-21. Kingdom Animalia: Mollusca. (**A**) *Anodonta* (freshwater clam), internal aspect showing general structural features. (**B**) *Loligo* (squid). (**C**) *Octopus* (octopus). (**D**) *Helix aspersa* (garden snail).

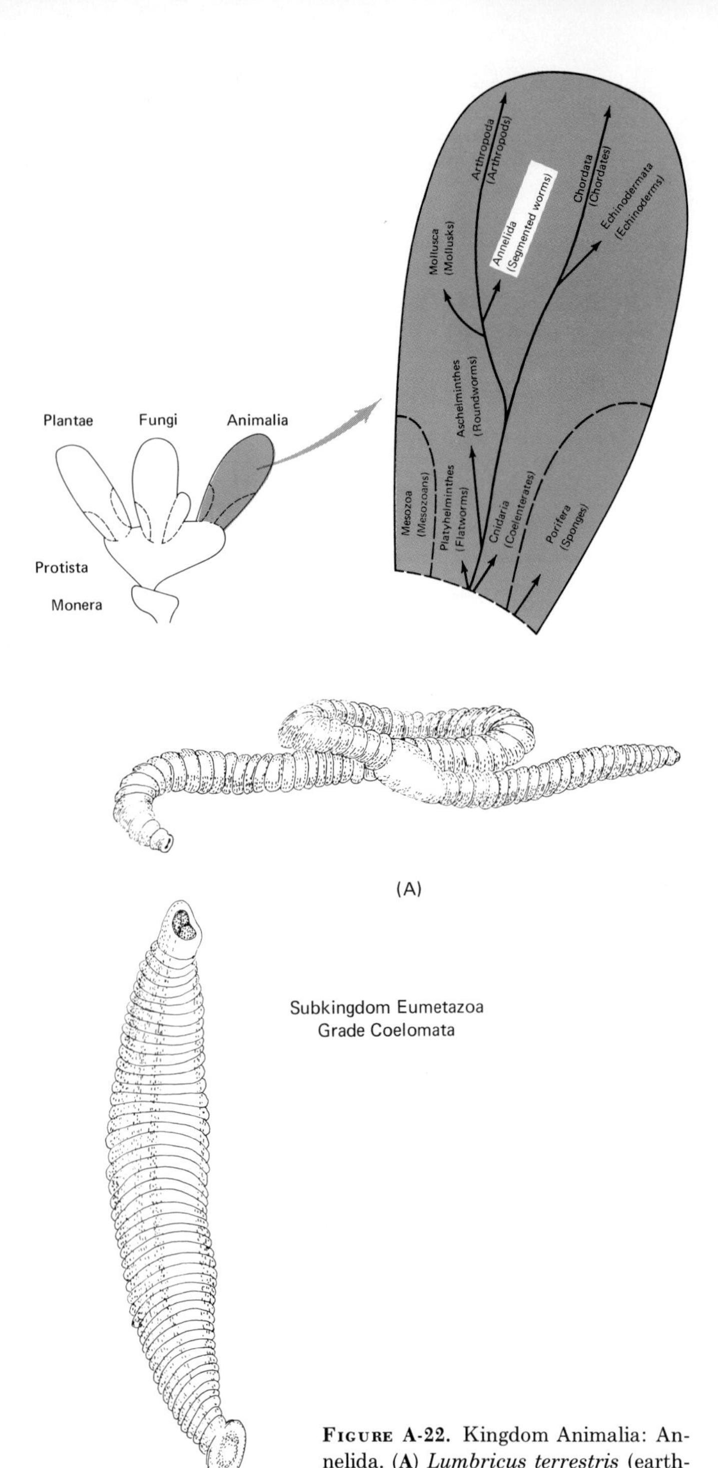

Figure A-22. Kingdom Animalia: Annelida. (**A**) *Lumbricus terrestris* (earthworm). (**B**) *Hirudo medicinalis* (medicinal leech).

but is returned through body spaces (hemocoel) to the heart.

Once again, as with the subgrade Schizocoela, only selected phyla of the subgrade Enterocoela will be discussed. These are Echinodermata and Chordata.

Included in the phylum *Echinodermata* are such familiar marine organisms as starfish, brittle stars, sea lilies, sea urchins, and sand dollars (Figure A-24). Although echinoderms have a predominantly radial symmetry, it is not as well developed as in the coelenterates. The skeleton of echinoderms is internal, consisting of calcareous plates embedded within the body wall. A unique system of water-filled channels represents a division of the coelom found in no other group of animals. It terminates externally in the form of hundreds of small tubular feet which aid in movement, food-getting, respiration, and even sensory perception.

The final phylum to be considered, the phylum *Chordata*, is perhaps the most familiar one. The phylum includes fishes, amphibians, reptiles, birds, and mammals, which belong to the subphylum *Vertebrata* (Figure A-25), and two less well known subphyla, the *Urochordata* and *Cephalochordata*, represented by the sea squirts and lancelets, respectively. Although these adult chordates do not look alike, they share certain unique features at some stage in their life history.

1. A dorsal, longitudinal rod known as the **notochord** is present in the embryos of all, and sometimes in the adult. The notochord is the first supporting structure of the chordate body and is replaced or surrounded by the vertebral column in vertebrates.

2. A longitudinal nerve cord lies dorsal to the notochord. In the vertebrates, the nerve cord differentiates into the brain and spinal cord.

3. Paired gill slits develop on the sides of the embryonic pharynx. Nearly all aquatic chordates up to amphibians respire by means of gills. In most amphibians, the gills are lost and replaced by lungs. The reptiles, birds, and mammals all develop several pairs of gill slits during early embryonic life, but they are never functional and soon close; all these animals later develop lungs for breathing air.

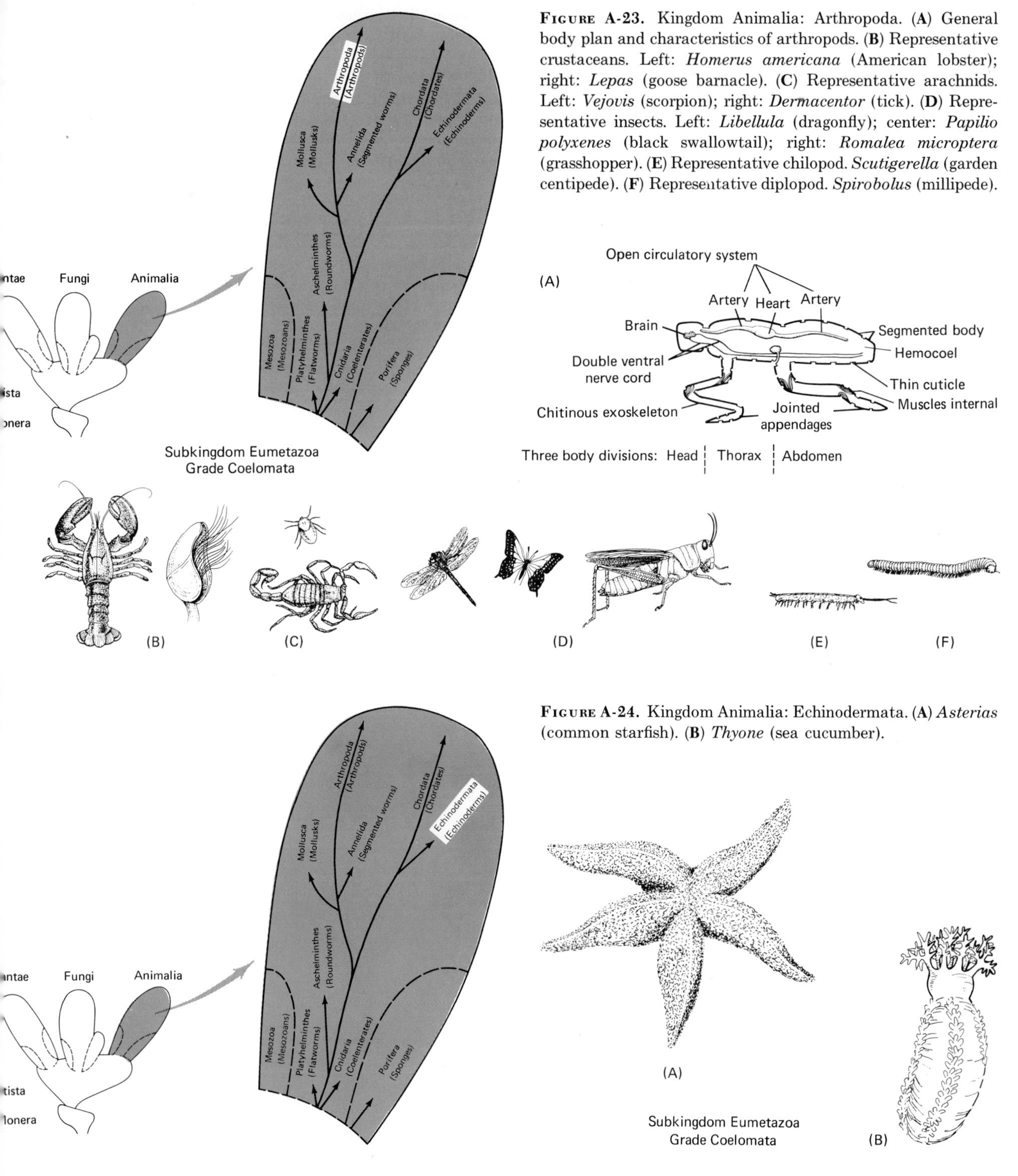

Figure A-23. Kingdom Animalia: Arthropoda. (**A**) General body plan and characteristics of arthropods. (**B**) Representative crustaceans. Left: *Homerus americana* (American lobster); right: *Lepas* (goose barnacle). (**C**) Representative arachnids. Left: *Vejovis* (scorpion); right: *Dermacentor* (tick). (**D**) Representative insects. Left: *Libellula* (dragonfly); center: *Papilio polyxenes* (black swallowtail); right: *Romalea microptera* (grasshopper). (**E**) Representative chilopod. *Scutigerella* (garden centipede). (**F**) Representative diplopod. *Spirobolus* (millipede).

Figure A-24. Kingdom Animalia: Echinodermata. (**A**) *Asterias* (common starfish). (**B**) *Thyone* (sea cucumber).

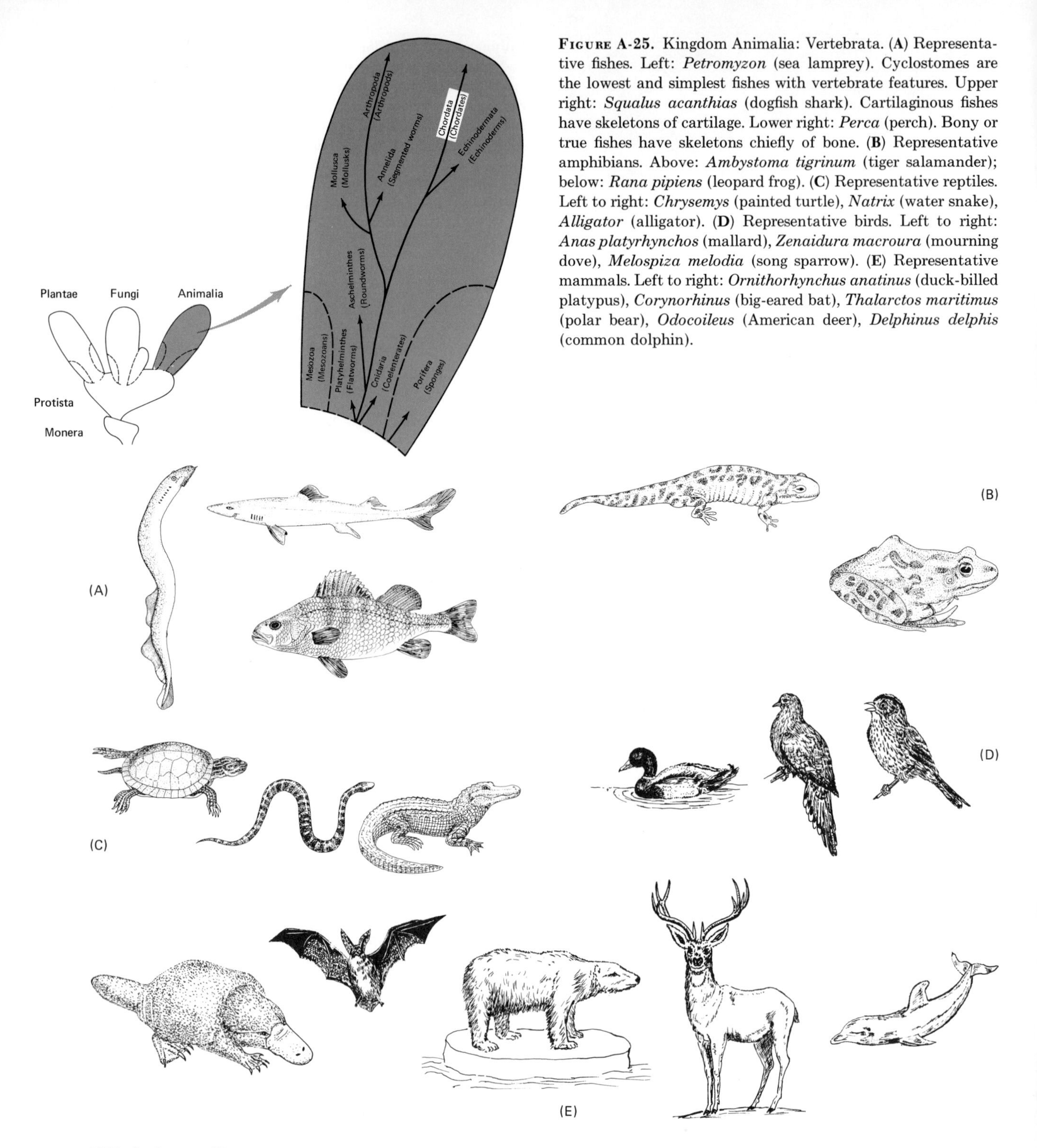

Figure A-25. Kingdom Animalia: Vertebrata. (**A**) Representative fishes. Left: *Petromyzon* (sea lamprey). Cyclostomes are the lowest and simplest fishes with vertebrate features. Upper right: *Squalus acanthias* (dogfish shark). Cartilaginous fishes have skeletons of cartilage. Lower right: *Perca* (perch). Bony or true fishes have skeletons chiefly of bone. (**B**) Representative amphibians. Above: *Ambystoma tigrinum* (tiger salamander); below: *Rana pipiens* (leopard frog). (**C**) Representative reptiles. Left to right: *Chrysemys* (painted turtle), *Natrix* (water snake), *Alligator* (alligator). (**D**) Representative birds. Left to right: *Anas platyrhynchos* (mallard), *Zenaidura macroura* (mourning dove), *Melospiza melodia* (song sparrow). (**E**) Representative mammals. Left to right: *Ornithorhynchus anatinus* (duck-billed platypus), *Corynorhinus* (big-eared bat), *Thalarctos maritimus* (polar bear), *Odocoileus* (American deer), *Delphinus delphis* (common dolphin).

Glossary

Abiogenesis Theory that living forms are derived from nonliving matter.

Abiotic environment Nonliving aspects of the environment; includes such factors as solar radiation, temperature, atmospheric pressure, topography, wind, precipitation, and inorganic and organic substances.

Abscission layer Layer of cells formed across the base of the petiole or stem; when modified by auxins, this layer brings about separation of an organ (leaf or fruit) from the plant body.

Absorption Uptake of water and dissolved substances by a cell or organism.

Absorption spectrum Plot of the efficiency with which a substance absorbs light as a function of wavelength.

Acetylcholine Chemical substance liberated at or near a nerve ending and thought to be involved in the transmission of nerve impulses across many synapses.

Acid Any substance that is a proton (H^+) donor or will accept an electron pair.

Acoelomate Animal in which a true internal cavity is lacking.

Acromegaly Abnormal development brought about by hypersecretion of somatotropin from the pituitary during adult life; characterized by excessive growth of the extremities, nose, jaw, fingers, and toes.

Acrosome Cap-like structure covering the head of a sperm.

Actin Protein arranged as thin filaments in a myofibril; together with myosin it is responsible for the shortening of muscle cells.

Action current Electric current that flows between a region of excitation and an adjacent unexcited region.

Activation energy Amount of energy required to initiate a chemical reaction.

Activator One of two kinds of prosthetic groups associated with enzymes; metal, usually an ion of a salt.

Active site Portion of an enzyme that binds the substrate during catalysis.

Active transport Movement of substances across a cell membrane at a constant rate and commonly against a concentration gradient; an energy-expending process in which the substance is moved via an enzyme-like carrier.

Adaptation Any change in an organism's structure, physiology, or behavior that results in its better adjustment to conditions in the environment; the acquisition of characteristics by an organism that make it better suited to function and reproduce in its environment.

Adaptive radiation Evolution and divergent development of a number of species adapted to ways of life different from a single ancestral species.

Adenosine triphosphate (ATP) Organic molecule consisting of adenine, ribose, and three phosphate groups; the major source of usable biological energy, which is stored in the high-energy linkages of the phosphate groups and liberated when one or two phosphate groups are split off from the ATP molecule; if one phosphate group is split off, the resulting compound is called adenosine diphosphate (ADP).

Adrenal gland Vertebrate endocrine gland located above the kidney; anatomically and physiologically subdivided into a cortex and medulla; among its functions are the hormonal regulation of heartbeat, blood pressure, respiration rate, mineral ions in the body fluid, carbohydrate metabolism, and development of sexual characteristics.

Aerobic respiration Type of respiration in which the H of NADH + H^+ ultimately combines with molecular oxygen.

Allele One of the alternate forms of the same functional gene that may occupy the same locus on homologous chromosomes.

Allograft Transplantation between individuals of the same species but with different genetic backgrounds.

Allometric growth Differential growth rate in various parts of an organism.

Allosteric interaction Change in the activity of an enzyme brought about by the selective binding of a compound at a second site on the enzyme that does not overlap the substrate binding site.

Alternation of generations In plants, a life cycle in which a haploid organism, the gametophyte, produces gametes that upon fusion, form a zygote that develops into a diploid sporophyte; following meiosis, spores produced by the sporophyte give rise to gametophytes.

Alveolus Small sac-like cavity within the lungs; site of gaseous exchange between air and blood.

Amino acid Organic acid consisting of an amino ($-NH_2$) group and a carboxyl ($-COOH$) group; subunit of a protein molecule in which it is joined to the next amino acid via a peptide linkage.

Amnion Extraembryonic membrane of reptiles, birds, and

mammals consisting of a double membrane fluid-filled sac that surrounds the embryo.

Anabolic reaction Enzymatic synthesis of relatively large molecules from simple precursor molecules in which an input of free energy is required.

Anabolism Sum total of chemical reactions concerned with the synthesis of molecules from simpler materials; typically endergonic.

Anaerobic respiration Type of respiration in which the H of NADH + H^+ is ultimately received by oxygen found as part of a chemical compound; carried on by relatively few bacteria.

Analogous Similar in function and/or appearance but not in evolutionary origin.

Androgen General term applied to one of several male sex hormones.

Ångström Microscopic unit of measurement (symbolized Å) equal to $\frac{1}{10,000}$ of a micron.

Annual (growth) ring Cylindrical deposition of xylem cells in stems of plants during one growing season.

Antibiotic Chemical substance that can destroy or selectively inhibit the growth of microorganisms.

Antibody Globular protein produced in response to the presence of some foreign substance (antigen) in the blood or tissues that reacts directly with the antigen.

Anticodon Three nucleotides in *t*RNA that are complementary to a codon of *m*RNA.

Antigen Foreign substance, usually a protein or protein–polysaccharide complex, that stimulates the production of specific antibodies.

Apical dominance Inhibition of lateral growth by apical buds.

Apical meristem Meristematic tissue located near the tips of roots and stems; also referred to as promeristem.

Apoenzyme Protein portion of an enzyme; requires a specific coenzyme to become a complete functional enzyme (holoenzyme).

Artery Kind of blood vessel that carries blood, usually oxygenated, away from the heart.

Articulation (joint) Point at which two or more bones, or a bone and cartilage, meet.

Ascus Sac-like structure in which ascospores are formed; characteristic of Ascomycota.

Atmosphere Layer of gases that surrounds the earth.

Atom Smallest whole unit of a chemical element; made up of protons, neutrons, as well as other particles, which form an atomic nucleus, and a specific number of electrons that orbit about the nucleus.

Atomic number Number of protons in the nucleus of an atom.

Autonomic nervous system Nervous coordinating system in vertebrates consisting of motor nerves and ganglia; fibers innervate smooth muscle and glands and elicit inhibition and/or acceleration of internal organs; structurally organized into sympathetic and parasympathetic divisions.

Autosome Chromosome other than a sex chromosome.

Autotroph Organism capable of synthesizing its own required nutrients from simple organic and inorganic substances.

Auxin Growth substance characterized by its capacity to induce elongation of plant cells.

Axon Process of a neuron that is single, often long, and slightly branched; conducts nerve impulses.

Bacteriophage Virus that infects bacteria.

Bark Collective designation for secondary phloem, cork cambium, and cork; tissue external to epidermis.

Base Any substance that is a proton (H^+) acceptor or an electron pair donor.

Basidium Club-shaped structure in which basidiospores are developed; characteristic of Basidiomycota.

Benthos Crawling, creeping, and sessile organisms inhabiting the benthic regions of oceans and lakes.

Biogenesis Theory that all living forms are derived from other living forms.

Biogeographical region Large region of the earth classified on the basis of characteristic fauna and flora.

Biological catalyst *See* Enzyme.

Biological clock Endogenous factor or factors that govern the biological rhythms of organisms.

Biological oxidation *See* Respiration.

Bioluminescence Ability of an organism to emit light.

Biomass Total weight of all living organisms in an ecosystem.

Biome Relatively large community easily recognized by characteristic fauna, flora, soil, and climate; many biomes may exist in a biogeographical region.

Biosphere Entire zone of water, land, and air at the surface of the earth that is occupied by living things.

Biosynthesis Formation of organic compounds by living organisms.

Biotic environment Portion of the environment that includes all living organisms.

Biotic potential Ability of a population to grow; reproductive potential under unlimited environmental conditions; overall population growth rate under unlimited environmental conditions.

Birth rate Number of individuals born to a given population within a given period of time.

Blastocoel Fluid-filled cavity of a blastula.

Blastomere Cell in blastula resulting from the cleavage of an egg.

Blastula Typically spherical structure composed of a single layer of cells surrounding a fluid-filled cavity; produced by cleavage and subsequent rearrangement of cells of a fertilized egg.

Bud Meristematic tissue enclosed by modified leaves; terminal and lateral buds of stems and floral buds; in some animals, an asexually produced swelling that develops into a new individual.

Buffer Substance that prevents significant changes in pH in solutions to which small quantities of acids or bases are added; usually a weak acid and its salt.

Cambium Cylindrical sheath of meristematic cells in stems and roots of many tracheophytes that divides to produce secondary phloem outwardly and secondary xylem inwardly.

Capillary Microscopic blood vessel that facilitates exchange of nutrients and wastes.

Carbohydrate Organic compound containing a chain of carbon atoms to which hydrogen and oxygen atoms are attached; polyhydroxy ketones or aldehydes.

Cardiac output Volume of blood pumped by one ventricle into the arteries per minute.

Carotene Yellow or orange pigment found in plants; precursor of vitamin A in the vertebrate liver.

Carpel Chamber in plant ovary in which ovules and seeds are produced.

Carrying capacity Number of organisms or biomass a portion of the ecosystem can support.

Castration Removal of the testes.

Catabolic reaction Enzymatic breakdown, usually by oxidative reactions, of relatively large nutrient molecules into smaller, simpler molecules; accompanied by a release of free energy.

Catabolism Chemical reactions concerned with the degradation of molecules into simpler substances; typically release energy and are exergonic.

Cell Basic structural and functional unit of a living system; microscopic unit of protoplasm surrounded by a membrane and containing genetic material.

Cell (plasma) membrane Thin, outermost membrane that surrounds and is differentiated from the cytoplasm of the cell.

Cell plate Dividing structure that appears during plant cell cytokinesis at the equator of the mitotic spindle; becomes impregnated with calcium pectate and forms a cell wall between daughter cells after division.

Cell sap Water and dissolved materials within vacuoles of plant cells.

Cell wall Rigid, outermost layer of a plant cell synthesized by the cytoplasm and consisting chiefly of cellulose.

Central nervous system Brain and spinal cord of vertebrates.

Centriole Cylinder-shaped organelle, usually found in pairs near the nucleus of animal cells and many protists, which doubles prior to mitosis; during mitotic division the centrioles move apart to form the spindle poles; also located at the base of a cilium.

Centromere Small spherical body joining chromatids together; point of spindle fiber attachment.

Cerebellum Portion of vertebrate brain that serves as the principal muscle-coordinating center.

Cerebrum Main portion of the vertebrate brain consisting of two cerebral hemispheres; functions principally in conscious sensations, learning, memory, reasoning, and will.

Chemical bond Attractive force that binds atoms together.

Chemical element One of about 100 different natural or man-made types of matter that, singly or in combination, comprise all materials of the universe; a substance composed of only one kind of atom.

Chemical reaction Change of one or more substances into different ones by rearrangement of atoms.

Chitin Insoluble protein–polysaccharide that forms the exoskeleton of arthropods and cell walls of some fungi.

Chloroplast Typically disc-shaped organelle of plant cells consisting of a double-layered external membrane and an internal membrane system; contains chlorophylls and accessory pigments and is the site of photosynthesis.

Chorion Outermost extraembryonic membrane of reptiles, birds, and mammals; in placental mammals, enters into the formation of the placenta.

Chromatin Deeply staining, diffuse hereditary material in the nucleus; composed of DNA; condenses into chromosomes; has RNA and proteins associated with it.

Chromatophore Pigmented cell responsible for color markings of animals.

Chromoplast A plastid that contains a pigment.

Chromosome Deeply staining structure visible in the nucleus of eucaryotic cells during division; consists mostly of DNA and associated proteins; carries hereditary determinants.

Cilium Short, hair-like protoplasmic structure present on the surface of certain types of cells; cilia beat in unison to move the cell or organism or to move food particles over the surface of the cell; each consists of a characteristic internal structure of two inner fibrils surrounded by nine pairs of outer fibrils.

Circadian rhythm Regular periods of growth and activity within organisms, usually on a 24-hour basis.

Cistron Segment of a chromosome that specifies one polypeptide chain (*see* Structural gene).

Cleavage Successive early cell divisions of a fertilized egg into smaller cells (blastomeres), leading to the development of a blastula.

Climax community Community that will not be replaced by another through normal succession; stable and terminal community in an ecological succession.

Coacervate Protein aggregate enveloped by water molecules owing to the effects of surface tension.

Codon Sequence of three nucleotides; unit of genetic information in DNA; a sequence of codons specifies the order of amino acids in a polypeptide chain.

Coelom Body cavity of triploblastic animals lying within and lined by the mesoderm.

Coenzyme Organic molecule, loosely bound to an enzyme, that aids the enzyme in catalyzing a particular reaction; often acts as an acceptor or donor of a reactant.

Coleoptile Cylindrical covering of the epicotyl of the embryo in a monocot seed.

Collaboration Phenomenon by which two different genes influencing the same trait interact so as to produce single trait phenotypes that neither gene by itself could produce.

Collenchyma Plant tissue occurring beneath the epidermis of stems and leaf stalks that provides mechanical support and strength; composed of cells with primary cell walls thickened at their corners.

Collision theory Assumption that particles must collide in order for a chemical reaction to occur.

Commensalism Two species living together in which one derives benefit from the association while the other is neither helped nor harmed.

Community Species of different kinds inhabiting a common environment.

Complementary gene One of several dominant genes that interact to produce a phenotype although neither gene produces a phenotypic effect by itself.

Compound Any substance formed by the chemical combination of the atoms of two or more elements.

Compound eye Complex eye common in arthropods and consisting of a number of separate units, each capable of forming an image or a portion thereof.

Cone Light-sensitive nerve cell in the retina responsible for the perception of color and daylight vision.

Conjugation Temporary union of two cells in which nuclear material is exchanged; common in ciliate protozoans and bacteria.

Connective tissue Kind of animal tissue that supports, protects, and binds together other tissues; comprised of cells that are irregularly distributed through a relatively large amount of fibrous intercellular substance.

Consumer Organism that cannot synthesize its own food; consumes preformed organic compounds.

Convergent evolution Independent evolution of similar adaptations in two or more organisms not closely related.

Corepressor End product of a metabolic pathway that inhibits the formation of a specific enzyme.

Cork Outer protective layer of woody plants composed of dead cells with thick walls impregnated with suberin.

Cornea Transparent anterior portion of the eye.

Cortex Storage tissue beneath the epidermis in roots and stems; outer area of an organ in animals.

Cotyledon Part of angiosperm seed; a food storage organ that absorbs nutrients from the endosperm and supplies them to the developing embryo; may function as a leaf after germination.

Covalent bond Chemical bond formed between atoms as a result of sharing a pair of electrons.

Cretinism Abnormality during youth resulting from a hyposecretion of thyroxine by the thyroid gland; characterized by stunted growth and retarded mental development.

Crossing-over Process that may occur during synapsis of meiosis in which homologous chromosomes undergo an exchange of segments.

Cutin Waxy, waterproofing material, secreted by plant epidermal cells, that forms the cuticle; found in leaves, some stems, and some fruits.

Cyclosis Circular flowing or streaming of cytoplasm within a plant cell.

Cytochrome Iron-containing pigment important in cellular respiration; alternately oxidized and reduced during biological oxidation by acting as an electron donor and acceptor.

Cytochrome system Group of iron-containing pigments (cytochromes) that transfer electrons to free oxygen in aerobic respiration.

Cytokinesis Division of the cytoplasm into two separate components; usually follows mitosis.

Cytokinin Growth substance produced by plants; active in inducing cell division.

Cytoplasm Living matter of the cell external to the nucleus and internal to the plasma membrane; ground substance of a cell containing organelles.

Death rate Decrease in the size of a given population during a given period of time.

Decomposer Organism that breaks down nonliving organic matter and converts it to simpler substances.

Degradable pollutant Any substance causing pollution that can be altered to a nonpollutant by naturally occurring biological, chemical, or physical processes.

Dehydration synthesis Joining together of smaller units into a single, larger molecule by the loss of water; one unit contributes H^+ ions, the other OH^- ions.

Dehydrogenase Class of enzymes that remove hydrogen atoms from a substrate molecule.

Dehydrogenation Removal of hydrogen atom(s) from a molecule.

Deletion (deficiency) Chromosomal aberration that is the loss of a part of a chromosome.

Demography Statistical analysis of human populations including such aspects as birth rates, death rates, growth, life expectancy, age distribution, and sex ratios.

Denaturation Alteration of the three-dimensional structure of a protein by mild treatment that does not break the primary structure; results in alteration of physical and functional properties of the protein.

Dendrite Process of a neuron that is multiple, relatively thin, and profusely branched; either receives sensory information from the environment or receives impulses from other neurons.

Deoxyribonucleic acid (DNA) Macromolecule that stores genetic information in cells; capable of self-replicating and of determining synthesis of proteins; composed of chains of phosphates, sugars (deoxyribose), and nitrogenous bases.

Depressants A class of drugs that can produce effects ranging from relief of anxiety and tension to sleep, anesthesia, coma, and death.

Desmosome Modification of the plasma membrane represented by a local thickening of opposing surfaces of adjacent cells from which fine filaments radiate into the cytoplasm; desmosomes bind cells together in epithelial tissue.

Development Complex series of events between fertilization and independent life, involving cell division and growth, morphogenesis, and differentiation, leading to the expression of phenotypic traits characteristic of an organism.

Diastole Relaxation of the chambers (atria and ventricles) of the heart that allows them to become filled with blood.

Dicotyledonae Subgroup of angiosperms possessing two seed leaves (cotyledons); also called dicots.

Differentiation Development of a relatively unspecialized cell or tissue into a more specialized form.

Differentially permeable Property of a membrane by which certain ions and molecules pass through, to the exclusion of others.

Diffusion New movement of molecules or ions from a region of higher concentration to a region of lower concentration as a result of molecular activity (kinetic energy); a tendency toward reaching an equilibrium in which there is an equal distribution of molecules or ions within a system.

Digestion Series of hydrolysis reactions in which insoluble nutrients are made soluble for use by cells.

Dioecious Organism possessing distinct male or female reproductive structures.

Diploblastic Two-layered body derived from ectoderm and endoderm tissues.

Diploid Chromosome state in which each type of chromosome occurs in pairs; a chromosome number twice that found in gametes; symbolized $2n$.

Disaccharide Sugar consisting of two monosaccharides joined together via dehydration synthesis.

Disinfection Process of treating an object, usually with a chemical, to remove or kill pathogens.

Divergent evolution Evolution of dissimilar characteristics in two or more lines descended from a common ancestor.

Dominant Term applied to a gene that exerts its complete phenotypic effect regardless of the presence of another allele; a gene that masks or prevents the expression of any of its alleles.

Duplication Chromosomal aberration in which a portion of a chromosome is duplicated.

Ecological equivalents Organisms that perform the same role (occupy the same niche) in different environments.

Ecological niche Specific position and role of an organism in an ecosystem.

Ecosystem Community of organisms interacting with abiotic factors of the environment; an area with arbitrary boundaries in which biotic and abiotic interactions can be studied.

Ectoderm Outer germ layer in the gastrula that gives rise to the nervous system, sense organs, and skin and its outgrowths.

Effector Cell, tissue, or organ (muscle or gland) that reacts to stimuli and elicits physiological responses.

Electrolyte Substance that dissociates in solution into ions and permits the conduction of an electric current through the solution.

Embryo Early stage of development of an organism; developing product of a fertilized egg.

Endergonic reaction Chemical reaction with a positive standard free energy; stores energy.

Endocrine gland Ductless gland whose hormones are discharged directly into the circulatory system of vertebrates.

Endoderm Inner germ layer of the gastrula stage; lines the archenteron; gives rise to the digestive tract and its outgrowths.

Endoplasmic reticulum (ER) Extensive system of double membranes considered to be continuous with the cell membrane; divides the cytoplasm into channels and compartments; serves for distribution of materials and is frequently associated with ribosomes; provides a large surface area for exchange of materials.

Endoskeleton Bony and/or cartilaginous structure that provides internal support for the body of an organism.

Endosperm In angiosperms, a food-storing tissue, that develops from the union of a male nucleus and two polar nuclei within the embryo sac.

Energy Capacity of a system to do work.

Energy transduction Change of energy from one form to another.

Environmental resistance Opposition to unlimited growth of a species coming from the abiotic and biotic environments in which the species exists.

Enzyme Highly specialized protein that expresses gene action by catalyzing a chemical reaction; a biological catalyst.

Epidermis Outermost layer of cells of an organism.

Epistatic gene Gene that, at one locus, influences the phenotypic expression of a gene at another locus; the phenomenon is called epistasis.

Epithelium Kind of animal tissue that covers external body surfaces and lines cavities; consists of cells arranged in one or more layers with little intercellular substance.

Erythrocyte Red blood cell; functions in the transportation of oxygen and carbon dioxide.

Estivation Dormancy during dry periods, generally in the warmer regions of the earth.

Estrogens General term applied to several female sex hormones.

Estrus Recurrent, restricted period of sexual receptivity of certain nonprimate female animals; characterized by intense sexual urge.

Ethology Biological approach to the study of animal behavior that stresses behavior that occurs in the natural environment in which the animal lives.

Etiolation Phenomenon by which plants grown in the dark exhibit certain abnormal features including lack of chlorophyll (and thus yellow leaves), spindly, elongated stems, and poorly developed vascular tissue.

Eucaryotic Applied to organisms whose cells have a true nucleus enclosed by a membrane and, usually, other organelles such as plastids, mitochondria, and Golgi apparatus.

Evolution Genetic and phenotypic change according to an orderly sequence of events over a long period of time; a change in a gene pool over a long period of time.

Excretion Elimination of metabolic wastes by an organism.

Exergonic reaction Chemical reaction with a negative standard free energy exchange; chemical reaction that yields energy.

Exobiology Science of the search for life in the universe other than on earth.

Exoskeleton Skeleton covering the outside of the body; common in arthropods.

Expressivity Manner in which a phenotype is expressed; condition in which individuals with the same genotype show different phenotypes.

Feedback inhibition Phenomenon by which the end product of an enzymatic reaction may inhibit the activity of one or more enzymes in a metabolic pathway.

Fermentation Type of respiration in which the H of NADH + H^+ is passed to pyruvic acid or an organic compound derived from pyruvic acid; breakdown of carbohydrates, usually by microorganisms, under anaerobic conditions.

Ferredoxin Iron-containing protein (coenzyme) that serves as an electron acceptor in photosynthesis.

Fertilization Union of two haploid gametes that initiates the development of a diploid zygote.

Fission Kind of asexual reproduction in which an organism divides into two or more equal (or nearly equal) parts.

Flagellum Long, whip-like protoplasmic process protruding from the cell body and utilized in locomotion and feeding; longer than a cilium, but identical in structure.

Flower In angiosperms, the reproductive structure composed of reproductive organs with or without associated parts; a specialized branched stem bearing lateral appendages.

Food chain Sequence of organisms through which energy is transferred from its ultimate source in an autotroph; each organism in the chain feeds on the one below and is eaten by the one above until there is an ultimate consumer.

Fossils Remains, impressions, petrifactions, or casts of organisms that lived in the distant past.

Fruit Structure composed of a ripened ovary, with or without seeds, together with accessory flower parts; may be associated with the ovary.

Gamete Functional haploid cell produced by meiosis that in sexual reproduction, unites with another gamete to establish a diploid zygote; egg or sperm.

Gametophyte In plants, the haploid, gamete-producing stage in the alternation of generations in the life cycle.

Ganglion Collection of nerve cell bodies outside of the central nervous system.

Gastrula Early embryonic stage, following blastula; initially consists of two germ layers (ectoderm and endoderm) and the archenteron opening to the exterior by way of a blastopore.

Gastrulation Process by which a young embryo develops into a gastrula consisting of two and then three layers of cells.

Gene Unit of hereditary material represented as a small section of a chromosome; it is capable of self-replication and transmission as part of a chromosome; a unit of function that specifies a polypeptide chain (cistron); a unit of mutation (muton); a unit capable of recombination (recon).

Gene pool Sum total of all the alleles of all the genes in a population.

Genetic drift Tendency for change in the genetic composition of a population due to random mating; may cause drastic changes in small populations.

Genetic equilibrium Maintenance of a fairly constant ratio between different alleles in a gene pool in successive generations; basic assumption in the Hardy–Weinberg law.

Genome Complete genetic complement contained in a haploid cell.

Genotype Genetic constitution of an organism deduced from data provided by breeding experiments; entire genetic makeup of an individual.

Germination Resumption of growth of an embryo (or spore) after a period of reduced metabolic activity.

Germ layer One of three primary embryonic layers of cells that give rise to all tissues of the adult organism.

Gestation Period of development of a fertilized egg within the uterus.

Gibberellin Growth substance in plants which typically brings about an increase in shoot length.

Gill Respiratory structure of aquatic organisms usually consisting of a thin-walled projection from some part of the external body surface.

Girdling (ringing) Removing the bark (including the phloem), but leaving the xylem intact.

Gland Cell or organ, consisting of modified epithelial cells, that produces one or more secretions.

Glycolysis Metabolic conversion of glucose into pyruvic acid without the utilization of free molecular oxygen.

Goiter Enlargement of the thyroid gland characterized by a swelling on the front of the neck.

Golgi body (dictyosome) System of membrane-bounded cavities in the cytoplasm of eucaryotic cells; may be related to the storage of secretory products.

Gonad In a multicellular animal, the gamete-producing structure.

Granum A layer in the chloroplast upon which is found chlorophyll.

Gray crescent Crescent-shaped region on the surface of the fertilized egg, between the animal and vegetal poles, caused by a shift of pigments.

Growth regulator Chemical substance manufactured primarily in meristems, transported in phloem tissue, and concerned largely with regulation of growth patterns.

Guard cell Modified epidermal cell, often kidney-shaped, that, in conjunction with another guard cell, regulates the size of a stoma.

Guttation Loss of liquid water by plants, usually through veins of leaves, as a result of increased internal pressure.

Habitat Physical location where an organism lives and interacts with the abiotic and biotic environments.

Half-life Amount of time required for half the atoms of a given radioactive substance to decay to a different form.

Hallucinogen A variety of psychedelic drugs that alter mood, perception, thinking, and ego.

Haploid Chromosome state in which each type of chromosome is represented only once; symbolized as n.

Hardy-Weinberg law Mathematical expression of the relationship between the relative frequencies in successive generations of two alleles in a population; described by the binomial equation $q^2 + 2pq + p^2$.

Hemoglobin Red, iron-containing protein pigment found in vertebrate erythrocytes; transports oxygen and carbon dioxide and aids in the maintenance of pH.

Hermaphrodite Organism containing both male and female gonads.

Heterotroph Organism that cannot synthesize its own organic nutritive substances and must feed on nutrients manufactured by autotrophs.

Heterozygous Having two different alleles at the same locus on homologous chromosomes.

Holoenzyme Complete enzyme molecule consisting of a protein portion (apoenzyme) and a nonprotein constituent (prosthetic group).

Homeostasis Condition that results in the maintenance of stability (within certain limits) in an organism; maintenance of a stable internal environment compatible with the continuance of life; mechanisms for maintaining a stable internal state.

Homologous chromosome One of a pair of chromosomes generally having the same gene locus; normally separate during meiosis.

Homology Similarity in structure as a result of common ancestry regardless of function.

Homozygous Having identical alleles at the same loci on homologous chromosomes.

Hormone Chemical substance produced on a regular basis in minute concentrations by specialized cells in one part of the animal that exerts its influence on another part of the animal.

Hydrogen bond Electrostatic interaction between the partial negative charge of an electron pair of certain atoms such as oxygen and nitrogen and the partial positive charge of a hydrogen atom; may form between two molecules or between two parts of the same molecule.

Hydrolysis Chemical reactions involving the addition of water molecules. Usually to change large molecules into smaller molecules.

Hydrosphere That portion of the abiotic environment made up of the water areas of the earth.

Hypersensitivity Ability to react to the presence of certain substances (antigens) in amounts innocuous to normal individuals.

Hypertonic Solution having a higher concentration of solute molecules and a lower concentration of solvent molecules (water) than that of a solution to which it is compared.

Hypha Microscopic tubular filament forming the body of a fungus that branches and rebranches in the substrate from which the organism obtains nourishment.

Hypothalamus Portion of the vertebrate forebrain below the cerebrum; contains centers for coordination and temperature control; main homeostatic control center of the body.

Hypotonic Solution having a lower concentration of solute molecules and a higher concentration of solvent (water) molecules than that of a solution to which it is compared.

Immunity Ability of a host to prevent or overcome invasion by virulent pathogenic organisms.

Implantation (nidation) Attachment of a developing embryo to the uterine endometrium.

Inclusions Inactive materials such as droplets or crystals, found within cells.

Inducer Substance that, when present, brings about an increased synthesis of an enzyme.

Inducible enzyme Enzyme that is synthesized in the presence of an inducer.

Induction Process in which one tissue or body part brings about the differentiation of another tissue or body part.

Inflammation Local reaction caused by an agent that injures tissues.

Ingestion Process of taking solid food into the body.

Integument Covering of the body; the skin.

Inversion Chromosomal aberration in which reversal of a segment of a chromosome results in a reversal of the linear arrangement of certain genes.

Ion Atom or group of atoms bearing an electric charge, either positive (cation) or negative (anion).

Ionic (electrostatic) bond Chemical bond formed between atoms of different electrostatic charge as a result of the transfer of one or more electrons from an atom of lower electron affinity to one of higher electron affinity; ions formed as a result of the transfer are attracted to each other.

Isograft Transplantation in which the donor and recipient have identical genetic backgrounds.

Isolating mechanism Morphological, behavioral, or physiological mechanism that prevents the exchange of genetic material between different species.

Isomer One of a group of compounds having the same atomic composition but a different structural arrangement.

Isotonic Solution having the concentrations of solute and solvent (water) molecules exactly equal to those in another solution to which it is compared.

Isotope One of a group of atoms having the same atomic number but different atomic masses because of different numbers of neutrons in the atomic nuclei.

Karyogamy In fungus organisms, the fusion of two nuclei resulting in a zygote nucleus that is diploid.

Keratin Horny, protein-like substance formed by certain epidermal cells; especially prominent in skin, claws, hair, feathers, and hooves.

Kinesis Movement or activity of an organism in which the direction of the response is not controlled by the direction of the stimulus.

Krill Shrimp-like organisms that feed upon plankton and play an important role in the ocean food chain.

Larva Immature stage in the life history of an organism; morphologically distinct from the adult form.

Laterite Hard, rock-like substance resulting from the exposure of soil to sun and air after removal of vegetation; common in tropical areas.

Leaf Characteristic photosynthetic organ of most vascular plants.

Lentic habitat Standing water habitat such as a lake, pond, swamp, and bog.

Leucoplast A colorless plastid.

Leukocyte White blood cell; involved in defense against microorganisms by phagocytosis and antibody production.

Linkage Phenomenon by which certain genes are inherited together because they are linked together on the same chromosome.

Lipid Fat or fat-like organic compound soluble in nonpolar solvents such as ether or chloroform and insoluble in water.

Lithosphere Solid component of the abiotic environment in which organisms live.

Littoral zone That area of the shoreline lying between extreme high and low tide lines.

Lotic habitat Running water habitat such as a spring, brook, creek, and river.

Lymph Colorless fluid present in lymph vessels, organs, and ducts derived from blood by filtration; quite similar to plasma in chemical composition.

Lysosome Spherical membrane-enclosed organelle containing hydrolytic enzymes.

Macromolecule Molecule of relatively high molecular weight; usually refers to polysaccharides, proteins, nucleic acids, and their complexes.

Macronutrient Essential mineral element needed in relatively large amounts by organisms.

Matter That which has mass and occupies space.

Medulla Portion of the vertebrate hindbrain that contains, among others, centers for heart action, respiration, coughing, and swallowing; inner part of an organ.

Meiosis Special cell division involved in the formation of gametes (in animals) and spores (in plants); mechanism by which chromosome number is reduced from $2n$ to n and in which segregation and independent assortment of genes occur.

Meristematic tissue Mitotically active, undifferentiated plant tissue from which all other tissues are derived.

Mesoderm Middle layer of the three primary germ layers located between the ectoderm and endoderm; gives rise to circulatory system, excretory system, most of the reproductive system, the skeleton, and the muscles.

Mesophyll Parenchyma tissue found in leaves; contains numerous chloroplasts.

Messenger RNA (mRNA) Kind of RNA that carries genetic information from DNA to the surface of ribosomes during protein synthesis; acts as a template for protein synthesis.

Metabolism Sum total of all matter and energy transformations within a living system.

Metamorphosis Abrupt transition from one developmental stage to another.

Micron Unit of microscopic measurement (symbolized μ) equal to $\frac{1}{1000}$ mm.

Micronutrient Essential mineral element needed in minute amounts by organisms.

Microorganism Microscopic or submicroscopic form of life.

Microsphere Spherical, cell-like structure that arises spontaneously when hot concentrated solutions of thermally formed proteinoids are allowed to cool slowly over a period of several weeks.

Microtubule Thin, tube-like structure often found near centrioles; forms the spindle apparatus of dividing cells, may help to maintain cell shape, and may be involved in the internal movements of cytoplasm.

Microvillus One of many minute, finger-like extensions of the cell membrane that, collectively, increase the area of cell surface.

Middle lamella Thin layer of intercellular material formed between adjacent plant cells during division; consists chiefly of calcium pectate.

Mimicry Adaptation for survival in which a particular organism resembles another living form or a nonliving object.

Mitochondrion Small spherical, rod-shaped, or filament-shaped organelle within the cytoplasm; bounded by a double unit membrane; functions in the process of cellular respiration; site of ATP production in the cell.

Mitosis Basic process of nuclear division in which the two members of each pair of daughter chromosomes move to opposite poles of the spindle; maintains chromosome number at $2n$; process of somatic cell division.

Modifying gene Gene that modifies the expression of other genes, usually by suppressing or enhancing their visible effects.

Molecule Smallest whole unit of a chemical compound; formed from the chemical combination of two or more atoms.

Molting Periodic shedding of all or part of the outer covering; in arthropods, shedding of the exoskeleton to accommodate increase in growth; in birds, shedding of feathers.

Monocotyledonae Subgroup of angiosperms possessing a single seed (cotyledon); also called monocots.

Monosaccharide Simplest of the carbohydrates; single sugar, cannot be hydrolyzed into smaller sugar units; may be classified according to whether it contains an aldehyde group (aldose) or a ketone group (ketose).

Morphogenesis Development of form, size, and other structural features of an organism.

Morphology Study of form and structure at various levels of organization.

Multiple allele Three or more alternative conditions of a single locus that produce different phenotypes.

Multiple genes (polygenes) Two or more independent pairs of genes that affect the same characteristic in the same way and in an additive fashion.

Mutation Any alteration of the genetic material; generally a change of a gene from one allelic form to another.

Muton Smallest portion of a chromosome capable of undergoing mutation.

Mutualism Nutritional relationship in which two organisms live together and both benefit.

Mycelium Entire system of fungal hyphae that produce spores.

Myelin sheath Multilayered membrane, fatty in nature, that covers some nerve fibers; functions by protecting the fiber and assisting in the transmission of nerve impulses.

Myofibril A long, contractile protein fibril of cardiac and skeletal muscle tissue.

Myosin Protein, arranged in thick filaments in a myofibril, responsible for some of the chemical reactions occurring during shortening of muscle cells; unlike actin, does not shorten.

Myxedema Abnormality resulting from a hypersecretion of thyroxin in adults; characterized by a low metabolic rate and decreased heat production.

Nasties Paratonic movements of plants; the movement is independent of the direction from which the stimulus is received.

Natural selection Selective survival due to differential reproductive success of certain genotypes best fitted to exist in a particular environment through successive generations.

Nekton Collective term for aquatic organisms that are active swimmers and are capable of moving from one location to another.

Nematocyst Minute thread-like structural adaptation typical of coelenterates; used for anchorage, defense, and capture of prey.

Nephridium Kind of excretory tubule typical of many invertebrates.

Nephron Functional and structural unit of the vertebrate kidney.

Neritic (sublittoral) zone That portion of the ocean lying over the continental shelf.

Nerve Cord-like structure composed of bundles of nerve fibers enclosed in a connective tissue sheath.

Nerve net Diffuse network of simple nerve cells with no differentiation between axons and dendrites; characteristic of coelenterates.

Neurohumor A chemical substance produced by nerve cells that causes an endocrine gland to secrete its hormone.

Neuroglia Cells comprising the interstitial tissue of the nervous system; support nerve cells and probably assume a role in normal metabolism.

Neuron Nerve cell; consists of a nucleated cell body, a dendritic zone, and an axon.

Niche The particular way in which a population utilizes the resources of its habitat; no two populations can occupy the same niche for a long period of time.

Nondegradable pollutant Polluting substance that cannot be altered by natural biological and physical processes.

Nondisjunction Failure of homologous chromosomes to separate during meiosis; results in one or more extra chromosomes in the cells of some of the offspring.

Notochord Cylindrical rod of supporting cells in the anteroposterior axis that serves as an internal skeleton in all chordate vertebrates; becomes either surrounded or largely replaced by vertebrae.

Nuclear membrane (envelope) Double unit membrane surrounding the nucleus within the cell.

Nucleic acid Organic acid containing nucleotides chemically joined to each other; principal kinds are DNA and RNA.

Nucleolus Spherical, RNA-containing body within the nucleus of most cells.

Nucleotide Single unit of a nucleic acid composed of phosphate, sugar (ribose or deoxyribose), and a nitrogenous base.

Nucleus Typically spherical or oval structure, present in most cells, that contains chromosomes and nucleoli and is enclosed by a nuclear membrane.

Nutrient General term applied to any substance that can be utilized in the metabolic processes of an organism.

Oceanic zone That portion of the open ocean beyond the continental shelf.

Ocellus Light-sensitive pigmented structure that serves as an eye in most invertebrates.

Oogenesis Origin and development of an egg within the ovary.

Operator gene Region of DNA at which transcription is initiated; controls structural genes of its operon.

Operon Group of adjacent genes under the control of a single operator gene; functional unit of transcription.

Organ Unit consisting of different tissues that performs a specific activity.

Organelle Supramolecular complex found in the cytoplasm of the cell with a characteristic structure and a definite, although not always clear, function.

Organism Complete living system, either unicellular or multicellular.

Organizer Portion of an embryo capable of inducing undifferentiated cells to follow a specific histological and morphological pattern of development.

Osmosis Movement of water molecules through a differentially permeable membrane from a region containing a higher concentration of water to a lower concentration of water.

Ossification Process by which membranous and cartilaginous precursors of bone are replaced by bone.

Osteoblast Bone-forming cell.
Ovary Egg-producing organ of many animals.
Overturn Top to bottom circulation in large bodies of water as a result of changes in climate and variability in the density of water at different temperatures.
Ovulation Release of a mature egg from the Graafian follicle of the ovary.
Oxidation-reduction Chemical reaction in which most cellular energy, in the form of ATP, is finally obtained; may involve the transfer of electrons, the transfer of hydrogen atoms, or the transfer of hydrogen ions and an electron.

Paleontology Study of life during past geologic eras, primarily by means of fossils.
Parallel evolution Evolution in which there is a similarity in adaptation among organisms as a result of similar natural selection forces.
Parasite Organism that lives in or on an organism of a different species and derives nourishment from it.
Parasitism Nutritional relationship in which one organism lives in or on its host for part or most of its life cycle; typically harmful to the host organism.
Parathyroid gland One of several small pea-shaped glands embedded within the posterior thyroid tissue; secretes parathormone, which controls the metabolism of calcium and phosphate in the body.
Parenchyma Plant tissue composed of thin-walled cells and well-developed vacuoles; functions in photosynthesis and storage.
Parthenocarpy Development of a fruit without pollination.
Parthenogenesis Development of an egg into an adult without fertilization.
Passive absorption Movement of materials into cells as a result of diffusion only; cell does not expend energy (ATP).
Pathogenic Disease-producing.
Pelagic zone That portion of the ocean comprising the open water of an ocean basin.
Pellicle Thin, transparent covering external to the plasma membrane in many protozoans.
Penetrance Percentage of individuals that actually express a phenotype as determined by a given genotype.
Peptide linkage Chemical bond formed between two amino acids by dehydration synthesis; the carboxyl group of one amino acid is attached to the amine group of the next.
Peristalsis Rhythmical contractions and relaxations of the muscles in the wall of hollow, tubular organs such as the ureter or parts of the digestive tract; serves to move the contents through the tube.
Peritoneum Thin membrane lining the coelom and forming the external covering of the viscera.
Permanent tissue Specialized plant tissue that typically does not undergo division.
Permeable Generally applied to membranes that permit substances to pass through them.
Petiole Stalk that attaches a leaf to a stem.
pH Negative logarithm of hydrogen ion concentration by which the degree of acidity or alkalinity of a solution may be expressed.
Phagocytosis Process of "cell eating" in which projections of cytoplasm engulf solid particles and invaginations of the membrane pinch off and form a vesicle or vacuole that floats freely in the cytoplasm.
Phenotype Physical appearance of an organism.
Pheromone Chemical substance secreted by one organism into the environment that influences the development or behavior of other members of the same species.
Phloem Type of vascular tissue in tracheophyte plant composed of sieve cells that conducts dissolved organic substances from one plant part to another.
Phosphorylation Addition of phosphate to an organic substrate molecule.
Photoperiodism Response of plants to length of the daily period of illumination.
Photophosphorylation Production of ATP under the direct influence of light; transduction of light energy into chemical energy.
Photosynthesis Transduction of solar energy into chemical energy by chlorophyll-containing cells; synthesis of carbohydrates from carbon dioxide and water accompanied by the release of oxygen.
Phytochrome Protein-like receptor pigment found in leaves of green plants that is related to absorption of light.
Pinocytosis Process of "cell drinking" in which the engulfed material consists of a liquid containing small particles that are absorbed on the cell surface.
Placenta Temporary organ of mammals formed partly from the inner lining of the uterus and partly from tissues of the embryo; serves as a connection between mother and embryo during pregnancy and facilitates passage of nutrients and wastes between mother and embryo.
Plankton Free-floating, mainly microscopic aquatic organisms.
Plasma Liquid portion of blood containing the formed elements as well as dissolved substances.
Plasma membrane The outer layer, or living membrane, surrounding the cytoplasm of a cell.
Plasmogamy Mechanism by which two compatible nuclei in a single protoplast are brought together during sexual reproduction; occurs in fungi.
Plasmodesma Thread-like cytoplasmic projection connecting adjacent plant cells.
Plasmolysis Shrinkage of cytoplasm away from the cell wall or cell membrane due to loss of water by osmosis.
Plastid Cytoplasmic, frequently pigmented organelle found only in plant cells.
Platelet A colorless enucleated cell fragment in the circulating blood of all mammals; important in the clotting of blood.
Pollination Deposition of pollen grains on the stigma of the carpel.
Polymorphism Phenomenon by which certain members of

the same species are structurally and physiologically different from other members.

Polypeptide Sequence of amino acids joined together by peptide linkages.

Polysaccharide Carbohydrate consisting of monosaccharides joined together in a long chain by dehydration synthesis.

Pons Portion of the vertebrate hindbrain just above the medulla; center for reflexes mediated by the V, VI, VII, and VIII cranial nerves; conduction pathway between brain and spinal cord.

Population Interbreeding group of organisms of the same species.

Primary cell wall Thin, extensible structure found in all plant cells; forms during early stages of growth; consists mostly of cellulose.

Procaryotic Applied to organisms, such as bacteria and blue-green algae, that lack a membrane-bound nucleus, Golgi apparatus, plastids, and mitochondria.

Producer An organism capable of producing its own food by synthesizing organic compounds—for example, photosynthetic plants.

Promeristem *See* Apical meristem.

Proprioceptor Internal sensory receptor within skeletal muscles, tendons, connective tissue, and joints that are sensitive to pressure; relays to the brain impulses that are related to movements, muscle stretch, and body position.

Prosthetic group Nonprotein constituent of a holoenzyme.

Protein High molecular weight organic compound consisting of one or more polypeptides; each polypeptide is made up of many amino acids joined via peptide linkages.

Pseudocoelomate Animal having a body cavity that is not lined with peritoneum.

Pseudopodium Temporary membrane-bound projection of cytoplasm of a cell; involved in locomotion and feeding.

Pulse Shock wave transmitted through a column of blood in an artery by each cardiac systole and which causes alternate expansion and contraction of an artery.

Pupa Nonmotile, nonfeeding developmental stage between the larva stage and the adult stage in the life history of certain insects.

Recapitulation Tendency for embryos to repeat, in abbreviated fashion, the sequence of stages in the embryonic development of ancestral forms.

Receptor Cell, tissue, or organ containing sensory nerve endings that respond to specific types of internal or external stimuli; transmits stimuli in the form of electrical impulses to the central nervous system.

Recessive Term applied to an allele whose phenotypic effect is prevented by a dominant allele; appears phenotypically only in the homozygous condition.

Recon Smallest portion of a chromosome capable or recombination.

Reflex Functional unit of the nervous system involving a sensory neuron, one or more association neurons, and one or more motor neurons; usually occurs below the level of consciousness.

Regeneration Regrowth of lost or injured tissue or part of an organism.

Regulator gene Gene that produces a repressor substance (protein) that inactivates structural genes.

Repressible enzyme Enzyme that is reduced in amount by the presence of the end product of a metabolic reaction.

Repressor Protein that recognizes operator genes; when associated with an operator gene, inhibits mRNA synthesis.

Reproductive potential The inherent ability of each species to reproduce at theoretical maximum.

Respiration (biological oxidation) Oxidation of organic compounds resulting in the release of chemical energy (ATP).

Retina Innermost layer of nervous tissue within the eyeball; surrounds the vitreous body; and is continuous posteriorly with the optic nerve; contains light-sensitive receptor cells, rods, and cones.

Rhodopsin Substance in the retina of the eye consisting of opsin, a protein, and retinal, a derivative of vitamin A (retinol); capable of undergoing, in the presence of light, a chemical reaction that results in the sensation of sight.

Ribonucleic acid (RNA) Nucleic acid involved in transcribing the genetic code into protein molecular structures; complex, single-stranded molecule composed of chains of phosphate, sugar (ribose), and nitrogenous bases.

Ribosomal RNA (rRNA) Kind of RNA formed by DNA transcription that constitutes the bulk of cellular RNA and occurs in the ribosomes.

Ribosome Small ribonucleoprotein particle found either on the surface of the endoplasmic reticulum or freely in the cytoplasm; site of protein synthesis.

Rod Light-sensitive nerve cell in the retina that is sensitive to very dim light.

Root An organ of most vascular plants; typically subterranean; functions chiefly in anchorage, absorption, and conduction; may also store food.

Root pressure Hydrostatic pressure within a root system due to the inward movement of water by osmosis.

Salt Substance that dissociates in water and releases oppositely charged ions other than H^+ and OH^- ions.

Saprophyte Organism that derives nutrients from nonliving organic matter.

Sclerenchyma Plant tissue consisting of cells with greatly thickened, lignin-impregnated secondary walls; a supporting tissue.

Secondary cell wall Relatively thick structure found only in certain mature plant cells; forms after the primary cell wall; after its formation, the living components of the cell die; consists primarily of cellulose and lignin and provides strength and support.

Seed Ripened ovule; reproductive structure consisting of an immature plant and stored food enclosed in a protective outer covering.

Semiconservative replication Replication of a DNA molecule in which half (one strand) of each parent molecule is conserved in each daughter molecule; the daughter molecule consists of one strand from the parent and one strand synthesized from the surrounding medium.

Sere One of a series of communities within an ecological succession.

Sex chromosome Sex-determining chromosome; number and shape in male differ from those in female.

Sex-influenced gene Autosomal gene that is expressed differently in males and females.

Sex-limited gene Autosomal gene whose effect is limited to one of the sexes.

Sex-linked trait Characteristic brought about by gene(s) carried on sex chromosomes.

Sieve cell Fundamental thick-walled cell in phloem tissue that, at maturity, retains its cytoplasm but loses its nucleus; when attached end to end, sieve cells form a continuous tubular structure, the sieve tube.

Somites Masses of mesoderm, occurring in pairs, in a longitudinal series alongside the embryonic neural tube; give rise to the vertebral column and dorsal muscles.

Speciation Evolutionary process by which a new species is produced; process in which, through geographical isolation over long periods of time, natural selection acts upon variation to bring about reproductive isolation.

Species Group of (actually or potentially) interbreeding organisms that is reproductively isolated from all other groups.

Species-specific behavior Behavior that characteristically develops in a species and is observable when the organism exists freely in its natural surroundings.

Spermatogenesis Origin and development of sperm within the testes.

Spinal cord One of two principal subdivisions of the vertebrate central nervous system; consists of longitudinal bundles of nerve fibers extending from the brain through the vertebral canal.

Spiracles External openings of the respiratory system in some arthropods.

Spore Asexual reproductive cell that is capable of developing into an adult organism without fertilization.

Sporophyte In plants, the diploid, asexual spore-producing stage in the alternation of generations in the life cycle.

Stamen Male reproductive structure of a flower that produces pollen grains.

Statocyst Organ of equilibrium that enables certain invertebrate organisms to maintain balance.

Stele (vascular cylinder) Central cylinder of a root; contains both xylem and phloem elements.

Stem An organ of most vascular plants; typically erect and aerial; functions in the mechanical support of leaves, flowers, and fruits and in the conduction of materials; together with the leaves comprises the shoot.

Sterilization (1) Process by which an object or material is rendered free of living organisms of any kind. (2) Rendering an organism incapable of producing offspring.

Stigma Uppermost portion of the carpel that serves as a receptive surface for the deposition of pollen; photosensitive receptor organelle of certain unicells.

Stimulus Any physical or chemical change, either external or internal, that is capable of eliciting a reaction from a living system.

Stoma Pore in the epidermis of a plant through which gaseous exchange and evaporation of water occur; size of pore is regulated by a pair of guard cells.

Structural gene Gene that specifies a polypeptide chain by transcription; also called a cistron.

Suberin Insoluble, waxy material present in the walls of cork cells and endodermis that makes them waterproof.

Substrate Molecule upon which an enzyme acts.

Succession Orderly process of community change in which one community replaces another in a given area until the climax community for the particular environment is attained; occurs in areas that are newly exposed, such as a newly emerged island, or areas from which the existing vegetation has been removed; proceeds from pioneer to climax communities.

Symbiosis Nutritional relationship in which two organisms live together.

Synapse Region of nerve impulse transfer between the axon of one neuron and the dendrite of the next or between a nerve and a muscle.

Synapsis Pairing of homologous chromosomes that occurs prior to the first meiotic division.

System Group of interacting organs that specializes as a functional unit.

Systole Contraction of the heart chambers (atria and ventricles); interval between first and second heart sounds during which blood is forced into the aorta and pulmonary artery.

Tautomer One of several alternate forms of purines and pyrimidines of DNA and RNA.

Taxis Movement of an organism toward or away from a stimulus; generally referred to in the behavior of simple motile organisms.

Taxonomy Science of classification of organisms.

Test cross Mating of an organism of unknown genotype with another that exhibits the recessive trait in order to ascertain the genotype.

Testis Male gamete-producing organ.

Thalamus Portion of the vertebrate forebrain that serves as intermediary between the cerebrum and all other parts of the nervous system.

Threshold Value at which a stimulus just produces a response.

Thrombocyte Platelet; aids in blood clotting.

Thyroid gland Endocrine gland of vertebrates, located in the neck region, that produces the hormone thyroxin; regulates the rate of metabolism.

Tissue Aggregation of similar cells organized into a structural and functional unit.

Toxin Poisonous substance produced by one organism that usually affects a particular part of another organism.

Tracheid Long, slender xylem cell, tapered at the ends, with well-developed, lignified secondary walls.

Transcription Transfer of genetic information from DNA to mRNA.

Transduction Conversion of energy from one form to another; transfer of genetic material from one bacterial cell to another by means of a bacteriophage.

Transfer RNA (tRNA) Kind of RNA synthesized on DNA templates; composed of about 80 ribonucleotides; serves as an adapter molecule in protein synthesis; combines with specific amino acids and transports them to mRNA; also known as soluble RNA (sRNA).

Transformation Process in which hereditary material released by nonliving bacterial cells (donors) is incorporated into other bacterial cells (recipients).

Translocation Transportation of dissolved organic materials through phloem tissue; transfer of a fragment of one chromosome to a different nonhomologous chromosome.

Transpiration Evaporation of water vapor through stomata of leaves or from other exposed surfaces.

Transplantation Replacement of diseased or injured tissues or organs with natural or artificial ones.

Triploblastic Having three germ layers—ectoderm, mesoderm, and endoderm—during development.

Tropism Involuntary movement or response of an organism; direction is determined by the source of the stimulus; generally referred to in the behavior of plants.

Tuber Enlarged portion of an underground stem (stolon).

Turgor pressure Internal pressure, developed in a plant cell by osmosis, causing the cytoplasm to be forced against the wall.

Ulcer A crater-like lesion in a membrane.

Unit membrane Trilaminar membrane composed of external and internal layers of protein molecules and a middle layer of lipid material; most cellular membranes are so constructed.

Vacuole Membrane-bound inclusion within the cytoplasm of mature plant cells and some unicells; contains water and dissolved substances.

Vasectomy Sterilization procedure in the male in which a portion of each vas deferens is removed.

Vein Kind of blood vessel that collects blood, usually deoxygenated, and returns it to the heart.

Ventricle Chamber of the heart that receives blood from an atrium and pumps it out of the heart; cavity in an organ such as the brain.

Vertebra One of the bones comprising the spinal column in higher vertebrates.

Vessel Series of xylem elements stacked end to end to form a cylindrical tube.

Villus One of many finger-like projections from the intestinal wall that increases surface area and facilitates absorption of nutrients; contains blood and lymph vessels for transportation of absorbed materials.

Virulence Disease-producing capacity of an organism.

Vitamin Chemical substance, organic in nature, ingested in minute quantities and necessary for normal metabolism.

Xenograft Transplantation between animals of different species.

X-linked gene Gene in a portion of the X chromosome that has no allele in the Y chromosome.

Xylem Tissue in tracheophytes that transports water and dissolved salts; consists of tracheids, fibers, and vessels; provides support; commonly referred to as wood.

Y-linked gene Gene in a portion of the Y chromosome that has no allele in the X chromosome.

Yolk Stored food material in eggs; consists of proteins, phospholipids, and neutral fats.

Zygote Diploid cell resulting from the union of gametes.

Index

Figures are represented by numbers in **boldface** type, tables by the letter t after the page number.

A

A band, of muscle, 165, **166**
Abiogenesis, 52–54, **54**
  Redi's experiment, 53–54, **54**
Abiotic environment, 561–581
  changes in, 582–583
  importance of, 562
Abortion, birth control, 665
Abscess, 380
Abscission
  auxin gradient, 235
  layer, 235
Absorption
  in gastrointestinal tract 333–341, **334–337,** 338t, **339,** 341t
  in root, 322–323, **323**
  spectrum, 133, **133**
Absorptive cells, 217
Acceptors, 137, 139
Accommodation, in vision, 292
Acer rubrum, 555
Acer saccharum, 555
Acetabularia crenulata, 107, **108**
  development, 477
Acetabularia mediterranea, 107, **108**
Acetaldehyde, 146, 147
Acetylcholine, in neuron, 163, 257–258
Acetyl CoA
  in fatty acid metabolism, 153–154, **153,** 170–173, **171–172**
  in carbohydrate metabolism, 147–149
Achilles reflex, 266–267
Acid, 24
Acid [*cont.*]
  Arrhenius, 24, **25**
  Brønsted-Lowry, 24, **25,** 36, 37
  organic, 30, **30**
  strong-weak, 26
Acid-base balance, 28–29, 412
Acoelomata, 76
  gaseous exchange in, 394–395
  structure, 766, **767**
Acrasin, 760
Acrasiomycota, 760, **761**
Acromegaly, 255, **256**
Acrosome, 444, **445,** 461
ACTH, 251, 254
  hormones, 254–255, **255**
Actin, 165, 167
Actinomycota, 749, **750**
Action current, 165
Activation energy, 16, **16,** 17, **17**
  glucose, 142–143
Activators, 46, 196
Active site, 45–46, 197
Active transport, 97–98, **97,** 159, 164
  in erythrocyte, 97
  of potassium ions, 97
Actomyosin, 126
  in muscle contraction, 165–166
Acupuncture, 287
Adaptation
  analogy, 75, **75**
  of behavior, 712–722
  of chemical responses, 714
  in development, 483
  in evolution, 531, 555–556, **556**
  of habitat, 519
  individual, 712–713
  natural selection and, 541, 543–544, **545**
  of plants, 561, 713–714
Adaptation [*cont.*]
  patterns of, 555–557, **556**
Adapter molecules, 187
Adaptive radiation, 556–557, **556**
Addison's disease, 253–254
Adenine, 179, 183
Adenohypophysis, **249,** 254–255
Adenosine diphosphate. *See* ADP; *also* ATP
Adenosine monophosphate. *See* AMP
Adenosine triphosphate. *See* ATP
Adipose tissue, 222t
ADP, 135
Adrenal gland, 248, **249,** 251t, 252–254
  cortex, **249,** 251t, 252
  functions, 252–254
  hormones, 251t, 253–254
  medulla, 251t, 252
  reptile, 252–253
  structure, 249, 253
Adrenaline, 251t, 253, **258**
  physiological responses, 253
Adrenocorticotropin. *See* ACTH
Advertising mimicry, 735
Aerobic respiration, 146, 147–151
  cytochrome system, 150–151, **150**
  in evolution, 62, 64
  fermentation and, compared, 151
  See also Respiration
Afferent neuron, 267
  See also Sensory neuron
African peach-faced love bird, 696–697, **697**
Agaricus, 763
Age distribution pyramids, 585, **585**
Aggression, 731–732, **732**
Aging, 484–485
  free radical theory, 484
  immunological theory, 485

Agnotoza, 764, **765**
Agranoff, B., 710
Agranular reticulum, 113, **113**
Agriculture
dust bowl, 623, **625**
in ecosystems, 622–627, **625**
insects and, 623
minerals and, 627
pollution and, 676, **676,** 680, **680**
Air
pollution in, 676–677, **676,** 681–682
in soil, 564–565
Alamine, 173
Albinism, 303, 518
Alcohol, 30, 30t, 343–345
absorption of, 344
fermentation, 146, **147**
and human behavior, 739
physiological effects, 344–345
Aleurone grains, 119
Alga(e)
blue-green, 749–751, **750**
brown, 755, **756**
disease-producing, 374, 374t
green, 755, **756**
in intertidal zone, 634, **635**
in mutualism, 594
in nitrogen cycle, 604–606, **605**
in pelagic zone, 635, 637
red, 755, **756**
and world food supply, 661
Alkaptonuria, 181–182, 524–525
Allantois, 470, **470**
Allele
in gene pool, 546, **547**
linkage, 505
multiple, 497, 521–524
Allergy, response, 386
Allium, **122, 123**
Allograft, 389, **389**
Allometric growth, 482, **483**
Allomyces macrogymus, 760, 762, **762**
Allosteric interaction, 195, **195**
Alpha cells, 248, **249,** 251t
Alternation of generations, 429–430, **429**
in Bryophytes, 432–433, **433,** 757, **758**
in Eumetozoa, 764
in ferns, 433–434, **434**
in Plantae, 430–433, **432**
in *Ulva,* 432, **432**
Alveoli, 400, **400**
Amine, 245
Amino acids, 30, 36–40, **37,** 38–40t
activation, 187, **187**
biosynthesis, 171–173
Amino acids [*cont.*]
catabolism, 154–156, **156**
coding, 190–191, 193, **194**
decarboxylation, 173
in digestion, 338t
essential, 173
in Kreb's cycle, 173
structure, 36–37, **37**
Ammonia
in excretion, 407
in nitrogen cycle, 604–606, **605**
in urea cycle, 155
Amniocentesis, 527
Amnion, 470, **470**
Amoeba, 96, 98, 107, 753, **754**
coverings, 299
digestion, 324–325, **326,** 350
excretion, 407
gaseous exchange, 394
movement, 168–169, **169,** 313
Amoebocyte, 350, 355, 356
AMP
allosteric reaction, 195
amino acid activation, 187, **187**
cyclic, 50, 248
hormone action, 246
Amphetamines, 296t, 297
Amphibians, 301
adrenal glands, 252
behavior, 702
brain, 264
embryo development, 464–465
respiratory system, 396–397
skin, 301, 302
stomach, 334
Amphioxus, 463, **463**
Amylopectin, 33
Amylose, 33
Anabolic reaction, 18–19
Anabolism, 130, **130**
Anaerobic respiration, 146–147
glycolysis, 142–145, 144
Anal pore, 326, **326,** 407, **408,** 754
Analogy, 75, **75**
Anaphase
in meiosis, **426,** 427
in mitosis, **120–123,** 124–125
Androgen, 248, 251t, 253
Aneuploid, 509–510
Angiosperm, 204, **531,** 759
flowering, 238–239
leaf, 393
life cycle, 434–440, **438–439**
Angiotensin, 251, 254
Angstrom, 88
Animalia, 80–81, 762–772
budding in, 420, **422**
Annelids (Annelida)
circulatory system, 356–357, **356**
coverings, 299, **300**
cuticle, 299, **300**
digestive system, 328–329, **330**
nervous system, 262–263
respiratory system, 394–395
Annual rings, 214
Anodonta (freshwater clam), **769**
Anopheles, 753
Ant, pheromone, 720, **720**
Antennae
in chemoreception, 716–717, **716**
crayfish, 265
in hearing, 718, **719**
Antheridium
in Bryophyta, 432, **433**
in Ferns, 434, **434**
in *Oedogonium,* 430, **431**
Anthoceros, structure, **758**
Anthocyanin, 119, 495
Antibiotic, 376–377
myostatin, 377t
penicillin, 377t
selection of, 376–377
streptomycin, 377t
tetracyclines, 377t
Antibody, 340, 359, 377t, 381–382, **383–384,** 485, 533
antibody reaction, 382–385, **383–384**
antigens, 381–382, **383,** 385–391
antitoxins, 382, **383**
DNA, 384
lysins, 382–383, **383**
opsonins, 382–384, **383**
in plasma, 357–358, 358t, 369
precipitin, 382, **383**
therapy, 376–377
theories of formation, 384–385, **384**
Anticodon, 186–187, **186, 191**
Antidiuretic hormone, 256
Antigen, 357, 381–383, **383,** 533
antibody reaction, 385–391
types, 382
Antipodals, 436, **437, 438**
Antitoxin, 382, **383**
Antlers, 302
Aorta, 355, 356, 361–363, **362, 365, 366**
Apoenzyme, 46, **46**
Apical dominance, 235–236, **238**
Apical meristem, 204, **209**
Arachnoidiscus, **753**
Arachnoid, 269, **269**

Arabidopsis thaliana, **106**
Archegonium
in Bryophytes, 432, **433**
in Ferns, 434, **434**
Aristotle, 52
Arm
bones, **305,** 307
muscles, 315, **315,** 317t
Arteriole, structure, 363
Arteriosclerosis, 372
Artery, 363
aorta, 355, 356, 361–363, **362, 365, 366**
principal, **365**
pulmonary, 364, 365, **362, 365, 366**
renal, 410, **411**
structure, 363
Arthropod, 768, **771**
behavior, 695
chemoreceptors, 283, 716
coverings, 299, **300, 301**
cuticle, 299, **301**
gaseous exchange, 395–396, **396**
movement, 313
nervous system, 262, 264, **265**
structure, 303, 770, **771**
Articulations. *See* Joints
Ascaris, 766, **768**
Aschelmenthes, 766, **768**
Ascomycota, 762–763, **763**
Ascorbic acid. *See* Vitamin C
Ascospore, 763, **763**
Ascus, 762, **763**
Asepsis, 374
Asexual reproduction, 420–422, **421–424**
budding, 420, **422**
fission, 420, **421**
in Plantae, 421–422, **424**
regeneration, 422, **425**
sexual compared, 428–429
Aspartic acid, 173
Association neurons, 161
Aster, in mitosis, 120–123, 124
Astigmatism, 292, 293
Atmosphere, 573–581
ancient, 57–58
in carbon cycle, 603–604, **603**
carbon dioxide in, 574–576
composition, 24, 403, 573–574, 576
in nitrogen cycle, 604–606, **605**
oxygen in, 24
pressure, in respiration, 401–402
in water cycle, 574, 576, 606, 607
wind, 576–579, 576, **577**
Atom, 6, 8–10, **9**
electronic configuration, 9–10, 10t
Atom [*cont.*]
half-life, 536, 536t
radioactive dating, 536
size, 8
structure, 8–9
Atomic number, 9
ATP (adenosine triphosphate) 48–49, **49,** 62–63, 114, 131t, 135, 137, 139, 140, 142, **143,** 169, 170
in active transport, 97–98
in alcoholic fermentation, 146–147
in amino acid activation, 187, **187**
in anabolic reactions, 19
in bioelectricity, 160
in cellular bioenergetics, 173
cycle, 143, **143**
in cytochrome system, 150–151, **150**
in energy transduction, 159
in enzyme activation, 46, 195
in evolution, 62–63
in fatty acid catabolism, 153–154, **153**
in fat synthesis, 171, **172**
in fermentation, 146–147
in glycolysis, net yield, 151
in Kreb's cycle, 148–150, **149**
in metabolism, 143–146, **144,** 150, 171
in movement, 167–168
in muscle contraction, 167–168
in nerve fiber, 162, 167
in photophosphorylation, 135, 137, 139
in photosynthesis, 140–142, **141**
in polypeptide synthesis, 191
structure, 48–50, **49**
Atrium of heart, 360–361, **360**
of fish, 361–362, **362**
of frog, 362, **362**
Auditory canal, 287, **288**
Auricles, *Planaria,* 262, **263**
Australian region, 647, **648**
Autonomic movement, in plants, 242, **242**
Autonomic nervous system, 266–267, 276–278, **277,** 302
functions, 276
organization, 276
parasympathetic, 276, **277**
sympathetic, 276, **277**
Autosome, 501, **501**
Autotrophs, 64, 77, 322
bacteria, 77–79
in carbon cycle, 603–604, **603**
in classification, 77
in ecosystem, 599
nutrition, 322–323
Protistans, 751–752
Auxins, 236, 237, 714
gradient, 235
in apical bud dominance, 235–236
in fruit drop, 235–236
influences of, 235
plant movement, 242–245, **242–244**
Avena sativa, 233, **233**
Avena test, 234
Avery, O., 175–176, **176**
Axillary bud, **215**
Axon, 98–101, **100,** 161, **161,** 228, **228,** 267
Azalea, 510, **511**
Azotobacter, **605,** 606

B

Bacillus, 749, **750**
Bacteria, 158, 374, 374t, 749, **750**
Actinomycota, 749, **750**
antibiotic therapy, 376–377, 377t
asexual reproduction, 420
bioluminescence, 158–159
denitrifying, 604–606, **605**
disease-producing, 373, 374, 374t
endospores, 420
Eubacteriae, 749, **750**
in evolution, 365
forms, 749, **750**
in genetic experiments, 176–177, **177**
genetic transformation, 175–176, **176**
Myxobacteriae, 749, **750**
nitrifying, 604–606, **605**
Spirochaetae, 749, **750**
sterilization, 374–375
Bacteriophage, 176–177, **177**
in genetic experiments, 176–177, **177**
in genetic recombination, 430
structure, **176**
Bacteriovirdin, 132
Balanus, 593–594, **594**
Balbiani rings, 478
Baldness, 524
Barbiturates, 295–297, 296t
Bark, 214
Basal body, 110, **111**
Basal ganglia, 270, **271**
Basal plate, 110, **111**
Base, 24, **25**
Brønsted-Lowry, 24, 37
weak, 26, 37
Basement membrane, 217, 219t
Basidiomycota, 763, **763**
Basophil, 225t

Bat, 702
Bates, H., 735
Baur, H., 477–478
Beadle, G. 182
Beadle-Tatum hypothesis, 182–183, **183**
  exceptions to, 193–194
Bear
  carnivorian lethargy, 702
  polar, **722**
Beavers, dam construction, 569
Bee. *See* Honeybee
Beerman, W., 478
Beggiatoa alba, **750**
Behavior
  adaptation, 712–713
  advertising mimicry, 735
  aggression, 731–732
  agnostic, 731
  amphibian, 702
  appetitive, 697
  avoidance, 734–736, **735, 736**
  body build and, 738–739
  brain, 740
  breathing, 716
  chemoreception, 714, 716–717
  communication, 723–724, **725**
  concealing coloration, 734–735, **734**
  conditioned response, 705–707, **706**
  courtship, 726–727, **727**
  cyclic, 699–705, **701,** 703–704
  deceptive mimicry, 735–736, **736**
  dominance, 729–730, 732–733
  drinking, 715–716
  and electricity, 719–720
  estivation, 703
  ethological approach, 691–692
  evolution of, 721–722
  family, **725,** 728–729
  feeding patterns, 715
  group, 729–730, **729**
  habituation, 705
  Hawthorne effect, 689–690
  hibernation, 702–703
  homing, 705
  hormones, effect of, 720–721, **720,** 741
  human, 737–746
    factors modifying, 739–740
    heredity, effect of, 738–739
    levels of social, 742–745
    personality, 744
    postnatal influences, 739
    prenatal influences, 739
    social, 741–745
  imprinting, 707, **707**
  individual, 715–721
Behavior [*cont.*]
  insight, 707–708, **708**
  instinctive, 691
  kinesis, 694
  learned, 705–711, **708**
  LSD and, 740
  mating, 726–728, **727**
  methods of study, 689–690
  neurophysiological approach, 962
  plants, 713–714
  plankton, 699–700
  problems in studying, 692–693
  programmed learning, 690–691
  psychological approach, 690–691
  radiation effects on, 739
  release mechanism, 733–734, **734**
  reproductive, 702, 703
  seasonal cycles, 702–705, **703, 704**
  stickleback, 697–702, **698, 702**
  simple, 693–695
  sleep, 721
  social, 723, 725–733
  sound, effect of, 718–719
  species-specific, 691–692, 695–699,
      696–699
  stimuli, 690–693
  taxis, 694
  teleological theory, 690
  territoriality, 730–731
  tropism, 694
  vision, effect of, 717–718
Belling, J., 510
Benthonic zone, 571, **571**
Benthos zone, 571, **571**
  consumers, **635,** 636
Benzor, S., 514
Beta cells, 248, 251t
Beta oxidation, 153–154, **153,** 156, **156;**
      see also Lipid metabolism
Bicarbonate ion, 359
  in respiration, 402
Bilateral symmetry, 76, 764, 766
  in Eumetazoa, 764, 766
Bile, 337–338, 338t
  duct, 337, **337**
  pigments, 337
Binding sites, hormones, 246
Biochemical cycle, 603
  calcium, 608–609, **608**
  carbon, 603–604, **603**
  energy in, 602–603
  gaseous, 603–606, **603, 605**
  mineral, 606–609, **608**
  nitrogen, 604–606, **605**
  oxygen, 603–604, **603**
Biochemical cycle [*cont.*]
  pesticides in, 609
  water, 606, **607**
Biochemical oxygen demand (BOD), 674
Bioelectricity, 160–164
  electrical potential, 160, **162**
  resting potential, 162, **162**
Biogenesis, 54–55
Biogeographical regions, 647–649
  Australian, 647, **648**
  Neotropic, 647–648, **648**
  World Continent, 648–649, **648**
Biological oxidation, 146, **146**
Bioluminescence, 158–160, **159**
  adaptive nature, 158
  organisms, 158–160
Biomass, 600–601
  agriculture, 622
Biome, 580, 640–646, **641–646**
  deciduous forest, 641–642, **643**
  desert, 643, **644,** 645
  grassland, 642–643
  indicator species, 616
  mountain zonation, 646, **646**
  in succession, 619
  taiga, 641, **642**
  tropical rain forest, 645, **645,** 646
  tundra, 640–641, **641**
Biosphere, 7, 561
  human population, effect of, 650, 669
  oxygen in, 169–170
  pollution, effects of, 673–684
Biosynthesis, 169–173
  amino acids, 171–173
  fat, 170–171, **172,** 173
  polysaccharide, 170, **171**
Biotic environment, 561, 582
Biotic potential, 583, **587**
Biotin, 172, **172**
Bird
  behavior, 696
  in calcium cycle, 608, **608**
  communication, 724
  heart of, **362,** 363
  hormones, 721
  sound, 719
  territoriality, 730
Birth
  human, 473–475, **475**
  labor, 473–474
  stages of, 473–474
Birth control, human, 453
  abortion, 665
  chemical methods, 667–668t
  condom, 663, 667t

Birth control [*cont.*]
contraceptive methods, 663–665, **664**
corrective methods, 665
douche, 668
education and, 665, 686
intrauterine device, 663–665, **665,** 666–667t
methods of contraception, 662–668, 666–667t
oral contraception, 666–668t
preventive methods, 662–663
rhythm method, 667t
tubal ligation, 663, **663**
vasectomy, 662, **663**
Birth rate
in ecosystem, 584
population control and, 662
Biston betularia, 544, **545**
Bladder, urinary, **410,** 411, **444,** 446
Blakeslee, A., 510
Blastocoel, 467, **468**
Blastocyst, development, 467, 468
Blastomer, 461
Blastopore, **463,** 464, 479
Blastula, formation, 461–462, **461, 462**
Blind spot, 291–292
Blood, 224, 232
antigen, 357, 381–383, **383, 533**
agglutination, 357–358t
buffering action, 359
cell, 22, 357–358
circulation, 364–366, **365**
clotting, 257, 358–359
components, 224, 225–226t, 357–359, **358,** 358t, 382
composition, 224, 225–226t, 357–359, **358,** 358t
earthworm, 356–357, **356**
erythrocytes, 160, 224, 225t, 357–358
function, 224
fibrinogen, 381
groups, antigens and antibodies, 521–522, **522**
hormones in, 359
ions in, 224
invertebrate, 355–357
infection, 379
leukocytes, 224, 225t, 357, 358
leucoplasts, 117
mollusk, 356
plasma, 357, 382
platelets, 357, 382
portal, 364–366, **365**
pressure, 367–368
diastolic, 367–368
Blood [*cont.*]
measurement, 367–368
systolic, 367
proteins, 172
Rh factor, 357, 521–522, **522**
systemic circulation, 361, 364
types, 357, 358t
vertebrate, 357–359, **358,** 358t
vessels, 363
arteriole, 363
artery, 363, 364, **365**
capillary, 270
of heart, 364, **365**
pressure, 367–368
of small intestine, 336–337, **336**
vein, 363, 364, 365, 370, 412
Body build, behavior and, 738–739
Body cavity. *See* Coelom
Body forms, 76
symmetry, 764–765
Bog, 569
Bond, chemical. *See* Molecule
Bone, 220–224, 223t
of arm, **305,** 307
articulation, 307–308, 309–311t
calcium control, 252
cancellous, 221
cell, 220–221, **221,** 223t
density, 221
of ear, 288, **288**
formation, 309
fracture, 308, 311, **312**
of head, 304, **306**
growth, 308–309
human, 304–313
of leg, **305,** 307
repair, 308, 311, **312,** 313
skeleton, 304–305, 306
structures, 308
sutures, 304
type, 221, 304
vertebrate, 305, **306**
Bonnemaisonia, **756**
Book lungs, of spider, **395,** 396
Bordered pits, 101, **103**
Bowman's capsule, 411–412, **411**
Boycott, B., 709
Boysen-Jensen, P., 233–234
Brachial plexus, 276, 276t
Brain, **267,** 269–270
of amphibian, 264
behavior and, 740
cerebral cortex, 291
cerebrum, 270, **270, 271,** 272
cerebellum, **270, 271,** 272
Brain [*cont.*]
cortex, 272
dura mater, 269, **269**
of fish, 264, **266**
fissures, 270, 271
forebrain, 270–272, **270, 271**
of frog, **266**
function of areas, 264
grey matter, 270, 272, **273,** 278
hemispheres, 270, **271,** 272
hindbrain, 264, 266, 270, 271–273
horse, 267
hypothalamus, 264, **270,** 271–272
medulla oblongata, 273
midbrain, 264, **266,** 272
motor areas, 271, **272**
pons, **270,** 273
primitive, 264
reptile, **266**
sensory areas, 271, **272**
simple, 261–262
in sleep, 721
spinal cord relationship, **269**
structure, 269–273, **269–271**
subdivisions, 270, **270**
types of fibers, 273, **273**
ventricles, 273
waves, 281, **282**
white matter, 270, **270**
Breathing, 392–403
behavior, 716
carbon dioxide in, 393
expiration, 401–402, **401**
inspiration, 401–402, **401**
in man, 401–402, **401**
mechanisms, 401–402, **401**
transportation of gases, 402
See also Gaseous exchange
Bridges, C., 502
Briggs, R., 474
Bronchiole, 400, **400**
Bronchus, 399–400, **399, 400**
Brook, 570
Brown, R., 87
Brownian movement, 87
Bryophyta, 757, **758**
life cycle, 432–433, **432**
protonema, 432, **433**
Bud, 212, **212**
apical dominance, 235–236, **236**
axillary, 212, **212,** 215
lateral, 212, **212**
scale, 212, **212**
scars, 212, **212**
terminal, 212, **212**

Budding, 420, **422**
in fungi, 420, **422**
in hydra, 420, **422**
Buffer, 32, 359
inorganic, 27–28
protein, 37

C

Calcium, 346t
in blood clotting, 358
in carbon cycle, 603–604, **603**
cycle, 608–609, **608**
and DDT, 626
metabolism, 252
Calcium carbonate, 303
control, 252
in muscle contraction, 166
Calcium oxalate, 119
Callophyllis, **756**
Calviceps purpurea, **762–763**
Calvin, M., 139
Calyx, 435, **435**
Cambium, 204
cork, 205, 211, **214,** 459
development, 213, **213**
procambium, 458
vascular, 204–205, 210, **210,** 213, 459
Cancer, lung, 405–407, **406**
Canciliculi, **221,** 222
Cannabis sativa, 295, 296t
Capillary, 270
and circulatory system, 363
in inflammation, 379
in lymphatics, 368, **369**
of small intestine, 336–336, **336**
Capillary water, 564
Carbaminohemoglobin, 357
Carbohydrate, 30–31
classification, 31t
composition, 31, **32**
dehydration synthesis, 32, **33**
digestion, 338t
metabolism, 151–152
synthesis, 32, **33**
Carbon, 7, 29, 139
bonding, 29
cycle, 603–604, **603**
in living matter, 28t
in organic compounds 29–30, **29**
skeleton, 29
structure, 29, **29**
Carbon dioxide, 13, **13,** 130–131, 139, 168
in atomosphere, 24, 574–576
Carbon dioxide [*cont.*]
in biosynthesis, 169, 170
in blood, 359, 363
in breathing, 393
in carbon cycle, 603–604, **603**
in chemoreception, 696
in control of respiration, 402
in evolution, 64
in fat synthesis, **172**
in fermentation, 146–147, **147**
fixation, 140
in guard cells, 393–394
in hibernation, 702–703
in human excretion, 409–410
in Kreb's cycle, 148–150
limiting factor, 614
in metabolism, 344
in pentose phosphate shunt, 151–152, **152**
in photosynthesis, 130–132, **132,** 137
in respiration, 398, 402, 403
in resuscitation, 403
on planets, 68
in plasma, 359
transportation, 395–396
in veins, 363–364
Cardiac cycle, 361
Cardiac muscle, 224, 226, **227,** 333, **334,** 360–361, **360**
Carnivore, 323, **600**
Carotenes, 132, 134, 303
Carotenoid, 117
Carple, 305, 307, 435, **435**
Carrying capacity, 586, 610
Cartilage, 220, 223t
elastic, 223t
fibrous, 223t
hyaline, 223t
joint, 307, 309t
types, 220
Catabolism, 18, 130, **130**
fatty acid, 153–154, **153**
Catalyst, 17–18
Cecropia moth, 246–247, **246, 247**
Cecum, 339–340, **339**
Cell, 6, **91**
absorptive, 217
alpha, 248, 249, 251t
basic portions, 90
beta, 248, 251t
blastomere, 461
blood, 357–358
body, 161
bone, 220–221, **221,** 223t
centriole, 110, **110,** 124
Cell [*cont.*]
chlorenchyma, 215, **215**
chloroplast, **90, 114, 116,** 117, 137, 139, **159**
chromoplast, 117
cilia, 110–111, **111,** 168–169, **169,** 219t
in classification, 75–76
coenocytic, 107, 125
collenchyma, 206, **206**
columnar, 217, 218t
companion, 208, **208**
concept, 85, 87
cone, 98, 291, **291**
cork, 206, 207
cuboidal, 217, 218t
cytoplasm, 105–107, **106**
definition, 88
differentiation, 455, 459
diploid, 423
discovery, 85, 87
division, 120–128, **120–123,** 126–127, 455, 459
chromosomes, 110, 124–127
cytokinesis, **121–123,** 125
duration of stages, 125–126, 126t, 127
influence of cytokinins, 237–238
exceptions to typical pattern, 128
significance, 128
See also Mitosis.
endoplasmic reticulum, 107, 111, **112,** 113, 117
enlargement, 459
erythrocytes, 160, 224, 225t
eucaryotic, 65–66, 75
extracellular materials, 101–104, **102, 103, 104**
follicles, 446–448, 447
fat, **230**
flagellum, 110, **111**
flame, 408, **408**
germ, 459, **460**
goblet, 217, 219t
Golgi apparatus, 75, **91,** 113–114, **113,** 125
gonial, 459–460
growth duration, 125–126, 126t
guard, 206, **206**
haploid, 423
inclusions, **91,** 118–119
intercellular substances, 104–105
interstitial, 368, 443
leucoplast, 117
leukocytes, 224, 225, 357, 358, 368
lumen, 102

Cell [*cont.*]
lymphocytes, 385
lysosome, 117, **118**
mast, 220, 258
membrane, 90, **91, 92,** 111
in active transport, 97, **97**
functions of, 93–98
modifications in, 98–101, **99, 100**
in osmosis, 95–96, **95**
nuclear, 107, **109**
in phagocytosis, 98, **98**
in pinocytosis, 98, **98**
structure, 90–93, **91, 92**
See also Membrane.
meiosis, 424, 427–429, **426**
microtubule, 118, **120–123**
mitochondria, **91,** 114, **115**
mitosis, 120–128, **120–123,** 126t
nucleolus, **91,** 107, **109,** 110
nucleus, **91,** 107, **109**
organelles, 107–118
phagocytic, 98
phloem, 208, 210, 213, **213,** 214, **232**
plasma, 220, 224, 357, 359, 368
plate, **122, 123,** 125
primary wall, 101, **102**
procaryotic, 65, 75–76, 107, 420
procell, 61–62, **62**
processes, 228, **228**
reproduction. *See* Mitosis; Ovum; Sperm
rod, 98, **100,** 291, **291,** 294
sap, 119
Schwann, 98, **100,** 101, **228,** 229
secondary wall, 101, **102**
sieve, 208, **208**
specialization, 204
sperm, 444–445, **445**
squamous, 217, 218t
structure, 88–128
surface, 90–104
tetraploid, 510, **511**
theory, 56–57
vacuole, **91,** 119
wall, 101, **102, 103,** 104
cellulose in, 101, 104, **104**
middle lamella, 101, **102, 103**
organization, 104, **104**
pits, 101, 103
primary, 101, **102**
secondary, 101, **102**
Cellular metabolites, 197–198, **198**
Cellulose, 32, 33
in cell wall, 101, 104, **104**
in ecosystems, 602
Cellulose [*cont.*]
in slime molds, 760
Centipedes, 158
Central nervous system, 264, 269–276, **269–274,** 275t
Centriole, 110, **110,** 124
function of, 110
Centromere, **120–123,** 214, 427, 523
Centipede, 771
Cephalization, 261
Cephalochordata, 770, **772**
Cerebellar, hemisphere, **270, 271,** 272
Cerebellum, **270, 271,** 272–273
functions of, 272–273
Cerebral
cortex, 291
hemispheres, 272
tracts, 272
palsy, 279–280
Cerebrospinal fluid, 269
Cerebrum, 270, **270,** 271, 272
Cervical plexus, 276, 276t
Changeux, J., 196
Chailakhian, 239–240, **239**
Charophyta, 755, 757, 758
Chase, M., 178
Chemical
bonding. *See* Molecule
control, 231–258
in animals, 245–258
in plants, 232–245
disinfectants, 375–376, 376t
elements, 7–8, 7t
in earth's crust, 8t
in living systems, 7–8, 8t, 21–50
energy, 4
equilibrium, 19–20, **20**
evolution, 51–52, 64
reactions, 14
anabolic, 18
in blood, 357–359
catabolic, 18–19
collision theory, 16
deamination, 154–155
decarboxylation, 148, **149,** 154–155
decomposition, 15
dehydration, 143, 148, **149**
displacement, 15
double displacement, 15
energy in, 18–19, **17,** 18
of eye, 294, **294**
factors influencing, 17
rate, 17–18
rearrangement, 14–15
receptors, 283–284, **284,** 285
Chemical [*cont.*]
responses, 714
reversible, 19–20
synthetic, 15, 18–19
types, 14–19
symbols, 7, 8t
Chemoreception, 283–285, **284, 285**
in arthropods, 283, 716
behavior, 716–717
in housefly, 283
in parasites, 717
pH in, 717
setae in, 716
Chemosynthesis. *See* Evolution
Chiasma, 427, **427**
Chickens
Andalusian, 494
embryo, 465–466, **466**
inheritance in, 494
pecking order in, 729, **729**
sex-limited traits in, 505, **505**
Chilopod, **771**
Chimpanzee, 707–708, **707**
Chitin, 32, 34
of arthropods, 299, 301
Chlamydomonas, 11, 755, **757**
chromosome number, 423
isogamy in, 430, **431**
life cycle, 429–431, **431**
Chlorella, 139
Chlorencyma, 215, **215**
Chlorine, 346t
Chlorophyll, 132–134, **132, 134,** 135, 137, **137,** 139, 173
a, 132, 134, 137, **137,** 139, 755
b, 132, 751
c, 132
d, 132, 755
e, 132
in light reactions, 134, 135, 137, 139
in photosynthesis, 134, 135, 137, 139
in Protista, 751
Chlorophyta, 81, 755, **757**
Chloroplast, **90, 91,** 114, **116,** 117, 137, 139, **159**
in energy transduction, **159**
granum lamella, **116,** 117
stroma, **116,** 117
Choanocyte, 326, **327,** 355
Cholecystokinin, **250,** 251t
pancreozymin, 248
Cholesterol, 35t, 359
in arteriosclerosis, 372
Chondroitin sulfate, 104–105
Chondrocytes, 220

Chordata, characteristics of, 770, 772
Chorid, 290, **291**
Chorion, 468, 470, **470**
Chorionic gonadotropin, 257
Chromatid, **120–123,** 124, 427, **427**
  formation, 506, **507**
Chromatin, 107, 109
Chromophore, 303
Chromoplast, 117
  in diatoms, 751–752
Chromosome, 110, **511**
  aberrations, 509–513
  alteration, 509–513, **514**
  abnormalities, 525–527
  aneuploids, 509–510
  autosomes, 501, **501**
  centromere, 120–123, 124
  chromatids, **120–123,** 124, 427, **427**
  chromatin, 107, 109
  crossing over, 427, **427**
  deletion, 513, **514**
  in Down's syndrome, 525–526, **526**
  duplication, 513, 514
  giant, 477–478, 508–509, **508**
  homologous, 426, 488
  human, **423,** 525–526
  linkage, 501–502, **501, 503**
  mapping, 175, 507–509, **508**
  in mitosis, 119, **120–123,** 124–127
  movement, 126–127
  nondisjunction, 502–504
  polyploidy, 510–513, **511**
  puffs, 478
  sex, 501, **501**
  sex linkage, 501–502, **502**
  in sexual reproduction, 423–424
  shape, 127
  synapse, 427, **427**
  theory of heredity, 499–507
  translocation, 513, **514**
  Turner's syndrome, 526, **527**
  X chromosome, 500–505, 509, 522–523, 526, **526,** 527, **527**
  Y chromosome, 500–505, 509, 522–523, **523**
Chrysophyta, 751, **753**
Chthalamus, 593–594, **594**
Chyme, 340
Chytridiomycota, 760, 762, **762**
Cilia, 110–111, **111,** 168–169, **169,** 261, **261**
  movement, 168–169, **169**
Ciliated cell, 219t
Ciliates
  covering, 299
Ciliates [*cont.*]
  nervous control, 261
Ciliophora, **754,** 755
  nervous control, 260
Circadian rhythm, 699–702, **701**
Circulatory system
  of annelids, 356–357, **356**
  closed, 350, 356–357, **356**
  components, 357–366
  disorders, 371–372
  of earthworm, 356–357, **356**
  fetal, 366–367, **366**
  of frog, 362, **362**
  hydra, 355
  and lymphatics, 368–369, **369**
  of man, 356–371
  of mollusk, 356
  open, 350, 355–356, **355**
  paramecium, 354
  physiology, 367–368
  placenta in, 366, **366**
  plant mechanisms, 350–354, **351, 352, 354**
  portal, 364, 366, **366**
  prenatal, 366, **366**
  pulmonary, 360–364
  routes, 363–367, **365, 366**
  systemic, 361, 364
  of unicells, 354
  of vertebrates, 357–369
Cirrhosis, 345
Cisterna, 111, **113**
Citric acid, 148, **149**
Cistron, 514
Clam, 301
  respiratory system, 395–396, **395**
  shell, 299
  structure, 768, **769**
Classification
  analogy, 75, **75**
  body cavity in, 76–77, **78**
  cellularity in, 76
  embryonic layers in, 77
  homology, 75, 532
  levels, 74
  natural, 73–74
  nucleus in, 75–76
  nutritional patterns in, 77–79
  of organisms, 73–81
  scheme, **80**
  systems, artificial, 73
  symmetry in, 76, **77**
  taxonomic criteria, 74–81
  types, 73–74
  units, 74
Claviceps purpurea, **762**
Cleavage, 461–462, **462**
  determinate, 461–462
  indeterminate, 461–462
  relationship of yolk, 463, **463**
Clever, U., 478
Climate, 579–581, **579, 580,** 581t
  changes in, 619
  in ecosystems, 637
  in tolerance levels, 614
  in pollution, 672–673
Climax, ecological, 617
Clone, in antibody formation, 385
Clostridium, 605, 606
Club fungi (Basidiomycota), 763, **763**
Club moss, **758,** 759
Cnidaria, 764, **765**
  digestion in, 326, **327**
Coacervate, 61; *see also* Procells
Cobalt, 346t
Coccus, 749, **750**
Cochlea, **288,** 289
Codon, 183–184, **184,** 190, **191,** 193, **193,** 514
  assignments, 194
Coelom, 76–77, **78**
  acoelomates, 76–77, **78**
  coelomates, 77, **78**
  development, 769, **769**
Coelomata, 766, 768, **769**
  gaseous exhange in, 395–396, **395**
Coenocytic, 107
Coenzyme, 46, 135
  in cytochrome system, 150, **150,** 151
  oxidation-reduction, 145
Coenzyme A, 171–172, **172;** *see also* Acetyl Co-A
Cofactor, 135
Cohesion tension theory, 351–353, 352t
Colchicine, 512
Coleoptile, 233, **233,** 457, **457**
Coleus, 518
Collaboration of genes, 496–497, **496**
Collagen, 105
Collagenous fiber, 220
Collenchyma, 206, **206**
Collision theory, 16
Colloid, 105
Colon, 339–340, **339**
Colorblindness, 523
Commensalism, 595, **595**
Communication
  bird song, 724
  crickets, 724

Communication [*cont.*]
honey bee, 724, **725**
types, 724
Community, 7
climax, 617
ecological, 593
pioneer, 617
in succession, 617–619, **618**
Companion cell, 208, **208**
Competition
interspecific, 593–594, **594**
intraspecific, 588–589
Complement, 382
compound, chemical, 21
types, 21
See also Organic compound
Concealing coloration, 734–735
Concentration gradient, 160
Conditioned reflex, 267–269, 706
Conditioning
classical, 706
instrumental, 706–707, **706**
operant, 706–707, **706**
Conducting tissue. *See* Xylem
Cone cells, 98, 291, **291**
Conjugation, 441–442, **441**
in Monera, 441–442, **441**
in paramecium, 441–442, **441**
Connell, J., 593
Connective tissue, 218, 220–224, 222t
elastic, 222t
fibrous, 222t
reticular, 22t
Conservation, 629–630
Consumer, **600,** 601–602
benthos, 635–636
in energy flow, 610
limnetic, **632,** 633
littoral, 632–633, **632,** 635, 636
nekton, **635,** 636
neritic, 636, **636**
terrestrial, 635, **638,** 639
Consummatory act, 697–699, **698, 699, 700**
Continental shelf, 570, **571**
Continental slope, 570, **571**
Contractile
protein, 169
vacuole, 407, **408**
Convergent evolution, **556,** 557, 735
Copper, 346t
Coral, 158
Coreolis effect, 576
in carbon cycle, **603,** 604
Cork, **206,** 207, **214,** 299, 459
Cork [*cont.*]
cambium, 205, 211, **214**
root, 210–211, **211**
stem, **213,** 214, **214**
Corm, 422, **424**
Corn, 540
albinism, 518
germination, 457, **457**
seed, 457
Cornea, 290, **291**
Corning, 710
Corepressor, 197–198, **198**
Corolla, 435, **435**
Corpora allata, 247, **247**
Corpus luteum, **255,** 256, 448
Cortex
adrenal gland, 251t, **249,** 252
brain, 272, **272**
role in pain, 286
root, **209,** 210
stem, 210, 213, **214**
Cortical hormones, types, 251t, 253
Corticoids, 251t, 253
Cotyledon, 455, 456
Cosmozoic hypothesis, 65
Courtship, 726–727
Covalent bonding, 11–12, **13,** 12–14, 29
Coverings
amoeba, 299
annelid, 299, **300**
arthropod, 299, **300, 301**
Ciliates, 299
crayfish, 264, **265**
crustaceans, 299
invertebrate, 299, **300**
Monera, 299
plant, 299
Protistan, 299
vertebrate, 299, 301–303
Cowper's gland, **444,** 446
Cranial nerve, 264, 267
function, 275t
Cranium, 304, **306**
Crayfish, 301
antennae, **265**
exoskeleton, **301**
nervous system, 264, **265**
Creatin, 167–168, **168**
phosphate, 168, **168**
Creatinine, 301
Cretinism. *See* Hypothyroidism
Crick, F., 179, 183
Crocus, 422, **424**
Crossing over, in meiosis, 427, **427**
Crustaceans, 158, 246
Crustaceans [*cont.*]
coverings, 299
regeneration, 422
Cryptorchidism, 444
Ctenophores, 158
Cutaneous sense
receptors, **286**
senses, 285–286, **286**
Cuticle, **215**
in *Ascaris,* 766, **768**
of annelid, 299, **300**
of arthropod, 299, **301**
layers, 299, **301**
of plant, **352**
of roundworm, 299, **300**
Cutin, 206, 299
Cuttings, vegetative reproduction, 421
Cyanophyta. *See* Algae, bluegreen
Cyclic photophosphorylation, 135, **135,** 139
Cyclosis, 105, 350
Cysteine, 173, 179, 183
Cytochrome, 46–47, 135, **135,** 137, 145
types, **150,** 151
Cytochrome system, 150–151, **150**
Cytokinesis, **120–123,** 124, 125, 127, 237–238
Cytokinin, 237, 238
influence of, 237–238
Cytopharynx, 326, **326**
Cytoplasm, 105–107, **106**
determinants, 477–478
division, 119, **120–123,** 125
in movement, 118, 165, 168
streaming, 105, **106,** 350
Cytopyge, 326, **326**

D

Danielli, J., 90
Dark reaction, 139–140, **141**
Darwin, C., 233, 538–539, 552, 690
Davenport, C., 519
Davson, H., 90
Deamination, 43, 154–155, 173
Death rate, 584–585, **584**
human, **584,** 656
Decarboxylation, 148, **149,** 154–155, 173
Deceptive mimicry, 735–736, **736**
Decomposition, 15
DDT, effect on ecosystem, 623, 624–627, 672
Deciduous forest, 641–642, **643**
Decomposer, 322, **600,** 601–602

Decomposer [*cont.*]
in carbon cycle, **603,** 604
in ecosystem, 599
littoral, 632, **632,** 635, 636
neritic, **635,** 636
profundal zone, 633
terrestrial, 638, 639–640
Deer, 772
white-tailed, **735**
Dehydration, 143, 148, **149**
amino acids, 37, **37**
fats, 34, **35**
monosaccharides, 32, **33**
demography. *See* Population
dendrite, 161–162, **161,** 228, **228**
Dental
caries, 342
plaque, 342
Deoxyrybonucleic acid. *See* DNA
Deoxyribose, 31, 48, **48**
Dependency ration, 654
Depolarization, 163
Depressants, 295, 296t, 297
Dermacentor, **771**
Dermis, **230**
Desert, 643–645, **644**
Desmosome, 98, **99**
of duodenum, **99**
de Soussure, N., 130–131
Development, 235, 474–477, **476**
in *Acetabularia,* **476,** 477
in ageing, 484
of animal embryo, 459–466
blastocyst, 467, **468**
blastula, 461–462, **461, 462**
cleavage, 461–462, **462**
closed, 455
continuity, 454
control of, 474–480
cytoplasmic determinants, 477–478
of digestive tract, 467, **467**
of egg, 460–461
embryonic, **456,** 464–466, 468–469, 532, **532**
human, **472, 473,** 473
external environment, 480
of eye, 464–465, **465**
feedback control, 479
of fern, 478–479
fruit, 235, 236
gamete, 459–461, **460**
germ layers, 77, 463–470, **463,** 466, **469,** 470t
growth, 210–211, **211,** 480–482, **481, 482**
Development [*cont.*]
hormones, in, 479–480
microenvironment, 478
neighboring cell determinants, 478–480
nervous system, 261
neural folds, 464–466, **465, 466**
organ, 468–469, **467,** 470t
pH, effect of, 480
processes, 454–485
root, 236–237, **237, 457,** 458, **458**
seed, 455–457, 456
shoot, 455, 456
tissue, 470t
Diaphram, in birth control, 663, **664,** 667t
Diaphysis, 220, **221**
Diarrhea, 340
Diastolic pressure, 367–368
Diathrosis joints, 307–308, 309t–311t
Diatom, 635, 751–752, **753**
Diatomaceous earth, 751
Dichlorodiphenyltrichloroethane. *See* DDT
Dicot, 204, **205**
flower, 436
germination, 457–458, **457**
monocot and, compared, 436
seed, 455, **456,** 457
Dicotyledonae. *See* Dicot
Dictostelium discoideum, **761**
Dictyosome, 113–114, **113**
Dicynema, 763, **765**
Diencephalon, 264
Differentiation
in animal embryonic layer, 77
in cell, 455
in plant growth, 209–210, 458–459, **458**
Diffusion, 232, 323, 350
in plants, 93–96, **93**
pressure, 94
Digestion, 321–344, 350
in annelidia, 328–329, **330**
of carbohydrates, 338t
in coelenterates, 327–329, **329**
control, 341–342, 341t
in earthworm, 328–329, **330**
enzymes, 337–338, 338t
of fats, 338t
gland secretions, 338, 338t, 341–342
hormone control, 248, **249**
human, 329–342
in hydra, 326–327, **327,** 328, 355
hydrolysis, 32
intestinal juices, 338
intracellular, 324, 350
Digestion [*cont.*]
in invertebrates, 326–329
in large intestine, 339–341, **339,** 342
lymphatic system, role of, 270
muscle action in, 333
in *Paramecium,* 325–326, 326, 329, 358
pendular movements, 339
physical processes, in, 341t
of protein, 172, 338t
in Protista, 324–326, **326**
in sponges, 326, **327,** 355
types, 323–324
Digestive system
of Annelida, 328–329, **330**
of Cnidaria, 326, 327
disorders of human, 342–343
of earthworm, 328–329, **330**
of hydra, 326–327, **327,** 328, 355
invertebrate, 326–329, **327–329**
of planaria, **329,** 358
of Platyhelminthes, 327–328, **328**
of Porifera, 255, 326, **327**
of Protozoa, 324–326, 326
of vertebrate, 329–342
Dihybrid cross, 492–493, **492**
Dilger, W., 696–697
Dinoflagellate, 635, 752, **753**
Diplococcus pneumoniae, in genetic studies, 175
Disaccharide, 32
Disc, embryonic, 469, **469**
Disease
alkaptonuria, 524
arteriosclerosis, 372
beriberi, 348t
cryptorchidism, 444
dermatitis, 347t
Down's syndrome, 525–526, **526**
emphysema, 405–407, **406**
endema, 380
of eye, 292, **293**
heart, 373
hereditary, 182, 524–528
Klinefelter's Syndrome, 527, **528**
lung cancer, 405–406, **406**
malaria, 742–753
meningitis, 269
multiple sclerosis, 283
nervous system, 278–283
nutritional, 347–348
Parkinsonism, 280
pellagra, 347t
peptic ulcer, 343
periodontal, 342–343
peritonitis, 343

Disease [*cont.*]
  poliomyelitis, 278–279
  prenatal detection of, 527
  rickets, 347t
  sickle cell anemia, 525
  sources of, 377
  syphilis, 279
  Turner's Syndrome, 526
  and world food supply, 659–660
Disinfectant, 374, 375–376, 376t
Dissociation, 23, 24–25
DMT, 295, 296t
DNA (deoxyribonucleic acid), 47, **49,** 110, 111, 119, 124, 125–126, **185,** 533
  in antibody formation, 384
  in bacterial transformation, 178
  in behavior, 697
  characteristics of, 47–48
  in cell reproduction, 119, 419–420
  in chloroplasts, 117
  cistron, 514
  deciphering, 190, 191, 193
  discovery, 175
  in enzyme synthesis, 197–198
  in eucaryotic cells, 75
  in evolution, 66
  function, 175
  in gene induction, 198, **198**
  genetic code, 173, 183
  genetic mutations, 515–516, 540
  helix, 179–180
  hereditary material, 175–177, **176**
  hydrogen bonding, 179
  in immunity, 384
  Meselson-Stahl experiment, 180–181, **181**
  in mitochondria, 114
  in mitosis, 119, 419–420
  muton, 515
  nature of genes, 513
  in operon, 198, **198**
  in protein synthesis, 184
  recon, 515
  replication, 125–126, 179–181, **180**
  in reproduction, 419–420
  in ribosomes, 184
  RNA and, 184
  structure, 179–181, **180**
  synthesis, 170
  in transcription, 185
  triplets, 190
  Watson-Crick model, 179–180, **180**
  See also RNA
Dog, in experiments, 267
Dolphin, **772**
Dove, **772**
Dominance, 487
  in behavior, 729–730, 732–733
  law of, **488,** 489
  social, 589–590
Dorsal aorta, 355, **355**
Dorsal lip, 479
Dorsal root. *See* Spinal cord
Double displacement reactions, 15
Doubleday, T., 651
Down's syndrome, 525–526, **526**
Drosophila, 477
  aneuploids, 509–510
  chromosomes, 501, **501**
  gene mutations in, 540–541
  giant chromosome, 508–509, **509**
  in genetic experiments, 182, 477, 519
  in mapping chromosomes, 507–509, **508**
  nondisjunctive heredity, 503–505, **504**
  sex chromosomes, 501–502, **502**
  survivorship curve, 584, **584**
Drugs
  amphetamines, 296t, 297
  barbiturates, 295–297, 296t
  behavior, effects of, 740
  depressants, 295–297, 296t
  DMT, 295, 296t
  effects of, 295–297, 296t
  hallucinogens, 295, 296t
  heroin, 297
  immune response, 388
  LSD, 295, 296t
  mescaline, 295, 296t
  and nervous system, 294–297
  opiates, 296t, 297
  opium, 297
  prenatal effects, 739
  psilocybin, 295, 296t
  STP, 295, 296t
  sulfonamide, 375, **375**
  tranquilizers, 296t, 297
Druse, **91,** 119
Duck, **772**
  imprinting, 707
Ductless glands. *See* Endocrine glands
Ductus deferens, **663**
Duodenum, 248, **249,** 251t
  ulcers in, 343
Dura mater, 269, **269**
Dutrochet, R., 56, 131
Dwarfism, 237, 256, **256**

E

E. coli. See Escherichia coli
Ear
  auditory canal, 287–288, **288**
  bones, 288, **288**
  cochlea, **288,** 289
  in equilibrium, 288–289
  human, 287–290, **288**
  inner, **288,** 289
  middle, 288–289, **288**
  outer, 287–288, **288**
  oval window, 288, **288**
  semicircular canals, 288, 289, **288**
  structure, 287–289, **288**
Earth, origin, of, 57
Earthworm
  blood, 356–357
  blood vessels, 356–357, **356**
  circulatory system, 356–357, **356**
  digestive system, 328–329, **330**
  epidermis, 299, 300
  execretory system, 408–409, **409**
  ganglion, 262, **263**
  gizzard, 329, **330**
  intestine, 329, **330**
  mouth, 329, **330**
  movement, 313–314
  nephridia, 408–409, **409**
  nervous system, 262, **263**
  structure, 768, **770**
Ecdysone, 247, **247**
Ecdysial glands, 247, **247**
Echinodermata, 158, 770, **771**
Ecological climax, 617–619, **618**
Ecological equivalent, 647
Ecological pyramids, 612–614, **615**
Ecosystem, 7, 562, 598–631
  age distribution in, 585, **585**
  agriculture, effect on, 622–623, **625,** 624–627
  aquatic, 631–637, **632, 635**
  artificial changes in, 622–630
  balance within, 630
  changes in, 617–629, **618, 621–629**
  climate, effect of, 619
  components, 598–599, **600,** 601–602
  commensalism in, 595, **595**
  community in, 562, 593
  conservation and, 629–630
  decomposition in, 621–622
  DDT, effect of, 623, 624–627, 672
  deposition in, 621–622
  energy in, 602–603, 609–612, **611,** 612–614, **615**

Ecosystem [*cont.*]
equivalents, 647
fire, effect of, 619, 621, **622, 624**
flowing water, 633–634
food chains, **611,** 612
habitat, 591–593, **592**
herbivores in, 599
heterotrophs in, 599
indicator species, 616–617
interaction in, 588–597, **594, 595, 596**
lintic habitat. *See* Standing water
intertidal, 634, **635**
mutualism in, 594
neritic, 634–636, **635**
niche, 591–593, **592,** 594
nutrients, 599, **600**
parasitism in, 595–597, **596**
pelagic, 635, 636–637
photosynthesis in, 599–601
populations in, 582–597
predators in, 597
sere, 617–618
standing freshwater, 631–633, **632**
social organization in, 589–590, **589**
succession in, 617, 618, 619
symbiotic interaction, 594–597
technology in, 629
terrestrial, 590–591, 637–640, **638**
trophic levels, **611,** 612, **614**
tundra, 612, **614**
types, 631–649
ubanization, effect of, 627–629
Ectocrines, 602
Ectoderm, 464–466, **463, 465,** 466, 468, **469,** 470t
Ectoplasm, 299
Edema, 381
Effector, 232, **232,** 260
neuron, **161**
Egg, 120, **121,** 246, 430, **431,** 436–437, **438, 439,** 460, 463, **463**
determinate, 461
development, 460–461, 474
fertilization, 461
indeterminate, 461–462
regions in, 477
types, 462
Elastic fibers, 220
Elastin, 105
Electrical potential, 160
Electricity in organisms, 164, 719–720
Electrocardiogram, 370–371, **371**
deflection waves, 371, **371**
P wave, 371, **371**
QRS wave, 373
Electroencephalogram, 281, **282**
Electromagnetic radiation, 72
Electron, 9, 134–135
acceptor, 135
configuration, 9–10, 10t
in cytochrome system, 150–151, **150**
donor, 135
energy levels, 9
excited, 134–135, **134,** 137
in photosynthesis, 137, **137,** 139
Elements
in living organisms, 28t
naturally occurring, 7t
Eliphidium, **754**
Embryonic development, 532, **532**
animal, 459–466, **466**
comparative, 532–533, **532**
disc, 469, **469**
germ layer, 77, 463–470, **463, 466, 469,** 470t
human, 466–473
membranes, 469–479, **469, 470**
of plant, 455–458
stages, 468–469
Embryonic induction, 479
Emphysema, 405–407, **406**
Endema, 380
Endocardium, **360,** 361
Endocrine gland, 247–248, **249,** 251t
Endoderm, 464–466, **463, 465, 466,** 468, **469,** 470t
Endoplasmic reticulum, 107, 111, **112,** 113, 117
membrane types, 111, **112,** 113
Endoskeleton, 303
Endosperm, 438, **439,** 455, **456**
Endosteum, 220, **221**
Energy, 2–5, 52
activation, 16, **16,** 17
and biogeochemical cycles, 602–603
in ecological pyramids, 612–614, **615**
in ecosystems, 609–612, **611**
endergonic, 18, **18,** 130
entropy, 18
exchange and chemical reactions, 4, **17,** 18–19
exergonic, 18, **18,** 130
first law of thermodynamics, 5
forms, 3–5
free, 48–49
herbivores and, 610, **611**
in Kreb's cycle, 148, **149,** 150
kinetic, 4
mechanical, 4–5, 164–169
in metabolism, 130, **130,** 142–151, **143**
Energy [*cont.*]
in photosynthesis, 135, 610
potential, 3, **4,** 610
in procells, 62
radiant, 5, 602, 631, 632, **632**
reactions of metabolism, 158–173
second law of thermodynamics, 6
storage, in photosynthesis, 137
transformation, 140, 142
transduction, 5–6, 5t, 159, **159**
utilization by cells, 159, **159**
Entamoeba histolytica, 753, **754**
Enterocolla, **768**
Enterogastrone, 248
Enthalpy, 18
Entropy, 18
Environment, 561–581
abiotic, 561–581
biotic, 561, 582
development and, 516, **516**
heredity and, 516–518, 527–529
human population and, 669–681
physical, 322
resistance in, 585–587, **586, 587**
social, 742–745
Environmental resistance, 583, 585–587, **587**
Enzyme, 16–17, **17,** 45–47, 130, 131t
actions of antibiotics, 375, **375**
action mechanisms of, 45, **45**
activator, 46
active site, 45
allosteric interaction, 195, **195**
amino acid activation, 187
apoenzyme, 46, **46**
in bioluminescence, 158–160, **159**
coenzymes, 46, 135, 145, 150–151, **150**
cofactor, 135
components, **46**
control, 194–196, **194**
corepressor, 197–198, **198**
in digestion, 338t
feedback inhibition, 194–196, **195, 196**
function, 196
and genes, 181–182
holoenzyme, 46, **46**
induced fit, 196
inducible system, 196, **198**
mechanisms of, 196
naming, 130
in polysaccharide synthesis, 170
prosthetic group, 46, **46**
reaction, 43, 45–47, **46**
regulation, 196
repressor, 196, 197–198, **198**

Enzyme [*cont.*]
specificity, 45, 46
substrate, 43
synthesis, control of, 189–190, 196–199
turnover number, 45
Eosinophil, 225t
Epiactis, 746, **766**
Epidermis, 205–206, **206, 209, 215**
invertebrate, 299, **300**
leaf, 214–216, **215**
origin, 210
plant, 299
root, 209–211, **209**
stem, 212–214, **212, 213,** 216
See also Skin
Epididymis, **444,** 445
Epiglottis, 332, **333**
Epilepsy, 280, 281, 282
Epinephrine, 251t, 253, 297
Epiphysis, 220, **221**
Epistasis, 495–496, **495**
Epithelium, 216–218, 217t
classification, 218–219, 219t
columnar cells, 217, 218t
cuboidal cells, 217, 218t
germinal, 446
olfactory, 285, **285**
pseudostratified, 217, 219t
simple, 217, 218t
squamous cells, 217, 218t
stratified, 217, 219t
transitional, 217, 219t
Epithemia, **753**
Equilibrium
chemical, 19–20, **20**
sense of, 288–289
Equivalents, ecological, 647
Equisetum, **759**
Erythrocruorin, 356
Erythrocyte, 160, 224, 225t, 357, 358
active transport in, 97
in infection, 379
Escherichia coli, 176, 180–181
in DNA research, 190–191
in enzyme research, 195, 196–197
in operon model, 197, **197**
in protein synthesis, 188–189
in RNA synthesis, 196–197
in vitamin synthesis, 359
Esophagus, 332–333, **333**
structure, **333**
Ester link, 34, **36**
Estivation, 703
Estradiol, 248, 251t
Estrogen, 248, 251t
Estrus cycle, 450–453, **451**
Ethiopian region, 648, **648**
Ethology, 691–692
Ethyl alcohol, 146, **147**
Etiolation, 517, **517**
Eubacteria, 749, **750**
Eucaryotic, 65–66, 75
DNA in, 75
evolution of, 65–66
in reproduction, 423–424
Euglena, 111, **261,** 290, 420, **421,** 751, **752**
behavior of, 694
binary fission, 420, **421**
eye spot, 290
nervous control, 260, **261**
Euglenophyta, 751, **752**
Eumetazoa, 764, 766, **766**
bilateral symmetry, 764, 766
movement, 313
nervous control, 260
radial symmetry, 764
Euphorbia, **556**
Euploidy, 510–513, **511**
Euplotes, 260, **261**
Eustachian tube, 288, **288, 399**
Evolution, 530–557
abiogenesis, 52–53
adaptation in, 531
adaptive radiation, 557
Aristotle's hypothesis, 52
autotroph hypothesis, 64–65
of behavior, 721–722
biogenesis, 53–57
biological, 63–65
chemical, 51–52, 64
chemosynthesis, 57–64
convergent, 556–557, **556**
cosmozoic hypothesis, 65
Darwinian theory, 538–539
divergent, 556, **556**
embryology and, 532–533, **532**
eucaryotic cell in, 65–66
evidence for, 531–538
genetic data, 533–534, 546–547, **546,** 548–555
genetic drift, 548–549
geographic distribution and, 534–538
geographical isolation and, 553–555, **554**
Hardy Weinberg Law and, 547
heterotroph hypothesis, 57, **59**
indeterminant, 548
inorganic, 531
Lamarck's theory, 538
Evolution [*cont.*]
meaning of, 530–531
mechanisms, 539–555
of Mesozoans, 764
migration and, 548
morphological, 531–532, 533
mutation and, 540–541, **539**
natural selection in, 538–539, 541–545, 549
nucleic acids in, 65
organic, 531–532, **533**
organismic diversity in, 539
paleontology and, 534–538
patterns of, 555–557, **556**
populations in, 549–550, **549, 550**
procaryotic cell in, 65–66
recapitulation in, 532, **532**
reproductive isolation and, 550–555, **551,** 553t, **554**
speciation and, 550–555, 551, 553t, **554**
in succession, 619
synthetic theory, 539
taxonomic data, 534
theories, 538–539
Excretion, 407–411
homeostasis, 412
pH in, 412
skin, role of, 410
types, 407
urea, 407
uric acid, 407
urine, 412
Excretory system
of *Amoeba,* 407
of earthworm, 408–409, **409**
excretion and, 407
flatworm, 407–408, **408**
human, 409–412, **410, 411**
of insects, 409, **410,** 768–770
of invertebrates, 407–409, **408, 409**
organs, 229t
of *Paramecium,* 407, **408**
of planaria, 408, **408**
processes involved in, 407
of *Protista,* 407, **408**
types of wastes, 407
Exergonic reactions, 18, **18**
Exobiology, 66–72
nature of, 66–67
Exocarp, 440, **440**
Exophthalmos, 252, **252**
Exoskeleton, 303
arthropod, 768
in movement, 313–314
Extinction, marsupials, 647

Extraterrestrial life, 66–67
  significance of discovery, 66–67
Eye
  blind spot, 291–292
  chemical reactions in, 294, **294**
  color, 520–521
  compound, 264, **265**
  cones, 291, **291**
  cornea, 290, **291**
  crayfish, 264, **265**
  development, 464–465, **465**
  disorders, 292–293, **293**
  function, 290–294, **291, 293, 294**
  human, 290–294, **291**
  image formation, 292–294, **293**
  iris, 291–293, **291**
  lens function, 291, **293**
  muscles, **291,** 292–294, 317t
  optic vesicles, 464–465, **465**
  photoreception, 290–294
  pupil, 291, **291**
  refraction in, 292, **293**
  retina, 291, **291**
  simple, 264, **265**
  spot
    Euglena, 290
    planaria, 262, **263**
  structure, 264, 265, 290–294, **291, 293, 294**

F

FAD (flavin adenine dinucleotide) 46, 135, **136,** 145
  in cytochrome system, 150–151, **150**
  in Kreb's cycle, 148–150, **149**
Fall overturn, **568,** 569
Fallopian tubes. *See* Uterine tube
Fat
  catabolism, 153, **153,** 154
  cells, **230**
  digestion of, **36,** 153–154, **153,** 338t
  metabolism, 152–154, **153**
  malonyl CoA in synthesis, 170–171, **172**
  synthesis, **36,** 170–171, **171, 172,** 173
Fatty acids, 34, **35, 36**
Faunal regions. *See* Biogeographical regions
Feedback
  control in development, 479
  inhibition, 194
  negative, 231–232
  role of pituitary, 254–255
Femur, 220, **305,** 307
Fermentation, 63
  aerobic respiration and, compared, 151
  alcoholic, 146–147, **147**
  bacteria, 147
  in evolution, 63
  lactic acid, 147, **147**
Fern, **759**
  development, 478–479
  fertilization, 434, **434**
  life cycle, 433–434, **434**
Ferredoxin, 135
Fertilization, 438, **438, 439,** 461, **462**
  in Angiosperms, 434, 438, **438**
  in bryophytes, 432–433, **433**
  double, 438
  in ferns, 434, **434**
  and growth, 483
  human, 466–467
  sperm in, 467
  types, 442
  in vertebrates, 443, 461–462
Fiber, 105, 207, **207**
  elastic, 220
  cartilage, 223t
  collagenous, 220
  elastic, 220
  muscle, 316
  phloem, 208
  Purkinje, 370, **371**
  reticular, 220
  unmyelinated, **228,** 229
  wood, 207–208
  See also Neuron
Fibrils, 110, 111, **111,** 126, **261**
  in flagellum, 168
  in movement, 313
  myonemes, 313, **314**
Fibrinogen, 381
Fibroblasts, 220
Fibula, **305,** 307
Fiddler crabs, behavior, 701
Finches, Darwin, **592**
Fire
  in ecosystem, 619, 621, **622, 624**
  in succession, 619, 622, **624**
First law of thermodynamics, 5, 610
Fish
  adrenal gland, 252
  behavior, 700–701
  brain, 264, **266**
  circulatory system, 361–362, 363
  electrical, 719–720
  in freshwater ecosystem, 631–634, **632**
  gill, 396–397, **397**
  hearing, 718
Fish [*cont.*]
  heart, 361–362, **362**
  in neritic zone, **635,** 636
  nervous system, 264
  scales, 302
  skin, 301, 302
  stomach, 334
  territoriality of, 730
  and thermal pollution, 675–676
  vision of, 717
Fission
  asexual reproduction, 420, **421**
  binary, 420
  in Monera, 420
  in *Paramecium,* 420, **421**
  in Protista, 420, **421**
Flagellates, 158
Flagellum, 110–111, **111,** 168–169, **169**
  movement, 168–169, **169**
  in Protista, 751
Flame cell, 408, **408**
Flatworm
  digestion, 327–328, **328**
  excretory system, 407–408, **408**
  mouth, 327, **328**
  movement, 313
  nervous system, 262, **263**
  pharynx, 327, **328**
Flavin mononucleotide. *See* FMN
Florigen, 240
Flower
  color, 428
  corolla, 435, **435**
  dicot, 436
  monocot, 436
  ovary, 435, **435**
  petal, 435, **435**
  placenta, 435, **435**
  receptacle, 435, **435**
  releasing mechanisms of, 733–734, **734**
  sperm, 438, **438**
  stamen, 435, **435**
  structure, 435–436, **435**
Flowering
  control, 238–242
  experiments, 239, 241
  plant, **216**
  wavelengths in, 241
Fluoride, 342, 346t
FMN (flavin mononucleotide), 46, 135, **136,** 145, 146, **146**
Follicle, 446–448, **447**
Follicle stimulating hormone. *See* FSH

Fontanel, 304, **306**
Food chain, **611,** 612, **613**
  DDT in, 625–626
Food vacuole, 325, **326**
Food web. *See* Food chain
Forebrain, 264, 266, **269,** 270–272, **270, 271**
  cerebral tracts, 270
  cerebrum, 271, **271**
  divisions, 264
  hypothalamus, 271, **271,** 272
  major areas, 270, 271
  portions of, 271–272, **271**
  thalamus, 270, 271
Forest
  deciduous, 641–642, 643
  tropical rain, 645, **645,** 646
Fossil, 534–536, **535,** 537t
Fox, S., 61–62
Free energy, 18, 48, 49–50
Free radical theory, 484
Freshwater ecosystem, 631–634
Frisch, K. von, 733
Frog
  brain, 266
  circulatory system, 362, **362**
  development experiments, 479
  heart, 362, **362**
  respiratory system, 397–398, **398**
  vision of, 717
Fructose, 31, **31, 32**
Fruit
  development, 235, 236
  drop, 235, **236**
  structure, 440, **440**
FSH (follicle stimulating hormone), 251t, **255,** 256, 448, **451,** 452, 453
Fucoxanthin, 755
Fucus, 480
Fumaric acid, 148, **149,** 150
Functional groups, organic compounds, 30t
Fungi, 80–81, 158, 374, 374t, 758, 760–763, **761, 762, 763**
  budding in, 420, **422**
  characteristics, 758, 760
  karygamy in, 760
  meiosis, 760
  in mutualism, 594
  sac, 762–763, **762**
  sporangium, 421, **423**
  spore production, 420–421, **423**
  true, 760, 762–763, **762, 763**

G

Galactose, 31
Gall bladder, **250,** 337, **337**
Gametangium, 762
Gamete, 420
  development, 459–461, **460**
  haploid, 506–507, **507**
  in inheritance, 488
Gametophyte
  in bryophyte, 757
  generation, 322, 333, 334
Ganglion, 261–262, **263**
  basal, 270, **271**
  dorsal root, 273, **273**
  earthworm, 262, **263**
  flatworm, 261, **263**
  neurons, 291
Garrod, A., 181–182
Gaseous exchange
  in amphibians, 396, 398, **398**
  in breathing, 398
  in clam, 395–396, **395**
  fish, 396, 397
  frog, 398
  in grasshopper, 395–396, **396**
  hydra, 396–397, **397**
  in invertebrates, 394–395
  in mollusk, 395
  oxygen in, 393
  in photosynthesis, 393
  physiology of, 394
  planaria, 395
  in plants, 393–394
  in Protista, 394
  in spider, **395,** 396
  in vertebrates, 396–403
Gastric juice, 333–334, 338t
Gastric secretion, phases, 341
Gastrin, **250,** 251t
Gastrula, 463, **463,** 479
Gastrulation, 463–464, **463**
Gates, R., 520
Gel, 105–106, **106**
Gene, 175
  action, 181–183
  in Beadle-Tatum hypothesis, 182–183, **183**
  chemistry of, 175–179, **176**
  chromosomes, relationship to, 499–500
  collaboration, 496–497, **496**
  complementary, 494–496, **494**
  control, 182–183, 199
  in enzymatic activity, 181–182
  epistasis, 495–496, **495**
Gene [*cont.*]
  equilibrium in population, 547–555
  in evolution, 533–534
  expressivity, 498–499
  function, 175, 184–190, 194
  induction, 198
  in inheritance, 488, 494–499
  interaction of, 494–499
  holandric, 523–524
  multiple, 497–498, **497,** 519–521, **520**
  multiple alleles, 487–488, **488,** 497
  mutation, 515–516, 540–541, 548
  nature of, 181–193, 513–516
  operon, 513
  operator, 198
  organizer, 479
  penetrance, 498–499
  polygene, 497
  pool, 546–547, **546, 547**
  in protein synthesis, 184–190, **185, 189**
  recombination, 428, **428**
  regulator, 197–198, **198,** 199
  sex-influenced, 504–505, **504,** 523
  sex-linked, 501–502, **502,** 504–505, **504,** 522–524, **523**
  sex limited, 504, 505
  structural, 197–198, **198**
  structure, 514–515
  Sutton's hypothesis, 499–500
Genetic code. *See* DNA
Genetics
  and behavior, 692
  data, 533–534
  dihybrid cross, 492–493, **492**
  drift, 548–549
  environment and, 549–550, **549, 550**
  equilibrium, 547–555
  euploidy, 510–513, **511**
  evolution, 533–534
  Hardy-Weinberg law, 557
  interaction of genes, 494–499
  since Mendel, 493–518
  Mendelian laws, 488–489
  monohybrid cross, 487–487, **488, 490**
  mutations, 540–541
  population, 541–550
  probability in, 489–490
  recombination, 539–540
  transformation, 176–177, **177,** 430
  test cross, 490–491, **491**
  variability, 549
Genotype, 488–489, 546, **547**
Genus, 74
Geologic time scale, 536, 537t
Geology and evolution, 534–538

Geotropism, 244–245, **244**
Germ layer, 77, 766
in classification, 463–464, **463,** 468–469, **469**
derivatives of, 470t
development of, 464–466, **465, 466**
Germination
of pollen, 437–438, **438**
of seed, 454–458, **457**
of spores, 533
Gestation
human, **472,** 473, **473**
Giantism, 259, 483
Gibberella fujikuroi, 236
Gibberellin, 237
in genetic dwarfism, 237
gibberellic acid, 237
influences of, 236–237
Gill
of fish, 395–396, **395**
of mollusk, 395–396, **395**
Gill slit, 466, 770
Gini, G., 651
Gizzard, earthworm, 329, **330**
Glands, 218
poison, 302
scent, 302
skin, 302–303
sweat, 302
Glomerulus, 412
Glottis, 397, **398**
Glucocorticoids, 251, 251t, 253–254, 257
Glucose, **18,** 31, **31, 32,** 33, 105, 131t, 142–143
control of, 248, **249**
in fermentation, **147**
in photosynthesis, **132**
in polysaccharide synthesis, 170, **171**
Glutamic acid, 155
Gluten, 119
Glycerol, 34, **35,** 154
in fat biosynthesis, 170, **171**
Glycine, 173
Glycogen, 32, 34, 168, 195
in muscle contraction, 168
synthesis, 170, **171**
Glycolysis, 142–145, **144,** 151, 168, 170–171
Goblet cells, 217, 219t
Goiter, 250
Golden plover, migration, 704, **704**
Goldfish, in memory experiments, 710
Golgi apparatus, 75, 113–114, **113,** 125
cisternae, 113, **113**
functions, 113–114
Golgi apparatus [*cont.*]
vesicles, 113, **113**
Golgi bodies. *See* Golgi apparatus
Golgi-Mazzoni corpuscle, **286**
Gonad, 248, **249,** 251t
control of reproduction, 248, 250
hormones, 248, 250, 251t
Grand mal, 280, **282**
Granular reticulum, 113, **113**
Granum lamellae, **116,** 117
Grasshopper, **771**
blood, 355
circulatory system, 355
excretory system, 409–410
metamorphosis, 484, **484**
nervous system, 264, **265**
respiratory system, 396, **396**
tracheae, 396, **396**
Grassland, 642–643
dustbowl, 623, **625**
in ecological pyramids, 612, **615**
Gray matter, 270, 272, **273,** 278
Grey crescent, 477, 479
Griffith, F., 175
Ground meristem, 212
Group roles, human, 743–744
Growth, 480–482, **481, 482**
allometric, 482, **483**
carrying capacity, 586
hormones, 483
movements in plants, 242
nature of, 480–482, **481, 482**
patterns, 586, **586, 587**
pituitary, role of, 255–256, **255, 256**
plant, 233–234, **233, 234, 457,** 458–459, 517–518, **517,** 713–714
population, 583–584, **587**
rate, 481–482, **481, 482**
regulators. *See* Hormones
rings, 213, **214**
root, 458, **458**
termination, 482–483
tissue. *See* Meristem
Grunion fish, 702, **703**
Guanine, 179, 183
triphosphate, 148, **149**
Guard cell, 206, **215,** 393, **393,** 394
Gull, behavior, 727, **727**
Gullet, 754
Gurdon, J., 474
Guttation, 351
Gymnodinum brevis, 599
Gymnosperms, 531, **759**
Gyri, 270, **271**

H

H zone, of muscle, 165, **166**
Habitat, 591–593, **592**
and vision, 717
Hair, **230**
follicle, 302
papilla, 302
structure, 302
tactile, 264
Hallucinogens, 295, 296t
effects, 296t
Hämmerling, J., 107
Haploid, gametes, 506–507, **507**
Hardy-Wineberg law, 457, 548
Harlow, H., 690–691, 728
Haversian system, 221, **221**
Hawthorne effect, 689–690
HCN (hydrogen cyanide)
in chemical evolution, **59,** 60
in organic synthesis, 60
Head
bones of, 304, **306**
muscles of, 317t
receptors in, 283–294
Hearing, 287–290, **288,** 718–719, **719**
antennae, 718, **719**
in fish, 718
insects, 718, **719**
physiology, 289–290
Heart, 359–363, **360, 362**
atrium, 360–361, **360**
of bird, **362,** 363
blood vessels, 364, **365**
cardiac compression, 404
cardiac cycle, 361
cardiac output, 367
conduction system, 370–371, **371**
contraction, 367
diastolic pressure, 367–368
disease, 373
endocardium, **360,** 361
evolution, 361–363
of fish, 361–362, **362**
of frog, 362, **362**
of human, 359–361, **360**
massage, 404
mechanism, 359–361, **360**
muscles, 314, 359–360, **360**
myocardium, 360, **360**
node, 370, **371**
pericardium, 359–360, **360**
of reptiles, 362–363, **362**
resuscitation, 404
in sleep, 721

Heart [*cont.*]
systolic pressure, 367–368
valves, **360,** 361
ventricle, **360,** 361
vertebrate types, 361–363, **362**
Hebb, D., 709
Heinroth, O., 707
Heitz, E., 477–478
Helix aspersa, **769**
Hemicellulose, 102, 104
Hemocyanin, 356
Hemodialysis, 413–414, **413**
Hemoglobin, 135, 173, 303, 357, 525
in plasma, 359
Hemophilia, 523
Henle, loop of, 411, **412**
Heparin, 220, 258
Herbivore, 323, **600,** 601
and DDT, 625
in ecosystems, 599
in energy flow, 610, **611**
in predation, 597
Heredity
and behavior, 738–741
and body build, 738–739
chromosome theory, 499–507
cross over, 505, 506, **507**
disease, 182, 524–528
environment and, 516–518, 527–529
human, 519–525
chromosome abnormalities in, 525–**526, 527,** 528
chromosome number, **523,** 525–526
defects in metabolism, 524–525
eye color, 520–521
holandric, 523–524
multiple alleles, 521, 522
multiple gene inheritance, 519, 520
sex, 522–524, **523, 524**
skin color, 519–521, **519**
nondisjunction, 503–505, **504**
sex influenced, 504–505, **504**
sex limited, 504
sex-linkage, 501–502, **502,** 504–505, **504,** 505–506, **506**
Sutton's hypothesis, 499–500
twin studies, 527–529
X-linked, 522–524, **523**
Y-linked, 522–524, **523**
Hermaphrodism, 442
Heroin, 297
Hershey, A., 178
Heterogamy, 430, **431**
Heterotroph, 63, 77, 322
in carbon cycle, 603–604, **603**
Heterotroph [*cont.*]
in classification, 77
in ecosystem, 599
evolution, 57, **59**
nutrition, 323–324
Protista, 752–755, **752, 754**
See also Consumer; Decomposer
Heterozygous, 489
Hexose, 140
monophosphate shunt, 151, 152, **152**
Hibernation, 702–703
High-energy phosphate, 48, 602
Hindbrain, 264, 266, **270,** 271–272
cerebellum, **270,** 272
medulla oblongata, **270,** 273
pons, **270,** 273
portions of, **270,** 271–273
Hirudo medicinalis, **770**
Histamine, 258
in allergy, 386
in inflammation, 379
Holandric inheritance, 523–524
Holarctic region, 648, **648**
Holley, R., 186
Holoenzyme, 46, **46**
Holotroph, 77
Homeostasis, 231–258
in excretion, 412
in nervous control, 259–297
pattern of control, 231–232
See also Chemical control
Homeotherms, 232
Homerus americana, **771**
Homing, 705
Homology, 75
Homozygous, 488–489
Honey bee
communication, 724, **725**
social organization, 589–590, **590**
Hooke, R., 87
Hormone, 232
activities of, 246
adrenal gland, 251t, 253, 254, 258
androgens, 248, 251t, 253
antidiuretic, 256
auxins, 235–236, 237
and behavior, 713, 720–721, **720,** 741
in blood, 359
composition, 245
cortical, 251t, 253, 254–255
in development, 479–480
in digestion, 248
female sex, 448, 450, 526
in growth, 233–235, 232, 483
in hibernation, 703
Hormone [*cont.*]
as insecticides, 672, **672**
invertebrate, 246–247, **246**
lactogenic, **451,** 452–453
luteinizing, **451,** 452–453
mechanisms, 245–246
metabolic, 248, 250, 251t, 252
in metamorphosis, 246–247
morphogenic, 247
ovary, 248–250, 446, 448, 450, **451, 452,** 453
parathyroid, 251, 252
pineal gland, 257
pituitary, 248, **249,** 254–257, 450–453, **451**
plant. *See* Auxin
progesterone, 250, 251t
regulatory, 257
roles, 245
sex, 444
thymus gland, 257
vertebrate, 247–258, 251t
in vertebrate control, 247–248
Hornworts, 757, **758**
Horse
brain, **266**
evolution, **542,** 543–544
Housefly, chemoreceptors, 283
Huange, S., 72
Human
behavior
affects of alcohol, 739
and body build, 738–739
group behavior, 742–744
group roles, 743–744
and heredity, 738–741
interpersonal roles, 743
oxygen, effects of, 739
psychological approach to, 737–738
social behavior, 742
and social institutions, 745–746
and social motivation, 744–745
social personality, 744
digestive system, 329–342
disorders of, 342–343
ear, 287–290, **288**
excretory system, 407–411, **412**
female reproductive cycle, 447, **448**
heart, 359–361, **360**
heredity, 519–529
insight, 745
kidney, 410–412, **410, 411**
lungs, 399–401, **400**
lymphatic system, 369
movement, 314–316, 317–320t

Human [*cont.*]
muscles, 317, 320t
nervous system, 264–278
nostrils, 399, **399**
population
age and sex, 654–656, 655t
and changing environment, 669–681
control, 662–668
dependency ratio, 654
growth, 653–654, **654**
growth rate, 656
theories, 650–652
and urbanization, 656, 657
respiratory system, 398–403, **399**
skin, 302
sperm, 444–445, **445**
stomach, 333–335, **334**
survivorship curve, 584, **584**
teeth, 331, **332**
thorax, 305, **305**
vocal cords, 399, **399**
Humerus, **305,** 307
Huxley, J., 166
Hyacinth, water, **628**
Hyden, H., 710
Hydra, **262,** 746, **766**
budding, 420, **422**
circulation, 355
digestion, 326–327, **327,** 328, 355
epidermis, 299, **300**
gaseous exchange, 395
gastrovascular cavity, 395
movement, 313, 314
nervous system, 260, 261, **262**
regeneration, 422, 425
structure, **766**
Hydrochloric acid, in stomach, 334
Hydrogen
acceptor, 151
bond, 14, **14,** 22, 179, 187
in cytochrome system, 150, **150**
Hydrogen cyanide. *See* HCN
Hydrolase, 131t
Hydrolysis, 154, 340
Hydrosphere, 566–573
bogs, 569
brooks, 570
climate, 579–581, **579, 580,** 581t
lakes, 567–569, **568**
oceans, 570–573, **571,** 572t, **573, 574, 575**
ponds, 569
rivers, 570
springs, 569–570
Hydrosphere [*cont.*]
streams, 570
in water cycle, 606, **607**
Hymen, **447,** 449
Hypertension, 372
Hypertonic solution, 95–96
Hypha, 421, **423,** 760, **763**
Hypocotyl, 455, **456**
Hypophysis. *See* Pituitary gland
Hypothalamus, 264
functions, **270,** 271–272
and sleep, 721
Hypothyrodism, 250
Hypotonic solution, 95
Hyscyamus, 240

I

I band of muscle, 165, **166**
IAA, 235, 236
Ilex aquifolium, 237
Immunity, 385–391
active, 385
allergy, 386
passive, 385
transplantation and, 386–391
Immunological theory of aging, 548
Implantation, 467, **468**
Imprinting, 707, **707**
Indicator species, 616–617
Indole-3-acetic acid. *See* IAA
Inducer, 479
Indusium, 434
Industrial pollution
air, 676, **676**
land, 677, **678,** 679–680
strip mining, 678–679, **679**
Infection
blood cells in, 379
defensive mechanisms, 378–381
inflammatory response, 379
microphages, 379
plasma in, 379
sources of, 377–378
specific resistance, 381–385, **383, 384**
Inflammatory response, 379–381, **380**
pain, 379
phagocytosis, 379
symptoms, 379
Infundibulum, 447, 448
Ingen-Housz, J., 130
Ingestion, 322
Inheritance
basic plan, 486–487
Inheritance [*cont.*]
in chickens, 494
collaboration, 496–497, **496**
complementary genes, 494–496, **494**
dihybrid cross, 492–493, **492**
epistasis, 495–496, **495**
holandric, 523–524
incomplete dominance, 493–494, **493**
intermediate, 494–494, **493**
law of dominance, **488,** 489
law of independent assortment, 492–493
interaction of genes, 494–499
law of segregation, 488
law of unit characters, 488
monohybrid cross, 487–488, **488**
polygenic trait, 521
probability in, 589–590
test cross, 490–491, **491**
Inner ear, **288,** 289
Inoue, S. 127
Insect, 58
in agriculture, 623
chemoreception, 716–717, **716**
excretory system, 409, 410
hearing, 718–719, **719**
social organization among, 589–590, **590**
structure, 771
territoriality, 730
See also Arthropod
Insect guide, 733, **734**
Insectivorous plant, 324, **325,** 597, 714
Insight, 707–708, **708**
human, 745
Integument. *See* Skin
Intercalated disc, 227, **227**
Intercellular substance, 104–105
Intermediary metabolism, 155–157, **156,** 173
Interpersonal roles, 743
Interphase
in meiosis, 426, **426,** 427
in mitosis, 124
Interspecific interactions, 593–597, **594**
commensalism, 595
facultative, 594
obligatory, 594
parasitism, 595–597, **596**
predation, 597
Intestinal juice, 338–342
in digestion, 338, 338t
function, 342
secretions, 338, 342

Intestine
large, 339–341, **339**
small, 335–339, **335, 336**
Intertidal ecosystems, 634, **635**
Intraspecific interactions, 588–593
Intrauterine device (IUD), 663–665, **665,** 666t
Invertebrate
blood, 355–356
coverings, 299, **300**
digestion, 326–329, **327, 328, 329**
epidermis, 229, **300**
excretory system, 407–409, **408, 409**
gaseous exchange, 394–395
nervous control, 260
respiratory system, 394–396, **395, 396**
skeleton, 303–304, **304**
iodine, 346t
in biogeochemical cyles, 609
Ionic bonding, 11, **12**
Ionization, **23,** 24
in active transport, 97–98
Ions, 11, 23, 24, 25, 135
in calcium cycle, 608
in muscle contraction, 165, 166
in nerve impulse, 162–164
nutrition, 323
in osmosis, 97
Iris, 290–291, **291,** 292–293
Iron, 346t
Irrigation, effects of, 680, **680**
Islets of Langerhans, 248, **249,** 337
Isogamy, 430, 431
Isograft, 338–339
Isolation
geographic, 552, 553–555, **554**
mechanism, 552, 553t
reproductive, 550–552, **551,** 553t
Isomerase, 131t
Isomerism, 31
Isotonic solution, 96
Isotope, 9, 132
in DNA determination, 180
radioactive, 536, 536t

J

Janssen, Z., 87
Jelly fish, **262,** 764, **766**
nervous system, 260-261, **262**
statocysts, 260, 262
Jennings, H., 690
John, E., 710
Joints, 307–308, 309–311t
cartilaginous, 307, 309t
fibrous, 307, 309t
types, 307, 308, 309-311t
See also Bones
Jupiter, 69, 71

K

Kalanchoe, vegetative reproduction, 422, **424**
Kamen, M., 132
Karpenchenko, G., 512
Karyogamy, 760, 763
Karyokinesis. *See* Mitosis
Karyolymph, **109,** 124
Keratin, 302, 379
α-keto-glutaric acid, 145, **149,** 173
Kettlewell, H., 544
Kidney, 301
in alcohol metabolism, 344
artifical, 413–414, **413**
Bowman's capsule, 411–412, **411**
hemodialysis, 413–414, **413**
nephron, 411–412, **411**
physiology, 412
transplant, 388
tubule, 410–413, **411**
Kinase, 143
Kinesis, 694
Kinetic energy, 4
King, T., 474
Klebsiella pneumoniae, 75, **750**
Klinefelter's syndrome, 527, **528**
Knoop, F., 153
Korn, E., 100
Kornberg, A., 180
Krause, end bulb of, 285, **286**
Kreb's cycle, 47, 148, **149,** 150, 168
amino acid metabolism, 155
fat biosynthesis, 171
intermediate metabolism, 156, **156**
Krech, D., 710–711
Krill, 637

L

Labor, stages of, 474
Labyrinthulomycota, **760**
Lack, D., 732
Lactic acid
fermentation, 147, **147**
muscle contraction, 168
Lactobacillus acidophilus, 342
Lactogenichormone, **451,** 452–453
Lactotropin (LTH), 452–453
Lacunae, 222, **222**
Lake
eutrophic, 617
oligotrophic, 617
overturn, 568–569, **568**
pollution, 674
succession, 617, **618**
types, 568–569, **568**
Lamarck, J., 538, 690
Lamellae, 221, **221**
Laminaria, **756**
Lamprey, reproductive system, 448
Land
bridges, 647, **648**
inhabitable areas, 652, **653**
Landsteiner, K., 521
Langerhans, islets of, 248, **249,** 337
Large intestine, 339–341, **339**
digestion, 340
function, 340–341, 341t
structure, 339–340, **339**
Larva, 246
Larynx, 339, **399**
Lashley, K., 709
Latent period, 165
Laterite, 658
Layering, 421–422, **424**
Leaf, 214–216
abscission, 235
angiosperm, 393
dicot, 214–216, **215**
fossil, 535
mosaic, 244, **244**
petiole, 215, **215, 216**
scar, 212, **212**
stipules, 215, **215, 216**
structure, 215–216, **215, 216**
tissue, 352–353, **352**
transpiration, 214, 352–353, **352**
Learning
conditioned response, 705–707, **706**
imprinting, 707, **707**
insight, 707–708, **708**
psychological approach, 690–691
Leeuwenhoek, A. van, 54
Leg
bones, **305,** 307
muscles, 318t
Lentic habitat, 631-633, **632**
Lepas, **771**
Leucine, 173
Leucoplast, 117

Leukocyte, 224, 225t, 357, 358, 368
agranular, 226t
basophils, 225t
eosinophils, 225t
in lymphatics, 368
types, 226-226t
Lewis, W., 479
LH (luteinizing hormone), 251t, **255,** 256, **451,** 452-453
Libellula, **771**
Lichen
in mutualism, 594
in Tundra, 612, **614**
Liebig, J., 614
Life, extraterrestrial, 66-67
Life cycle
alternation of generation, **429,** 430
of bryophytes, 432-433, **433**
of chlamydomonas, 429, 431
diplontic, 429-430, **429**
diplohaplontic, **429,** 430
of fern, 433-434, **434**
gametophyte generation in, **429,** 430
haplontic, 429, **429**
of liver fluke, 596-597, **597**
and meiosis, 429-430, **429**
of Plantae, 430-433, **432**
of *Polytrichum,* 432-433, **433**
sporophytic generation in, **429,** 430
of *Ulva,* 432, **432**
Ligament, 305
Ligase, 313t
Light
in flowering, 238-241
in photosynthesis, 132-139, **134, 135, 137**
Lignin, 102, 104, 206, 207
Lily, insect guide, **734**
Lima bean
germination, 457-458, **457**
seed structure, 455-457, **456**
Limiting factors, law of, 614
Limnetic
consumers, **632,** 633
decomposers, **632,** 633
producers, **632,** 633
Linkage, sex, 501-502, **502,** 504-505, **504,** 522-524, **523**
Linnaeus, C., 73
Lipid, 34, 35t, 102, 105
classification, 35t
biological importance, 35t
biosynthesis, 170-171, **172, 173**
in cells, **92,** 93
characteristics, 34, 35t
Lipid [*cont.*]
function, 34
Kreb's cycle, 171
in membrane, **92,** 93
metabolism, 170-171
Lithosphere, 562-566
in sedimentary cycles, 606
in water cycle, 606, **607**
Littoral, 631, **632**
consumers, 632-633, **632, 635,** 636
decomposers, 632, **632,** 635, **635**
pollution in, 628
producers, 631-632, **632**
Liver, 337, **337**
in digestion, 337-338, **337**
function, 337-338, 412-413
in homeostasis, 412-413
and lymphatic system, 370
protein biosynthesis, 172-173
structure, 337-338, **337**
Liver fluke, life cycle, 596-597, **596**
Liverworts, 757, **758**
Locomotion. *See* Movement
Loeb, J., 690
Loligo, **769**
Long day plants, 238, **238**
Longitudinal fissures, 270, **271**
Loop of Henle, **411,** 412
Lorenz, K., 691, 707
Lotic habitat, 633-634
LSD, 295, 296t, 763
effect on behavior, 740
LTH (lactotropin) 452-453
Luciferase, 160
Luciferin, 160
Lumbricus terrestris. See Earthworm
Lumen, 207, **207**
Lung
alveoli, 400, **400**
cancer, 405-407, **406**
capacity, 402
diaphram, 359, **360,** 401-402
function, 402-403
human, 399-401, **400**
physiology, 399-401, **400**
resuscitation, 403-404
structure, 400-401, **400**
Lutein, 448
Luteinizing hormone. *See* LH
Lyase, 131t
Lycoperdon, **763**
Lycopodium, **759**
Lymph, 368-369, **369**
nodes, 368
Lymphatic system, 368-369, **369**
in digestion, 279
function, 368
in infection, 369, **369,** 379
lacteal, 270
liver, role of, 370
microphages in, 369
organs, 229t
of thoracic duct, 368
Lymphocytes
B cells, 385
T cells, 385
Lysin, 382, **383**
Lysosome, 113, 117, **118**
Lysozyme, 379

M

M line, of muscle, 165
McConnell, J., 710
Macromolecule, 6
formation, 60-61, 170
Macrophage, 220
in lymphatic system, 369
Magnesium, 346t
Malleus, 288, **288**
Mallomonas, **753**
Malonyl Co-A, in lipid synthesis, 172, **172;** *see also* Lipid, metabolism
Malpighi, M., 87
Malpighian tubule, 409, **410**
Malthus, T., 650-651
Maltose, 33
Mammary gland, 301, 449-450
Mangold, H., 479
Marchantia, **758**
Marcker, A., 188
Marijuana, 295, 296t
Marrow, 221
Marsupial, 647
Mass, 3
Mass flow hypothesis, 354, **354**
Mast cells, 220, 258
Mating, 158-159
Matter
in biological systems, 7-8
energy and, 3-20
structural organization in living systems, 6-7
Matthaei, J., 190, 193
Mazia, D., 127
Measurement, metric system, 87-88t
Mechanical receptors, 285-287, **286**
cutaneous senses, 285-286

Mechanical receptors [*cont.*]
hearing, 287–290, **288**
pain, 286–287
proprioceptors, 287
Medulla oblongata, 273
Megagametophyte, 436, **437**
Megasporangium, 436, **437, 439**
Megaspore, 436, **437, 439**
Megasporocyte, 436, **437, 439**
Meiosis, 423–428, **426, 428,** 460
in fungi, 760
life cycles and, 429–430, **429**
results of, 428
Meiosporangium, 436, **437**
Meiospore, 436, 760
Meissner's corpuscle, 285, **286**
Melanin, 119, 303
Melanism, industrial, 544, **545**
Melatonin, 278
Membrane, 93–98, 111
basement, 217, 219t
mitochondrial, 114, **115**
mucous, infection in, 378–379
neuron, 98, **100,** 101, 162–164
nuclear, 107, **109**
permeability, 93–98
plasma, 90, **92,** 98–104
potential, 163
action, 162, **162**
resting, 162, **162**
reversal, **162,** 163
sarcoplasmic, 165, **166**
See also Cell, membrane
Memory, 709–711, 745
long-term, 709–710
RNA in, 709–710
types, 709
traces, 709
Mendel, G., 175, 487–500
Meninges, 269
Meningitis, 269
Menstrual cycle, 250, 256–257, 450–453, **451**
birth control and, 453
phases, 450–452, **451**
Menstruation, 449
Mercury, 68
Meristem, 204, **209**
apical, 458, **458, 459**
ground, **209,** 210, **212,** 458
region, 210
Mescaline, 295, 296t
Meselson, M., 180
Mesocarp, 440, **440**
Mesophyll, 206, 215, **215,** 517, **517**
palisade, 215, **215**
role in transpiration, 352–353, **352**
spongy, 215, **215**
Mesozoa, **764**
Metabolism, 140, 345
anabolism, 130, **130**
catabolism, 130, **130**
control of, 174–199, 250, 252
energy reactions, 130, **130,** 142–151, **143**
fatty acid, 152–154, **153**
in hibernation, 702
inheritable defects in, 524–525
intermediary, 156, **156**
Kreb's cycle, 47, 148, **149,** 150, 156, **156**
metabolites, integration of, **156**
parathyroid, role of, 252
phosphates, 251t, 252
thyroid, role of, 250, 252, **252**
Metamorphosis, 117, 246, 250, 483–484, **484**
in *Cecropia* moth, 246–247, **246**
complete, 483–484
gradual, 484
in grasshopper, 484, **484**
hormones in, 246–247
lysosome, role of, 117
stages, 246–247, **246**
Metaphase
in meiosis, **426,** 427
in mitosis, 120–**123,** 124
Methane, 13, **13**
Methionine, 188
Micelle, 104, **104**
Microbe
characteristics, 374t
control, 374–375
host relationship, 377–385
pathogenecity, 378
relationship to humans, 374t
toxins, 378
types, 373–374, 374t
Microenvironment, 478–479
Microfibril, 104, **104**
Micron, 87
Microphages, in infection, 379
Micropyle, 436–437, **438**
in seed, 455–457, **456**
Microscope, 85, **86,** 87
compound, 87, **89**
electron, 88, **89**
Microsphere, 61–62, **62**
Microsporangia, 436, 437
Microspore, 436, **437**
Microsporocyte, 436, **437**
Microtubule, 118, **120, 122**
Microvillus, 98, **99**
Midbrain, 264, **266,** 272
Migration, 703–705, **704**
in evolution, 548
types, 703
Miller, L., 60
Millipede, 158
Mimicry
advertising, 735
Batesian, 735–736, **736**
deceptive, 735–736, **736**
Mullerian, 735
Mimosa pudica, **243**
Minerals, 345, 346t
in agriculture, 627
concentration in organisms, 609
cycle, 606–609, **608**
in ecosystems, **638,** 639
in food chains, 612, **613**
in soil water, 564
Mirsky, A., 178
Mitochondria, 75, 114, **115,** 159, **164**
cristae, 114, **115**
matrix, 114, **115**
membrane, 114, **115**
Mitosis, 119–128, **120–123**
centromere in, **110–123,** 124, 427, **523**
chromosomes in, 126–127
cytokinesis, **120–123,** 124, 125, 127, 237–238
features, 125
significance, 128
stages, **120–123,** 124–128
duration, 127
Mitospores, 429, **429**
Modulator, 232, **232**
Molecule, 6, 11, 21
covalent bonding, 11–14, **13,** 29
formation, 11–14
hydrogen bonding, 14, **14,** 22, 179, 187
polar, 22, **23**
Mollusca. *See* Mollusk
Mollusk, 299
circulatory system, 356
gaseous exchange, 395–396, **395**
gill, 395–396, **395**
in neritic zone, **635,** 636
respiratory system, 395–396, **395**
shell, 299
structure, 768, **769**
Monarch butterfly, 733
Mongoloid idiocy, 525

Monera, 79–81, 749–751, **750**
asexual reproduction, 420, **421**
cell wall, 299
coverings, 299
in ecosystems, 638, 639
movement, 749
terrestrial consumers, 639
types of reproduction, 430
Monocot, 204, **205**
development, 455, **456**, 457, **457**
flower, 456
germination, 457, **457**
seed, **456**
Monocotyledonae. *See* Monocot
Monocytes, 226t
Monod, J., 196
Monohybrid cross, 487–489, **488, 490**
Monoploidy, 510
Monosaccharide, 31, **31**
biologically important, 30–32
characteristics, 31
classification, 31t
ring form, 31, **32**
Moon, effect on tides, 672
Morgan, L., 690
Morphogenesis, 455
Mosquito, in predation, 597
Moss, 432–433, **433**
club, 758, **759**
structure, **433, 758**
Motion. *See* Energy
Motor
neurons, 161, **161**
pathway, 232, **232**
Mountain zonation, 646, **646**
Mouth
earthworm, 329, **330**
flatworm, 327, **328**
human, 329–330, **333**
Movement, 311, 313–320, **314, 315,** 317–320t
of amoeba, 313, **314**
of arthropods, 313
of chromosomes, 126, 127
of cilium, 168–169, **169**
of earthworm, 313, **314**
of Eumetazoa, 313
of flagellum, 168–169, **169**
human, 314–316, 317–320t
involuntary, 224
of monera, 749
of planaria, 313
of plants, 242–244, **243,** 714
of Protista, 313, **314**
of Vertebrates, 314–316, 317–320t
Movement [*cont.*]
voluntary, 224
of *Vorticella,* 313, **314**
Mucus, 217
Muller, F., 735
Multiple sclerosis, 283
Muscle
of abdomen, 318t
action 315–316, **315**
adaptation, 224
in breathing, 401–402, **401**
cardiac, 224, 226, **227,** 359–361, **360**
intercalated disc, 225, **228**
classification, 224–225
contraction, 165–168, **166, 167, 168,** 224, 315, 319
actomysin in, 165–167
biochemistry of, 165–168, **168**
creatin in, 167–168, **168**
in respiration, 401–403, **401**
sliding filament hypothesis, 166, **167**
of eye, **291,** 292–294, 317t
fiber, 165, 166, **166,** 316
depolarization, 165
elasticity, 224
myofibrils, 165, **166**
sarcoplasmic reticulum, 165, **166**
size, 165
function, 314
of head, 317, 317t
of heart. *See* Muscle, cardiac
human, 317–320t
involuntary movements, 224
latent period, 165
naming, 316, 317–320t
nervous control of, 224
of shoulder, 317t
smooth, 224, 226–227, **227**
sodium ion in, 165
striated, 165, **166**
system, 314–316, 317–320t
tissue, 224, 226–227
voluntary movements, 224
Mushroom, 763, **763**
Mutase, 143
Mutation, 237, 515–516, 540–541
dwarfism, 237
in evolution, 539–541
induced, 182
in populations, 541–543
pressure, 548
Muton, 515
Mutualism, 594
in nitrogen cycle, **605,** 606
Mycelium, 760
Myelin sheath, 98, **100,** 101, **228,** 229
Myelinated fibers, **228,** 229
Myocardium, 360, **360**
Myofibril, 165, **166,** 226, **227**
Myopia, 292, **293**
Myosin, 165
Myostatin, 377t
Myxomycota, **760**
Myxomysin, 169

N

NAD (nicotinamide adenine dinucleotide), 46, 136, **138,** 146, **146**
in alcohol metabolism, 344
in cytochrome system, 150–151, **150**
in deamination, 155
in fatty acid catabolism, **153,** 154
in fermentation, 146–147, **147**
in glycolysis, **144,** 145
in hexose monophosphate shunt, **152**
in Kreb's cycle, 148, **149,** 150
in respiration, 145–146, **146**
NADP (nicotinamide adenine dinucleotide phosphate), 46
in photophosphorylation, 137, **137, 138,** 140
Nare, 397, **398,** 399, **399**
Nasal cavity, 399, **399**
Nasties, 242–243, **243**
Natural selection, 538–545, 549
in evolution, 538–539
in population genetics, 541–544, **545**
Nearctic region, 623, 648, **648**
Necturus, 397
Needham, J., 54
Nekton, **635,** 636
consumers, **635,** 636
Nematocysts, 326, **327**
Neotenin, 247, **247**
Neotropic region, 647–648, **648**
Nephridian, 408–409, **409**
Nephron, 411–412, **411**
Neptune, 69
Nereocystis, **756**
Neris, 263
Neritic zone, 634–636, **635**
consumers, **635,** 636
decomposers, **635,** 636
phytoplankon, 634–635, **635**
zooplankon, **635,** 636
Nerve
auditory, **288,** 290

Nerve [*cont.*]
autonomic, 276–278, **277**
axon, 98, **100,** 101, 161, **161,** 228, **228,** 267
cell. *See* Neuron
cord, 261, **263**
cranial, 264, **267,** 275t
cutaneous, 285–286, **286**
impulse, 162–164, **162**
action current, 165
action potential, 163
depolarization, 163
electrical potential, 160
electrochemical change, 162–164
ions, role of, 162
resting potential, 160, 162, **162**
synapse, 162, 163, **164**
net, 260, **262**
olfactory, 275t
optic, 275t
peripheral, 264, 266, 276, 275–276t
in reflex arc, 266–269, **268**
ring, 260, **262**
spinal, 264, 265, 276, 276t
tissue, 227–229, 275t, 284
Nervous system, 229
activities, 260
of annelids, 262–263
of amphibians, 264
of arthropods, 262, 264, 265
autonomic, 266, **267,** 276, **277,** 278, 302
in behavior, 693
central, 264, 269–276, **269, 270, 271, 272, 273, 273,** 275t
of ciliates, 261
of coelenterates, 260–261
components of, 260
conduction system of, 370–371, **371**
control, 259–297
of crayfish, 264, **265**
development of, 261
disorders of, 278–283
drugs and, 294–297, 296t
of earthworm, 262, **263**
endocrine system integration, 257
of *Euglena,* 260, **261**
of Euplotes, 260, **261**
of fish, 264
of flatworm, 261–262, **263**
of grasshopper, 264, **265**
of human, 265–278, 275–276t
of hydra, 260, **262**
of invertebrates, 260
of jelly fish, 260–261, **262**
Nervous system [*cont.*]
of *Neris,* **263**
parasympathetic, 266, **267,** 276, **276**
peripheral, 264
of planaria, 262, **263**
of paramecium, 260, **261**
reflex, 266–269, **268**
of reptiles, 264
sense organs and, 283–294
somatic, 267
of sponge, 260, **262**
sympathetic, 266, **267,** 276, **277**
Neural fold, 464–466, **465, 466**
Neural plate, 464, **465**
Neural tube, 464–466, **465, 466**
Neurilemma, 228–229, **228**
Neurofibril, 228, **228**
Neuroglia, 227
Neurohumor, 257
Neurohypophysis, **249,** 254, **255**
Neuron 160–164, **161, 162,** 227
afferent, 267, 269
axon, 225, **228**
cell body, 228, **228**
classification, 227–228, 229t
dendrite, 161, **161,** 162, 228, **228**
depolarization of, 163
effector, 161–162, **161,** 232, **232,** 260
function, 229t
impulse conduction, 160–164, **162, 164**
membrane, 98, **100,** 101
morphology, 160–164
motor, 161, **161,** 225, 229t, 267
neurofibrils, 228, **228**
perikaryon, **228**
receptor, 161–162, **161**
respiration and, 402–403
sensory, 225, 229t, 267, 269
spinal, 264–265, 276, 276t
structure, 225, 227–229, 228t
synapse, 163, **164**
Neutrons, 9
Niche, 713, 715
ecological, 591–593, **592,** 594
food, 592–593, **592**
Nicotinamide adenine dinucleotide. *See* NAD
Nirenberg, M., 190, 193
Nissl granule, 225, **228**
Nitella, 97, **756**
Nitrates in nitrogen cycle, 604–606, **605**
Nitrogen
in atmosphere, 24
cycle, 604–606, **605**
fixation, 605–606, **605**
Nitrogeneous base, 183
Nitrosomonas, 605–606, **605**
Node, 212, **212**
of heart, 370, **371**
lymph, 368
of Ranvier, 228, **229**
Nomura, M., 189
Noncyclic photophosphorylation, 137–139, **137**
Norepinephrine, 251t, 253, 258
Nose, 399, **399**
Nostoc, 128, 755, **757**
Nostril. *See* Nare
Notochord, 464–466, **465,** 770
Nuclear membrane, 107, **109**
matrix, 107
Nucleic acid, 47–49, 171
in evolution, 65
ribosome and, 184
Nucleolus, 107, **109**
Nucleotide, 179, 180, 183, 514
Nucleus, 105, 107, 108, **109**
activities, 126, 474, 477
antipodal, 436, **437, 438**
appearance, 75–76
in conjugation, 441, **441**
control of, 107, **107**
function in development, 474–477
generative, 436, **437**
membrane, 107, **109**
nucleolus in, 107, **109,** 110
polar, 436, **437**
pronucleus, 442, 461
segmentation, 467
tube, 436, **437,** 438, **438**
Nutation, 242
Nutrient, 322, 345
absorption of, 270
autotophic, 322–323
in ecosystem, 599, **600**
heterotrophic, 323–324
transportation in organisms, 349–370
transportation in plants, 350–354, **351, 352, 354**
Nutritional pattern, 77–79

O

Ocean, 570–573
abyssal plain, 570, **571**
basin, 570–571, **571**
in calcium cycle, 608–609, **608**
composition, 571–572, **572**
currents, 572–573, **572**

Ocean [*cont.*]
ecosystem, 634–637, **635**
nutrients in, 572, 572t
in sedimentary cycles, 606–609, **608**
thermocline, 572, **573**
tides, 572–573, **575**
world food supply, 660–661
zones, 570–571, **571**
Ocelli, 260, **262**
Ochoa, S., 190–191, 193
Octopus, 768, **769**
eye, 717
Oedogonium, 430, **431**
heterogamy, **431**
Olifaction, 285
Olifactory epithelium, 285, **285**
Omnivores, 323, **600**
Oogenesis, 460–461, **460**
Oogonium, in *Oedogonium,* 430, **431**
Oomycota, 762, **762**
Oophorectomy, 622
Oparin, A., 57, 58–59
Operon, 197, 513
Opiates, 296t, 297
Opium, 296t, 297
Opossum, 647
Opsonin, 294
Optic vesicle, 464–465, **464**
Organ, 6, 204
of Corti, 288, 289–290
development, 468–469, **467,** 470t
evolution of, 532, **533**
excretory, 229t
functions, 229t
homologous, 532, **532**
human, 229t
lymphatic, 229t
plant, 208–216
of Ruffini, 276–276, **276**
sense, 283–294
system concept, 228–229, 229t
transplant, 386–391
Organelle, 6, 107–118, 755
in cell evolution, 65–66
See also Cell
Organic compound
formation 58–60, **59, 60**
functional groups, 29–30, 30t
in organisms, 30
structural formulas, 29–30
synthesis experiment, 60, **60**
Organic matter
decomposition, 621–622
deposition, 621–622
Organic matter [*cont.*]
in ecosystems, 621–622
in soil, 565–566
Organism
classification, 73–81
diversity, 79–81
multicellular, 204
Organizer, 479
Oriental region, 648, **648**
Osculum, 764, **765**
Osmosis, 94–96, **94, 95,** 323, 353
cell membrane, 96
Osmotic pressure, 95–96
Ossification, 308
Osteoblast, **221,** 224, 309
Osteocyte, 222
Otolith, 289
Outer ear, 287–288, **288**
Ovary, 248, **249,** 446–447, **447**
Overturn, 568–569, **568**
Ovulation, 250, 448
Ovule, **435,** 436–437, **438, 439**
Ovum, 461, **461,** 467, 468
human, 461, **461**
Oxaloacetic acid, 148, **149**
Oxidation-reduction, 145
biological, 146, **146**
Oxygen, 7t, 301
in atmosphere, 24
in biosynthesis, 169, 170
in blood, 357
cycle, 603–604, **603**
debt, 168
in ecosystems, 600, 633–634
in gaseous exchange, 393, 394
in living matter, 28t
in photosynthesis, 131, 139
in plasma, 359
in respiration, 398, 402
in transportation, 395
Oxytocin, 251t, 256–257
Oysters
population distribution, 587
structure, 768
survivorship curves, 584, 585
tolerance levels, 615
Ozone, 64

P

Pacemaker, 389
Pacinian corpusle, 285, **286**
Pain, 286–288
Palaearctic region, 648, **648**
Palate, 399, **399**
Paleontology, 534–538
Palisade layer, 215, **215**
Pancreas, 248, **249,** 251t, **335,** 337, 341–342
in digestion, 335–337, 338t, 341–342
hormones, 248, 249, 251t
Pancreatic juice, 337, 338t
Pancreozymin, 250, 251t
Papilio polyxenes, **771**
Papillae, **230,** 330, **333**
Paramecium, 96, 111, 261
anal pore in, 407, **408**
behavior, 694, **694**
binary fission, 420, **421**
circulation, 354
conjugation, 441–442, **441**
cytopyge, 326, **326**
digestion, 325–326, **326**
excretion, 407, **408**
fertilization, **441,** 442
food vacuole in, **326,** 354
nervous system, 260, **261**
reproduction, 420, **421**
Parasite, 323, 595–597, **596, 600,** 760–762
in classification, 79
chemoreception of, 717
in ecosystem, 595–597, **596,** 601–602
in food chain, 612, **613**
external, 595
facultative, 594
interspecific interactions of, 595–597, **596**
microbes, 377–385
obligatory, 594
types, 595
world food supply and, 659–600
Parasitic food chain, 612, 613
Parasympathetic nervous system, 266, **267,** 276, **277**
Parathyroid, 248, **249,** 251, 252
Paratonic movement, 242, **243**
Parazoa, 764, **765**
Parenchyma, 206, **206,** 208, 216, 352–353, **352**
Parkinsonism, 280
Parthenocarpy, 235, 440
Parthenogenesis, 442
Passive transport, 96–97
Pasteur, L., 55–56
biogenesis experiments, 55–56, **56**
Patella, **305,** 307
reflex, 266–267, **268**
Pathogens. *See* Microbe
Pavlov, I., 267

Pecking order, 729, **729**
Pedicel, 435, **435**
Pelagic zone, **635,** 636–637
  components, **635,** 636–637
  krill, 637
  zooplankton, 637
Pellicle, **261,** 299
Pelvic girdle, 307
Penicillin, 375, 377t
Penis, human, **444,** 446
Pentose phosphate shunt, 151–152, **152**
Pepsin, 334, 338t
Peptic ulcers, 343
Peptide link, 188
Peranema, **752**
Perca, **772**
Perianth, 435, **435**
Pericarp, 440, **440**
Pericardium, 359–360, **360**
Perichondrium, 220
Pericycle, 210
Periderm, 211, **211**
Peridinium, **735**
Perikaryon, 161
Perineum, **447,** 449
Perinuclear cisterna, 107, **109**
Periosteum, 220, **221,** 223
Peripheral nervous system, 264, **267,** 276
Peristalsis, 333
  in esophagus, 333
  in large intestine, 340
  in small intestine, 338–339
Peritonitis, 343
Pernicious anemia, 348t
Pesticides
  in biogeochemical cycles, 609
  DDT, 623, 624–627
Petit mal, 281, **282**
PGAL (phosphoglyceric acid), 139, 140, **141,** 170
  in glycolysis, 143–145, **144**
  in metabolism, 143–145
  in photosynthesis, 139, **141**
pH, 24, 26–27, **27,** 37
  in chemoreception, 717
  in dental caries, 342
  in excretion, 412
  in respiration, 403
  saliva, 332
  stomach, 334
Phaeophyta, 755, **756**
Phagocytosis, 98, **98,** 325
  in infection, 379
  of leukocyte, 98
  and virulence, 378
Pharynx, 332–333, **333**
  flatworms, 327, **328**
  in respiration, 399, **399**
  structure, 399, **399**
Phaseolus multiflorus, 237
Phaseolus vulgaris, **243**
Phellem, 211
Phelloderm, 211
Phellogen. *See* Cork, cambium
Phenotype, 489
Phenylalanine, polypeptide synthesis, 193, **193,** 524
Phenylketonuria, 524
Pheromone, 720, **720**
Phleum pratense
  chloroplast, **116**
  endoplasmic reticulum, **112**
Phloem, 208, 210, 213, **213, 214, 232**
  fiber, 208
  mass flow hypothesis, 354, **354**
  protoplasmic streaming hypothesis, 354
  primary, **209,** 210, **211,** 213
  ray, 213, **213,** 214
  root, **209,** 210, 211, **211**
  secondary, 211, **211,** 213, **213**
  sieve tube elements, 353, **354**
  stem, 213, 214, **214**
  transportation of nutrients, 353–354, **354**
Phosphate metabolism, 251t, 252
Phosphoglyceraldehyde, 137, **141**
Phosphoglyceric acid. *See* PGAL
Phospholipids, 35t
Phosphorus, 346t
Phosphorus in biogeochemical cycles, 609
Phosphorylation, 140, 143
Photic zone, 751, **751**
Photon, 133, 134, **134**
  theory, 133, 134
Photoperiodism, 238–242, **238**
Photophosphorylation, 135
  cyclic, 135, **135**
  noncyclic, 137–139, **137**
  oxidation-reduction in, 137, **138**
Photoreception, 290–294, **291, 293, 294**
  process, 292
Photosynthesis, 64, 130–140
  carbon dioxide in, 130–132, **132,** 137
  cyclic photophosphorylation, 135, **135**
  dark reaction, 134–140, **134, 135, 136,** 139–140, **141**
  in diatoms, 751–752
  in ecosystem, 599–601
Photosynthesis [*cont.*]
  electrontransfer in, 135–139, **135, 137**
  energy flow and, 610
  in evolution, 64
  gaseous exchange and, 393
  historical development, 130–131
  light, role of, 132
  light reaction, 132, 134–139, **135, 137**
  in Neritic zone, 634–635, **635**
  oxidation-reduction in, 137, **138**
  noncyclic photophosphorylation, 137–139, **137**
  oxygen in, 131, 139
  rate factors, 614
  raw materials, 131–132
  reactions, 134–140, **134, 135, 136, 141**
  and respiration compared, 142, 142t
  significance, 130
  water in, 131–132
Phototropism, 244, **244**
Phragmoplast, **123,** 125
Phycobilin, 132
Phycoerythrin, 755
Phylum, in classification, 74
Physalia, **764,** 766
Physical change, 3
Physical environment, 322
Phytophthora infestans, 762, **762**
Phytoplankton, 751
  blooms, 633
  in Limnetic zone, **632,** 633
  in Neritic zone, 634–635, **635**
Pia mater, 269, **269, 273**
Pigeon, homing in, 705
Pigment I. *See* Chlorophyll *a*
Pigment II. *See* Chlorophyll *b*
Pineal gland, **249,** 257
Pinna, 287, **288, 759**
Pinocytosis, 98, **98**
Pinus ponderosa, 555
Pinus sabiniana, 555
Pistil, 435, **435**
Pisum sativum, 237
Pith, 213, **213, 214**
Pit, 101, **103,** 207
  border, 101, **103**
  primary field, 101, **102, 103**
  simple, 101, **103**
  types, 101, **103**
Pituitary gland, 248, **249,** 448
  hormones, 254–257, 251t, **255,** 470–473, **470, 471, 472**
    control, 257
  posterior, 264
  role of, 256, **256**

Placenta, 257–258
  in circulation, 366, **366**
  in development, 470, **471**
  in flower, 435, **435**
  hormones, 257–258
Placental lactogen, 257
Planaria
  auricles, 262, **263**
  cell layers, 355
  digestive system, **329,** 358
  excretory system, 408, **408**
  eye spot, 262, **263**
  gaseous exchange, 395
  movement, 313
  nervous system, 262, **263**
  regeneration in, 422, **425**
  structure, **263, 329**
Planet, 67–71
Plankton, behavior of, 699–700
Plant, 80–81, 216, 257–258, 755–758, **756, 758**
  adaptation of, 561–562
  adaptive behavior, 731–732
  asexual reproduction, 422, **425**
  auxin, 235–237, **237,** 242–245, **242, 243, 244**
  coverings, 299
  cyclosis, 105, 350
  development, 236–237, **237,** 455–458, **456, 457, 458**
  diffusion in, 350
  epidermis, 299
  fertilization, 432–434, **433,** 438, **438**
  flowering
    day neutral, 238–239, **238**
    light, effect on, 238–242, **240**
    long-day plant, 238–242, **238, 239, 240**
    short-day plant, 238–242, **238, 239, 240**
  gaseous exchange, 393–394
  genetic dwarfism, 237
  growth, 204–205, 233–234, **233, 234,** 459
    autonomic, 713
    auxin, effect of, 235–236
    cytokinin, 237–238
    distribution of, 459
    experiment, 233–235
    embryonic development, 455, **456**
    gibberellin effects, 236–237
    movements, 242–245, **243, 244,** 713–714
    regulators. *See* Auxins
Plant [*cont.*]
    stages, 236–237, **237,** 455–458, **456, 457, 458**
  insectivorous, 324, **325,** 597, 714
  mass flow hypothesis, 354, **354**
  metabolism, 322–323
  movement
    autonomic, 242, **242**
    paratonic, 242–245, **243, 244**
  organ system concept, 208–216
  nutrient, transportation of, 350–354, **351, 352, 354**
  protoplasmic streaming hypothesis, 350, 354
  sexual reproduction, 430–441
  spore production, 420–421, **423**
  support, 303
  and terrestrial consumers, 635, 638
  tissues, 204–216
  vegetative reproduction, 421–422, **424**
Plasma, 220, 224, 259, 268
  cells, 220, 378, 384–385, **384**
  complement, 382
  function, 359
  hemoglobin in, 359
  in infection, 379
  oxygen in, 359
Plasma membrane, **92,** 98–104
Plasmodesmata, 101, **103**
Plasmodium vivax, 752–753, **754**
Plasmogamy, 760
Plasmolysis, **95,** 96
Plastid, 114–117, **115, 116**
Plastoquinone, 137, **137**
Platelet, 357, 358
Platyhelminthes, 766, **767**
  digestion in, 327–328
Platypus, duck-billed, **772**
Pleura, 400
Pleuropneumonia-like organism (PPLO), 374, 374t
Plexus, 276, 276t
Pneumonia bacterium, 75
Polar body, 460, **460, 461**
Polar molecule, 22, **23**
Polar nucleus, 436, **437**
Poliomyelitis, 278–279
Pollen
  development, 436, **437, 438**
  grain, **511**
  sacs, 436, 437
  tube, 438, **438**
Pollination, 436, **438**
  cross, 435, **438**
  self, 436, **438**
Pollution, 673–684, **629, 671**
  by agriculture, 675–676, **676,** 677–680
  air, **671,** 673, **673,** 676–677, **676**
  climatic effects, 672–673
  degradable, 674
  domestic, 674–675, 676, 677, 680–681
  effects of, 673–684, **682**
  indicator species and, 616–617
  industrial, 669–672, 675–676, **676,** 678–679, **679**
  and insecticides, 672
  land, 677–681, **678,** 680, **680**
  in littoral ecosystem, 628
  nondegradable, 674
  population and, 669
  radioactive, 677, **678**
  reduction of, 683–684
  solid wastes, **671,** 674
  technological change and, 669–672
  thermal, 672
  water, 674–676, **675,** 681–682, **683**
Polymerization, 33
Polymorphism, 589–590, **590**
Polypeptide synthesis. *See* Protein, synthesis
Polyphemus moth, 733–734, **735**
Polyploidy, 510–513, **511**
Polyribosomes, 113, **113,** 189
Polysaccharide, 32–33, **35**
  synthesis, 170, **171**
Polytrichum, 432–433, **433,** 582
Pond, 569, 631, **632**
Pons, **270**
  function, 273
Population, 6
  in biosphere, 650
  birthrate, 584
  characteristics, 582–583
  crash, 586, **586**
  death rate, 584–585, **584**
  density, 587–588
  distribution, 587
  in ecological pyramids, 612, **614**
  in ecosystems, 582–597
  emigration, 585
  environmental resistance, 585–587, **586, 587**
  in evolution, 549–550, **549, 550**
  genetics, 541–550
  growth, 585–587, **586, 587**
    in ecosystems, 582–597
    rate, 583–584, **587**
  human
    control of, 662–668
    implications, 661–662

Population [*cont.*]
types, 652
and world food supply, 657–662
See also Human, population; Population, world
immigration, 585
interaction in ecosystem, 585–588
in natural selection, 541–545
regulation, 585–588
size, 583
social organization, 589–590
species and, 74, 550–551
tolerance levels, 614
world
age and sex, 654–656, 655t
birth rates, 655–656
distribution, 652, **653**
growth rates, 653–654, **654,** 656
urban, 656–657
Pore, **109**
Porifera. *See* Sponge
Porphyrin, 35t, **156,** 573
Potamogeton pectinatus, Golgi complex, **113**
Potassium, 346t
ions in bioelectricity, 160
ions in muscle contraction, 165
Potato
sweet, 422, **424**
white, 422, **424**
Potential energy, 3, 4
Predation, 597
in ecosystems, 597
in food chain, 612, **613**
releasing mechanisms, 733–734
Predator chain, 612, **613**
Pressure
atmospheric, in breathing, 401–402
blood, 367–368
diastolic, 367–368
diffusion, 94
osmotic, 94, 95–96
systolic, 367–368
turgor, 95, **95**
Priestley, J., 130
Procambium, 209–210, **209,** 212
Procaryotic cell, 65, 75–76, 107
Procell, 61–62, **62**
Producer, 600–601, **600**
limnetic, **632,** 633
littoral, 632, **632,** 634
terrestrial, 638–639, **638**
Profundal zone, components of, **632,** 633
Progesterone, 250, 251t
Prolactin, 251t, 254
Promeristem, 210
Pronucleus, 467
Prophase
in meiosis, 426–427, **426**
in mitosis, 121, **123,** 124
Proprioceptors, 287, **287**
Prostate gland, **444,** 446
Prosthetic group, 45, **45**
Protection, 298–303; *see also* Coverings
Protein, 34–47, 171, 245
in cell, **92,** 93
chemical reaction of, 43
classification, 42–43, 44t
contractile, 167
denaturation, 43, 375
digestion, 172
formation, 37
crops high in, 659
levels of organization, 40–42, **41, 43**
in membrane, **92,** 93
in nitrogen cycle, 604
structure, 40–42, **41, 43**
synthesis, 172–173, 184–190, **189, 192**
anticodon in, 185–190
codon in, 184
deamination, 43, 155, 173
gene, role of, 185–190
ribosome, role of, 185–190
transamination, 154–155, 173
transcription, 185–186, **185**
translation, 186–187
Proteinoid microspheres. *See* Procells
Prothoracotropic hormone, 247, **247**
Prothrombin, 359
Protista
asexual reproduction, 420, **420**
autotrophic, 751–752, **752**
behavior, 694, **694**
cell wall, 299
chlorophyll, 751
chromosomes, 751
circulation, 354
conjugation, 441–442, **441**
coordination, 260
coverings, 299
digestion, 324–326, **326**
in ecosystem, 638, 639
excretion, 407, **408**
flagella, 751
framework, 303, 304
gaseous exchange, 394
heterotrophic, 752–755, **752, 754**
in limnetic zone, **632,** 633
movement, 313, **314**
Protista [*cont.*]
reproduction, 441–442, **441,** 751
structure, 303, **304**
Protoderm, **209,** 210, 458
Proton, 9
in cytochrome system, 150, **150**
Protoplasmic streaming hypothesis, 350, 354
Protozoa, 374, 374t
digestion, 324–326, **326**
disease-producing, 374, 374t
in mutualism, 594
Pseudocoelomata, 77, **78,** 250, 766, **768**
Pseudoplasmodium, 760, **761**
Pseudopodium, 98, 168–169, **169,** 325, **326,** 753
Psilocybin, 295, 296t
Pulmonary circulation, 361–362, **362**
Pulse, 368
Pupa, **246,** 484
Pupil, 291, **291**
Purine, 47–48, **48,** 195–196
Purkinje fibers, 370, **371**
Pus, 358, 380, **380**
Pyloric mucosa, 251t
Pyramids, ecological, 612–614, **615**
Pyrimidine, 48, **48,** 156, 195
Pyrrophyta, 752, **753**
Pyruvic acid, 142, **144,** 145, 171, 173, 196
glycolysis, 142, **144,** 145
in metabolism, 171, 173, 196

Q

Quantasomes, **116,** 117
Quercus, 74, **75**
Quercus suber, 207

R

Rabbit
in evolution, 540
Himalayan, 516, **516**
snowshoe, 734
Radial symmetry, 76, 260–261
in Eumetazoa, 764
Radiant energy, 5
Radiation
adaptive, 557
effects on behavior, 739
electromagnetic, 72
and mutation, 515

Radioactive fallout, 677, **678**
radioactive isotopes, 536
tracers, 353–354
Radioactivity, 536, **536**
Radiolarian, 158
Radio telescope, 71, **71**
Ranna pipens. See Frog
Ranvier, nodes of, 228, **229**
Raphanobrassica, 512–513, **512,** 533
Raphid, 119
Rearrangement reaction, 14–15
Recapitulation, 532–533, **532**
Receptor, 232, **232,** 260
chemical, 283–285, **284**
cutaneous, **286**
electrical, 717
mechanical, 285–287, **286**
neurons, 161–162, **161**
pain, 286–288
photoreceptor, 290–294, **291, 293, 294**
proprioceptors, 287, **287**
sesory, types, of, 283
Recessive character, 487
Recon, 515
Rectum, 339–340, **339**
Redi, F., 53–54, **53**
Reduction, in photosynthesis, 136; *see also* Oxidation-reduction
Reflex, 266–269, **268,** 695
Achilles, 266–267
arc, 266–269, **268**
conditioned, 267–269, 706
monosynaptic, 268
simple, 266–267, **268,** 695
Refraction, in eye, 292, **293**
Regeneration, 422, **425**
Regulation genes, 197–198, **198**
Releasing mechanism, in behavior, 697–699, **699, 700,** 733–734, **734**
Replication, semiconservative, 176–180
Reproduction, 229t
Angiosperms, 434–440, **435, 437, 438, 439**
asexual, 420–422, 425, 430, **431**
in bacteria, 420
in Fungi, 760
in Monera, 420, **421**
in plants, 421–422, **424**
in Protista, 420, **421,** 751
behavior, 702–703, **703**
of cell, 423–424
DNA in, 419–420
human, 446–450, **447**
hermaphroditism, 442
heterogamy, 430, **431**
Reproduction [*cont.*]
isolation, 550–555, **551,** 553t, **554**
parthenogenesis, 442
patterns, 419–453
potential, 583
random, 548–549
sexual, 422–453
in animals, 441–453
in Monera, 430
in Fungi, 760
in plants, 430–441
in Protista, 441–442, **441**
sexual vs. asexual, 428–429
stages in Fungi, 760
types, 419–420
vegetative, 421–422, **424**
Reproductive cycle, 421–422, **424**
and behavior, 702, 703
control, 452–453
female, 450–453
Reproductive system
human female, 446–450, **447**
human male, 443–446, **444**
Reptile
adrenal gland, 252–253
brain, 266, **266**
heart, 362–363, **362**
poison gland, 302
skin, 302
Respiration, 114, 140, 142
aerobic, 64, 145–146, 147–150, **149, 150**
anaerobic, 145–146, **147**
carbon dioxide in, 398, 402, 403
control, 402, 403
in evolution, 64
external 394
factors influencing, 403
internal, 394
Kreb's cycle, 148, **149**
lungs in, 399–401, **400**
in mitochondria, 114
pH in, 403
photosynthesis and, compared, 142, 142t
pigments, 356
types, 145–147
See also Gaseous exchange
Respiratory system
anatomy, 398–401, **399, 400**
of amphibian, 396–397
of clam, 395–396, **395**
of fish, 396, **397**
of frog, 397–398
of grasshopper, 396, **396**
of human, 398–403, **399, 400**
Respiratory system [*cont.*]
of mollusks, 395, **395**
of spider, 396, **396**
Response
allergy, 386
avoidance, 694
conditioned, 705–707, **706**
phototactic, 694
Resting potential, 160, 162, **162**
Reticular fibers, 220
Reticulin, 105
Reticuloendothelial system (RES), 712
in infection, 379
Retina, 291, **291,** 464–465, **465**
Retinene, 294
Rh factor, 357, 521–522, **522**
Rhagoletis pomonella, 555
Rhesus monkey, 690–691, **691,** 728–729
Rhizobium, **605,** 606
Rhizoid, 421
Rhizome, 211, **759**
Rhizopus stolonifer, 421, **423,** 762, **763**
Rhodophyta, 755, **756**
Rhodopsin, 294
Rhynia, **759**
Rhythmic behavior, 716
Ribitol, **136**
Ribonucleic acid. *See* RNA
Ribose, 31, **31**
Ribosome, 111, **112,** 113, **191,** 192
components, 187–188, **188**
nucleic acid and, 184
protein synthesis and, 187–190
RNA and, 184
structure, 187–188, **188**
subunits, 187–188
Ribs, 305
Ribulose phosphate, 140, **141,** 151, **152**
Rich, A., 189
Rickettsiae, 374, 374t
Rivers, 570
RNA (ribonucleic acid), 47–48, 111, 117, 184, 533
in antibody formation, 384
in brain, 740
in choroplast, 117
chromosome puffs, 478
in enzyme synthesis, 184
in memory, 709–710
messenger, 48, 185–186, 189, 190
in mitochondria, 114
in nucleoli, 107, 110
ribosomal, 48, 184, 187–188, **190**
structure, 184
subgroups, 48

RNA [*cont.*]
synthesis, 191, 193
transfer, 48, 186, **186,** 188, 189, 190, 191, **192, 193**
types, 49
See also DNA
Robertson, J., 90
Robina pseudoacacia, **101**
Rods, 98, 100, 294
Romalea microptera. See Grasshopper
Root, 216, **457**
absorption, 322, **323**
cap, 209, **209,** 458, **458**
cork cambium, 210–211, **211**
development, 236–237, **237, 457,** 458, **458**
differentiation region, **209,** 210, 458
enlargement region, **209,** 210, 458
function, 323, **323**
geotropism, 244, **244**
growth, **209,** 210, 458, **458**
hairs, 201, **209,** 323, **323**
pressure, 351–352, **351**
structure, 209–211, **216,** 323, **323**
tip, 209–210, **209**
tissue
primary, 210
secondary, 211
Roundworm, cuticle, 299, **300**
Ruffini, end organs of, 275–276, **276**

S

Sac fungi (*Ascomycota*), 762–763, **762**
Saccharomyces, **763**
Sacral, 276, 276t
Sadler, M., 651
Salamander, **772**
Saliva, 338t, 341
in digestion, 331–332
secretion, 331–332
Salivary gland, 331–332, 341
Salmon, behavior of, 703
Salt, 23, 24, **24**
in excretion, 410
Sanger, F., 188
Sap, 119, 350–351, 353
Saprolegnia parasitica, **762**
Saprophyte, 77–78, 323, 599, **600**
types, 760
Saprophytic chain, 612, **613**
Sarcodina, 753, **754**
Sarcolemma, **164,** 165, 226, **227**
Sarcomeres, 165, 166, 167
Sarcoplasm, 226–227, **227**
Sarcoplasmic membrane, 165, **166**
Sarcoplasmic reticulum, 165, **166,** 167
Saturn, 69
Saussure, N. de, 130–131
Savannas, 642
Scavanger, 601
Schizocoela, 766, 768, **770**
Schleiden, M., 56, 87
Schotte, O., 479
Schwann, T., 56, 87
Schwann cell, 98, **100,** 101, **228,** 229
Sclereids, 207
Sclerenchyma, 206–207, **206**
Scrotum, 443, **444,** 663
Scypha, **765**
Seal, Alaskan, 558–589, **589**
Sebaceous glands, **203,** 378
Second law of thermodynamics, 6, 602
Secretin, 248, **250,** 251t, 342
Sedimentary cycles, 606–609, **608**
Seed, 438–440, **439**
development, 455–457, **456**
dissemination, 439–440
endosperm, 438, **439**
germination, 457–458, **457**
micropyle, 436, 437, **438**
structure, **439,** 455, **456,** 457
viability, 440
Selective theory, in antibody formation, 384–385, **384**
Semicircular canal, **288,** 289
in equilibrium, 289
Seminal fluid, 446
Seminal vesicles, **444,** 446
Seminiferous tubule, 443, **444**
Sense organs, 283–294, 470t
Sensory neuron, 161, 267, 269
Sensory pathway, 232, **232**
Sepal, 435–436, **435**
Seral stage, 617, **618**
Serine, 173
Serotonin, 258, 721
Setae, 313, **314**
in chemoreception, 716
Sex
characteristics, 241, 249, 253
chromosomes, 501, **501**
determination, 501, **502**
and inheretance, 522–523
See also Genetics; Heredity
Sexual intercourse, 453
Shell, 299
Sheldon, W., 738
Shelford, V., 615
Shoot, development, **216,** 457–459, **457**
Short-day plant, 238–242, **238, 239, 240**
Sickle-cell anemia, 525
Sieve plate, 208, **208**
Sieve tube, 208, **208,** 353
Simple reflex, 267–268, **268**
Sign stimuli, 731, 732, 742
Siren, 297
Skeleton
appendicular, 304, 305–307, **305**
axial, 304–305, **306**
exoskeleton, 303
functions, 304
human, 304–308, **305, 306,** 309–311t
articulation, 307, 309–311t
components, 304
muscles, 315, 317–320t
of protista, 303–304, **304**
of sponge, 303
vertebrates, 304–308, **305,** 306
Skin, 217, **230,** 232, **232,** 298, 299, 300–303
color, 303, 519–520, **519**
defense mechanisms, 378–379
in excretion, 410
of fish, 302
function of, 299, 300, 311–312
horny, 302
human, 302
in infection, 378–379
of land vertebrates, 302
modification 302–303
regeneration, 422
in respiration, 301
reptile, 302
sense organs, 285–287, **286**
structure, 302
Skinner, B., 690, 760
Skinner box, 706, **706**
Skull, **305,** 306
Sleep, 721
Sliding filament hypothesis, 166, **167**
Slime mold, 760, **761**
Small intestine, 335–339, **335, 336**
function, 269–270, 335–337, 338–339
peristalsis in, 338, 339
regions, **335,** 336
secretions, 338, 338t
structure, 336–337
Smell, sense of, 264, 283, 284–285, **285**
in mating behavior, 727–728
See also Chemoreception
Smog, 672–673, **673**
Snail, 158
structure, 768, **769**

Snake, water, **772**
Social behavior
aggression, 731–732, **731**
communication, 723–724
dominance in, 731, 732–733
of family, 728, 729
of group, 729–730, **729**
human, 741–746
environment, effect on, 742–743
levels, 742, 743–744
motivation in, 744–745
mating, 726–728, **727**
nature, of, 725–733
releasing mechanism, 733–736, **734, 735, 736**
territoriality in, 730–731
Social institutions, 745–746
Social organization, insects, 589–590, **590**
Sodium chloride, **22,** 105, 346t
ion, 160–161, 162–163
in sea water, 571–572, 572t
Soil, 562–566, **563, 565**
air in, 393, 564–565
in ecosystems, 637–638
organic matter, in 565–566
organisms in, 566
profile, 562, **563**
rock and mineral matter, 562–564, **563**
types, 563–564
water, 564
Sol, 105, **106**
Solar radiation, in photosynthesis, 133–134
Solution, 22, **23**
hypertonic, 95–96, **95**
hypotonic, 95, **95**
isotonic, **95,** 96
Somatic nervous system, **267**
Somatotropin. *See* STH
Somite, 466, **466**
Sorocarps, 760
Sorus, 433–434, **434**
Sound and behavior, 718–719
Space
exploration, 69–70
extraterrestrial life, 66–67
vehicles, 67, 68
Space biology. *See* Exobiology
Spallanzani, L., 54–55
Spanish moss, 595, **595**
Sparrow, **772**
Speciation, 550–555, **551,** 553t, **554**
Species, 74
alien, 627, **628**
indicator, 616–617
See also Speciation
Species-specific behavior, 695–699, **696, 697, 698**
between species, 733–734, **734**
components of, 697–699
consummatory act, 697–699, **698, 699, 700**
criteria, 696–697
experiences, 711
nature of, 695–699
and vision, 718
Spectrum, visible, 133–134, **133**
Spemann, H., 479
Spencer, H., 651
Sperm, 444–445, **445**
in fertilization, 467
human, 444–445
in pollination, 438, **438**
structure, 444–445, **445**
types, **445**
Spermatocyte, 460, **460**
Spermatogenesis, in animal development, 443–44, 460, **460**
Spermatozoa. *See* Sperm
Spherical symmetry, 76, **77**
Spider
behavior, 696, **696**
book lung, 396, **396**
respiratory system, 396
web of, 696, **696**
Spinal cord, 264, 267, **267,** 273, **273, 274,** 275t, 276
dorsal root, 273, **273**
function, 273, 274
ganglion, 273, **273**
grey matter, 270, 272, **273,** 278
structure, 273, **273**
white matter, 273, **273,** 276, 276t
Spinal nerve, 264–273, 276, 276t
Spindle
apparatus, **120–123,** 124
in cell division, **120–123,** 124, 126
origin and nature, 126
protein, 126
Spiracle, in grasshopper, 396, **396**
Spirillum, 749, **750**
Spirochaetae, 749, **750**
Spirogyra, 755, **757**
Sponge, 158, **764,** 765
amoebocytes, 355
choanocytes, 355
digestion, 326, **327,** 377
Sponge [*cont.*]
coordination, 260, **262**
gaseous exchange, 394
structure, 303, **304**
Spongocoel, 764, **765**
Spontaneous generation. *See* Abiogenesis; evolution
Sporangium, 421, **423,** 760, 762, **762,** 763
in Fungi, 420, 760
in Plantae, 420, 423, 430, **431,** 433–434, **434**
Spore
endospore, 420
formation, 420–421, **423**
megaspore, 436, **437**
meiospore, 760
microspore, 436, **437**
mitospore, 429, **429**
production, 420–421, **423**
in Protista, 752
in slime molds, 760
zoospore, 420–421, 423
zygospore, 429–430, **429, 430**
Sporophyte generation, 430–432, **432, 433–434, 434**
Sporozoa, 752–753, **754**
Spring overturn, 568, 569
Springs, 569–570
Squid, 158
Squirrel
Abert, 553–555, **554**
Kaibab, 553, 555, **554**
Stahl, P., 180
Stamen, 435, 435
Standing water ecosystem, 631–633, **632**
Stapes, 288, **288**
Star, 71–72
Starch, 32–33
digestion of, 331–332
grain, 33, 119
Starfish, regeneration in, 422, **425**
Statice sinuata
mitochondrion of, **115**
nucleus of, **109**
Statocyst, 260, **262**
of crayfish, 264–265
of jelly fish, 260, **262**
Steady state. *See* Chemical equilibrium
Stele, 210
Stem, 211–214, **212**
cork cambium, 210, 213, **214**
cortex, **212**
epidermis, 212
structure, 212, **212**
tissue

Stem [*cont.*]
functions, 211
primary, 212, **212**
secondary, 213–214, **213**
types, 211–212
Steppe, 643
Sterilization
in birth control, 662–663
of microbes, 374–375, **375**
Steroid, 35t, 245
Steward, F., 474, 477
STH (somatotropin), 251t, 255–256, **255,** 257, 479
Stickleback, behavior of, 697–699, **698, 699, 700,** 732
Stigma, 261, 434–437, **435**
in *Euglena,* 260
Stimulus, 232, **232**
external, 693
internal, 693
releasing, 697–699, **699**
response, 692
sign, 697–699, **698, 699, 700**
subnormal, 699
Stomach, 248, **249**
of cow, 334–334, **334**
digestion in, 334
fish, 334
function, 335
gastric juice, 334, 338t
human, 333–335, **334**
secretions, 341, 334–335, **334**
structure, 334–335
stomata, 206, **215, 352,** 393, **393, 511**
STP, 295, 296t
Stream, 570
Streptomyces, **750**
Streptomycin, 377t
Strip mining, 677, 679–680, **679**
Strobilus, **759**
Stroma, in chloroplast, 116, **117**
Structural gene, 197–198, **198**
Sturtevant, A., 507
Subcutaneous layer, 230
Suberin, 207
Subspecies, 74
Succession
characteristics, 617–619
lake, 617, 618
primary, 619
secondary, 619, **621**
terrestrial, 619, **620**
Succinic acid, 148, **149**
Succinyl Co-A, 173
Sucrose, **32**
Sulci, 270, **271**
Sulfonamine, 375, **375**
Sulfur, 346t
Sundew, 324, 325
Support, 303–311, **304, 305, 306,** 309–311t, **312, 313**
Surveyor III, **67**
Sutton, W., 499–500
Sweat gland, **230,** 232, **232,** 379
Symbiosis, 594–597
Symmetry, 76, **77,** 764, 766
bilateral, 76, **77,** 764
radial, 76, **77,** 764, 766
spherical, 76, **77**
Sympathetic nervous system, 266, **267,** 276, 277
Symphysis joint, 307, 309t
Synapse, 162, 163, **164**
potential, 163
Synchytrium endobioticium, 760, **762**
Syphilis, 279
Systemic circulation, 361, 364
Systems, 204, 229–230

T

Tactile hairs, in crayfish, 264, **265**
Taenia, **767**
Taiga, 614, **642**
Tapeworm, 327–328, 595, **767**
Taste, 283, 284, **284**
bud, 284, **284,** 330, **333,** 716
See also Chemoreception
Tatum, E., 182
Taxis, 694, 713
Taxonomy, 73; *see also* Classification systems
Teeth
in digestion, 331
disorders, 342–343
structure, 331, **332**
types, 331
Telodendria, 162
Telophase
in meiosis, **426,** 427–428
in mitosis, **120–123,** 125
Temperature
body, 232
inversion, 672–673
in metamorphosis, 247
Termites in mutualism, 594
Terrestrial ecosystem, 637–640, **638**
Territoriality, 590–591
function of, 730–731
Territoriality [*cont.*]
sign stimulus in, 731
succession, 619, **620**
Testa, 455
Ts, **444, 663**
function, 443–444
structure, 443–444, **444**
Testosterone, 248, 251t, **444**
Tetany, 252
Tetracycline, 377t
Thalamus, 264, **270,** 271
function, 271
Thermal pollution, 675–676
Thermocline, 572–573, **573**
Thiamine. *See* Vitamin B
Thigmotropism, 243, **244,** 245
Thoracic duct, 368
Thorax
human, 305, **305**
insects, 768, **777**
Thorpe, W., 707
Thrombin, in blood clotting, 358–359
Thrombocyte, 224, 226t
Thromboplastin, 358–359
Thylakoids, **116,** 117
Thymine, 175, 183
Thymus gland, **249,** 251t, 257, 258
Thyrocalcitonin, 251, 252
Thyroglobin, 250
Thyroid gland, **249,** 250, 251t, 252
hormones, 250, 251t, 254, 255
Thyrotropin. *See* TSH
Thyroxin, 250, 251t, 479
Tibia, **305, 307**
Tick, 771
Tides, 572–573, **575**
affect on ecosystems, 594
red, 599
Tinbergen, N., 691
Tissue, 2, 204
adipose, 34, 386
animal, 216–230
animal organs and, 229–30, 229t, **230**
chlorenchyma, 215, **215**
collenchyma, 206, **206**
complex permanent, 207–208, **207, 208**
connective, 218, 220–224, 222t
cork, 207, **206,** 299
development of, 470t
embryonic layers, 77, 463–470, **463, 466, 469,** 470t
epithelial, 216–218, 217–218t
meristematic, **209,** 210, **212,** 254, 458, **458, 459**
mesophyll, 352–353, **352**

Tissue [*cont.*]
palisade, 215, **215**
parenchyma, 206, 208, 216, 352–353, **352**
permanent, 205–216, **209, 210**
phloem, 208, **209,** 210, **210,** 211, **211,** 213, **213,** 214, 232
plant, 204–216
primary, 210, 212–213, 458
procambium, 458
protoderm, 458
sclernchyma, 206–207, 206
typing, 388, **388**
vascular, 207–208, 249
xylem, 207–208, **207,** 459
Tobacco, effects of smoking, 404–405
Tobacco mosaic virus, 176, 177
Tolerence levels, 614
Tolerence, law of, 615–616
Tongue
structure, 330–331, **333**
papillae, 284, **284**
taste bud, 284, **284**
Tonoplast, 91, 119; *see also* Vacuole
Toxin, 378
Trachea, 399, **399**
in grasshopper, 396, **396**
Tracheid, 207, 351
Tracheophyta, 758, **759;** *see also* Plant
Tranquilizers, 296t, 297
Transamination, 154–155
Transcription, 185–186, 192, 198
Transduction, 5–6, 5t, 159, **159,** 430
in ecosystems, 609–612, **611**
See also Energy
Transferase, 131t
Transformation, genetic, 176–177, **177,** 430
Translation, 186–187
Translocation, 207, 354
chromosome, 513, **514**
mass flow hypothesis, 354, **354**
protoplasmic streaming hypothesis, 354
Transpiration, 214, 352–353, **352**
Transplantation, 386–391
allograft, 389, **389**
artificial replacements vs., 389–390
implication, 390–391
isograft, 388–389
organ, 386–391
types, 388–389
xenograft, 389
Trechinympha, 753, **754**
Treponema pallidum, 297
Trial and error, 707, **708**
Triplet in genetic code, 183, **184,** 193, **194**
Trophic levels, 633, 639
in ecological pyramids, 612–614, **615**
in ecosystems, **611,** 612, **614**
in energy flow, 610
Tropical rain forest, 645, **645,** 646
Tropism, 242, 244–245, 694
geotropism, 244–245, **244**
nasties, 242–244, **243**
paratonic movement, 242, **243**
phototropism, 244, **244**
thigmotropism, **244,** 245
True fungi, 760–763, **762, 763**
Trypanosoma gambiense, 753, **754**
Tryptophane, 173
TSH (thyrotropin), 257
Tuber, 422, **424**
Tubule, 166, **166**
kidney, 410–413, **411**
Tundra, 640–641, **641**
indicator species in, 616
Turgor movement, 242
Turgor pressure, 95
Turner's syndrome, 526, **527**
Turtle, shell, 302
Twins, 527–529
in experiments, 528–529
fraternal, 462, 528
identical, 462, 528
Tympanic membrane, 287–288, **288**
Tympanum, 718, **719**
Typhlosole, 329, **330**

U

Ulcer, 380
Ultraviolet radiation, 301
Ulothrix, 421, **423**
Ulva, **305,** 307, 755, **757**
life cycle, 432, **432**
meiospores, 432, **432**
Umbarger, H., 194
Umbilical cord, **366,** 367, 471, **471**
Unit membrane hypothesis, 90, **92**
Unmyelinated fibers, **228,** 229
Uracil, 184, 191
Uranus, 69
Urbanization, 627–629, **629,** 656–657
Urea, 301
cycle, 155
in excretion, 407
Ureter, **410,** 411, **411**
Urey, H., 60
Uric acid, 407
Uridine triphosphate, 170, 171
Urinary bladder, **410,** 411
Urochordata, 770
Uterine tube, **447,** 448, 662–663, **663**
Uterus, **447,** 448–449

V

Vacuole, **91,** 119
contractile, 119
food, 119, 325–326, **326**
Vagina, **447,** 449
Valence, 11–14
Valine, 173
Valve, heart, **360,** 361
Vas deferens, **444,** 445–446
Vascular
bundle, 212, **212,** 215, **215**
cambium, 204–205, 210, **210,** 213
conduction of nutrients, 350–354, **351, 352, 354**
cylinder, 210
rays, 214
tissue, 224
animal, 216–230
plant, 204–216
Vasectomy, 662, **663**
Vasopressin, 251t, 256–257
Vegetative reproduction. *See* Reproduction
Vein
function, 363–364
inferior vena cava, 363–364, **365**
major groups, **364, 365**
portal, 370, 412
pulmonary, 362, 364, **365**
structure, 363
superior vena cava, 363, **365**
Vejovis, **771**
Ventral root, 267, 273, **273**
Ventricle, 269, **269**
of brain, 273
of heart, **360,** 361
Venus, 68
Venus's flytrap, 324, **325,** 714
Vertebra, 305, **306**
Vertebrata, 770, **772;** *see also* Vertebrates
Vertebrates, 78
blood, 357–359, **358,** 358t
care of young, 443
circulatory system, 357–369

Vertebrates [*cont.*]
digestive system, 329–342
excretory system, 409–413
fertilization in, 443, 461–462
heart, 361–363, **362**
hormones, 251t
movement, 314–316, 317–320t
reproductive system, 443–450
respiratory system, 396–398, **397, 398**
skeleton, 304–308, **305, 306**
skin, 299, 301–303
Vessel elements, 208, **208**
annular, 207
Ventral root, 267, 273, **273**
Viceroy butterfly, 736, **736**
Vici faba, 327
Villus
chorionic, 471, **471**
in digestion, 336, **336**
structure, 336, **336**
Virchow, R., 56
Virus, 176, **177,** 374, 374t
in evolution, 65
in genetic experiments, 175–178, **176, 177, 179**
Vision
accommodation in, 294
behavior and, 717–718
chemical reaction of, 294, **294**
of fish, 717
of frog, 717
of human, 292, **293,** 294
in mating behavior, 727–729
pigments, 98
and species specific behavior, 717
See also Photoreception
Vitamin, 345, 346–348t, 348
A, 347t, 413
B_1, 348t
B_2, 348t
B_6, 348t
B_{12}, 348t
B complex, 340
C, 348t
D, 342, 347t, 413
deficiency, 347, 348t
E, 347t
folic acid, 348t
functions of, 347–348t
K, 337–338, 340, 347t, 359, 413
in liver, 413
niacin, 348t
pantothenic acid, 348t
sources of, 347–348t, 348
Vocal cord, 396–397
Volkmann's canal, **221,** 224
Volvox, 128, 755, 757
Vorticella, movement, 313, **314**
Vulva, **447,** 449

W

Wasp, chemoreception in, 717
Waste disposal. *See* Excretory system; Pollution
Water, 22–24, **23,** 168, 323
absorption, 340
annual rainfall, 579, 581, 581t
in atmosphere, 574, 576
in aquatic ecosystems, 631–637, **632, 635**
in biological oxidations, 146, **146**
capillary, 564, **565**
conduction in plants, 351–353, **351**
cycle, 606, **607**
in cytoplasm, 105, 224
density, 566–567, **567**
in ecosystems, 633–634, 637
ground runoff, 569
hyacinth, 627, **628**
hydrogen bonding in, 14, 22
hydrosphere, 566–573
in ionization, 25
in Kreb's cycle, 148–149, **149**
in living organisms, 22
in osmosis, 94–96
pH, 27
in photosynthesis, 132
physical properties, 566–567
in plants, 351
polarity, 22, **23**
pollution, 674–676, **675,** 681–682, **683**
properties, 22–24, **23,** 566–567, **567**
sea, 571–572, 572t
in soil, 562–565, **565**
solvating property, 22, **23**
Watson, J., 47, **49,** 179
Watson-Crick model of DNA, 47, **49,** 513
Wave theory, 133
Waves, brain, 281, **282**
Waxes, 35t
Went, F., 234
Whitaker, D., 480
White matter, 270, **270,** 273, **273**
Whitman, C., 707
Whittaker, R., 79–81
Wiener, A., 521
Wind
belts, 578, **579**
currents, 576–579, **578**
in transpiration, 373
World continent, 648–649, **648**
Wort (*Charophyta*), 755–757, **757**
Wyman, J., 196

X

Xanthium, 241
Xanthophyll, 132, 134
X chromosome
Kleinfelter's syndrome, 527, **528**
linkage, 523
in man, 500, 501, 502, 503, 504, 505, 509
Turner's syndrome, 526, **527**
Xenograft, 389
Xenopus laevis, 474
X-linked inheritance, 523
X ray
diffection, 42–43
mutations induced by, 515
Xylem, **207,** 210, 214, **215**
fibers, 207, **207**
lignin, 102, 104, 206–207
nutrient transport, 351
primary, 210, 213
ray cell, 214, **215**
of root, **209,** 210, 211
secondary, 211, **211,** 214
of stem, 207–208, **207, 212,** 213–214, **213**
water conduction, 351–353, **351, 352**

Y

Y chromosome, 500, 501, 502, 503, 504, 505, 509, 522–523, **523**
in man, 523–524
Yeast, 763, **763**
Y-linked inheritance, 523–524, **523,** 527, **527**
Yolk, 460, 463, **463**
Young, J., 709

Z

Z line of muscle, 165–166
Zea mays, **102**
Zinc, 346t, 609

Zoomastigina, **753**
Zonation
in flowing water, 634
in standing water, 631–632
Zooplankton, 601, **635,** 636
Littoral consumer, **632,** 633
Neritic zone, **635,** 636
in Plagic zone, 637
Zoosporangium, 420–421, 423
Zoospore, 420–421, **423**
in meiosis, 430, **431**
in *Oedogonium,* **431**
Zygomycota, 762, **763**
Zygospore, 429, **429, 763**
Zygote, 420, 422, 429, **429,** 455, 461–462, 467
Zygote [*cont.*]
development, 438, **439**
polarization in, 461
Zymogen, 113